QUANTUM COMPUTATION

AND

QUANTUM INFORMATION THEORY

QUANTUM COMPUTATION

AND

QUANTUM INFORMATION THEORY

Reprint volume with introductory notes for
ISI TMR Network School

12–23 July 1999 Villa Gualino, Torino, Italy

Editors

C. Macchiavello
University of Pavia, Italy

G. M. Palma
University of Palermo, Italy

A. Zeilinger
University of Vienna, Austria

World Scientific
Singapore • New Jersey • London • Hong Kong

Published by

World Scientific Publishing Co. Pte. Ltd.
P O Box 128, Farrer Road, Singapore 912805
USA office: Suite 1B, 1060 Main Street, River Edge, NJ 07661
UK office: 57 Shelton Street, Covent Garden, London WC2H 9HE

British Library Cataloguing-in-Publication Data
A catalogue record for this book is available from the British Library.

The editors and the publisher would like to thank the authors and the following organizations for their assistance and their permission to reproduce the articles found in this volume:

American Association for Advancement of Science
American Institute of Physics
American Physical Society
Elsevier Science Publishers
Institute of Physics Publishing
Kluwer Academic Publishers
National Academy of Sciences
Nature Publishing Group
Plenum Publishing Corporation
Royal Swedish Academy of Sciences
Scientific American
Springer Verlag

QUANTUM COMPUTATION AND QUANTUM INFORMATION THEORY

ISBN 981-02-4117-8

Printed in Singapore.

Preface

What can you do when you write and read your information on single atoms? How secret are your messages if you conceal them in the state of a single photon? Can you teleport at a distant point the unknown state of a particle? These questions, which would have seemed idle - if not meaningless - just a decade ago have now become hot topic of research. Quantum Information Theory has indeed revolutionised our view of what is the nature of information, imposing itself as a new paradigm. Since the early seminal papers in the 80' s the rate at which results have appeared in the scientific literature, mirroring the rapid progresses in this - now mature - discipline, has been astonishing. As the field still lacks of a comprehensive textbook it is a hard job for the newcomer to find his way in the existing literature. The job being made harder by the very interdisciplinary nature of this new area of research, with deep roots both in the foundations of Quantum Mechanics as well as of Information Theory and Computer Science. In this volume we have asked some of the leading researchers in the field to select and annotate a choice of articles in their area in order to give the reader a broad perspective of the present stage of research in all the various aspects of quantum information processing. The articles have been chosen with particular attention to their tutorial value. Each expert has been the curator of one section of the book. This initiative has been taken as part of the summer school on "Quantum Computation and Quantum Information Theory" organised by the Institute for Scientific Interchange Foundation in Villa Gualino, Torino, July '99.

In our choice of topics we have tried to give a broad overview of both the theoretical foundation of the field as well as of its practical design and experimental aspects.

The opening chapter is dedicated to the physical ingredients of the theory, introducing key concepts like entanglement and non locality which are the core of the quantum world, followed by a chapter dealing with the experimental aspects of entanglement measurement and manipulation. Quantum world manifests itself with a puzzling weird looking behavior which confuses our classical perception of daily world. Indeed it is this "weirdness" which, when properly harnessed, allows for new efficient forms of information coding, transmission and manipulation.

Three sections are then devoted to clarifying the computer - science aspects of the theory. The first of these chapters is an overview of quantum algorithms. The aim is to clarify which features of quantum theory make possible an exponential gain of resource in computation and to provide a unified description of the known quantum algorithms. This is followed by a chapter specially written for this book devoted to the theory of quantum complexity, in which the complexity classes of quantum algorithms are analysed with respect to the resources needed to implement them. The discussion of this part of the book ends with a chapter dedicated to quantum error correction. The extention of the powerful techniques developed for the reliable transmission and storage of classical information are far from being straightforwardly transferable in the quantum domain. Quantum error correction is an example of area of research in which, in the time span of less than two years the theory has reached a high level of sophistication, starting basically from sctatch.

The three following chapters overview three different aspects of information transmission. The Quantum Channels section provides an introduction to the tools used to describe noisy quantum channels and to measure the amount of reliable information sent through them via appropriate encoding and decoding. Related to this is the section on Long distance Quantum Communication, which describes protocols aiming at achieving the transmission of coherent quantum information over long distances overcoming the effects of the external environment. This is followed by a section devoted Quantum Cryptography, the art of sharing secret keys with the help of quantum mechanics. One of the oldest branch of quantum information theory, it is now in the realm of technological applications, with existing pre industrial prototypes. The related chapter gives an overview of both the theoretical and the experimental aspects of the subject.

The remaining part of the book reviews all the candidate technologies for the practical implementation of quantum computing devices. These can be broadly divided into two categories: atomic systems (like cavity Q.E.D., cold trapped ions) and mesoscopic ones (like Josephson junctions, quantum dots), with some technologies at the borderline between the two, as NMR and cold atoms in optical lattices. While atomic systems are characterised by a relatively easily controllable dynamics their scalability remains an open problem. On the other hand mesoscopic systems, although vulnerable to the spoiling action of the environment, open the possibility of large scale integration.

At the end of the book a bibliography is appended, with all the papers reproduced or quoted in the introduction plus a few others chosen directly by the editors. We have made no attempt neither to be comprehensive of all contributions to the field nor to be exhaustive of all the research going on in Quantum Information Theory. We apologise to whoever feels neglected in this list.

The limited space at our disposal has left no room for a description of many other recent developments of the theory, like the thermodynamics of entanglement, quantum cloning, topological quantum computation, just to mention some. As these topics are at the front line of such a rapidly expanding field it would have been extremely difficult to make a choice of papers which could be considered as reference. This, we feel, will be probably possible in the next few years.

Our thanks go to all the lecturers to the summer school on "Quantum Computation and Quantum Information Theory" organised by the Institute for Scientific Interchange Foundation, Villa Gualino, Torino, which have patiently helped us in selecting the papers of each section and which have written the relative introductory notes.

C. Macchiavello
G. M. Palma
A. Zeilinger

CONTENTS

Preface v

Introductory Concepts 1

1. Introductory Concepts 3
A. Zeilinger (University of Vienna)
   - 1.1 D. M. Greenberger, M. A. Horne and A. Zeilinger
     Multiparticle Interferometry and the Superposition Principle
     Physics Today **46**, No. 8, 22–29 (August 93) 4
   - 1.2 A. Zeilinger
     Quantum Entanglement: A Fundamental Concept Finding its Applications
     Physica Scripta, **T76**, 203–209 (1998) 12
   - 1.3 A. Zeilinger
     Experiment and the Foundations of Quantum Physics
     Rev. Mod. Phys. **71**(2), S288–S297 (1999) 19

Quantum Entanglement Manipulation 29

2. Quantum Entanglement Manipulation 31
D. Bouwmeester (Center of Quantum Computation, University of Oxford)
   - 2.1 C. H. Bennett, G. Brassard, C. Crépeau, R. Jozsa, A. Peres and W. K. Wootters,
     Teleporting an Unknown Quantum State via Dual Classic and Einstein–Podolsky–Rosen Channels
     Phys. Rev. Lett. **70**(13), 1895–1899 (1993) 35
   - 2.2 D. Bouwmeester, J.-W. Pan, K. Mattle, M. Eibl, H. Weinfurter and A. Zeilinger
     Experimental Quantum Teleportation
     Nature **390**, 575–579 (1997) 40
   - 2.3 S. L. Braunstein and H. J. Kimble
     Teleportation of Continuous Quantum Variables
     Phys. Rev. Lett. **80**(4), 869–972 (1998) 45
   - 2.4 D. M. Greenberger, M. A. Horne and A. Zeilinger
     Going Beyond Bell's Theorem
     in *Bell's Theorem, Quantum Theory and Conception of the Universe*
     M. Kafatos (Ed.) (Kluwer, Dodrecht, 1989) 49
   - 2.5 N. David Mermin
     What is Wrong with These Elements of Reality
     Physics Today, **43**, No. 6, 9 & 11 (June 1990) 53
   - 2.6 D. Bouwmeester, J.-W. Pan, M. Daniell, H. Weinfurter and A. Zeilinger
     Observation of Three-Photon Greenberger–Horne–Zeilinger Entanglement
     Phys. Rev. Lett. **82**(7), 1345–1349 (1999) 55

Quantum Algorithms 61

3. Quantum Algorithms 63
A. Ekert (Center for Quantum Computation, University of Oxford)
3.1 A. Ekert and C. Macchiavello
An Overview of Quantum Computing
in *Unconventional Models of Computation*, 19–44, C. S. Calude, J. Casti and M. J. Dinneen Eds., Springer Series in Discrete Mathematics and Theoretical Computer Science (Springer, Singapore, 1998) 66
3.2 R. Cleve, A. Ekert, L. Henderson, C. Macchiavello and M. Mosca
On Quantum Algorithms
Complexity **4**(1), 33–42 (Sept/Oct 1998) also preprint quant-ph/9903061 86

Quantum Complexity 101

4. *An Introduction to Quantum Complexity Theory*
R. Cleve (University of Calgary) 103

Quantum Error Correction 129

5. Quantum Error Correction 131
D. DiVincenzo (IBM Research Laboratories)
5.1 P. W. Shor
Scheme for Reducing Decoherence in Quantum Computer Memory
Phys. Rev. A **52**(4), R2493–R2496 (1995) 134
5.2 A. M. Steane
Error Correcting Codes in Quantum Theory
Phys. Rev. Lett. **77**(5), 793–797 (1996) 138
5.3 D. Gottesman
Class of Quantum Error-Correcting Codes Saturating the Quantum Hamming Bound
Phys. Rev. A **54**(3), 1862–1868 (1996) 143
5.4 D. P. DiVincenzo and P. W. Shor
Fault-Tolerant Error Correction with Efficient Quantum Codes
Phys. Rev. Lett. **77**(15), 3260–3263 (1996) 150

Quantum Channels 155

6. Quantum Channels 157
C. A. Fuchs (California Institute of Technology)
6.1 K.-E. Hellwig
General Scheme of Measurement Processes
Int. J. Theor. Phys. **34**(8), 1467–1479 (1995) 161
6.2 M.-D. Choi
Completely Positive Linear Maps on Complex Matrices
Lin. Alg. Appl. **10**, 285–289 (1975) 174

6.3 B. Schumacher
Sending Entanglement Through Noisy Quantum Channels
Phys. Rev. A **54**(4), 2614–2628 (1996) 180

6.4 H. Barnum, C. M. Caves, C. A. Fuchs, R. Jozsa and B. Schumacher
Noncommuting Mixed States Cannot Be Broadcast
Phys. Rev. Lett. **76**(15), 2818–2821 (1996) 195

6.5 B. Schumacher and M. D. Westmoreland
Sending Classical Information via Noisy Quantum Channels
Phys. Rev. A **56**(1), 131–138 (1997) 199

6.6 C. A. Fuchs
Nonorthogonal Quantum States Maximize Classical Information Capacity
Phys. Rev. Lett. **79**(6), 1163–1166 (1997) 207

Entanglement Purification and Long-Distance Quantum Communication 211

7. Long-Distance Quantum Communication 213
H. Briegel (Ludwig-Maximilians-Universitaet, Munich)

7.1 H.-J. Briegel, W. Dür, J. I. Cirac and P. Zoller
Quantum Repeaters: The Role of Imperfect Local Operations in Quantum Communication
Phys. Rev. Lett. **81**(26), 5932–5935 (1998) 217

7.2 C. H. Bennett, G. Brassard, S. Popescu, B. Schumacher, J. A. Smolin and W. K. Wootters
Purification of Noisy Entanglement and Faithful Teleportation via Noisy Channels
Phys. Rev. Lett. **76**(5), 722–725 (1996) 221

7.3 D. Deutsch, A. Ekert, R. Jozsa, C. Macchiavello, S. Popescu and A. Sanpera
Quantum Privacy Amplification and the Security of Quantum Cryptography over Noisy Channels
Phys. Rev. Lett. **77**(13), 2818–2821 (1996) 225

7.4 S. J. van Enk, J. I. Cirac and P. Zoller
Photonic Channels for Quantum Communication
Science **279**, 205–208 (1998) 229

Quantum Key Distribution 233

8. Quantum Key Distribution 235
G. Ribordy, N. Gisin and H. Zbinden (University of Geneva)

8.1 W. Tittel, G. Ribordy and N. Gisin
Quantum Cryptography
Physics World, pp. 41–45 (March 1998) 240

8.2 H. Zbinden, N. Gisin, B. Huttner, A. Muller and W. Tittel
Practical Aspects of Quantum Cryptographic Key Distribution
J. Cryptology **11**, 1–14 (1998) 245

8.3 D. Bruß and N. Lütkenhaus
Quantum Key Distribution: From Principles to Practicalities
in *Applicable Algebra in Engineering, Communication & Communication* 259

Cavity Quantum Electrodynamics 275

9. Cavity Quantum Electrodynamics 277
H. Mabuchi (California Institute of Technology)

9.1 S. Haroche and J. M. Raimond
Cavity Quantum Electrodynamics
Scientific American (April 1993), p. 26–33 282

9.2 Q. A. Turchette, C. J. Hood, W. Lange, H. Mabuchi and H. J. Kimble
Measurement of Conditional Phase Shifts for Quantum Logic
Phys. Rev. Lett. **75**(25), 4710–4713 (1995) 290

9.3 C. J. Hood, M. S. Chapman, T. W. Lynn and H. J. Kimble
Real-Time Cavity QED with Single Atoms
Phys. Rev. Lett. **80**(19), 4157–4160 (1998) 294

9.4 X. Maître, E. Hagley, G. Nogues, C. Wunderklich, P. Goy, M. Brune, J. M. Raimond and S. Haroche
Quantum Memory with a Single Photon in a Cavity
Phys. Rev. Lett. **79**(4), 769–772 (1997) 298

9.5 M. Brune, E. Hagley, J. Dreyer, X. Maître, A. Maali, C. Wunderlich, J. M. Raimond and S. Haroche
Observing the Progressive Decoherence of the "Meter" in a Quantum Measurement
Phys. Rev. Lett. **77**(24), 4887–4890 (1996) 302

9.6 H. Mabuchi and P. Zoller
Inversion of Quantum Jumps in Quantum Optical Systems under Continuous Observation
Phys. Rev. Lett. **76**(17), 3108–3111 (1996) 306

Quantum Computation with Ion Traps 311

10. Quantum Computation with Ion Traps 313
R. Blatt (University of Innsbruck) & W. Lange (Max-Planck-Instut für Quantenoptik, Munich)

10.1 W. Paul
Electromagnetic Traps for Charged and Neutral Particles
Rev. Mod. Phys. **62**(3), 531–540 (1990) 320

10.2 J. I. Cirac and P. Zoller
Quantum Computations with Cold Trapped Ions
Phys. Rev. Lett. **74**(20), 4091–4094 (1995) 330

10.3 J. D. Prestage, G. J. Dick and L. Maleki
New Ion Trap for Frequency Standard Applications
J. Appl. Phys. **66**(3), 1013–1017 (1989) 334

10.4 H. C. Nägerl, W. Bechter, J. Eschner, F. Schmidt-Kaler and R. Blatt
Ion Strings for Quantum Gates
Appl. Phys. B – Lasers Opt. **66**, 603–608 (1998) 339

10.5 D. F. V. James
Quantum Dynamics of Cold Trapped Ions with Application to Quantum Computation
Appl. Phys. B – Lasers Opt. **66**, 181–190 (1998) 345

10.6 C. Monroe, D. M. Meekhof, B. E. King, W. M. Itano and D. J. Wineland
Demonstration of a Fundamental Quantum Logic Gate
Phys. Rev. Lett. **75**(25), 4714–4717 (1995) 355

10.7 Q. A. Turchette, C. S. Wood, B. E. King, C. J. Myatt, D. Leibfried, W. M. Itano, C. Monroe and D. J. Wineland
Deterministic Entanglement of Two Trapped Ions
Phys. Rev. Lett. **81**(17), 3631–3614 (1998) 359

Josephson Junctions and Quantum Computation 363

11. Josephson Junctions and Quantum Computation 365
R. Fazio (University of Catania) & G. Schön (University of Karlsruhe)

11.1 R. Fazio and G. Schön
Mesoscopic Effects in Superconductivity
in *Mesoscopic Electron Transport*, NATO ASI series E, **345**, 407, Kluver (1997) 368

11.2 A. Shnirman, G. Schön and Ziv Hermon
Quantum Manipulations of Small Josephson Junctions
Phys. Rev. Lett. **79**(12), 2371–2374 (1997) 391

11.3 Y. Makhlin, G. Schön and A. Shnirman
Josephson–Junction Qubits with Controlled Couplings
Nature, **398**, 305–307 (1999) 395

Quantum Computing in Optical Lattices 399

12. Quantum Information in Optical Lattices 401
H. Briegel (Ludwig-Maximilians-Universitaet, Munich)

12.1 H.-J. Briegel, T. Calarco, D. Jaksch, J. I. Cirac and P. Zoller
Quantum Computing with Neutral Atoms
Reprint, to appear in the special issue of J. Mod. Opt. (2000) on *The Physics of Quantum Information* 404

Quantum Computation and Quantum Communication with Electrons 425

13. Quantum Computation and Quantum Communication with Electrons 427
D. Loss (University of Basel)

13.1 D. Loss and D. P. DiVincenzo
Quantum Computation with Quantum Dots
Phys. Rev. A **57**(1), 120–126 (1998) 433

13.2 G. Burkard, D. Loss and D. P. DiVincenzo
Coupled Quantum Dots as Quantum Gates
Phys. Rev. B **59**(3), 2070–2078 (1999) 440

13.3 D. P. DiVincenzo and D. Loss
Quantum Computers and Quantum Coherence
J. Magn. Magn. Matler. **200** special issue on *Magnetism beyond 2000*
also cond-mat/9901137 449

NMR Quantum Computing 463

14. Quantum Computing with NMR 465
J. Jones (Oxford University)

14.1 D. G. Cory, A. F. Fahmy and T. F. Havel
Nuclear Magnetic Resonance Spectroscopy: An Experimentally Accessible Paradigm for Quantum Computing
Proceedings of PhysComp '96 (eds. T. Toffoli, M. Biafore and J. Leão), (New England Complex Systems Institute, 1996) pp. 87–91 471

14.2 J. A. Jones and M. Mosca
Implementation of a Quantum Algorithm on a Nuclear Magnetic Resonance Quantum Computer
J. Chem. Phys. **109**(5), 1648–1653 (1998) 476

14.3 I. L. Chuang, N. Gershenfeld and M. Kubinec
Experimental Implementation of Fast Quantum Searching
Phys. Rev. Lett. **80**(15), 3408–3411 (1998) 482

14.4 N. Linden, H. Barjat and R. Freeman
An Implementation of the Deutsch–Jozsa Algorithm on a Three-Qubit NMR Quantum Computer
Chem. Phys. Lett. **296**, 61–67 (1998) 486

Selected Bibliography 493

Introductory Concepts

Introductory Concepts

Anton Zeilinger
University of Vienna

The first three papers selected provide an elementary introduction to the fundamental issues in quantum physics related to quantum information theory. The most basic concept is the superposition principle of quantum mechanics which says that whenever we have two quantum states admissible in a given experimental situation, a linear superposition of the two is also an admissible quantum state. This leads already to the quantum interference phenomena for individual partuicles as exemplified by the famous double slit experiment.

Superposition of multiparticle quantum states then leads to the very basic notion of quantum entanglement. While from a fundamental point of view entanglement implies quantum nonlocality (or nonseparability) as signified by Bell's theorem, it also forms the essential basis for some quantum crytography protocols, for all nontrivial quantum computation algorithms and for novel communication concepts like quantum teleportation.

All papers chosen are of a review character and were written for a broader audience. The first paper adresses the fundamental issues in the quantum superposition of multiparticle states. Building on these notions, the second paper then explains in more detail quantum entanglement and the basic ideas behind its applications in quantum information concepts. The last paper then addresses the experimental side, particularly the status of current experimentation and its relation to the foundations of quantum physics.

Selected bibliography

1. D.M. Greenberger, M.A. Horne and A. Zeilinger
   Multiparticle Interferometry and the Superposition Principle
   Physics Today **46**, 8 (Aug. 93), 22-29, 8 pages

2. A. Zeilinger
   Quantum Entanglement: A Fundamental Concept Finding its Applications
   Physica Scripta, **T76**, 203-209, (1998), 7 pages

3. A. Zeilinger
   Experiment and the Foundations of Quantum Physics
   Rev.Mod.Phys.**71** (2) , p. S 288 - S 297 (1999) 11 pages

MULTIPARTICLE INTERFEROMETRY AND THE SUPERPOSITION PRINCIPLE

We're just beginning to understand the ramifications of the superposition principle at the heart of quantum mechanics. Multiparticle interference experiments can exhibit wonderful new phenomena.

Daniel M. Greenberger, Michael A. Horne and Anton Zeilinger

Discussing the particle analog of Thomas Young's classic double-slit experiment, Richard Feynman wrote in 1964 that it "has in it the heart of quantum mechanics. In reality, it contains the *only* mystery."[1] That mystery is the one-particle superposition principle. But Feynman's discussion and statement have to be generalized. Superposition may be the only true quantum mystery, but in multiparticle systems the principle yields phenomena that are much richer and more interesting than anything that can be seen in one-particle systems.

The famous 1935 paper by Albert Einstein, Boris Podolsky and Nathan Rosen pointed out some startling features of two-particle quantum theory.[2] Erwin Schrödinger emphasized that these features are due to the existence of what he called "entangled states," which are two-particle states that cannot be factored into products of two single-particle states in any representation. "Entanglement" is simply Schrödinger's name for superposition in a multiparticle system. Schrödinger was so taken with the significance of multiparticle superposition that he said entanglement is "not *one* but rather *the* characteristic trait of quantum mechanics."

Until the mid-1980s, the quintessential example of an entangled state was the singlet state of two spin-½ particles,

$$|\psi\rangle = (1/\sqrt{2})(|+\rangle_1|-\rangle_2 - |-\rangle_1|+\rangle_2)$$

or its photon analog. The subscripts 1 and 2 refer to the two particles (distinguished, for example, by their flight directions), and the plus and minus signs refer to spin up or down with respect to any specified axis. This state of two spatially separated particles was introduced into the Einstein–Podolsky–Rosen discussion by David Bohm[3] in 1951. It inspired a spate of experiments in the 1970s and '80s.

Since the mid-1980s there has been a revolution in the laboratory preparation of new types of two-particle entanglements. Various experimental groups started doing interferometry with down-conversion photon pairs. Down-conversion is a process in which one ultraviolet photon converts into two photons inside a nonlinear crystal.[4] This process allows one to construct "two-particle interferometers" that entangle the two photons in a way that needn't involve polarization at all. Many experimental groups independently came up with this idea, but the first explicit proposal was made by two of us.[5]

Real experiments commenced when Carroll Alley and Yan Hua Shih[6] at the University of Maryland first used down-conversion to produce an entangled state and when Ruba Ghosh and Leonard Mandel[7] at the University of Rochester first produced two-particle fringes without using polarizers. Since these pioneering efforts, many increasingly sophisticated experiments have been performed, with important lessons for quantum theory.

In all of these two-particle experiments, the source of the entanglement has been down-conversion. We will discuss a small sampling of recent developments, with particular emphasis on the fundamental ideas. Three-particle interferometry is even richer, and we shall say something about it. But it is mostly unexplored territory, both experimentally and theoretically.

One of our motivations for writing this article was to make the point that one doesn't have to be a quantum optics expert to understand or analyze such experiments. They illustrate beautifully the general principles of quantum mechanics, and they can be understood, both qualitatively and quantitatively, in those terms. Calculations based on detailed nonlinear quantum optics Hamiltonians do describe specific mechanisms. But they tend to obscure the generality of the conclusions, which depend primarily on the fact that if one cannot distinguish (even in principle) between different paths from source to detector, the amplitudes for these alternative paths will add coherently.

Two-particle double-slit interferometry

Figure 1 is a sketch of an idealized two-particle interference experiment that nonetheless exhibits some intriguing phenomena which have been verified experimentally. Consider a particle Ω near the center that can decay into two daughter particles. If the original particle is essentially at rest, then the momenta of its two daughters will be approximately equal and opposite. Now imagine that there are screens on both sides of the center, each with

Daniel Greenberger is a professor of physics at the City College of New York. **Michael Horne** is a professor of physics at Stonehill College, in North Easton, Massachusetts. **Anton Zeilinger** is a professor of physics at the Institute for Experimental Physics of the University of Innsbruck, in Austria.

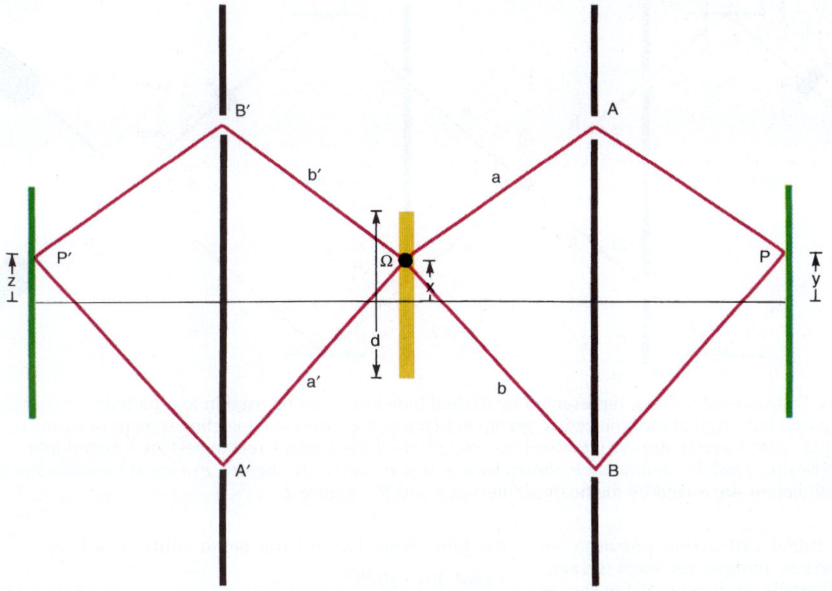

Idealized two-particle interferometer has two detecting screens (green) flanking two collimating screens, each with a pair of holes, that bracket a source (orange) of decaying particles at the center. The vertical extension of the source is *d*. A particle Ω decays at height *x* above the center line into two particles (trajectories in red) that hit their respective detecting screens at heights *y* and *z*. Because one can't tell whether the pair passes through holes *A* and *A′* or *B* and *B′*, these alternatives can interfere. **Figure 1**

two holes in it, as shown in the figure. These holes confine the escaping decay particles to either of a pair of opposite directions. The decay particles can pass either through holes *A* and *A′* or through holes *B* and *B′*. We can then write the state of the two-particle system as

$$|\psi\rangle = (1/\sqrt{2})(|a\rangle_1\, |a'\rangle_2 + |b\rangle_1\, |b'\rangle_2)$$

where the letter in the state-vector bracket denotes the escape direction defined by the corresponding holes. Beyond the two perforated screens are two scintillation screens that record the positions of particles landing on them.

Because each of the particles can reach its detecting screen by way of two different paths, one might expect these screens to show interference patterns. But they don't. That's where the two-particle interferometer differs from a single-particle interferometer based on Young's classic double-slit experiment. There is no interfence pattern at either screen in these two-particle experiments because, for reasons that will become clear, the vertical position of the decaying particle is unknown to within some source size d that is considerably larger than λ/θ, where θ is the angle subtended by the hole pairs at the source, and λ is the relevant wavelength—optical or de Broglie, as the case may be. Thus the initial position uncertainty d washes out any interference fringes.

In the single-particle case, by contrast, the geometry is so determined, usually by lenses, that the source is effectively a point. Its positional uncertainty is much smaller than λ/θ, so that the waves arrive at the two holes with a definite phase relation and therefore interfere. The experimenter produces a diffraction pattern by controlling the geometry of the emission process.

In the two-particle case, something like the opposite of this process happens. With a large source, if one looks at either particle *separately*, one sees no interference pattern. But there is a *two*-particle interference pattern! If one monitors the arrival positions *P* and *P′* at the two scintillation screens *in coincidence*, then one sees that the two particles are much more likely to land where the alternative paths *PAΩA′P′* and *PBΩB′P′* differ in length by an integral multiple of the wavelength and so interfere constructively. Note that one cannot catch such two-particle interference patterns on film at either screen; one has to record coincidences.

One may well ask how one particle could have any knowledge of where the other can land, especially when the source position is unknown and therefore the first particle doesn't even know where it itself will land. The answer is that by landing at a specific point, one particle actually creates a sinusoidal amplitude of possible positions where the source is likely to have been—a sort of "virtual crystal." This crystal in turn creates the two-particle diffraction pattern. Virtual slit systems can be exploited in actual experiments.[8]

To see how this works here, let us for simplicity consider only the vertical degree of freedom x for the source position relative to the horizontal center line in figure 1. And let us say that the decay particles are photons. If y and z are the corresponding vertical distances above center of the landing points P and P', respectively, then the quantum mechanical amplitude for landing at P is

$$\psi \sim \exp(ikL_a) + \exp(ikL_b) \sim \cos\frac{2\pi\theta}{\lambda}(y+x)$$

where k is $2\pi/\lambda$ and the L's are the alternative path lengths for the photon on the right side of the apparatus. Similarly the amplitude for the other photon to land at P' is

$$\psi' \sim \cos\frac{2\pi\theta}{\lambda}(z+x)$$

Then the total amplitude for the two photons to land at heights of x and y, respectively, above center will be

$$\psi(y,z) \sim \frac{1}{d}\int_{-d/2}^{d/2} dx\, \cos\frac{2\pi\theta}{\lambda}(y+x)\cos\frac{2\pi\theta}{\lambda}(z+x)$$

If d is much larger than λ/θ, this integral becomes $\frac{1}{2}\cos(2\pi\theta(z-y)/\lambda)$, and one gets 100% visibility for "conditional fringes" between the two photons on opposite sides.

At the other extreme, if d is much smaller than λ/θ, the integral gives $\cos(2\pi\theta y/\lambda) \times \cos(2\pi\theta z/\lambda)$.

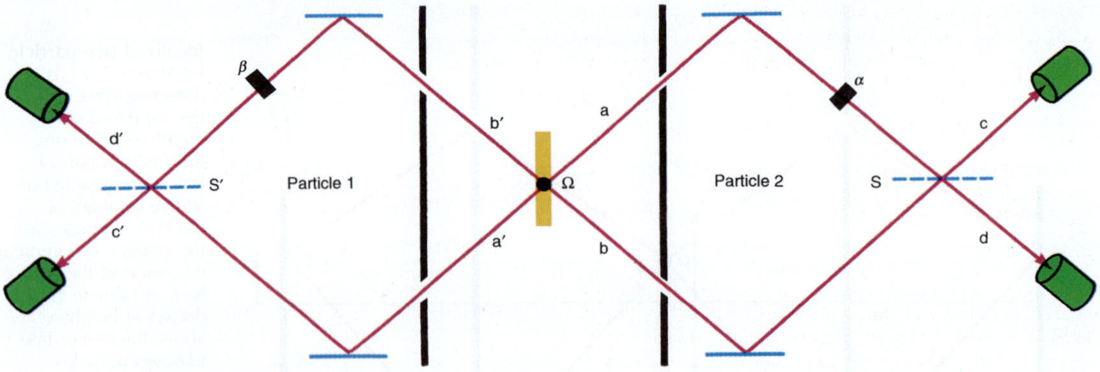

Beam splitters (half-silvered mirrors, represented by dashed blue lines) can be used in two-particle interferometry with localized photon detectors (green) in place of the extended detecting screens of figure 1. Particle Ω in the central source decays into two photons. Phase shifters (black rectangles) are inserted into decay-photon beams *a* and *b′*, shifting their phases by α and β, respectively. Before arriving at the detectors the alternative-path beams are mixed by the beam splitters at *S* and *S′*. **Figure 2**

That's a product of independent diffraction patterns, so we actually see single-particle fringes on each screen. The condition $d \ll \lambda/\theta$ is just the requirement for seeing fringes in a usual single-particle diffraction experiment. So there is a sort of complementarity between one- and two-particle fringes: The conditions for seeing one *preclude* the possibility of seeing the other.

One can make the same argument in momentum space. Momentum p is related to wavenumber by $k = p/\hbar$. If $d \gg \lambda/\theta$, the uncertainty principle tells us that the fractional transverse momentum spread $\delta k/k$ of each emitted photon will be much less than θ. That's too little to illuminate both pinholes simultaneously, so there can be no single-particle interference. On the other hand, if the source is small, then $\delta k/k \gg \theta$ and the particle can pass through either hole. The two paths can then interfere, and one will see fringes at the individual screens. But then one can no longer guarantee that if one photon goes through pinhole A, the other will go through A'. That destroys the two-particle entangled state. Once again, the two conditions are mutually exclusive.

The gedankenexperiment we have just described is essentially what Ghosh and Mandel did in their pioneering down-conversion experiment.[7] Its main difference from our description is that their originating ultraviolet photon strikes the nonlinear crystal with a substantial momentum, so that the two down-converted photons it generates both emerge together from the back of the crystal. Thus Ghosh and Mandel were able to catch both photons on a single screen.

Beam splitters

Most subsequent down-conversion experiments have used a different technique, replacing the scintillation screens with beam splitters. Figure 2 is a schematic illustration of such an interferometer. Beam splitters at S and S' have replaced the two detector screens of figure 1. Small detectors beyond the beam splitters monitor the counts in the four photon beams labeled c, d, c' and d'. To detect interference, one inserts a phase shifter of phase α into beam a, and one of phase β into b'.

We assume that each beam splitter transmits precisely half of each incident beam and reflects the other half. (Without loss of generality we can take the transmitted and reflected beams to be 90° out of phase.) Then the joint state beyond the beam splitters will be

$$\frac{i\exp(-i(\alpha+\beta)/2)}{\sqrt{2}}\,[\sin(\Delta/2)\,|c\rangle_1\,|c'\rangle_2 + \cos(\Delta/2)\,|c\rangle_1\,|d'\rangle_2 + \cos(\Delta/2)\,|d\rangle_1\,|c'\rangle_2 - \sin(\Delta/2)\,|d\rangle_1\,|d'\rangle_2]$$

where $\Delta = \alpha - \beta$ is the difference between the parameters of the two phase shifters. To see two-particle interference effects one must simultaneously monitor beam detectors on the left and right sides of the apparatus (c and c', for example) for coincident counts while varying Δ. In any single detector, by contrast, one sees no interference; the counting rate is a constant independent of the variable phase shifters. Each detector on its own is seen to record at random half of all events. For example,

$$P(c) = P(c,c') + P(c,d') = \frac{1}{2}\sin^2\frac{\Delta}{2} + \frac{1}{2}\cos^2\frac{\Delta}{2} = \frac{1}{2}$$

independent of Δ, where $P(c,c')$ is the joint probability of simultaneous counts in c *and* c'. One can also use this setup to perform an Einstein–Podolsky–Rosen experiment. We assign a value of +1 to a detection at either detector c or c', and −1 to d or d', and take the product of the appropriate values for a pair of simultaneous counts at left and right. Simultaneous counts in detectors c and d', for example, get a score of $(+1)\times(-1) = -1$. One can then take the expectation value over a long series of counts to confirm the quantum mechanical prediction:

$$E(\alpha,\beta) = P(c,c') - P(c,d') - P(d,c') + P(d,d') = -\cos\Delta$$

This cosine form is identical to what one gets for Bohm's version of the Einstein–Podolsky–Rosen experiment, with its two spin-½ particles in the singlet state.[9] The phase shifters play the role of the spin polarizer angles in that experiment. Variants of Bohm's version have been performed many times, generally using photon polarization rather than the spin of massive particles. The first such experiment was done by John Clauser and Stuart Freedman at Berkeley, and the most famous is the experiment of Alain Aspect and his coworkers at Orsay, near Paris.[10] With the advent of the parametric down-converter, the new version without polarization that we've been describing has also been done.[11]

The fact that these experiments exhibit two-particle correlations but not single-particle interference has some important (if poorly understood) ramifications, of which

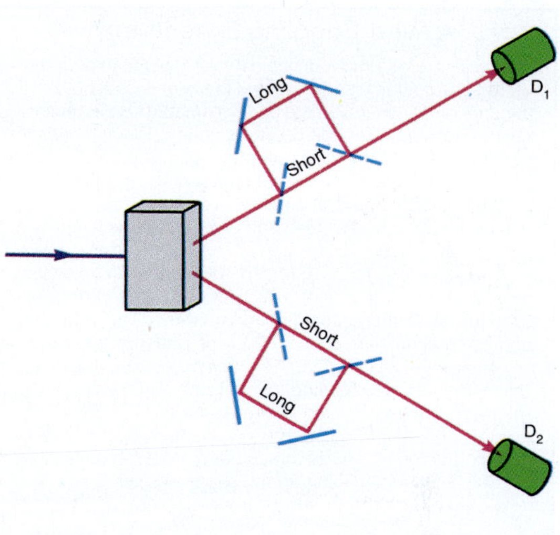

Temporal double-slit experiment proposed by James Franson[13] produces two-photon interference because one doesn't know *when* the pair was produced by down-conversion of an incident ultraviolet photon in the crystal (gray). The down-conversion photons (red trajectories) arrive simultaneously at their respective detectors (green), so one knows that both took paths of equal length. But one doesn't know whether both took the long or the short alternative paths offered by the beam splitters and mirrors in each arm of the apparatus. **Figure 3**

we shall mention two. One is that it is impossible to use such a system to communicate faster than the speed of light. If the value of α had any effect on the counting statistics at c' and d', that would clearly violate special relativity. Why quantum theory, a specifically nonrelativistic theory, should conspire to be consistent with relativity in this way is a deep mystery.

To illustrate the other particularly interesting feature of the experiment sketched in figure 2, consider keeping a record of the results of repeated outcomes for particle 1, the decay product that goes to the right. Such a string of 1's and −1's (depending on whether detector c or d clicked) would be useless by itself, because the numbers would be random. It is not until the lists for the left and right decay particles are brought to the same place for comparison, possibly weeks later, that the correlations between them can be seen. So these correlations, necessarily nonlocal in character, are worthless until they are locally compared.

In England, John Rarity and Paul Tapster[11] have performed such an experiment by means of down-conversion. The converted photons emerge from the crystal with a broad range of colors and directions, but they can be accurately selected by filters and other optical devices. Rarity and Tapster employed a folded version of the configuration in figure 2, with both photons coming out on the same side of the crystal. Although several other groups[6,7] have done tests of Bell's inequality with the new techniques, the experiment of Rarity and Tapster was the first one that did not rely at all on the polarization of the photons. Their data reflect the transverse correlation of the two photons beyond the beam splitters.

Interference of emission times

Ten years ago Mandel pointed out[12] that one could get two-photon fringes when two independent and spatially separate single-atom sources produce coincident photon counts in a pair of detectors. He traced this idea back to the 1950s. Interference occurs because, as Mandel put it, "one photon must have come from one source and one from the other, but we cannot tell which came from which."

An ingenious alternative was proposed by James Franson at Johns Hopkins.[13] He pointed out that two-particle fringes can also arise because we don't know *when* the particles were produced. Several groups[14] have successfully produced interference fringes using Franson's scheme, and it is becoming an efficient way to produce such fringes. We will describe a recent experiment by Raymond Chiao and colleagues at Berkeley,[15] which is the first to produce high-visibility fringes in this way.

Figure 3 illustrates the novel type of superposition in Franson's proposal. A pair of down-conversion photons register at detectors D_1 and D_2 within a coincident-time window (1 nsec in the newest experiment) that is small compared with the travel-time difference (4 nsec) between the short and long alternative routes in each arm of the interferometer. Which route did each photon take? Because the down-conversion photon pairs are produced together and arrive together (within the coincidence window) at detectors D_1 and D_2, they must *both* have taken either the long way or the short way. But because we can't know at what time the down-conversion took place, we must use the quantum-state superposition

$$(1/\sqrt{2})(|s\rangle_1 |s\rangle_2 + |l\rangle_1 |l\rangle_2)$$

where, for example, $|s\rangle_1$ denotes the short route for particle 1. Because we don't know whether the photon pair was produced at the earlier or later time consistent with the coincidence observation, this device is in effect a "temporal double slit." It is easy to see how this kind of arrangement could be generalized from a two-time slit to a multitime grating.

By changing the length of one of the long paths and thus altering the relative phase of the two terms, one can produce sinusoidal oscillations ("fringes") in the coincident count rates. But the singles rate in each separate detector is constant.

Why is there no interference in the individual detectors, even though each particle passes through a self-contained interferometer? Well, one could completely remove the half-silvered mirrors from one of these interferometers and monitor the short route, thereby ascertaining whether the particle in the other one took the short or long route. That is, if the counts are nearly simultaneous, both must have taken the short (straight) route, but if the photon in the intact interferometer arrives 4 nsec late, it must have taken the long route. Thus one can use one particle to obtain path information about the other one, even though the latter passes through an interferometer, and hence no single-detector oscillations are possible. But when both interferometers are intact and both particles have been detected within a 1-nsec window, the opportunity to obtain path information is lost forever. Then the two-particle oscillations appear!

This explanation stresses an aspect of the interpretation of the wavefunction that is not often emphasized: The wavefunction contains all information about the system that is potentially available, not just the information actually in hand. It is the mere possibility of obtaining path information for the individual photon in a particular experimental configuration that guarantees that the amplitudes along those paths will not interfere. It is what the experimenter can do, not what he bothers to do, that is important. Changing the configuration to eliminate the path information can restore the interference fringes.

Path distinguishability and coherence

To show some of the possibilities inherent in two-particle superpositions, we will describe a truly mind-boggling experiment by Mandel and his Rochester colleagues Xin Yu Zou and Li Jun Wang.[16] Their experiment, schematically illustrated in figure 4a, marvelously vindicates Feynman's dictum that states interfere with each other only when they cannot physically be distinguished in a particular experimental setup.

Consider a single photon in beam a entering the Zou–Wang–Mandel apparatus. After the beam splitter A, this particle's wavefunction illuminates *both* of the down-conversion crystals X_1 and X_2. But because there is only one photon, it can down-convert at only one of the two crystals, creating the entangled state $(1/\sqrt{2})(|d\rangle_1\,|e\rangle_2 + |h\rangle_1\,|k\rangle_2)$. If one recombines the beams h and d at beam splitter C, will they interfere? Normally they will not, because one could catch the companion photon (in beam e or k) and thus know in which crystal the down-conversion occurred. But Zou and company, following a suggestion of their colleague Zhe Yu Ou, cleverly overlapped the beams e and k (the crystals being transparent) to erase that potential information. Thus the state $|e\rangle$ evolves into the state $|k\rangle$, and the entangled state now becomes $(1/\sqrt{2})(|d\rangle + |h\rangle)_1\,|k\rangle_2$, which is no longer entangled. Consequently one can get ordinary single-particle interference fringes at detector D_1 by varying the phase φ at the phase shifter P.

It is important to note several things here. First, as opposed to the two-particle fringes discussed earlier, in this experiment the second photon does not actually need to be detected. The fringes at the detector looking at photon 1 are there in any case because the detection of photon 2 in beam k cannot provide information about where photon 1 was created. Second, if beam e is blocked, the fringes at detector D_1 will *disappear*, because in that case the detection of a photon in beam k would tell you that it was created in crystal X_2. Zou and his collaborators demonstrated this by inserting a beam splitter at B and showing that the visibility of the fringes in D_1 varied with the transmission coefficient of the splitter.

The mind-boggling feature of this experiment is that the beams e and k are not even in the two alternative coherence paths that run from A to D_1! The only role they play in the experiment is that, by overlapping, they prevent us from determining in which crystal the down-conversion occurs.

One might be tempted to think that beam e contributes to the amplitude leading to beam h through some nonlinear coupling. That would make the effect seem less amazing. Be that as it may, note that if we placed an additional phase shifter γ into beam e, the state created in crystal X_1 would become $e^{i\gamma}|d\rangle_1\,|e\rangle_2$. The consequent phase shift in beam d would produce a modulation of the form $\cos(\varphi - \gamma)$ at detector D_1. So it is a mistake to think that beam e is contributing to beam h in any direct way, in which case one might have expected $\cos(\varphi + \gamma)$.

This experiment is a prime example of how one could certainly interpret the same result in terms of some nonlinear Hamiltonian mechanism that couples the two-photon state to the vacuum. But one is then in danger of missing the point that *any* such mechanism will produce the effect, if and only if the beams are experimentally indistinguishable.

The experiment also reinforces the point that it is potential information, not actual information, that destroys coherence. Furthermore we have to generalize Dirac's famous dictum that a photon can only interfere with itself. In this experiment the original photon in beam a is not even present to interfere at D_1. Its down-converted progeny are doing the interfering. We prefer to think of the down-converted pair as a single entity, a "two-photon." It is this two-photon, created at either X_1 or X_2, that is interfering with itself. More generally, it is the time-evolved continuation of the photon state that is interfering with itself. (See the box at left.)

Figure 4b is a photograph of an interesting variant of this experiment very recently carried out by Thomas Herzog and coworkers in Zeilinger's Innsbruck laboratory.[16] They use only one crystal, but the continuation of the original beam and its down-converted progeny are reflected back through the crystal so that one can't tell whether the down-converted pair was created during the the first pass of the incident photon or on its return. Therefore these two possibilities interfere and one can, in

A Mind-Boggling Experiment

We present here a more detailed analysis of the experiment of Zou and coworkers,[16] illustrated in figure 4a, to show that the phenomena involved are understandable by elementary quantum mechanics; you don't need the full machinery of quantum optics. (In our description, $|a\rangle_1$ simply denotes particle 1 in beam a. It does not represent the field of beam a. The particle could just as well be an electron.[24]) We want to encourage nonexperts to enter the game.

At A, the first beam splitter, we have $|a\rangle \to (|b\rangle + i|c\rangle)/\sqrt{2}$. At B, $|e\rangle \to (T|g\rangle + iR|f\rangle)$, where T and R are the (real) transmission and reflection amplitudes at the second beam splitter. (The reflected and transmitted parts are 90° out of phase to conserve probability.) At C, the third beam splitter, $|h\rangle \to (|l\rangle + i|m\rangle)/\sqrt{2}$ and $|d\rangle \to (|m\rangle + i|l\rangle)/\sqrt{2}$. At P, the phase shifter, $|h\rangle \to e^{i\varphi}\,|h\rangle$. At the down-conversion crystals, $|b\rangle \to \eta|d\rangle_1|e\rangle_2$ and $|c\rangle \to \eta|h\rangle_1|k\rangle_2$, where η, on the order of 10^{-6}, is the amplitude for down-conversion. Finally, by perfectly lining up the beams we get $|g\rangle \to |k\rangle$.

Combining all these terms gives

$$|a\rangle \to \frac{1}{\sqrt{2}}(|b\rangle + i|c\rangle) \to \frac{\eta}{\sqrt{2}}(|d\rangle_1\,|e\rangle_2 + i|h\rangle_1\,|k\rangle_2)$$

$$\to \frac{\eta}{2}[(T - e^{i\varphi})(|m\rangle + i(T + e^{i\varphi})|l\rangle]_1\,|k\rangle_2$$

$$+ i\frac{\eta}{2}R(|m\rangle + i|l\rangle)_1\,|f\rangle_2$$

By counting coincidences at detectors D_1 and D_2 one measures the square of the amplitude of the $|l\rangle_1\,|k\rangle_2$ term,

$$\frac{\eta}{2}(1 + T^2 + 2T\cos\varphi)$$

That gives a fringe contrast $\nu = 2T/(1 + T^2)$ for coincident counts at D_1 and D_2. This contrast increases with T, which is the amplitude ratio of the two contributions to beam $|k\rangle$. If instead of recording coincidences one only monitors beam $|l\rangle$ at D_1, there are contributions from the $|l\rangle_1\,|k\rangle_2$ and $|l\rangle_1\,|f\rangle_2$ terms, yielding

$$\frac{\eta^2}{2}(1 + T\cos\varphi)$$

so that the contrast is reduced to $\nu = T$. Thus the degree of coherence of beam 1 is controlled by beam 2, even though beam 2 isn't even in the coherence path. The only purpose the second beam serves is to make it impossible to tell in which crystal the down-conversion occurred.

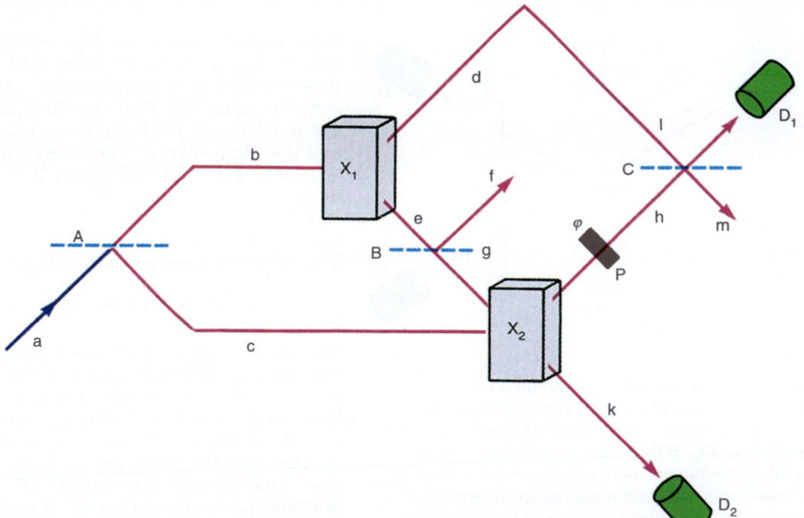

a

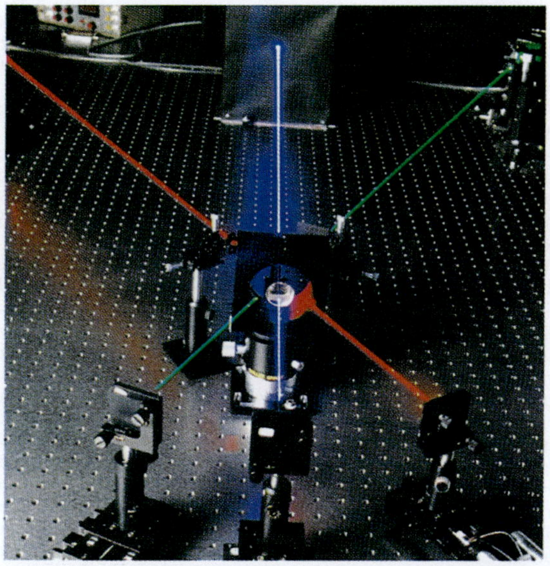

b

Manipulating one photon can alter the interference pattern of another. The arrangement[16] sketched in **a** can produce an interference pattern at detector D_1 when the phase shifter P is varied. An entering ultraviolet photon a is split at beam splitter A so that both down-conversion crystals (X_1 and X_2) are illuminated. One of the resulting pair of down-conversion photons can reach D_1 by way of beam path d or h. If one could monitor beams e and k separately, one would know in which crystal the down-conversion occurred, and there would be no interference. But merging beams e and k in this configuration lets the alternative paths of the other photon interfere. A new variant of this scheme, shown in **b**, uses only one crystal (at the center). The ambiguity here is created by reflecting the originating beam (blue, from the top of the photo) and its down-converted progeny (red and green) back through the crystal from mirrors (at the bottom of the photo), so one can't know on which pass the down-conversion occurred. **Figure 4**

effect, enhance or suppress the atomic emission process in the crystal by small movements of mirrors that are several feet away!

The quantum eraser

Figure 5 shows an experimental arrangement first used by Alley and Shih,[6] and recently exploited by Paul Kwiat and coworkers at Berkeley[17] to demonstrate Marlan Scully's notion of a "quantum eraser." Others have called it "haunted" or "phantom" measurement.[18] To appreciate the basic idea, first suppose that only the beam splitter is in the path of the two beams emerging from the down-converter crystal. This arrangement produces an interesting and very basic two-particle effect: Both particles must end up in the same detector.[19]

The reason is simple. To get coincident counts at the two detectors, either particle 1 (the photon that ultimately lands in detector 1) takes route a and particle 2 takes route b, or vice versa. In the former alternative, both photons are reflected by the beam splitter, each reflection contributing a 90° phase shift to the overall amplitude. The latter alternative, by contrast, involves *no* phase-shifting reflections. Thus the two alternative amplitudes are 180° out of phase with each other, and they cancel.

So if both beams are, for example, horizontally polarized, they will remain so, and there will be no coincident counts in detectors 1 and 2. But if we now insert a 90° polarization rotator into beam b, it will become vertically polarized and the two amplitudes will no longer interfere because one can tell which path a photon took to its detector by measuring its polarization. Thus there will now be no superposition of the two amplitudes, and therefore coincidence counts between the two detectors will be observed.

But it is still possible to "erase" this path information and recover the interference by placing a linear polarizer in each beam, as shown in the figure. If both polarizers are either horizontal or vertical there will be path information present. But if they are oriented at 45° to the horizontal, either route, a or b, can now lead to either detector. The coherence is restored and there are, once again, no coincident counts.

The idea here is that one can seem to destroy information about a system without actually doing so. The information remains subliminally present; it can be recaptured. Or, as Chiao has put it, "Nothing has really been erased here, only scrambled!" Helmut Rauch's Vienna group performed a conceptually identical experiment in 1982 with a neutron interferometer, but that involved only single-particle interference.[17]

Generalization to many states

Experiments with entangled particles have generally been confined to two-state systems, either spin-½ particles or photons. One way to generate superpositions

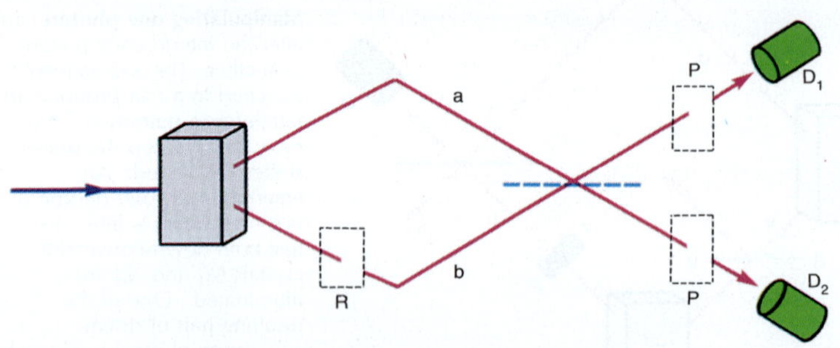

Polarizers can serve as quantum erasers.[17] A single ultraviolet photon entering a down-conversion crystal (gray) produces two optical photons that mix at a beam splitter (dashed blue line) before arriving at detectors D_1 and D_2. If there are no polarizers (*P*) or polarization rotator (*R*) in the beams, both photons must end up in the same detector. Inserting a 90° rotator into beam *b* provides information that renders the beams incoherent and thus produces coincidence counts in the two detectors. Inserting the polarizers oriented 45° to the horizontal after the beam splitter erases that information and recovers the coherence that prevents coincidence counts. **Figure 5**

involving more than two states is to use systems of higher spin. An easier alternative is to generalize the beam splitter to provide more than two paths for each photon. We call such systems "multiports."[20] The half-silvered mirror that serves as a beam splitter is a four-port; it has two input ports and two output ports. The output beams in such devices are mathematically related to the input beams by a unitary transformation.

A simple generalization of the beam splitter, shown in figure 6, is the six-port, with three input beams and three output beams. We call it a "tritter." (An eight-port would be called a "quitter.") If the beam splitters A, B and C in the figure are chosen to have reflectivities $1/\sqrt{2}$, $1/\sqrt{3}$, $1/\sqrt{2}$, respectively, and the phase shifters α, β and γ are chosen appropriately, this device will yield three equally intense output beams if any one of the three input ports is illuminated. That's a straightforward generalization of the symmetric beam splitter. But by varying all these parameters one can induce a range of unitary transformations. Multiports provide a practical method for investigating such transformations in an N-dimensional Hilbert space.

Three-particle interferometry

Figure 7 shows an idealized three-particle interferometer.[21] No such device has been been built, but a number of models have been proposed.[22] This particular configuration was inspired by David Mermin's Reference Frame column in PHYSICS TODAY, June 1990, page 9. Imagine a three-particle decay at the center. (Alternatively one could start with a three-particle down-conversion.) Assume for simplicity that the three decay particles all have the same mass and the same energy. Then they will, of course, come off 120° apart in the decay plane. A suitable placement of screens with holes restricts them to two possible states: $|abc\rangle$ or $|a'b'c'\rangle$. The coherent superposition will be

$$\psi = \frac{1}{\sqrt{2}}\left(|abc\rangle + |a'b'c'\rangle\right)$$

The beams a', b' and c' pass through the phase shifters with phase shifts α, β and γ, after which the appropriate beams are recombined at beam splitters A, B and C. Beyond the beam splitters are the three pairs of counters. One records only triplets of simultaneous counts at G or G', H or H' and K or K'. If one assigns a +1 (or −1) to a count in an unprimed (or primed) counter, then the probability that a triplet of counts will give the product +1 (for example, GHK or $GH'K'$) will be $(1+\sin\Delta)/2$, where $\Delta = \alpha+\beta+\gamma$. The probability for a product −1 (for example, $G'H'K'$ or $G'HK$) will be $(1-\sin\Delta)/2$. Then the expectation value over a large number of counts is simply $\sin\Delta$.

This result is remarkably similar to the case of the two-particle interferometer, but the implications are entirely different. In the three-particle case a perfect correlation occurs when $\Delta = \pi/2$ or $3\pi/2$. Then if one knows at which counter two of the particles have landed, one can predict with certainty the counter at which the third particle will land, without having disturbed it at all. Hence there exists what Einstein, Podolsky and Rosen called an "element of reality" associated with the path from the beam splitter to the counter. Elements of reality are those entities to which, from the Einstein–Podolsky–Rosen point of view, the concept of an objective reality most clearly applies. They are natural entities for discussions of local, realistic descriptions of quantum events.

In the two-particle interferometer, measurements involving only perfect correlations are uninteresting, in the sense that they cannot violate Bell's inequality. (This famous inequality is a statement of the limits of correlation allowable between separated events in any theory that preserves local reality.) But from the Einstein–Podolsky–Rosen viewpoint these perfect correlations are essential for the introduction of elements of reality.

In the three-particle interferometer, however, if one assumes the existence of these elements of reality, one already runs into a contradiction even if the correlations are perfect.[21] If the particles are perfectly correlated in the two-particle case, their spin directions are precisely opposite. But perfect correlation in the three-particle case yields a continuum of possibilities, a condition that is impossible to satisfy within the classical restrictions.

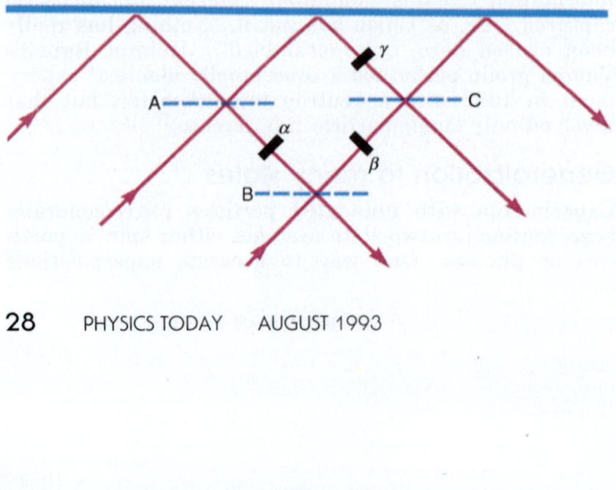

A six-port, or 'tritter,' is made with three beam splitters. If the reflectivities of the beam splitters *A*, *B* and *C* and the phase angles α, β and γ of the phase shifters (black) are properly chosen, this configuration yields three output beams of equal intensity when any one of the three input ports is illuminated. **Figure 6**

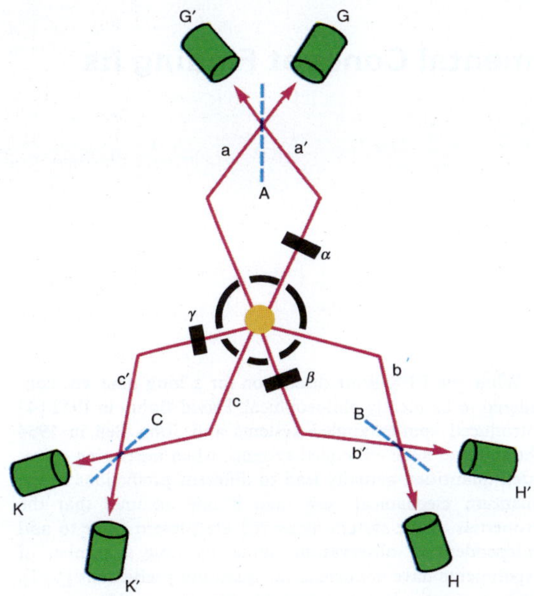

Three-particle interferometer. A particle at the center decays into three daughters of equal mass and momentum. Collimation restricts the decay state to beams *a*, *b* and *c* or *a*′, *b*′ and *c*′. The primed beams pass through phase shifters (black rectangles) with phase shifts α, β and γ, respectively, after which the alternative paths are mixed at beam splitters *A*, *B* and *C* before arriving at three pairs of detectors (green). **Figure 7**

Therefore one can disprove the existence of elements of reality by observing just one event. (Of course one can never attain perfect correlations, so the situation will be more complicated in any realistic experiment.) Clearly correlations among three particles are even richer than two-particle correlations.

Another interesting feature of the three-particle interferometer is that *only* three-particle correlations show up. When one looks at only one particle or at coincidence counts between two particles, one gets random results. One has to look at all three particles to see any correlations.

We have tried to point out here that the superposition principle, the source of much of the strangeness in one-particle quantum theory, has proved to contain even more mysteries when several particles are involved. In addition to the new experimental techniques we have discussed, other new directions are being explored. One of these goes by the name of quantum cryptography. That's rather a poor name, because this new field has very general implications for quantum theory. (Reference 23 gives a sampling of recent papers; see also PHYSICS TODAY, November 1992, page 21.) For example, the paper by Charles Bennett and collaborators points out that quantum cryptographic techniques can be used to "teleport" a quantum state from one observer to another. This futuristic scheme does not violate relativity, because the receiver cannot decipher the quantum state without additional information that comes through conventional channels.

The superposition principle is at the very heart of quantum theory. It seems funny, therefore, to say that this central idea is just beginning to be explored in depth. Our guess is that many more surprises await us in multiparticle interferometry.

References

1. R. P. Feynman, R. B. Leighton, M. Sands, *The Feynman Lectures on Physics,* Addison–Wesley, Reading, Mass. (1963).
2. A. Einstein, B. Podolsky, N. Rosen, Phys. Rev. **47**, 777 (1935); reprinted in J. A. Wheeler, W. H. Zurek, *Quantum Measurement Theory*, Princeton U. P., Princeton (1983).
3. D. Bohm, *Quantum Theory*, Prentice–Hall, New York (1951).
4. D. C. Burnham, D. L. Weinberg, Phys. Rev. Lett. **25**, 84 (1970).
5. M. A. Horne, A. Zeilinger, in *Proc. Symp. on Foundations of Modern Physics*, P. Lahti, P. Mittelstaedt, eds., World Scientific, Singapore (1985), p. 435. M. A. Horne, A. Shimony, A. Zeilinger, Phys. Rev. Lett. **62**, 2209 (1989).
6. C. O. Alley, Y. H. Shih, *Proc. 2nd Int. Symp. on Foundations of Quantum Mechanics in the Light of New Technology*, M. Namiki *et al.*, eds., Phys. Soc. Jpn., Tokyo (1986), p. 47. Y. H. Shih, C. O. Alley, Phys. Rev. Lett. **61**, 2921 (1988).
7. R. Ghosh, L. Mandel, Phys. Rev. Lett. **59**, 1903 (1987).
8. P. Storey, M. Collett, D. Walls, Phys. Rev. Lett. **68**, 472 (1992).
9. J. S. Bell, Physics **1**, 195 (1964). All of Bell's basic writings on the subject are reprinted in J. S. Bell, *Speakable and Unspeakable in Quantum Mechanics*, Cambridge U. P., New York (1987).
10. J. Clauser, S. Freedman, Phys. Rev. Lett. **28**, 938 (1972). A. Aspect, J. Dalibard, G. Roger, Phys. Rev. Lett. **47**, 1804 (1982). For an analysis, see F. Selleri, *Quantum Paradoxes and Physical Reality*, Kluwer Academic, Dordrecht, The Netherlands (1990).
11. J. G. Rarity, P. R. Tapster, Phys. Rev. Lett. **64**, 2495 (1990).
12. L. Mandel, Phys. Rev. A **28**, 929 (1983).
13. J. D. Franson, Phys. Rev. Lett. **62**, 2205 (1989).
14. Z. Y. Ou, L. J. Wang, L. Mandel, Phys. Rev. Lett. **65**, 321 (1990). P. G. Kwiat, W. A. Vareka, C. K. Hong, H. Nathel, R. Y. Chiao, Phys. Rev. A **41**, 2910 (1990).
15. P. G. Kwiat, A. M. Steinberg, R. Y. Chiao, Phys. Rev. A **47**, 2472 (1993).
16. X. Y. Zou, L. J. Wang, L. Mandel, Phys. Rev. Lett. **67**, 318 (1991). L. J. Wang, X. Y. Zou, L. Mandel, Phys. Rev. A **44**, 4614 (1991). T. Herzog, J. Rarity, H. Wienfurter, A. Zeilinger, in Proc. Workshop on Quantum Optics, Ulm, April 1993 (to be published).
17. P. G. Kwiat, A. M. Steinberg, R. Y. Chiao, Phys. Rev. A **45**, 7729 (1992). J. Summhammer, G. Badurek, H. Rauch, U. Kischico, Phys. Lett. A **90**, 110 (1982).
18. M. O. Scully, R. Shea, J. D. McCullen, Phys. Rep. **43**, 485 (1978). M. O. Scully, B.-G. Englert, H. Walther, Nature **351**, 111 (1991). D. M. Greenberger, A. YaSin, Found. Phys. **19**, 679 (1989).
19. C. K. Hong, Z. Y. Ou, L. Mandel, Phys. Rev. Lett. **59**, 2044 (1987).
20. A. Zeilinger, H. J. Bernstein, D. M. Greenberger, M. A. Horne, M. Zukowski, in *Proc. 4th Int. Symp. on Foundations of Quantum Mechanics in the Light of New Technology*, H. Ezawa *et al.*, eds., World Scientific, Singapore (1993).
21. D. M. Greenberger, M. A. Horne, A. Zeilinger, in *Bell's Theorem, Quantum Theory, and Conceptions of the Universe*, M. Katafos, ed., Kluwer Academic, Dordrecht, The Netherlands (1989), p. 173. D. M. Greenberger, M. A. Horne, A. Shimony, A. Zeilinger, Am. J. Phys. **58**, 1131 (1990). N. D. Mermin, Am. J. Phys. **58**, 731 (1990).
22. D. M. Greenberger, H. J. Bernstein, M. A. Horne, A. Zeilinger, in *Proc. 4th Int. Symp. on Foundations of Quantum Mechanics in the Light of New Technology*, H. Ezawa *et al.*, eds., World Scientific, Singapore (1993). B. Yurke, D. Stoler, Phys. Rev. Lett. **68**, 1251 (1992). L. Wang, K. Wodkiewicz, J. H. Eberly, in *Tech. Digest Annu. Mtg. of the OSA*, Albuquerque, September 1992, abstr. MUU4.
23. A. K. Ekert, J. G. Rarity, P. R. Tapster, G. M. Palme, Phys. Rev. Lett. **69**, 1293 (1992). C. H. Bennett, G. Brassard, N. D. Mermin, Phys. Rev. Lett. **68**, 557 (1992). C. H. Bennett, G. Brassard, C. Crepeau, R. Jozsa, A. Peres, W. Wooters, Phys. Rev. Lett. **70**, 1895 (1993).
24. R. J. Glauber, Phys. Rev. **130**, 2529 (1963); **131**, 2766 (1963). E. C. G. Sudarshan, T. Rothman, Am. J. Phys. **59**, 592 (1991). D. Walls, Am. J. Phys. **45**, 952 (1977). ■

Physica Scripta. Vol. T76, 203–209, 1998

Quantum Entanglement: A Fundamental Concept Finding its Applications

Anton Zeilinger*

Institut für Experimentalphysik, Universität Innsbruck, A-6020 Innsbruck, Austria

Received November 4, 1997; accepted December 17, 1997

PACS Ref: 03.67.−a, 42.50.−p

Abstract

Entanglement, according to the Austrian physicist Erwin Schrödinger *the Essence* of Quantum Mechanics, has been known for a long time now to be the source of a number of paradoxical and counterintuitive phenomena. Of those the most remarkable one is usually called non-locality and it is at the heart of the Einstein-Podolsky-Rosen Paradox and of the fact that Quantum Mechanics violates Bell's inequalities. Recent years saw an emergence of novel ideas in entanglement of three or more particles. Most recently it turned out that entanglement is an important concept in the development of quantum communication, quantum cryptography and quantum computation. First explicit experimental realizations with two or more photons include quantum dense coding and quantum teleportation.

1. Introduction

Immediately after the discovery of modern quantum mechanics it was realized that it contains novel counterintuitive features, as witnessed most remarkably in the famous dialogue between Niels Bohr and Albert Einstein [1]. While Einstein initially tried to argue that quantum mechanics is inconsistent, he later turned his argument towards aiming at demonstrating that quantum mechanics is incomplete. In his seminal paper [2] where Einstein – together with Podolsky and Rosen – considers quantum systems consisting of two particles such that, while neither position nor momentum of either particle is well defined, the sum of their positions, that is their center of mass, and the difference of their momenta, that is their individual momenta in the center of mass system, are both precisely defined. It then follows that measurement of either position or momentum performed on, say, particle 1 immediately implies for particle 2 a precise position or momentum respectively without interacting with that particle. Assuming that the two particles can be separated by arbitrary distances, EPR suggest that measurement on particle 1 cannot have any actual influence on particle 2 (locality condition); thus the property of particle 2 must be independent of the measurement performed on particle 1. To them, it then follows that both position and momentum can simultaneously be well defined properties of a quantum system.

In his famous reply [3] Niels Bohr argues that the two particles in the EPR case are always parts of one quantum system and thus measurement on one particle changes the possible predictions that can be made for the whole system and therefore for the other particle. This is usually interpreted as implying a non-local feature of quantum mechanics, though that is by no means the only possible interpretation of the situation.

* electronic mail address: Anton.Zeilinger@uibk.ac.at

While the EPR-Bohr discussion for a long time was considered to be merely philosophical, David Bohm in 1952 [4] introduced spin-entangled systems and John Bell in 1964 showed that such entangled systems, when measuring correlated quantities, actually lead to different predictions in the quantum mechanical case than if one assumes that the properties of the system measured are present prior to and independent of observation. While by now a number of experiments have confirmed the quantum predictions [5–7], from a strictly logical point of view the problem is not closed yet, because some loopholes in the existing experiments still make it at least logically possible to uphold a local realist world view [8].

In recent years entanglement has become a new focus of activity in quantum physics, because of (1) immense technological and experimental progress (2) the discovery of novel non-classical features in multi-particle situations [9] and (3) development of quantum information physics which heavily draws on multi-particle entangled systems.

2. Quantum coding

Let us consider a digital encoding scheme which is equivalent to encoding *yes/no* answers to individual questions or the truth value of elementary propositions. Usually these are expressed by the bit value "1" for *yes* and "0" for *no*. A good coding scheme should then be such that "0" or "1" can be distinguished clearly. Physical realizations of such coding schemes will therefore use systems which have two well-defined states. In classical coding these could be, for example, the two positions of a switch as in a simple relay, they could be two different voltage levels in an electronic circuit, two different colours of beads, a knot in my handkerchief, etc.

Considering quantum coding schemes it is then natural to associate two different orthogonal states to "0" and "1". Given therefore a two-state system we have to choose a basis in which we wish to encode our information. Then any two orthogonal base states can be used to implement "0" and "1". Let us therefore designate the chosen orthogonal basis states as $|0\rangle$ and $|1\rangle$ respectively. Evidently a novel feature of quantum information is that it can be in a superposition of these two states. A general superposition is called a qubit.

Considering then the encoding of a large number of bits of information it is evident that this could be done by simply assigning each bit its own two-state system. In that case the information will be encoded into orthogonal product states, e.g. for three bits

$$|0\rangle|0\rangle|0\rangle$$
$$|0\rangle|0\rangle|1\rangle$$
$$|0\rangle|1\rangle|1\rangle$$
$$|1\rangle|0\rangle|0\rangle$$
$$|1\rangle|1\rangle|0\rangle$$
$$|1\rangle|1\rangle|1\rangle. \quad (1)$$

The corresponding, in general 2^n, states form a complete orthonormal basis for the n-qubit space. Yet, alternatively, in such a space we could also choose very different bases which even could be entangled. A maximally entangled basis for two independent particles, two qubits, is

$$|\Psi^+\rangle = \frac{1}{\sqrt{2}}(|0\rangle|1\rangle + |1\rangle|0\rangle),$$

$$|\Psi^-\rangle = \frac{1}{\sqrt{2}}(|0\rangle|1\rangle - |1\rangle|0\rangle),$$

$$|\Phi^+\rangle = \frac{1}{\sqrt{2}}(|0\rangle|0\rangle + |1\rangle|1\rangle),$$

$$|\Phi^-\rangle = \frac{1}{\sqrt{2}}(|0\rangle|0\rangle - |1\rangle|1\rangle). \quad (2)$$

This is the so-called Bell basis. It is important to notice that here we can still encode two bits of information, that is we have four different possibilities, but now this encoding is done in such a way that none of the bits carries any well-defined information on its own. All information is encoded into relational properties of the two qubits. It thus follows immediately that in order to read out the information one has to have access to both qubits. The corresponding measurement is called a Bell-state measurement. This is to be compared with the classical case where access to one qubit is simply enough to determine the answer to one *yes/no* question. In contrast, in the case of the maximally entangled basis access to an individual qubit does not provide any information.

3. Quantum communication and dense coding

Whenever, say, two parties A (Alice) and B (Bob) wish to communicate with each other they have to agree first on a coding procedure, that is they have to agree which symbol means what. In classical coding the situation is very simple. Restricting ourselves to binary information, that is to bits, we need some information carrier which has two states. A famous historic example from the American revolution was when Paul Revere informed the Revolutionaries about the path taken by the Royals by displaying one or two lamps in the steeple of Old North Church in Boston. In quantum physics again we can have information encoding in a novel way using entangled states and thus encode information into joint properties of elementary systems. Then the elementary systems themselves do not carry any information. A first elementary case where this is clearly demonstrated is quantum dense coding.

The maximally entangled Bell basis of eq. (2) has a very important and interesting property which was exploited by Bennett and Wiesner [10] in their proposal for quantum

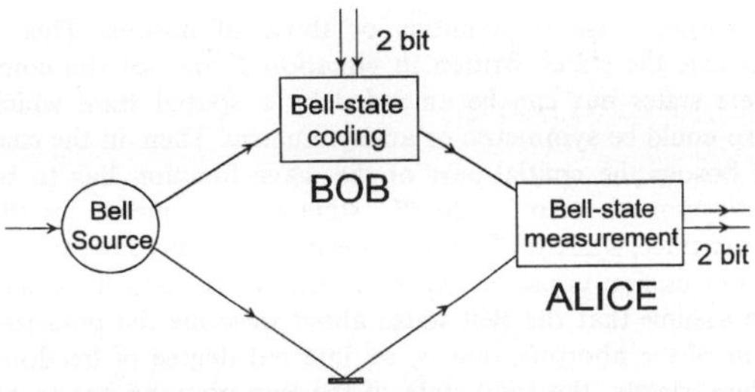

Fig. 1. In quantum dense coding Bob, having access to one of the two entangled qubits only, can actually encode two bits of information, because he can switch around in the full Bell basis. Alice can identify the information encoded by performing a complete Bell-state measurement on the entangled pair.

dense coding. This is the property that in order to switch from any one of the four Bell states to all other four it is sufficient to manipulate only one of the two qubits while in the classical case one has to manipulate both. Thus, the sender Bob (Fig. 1) can actually encode two bits of information into the whole entangled system by just manipulating one of the two qubits. Let us, for example, assume that we start from the state $|\Psi^+\rangle$ then we can obtain $|\Psi^-\rangle$ by just introducing a phase shift of π onto, say, the second qubit, $|\Phi^+\rangle$ is obtained by flipping the second qubit and the last state $|\Phi^-\rangle$ is obtained by a combination of both.

In order to read out this information the receiver, Alice, needs to be able to identify the four Bell states, that is she needs a Bell-state analyzer. It can be shown that, in order to identify all four Bell states, one needs some non-linear interaction between the two qubits. In our experimental realization [11] of the dense coding scheme it was possible by just using linear elements like beam splitters and polarizers to identify three of the four possibilities and thus to encode and identify $\log_2 3 = 1{,}58$ bits of information per photon manipulated.

4. Photon statistics at a beam splitter and Bell-state analysis

Formally speaking Bell-state analysis is not a problem. All you have to do is project any incoming state onto the Bell basis and you will find out by repeating the experiment with which probability the original state can be found in one of the Bell states. The experimental problem is that thus far no complete Bell-state analyser exists in the laboratory for any kind of quantum systems. This is due to the fact that complete Bell-state analysis requires non-linear interaction between the two qubits [12] which has not been realized so far.

Interestingly, partial Bell-state analysis is possible exploiting the statistics of two qubits at a beam splitter. The basic principle of that Bell-state analyser rests on the observation that of the states in equation 2 one state is antisymmetric. This is the $|\Psi^-\rangle$ state which clearly changes sign upon exchange between the two particles. The other three states are symmetric. We thus observe that the qubit obeys fermionic symmetry in the case of $|\Psi^-\rangle$ and bosonic symmetry in case of the other three states. Thus far we have not identified whether we use bosons or fermions in our experimental scheme. In fact, the four Bell states could very well be either those of fermions or those of bosons. This is because the states written in equation 2 are not the complete states but can be amended by a spatial state which also could be symmetric or antisymmetric. Then, in the case of bosons the spatial part of the wave function has to be antisymmetric also for the Ψ^- state and symmetric for the other three, while for fermions this has to be just reverse.

Let us first consider two photons, clearly bosons, where we assume that the Bell states above describe the polarization of the photons, that is, an internal degree of freedom. Then, clearly, the total state of the two photons has to be symmetric. For the case of the two particles incident symmetrically onto a beam splitter (Fig. 2), one from each mode $|a\rangle$ and $|b\rangle$, the possible external states are

$$|\Psi_A\rangle = \frac{1}{\sqrt{2}}(|a\rangle|b\rangle - |b\rangle|a\rangle),$$

$$|\Psi_S\rangle = \frac{1}{\sqrt{2}}(|a\rangle|b\rangle + |b\rangle|a\rangle), \quad (3)$$

where $|\Psi_A\rangle$ and $|\Psi_S\rangle$ are anti-symmetric or symmetric respectively. Because of the requirement of symmetry the total two-photon states are

$$|\Psi^+\rangle|\Psi_S\rangle,$$

$$|\Psi^-\rangle|\Psi_A\rangle,$$

$$|\Phi^+\rangle|\Psi_S\rangle,$$

$$|\Phi^-\rangle|\Psi_S\rangle. \quad (4)$$

We note that only the state anti-symmetric in external variables is also anti-symmetric in internal variables. It is this state which also emerges from the beam splitter in an external anti-symmetric state. This can easily be found by assuming firstly that the beam splitter does not influence the internal state and secondly that at the beam splitter we have a phase shift of $\pi/2$ upon reflection. Actually it can easily be seen [13] that the spatially anti-symmetric state is an eigenstate to any beam splitter operator. In contrast, in all three cases of the symmetric external state $|\Psi_S\rangle$ the two photons emerge together in one of the two outputs of the beam splitter. This behaviour of symmetric states can readily be demonstrated even using photons in an unentangled internal state [14]. Figure 3 and 4 show the consequent experimental behaviour [15] for the entangled internal state $|\Psi^+\rangle$ and that for the internally antisymmetric state $|\Psi^-\rangle$.

It is therefore evident that the state $|\Psi^-\rangle$ can clearly be discriminated from all the other states. It is the only one of the four Bell states which leads to coincidences between

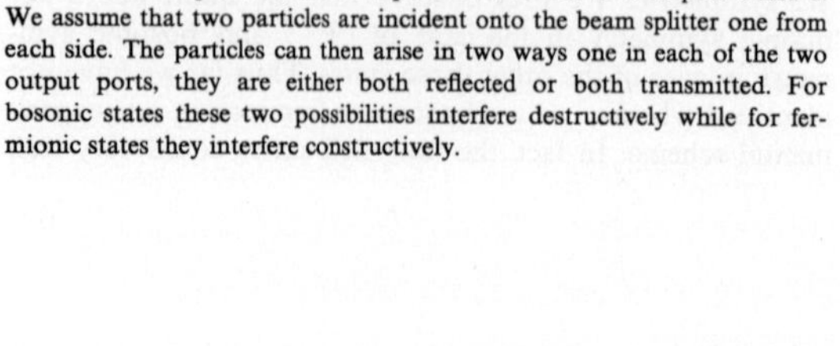

Fig. 2. A beam splitter with two inputs a and b and two outputs c and d. We assume that two particles are incident onto the beam splitter one from each side. The particles can then arise in two ways one in each of the two output ports, they are either both reflected or both transmitted. For bosonic states these two possibilities interfere destructively while for fermionic states they interfere constructively.

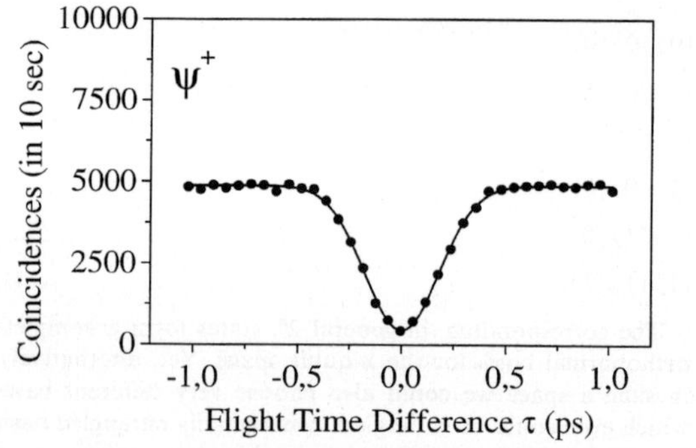

Fig. 3. Coincidence counts behind a beam splitter as a function of the difference of arrival time of two photons in an internally symmetric entangled state. For large time differences the two photons are distinguishable and therefore they behave independently and a certain coincidence rate equal to the classical rate is observed. For zero time difference there are no coincidences because both photons always leave the beam splitter together in the same output port.

detectors placed on each side after a beam splitter. How can we then identify the other three states? It turns out that distinction between $|\Psi^+\rangle$ on the one hand and $|\Phi^+\rangle$ and $|\Phi^-\rangle$ on the other hand can be based on the fact that only in $|\Psi^+\rangle$ the two photons have different polarization while in the other two they have the same polarization. Thus performing polarization measurements and observing the photons on the same side of the beam splitter one can determine whether they are in the state $|\Psi^+\rangle$ or in one of the states $|\Phi^+\rangle$ and $|\Phi^-\rangle$. It should be remarked that a generalization of this procedure implies that any three orthogonal maximally entangled states can be distinguished from each other in the same way, because by local unitary transformations one can perform rotations in the two-dimensional Hilbert-space.

Consider now the same experiment with fermions where again the Bell states describe the internal states, for example if the two qubits are entangled in spin. We then find that

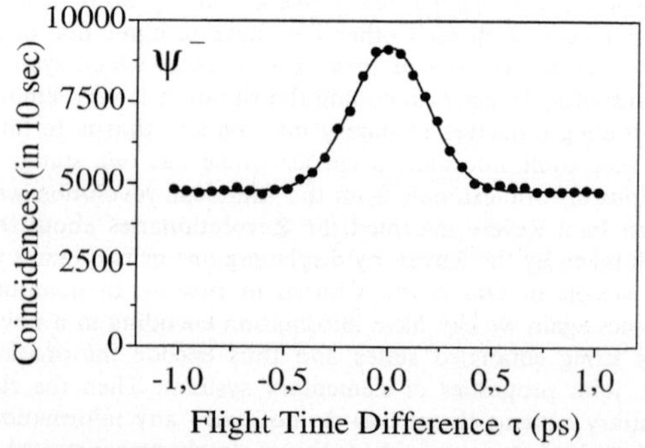

Fig. 4. Coincidence rate for two photons in the internally antisymmetric entangled state. Here again for large time differences we obtain the classical behaviour. When the time difference approaches zero we get twice the coincidence rate implying that the two photons never leave the beam splitter together in the same output port.

the four possible states now are

$$|\Psi^+\rangle|\Psi_A\rangle,$$
$$|\Psi^-\rangle|\Psi_S\rangle,$$
$$|\Phi^+\rangle|\Psi_A\rangle,$$
$$|\Phi^-\rangle|\Psi_A\rangle, \quad (5)$$

because of the antisymmetry requirement of the total state.

For fermions therefore only one of the states is spatially symmetric, the other three are spatially anti-symmetric. It will thus be in only one of the cases, namely for $|\Psi^-\rangle$ that both fermions will emerge together from the beam splitter. In all other three cases they will emerge from different sides [16]. Yet, remarkably, it is again this state which can immediately be distinguished from the other three because of its distinct symmetry properties.

5. Entanglement swapping and the nature of quantum information

Entanglement used to be considered as a consequence of the fact that the entangled particles interacted in their past or that they came from a common source. That this is too restricted a view is witnessed by the concept of entanglement swapping [17]. In the simplest case of entanglement swapping we take two entangled pairs (Fig. 5) and subject two particles, one from each source, to a Bell-state measurement. Then the other two particles which have never interacted in the past and also did not come from a common source are projected into an entangled state.

This experiment, whose realization is presently proceeding in Innsbruck [31], has some deep implications for the meaning of information in quantum physics. As mentioned above, entanglement between two quantum systems implies that at least some information is stored only in joint or relational properties of the two systems. In the case of maximal entanglement there is actually no information stored in any individual particle, all information is carried by all the particles jointly. This fact in itself poses already a challenge for a naive realist position. How can it be that particles have well-defined joint properties – for example when we know that their polarizations in a certain basis are equal – without carrying any properties by themselves? For a Bell state each of the systems is completely unpolarized on its own? In the case of entanglement swapping this interpretive problem is even more striking.

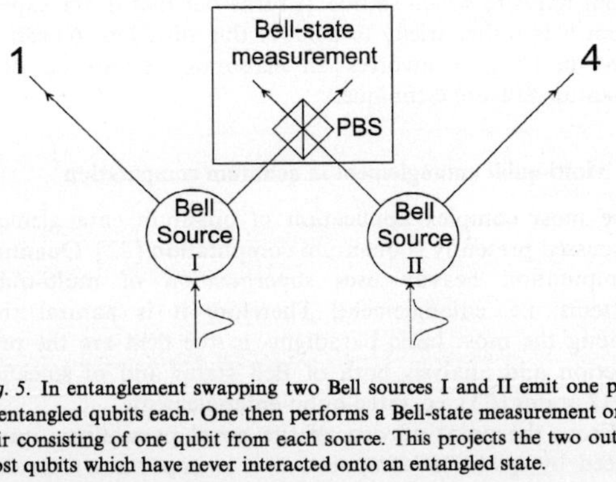

Fig. 5. In entanglement swapping two Bell sources I and II emit one pair of entangled qubits each. One then performs a Bell-state measurement on a pair consisting of one qubit from each source. This projects the two outermost qubits which have never interacted onto an entangled state.

Let us, for simplicity of discussion, just consider the case where we have sources that produce our two qubits in the anti-symmetric state $|\Psi^-\rangle$. This state has the unique feature that it is anti-symmetric in any basis. Thus, in terms of its information content, the statement is that we know the two qubits are different whatever basis we choose[1]. We thus know simply by the choice of preparation that in each of the two entangled pairs to be used in entanglement swapping the two qubits are completely different.

We now have to discuss the information content of Bell-state analysis. At first we note that, whichever states we would produce at the sources, a fair Bell-state analyser will return any of the four possible answers with equal probability of 25%. That is, the action of the Bell-state analyser is such that it projects the two photons onto an entangled state and, since in our case the two qubits are themselves members of maximally entangled pairs and therefore carry no information, this has to happen with equal probability for all four Bell states measured. In fact, the Bell-state measurement does not reveal any information about any of the qubits emitted by one of the two sources nor any joint information about each source. Yet, what we gain is joint (or relative) information about the two sources.

Suppose, specifically, that in a certain experimental run we obtain the result $|\Psi^-\rangle$ for the Bell-state measurement. We then know that qubits 2 and 3 have been projected by the measurement onto a state which is characterized by the fact that these two qubits are different in whatever basis. Interestingly, and again most remarkably, this statement is even true as none of the two qubits themselves are yet well-defined in any basis and have no properties by themselves.

Now we are in a position to complete our chain of reasoning. By the properties of source 1 we know that qubit 1 and qubit 2 are different. By the result of the specific Bell-state measurement we know that qubit 2 and 3 are different and, finally, from the property of source 2 we know that qubit 3 and 4 are different. Therefore, since our qubits are defined in a Hilbert-space of dimension 2 only, we conclude that qubits 1 and 4 also have to be different in any basis. Therefore they emerge in the anti-symmetric state $|\Psi^-\rangle$. Analogous reasoning can be built up for the other three possible Bell-state measurement results.

This analysis supports the interpretation that quantum mechanics is just a formulation of what can be said about quantum systems very much in the spirit of Bohr [18]. The question as to what "really" the properties of these quantum systems are is empty and devoid of any meaning.

We could go even one step further realising that in quantum physics information has a novel quality which goes beyond its nature in classical physics. In classical physics it is natural to assume that the objects we observe

[1] We remark that for the three symmetric states we can make analogous statements but the situation is slightly more complicated. It turns out in the end that, since the Hilbert-space of the four Bell-states is four-dimensional, we can encode two independent bits of information into these four states. Therefore the four states can be characterised by *yes/no* answers to two distinct questions. These two questions are questions about the identity of the two qubits in two different bases conjugate to each other.

exist prior to and independent of any measurement and the information we gain by measurement is an entity of secondary quality while the objects are primary. In contrast it is suggestive that in quantum physics the situation is that information can be a primary quality and the objects with their properties have a secondary quality, they can be constructed in retrospect after the information has been gained in a measurement. This is consistent with Bohr's Caveat that we should be very cautious about making mental pictures of quantum systems. Such pictures are only allowed in the context of a specific experimental setup. In fact, most of the misunderstandings and misinterpretations of quantum mechanics result from assuming that quantum objects have properties prior to and independent of the specific apparatus chosen to observe them.

6. Quantum teleportation and the non-locality of information

A most remarkable application of the concept of entanglement is quantum teleportation. Consider first the problem. Suppose that Alice has an object which Bob, who could be anywhere, might need at a certain time. In classical physics what she can do is perform many precise measurements on the object and send the information to Bob who then can reconstitute the object. Within classical physics the accuracy of this is only limited by the precision with which Alice can measure and by the technical abilities of Bob. Yet, we know that in the end any measurement will run into the limitations imposed by quantum mechanics. It is evident that no measurement whatsoever performed by Alice can reveal the full quantum state of the object. We therefore ask which strategy can Alice pursue in order for Bob to obtain the object in its full quantum state when he needs it.

A strategy proposed by Bennett *et al.* [19] uses exactly the information-theoretic features of entanglement mentioned above. According to that protocol, Alice and Bob have to share in advance an entangled pair of quantum states (Fig. 6). Let us consider for simplicity that the object is simply a two-state system, a qubit. Then Alice and Bob share from the beginning an ancillary entangled pair which for convenience we again consider to be in the state $|\Psi^-\rangle$. Subsequently Alice performs a Bell-state measurement on her qubit and one of the two ancillaries. As discussed above, for

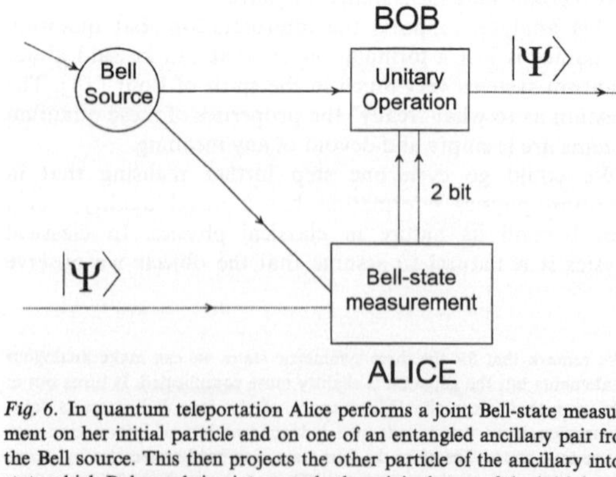

Fig. 6. In quantum teleportation Alice performs a joint Bell-state measurement on her initial particle and on one of an entangled ancillary pair from the Bell source. This then projects the other particle of the ancillary into a state which Bob can bring into exactly the original state of the initial particle using one of the four unitary transformations according to Alice's Bell-state measurement result.

the case of entanglement swapping, Alice will obtain each one of the four possible answers with equal probability, that is her original qubit and her qubit from the ancillary pair will be projected onto any one of the four Bell-states each with probability 25%. We note again that this measurement does not reveal any information, neither about the properties of the original qubit nor about the properties of the ancillary pair. So Alice obtains one of four possible results, "Ψ^+", "Ψ^-", "Φ^+", or "Φ^-". She then broadcasts this information, that is two classical bits, such that Bob can receive them. By now Bob is in posssession of a specific state as a consequence of Alice's Bell-state measurement. Performing one of four unitary transformations depending on Alice's specific result Bob can transform his particle into the original qubit. We also note that the original qubit disappears during the Bell-state measurement, it looses its identity, and thus Bob's qubit is not a copy but really a teleported reappearance of the original.

A remarkable property of the scheme is that the four unitary operations Bob has to perform are completely unrelated to the state Alice wants to teleport. This is because Alice's Bell-state measurement does not convey any information about the original state.

The experiment can most easily be understood again on the basis of the fact that the meaning of entanglement is that it represents relational statements between quantum systems. So, in the case of teleportation we have to use entanglement twice, firstly when the ancillary pair is produced and secondly in the Bell-state measurement. It thus turns out that the entanglement of the ancillary pair implies a well-defined relation between these two qubits without defining their individual properties, and the Bell-state measurement implies a well-defined relation between the qubit to be teleported and one of the ancillaries. Therefore, the second ancillary, by the same chain of reasoning as above in our discussion of entanglement swapping, is in a well-defined relationship to the original qubits to be teleported. It is again this remarkable feature of information in quantum mechanics that we have a closed chain of logical reasoning about relational properties without any statement on the properties of the individual systems.

The major problem in the experimental verification of quantum teleportation [20] is the Bell-state measurement of two independently created quantum systems. This means that these two qubits have to be measured such that their identity is lost, that it is not possible to infer which detection event refers to which source. It turns out that in the experiment it is rather tricky to achieve this situation. As can be seen in [21], it involves an elaborate application of a quantum erasure technique.

7. Multi-qubit entanglement in quantum computation

The most complex application of quantum entanglement discussed presently is quantum computation [22]. Quantum computation heavily uses superposition of multi-qubit systems, i.e. entanglement. Therefore it is natural that among the most basic paradigms in the field are the production and analysis both of Bell states and of so-called GHZ states [23], i.e. three-qubit entanglements.

From the point of view of the novel possibilities introduced by quantum computation it is important to realize that superposition of multi-qubit states makes massive quantum parallelism possible [24]. This is that fact that a quantum computer can produce the result of various different inputs at the same time if the input state is a superposition of the individual informations. Consider, for example, that using standard binary encoding the most simple GHZ state can be interpreted as a superposition of two numbers

$$\frac{1}{\sqrt{2}}(|000\rangle + |111\rangle = \frac{1}{\sqrt{2}}(|\text{“0”}\rangle + |\text{“7”}\rangle) \quad (6)$$

that is, it represents a superposition of the numbers “0” and “7”. Taking this as the input to a linear quantum computer it will produce in the output a superposition of the results corresponding to the two inputs.

It is well known that a most important practical problem a future quantum computer faces is decoherence due to coupling to the environment. Somewhat analogous to classical computers decoherence can be viewed as a quantum analogue of classical noise and it can be overcome using redundant information. Therefore, quantum error correction codes [25] utilize the possibility of encoding n qubit into the states of a Hilbert-space of higher dimension N. One then exploits the additional degrees of freedom to eliminate the quantum noise. This, naturally, implies entangled qubits in a Hilbert-space of high dimension.

While the field of quantum computation is signified presently by development of a vast number of different codes and theoretical approaches, the status of experiments is to verify the individual elementary steps including two-qubit quantum gates. It is presently completely open whether future development will utilize multi-photon states as discussed in this paper, quantum encoding into individual atoms [26] or individual ions [27] in trap configurations or bulk nuclear magnetic resonance techniques [28]. Yet, the fundamental individual steps in all these techniques are all the same from a quantum information science point of view.

8. Concluding comments

There are several further applications of entanglement in information science. One of them is the use of EPR pairs in quantum cryptography as proposed by Ekert [29]. There one actually uses a test of Bell's inequality to find out whether or not an eavesdropper has gained any access to the qubits sent.

Another concept utilizing entanglement is quantum computation in distributed networks [30]. This concept presently is in the stage of ideas and has not yet found experimental verification. It would imply extensive use of quantum teleportation, of Bell-state measurements, of entanglement swapping, and of multi-particle entanglements.

Taken together, all the examples given above demonstrate already now that quantum entanglement is one of the most important novel concepts in information science. It is furthermore not unreasonable to even expect future applications of entanglement in information technology itself. This is supported by the observation that the present development of information technology towards further and further miniturization will reach the single quantum limit sometime in the first decades of the next century. Once that happens information technology will by necessity be quantum. The radically new features of quantum entanglement let us expect that this will not just be a limitation but that it will open up completely novel possibilities for communication and computation.

The intricate interpretive questions raised by the multiple use of entanglement in the examples given here and as discussed above underline once more the deep conceptual problems of quantum reality. It is hard to believe that there is a possibility to escape Niels Bohr's verdict “There is no quantum world, there is only an abstract quantum description”.

Acknowledgements

The author thanks the Nobel Committee for its hospitality. This work was supported by the Austrian Science Foundation (FWF), project S65/04 and by the US National Science Foundation (NSF), grant number PHY97-22614. I also wish to acknowledge the collaboration with all members of my group at the University of Innsbruck.

References

1. Bohr, N., “Discussions with Einstein on Epistemological Problems in Atomic Physics” in Albert Einstein: Philosopher-Scientist“ (Edited by P. A. Schilpp) (The Library of Living Philosophers, Evanston 1949), p. 200.
2. Einstein, A., Podolsky, B., and Rosen, N., Phys. Rev. **47**, 777 (1935).
3. Bohr N., Phys. Rev. **48**, 696 (1935).
4. Bohm, D., ”Quantum Theory“ (Prentice-Hall, Englewood Cliffs, NJ, 1951) p. 614.
5. Freedman, S. J. and Clauser, J. S., Phys. Rev. Lett. **28**, 938-941 (1972).
6. Aspect, A., Grangier, P. and Roger, G., Phys. Rev. Lett. **47**, 460 (1981); **49**, 91 (1982); Aspect, A., Dalibard, J. and Roger, G., ibid. **49**, 1804 (1982).
7. Kwiat, P. G., *et al.* Phys. Rev. Lett. **75**, 4337 (1995).
8. Pearle, P. M., Phys. Rev. **D2**, 1418 (1970); Clauser, J. F. and Shimony, A., Rep. Prog. Phys. **41**, 1881 (1978).
9. Greenberger, D., Horne, M. A. and Zeilinger, A., ”Bell's Theorem, Quantum Theory, and Conceptions of the Universe“, (Edited by M. Kafatos) (Kluwer, Dordrecht, 1989) p. 69; Greenberger, D. M., Horne, M. A., Shimony, A. and Zeilinger, A., Amer. J. Phys. **58**, 1131 (1990).
10. Bennett, C. H. and Wiesner, S. J., Phys. Rev. Lett. **69**, 2881 (1992).
11. Mattle, K., Weinfurter, H., Kwiat, P. G. and Zeilinger, A., Phys. Rev. Lett. **76**, 4656 (1996).
12. Michael Reck, unpublished.
13. Zeilinger, A., Bernstein, H. J. and Horne, M. A., J. Modern Opt. **41**, 2375 (1994).
14. Hong, C. K., Ou, Z. Y. and Mandel, L., Phys. Rev. Lett. **59**, 2044 (1987).
15. Oberparleiter, M., Diploma-thesis, University of Innsbruck, 1997, unpublished.
16. Loudon, R., ”Coherence and Quantum Optics VI“ (Edited by J. H. Eberly and L. Mandel) (New York: Plenum, 1990), p. 703; Loudon, R., ”Disorder and Condensed Matter Physics“ (Edited by J. A. Blackman and J. Taguena) (Oxford: Clarendon, 1991), p. 441.
17. Zukowski, M., Zeilinger, A., Horne, M. A. and Ekert, A. K., Phys. Rev. Lett. **71**, 4287 (1993).
18. Nagel, B., in ”Possible Worlds in Humanities, Arts and Sciences“. (Proceedings of Nobel Symposium **65**) (Edited by Sture Allen) (Gruyter, Berlin 1989), p. 388.
19. Bennett, C. H. *et al.*, Phys. Rev. Lett. **70**, 1895 (1993).
20. Bouwmeester, D. *et al.*, Experimental Quantum Teleportation, Nature 390, 575 (1997).
21. Zukowski, M., Zeilinger, A. and Weinfurter, H., ”Fundamental Problems in Quantum Theory“ (Edited by D. M. Greenberger, A. Zeilinger) (Annals of the New York Academy of Sciences, 1995) vol. 755, p. 91.

22. C. H. Bennett, Phys. Today **48**, 24 (1995).
23. Bruss, D., Ekert, A., Huelga, S. F., Pan, J.-W. and Zeilinger, A., Phil. Trans. R. Soc. Lond. A (1997) in press.
24. Shor, P. W., in Proc. of the 35th Ann. Symp. on the Foundations of Computer Science, Los Alamitos, CA, (IEEE Press, New York 1994); Ekert, A., Jozsa, R., Rev. Mod. Phys. **68**, 733 (1996).
25. Shor, P. W., Phys. Rev. **A52**, 2493 (1995); Steane, A. M., Proc. Roy. Soc. Lond. **A452**, 2551 (1996).
26. Hagley, E. *et al.*, Phys. Rev. **79**, 1, (1997).
27. Monroe, C. *et al.*, Phys. Rev. Lett. **75**, 4714, (1995); Wineland, D., this volume.
28. Gershenfeld, N. A. and Chuang, I. L., Science **275**, 350 (1997).
29. Ekert, A. K., Phys. Rev. Lett. **67**, 661 (1991).
30. Cirac, I., Zoller, P., Kimble, J. and Mabuchi, H., Phys. Rev. Lett. **78**, 3221 (1997); Bose, S., Vedral, V. and Knight, P. L., Phys. Rev. **A57**, 822 (1998).
31. Pan, J.-W., Bouwmeester, D., Weinfurter, H. and Zeilinger, A., Phys. Rev. Lett. **80**, 3891 (1998).

Experiment and the foundations of quantum physics

Anton Zeilinger*

Institut für Experimentalphysik, University of Vienna, Boltzmanngasse 5, A-1090 Vienna, Austria

Instead of having to rely on gedanken (thought) experiments, it is possible to base this discussion of the foundations of quantum physics on actually performed experiments because of the enormous experimental progress in recent years. For reasons of space, the author discusses mainly experiments related to the Einstein-Podolsky-Rosen paradox and Bell's theorem, that is, to quantum entanglement. Not only have such fundamental experiments realized many historic proposals, they also helped to sharpen our quantum intuition. This recently led to the development of a new field, quantum information, where quantum teleportation and quantum computation are some of the most fascinating topics. Finally the author ventures into a discussion of future prospects in experiment and theory. [S0034-6861(99)03602-8]

CONTENTS

I. The Background S288
II. A Double Slit and One Particle S288
III. A Double Slit and Two Particles S289
IV. Quantum Complementarity S291
V. Einstein-Podolsky-Rosen and Bell's Inequality S292
VI. Quantum Information and Entanglement S295
VII. Final Remarks and Outlook S296
Acknowledgments S296
References S297

I. THE BACKGROUND

Quantum physics, a child of the early 20th century, is probably the most successful description of nature ever invented by man. The range of phenomena it has been applied to is enormous. It covers phenomena from the elementary-particle level all the way to the physics of the early universe. Many modern technologies would be impossible without quantum physics—witness, for example, that all information technologies are based on a quantum understanding of solids, particularly of semiconductors, or that the operation of lasers is based on a quantum understanding of atomic and molecular phenomena.

So, where is the problem? The problem arises when one realizes that quantum physics implies a number of very counterintuitive concepts and notions. This has led, for example, R. P. Feynman to remark, "I think I can safely say that nobody today understands quantum physics," or Roger Penrose (1986) to comment that the theory "makes absolutely no sense."

From the beginning, gedanken (thought) experiments were used to discuss fundamental issues in quantum physics. At that time, Heisenberg invented his gedanken gamma-ray microscope to demonstrate the uncertainty principle while Niels Bohr and Albert Einstein in their famous dialogue on epistemological problems in what was then called atomic physics made extensive use of gedanken experiments to make their points.

Now, at the end of the 20th century, the situation has changed dramatically. Real experiments on the foundations of quantum physics abound. This has not only given dramatic support to the early views, it has also helped to sharpen our intuition with respect to quantum phenomena. Most recently, experimentation is already applying some of the fundamental phenomena in completely novel ways. For example, quantum cryptography is a direct application of quantum uncertainty and both quantum teleportation and quantum computation are direct applications of quantum entanglement, the concept underlying quantum nonlocality (Schrödinger, 1935).

I will discuss a number of fundamental concepts in quantum physics with direct reference to experiments. For the sake of the consistency of the discussion and because I know them best I will mainly present experiments performed by my group. In view of the limited space available my aim can neither be completeness, nor a historical overview. Rather, I will focus on those issues I consider most fundamental.

II. A DOUBLE SLIT AND ONE PARTICLE

Feynman (1965) has said that the double-slit "has in it the heart of quantum mechanics. In reality, it contains the only mystery." As we shall see, entangled states of two or more particles imply that there are further mysteries (Silverman, 1995). Nevertheless, the two-slit experiment merits our attention, and we show the results of a typical two-slit experiment done with neutrons in Fig. 1 (Zeilinger *et al.*, 1988). The measured distribution of the neutrons has two remarkable features. First, the observed interference pattern showing the expected fringes agrees perfectly well with theoretical prediction (solid line), taking into account all features of the experimental setup. Assuming symmetric illumination the neutron state at the double slit can be symbolized as

$$|\psi\rangle = \frac{1}{\sqrt{2}}(|\text{passage through slit } a\rangle + |\text{passage through slit } b\rangle). \quad (1)$$

*Electronic address: anton.zeilinger@uibk.ac.at

 0034-6861/99/71(2)/288(10)/$17.00

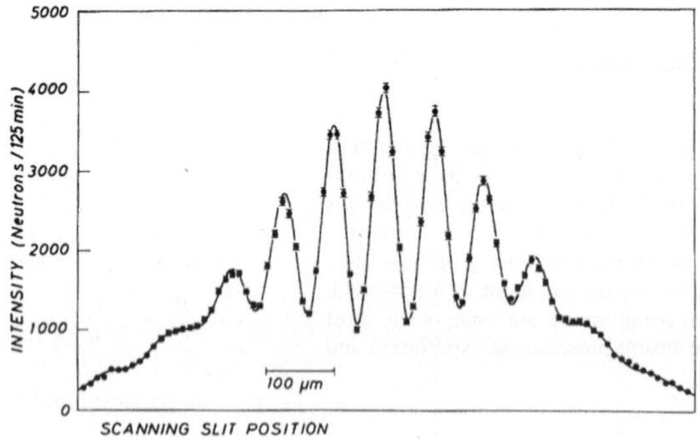

FIG. 1. A double-slit diffraction pattern measured with very cold neutrons with a wavelength of 2 nm corresponding to a velocity of 200 ms^{-1}. The two slits were 22 μm and 23 μm wide, respectively, separated by a distance of 104 μm. The resulting diffraction angles were only of the order of 10 μrad, hence the observation plane was located 5 m downstream from the double slit in order to resolve the interference pattern. (For experimental details see Zeilinger *et al.*, 1988.) The solid line represents first-principles prediction from quantum mechanics, including all features of the experimental apparatus. For example, the fact that the modulation of the interference pattern was not perfect can fully be understood on the basis that a broad wavelength band had to be used for intensity reasons and the experiment was not operated in the Fraunhofer regime.

The interference pattern is then obtained as the superposition of two probability amplitudes. The particle could have arrived at a given observation point $\vec{r}$ either via slit 1 with probability amplitude $a(\vec{r})$ or via slit 2 with probability amplitude $b(\vec{r})$. The total probability density to find the particle at point $\vec{r}$ is then simply given as

$$p(\vec{r})=|a(\vec{r})+b(\vec{r})|^2. \tag{2}$$

This picture suggests that the pattern be interpreted as a wave phenomenon.

Yet, second, we note that the maximum observed intensity is of the order of one neutron every two seconds. This means that, while one neutron is being registered, the next one to be registered usually is still confined to its uranium nucleus inside the nuclear reactor, waiting for nuclear fission to release it to freedom!

This feature of very low-intensity interference is shared by all existing neutron interferometer experiments (Rauch and Werner, in press). These pioneering matter-wave experiments led to the realization of a number of very basic experiments in quantum mechanics including the change of the sign of a spinor under a full rotation, the effect of gravity on the phase of a neutron wave, a number of experiments related to quantum complementarity, and many others.

Thus the interference pattern is really collected one by one and this suggests the particle nature. Then the famous question can be posed: through which of the two slits did the particle actually pass on its way from source to detector? The well-known answer according to standard quantum physics is that such a question only makes sense when the experiment is such that the path taken can actually be determined for each particle. In other words, the superposition of amplitudes in Eq. (1) is only valid if there is no way to know, even in principle, which path the particle took. It is important to realize that this does not imply that an observer actually takes note of what happens. It is sufficient to destroy the interference pattern, if the path information is accessible in principle from the experiment or even if it is dispersed in the environment and beyond any technical possibility to be recovered, but in principle still "out there." The absence of any such information is *the essential criterion* for quantum interference to appear. For a parallel discussion, see the accompanying article by Mandel (1999) in this volume.

To emphasize this point, let us consider now a gedanken experiment where a second, probe, particle is scattered by the neutron while it passes through the double slit. Then the state will be

$$|\psi\rangle=\frac{1}{\sqrt{2}}\begin{array}{l}(|\text{passage through slit } a\rangle_1|\text{scattered in region } a\rangle_2\\ +|\text{passage through slit } b\rangle_1|\text{scattered in region } b\rangle_2)\end{array}. \tag{3}$$

There the subscripts 1 and 2 refer to the neutron and the probe particle, respectively. The state (3) is entangled and if the two states for particle 2 are orthogonal, no interference for particle 1 can arise. Yet, if particle 2 is measured such that this measurement is not able, *even in principle*, to reveal any information about the slit particle 1 passes, then particle 1 will show interference. Obviously, there is a continuous transition between these two extreme situations.

We thus have seen that one can either observe a wavelike feature (the interference pattern) or a particle feature (the path a particle takes through the apparatus) depending on which experiment one chooses. Yet one could still have a naive picture in one's mind essentially assuming waves propagating through the apparatus which can only be observed in quanta. That such a picture is not possible is demonstrated by two-particle interferences, as we will discuss now.

III. A DOUBLE SLIT AND TWO PARTICLES

The situation is strikingly illustrated if one employs pairs of particles which are strongly correlated ("en-

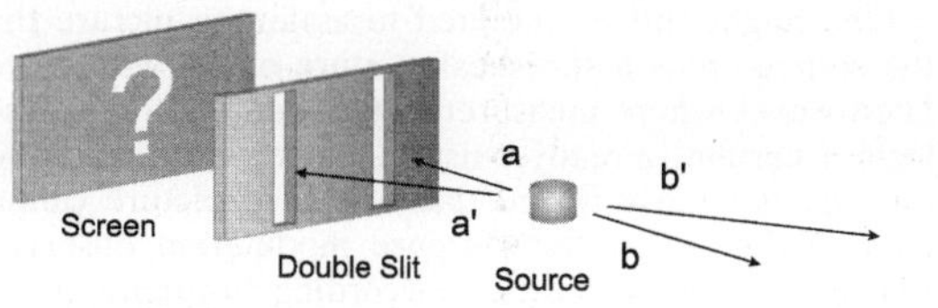

FIG. 2. A source emits pairs of particles with total zero momentum. Particle 1 is either emitted into beams a or a' and particle 2 into beams b or b' with perfect correlations between a and b and a' and b', respectively. The beams of particle 1 then pass a double-slit assembly. Because of the perfect correlation between the two particles, particle 2 can serve to find out which slit particle 1 passed and therefore no interference pattern arises.

tangled") such that either particle carries information about the other (Horne and Zeilinger, 1985; Greenberger, Horne, and Zeilinger, 1993). Consider a setup where a source emits two particles with antiparallel momenta (Fig. 2). Then, whenever particle 1 is found in beam a, particle 2 is found in beam b and whenever particle 1 is found in beam a', particle 2 is found in beam b'. The quantum state is

$$|\psi\rangle = \frac{1}{\sqrt{2}}(|a\rangle_1|b\rangle_2 + |a'\rangle_1|b'\rangle_2). \qquad (4)$$

Will we now observe an interference pattern for particle 1 behind its double slit? The answer has again to be negative because by simply placing detectors in the beams b and b' of particle 2 we can determine which path particle 1 took. Formally speaking, the states $|a\rangle_1$ and $|a'\rangle_1$ again cannot be coherently superposed because they are entangled with the two orthogonal states $|b'\rangle_2$ and $|b'\rangle_2$.

Obviously, the interference pattern can be obtained if one applies a so-called quantum eraser which completely erases the path information carried by particle 2. That is, one has to measure particle 2 in such a way that it is not possible, even in principle, to know from the measurement which path it took, a' or b'.

A recent experiment (Dopfer, 1998) used the so-called process of parametric down conversion to create entangled pairs of photons (Fig. 3) where a UV beam entering a nonlinear optical crystal spontaneously creates pairs of photons such that the sum of their linear momenta is constant. In type-I parametric down conversion, the two photons carry equal polarization. Parametric down conversion is discussed in somewhat more detail below. Although the experimental situations are different, conceptually this is equivalent to the case discussed above. In this experiment, photon 2 passes a double slit while the other, photon 1, can be observed by a detector placed at various distances behind the Heisenberg lens which plays exactly the same role as the lens in the gamma-ray microscope discussed by Heisenberg (1928) and extended by Weizsächer (1931). If the detector is placed at the focal plane of the lens, then registration of a photon there provides information about its direction, i.e., momentum, before entering the lens. Thus, because of the strict momentum correlation, the momentum of the other photon incident on the double slit and registered in coincidence is also well defined. A momentum eigenstate cannot carry any position information, i.e., no information about which slit the particle passes through. Therefore, a double-slit interference pattern for photon 2 is registered conditioned on registration of photon 1 in the focal plane of the lens. It is important to note that it is actually necessary to register photon 1 at the focal plane because without registration one could always, at least in principle, reconstruct the state in front of the lens. Most strikingly, therefore, one can find out the slit photon 2 passed by placing the detector for photon 1 into the imaging plane of the lens. The imaging plane is simply obtained by taking the object distance as the sum of the distances from the lens to the crystal and from the crystal to the double slit. Then, as has also been demonstrated in the experiment, a one-to-one relationship exists between positions in the plane of the double slit and in the imaging plane and thus, the slit particle 2 passes through can readily be determined by observing photon 1 in the imaging plane. Only after registration of photon 1 in the

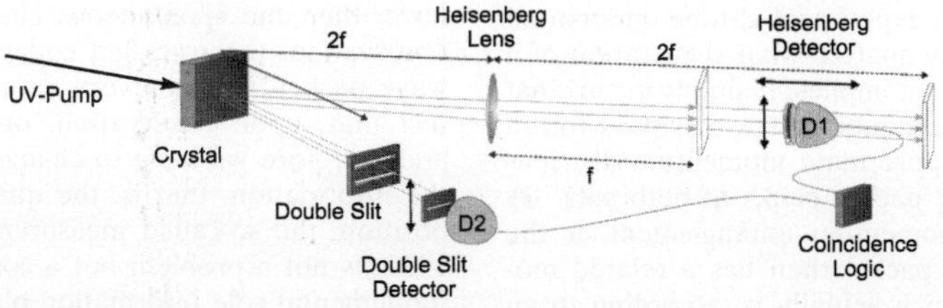

FIG. 3. Two photons and one double slit. A pair of momentum-entangled photons is created by type-*I* parametric down conversion. Photon 2 enters a double-slit assembly and photon 1 is registered by the Heisenberg detector arranged behind the Heisenberg lens. If the Heisenberg detector is placed in the focal plane of the lens, it projects the state of the second photon into a momentum eigenstate which cannot reveal any position information and hence no information about slit passage. Therefore, in coincidence with a registration of photon 1 in the focal plane, photon 2 exhibits an interference pattern. On the other hand, if the Heisenberg detector is placed in the imaging plane at $2f$, it can reveal the path the second photon takes through the slit assembly which therefore connot show the interference pattern (Dopfer, 1998).

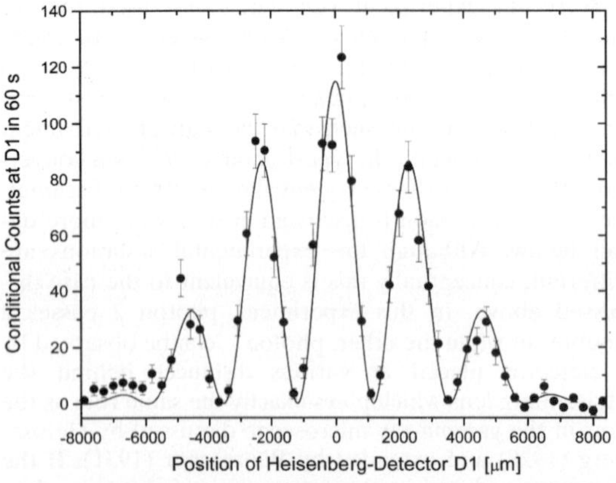

FIG. 4. Double-slit pattern registered by the Heisenberg detector of photon 1 (Fig. 3). The graph shows the counts registered by that detector as a function of its lateral position, if that detector is arranged in the focal plane of the lens. The counts are conditioned on registration of the second photon behind its double slit. Note that the photons registered in detector $D1$ exhibit a double-slit pattern even though they never pass through a double-slit assembly. Note also the low intensity which indicates that the interference pattern is collected photon by photon.

focal plane of the lens is any possibility to obtain any path information from photon 1 irrecoverably destroyed.

We note that the distribution of photons behind the double slit without registration of the other photon is just an incoherent sum of probabilities having passed through either slit and, as shown in the experiment, no interference pattern arises if one does not look at the other photon. This is again a result of the fact that, indeed, path information is still present and can easily be extracted by placing the detector of photon 1 into the imaging plane of the lens.

Likewise, registration of photon 2 behind its double slit destroys any path information it may carry and thus, by symmetry, a Fraunhofer double-slit pattern is obtained for the distribution of photon 1 in the focal plane behind its lens, even though that photon never passed a double slit (Fig. 4)! This experiment can be understood intuitively if we carefully analyze what registration of a photon behind a double slit implies. It simply means that the state incident on the double slit is collapsed into a wave packet with the appropriate momentum distribution such that the wave packet peaks at both slits. By virtue of the strong momentum entanglement at the source, the other wave packet then has a related momentum distribution which actually is, according to an argument put forward by Klyshko (1988), the time reversal of the other wave packet. Thus, photon 1 appears to originate backwards from the double slit assembly and is then considered to be reflected by the wave fronts of the pump beam into the beam towards the lens which then simply realizes the standard Fraunhofer observation conditions.

One might still be tempted to assume a picture that the source emits a statistical mixture of pairwise correlated waves where measurement of one photon just selects a certain, already existing, wavelet for the other photon. It is easy to see that any such picture cannot lead to the perfect interference modulation observed. The most sensible position, according to quantum mechanics, is to assume that no such waves preexist before any measurement.

IV. QUANTUM COMPLEMENTARITY

The observation that particle path and interference pattern mutually exclude each other is one specific manifestation of the general concept of complementary in quantum physics. Other examples are position and linear momentum as highlighted in Heisenberg's uncertainty relation, or the different components of angular momentum. It is often said that complementarity is due to an unavoidable disturbance during observation. This is suggested if, as in our example in Sec. II, we consider determining the path a particle takes through the double-slit assembly by scattering some other particle from it. That this is too limited a view is brought out by the experiment discussed in the preceding section.

The absence of the interference pattern for photon 2 if no measurement is performed on photon 1, is not due to it being disturbed by observation; rather, it can be understood if we consider the complete set of possible statements which can be made about the experiment as a whole (Bohr, 1935) including the other photon.

As long as no observation whatsoever is made on the complete quantum system comprised of both photons our description of the situation has to encompass all possible experimental results. The quantum state is exactly that representation of our knowledge of the complete situation which enables the maximal set of (probabilistic) predictions for any possible future observation. What comes new in quantum mechanics is that, instead of just listing the various experimental possibilities with the individual probabilities, we have to represent our knowledge of the situation by the quantum state using complex amplitudes. If we accept that the quantum state is no more than a representation of the information we have, then the spontaneous change of the state upon observation, the so-called collapse or reduction of the wave packet, is just a very natural consequence of the fact that, upon observation, our information changes and therefore we have to change our representation of the information, that is, the quantum state. From that position, the so-called measurement problem (Wigner, 1970) is not a problem but a consequence of the more fundamental role information plays in quantum physics as compared to classical physics (Zeilinger, 1999).

Quantum complementarity then is simply an expression of the fact that in order to measure two complementary quantities, we would have to use apparatuses which mutually exclude each other. In the example of our experiment, interference pattern and path information for photon 2 are mutually exclusive, i.e., complementary, because it is not possible to position the detector for photon 1 simultaneously in the focal plane and in the image plane of the lens. Yet the complete quantum state encompasses both possible experiments.

We finally note two corollaries of our analysis. First, it is clearly possible to have a concept of continuous complementarity. In our case, placing the detector of photon 1 somewhere in between the two extreme positions mentioned will reveal partial path information and thus an interference pattern of reduced visibility. And second, the choice whether or not path information or the interference pattern become manifest for photon 2 can be delayed to arbitrary times after that photon has been registered. In the experiment discussed, the choice where detector D_1 is placed can be delayed until after photon 2 has been detected behind its double slit. While we note that in the experiment, the lens was already arranged at a larger distance from the crystal than the double slit, a future experiment will actually employ a rapidly switched mirror sending photon 1 either to a detector placed in the focal plane of the lens or to a detector placed in the imaging plane.

This possibility of deciding long after registration of the photon whether a wave feature or a particle feature manifests itself is another warning that one should not have any realistic pictures in one's mind when considering a quantum phenomenon. Any detailed picture of what goes on in a specific individual observation of one photon has to take into account the whole experimental apparatus of the complete quantum system consisting of both photons and it can only make sense after the fact, i.e., after all information concerning complementary variables has irrecoverably been erased.

V. EINSTEIN-PODOLSKY-ROSEN AND BELL'S INEQUALITY

In 1935 Einstein, Podolsky, and Rosen (EPR) studied entangled states of the general type used in the two-photon experiment discussed above. They realized that in many such states, when measuring either linear momentum or position of one of the two particles, one can infer precisely either momentum or position of the other. As the two particles might be widely separated, it is natural to assume validity of the locality condition suggested by EPR: "Since at the time of measurement the two systems no longer interact, no real change can take place in the second system in consequence of anything that may be done to the first system." Then, whether or not momentum or position can be assigned to particle (system) 2 must be independent of what measurement is performed on particle 1 or even whether any measurement is performed on it at all. The question therefore arises whether the specific results obtained for either particle can be understood without reference to which measurement is actually performed on the other particle. Such a picture would imply a theory, underlying quantum physics, which provides a more detailed account of individual measurements. Specifically, following Bell, it might explain "why events happen" (Bell, 1990; Gottfried, 1991).

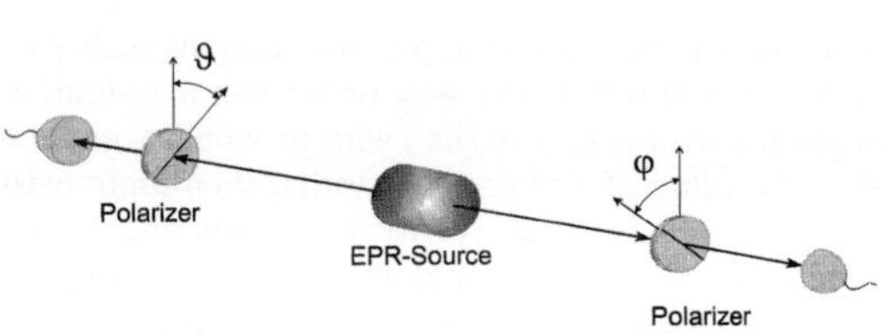

FIG. 5. Typical experimental arrangement to test Bell's inequality. A source emits, say, polarization-entangled pairs of photons. Each photon is sent through a polarizer whose orientation can be varied. Finally behind each polarizer, the transmitted photons are registered. Quantum mechanics predicts a sinusoidal variation of the coincidence count rate as a function of the relative angular orientation of the polarizers. Any such variation violates local realism as expressed in Bell's inequality.

In the sixties, two different developments started, which nicely complement each other. First, it was initially argued by Specker (1960) for Hibbert spaces of dimension larger than two that quantum mechanics cannot be supplemented by additional variables. Later it was shown by Kochen and Specker (1967) and by Bell (1966; for a review see Mermin, 1993), that for the specific case of a spin-1 particle, it is not possible to assign in a consistent way measurement values to the squares of any three orthogonal spin projections, despite the fact that the three measurements commute with each other. This is a purely geometric argument which only makes use of some very basic geometric considerations. The conclusion here is very important. The quantum system cannot be assigned properties independent of the context of the complete experimental arrangement. This is just in the spirit of Bohr's interpretation. This so-called contextuality of quantum physics is another central and subtle feature of quantum mechanics.

Second, a most important development was due to John Bell (1964) who continued the EPR line of reasoning and demonstrated that a contradiction arises between the EPR assumptions and quantum physics. The most essential assumptions are realism and locality. This contradiction is called Bell's theorem.

To be specific, and in anticipation of experiments we will discuss below, let us assume we have a pair of photons in the state:

$$|\psi\rangle = \frac{1}{\sqrt{2}}(|H\rangle_1|V\rangle_2 - |V\rangle_1|H\rangle_2). \quad (5)$$

This polarization-entangled state implies that whenever (Fig. 5) photon 1 is measured and found to have horizontal (H) polarization, the polarization of photon 2 will be vertical (V) and vice versa. Actually, the state of Eq. (5) has the same form in any basis. This means whichever state photon 1 will be found in, photon 2 can definitely be predicted to be found in the orthogonal state if measured.

Following EPR one can apply their famous reality criterion, "If, without in any way disturbing a system, we can predict with certainty (i.e., with probability equal to unity) the value of a physical quantity, then there exists an element of physical reality corresponding to this physical quantity." This would imply that to any possible polarization measurement on any one of our photons we can assign such an element of physical reality on the basis of a corresponding measurement on the other photon of any given pair.

The next step then is to assume the two photons (systems) to be widely separated so that we can invoke EPR's locality assumption as given above. Within this line of reasoning, whether or not we can assign an element of reality to a specific polarization of one of the systems must be independent of which measurement we actually perform on the other system and even independent of whether we care to perform any measurement at all on that system. To put it dramatically, one experiment could be performed here on earth and the other on a planet of another star a couple of light years away. It is this very independence of a measurement result on one side from what may be done on the other side, as assumed by EPR, which is at variance with quantum mechanics. Indeed, this assumption implies that certain combinations of expectation values have definite bounds. The mathematical expression of that bound is called Bell's inequality, of which many variants exist. For example, a version given by Clauser, Horne, Shimony, and Holt (1969) is

$$|E(\alpha,\beta)-E(\alpha',\beta)|+|E(\alpha,\beta')+E(\alpha',\beta')|\leq 2, \quad (6)$$

where

$$E(\alpha,\beta)=\frac{1}{N}[C_{++}(\alpha,\beta)+C_{--}(\alpha,\beta)-C_{+-}(\alpha,\beta)-C_{-+}(\alpha,\beta)]. \quad (7)$$

Here we assume that each photon is subject to a measurement of linear polarization with a two-channel polarizer whose outputs are + and −. Then, e.g., $C_{++}(\alpha,\beta)$ is the number of coincidences between the + output port of the polarizer measuring photon 1 along α and the + output port of the polarizer measuring photon 2 along β. Maximal violation occurs for $\alpha=0°$, $\beta=22.5°$, $\alpha'=45°$, $\beta'=67.5°$. Then the left-hand side of Eq. (6) will be $2\sqrt{2}$ in clear violation of the inequality. Thus Bell discovered that the assumption of local realism is in conflict with quantum physics itself and it became a matter of experiment to find out which of the two world views is correct.

Interestingly, at the time of Bell's discovery no experimental evidence existed which was able to decide between quantum physics and local realism as defined in Bell's derivation. An earlier experiment by Wu and Shaknov (1950) had demonstrated the existence of spatially separated entangled states, yet failed to give data for nonorthogonal measurement directions. After the realization that the polarization entangled state of photons emitted in atomic cascades can be used to test

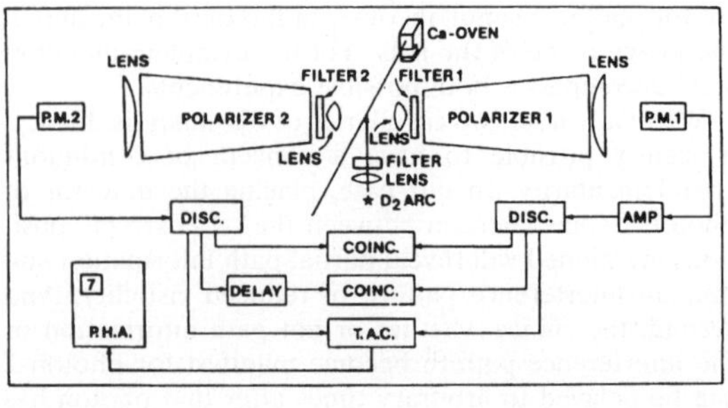

FIG. 6. Sketch of the experimental setup used in the first experiment demonstrating a violation of Bell's inequality (Freedman and Clauser, 1972). The two photons emitted in an atomic cascade in Ca are collected with lenses and, after passage through adjustable polarizers, coincidences are registered using photomultiplier detectors and suitable discriminators and coincidence logic. The observed coincidence counts violate an inequality derived from Bell's inequality under the fair sampling assumption.

Bell's inequalities, the first experiment was performed by Freedman and Clauser in 1972 (Fig. 6). By now, there exists a large number of such experiments. The ones showing the largest violation of a Bell-type inequality have for a long time been the experiments by Aspect, Grangier, and Roger (1981, 1982) in the early eighties. Aside from two early experiments, all agreed with the predictions of quantum mechanics and violated inequalities derived from Bell's original version using certain additional assumptions. Actually, while the experimental evidence strongly favors quantum mechanics, there remained two possible mechanisms for which a local realistic view could still be maintained.

One problem in all experimental situations thus far is due to technical insufficiencies, namely that only a small fraction of all pairs emitted by the source is registered. This is a standard problem in experimental work and experimentalists take great care to ensure that it is reasonable to assume that the detected pairs are a faithful representative of all pairs emitted. Yet, at least in principle, it is certainly thinkable that this is not the case and that, should we once be able to detect all pairs, a violation of quantum mechanics and data in agreement with local realism would be observed. While this is in principle possible, I would agree with Bell's judgment (1981) that "although there is an escape route there, it is hard for me to believe that quantum mechanics works so nicely for inefficient practical set-ups, and is yet going to fail badly when sufficient refinements are made. Of more importance, in my opinion, is the complete absence of the vital *time* factor in existing experiments. The analyzers are not rotated during the flight of the particles. Even if one is obliged to admit some long-range influence, it need not travel faster than light—and so would be much less indigestible." Until recently, there has been only one experiment where the time factor played a role. In that experiment (Aspect, Dalibard,

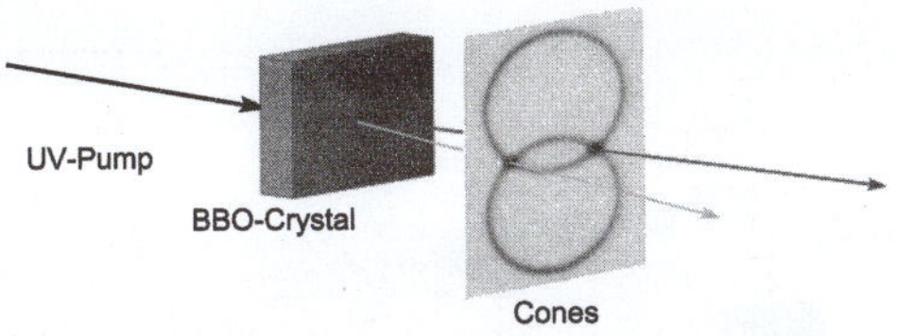

FIG. 7. Principle of type-II parametric down conversion to produce directed beams of polarization entangled photons (Kwiat *et al.*, 1995). An incident pump photon can spontaneously decay into two photons which are entangled in momentum and energy. Each photon can be emitted along a cone in such a way that two photons of a pair are found opposite to each other on the respective cones. The two photons are orthogonally polarized. Along the directions where the two cones overlap, one obtains polarization-entangled pairs. In the figure, it is assumed that a filter already selects those photons which have exactly half the energy of a pump photon.

and Roger, 1982) each of the two photons could be switched between two different polarizers on a time scale which was small compared to the flight time of the photons. Due to technical limitations at the time of the experiment and because this switching back and forth between two different polarizations was periodic, the experiment does not completely fulfill Bell's desideratum, but it is an important step.

Experimental development in the last decade is marked by two new features. First, it was realized initially by Horne and Zeilinger (1985, 1988) for momentum and position, and then by Franson (1989) for time and energy, that situations can arise where Bell's inequality is violated not just for internal variables, like spin, but also for external ones. This observation put Bell's theorem in a much broader perspective than before. Second, a new type of source was employed (Burnham and Weinberg, 1970), based on the process of spontaneous parametric down conversion. The first to use such a source in a Bell-inequality experiment were Alley and Shih in 1986. In such experiments, a nonlinear optical crystal is pumped by a sufficiently strong laser beam. Then, with a certain very small probability, a photon in the laser beam can spontaneously decay into two photons. The propagation directions of the photons and the polarization are determined by the dispersion surfaces inside the medium. The so-called phase-matching conditions of quantum optics, which for sufficiently large crystals are practically equivalent to energy and momentum conservation, imply that the momenta and the energies of the two created photons have to sum up to the corresponding value of the original pump photon inside the crystal. In effect, a very rich entangled state results. The two emerging photons are entangled both in energy and in momentum. In type-I down conversion, these two photons have the same polarization while in type-II down conversion, they have different polarization.

A recent experiment utilized type-II down conversion (Figs. 7 and 8) such that the two emerging photons having orthogonal polarizations effectively emerge in a polarization-entangled state as discussed above [see Eq. (5)]. In the experiment (Weihs *et al.*, 1998), the photons were coupled into long glass fibers and the polarization correlations over a distance of the order of 400 m was measured. The important feature of that experiment is that the polarization of the photons could be rotated in the last instant, thus effectively realizing the rotatable polarizers suggested by Bell. The decision whether or not to rotate the polarization was made by a physical random-number generator on a time scale short compared to the flight time of the photons. Figure 9 shows the principle of the experimental setup. Due to technological progress it is possible now in such experiments to violate Bell's inequality by many standard deviations in a very short time: in this experiment by about 100 standard deviations in measurement times of the order of a minute. In a related experiment (Tittel *et al.*, 1998), en-

FIG. 8. (Color) A more complete representation of the radiation produced in type-II parametric down-conversion (photo: Paul Kwiat and Michael Reck). Three photographs taken with different color filters have been superposed here. The colors are actually false colors for clarity of presentation. The photons emitted from the source are momentum and energy entangled in such a way that each photon can be emitted with a variety of different momenta and frequencies, each frequency defining a cone of emission for each photon. The whole quantum state is then a superposition of many different pairs. For example, if measurement reveals a photon to be found somewhere on the red small circle in the figure, its brother photon is found exactly opposite on the blue small circle. The green circles represent the case where the two colors are identical.

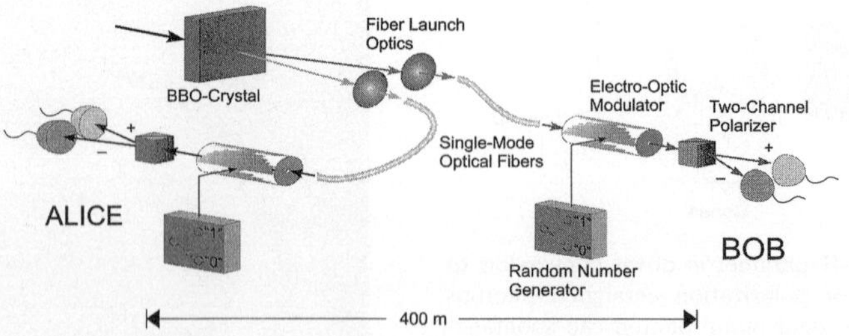

FIG. 9. Long-distance Bell-inequality experiment with independent observers (Weihs *et al.*, 1998). The two entangled photons are individually launched into optical fibers and sent to the measurement stations of the experimenters Alice and Bob which are separated from each other by a distance of 400 m. At each of the measurement stations, an independent, very fast, random-number generator decides, while the photons are really in flight, the direction along which the polarization will be measured. Finally, events are registered independently on both sides and coincidences are only identified long after the experiment is finished.

tanglement could be demonstrated over distances of more than 10 km but without random switching.

A few points deserve consideration regarding future experiments. On the one hand experiments must be improved to high enough pair-collection efficiencies in order to finally prove that the fair sampling hypothesis used in all existing experiments was justified. On the other hand, and more interesting from a fundamental point of view, one could still assume that both random-number generators are influenced by joint events in their common past. This suggests a final experiment in which two experimenters exercise their free will and choose independently the measurement directions. Such an experiment would require distances of order of a few light seconds and thus can only be performed in outer space.

Another future direction of research will certainly be directed at quantum entanglement employing more systems, or at larger, specifically more massive, systems. A first experiment in Paris was able to demonstrate entanglement between atoms (Hagley *et al.*, 1997).

VI. QUANTUM INFORMATION AND ENTANGLEMENT

While most work on the foundations of quantum physics was initially motivated by curiosity and even by philosophical considerations, this has recently led to the emergence of novel ideas in information science. A significant result is already a new perspective on information itself. Eventually, applications might include quantum communication, quantum cryptography, possibly even quantum computation.

Some of the basic novel features are contained in quantum teleportation involving two distant experimenters, conventionally called Alice and Bob (Fig. 10). Here, Alice initially has a single particle in the quantum state $|\psi\rangle$ (the "teleportee"). The state may be unknown to her or possibly even undefined. The aim is that the distant experimenter Bob obtains an exact replica of that particle. It is evident that no measurement whatsoever Alice might perform on the particle could reveal all necessary information to enable Bob to reconstruct its state. The quantum teleportation protocol (Bennett *et al.*, 1993) proceeds by Alice and Bob agreeing to share initially an entangled pair of "ancillary" photons. Alice then performs a joint Bell-state measurement on the teleportee and her ancillary photon, and obtains one of the four possible Bell results. The four possible Bell states (Braunstein *et al.*, 1992) are

$$|\psi^{\pm}\rangle=\frac{1}{\sqrt{2}}(|H\rangle_1|V\rangle_2\pm|V\rangle_1|H\rangle_2),$$

$$|\phi^{\pm}\rangle=\frac{1}{\sqrt{2}}(|H\rangle_1|H\rangle_2\pm|V\rangle_1|V\rangle_2). \tag{8}$$

They form a maximally entangled basis for the two-photon four-dimensional spin Hilbert space. These Bell states are essential in many quantum information scenarios. Alice's measurement also projects Bob's ancillary photon into a well-defined quantum state. Alice then transmits her result as a classical two-bit message to Bob, who performs one of four unitary operations, independent of the state $|\psi\rangle$, to obtain the original state. In the experiment (Bouwmeester *et al.*, 1997), femtosecond pulse technology had to be used in order to obtain the necessary nontrivial coherence conditions for the Bell-state measurements.

While teleportation presently might sound like a strange name conjuring up futuristic images, it is appropriate. The reader should be reminded of the strange connotations of the notion of magnetism before its clear definition by physicists. Quantum teleportation actually demonstrates some of the salient features of entanglement and quantum information. It also raises deep questions about the nature of reality in the quantum world.

Most important for the understanding of the quantum teleportation scheme is the realization that maximally entangled states such as the Bell basis are characterized by the fact that none of the individual members of the entangled state, in our case, the two photons, carries any information on its own. All information is only encoded in joint properties. Thus, an entangled state is a repre-

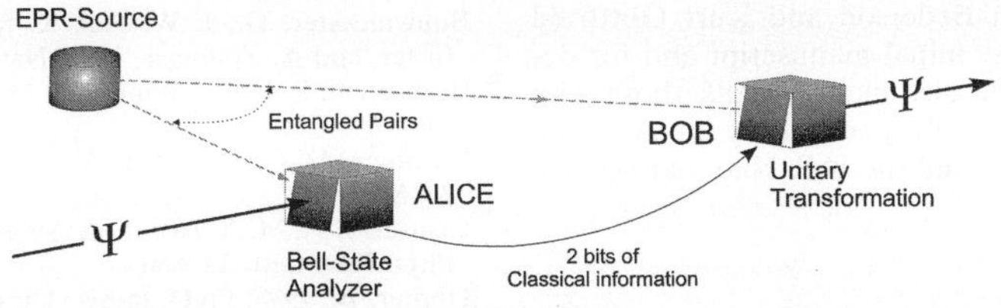

FIG. 10. Principle of quantum teleportation (Bennett *et al.*, 1993). In order to teleport her, possibly even unknown, quantum state $|\psi\rangle$ to Bob, Alice shares with him initially an entangled pair. She then performs a Bell-state analysis and, after receipt of Alice's measurement result, Bob can turn his member of the entangled pair into the original state by applying a unitary transformation which only depends on the specific Bell state result obtained by Alice and is independent of any properties of the teleportee state $|\psi\rangle$.

sentation of the relations between two possible measurements on the two members of the entangled pair. In the most simple case, the state $|\psi^-\rangle$ is a representation of the prediction that in any basis whatsoever, the two photons will be found to have orthogonal states with none of the photons having any well-defined state before measurement. The teleportation scheme then simply means that Alice's Bell-state measurement results in a well-defined relational statement between the original $|\psi\rangle$ and one of the two photons emerging from the EPR source. The specific entangled state emitted by the source then implies another relational statement with Bob's photon, and thus, by this line of reasoning, we have a clear relational statement connecting his photon with Alice's original. That statement is independent of the properties of $|\psi\rangle$, and Bob just has to apply the proper unitary transformation defined by the specific one of the four Bell states Alice happened to obtain randomly. In the most simple case, suppose Alice's Bell-state measurement happens to give the same result as the state emitted by the source. Then, Bob's particle is immediately identical to the original, and his unitary transformation is the identity. Even more striking is the possibility to teleport a quantum state which itself is entangled to another particle. Then, the teleported state is not just unknown but undefined. This possibility results in entanglement swapping (Zukowski *et al.*, 1993; Pan *et al.*, 1998), that is, in entangling two particles which were created completely independently and which never interacted.

The essential feature in all these schemes is again entanglement. Information can be shared by two photons in a way where none of the individuals carries any information on its own.

As a most striking example consider entangled superpositions of three quanta, e.g.,

$$|\psi\rangle = \frac{1}{\sqrt{2}}(|H\rangle|H\rangle|H\rangle + |V\rangle|V\rangle|V\rangle). \tag{9}$$

Such states, usually called Greenberger-Horne-Zeilinger states (Greenberger *et al.*, 1989; Greenberger *et al.*, 1990), exhibit very rich perfect correlations. For such states, these perfect correlations lead to a dramatic conflict with local realism on an event-by-event basis and not just on a statistical basis as in experiments testing Bell's inequality. Such states and their multiquanta generalizations are essential ingredients in many quantum communication and quantum computation schemes (*Physics World*, 1998).

VII. FINAL REMARKS AND OUTLOOK

I hope that the reader can sympathize now with my viewpoint that quantum physics goes beyond Wittgenstein, who starts his Tractatus Logico-Philosophicus with the sentence, "The world is everything that is the case." This is a classical viewpoint, a quantum state goes beyond. It represents all possibilities of everything that could be the case.

In any case, it will be interesting in the future to see more and more quantum experiments realized with increasingly larger objects. Another very promising future avenue of development is to realize entanglements of increasing complexity, either by entangling more and more systems with each other, or by entangling systems with a larger number of degrees of freedom. Eventually, all these developments will push the realm of quantum physics well into the macroscopic world. I expect that they will further elucidate Bohr's viewpoint that over a very large range the classical-quantum boundary is at the whim of the experimenter. Which parts we can talk about using our classical language and which parts are the quantum system depends on the specific experimental setup.

In the present brief overview I avoided all discussion of various alternative interpretations of quantum physics. I also did not venture into analyzing possible suggested alternatives to quantum mechanics. All these topics are quite important, interesting and in lively development. I hope my omissions are justified by the lack of space. It is my personal expectation that new insight and any progress in the interpretive discussion of quantum mechanics will bring along fundamentally new assessment of our humble role in the Universe.

ACKNOWLEDGMENTS

I would like to thank all my colleagues and collaborators who over many years shared with me the joy of working in the foundations of quantum physics. I am deeply indebted to Ben Bederson and Kurt Gottfried for careful reading of my initial manuscript and for detailed comments. I would also like to thank those who supported my work financially, particularly the Austrian Science Foundation FWF and the U.S. National Science Foundation NSF.

REFERENCES

As impossible as it is to do justice to all the work done in the field, it is equally impossible to attempt to quote even a minute fraction only of the relevant papers published. The general references below should help the reader to delve deeper into the subject:

Bell, J. S., 1987, *Speakable and Unspeakable in Quantum Mechanics* (Cambridge University, Cambridge).

Feynman, R. P., R. B. Leighton, and M. Sands, 1965, *The Feynman Lectures of Physics*, Vol. III, Quantum Mechanics (Addison-Wesley, Reading).

Greenberger, D. M., and A. Zeilinger, 1995, Eds., *Fundamental Problems in Quantum Theory* (Annals of the New York Academy of Sciences, Vol. 755, New York).

Greenstein, G. and A. G. Zajonc, 1997, *The Quantum Challenge: Modern Research on the Foundations of Quantum Mechanics* (Jones and Bartlett, New York).

Klyshko D., 1988, *Photons and Nonlinear Optics* (Gordon and Breech, New York).

Quantum Information, 1998, Phys. World **11** (3).

Rauch, H., and S. A. Werner, *Neutron Interferometry. Lessons in Experimental Quantum Mechanics* (Oxford University, Oxford) (in press).

Silverman, M. P., 1995, *More than One Mystery: Explorations in Quantum Interference* (Springer, Berlin).

Wheeler J. A., and W. H. Zurek, 1983, Eds. *Quantum Theory and Measurement* (Princeton University Princeton).

Specific papers quoted in the present paper are:

Alley, C. O., and Y. H. Shih, 1986, in *Proceedings of the Second International Symposium on Foundations of Quantum Mechanics in the Light of New Technology*, Tokyo, 1986, edited by M. Namiki *et al.* (Physical Society of Japan), p. 47.

Aspect, A., J. Dalibard, and G. Roger, 1982, Phys. Rev. Lett. **49**, 1804.

Aspect, A., P. Grangier, and G. Roger, 1981, Phys. Rev. Lett. **47**, 460.

Aspect, A., P. Grangier, and G. Roger, 1982, Phys. Rev. Lett. **49**, 91.

Bell, J. S., 1964, Physics (Long Island City, N.Y.) **1**, reprinted in J. S. Bell, 1987, *Speakable and Unspeakable in Quantum Mechanics* (Cambridge University, Cambridge).

Bell, J. S., 1966, Rev. Mod. Phys. **38**, 447.

Bell, J. S., 1981, J. Phys. C2 **42**, 41, reprinted in J. S. Bell, 1987, *Speakable and Unspeakable in Quantum Mechanics* (Cambridge University, Cambridge).

Bell, J. S., 1990, Phys. World **3**(August), 33.

Bennett, C. H., G. Brassard, C. Crépeau, R. Josza, A. Peres, and W. K. Wootters, 1993, Phys. Rev. Lett. **70**, 1895.

Bohr, N., 1935, Phys. Rev. **48**, 696.

Bouwmeester, D., J. W. Pan, K. Mattle, M. Eibl, H. Weinfurter, and A. Zeilinger, 1997, Nature (London) **390**, 575.

Braunstein, S. L., A. Mann, and M. Revzen, 1992, Phys. Rev. Lett. **68**, 3259.

Burnham, D. C., and D. L. Weinberg, 1970, Phys. Rev. Lett. **25**, 84.

Clauser, J. F., M. A. Horne, A. Shimony, and R. A. Holt, 1969, Phys. Rev. Lett. **23**, 880.

Dopfer, B., 1998, Ph.D. thesis (University of Innsbruck).

Einstein, A., B. Podolsky, and N. Rosen, 1935, Phys. Rev. **47**, 777.

Franson, J. D., 1989, Phys. Rev. Lett. **62**, 2205.

Freedman, S. J., and J. S. Clauser, 1972, Phys. Rev. Lett. **28**, 938.

Gottfried, K., 1991, Phys. World **4**(October), 41.

Greenberger, D. M., M. A. Horne, and A. Zeilinger, 1993, Phys. Today (8), 22.

Greenberger, D. M., M. A. Horne, and A. Zeilinger, 1989, in *Bell's Theorem, Quantum Theory, and Conceptions of the Universe*, edited by M. Kafatos (Kluwer, Dordrecht), p. 74.

Greenberger, D. M., M. A. Horne, A. Shimony, and A. Zeilinger, 1990, Am. J. Phys. **58**, 1131.

Hagley, E., X. Maître, G. Nogues, C. Wunderlich, M. Brune, J. M. Raimond, and S. Haroche, 1997, Phys. Rev. Lett. **79**, .

Heisenberg, W., 1927, Z. Phys. (Leipzig) **43**, 172, reprinted in English in Wheeler and Zurek (above).

Horne, M. A., and A. Zeilinger, 1985, in *Proceedings of the Symposium Foundations of Modern Physics*, edited by P. Lahti and P. Mittelstaedt (World Scientific, Singapore), p. 435.

Horne, M. A., and A. Zeilinger, 1988, in *Microphysical Reality and Quantum Formalism*, edited by A. van der Merwe, F. Seller, and G. Tarozzi (Kluwer, Dordrecht), p. 401.

Kochen, S. and E. Specker, 1967, J. Math. Mech. **17**, 59.

Kwiat, P., H. Weinfurter, T. Herzog, and A. Zeilinger, 1995, Phys. Rev. Lett. **74**, 4763.

Mandel, L., 1999, Rev. Mod. Phys. **71** (this issue).

Mermin, N. D., 1993, Rev. Mod. Phys. **65**, 803.

Pan, J. W., D. Bouwmeester, H. Weinfurter, and A. Zeilinger, 1998, Phys. Rev. Lett. **80**, 3891.

Penrose, R., 1986, in *Quantum Concepts in Space and Time*, edited by R. Penrose and C. J. Isham (Clarendon Press, Oxford), p. 139.

Schrödinger, E., 1935, Naturwissenschaften **23**, English translation in *Proceedings of the American Philosophical Society*, 124 (1980) reprinted in Wheeler & Zurek, above.

Specker, F., 1960, Dialectica **14**, 239.

Tittel, W., J. Brendel, B. Gisin, T. Herzog, H. Zbinden, and N. Gisin, 1998, Phys. Rev. A **57**, 3229.

Weihs, G., T. Jenewein, C. Simon, H. Weinfurter, and A. Zeilinger, 1998, Phys. Rev. Lett **81**, 5039.

Weizsächer, K. F., 1931, Z. Phys. **40**, 114.

Wigner, E. P., 1970, Am. J. Phys. **38**, 1005.

Wu, C. S., and I. Shaknov, 1950, Phys. Rev. **77**, 136.

Zeilinger, A., 1999, Found. Phys. (in press).

Zeilinger, A., R. Gähler, C. G. Shull, W. Treimer, and W. Hampe, 1988, Rev. Mod. Phys. **60**, 1067.

Zukowski, M., A. Zeilinger, M. A. Horne, and A. K. Ekert, 1993, Phys. Rev. Lett. **71**, 4287.

Quantum Entanglement Manipulation

Quantum Entanglement Manipulation

Dik Bouwmeester
Center for Quantum Computation, Oxford

1 Quantum Teleportation

The scheme for quantum teleportation – the transmission and reconstruction over arbitrary distances of the state of a quantum system – has been proposed by Bennett, Brassard, Crépeau, Jozsa, Peres, and Wootters [1]. This publication is the first selected paper of this Chapter. Quantum optics has proven very successful for the experimental implementation of the quantum teleportation scheme. The second selected paper presents the quantum teleportation experiment performed in Innsbruck [2]. In this experiment the polarization state of a single photon is teleported using an auxiliary pair of entangled photons. Another experiment that demonstrates the transfer of the polarization state of a single photon onto another photon has been proposed by Popescu [3] and has been performed in Rome [4].

The third selected paper is by Braunstein and Kimble [5] and explains teleportation of continuous quantum variables, which has initially been proposed by Vaidman [6]. The continuous quantum variables under consideration are the quadrature amplitude components of optical fields. An experimental demonstration of teleportation of these continuous quantum variables has been performed at Caltech [7].

Each of the three quantum teleportation experiments mentioned above has its own advantages and disadvantages and a comparison between the various methods can be found in the literature [8, 9, 10].

If the initial quantum state of the teleportation protocol is part of an entangled state, the result of the teleportation process is that two systems become entangled that did not directly interact with one another. This process, referred to as "Entanglement Swapping" [11] has experimentally been realized [12] and plays an important role in several quantum communication schemes [13, 14].

2 GHZ States

Ever since the seminal work of Einstein, Podolsky and Rosen [15] there has been a quest for generating entanglement between quantum particles. Two-particle entanglements have long been demonstrated experimentally [16], and only very recently the creation of entanglement between three spatial separated photons [17] has been achieved [18]. Proposals exist, and experiments are in progress, to generate entanglement between three atoms [19], and three nuclear spins within a single molecule have been prepared such that they locally exhibit three-particle correlations [20].

The original motivation to prepare three-particle entanglements stems from the observation by Greenberger, Horne and Zeilinger (GHZ) that three-particle entanglement leads to a conflict with local realism for non-statistical predictions of quantum mechanics. One of the initial publications on this topic is the fourth and fifth selected papers of this chapter [21] and additional references are given in [22]. This non-statistical nature of the conflict is in contrast to the case of Einstein, Podolsky and Rosen experiments with two entangled particles testing Bellś inequalities, where the conflict only arises for statistical predictions [23]. This issue is experimentally address in the final reprinted paper Ref. [24].

The incentive to produce GHZ states has been significantly increased by the advance of the field of quantum communication and quantum information processing. Entanglement between several particles is the most important feature of many such quantum communication and computation protocols [25, 26]. Additional information on the topics mentioned above and on most topics presented in other Chapters can be found in Ref. [27]

References

[1] Bennett, C. H., Brassard, G., Crépeau, C., Jozsa, R., Peres, A. and Wootters, W. K. Teleporting an unknown quantum state via dual classic and Einstein-Podolsky-Rosen channels. Phys. Rev. Lett. **70**, 1895-1899 (1993).

[2] D. Bouwmeester, J.-W. Pan, K. Mattle, M. Eibl, H. Weinfurter, and A. Zeilinger, Experimental Quantum Teleportation. Nature **390**, 575 (1997).

[3] S. Popescu, (quant-ph 9501020).

[4] D. Boschi, S. Branca, F. De Martini, L. Hardy, and S. Popescu, Phys. Rev. Lett. **80**, 1121 (1998).

[5] S.L. Braunstein and H.J. Kimble, Phys. Rev. Lett. **80**, 869 (1998).

[6] L. Vaidman, Phys. Rev. A **49**, 1473 (1994).

[7] A. Furusawa, J.Ł. Sørenson, S.L. Braunstein, C.A. Fuchs, H.J. Kimble, and E.S. Polzik, Science Oct23 1998 pp 706-709.

[8] Comment by F. De Martini, and Reply by A. Zeilinger, Physics World **11**, nr.3, 23-24 (March 1998).

[9] Comment by S.L. Braunstein and H.J. Kimble, and Reply by D. Bouwmeester, J-W. Pan, M. Daniell, H. Weinfurter, M. Zukowski, and A. Zeilinger, Nature (London) **394**, 840-841 (1998).

[10] D. Bouwmeester, J.-W. Pan, H. Weinfurter, and A. Zeilinger, High-fidelity teleportation of qubits, J. Mod. Opt.XX.

[11] M. Zukowski, A. Zeilinger, M.A. Horne, and A. Ekert, "Event-ready-detectors" Bell experiment via entanglement swapping. Phys. Rev. Lett. **71**, 4287-4290 (1993).

[12] J.-W. Pan, D. Bouwmeester, H. Weinfurter, and A. Zeilinger, Experimental entanglement swapping: entangling photons that never interacted. Phys. Rev. Lett. **80**, 3891 (1998).

[13] S. Bose, V. Vedral, and P.L. Knight, *A multiparticle generalization of entanglement swapping* Phys. Rev. A **57**, 822 (1998).

[14] H-J. Briegel, W. Dür, J.I. Cirac, and P. Zoller, Phys. Rev. Lett. **81**, 5932 (1998).

[15] A. Einstein, B. Podolsky, N. Rosen, Phys. Rev. **47**, 777 (1935).

[16] C.S. Wu, I. Shaknov, Phys. Rev. **77**, 136 (1950); S.J. Freedman, J.S. Clauser, Phys. Rev. Lett. **28**, 938 (1972); A. Aspect, J. Dalibard, G. Roger, Phys. Rev. Lett. **47**, 1804 (1982); E. Hagley, X. Maître, G. Nogues, C. Wunderlich, M. Brune, J.M. Raimond, S. Haroche, Phys. Rev. Lett. **79**, 1 (1997); P.G. Kwiat, K. Mattle, H. Weinfurter, A. Zeilinger, A.V. Sergienko, Y.H. Shih, Phys. Rev. Lett. **75**, 4337 (1995); G. Weish, T. Jennewein, C. Simon, H. Weinfurter, and A. Zeilinger, Phys. Rev. Lett. **81**, 5039 (1998).

[17] A. Zeilinger, M.A. Horne, H. Weinfurter, M. Zukowski, Phys. Rev. Lett. **78**, 3031 (1997).

[18] D. Bouwmeester, J.-W. Pan, M. Daniell, H. Weinfurter, and A. Zeilinger, Phys. Rev. Lett. **82**, 1345 (1999).

[19] S. Haroche, Ann. N. Y. Acad. Sci. **755**, 73 (1995); J.I. Cirac, P. Zoller, Phys. Rev. A **50**, R2799 (1994).

[20] S. Lloyd, Phys. Rev. A **57**, R1473 (1998); R. Laflamme, E. Knill, W.H. Zurek, P. Catasti, S.V.S. Mariappan, Phil.Trans. R. Soc. Lond. A **356**, 1941 (1998).

[21] D.M. Greenberger, M.A. Horne, A. Zeilinger, A. Going beyond Bell's theorem, in *Bell's Theorem, Quantum Theory, and Conceptions of the Universe,* edited by M. Kafatos, (Kluwer Academics, Dordrecht, The Netherlands, 1989), pp. 73-76.
N.D. Mermin, Physics Today, 9, June 1990.

[22] D.M. Greenberger, M.A. Horne,A. Shimony, A. Zeilinger, Am. J. Phys. **58**, 1131 (1990). N.D. Mermin, Am. J. Phys. **58**, 731 (1990)D.M. Greenberger, M.A. Horne, A. Zeilinger, Physics Today, 22, August 1993.

[23] Bell, J. S. *On the Einstein-Podolsky-Rosen paradox*, Physics **1**, 195 (1964), reprinted in J. S. Bell, *Speakable and Unspeakable in Quantum Mechanics*, Cambridge U.P., Cambridge, (1987).

[24] D. Bouwmeester,J.-W. Pan, M. Daniell, H. Weinfurter, and A. Zeilinger,*Observation of three-photon Greenberger-Horne-Zeilinger Entanglement.* Phys. Rev. Lett. **82**, 1345-1349

[25] C.H. Bennett, Physics Today, 24, October 1995; Feature issue Quantum Information, Physics World, March 1998.

[26] R. Cleve and H. Buhrman, Phys. Rev. A **56**, 1201 (1997); D. Bruss, D. DiVincenzo, A. Ekert, C. Fuchs, C. Mecchiavello, J. Smolin, Phys. Rev. A **57**, 2368 (1998).

[27] *The Physics of Quantum Information: quantum cryptography, quantum teleportation and quantum computation*, edited by D. Bouwmeester, A. Ekert, and A. Zeilinger, Springer-Verlag, Berlin, to be published.

PHYSICAL REVIEW LETTERS

VOLUME 70 | 29 MARCH 1993 | NUMBER 13

Teleporting an Unknown Quantum State via Dual Classical and Einstein-Podolsky-Rosen Channels

Charles H. Bennett,[1] Gilles Brassard,[2] Claude Crépeau,[2],[3] Richard Jozsa,[2] Asher Peres,[4] and William K. Wootters[5]

[1] *IBM Research Division, T.J. Watson Research Center, Yorktown Heights, New York 10598*

[2] *Département IRO, Université de Montréal, C.P. 6128, Succursale "A", Montréal, Québec, Canada H3C 3J7*

[3] *Laboratoire d'Informatique de l'École Normale Supérieure, 45 rue d'Ulm, 75230 Paris CEDEX 05, France*[a]

[4] *Department of Physics, Technion–Israel Institute of Technology, 32000 Haifa, Israel*

[5] *Department of Physics, Williams College, Williamstown, Massachusetts 01267*

(Received 2 December 1992)

An unknown quantum state $|\phi\rangle$ can be disassembled into, then later reconstructed from, purely classical information and purely nonclassical Einstein-Podolsky-Rosen (EPR) correlations. To do so the sender, "Alice," and the receiver, "Bob," must prearrange the sharing of an EPR-correlated pair of particles. Alice makes a joint measurement on her EPR particle and the unknown quantum system, and sends Bob the classical result of this measurement. Knowing this, Bob can convert the state of his EPR particle into an exact replica of the unknown state $|\phi\rangle$ which Alice destroyed.

PACS numbers: 03.65.Bz, 42.50.Dv, 89.70.+c

The existence of long range correlations between Einstein-Podolsky-Rosen (EPR) [1] pairs of particles raises the question of their use for information transfer. Einstein himself used the word "telepathically" in this context [2]. It is known that *instantaneous* information transfer is definitely impossible [3]. Here, we show that EPR correlations can nevertheless assist in the "teleportation" of an intact quantum state from one place to another, by a sender who knows neither the state to be teleported nor the location of the intended receiver.

Suppose one observer, whom we shall call "Alice," has been given a quantum system such as a photon or spin-$\frac{1}{2}$ particle, prepared in a state $|\phi\rangle$ unknown to her, and she wishes to communicate to another observer, "Bob," sufficient information about the quantum system for him to make an accurate copy of it. Knowing the state vector $|\phi\rangle$ itself would be sufficient information, but in general there is no way to learn it. Only if Alice knows beforehand that $|\phi\rangle$ belongs to a given orthonormal set can she make a measurement whose result will allow her to make an accurate copy of $|\phi\rangle$. Conversely, if the possibilities for $|\phi\rangle$ include two or more nonorthogonal states, then no measurement will yield sufficient information to prepare a perfectly accurate copy.

A trivial way for Alice to provide Bob with all the information in $|\phi\rangle$ would be to send the particle itself. If she wants to avoid transferring the original particle, she can make it interact unitarily with another system, or "ancilla," initially in a known state $|a_0\rangle$, in such a way that after the interaction the original particle is left in a standard state $|\phi_0\rangle$ and the ancilla is in an unknown state $|a\rangle$ containing complete information about $|\phi\rangle$. If Alice now sends Bob the ancilla (perhaps technically easier than sending the original particle), Bob can reverse her actions to prepare a replica of her original state $|\phi\rangle$. This "spin-exchange measurement" [4] illustrates an essential feature of quantum information: it can be swapped from one system to another, but it cannot be duplicated or "cloned" [5]. In this regard it is quite unlike classical information, which can be duplicated at will. The most tangible manifestation of the nonclassicality of quantum information is the violation of Bell's inequalities [6] observed [7] in experiments on EPR states. Other manifestations include the possibility of quantum cryptography [8], quantum parallel computation [9], and the superiority of interactive measurements for extracting information from a pair of identically prepared particles [10].

The spin-exchange method of sending full information to Bob still lumps classical and nonclassical information together in a single transmission. Below, we show how Alice can divide the full information encoded in $|\phi\rangle$ into two parts, one purely classical and the other purely nonclassical, and send them to Bob through two different channels. Having received these two transmissions, Bob can construct an accurate replica of $|\phi\rangle$. Of course Alice's original $|\phi\rangle$ is destroyed in the process, as it must be to obey the no-cloning theorem. We call the process we are about to describe teleportation, a term from science fiction meaning to make a person or object disappear while an exact replica appears somewhere else. It must be emphasized that our teleportation, unlike some science fiction versions, defies no physical laws. In particular, it cannot take place instantaneously or over a spacelike interval, because it requires, among other things, sending a classical message from Alice to Bob. The net result of teleportation is completely prosaic: the removal of $|\phi\rangle$ from Alice's hands and its appearance in Bob's hands a suitable time later. The only remarkable feature is that, in the interim, the information in $|\phi\rangle$ has been cleanly separated into classical and nonclassical parts. First we shall show how to teleport the quantum state $|\phi\rangle$ of a spin-$\frac{1}{2}$ particle. Later we discuss teleportation of more complicated states.

The nonclassical part is transmitted first. To do so, two spin-$\frac{1}{2}$ particles are prepared in an EPR singlet state

$$|\Psi_{23}^{(-)}\rangle = \sqrt{\tfrac{1}{2}}\,(|\uparrow_2\rangle|\downarrow_3\rangle - |\downarrow_2\rangle|\uparrow_3\rangle). \qquad (1)$$

The subscripts 2 and 3 label the particles in this EPR pair. Alice's original particle, whose unknown state $|\phi\rangle$ she seeks to teleport to Bob, will be designated by a subscript 1 when necessary. These three particles may be of different kinds, e.g., one or more may be photons, the polarization degree of freedom having the same algebra as a spin.

One EPR particle (particle 2) is given to Alice, while the other (particle 3) is given to Bob. Although this establishes the possibility of nonclassical correlations between Alice and Bob, the EPR pair at this stage contains no information about $|\phi\rangle$. Indeed the entire system, comprising Alice's unknown particle 1 and the EPR pair, is in a pure product state, $|\phi_1\rangle\,|\Psi_{23}^{(-)}\rangle$, involving neither classical correlation nor quantum entanglement between the unknown particle and the EPR pair. Therefore no measurement on either member of the EPR pair, or both together, can yield any information about $|\phi\rangle$. An entanglement between these two subsystems is brought about in the next step.

To couple the first particle with the EPR pair, Alice performs a complete measurement of the von Neumann type on the joint system consisting of particle 1 and particle 2 (her EPR particle). This measurement is performed in the Bell operator basis [11] consisting of $|\Psi_{12}^{(-)}\rangle$ and

$$|\Psi_{12}^{(+)}\rangle = \sqrt{\tfrac{1}{2}}\,(|\uparrow_1\rangle|\downarrow_2\rangle + |\downarrow_1\rangle|\uparrow_2\rangle),$$
$$|\Phi_{12}^{(\pm)}\rangle = \sqrt{\tfrac{1}{2}}\,(|\uparrow_1\rangle|\uparrow_2\rangle \pm |\downarrow_1\rangle|\downarrow_2\rangle). \qquad (2)$$

Note that these four states are a complete orthonormal basis for particles 1 and 2.

It is convenient to write the unknown state of the first particle as

$$|\phi_1\rangle = a\,|\uparrow_1\rangle + b\,|\downarrow_1\rangle, \qquad (3)$$

with $|a|^2 + |b|^2 = 1$. The complete state of the three particles before Alice's measurement is thus

$$|\Psi_{123}\rangle = \frac{a}{\sqrt{2}}\,(|\uparrow_1\rangle|\uparrow_2\rangle|\downarrow_3\rangle - |\uparrow_1\rangle|\downarrow_2\rangle|\uparrow_3\rangle) + \frac{b}{\sqrt{2}}\,(|\downarrow_1\rangle|\uparrow_2\rangle|\downarrow_3\rangle - |\downarrow_1\rangle|\downarrow_2\rangle|\uparrow_3\rangle). \qquad (4)$$

In this equation, each direct product $|\ _1\rangle|\ _2\rangle$ can be expressed in terms of the Bell operator basis vectors $|\Phi_{12}^{(\pm)}\rangle$ and $|\Psi_{12}^{(\pm)}\rangle$, and we obtain

$$|\Psi_{123}\rangle = \tfrac{1}{2}\,[|\Psi_{12}^{(-)}\rangle\,(-a|\uparrow_3\rangle - b|\downarrow_3\rangle) + |\Psi_{12}^{(+)}\rangle\,(-a|\uparrow_3\rangle + b|\downarrow_3\rangle) + |\Phi_{12}^{(-)}\rangle\,(a|\downarrow_3\rangle + b|\uparrow_3\rangle) + |\Phi_{12}^{(+)}\rangle\,(a|\downarrow_3\rangle - b|\uparrow_3\rangle)]. \qquad (5)$$

It follows that, regardless of the unknown state $|\phi_1\rangle$, the four measurement outcomes are equally likely, each occurring with probability 1/4. Furthermore, after Alice's measurement, Bob's particle 3 will have been projected into one of the four pure states superposed in Eq. (5), according to the measurement outcome. These are, respectively,

$$-|\phi_3\rangle \equiv -\begin{pmatrix} a \\ b \end{pmatrix}, \quad \begin{pmatrix} -1 & 0 \\ 0 & 1 \end{pmatrix}|\phi_3\rangle,$$
$$\begin{pmatrix} 0 & 1 \\ 1 & 0 \end{pmatrix}|\phi_3\rangle, \quad \begin{pmatrix} 0 & -1 \\ 1 & 0 \end{pmatrix}|\phi_3\rangle. \qquad (6)$$

Each of these possible resultant states for Bob's EPR particle is related in a simple way to the original state $|\phi\rangle$ which Alice sought to teleport. In the case of the first (singlet) outcome, Bob's state is the same except for an irrelevant phase factor, so Bob need do nothing further to produce a replica of Alice's spin. In the three other cases, Bob must apply one of the unitary operators in Eq. (6), corresponding, respectively, to 180° rotations around the z, x, and y axes, in order to convert his EPR particle into a replica of Alice's original state $|\phi\rangle$. (If $|\phi\rangle$ represents a photon polarization state, a suitable combination of half-wave plates will perform these unitary operations.) Thus an accurate teleportation can be achieved in all cases by having Alice tell Bob the classical outcome of her measurement, after which Bob applies the required rotation to transform the state of his particle into a replica of $|\phi\rangle$. Alice, on the other hand, is left with particles 1 and 2 in one of the states $|\Psi_{12}^{(\pm)}\rangle$ or $|\Phi_{12}^{(\pm)}\rangle$, without any trace of the original state $|\phi\rangle$.

Unlike the quantum correlation of Bob's EPR particle 3 to Alice's particle 2, the result of Alice's measurement is purely classical information, which can be transmitted, copied, and stored at will in any suitable physical medium. In particular, this information need not be destroyed or canceled to bring the teleportation process to a successful conclusion: The teleportation of $|\phi\rangle$ from Alice to Bob has the side effect of producing two bits of random classical information, uncorrelated to $|\phi\rangle$, which are left behind at the end of the process.

Since teleportation is a *linear* operation applied to the quantum state $|\phi\rangle$, it will work not only with pure states, but also with mixed or entangled states. For example, let Alice's original particle 1 be itself part of an EPR singlet with another particle, labeled 0, which may be far away from both Alice and Bob. Then, after teleportation, particles 0 and 3 would be left in a singlet state, even though they had originally belonged to separate EPR pairs.

All of what we have said above can be generalized to systems having $N > 2$ orthogonal states. In place of an EPR spin pair in the singlet state, Alice would use a pair of N-state particles in a completely entangled state. For definiteness let us write this entangled state as $\sum_j |j\rangle \otimes |j\rangle/\sqrt{N}$, where $j = 0, 1, \ldots, N-1$ labels the N elements of an orthonormal basis for each of the N-state systems. As before, Alice performs a joint measurement on particles 1 and 2. One such measurement that has the desired effect is the one whose eigenstates are $|\psi_{nm}\rangle$, defined by

$$|\psi_{nm}\rangle = \sum_j e^{2\pi i j n/N}\, |j\rangle \otimes |(j+m) \bmod N\rangle \,/\, \sqrt{N}. \tag{7}$$

Once Bob learns from Alice that she has obtained the result nm, he performs on his previously entangled particle (particle 3) the unitary transformation

$$U_{nm} = \sum_k e^{2\pi i k n/N}\, |k\rangle\, \langle (k+m) \bmod N| \,. \tag{8}$$

This transformation brings Bob's particle to the original state of Alice's particle 1, and the teleportation is complete.

The classical message plays an essential role in teleportation. To see why, suppose that Bob is impatient, and tries to complete the teleportation by guessing Alice's classical message before it arrives. Then Alice's expected $|\phi\rangle$ will be reconstructed (in the spin-$\frac{1}{2}$ case) as a random mixture of the four states of Eq. (6). For any $|\phi\rangle$, this is a maximally mixed state, giving no information about the input state $|\phi\rangle$. It could not be otherwise, because any correlation between the input and the guessed output could be used to send a superluminal signal.

One may still inquire whether accurate teleportation of a two-state particle requires a full two bits of classical information. Could it be done, for example, using only two or three distinct classical messages instead of four, or four messages of unequal probability? Later we show that a full two bits of classical channel capacity are necessary. Accurate teleportation using a classical channel of any lesser capacity would allow Bob to send superluminal messages through the teleported particle, by guessing the classical message before it arrived (cf. Fig. 2).

Conversely one may inquire whether other states besides an EPR singlet can be used as the nonclassical channel of the teleportation process. Clearly any direct product state of particles 2 and 3 is useless, because for such states manipulation of particle 2 has no effect on what can be predicted about particle 3. Consider now a nonfactorable state $|\Upsilon_{23}\rangle$. It can readily be seen that after Alice's measurement, Bob's particle 3 will be related to $|\phi_1\rangle$ by four fixed unitary operations if and only if $|\Upsilon_{23}\rangle$ has the form

$$\sqrt{\tfrac{1}{2}}(|u_2\rangle\,|p_3\rangle + |v_2\rangle\,|q_3\rangle), \tag{9}$$

where $\{|u\rangle, |v\rangle\}$ and $\{|p\rangle, |q\rangle\}$ are any two pairs of orthonormal states. These are maximally entangled states [11], having maximally random marginal statistics for measurements on either particle separately. States which are less entangled reduce the fidelity of teleportation, and/or the range of states $|\phi\rangle$ that can be accurately teleported. The states in Eq. (9) are also precisely those obtainable from the EPR singlet by a local one-particle unitary operation [12]. Their use for the nonclassical channel is entirely equivalent to that of the singlet (1). Maximal entanglement is necessary and sufficient for faithful tele-

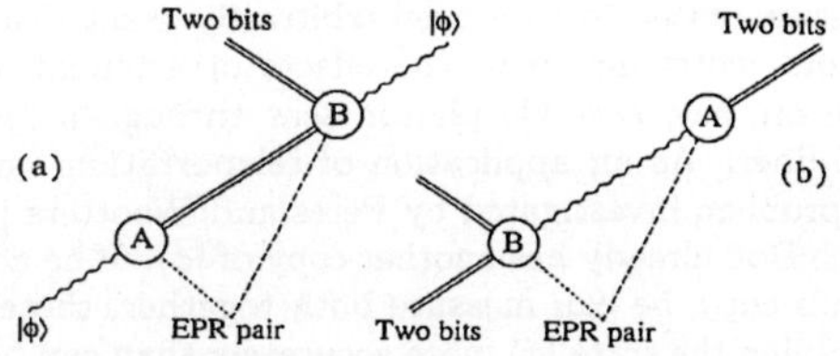

FIG. 1. Spacetime diagrams for (a) quantum teleportation, and (b) 4-way coding [12]. As usual, time increases from bottom to top. The solid lines represent a classical pair of bits, the dashed lines an EPR pair of particles (which may be of different types), and the wavy line a quantum particle in an unknown state $|\phi\rangle$. Alice (A) performs a quantum measurement, and Bob (B) a unitary operation.

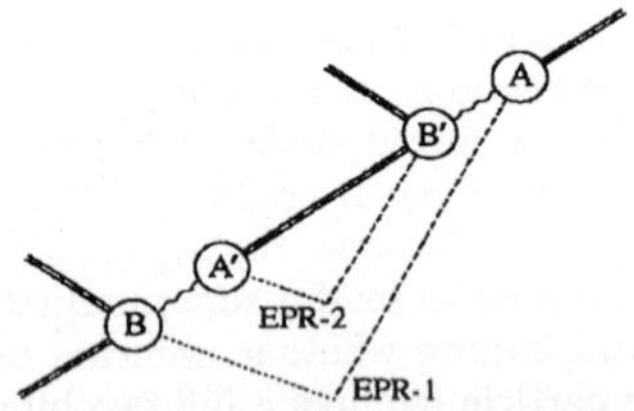

FIG. 2. Spacetime diagram of a more complex 4-way coding scheme in which the modulated EPR particle (wavy line) is teleported rather than being transmitted directly. This diagram can be used to prove that a classical channel of two bits of capacity is necessary for teleportation. To do so, assume on the contrary that the teleportation from A' to B' uses an internal classical channel of capacity $C < 2$ bits, but is still able to transmit the wavy particle's state accurately from A' to B', and therefore still transmit the external two-bit message accurately from B to A. The assumed lower capacity $C < 2$ of the internal channel means that if B' were to guess the internal classical message superluminally instead of waiting for it to arrive, his probability 2^{-C} of guessing correctly would exceed 1/4, resulting in a probability greater than 1/4 for successful superluminal transmission of the external two-bit message from B to A. This in turn entails the existence of two distinct external two-bit messages, r and s, such that $P(r|s)$, the probability of superluminally receiving r if s was sent, is less than 1/4, while $P(r|r)$, the probability of superluminally receiving r if r was sent, is greater than 1/4. By redundant coding, even this statistical difference between r and s could be used to send reliable superluminal messages; therefore reliable teleportation of a two-state particle cannot be achieved with a classical channel of less than two bits of capacity. By the same argument, reliable teleportation of an N-state particle requires a classical channel of $2\log_2(N)$ bits capacity.

portation.

Although it is currently unfeasible to store separated EPR particles for more than a brief time, if it becomes feasible to do so, quantum teleportation could be quite useful. Alice and Bob would only need a stockpile of EPR pairs (whose reliability can be tested by violations of Bell's inequality [7]) and a channel capable of carrying robust classical messages. Alice could then teleport quantum states to Bob over arbitrarily great distances, without worrying about the effects of attenuation and noise on, say, a single photon sent through a long optical fiber. As an application of teleportation, consider the problem investigated by Peres and Wootters [10], in which Bob already has another copy of $|\phi\rangle$. If he acquires Alice's copy, he can measure both together, thereby determining the state $|\phi\rangle$ more accurately than can be done by making a separate measurement on each one. Finally, teleportation has the advantage of still being possible in situations where Alice and Bob, after sharing their EPR pairs, have wandered about independently and no longer know each others' locations. Alice cannot reliably send Bob the original quantum particle, or a spin-exchanged version of it, if she does not know where he is; but she can still teleport the quantum state to him, by broadcasting the classical information to all places where he might be.

Teleportation resembles another recent scheme for using EPR correlations to help transmit useful information. In "4-way coding" [12] modulation of one member of an EPR pair serves to reliably encode a 2-bit message in the joint state of the complete pair. Teleportation and 4-way coding can be seen as variations on the same underlying process, illustrated by the spacetime diagrams in Fig. 1. Note that *closed loops* are involved for both processes. Trying to draw similar "Feynman diagrams" with tree structure, rather than loops, would lead to physically impossible processes.

On the other hand, more complicated closed-loop diagrams are possible, such as Fig. 2, obtained by substituting Fig. 1(a) into the wavy line of Fig. 1(b). This represents a 4-way coding scheme in which the modulated EPR particle is teleported instead of being transmitted directly. Two incoming classical bits on the lower left are reproduced reliably on the upper right, with the assistance of two shared EPR pairs and two other classical bits, uncorrelated with the external bits, in an internal channel from A' to B'. This diagram is of interest because it can be used to show that a full two bits of classical channel capacity are necessary for accurate teleportation of a two-state particle (cf. caption).

Work by G.B. is supported by NSERC's E. W. R. Steacie Memorial Fellowship and Québec's FCAR. A.P. was supported by the Gerard Swope Fund and the Fund for Encouragement of Research. Laboratoire d'Informatique de l'Ecole Normale Supérieure is associée au CNRS URA 1327.

(a) Permanent address.

[1] A. Einstein, B. Podolsky, and N. Rosen, Phys. Rev. **47**, 777 (1935); D. Bohm, *Quantum Theory* (Prentice Hall, Englewood Cliffs, NJ, 1951).
[2] A. Einstein, in *Albert Einstein, Philosopher-Scientist*, edited by P. A. Schilpp (Library of Living Philosophers, Evanston, 1949) p. 85.
[3] A. Shimony, in *Proceedings of the International Symposium on Foundations of Quantum Theory* (Physical Society of Japan, Tokyo, 1984).
[4] J. L. Park, Found. Phys. **1**, 23 (1970).
[5] W. K. Wootters and W. H. Zurek, Nature (London) **299**, 802 (1982).
[6] J. S. Bell, Physics (Long Island City, N.Y.) **1**, 195 (1964); J. F. Clauser, M. A. Horne, A. Shimony, and R. A. Holt, Phys. Rev. Lett. **23**, 880 (1969).
[7] A. Aspect, J. Dalibard, and G. Roger, Phys. Rev. Lett. **49**, 1804 (1982); Y. H. Shih and C. O. Alley, Phys. Rev. Lett. **61**, 2921 (1988).
[8] S. Wiesner, Sigact News **15**, 78 (1983); C. H. Bennett and G. Brassard, in *Proceedings of IEEE International*

Conference on Computers, Systems, and Signal Processing, Bangalore, India (IEEE, New York, 1984), pp. 175–179; A. K. Ekert, Phys. Rev. Lett. **67**, 661 (1991); C. H. Bennett, G. Brassard, and N. D. Mermin, Phys. Rev. Lett. **68**, 557–559 (1992); C. H. Bennett, Phys. Rev. Lett. **68**, 3121 (1992); A. K. Ekert, J. G. Rarity, P. R. Tapster, and G. M. Palma, Phys. Rev. Lett. **69**, 1293 (1992); C. H. Bennett, G. Brassard, C. Crépeau, and M.-H. Skubiszewska, *Advances in Cryptology—Crypto '91 Proceedings, August 1991* (Springer, New York, 1992), pp. 351–366; G. Brassard and C. Crépeau, *Advances in Cryptology—Crypto '90 Proceedings, August 1990* (Springer, New York, 1991), pp. 49–61.

[9] D. Deutsch Proc. R. Soc. London A **400**, 97 (1985); D. Deutsch and R. Jozsa, Proc. R. Soc. London A **439**, 553–558 (1992); A. Berthiaume and G. Brassard, in *Proceedings of the Seventh Annual IEEE Conference on Structure in Complexity Theory, Boston, June 1992*, (IEEE, New York, 1989), pp. 132–137; "Oracle Quantum Computing," Proceedings of the Workshop on Physics and Computation, PhysComp 92, (IEEE, Dallas, to be published).

[10] A. Peres and W. K. Wootters, Phys. Rev. Lett. **66**, 1119 (1991).

[11] S. L. Braunstein, A. Mann, and M. Revzen, Phys. Rev. Lett. **68**, 3259 (1992).

[12] C. H. Bennett and S. J. Wiesner, Phys. Rev. Lett. **69**, 2881 (1992).

Experimental quantum teleportation

Dik Bouwmeester, Jian-Wei Pan, Klaus Mattle, Manfred Eibl, Harald Weinfurter & Anton Zeilinger

Institut für Experimentalphysik, Universität Innsbruck, Technikerstr. 25, A-6020 Innsbruck, Austria

Quantum teleportation—the transmission and reconstruction over arbitrary distances of the state of a quantum system—is demonstrated experimentally. During teleportation, an initial photon which carries the polarization that is to be transferred and one of a pair of entangled photons are subjected to a measurement such that the second photon of the entangled pair acquires the polarization of the initial photon. This latter photon can be arbitrarily far away from the initial one. Quantum teleportation will be a critical ingredient for quantum computation networks.

The dream of teleportation is to be able to travel by simply reappearing at some distant location. An object to be teleported can be fully characterized by its properties, which in classical physics can be determined by measurement. To make a copy of that object at a distant location one does not need the original parts and pieces—all that is needed is to send the scanned information so that it can be used for reconstructing the object. But how precisely can this be a true copy of the original? What if these parts and pieces are electrons, atoms and molecules? What happens to their individual quantum properties, which according to the Heisenberg's uncertainty principle cannot be measured with arbitrary precision?

Bennett *et al.*[1] have suggested that it is possible to transfer the quantum state of a particle onto another particle—the process of quantum teleportation—provided one does not get any information about the state in the course of this transformation. This requirement can be fulfilled by using entanglement, the essential feature of quantum mechanics[2]. It describes correlations between quantum systems much stronger than any classical correlation could be.

The possibility of transferring quantum information is one of the cornerstones of the emerging field of quantum communication and quantum computation[3]. Although there is fast progress in the theoretical description of quantum information processing, the difficulties in handling quantum systems have not allowed an equal advance in the experimental realization of the new proposals. Besides the promising developments of quantum cryptography[4] (the first provably secure way to send secret messages), we have only recently succeeded in demonstrating the possibility of quantum dense coding[5], a way to quantum mechanically enhance data compression. The main reason for this slow experimental progress is that, although there exist methods to produce pairs of entangled photons[6], entanglement has been demonstrated for atoms only very recently[7] and it has not been possible thus far to produce entangled states of more than two quanta.

Here we report the first experimental verification of quantum teleportation. By producing pairs of entangled photons by the process of parametric down-conversion and using two-photon interferometry for analysing entanglement, we could transfer a quantum property (in our case the polarization state) from one photon to another. The methods developed for this experiment will be of great importance both for exploring the field of quantum communication and for future experiments on the foundations of quantum mechanics.

The problem

To make the problem of transferring quantum information clearer, suppose that Alice has some particle in a certain quantum state $|\psi\rangle$ and she wants Bob, at a distant location, to have a particle in that state. There is certainly the possibility of sending Bob the particle directly. But suppose that the communication channel between Alice and Bob is not good enough to preserve the necessary quantum coherence or suppose that this would take too much time, which could easily be the case if $|\psi\rangle$ is the state of a more complicated or massive object. Then, what strategy can Alice and Bob pursue?

As mentioned above, no measurement that Alice can perform on $|\psi\rangle$ will be sufficient for Bob to reconstruct the state because the state of a quantum system cannot be fully determined by measurements. Quantum systems are so evasive because they can be in a superposition of several states at the same time. A measurement on the quantum system will force it into only one of these states—this is often referred to as the projection postulate. We can illustrate this important quantum feature by taking a single photon, which can be horizontally or vertically polarized, indicated by the states $|\leftrightarrow\rangle$ and $|\updownarrow\rangle$. It can even be polarized in the general superposition of these two states

$$|\psi\rangle = \alpha|\leftrightarrow\rangle + \beta|\updownarrow\rangle \quad (1)$$

where α and β are two complex numbers satisfying $|\alpha|^2 + |\beta|^2 = 1$. To place this example in a more general setting we can replace the states $|\leftrightarrow\rangle$ and $|\updownarrow\rangle$ in equation (1) by $|0\rangle$ and $|1\rangle$, which refer to the states of any two-state quantum system. Superpositions of $|0\rangle$ and $|1\rangle$ are called qubits to signify the new possibilities introduced by quantum physics into information science[8].

If a photon in state $|\psi\rangle$ passes through a polarizing beamsplitter—a device that reflects (transmits) horizontally (vertically) polarized photons—it will be found in the reflected (transmitted) beam with probability $|\alpha|^2$ ($|\beta|^2$). Then the general state $|\psi\rangle$ has been projected either onto $|\leftrightarrow\rangle$ or onto $|\updownarrow\rangle$ by the action of the measurement. We conclude that the rules of quantum mechanics, in particular the projection postulate, make it impossible for Alice to perform a measurement on $|\psi\rangle$ by which she would obtain all the information necessary to reconstruct the state.

The concept of quantum teleportation

Although the projection postulate in quantum mechanics seems to bring Alice's attempts to provide Bob with the state $|\psi\rangle$ to a halt, it was realised by Bennett *et al.*[1] that precisely this projection postulate enables teleportation of $|\psi\rangle$ from Alice to Bob. During teleportation Alice will destroy the quantum state at hand while Bob receives the quantum state, with neither Alice nor Bob obtaining information about the state $|\psi\rangle$. A key role in the teleportation scheme is played by an entangled ancillary pair of particles which will be initially shared by Alice and Bob.

Suppose particle 1 which Alice wants to teleport is in the initial state $|\psi\rangle_1 = \alpha|\leftrightarrow\rangle_1 + \beta|\updownarrow\rangle_1$ (Fig. 1a), and the entangled pair of particles 2 and 3 shared by Alice and Bob is in the state:

$$|\Psi^-\rangle_{23} = \frac{1}{\sqrt{2}}\left(|\leftrightarrow\rangle_2|\updownarrow\rangle_3 - |\updownarrow\rangle_2|\leftrightarrow\rangle_3\right) \qquad (2)$$

That entangled pair is a single quantum system in an equal superposition of the states $|\leftrightarrow\rangle_2|\updownarrow\rangle_3$ and $|\updownarrow\rangle_2|\leftrightarrow\rangle_3$. The entangled state contains no information on the individual particles; it only indicates that the two particles will be in opposite states. The important property of an entangled pair is that as soon as a measurement on one of the particles projects it, say, onto $|\leftrightarrow\rangle$ the state of the other one is determined to be $|\updownarrow\rangle$, and vice versa. How could a measurement on one of the particles instantaneously influence the state of the other particle, which can be arbitrarily far away? Einstein, among many other distinguished physicists, could simply not accept this "spooky action at a distance". But this property of entangled states has now been demonstrated by numerous experiments (for reviews, see refs 9, 10).

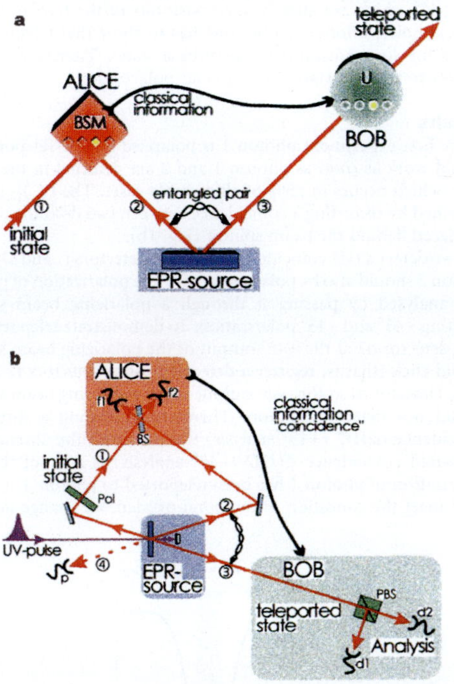

Figure 1 Scheme showing principles involved in quantum teleportation (**a**) and the experimental set-up (**b**). **a**, Alice has a quantum system, particle 1, in an initial state which she wants to teleport to Bob. Alice and Bob also share an ancillary entangled pair of particles 2 and 3 emitted by an Einstein-Podolsky-Rosen (EPR) source. Alice then performs a joint Bell-state measurement (BSM) on the initial particle and one of the ancillaries, projecting them also onto an entangled state. After she has sent the result of her measurement as classical information to Bob, he can perform a unitary transformation (U) on the other ancillary particle resulting in it being in the state of the original particle. **b**, A pulse of ultraviolet radiation passing through a nonlinear crystal creates the ancillary pair of photons 2 and 3. After retroflection during its second passage through the crystal the ultraviolet pulse creates another pair of photons, one of which will be prepared in the initial state of photon 1 to be teleported, the other one serving as a trigger indicating that a photon to be teleported is under way. Alice then looks for coincidences after a beam splitter BS where the initial photon and one of the ancillaries are superposed. Bob, after receiving the classical information that Alice obtained a coincidence count in detectors f1 and f2 identifying the $|\Psi^-\rangle_{12}$ Bell state, knows that his photon 3 is in the initial state of photon 1 which he then can check using polarization analysis with the polarizing beam splitter PBS and the detectors d1 and d2. The detector p provides the information that photon 1 is under way.

The teleportation scheme works as follows. Alice has the particle 1 in the initial state $|\psi\rangle_1$ and particle 2. Particle 2 is entangled with particle 3 in the hands of Bob. The essential point is to perform a specific measurement on particles 1 and 2 which projects them onto the entangled state:

$$|\Psi^-\rangle_{12} = \frac{1}{\sqrt{2}}\left(|\leftrightarrow\rangle_1|\updownarrow\rangle_2 - |\updownarrow\rangle_1|\leftrightarrow\rangle_2\right) \qquad (3)$$

This is only one of four possible maximally entangled states into which any state of two particles can be decomposed. The projection of an arbitrary state of two particles onto the basis of the four states is called a Bell-state measurement. The state given in equation (3) distinguishes itself from the three other maximally entangled states by the fact that it changes sign upon interchanging particle 1 and particle 2. This unique antisymmetric feature of $|\Psi^-\rangle_{12}$ will play an important role in the experimental identification, that is, in measurements of this state.

Quantum physics predicts[1] that once particles 1 and 2 are projected into $|\Psi^-\rangle_{12}$, particle 3 is instantaneously projected into the initial state of particle 1. The reason for this is as follows. Because we observe particles 1 and 2 in the state $|\Psi^-\rangle_{12}$ we know that whatever the state of particle 1 is, particle 2 must be in the opposite state, that is, in the state orthogonal to the state of particle 1. But we had initially prepared particle 2 and 3 in the state $|\Psi^-\rangle_{23}$, which means that particle 2 is also orthogonal to particle 3. This is only possible if particle 3 is in the same state as particle 1 was initially. The final state of particle 3 is therefore:

$$|\psi\rangle_3 = \alpha|\leftrightarrow\rangle_3 + \beta|\updownarrow\rangle_3 \qquad (4)$$

We note that during the Bell-state measurement particle 1 loses its identity because it becomes entangled with particle 2. Therefore the state $|\psi\rangle_1$ is destroyed on Alice's side during teleportation.

This result (equation (4)) deserves some further comments. The transfer of quantum information from particle 1 to particle 3 can happen over arbitrary distances, hence the name teleportation. Experimentally, quantum entanglement has been shown[11] to survive over distances of the order of 10 km. We note that in the teleportation scheme it is not necessary for Alice to know where Bob is. Furthermore, the initial state of particle 1 can be completely unknown not only to Alice but to anyone. It could even be quantum mechanically completely undefined at the time the Bell-state measurement takes place. This is the case when, as already remarked by Bennett *et al.*[1], particle 1 itself is a member of an entangled pair and therefore has no well-defined properties on its own. This ultimately leads to entanglement swapping[12,13].

It is also important to notice that the Bell-state measurement does not reveal any information on the properties of any of the particles. This is the very reason why quantum teleportation using coherent two-particle superpositions works, while any measurement on one-particle superpositions would fail. The fact that no information whatsoever is gained on either particle is also the reason why quantum teleportation escapes the verdict of the no-cloning theorem[14]. After successful teleportation particle 1 is not available in its original state any more, and therefore particle 3 is not a clone but is really the result of teleportation.

A complete Bell-state measurement can not only give the result that the two particles 1 and 2 are in the antisymmetric state, but with equal probabilities of 25% we could find them in any one of the three other entangled states. When this happens, particle 3 is left in one of three different states. It can then be brought by Bob into the original state of particle 1 by an accordingly chosen transformation, independent of the state of particle 1, after receiving via a classical communication channel the information on which of the Bell-state results was obtained by Alice. Yet we note, with emphasis, that even if we chose to identify only one of the four Bell states as discussed above, teleportation is successfully achieved, albeit only in a quarter of the cases.

Experimental realization

Teleportation necessitates both production and measurement of entangled states; these are the two most challenging tasks for any experimental realization. Thus far there are only a few experimental techniques by which one can prepare entangled states, and there exist no experimentally realized procedures to identify all four Bell states for any kind of quantum system. However, entangled pairs of photons can readily be generated and they can be projected onto at least two of the four Bell states.

We produced the entangled photons 2 and 3 by parametric down-conversion. In this technique, inside a nonlinear crystal, an incoming pump photon can decay spontaneously into two photons which, in the case of type II parametric down-conversion, are in the state given by equation (2) (Fig. 2)[6].

To achieve projection of photons 1 and 2 into a Bell state we have to make them indistinguishable. To achieve this indistinguishability we superpose the two photons at a beam splitter (Fig. 1b). Then if they are incident one from each side, how can it happen that they emerge still one on each side? Clearly this can happen if they are either both reflected or both transmitted. In quantum physics we have to superimpose the amplitudes for these two possibilities. Unitarity implies that the amplitude for both photons being reflected obtains an additional minus sign. Therefore, it seems that the two processes cancel each other. This is, however, only true for a symmetric input state. For an antisymmetric state, the two possibilities obtain another relative minus sign, and therefore they constructively interfere[15,16]. It is thus sufficient for projecting photons 1 and 2 onto the antisymmetric state $|\psi^-\rangle_{12}$ to place detectors in each of the outputs of the beam splitter and to register simultaneous detections (coincidence)[17–19].

To make sure that photons 1 and 2 cannot be distinguished by their arrival times, they were generated using a pulsed pump beam and sent through narrow-bandwidth filters producing a coherence time much longer than the pump pulse length[20]. In the experiment, the pump pulses had a duration of 200 fs at a repetition rate of 76 MHz. Observing the down-converted photons at a wavelength of 788 nm and a bandwidth of 4 nm results in a coherence time of 520 fs. It should be mentioned that, because photon 1 is also produced as part of an entangled pair, its partner can serve to indicate that it was emitted.

How can one experimentally prove that an unknown quantum state can be teleported? First, one has to show that teleportation works for a (complete) basis, a set of known states into which any other state can be decomposed. A basis for polarization states has just two components, and in principle we could choose as the basis horizontal and vertical polarization as emitted by the source. Yet this would not demonstrate that teleportation works for any general superposition, because these two directions are preferred directions in our experiment. Therefore, in the first demonstration we choose as the basis for teleportation the two states linearly polarized at −45° and +45° which are already superpositions of the horizontal and vertical polarizations. Second, one has to show that teleportation works for superpositions of these base states. Therefore we also demonstrate teleportation for circular polarization.

Results

In the first experiment photon 1 is polarized at 45°. Teleportation should work as soon as photon 1 and 2 are detected in the $|\psi^-\rangle_{12}$ state, which occurs in 25% of all possible cases. The $|\psi^-\rangle_{12}$ state is identified by recording a coincidence between two detectors, f1 and f2, placed behind the beam splitter (Fig. 1b).

If we detect a f1f2 coincidence (between detectors f1 and f2), then photon 3 should also be polarized at 45°. The polarization of photon 3 is analysed by passing it through a polarizing beam splitter selecting +45° and −45° polarization. To demonstrate teleportation, only detector d2 at the +45° output of the polarizing beam splitter should click (that is, register a detection) once detectors f1 and f2 click. Detector d1 at the −45° output of the polarizing beam splitter should not detect a photon. Therefore, recording a three-fold coincidence d2f1f2 (+45° analysis) together with the absence of a three-fold coincidence d1f1f2 (−45° analysis) is a proof that the polarization of photon 1 has been teleported to photon 3.

To meet the condition of temporal overlap, we change in small

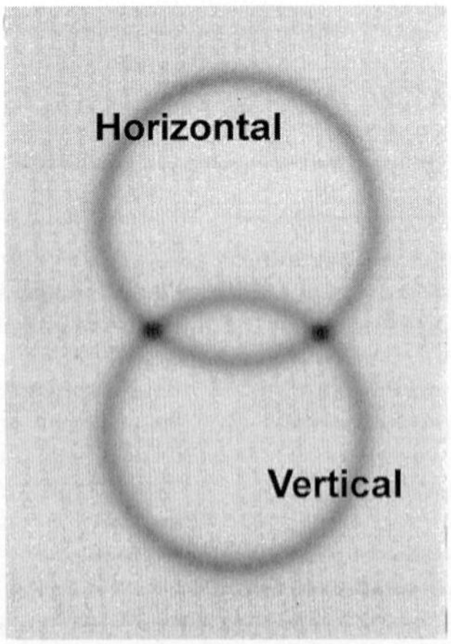

Figure 2 Photons emerging from type II down-conversion (see text). Photograph taken perpendicular to the propagation direction. Photons are produced in pairs. A photon on the top circle is horizontally polarized while its exactly opposite partner in the bottom circle is vertically polarized. At the intersection points their polarizations are undefined; all that is known is that they have to be different, which results in entanglement.

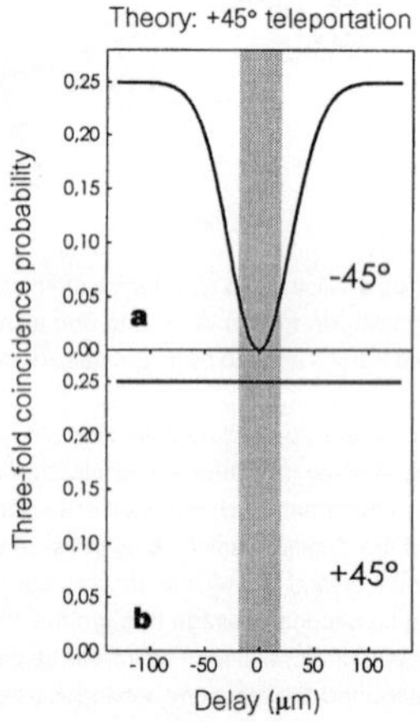

Figure 3 Theoretical prediction for the three-fold coincidence probability between the two Bell-state detectors (f1, f2) and one of the detectors analysing the teleported state. The signature of teleportation of a photon polarization state at +45° is a dip to zero at zero delay in the three-fold coincidence rate with the detector analysing −45° (d1f1f2) (**a**) and a constant value for the detector analysis +45° (d2f1f2) (**b**). The shaded area indicates the region of teleportation.

steps the arrival time of photon 2 by changing the delay between the first and second down-conversion by translating the retroflection mirror (Fig. 1b). In this way we scan into the region of temporal overlap at the beam splitter so that teleportation should occur.

Outside the region of teleportation, photon 1 and 2 each will go either to f1 or to f2 independent of one another. The probability of having a coincidence between f1 and f2 is therefore 50%, which is twice as high as inside the region of teleportation. Photon 3 should not have a well-defined polarization because it is part of an entangled pair. Therefore, d1 and d2 have both a 50% chance of receiving photon 3. This simple argument yields a 25% probability both for the −45° analysis (d1f1f2 coincidences) and for the +45° analysis (d2f1f2 coincidences) outside the region of teleportation. Figure 3 summarizes the predictions as a function of the delay. Successful teleportation of the +45° polarization state is then characterized by a decrease to zero in the −45° analysis (Fig. 3a), and by a constant value for the +45° analysis (Fig. 3b).

The theoretical prediction of Fig. 3 may easily be understood by realizing that at zero delay there is a decrease to half in the coincidence rate for the two detectors of the Bell-state analyser, f1 and f2, compared with outside the region of teleportation. Therefore, if the polarization of photon 3 were completely uncorrelated to the others the three-fold coincidence should also show this dip to half. That the right state is teleported is indicated by the fact that the dip goes to zero in Fig. 3a and that it is filled to a flat curve in Fig. 3b.

We note that equally as likely as the production of photons 1, 2 and 3 is the emission of two pairs of down-converted photons by a single source. Although there is no photon coming from the first source (photon 1 is absent), there will still be a significant contribution to the three-fold coincidence rates. These coincidences have nothing to do with teleportation and can be identified by blocking the path of photon 1.

The probability for this process to yield spurious two- and three-fold coincidences can be estimated by taking into account the experimental parameters. The experimentally determined value

Table 1 Visibility of teleportation in three fold coincidences

Polarization	Visibility
+45°	0.63 ± 0.02
−45°	0.64 ± 0.02
0°	0.66 ± 0.02
90°	0.61 ± 0.02
Circular	0.57 ± 0.02

for the percentage of spurious three-fold coincidences is 68% ± 1%. In the experimental graphs of Fig. 4 we have subtracted the experimentally determined spurious coincidences.

The experimental results for teleportation of photons polarized under +45° are shown in the left-hand column of Fig. 4; Fig. 4a and b should be compared with the theoretical predictions shown in Fig. 3. The strong decrease in the −45° analysis, and the constant signal for the +45° analysis, indicate that photon 3 is polarized along the direction of photon 1, confirming teleportation.

The results for photon 1 polarized at −45° demonstrate that teleportation works for a complete basis for polarization states (right-hand column of Fig. 4). To rule out any classical explanation for the experimental results, we have produced further confirmation that our procedure works by additional experiments. In these experiments we teleported photons linearly polarized at 0° and at 90°, and also teleported circularly polarized photons. The experimental results are summarized in Table 1, where we list the visibility of the dip in three-fold coincidences, which occurs for analysis orthogonal to the input polarization.

As mentioned above, the values for the visibilities are obtained after subtracting the offset caused by spurious three-fold coincidences. These can experimentally be excluded by conditioning the three-fold coincidences on the detection of photon 4, which effectively projects photon 1 into a single-particle state. We have performed this four-fold coincidence measurement for the case of teleportation of the +45° and +90° polarization states, that is, for two non-orthogonal

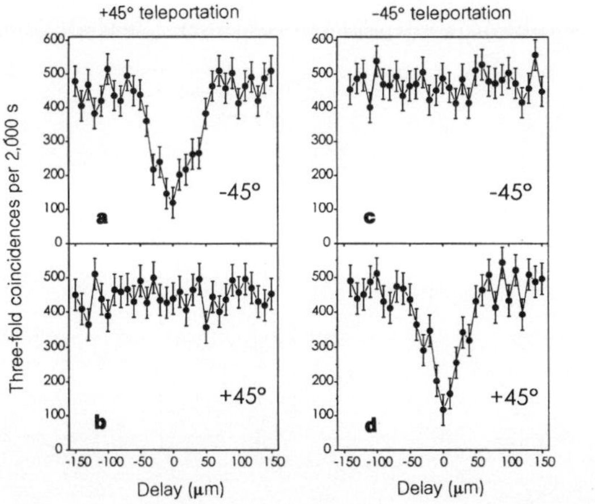

Figure 4 Experimental results. Measured three-fold coincidence rates d1f1f2 (−45°) and d2f1f2 (+45°) in the case that the photon state to be teleported is polarized at +45° (**a** and **b**) or at −45° (**c** and **d**). The coincidence rates are plotted as function of the delay between the arrival of photon 1 and 2 at Alice's beam splitter (see Fig. 1b). The three-fold coincidence rates are plotted after subtracting the spurious three-fold contribution (see text). These data, compared with Fig. 3, together with similar ones for other polarizations (Table 1) confirm teleportation for an arbitrary state.

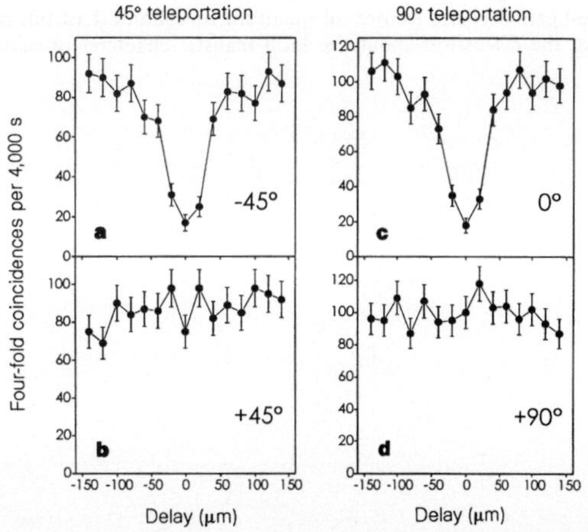

Figure 5 Four-fold coincidence rates (without background subtraction). Conditioning the three-fold coincidences as shown in Fig. 4 on the registration of photon 4 (see Fig. 1b) eliminates the spurious three-fold background. **a** and **b** show the four-fold coincidence measurements for the case of teleportation of the +45° polarization state; **c** and **d** show the results for the +90° polarization state. The visibilities, and thus the polarizations of the teleported photons, obtained without any background subtraction are 70% ± 3%. These results for teleportation of two non-orthogonal states prove that we have demonstrated teleportation of the quantum state of a single photon.

states. The experimental results are shown in Fig. 5. Visibilities of 70% ± 3% are obtained for the dips in the orthogonal polarization states. Here, these visibilities are directly the degree of polarization of the teleported photon in the right state. This proves that we have demonstrated teleportation of the quantum state of a single photon.

The next steps

In our experiment, we used pairs of polarization entangled photons as produced by pulsed down-conversion and two-photon interferometric methods to transfer the polarization state of one photon onto another one. But teleportation is by no means restricted to this system. In addition to pairs of entangled photons or entangled atoms[7,21], one could imagine entangling photons with atoms, or phonons with ions, and so on. Then teleportation would allow us to transfer the state of, for example, fast-decohering, short-lived particles, onto some more stable systems. This opens the possibility of quantum memories, where the information of incoming photons is stored on trapped ions, carefully shielded from the environment.

Furthermore, by using entanglement purification[22]—a scheme of improving the quality of entanglement if it was degraded by decoherence during storage or transmission of the particles over noisy channels—it becomes possible to teleport the quantum state of a particle to some place, even if the available quantum channels are of very poor quality and thus sending the particle itself would very probably destroy the fragile quantum state. The feasibility of preserving quantum states in a hostile environment will have great advantages in the realm of quantum computation. The teleportation scheme could also be used to provide links between quantum computers.

Quantum teleportation is not only an important ingredient in quantum information tasks; it also allows new types of experiments and investigations of the foundations of quantum mechanics. As any arbitrary state can be teleported, so can the fully undetermined state of a particle which is member of an entangled pair. Doing so, one transfers the entanglement between particles. This allows us not only to chain the transmission of quantum states over distances, where decoherence would have already destroyed the state completely, but it also enables us to perform a test of Bell's theorem on particles which do not share any common past, a new step in the investigation of the features of quantum mechanics. Last but not least, the discussion about the local realistic character of nature could be settled firmly if one used features of the experiment presented here to generate entanglement between more than two spatially separated particles[23,24]. □

Received 16 October; accepted 18 November 1997.

1. Bennett, C. H. *et al.* Teleporting an unknown quantum state via dual classic and Einstein-Podolsky-Rosen channels. *Phys. Rev. Lett.* **70**, 1895–1899 (1993).
2. Schrödinger, E. Die gegenwärtige Situation in der Quantenmechanik. *Naturwissenschaften* **23**, 807–812; 823–828; 844–849 (1935).
3. Bennett, C. H. Quantum information and computation. *Phys. Today* **48(10)**, 24–30, October (1995).
4. Bennett, C. H., Brassard, G. & Ekert, A. K. Quantum Cryptography. *Sci. Am.* **267(4)**, 50–57, October (1992).
5. Mattle, K., Weinfurter, H., Kwiat, P. G. & Zeilinger, A. Dense coding in experimental quantum communication. *Phys. Rev. Lett.* **76**, 4656–4659 (1996).
6. Kwiat, P. G. *et al.* New high intensity source of polarization-entangled photon pairs. *Phys. Rev. Lett.* **75**, 4337–4341 (1995).
7. Hagley, E. *et al.* Generation of Einstein-Podolsky-Rosen pairs of atoms. *Phys. Rev. Lett.* **79**, 1–5 (1997).
8. Schumacher, B. Quantum coding. *Phys. Rev. A* **51**, 2738–2747 (1995).
9. Clauser, J. F. & Shimony, A. Bell's theorem: experimental tests and implications. *Rep. Prog. Phys.* **41**, 1881–1927 (1978).
10. Greenberger, D. M., Horne, M. A. & Zeilinger, A. Multiparticle interferometry and the superposition principle. *Phys. Today* August, 22–29 (1993).
11. Tittel, W. *et al.* Experimental demonstration of quantum-correlations over more than 10 kilometers. *Phys. Rev. Lett.* (submitted).
12. Zukowski, M., Zeilinger, A., Horne, M. A. & Ekert, A. "Event-ready-detectors" Bell experiment via entanglement swapping. *Phys. Rev. Lett.* **71**, 4287–4290 (1993).
13. Bose, S., Vedral, V. & Knight, P. L. A multiparticle generalization of entanglement swapping. preprint.
14. Wootters, W. K. & Zurek, W. H. A single quantum cannot be cloned. *Nature* **299**, 802–803 (1982).
15. Loudon, R. *Coherence and Quantum Optics VI* (eds Everly, J. H. & Mandel, L.) 703–708 (Plenum, New York, 1990).
16. Zeilinger, A., Bernstein, H. J. & Horne, M. A. Information transfer with two-state two-particle quantum systems. *J. Mod. Optics* **41**, 2375–2384 (1994).
17. Weinfurter, H. Experimental Bell-state analysis. *Europhys. Lett.* **25**, 559–564 (1994).
18. Braunstein, S. L. & Mann, A. Measurement of the Bell operator and quantum teleportation. *Phys. Rev. A* **51**, R1727–R1730 (1995).
19. Michler, M., Mattle, K., Weinfurter, H. & Zeilinger, A. Interferometric Bell-state analysis. *Phys. Rev. A* **53**, R1209–R1212 (1996).
20. Zukowski, M., Zeilinger, A. & Weinfurter, H. Entangling photons radiated by independent pulsed sources. *Ann. NY Acad. Sci.* **755**, 91–102 (1995).
21. Fry, E. S., Walther, T. & Li, S. Proposal for a loophole-free test of the Bell inequalities. *Phys. Rev. A* **52**, 4381–4395 (1995).
22. Bennett, C. H. *et al.* Purification of noisy entanglement and faithful teleportation via noisy channels. *Phys. Rev. Lett.* **76**, 722–725 (1996).
23. Greenberger, D. M., Horne, M. A., Shimony, A. & Zeilinger, A. Bell's theorem without inequalities. *Am. J. Phys.* **58**, 1131–1143 (1990).
24. Zeilinger, A., Horne, M. A., Weinfurter, H. & Zukowski, M. Three particle entanglements from two entangled pairs. *Phys. Rev. Lett.* **78**, 3031–3034 (1997).

Acknowledgements. We thank C. Bennett, I. Cirac, J. Rarity, W. Wootters and P. Zoller for discussions, and M. Zukowski for suggestions about various aspects of the experiments. This work was supported by the Austrian Science Foundation FWF, the Austrian Academy of Sciences, the TMR program of the European Union and the US NSF.

Correspondence and requests for materials should be addressed to D.B. (e-mail: Dik.Bouwmeester@uibk.ac.at).

VOLUME 80, NUMBER 4 PHYSICAL REVIEW LETTERS 26 JANUARY 1998

Teleportation of Continuous Quantum Variables

Samuel L. Braunstein
SEECS, University of Wales, Bangor LL57 1UT, United Kingdom

H. J. Kimble
Norman Bridge Laboratory of Physics 12-33, California Institute of Technology, Pasadena, California 91125
(Received 8 September 1997)

Quantum teleportation is analyzed for states of dynamical variables with continuous spectra, in contrast to previous work with discrete (spin) variables. The entanglement fidelity of the scheme is computed, including the roles of finite quantum correlation and nonideal detection efficiency. A protocol is presented for teleporting the wave function of a single mode of the electromagnetic field with high fidelity using squeezed-state entanglement and current experimental capability. [S0031-9007(97)05114-4]

PACS numbers: 03.67.–a, 03.65.Bz, 42.50.Dv

Quantum mechanics offers certain unique capabilities for the processing of information, whether for computation or communication [1]. A particularly startling discovery by Bennett *et al.* is the possibility for teleportation of a quantum state, whereby an *unknown* state of a spin-$\frac{1}{2}$ particle is transported by "Alice" from a sending station to "Bob" at a receiving terminal by conveying 2 bits of classical information [2]. The enabling capability for this remarkable process is what Bell termed the irreducible nonlocal content of quantum mechanics, namely that Alice and Bob share an entangled quantum state and exploit its nonlocal characteristics for the teleportation process. For spin-$\frac{1}{2}$ particles, this entangled state is a pair of spins in a Bell state as in Bohm's version of the Einstein, Podolsky, and Rosen (EPR) paradox [3] and for which Bell formulated his famous inequalities [4].

Beyond the context of dichotomic variables, Vaidman has analyzed teleportation of the wave function of a one-dimensional particle in a beautiful variation of the original EPR paradox [5]. In this case, the nonlocal resource shared by Alice and Bob is the EPR state with perfect correlations in both position and momentum. The goal of this Letter is to extend Vaidman's analysis to incorporate finite (nonsingular) degrees of correlation among the relevant particles and to include inefficiencies in the measurement process. The "quality" of the resulting protocol for teleportation is quantified with the first explicit computation of the fidelity of entanglement for a process acting on an infinite dimensional Hilbert space. We further describe a realistic implementation for the quantum teleportation of states of continuous variables, where now the entangled state shared by Alice and Bob is a highly squeezed two-mode state of the electromagnetic field, with the quadrature amplitudes of the field playing the roles of position and momentum. Indeed, an experimental demonstration of the original EPR paradox for variables with a continuous spectrum has previously been carried out [6,7], which when combined with our analysis, forms the basis of a realizable experiment to teleport the complete quantum state of a single mode of the electromagnetic field.

Note that up until now, all experimental proposals for teleportation have involved dichotomic variables in $SU(2)$ [2,8–11], with optical schemes accomplishing the Bell-operator measurement with low efficiency. Indeed, the recent report of teleportation via parametric down conversion [12] succeeds only *a posteriori* with rare post-selected detection events. By contrast, our scheme employs linear elements corresponding to operations in SU(1,1) [13] for Bell-state detection and thus should operate at near unit absolute efficiency, enabling *a priori* teleportation as originally envisionaged in Ref. [2].

As shown schematically in Fig. 1, an unknown input state described by the Wigner function $W_{\text{in}}(\alpha)$ is to be teleported to a remote station, with the teleported (output) state denoted by $W_{\text{out}}(\alpha)$. In analogy with the previously proposed scheme for teleportation of the state of a spin-$\frac{1}{2}$ particle, Alice (at the sending station) and Bob (at the receiving terminal) have previously arranged to share an entangled state which is sent along paths 1 and 2. Within the context of our scheme in SU(1,1), the entangled state distributed to Alice and Bob is described by the Wigner function $W_{\text{EPR}}(\alpha_1, \alpha_2)$ [4]

$$W_{\text{EPR}}(\alpha_1;\alpha_2) = \frac{4}{\pi^2}\exp\{-e^{-2r}[(x_1 - x_2)^2 + (p_1 + p_2)^2] - e^{+2r}[(x_1 + x_2)^2 + (p_1 - p_2)^2]\}$$
$$\to C\delta(x_1 + x_2)\delta(p_1 - p_2), \tag{1}$$

where $\alpha_j = x_j + ip_j$. Here, the real quantities (x_j, p_j) correspond to canonically conjugate variables for the relevant pathways and describe, for example, position and momentum for a massive particle, and quadrature amplitudes for the

0031-9007/98/80(4)/869(4)$15.00

VOLUME 80, NUMBER 4 PHYSICAL REVIEW LETTERS 26 JANUARY 1998

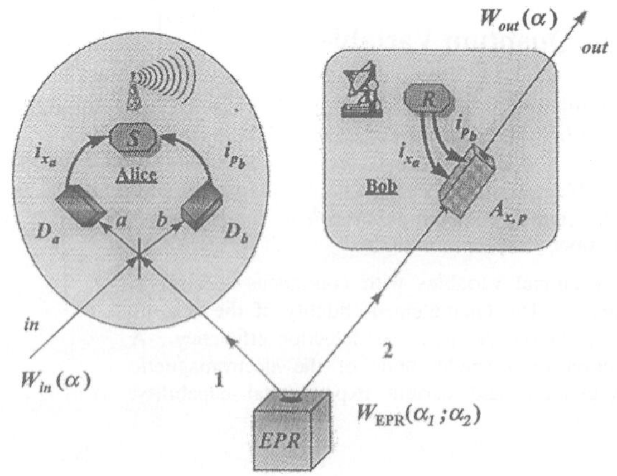

FIG. 1. Scheme for quantum teleportation of an (unknown) input state $W_{\text{in}}(\alpha)$ from Alice's sending station S to Bob's remote receiving terminal R, resulting in the teleported output state $W_{\text{out}}(\alpha)$.

electromagnetic field. Note that for $r \to \infty$, the state described by Eq. (1) becomes precisely the EPR state of Ref. [3] employed by Vaidman [5] and provides an ideal entangled "pair" shared between the teleportation sending and receiving stations, albeit with divergent energy in this limit.

As for the protocol itself, the first step in teleporting the (unknown) state $W_{\text{in}}(\alpha_{\text{in}})$ is to form new variables $\beta_{a,b}$ along paths (a, b) which are linear superpositions of those of the initially independent pathways *in* and 1 at the sending station S of Fig. 1, namely $\beta_{a,b} = \frac{1}{\sqrt{2}}(\alpha_1 \pm \alpha_{\text{in}})$. The resulting Wigner function in the variables $(\beta_a; \beta_b; \alpha_2)$ exhibits "entanglement" between the paths (a, b) and the remote path 2. Step 2 at S is then to measure the observables corresponding to $\text{Re}\,\beta_a = \frac{1}{\sqrt{2}}(x_1 + x_{\text{in}}) \equiv x_a$ and $\text{Im}\,\beta_b = \frac{1}{\sqrt{2}}(p_1 - p_{\text{in}}) \equiv p_b$ at the detectors (D_a, D_b) shown in Fig. 1, with the resulting classical outcomes denoted by (i_{x_z}, i_{p_b}), respectively. We define ideal measurement of (x_a, p_b) to be that for which the distribution $P_{ab}(i_{x_a}; i_{p_b})$ is identical to the associated Wigner function $W_{ab}(x_a; p_b)$. With the entangled state of paths (1,2) given by Eq. (1), we find

$$P_{ab}(i_{x_a}; i_{p_b}) = 2 \int d^2\alpha\, W_{\text{in}}(\alpha) G_\nu[\sqrt{2}(i_{x_a} - i i_{p_b}) - \alpha] \equiv 2[W_{\text{in}} \circ G_\nu][\sqrt{2}(i_{x_a} - i i_{p_b})], \quad (2)$$

with $\circ$ denoting convolution and G_ν as a complex Gaussian distribution with variance $\nu = \cosh 2r/2$. Note that such ideal detectors provide "perfect" information about (x_a, p_b) via (i_{x_a}, i_{p_b}), while all information about $(p_a, x_b) \equiv (\text{Im}\,\beta_a = \frac{1}{\sqrt{2}}(p_1 + p_{\text{in}}), \text{Re}\,\beta_b = \frac{1}{\sqrt{2}}(x_1 - x_{\text{in}}))$ is lost. Furthermore, although (i_{x_a}, i_{p_b}) contains a small amount of information about the fiducial state $W_{\text{in}}(\alpha) = W_{\text{in}}(x_{\text{in}}, p_{\text{in}})$, this information goes to zero for $r \to \infty$. Nonetheless, the third and final step at the sending station is to transmit this *classical* information to the receiving terminal.

As illustrated in Fig. 1, receipt of (i_{x_a}, i_{p_b}) allows Bob to construct the teleported state $W_{\text{out}}(\alpha_2)$ from component 2 of the EPR state. That this resurrection is possible can be understood by examining the (unnormalized) Wigner function for the system obtained by integrating out (p_a, x_b) in correspondence to Alice's detection of (x_a, p_b), namely

$$G_\nu(\alpha_2)[W_{\text{in}} \circ G_\tau](\sqrt{2}(i_{x_a} - i i_{p_b}) + \tanh 2r\alpha_2), \quad (3)$$

where the variance $\tau = \text{sech}\, 2r/2$. Note that as $r \to \infty$, $G_\tau(\alpha)$ quickly approaches a delta function, while $G_\nu(\alpha)$ describes a broad background state. Thus, for large r, the reduced state of mode 2 is described by a broad pedestal with negligible probability upon which sits a randomly located peak at $\alpha_2 \approx \sqrt{2}(i_{x_a} - i i_{p_b})$ closely mimicing the incoming state $W_{\text{in}}(\alpha)$. The location of this random "displacement" is distributed according to Eq. (2), and is the classical information that Alice sends to Bob.

By way of the actuator $A_{x,p}$ shown in Fig. 1, Bob thus performs linear displacements of the real and imaginery components of the complex amplitude α_2 to produce $\alpha_{\text{out}} = \alpha_2 + \sqrt{2}(i_{x_a} - i i_{p_b})$, where the quantities (i_{x_a}, i_{p_b}) are scaled to (x_a, p_b). Integrating out i_{x_a} and i_{p_b} yields the ensemble description of states produced at the output of the teleportation device on an ensemble of input states W_{in}, namely

$$W_{\text{out}} = W_{\text{in}} \circ G_\sigma, \quad (4)$$

where $\sigma = e^{-2r}$ is the variance of the complex Gaussian G_σ, thus completing the teleportation process.

Clearly, for $r \to \infty$ the teleported state of Eq. (4) reproduces the original unknown state W_{in} [5]. However, note that as $r \to 0$, W_{out} also mimics W_{in}, now with *two* extra units of vacuum noise (i.e., $\sigma = \frac{1}{2} + \frac{1}{2}$). One of these noise contributions arises from Alice's attempt to measure both $(x_{\text{in}}, p_{\text{in}})$ [14], while the second comes from Bob's use of this necessarily noisy information to generate a coherent state at $\sqrt{2}(i_{x_a} - i i_{p_b})$. In this way *quantum mechanics extracts two tariffs* (one at each instance of the border crossing between quantum and classical domains), each of which we term the quantum duty or *quduty*). Note that the limit $r = 0$ corresponds to what might be considered "classical" teleportation for which the "best measurement" of the coherent amplitude of the unknown state is made [14] and sent to the receiving station, where it is used to produce a coherent state of that classical amplitude. For any $r > 0$, our quantum teleportation protocol beats this classical scheme.

Before calculating an actual figure of merit for our protocol, we now specialize from general continuous variables to the case of a single mode of the electromagnetic field and thereby to actual physical implementations of the various transformations shown in Fig. 1. Beginning with the EPR state itself, we note that such a state can be generated by nondegenerate parametric amplification with the quantities (x_j, p_j) as the quadrature-phase amplitudes of the field [6], as has been experimentally confirmed via type-II down-conversion [7]. The linear transformation $\beta_{a,b} = \frac{1}{\sqrt{2}}(\alpha_1 \pm \alpha_{\rm in})$ is accomplished by the simple superposition of modes *in* and 1 at a 50/50 beam splitter. The detectors (D_a, D_b) of Fig. 1 are now just balanced homodyne detectors with the phases of their respective local oscillators set to record (x_a, p_b) in the observed photocurrents (i_{x_a}, i_{p_b}). Note that for unit efficiency, homodyne detection provides an ideal quantum measurement of the quadrature amplitudes required for our protocol [15–17].

Nonideal detectors, each having (amplitude) efficiency η, may be modeled by using a pair of auxiliary beam splitters at (D_a, D_b) to introduce noise from a pair of vacuum modes described by annihilation operators $(\hat{c}_{a,b}, \hat{d}_{a,b})$ [15,18]. It is then convenient to introduce annihiliation operators corresponding to the "modes" of the photocurrents described by

$$\hat{i}_{a,b} = \eta \hat{\beta}_{a,b} + \sqrt{\frac{1-\eta^2}{2}}(\hat{c}_{a,b} + \hat{d}_{a,b}), \qquad (5)$$

where these fictitious objects allow us to apply an analog of the Wigner-function formalism to the photocurrents and to incorporate the effects of nonideal photodetection in a straightforward fashion. For example, loss in the response of Alice's detectors [Eq. (2)] leads to the convolution

$$\bar{P}_{ab}(i_{x_a}, i_{p_b}) = \frac{1}{\eta^2}[P_{ab} \circ G_\zeta][(i_{x_a} + i i_{p_b})/\eta], \qquad (6)$$

where G_ζ has variance $\zeta = (1-\eta^2)/2\eta^2$, which goes to zero for $\eta \to 1$ in correspondence with the ideal character of homodyne detection. Substituting for P_{ab} from Eq. (2) then gives

$$\bar{P}_{ab}(i_{x_a}, i_{p_b}) = \frac{2}{\eta^2}[W_{\rm in} \circ G_{\bar{\nu}}]\left[\frac{\sqrt{2}}{\eta}(i_{x_a} - i i_{p_b})\right], \qquad (7)$$

where $\bar{\nu} = \frac{1}{2}\cosh 2r + (1-\eta^2)/\eta^2$.

Within the context of the electromagnetic field, Bob can efficiently perform the required phase-space displacement of mode 2 based upon the classical information (i_{x_a}, i_{p_b}) received from Alice by combining the field of mode 2 with a (classical) coherent state of mean amplitude E/t, where $E = \sqrt{2}(i_{x_a} - i i_{p_b})/\eta$, at a highly reflecting mirror of transmissivity $t \to 0$. The mean state after this shift is the final teleported state, namely

$$W_{\rm out} = W_{\rm in} \circ G_{\bar{\sigma}}, \qquad (8)$$

where $G_{\bar{\sigma}}(\alpha) = \frac{1}{\pi\bar{\sigma}}\exp(\frac{-|\alpha|^2}{\bar{\sigma}})$ with $\bar{\sigma} = e^{-2r} + \frac{1-\eta^2}{\eta^2}$.

The teleportation evolution described by Eq. (8) may be written in density matrix form as

$$\hat{\rho}_{\rm out} = \int d^2\xi G_{\bar{\sigma}}(\xi)\hat{D}(i\xi)\hat{\rho}_{\rm in}\hat{D}^\dagger(i\xi), \qquad (9)$$

where $\hat{\rho}_{\rm in}$ is the original state being teleported and $\hat{D}(\alpha)$ is the displacement operator. The dynamics associated with Eq. (9) were first studied by Glauber [19] and Lachs [20] for an "incoming" vacuum state $\hat{\rho} = |0\rangle\langle 0|$ and for squeezed vacuum by Vourdas and Weiner [21]. The detailed behavior of the photocount statistics under this dynamics was investigated by Musslimani *et al.* [22]. These references also relate the development of the convolutional formalism used here (see also Refs. [23,24]).

To illustrate the protocol, consider teleportation of the coherent superposition state

$$|\psi\rangle \propto |+\alpha\rangle + e^{i\phi}|-\alpha\rangle, \qquad (10)$$

with corresponding Wigner function $W_{\rm in}(\alpha)$ illustrated in Fig. 2(a). The teleported Wigner function $W_{\rm out}(\alpha)$ as computed from Eq. (8) is shown for Fig. 2(b) for parameters corresponding to -10 dB of squeezing (i.e., $r = 1.15$) with efficiency $\eta^2 = 0.99$, which should be compared to the parameters of Ref. [25] [namely squeezing $r = 0.69$ (i.e., 6 dB of squeezing), and detectors with absolute quantum efficiency $\eta^2 = 0.99 \pm 0.02$]. Note that the quantum character of the state survives teleportation, including negative values for $W_{\rm out}$ associated with quantum interference for the off-diagonal components of $\hat{\rho}_{\rm in}$. For comparison, note that for classical teleportation (i.e., $r = 0$), $W^{\rm cl}_{\rm out}$ consists of the (incoherent) superposition of two distributions centered at $\pm\alpha$, each of which is broadened by the *quduty*.

To provide a quantitative measure of the "quality" of the output state, we note that the strongest measure of fidelity of a teleported state relative to the input state is given by the *entanglement fidelity* [26]. For processes described by Eq. (9), it is given by

$$F_e = \int d^2\xi G_{\bar{\sigma}}(\xi)|\chi_{W_{\rm in}}(\xi)|^2, \qquad (11)$$

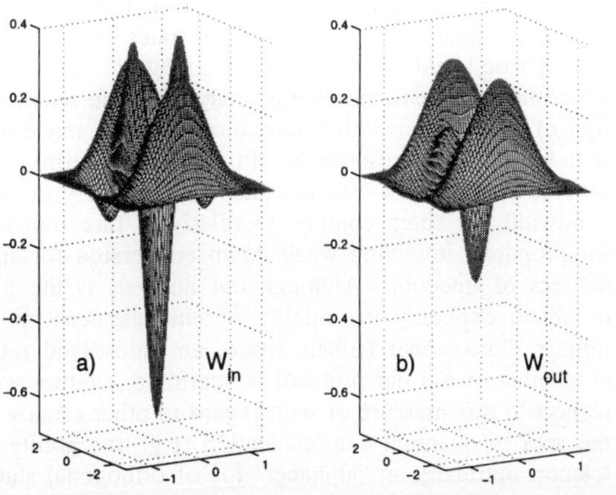

FIG. 2. (a) Wigner function $W_{\rm in}(\alpha)$ for the input state of Eq. (10) with $\alpha = 1.5i$ and $\phi = \pi$. (b) Teleported output state $W_{\rm out}(\alpha)$ for $r = 1.15$ and $\eta^2 = 0.99$.

where $\chi_{W_{\text{in}}}(\xi) = \text{tr}\, \hat{D}(i\xi)\hat{\rho}_{\text{in}}$ is the characteristic function for the incoming state's Wigner function.

For the coherent superposition of Eq. (10) direct substitution yields a fidelity of entanglement F_e of

$$\frac{1}{1+\bar{\sigma}} - \frac{1 + e^{-4|\alpha|^2} - \exp(\frac{-4\bar{\sigma}|\alpha|^2}{1+\bar{\sigma}}) - \exp(\frac{-4|\alpha|^2}{1+\bar{\sigma}})}{2(1+\bar{\sigma})(1+e^{-2|\alpha|^2}\cos\phi)^2}. \tag{12}$$

For the state shown in Fig. 2(b) this fidelity is 0.6285 for $r = 1.15$ and $\eta^2 = 0.99$ compared to 0.2487 for $r = 0$ and the same detector efficiency. This latter fidelity precludes observation of any quantum features in the classically teleported state, while the former case yields observable quantum characteristics as seen in Fig. 2.

Beyond any one particular state, let us now concentrate on high fidelity teleportation in general. In this case the Gaussian weighting described by $G_{\bar{\sigma}}$ is sufficiently narrow so that only the lowest terms in an expansion about $\xi = 0$ of $\chi_{W_{\text{in}}}$ will contribute. That is, $|\chi_{W_{\text{in}}}(\xi)|^2$ may be approximated by

$$1 - \xi^{*2}(\Delta\alpha)^2 - \xi^2(\Delta\alpha^*)^2 - 2|\xi|^2|\Delta\alpha|^2, \tag{13}$$

where $|\Delta\alpha|^2 \equiv \langle|\alpha|^2\rangle - |\langle\alpha\rangle|^2$ averaged over $W_{\text{in}}(\alpha)$. Thus, the condition for high fidelity teleportation (i.e., $1 - F_e \ll 1$) becomes $1/|\Delta\alpha|^2 \gg \bar{\sigma}$. Now $|\Delta\alpha|^2$ is just the number of photons (plus $\frac{1}{2}$) in the incoming state *after* it has been shifted so as to have *no* coherent amplitude. Roughly speaking it is the maximal rms spread of the Wigner function of the unknown quantum state being teleported, and so its reciprocal bounds the size of "important" small scale features in that state, though there can indeed be smaller features. Apparently then the condition for high entanglement fidelity says that features in the Wigner function smaller than $1/|\Delta\alpha|$ do not give a significant contribution to the state's identity.

In conclusion, our analysis suggests that existing experimental capabilities should suffice to teleport manifestly quantum or nonclassical states of the electromagnetic field with reasonable fidelity. For such experiments, extensions of our analysis to the teleportation of broad bandwidth information must be made and will be discussed elsewhere. In qualitative terms, our scheme should allow efficient teleportation every inverse bandwidth, in sharp contrast to relatively rare transfers for proposals involving weak down conversion for spin degrees of freedom. Although our analysis is the first to obtain explicitly the fidelity of entanglement on an infinite dimensional Hilbert space, an unresolved issue is whether or not our protocol is "optimum," either with respect to this measure or with regard to other criteria in the area of quantum communication (e.g., the ability to teleport optimally an "alphabet" $\{j\}$ of orthogonal states W^j_{in}). More generally, the work presented here is part of a larger program to extend classical communication with complex amplitutes into the quantum domain.

S. L. B. was funded in part by EPSRC Grant No. GR/L91344 and by a Humboldt Fellowship. H. J. K. acknowledges support from DARPA via the QUIC Institute administered by ARO, from the Office of Naval Research, and from the National Science Foundation. Both appreciate the hospitality of the Institute for Theoretical Physics under National Science Foundation Grant No. PHY94-07194.

[1] A. Steane, LANL Report No. quant-ph/9708022; A. S. Holevo, LANL Report No. quant-ph/9708046.
[2] C. H. Bennett, G. Brassard, C. Crepeau, R. Jozsa, A. Peres, and W. K. Wootters, Phys. Rev. Lett. **70**, 1895 (1993).
[3] A. Einstein, B. Podolsky, and N. Rosen, Phys. Rev. **47**, 777 (1935).
[4] J. S. Bell, *Speakable and Unspeakable in Quantum Mechanics* (Cambridge Univ. Press, Cambridge, England, 1988), p. 196.
[5] L. Vaidman, Phys. Rev. A **49**, 1473 (1994).
[6] M. D. Reid and P. D. Drummond, Phys. Rev. Lett. **60**, 2731 (1988); M. D. Reid, Phys. Rev. A **40**, 913 (1989).
[7] (a) Z. Y. Ou, S. F. Pereira, H. J. Kimble, and K. C. Peng, Phys. Rev. Lett. **68**, 3663 (1992); (b) Appl. Phys. B **55**, 265 (1992).
[8] L. Davidovich, N. Zagury, M. Brune, J. M. Raimond, and S. Haroche, Phys. Rev. A **50**, R895 (1994).
[9] J. I. Cirac and A. S. Parkins, Phys. Rev. A **50**, R4441 (1994).
[10] T. Sleator and H. Weinfurter, Ann. N. Y. Acad. Sci. **755**, 715 (1995).
[11] S. L. Braunstein and A. Mann, Phys. Rev. A **51**, R1727 (1995); **53**, 630(E) (1996).
[12] D. Boumeester *et al.,* Nature (London) **390**, 575 (1997).
[13] B. Yurke, S. L. McCall, and J. R. Klauder, Phys. Rev. A **33**, 4033 (1986).
[14] E. Arthurs and J. L. Kelly, Jr., Bell. Syst. Tech. J. **44**, 725 (1965).
[15] H. P. Yuen and J. H. Shapiro, IEEE Trans. Inf. Theory **26**, 78 (1980).
[16] S. L. Braunstein, Phys. Rev. A **42**, 474 (1990).
[17] Z. Y. Ou and H. J. Kimble, Phys. Rev. A **52**, 3126 (1995).
[18] K. Banaszek and K. Wódkiewicz, Phys. Phys. A **55**, 3117 (1997).
[19] R. J. Glauber, Phys. Rev. **131**, 2766 (1963).
[20] G. Lachs, Phys. Rev. **138**, B1012 (1965).
[21] A. Vourdas and R. M. Weiner, Phys. Rev. A **36**, 5866 (1987).
[22] Z. H. Musslimani, S. L. Braunstein, A. Mann, and M. Revzen, Phys. Rev. A **51**, 4967 (1995).
[23] M. S. Kim and N. Imoto, Phys. Rev. A **52**, 2401 (1995).
[24] K. Banaszek and K. Wódkiewicz, Phys. Rev. Lett. **76**, 4344 (1996).
[25] E. S. Polzik, J. Carri, and H. J. Kimble, Phys. Rev. Lett. **68**, 3020 (1992); (b) Appl. Phys. B **55**, 279 (1992).
[26] B. Schumacher, Phys. Rev. A **54**, 2614 (1996).

GOING BEYOND BELL'S THEOREM

Daniel M.Greenberger[1], Michael A.Horne[2] and Anton Zeilinger[3]
[1]*City College of the City University of New York, New York, New York*
[2]*Stonehill College, North Easton, Massachussetts*
[3]*Atominstitut der Oesterreichischen Universitaeten, Wien, Austria*

ABSTRACT. Bell's Theorem proved that one cannot in general reproduce the results of quantum theory with a classical, deterministic model. However, Einstein originally considered the case where one could define an "element of reality", namely for the much simpler case where one could predict with certainty a definite outcome for an experiment. For this simple case, Bell's theorem says nothing. But by using a slightly more complicated model than Bell, one can show that even in this simple case where one can make definite predictions, one still cannot generally introduce deterministic, local models to explain the results.

In 1935 Einstein, Poldosky and Rosen (1) wrote their classic paper (EPR) which pointed directly to the Achilles' Heel of quantum theory. They pointed out that if quantum theory were true, it would have to defy common sense in a manner which was very distasteful to a classically oriented mind. Bohr's answer (2) was not a refutation of their logic, but rather an affirmation of the fact that quantum theory does just that. The subsequent history of the subject, which has vindicated Bohr, is not to be taken as a refutation of EPR, but rather as a confirmation of just exactly how counter-intuitive a theory quantum theory is. An indication of how expertly they zeroed in on the most troubling aspect of the subject is the fact that in 1985 alone, 50 years after their paper was written, there were still 48 journal citation of their original article.

They were interested in the completeness of the theory, and they defined a complete theory as one in which "Every element of the Physical Reality must have a counterpart in the physical theory" . As to the phrase "Physical Reality" that occurs here, they made no claim to be able to define it in general. Rather, they gave what they thought should be one minimal requirement that an element of physical reality should exhibit. It is this requirement, which seems so necessary and obvious, that quantum theory violates. They proposed that "if, without in any way disturbing a system, we can predict with certainty (i.e. a probability equal to unity) the value of a physical quantity, then there exists an element of physical reality corresponding to this physical quantity."

They gave an example, but most subsequent discussion has used a different example given by Bohm (3). Consider a spin-0 system which decays into two spin 1/2 particles. The wave function will be

$$|\psi\rangle = (|\uparrow\downarrow\rangle - |\downarrow\uparrow\rangle)/\sqrt{2}$$

so that if particle 1 comes off with spin up (↑), particle 2 will have spin down (↓), and vice versa. The two particles will come off in opposite directions to conserve momentum. If one measures the spin of particle 1, far from the decay point, and finds spin up, say, then one knows with certainty that particle 2, which is far away, has spin down.

According to the EPR argument, since one has to in no way disturbed particle 2, then this feature

Originally printed in: M.Kafatos (ed.), Bell's Theorem, Quantum Theory and Conceptions of the Universe, 69-72
1989, Kluwer Academic Publisher

feature, spin down, must be an element of physical reality. Therefore having spin down is a property of the particle itself, and cannot have been produced by any measurement we made on particle 1. It must have come away from the point of interaction, the decay point with spin down.

Quantum mechanics denies this simple point. It says that the spin of particle 2 is indeterminate until the spin of particle 1 is measured, as until then it was in a superposition of states up and down, and one could in principle have interference between the possibilities. This was the crux of the dispute between Einstein and Bohr, but it was thought until 1965 that the difference between the two point of view had no experimental consequences. Only then did Bell prove his famous theorem (4) that in fact the assumption of the reality of the spin places severe restrictions on the possible correlations that can exist between the particles, if one makes spin measurements in arbitrary directions. Many experiments done since then have confirmed the results of quantum theory.

But it is interesting that Bell's results say nothing in the special case covered directly by the EPR argument, namely the case where a measurement on one particle allows one to predict what happens to the other particle with 100% certainty. This is the case where one measures the spin of one particle, and then measures the other either in the same or opposite direction. Not only does this case yield certainty in its measurement, but in fact one can arrive at a classical model of the system which gives the same result. It is only in the general case of an arbitrary angle between the particles, where one does not have certain knowledge, that quantum theory yields results that contradict the classical ones.

Specifically, with the wave function above, if one measures the spin of particle 1 in some direction **n**, while one measures the spin of particle 2 in a different direction **l**, then the expectation of the correlation between the two particles quantum mechanically will be

$$E(\mathbf{n}\cdot\mathbf{l}) = <\Psi|(\sigma\cdot\mathbf{n})(\sigma\cdot\mathbf{l})|\Psi> = -cos(\mathbf{n}\cdot\mathbf{l})$$

and in the case where the particles are moving along the ±z direction while **n** and **l** are in the x-y plane at angles α and β, this becomes $\cos(\alpha\cdot\beta)$. The cases where a definite prediction is possible are given by those mentioned above, where the measurement directions differ by 0° or 180°. We call this case the "super-classical" case, where an element of reality exists by virtue of perfect predictability, according to the EPR criterion.

In constructing a model for correlations in the case of a deterministic and local theory, Bell assumed that the spin of the particles were determined at the point they separated, according to EPR. Since the measurement of the spin in a given direction can only give two possible values, he assigned a value ±1 to the result. Thus he gave as the result of a measurement of the spin of both particles, one along **n** and the other along **l**, the value

$$A_\lambda(\mathbf{n})\cdot B_\lambda(\mathbf{l})$$

where both A and B could have the values only ±1 which in a particular case were determined by some internal, hidden variable λ. The only limitation on the product was that, as stated above, if **l** ± **n**, than one had

$$A_\lambda(\mathbf{n})\cdot B_\lambda(\mathbf{n}) = -1 \qquad A_\lambda(\mathbf{n})\cdot B_\lambda(-\mathbf{n}) = +1$$

Finally, the expectation value of the measurement represented the weighted sum over all possibilities λ,

$$E(\mathbf{n},\mathbf{l}) = \int d\lambda\, \rho(\lambda)\, A_\lambda(\mathbf{n}) \cdot B_\lambda(\mathbf{l})$$

Because of the factorable nature of the probabilities, he was able to derive the inequality

$$| E(\mathbf{n},\mathbf{l}) - E(\mathbf{n},\mathbf{k}) | \leq 1 + E(\mathbf{l},\mathbf{k})$$

This inequality is broken by the quantum mechanical result for most angles. But unfortunately, it gives no information at all when $\mathbf{l} \pm \mathbf{n}$.

In the super-classical case, where one can make a definite prediction, but where the Bell inequality above gives no information, one can make a simple deterministic model to explain the result, but of course, the Bell inequality shows that this is not possible for general angles. However it leads naturally to the question "Can one always find a classical model for the superclassical case?" While that is the trivial case from the point of view of Bell's inequality, it is the most interesting from the point of view of reality. In other words, Bell's theorem answers the question of whether one can make a classical, local, deterministic model to duplicate the results of quantum theory in general, and the answer is no. But it does not address the question of whether one can make such a model in the special case in which one can make definite predictions, the EPR case.

The answer to this question is also no. But in order to answer it one needs a more complicated model than the two-body decay above. We have constructed a simple generalization of the Bohm model, which provides a greater restriction on the possibilities for the various particles. In fact, the restriction is so great that even in the super classical case, one cannot make a derministic, local model, and one does not even need inequalities. For any given value of the hidden variable λ one can show that it is impossible to construct such a model.

Our general attitude is that we assume that quantum mechanics gives correct answers, and the question is whether a classical model can reproduce these answers. The example we have chosen is of a particle of spin 1, initially in the state m=0 in a magnetic field along the z axis. The particle then decays into two particles, each of spin 1, one along the +z axis and one along the -z axis, Subsequently, each of these two particles decays into two spin 1/2 particles, the first two also moving along the +z axis, and the other two along the -z axis. We introduce this restriction so as not to introduce any spin-orbit type considerations into the problem.

In this problem the initial particle decays into four spin 1/2 particles and its wave function can be expressed as

$$|1,0\rangle = (|\uparrow\uparrow\downarrow\downarrow\rangle - |\downarrow\downarrow\uparrow\uparrow\rangle)/\sqrt{2}$$

The quantum mechanical expectation value for the spins in four given directions is

$$E(\mathbf{n}_1,\mathbf{n}_2,\mathbf{n}_3,\mathbf{n}_4) = -\cos(\alpha+\beta-\gamma-\delta)$$

where each of the directions $\mathbf{n}_i$ is assumed to be in the x-y plane at angles α,β,γ,δ, respectively. Note that if any two of the angles are fixed, the other two obey the same law as the two body decay before, so that they will obey the Bell inequality. But the important point for us is that if

$$\alpha + \beta - \gamma - \delta = 0\,,\, \pi$$

than the *cos* term will equal ±1, and so if we measure three of the angles, we can predict the fourth with 100% certainty. This is exactly the super-classical state EPR case!

Thus we have again a general case where for most angles, we can make no specific classical prediction, however there is a range of parameters, given by the equation above, for which we can indeed make a definite prediction. The next question is whether we can find a classical, deterministic, local model for it. Since as before, if we measure the spin in any specific direction, we can get only two answers, we can use the same type of parametrization as before. We then get

$$A_\lambda(\alpha)B_\lambda(\beta)C_\lambda(\gamma)D_\lambda(\delta)$$

as our measure of the four particle landing at different angles. (The hidden variable λ can stand for any configuration of hidden parameters. But in this case they are all determined back at the original decay. The subsequent decay hidden variables will be determined by the original hidden variables of the first decay.) The condition corresponding to the superclassical case is

$$A_\lambda(\alpha)B_\lambda(\beta)C_\lambda(\gamma)D_\lambda(\delta) = \pm 1$$
$$\text{for} \quad \alpha + \beta + \gamma + \delta = 0, \pi$$

But it turns out that there is no way to satisfy this condition. It is too restrictive, because we can continuously vary two of the parameters while keeping the other two constant. This leads to the conclusion that A=B=C=D= constant. But this is impossible, since the product sometimes equals +1 and sometimes -1. This is true for any value of λ so that there is no need to integrate over it. Thus we reach the general conclusion that not only is there no way to form a classical, deterministic, local theory that reproduces quantum theory in general, but that even in the simpler case that one can make definite predictions in the EPR sense, it is impossible to do so with such a model. However one must go beyond the Bell theorem in order to prove this. A further conclusion is that with the appropriate 4-particle (or even 3-particle) system, all one must do is prove that quantum theory holds experimentally, and then we know that it cannot be classically duplicated, so that it will be much easier to disprove the classical type of loop-holes that are constantly being sought to explain the results of 2-particle experiments which verify quantum theory.

ACKNOWLEDGMENT. We would like to thank the National Science Foundation and the Humbold Stiftung (DMG) for providing partial support for this work.

REFERENCES:

1. A.Einstein, B.Poldosky and N.Rosen, (1935) *Phys.Rev.*, **47**, 777
2. N.Bohr, (1935) *Phys.Rev.*, **48**, 696
3. D.Bohm, (1951)*"Quantum Theory"* Pretience - Hall, New York
4. J.S.Bell, (1965) *Physics* (N.Y.) **1**, 195

REFERENCE FRAME

WHAT'S WRONG WITH THESE ELEMENTS OF REALITY?

N. David Mermin

The subject of Einstein–Podolsky–Rosen correlations—those strong quantum correlations that seem to imply "spooky actions at a distance"—has just been given a new and beautiful twist. Daniel Greenberger, Michael Horne and Anton Zeilinger have found a clever and powerful extension of the two-particle EPR experiment to *gedanken* decays that produce more than two particles.[1] In the GHZ experiment the spookiness assumes an even more vivid form than it acquired in John Bell's celebrated analysis of the EPR experiment, given over 25 years ago.[2] The argument that follows is my attempt to simplify a refinement of the GHZ argument given by the philosophers Robert Clifton, Michael Redhead and Jeremy Butterfield.[3]

Consider *three* spin-½ particles, named 1, 2 and 3. They have originated in a spin-conserving *gedanken* decay and are now *gedanken* flying apart along three different straight lines in the horizontal plane. (It's not essential for the *gedanken* trajectories to be coplanar, but it makes it easier to describe the rest of the geometry.) I specify the spin state Ψ of the three particles in a time-honored manner, giving you a complete set of commuting Hermitian spin-space operators of which Ψ is an eigenstate.

Those operators are assembled out of the following pieces (measuring all spins in units of $\frac{1}{2}\hbar$): σ_z^i, the operator for the spin of particle i along its direction of motion; σ_x^i, the spin along the vertical direction; and σ_y^i, the spin along the horizontal direction orthogonal to the trajectory. (Any three orthogonal directions independently chosen for each particle would do. But we're going to be *gedanken* measuring x and y components of each particle's spin, so it's nice to think of the x and y directions as orthogonal to the direction of motion, since the components can then be straightforwardly measured by passage through a conventional Stern–Gerlach magnet.) The complete set of commuting Hermitian operators consists of

$$\sigma_x^1\sigma_y^2\sigma_y^3, \quad \sigma_y^1\sigma_x^2\sigma_y^3, \quad \sigma_y^1\sigma_y^2\sigma_x^3. \qquad (1)$$

Even though the x and y components of a given particle's spin anticommute—a fact of paramount importance in what follows—all three of the operators in (1) do indeed commute with one another, because the product of any two of them differs from the product in the reverse order by an even number of such anticommutations. Because they all commute, the three operators can be provided with simultaneous eigenstates. Since the square of each of the three is unity, the eigenvalues of each are $+1$ or -1, and the 2^3 possible choices are indeed just what we need to span the eight-dimensional space of three spins-½.

For simplicity of exposition let's focus our attention on the symmetric eigenstate in which each of the operators (1) has the eigenvalue $+1$. (Its state vector is $\Psi = (1/\sqrt{2})(|1,1,1\rangle - |-1,-1,-1\rangle)$, where 1 or -1 specifies spin up or down along the appropriate z axis, but you don't need to know this. I'm only telling you because discussions of EPR always write down an explicit form for the state vector and I wouldn't want you to think you were missing anything.) Because the spin vectors of distinct particles commute component by component, we can simultaneously measure the x component of one particle and the y components of the other two (using three Stern–Gerlach magnets in three remote regions of space). Since the three particles are in an eigenstate of all three operators (1) with eigenvalue unity, the product of the results of the three spin measurements has to be $+1$, regardless of which particle we single out for the x-spin measurement.

This affords an immediate application of the EPR reality criterion[4]: "If, without in any way disturbing a system, we can predict with certainty the value of a physical quantity, then there exists an element of physical reality corresponding to this physical quantity." The "element of physical reality" is that predictable value, and it ought exist whether or not we actually carry out the procedure necessary for its prediction, since that procedure in no way disturbs it. Because the product of the results of measuring one x component and two y components is unity in the state Ψ, we can predict with certainty the result of measuring the x component of the spin of any one of the three particles by measuring the y components of the two other, far away particles. For if both y components turn out to be the same then the x component, when measured, must yield the value $+1$; if the two y components turn out to be different, the subsequently measured x component will necessarily yield the value -1. In the absence of spooky actions at a distance or the metaphysical cunning of a Niels Bohr, the two far away y-component measurements cannot "disturb" the particle whose x component is subsequently to be measured. The EPR reality criterion therefore asserts the existence of elements of reality m_x^1, m_x^2 and m_x^3, each having the value $+1$ or -1, each waiting to be revealed by the appropriate pair of far away y-component measurements.

In much the same way, we can also predict the result of measuring the y component of the spin of any particle with certainty, by measuring one x component and one y component of the spins of the other two. There are thus elements of reality m_y^1, m_y^2 and m_y^3, with values $+1$ or -1, also waiting to be revealed by far away measurements. All six of the elements of reality m_x^i and m_y^i have to be there, because we can predict in advance what any one of the six values will be by measurements made

David Mermin is a professor at Cornell University and director of the Laboratory of Atomic and Solid-State Physics. His thoughts about physics and physicists have just been collected together and published under the preposterous title *Boojums All The Way Through* (Cambridge U. P., New York, 1990).

REFERENCE FRAME

so far away that they cannot disturb the particle that subsequently does indeed display the predicted value.

This conclusion is, of course, highly heretical, because σ_x^i does not commute with σ_y^i—in fact the two *anti*-commute—and therefore they cannot have simultaneous values. (The operators (1) are nicely chosen to hide this failure to commute, since the anticommutations always occur in pairs.) But heresy or not, since the result of either measurement can be predicted with probability 1 from the results of other measurements made arbitrarily far away, an open-minded person might be sorely tempted to renounce quantum theology in favor of an interpretation less hostile to the elements of reality.

In the GHZ experiment, however, as in Bell's version of the EPR, the elements of reality are demolished by the straightforward quantum mechanical predictions for some additional experiments, entirely unencumbered by accompanying metaphysical baggage.

In the GHZ case the demolition is spectacularly more efficient. Suppose, heretically, that the elements of reality really do exist in each run of the experiment. While we cannot know all six of their values, those values are constrained by the fact that the values of $\sigma_x^1\sigma_y^2\sigma_y^3$, $\sigma_y^1\sigma_x^2\sigma_y^3$ and $\sigma_y^1\sigma_y^2\sigma_x^3$, all unity in the state Ψ, are given by the values of the corresponding products $m_x^1m_y^2m_y^3$, $m_y^1m_x^2m_y^3$ and $m_y^1m_y^2m_x^3$. But if these latter three quantities are unity, so is their combined product. Since each individual m_y^i is either $+1$ or -1 and each occurs twice in the combined product, that combined product is just $m_x^1m_x^2m_x^3$. So the existence of the elements of reality implies that should we choose to measure the x components of all three spins in the state Ψ, the product of the three resulting values must once again be $+1$.

The value of that product can also be determined without invoking disreputable elements of reality by a simple quantum mechanical calculation, since it is just the result of measuring the Hermitian operator

$$\sigma_x^1\sigma_x^2\sigma_x^3. \qquad (2)$$

You can easily check that this operator also commutes with all of the operators (1): Once again the number of anticommutations is always even. This is encouraging, for if the value of the operator (2) in the state Ψ is invariably to be $+1$, it had better also have Ψ for an eigenstate, a requirement that is guaranteed by its commuting with all three members of the complete set of commuting operators (1) whose eigenvalues define Ψ.

However:

Not only does (2) commute with each of the operators (1), but you can easily check that it is a simple explicit function of them, namely, *minus* the product of all three. The (crucial) minus sign arises because here, at last, in bringing the pairs of operators σ_y^i together to produce unity, one runs up against an *odd* number of anticommutations of σ_y^i's with σ_x^i's. Since Ψ is an eigenstate with eigenvalue $+1$ of each of the operators (1), it is therefore indeed an eigenstate of the operator (2), but with the *wrong* eigenvalue, opposite in sign to the one required by the existence of the elements of reality.

So farewell elements of reality! And farewell in a hurry. The compelling hypothesis that they exist can be refuted by a *single* measurement of the three x components: The elements of reality require the product of the three outcomes *invariably* to be $+1$; but *invariably* the product of the three outcomes is -1.

This is an altogether more powerful refutation of the existence of elements of reality than the one provided by Bell's theorem for the two-particle EPR experiment. Bell showed that the elements of reality inferred from one group of measurements are incompatible with the *statistics* produced by a second group of measurements. Such a refutation cannot be accomplished in a single run, but is built up with increasing confidence as the number of runs increases. Thus in one simple version of the two-particle EPR experiment (which I described in PHYSICS TODAY, April 1985, page 38) the hypothesis of elements of reality requires a class of outcomes to occur at least 55.5% of the time, while quantum mechanics allows them to occur only 50% of the time. In the GHZ experiment, on the other hand, the elements of reality require a class of outcomes to occur *all* of the time, while quantum mechanics *never* allows them to occur.

It is also appealing to see the failure of the EPR reality criterion emerge quite directly from the one crucial difference between the elements of reality (which, being ordinary numbers, necessarily commute) and the precisely corresponding quantum mechanical observables (which sometimes anticommute).

I was surprised to learn of this always-vs-never refutation of Einstein, Podolsky and Rosen. After all, quantum magic generally flows from the fact that it is the amplitudes that combine like probabilities rather than the probabilities themselves. But when the probabilities are zero, so are the amplitudes. Guided by such woolly thinking, and the failure of anybody to strengthen Bell's result in this direction in the ensuing 25 years, I recently declared in writing[5] that no set of experiments, real or *gedanken*, was known that could produce such an all-or-nothing demolition of the elements of reality. With a bow of admiration to Greenberger, Horne and Zeilinger, I hereby recant.

References

1. D. M. Greenberger, M. Horne, A. Zeilinger, in *Bell's Theorem, Quantum Theory, and Conceptions of the Universe*, M. Kafatos, ed., Kluwer, Dordrecht, The Netherlands (1989), p. 69.
2. J. S. Bell, Physics 1, 195 (1964).
3. R. K. Clifton, M. L. G. Redhead, J. M. Butterfield, "Generalization of the Greenberger–Horne–Zeilinger Algebraic Proof of Nonlocality," submitted to Found. Phys.
4. A. Einstein, B. Podolsky, N. Rosen, Phys. Rev. **47**, 777 (1935).
5. N. D. Mermin, in *Philosophical Consequences of Quantum Theory*, J. T. Cushing, E. McMullin, eds., Notre Dame U. P., Notre Dame, Ind. (1989), p. 48. ■

PHYSICAL REVIEW LETTERS

VOLUME 82 15 FEBRUARY 1999 NUMBER 7

Observation of Three-Photon Greenberger-Horne-Zeilinger Entanglement

Dik Bouwmeester, Jian-Wei Pan, Matthew Daniell, Harald Weinfurter, and Anton Zeilinger

Institut für Experimentalphysik, Universität Innsbruck, Technikerstrasse 25, A-6020 Innsbruck, Austria

(Received 6 October 1998)

We present the experimental observation of polarization entanglement for three spatially separated photons. Such states of more than two entangled particles, known as Greenberger-Horne-Zeilinger (GHZ) states, play a crucial role in fundamental tests of quantum mechanics versus local realism and in many quantum information and quantum computation schemes. Our experimental arrangement is such that we start with two pairs of entangled photons and register the photons in a way that any information as to which pair each photon belongs to is erased. After detecting a trigger photon, the registered events at the detectors for the remaining three photons exhibit the desired GHZ correlations. [S0031-9007(98)08348-3]

PACS numbers: 03.65.Bz, 03.67.–a, 42.50.Ar

Since the seminal work of Einstein, Podolsky, and Rosen [1], there has been a quest for generating entanglement between quantum particles. Although two-particle entanglements have long been demonstrated experimentally [2,3], the preparation of entanglement between three or more particles remains an experimental challenge. Proposals have been made for experiments with photons [4] and atoms [5], and three nuclear spins within a single molecule have been prepared such that they locally exhibit three-particle correlations [6]. However, until now there has been no experiment which demonstrates the existence of entanglement of more than two spatially separated particles. Here we report the experimental observation of polarization entanglement of three spatially separated photons.

The original motivation to prepare three-particle entanglements stems from the observation by Greenberger, Horne, and Zeilinger (GHZ) that entanglement of more than two particles leads to a conflict with local realism for nonstatistical predictions of quantum mechanics [7]. This is in contrast to the case of experiments with two entangled particles testing Bell's inequalities, where the conflict only arises for statistical predictions [8].

The incentive to produce GHZ states has been significantly increased by the advance of the fields of quantum communication and quantum information processing. Entanglement between several particles is the most important feature of many such quantum communication and computation protocols [9,10].

The experiment described here is based on techniques that have been developed for our previous experiments on quantum teleportation [11] and entanglement swapping [12]. In fact, one of the main complications in those experiments, namely, the creation of two pairs of photons by a single source, is here turned into a virtue.

The main idea, as was put forward in Ref. [4], is to transform two pairs of polarization entangled photons into three entangled photons and a fourth independent photon. In our experiment the GHZ entanglement is observed only under the condition that both the trigger photon and the three entangled photons are actually detected. Thus, detection plays the double role of both projecting into the GHZ state and performing a specific measurement on the state. This, we submit, in practice will not be a severe limitation because, on the one hand, in any realistic scheme one always has losses, and information is only obtained if the photons are actually observed, as, for instance, in third-man quantum cryptography. On the other hand, many applications explicitly use specific measurement results. For example, the GHZ argument for testing local realism is based on detection events, and knowledge of the underlying quantum state is not even necessary.

Figure 1 is a schematic drawing of our experimental setup. Pairs of polarization entangled photons are

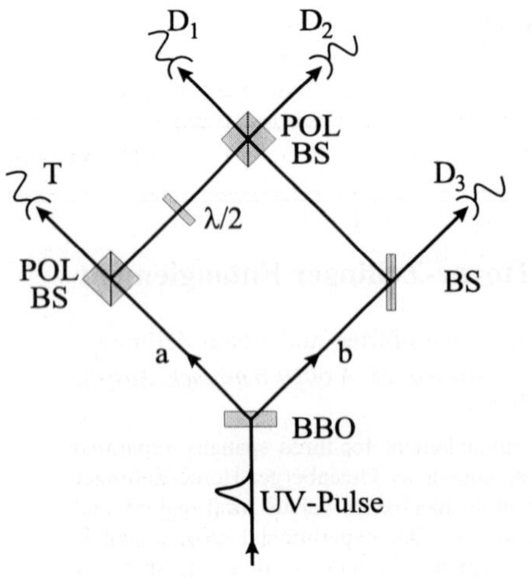

FIG. 1. Schematic drawing of the experimental setup for the demomstration of the Greenberger-Horne-Zeilinger entanglement for spatially separated photons. Conditioned on the registration of one photon at the trigger detector T, the three photons registered at D_1, D_2, and D_3 exhibit the desired GHZ correlations.

generated by a short pulse of ultraviolet (UV) light ($\approx$200 fs, $\lambda = 394$ nm from a frequency-doubled, mode-locked Ti-sapphire laser), which passes through a nonlinear crystal (here, β-barium-borate, BBO). The probability per pulse to create a single pair in the desired modes, selected by irises, about 1.5 mm wide and 25 cm behind the crystal, is low and of the order of a few 10^{-4}. The pair creation is such that the following polarization entangled state is obtained [3]:

$$\frac{1}{\sqrt{2}}(|H\rangle_a|V\rangle_b - |V\rangle_a|H\rangle_b). \qquad (1)$$

This state indicates a superposition of the possibility that the photon in arm a is horizontally polarized and the one in arm b is vertically polarized ($|H\rangle_a|V\rangle_b$), and the opposite possibility ($|V\rangle_a|H\rangle_b$). The minus sign indicates that there is a fixed phase difference of π between the two possibilities. For our GHZ experiment this phase factor is actually allowed to have any value, as long as it is fixed for all pair creations.

The setup is such that arm a continues towards a polarizing beam splitter, where V photons are reflected and H photons are transmitted towards detector T (behind an interference filter $\delta\lambda = 4.6$ nm at 788 nm). Arm b continues towards a 50/50 polarization-independent beam splitter. From each beam splitter, one output is directed to a final polarizing beam splitter. In between the two polarizing beam splitters, vertical polarization is rotated to 45° polarization using a $\lambda/2$ plate. The remaining three output arms continue through interference filters ($\delta\lambda =$ 3.6 nm) and single-mode fibers towards the single-photon detectors D_1, D_2, and D_3. Including filter losses, coupling into single-mode fibers, and the Si-avalanche detector efficiency, the total collection and detection probability of a photon is about 10%.

Consider now the case that *two* pairs are generated by a single UV pulse, and that the four photons are all detected, one by each detector T, D_1, D_2, and D_3. Our claim is that, by the coincident detection of the four photons and because of the brief duration of the UV pulse and the narrowness of the filters, one can conclude that a three-photon GHZ state has been recorded by detectors D_1, D_2, and D_3. The reasoning is as follows. When a fourfold coincidence recording is obtained, one photon in path a must have been horizontally polarized and detected by the trigger detector T. Its companion photon in path b must then be vertically polarized, and it has a 50% chance to be transmitted by the beam splitter (see Fig. 1) towards detector D_3 and a 50% chance to be reflected by the beam splitter towards the final polarizing beam splitter, where it will be reflected to D_2. Consider the first possibility, i.e., the companion of the photon detected at T is detected by D_3 and necessarily carried polarization V. Then the counts at detectors D_1 and D_2 were due to a second pair, one photon traveling via path a and the other one via path b. The photon traveling via path a must necessarily be V polarized in order to be reflected by the polarizing beam splitter in path a; thus its companion, taking path b, must be H polarized and, after reflection at the beam spliter in path b, it will be transmitted by the final polarizing beam splitter and arrive at detector D_1. The photon detected by D_2 therefore must be H polarized since it came via path a and had to transit the last polarizing beam splitter. Note that this latter photon was V polarized but after passing the $\lambda/2$ plate it became polarized at 45° which gave it a 50% chance to arrive as an H polarized photon at detector D_2. Thus we conclude that, if the photon detected by D_3 is the companion of the T photon, the coincidence detection by D_1, D_2, and D_3 then corresponds to the detection of the state

$$|H\rangle_1|H\rangle_2|V\rangle_3. \qquad (2)$$

By a similar argument one can show that, if the photon detected by D_2 is the companion of the T photon, the coincidence detection by D_1, D_2, and D_3 corresponds to the detection of the state

$$|V\rangle_1|V\rangle_2|H\rangle_3. \qquad (3)$$

In general, the two possible states (2) and (3), corresponding to a fourfold coincidence recording, will not form a coherent superposition, i.e., a GHZ state, because they could, in principle, be distinguishable. Besides the possible lack of mode overlap at the detectors, the exact detection time of each photon can reveal which state is present. For example, state (2) is identified by noting that T and D_3, or D_1 and D_2, fire nearly simultaneously. To erase this information it is necessary that the coherence time of the photons is substantially longer than the duration of the UV pulse (approximately 200 fs) [13]. We achieved this by detecting the photons behind narrow bandwidth filters which yields a coherence time of approximately 500 fs. Thus, the possibility to distinguish between states (2) and (3) is no longer present, and, by a basic rule of quantum mechanics, the state detected by a coincidence recording of D_1, D_2, and D_3, conditioned on the trigger T, is the quantum superposition

$$\frac{1}{\sqrt{2}}(|H\rangle_1|H\rangle_2|V\rangle_3 + |V\rangle_1|V\rangle_2|H\rangle_3), \quad (4)$$

which is a GHZ state [14].

The plus sign in Eq. (4) follows from the following more formal derivation. Consider two down-conversions producing the product state

$$\tfrac{1}{2}(|H\rangle_a|V\rangle_b - |V\rangle_a|H\rangle_b)(|H\rangle'_a|V\rangle'_b - |V\rangle'_a|H\rangle'_b). \quad (5)$$

Initially, we assume that the components $|H\rangle_{a,b}$ and $|V\rangle_{a,b}$ created in one down-conversion might be distinguishable from the components $|H\rangle'_{a,b}$ and $|V\rangle'_{a,b}$ created in the other one. The evolution of the individual components of state (5) through the apparatus towards the detectors T, D_1, D_2, and D_3 is given by

$$|H\rangle_a \rightarrow |H\rangle_T, \quad (6)$$

$$|V\rangle_b \rightarrow \frac{1}{\sqrt{2}}(|V\rangle_2 + |V\rangle_3), \quad (7)$$

$$|V\rangle_a \rightarrow \frac{1}{\sqrt{2}}(|V\rangle_1 + |H\rangle_2), \quad (8)$$

$$|H\rangle_b \rightarrow \frac{1}{\sqrt{2}}(|H\rangle_1 + |H\rangle_3). \quad (9)$$

Identical expressions hold for the primed components. Inserting these expressions into state (5) and restricting ourselves to those terms where only one photon is found in each output we obtain, after normalization,

$$\tfrac{1}{2}\{|H\rangle_T(|H\rangle'_1|H\rangle'_2|V\rangle_3 + |V\rangle'_1|V\rangle_2|H\rangle'_3) + |H\rangle'_T(|H\rangle_1|H\rangle_2|V\rangle'_3 + |V\rangle_1|V\rangle'_2|H\rangle_3)\}. \quad (10)$$

If now the experiment is performed such that the photon states from the two down-conversions are indistinguishable, we finally obtain the desired state

$$\frac{1}{\sqrt{2}}|H\rangle_T(|H\rangle_1|H\rangle_2|V\rangle_3 + |V\rangle_1|V\rangle_2|H\rangle_3). \quad (11)$$

Note that the total photon state produced by our setup, i.e., the state before detection, also contains terms in which, for example, two photons enter the same detector. In addition, the total state contains contributions from single down-conversions. The fourfold coincidence detection acts as a projection measurement onto the desired GHZ state (11) and filters out these undesirable terms. The efficiency for one UV pump pulse to yield such a fourfold coincidence detection is very low (of the order of 10^{-10}). Fortunately, 7.6×10^7 UV pulses are generated per second, which yields about one double pair creation and detection per 150 seconds, which is just enough to perform our experiments [15]. Triple pair creations can be completely neglected since they can give rise to a fourfold coincidence detection only about once each day.

To experimentally demonstrate that GHZ entanglement has been obtained by the method described above, we first verified that, conditioned on a photon detection by the trigger T, both the $H_1H_2V_3$ and the $V_1V_2H_3$ components can be observed, but no others. This was done by comparing the count rates of the eight possible combinations of polarization measurements, $H_1H_2H_3$, $H_1H_2V_3, \ldots, V_1V_2V_3$. The observed intensity ratio between the desired and undesired states was 12:1. Existence of the two terms as just demonstrated is a necessary but not yet sufficient condition for demonstrating GHZ entanglement. In fact, there could, in principle, be just a statistical mixture of those two states. Therefore, one has to prove that the two terms coherently superpose. This we did by a measurement of linear polarization of photon 1 along +45°, bisecting the H and V directions. Such a measurement projects photon 1 into the superposition $|+45°\rangle_1 = \frac{1}{\sqrt{2}}(|H\rangle_1 + |V\rangle_1)$, implying that the state (11) is projected into

$$\frac{1}{\sqrt{2}}|H\rangle_T|+45°\rangle_1(|H\rangle_2|V\rangle_3 + |V\rangle_2|H\rangle_3). \quad (12)$$

Thus photons 2 and 3 end up entangled as predicted under the notion of "entangled entanglement" [16]. Rewriting the state of photons 2 and 3 in the 45° basis results in the state

$$\frac{1}{\sqrt{2}}(|+45°\rangle_2|+45°\rangle_3 - |-45°\rangle_2|-45°\rangle_3), \quad (13)$$

which implies that if photon 2 is found to be polarized along −45° (or along +45°), photon 3 is polarized along the same direction. The absence of the terms $|+45°\rangle_2|-45°\rangle_3$ and $|-45°\rangle_2|+45°\rangle_3$ is due to destructive interference and thus indicates the desired coherent superposition of the terms in the GHZ state (11). The experiment therefore consisted of measuring fourfold coincidences between the detector T, detector 1 behind a +45° polarizer, detector 2 behind a −45° polarizer, and measuring photon 3 behind either a +45° polarizer or a −45° polarizer. In the experiment, the difference in arrival time of the photons at the final polarizer or, more specifically, at the detectors D_1 and D_2 was varied.

The data points in Fig. 2(a) are the experimental results obtained for the polarization analysis of the photon at D_3, conditioned on the trigger and on detection of two photons polarized at 45° and −45° by the two detectors D_1 and D_2, respectively. The two curves show the fourfold coincidences for a polarizer oriented at −45° (squares)

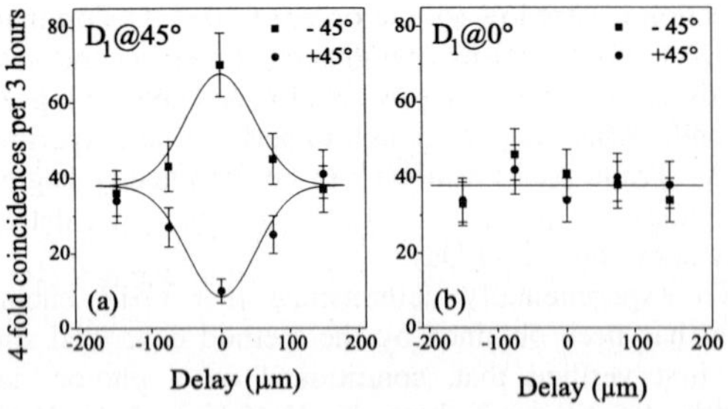

FIG. 2. Experimental confirmation of GHZ entanglement. Graph (a) shows the results obtained for polarization analysis of the photon at D_3, conditioned on the trigger, and the detection of one photon at D_1 polarized at 45° and one photon at detector D_2 polarized at $-45°$. The two curves show the fourfold coincidences for a polarizer oriented at $-45°$ and 45°, respectively, in front of detector D_3 as a function of the spatial delay in path a. The difference between the two curves at zero delay confirms the GHZ entanglement. By comparison [graph (b)] no such intensity difference is predicted if the polarizer in front of detector D_1 is set at 0°. Error bars are given by the square root of the coincidence counts.

and $+45°$ (circles) in front of detector D_3 as a function of the spatial delay in path a. From the two curves it follows that for zero delay the polarization of the photon at D_3 is oriented along $-45°$, in accordance with the quantum-mechanical predictions for the GHZ state. For nonzero delay, the photons traveling via path a towards the second polarizing beam splitter and those traveling via path b become distinguishable. Therefore increasing the delay gradually destroys the quantum superposition in the three-particle state.

Note that one can equally well conclude from the data that, at zero delay, the photons at D_1 and D_3 have been projected onto a two-particle entangled state by the projection of the photon at D_2 onto $-45°$. The two conclusions are only compatible for a genuine GHZ state. We note that the observed visibility was as high as 75% [17].

For an additional confirmation of state (11) we performed measurements conditioned on the detection of the photon at D_1 under 0° polarization (i.e., V polarization). For the GHZ state $(1/\sqrt{2})(|H\rangle_1|H\rangle_2|V\rangle_3 + |V\rangle_1|V\rangle_2|H\rangle_3)$ this implies that the remaining two photons should be in the state $|V\rangle_2|H\rangle_3$ which cannot give rise to any correlation between these two photons in the 45° detection basis. The experimental results of these measurement are presented in Fig. 2(b). The data clearly indicate the absence of two-photon correlations and thereby confirm our claim of the observation of GHZ entanglement between three spatially separated photons.

Although the extension from two to three entangled particles might seem to be only a modest step forward, the implications are rather profound. First, GHZ entanglements allow for novel tests of quantum mechanics versus local realistic models [7,18]. Second, three-particle GHZ states might find a direct application, for example, in third-man quantum cryptography. Third, the method developed to obtain three-particle entanglement from a source of pairs of entangled particles can be extended to obtain entanglement between many more particles [19], which is the basis of many quantum communication and computation protocols. Finally, we note that our experiment, together with our earlier realization of quantum teleportation [11] and entanglement swapping [12], provides necessary tools to implement a number of novel entanglement distribution and network ideas as recently proposed [20].

We are very grateful to D.M. Greenberger for useful discussions and criticism, and also M.A. Horne for detailed improvements of our initial manuscript. This work was supported by the Austrian Science Foundation FWF (Project No. S6502), the Austrian Academy of Sciences, the TMR program of the European Union (Contract No. ERBFMRXCT96-0087), and by NSF (Grant No. PHY 97-22614).

[1] A. Einstein, B. Podolsky, and N. Rosen, Phys. Rev. **47**, 777 (1935).

[2] C.S. Wu, I. Shaknov, Phys. Rev. **77**, 136 (1950); S.J. Freedman, and J.S. Clauser, Phys. Rev. Lett. **28**, 938 (1972); A. Aspect, J. Dalibard, and G. Roger, Phys. Rev. Lett. **49**, 1804 (1982); E. Hagley, X. Maître, G. Nogues, C. Wunderlich, M. Brune, J.M. Raimond, and S. Haroche, Phys. Rev. Lett. **79**, 1 (1997).

[3] P.G. Kwiat, K. Mattle, H. Weinfurter, A. Zeilinger, A.V. Sergienko, and Y.H. Shih, Phys. Rev. Lett. **75**, 4337 (1995).

[4] A. Zeilinger, M.A. Horne, H. Weinfurter, and M. Zukowski, Phys. Rev. Lett. **78**, 3031 (1997).

[5] S. Haroche, Ann. N.Y. Acad. Sci. **755**, 73 (1995); J.I. Cirac and P. Zoller, Phys. Rev. A **50**, R2799 (1994).

[6] S. Lloyd, Phys. Rev. A **57**, R1473 (1998); R. Laflamme, E. Knill, W.H. Zurek, P. Catasti, and S.V.S. Mariappan, Philos. Trans. R. Soc. London A **356**, 1941 (1998).

[7] D.M. Greenberger, M.A. Horne, and A. Zeilinger, in *Bell's Theorem, Quantum Theory, and Conceptions of the Universe,* edited by M. Kafatos (Kluwer Academics, Dordrecht, The Netherlands 1989), pp. 73–76; D.M. Greenberger, M.A. Horne, A. Shimony, and A. Zeilinger, Am. J. Phys. **58**, 1131 (1990); D.M. Greenberger, M.A. Horne, and A. Zeilinger, Phys. Today **46**, No. 8, 22 (1993); N.D. Mermin, Am. J. Phys. **58**, 731 (1990); N.D. Mermin, Phys. Today **43**, No. 6, 9 (1990).

[8] J.S. Bell, Phys. **1**, 195 (1964); reprinted in J.S. Bell, *Speakable and Unspeakable in Quantum Mechanics* (Cambridge University Press, Cambridge, England, 1987).

[9] C.H. Bennett, Phys. Today **48**, No. 10, 24 (1995); Special issue on Quantum Information, Phys. World **11** (1998).

[10] R. Cleve and H. Buhrman, Phys. Rev. A **56**, 1201 (1997); D. Bruss, D. DiVincenzo, A. Ekert, C. Fuchs, C. Macchiavello, and J. Smolin, Phys. Rev. A **57**, 2368 (1998).

[11] D. Bouwmeester, J.-W. Pan, K. Mattle, M. Eibl, H. Weinfurter, and A. Zeilinger, Nature (London) **390**, 575 (1997).

[12] J.-W. Pan, D. Bouwmeester, H. Weinfurter, and A. Zeilinger, Phys. Rev. Lett. **80**, 3891 (1998).

[13] M. Zukowski, A. Zeilinger, and H. Weinfurter, Ann. N.Y. Acad. Sci. **755**, 91 (1995).

[14] Rigorously speaking, this erasure technique is perfect, hence produces a pure GHZ state, only in the limit of infinitesimal pulse duration and infinitesimal filter bandwidth, but detailed calculations [See Ref. [13] and M. A. Horne, Fortschr. Phys. **46**, 6 (1998)] reveal that our pulse and filter values are sufficient to create a clearly observable entanglement, as confirmed by our experimental data.

[15] The singles detection rate at detectors D_1, D_2, and D_3, is about 15.000 counts per second, and at the trigger detector T about 100.000 counts per second, due to the larger filter bandwidth and mode acceptance. Fourfold coincidence is registered with logic AND circuitry with a coincidence time of 6 ns.

[16] G. Krenn and A. Zeilinger, Phys. Rev. A **54**, 1793 (1996).

[17] The limited visibility is due mainly to the finite width of the interference filters, the finite pulse duration [14], and the limited quality of the polarization optics. Detector noise or accidental coincidences do not play any role.

[18] M. Zukowski, quant-ph/9811013.

[19] S. Bose, V. Vedral, and P. L. Knight, Phys. Rev. A **57**, 822 (1998).

[20] L. K. Grover, quant-ph/9704012.; W. Dür, H.-J. Briegel, J. I. Cirac, and P. Zoller, Phys. Rev. A **59**, 169 (1999).

Quantum Algorithms

Quantum Algorithms

Artur K. Ekert
Center for Quantum Computation, Oxford

Needless to say there is no universal way to learn about quantum computation. If you want to understand quantum algorithms you should pick up some quantum mechanics, at least at the level of the first few chapters of Feynman Lectures on Physics vol. III, and some elements of computer science including computational complexity. I recommend the following five books

1. Feynman, R. P., Leighton, R. B. and Sands, M. *The Feynman Lectures on Physics*, Vol. III Addison-Wesley, Reading, MA 1966.
2. Feynman, R.P., *Feynman Lectures on Computation*, (Edited by A.J.G. Hey and R.W. Allen)Addison-Wesley, 1996.
3. Papadimitriou, C.H., *Computational Complexity*, Addison-Wesley, 1994.
4. Welsh, D. "Codes & Cryptography" Clarendon Press, Oxford, 1988.
5. Some of the deeper implications of quantum computing are discussed at length in *The Fabric of Reality* by David Deutsch (Allen Lane, The Penguin Press, 1997).

If you have only a casual interest in the field then try some popular papers such as

- D.Deutsch and A.Ekert, "Quantum computation" Physics World , Vol. 11 No.3, pp.47-52 (March 1998).
- Lloyd, S. "Quantum-Mechanical Computers" Scientific American, October 1995, pp. 140-145.
- A.Barenco, A. Ekert, C. Macchiavello, and A. Sanpera, "Un Saut Quantique Pour Les Calculateurs", La Recherche, No. 292, pp. 52-58 (November 1996).

or simply surf the Web. For example, The Centre for Quantum Computation at the University of Oxford (http://www.qubit.org)has several WWW pages and links devoted to quantum computation and cryptography. For a more technical overview try

- Ekert, A., Josza, R. "Quantum Computation and Shor's Factoring Algorithm" Reviews of Modern Physics, Vol. 68 (July 1996) pp. 733-753.

Regarding the original work on quantum algorithms one should start with the two historic papers

- Feynman, R. P. "Simulating Physics with Computers" International Journal of Theoretical Physics, Vol. 21 (1982) pp. 467-488.
- Deutsch, D. "Quantum Theory, the Church-Turing Principle, and the Universal Quantum Computer" Proc. Roy. Soc. Lond. A400 (1985) pp. 97-117.

Deutsch's paper described the first quantum algorithm. His result laid the foundation for the new field of quantum computation, and was followed by a sequence of steadily improved quantum algorithms:

- Deutsch, D. and Jozsa, R. "Rapid Solution of Problems by Quantum Computation" Proceedings of the Royal Society of London, Vol. 439A (1992) pp. 553-558.
- Bernstein, E., Vazirani, U. "Quantum Complexity Theory" Proceedings of the 25th Annual ACM Symposium on the Theory of Computing" (1993) pp. 11-20.
- Simon, D. "On the Power of Quantum Computation" Proceedings of the 35th Annual IEEE Symposium on Foundations of Computer Science (1994) pp. 116-123.

Simon's paper, in which he described quantum periodicity estimation, led to a sudden change in 1994 when Peter Shor devised the first quantum algorithm that, in principle, can perform efficient factorisation.

- Shor, P. "Algorithms for quantum computation: discrete logarithms and factoring" Proceedings 35th Annual Symposium on Foundations of Computer Science, Santa Fe, NM, USA, 20-22 Nov. 1994, IEEE Comput. Soc. Press (1994) pp. 124-134.

Difficulty of factorisation underpins security of many common methods of encryption, such as the most popular public key cryptosystem RSA. Thus factoring became very quickly a 'killer application' for quantum computers.

Subsequent developments include:

- Coppersmith, D., "An Approximate Fourier Transform Useful in Quantum Factoring", IBM Research Report No. RC19642 (1994).
- Grover, L. K. "A Fast Quantum Mechanical Algorithm for Database Search" Proceedings of the 28th Annual ACM Symposium on the Theory of Computing, Philadelphia, (1996) pp. 212-219.
- A method for estimating eigenvalues of unitary operators in Kitaev, A. Yu., "Quantum measurements and the Abelian stabilizer problem", November 1995. Available at Los Alamos e-Print achive (`http://xxx.lanl.gov`) as quant-ph/9511026.

- Several further generalisations and applications of the method laid out by Simon and Shor, summarised in Mosca, M. and Ekert, A., "The Hidden Subgroup Problem and Eigenvalue Estimation on a Quantum Computer", to appear in Proceedings of the 1st NASA International Conference on Quantum Computing and Quantum Communication (Springer-Verlag).

- Work on quantum simulations inspired by Feynman's 1982 paper: Wiesner, S., "Simulation of Many-Body Quantum Systems by a Quantum Computer" available at Los Alamos e-Print achive (`http://xxx.lanl.gov`) as quant-ph/9603028; Zalka, C., "Efficient Simulation of Quantum Systems by Quantum Computers, quant-ph/9603026; and Lloyd, S. "Universal Quantum Simulators" Science, Vol. 273, 23 August, 1996, pp. 1073-1078.

Finally, a unified approach to quantum algorithms, in terms of multiparticle interferometry and the phase estimation, is presented in

- R.Cleve, A.Ekert, C.Macchiavello, and M. Mosca, "Quantum Algorithms Revisited", Proc. Roy. Soc. A vol. 454, pp. 339-354 (1998).

An Overview of Quantum Computing

Artur Ekert[a] and Chiara Macchiavello[a,b]

a) Clarendon Laboratory, University of Oxford, Oxford OX1 3PU, U.K.
b) Dipartimento di Fisica "A. Volta", Via Bassi 6, 27100 Pavia, Italy

Abstract

Information is physical and any processing of information is always performed by physical means - an innocent-sounding statement, but its consequences are anything but trivial. When quantum effects become important, for example at the level of single atoms and photons, the existing, purely abstract, classical theory of computation becomes fundamentally inadequate. Entirely new modes of computation and information processing become possible. In the last few years there has been an explosion of theoretical and experimental research in quantum computation. In this brief review we describe some of these new developments.

1 Introduction

The classical theory of computation is essentially the theory of the universal Turing machine - the most popular mathematical model of computation. Its significance relies on the fact that, given a large but finite amount of time, the universal Turing machine is capable of any computation that can be done by any modern classical digital computer, no matter how powerful. One of the strengths of the classical theory of computation is that it abstracts away the physics of the machines that perform computation. As such it becomes a branch of mathematics, and the study of computation requires no experimentations and can be done by pure thought. Pioneers such as Turing, Church, Post and Gödel managed to capture the correct classical theory by intuition alone and as the result it is often falsely assumed that its foundations are self-evident and purely abstract. However, the concepts of information and computation can be properly formulated only in the context of a physical theory – information is stored, transmitted and processed always by *physical* means. This approach, pioneered by Rolf Landauer and Charles H. Bennett in the sixties, led to questions about the physical limits of computation including the limits of miniaturisation of computing devices. It became clear that if computers are to become much smaller in the future their description must be given by quantum mechanics. In the early 1980s Paul Benioff described a (classical) Turing machine made of quantum components and showed that any (classical) computation could in principle be supported by a quantum hardware [1]. However, when quantum effects such as interference and entanglement become important, for example at the level of single atoms and photons, the existing, purely abstract, classical theory of computation becomes fundamentally inadequate. Entirely new modes of computation and information processing become possible!

The unusual power of quantum computers was first anticipated by Richard Feynman [2] and demonstrated and discussed in detail by David Deutsch in his seminal paper [3] which laid the foundation for the new field of quantum computation.

Feynman [2] in his talk during the First Conference on the Physics of Computation held at MIT in 1981 observed that it appears to be impossible to simulate a general quantum evolution on a classical probabilistic computer in an *efficient* way i.e. any classical simulation of quantum evolution appears to involve an exponential slowdown in time as compared to the natural evolution since the amount of information required to describe the evolving quantum state in classical terms generally grows exponentially in time. However, instead of viewing this fact as an obstacle, Feynman regarded it as an opportunity. If it requires so much computation to work out what will happen in a complicated multiparticle interference experiment then, he argued, the very act of setting up such an experiment and measuring the outcome is tantamount to performing a complex computation. Thus, Feynman suggested that it might be possible to simulate a quantum evolution efficiently after all, provided that the simulator itself is a quantum mechanical device. Furthermore, he conjectured that if one wanted to simulate a different quantum evolution, it would not be necessary to construct a new simulator from scratch. It should be possible to choose the simulator so that minor systematic modifications of it would suffice to give it any desired interference properties. He called such a device a universal quantum simulator. In 1985 Deutsch proved that such a universal simulator or a universal quantum computer does exist and that it could perform any computation that any other quantum computer (or any Turing-type computer) could perform. Moreover, it has since been shown that the time and other resources that it would need to do these things would not increase exponentially with the size or detail of the physical system being simulated, so the relevant computations would be tractable by the standards of complexity theory [4].

After the Deutsch paper, the hunt was on for something interesting for quantum computers to do. A sequence of steadily improved quantum algorithms by Deutsch and Jozsa [5], Bernstein and Vazirani [4], and by Simon [6] led to a sudden change in 1994 when Peter Shor devised the first quantum algorithm that, in principle, can perform efficient factorisation [7]. Difficulty of factorisation underpins security of many common methods of encryption, such as the most popular public key cryptosystem RSA (see, for example [8]). Thus factoring became very quickly a 'killer application' — something very useful that only a quantum computer could do. Shor's algorithm appeared right after quantum cryptography established itself as a respectable branch of physics and cryptology [9, 10, 11] stimulating a rapid development of quantum communication technology and right after Seth Lloyd's description of physical systems that, in principle, might act as basic units of quantum computers [12]. Thus the timing was right and the whole field of quantum computation very quickly became fashionable among physicists and computer scientists. Today, with only few more algorithms proposed (e.g. Grover's "data base search" [13] and Lloyd's quantum simulations [14]) but with much better understanding of their underlying structure [15] the hunt is on for a good technology which can support quantum computers and make them practical. Research in quantum computation is so vigorously active these days that any comprehensive review of the field must be obsolete as soon as it is written. Thus we have decided to provide only a very basic outline of quantum computation, hoping that it will serve as a good starting point for all those who want to enter the field.

2 Computational complexity

In order to solve a particular problem computers follow a precise set of instructions that can be mechanically applied to yield the solution to any given instance of the problem. A specification of this set of instructions is called an algorithm. Examples of algorithms are the procedures taught in elementary schools for adding and multiplying whole numbers; when these procedures are mechanically applied, they always yield the correct result for any pair of whole numbers. Some algorithms are fast (e.g. multiplication), others are very slow (e.g. factorisation, playing chess). An algorithm is said to be fast or efficient if the time taken to execute it increases no faster than a polynomial function of the size of the input [16]. We generally take the input size to be the total number of bits needed to specify the input (for example, a number n requires $\log_2 n$ bits of binary storage in a computer) and measure the execution time as the number of computational steps. Thus an efficient algorithm on a general input n runs in $poly(\log n)$ time. In 1965 A.Cobham proposed that the two crucial cases roughly corresponding to what we would loosely call "good" or "bad" algorithms are respectively polynomial and exponential time algorithms [17]. As an example of a "bad" algorithm consider the most naive factoring method: try dividing n by each number from 1 to $\sqrt{n}$ (as any composite n must have a factor in this range). This requires at least $\sqrt{n}$ steps (at least one step for each tried factor). However $\sqrt{n} = 2^{\frac{1}{2}\log n}$ is *exponential* in $\log n$ so this is not an efficient algorithm.

Computational complexity theory classifies problems according to the efficiency of algorithms required to solve them. The theory looks at the minimum time and space (memory) required to solve the hardest instance of the problem. Fig.1 shows some of the complexity classes and their presumed relationships

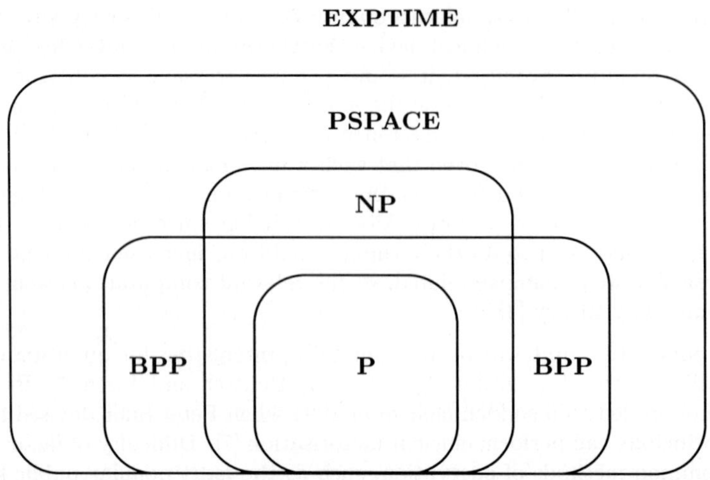

Figure 1: Simplified diagram showing some of the complexity classes and their presumed relationships. The diagram refers to classical computation — quantum complexity classes could be quite different.

(not much about this classification has been proved mathematically). The class **P** consists of all problems that can be solved in polynomial time (such as addition and multiplication). A presumably more general class called **NP** contains problems which cannot be solved (or we do not know how to solve them) in polynomial time but verifying that an attempted solution is indeed a solution can be performed in polynomial time (e.g. factoring belongs to **NP** because we do not know how to factor in polynomial time but we can easily check any proposed solution by multiplication). The next important class in the diagram, known in the literature as **BPP** [16], is somewhat different. It refers to randomised computation and concerns decision problems i.e. problems for which the output is just "yes" or "no" (problems in other complexity classes can also be phrased this way, e.g. given n and $m < n$, is there a factor of n less than m?). A **BPP** algorithm $\mathcal{A}$ is an efficient algorithm providing an answer which for any input is correct with a probability greater than some constant $\delta > 1/2$ (e.g. greater than $2/3$). We cannot check easily if the answer is correct or not but we may repeat $\mathcal{A}$ some fixed number k times and then take a majority vote of all the k answers. For sufficiently large k the majority answer will be correct with probability as close to 1 as desired (in fact the probability even converges to 1 *exponentially* fast in k [16]). In computational complexity theory it is customary to view problems in **BPP** as being "tractable" or "solvable in practice" and problems not in **BPP** as "intractable" or " unsolvable in practice on a computer". Further out in the complexity hierarchy is **PSPACE**. Problems in this class can be solved using only polynomial memory space, but not necessarily polynomial time. **PSPACE** includes **NP** but there are problems in **PSPACE** that are thought to be harder than **NP** (needless to say, this isn't proven either). Finally there is the class of problems called **EXPTIME** which contains problems solvable in exponential time, some of them can actually be proven not to be solvable in deterministic polynomial time. A rich source of **PSPACE** and **EXPTIME** problems has been the area of combinatorial games e.g. deciding whether or not the first player in a game such as Go or Chess on $l \times l$ board has a winning strategy [16, 18, 8].

It is worth pointing out that the definitions of efficient and inefficient algorithms have been carefully constructed to avoid any reference to a physical hardware. Indeed, they do not capture, for example, differences between classical and quantum algorithms and yet Feynman's observation that it appears impossible to simulate a general quantum evolution on a classical probabilistic computer in an *efficient* way suggests that the quantum theory allows Nature to efficiently keep track of exponentially many branching amplitudes in a way that we cannot simulate classically! Thus we may suspect that the computing power of a quantum device can exceed that of any classical device and new quantum complexity classes are needed to assess 'difficulty' of some mathematical operations such as factoring. This provides a fundamental impetus for the study of quantum computation and its possible experimental realisation.

3 Quantum Algorithms

Quantum computers can compute faster because they can accept as the input not a single number but a coherent superposition of many different numbers and subsequently perform a computation (a sequence of unitary operations) on all of these numbers simultaneously. This can be viewed as a massive parallel computation, but instead of having many processors working in parallel we have only one quantum processor performing a computation that affects all numbers in a superposition i.e. all components of the input state vector.

The exponential speed-up of quantum computers takes place at the very beginning of their computation. Qubits, i.e. physical systems which can be prepared in one of the two orthogonal states labeled as $|0\rangle$ and $|1\rangle$ or in a superposition of the two, can store superpositions of many 'classical' inputs. For example, the equally weighted superposition of $|0\rangle$ and $|1\rangle$ can be prepared by taking a qubit initially in state $|0\rangle$ and applying to it transformation $\mathbf{H}$ (also known as the Hadamard transform) which maps

$$|0\rangle \longrightarrow \frac{1}{\sqrt{2}}(|0\rangle + |1\rangle), \tag{1}$$

$$|1\rangle \longrightarrow \frac{1}{\sqrt{2}}(|0\rangle - |1\rangle). \tag{2}$$

If this transformation is applied to each qubit in a register composed of two qubits it will generate the superposition of four numbers

$$|0\rangle|0\rangle \longrightarrow \frac{1}{\sqrt{2}}(|0\rangle + |1\rangle)\frac{1}{\sqrt{2}}(|0\rangle + |1\rangle) \tag{3}$$

$$= \tfrac{1}{2}(|00\rangle + |01\rangle + |10\rangle + |11\rangle). \tag{4}$$

In general a quantum register composed of l qubits can be prepared in a superposition of 2^l different numbers (inputs) with only l elementary operations. This can be written as

$$|0\rangle \longrightarrow 2^{-l/2} \sum_{x\in\{0,1\}^l} |x\rangle\,. \tag{5}$$

Operation $\mathbf{H}$ applied to each of the l qubits in the register is referred to as the l-qubit Hadamard transform or simply the Hadamard transform. If the register is initially in some state $|u\rangle$ which belongs to the computational basis, i.e. it represents one particular binary number $u \in \{0,1\}^l$ rather than a superposition of numbers, then the action of the Hadamard transform can be described as

$$|u\rangle \longrightarrow 2^{-l/2} \sum_{x\in\{0,1\}^l} (-1)^{u\cdot x}|x\rangle\,, \tag{6}$$

where the product of $u = (u_1, \ldots, u_l)$ and $x = (x_1, \ldots, x_l)$ is defined as

$$u \cdot x = (u_1x_1 + u_2x_2 + \ldots + u_lx_l) \in \{0,1\}, \tag{7}$$

with additions and multiplications performed modulo 2.

Thus the l-qubit Hadamard transform, which involves only l elementary operations, generates exponentially many, that is 2^l, different binary numbers at the input!

The next task is to process all the numbers in parallel within the superposition by a sequence of unitary operations. This is how quantum devices must compute functions. Consider a function

$$f: \{0,1,\ldots 2^l - 1\} \longrightarrow \{0,1,\ldots 2^k - 1\}. \tag{8}$$

A classical computer computes f by evolving each labeled input, $0, 1, \ldots, 2^l - 1$ into a respective labeled output, $f(0), f(1), \ldots, f(2^l - 1)$. Quantum computers, due to the unitary (and therefore reversible) nature of their evolution, compute functions in a slightly different way. In order to compute functions which are not one-to-one and to preserve the reversibility of computation, quantum computers have to keep the record of the input. Here is how it is done. We need two quantum registers of length l and k. The first register is loaded with value x, i.e. it is prepared in state $|\,x\rangle$, the second register may initially contain an arbitrary number y. The function evaluation is then determined by an appropriate unitary evolution of the two registers,

$$|\,x\rangle\,|\,y\rangle \xrightarrow{U_f} |\,x\rangle\,|\,y+f(x)\rangle\,. \tag{9}$$

Here $y+f(x)$ means addition modulo the maximum number of configurations of the second register, i.e. 2^k in our case.

The computation we are considering here is not only reversible but also quantum and we can do much more than computing values of $f(x)$ one by one. We can prepare a superposition of all input values as a single state and by running the computation U_f *only once*, we can compute *all* of the 2^l values $f(0),\ldots,f(2^l-1)$ (here and in the following we ignore the normalisation constants),

$$\sum_{x=0}^{2^l-1} |\,x\rangle\,|\,y\rangle \xrightarrow{U_f} |\,y+f\rangle = \sum_{x=0}^{2^l-1} |\,x\rangle\,|\,y+f(x)\rangle\,. \tag{10}$$

It looks too good to be true so where is the catch? How much information about f does the state

$$|\,f\rangle = |\,0\rangle\,|\,f(0)\rangle + |\,1\rangle\,|\,f(1)\rangle + \ldots + \left|\,2^l-1\right\rangle \left|\,f(2^l-1)\right\rangle \tag{11}$$

really contain?

Unfortunately no quantum measurement can extract all of the 2^l values $f(0), f(1),\ldots,f(2^l-1)$ from $|\,f\rangle$. If we measure the two registers after the computation U_f we register one output $|\,x\rangle\,|\,y+f(x)\rangle$ for some value x. However, there are measurements that provide us with information about joint properties of all the output values $f(x)$, such as, for example, periodicity, without providing any information about particular values of $f(x)$. Let us illustrate this with a simple example.

Consider a Boolean function f which maps $\{0,1\}\to\{0,1\}$. There are exactly four functions of this type: two constant functions ($f(0)=f(1)=0$ and $f(0)=f(1)=1$) and two balanced functions ($f(0)=0, f(1)=1$ and $f(0)=1, f(1)=0$). Is it possible to compute function f *only once* and to find out whether it is constant or balanced, i.e. whether the binary numbers $f(0)$ and $f(1)$ are the same or different? N.B. we are not asking for particular values $f(0)$ and $f(0)$ but for a global property of f.

Classical intuition tells us that we have to evaluate both $f(0)$ and $f(1)$, that is to compute f twice. This is not so. Quantum mechanics allows us to perform the trick with a single function evaluation. We simply take two qubits, each qubit serves as a single qubit register, prepare the first qubit in state $|\,0\rangle$ and the second in state $|\,1\rangle$ and compute

$$\begin{aligned} |\,0\rangle\,|\,1\rangle &\longrightarrow (|\,0\rangle + |\,1\rangle)(|\,0\rangle - |\,1\rangle) \longrightarrow \\ &\longrightarrow |\,0\rangle\,(|\,f(0)\rangle - |\,1+f(0)\rangle) + |\,1\rangle\,(|\,f(1)\rangle - |\,1+f(1)\rangle). \end{aligned} \tag{12}$$

We start with transformation $\mathbf{H}$ applied both to the first and the second qubit, followed by the function evaluation. Here $1+f(0)$ denotes addition modulo 2 and simply means taking the negation of $f(0)$. At this stage, depending on values $f(0)$ and $f(1)$, we have one of the four possible states of the two qubits. We apply $\mathbf{H}$ again to the first and the second qubit and evolve the four states as follows

$$|\,0\rangle\,(|\,0\rangle - |\,1\rangle) + |\,1\rangle\,(|\,0\rangle - |\,1\rangle) \longrightarrow +|\,0\rangle\,|\,1\rangle\,, \tag{13}$$

$$|\,0\rangle\,(|\,1\rangle - |\,0\rangle) + |\,1\rangle\,(|\,1\rangle - |\,0\rangle) \longrightarrow -|\,0\rangle\,|\,1\rangle\,, \tag{14}$$

$$|\,0\rangle\,(|\,0\rangle - |\,1\rangle) + |\,1\rangle\,(|\,1\rangle - |\,0\rangle) \longrightarrow +|\,1\rangle\,|\,1\rangle\,, \tag{15}$$

$$|\,0\rangle\,(|\,1\rangle - |\,0\rangle) + |\,1\rangle\,(|\,0\rangle - |\,1\rangle) \longrightarrow -|\,1\rangle\,|\,1\rangle\,. \tag{16}$$

The second qubit returns to its initial state $|1\rangle$ but the first qubit contains the relevant information. We measure its bit value — if we register '0' the function is constant, if we register '1' the function is balanced !

This example [15] is an improved version of the first quantum algorithm proposed by Deutsch [3]. (The original Deutsch algorithm provides the correct answer with probability 50% .) Deutsch's result laid the foundation for the new field of quantum computation, and was followed by several other quantum algorithms for various problems.

Deutsch's original problem was subsequently generalised by Deutsch and Jozsa [5] for Boolean functions $f : \{0,1\}^l \to \{0,1\}$ in the following way. Assume that, for one of these functions (which are computed by "black boxes" or "oracles" so that each evaluation of f counts as one computational step), it is "promised" that it is either constant or balanced (i.e. has an equal number of 0 outputs as 1's), and consider the goal of determining which of the two properties the function actually has. How many evaluations of f are required to do this? Any classical algorithm for this problem would, in the worst-case, require $2^{l-1}+1$ evaluations of f before determining the answer with certainty. There is, however, a quantum algorithm that solves this problem with a single evaluation of f. It works as follows.

- Start with the first register (l qubits) in state $|0\rangle$ and the second one (one qubit) in state $|1\rangle$, and apply the Hadamard transform to the two registers. This gives

$$\sum_{x\in\{0,1\}^l} |x\rangle\,(|0\rangle - |1\rangle). \tag{17}$$

- Evaluate the function f. This generates the state

$$\sum_{x\in\{0,1\}^l} |x\rangle\,(|f(x)\rangle - |f(x)+1\rangle) = \sum_{x\in\{0,1\}^l} (-1)^{f(x)}\,|x\rangle\,(|0\rangle - |1\rangle). \tag{18}$$

- Apply the Hadamard transform to the first register. The state now is

$$\sum_{x,y\in\{0,1\}^l} (-1)^{f(x)+x\cdot y}\,|y\rangle\,(|0\rangle - |1\rangle). \tag{19}$$

- Measure the first register. If $y = 0$ (i.e. $y = (0,0,\ldots,0)$), then the function is constant. If $y \neq 0$ the function is balanced.

The last part follows from the fact that $\sum_{x\in\{0,1\}^l}(-1)^{x\cdot y} = 0$ for all $y \neq 0$. (Deutsch and Jozsa's algorithm is similar to the above, except that it employs two function evaluations instead of one).

It was the first quantum algorithm which indicated the exponential power of quantum speed-up – $2^{l-1}+1$ versus 1 function evaluations, however, when compared with classical probabilistic computation the difference in the performance was less dramatic.

An interesting variation of this problem has been discussed by Ethan Bernstein and Umesh Vazirani [4]. Suppose that $f : \{0,1\}^l \to \{0,1\}$ is of the form

$$f(x) = a \cdot x, \tag{20}$$

where $a \in \{0,1\}^l$ and consider the goal of determining a. The classical determination of a requires at least l evaluations of f whereas the quantum solution involves computation of f only once. The quantum algorithm follows the same sequence of operations as in the Deutsch-Jozsa algorithm; the final measurement on the first register gives value a.

The next major progress in quantum algorithms was due to Dan Simon and his quantum periodicity estimation [6]. Consider a black box (or oracle) which computes function $f : \{0,1\}^l \to \{0,1\}^l$. The function is guaranteed to be a two-to-one function with some periodicity $r \in \{0,1\}^l$ i.e. $f(x) = f(y)$ iff $y = x + r$ (this is a bit by bit addition mod2) for all $x, y \in \{0,1\}^l$. The goal is to determine r.

The classical (probabilistic) algorithm which gives r with some fixed probability requires exponential number of f-evaluations. The quantum algorithm computes f only m times where m is a number which is of the order of l. The algorithm proceeds as follows.

- Start with the first and the second register (both with l qubits) in state $|\,0\rangle$ and apply the Hadamard transform to the first register to get

$$\sum_{x\in\{0,1\}^l} |\,x\rangle\,|\,0\rangle\,. \tag{21}$$

- Evaluate f. This gives

$$\sum_{x\in\{0,1\}^l} |\,x\rangle\,|\,f(x)\rangle\,. \tag{22}$$

- Measure the second register to get

$$(|\,k\rangle + |\,k+r\rangle)\,|\,f(k)\rangle \tag{23}$$

  for some k.

- Apply the Hadamard transform to the first register to get

$$\sum_{y\in\{0,1\}^l} ((-1)^{k\cdot y} + (-1)^{(k+r)\cdot y})\,|\,y\rangle\,|\,f(k)\rangle = \sum_{y\in\{0,1\}^l} (-1)^{k\cdot y}(1+(-1)^{r\cdot y})\,|\,y\rangle\,|\,f(k)\rangle \tag{24}$$

- Measure the first register. From the last equation above it follows that if $r \cdot y = 1$ then $(1 + (-1)^{r\cdot y}) = 0$ and the probability amplitude of state $|\,y\rangle$ is zero. Thus the measurement must give y such that $y \cdot r = 0$.

- Repeat the above to find enough different y_i's so that r can be determined by solving the system of linear equations $y_1 \cdot r = 0, \ldots y_m \cdot r = 0$.

The quantum factoring algorithm [7] was inspired by Simon's quantum periodicity estimation. Shor's quantum factoring of an integer n is based on calculating the period of the function $F_n(x) = a^x \bmod n$ for a randomly selected integer a between 0 and n. It turns out that for increasing powers of a, the remainders form a repeating sequence with a period which we denote r. Once r is known the factors of n are obtained by calculating the greatest common divisor of n and $a^{r/2} \pm 1$.

Suppose we want to factor 15 using this method. Let $a = 11$. For increasing x the function $11^x \bmod 15$ forms a repeating sequence $1, 11, 1, 11, 1, 11, \ldots$. The period is $r = 2$, and $a^{r/2} \bmod 15 = 11$. Then we take the greatest common divisor of 10 and 15, and of 12 and 15 which gives us respectively 5 and 3, the two factors of 15. Classically calculating r is at least as difficult as trying to factor n; the execution time of calculations grows exponentially with the number of digits in n. Quantum computers can find r in time which grows only as a cubic function of the number of digits in n.

The structure of Shor's algorithm is very similar to Simon's periodicity estimations but unlike its predecessors Shor's algorithm does not involve any "black boxes" (or "oracles"). To estimate the period r we prepare two quantum registers; the first register, with l qubits, in the equally weighted superposition of all numbers it can contain, and the second register in state zero. Then we perform an arithmetical operation that takes advantage of quantum parallelism by computing the function $F_n(x)$ for each number x in the superposition. The values of $F_n(x)$ are placed in the second register so that after the computation the two registers become entangled:

$$\sum_x |\,x\rangle\,|\,0\rangle \longrightarrow \sum_x |\,x\rangle\,|\,F_n(x)\rangle\;. \tag{25}$$

Now we perform a measurement on the second register. We measure each qubit and obtain either "0" or "1" . This measurement yields value $F_n(k)$ (in binary notation) for some randomly selected k. The state of the first register right after the measurement, due to the periodicity of $F_n(x)$, is a coherent superposition of all states $|\,x\rangle$ such that $x = k, k+r, k+2r, \ldots$, i.e. all x for which $F_n(x) = F_n(k)$. The periodicity in the probability amplitudes in the first register cannot be simply measured because the offset i.e. the value k is randomly selected by the measurement. However the state of the first register can be subsequently transformed via a unitary operation which effectively removes the offset and modifies the period in the probability amplitudes from r to a multiple of $2^l/r$. This operation is known as the quantum Fourier transform (QFT) and can be written as

$$\mathrm{QFT} : |\,x\rangle \longmapsto 2^{-l/2} \sum_{y=0}^{2^l-1} \exp(2\pi i x y/2^l)\ |\,y\rangle\,. \tag{26}$$

After QFT the first register is ready for the final measurement which yields with high probability an integer which is the best whole approximation of a multiple of $2^l/r$ i.e. $x = m2^l/r$ for some integer m. We know the measured value x and the size of the register l hence if m and r are coprime we can determine r by canceling $x/2^l$ down to an irreducible fraction and taking its denominator. Since the probability that m and r are coprime is sufficiently large (greater than $1/\log r$ for large r) this gives an efficient randomised algorithm for determination of r. A more detailed description of Shor's algorithm can be found in [7, 19].

Perhaps the most important recent development in quantum algorithms is Grover's 'quantum database search' [13]. Suppose we are given (as an oracle) a function f_k which maps $\{0,1\}^l$ to $\{0,1\}$ such that $f_k(x) = \delta_{xk}$ for some k. Our task is to find k. Thus in a set of numbers from 0 to $2^l - 1$ one element has been "tagged" and by evaluating f_k we have to find which one. To find k with probability of 50% any classical algorithm, be it deterministic or randomised, will need to evaluate f_k a minimum of 2^{l-1} times. In contrast, Grover's quantum algorithm needs only $O(2^{l/2})$ evaluations. Again, the first register is prepared in state $|\,0\rangle$ and the second one (containing only one qubit) in state $|\,1\rangle$. The the sequence of operations $\mathbf{H}U_{f_k}\mathbf{H}\ U_{f_0}$, where $\mathbf{H}$ is the Hadamard transform performed both on the first and the second register and U_{f_k} is the quantum evaluation of f_k, is repeated roughly $2^{l/2}$ times and the first register is measured. The result of the measurement is, with probability greater than half, the sought after value k. Although the speed-up the data base search remains computationally hard even for quantum computers (the speed-up is only by the square root factor), but it is truly remarkable that such a difficult problem as searching an unstructured data base can be improved at all. It seems that Grover's algorithm probes the very limits of power of quantum computation (see, for example, discussion by Bennett *et al.* [20]).

4 Quantum Networks

Our description of quantum algorithms, as presented in the previous section, is of little help to those who would like to make a practical use of the awesome computational power of quantum devices. Clearly we cannot just assume that any unitary transformation U representing a mathematical operation, e.g. a function evaluation

$$\sum_x |\,x\rangle\,|\,y\rangle \xrightarrow{U_f} |\,y+f\rangle = \sum_x |\,x\rangle\,|\,y+f(x)\rangle \tag{27}$$

or the quantum Fourier transform (Eq. 26), may be efficiently implemented. We have to show how to *construct* U using some *finite* basic set of transformations. In classical computation any Boolean function can be constructed as a Boolean network composed of elementary logic gates. Extending this idea to the quantum domain we will now express quantum algorithms in terms of quantum computational networks. A quantum network, introduced by Deutsch [21] in 1989, is a quantum computing device consisting of quantum logic gates whose computational steps are synchronised in time. The outputs of some of the gates are connected by wires to the inputs of others. Quantum gates are the active components of quantum networks – in an n-bit quantum gate, n qubits undergo a coherent interaction. Wires are the passive components which allow to carry quantum states from one computational step to another. However, the distinction between "active" and "passive" components can only be made relative to a given physical and technological implementation. Different technologies would lead us to draw the line differently. In most current quantum technologies there are no physical "wires" to move qubits into adjacent positions so that they can undergo a gate-type interaction – a "wire" is just a convenient symbol which acts as a time-separator between two subsequent computational steps.

So how many quantum gates do we need? In fact only one and almost any two-qubit gate can act as a universal quantum gate [22, 23]! Of course, since there is a continuum of possible unitary transformations quantum networks in general will only approximate these transformations, but any desired degree of approximation can be obtained using sufficiently long networks.

Any quantum algorithm can be represented as a family of networks, for example, quantum addition is represented by a family of quantum adders, each adder in the family acts on a different number of input qubits. The computational complexity of a family of networks $\{\mathcal{Q}_l\}$ may now be defined in terms of the *size* of $\mathcal{Q}_l$ i.e. the number of gates involved in $\mathcal{Q}_l$. An algorithm will be called efficient if it has a (uniform) family of *polynomial-size* networks i.e. a family $\{\mathcal{Q}_l\}$ such that the size of $\mathcal{Q}_l$ grows only polynomially with l. "Uniform" means that it is easy to construct the whole family of networks e.g. the network has a pattern which allows to build $\mathcal{Q}_{l+1}$ as a simple extension of $\mathcal{Q}_l$ (c.f. the QFT network below).

Let us introduce some elementary gates. The Hadamard gate implements the Hadamard transform (Eq.2). It is the single qubit gate **H** performing the unitary transformation

$$\mathbf{H} = \frac{1}{\sqrt{2}}\begin{pmatrix} 1 & 1 \\ 1 & -1 \end{pmatrix} \tag{28}$$

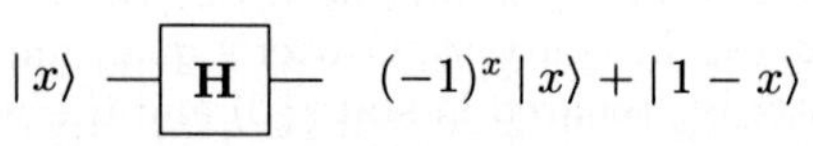

where the diagram on the right provides a schematic representation of the gate **H** acting on a qubit in state $|\,x\rangle$, with $x = 0, 1$. The conditional phase shift is the two–bit gate $\mathbf{B}(\phi)$ defined as

$$\mathbf{B}(\phi) = \begin{pmatrix} 1 & 0 & 0 & 0 \\ 0 & 1 & 0 & 0 \\ 0 & 0 & 1 & 0 \\ 0 & 0 & 0 & e^{i\phi} \end{pmatrix} \tag{29}$$

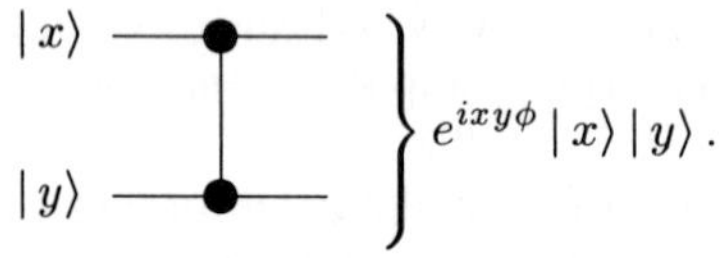

The matrix is written in the basis $\{|\,0\rangle\,|\,0\rangle\,,|\,0\rangle\,|\,1\rangle\,,|\,1\rangle\,|\,0\rangle\,,|\,1\rangle\,|\,1\rangle\}$ (the diagram on the right shows the structure of the gate). Another important two-qubit gate is the quantum controlled-NOT (or XOR) operation defined as

$$C = \begin{pmatrix} 1 & 0 & 0 & 0 \\ 0 & 1 & 0 & 0 \\ 0 & 0 & 0 & 1 \\ 0 & 0 & 1 & 0 \end{pmatrix} \tag{30}$$

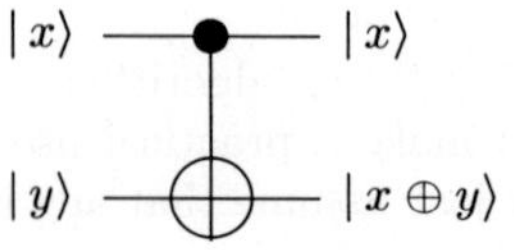

where $x, y = 0$ or 1 and $\oplus$ denotes XOR or addition modulo 2. The quantum controlled-NOT gate is not a universal gate but a universal quantum gate can be constructed by a combination of the controlled-NOT and simple unitary operations on a single qubit (for more details about the relevance and possible implementations of this gate see [24, 25]).

Suppose we want to implement Shor's factoring algorithm, i.e. we want to build a dedicated quantum device to factor large integers, how shall we start?

Firstly we notice that quantum factorisation contains two major operations: quantum exponentiation (computing $a^x \bmod n$) followed by the quantum Fourier transform. Quantum exponentiation can be

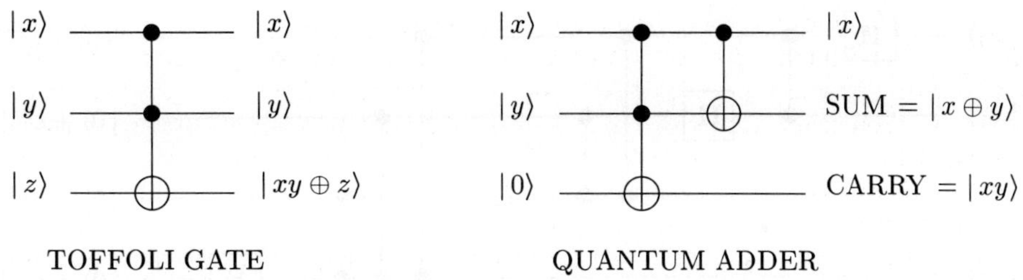

Figure 2: Diagrammatic representation of the controlled-controlled-NOT (Toffoli) gate and a simplified quantum adder. States $|x\rangle, |y\rangle$, and $|z\rangle$ belong to the computational basis $x, y, z = 0$ or 1 and both addition $\oplus$ and multiplication xy are performed modulo 2. The Toffoli gate is a very handy basic unit which features prominently in the network implementing elementary quantum arithmetic i.e. in quantum adders, multipliers etc. It can be decomposed and written as a quantum network of elementary two-qubit and one-qubit gates. A simplified quantum adder is a starting point for constructing more elaborate networks.

decomposed into a sequence of squaring,

$$a^x = a^{2^0 x_0} \cdot a^{2^1 x_1} \cdot \ldots a^{2^{l-1} x_{l-1}}, \tag{31}$$

where $x_0, x_1 \ldots$ are the binary digits of x. Squaring is achieved by multiplication and multiplication by a sequence of additions. Following this reduction procedure we end up with a quantum adder as a basic unit for the whole network. However, a quantum adder is different from a classical adder. Any unitary operation is reversible which is why quantum networks for basic arithmetic cannot be directly deduced from their classical Boolean counterparts (classical logic gates such as AND or OR are clearly irreversible - reading 1 at the output of the OR gate does not provide enough information to determine the input which could be either $(0,1)$ or $(1,0)$ or $(1,1)$). Quantum arithmetic must be build *ab initio* from the reversible logical components. A good starting point is a simplified quantum adder shown in Fig.2. Explicit constructions of more elaborate quantum networks leading to modular exponentiation have been described in detail by Vedral et al. [26] and Beckman et al. [27].

The second part of the Shor algorithm, i.e. the discrete quantum Fourier transform (Eq. 26), is much easier to implement. The QFT network will consist of only one–qubit and two–qubit gates, which are: the Hadamard gate $\mathbf{H}$ and the conditional phase shift $\mathbf{B}(\phi)$. In Fig.3 we show the network which performs the quantum Fourier transform for $l = 4$ (for more details on the implementations of QFT see [28]).

A general case of l qubits requires a trivial extension of the network following the same sequence pattern of gates $\mathbf{H}$ and $\mathbf{B}$. The QFT network operating on l qubits contains l Hadamard gates $\mathbf{H}$ and $l(l-1)/2$ phase shifts $\mathbf{B}$, in total $l(l+1)/2$ elementary gates. Thus the quantum Fourier transform can be performed in an efficient way, the network size grows only as a quadratic function of the size of the input.

5 Conditional quantum dynamics

In the past three years researchers from different walks of physics have proposed many technologies for quantum logic gates ranging from the cavity QED [29] via ions in linear traps [30, 31] to the NMR based bulk spin computation [32, 33]. Single qubit quantum gates are regarded as relatively easy to implement. For example, a typical quantum optical realisation uses atoms as qubits and controls their states with laser light pulses of carefully selected frequency, intensity and duration; any prescribed

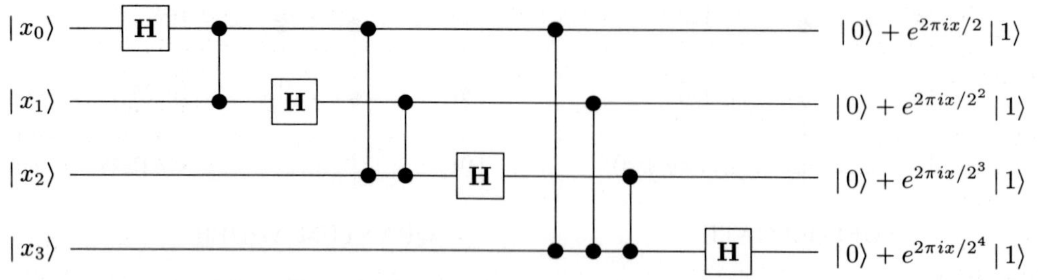

$$\mathbf{H}\,\mathbf{B}(\pi)\,\mathbf{H}\,\mathbf{B}(\pi/2)\mathbf{B}(\pi)\,\mathbf{H}\,\mathbf{B}(\pi/4)\mathbf{B}(\pi/2)\mathbf{B}(\pi)\,\mathbf{H}$$

Figure 3: The quantum Fourier transform (QFT) network operating on four qubits. If the input state represents number $x = \sum_k 2^k x_k$ the output state of each qubit is of the form $|0\rangle + e^{i\phi_k}|1\rangle$, where $\phi_k = 2\pi/2^k$ and $k = 0, 1, 3 \ldots$. N.B. there are three different types of the $\mathbf{B}(\phi)$ gate in the network above: $\mathbf{B}(\pi)$, $\mathbf{B}(\pi/2)$ and $\mathbf{B}(\pi/4)$.

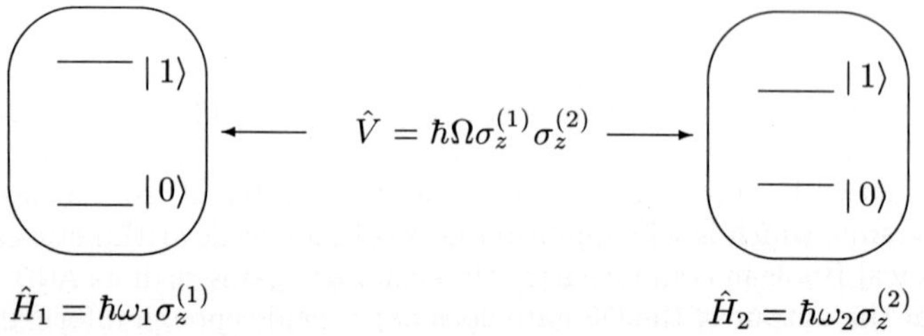

Figure 4: The control qubit of resonant frequency ω_1 interacts via $\hat{V}$ with the target qubit of resonant frequency ω_2. Due to the interaction the two resonant frequencies are modified and the combined system of the two qubits has four different resonant frequencies $\omega_1 \pm \Omega$ and $\omega_2 \pm \Omega$. A π-pulse at frequency $\omega_2 + \Omega$ causes the transition $|0\rangle \leftrightarrow |1\rangle$ in the second qubit only if the first qubit is in state $|1\rangle$. This is one possible realisation of the quantum controlled-NOT gate.

superposition of two selected atomic states can be prepared this way. Two-qubit gates are much more difficult to build.

In order to implement two-qubit quantum logic gates it is sufficient, from the experimental point of view, to induce a conditional dynamics of physical bits, i.e. to perform a unitary transformation on one physical subsystem conditioned upon the quantum state of another subsystem,

$$U = |0\rangle\langle 0| \otimes U_0 + |1\rangle\langle 1| \otimes U_1 + \ldots + |k\rangle\langle k| \otimes U_k, \tag{32}$$

where the projectors refer to quantum states of the control subsystem and the unitary operations U_i are performed on the target subsystem [24]. The simplest non-trivial operation of this sort is probably a conditional phase shift such as $\mathbf{B}(\phi)$ which we used to implement the quantum Fourier transform and the quantum controlled-NOT (or XOR) gate.

Let us illustrate the notion of the conditional quantum dynamics with a simple example (see Fig.4). Consider two qubits, e.g. two spins, atoms, single-electron quantum dots, which are coupled via $\sigma_z^{(1)}\sigma_z^{(2)}$ interaction (e.g. a dipole-dipole interaction). The first qubit with the resonant frequency ω_1

will act as the control qubit and the second one, with the resonant frequency ω_2, as the target qubit. Due to the coupling V the resonant frequency for transitions between the states $|0\rangle$ and $|1\rangle$ of one qubit ***depends on the neighbour's state.*** The resonant frequency for the first qubit becomes $\omega_1 \pm \Omega$ depending on whether the second qubit is in state $|0\rangle$ or $|1\rangle$. Similarly the second qubit's resonant frequency becomes $\omega_2 \pm \Omega$, depending on the state of the first qubit. Thus a π-pulse at frequency $\omega_2 + \Omega$ causes the transition $|0\rangle \leftrightarrow |1\rangle$ in the second qubit only if the first qubit is in $|1\rangle$ state. This way we can implement the quantum controlled-NOT gate.

Thus in principle we know how to build a quantum computer; we can start with simple quantum logic gates and try to integrate them together into quantum networks. However if we keep on putting quantum gates together into networks we will quickly run into some serious practical problems. The more interacting qubits are involved the harder it tends to be to engineer the interaction that would display the quantum interference. Apart from the technical difficulties of working at single-atom and single-photon scales, one of the most important problems is that of preventing the surrounding environment from being affected by the interactions that generate quantum superpositions. The more components the more likely it is that quantum computation will spread outside the computational unit and will irreversibly dissipate useful information to the environment. This degrading effect of the computer-environment interaction on the computer is generally known as decoherence [34]. Thus the race is to engineer sub-microscopic systems in which qubits interact only with themselves but not with the environment.

6 Stability of Quantum Computation

When we analyse ***physically realisable*** computations we have to consider errors which are due to the computer-environment coupling and from the computational complexity point of view we need to assess how these errors scale with the input size l. If the probability of an error in a single run, $\epsilon(l)$, grows exponentially with l, i.e. if $\epsilon(l) = 1 - A\exp(-\alpha l)$, where A and α are positive constants, then the randomised algorithm cannot technically be regarded as efficient any more regardless of how weak the coupling to the environment may be. Unfortunately the computer-environment interaction leads to just such an unwelcome exponential increase of the error rate with the input size [35]. It is clear that for quantum computation of any reasonable length to ever be physically feasible it will be necessary to incorporate some efficiently realisable stabilisation scheme to combat the effects of decoherence. Deutsch discussed this problem during the Rank Prize Funds Mini–Symposium on Quantum Communication and Cryptography, Broadway, England in 1993 and proposed 'recoherence' based on a symmetrisation procedure (for details see [36]).

The basic idea is as follows. Suppose we have a quantum system, we prepare it in some initial state $|\Psi_i\rangle$ and we want to implement a prescribed unitary evolution $|\Psi(t)\rangle$ or just preserve $|\Psi_i\rangle$ for some period of time t. Now, suppose that instead of a single system we can prepare R copies of $|\Psi_i\rangle$ and subsequently we can project the state of the combined system into the symmetric subspace i.e. the subspace containing all states which are invariant under any permutation of the sub-systems. The claim is that frequent projections into the symmetric subspace will reduce errors induced by the environment. The intuition behind this concept is based on the observation that a prescribed error-free storage or evolution of the R independent copies starts in the symmetric sub-space and should remain in that sub-space. Therefore, since the error-free component of any state always lies in the symmetric subspace, upon successful projection it will be unchanged and part of the error will have been removed. Note however that the projected state is generally not error–free since the symmetric subspace contains states which are not of the simple product form $|\Psi\rangle|\Psi\rangle\ldots|\Psi\rangle$. Nevertheless it has been shown that the error probability will be suppressed by a factor of $1/R$ [36]. More recently projections on symmetric subspaces were replaced by more complicated projections on carefully selected subspaces. These clever projections, proposed by Shor [37], Calderbank and Shor [38], Steane [39] and others [40, 41, 42, 43, 44], are constructed on the basis of classical error-correcting methods but represent intrinsically new quantum error-correction and stabilisation schemes;

they are the subject of much current study. There are also other approaches to recoherence [45, 46, 47] but here we illustrate the main idea with the simplest three-qubit quantum error-correcting code.

Within a *simplified* model of the computer-environment interaction, known as decoherence [34], it is assumed that the register in the computer and the environment undergo the following unitary evolution

$$|\, i\rangle\, |\, R\rangle \longrightarrow |\, i\rangle\, |\, R_i(t)\rangle\,, \tag{33}$$

where $|\, i\rangle$ is the state from the computational basis and $|\, R\rangle$ is the initial state of the environment. States $|\, R_i(t)\rangle$ are normalised but not necessarily orthogonal to each other. Now, consider the following initial state of the computer and the reservoir

$$|\, \Psi(0)\rangle = \sum_i c_i\, |\, i\rangle \otimes |\, R\rangle\,. \tag{34}$$

The unitary evolution of the composed system results in an entangled computer-reservoir state which can be written as

$$|\, \Psi(t)\rangle = \sum_i c_i\, |\, i\rangle \otimes |\, R_i(t)\rangle\,, \tag{35}$$

where, in general, $\langle R_i \,|\, R_j\rangle \neq 0$ for $i \neq j$. The elements of the density matrix evolve as

$$\rho_{ij}(0) = c_i\, c_j^* \longrightarrow \rho_{ij}(t) = c_i\, c_j^*\, \langle R_i(t) \,|\, R_j(t)\rangle. \tag{36}$$

In this popular model of decoherence [34] the environment effectively acts as a measuring apparatus; in time the reservoir states $\{|\, R_i\rangle\}$ become more and more orthogonal to each other whilst the coefficients $\{c_i\}$ remain unchanged. Consequently, after some period of time known as the decoherence time, the off-diagonal elements ρ_{ij} disappear due to the $\langle R_i(t) \,|\, R_j(t)\rangle$ factors. If the computing device contains l qubits the typical decoherence time for the computer will be of the order t_d/l, where t_d is a decoherence time of a single qubit. Clearly decoherence puts an upper bound on the length of any feasible quantum computation. For if the elementary computational step takes time τ then the requirement that a coherent computation of K steps involving l qubits be completed within the decoherence time of the computer can be written as

$$\tau K < t_d/l. \tag{37}$$

In the case of Shor's algorithm for factorisation of an l-bit number we may take $K = l^3$ and (37) provides an upper bound

$$l < \left(\frac{t_d}{\tau}\right)^{1/4}. \tag{38}$$

Thus the ratio t_d/τ, which depends on the technology employed, determines the limits of the algorithm and it is unrealistic to assume that this ratio can be made infinite. This does not look very promising at first glance and indicates that quantum algorithms have to incorporate some additional recoherence-type operations. The good news is that recoherence is perfectly possible and several methods for protecting quantum states have been recently proposed.

Let us illustrate the main idea of recoherence by describing a simple method for protecting an unknown state of a single qubit in a noisy quantum register. Consider the following scenario: we want to store in a computer memory one qubit in an unknown quantum state of the form $|\, \phi\rangle = c_0\, |\, 0\rangle + c_1\, |\, 1\rangle$ and we know that any single qubit which is stored in a register undergoes a decoherence type entanglement with an environment (Eq. 36); in particular

$$(c_0\, |\, 0\rangle + c_1\, |\, 1\rangle)\, |\, R\rangle \longrightarrow c_0\, |\, 0\rangle\, |\, R_0(t)\rangle + c_1\, |\, 1\rangle\, |\, R_1(t)\rangle\,. \tag{39}$$

To see how the state of the qubit is affected by the environment, we calculate the fidelity of the decohered state at time t with respect to the initial state $|\phi\rangle$

$$F(t) = \langle\phi\,|\,\rho(t)\,|\,\phi\rangle \ , \tag{40}$$

where $\rho(t)$ is given by Eq. (36). It follows that

$$F(t) = |c_0|^4 + |c_1|^4 + 2|c_0|^2|c_1|^2 \mathrm{Re}[\langle R_0(t)\,|\,R_1(t)\rangle] \ . \tag{41}$$

The expression above depends on the initial state $|\phi\rangle$ and clearly indicates that some states are more vulnerable to decoherence than others. In order to get rid of this dependence we consider the average fidelity, calculated under the assumption that any initial state $|\phi\rangle$ is equally probable. Taking into account the normalisation constraint the average fidelity is given by

$$\bar{F}(t) = \int_0^1 F(t)\, d\,|c_0|^2 = \frac{1}{3}(2 + \mathrm{Re}[\langle R_0(t)\,|\,R_1(t)\rangle]) \ . \tag{42}$$

If we assume an exponential-type decoherence, where $\langle R_0(t)\,|\,R_1(t)\rangle = e^{-\gamma t}$, the average fidelity takes the simple form

$$\bar{F}(t) = \frac{1}{3}(2 + e^{-\gamma t}) \ . \tag{43}$$

In particular, for times much shorter than the decoherence time $t_d = 1/\gamma$, the above fidelity can be approximated as

$$\bar{F}(t) \simeq 1 - \frac{2}{3}\gamma t + O(\gamma^2 t^2) \ . \tag{44}$$

Let us now show how to improve the average fidelity by quantum encoding.

Before we place the qubit in the memory register we *encode* it: we can add two qubits, initially both in state $|0\rangle$, to the original qubit and then perform an encoding unitary transformation

$$|000\rangle \longrightarrow |\bar{0}\bar{0}\bar{0}\rangle = (|0\rangle + |1\rangle)(|0\rangle + |1\rangle)(|0\rangle + |1\rangle), \tag{45}$$
$$|100\rangle \longrightarrow |\bar{1}\bar{1}\bar{1}\rangle = (|0\rangle - |1\rangle)(|0\rangle - |1\rangle)(|0\rangle - |1\rangle), \tag{46}$$

generating state $c_0\,|\bar{0}\bar{0}\bar{0}\rangle + c_1\,|\bar{1}\bar{1}\bar{1}\rangle$, where $|\bar{0}\rangle = |0\rangle + |1\rangle$ and $|\bar{1}\rangle = |0\rangle - |1\rangle$. Now, suppose that only the second stored qubit was affected by decoherence and became entangled with the environment:

$$\begin{aligned} &c_0(|0\rangle + |1\rangle)(|0\rangle\,|R_0\rangle + |1\rangle\,|R_1\rangle)(|0\rangle + |1\rangle) + \\ &c_1(|0\rangle - |1\rangle)(|0\rangle\,|R_0\rangle - |1\rangle\,|R_1\rangle)(|0\rangle - |1\rangle), \end{aligned} \tag{47}$$

which can also be written as

$$(c_0\,|\bar{0}\bar{0}\bar{0}\rangle + c_1\,|\bar{1}\bar{1}\bar{1}\rangle)(|R_0\rangle + |R_1\rangle) + (c_0\,|\bar{0}\bar{1}\bar{0}\rangle + c_1\,|\bar{1}\bar{0}\bar{1}\rangle)(|R_0\rangle - |R_1\rangle). \tag{48}$$

The decoding unitary transformation can be constructed using a couple of quantum controlled-NOT gates and the Toffoli gate as shown in Fig.5. More careful inspection of the network in Fig.5 shows that any single phase-flip $|\bar{0}\rangle \leftrightarrow |\bar{1}\rangle$ will be corrected and the environment will be effectively disentangled from the qubits. In our particular case we obtain

$$(c_0\,|0\rangle + c_1\,|1\rangle)\,[|00\rangle\,(|R_0\rangle + |R_1\rangle) + |10\rangle\,(|R_0\rangle - |R_1\rangle)]. \tag{49}$$

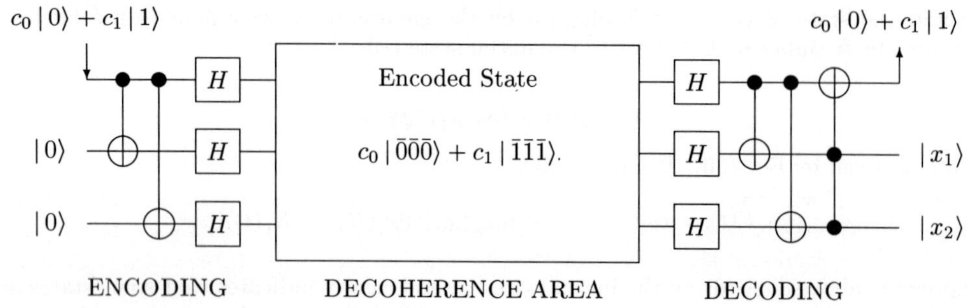

Figure 5: An encoding and decoding (and correcting) network for the three-qubit code. Any state of the form $c_0\,|\,0\rangle + c_1\,|\,1\rangle$ can be encoded into $c_0\,|\,\bar{0}\bar{0}\bar{0}\rangle + c_1\,|\,\bar{1}\bar{1}\bar{1}\rangle$, where $|\,\bar{0}\rangle = |\,0\rangle + |\,1\rangle$ and $|\,\bar{1}\rangle = |\,0\rangle - |\,1\rangle$. The decoding unitary transformation can be constructed using a couple of quantum controlled-NOT gates and the Toffoli gate. This network corrects up to one phase-flip error in any location. The two auxiliary outputs x_1 and x_2 carry information about the error syndrome - 00 means no error, 01 means the phase-flip occurred in the bottom qubit,10 means the phase-flip in the middle qubit and 11 signals the phase flip in the top qubit.

The two auxiliary outputs carry information about the error syndrome - 00 means no error, 01 means the phase-flip occurred in the third qubit, 10 means the phase-flip in the second qubit and 11 signals the phase flip in the first qubit (c.f. Fig. 5).

Thus if only one qubit in the encoded triplet decoheres we can recover the original state perfectly. In reality all three qubits decohere simultaneously and, as the result, only partial recovery of the original state is possible. In this case lengthy but straightforward calculations show that the average fidelity of the reconstructed state after the decoding operation for an exponential-type decoherence is

$$\bar{F}_{\mathrm{ec}}(t) = \frac{1}{6}[4 + 3e^{-\gamma t} - e^{-3\gamma t}]\,. \tag{50}$$

For short times this can be written as

$$\bar{F}_{\mathrm{ec}}(t) \simeq 1 - \frac{1}{2}\gamma^2 t^2 + O(\gamma^3 t^3). \tag{51}$$

Comparing Eq. (43) with Eq. (50), we can easily see that for all times t,

$$\bar{F}_{\mathrm{ec}}(t) \geq \bar{F}(t). \tag{52}$$

This is the essence of recoherence via encoding and decoding.

More general qubit–environment interaction leads to the qubit–environment entanglement given by

$$|\,0\rangle\,|\,R\rangle \;\longrightarrow\; |\,0\rangle\,|\,R_{00}(t)\rangle + |\,1\rangle\,|\,R_{01}(t)\rangle\,, \tag{53}$$
$$|\,1\rangle\,|\,R\rangle \;\longrightarrow\; |\,0\rangle\,|\,R_{10}(t)\rangle + |\,1\rangle\,|\,R_{11}(t)\rangle\,, \tag{54}$$

where the states of the environment $|\,R\rangle$ and $|\,R_{ij}\rangle$ are still normalised but not necessarily orthogonal to each other. The r.h.s. of the formulae above can also be written in a matrix form as

$$\begin{pmatrix} |\,R_{00}\rangle & |\,R_{01}\rangle \\ |\,R_{10}\rangle & |\,R_{11}\rangle \end{pmatrix} \begin{pmatrix} |\,0\rangle \\ |\,1\rangle \end{pmatrix}, \tag{55}$$

and the 2 × 2 matrix can be subsequently decomposed into some basis matrices e.g. into the unity and the Pauli matrices

$$|\,R_0\rangle\,1 + |\,R_1\rangle\,\sigma_x + i\,|\,R_2\rangle\,\sigma_y + |\,R_3\rangle\,\sigma_z, \tag{56}$$

where $|R_0\rangle = (|R_{00}\rangle + |R_{11}\rangle)/2$, $|R_3\rangle = (|R_{00}\rangle - |R_{11}\rangle)/2$, $|R_1\rangle = (|R_{01}\rangle + |R_{10}\rangle)/2$, and $|R_2\rangle = (|R_{01}\rangle - |R_{10}\rangle)/2$. Thus the qubit initially in state $|\phi\rangle$ will evolve as

$$|\phi\rangle\,|R\rangle \longrightarrow \sum_{i=0}^{3} \sigma_i\,|\phi\rangle\,|R_i\rangle \tag{57}$$

becoming entangled with the environment (we have relabeled the unity operator and the Pauli matrices $\{1, \sigma_x, \sigma_y, \sigma_z\}$ respectively as $\{\sigma_0, \sigma_1, \sigma_2, \sigma_3\}$). Formula (57) shows that a general qubit–environment interaction can be expressed as a superposition of unity and Pauli operators acting on the qubit. In the language of error correcting codes this means that the qubit state is evolved into a superposition of an error-free component and three erroneous components, with errors of the σ_x, σ_y and σ_z type. The fidelity of the evolved state with respect to the initial state $|\phi\rangle$ is

$$F(t) = \sum_{i,j} \langle\phi\,|\,\sigma_i\,|\,\phi\rangle\,\langle\phi\,|\,\sigma_j\,|\,\phi\rangle\,\langle R_j(t)\,|\,R_i(t)\rangle. \tag{58}$$

We can carry on this description even if the qubit itself is not in a pure state $|\phi\rangle$ but is entangled with some other qubits. For example, if in a three qubit register initially in state $\left|\tilde{\phi}\right\rangle = |0\rangle\,|0\rangle\,|0\rangle - |1\rangle\,|1\rangle\,|1\rangle$ the second qubit interacted with its environment then the state of the register at some time t is given by

$$\sum_{i=0}^{3} \sigma_i^{(2)} \left|\tilde{\phi}\right\rangle |R_i(t)\rangle = \sum_{i=0}^{3} (|0\rangle\,(\sigma_i\,|0\rangle)\,|0\rangle - |1\rangle\,(\sigma_i\,|1\rangle)\,|1\rangle)\,|R_i(t)\rangle\,, \tag{59}$$

where the superscript (2) reminds us that the Pauli operators act only on the second qubit. We can then say that the second qubit was affected by quantum errors which are represented by the Pauli operators σ_i. Unlike errors affecting classical bits, which can only change their binary values ($0 \leftrightarrow 1$), quantum errors operators σ_i acting on qubits can change their binary values (σ_x), their phases (σ_z) or both simultaneously (σ_y).

In general, a batch of n qubits initially in some state $\left|\tilde{\phi}\right\rangle$, each of them interacting with different environments, will evolve as

$$\prod_{k=1}^{n} \sum_{i=0}^{3} \sigma_i^{(k)} \left|\tilde{\phi}\right\rangle \left|R_i^{(k)}(t)\right\rangle\,, \tag{60}$$

namely multiple errors of the form $\sigma_i \otimes \sigma_j \cdots \otimes \sigma_k$ may occur, affecting several qubits at the same time.

Let us now specify the conditions for the existence of quantum error-correcting codes.

We say we can correct a single error $\sigma_i^{(k)}$ (where $i = 0 \ldots 3$ refers to the type of error) if we can find a transformation such that it maps all states with a single error $\sigma_i^{(k)} \left|\tilde{\phi}\right\rangle$ into the proper error free state $\left|\tilde{\phi}\right\rangle$:

$$\sigma_i^{(k)} \left|\tilde{\phi}\right\rangle \longrightarrow \left|\tilde{\phi}\right\rangle \tag{61}$$

To make it unitary we may need an ancilla

$$\sigma_i^{(k)} \left|\tilde{\phi}\right\rangle |0\rangle \longrightarrow \left|\tilde{\phi}\right\rangle |a_i^k\rangle\,. \tag{62}$$

For encoded basis states of a single qubit $|C_0\rangle$ and $|C_1\rangle$ this implies [47]

$$A_k\,|C_0\rangle\,|0\rangle \quad\longrightarrow\quad |C_0\rangle\,|a_k\rangle\,, \tag{63}$$
$$A_k\,|C_1\rangle\,|0\rangle \quad\longrightarrow\quad |C_1\rangle\,|a_k\rangle\,, \tag{64}$$

where A_k denotes all the possible types of independent errors affecting at most one of the qubits. The above requirement leads to the following unitarity conditions

$$\langle C_0 \,|\, A_k^\dagger A_l \,|\, C_0 \rangle \;=\; \langle C_1 \,|\, A_k^\dagger A_l \,|\, C_1 \rangle = \langle a_k \,|\, a_l \rangle \;, \tag{65}$$

$$\langle C_0 \,|\, A_k^\dagger A_l \,|\, C_1 \rangle \;=\; 0 \,. \tag{66}$$

The above conditions can be easily generalised to an arbitrary t error correcting code, which corrects any kind of transformations affecting up to t qubits in the encoded state. In this case the operators A_k are all the possible independent errors affecting up to t qubits, namely operators of the form $\Pi_{i=1}^{t}\sigma_i$ acting on t different qubits. In the case of the so-called "nondegenerate codes" Eq. (65) takes the simple form [40],

$$\langle C_0 \,|\, A_k^\dagger A_l \,|\, C_0 \rangle = \langle C_1 \,|\, A_k^\dagger A_l \,|\, C_1 \rangle = 0 \,. \tag{67}$$

This condition requires that all states which are obtained by affecting up to t qubits in the encoded states are all orthogonal to each other, and therefore distinguishable. This ensures that by performing suitable projections of the encoded state we are able to detect the kind of error which occurred and "undo" it to recover the desired error free state. Condition (67), even though more restrictive than (65), is quite useful because it allows to establish bounds on the resources needed in constructing efficient nondegenerate codes. Let us assume that the initial state of l qubits is encoded in a redundant Hilbert space of n qubits. If we want to encode 2^l input basis states and correct up to t errors we must choose the dimension of the encoding Hilbert space 2^n such that all the necessary orthogonal states can be accommodated. According to Eq. (67), the total number of orthogonal states that we need in order to be able to correct i errors of the three types σ_x, σ_y and σ_z in an n-qubit state is $3^i \begin{pmatrix} n \\ i \end{pmatrix}$ (this is the number of different ways in which the errors can occur). The argument based on counting orthogonal states then leads to the following condition

$$2^l \sum_{i=0}^{t} 3^i \begin{pmatrix} n \\ i \end{pmatrix} \le 2^n. \tag{68}$$

Eq. (68) is the quantum version of the Hamming bound for classical error-correcting codes [48]; given l and t it provides a lower bound on the dimension of the encoding Hilbert space for nondegenerate codes.

The quantum version of the classical Gilbert-Varshamov bound [48] can be also obtained, which gives an upper bound on the dimension of the encoding Hilbert space for optimal non degenerate codes:

$$2^l \sum_{i=0}^{2t} 3^i \begin{pmatrix} n \\ i \end{pmatrix} \ge 2^n. \tag{69}$$

This expression can be proved from the observation that in the 2^n dimensional Hilbert space with a maximum number of encoded basis vectors (or code-vectors) $|\,C^k\rangle$ any vector which is orthogonal to $|\,C^k\rangle$ (for any k) can be reached by applying up to $2t$ error operations of σ_x, σ_y, and σ_z type to any of the 2^l encoded basis vectors. Clearly all vectors which cannot be reached in the $2t$ operations can be added to the encoded basis states $|\,C^k\rangle$ as all the vectors into which they can be transformed by applying up to t amplitude and/or phase transformations are orthogonal to all the others. This situation cannot happen because we have assumed that the number of code-vectors is maximal. Thus the number of orthogonal vectors that can be obtained by performing up to $2t$ transformations on the code-vectors must be at least equal to the dimension of the encoding Hilbert space.

It follows from Eq. (68) that a one-bit quantum error correcting code to protect a single qubit ($l = 1$, $t = 1$) requires at least 5 encoding qubits and, according to Eq. (69), this can be achieved with less than 10 qubits. Explicit constructions for an optimal five qubit code were first given in [41, 47].

The asymptotic form of the quantum Hamming bound (68) in the limit of large n is given by

$$\frac{l}{n} \le 1 - \frac{t}{n}\log_2 3 - H(\frac{t}{n}), \tag{70}$$

where H is the entropy function $H(x) = -x\log_2 x - (1-x)\log_2(1-x)$. The corresponding asymptotic form for the quantum Gilbert-Varshamov bound (69) is

$$\frac{l}{n} \geq 1 - \frac{2t}{n}\log_2 3 - H(\frac{2t}{n}). \qquad (71)$$

As we can see from eq. (70), in quantum error correction there is an upper bound on the error rate t/n which a code can tolerate. In fact, differently from the classical case, where any arbitrary error rate can be corrected by a suitable code, in the quantum world the ratio t/n cannot be larger than 0.18929 for nondegenerate codes. This bound corresponds to the value of t/n for which the r.h.s. of Eq. (70) equals zero. For other bounds regarding quantum error correcting codes see [49].

There is much more to say (and write) about quantum codes and the reader should be warned that we have barely scratched the surface of the current activities in quantum error correction, neglecting topics such as group theoretical ways of constructing good quantum codes [42, 43], concatenated codes [44], quantum fault tolerant computation [50] and many others.

7 Concluding remarks

It is not clear at present which technology, if any, will support quantum computation in the future. Nevertheless both experimental and theoretical research in quantum computation is accelerating world–wide. New technologies for realising quantum computers are being proposed, and new types of quantum computation with various advantages over classical computation are continually being discovered and analysed and we believe some of them will bear technological fruit. From a fundamental standpoint, however, it does not matter how useful quantum computation turns out to be, nor does it matter whether we build the first quantum computer tomorrow, next year or centuries from now. The quantum theory of computation must in any case be an integral part of the world view of anyone who seeks a fundamental understanding of the quantum theory and the processing of information [51].

8 Acknowledgments

This work was supported in part by the European TMR Research Network ERP-4061PL95-1412, the TMR Marie Curie Fellowship Programme, Hewlett-Packard, The Royal Society London and Elsag-Bailey, a Finmeccanica Company.

References

[1] P. Benioff, J. Stat. Phys. **22**, 563 (1980).

[2] R. Feynman, Int. J. Theor. Phys. **21**, 467 (1982).

[3] D. Deutsch, Proc. R. Soc. Lond. A **400**, 97 (1985).

[4] E. Bernstein and U. Vazirani, *Proc. 25th ACM Symposium on the Theory of Computation*, p. 11 (1993).

[5] D. Deutsch and R. Jozsa, Proc. R. Soc. Lond. A **439**, 553 (1992).

[6] D. S. Simon, *Proc. 35th Annual Symposium on the Foundations of Computer Science*, Edited by S. Goldwasser, IEEE Computer Society Press, Los Alamitos, CA (1994), p. 116.

[7] P.W. Shor, *Proc. 35th Annual Symposium on the Foundations of Computer Science*, Edited by S. Goldwasser, IEEE Computer Society Press, Los Alamitos, CA (1994), p. 124. Expanded version of this paper is available at LANL quant-ph archive.

[8] D. Welsh, *Codes and Cryptography*, Clarendon Press, Oxford (1988).

[9] S. Wiesner, *SIGACT News*, **15**, 78 (1983); original manuscript written *circa* 1970.

[10] C.H. Bennett and G. Brassard, in *Proceedings of IEEE International Conference on Computers, Systems and Signal Processing*, Bangalore, India (IEEE, 1984) p. 175.

[11] A.K. Ekert, Phys. Rev. Lett. **68**, 661 (1991).

[12] S. Lloyd, Science **261**, 1569 (1993).

[13] L. Grover, *Proc. 28th Annual ACM Symposium on the Theory of Computing*, ACM Press, New York (1996), p. 212.

[14] S. Lloyd, Science **273**, 1073 (1996).

[15] R. Cleve, A. Ekert, C. Macchiavello and M. Mosca, *Quantum Algorithms Revisited*, Proc. R. Soc. Lond. A (in print).

[16] C.H. Papadimitriou, *Computational Complexity*, Addison-Wesley, (1994).

[17] A. Cobham in *Proc. 1964 International Congress for Logic, Methodology, and Philosophy of Sciences* ed. Y.Bar-Hellel (North Holland, Amsterdam, 1965) p.24.

[18] M.R. Garey and D.S.Johnson *Computers and Intractability; A guide to the theory of NP-completeness*, W.H.Freeman and Company, New York (1979).

[19] A. Ekert and R. Jozsa, Rev. Mod. Phys. **68**, 733 (1996).

[20] C.H. Bennett, E.Bernstein, G.Brassard, and U.Vazirani *Strentgh and weaknesses of quantum computing* e-print quant-ph/9701001 (1997).

[21] D. Deutsch, Proc. R. Soc. Lond. A **425**, 73 (1989).

[22] D. Deutsch, A. Barenco and A. Ekert, Proc. R. Soc. Lond. A **449**, 669 (1995).

[23] S. Lloyd, Science **261**, 1569 (1993).

[24] A. Barenco, D. Deutsch, A. Ekert, R. and Jozsa, Phys. Rev. Lett. **74**, 4083 (1995).

[25] A. Barenco, C.H. Bennett, R. Cleve, D.P. DiVincenzo, N. Margolus, P.W. Shor, T. Sleator, J.A. Smolin and H. Weinfurter, Phys. Rev. A **52**, 3457 (1995).

[26] V. Vedral, A. Barenco and A. Ekert, Phys. Rev. A **54**, 139 (1996).

[27] D. Beckman, A. Chari, S. Devabhaktuni, and J. Preskill, Phys. Rev. A **54** 1034 (1996).

[28] D. Coppersmith, *An Approximate Fourier Transform Useful in Quantum Factoring*, IBM Research Report No. RC19642 (1994).

[29] Q.A. Turchette, C.J. Hood, W. Lange, H. Mabuchi and H.J. Kimble, Phys. Rev. Lett. **75**, 4710 (1995).

[30] J.I. Cirac and P. Zoller, Phys. Rev. Lett. **74**, 4091 (1995).

[31] C. Monroe, D.M. Meekhof, B.E. King, W.M. Itano and D.J. Wineland, Phys. Rev. Lett. **75**, 4714 (1995).

[32] N. Gershenfeld and I. Chuang, Science **275**, 350 (1997).

[33] D.G. Cory, A.F. Fahmy and T.F. Havel, Proc. of the 4th Workshop on Physics and Computation, Boston, New England Complex Systems Institute (1996), p. 87.

[34] W.H. Zurek, Phys. Today, October p. 36 (1991).

[35] G.M. Palma, K.-A. Suominen and A. Ekert, Proc. R. Soc. Lond. A **452**, 567 (1996).

[36] A. Berthiaume, D. Deutsch and R. Jozsa, Proceedings of the Workshop on the Physics and Computation—PhysComp '94, IEEE Computer Society Press, Dallas, Texas (1994); A. Barenco, A. Berthiaume, D. Deutsch, A. Ekert, R. Jozsa and C. Macchiavello, SIAM J. Comput. **26**, in press.

[37] P.W. Shor, Phys. Rev. A **52**, R2493 (1995).

[38] R. Calderbank and P.W. Shor, Phys. Rev. A **54**, 1098 (1996).

[39] A. Steane, Phys. Rev. Lett. **77**, 793 (1996); A. Steane, Proc. R. Soc. Lond. A **452**, 2551 (1996).

[40] A. Ekert and C. Macchiavello, Phys. Rev. Lett. **77**, 2585 (1996).

[41] R. Laflamme, C. Miquel, J.P. Paz and W.H. Zurek, Phys. Rev. Lett. **77**, 198 (1996).

[42] D. Gottesman, Phys. Rev. A **54**, 1862 (1996).

[43] A.R. Calderbank, E.M. Rains, P.W. Shor and N.J.A. Sloane, Phys. Rev. Lett. **78**, 405 (1997).

[44] E. Knill and R. Laflamme, e-print quant-ph/9608012 (1996).

[45] I. Chuang and R. Laflamme, e-print quant-ph/9511003 (1995).

[46] C.H. Bennett, G. Brassard, S. Popescu, B. Schumacher, J.A. Smolin and W.K. Wootters, Phys. Rev. Lett. **76**, 722 (1996).

[47] C.H. Bennett, D.P. DiVincenzo, J.A. Smolin and W.K. Wootters, Phys. Rev. A **54**, 3824 (1996).

[48] F.J. MacWilliams and N.J.A. Sloane, *The theory of error correcting codes*, Amsterdam: North Holland (1977).

[49] R. Cleve, Phys. Rev. A **55**, 4054 (1997).

[50] P.W. Shor, e-print quant-ph/9605011 (1996); D.P. DiVincenzo and P.W. Shor, Phys. Rev. Lett. **77**, 3260 (1996).

[51] D. Deutsch, *The Fabric of Reality*, The Penguin Press (1997).

On quantum algorithms

Richard Cleve, Artur Ekert, Leah Henderson, Chiara Macchiavello and Michele Mosca

Centre for Quantum Computation
Clarendon Laboratory, University of Oxford, Oxford OX1 3PU, U.K.
Department of Computer Science
University of Calgary, Calgary, Alberta, Canada T2N 1N4
Theoretical Quantum Optics Group
Dipartimento di Fisica "A. Volta" and I.N.F.M. - Unità di Pavia
Via Bassi 6, I-27100 Pavia, Italy

Abstract

Quantum computers use the quantum interference of different computational paths to enhance correct outcomes and suppress erroneous outcomes of computations. In effect, they follow the same logical paradigm as (multi-particle) interferometers. We show how most known quantum algorithms, including quantum algorithms for factorising and counting, may be cast in this manner. Quantum searching is described as inducing a desired relative phase between two eigenvectors to yield constructive interference on the sought elements and destructive interference on the remaining terms.

1 From Interferometers to Computers

Richard Feynman [1] in his talk during the First Conference on the Physics of Computation held at MIT in 1981 observed that it appears to be impossible to simulate a general quantum evolution on a classical probabilistic computer in an *efficient* way. He pointed out that any classical simulation of quantum evolution appears to involve an exponential slowdown in time as compared to the natural evolution since the amount of information required to describe the evolving quantum state in classical terms generally grows exponentially in time. However, instead of viewing this as an obstacle, Feynman regarded it as an opportunity. If it requires so much computation to work out what will happen in a complicated multiparticle interference experiment then, he argued, the very act of setting up such an experiment and measuring the outcome is tantamount to performing a complex computation. Indeed, all quantum multiparticle interferometers *are* quantum computers and some interesting computational problems can be based on estimating internal phase shifts in these interferometers. This approach leads to a unified picture of quantum algorithms and has been recently discussed in detail by Cleve *et al.* [2].

Let us start with the textbook example of quantum interference, namely the double-slit experiment, which, in a more modern version, can be rephrased in terms of Mach-Zehnder interferometry (see Fig. 1).

A particle, say a photon, impinges on a beam-splitter (BS1), and, with some probability amplitudes, propagates via two different paths to another beam-splitter (BS2) which directs the particle to one of the two detectors. Along each path between the two beam-splitters, is a phase shifter (PS). If the lower path is labelled as state $|0\rangle$ and the upper one as state $|1\rangle$ then the particle, initially in path $|0\rangle$, undergoes the following sequence of transformations

$$\begin{aligned} |0\rangle &\xrightarrow{\text{BS1}} \frac{1}{\sqrt{2}}(|0\rangle + |1\rangle) \\ &\xrightarrow{\text{PS}} \frac{1}{\sqrt{2}}(e^{i\phi_0}|0\rangle + e^{i\phi_1}|1\rangle) = e^{i\frac{\phi_0+\phi_1}{2}}\frac{1}{\sqrt{2}}(e^{i\frac{\phi_0-\phi_1}{2}}|0\rangle + e^{-i\frac{\phi_0-\phi_1}{2}}|1\rangle) \\ &\xrightarrow{\text{BS2}} e^{i\frac{\phi_1+\phi_2}{2}}(\cos\tfrac{1}{2}(\phi_0-\phi_1)|0\rangle + i\sin\tfrac{1}{2}(\phi_0-\phi_1)|1\rangle), \end{aligned} \quad (1)$$

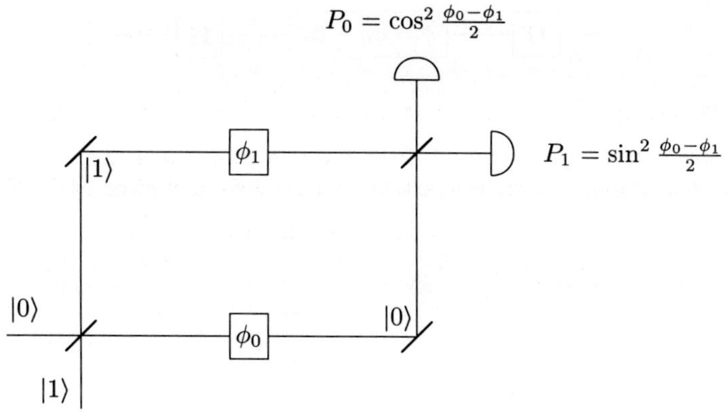

Figure 1: A Mach-Zehnder interferometer with two phase shifters. The interference pattern depends on the difference between the phase shifts in different arms of the interferometer.

where ϕ_0 and ϕ_1 are the settings of the two phase shifters and the action of the beam-splitters is defined as

$$\begin{aligned} |\,0\rangle &\longrightarrow \tfrac{1}{\sqrt{2}}(|\,0\rangle + |\,1\rangle) \\ |\,1\rangle &\longrightarrow \tfrac{1}{\sqrt{2}}(|\,0\rangle - |\,1\rangle) \end{aligned} \tag{2}$$

(and extends by linearity to states of the form $\alpha\,|\,0\rangle + \beta\,|\,1\rangle$). Here, we have ignored the $e^{i\frac{\phi_0+\phi_0}{2}}$ phase shift in the reflected beam, which is irrelevant because the interference pattern depends only on the *difference* between the phase shifts in different arms of the interferometer. The phase shifters in the two paths can be tuned to effect any prescribed relative phase shift $\phi = \phi_0 - \phi_1$ and to direct the particle with probabilities $\cos^2\left(\frac{\phi}{2}\right)$ and $\sin^2\left(\frac{\phi}{2}\right)$ respectively to detectors "0" and "1".

The roles of the three key ingredients in this experiment are clear. The first beam splitter prepares a superposition of possible paths, the phase shifters modify quantum phases in different paths and the second beam-splitter combines all the paths together. As we shall see in the following sections, quantum algorithms follow this interferometry paradigm: a superposition of computational paths is prepared by the Hadamard (or the Fourier) transform, followed by a quantum function evaluation which effectively introduces phase shifts into different computational paths, followed by the Hadamard or the Fourier transform which acts somewhat in reverse to the first Hadamard/Fourier transform and combines the computational paths together. To see this, let us start by rephrasing Mach-Zehnder interferometry in terms of quantum networks.

2 Quantum gates & networks

In order to avoid references to specific technological choices (hardware), let us now describe our Mach-Zehnder interference experiment in more abstract terms. It is convenient to view this experiment as a *quantum network* with three quantum logic gates (elementary unitary transformations) operating on a qubit (a generic two-state system with a prescribed computational basis $\{|\,0\rangle\,, |\,1\rangle\}$). The beam-splitters will be now called the Hadamard gates and the phase shifters the phase shift gates (see Fig. 2).

The Hadamard gate is the single qubit gate **H** performing the unitary transformation known as the

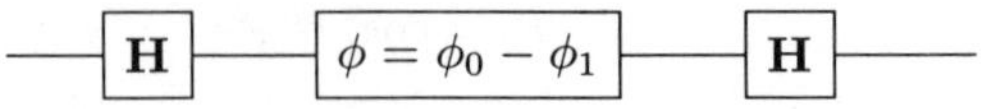

Figure 2: A quantum network composed of three single qubit gates. This network provides a hardware-independent description of any single-particle interference, including Mach-Zehnder interferometry.

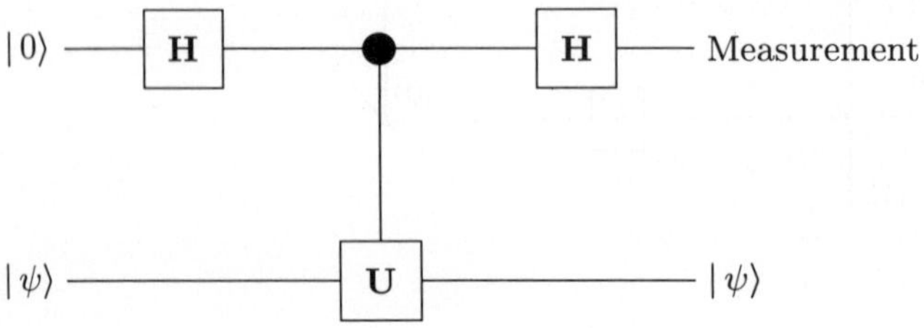

Figure 3: Phase factors can be introduced into different computational paths via the controlled-U operations. The controlled-U means that the form of U depends on the logical value of the control qubit (the upper qubit). Here, we apply the identity transformation to the auxiliary (lower) qubits (i.e. do nothing) when the control qubit is in state $|0\rangle$ and apply a prescribed U when the control qubit is in state $|1\rangle$. The auxiliary or the target qubit is initially prepared in state $|\psi\rangle$ which is one of the eigenstates of U.

Hadamard transform given by (Eq. 2)

$$\mathbf{H} = \frac{1}{\sqrt{2}}\begin{pmatrix} 1 & 1 \\ 1 & -1 \end{pmatrix} \tag{3}$$

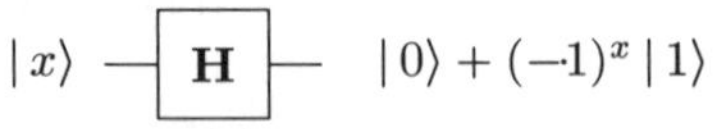

The matrix is written in the basis $\{|0\rangle, |1\rangle\}$ and the diagram on the right provides a schematic representation of the gate $\mathbf{H}$ acting on a qubit in state $|x\rangle$, with $x = 0, 1$. Using the same notation we define the phase shift gate ϕ as a single qubit gate such that $|0\rangle \mapsto |0\rangle$ and $|1\rangle \mapsto e^{i\phi}|1\rangle$,

$$\phi = \begin{pmatrix} 1 & 0 \\ 0 & e^{i\phi} \end{pmatrix} \tag{4}$$

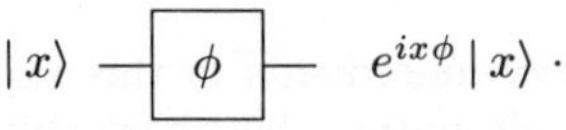

Let us explain now how the phase shift ϕ can be "computed" with the help of an auxiliary qubit (or a set of qubits) in a prescribed state $|\psi\rangle$ and some controlled-U transformation where $U|\psi\rangle = e^{i\phi}|\psi\rangle$ (see Fig. 3).

Here the controlled-U is a transformation involving two qubits, where the form of U applied to the auxiliary or target qubit depends on the logical value of the control qubit. For example, we can apply the identity transformation to the auxiliary qubits (i.e. do nothing) when the control qubit is in state $|0\rangle$ and apply a prescribed U when the control qubit is in state $|1\rangle$. In our example shown in Fig. 3, we obtain the following sequence of transformations on the two qubits

$$\begin{aligned} |0\rangle|\psi\rangle \xrightarrow{H} \tfrac{1}{\sqrt{2}}(|0\rangle + |1\rangle)|\psi\rangle &\xrightarrow{c-U} \tfrac{1}{\sqrt{2}}(|0\rangle + e^{i\phi}|1\rangle)|\psi\rangle \\ &\xrightarrow{H} e^{(i\frac{\phi}{2})}(\cos\tfrac{\phi}{2}|0\rangle + i\sin\tfrac{\phi}{2}|1\rangle)|\psi\rangle \,. \end{aligned} \tag{5}$$

We note that the state of the auxiliary register $|\psi\rangle$, being an eigenstate of U, is not altered along this network, but its eigenvalue $e^{i\phi}$ is "kicked back" in front of the $|1\rangle$ component in the first qubit. The sequence (5) is equivalent to the steps of the Mach-Zehnder interferometer (1) and, as was shown in [2], the kernel of most known quantum algorithms.

3 The first quantum algorithm

Since quantum phases in interferometers can be introduced by some controlled-U operations, it is natural to ask whether effecting these operations can be described as an interesting computational problem.

Suppose an experimentalist, Alice, who runs the Mach-Zehnder interferometer delegates the control of the phase shifters to her colleague, Bob. Bob is allowed to set up any value $\phi = \phi_0 - \phi_1$ and Alice's task is to estimate ϕ. Clearly for general ϕ this involves running the device several times until Alice accumulates enough data to estimate probabilities P_0 and P_1, however, if Bob promises to set up ϕ either at 0 or at π then a single-shot experiment can deliver the conclusive outcome (click in detector "0" corresponds to $\phi = 0$ and in detector "1" corresponds to $\phi = \pi$). The first quantum algorithm proposed by David Deutsch in 1985 [3] is related to this effect.

We have seen in the previous section that a controlled-U transformation can be used to produce a particular phase shift on the control qubit corresponding to its eigenvalue on the auxiliary qubit. If two eigenvalues of the controlled-U transformation lead to different orthogonal states in the control qubit, a single measurement on this qubit will suffice to distinguish the two cases.

For example consider the Boolean functions f that map $\{0,1\}$ to $\{0,1\}$. There are exactly four such functions: two constant functions ($f(0) = f(1) = 0$ and $f(0) = f(1) = 1$) and two "balanced" functions ($f(0) = 0, f(1) = 1$ and $f(0) = 1, f(1) = 0$). It turns out that it is possible to construct a controlled function evaluation such that two possible eigenvalues are produced which may be used to determine whether the function is constant or balanced. This is done in the following way.

Let us formally define the operation of "evaluating" f in terms of the *f-controlled-NOT* operation on two bits: the first contains the input value and the second contains the output value. If the second bit is initialised to 0, the f-controlled-NOT maps $(x, 0)$ to $(x, f(x))$. This is clearly just a formalization of the operation of computing f. In order to make the operation reversible, the mapping is defined for *all* initial settings of the two bits, taking (x, y) to $(x, y \oplus f(x))$, where $\oplus$ denotes addition modulo two.

A single evaluation of the f-controlled-NOT on quantum superpositions suffices to classify f as constant or balanced. This is the real advantage of the quantum method over the classical. Classically if the f-controlled-NOT operation may be performed only once then it is *impossible* to distinguish between balanced and constant functions. Whatever the outcome, both possibilities (balanced and constant) remain for f. This corresponds to our classical intuition about the problem since it involves determining not particular values of $f(0)$ and $f(1)$, but a global property of f. Classically to determine this global property of f, we have to evaluate both $f(0)$ and $f(1)$, which involves evaluating f twice.

Deutsch's quantum algorithm has the same mathematical structure as the Mach-Zehnder interferometer, with the two phase settings $\phi = 0, \pi$. It is best represented as the quantum network shown in Fig. 4, where the middle operation is the f-controlled-NOT, which can be defined as:

$$|x\rangle\,|y\rangle \stackrel{f-c-N}{\longrightarrow} |x\rangle\,|y \oplus f(x)\rangle\ . \tag{6}$$

The initial state of the qubits in the quantum network is $|0\rangle\,(|0\rangle - |1\rangle)$ (apart from a normalization factor, which will be omitted in the following). After the first Hadamard transform, the state of the two qubits has the form $(|0\rangle + |1\rangle)(|0\rangle - |1\rangle)$. To determine the effect of the f-controlled-NOT on

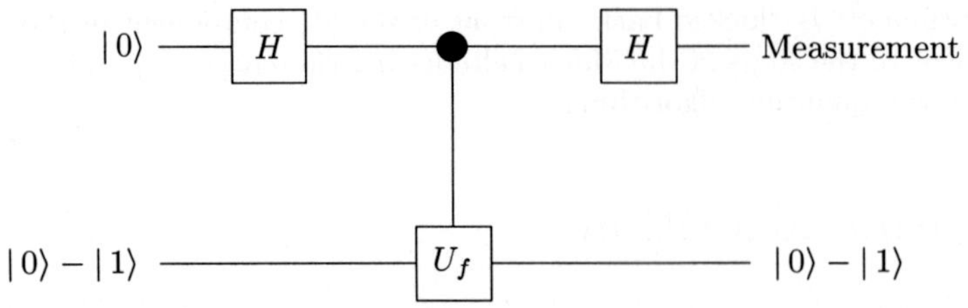

Figure 4: Quantum network which implements Deutsch's algorithm. The middle gate is the f-controlled-NOT which evaluates one of the four functions $f:\ \{0,1\} \mapsto \{0,1\}$. If the first qubit is measured to be $|0\rangle$, then the function is constant, and if $|1\rangle$, the function is balanced.

this state, first note that, for each $x \in \{0,1\}$,

$$|x\rangle(|0\rangle - |1\rangle) \stackrel{f-c-N}{\longrightarrow} |x\rangle(|0 \oplus f(x)\rangle - |1 \oplus f(x)\rangle) = (-1)^{f(x)}|x\rangle(|0\rangle - |1\rangle)\,. \tag{7}$$

Therefore, the state after the f-controlled-NOT is

$$((-1)^{f(0)}|0\rangle + (-1)^{f(1)}|1\rangle)(|0\rangle - |1\rangle)\,. \tag{8}$$

That is, for each x, the $|x\rangle$ term acquires a phase factor of $(-1)^{f(x)}$, which corresponds to the eigenvalue of the state of the auxiliary qubit under the action of the operator that sends $|y\rangle$ to $|y \oplus f(x)\rangle$.

This state can also be written as

$$(-1)^{f(0)}(|0\rangle + (-1)^{f(0)\oplus f(1)}|1\rangle)(|0\rangle - |1\rangle)\,, \tag{9}$$

which, after applying the second Hadamard transform to the first qubit, becomes

$$(-1)^{f(0)}|f(0) \oplus f(1)\rangle(|0\rangle - |1\rangle)\,. \tag{10}$$

Therefore, the first qubit is finally in state $|0\rangle$ if the function f is constant and in state $|1\rangle$ if the function is balanced, and a measurement of this qubit distinguishes these cases with certainty.

The Mach-Zehnder interferometer with phases ϕ_0 and ϕ_1 each set to either 0 or π can be regarded as an implementation of the above algorithm. In this case, ϕ_0 and ϕ_1 respectively encode $f(0)$ and $f(1)$ (with π representing 1), and a single photon can query both phase shifters (i.e. $f(0)$ and $f(1)$) in superposition. More recently, this algorithm (Fig. 4) has been implemented using a very different quantum physical technology, nuclear magnetic resonance [4, 5].

More general algorithms may operate not just on single qubits, as in Deutsch's case, but on sets of qubits or 'registers'. The second qubit becomes an auxiliary register $|\psi\rangle$ prepared in a superposition of basis states, each weighted by a different phase factor,

$$|\psi\rangle = \sum_{y=0}^{2^m-1} e^{-2\pi i y/2^m}|y\rangle\,. \tag{11}$$

In general, the middle gate which produces the phase shift is some controlled function evaluation. A controlled function evaluation operates on its second input, the 'target', according to the state of the first input, the 'control'. A controlled function f applied to a control state $|x\rangle$, and a target state $|\psi\rangle$ gives

$$|x\rangle|\psi\rangle \longrightarrow |x\rangle|\psi + f(x)\rangle\,. \tag{12}$$

where the addition is mod 2^m. Hence for the register in state (11)

$$|\,x\rangle \sum_{y=0}^{2^m-1} e^{-2\pi i y/2^m}\,|\,y\rangle \longrightarrow e^{2\pi i f(x)/2^m}\,|\,x\rangle \sum_{y=0}^{2^m-1} e^{-2\pi i(y+f(x))/2^m}\,|\,y+f(x)\rangle = e^{2\pi i f(x)/2^m}\,|\,x\rangle\,|\,\psi\rangle\,. \tag{13}$$

Effectively a phase shift proportional to the value of $f(x)$ is produced on the first input.

We will now see how phase estimation on registers may be carried out by networks consisting of only two types of quantum gates: the Hadamard gate $\mathbf{H}$ and the conditional phase shift $\mathbf{R}(\phi)$. The conditional phase shift is the two-qubit gate $\mathbf{R}(\phi)$ defined as

$$\mathbf{R}(\phi) = \begin{pmatrix} 1 & 0 & 0 & 0 \\ 0 & 1 & 0 & 0 \\ 0 & 0 & 1 & 0 \\ 0 & 0 & 0 & e^{i\phi} \end{pmatrix}$$

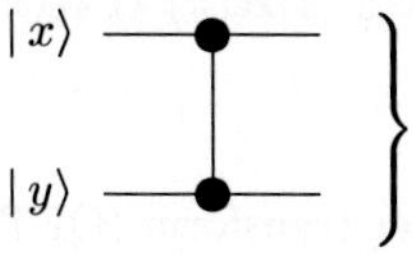

$$e^{ixy\phi}\,|\,x\rangle\,|\,y\rangle\,. \tag{14}$$

The matrix is written in the basis $\{|\,0\rangle\,|\,0\rangle\,, |\,0\rangle\,|\,1\rangle\,, |\,1\rangle\,|\,0\rangle\,, |\,1\rangle\,|\,1\rangle\}$, (the diagram on the right shows the structure of the gate). For some of the known quantum algorithms, when working with registers, the Hadamard transformation, corresponding to the beamsplitters in the interferometer, is generalised to a quantum Fourier transform.

4 Quantum Fourier transform and computing phase shifts

The discrete Fourier transform is a unitary transformation of a s–dimensional vector

$$(f(0), f(1), f(2), \ldots, f(s-1)) \to (\tilde{f}(0), \tilde{f}(1), \tilde{f}(2), \ldots, \tilde{f}(s-1)) \tag{15}$$

defined by:

$$\tilde{f}(y) = \frac{1}{\sqrt{s}} \sum_{x=0}^{s-1} e^{2\pi i x y/s} f(x), \tag{16}$$

where $f(x)$ and $\tilde{f}(y)$ are in general complex numbers. In the following, we assume that s is a power of 2, i.e., $s = 2^n$ for some n; this is a natural choice when binary coding is used.

The quantum version of the discrete Fourier transform (QFT) is a unitary transformation which can be written in a chosen computational basis $\{|0\rangle, |1\rangle, \ldots, |2^n-1\rangle\}$ as,

$$|x\rangle \longmapsto \frac{1}{\sqrt{s}} \sum_{y=0}^{s-1} \exp(2\pi i x y/s)\,|y\rangle. \tag{17}$$

More generally, the QFT effects the discrete Fourier transform of the input amplitudes. If

$$\mathrm{QFT}: \sum_x f(x)|x\rangle \longmapsto \sum_y \tilde{f}(y)|y\rangle, \tag{18}$$

then the coefficients $\tilde{f}(y)$ are the discrete Fourier transforms of the $f(x)$'s.

A given phase $\phi_x = 2\pi x/2^n$ can be encoded by a QFT. In this process the information about ϕ_x is distributed between states of a register. Let x be represented in binary as $x_0 \ldots x_{n-1} \in \{0,1\}^n$, where $x = \sum_{i=0}^{n-1} x_i 2^i$ (and similarly for y). An important observation is that the QFT of x, $\sum_{y=0}^{s-1} \exp(2\pi i x y/s)\,|y\rangle$, is unentangled, and can in fact be factorised as

$$(|\,0\rangle + e^{i\phi_x}\,|\,1\rangle)(|\,0\rangle + e^{i2\phi_x}\,|\,1\rangle)\cdots(|\,0\rangle + e^{i2^{n-1}\phi_x}\,|\,1\rangle)\,. \tag{19}$$

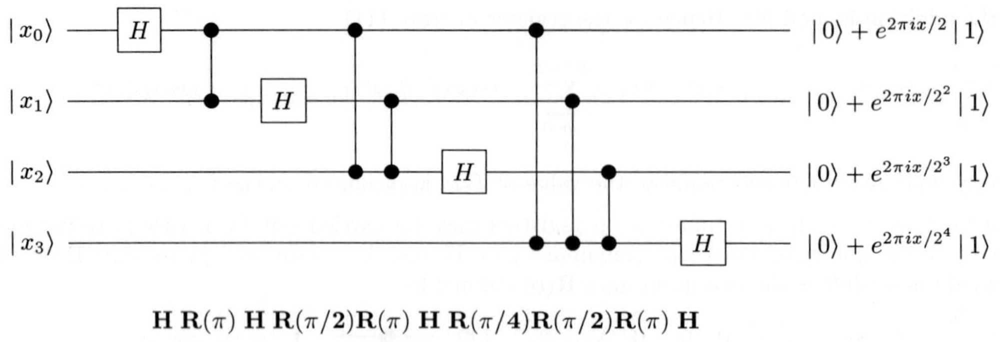

Figure 5: The quantum Fourier transform (QFT) network operating on four qubits. If the input state represents number $x = \sum_k 2^k x_k$ the output state of each qubit is of the form $|0\rangle + e^{i2^{n-1-k}\phi_x}|1\rangle$, where $\phi_x = 2\pi x/2^n$ and $k = 0, 1, 2 \ldots n-1$. N.B. there are three different types of the $R(\phi)$ gate in the network above: $R(\pi)$, $R(\pi/2)$ and $R(\pi/4)$. The size of the rotation is indicated by the distance between the 'wires'.

The network for performing the QFT is shown in Fig. 5. The input qubits are initially in some state $|x\rangle = |x_0\rangle|x_1\rangle|x_2\rangle|x_3\rangle$ where $x_0x_1x_2x_3$ is the binary representation of x, that is, $x = \sum_{i=0}^{3} x_i 2^i$. As the number of qubits becomes large, the rotations $R(\pi/2^n)$ will require exponential precision, which is impractical. Fortunately, the algorithm will work even if we omit the small rotations, [6, 7].

The general case of n qubits requires a simple extension of the network following the same pattern of **H** and **R** gates.

States of the form (19) are produced by function evaluation in a quantum computer. Suppose that U is any unitary transformation on m qubits and $|\psi\rangle$ is an eigenvector of U with eigenvalue $e^{i\phi}$. The scenario is that we do not explicitly know U or $|\psi\rangle$ or $e^{i\phi}$, but instead are given devices that perform controlled-U, controlled-U^{2^1}, controlled-U^{2^2} and so on until we reach controlled-$U^{2^{n-1}}$. Also, assume that we are given a single preparation of the state $|\psi\rangle$. From this, our goal is to obtain an n-bit estimator of ϕ.

In a quantum algorithm a quantum state of the form

$$(|0\rangle + e^{i2^{n-1}\phi}|1\rangle)(|0\rangle + e^{i2^{n-2}\phi}|1\rangle)\cdots(|0\rangle + e^{i\phi}|1\rangle) \tag{20}$$

is created by applying the network of Fig. 6.

Then, in the special case where $\phi = 2\pi x/2^n$, the state $|x_0 \cdots x_{n-1}\rangle$ (and hence ϕ) can be obtained by just applying the inverse of the QFT (which is the network of Fig. 5 in the backwards direction and with the qubits in reverse order). If x is an n-bit number this will produce the state $|x_0 \cdots x_{n-1}\rangle$ exactly (and hence the exact value ϕ).

However, ϕ is not in general a fraction of a power of two (and may not even be a rational number). For such a $\phi = 2\pi\omega$, it turns out that applying the inverse of the QFT produces the best n-bit approximation of ω with probability at least $4/\pi^2 \approx 0.41$ [2]. The probability of obtaining the best[1] estimate can be made $1-\delta$ for any δ, $0 < \delta < 1$, by creating the state in equation (20) but with $n + O(\log(1/\delta))$ qubits and rounding the answer off to the nearest n bits [2].

[1]Though this process produces the best estimate of ω with significant probability, it is not necessarily the best estimator of ω, since, for example, we might be able to to obtain as close an estimate with higher probability. See [8] for details.

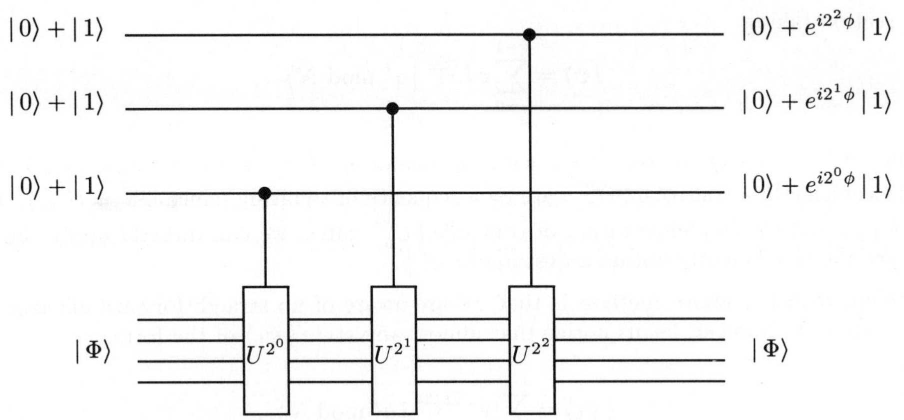

Figure 6: The network which computes phase shifts in Shor's algorithms; it also implements the modular exponentiation function via repeated squarings.

5 Examples

We will now illustrate the general framework described in the preceding section by showing how some of the most important quantum algorithms can be viewed in this light. We start with Shor's quantum algorithm for efficient factorisation (for a comprehensive discussion of quantum factoring see [9, 10, 2]).

5.1 Quantum Factoring

Shor's quantum factoring of an integer N is based on calculating the period of the function $f(x) = a^x \bmod N$ for a randomly selected integer a between 1 and N. For any positive integer y, we define $y \bmod N$ to be the remainder (between 0 and $N-1$) when we divide y by N. More generally, $y \bmod N$ is the unique positive integer $\overline{y}$ between 0 and $N-1$ such that N evenly divides $y - \overline{y}$. For example, $2 \bmod 35 = 2$, $107 \bmod 35 = 2$, and $-3 \bmod 35 = 32$. We can test if a is relatively prime to N using the Euclidean algorithm. If it is not, we can compute the greatest common divisor of a and N using the extended Euclidean algorithm. This will factor N into two factors N_1 and N_2 (this is called *splitting* N). We can then test if N_1 and N_2 are powers of primes, and otherwise proceed to split them if they are composite. We will require at most $\log_2(N)$ splittings before we factor N into its prime factors. These techniques are summarised in [11].

It turns out that for increasing powers of a, the remainders form a repeating sequence with a period r. We can also call r the *order* of a since $a^r = 1 \bmod N$. Once r is known, factors of N are obtained by calculating the greatest common divisor of N and $a^{r/2} \pm 1$.

Suppose we want to factor 35 using this method. Let $a = 4$. For increasing x the function $4^x \bmod 35$ forms a repeating sequence $4, 16, 32, 29, 9, 1, 4, 16, 29, 32, 9, 1, \ldots$. The period is $r = 6$, and $a^{r/2} \bmod 35 = 29$. Then we take the greatest common divisor of 28 and 35, and of 30 and 35, which gives us 7 and 5, respectively, the two factors of 35. Classically, calculating r is at least as difficult as trying to factor N; the execution time of the best currently-known algorithms grows exponentially with the number of digits in N. Quantum computers can find r very efficiently.

Consider the unitary transformation U_a that maps $|x\rangle$ to $|ax \bmod N\rangle$. Such a transformation is realised by simply implementing the reversible classical network for multiplication by a modulo N using quantum gates. The transformation U_a, like the element a, has order r, that is, $U_a^r = I$, the identity operator. Such an operator has eigenvalues of the form $e^{\frac{2\pi i k}{r}}$ for $k = 0, 1, 2, \ldots, r-1$. In order to formulate Shor's algorithm in terms of phase estimation let us apply the construction from the last section taking

$$|\psi\rangle = \sum_{j=0}^{r-1} e^{\frac{-2\pi ij}{r}} |a^j \bmod N\rangle \ . \tag{21}$$

Note that $|\psi\rangle$ is an eigenvector of U_a with eigenvalue $e^{2\pi i(\frac{1}{r})}$. Also, for any j, it is possible to implement efficiently a controlled-$U_a^{2^j}$ gate by a sequence of squaring (since $U_a^{2^j} = U_{a^{2^j}}$). Thus, using the state $|\psi\rangle$ and the implementation of controlled-$U_a^{2^j}$ gates, we can directly apply the method of the last section to efficiently obtain an estimator of $\frac{1}{r}$.

The problem with the above method is that we are aware of no straightforward efficient method to prepare state $|\psi\rangle$, however, let us notice that almost any state $|\psi_k\rangle$ of the form

$$|\psi_k\rangle = \sum_{j=0}^{r-1} e^{-\frac{2\pi ikj}{r}} |a^j \bmod N\rangle \ , \tag{22}$$

where k is from $\{0,\ldots,r-1\}$ would also do the job. For each $k \in \{0,1,\ldots,r-1\}$, the eigenvalue of state $|\psi_k\rangle$ is $e^{2\pi i(\frac{k}{r})}$. We can again use the technique from the last section to efficiently determine $\frac{k}{r}$ and if k and r are coprime then this yields [2] r. Now the key observation is that

$$|1\rangle = \sum_{k=1}^{r} |\psi_k\rangle \ , \tag{23}$$

and $|1\rangle$ *is* an easy state to prepare.

If we substituted $|1\rangle$ in place of $|\psi\rangle$ in the last section then effectively we would be estimating one of the r, randomly chosen, eigenvalues $e^{2\pi i(\frac{k}{r})}$. This demonstrates that Shor's algorithm, in effect, estimates the eigenvalue corresponding to an eigenstate of the operation U_a that maps $|x\rangle$ to $|ax \bmod N\rangle$. A classical procedure - the continued fractions algorithm - can be employed to estimate r from these results. The value of r is then used to factorise the integer.

5.2 Finding hidden subgroups

A number of algorithms can be generalised in terms of group theory as examples of finding hidden subgroups. For any $g \in G$, the coset gK, of the subgroup K is defined as $\{gK | g \in G\}$. Say we have a function f which maps a group G to a set X, and f is constant on each coset of the subgroup K, and distinct on each coset, as illustrated in Figure 7. In other words, $f(x) = f(y)$ if and only if $x - y$ is an element of K.

In Deutsch's case, $G = \{0,1\}$ with addition mod 2 as the group operation, and X is also $\{0,1\}$. There are two possible subgroups K: $|0\rangle$, and G itself. We are given a black-box U_f for computing f

$$|x\rangle|y\rangle \rightarrow |x\rangle|y \oplus f(x)\rangle \ .$$

There are two cosets of the subgroup $\{0\}$: $\{0\}$ and $\{1\}$. If the function is defined to be constant and distinct on each coset, it must be balanced. On the other hand, there is only one coset of the

[2] If the estimate $y/2^m$ of k/r satisfies

$$\left|\frac{y}{2^m} - \frac{k}{r}\right| < \frac{1}{2N^2},$$

then there is a unique rational of the form $\frac{a}{b}$ with $0 < b \leq N$ satisfying

$$\left|\frac{y}{2^m} - \frac{a}{b}\right| < \frac{1}{2N^2}.$$

Consequently, $a/b = k/r$, and the continued fractions algorithm will find the fraction for us. We might be unlucky and get a k like 0, but with even 2 repetitions with random k we can find r with probability at least 0.54 [2].

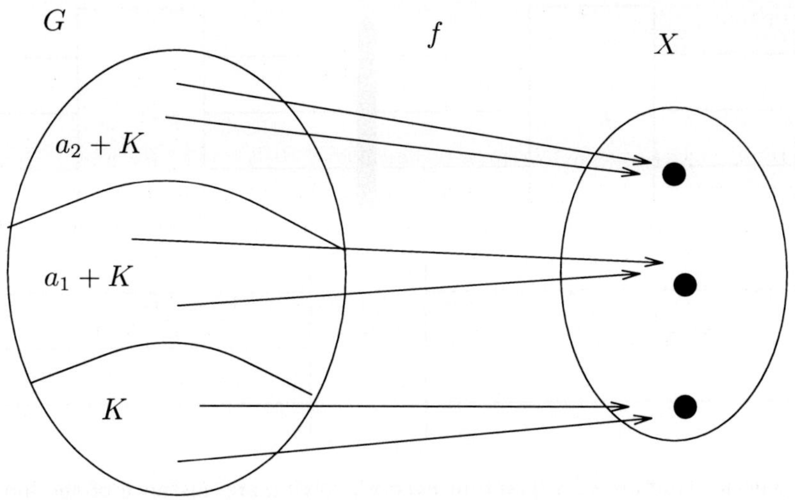

Figure 7: A function f mapping elements of a group G to a set X with a hidden subgroup K. This means that $f(g_1) = f(g_2)$ if and only if g_1 and g_2 are in the same coset of K.

other subgroup G, the group itself. In this case the function is constant. With our specially chosen eigenvector $|\,0\rangle - |\,1\rangle$ the algorithm always outputs $|\,0\rangle$ if $K = \{0,1\}$ (f is constant), and $|\,1\rangle$ if $K = \{0\}$, (f is balanced). Therefore we can view Deutsch's algorithm as distinguishing between the 'hidden subgroups'.

The hidden subgroup problem also encompasses the problem of finding orders of elements in a group, of which the factoring algorithm is a special case. In quantum factoring, we wish to find the order r of the element a in some group represented by X. Here G is the group of integers $\mathbf{Z}$ and K is the additive subgroup $r\mathbf{Z}$ of integer multiples of r, where r is the order of a, and a is from the multiplicative group of integers modulo N. The function f maps x to $a^x \bmod N$.

The output $|\,y\rangle$ in this case estimates an element which is orthogonal [3] to the subgroup K. The output $|\,z\rangle$ corresponds to the estimate $z/2^n$ of the eigenvalue k/r of the operator U_a which maps $|\,x\rangle$ to $|\,ax\rangle$ (that is, the operator which maps $|\,f(g)\rangle$ to $|\,f(g+1)\rangle$) on the eigenvector $|\,\psi_k\rangle$. In general, for any function f mapping a finitely generated Abelian group G to a finite set X, the quantum network shown in figure 8 will output an estimate of a random element orthogonal to the hidden subgroup K. With enough such elements, we can easily determine K using linear algebra.

By framing algorithms in terms of hidden subgroups, it may be possible to think of other problems associated with this structure in groups which we can treat with quantum algorithms. A number of algorithms have already been cast in this language, including Deutsch's problem [3, 2], Simon's problem [12], factoring integers [9], finding discrete logarithms [9], Abelian stabilisers [13], self-shift-equivalences [14], and others [15] (see [16] and [17] for details).

5.3 Quantum Counting and Searching

The first quantum algorithm for searching was constructed by Grover [18]. This has led to a large class of searching and counting algorithms.

We again consider a function f, this time mapping us from a set X to the set $\{0,1\}$.

[3]By *orthogonal* here, we are not referring to the orthogonality of states in our computational Hilbert space. When we say k/r is orthogonal to $K = r\mathbf{Z}$, we mean that $\exp(2\pi i z\frac{k}{r}) = 1$ for every $z \in K$. This notion of orthogonality generalises to groups with several generators as well.

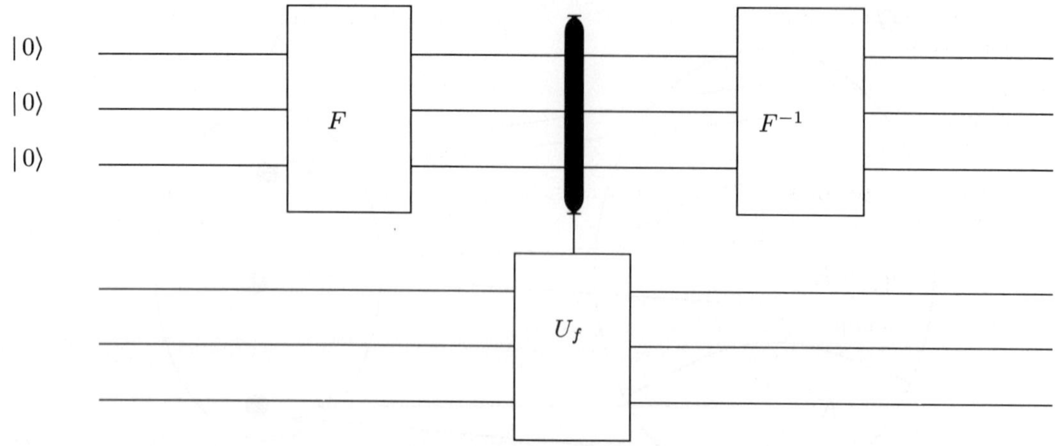

Figure 8: The generic structure of a quantum network solving any instance of the hidden subgroup problem. The first register contains tuples of integers corresponding to the Abelian group G. The role of the first Fourier transform is to create a superposition of many computational paths corresponding to different elements of G. The evaluation of the function simply kicks phases back into the control register states, and the final inverse Fourier transform produces the estimates of the eigenvalues of operators related to the function f. The set of eigenvalues corresponding to a particular eigenvector produces an element orthogonal to K. By collecting enough such orthogonal elements we can efficiently find a generating set for K.

We might wish to decide if there is a solution to $f(x) = 1$ (the *decision* problem) , or to actually find a solution to $f(x) = 1$ (the *searching* problem). We might be more demanding and want to know how many solutions x there are to $f(x) = 1$ (*counting problem*). Small cases of the searching [19, 20] and counting [21] algorithms have been implemented using NMR technology.

In this section we will show how approximate quantum counting can easily be phrased as an instance of phase estimation, and quantum searching as an instance of inducing a desired relative phase between two eigenvectors.

In the following sections analysing quantum counting and searching, we will be considering the *Grover iterate*

$$G = -A U_0 A^{-1} U_f \tag{24}$$

which was defined in [18] with A as the Hadamard transform. It was later generalised in [22], [23], [24] and [25] with A being any transformation such that $A\,|\,0\rangle$ contains a solution to $f(x) = 1$ with non-zero amplitude, i.e. $|\,\langle x\,|\,A\,|\,0\rangle\,|^2 > 0$ for some x with $f(x) = 1$. The operator U_f maps

$$|\,x\rangle \rightarrow -\,|\,x\rangle$$

for all x satisfying $f(x) = 1$, and the operator U_0 maps

$$|\,0\rangle \rightarrow -\,|\,0\rangle$$

leaving the remaining basis states alone. Note that this U_f is slightly different than the standard U_f which maps $|\,x\rangle\,|\,b\rangle$ to $|\,x\rangle\,|\,b \oplus f(x)\rangle$, but can be easily obtained from it by setting $|\,b\rangle$ to $|\,0\rangle - |\,1\rangle$.

5.3.1 Quantum Counting

Quantum counting was first discussed in [25], where it was observed that the Grover iterate is almost periodic with a period dependent on the number of solutions. Therefore the techniques of period-finding, as in Shor's algorithm, were applied [24]. It is also possible to think of the problem as a phase estimation (see [26]).

We simply observe that the eigenvalues[4] of G are 1, -1, $e^{2\pi i\omega_j}$, and $e^{-2\pi i\omega_j}$ where $f(x) = 1$ has j solutions and

$$e^{2\pi i\omega_j} = 1 - 2j/N + 2i\sqrt{j/N - (j/N)^2}.$$

Let X_1 denote the set of solutions to $f(x) = 1$, and X_0 denote the set of solutions to $f(x) = 0$. Estimating ω_j (or $-\omega_j$) will give us information about the number of solutions to $f(x) = 1$. For example, for small ω_j, the number of solutions, j, is roughly $N\pi^2\omega_j^2$ since $\cos(2\pi\omega_j) = 1 - 2j/N \approx 1 - 2\pi^2\omega_j^2$ for small ω_j.

We can use the techniques of the previous sections to estimate this phase ω_j provided we know how to create a starting state containing the eigenvectors with eigenvalues $e^{2\pi i\omega_j}$ and $e^{-2\pi i\omega_j}$. For non-trivial j, these eigenvectors are given by

$$|\psi_+\rangle = \frac{1}{\sqrt{2}}(|X_1\rangle + i\,|X_0\rangle) \tag{25}$$

$$|\psi_-\rangle = \frac{1}{\sqrt{2}}(|X_1\rangle - i\,|X_0\rangle) \tag{26}$$

where

$$|X_1\rangle = \frac{1}{\sqrt{j}} \sum_{f(x)=1} |x\rangle \tag{27}$$

$$|X_0\rangle = \frac{1}{\sqrt{N-j}} \sum_{f(x)=0} |x\rangle\,. \tag{28}$$

Fortunately, the starting state

$$A\,|0\rangle = \frac{1}{\sqrt{N}} \sum_{x=0}^{N-1} |x\rangle$$

is equal to

$$\frac{1}{\sqrt{2}}(e^{-2\pi i\theta_j}\,|\psi_+\rangle + e^{2\pi i\theta_j}\,|\psi_-\rangle) \tag{29}$$

for some real number θ_j, which is not important as far as counting is concerned, since all that is required for the phase estimation procedure is any superposition of these two eigenvectors of G.

Thus using a controlled-G, controlled-G^2, ..., and a controlled-G^{2^n}, (as done with controlled-Us in Figure 6) and applying a quantum Fourier transform, we can get an n-bit estimate of either ω_j or $-\omega_j$. This gives us an estimate of j, the number of solutions. Note that, unlike in the case of finding orders, there are in general no short-cuts for computing higher powers of G. That is, computing G^{2^n} requires 2^n repetitions of G.

Quantum algorithms for approximate counting require roughly only square root of the number of calls a classical algorithm would require.

5.3.2 Quantum searching

While estimating the number of solutions to $f(x) = 1$ is a special case of quantum phase estimation, the algorithm for searching for these solutions can be viewed as a clever use of the phase kick-back

[4]The eigenvalue -1 has multiplicity $j-1$, 1 has multiplicity $N-j-1$, and $e^{2\pi i\omega}$ and $e^{-2\pi i\omega}$ each have multiplicity 1. If $j = 0$, then 1 has multiplicity N, (note that $e^{2\pi i\omega_0} = e^{-2\pi i\omega_0} = 1$), if $j = N$, then -1 has multiplicity N, ($e^{2\pi i\omega_N} = e^{-2\pi i\omega_N} = -1$).

technique to induce a desired relative phase between two eigenvectors of G. The state $|X_1\rangle$ is a superposition of solutions to $f(x) = 1$, so it is itself a solution which it is possible for us to construct. We note that

$$|X_1\rangle = |\psi_+\rangle + |\psi_-\rangle \tag{30}$$

and our starting state for quantum searching is

$$A|0\rangle = e^{-2\pi i\theta_j}|\psi_+\rangle + e^{2\pi i\theta_j}|\psi_-\rangle\,. \tag{31}$$

Each iteration of G kicks back a phase of $e^{2\pi i\omega_j}$ in front of $|\psi_+\rangle$ and $e^{-2\pi i\omega_j}$ in front of $|\psi_-\rangle$. So k iterations of G produces the state

$$A|0\rangle = \frac{1}{\sqrt{2}}(e^{2\pi i(k\omega_j-\theta_j)})|\psi_+\rangle + e^{-2\pi i(k\omega_j-\theta_j)}|\psi_-\rangle\,. \tag{32}$$

Since we seek

$$|X_1\rangle = \frac{1}{\sqrt{2}}(|\psi_+\rangle + |\psi_-\rangle)$$

we want to choose the number of iterations k so that

$$k\omega_j - \theta_j \tag{33}$$

is as close to an integer as possible. When j is small, this means selecting the number of iterations close to

$$\frac{\pi}{4}\sqrt{N/j}. \tag{34}$$

Note that any classical algorithm would require N/j evaluations of f before finding a solution to $f(x) = 1$ with high probability.

6 Concluding remarks

Multi-particle interferometers can be viewed as quantum computers and any quantum algorithm follows the typical structure of a multi-particle interferometry sequence of operations. This approach seems to provide an additional insight into the nature of quantum computation and, we believe, will help to unify all quantum algorithms and relate them to different instances of quantum phase estimation.

7 Acknowledgements

This work was supported in part by the European TMR Research Network ERP-4061PL95-1412, Hewlett-Packard and Elsag-Bailey, The Royal Society, CESG and the Rhodes Trust. R.C. is partially supported by Canada's NSERC.

References

[1] R. Feynman: Simulating physics with computers. Int. J. Theor. Phys. **21**, 1982, pp. 467-488.

[2] R. Cleve, A. Ekert, C. Macchiavello, and M. Mosca: Quantum Algorithms Revisited, Proc. R. Soc. Lond. A **454**, 1998, pp. 339–354. See also LANL preprint/quant-ph/9708016.

[3] D. Deutsch: Quantum-theory, the Church-Turing principle and the universal quantum computer. Proc. R. Soc. Lond. A **400**,1985, pp. 97-117.

[4] J. Jones and M. Mosca: Implementation of a quantum algorithm on a nuclear-magnetic resonance quantum computer. J. Chem. Phys. **109**, pp. 1648-1653. See also LANL preprint quant-ph/9801027.

[5] I. Chuang, L. Vandersypen, X. Zhou, D. Leung and S. Lloyd: Experimental realisation of a quantum algorithm. Nature, **393**, 1998, pp. 143-146. See also LANL preprint quant-ph/9801037.

[6] D. Coppersmith: An Approximate Fourier Transform Useful in Quantum Factoring, IBM Research Report No. RC19642, 1994.

[7] A. Barenco, A. Ekert, K. Suominen and P. Törma: Approximate quantum Fourier-transform and decoherence. Phys. Rev. A **54**, 1996, pp. 139-146. See also LANL preprint quant-ph/9601018.

[8] W. van Dam, G. D'Ariano, A. Ekert, C. Macchiavello and M. Mosca: Estimating Phase Rotations on a Quantum Computer, preprint.

[9] P.Shor: Algorithms for quantum computation: Discrete logarithms and factoring. Proc. 35th Annual Symposium on Foundations of Computer Science, 1994, pp. 124–134. See also LANL preprint quant-ph/9508027.

[10] A. Ekert and R. Jozsa: Quantum computation and Shor's factoring algorithm, Rev. Mod. Phys. **68**, 733, 1996, pp. 733-753.

[11] A. Menezes, P. van Oorschot, and S. Vanstone: Handbook of Applied Cryptography, CRC Press, London, 1996.

[12] D. Simon: On the Power of Quantum Computation. Proc. 35th Annual Symposium on Foundations of Computer Science, 1994, pp. 116-123.

[13] A. Kitaev: Quantum measurements and the Abelian stabiliser problem. LANL preprint quant-ph/9511026, 1995.

[14] D. Grigoriev,: Testing the shift-equivalence of polynomials by deterministic, probabilistic and quantum machines. Theoretical Computer Science, **180**, 1997, pp. 217-228.

[15] D. Boneh, and R. Lipton: Quantum cryptanalysis of hidden linear functions (Extended abstract). Lecture Notes on Computer Science, **963**, 1995, pp.424-437.

[16] M. Mosca and A. Ekert: Hidden subgroups and estimation of eigenvalues on a quantum computer. To appear in the Proc. of the 1st International NASA Conference on Quantum Computing and Quantum Information Processing, Lecture Notes on Computer Science, 1998.

[17] P. Høyer: Conjugated Operators in Quantum Algorithms. preprint, 1997.

[18] L. Grover: A fast quantum mechanical algorithm for database search, Proc. 28 Annual ACM Symposium on the Theory of Computing, ACM Press New York, 1996, pp. 212-219. Journal version, "Quantum Mechanics helps in searching for a needle in a haystack", appeared in *Physical Review Letters*, **79** (1997) 325-328. See also LANL preprint quant-ph/9706033.

[19] N. Gershenfeld, I. Chuang and M. Kubinec: Experimental implementation of fast quantum searching. Phys. Rev. Lett., **80**, 1998, pp. 3408-3411.

[20] J. Jones, R. Hansen and M. Mosca: Implementation of a quantum search algorithm on a quantum computer. Nature, **393**, 1998, pp. 344-346. See also LANL preprint quant-ph/9805069.

[21] J. Jones and M. Mosca: Approximate quantum computing on an NMR ensemble quantum computer. Submitted. See LANL preprint quant-ph/quant-ph/9808056.

[22] G. Brassard and P. Høyer: An exact quantum polynomial-time algorithm for Simon's problem. Proceedings of the Fifth Israeli Symposium on Theory of Computing and Systems, IEEE Computer Society Press, 1997, pp.12-23. See also LANL preprint quant-ph/9704027.

[23] L. Grover: A framework for fast quantum mechanical algorithms. Proc. 30th Annual ACM Symposium on the Theory of Computing, 1998. See also LANL preprint quant-ph/9711043.

[24] G. Brassard, P. Høyer and A. Tapp: Quantum Counting, Proc. 25th International Colloquium on Automata, Languages and Programming, Lecture Notes on Computer Science, **1443**, pp. 820-831, 1998. See also LANL preprint quant-ph/9805082.

[25] M. Boyer, G. Brassard, P. Høyer and A. Tapp: Tight bounds on quantum searching, Proceedings of the Fourth Workshop on Physics and Computation, 1996, pp. 36-43. Forschritte Der Physik, Special issue on quantum computing and quantum cryptography, **4**, pp. 493-505, 1998. See also LANL preprint quant-ph/9605034.

[26] M. Mosca: Quantum Searching and Counting by Eigenvector Analysis. Proceedings of Randomized Algorithms, satellite workshop of MFCS '98. Available at www.eccc.uni-trier.de/eccc-local/ECCC-LectureNotes/randalg/.

Quantum Complexity

An Introduction to Quantum Complexity Theory

Richard Cleve
*University of Calgary**

Abstract

We give an overview of basic quantum complexity theory. We focus on three fundamental models: computational complexity, query complexity, and communication complexity, highlighting the relationships between them.

Complexity theory is concerned with the inherent cost required to solve information processing problems, where the cost is measured in terms of various well-defined resources. In this context, a *problem* can usually be thought of as a function whose input is a *problem instance* and whose corresponding output is the *solution* to it. Sometimes the solution is not unique, in which case the problem can be thought of as a relation, rather than a function. *Resources* are usually measured in terms of: some designated elementary operations, memory usage, or communication. We consider three specific complexity scenarios, which highlight different advantages of working with quantum information:

1. **Computational complexity**
2. **Query complexity**
3. **Communication complexity**.

Despite the differences between these models, there are some intimate relationships among them. The usefulness of many currently-known quantum algorithms is ultimately best expressed in the computational complexity model; however, virtually all of these algorithms evolved from algorithms in the query complexity model. The query complexity model is a natural setting for discovering interesting quantum algorithms, which frequently have interesting counterparts in the computational complexity model. Quantum algorithms in the query complexity model can also be transformed into protocols in the communication complexity model that use quantum information (and sometimes these are more efficient than any classical protocol can be). Also, this latter relationship, taken in its contrapositive form, can be used to prove that some problems are inherently difficult in the query complexity model.

*Department of Computer Science, University of Calgary, Calgary, Alberta, Canada T2N 1N4. Email: `cleve@cpsc.ucalgary.ca`.

1 Computational complexity

In the *computational complexity* scenario, an *input* is encoded as a binary string (say) and supplied to an algorithm, which must compute an *output* string corresponding to the input. For example, in the case of the **factoring problem**, for input 100011 (representing 35 in binary), the valid outputs might be 000101 or 000111. The algorithm must produce the required output by a series of *local* operations. By "local", we do not necessarily mean "local in space", but, rather, that each operation involves a small portion of the data. In other words, a local operation is a transformation that is confined to a small number of bits or qubits (such as two or three). The above property is satisfied by Turing machines and circuits, and also by quantum Turing machines [7, 20] and quantum circuits [21, 48]. We shall find it most convenient to work with circuit models here.

1.1 Classical circuits

For classical circuits, the basic operations can be taken as the binary $\wedge$ (AND) gate, the binary $\vee$ (OR) gate, and the unary $\neg$ (NOT) gate. In Fig. 1 is a boolean circuit consisting of five gates that computes the parity of two bits. The

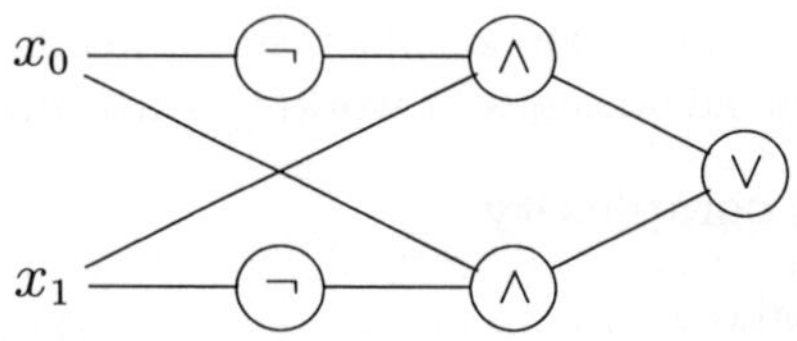

Figure 1: A classical circuit for computing the parity of two bits.

inputs are denoted as x_0 and x_1, and the "data-flow" is from left to right. The rightmost gate is designated as the output, whose value is $x_0 \oplus x_1$, as required. This is the smallest circuit consisting of $\wedge$, $\vee$, and $\neg$ gates that computes the parity. Based on this fact, we could say that the computational complexity of the binary parity function is five. But note that this value is highly dependent on the specific set of basic operations that we started with. If we included the binary $\oplus$ (EXCLUSIVE-OR) gate as a basic operation then a single gate suffices to compute the parity of two bits (Fig. 2).

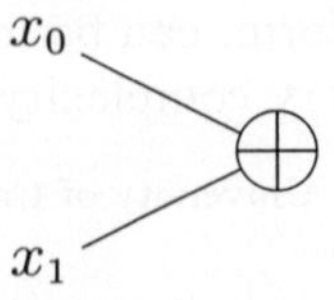

Figure 2: An alternative circuit for parity with one exclusive-or gate.

This illustrates a feature of the computational complexity model: the *exact* number of operations required to compute functions is quite sensitive to the technical choice of which basic operations to allow. The exact computational complexity of simple problems involving a small number of bits is somewhat arbitrary.

Computational complexity is more meaningful when larger problems that scale up are considered, such as the problem of computing the parity of n bits, $x_0, x_1, \ldots, x_{n-1}$. Using $\oplus$ gates, one can construct a tree with $n-1$ such gates that computes this parity. On the other hand, if only $\wedge$, $\vee$, and $\neg$ gates are available then it appears that something like $5(n-1)$ gates are needed. In both cases, the number of gates is $O(n)$, and it is also straightforward to prove that a constant times n gates are *necessary* for both cases. A similar property holds for *any* computational complexity problem: changing from one set of gates to any other set of gates (assuming that both sets are local and universal) can only affect computational complexity by a multiplicative constant. Thus, for any $f : \{0,1\}^* \to \{0,1\}$, its computational complexity is a well-defined function (of n, the length of the input to f) up to a multiplicative constant.

This is one reason why it is common to denote the computational complexity of functions using asymptotic notation that suppresses multiplicative constants. $O(T(n))$ means bounded above by $c\,T(n)$ for some constant $c > 0$ (for sufficiently large n). Also, $\Omega(T(n))$ means bounded below by $c\,T(n)$ for some constant $c > 0$, and $\Theta(T(n))$ means both $O(T(n))$ and $\Omega(T(n))$. A circuit is *polynomially-bounded* in size if its size is $O(n^d)$ for some constant d.

A matter that we have so far obscured concerns the treatment of the parameter n (denoting the input size). Although each circuit is for some *fixed* value of n, we are also speaking of n as a freely varying parameter. For problems where n is a variable (such as the problem of computing the parity of n bits), an algorithm in the circuit model must actually be a *circuit family* $(C_1, C_2, C_3, \ldots)$, where circuit C_n is responsible for all input instances of size n. To be meaningful, a circuit family has to be *uniform* in that it can somehow be finitely specified. For example, for the aforementioned parity problem, a finite specification of a circuit family can be informally: "for input size n, C_n is a binary tree of $\oplus$-gates with $x_0, \ldots, x_{n-1}$ at the leaves". Formally, a *specification* of a circuit family is an algorithm that maps each n to an explicit description of C_n. Technically, we ought to include the efficiency of the specification algorithm as part of the computational cost of a circuit family. This raises the question of what formalism one uses to describe the specification algorithm. Note that if we try to use another circuit family for this then *it* requires its own specification algorithm (and so on!), so this approach will not work. There are sophisticated ways of dealing with uniformity; a very simple way is to just use some non-circuit model, such as a Turing machine for the circuit specification algorithm. At this point, the reader may wonder why one does not just use the Turing machine model to begin with. A big advantage of circuits is that their structural elements are simple and easy to work with—and this appears to hold for quantum circuits as well. Uniformity tends to be a straightforward technicality, that can be worked out after a circuit family is discovered; the discovery of the circuit family is usually the interesting part of the algorithm design process.

Let us now consider the problem of **primality testing**, where the input is a number x represented as an n-bit binary string, and the output is (say) 1 if x is prime and 0 if x is composite. Notice that, in the cases where x is composite, there is no requirement here that a factor of x be produced. It turns out that the smallest currently-known uniform circuit family for this problem has size $O(n^{d \log\log n})$ (for some constant d), which is shy of being polynomially-bounded [2].

There exist *probabilistic* circuit families that solve primality testing more efficiently. A *probabilistic circuit* is one that can flip coins during its execution, and the evolution of the computation can depend on the outcomes. Formally, a $\not{c}$ (COIN-FLIP) gate, has no input and is understood to emit one uniformly-distributed random bit when executed during a computation. If m random bits are required then m $\not{c}$-gates can be inserted into a circuit. Solovay and Strassen [44] discovered a remarkable probabilistic algorithm for primality testing that can be expressed in terms of probabilistic circuits. For any $\varepsilon > 0$, there is a probabilistic circuit of size $O(n^3 \log(1/\varepsilon))$ that errs with probability at most ε. That is, given any $x \in \{0,1\}^n$ as input, the circuit correctly decides the primality of x with probability at least $1 - \varepsilon$ (note that the error probability is with respect to the $\not{c}$-gates, and not with respect to any assumed probability distribution on the input x). The circuit family is highly uniform, and there are versions of the algorithm that are quite efficient in practice, even when ε is very small (such as one billionth).

As an aside, we note that probabilistic circuit families can be translated into standard (deterministic) circuit families if one is willing to forfeit uniformity. For each n, by setting $\varepsilon = 1/(2^n + 1)$, we obtain a probabilistic circuit C_n of size $O(n^4)$ for primality testing that errs with probability less than $1/2^n$ for any input. Now consider the circuit C_n' that results if, for each $\not{c}$-gate in C_n, a uniformly distributed random bit is independently generated and substituted for that gate. This is a probabilistic construction that yields a deterministic circuit C_n'. For $x \in \{0,1\}^n$, let p_x be the probability that the resulting C_n' errs on input x. Then, for each x, $p_x < 1/2^n$, so the probability that C_n' errs for *any* $x \in \{0,1\}^n$ is strictly less than $\sum_{x\in\{0,1\}^n} 1/2^n = 1$. Therefore, with probability greater than 0, C_n' is correct for all of its 2^n possible input values. It follows that, for any n, a deterministic circuit of size $O(n^4)$ must exist for primality testing. The problem is that there is no known efficient way to explicitly construct the coin flips which yield a correct circuit. Thus, the implied $O(n^4)$-size circuit family for primality testing is merely established by an existence proof; this is an example of a *non-uniform* circuit family. The fact that uniform probabilistic circuit families can be converted into non-uniform deterministic circuit families is theoretically noteworthy, but not practical.

Let us now consider the **factoring problem**, where the input is an n-bit number x, and the output is a list of the prime factors of x. This is related to—but different from—primality testing, and is apparently much harder. The smallest currently-known circuit family for this problem is probabilistic and has size $O(2^{d\sqrt{n \log n}})$ (where d is a constant), which is far from being polynomially-bounded [33, 37]. One of the reasons why quantum algorithms are of interest is that there exists a *quantum* circuit family of polynomial-size that solves the factoring problem (this will be discussed later).

A problem related to the factoring problem is the **order-finding problem**, where the input is a pair of natural numbers a and N that are coprime (i.e. such that $\gcd(a, N) = 1$), and the goal is to find the smallest positive r such that $a^r \bmod N = 1$ (there always exists such an $r \in \{1, \ldots, N-1\}$). It turns out that the order-finding problem and the factoring problem are closely related in that a polynomial-size circuit family for either one of them can be converted into a polynomial-size probabilistic circuit family for the other one. In fact, the quantum circuit for factoring is actually obtained from a quantum circuit that solves the order-finding problem.

Although we have represented circuits pictorially as data-flow diagrams, it is useful to be able to encode circuits as binary strings. There are several reasonable encoding schemes. One such scheme encodes the graphical structure of a circuit C as an $m \times m$ adjacency matrix (where m is the number of gates plus the number of inputs in C), and then follows this by more bits that specify the labels of the nodes (e.g. $\wedge$, $\vee$, $\neg$, $x_0, \ldots, x_{n-1}$). Note that, using this encoding scheme, a circuit of size m has an encoding of $O(m^2)$ bits. There are more efficient encoding schemes, where the encodings are of length $O(m \log m)$, and, for any "reasonable" encoding scheme, the length of the string that encodes C is polynomially-related to the size of C. Let $e(C)$ denote a binary string that encodes the circuit C (relative to some reasonable encoding scheme).

A fundamental problem in classical computational complexity theory is the **circuit satisfiability problem**, which is defined as follows. Call a circuit *satisfiable* if there exists an input string to the circuit for which the corresponding output value of the circuit is 1. For example, the circuit in Fig. 1 is satisfiable. The input to the circuit satisfiability problem is a binary string $x = e(C)$ that encodes some boolean circuit C, and the output is 1 if C is satisfiable, and 0 otherwise. The best currently-known (deterministic or probabilistic) algorithm for circuit satisfiability is to simply try all possible inputs to C. When $e(C)$ encodes a circuit C with n inputs and m gates, this procedure takes $O(2^n m^d)$ steps, where d is a constant that depends on the encoding scheme used ($d = 2$ suffices for most reasonable encoding schemes). In interesting cases, m is typically polynomial in n, so the dominant factor in this quantity is 2^n. It is not known whether or not there is a polynomially-bounded circuit family for circuit satisfiability. In fact, circuit satisfiability is one of the so-called *NP*-complete problems [18, 25], for which a polynomially-bounded circuit family would yield polynomially-bounded circuits for *all* problems in *NP*.

1.2 Quantum circuits

To develop a theory of computational complexity for quantum information, it is natural to extend the notion of a circuit to a composition gates which perform *quantum* operations on *quantum bits* (called *qubits*). The most general quantum operations subsume all classical operations, which are frequently not reversible. It turns out that the quantum operations that seem to be the most useful in the context of quantum computation are those that are unitary (and hence reversible), as well as the von Neumann measurements.

Let us begin by recalling that the state of a system of m qubits can be described by associating an *amplitide* α_x with each $x \in \{0,1\}^m$ (we restrict our attention to *pure* quantum states). Each amplitude is a complex number and there is a condition that $\sum_{x\in\{0,1\}^m} |\alpha_x|^2 = 1$. Taken together, these amplitudes constitute a point in a 2^m-dimensional vector space. The *computational basis* associated with this vector space is $\{|x\rangle : x \in \{0,1\}^m\}$, and we follow the convention of writing states as linear combinations of these basis elements:

$$\sum_{x\in\{0,1\}^m} \alpha_x |x\rangle . \tag{1}$$

Given a quantum state, it is impossible to access the values of the amplitudes directly. What one can do is perform a (von Neumann) measurement on each qubit. If such an operation is performed then the result is a random m-bit string y, distributed as $\Pr[y = x] = |\alpha_x|^2$, for each $x \in \{0,1\}^m$. After this measurement, the original quantum state is destroyed.

One can also perform a unitary operation on an m-qubit system, which is a linear transformation U for which $UU^\dagger = I$, where $U^\dagger$ is the conjugate transpose of U. Such a unitary transformation can be represented by a $2^m \times 2^m$ matrix and will, in general, affect all of the m qubits. For the purposes of quantum computation, we restrict the basic operations to local unitary transformations that only involve a small number (say, one or two) of the qubits. A one-qubit unitary operation can be described by 2×2 unitary matrix U.

In the case where $m = 1$, this U transforms the state $\alpha|0\rangle + \beta|1\rangle$ to the state $\alpha'|0\rangle + \beta'|1\rangle$, where

$$\begin{pmatrix}\alpha'\\ \beta'\end{pmatrix} = U\begin{pmatrix}\alpha\\ \beta\end{pmatrix}. \tag{2}$$

In order to define the semantics of applying a one-qubit gate in the context of an m-qubit system for $m > 1$, we introduce a tensor product operation. Suppose that an m-qubit system is in state $\sum_{x\in\{0,1\}^m} \alpha_x |x\rangle$ and an n-qubit system is in state $\sum_{y\in\{0,1\}^n} \beta_y |y\rangle$. Then the state of the combined system (consisting of $m+n$ qubits) is defined to be the *tensor product* of the states of the individual systems, which is

$$\left(\sum_{x\in\{0,1\}^m} \alpha_x |x\rangle\right)\left(\sum_{y\in\{0,1\}^n} \beta_y |y\rangle\right) = \sum_{\substack{x\in\{0,1\}^m\\ y\in\{0,1\}^n}} \alpha_x\beta_y |xy\rangle . \tag{3}$$

For example, $(\frac{1}{\sqrt{2}}|0\rangle - \frac{1}{\sqrt{2}}|1\rangle)(\frac{1}{\sqrt{2}}|0\rangle - \frac{1}{\sqrt{2}}|1\rangle) = \frac{1}{2}|00\rangle - \frac{1}{2}|01\rangle - \frac{1}{2}|10\rangle + \frac{1}{2}|11\rangle$.

Now, applying a one-qubit U to the kth qubit of an m-qubit system is defined to be the unitary operation that maps each basis state

$$|x_0 \cdots x_{m-1}\rangle = |x_0 \cdots x_{k-2}\rangle\, |x_{k-1}\rangle\, |x_k \cdots x_{m-1}\rangle$$

to the state

$$|x_0 \cdots x_{k-2}\rangle \, (U \, |x_{k-1}\rangle) \, |x_k \cdots x_{m-1}\rangle$$

(for each $x \in \{0,1\}^m$). Note that, by linearity, this completely defines a unitary operation on an m-qubit system.

For example, the one-qubit *Hadamard* gate corresponds to the matrix

$$H = \frac{1}{\sqrt{2}} \begin{pmatrix} 1 & 1 \\ 1 & -1 \end{pmatrix}, \tag{4}$$

and, when it is applied to the second qubit of a two-qubit system, the resulting operation is

$$\frac{1}{\sqrt{2}} \begin{pmatrix} 1 & 1 & 0 & 0 \\ 1 & -1 & 0 & 0 \\ 0 & 0 & 1 & 1 \\ 0 & 0 & 1 & -1 \end{pmatrix} \tag{5}$$

(with respect to the ordering of basis states $|00\rangle$, $|01\rangle$, $|10\rangle$, $|11\rangle$). A quantum circuit corresponding to such an operation is in Fig. 3, which denotes that the first (top) qubit is left unaltered and H is applied to the second qubit.

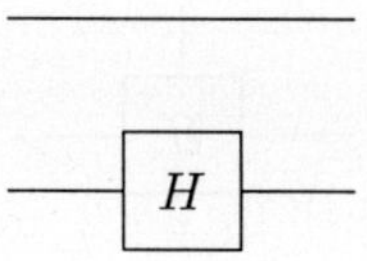

Figure 3: Quantum circuit applying a Hadamard transform to one of two qubits.

To construct nontrivial quantum circuits, it is necessary to include two-qubit unitary operations. A simple but quite useful two-qubit operation is the CONTROLLED-NOT gate (C-NOT, for short), which, for all $x, y \in \{0,1\}$, transforms the basis state $|x\rangle\,|y\rangle$ to the basis state $|x\rangle\,|y \oplus x\rangle$ (and this extends to arbitrary quantum states by linearity). The notation for the C-NOT gate in quantum circuits is indicated in Fig. 4 (it is also known as the "reversible exclusive-or" gate).

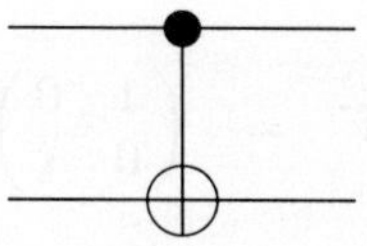

Figure 4: Notation for the CONTROLLED-NOT (C-NOT) gate.

Note that the C-NOT gate corresponds to the unitary transformation

$$\begin{pmatrix} 1 & 0 & 0 & 0 \\ 0 & 1 & 0 & 0 \\ 0 & 0 & 0 & 1 \\ 0 & 0 & 1 & 0 \end{pmatrix}. \tag{6}$$

The semantics of the C-NOT gate extends to the context of m-qubit systems with $m > 2$ in a manner similar to that of the one-qubit gates.

For basis states $|x\rangle|y\rangle$, the effect of the C-NOT gate is effectively the same as the classical two-bit gate that maps (x, y) to $(x, x \oplus y)$ (for all $x, y \in \{0, 1\}$). This gate negates the second bit conditional on the first bit being 1. For arbitrary quantum states, the behavior of this gate is more subtle. For example, although the classical gate never changes the value of its first "control" bit, the quantum gate sometimes does: applying the C-NOT gate to state $(\frac{1}{\sqrt{2}}|0\rangle - \frac{1}{\sqrt{2}}|1\rangle)(\frac{1}{\sqrt{2}}|0\rangle - \frac{1}{\sqrt{2}}|1\rangle)$ yields the state $(\frac{1}{\sqrt{2}}|0\rangle + \frac{1}{\sqrt{2}}|1\rangle)(\frac{1}{\sqrt{2}}|0\rangle - \frac{1}{\sqrt{2}}|1\rangle)$.

A more general kind of two-qubit gate is the CONTROLLED-U gate, where U is a 2×2 unitary matrix. This gate maps $|0\rangle|y\rangle$ to $|0\rangle|y\rangle$ and $|1\rangle|y\rangle$ to $|1\rangle(U|y\rangle)$ (for all $y \in \{0, 1\}$), and is denoted in Fig. 5.

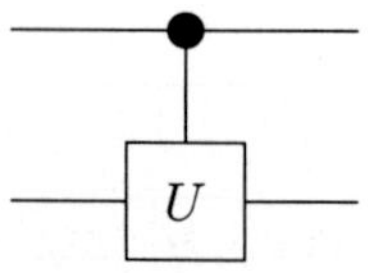

Figure 5: Notation for a CONTROLLED-U gate.

Note that the C-NOT gate is a special case of a CONTROLLED-U gate with

$$U = \begin{pmatrix} 0 & 1 \\ 1 & 0 \end{pmatrix} \tag{7}$$

(and this U itself is equivalent to a NOT gate).

Now, suppose that we want to compute the AND of two bits (i.e. take x_0 and x_1 as input and produce $x_0 \wedge x_1$ as output) using only the one- and two-qubit gates of the above form. This can be done in a manner that avoids irreversible operations via the quantum circuit in Fig. 6, where H is the Hadamard gate (Eq. 4) and

$$V = \begin{pmatrix} 1 & 0 \\ 0 & i \end{pmatrix} \tag{8}$$

(where $i = \sqrt{-1}$). For any $x_0, x_1, y \in \{0, 1\}$, setting the initial state of the qubits to $|x_0\rangle|x_1\rangle|y\rangle$ and tracing through the execution of this circuit reveals that the final state is $|x_0\rangle|x_1\rangle|y \oplus (x_0 \wedge x_1)\rangle$. Thus, when y is initialized to 0, the final state of the third qubit is $|x_0 \wedge x_1\rangle$ (and the explicit classical data,

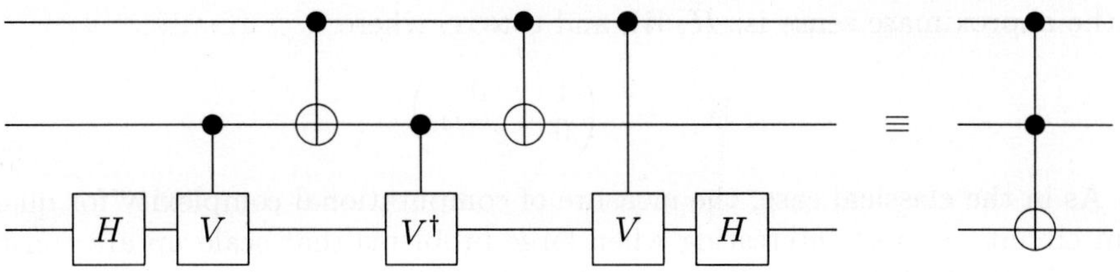

Figure 6: Quantum circuit simulating a C^2-NOT (Toffoli) gate.

$x_0 \wedge x_1$, can be extracted from this quantum state by a measurement). The three-qubit operation that is simulated in Fig. 6 is a so-called Toffoli gate (also called a CONTROLLED-CONTROLLED-NOT, or C^2-NOT for short). See [3, 22, 43] for some similar constructions.

For classical circuits, there are finite sets of gates that are universal in the sense that they can be used to simulate any other set of gates. For quantum circuits, the situation is different, since the set of all unitary operations is continuous, and hence uncountable—even when restricted to one-qubit gates. If one starts with any finite set of quantum gates then the set of all unitary operations that can implemented is limited to some countable subset of all the unitary operations. In spite of this, there are meaningful ways to capture the important features associated with universal sets of gates.

First, let us note that there are *infinite* sets consisting of one- and two-qubit of gates that are universal in the exact sense. For example, if the C-NOT gate as well as all unitary one-qubit gates are available then any k-qubit unitary operation can be simulated with $O(4^k k)$ such gates [3, 29]. Therefore, the overhead is constant when switching between different universal sets of local unitary gates (such as the set of all two-qubit gates and the set of all three-qubit gates).

Moreover, there are *finite* sets of one- and two-qubit gates that are universal in an *approximate* sense. For example, with the aforementioned one-qubit Hadamard gate H (Eq. 4) and the two-qubit CONTROLLED-V gate (where V is defined in Eq. 8), any two-qubit unitary operation can be simulated[1] within accuracy $\varepsilon > 0$ using $O(\log^d(1/\varepsilon))$ of these gates (for some constant d) [45]. The construction exploits the fact that the commutator of two unitary operators is not generally I (the identity operator), but it can converge very quickly to I. By *accuracy* ε, we mean with respect to the norm induced by the Euclidean distance between quantum state vectors. Thus, if U' approximates U within accuracy ε, and then U' is substituted for U in some quantum circuit, the final state $\sum_{x\in\{0,1\}^n} \alpha'_x |x\rangle$ approximates the final state of the original circuit $\sum_{x\in\{0,1\}^n} \alpha_x |x\rangle$ in the sense that $\sqrt{\sum_x |\alpha'_x - \alpha_x|^2} \leq \varepsilon$. This implies that if the final state is measured then the probability of any event among the possible outcomes is affected by at most ε. Another finite set of gates that is universal

[1]The simulation is up to a global phase factor, and such factors are irrelevant.

in the approximate sense is: H, W, and C-NOT, where

$$W = \begin{pmatrix} 1 & 0 \\ 0 & e^{i\pi/4} \end{pmatrix}. \tag{9}$$

As in the classical case, the measure of computational complexity for quantum circuits is most interesting when large problems that scale up are considered. Using sets of gates that are universal in the exact sense, computational complexity can vary only by constant factors. On the other hand, using sets of gates that are universal in the approximate sense, computational complexity can vary by at most polylogarithmic factors: any circuit with m gates can be simulated within accuracy ε by a circuit in terms of a different set of basic operations with $O(m \log^d(m/\varepsilon))$ gates. This is accomplished by simulating each of the m gates of the original circuit within accuracy ε/m, which results in a total accumulated error bounded by ε.

For example, computing the AND or the PARITY of n bits has quantum complexity $\Theta(n)$ in terms of the gates H, W, and C-NOT, and, with another set of gates, the complexity may be different, but it will remain between $\Omega(n)$ and $O(n \log^d(n/\varepsilon))$ (where d is some constant and ε is the accuracy level required).

Since it seems inconceivable that it would ever be possible to physically implement quantum gates with perfect accuracy, the need to ultimately work with approximations of quantum gates is inevitable. Fortunately, due the unitarity of quantum operations, inaccuracies only scale up linearly with the number of gates involved in a circuit. And, if one employs quantum error-correcting codes and fault-tolerant techniques then even gates with constant inaccuracies (and that are subject to "decoherence") can in principle be employed in arbitrarily large quantum circuits [1, 30, 41] (see [38] for an extensive review).

A convenient practice is to allow perfect universal sets of gates, bearing in mind that they can always be approximated using any finite set of gates that is universal in the approximate sense with only a polylogarithmic penalty in the circuit size (even if the implementations of these gates are approximate). It is also frequently convenient to disregard issues of uniformity, though, in principle, any legitimate quantum circuit family should be uniform (in the sense that it can be finitely specified in a computationally efficient way). Uniformity for quantum circuits can be defined as a straightforward extension of the uniformity definitions for classical circuit families, where the specification algorithm is *classical* and a finite set of gates that is universal in the approximate sense is used. All quantum algorithms proposed to date can be expressed as circuit families that are uniform in this sense.

Perhaps the most remarkable quantum algorithm that has been discovered to date is the factoring algorithm, due to Peter Shor [40].

Theorem 1 ([40]) *There exists a $O(n^2 \log n \log\log n \log(1/\varepsilon))$-size quantum circuit for the factoring problem, that errs with probability at most ε.*

Note that this circuit size is essentially exponentially smaller than the most efficient known classical probabilistic circuit for factoring (whose size is $O(2^{d\sqrt{n \log n}})$).

The quantum factoring algorithm actually arises from an algorithm for the order-finding problem, which in turn evolved from an algorithm in the query complexity model (explained in the next section).

The above result shows that, based on our current state of knowledge, quantum algorithms may be exponentially more efficient than classical algorithms for some problems. The next result shows that the gain in computational efficiency obtainable by quantum algorithms over classical algorithms can never exceed one exponential.

Theorem 2 *Any $S(n)$-qubit quantum circuit with $O(T(n))$ gates can be simulated by a classical circuit with $O(2^{S(n)}T(n)^3)$ gates.*

The idea behind the proof of Theorem 2 is to store the values of all $2^{S(n)}$ amplitudes associated with an $S(n)$-qubit quantum system in classical bits. Then these amplitudes are updated to reflect the effect of each of the $T(n)$ gates. It suffices store each amplitude with $O(T(n))$ bits of precision, which requires $O(2^{S(n)}T(n))$ bits in all. Since the effect of each quantum gate corresponds to multiplying the amplitude vector by a *sparse* $2^{S(n)} \times 2^{S(n)}$ matrix, this entails $O(2^{S(n)})$ arithmetic operations, which translates into $O(2^{S(n)}T(n)^2)$ bit operations per quantum gate. Thus, the total number of bit operations is $O(2^{S(n)}T(n)^3)$. A measurement step can also be simulated with $O(2^{S(n)}T(n)^2)$ classical gates, by first calculating the squares of the amplitudes and then sampling with respect to the appropriate probability distribution via $\not{c}$-gates.[2]

A more refined argument than the one above can be used to show that an $S(n)$-qubit circuit with $T(n)$ gates can be simulated using *space* that is polynomial in $S(n)$ and $T(n)$ (but still with an exponential number of operations), and there are also more esoteric computational models that subsume the power of quantum circuit families [24].

Regarding the circuit satisfiability problem, it is currently unknown whether or not there exists a polynomially-bounded quantum circuit family that solves it. What is known is that quantum algorithms can solve this problem quadratically faster than the best currently-known classical algorithms for this problem.

Theorem 3 *There exists a quantum circuit family of size $O(\sqrt{2^n}\log(1/\varepsilon)m^d)$ that solves the circuit satisfiability problem within accuracy ε (for some constant d). Here, n and m measure the size of the input instance: n is the number of inputs to circuit C and m is the number of gates of C.*

Note how this compares with the best currently-known classical circuit family for the circuit satisfiability problem, which has size $O(2^n m^d)$. Both quantities are exponential, but $\sqrt{2^n}$ is nevertheless considerably smaller than 2^n for large values of n. The quantum algorithm is a consequence of a remarkable algorithm in the query complexity model that was discovered by Lov Grover [26] (explained in the next section).

[2]The simulation is not exact, but is a good approximation: the error probability is exponentially small in $T(n)$.

2 Query complexity

This is an abstract model which can be thought of as a game, like "twenty questions". The goal is to determine some information by asking as few questions as possible. This differs from the computational complexity model in that the "input" is not presented as a binary string at the beginning of the computation. Rather, the input can be thought of as a "black box" computing a *function* $f : S \to T$, and the basic operations are *queries*, in which the algorithm specifies a t from the domain of the function and receives the value $f(t)$ in response.

A natural example is that of "polynomial interpolation", where f is an arbitrary polynomial of degree d

$$f(t) \quad = \quad c_0 + c_1 t + \cdots + c_d t^d \tag{10}$$

and the goal is to determine the coefficients $c_0, c_1, \ldots, c_d$. It is well known that $d+1$ queries to f are necessary and sufficient to accomplish this.

In the classical case, an algorithm in this model can be represented by a circuit consisting of gates from some standard universal set (e.g. $\wedge$, $\vee$, $\neg$), as well as additional gates that perform queries. For $f : S \to T$, an *f-query gate* takes $t \in S$ as input and produces $f(t)$ as output. In this scenario, since there are no input bits related to the problem instance (the problem instance is embodied in f), the inputs to the circuit are all set to constant values (such as 0).

In order to be able to adapt this model to settings involving quantum information, we slightly modify the form of the query gates so that they are reversible. For example, for $f : \{0,1\}^n \to \{0,1\}$, we define a *reversible f-query gate* as the mapping $\widetilde{f} : \{0,1\}^n \times \{0,1\} \to \{0,1\}^n \times \{0,1\}$, where $\widetilde{f}(x,y) = (x, y \oplus f(x))$ (for $x \in \{0,1\}^n$ and $y \in \{0,1\}$). Note that, for classical algorithms, reversible f-queries yield exactly the same information as the non-reversible kind. Any circuit that makes reversible f-queries can be converted into one that makes exactly the same number of non-reversible f-queries (and vice versa). Henceforth, all queries will be assumed to be in reversible form.

In the quantum case, an f-query is a unitary transformation that permutes the basis states according to the classical mapping determined by f (in reversible form). For example, for $f : \{0,1\}^n \to \{0,1\}$, an f-query gate is the unitary transformation that maps $|x\rangle\,|y\rangle$ to $|x\rangle\,|y \oplus f(x)\rangle$ (for all $x \in \{0,1\}^n$ and $y \in \{0,1\}$). One way of denoting f-queries in both classical and quantum circuits is shown in Fig. 7 (for the case where $f : \{0,1\}^2 \to \{0,1\}$).

Consider **Deutsch's problem** [20], where $f : \{0,1\} \to \{0,1\}$ and $f(t) = (c_0 + c_1 t) \bmod 2$, and the goal is to determine the value of c_1 (note that $c_1 = f(0) \oplus f(1)$). A classical circuit (in reversible form) that computes c_1 with two f-queries is shown in Fig. 8. The inputs to the circuit are both initialized to 0, and the unary $\oplus$ operation between the two f-queries is a NOT gate. It is easy to see that the final values of the two bits are 1 and c_1. It can also be shown that no classical algorithm exists that computes c_1 with a single f-query (since it is impossible to determine $f(0) \oplus f(1)$ from just $f(0)$ or $f(1)$ alone).

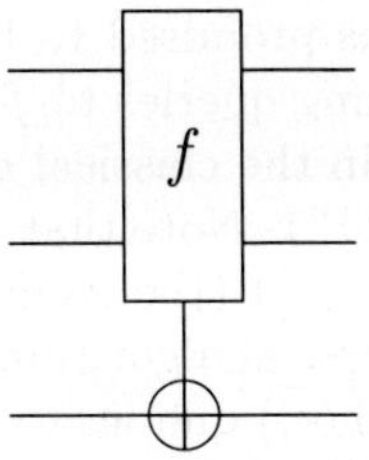

Figure 7: Notation for an f-query, when $f : \{0,1\}^2 \to \{0,1\}$.

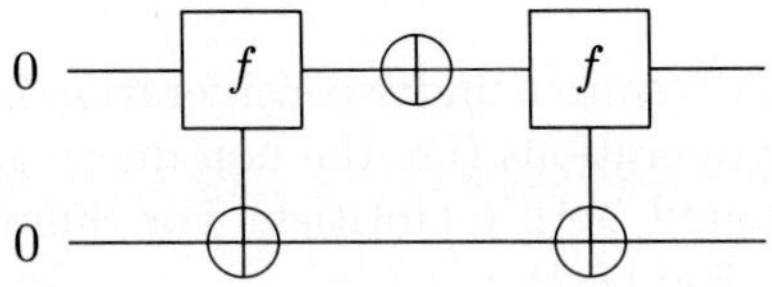

Figure 8: Classical circuit for Deutsch's problem using two queries.

But the quantum circuit in Fig. 9 [17, 20] computes c_1 with a single f-query gate. Here the initial state of the two-qubit system is $|0\rangle\,|1\rangle$ and its final state

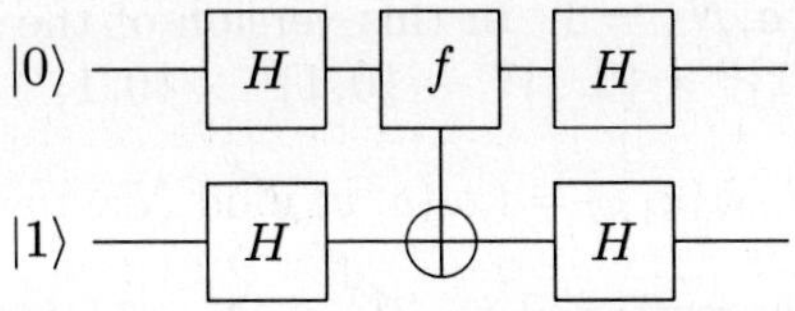

Figure 9: Quantum circuit for Deutsch's problem using one query.

is $(-1)^{c_0}\,|c_1\rangle\,|1\rangle$, which yields c_1 when the first qubit is measured.

Query complexity can be pinned down more precisely than computational complexity in that the "number of f-queries" is not sensitive to arbitrary technical conventions. So, it makes sense to consider the exact query complexity of a problem independent of linear factors, and to say that the classical query complexity of Deutsch's problem is two, whereas its quantum query complexity is one.

Although the above advantage is small, there are generalizations of Deutsch's problem for which the discrepancy between classical and quantum query complexity is much larger. One of these is **Simon's problem** [42], which is defined as follows. For a function $f : \{0,1\}^n \to \{0,1\}^n$, define $s \in \{0,1\}^n$ to be an *XOR-mask* of f if: $f(x) = f(y)$ if and only if $x \oplus y \in \{0^n, s\}$ (where $\oplus$ is defined over $\{0,1\}^n \times \{0,1\}^n$ bitwise). When $s = 0^n$, f is a bijection, and when $s \neq 0^n$, f is a two-to-one function with a special structure related to s. In Simon's problem, $f : \{0,1\}^n \to \{0,1\}^n$ is promised to have an XOR-mask $s \in \{0,1\}^n$, and the goal is to find s by making queries to f. In this case, an f-query is the mapping $(x,y) \mapsto (x, y \oplus f(x))$ in the classical case and $|x\rangle\,|y\rangle \mapsto |x\rangle\,|y \oplus f(x)\rangle$ in the quantum case ($x, y \in \{0,1\}^n$). Note that Deutsch's problem is the special case of Simon's problem where $n = 1$ (the XOR-mask is $\neg c_1$ in this case).

It can be proven that any classical algorithm in the query model for Simon's problem must make $\Omega(\sqrt{2^n \log(1/\varepsilon)})$ queries to f, even for probabilistic circuits with query gates that are permitted to err with probability up to ε. On the other hand, there is a simple quantum circuit that solves this problem with only $O(n \log(1/\varepsilon))$ queries to f [42]. (There is also a refinement of this algorithm [10] that makes a polynomial number of queries and solves Simon's problem *exactly.*)

Although the primary resource under consideration is the number of queries, the number of auxiliary operations (i.e. the non-query gates) is also of interest, and it is desirable to bound both quantities. For Simon's algorithm the total number of gates is $O(n^2 \log(1/\varepsilon))$.

Simon's problem demonstrates that, in the query complexity setting, there are quantum algorithms that are exponentially more efficient than any classical algorithm. Although the query complexity scenario is somewhat abstract, the significance of algorithms in this model will become apparent when the consequences of the next example are examined.

Consider the following version of the **order-finding problem** in the query complexity setting. Let N be an n-bit integer and $a \in \{1, \ldots, N-1\}$ be a number such that $\gcd(a, N) = 1$. In this version of the order-finding problem, the function $f_{a,N} : \{0,1\}^n \times \{0,1\}^n \to \{0,1\}^n \times \{0,1\}^n$ is defined as

$$f_{a,N}(x, y) = (x, (a^x y) \bmod N). \tag{11}$$

This is invertible if y is restricted to $\{0, \ldots, N-1\}$ (and can be extended to be invertible over its full domain by defining $f_{a,N}(x,y) = (x,y)$ for the case where $N \le y < 2^n$). The goal is to find the minimum $r \in \{1, \ldots, N-1\}$ such that $a^r \bmod N = 1$ by making queries to $f_{a,N}$ (in this case, $f_{a,N}$ is already in reversible form). Although there is no polynomially-bounded classical circuit that solves this problem, Shor [40] discovered a quantum circuit that solves it with probability $1-\varepsilon$ using only $O(\log(1/\varepsilon))$ queries to $f_{a,N}$ and $O(n^2 \log(1/\varepsilon))$ auxiliary gates.

A significant property of the function $f_{a,N}$ is that there exists a classical circuit of size $O(n^2 \log n \log\log n)$ that takes N (an n-bit number), $a \in \{1, \ldots, N-1\}$ (such that $\gcd(a, N) = 1$), and $x, y \in \{0,1\}^n$ as input, and produces $f_{a,N}(x,y)$ as output. In other words, given a and N, one can efficiently *simulate* an $f_{a,N}$-query gate. Moreover, this simulation can be implemented in terms of quantum gates, such as NOT, C-NOT, and C^2-NOT (using techniques for reversible classical computation [5]). By doing this simulation for each $f_{a,N}$-query gate in the quantum circuit for the order-finding problem, one obtains a quantum circuit of size $O(n^2 \log n \log\log n \log(1/\varepsilon))$ that takes a and N as input and produces the minimum positive r such that $a^r \bmod N = 1$ as output with probability $1 - \varepsilon$. Thus, the algorithm in the query complexity model yields an algorithm in the computational complexity model for order-finding— and hence also for factoring. This is a specific instance of the following general result relating algorithms in the query complexity model to algorithms in the computational complexity model.

Theorem 4 *Suppose that a function $f_z : \{0,1\}^m \to \{0,1\}^k$ is associated with each $z \in \{0,1\}^n$ (where m and k are functions of z), and that the classical computational complexity of the function that maps (z, x) to $f_z(x)$ is bounded above by $R(n)$. Suppose also that there is a problem in the query complexity model where some property $P(f_z)$ is to be determined in terms of f_z-queries, and that there is a quantum circuit that solves this problem using $S(n)$ queries to f_z and $T(n)$ auxiliary operations. Then the quantum computational complexity of the problem where the input is $z \in \{0,1\}^n$ and the output is the value of the property $P(f_z)$ is $O(R(n)S(n) + T(n))$.*

The circuit for the computational complexity problem is merely the circuit for the query complexity problem with a circuit simulating each f_z-query gate substituted for that f_z-query gate.

Let us now consider the **search problem** [26] in the query complexity model, where $f : \{0,1\}^n \to \{0,1\}$, and the goal is to find an $x \in \{0,1\}^n$ such that $f(x) = 1$ (or to indicate that no such x exists). Any classical algorithm for this problem must make $\Omega(2^n)$ f-queries, even if it is allowed to err with probability (say) $\frac{1}{3}$. Lov Grover [26] discovered a remarkable quantum algorithm that accomplishes this with $O(\sqrt{2^n})$ queries. Grover's result, with some later refinements [8, 9, 13, 34, 50] incorporated into it, is summarized as follows.

Theorem 5 ([26]) *There is a quantum algorithm that solves the search problem for $f : \{0,1\}^n \to \{0,1\}$ with $O(\sqrt{2^n \log(1/\varepsilon)})$ queries to f, and errs with probability at most ε.*

The efficiency of the above algorithm has been shown to be optimal [6, 8, 13, 49].

Clearly, Grover's algorithm can solve an existential version of the search problem, where the goal is just to determine whether or not there *exists* an $x \in \{0,1\}^n$ such that $f(x) = 1$ (a problem that also requires $\Omega(2^n)$ queries in the classical case). Note the similarity between this existential search problem and the circuit satisfiability problem. In fact, using Theorem 4, this algorithm in the query model translates into the algorithm for the circuit satisfiability problem that is claimed in Theorem 3. The input is $e(C)$, an encoding of a circuit C with m gates and n inputs that computes a mapping $C : \{0,1\}^n \to \{0,1\}$, and the output should be 1 if there exists an $x \in \{0,1\}^n$ such that $C(x) = 1$, and 0 otherwise. The mapping that takes $(e(C), x)$ to $C(x)$ can be computed by a classical circuit with $O(m^d)$ gates (where d is a constant that depends on the encoding scheme, and is usually small). Also, the algorithm in Theorem 5 makes $O(\sqrt{2^n \log(1/\varepsilon)}\, n)$ auxiliary operations. Therefore, applying Theorem 4, one obtains a quantum circuit of size $O(\sqrt{2^n \log(1/\varepsilon)}\, m^d)$ for the circuit satisfiability problem.

Let us now consider some variations and extensions of the existential search problem in the query model. We shall henceforth refer to the existential search problem as *OR*, defined as

$$OR(f) \quad = \quad (\exists x) f(x), \tag{12}$$

where $f : \{0,1\}^n \to \{0,1\}$ is accessed through f-queries. The name *OR* seems natural since

$$OR(f) \quad = \quad f(00\cdots 0) \vee f(00\cdots 1) \vee \cdots \vee f(11\cdots 1). \tag{13}$$

Note that the complementary problem $AND(f) = (\forall x) f(x)$ has computational complexity similar to that of *OR*, since $(\forall x) f(x) = \neg(\exists x)\neg f(x)$.

A non-trivial extension of *OR* and *AND* is the problem *OR-AND*, where there are two alternating quantifiers:

$$OR\text{-}AND(f) \quad = \quad (\exists x_1)(\forall x_2) f(x_1, x_2). \tag{14}$$

Here $f : \{0,1\}^{n_1} \times \{0,1\}^{n_2} \to \{0,1\}$, and n_1, n_2 are implicit parameters satisfying $n_1 + n_2 = n$. By a suitable recursive application of Grover's algorithm for *OR* [11], this problem can be solved with $O(\sqrt{2^n}\, n \log(1/\varepsilon))$ queries to f (the extra factor of $\sqrt{n}$ is to amplify the accuracy of the bottom level algorithm for *AND*).

In fact, one can extend the above to k alternations of quantifiers:

$$OR\text{-}AND\text{-}\cdots\text{-}Q(f) \quad = \quad (\exists x_1)(\forall x_2)\cdots(\mathrm{Q} x_k) f(x_1, x_2, \ldots, x_k) \tag{15}$$

where $Q \in \{OR, AND\}$ and $\mathrm{Q} \in \{\exists, \forall\}$, depending on whether k is even or odd, and $f : \{0,1\}^{n_1} \times \cdots \times \{0,1\}^{n_k} \to \{0,1\}$ with $n_1 + \cdots n_k = n$. The recursive application of Grover's technique in [11] also extends to k alternations with $O(\sqrt{2^n}\, n^{k-1} \log(1/\varepsilon))$ queries to f (see [36] for a related result).

For all of these variations of *OR* and *AND*, it can be shown that any classical algorithm for one of these problems must make $\Omega(2^n)$ queries, and the quantum algorithms for these problems are all nearly quadratically more efficient than this in the sense that they make $O((2^n)^{1/2+\delta})$ queries, for any $\delta > 0$ and $\varepsilon > 0$. In fact, even if k, the number of alternations of *OR* and *AND*, is set to $\delta n / 2 \log n$ (instead of being held constant), the quantum algorithms make $O((2^n)^{1/2+\delta})$ queries. All of these quantum algorithms also have counterparts for the corresponding problems in the computational model, where the function is specified by an encoding $e(C)$ of a circuit C.

Another problem that has a similar flavor to these problems is

$$PARITY(f) \quad = \quad \left(\sum_{x \in \{0,1\}^m} f(x) \right) \bmod 2. \tag{16}$$

It can be shown that any classical algorithm requires $\Omega(2^n)$ queries to solve *PARITY*, and it is natural to ask whether quantum algorithms can be quadratically more efficient—or even $O((2^n)^r)$, for some $r < 1$. One of the applications of the communication complexity model (explained in the next section) is to show that this is not possible: at least $\Omega(2^n/n)$ queries must be made by any quantum algorithm. In fact, a stronger lower bound of $\frac{1}{2}2^n$ is also known [4, 23] (using different methods).

It is important to note that, although upper bounds in the query model translate into upper bounds in the computational model, the converse of this need not be true: it is conceivable that there is a polynomially-bounded circuit that solves the **computational parity problem**, where the input is $e(C)$, an encoding of a circuit C that computes a function f and the output is $PARITY(f)$.

3 Communication complexity

In this model, there are two parties, traditionally referred to as Alice and Bob, who each receive an n-bit binary string as input ($x = x_0x_1 \ldots x_{n-1}$ for Alice and $y = y_0y_1 \ldots y_{n-1}$ for Bob) and the goal is for them to determine the value of some function of the of these $2n$ bits. The resource under consideration here is the *communication* between the two parties, and an algorithm is a *protocol*, where the parties send information to each other (possibly in both directions and over several rounds) until one of them (say, Bob) obtains the answer. This model was introduced by Yao [47] and has been widely studied in the classical context (see [32] for a survey).

An interesting example is the *equality* function EQ, defined as

$$EQ(x, y) = \begin{cases} 1 & \text{if } x = y \\ 0 & \text{if } x \neq y. \end{cases}$$

A simple n-bit protocol for EQ is for Alice to just send her bits $x_0, \ldots, x_{n-1}$ to Bob, after which Bob can evaluate the function by himself (in fact, there is a similar n-bit protocol for *any* function). The interesting question is whether or not the EQ function can be evaluated with fewer than n bits of communication—after all, the goal here is only for Bob to acquire one bit. The answer depends on whether or not any error probability is permitted.

If Bob must acquire the value of $EQ(x, y)$ with certainty then it turns out that n bits of communication are necessary. Note that Alice sending the first $n - 1$ bits of x will clearly *not* work, since the answer could critically depend on whether or not $x_{n-1} = y_{n-1}$. The number of possible protocols to consider is quite large and an actual *proof* that n bits communication are necessary is nontrivial. Such a proof uses methodologies that are beyond the scope of this paper. The interested reader is referred to [32] for a proof.

On the other hand, for probabilistic protocols (where Alice and Bob can flip coins and base their behavior on the outcomes), if an error probability of $\varepsilon > 0$ is permitted then $O(\log(n)\log(1/\varepsilon))$ bits of communication are sufficient. As usual, we are not assuming anything about a probability distribution on the input strings; the error probability is with respect to the random choices made by Alice and Bob, and it applies regardless of what x and y are.

We now describe an $O(\log(n)\log(1/\varepsilon))$-bit protocol for *EQ*. First of all, Alice and Bob agree on a finite field whose size is between $2n$ and $4n$ (such a field always exists, and its elements can be represented as $O(\log(n))$-bit strings). Now, consider the two polynomials

$$\begin{aligned} p_x(t) &= x_0 + x_1 t + \cdots + x_{n-1} t^{n-1} \quad (17) \\ p_y(t) &= y_0 + y_1 t + \cdots + y_{n-1} t^{n-1}. \quad (18) \end{aligned}$$

For any value of t in the field, Alice can evaluate $p_x(t)$ and Bob can evaluate $p_y(t)$. If $x = y$ then the two polynomials are identical, so $p_x(t) = p_y(t)$ for any value of t. But, if $x \neq y$ then, since $p_x(t)$ and $p_y(t)$ are polynomials of degree $n-1$, there can be at most $n-1$ distinct values of t for which $p_x(t) = p_y(t)$. Therefore, if a value of t is chosen randomly from the field then the probability that $p_x(t) = p_y(t)$ is at most $\frac{1}{2}$. Now, the protocol proceeds as follows. Alice chooses $k = \log(1/\varepsilon)$ independent random elements of the field, $t_1, \ldots, t_k$, and then sends $t_1, \ldots, t_k$ and $p_x(t_1), \ldots, p_x(t_k)$ to Bob (this consists of $O(\log(n)\log(1/\varepsilon))$ bits). Then Bob outputs 1 if and only if $p_x(t_i) = p_y(t_i)$ for all $i \in \{1, \ldots, k\}$. The probability that Bob erroneously outputs 1 when $x \neq y$ is at most $1/2^k = \varepsilon$.

Two other interesting communication complexity problems are the *intersection* function

$$IN(x,y) = (x_0 \wedge y_0) \vee (x_1 \wedge y_1) \vee \cdots \vee (x_{n-1} \wedge y_{n-1}) \quad (19)$$

and the *inner product* function

$$IP(x,y) = (x_0 \wedge y_0) \oplus (x_1 \wedge y_1) \oplus \cdots \oplus (x_{n-1} \wedge y_{n-1}). \quad (20)$$

Intuitively, for *IN*, the inputs x and y can be thought as encodings of two subsets of $\{0, \ldots, n-1\}$ and the output is a bit indicating whether or not they intersect. Also, *IP* is the inner product of x and y as bit vectors in modulo two arithmetic. The deterministic communication complexity of each of these problems is the same as that of *EQ*: any deterministic protocol requires n bits of communication. Also, it has been shown that both of these problems are more difficult than *EQ* when probabilistic protocols are considered: any probabilistic protocol with error probability up to (say) $\frac{1}{3}$ requires $\Omega(n)$ bits of communication (see [14] for *IP*, and [28] for *IN*; [32]).

It is natural to ask whether any reduction in communication can be obtained by somehow using *quantum* information. Define a *quantum* communication protocol as one where Alice and Bob can exchange messages that consist of qubits. In a more formal definition of this model, there is an *a priori* system of m qubits, some of them in Alice's possession and some of them in Bob's possession. The initial state of all of these qubits can be assumed to be $|0\rangle$, and Alice and Bob can each perform unitary transformations on those qubits that are in their possession and they can also send qubits between themselves (thereby changing the ownership of qubits). The output is then taken as the outcome of some measurement of Bob's qubits. Various preliminary results about communication complexity with quantum information occurred in [12, 15, 19, 31, 48].

There are fundamental results in quantum information theory which imply that classical information cannot be "compressed" within quantum information [27]. For example, Alice cannot convey more than r classical bits of information to Bob by sending him an r-qubit message. Based on this, one might mistakenly think that there is no advantage to using quantum information in the communication complexity context. In fact, there exists a quantum communication protocol that solves *IN* whose qubit communication is approximately the square root of the bit communication of the best possible classical probabilistic protocol.

Theorem 6 ([11]) *There exists a quantum protocol for the intersection function (IN) that uses* $O(\sqrt{n\log(1/\varepsilon)}\log(n))$ *qubits of communication and errs with probability at most* ε.

Moreover, the quantum protocol can be adapted to actually find a point in the intersection in the cases where $IN(x, y) = 1$. That is, to produce an $i \in \{0, \ldots, n-1\}$ such that $x_i \wedge y_i = 1$. This problem, like *IN*, has classical probabilistic communication complexity $\Omega(n)$.

To understand the protocol in Theorem 6, it is helpful to think of the inputs x and y as functions rather than strings, and we introduce some notation that makes this explicit. For convenience, assume that $n = 2^k$ for some k (if not then x and y can lengthened by padding them with zeroes), and define the functions $f_x, f_y : \{0,1\}^k \rightarrow \{0,1\}$ as

$$f_x(i) = x_i \tag{21}$$

$$f_y(i) = y_i \tag{22}$$

where $\{0,1\}^k$ and $\{0, 1, \ldots, 2^k - 1\}$ are identified in the natural way. Alice and Bob's input data can be thought of as f_x and f_y, rather than x and y (respectively). In particular, given x, Alice can simulate an f_x-query that maps $|i\rangle\,|j\rangle$ to $|i\rangle\,|j \oplus f_x(i)\rangle$ (for all $i \in \{0,1\}^k$ and $j \in \{0,1\}$), and Bob can simulate f_y-queries. (Although the resource that is of interest in this model is not the number of basic operations that Alice and Bob perform, it is worth noting that, Alice and Bob's simulations of these queries can be explicitly implemented by reversible circuits with $O(2^k k) = O(n\log(n))$ basic operations).

To construct an efficient quantum protocol for *IN*, define the function $f_x \wedge f_y : \{0,1\}^k \rightarrow \{0,1\}$ as $(f_x \wedge f_y)(i) = f_x(i) \wedge f_y(i)$ (for $i \in \{0,1\}^k$), and note that $IN(x, y) = OR(f_x \wedge f_y)$. Therefore, if Alice and Bob can somehow perform $(f_x \wedge f_y)$-queries then the value of $IN(x, y)$ can be determined by making $O(\sqrt{2^k\log(1/\varepsilon)}) = O(\sqrt{n\log(1/\varepsilon)})$ such queries. The problem is that neither Alice nor Bob individually have enough information to perform an $(f_x \wedge f_y)$-query (since this depends on both x and y). If Alice were to begin by sending x to Bob then Bob could make $(f_x \wedge f_y)$-queries on his own, but note that this entails n bits of communication to begin with. Another, more efficient, approach is for Alice and Bob to collectively simulate $(f_x \wedge f_y)$-queries by combining f_x-queries (which Alice can perform) with f_y-queries (which Bob can perform), and a small amount of communication. To see how this is accomplished, consider the circuit in Fig. 10.

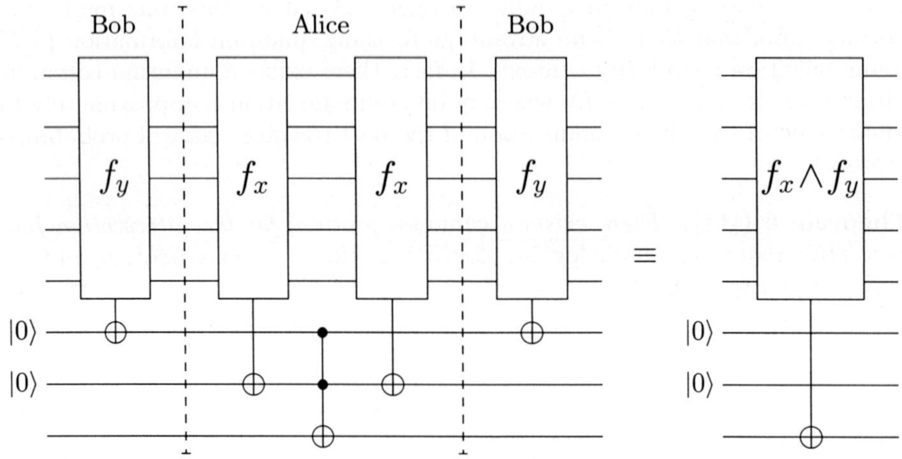

Figure 10: Simulation of an $(f_x \wedge f_y)$-query in terms of f_x-queries and f_y-queries.

First, ignoring the broken vertical lines, note that the quantum circuit (composed of two f_x-queries, two f_y-queries, and one Toffoli gate) is equivalent to an $(f_x \wedge f_y)$-query. That is, it implements the unitary transformation that maps the state $|i\rangle\,|0\rangle\,|0\rangle\,|j\rangle$ to the state $|i\rangle\,|0\rangle\,|0\rangle\,|j \oplus (f_x \wedge f_y)(i)\rangle$ (for all $i \in \{0,1\}^k$, $j \in \{0,1\}$). This circuit uses two extra qubits that are each initialized in state $|0\rangle$ and which incur no net change.

Now, the protocol for *IN* can be thought of as Bob executing the algorithm in the query model for *OR* with the function $f_x \wedge f_y$, except that, whenever an $(f_x \wedge f_y)$-query gate arises, he interacts with Alice to simulate the circuit in Fig. 10: first Bob performs an f_y-query gate, then he sends the $k+3$ qubits to Alice who performs some actions involving f_x-queries and a Toffoli gate (shown between the two broken lines) and sends the qubits back to Bob, who performs another f_y-query. Note that the total amount of communication that this entails is $2(k+3) \in O(\log n)$ qubits. Therefore, the total communication for Bob's simulation of the $O(\sqrt{n \log(1/\varepsilon)})$ queries to $(f_x \wedge f_y)$ is $O(\sqrt{n \log(1/\varepsilon)}\,\log(n))$, as claimed in Theorem 6.

More recently, Ran Raz has given an example of a communication complexity problem which a quantum protocol can solve with *exponentially* less communication than the best classical probabilistic protocol. The description of the problem is more complicated than *EQ*, *IN*, and *IP*, and the reader is referred to [39] for the details.

The methodology used to establish Theorem 6 involved the conversion of an algorithm in the query model (for OR) to a communication protocol (for $IN(x,y) = OR(f_x \wedge f_y)$). This conversion can be stated in a more general form.

Theorem 7 ([11]) *Suppose that there is a quantum algorithm in the query model that computes $P(f)$ in terms of $T(k,\varepsilon)$ queries to f, where $f : \{0,1\}^k \rightarrow \{0,1\}$, and ε is a bound on the error probability. For $n = 2^k$, define the communication problem $P_\wedge : \{0,1\}^n \times \{0,1\}^n \rightarrow \{0,1\}$ as $P_\wedge(x,y) = P(f_x \wedge f_y)$. Then there is a quantum protocol that solves $P_\wedge$ with $O(T(\log(n),\varepsilon)\log(n))$ qubits of communication. And a similar result holds for $P_\vee(x,y) = P(f_x \vee f_y)$ and $P_\oplus(x,y) = P(f_x \oplus f_y)$.*

We conclude with a discussion of the quantum communication complexity of the inner product function IP. It has been shown [31] (see also [16]) that even quantum protocols require communication $\Omega(n)$ for this problem, even when the error probability is permitted to be as large as (say) $\frac{1}{3}$. This fact, combined with Theorem 7 applied in its contrapositive form, can be used to establish a lower bound for a problem in the query model: $PARITY$ (defined in Eq. 16). The main observation is that $IP(x,y) = PARITY(f_x \wedge f_y)$. Suppose that there is a quantum algorithm that computes $PARITY(f)$ for $f : \{0,1\}^k \rightarrow \{0,1\}$ by making $T(k)$ f-queries (assume that the error probability is bounded by $\frac{1}{3}$). Then, by Theorem 7, there exists a quantum protocol that solves IP with $O(T(k)k)$ qubits of communication, where $n = 2^k$ is the size of the input instance to IP. Since there is a lower bound of $\Omega(n) = \Omega(2^k)$ for the communication complexity of IP, we must have $T(k)k \in \Omega(2^k)$, which implies that $T(k) \in \Omega(2^k/k)$. This is an easy way to get a "ball park" lower bound for the query complexity of $PARITY$, whose exact value is known to be $\frac{1}{2}2^k$ by other methods [4, 23].

Acknowledgments

I would like to thank Michele Mosca for providing some references to work in computational number theory. This research was partially supported by the Natural Sciences and Engineering Research Council of Canada.

References

[1] D. Aharonov and M. Ben-Or, "Fault-tolerant quantum computation with constant error", *Proc. 29th Ann. ACM Symp. on Theory of Computing (STOC '97)*, pp. 176–188, 1997.

[2] L. Adleman, C. Pomerance, R. Rumely, "On distinguishing prime numbers from composite numbers", *Annals of Mathematics*, Vol. 117, pp. 173–206, 1983.

[3] A. Barenco, C.H. Bennett, R. Cleve, D.P. DiVincenzo, N. Margolus, P. Shor, T. Sleator, J. Smolin, and H. Weinfurter, "Elementary gates for quantum computation", *Phys. Rev. A*, Vol. 52, pp. 3457–3467, 1995.

[4] R. Beals, H. Buhrman, R. Cleve, M. Mosca, R. de Wolf, "Quantum lower bounds by polynomials", *Proc. 39th Ann. IEEE Symp. on Foundations of Computer Science (FOCS '98)*, pp. 352–361, 1998.

[5] C.H. Bennett, "Logical reversibility of computation", *IBM J. of Research and Development*, Vol. 17, pp. 525–532, 1973.

[6] C.H. Bennett, E. Bernstein, G. Brassard, and U. Vazirani, "Strengths and weaknesses of quantum computing", *SIAM J. on Computing*, Vol. 26, No. 5, pp. 1510–1523, 1997.

[7] E. Bernstein and U.V. Vazirani, "Quantum complexity theory", *SIAM J. on Comput.*, Vol. 26, No. 5, pp. 1411–1473, 1997.

[8] M. Boyer, G. Brassard, P. Høyer, and A. Tapp, "Tight bounds on quantum searching", *Fortschritte der Physik*, Vol. 46, pp. 493–505, 1998. (An earlier version appeared in *Physcomp '96*.)

[9] G. Brassard, P. Høyer, and A. Tapp, "Quantum counting", *Proc. 25th ICALP*, Vol. 1443 of *Lecture Notes in Computer Science*, pp. 820–831 (Springer), 1998.

[10] G. Brassard and P. Høyer, "An exact quantum polynomial-time algorithm for Simon's problem", *Proc. 5th Israeli Symp. on Theory of Computing and Systems (ISTCS '97)*, pp. 12–23, 1997.

[11] H. Buhrman, R. Cleve, and A. Wigderson, "Quantum vs. classical communication and computation", *Proc. 30th Ann. ACM Symp. on Theory of Computing (STOC '98)*, pp. 63-68, 1998.

[12] H. Buhrman, R. Cleve, and W. van Dam, "Quantum entanglement and communication complexity", preprint quant-ph/9705033, 1997.

[13] H. Buhrman, R. de Wolf, "Lower bounds for quantum search and derandomization", preprint quant-ph/9811046, 1998.

[14] B. Chor and O. Goldreich, "Unbiased bits from sources of weak randomness and probabilistic communication complexity", *SIAM J. Comput.*, Vol. 17, No. 2, pp. 230–261, 1988.

[15] R. Cleve and H. Buhrman, "Substituting quantum entanglement for communication", *Phys. Rev. A*, Vol. 56, No. 2, pp. 1201–1204, 1997.

[16] R. Cleve, W. van Dam, M. Nielsen, and A. Tapp, "Quantum entanglement and the communication complexity of the inner product function", to appear in *Proc. 1st NASA Intl. Conf. on Quantum Computing and Quantum Communications*, Vol. 1509 of *Lecture Notes in Computer Science* (Springer), 1998. preprint quant-ph/9708019, 1997.

[17] R. Cleve, A. Ekert, C. Macchiavello, and M. Mosca, "Quantum algorithms revisited", *Proc. of the Royal Society of London*, Vol. A454, pp. 339–354, 1998.

[18] S.A. Cook, "The complexity of theorem proving procedures", *Proc. 3rd Ann. ACM Symp. on Theory of Computing (STOC '71)*, pp. 151–158, 1971.

[19] W. van Dam, P. Høyer, A. Tapp, "Multiparty quantum communication complexity", preprint quant-ph/9710054, 1997.

[20] D. Deutsch, "Quantum theory, the Church-Turing principle and the universal quantum computer", *Proc. of the Royal Society of London*, Vol. A400, pp. 96–117, 1985.

[21] D. Deutsch, "Quantum computational networks", *Proc. of the Royal Society of London*, Vol. A425, pp. 73–90, 1989.

[22] D.P. DiVincenzo, "Quantum gates and circuits", *Proc. of the Royal Society of London*, Vol. A454, pp. 261–276, 1998.

[23] E. Farhi, J. Goldstone, S. Gutmann, and M. Sipser, "A limit on the speed of quantum computation in determining parity." *Phys. Rev. Lett.*, Vol. 81, pp. 5442–5444, 1998.

[24] L. Fortnow and J. Rogers, "Complexity limitations on quantum computation", *Proc. 13th IEEE Conf. on Computational Complexity*, pp. 202–209, 1998.

[25] M.R. Garey and D.S. Johnson, *Computers and Intractibility: A Guide to the Theory of NP-Completeness*, W.H. Freeman, 1979.

[26] L.K. Grover, "A fast quantum mechanical algorithm for database search", *Proc. 28th Ann. ACM Symp. on Theory of Computing (STOC '96)*, pp. 212–219, 1996.

[27] A.S. Holevo, "Some estimates of the information transmitted by quantum communication channels", *Problemy Peredachi Informatsii*, Vol. 9, pp. 3–11, 1973. English translation in *Problems of Information Transmission (USSR)*, Vol. 9, pp. 177–183, 1973.

[28] B. Kalyanasundaram and G. Schnitger, "The probabilistic communication complexity of set intersection", *Proc. 2nd Conf. on Structure in Complexity Theory*, pp. 41–49, 1987.

[29] E. Knill, personal communication, 1996.

[30] E. Knill, R. Laflamme, W. Zurek, "Threshold accuracy for quantum computation", preprint quant-ph/9610011, 1996.

[31] I. Kremer, *Quantum Communication*, MSc Thesis, Computer Science Department, The Hebrew University, 1995.

[32] E. Kushilevitz and N. Nisan, *Communication Complexity*, (Cambridge University Press), 1998.

[33] H. Lenstra and C. Pomerance, "A rigorous time bound for factoring integers", *J. of the AMS*, Vol. 5, No. 2, pp. 482–516, 1992.

[34] M. Mosca, "Quantum searching, counting and amplitude amplification by eigenvector analysis", *MFCS '98 workshop on Randomized Algorithms*, 1998.

[35] M. Mosca and A. Ekert, "The hidden subgroup problem and eigenvalue estimation on a quantum computer", to appear in *Proc. 1st NASA Intl. Conf. on Quantum Computing and Quantum Communications*, Vol. 1509 of *Lecture Notes in Computer Science* (Springer), 1998. preprint quant-ph/9903071, 1998.

[36] Y. Ozhigov, "Fast quantum verification for the formulas of predicate calculus", preprint quant-ph/9809015, 1998.

[37] C. Pomerance, "Fast rigorous factorization and discrete logarithm algorithms", *Discrete Algorithms and Complexity (Proc. Japan-US Joint Seminar on Discrete Algorithms and Complexity theory)*, pp. 119–143, 1987.

[38] J. Preskill, "Fault-tolerant quantum computation", in *Introduction to Quantum Computation and Information*, edited by H.-K. Lo, S. Popescu, and T.P. Spiller (World Scientific), pp. 213–269, 1998.

[39] R. Raz, "Exponential separation of quantum and classical communication complexity", To appear in *Proc. 31st Ann. ACM Symp. on Theory of Computing (STOC '99)*, 1999.

[40] P.W. Shor, "Polynomial-time algorithms for prime factorization and discrete logarithms on a quantum computer", *SIAM J. on Computing*, Vol. 26, No. 5, pp. 1484–1509, 1997. (An earlier version appeared in *FOCS '94*.)

[41] P.W. Shor, "Fault-tolerant quantum computation", *Proc. 37th Ann. IEEE Symp. on Foundations of Computer Science (FOCS '96)*, pp. 56-65, 1996.

[42] D. Simon, "On the power of quantum computation", *SIAM J. on Computing*, Vol. 26, No. 5, pp. 1474–1483, 1997. (An earlier version appeared in *FOCS '94*.)

[43] T. Sleator and H. Weinfurter, "Realizable universal quantum logic gates", *Phys. Rev. Lett.*, Vol. 74, pp. 4087–4090, 1995.

[44] R. Solovay and V. Strassen, "A fast Monte Carlo test for primality", *SIAM J. on Computing*, Vol. 6, pp. 84–85, 1977. (Erratum in Vol. 7, p. 118, 1978.)

[45] R. Solovay, Lecture at the Institute for Theoretical Physics, Santa Barbara, U.S.A., 1996.

[46] U.V. Vazirani, "Strong communication complexity or generating quasi-random sequences from two communicating slightly-random sources", *Combinatorica*, Vol. 7, No. 4, pp. 375–392, 1987.

[47] A. C.-C. Yao, "Some complexity questions related to distributive computing", *Proc. 11th Ann. ACM Symp. on Theory of Computing (STOC '79)*, pp. 209-213, 1979.

[48] A. C.-C. Yao, "Quantum circuit complexity", *Proc. 34th Ann. IEEE Symp. on Foundations of Computer Science (FOCS '93)*, pp. 352–361, 1993.

[49] C. Zalka, "Grover's quantum searching algorithm is optimal", preprint quant-ph/9711070, 1997.

[50] C. Zalka, "A Grover-based quantum search of optimal order for an unknown number of marked elements", preprint quant-ph/9902049, 1999.

Quantum Error Correction

Quantum Error Correction

David P. DiVincenzo
IBM Research Division, T. J. Watson Research Center

Quantum error correction (QEC) prescribes a set of protocols that permit the coherence of quantum states to be protected from the decohering effects of the quantum environment. QEC can be used to make reliable quantum computation possible with imperfect apparatus, and to permit long-distance transmission of quantum states and of secure quantum cryptographic keys [see the contribution of Briegel to this volume].

During the early years of the development of quantum computing, there was considerable doubt that quantum error correction was possible at all[1]. But this doubt was dispelled by the discoveries of Shor[2] and then Steane[3]. This work introduced two crucial items: a workable description of the quantum noise process, and the first simple quantum error-correcting codes. Following the parallel developments in the theory of entangled mixed states [4, 5], it was noted that, if an independent environment is coupled to each qubit of the system and the correlation time of the environment is short, then any arbitrary effect of that environment is captured by a single universal error model, in which an ensemble of four discrete error operations can occur on the qubit: nothing, a bit flip (represented by the unitary operation of the σ_x Pauli operator), a π phase change (σ_z), or both (σ_y).

The identification of this universal, discrete error model permitted the reasoning of ordinary error correction codes to be applied to the quantum problem: the task became to devise a quantum state with enough redundancy that an incomplete quantum measurement on the state could reveal the identity of the error without disturbing the coded state. Shor[2] identified a two-dimensional subspace of the state of 9 qubits for which this is possible: that is, he found a 1-qubit-in-9 code which corrected all single-qubit errors. Steane[3] found a 1-in-7 code, other workers[6, 5] found a 1-in-5 code. At the same time, mathematical conditions for a general code were formulated[5, 7]; a large subset of the quantum codes, the "stabilizer codes," could be obtained immediately from these conditions and a knowledge of ordinary bit error correction codes[8, 9] and classical codes over 4-state bits[10]. A large family of random codes, and methods of improving error-correction coding for communication by exploiting classical communication between the parties, emerged from studies of entanglement purification[5].

These developments immediately raised the question of whether quantum computations could be performed on coded states, and whether the process of error correction could be performed successfully even with less than 100% fidelity quantum measurements and quantum gates. Shor[11] introduced a set of techniques for achieving fault tolerant quantum computation, and several other workers[12] have analyzed the application of these techniques thoroughly, showing that so long as the fidelity of quantum operations exceeds some finite value, efficient, reliable quantum computation can be achieved.

Finally, I should mention that some work has been done on the problem of error correction in the face of correlated errors. Some results have been obtained in the opposite limit, where the errors are completely *correlated* across a set of qubits, because they arise from coupling with a common environmental mode. Zanardi and Rasetti [13] showed that in the limit of completely correlated noise of this type, certain entangled quantum states (in the generalized singlet sector) do not have to be error corrected, because they are completely unaffected by this kind of error. Lidar and coworkers [14] have given further analysis of this situation. Finally, Viola *et al.* [15] have looked at a somewhat different case, in which the bath is *non-Markovian*, that is, the bath correlation time is longer than the clock-cycle time of the quantum computer. They have found that in this regime other approaches, involving the generalization of "refocusing" techniques of spin spectroscopies, become capable of warding off the effects of noise.

I thank the Army Research Office for support under contract DAAG55-98-C-0041.

References

[1] W. G. Unruh, Phys. Rev. A **51**, 992 (1995).

[2] P. W. Shor, Phys. Rev. A **52**, 2493 (1995).

[3] A. M. Steane, Phys. Rev. Lett. **77**, 793 (1996); Proc. R. Soc. Lond. **A452**, 2551 (1996). Phys. Rev. A **54**, 4741 (1996).

[4] C. H. Bennett, G. Brassard, S. Popescu, B. Schumacher, J. A. Smolin, and W. K. Wootters, Phys. Rev. Lett. **76**, 722 (1996).

[5] C. H. Bennett, D. P. DiVincenzo, J. A. Smolin, and W. K. Wootters, Phys. Rev. A **54**, 3825 (1996).

[6] R. Laflamme, C. Miquel, J. P. Paz, and W. H. Zurek, Phys. Rev. Lett. **77**, 198 (1996).

[7] E. Knill and R. Laflamme, Phys. Rev. A **55**, 900 (1997).

[8] D. Gottesman, Phys. Rev. A **54**, 1862 (1996).

[9] A. R. Calderbank, E. M. Rains, P. W. Shor, and N. J. A. Sloane, Phys. Rev. Lett. **78**, 405 (1997).

[10] A. R. Calderbank, E. M. Rains, P. W. Shor, and N. J. A. Sloane, IEEE Trans. Inf. Theory **44**, 1369 (1998).

[11] P. W. Shor, in *Proc. 37th Symp. on the Foundations of Computer Science* (IEEE Computer Society Press, 1996), p. 56; quant-ph/9605011.

[12] J. Preskill, Proc. R. Soc. Lond. A **454**, 385 (1998); see also D. P. DiVincenzo and P. Shor, Phys. Rev. Lett. **77**, 3260 (1996).

[13] P. Zanardi and M. Rasetti, "Noiseless quantum codes," Phys. Rev. Lett. **79**, 3306 (1997); quant-ph/9705044.

[14] D. A. Lidar, I. L. Chuang, and K. B. Whaley, "Decoherence free subspaces for quantum computation," Phys. Rev. Lett. **81**, 2594 (1998); quant-ph/9807004.

[15] L. Viola and S. Lloyd, "Dynamical suppression of decoherence in two-state quantum systems," Phys. Rev. A **58**, 2733 (1998) (quant-ph/9803057); L. Viola, E. Knill, and S. Lloyd, "Dynamical decoupling of open quantum systems," Phys. Rev. Lett., in press, quant-ph/9809071.

PHYSICAL REVIEW A

ATOMIC, MOLECULAR, AND OPTICAL PHYSICS

THIRD SERIES, VOLUME 52, NUMBER 4 OCTOBER 1995

RAPID COMMUNICATIONS

The Rapid Communications section is intended for the accelerated publication of important new results. Since manuscripts submitted to this section are given priority treatment both in the editorial office and in production, authors should explain in their submittal letter why the work justifies this special handling. A Rapid Communication should be no longer than 4 printed pages and must be accompanied by an abstract. Page proofs are sent to authors.

Scheme for reducing decoherence in quantum computer memory

Peter W. Shor*

AT&T Bell Laboratories, Room 2D-149, 600 Mountain Avenue, Murray Hill, New Jersey 07974

(Received 17 May 1995)

Recently, it was realized that use of the properties of quantum mechanics might speed up certain computations dramatically. Interest has since been growing in the area of quantum computation. One of the main difficulties of quantum computation is that decoherence destroys the information in a superposition of states contained in a quantum computer, thus making long computations impossible. It is shown how to reduce the effects of decoherence for information stored in quantum memory, assuming that the decoherence process acts independently on each of the bits stored in memory. This involves the use of a quantum analog of error-correcting codes.

PACS number(s): 03.65.Bz, 89.70.+c

I. INTRODUCTION

Recently, interest has been growing in an area called *quantum computation*, which involves computers that use the ability of quantum systems to be in a superposition of many states. These computations can be modeled formally by defining a quantum Turing machine [1,5], which is able to be in the superposition of many states. Instead of considering the computer itself to be in a superposition of states, it is sufficient to assume that the contents of the memory cells are in a superposition of different states and that the computer performs deterministic unitary transformations on the quantum states of these memory cells [2]. This model resembles a quantum circuit [3] more than a quantum Turing machine. After Schumacher [4], we will call a two-state memory cell that can be part of such a superposition a quantum bit, or *qubit*. Whereas k classical two-state memory cells can take on 2^k states, thereby requiring k bits to describe them, k quantum bits require 2^k-1 complex numbers to completely represent their state. Even though most of these numbers must be small, and only the most significant digits of these numbers are important, there still appears to be too much information contained in k qubits to represent in a polynomial number of classical bits. Although only k bits of classical information can be extracted from k qubits, the presence of extra unextractable quantum information is a barrier to efficient simulation of a quantum computer on a classical computer.

It now appears that, at least theoretically, quantum computation may be much faster than classical computation for solving certain problems [5–7], including prime factorization. However, it is not yet clear whether quantum computers are feasible to build. One reason that quantum computers will be difficult, if not impossible, to build is *decoherence*. In the process of decoherence, some qubit or qubits of the computation become entangled with the environment, thus in effect "collapsing" the state of the quantum computer. The conventional assumption has been that once one qubit has decohered, the entire computation of the quantum computer is corrupted, and the result of the computation will no longer be correct [8,9]. We believe that this may be too conservative an assumption. This paper gives a way to use software to reduce the rate of decoherence in quantum memory. Berthiaume, Deutsch, and Jozsa [10] have similarly proposed a way of partially correcting errors in a quantum computer by taking many copies of the computation and continually projecting the computation into the symmetric subspace of these many copies. The degree to which their method corrects errors will depend on the type of errors that the computers are likely to make. Unfortunately, a mathematical analysis of the efficiency of their error-correction scheme has not yet been accomplished.

Assuming that the decoherence process affects the differ-

*Electronic address: shor@research.att.com

1050-2947/95/52(4)/2493(4)/$06.00 52 R2493

ent qubits in memory independently, we show how to store an arbitrary state of n qubits using $9n$ qubits in a decoherence-resistant way. That is, even if one of the qubits decoheres, the original superposition can be reconstructed perfectly. In fact, we map each qubit of the original n qubits into nine qubits, and our process will reconstruct the original superposition if at most one qubit decoheres in each of these groups of nine qubits. We thus show that the identity computation can be performed in a more decoherence resistant manner than the naive implementation.

The classical analog of our problem is the transmission of information over a noisy channel; in this situation, error-correcting codes can be applied so as to recover with high probability the transmitted information even after corruption of some percentage of the transmitted bits, where the percentage depends on the Shannon entropy of the channel. We give a quantum analog of the most trivial classical coding scheme: the repetition code, which provides redundancy by duplicating each bit several times [11]. This encoding scheme might be useful when storing qubits in the internal memory of the quantum computer; so that while qubits are in storage they avoid (or at least undergo reduced) decoherence, leaving decoherence to occur mainly in qubits actively involved in the computation.

II. QUANTUM COMPUTATION

In our model of quantum computation (the *gate array* model) we assume that we have s qubits. These qubits start in some specified initial configuration, which we may take to be $|0,0,0,\ldots,0\rangle$. They are then acted on by a sequence of the following operations, which manipulate the state of these s qubits in the corresponding 2^s-dimensional Hilbert space.

(1) Measurement. One qubit is measured in some basis, and the result is recorded classically. This corresponds to a projection operation in the Hilbert space.

(2) Entanglement. Two qubits are entangled according to some four-by-four unitary matrix. The corresponding Hilbert space operation describing the entire s qubits is the tensor product of this four by four unitary matrix with the 2^{s-2} by 2^{s-2} identity matrix. Entanglements of three or more qubits can always be accomplished by a sequence of two-bit entanglements [12].

The sequence of operations can be arbitrary, and there is no reason to assume that it does not depend on the input to the computer. However, in comparing quantum computation with classical computation, in order to prevent the programmer from "cheating" by using the sequence of operations to give the computer information which might otherwise be impossible or difficult to compute, we require that this sequence of operations be generated by a classical computer in polynomial time (in computer science terminology, this keeps the class of problems solvable by a quantum computer *uniform)*. The result of the computation is extracted from the computer by measuring the values of the qubits.

We must also initialize the computer by putting its memory in some known state. This could be done by postulating a separate operation, initialize, which sets a qubit to a predetermined value. However, we can also initialize a qubit by first measuring it and then performing a rotation to put it in the proper state [rotations are a special case of operation (2), even though there is no actual entanglement of different qubits taking place].

III. ENCODING

Our encoding is as follows. Suppose we have k qubits that we wish to store. We have our quantum computer encode each of these qubits into nine qubits as follows:

$$|0\rangle \rightarrow \frac{1}{2\sqrt{2}}(|000\rangle+|111\rangle)(|000\rangle+|111\rangle)(|000\rangle+|111\rangle),$$

$$|1\rangle \rightarrow \frac{1}{2\sqrt{2}}(|000\rangle-|111\rangle)(|000\rangle-|111\rangle)(|000\rangle-|111\rangle). \tag{3.1}$$

Consider what happens when the nine qubits containing the encoding are read. We will actually read them in a quantum fashion using an ancilla, i.e., by entangling them with other qubits [as in process (2) above] and then measuring some of these other qubits, and not by measuring them directly [as in process (1)]. However, for explanatory purposes, it is best to first consider what happens if the qubits are measured using a Bell basis. This same basis will later be used to "read" them by entangling them with qubits that will then be measured.

Suppose that no decoherence has occurred, and that the first three qubits are in state $|000\rangle+|111\rangle$. This means that the other two sets are also in state $|000\rangle+|111\rangle$. Similarly, if the first three are in state $|000\rangle-|111\rangle$, the other two triplets must also be in this state. Thus, even if one qubit has decohered, when we measure the nine qubits (measuring each triple in the Bell basis $|000\rangle\pm|111\rangle,|001\rangle\pm|110\rangle,|010\rangle\pm|101\rangle,|100\rangle\pm|011\rangle$) we can deduce what the measurement should have been by taking the majority of the three triples, and thus we can tell whether the encoded bit was 0 or a 1. This, however, does not let us restore the first qubit to any superposition or entanglement that it may have been in, because by measuring the nine qubits, we are in effect projecting all the qubits onto a subspace. This process also projects the original encoded bit onto a subspace, and so does not let us recover a superposition of an encoded 0 and an encoded 1.

To preserve the state of superposition of the encoded qubit, what we do in effect is to measure the decoherence without measuring the state of the qubits. This allows us in effect to reverse the decoherence. To explain in detail, we must first examine the decoherence process more fully. The critical assumption here is that decoherence only affects one qubit of our superposition, while the other qubits remain *unchanged.* It is not clear how reasonable this assumption is physically, but it corresponds to the assumption in classical information theory of the independence of noise.

By quantum mechanics, decoherence must be a unitary process that entangles a qubit with the environment. We can describe this process by describing what happens to two basis states of the qubit undergoing decoherence, $|0\rangle$ and $|1\rangle$. Considering the environment as well, these must be taken to orthogonal states, but if the environment is neglected, they can get taken to any combination of states.

Let us describe more precisely what happens to the qubit that decoheres. We assume that this is the first qubit in the encoding, but the procedure works equally well if any of the other eight qubits is the one that decoheres. Because $|0\rangle$ and $|1\rangle$ form a basis for the first qubit, we need only consider what happens to these two states. In general, the decoherence process must be

$$|e_0\rangle|0\rangle \rightarrow |a_0\rangle|0\rangle + |a_1\rangle|1\rangle$$
$$|e_0\rangle|1\rangle \rightarrow |a_2\rangle|0\rangle + |a_3\rangle|1\rangle, \tag{3.2}$$

where $|a_0\rangle$, $|a_1\rangle$, $|a_2\rangle$, and $|a_3\rangle$ are states of the environment (not generally orthogonal or normalized). Let us now see what happens to an encoded 0, that is, to the superposition $(1/\sqrt{2})(|000\rangle + |111\rangle)$. After decoherence, it goes to the superposition

$$(1/\sqrt{2})[(|a_0\rangle|0\rangle + |a_1\rangle|1\rangle)|00\rangle + (|a_2\rangle|0\rangle + |a_3\rangle|1\rangle)|11\rangle]. \tag{3.3}$$

We now write this in terms of a Bell basis, obtaining

$$\frac{1}{2\sqrt{2}}(|a_0\rangle + |a_e\rangle)(|000\rangle + |111\rangle)$$
$$+ \frac{1}{2\sqrt{2}}(|a_0\rangle - |a_3\rangle)(|000\rangle - |111\rangle)$$
$$+ \frac{1}{2\sqrt{2}}(|a_1\rangle + |a_2\rangle)(|100\rangle + |011\rangle)$$
$$+ \frac{1}{2\sqrt{2}}(|a_1\rangle - |a_2\rangle)(|100\rangle - |011\rangle). \tag{3.4}$$

Similarly, the vector $|000\rangle - |111\rangle$ goes to

$$\frac{1}{2\sqrt{2}}(|a_0\rangle + |a_3\rangle)(|000\rangle - |111\rangle)$$
$$+ \frac{1}{2\sqrt{2}}(|a_0\rangle - |a_3\rangle)(|000\rangle + |111\rangle)$$
$$+ \frac{1}{2\sqrt{2}}(|a_1\rangle + |a_2\rangle)(|100\rangle - |011\rangle)$$
$$+ \frac{1}{2\sqrt{2}}:(|a_1\rangle - |a_2\rangle)(|100\rangle + |011\rangle). \tag{3.5}$$

The important thing to note is that the state of the environment is the same for corresponding vectors from the decoherence of the two quantum states encoding 0 and encoding 1. Further, we know what the original state of the encoded vector was by looking at the other two triples. Thus, we can restore the original state of the encoded vector and also keep the evolution unitary by creating a few ancillary qubits which tell which qubit decohered and whether the sign on the Bell superposition changed. By measuring these ancillary qubits, we can restore the original state. We still maintain any existing superposition of basis Bell states because the coefficients are the same whether the original vector decohered from the state $|000\rangle + |111\rangle$ or $|000\rangle - |111\rangle$. By measuring the ancillary qubits that tell which qubit was decohered, we in effect restore the original state.

We now describe this restoration process more fully. This restoration consists first of a unitary transformation which we can regard as being performed by a quantum computer, and then a measurement of some of the qubits of the outcome. What the computer first does is compare all three triples in the Bell basis. If these triples are the same, it outputs (in the ancillary qubits) "no decoherence," and leaves the encoded qubits alone. If these triplets are not the same, it outputs which triple is different, and how it is different. The computer must then restore the encoded qubits to their original state. For example, in Eqs. (3.4) and (3.5), the output corresponding to the second line would mean "wrong sign," and the output corresponding to the third line would mean "first qubit wrong, but right sign." These outputs are expressed by some quantum state of the ancilla, which then is measured. Because the coefficients on the corresponding vectors in Eqs. (3.4) and (3.5) are the same, the superposition of states after the measurement and the subsequent corrections will be the same as the original superposition of states before the decoherence. Further, the correction of errors is now a unitary transformation because we are not just correcting the error, but also "measuring" the error, in that we measure what and where the error was, so we do not have to combine two quantum states into one.

If more than one qubit of a nine-tuple decoheres, the encoding scheme does not work. However, the probability that this happens is proportional to the square of the probability that one qubit decoheres. That is, if each qubit decoheres with a probability p, then the probability that k qubits do not decohere is probability $(1-p)^k$. In our scheme, we replace each qubit by nine. The probability that at least two qubits in any particular nine-tuple decohere is $1-(1+8p)\times(1-p)^8 \approx 36p^2$, and the probability that our $9k$ qubits can be decoded to give the original quantum state is approximately $(1-36p^2)^k$. Thus, for a probability of decoherence less than $\frac{1}{36}$, we have an improved storage method for quantum-coherent states of large numbers of qubits. Since p generally increases with storage time the watchdog effect could be used to store quantum information over long periods by using the decoherence restoration scheme to frequently reset the quantum state. If the decoherence time for a qubit is t_d, the above analysis implies that use of the watchdog effect will be advantageous if the quantum state is reset at time intervals $t_r \ll \frac{1}{36} t_d$.

It seems that we are getting something for nothing, in that we are restoring the state of the superposition to the exact original predecoherence state, even though some of the information was destroyed. The reason we can do this is that we expand one qubit to nine encoded qubits, providing some sort of redundancy. Our encoding scheme stores information in the entanglement between qubits, so that no information is stored in any one specific qubit; i.e., measuring any one of the qubits gives no information about the encoded state. Essentially, what we are doing is putting all of the information in the signal into dimensions of the signal space that are unlikely to be affected by decoherence. We can then measure the effect of the decoherence in the other dimensions of this space, which contain no information about our signal, and use this measurement to restore the original signal.

There is a cost for using this scheme. First, the number of qubits is expanded from k to $9k$. Second, the machinery that implements the unitary transformations will not be exact. Thus, getting rid of the decoherence will introduce a small extra amount of error. This may cause problems if we wish to store quantum information for long periods of time by repeatedly using this decoherence reduction technique. If our unitary transformations were perfect, we could keep the information for large times using the watchdog effect by repeatedly measuring the state to eliminate the decoherence. However, each time we get rid of the decoherence (or even check whether there was decoherence) we introduce a small extra amount of error. We must therefore choose the rate at which we measure the state so as to balance the error introduced by decoherence with the error introduced by the restoration of decoherence.

The assumption that the qubits decohere independently is crucial. This is not completely unreasonable physically, and may in many cases be a good approximation of reality, but the effects of changing this assumption on the accuracy of the encoding must be investigated. This assumption corresponds to independence of errors between different bits in classical information theory; even though this does not always hold in practice for classical channels, classical error-correcting codes can still be made to work very well. This is done by exploiting the fact that errors in bits far enough apart from each other are, in fact, nearly independent. It is not clear what the corresponding property would be in a quantum channel, or whether it would hold in practice.

There are clearly improvements that can be made to the above scheme. What this scheme does is use the three-repetition code twice: once in an outer layer (repeating the triplet of qubits three times) and once in an inner layer (using $|000\rangle \pm |111\rangle$ for each triplet). In classical information theory, repetition codes are extremely inefficient. The outer layer of our quantum code can be replaced by a classical error-correcting code to produce a more efficient scheme; this reduces the cost of encoding k qubits from $9k$ qubits to a function that approaches $3k$ qubits asymptotically as k grows. The inner layer of our quantum code, however, needs to have more properties than a classical error-correcting code because it needs to be able to correct errors coherently. While longer repetition codes can be used for this inner layer, it is not immediately clear how to improve on repetition codes for this mechanism, but I believe it should be possible.

This scheme is a step toward the quantum analog of channel coding in classical information theory. Whereas the quantum analog of Shannon's source coding theorem is already known [4,13], it is not even clear how a noisy quantum channel should properly be defined. Other steps in this direction have also recently been taken in [14,15], which deal with transmitting classical information over a quantum channel, and in [16], which deals with transmitting quantum information over a quantum channel, given an auxiliary two-way classical channel. The ultimate goal would be to define the quantum analog of the Shannon capacity for a quantum channel, and find encoding schemes which approach this capacity. An intermediate goal would be to find schemes for faithfully encoding k qubits that use $k+\epsilon k$ qubits, where ϵ approaches 0 as the channel's error rate goes to 0, as in classical information theory.

[1] D. Deutsch, Proc. R. Soc. London Ser. A **400**, 96 (1985).

[2] A. Yao, in *Proceedings of the 34th Annual Symposium on Foundations of Computer Science* (IEEE Computer Society, Los Alamitos, CA, 1993), p. 352.

[3] D. Deutsch, Proc. R. Soc. London Ser. A **425**, 73 (1989).

[4] B. Schumacher, Phys. Rev. A **51**, 2738 (1995).

[5] E. Bernstein and U. Vazirani, in *Proceedings of the 25th Annual ACM Symposium on the Theory of Computing* (Association for Computing Machinery, New York, 1993), p. 11.

[6] D. Simon, in *Proceedings of the 35th Annual Symposium on Foundations of Computer Science,* edited by S. Goldwasser (IEEE Computer Society, Los Alamitos, CA, 1994), p. 116.

[7] P. W. Shor, in *Proceedings of the 35th Annual Symposium on Foundations of Computer Science* (Ref. [6]), p. 124.

[8] R. Landauer (unpublished).

[9] W. G. Unruh, Phys. Rev. A **51**, 992 (1995).

[10] A. Berthiaume, D. Deutsch, and R. Jozsa, in *Proceedings of the Workshop on Physics and Computation, PhysComp 94* (IEEE Computer Society, Los Alamitos, CA, 1994), p. 60.

[11] See any information theory textbook for these concepts; for example, T. M. Cover and J. A. Thomas, *Elements of Information Theory* (Wiley, New York, 1991).

[12] D. P. DiVincenzo, Phys. Rev. A **51**, 1015 (1995).

[13] R. Jozsa and B. Schumacher, J. Mod. Opt. **41**, 2343 (1994).

[14] A. Fujiwara and H. Nagaoka, Mathematical Engineering Technical Report No. 94-22, University of Tokyo, 1994 (unpublished).

[15] P. Hausladen, B. Schumacher, M. Westmoreland, and W. K. Wooters (unpublished).

[16] C. H. Bennett, G. Brassard, B. Schumacher, J. Smolin, and W. K. Wooters (unpublished).

PHYSICAL REVIEW
LETTERS

VOLUME 77 29 JULY 1996 NUMBER 5

Error Correcting Codes in Quantum Theory

A. M. Steane
Clarendon Laboratory, Parks Road, Oxford, OX1 3PU, England
(Received 4 October 1995)

A new type of uncertainty relation is presented, concerning the information-bearing properties of a discrete quantum system. A natural link is then revealed between basic quantum theory and the linear error correcting codes of classical information theory. A subset of the known codes is described, having properties which are important for error correction in quantum communication. It is shown that a pair of states which are, in a certain sense, "macroscopically different," can form a superposition in which the interference phase between the two parts is measurable. This provides a highly stabilized "Schrödinger cat" state. [S0031-9007(96)00779-X]

PACS numbers: 03.65.Bz, 03.75.Dg, 89.70.+c

This Letter discusses fundamental questions concerning quantum interference among many particles in a group. It will be shown that such questions are linked with the properties of the error correcting codes arising in classical information theory [1]. The possibility of error correction in quantum systems has been considered recently because of its importance in the theory of quantum computation [2] and quantum cryptography [3]. The present work provides the answers to fundamental questions in this area. First, a new way of expressing the Heisenberg uncertainty principle is presented. Here it describes a limit on the degree of robustness with which information can be encoded in a quantum state which is to be analyzed in either of two mutually rotated bases. In brief, if multiple error correction is possible in one basis, then it is ruled out in the other. The precise meaning of this sentence will be elucidated below. This gives a simple way of understanding the well-known instability of the phase relationship between quantum states expressing macroscopically different physical situations. Next, the *linear codes* of classical information theory are shown to have a remarkable property (Theorem 3 below) in the quantum mechanical context. This establishes a previously unremarked link between these two mathematical edifices. The new insights gained enable one to construct states which are both macroscopically distinguishable, in a technical sense to be described, and which also can be observed to show stable quantum mechanical interference. This has important implications for the possibility of quantum computation and is a new development in the understanding of the famous "Schrödinger's cat" experiment [4].

Consider a quantum system having a Hilbert space of 2^n dimensions (with positive integer n). For example, this could be a set of n two-state systems, such as n spin one-half particles, or n two-level atoms. Such systems can model the behavior of any other quantum system [5], including macroscopic objects such as measuring devices.

The two orthogonal states of each particle are written $|0\rangle$ and $|1\rangle$, and a product state such as $|0\rangle \otimes |0\rangle \otimes |1\rangle$ is written $|001\rangle$, where it is understood that the first binary digit (0 or 1) refers to the state of the first particle, the second digit the second particle, and so on. A general state of n particles can be written as a sum (entanglement) of product states. The singlet state of two particles, for example, is $(|10\rangle - |01\rangle)/\sqrt{2}$. In what follows, the notation will be simplified by omitting the overall normalization factor in such expressions. This will not affect the argument, and the factor can be reintroduced easily if necessary.

The states $|0\rangle$ and $|1\rangle$ form a basis, hereafter called "basis 1." We will be concerned with the state of the system as expressed using the states of basis 1, and also those of a rotated basis, "basis 2." For example, the two bases could be those corresponding to a vertical or horizontal choice of quantization axis, in the case of the spin state of spin-half particles. The basis states of basis

0031-9007/96/77(5)/793(5)$10.00

1 will be written using a plain $|0\rangle$ and $|1\rangle$; those of basis 2 will be written using bold fond $|\mathbf{0}\rangle$ and $|\mathbf{1}\rangle$. Thus ignoring normalization as already remarked, $|0\rangle \equiv |\mathbf{0}\rangle + |\mathbf{1}\rangle$, $|1\rangle \equiv |\mathbf{0}\rangle - |\mathbf{1}\rangle$, $|00\rangle \equiv |\mathbf{00}\rangle + |\mathbf{01}\rangle + |\mathbf{10}\rangle + |\mathbf{11}\rangle$, and so on. It will be convenient to have a shorthand for referring to the individual product states making up a superposition. Since a product state is identified by a unique string of bits, it will be referred to as a *word.*

A state which is equal to a superposition of words in basis 1 is equal to some other superposition in basis 2. Some basic relationships between the two bases will now be stated.

Theorem 1. *The word $|000\cdots 0\rangle$ consisting of all zeros in basis 1 is equal to a superposition of all 2^n possible words in basis 2, with equal coefficients.*

Theorem 2. *If the* jth *bit of each word is complemented* $(0 \leftrightarrow 1)$ *in basis 1, then all words in basis 2 in which the* jth *bit is set (is a* $\mathbf{1}$*) change sign.* For example,

$$|000\rangle + |111\rangle \equiv |\mathbf{000}\rangle + |\mathbf{011}\rangle + |\mathbf{101}\rangle + |\mathbf{110}\rangle,$$

$$|001\rangle + |110\rangle \equiv |\mathbf{000}\rangle - |\mathbf{011}\rangle - |\mathbf{101}\rangle + |\mathbf{110}\rangle.$$

Corollary. *If all the words are complemented in basis 1, then all words of odd parity change sign in basis 2, and vice versa.* (Odd parity means having an odd number of 1's.)

These theorems are easy to prove by writing each word in basis 1 as a product of bits, converting each bit to the form $(|\mathbf{0}\rangle \pm |\mathbf{1}\rangle)$, and multiplying out the products.

Next some of the standard results and notation of coding theory will be described. This is very basic material but is necessary in order to make the argument widely accessible.

In coding theory, information takes the form of a string of bits, or "words." A *code* is a set of words, all of the same *length* (number of bits). Words in the code are *code words.* The Hamming *distance* between two words (of the same length) is the number of places where they differ, i.e., the number of positions where one has a 0 and the other a 1. The *minimum distance* of a code is the smallest Hamming distance between any two code words in the code. A single error is the erroneous complementing of a single bit of a word, for example, when the word is transmitted or stored. *A code of minimum distance d allows $\lfloor (d-1)/2 \rfloor$ errors to be corrected.* This is because if less than $d/2$ errors occur, then the correct original code word, which gave rise to the erroneous received word, can be identified as the only code word at a distance less than $d/2$ from the received word. The price of this error correction is that only code words (i.e., a subset of the 2^n possible n-bit words) may be transmitted.

The fundamental problem of coding theory is to find codes having the maximum number of code words for given length n and minimum distance d. Let $A(n,d)$ be defined as this maximum number of words. The problem is notoriously difficult and has no general solution.

The notation $[n,k,d]$ refers to a set of 2^k code words, each of length n, with minimum distance d, and having the property of being a *linear code.* This means that if the EXCLUSIVE-OR operation is carried out bitwise between any two code words, then the resulting word is also a member of the code. (Not all codes are linear.) If C is a code, then the *dual* code $C^\perp$ is the set of all words u for which $u \cdot v$ has even parity for all $v \in C$, where the dot signifies the bitwise AND operation. The dual of a linear $[n,k,d]$ code is a linear $[n, n-k, d^\perp]$ code. In general, there is no simple precise relationship between d and $d^\perp$, though they are related indirectly through a theorem due to MacWilliams [1]. If d is large, then $d^\perp$ is small.

A linear code C is completely specified by its $n \times (n-k)$ *parity check matrix H,* or equivalently by its $n \times k$ *generator matrix G.* The rows of these matrices are n-bit words. The code C is the set of all words u for which $H_i \cdot u$ has even parity, for all rows H_i of H. Also the code C is the set of all linear combinations (by bitwise EXCLUSIVE-OR) of the rows of G. It can be shown that *the parity check matrix of a code C is the generator matrix of the dual code $C^\perp$.* This property will be used below and ends the present list of standard results.

In the context of the set of n quantum bits (two-state systems), the sets of words which express a given state in bases 1 and 2 are related through the basis rotation operation, which is a Hadamard transform. Just as the properties of the continuous Fourier transform lead to the Heisenberg uncertainty principle $\Delta x \Delta p \geq \hbar/2$ where x and p are conjugate continuous variables, so also for the discrete case the basis rotation operation implies a limit on the way a given state can be expressed in two mutually rotated bases.

Suppose a state can be written as a superposition of m_1 of the product states of basis 1 and as a superposition of m_2 of the product states of basis 2. Then

$$m_1 m_2 \geq 2^n. \tag{1}$$

Proof. Inequality (1) is subsumed by the "entropic uncertainty relation" introduced by Bialynicki-Birula and Mycielski [6] and by Deutsch [7], as improved by Maassen and Uffink [8].

Now suppose we wish to find a state which is expressed in basis 1 by a set of words of minimum Hamming distance d_1, and simultaneously in basis 2 by a set of words of minimum Hamming distance d_2. By definition, $m_1 \leq A(n, d_1)$ and $m_2 \leq A(n, d_2)$; therefore, using inequality (1), we have

$$A(n, d_1) A(n, d_2) \geq 2^n. \tag{2}$$

This "error correction uncertainty relation" places a limit on the highest minimum distance simultaneously achievable in bases 1 and 2. If d_1 is large, then $A(n, d_1)$ is necessarily small, which means, by (2), that $A(n, d_2)$ must be large, which in turn means that d_2 must be small. Thus we have a complementarity between d_1 and d_2. Its implications will be described below.

For odd d, Hamming [9] derived the Hamming or "sphere-packing bound" $A(n,d) \leq 2^n / \sum_{i=0}^{(d-1)/2} \binom{n}{i}$, where $\binom{n}{i}$ is the binomial coefficient $n!/i!(n-i)!$. Substituting in (2), one obtains, for odd d_1 and d_2,

$$\sum_{i=0}^{(d_1-1)/2} \binom{n}{i} \sum_{i=0}^{(d_2-1)/2} \binom{n}{i} \leq 2^n. \tag{3}$$

It is not generally possible to find codes which satisfy the upper limit of the Hamming bound, but it can be shown that for large enough n, codes exist which allow d_2 to exceed any value for any given d_1.

We will now consider the state

$$|\psi\rangle = |000\cdots 0\rangle + e^{i\phi}|111\cdots 1\rangle, \tag{4}$$

where the two words are those of all zeros or all ones, in basis 1. Such a state can be shown to violate a Bell-type inequality by an amount that grows exponentially with the number of bits n [10]. If n is large, then we have a superposition of two states representing macroscopically different situations (somewhat like a cat alive or dead [11]). However, the presence of both parts of the superposition, rather than simply of one part *or* the other, can be revealed only in experiments whose outcome depends on the value of ϕ. In practice, technological difficulties make ϕ extremely difficult to measure with an experimental uncertainty less than $\pm\pi$. In other words, experimental observation of the quantum interference is prevented by the sensitivity of ϕ to random errors.

A simple way to understand the state $|\psi\rangle$ is revealed by Theorem 1 and the corollary to Theorem 2. When $\phi = 0$, it is easy to see that in basis 2, $|\psi\rangle$ is equal to a superposition of all words having even parity (even number of **1**'s), while if $\phi = \pi$, the state is a superposition of all words having odd parity. Therefore to distinguish the cases $\phi = 0$ and $\phi = \pi$ experimentally one must find out whether the state in basis 2 has even or odd parity. However, a single error (complementing of a bit) in basis 2 is sufficient to destroy this information. If the probability of an error in any one bit is p, then the probability that no errors occur, enabling ϕ to be deduced, is $(1-p)^n$, which falls off exponentially with n [12]. For example, if $n = 1001$, $p = 0.02$, then $(1-p)^n \sim 10^{-9}$.

In the state just discussed, the code obtained in basis 2 (that consisting of all words of even parity) is the dual of the code appearing in basis 1 [Eq. (4)]. This is an example of a more general property which will now be stated.

Theorem 3. *When the quantum state of the system forms a linear code C in basis 1, in a superposition with equal coefficients, then in basis 2 the words appearing in the superposition are those of the dual code $C^\perp$.*

Proof. We will construct a code in basis 1 having generator matrix G and show that in basis 2 the code of which G is the parity check matrix appears.

Consider first the state $|\{0\}\rangle = |000\cdots 0\rangle$ consisting of all zeros in basis 1. In basis 2, all 2^n possible words are superposed (Theorem 1), each with positive sign. Let G be a generator matrix, that is, a matrix of k_1 rows, each row being a word n bits long. Take the first row G_1 of G and form the corresponding word $|G_1\rangle$ in basis 1 by starting from $|\{0\}\rangle$ and applying Theorem 2 once for each nonzero bit in G_1. These successive applications of Theorem 2 show that the state $|G_1\rangle$ is one for which all 2^n possible words appear in basis 2, and all those, and only those, words in basis 2 change sign which do not satisfy the parity check G_1.

Now form the state $|\{0\}\rangle + |G_1\rangle$. By the argument just given, when the sum is formed, all words in basis 2 which do not satisfy the parity check G_1 disappear. Therefore at this stage of the argument, G_1 is the (single-row) generator matrix of the code in basis 1 and also the parity check matrix of the code in basis 2.

Now take the next row G_2 of G, and form the pair of words $|G_2\rangle + |G_1 \oplus G_2\rangle$ by applying Theorem 2 the necessary number of times to the state $|\{0\}\rangle + |G_1\rangle$. Here $\oplus$ signifies the bitwise addition modulo 2 (EXCLUSIVE-OR) operation. By Theorem 2 again, all those and only those words in basis 2 change sign which do not satisfy the parity check G_2. Therefore the state $|\{0\}\rangle + |G_1\rangle + |G_2\rangle + |G_1 \oplus G_2\rangle$ has the property that the first two rows of G form the generator matrix of the code in basis 1 and also the parity check matrix of the code in basis 2.

The above process is continued for the rest of the rows of G, and the theorem is proved.

Since the dual of an $[n,k,d]$ code is an $[n, n-k, d^\perp]$ code, Theorem 3 shows that the linear codes satisfy the lower bound of inequality (1). However, it seems unlikely that nonlinear codes should do so; therefore we may conjecture that the linear codes approach the lower bound of the error correction uncertainty relation (2) more closely. In this case the MacWilliams theorem also yields a limit on d_1 and d_2, though often one must use tables of known codes to find the smallest length n which permits the distances d_1 and d_2 to attain given sizes simultaneously [1].

Error correction in quantum computation might be suggested through the use of the simple repetition code. A bit value 0 is represented by $|000\rangle$, and a value 1 by $|111\rangle$, for example. This allows single error correction in basis 1. However, the possibility of superpositions such as $|000\rangle \pm |111\rangle$ is fundamentally important to quantum computation, and as we have just seen, the sign in such superpositions is highly sensitive to errors in basis 2. It will now be shown how to find state $|a\rangle$ and $|b\rangle$ such that error correction is possible in both bases, in the following sense. In basis 1, the Hamming distance between $|a\rangle$ and $|b\rangle$ will be greater than 2, while in basis 2 the Hamming distance between $|c\rangle = |a\rangle + |b\rangle$ and $|d\rangle = |a\rangle - |b\rangle$ will be greater than 2. The Hamming distance between two states, in a given basis, is here defined to be the smallest distance between any word appearing in the first state and any word appearing in the second.

Let $|a\rangle$ be expressed by the [7, 3, 4] simplex code in basis 1:

$$|a\rangle = |0000000\rangle + |1010101\rangle + |0110011\rangle + |1100110\rangle + |0001111\rangle + |1011010\rangle + |0111100\rangle + |1101001\rangle .$$

This code has the following properties: It can be augmented to produce a code of minimum distance 3, and its dual code (the [7, 4, 3] Hamming code) has minimum distance 3. The process of augmentation consists of adding to the code the complement of each of its words (equivalent to adding a row of 1's to the generator matrix). Therefore if we let $|b\rangle$ be the complement of $|a\rangle$ in basis 1,

$$|b\rangle = |1111111\rangle + |0101010\rangle + |1001100\rangle + |0011001\rangle + |1110000\rangle + |0100101\rangle + |1000011\rangle + |0010110\rangle ,$$

then the desired properties are obtained. For in basis 1, $|a\rangle$ and $|b\rangle$ are nonoverlapping subsets of a distance 3 code, which means the distance between them is at least 3, and in basis 2, $|c\rangle = |a\rangle + |b\rangle$ contains just the even parity words of a [7, 4, 3] code, while $|d\rangle = |a\rangle - |b\rangle$ contains just the odd parity words of the same code. Since these are nonoverlapping subsets of a distance 3 code, the distance in basis 2 between $|c\rangle$ and $|d\rangle$ is at least 3.

Thus a method for error correction of quantum bits has been found, which enables both the bits themselves to be encoded robustly in basis 1 and the values of the signs appearing in superpositions in basis 1 to be encoded robustly.

The above argument can be extended to higher Hamming distances, which leads to the possibility of macroscopic—or at least mesoscopic—superpositions with measurable interference phase. For example, the case was considered of the Schrödinger cat state $|\psi\rangle$ of Eq. (4) involving $n = 1001$ two-state systems. The two parts of the superposition were "macroscopically different" in the sense that any property proportional to the sum of the bits in basis 1 would have a mean value in the state $|000\cdots0\rangle$ very different from its mean value in the state $|111\cdots1\rangle$. However, the spirit of Schrödinger's thought experiment can also be retained by arguing that two states are macroscopically different if a macroscopic number of errors would have to occur in order to make it possible to mistake one state for the other. Now suppose we use $n = 5000$ and seek two states $|a\rangle$ and $|b\rangle$ separated by Hamming distance $d_1 = 1001$ in basis 1. The uncertainty relation (3) then implies $d_2 \le 1213$, and it should be possible to find a dual pair of linear codes of which one is capable of augmentation and $d_1 = 1002$, $d_2 \ge 241$. If so, then subcodes of the augmented code are used to produce $|a\rangle$ and $|b\rangle$ as before, and we consider the superposition $|a\rangle + |b\rangle$. Quantum interference between $|a\rangle$ and $|b\rangle$ can be demonstrated if it can be shown experimentally that the sign in this superposition state is positive and not negative. To do this, measurements are carried out in basis 2. This measurement is the experimental method by which quantum interference between $|a\rangle$ and $|b\rangle$ is observed. Now, by construction, the state $|a\rangle + |b\rangle$ will be mistaken for only $|a\rangle - |b\rangle$ if at least $(d_2 - 1)/2 \ge 120$ errors occur. If these errors are independent, then the probability that the sign is revealed correctly in each experimental run (in which all the bits are measured) is

$$\sum_{i=0}^{120} \binom{n}{i} p^i (1 - p)^{n-i} \simeq 0.98 , \qquad (5)$$

where the error per bit $p = 0.02$ as before. This is to be compared with the result of order 10^{-9} obtained for the Schrödinger cat state of the type given in Eq. (4), having the same Hamming distance between its two parts in basis 1. In fact, the error per bit in a real experiment is likely to increase somewhat with n, but as long as $p < 0.055$, then a number n can always be found with makes the interference observable between states separated by a given distance d_1; this is proved in [13,14]. Also it is not always true that errors in different quantum bits are independent. However, situations can be found in which the errors are independent, and in such cases the above argument applies.

In conclusion, a new type of uncertainty relation has been presented in which a discrete quantum system is regarded as an information-bearing entity, with limitations on the degree to which it can store information robustly. The interference phase between two product states separated by a large Hamming distance in one basis is a particularly fragile piece of information because it is expressed by the value of a parity check covering a large number of bits in the rotated basis.

A method has been presented for finding codes which enable error correction in both of two mutually rotated bases. This type of correction does not arise in the classical context, but is important for quantum bits. The argument enables states to be identified in which interferences involving a macroscopic number of particles may be observable. The experimental production of such states is, however, a demanding task which remains to be addressed.

The author is supported by the Royal Society.

Note Added.—During resubmission of this Letter, related work [15] on quantum coding has become known to me. In addition, the coding method introduced in this letter has now been generalized and shown to be fully applicable to quantum communication in that general errors affecting general states of many information qubits can be corrected [13,14].

[1] F. J. MacWilliams and N. J. A. Sloane, *The Theory of Error-Correcting Codes* (North-Holland, Amsterdam, 1977).
[2] For reviews and references, see, e.g., A. Ekert, in *Atomic Physics 14,* edited by D. J. Wineland, C. E. Wieman, and S. J. Smith (AIP Press, New York, 1995); A. Ekert and R. Jozsa (to be published).
[3] For reviews, see, e.g., R. J. Hughes, D. M. Alde, P. Dyer, G. G. Luther, G. L. Morgan, and M. Schauer, Contemp. Phys. **36**, 149 (1995); S. J. D. Phoenix and P. D. Townsend, Contemp. Phys. **36**, 165 (1995).
[4] E. Schrödinger, Naturwissenschaften **23**, 807 (1935); translated in *Quantum Theory and Measurement,* edited by J. A. Wheeler and W. H. Zurek (Princeton University, Princeton, NJ, 1983).
[5] D. Deutsch, Proc. R. Soc. London A **400**, 97 (1985).
[6] I. Bialynicki and J. Mycielski, Commun. Math. Phys. **44**, 129 (1975).
[7] D. Deutsch, Phys. Rev. Lett. **50**, 631 (1983).
[8] H. Maassen and J. B. M. Uffink, Phys. Rev. Lett. **60**, 1103 (1988).
[9] R. W. Hamming, Bell Syst. Tech. J **29**, 147 (1950).
[10] N. David Mermin, Phys. Rev. Lett. **65**, 1838 (1990).
[11] A. Peres, *Quantum Theory: Concepts and Methods* (Kluwer Academic, Dordrecht, 1993).
[12] More precisely, the probability that ϕ is revealed in the correct one of the two ranges $(-\pi/2,\ldots,\pi/2)$, $(\pi/2,\ldots,3\pi/2)$ in a given experimental run is $P = \sum_{i=0}^{\lfloor n/2 \rfloor} \binom{n}{2i} p^{2i}(1-p)^{n-2i}$ (the probability that zero or an even number of errors occur), which approaches $1/2$ very rapidly as n increases. For example, if $p = 0.02$, $n = 1001$ then $|P - 1/2| \sim 10^{-14}$. When $P = 1/2$ the experimental results bear no correlation to ϕ.
[13] A. R. Calderbank and P. W. Shor, Phys. Rev. A (to be published).
[14] A. M. Steane, Proc. R. Soc. London A (to be published).
[15] P. W. Shor, Phys. Rev. A **52**, R2493 (1995).

PHYSICAL REVIEW A VOLUME 54, NUMBER 3 SEPTEMBER 1996

Class of quantum error-correcting codes saturating the quantum Hamming bound

Daniel Gottesman*
California Institute of Technology, Pasadena, California 91125
(Received 29 April 1996)

I develop methods for analyzing quantum error-correcting codes, and use these methods to construct an infinite class of codes saturating the quantum Hamming bound. These codes encode $k=n-j-2$ quantum bits (qubits) in $n=2^j$ qubits and correct $t=1$ error. [S1050-2947(96)09309-2]

PACS number(s): 03.65.Bz, 89.80.+h

I. INTRODUCTION

Since Shor [1] showed that it was possible to create quantum error-correcting codes, there has been a great deal of work on trying to create efficient codes. Calderbank and Shor [2] and Steane [3] demonstrated a method of converting certain classical error-correcting codes into quantum ones, and Laflamme *et al.* [4] and Bennett *et al.* [5] produced codes to correct one error that encode 1 qubit in 5 qubits.

Suppose we want to encode k qubits in n qubits. The space of code words is then some 2^k-dimensional subspace of the full 2^n-dimensional Hilbert space. The encodings $|\psi_i\rangle$ of the original 2^k basis states form a basis for the space of code words. When a coherent error occurs, the code states are altered by some linear transformation M:

$$|\psi_i\rangle \mapsto M|\psi_i\rangle. \tag{1}$$

We do not require that M be unitary, which will allow us to also correct incoherent errors.

Typically, we only consider the possibility of errors that act on no more than t qubits. An error that acts nontrivially on exactly t qubits will be said to have *length t*. An error of length 1 only acts on a two-dimensional Hilbert space, so the space of 1-qubit errors is $\mathcal{M}_2$, the space of 2×2 matrices.

An error-correction process can be modeled by a unitary linear transformation that entangles the erroneous states $M|\psi_i\rangle$ with an ancilla $|A\rangle$ and transforms the combination to a corrected state

$$(M|\psi_i\rangle)\otimes|A\rangle \mapsto |\psi_i\rangle\otimes|A_M\rangle. \tag{2}$$

Note that the map $M\mapsto|A_M\rangle$ must be linear, but not necessarily one-to-one. If the map is injective, I will call the code *nondegenerate*, and if it is not, I will call the code *degenerate*. A degenerate code has linearly independent matrices that act in a linearly dependent way on the code words, while in a nondegenerate code, all of the errors acting on the code words produce linearly independent states. Note that Shor's original code [1] is a degenerate code (phase errors within a group of 3 qubits act the same way), while the $k=1, n=5$ codes [4,5] are nondegenerate.

At this point, we can measure the ancilla preparatory to restoring it to its original state without disturbing the states $|\psi_i\rangle$. This process will correct the error even if the original state is a superposition of the basis states:

$$\left(M\sum_{i=1}^{2^k} c_i|\psi_i\rangle\right)\otimes|A\rangle \mapsto \left(\sum_{i=1}^{2^k} c_i|\psi_i\rangle\right)\otimes|A_M\rangle. \tag{3}$$

An incoherent error can be modeled as an ensemble of coherent errors. Since the above process corrects all coherent errors, it will therefore also correct incoherent errors. After the ancilla is measured and restored to its original state, the system will once again be in a pure state. Sufficient and necessary conditions for the system to form a quantum error-correcting code are given in [5] and [6]. While errors acting on different code words must produce orthogonal results, different errors acting on the same code word can produce nonorthogonal states, even in the nondegenerate case.

We can use the definition of nondegenerate quantum error-correcting codes to derive the quantum Hamming bound [7] on their possible efficiency. It is not known whether the quantum Hamming bound applies to degenerate codes, although some recent evidence suggests that it does not [8,9]. However, the breeding and hashing protocol presented by Shor and Smolin [8] and the random matrix encodings mentioned by Lloyd [9] do not give a 100% chance of successful decoding, even if only a fixed finite number of errors occurs. There are no known degenerate codes that guarantee success that violate the quantum Hamming bound. I show in Appendix A that a certain class of degenerate codes to correct one error are, in fact, limited by the quantum Hamming bound. The question for fully general degenerate codes remains open, although Knill and Laflamme [6] showed that at least five qubits are necessary to correct one error. Below, I will assume the code is nondegenerate.

Since there are three possible nontrivial 1-qubit errors, the number of possible errors M of length l on an n-qubit code is $3^l\binom{n}{l}$. Each of the states $M|\psi_i\rangle$ must be linearly independent, and all of these different errors must fit into the 2^n-dimensional Hilbert space of the n qubits. Thus, for a code that can correct up to t errors,

$$2^k\sum_{l=0}^{t} 3^l\binom{n}{l} \leq 2^n. \tag{4}$$

For large n, this becomes

$$\frac{k}{n} \leq 1 - \frac{t}{n}\log_2 3 - H(t/n), \tag{5}$$

*Electronic address: gottesma@theory.caltech.edu

where $H(x)=-x\log_2 x-(1-x)\log_2(1-x)$.

It is an interesting question whether it is generally possible to attain this bound, or whether some more restrictive upper bound holds. Breeding and hashing methods [10,5] can asymptotically saturate the quantum Hamming bound for large blocks, but have a small but nonzero probability of failure, even for only one error. For $t=1$ and $k=1$, the quantum Hamming bound (4) implies $n\geq 5$, so the known 5-qubit code does saturate the bound. Below, in Sec. III, I will give a class of codes saturating the bound for $t=1$ and $n=2^j$ (so $k=n-j-2$). For large n, the efficiency k/n of these codes approaches 1. In this sense, they are the analog of the classical Hamming codes. To aid in the construction, in Sec. II I will present some methods for analyzing quantum error-correcting codes. The method I present of using code stabilizers to describe codes is also given, using slightly different language, in [11].

Throughout this paper, I will assume the basis of $\mathcal{M}_2$ is

$$I=\begin{pmatrix}1 & 0\\ 0 & 1\end{pmatrix},\ X=\begin{pmatrix}0 & 1\\ 1 & 0\end{pmatrix},\ Y=\begin{pmatrix}0 & -1\\ 1 & 0\end{pmatrix},\ Z=\begin{pmatrix}1 & 0\\ 0 & -1\end{pmatrix}. \tag{6}$$

Some of the results will hold for other bases, but many will not. This basis has two important properties: all of the matrices either commute or anticommute, and $X^2=-Y^2=Z^2=I$.

II. CODE STABILIZERS

Suppose we have an n-qubit system. Let us write the matrices X, Y, and Z as X_i, Y_i, and Z_i when they act on the ith qubit. Let $\mathcal{G}$ be the group generated by all $3n$ of these matrices.[1] Since $(X_i)^2=(Z_i)^2=I$ and $Y_i=Z_iX_i=-X_iZ_i$, $\mathcal{G}$ has order 2^{2n+1} (for each i, we can have I, X_i, Y_i, or Z_i, plus a possible overall factor of -1). The group $\mathcal{G}$ has a few other useful features: every element in $\mathcal{G}$ squares to ± 1 and if $A,B\in\mathcal{G}$, then either $[A,B]=0$ or $\{A,B\}=0$.

The code words of the quantum error-correcting code span a subspace T of the Hilbert space. The group $\mathcal{G}$ acts on the vectors in T. Let $\mathcal{H}$ be the stabilizer of T — i.e.,

$$\mathcal{H}=\{M\in\mathcal{G}\text{ s.t. } M|\psi\rangle=|\psi\rangle\forall|\psi\rangle\in T\}. \tag{7}$$

Now suppose $E\in\mathcal{G}$ and $\exists M\in\mathcal{H}$ s.t. $\{E,M\}=0$. Then $\forall|\psi\rangle,|\phi\rangle\in T$,

$$\langle\phi|E|\psi\rangle=\langle\phi|EM|\psi\rangle=-\langle\phi|ME|\psi\rangle=-\langle\phi|E|\psi\rangle, \tag{8}$$

so $\langle\phi|E|\psi\rangle=0$.

The implications of this are profound. Suppose E and F are two errors, both of length t or less. Then $E|\psi\rangle$ and $F|\phi\rangle$ are orthogonal for all $|\psi\rangle,|\phi\rangle\in T$ whenever $F^\dagger E$ anticommutes with anything in $\mathcal{H}$. This is the requirement for a nondegenerate code, so to find such a code, we just need to pick T and corresponding $\mathcal{H}$ so that every nontrivial matrix in $\mathcal{G}$ of length less than or equal to $2t$ anticommutes with some member of $\mathcal{H}$.

It is unclear whether every quantum error-correcting code in the X, Y, Z basis can be completely described by its stabilizer $\mathcal{H}$. Certainly, a large class of codes can be described in this way, and I do not know of any quantum error-correcting codes that cannot be so described.

Given T, we can figure out $\mathcal{H}$, but it will be much easier to find codes using the above property if we can pick $\mathcal{H}$ and deduce a space T of code words. First I will discuss what properties $\mathcal{H}$ must have in order for it to be the stabilizer of a space T, then I will discuss how to choose $\mathcal{H}$ so that the matrices of length $2t$ or less anticommute with one of its elements.

Clearly, $\mathcal{H}$ must be a subgroup of $\mathcal{G}$. Also, if $M\in\mathcal{H}$, then $M^2|\psi\rangle=M|\psi\rangle=|\psi\rangle$ for $|\psi\rangle\in T$, so M cannot square to -1. Finally, if $M,N\in\mathcal{H}$, then

$$MN|\psi\rangle=|\psi\rangle, \tag{9}$$

$$NM|\psi\rangle=|\psi\rangle, \tag{10}$$

$$[M,N]|\psi\rangle=0. \tag{11}$$

If $\{M,N\}=0$, then $[M,N]=2MN$, but M and N are unitary, and cannot have 0 eigenvalues. Thus, $[M,N]=0$, and $\mathcal{H}$ must be Abelian.

Thus, $\mathcal{H}$ must be Abelian and every element of $\mathcal{H}$ must square to 1, so $\mathcal{H}$ is isomorphic to $(\mathbf{Z}_2)^a$ for some a. It turns out that these are sufficient conditions for there to exist nontrivial T with stabilizer $\mathcal{H}$, as long as $\mathcal{H}$ is not too big. The largest subspace T with stabilizer $\mathcal{H}$ will have dimension 2^{n-a}. To show this, I will give an algorithm for constructing a basis for T. Intuitively, it is unsurprising that this should be the dimension of T, since each generator of $\mathcal{H}$ has eigenvalues ± 1 and splits the Hilbert space in half.

Consider a state that can be written as a tensor product of 0's and 1's. This sort of state is analogous to one word of a classical code, so I will call it a *quasiclassical* state. Sometimes I will distinguish between quasiclassical states that differ by a phase and sometimes I will not. Now, given a quasiclassical state $|\phi\rangle$, then

$$|\psi\rangle=\sum_{M\in\mathcal{H}}M|\phi\rangle \tag{12}$$

is in T,[2] since applying an element of $\mathcal{H}$ to it will just rearrange the sum. I will call $|\phi\rangle$ the *seed* of the code word $|\psi\rangle$. By the same argument, if $M\in\mathcal{H}$, $M|\phi\rangle$ acts as the seed for the same quantum code word as $|\phi\rangle$. Not every possible seed will produce a nonzero code word. For instance, suppose $\mathcal{H}=\{I,Z_1Z_2\}$ and we use $|01\rangle$ as our seed. Then $|\psi\rangle=I|01\rangle+Z_1Z_2|01\rangle=0$.

To find elements of T, we try quasiclassical states until we get one that produces nonzero $|\psi\rangle$, call it $|\psi_1\rangle$. I will show later that such a state will always exist. We can write $|\psi_1\rangle$ as a sum of quasiclassical states, any of which could act as its seed. Pick a quasiclassical state that does not appear in

[1]For $n=1$, $\mathcal{G}$ is just D_4, the symmetry group of a square. For larger n, $\mathcal{G}$ is $(D_4)^n/(\mathbf{Z}_2)^{n-1}$.

[2]In fact, $|\phi\rangle$ does not need to be a quasiclassical state for $|\psi\rangle$ to be in T. Any state will do, but it is easiest to use quasiclassical states.

$|\psi_1\rangle$ and does not produce 0, and use it as the seed for a second state $|\psi_2\rangle$. Continue this process for all possible quasiclassical states. The states $|\psi_i\rangle$ will then form a basis for T. None of them share a quasiclassical state.

To see that $\{|\psi_i\rangle\}$ is a basis, imagine building up the elements of $\mathcal{H}$ by adding generators one by one. Suppose $\mathcal{H}=\langle M_1,M_2,\ldots,M_a\rangle$ (i.e., $\mathcal{H}$ is generated by M_1 through M_a). Let $\mathcal{H}_r$ be the group generated by M_1 through M_r, and look at the set S_r of quasiclassical states produced by acting with the elements of $\mathcal{H}_r$ on some given quasiclassical seed $|\phi\rangle$. The phases of these quasiclassical states will matter. The next generator M_{r+1} can do one of three things: (i) it can map the seed to some new quasiclassical state not in S_r, (ii) it can map the seed to plus or minus itself, or (iii) it can map the seed to plus or minus times some state in S_r other than the seed. I will call a generator that satisfies case (i) a type 1 generator, and so on.

In the first case, all of the elements of $\mathcal{H}_{r+1}-\mathcal{H}_r$ will also map the seed outside of S_r: If $N\in\mathcal{H}_{r+1}-\mathcal{H}_r$, then $N=MM_{r+1}$ for some $M\in\mathcal{H}_r$. Then if $\pm N|\phi\rangle\in S_r$, $N|\phi\rangle=\pm M'|\phi\rangle$ for some $M'\in\mathcal{H}_r$. Then $M_{r+1}|\phi\rangle=\pm M^{-1}M'|\phi\rangle\in S_r$, which contradicts the assumption. Thus, $S=S_a$ will always have size 2^b, where b is the number of type 1 generators.

In the second case, the new generator must act on each qubit as the identity I, as $-I$, or as Z_i, so type 2 generators can be written as the product of Z's. In principle, a type 2 generator could be -1 times the product of Z's, but the factor of -1 slightly complicates the process of picking seeds, so for simplicity I will assume it is not present. The method of choosing $\mathcal{H}$ that I give below will always create generators without such factors of -1.

In the third case, when $|\phi'\rangle=\pm M_{r+1}|\phi\rangle$ is already in S_r, then there exists $N\in\mathcal{H}_r$ with $N|\phi\rangle=|\phi'\rangle$. We can then use $N^{-1}M_{r+1}$ as a new generator instead of M_{r+1}, and since $N^{-1}M_{r+1}|\phi\rangle=\pm|\phi\rangle$, we are back to case (ii). After adding all of the generators, changing any of type 3 into type 2, we are left with b generators of type 1 and $a-b$ generators of type 2.

If one of the type 2 generators M_i gives a factor of -1 acting on the seed, the final state is 0:

$$\sum_{M\in\mathcal{H}} M|\phi\rangle=\left(\sum_{M\in\mathcal{H}} M\right)M_i|\phi\rangle=-\sum_{M\in\mathcal{H}} M|\phi\rangle=0. \tag{13}$$

Otherwise $|\psi\rangle$ is nonzero. We can simplify the computation of $|\psi\rangle$ by only summing over products of the type 1 generators, since the type 2 generators will only give us additional copies of the same sum. Then $|\psi\rangle$ will be the sum of 2^b quasiclassical states (with the appropriate signs).

Is this classification of generators going to be the same for all possible seeds? Anything that is a product of Z's has all quasiclassical states as eigenstates, and anything that is not a product of Z's has no quasiclassical states as eigenstates. Thus if a generator is type 2 for one seed, it is type 2 for all seeds. Type 1 generators cannot become type 3 generators because then the matrix $M^{-1}N$ would be type 2 for some states but not others. Thus, all of the states $|\psi_i\rangle$ are the sum of 2^b quasiclassical states, and $a-b$ of the generators of $\mathcal{H}$ are the product of Z's. Note that this also shows that the classification of generators into type 1 and type 2 generators does not depend on their order.

Since a seed produces a nontrivial final state if and only if it has an eigenvalue of $+1$ for all of the type 2 generators, all of the states $|\psi_i\rangle$ live in the joint $+1$ eigenspace of the $a-b$ type 2 generators, which has dimension $2^{n-(a-b)}$. We can partition the quasiclassical basis states of this eigenspace into classes based on the $|\psi_i\rangle$ in which they appear. Each partition has size 2^b, so there are 2^{n-a} partitions, proving the claimed dimension of T. The states $|\psi_i\rangle$ form a basis of T.

We can simplify the task of finding seeds for a basis of quantum code words. First, note that $|\mathbf{0}\rangle=|00\ldots0\rangle$ is always in the $+1$ eigenspace of any type 2 generator, so it can always provide our first seed. Any other quasiclassical seed $|\phi\rangle$ can be produced from $|\mathbf{0}\rangle$ by operating with some $N\in\mathcal{G}$ that is a product of X's. For $N|\mathbf{0}\rangle$ to act as the seed for a nontrivial state, N must commute with every type 2 generator in $\mathcal{H}$: If M_i is a type 2 generator, and $\{N,M_i\}=0$, then

$$M_i(N|\mathbf{0}\rangle)=-NM_i|\mathbf{0}\rangle=-N|\mathbf{0}\rangle. \tag{14}$$

But only quasiclassical states which have eigenvalue $+1$ give nontrivial code words, so N must commute with the type 2 generators. Two such operators N and N' will produce seeds for the same quantum code word iff they differ by an element of $\mathcal{H}$—i.e., $N^{-1}N'\in\mathcal{H}$. This provides a test for when two seeds will produce different code words, and also implies that the product of two operators producing different code words will also be a new code word. Thus, we can get a full set of 2^{n-a} seeds by taking products of $n-a$ operators $N_1,\ldots,N_{n-a}$. I will call the N_i *seed generators*. I do not know of any efficient method for determining the N_i.

Once we have determined the generators M_i of $\mathcal{H}$ and the seed generators N_i, we can define a unitary transformation to perform the encoding by

$$|c_1c_2,\ldots,c_k\rangle\mapsto\frac{1}{2^{b/2}}\prod_{M_i\text{ type }1}(I+M_i)N_1^{c_1}N_2^{c_2},\ldots,N_k^{c_k}|\mathbf{0}\rangle. \tag{15}$$

However, I do not know of an efficient way to implement this transformation.

Now I turn to the next question: how can we pick $\mathcal{H}$ so that all of the errors up to length $2t$ anticommute with some element of it? Given $M\in\mathcal{G}$, consider the function $f_M:\mathcal{G}\to\mathbf{Z}_2$,

$$f_M(N)=\begin{cases}0 & \text{if } [M,N]=0,\\ 1 & \text{if } \{M,N\}=0.\end{cases} \tag{16}$$

Then f_M is a homomorphism. If $\mathcal{H}=\langle M_1,M_2,\ldots,M_a\rangle$, then define a homomorphism $f:\mathcal{G}\to(\mathbf{Z}_2)^a$ by

$$f(N)=(f_{M_1}(N),f_{M_2}(N),\ldots,f_{M_a}(N)). \tag{17}$$

Below, I will actually write $f(N)$ as an a-bit binary string. With this definition of f, $f(N)=00\ldots0$ iff N commutes with everything in $\mathcal{H}$. We therefore wish to pick $\mathcal{H}$ so that $f(E)$ is nonzero for all E up to length $2t$. We can write any such E as the product of F and G, each of length t or less,

TABLE I. The values of $f(X_i)$, $f(Y_i)$, and $f(Z_i)$ for $n=8$.

X_1	01000	X_2	01001	X_3	01010	X_4	01011
Z_1	10111	Z_2	10000	Z_3	10110	Z_4	10001
Y_1	11111	Y_2	11001	Y_3	11100	Y_4	11010
X_5	01100	X_6	01101	X_7	01110	X_8	01111
Z_5	10010	Z_6	10101	Z_7	10011	Z_8	10100
Y_5	11110	Y_6	11000	Y_7	11101	Y_8	11011

and $f(E)\neq 0$ iff $f(F)\neq f(G)$. Therefore, we need to pick $\mathcal{H}$ so that $f(F)$ is different for each F of length t or less.

We can thus find a quantum error-correcting code by first choosing a different a-bit binary number for each X_i and Z_i. These numbers will be the values of $f(X_i)$ and $f(Z_i)$ for some $\mathcal{H}$ which we can then determine. We want to pick these binary numbers so that the corresponding values of $f(Y_i)$ and errors of length 2 or more (if $t>1$) are all different. While this task is difficult in general, it is tractable for $t=1$. In addition, even if all of the $f(E)$ are different, we still need to make sure that $\mathcal{H}$ fixes a nontrivial space of code words T by checking that $\mathcal{H}$ is Abelian and that its elements square to $+1$.

III. THE CODES

Now I will use the method described in Sec. II to construct an optimal nondegenerate quantum error-correcting code for $n=2^j$. The quantum Hamming bound (4) tells us that $k\leq n-j-2$, so we take $a=j+2$ and $j\geq 3$. I will also show explicitly the construction for $n=8$. Steane [12] has found the same $k=3$, $n=8$ code following inspiration from classical error-correcting codes, and Calderbank *et al.* [11] have found a different $k=3$, $n=8$ code.

We want to pick different $(j+2)$-bit binary numbers for X_i and Z_i $(i=1,\ldots,n)$ so that the numbers for Y_i, which are given by the bitwise XOR of the numbers for X_i and Z_i, are also all different. The numbers for $n=8$ are shown in Table I. In order to distinguish between the X's, the Y's, and the Z's, we will devote the first two bits to encoding which of the three it is, and the remaining j bits will encode which qubit i the error acts on (although this encoding will depend on whether it is an X, a Y, or a Z).

The first two bits are 01 for an X, 10 for a Z, and 11 for a Y, as required to make f a homomorphism. For the X_i's, the last j bits will just form the binary number for $i-1$, so X_1 is $0100\ldots0$, and X_n is $0111\ldots1$. The encoding for the last j bits for the Z_i's is more complicated. We cannot use the same pattern, or all of the Y_i's would just have all 0's for the last j bits. Instead of counting 0, 1, 2, 3, ..., we instead count 0, 0, 1, 1, 2, 2, ... Writing this in binary will not make all of the numbers for the Z's different, so what we do instead is to write them in binary and then take the bitwise NOT of one of each pair. This does make all of the Z's different. We then determine what the numbers for Y_i are.

How we pick which member of the pair to invert will determine whether all of the numbers for Y_i are different. For even j, we can just take the NOT for all odd i; but for odd j, we must take the NOT for odd i when $i\leq 2^{j-1}$ and for even i when $i>2^{j-1}$. A general proof that this method will give different numbers for all the Y_i's is given in Appendix B.

TABLE II. The generators of $\mathcal{H}$ and seed generators for $n=8$.

M_1	X_1	X_2	X_3	X_4	X_5	X_6	X_7	X_8
M_2	Z_1	Z_2	Z_3	Z_4	Z_5	Z_6	Z_7	Z_8
M_3	X_1	I	X_3	I	Z_5	Y_6	Z_7	Y_8
M_4	X_1	I	Y_3	Z_4	X_5	I	Y_7	Z_8
M_5	X_1	Z_2	I	Y_4	I	Y_6	X_7	Z_8
N_1	X_1	X_2	I	I	I	I	I	I
N_2	X_1	I	X_3	I	I	I	I	I
N_3	X_1	I	I	I	X_5	I	I	I

Now that we have the numbers for all of the 1-qubit errors, we need to determine the generators $M_1,\ldots,M_a$ of $\mathcal{H}$. Recall that the first digit of the binary numbers corresponds to the first generator. Since the first digit of the number for X_1 is 0, M_1 commutes with X_1; the first digits of the numbers for Y_1 and Z_1 are both 1, so M_1 anticommutes with Y_1 and Z_1. Therefore, M_1 is X_1 times the product of matrices which only act on the other qubits. Similarly, the first digit of the number for each X_i is 0 and the first digits for Y_i and Z_i are both 1, so $M_1=X_1X_2,\ldots,X_n$ (this is true even for $j>3$). Using the same principle, we can work out all of the generators.

The results for $n=8$ are summarized in Table II. Note that all of these generators square to $+1$ and that they all commute with each other. A proof of this fact for $j>3$ is given in Appendix C. Thus we have a code that encodes 3 qubits in 8 qubits, or more generally $n-j-2$ qubits in 2^j qubits. For these codes, there is 1 type 2 generator M_2. The remaining $j+1$ generators are type 1.

Table II also gives seed generators for $n=8$. We can see immediately that they all commute with M_2, the type 2 generator. It is less obvious that they all produce seeds for different states, but using them produces eight different quantum code words, listed in Table III, so they do, in fact, form a complete list of seed generators. This partly answers the question of how often we can saturate the quantum Hamming bound by showing that for one error, it can be saturated for arbitrarily large n. Although the methods given above may help somewhat, finding optimal codes to correct more than one error remains a difficult task.

ACKNOWLEDGMENTS

I would like to thank John Preskill for helpful discussions. This work was supported in part by the U.S. Department of Energy under Grant No. DE-FG03-92-ER40701.

APPENDIX A: PROOF THAT CERTAIN DEGENERATE CODES CANNOT DEFEAT THE QUANTUM HAMMING BOUND FOR $t=1$

While there is no known proof that degenerate quantum error-correcting codes cannot beat the quantum Hamming bound for arbitrary t and n, I will present a proof that codes to correct just one error are, in fact, limited by that bound, so long as the only source of degeneracies is when linearly in-

TABLE III. The quantum code words for the $n=8$ code.

$$|\psi_0\rangle=|00000000\rangle+|11111111\rangle+|10100101\rangle+|10101010\rangle+|10010110\rangle+|01011010\rangle$$
$$+|01010101\rangle+|01101001\rangle+|00001111\rangle+|00110011\rangle+|00111100\rangle$$
$$+|11110000\rangle+|11001100\rangle+|11000011\rangle+|10011001\rangle+|01100110\rangle$$
$$|\psi_1\rangle=|11000000\rangle+|00111111\rangle+|01100101\rangle+|01101010\rangle-|01010110\rangle$$
$$+|10011010\rangle+|10010101\rangle-|10101001\rangle+|11001111\rangle-|11110011\rangle$$
$$-|11111100\rangle+|00110000\rangle-|00001100\rangle-|00000011\rangle-|01011001\rangle-|10100110\rangle$$
$$|\psi_2\rangle=|10100000\rangle+|01011111\rangle+|00000101\rangle-|00001010\rangle+|00110110\rangle$$
$$+|11111010\rangle-|11110101\rangle+|11001001\rangle-|10101111\rangle+|10010011\rangle$$
$$-|10011100\rangle-|01010000\rangle+|01101100\rangle-|01100011\rangle-|00111001\rangle-|11000110\rangle$$
$$|\psi_3\rangle=|01100000\rangle+|10011111\rangle+|11000101\rangle-|11001010\rangle-|11110110\rangle$$
$$+|00111010\rangle-|00110101\rangle-|00001001\rangle-|01101111\rangle-|01010011\rangle$$
$$+|01011100\rangle-|10010000\rangle-|10101100\rangle+|10100011\rangle+|11111001\rangle+|00000110\rangle$$
$$|\psi_4\rangle=|10001000\rangle+|01110111\rangle-|00101101\rangle+|00100010\rangle+|00011110\rangle$$
$$-|11010010\rangle+|11011101\rangle+|11100001\rangle-|10000111\rangle-|10111011\rangle$$
$$+|10110100\rangle-|01111000\rangle-|01000100\rangle+|01001011\rangle-|00010001\rangle-|11101110\rangle$$
$$|\psi_5\rangle=|01001000\rangle+|10110111\rangle-|11101101\rangle+|11100010\rangle-|11011110\rangle$$
$$-|00010010\rangle+|00011101\rangle-|00100001\rangle-|01000111\rangle+|01111011\rangle$$
$$-|01110100\rangle-|10111000\rangle+|10000100\rangle-|10001011\rangle+|11010001\rangle+|00101110\rangle$$
$$|\psi_6\rangle=|00101000\rangle+|11010111\rangle-|10001101\rangle-|10000010\rangle+|10111110\rangle$$
$$-|01110010\rangle-|01111101\rangle+|01000001\rangle+|00100111\rangle-|00011011\rangle$$
$$-|00010100\rangle+|11011000\rangle-|11100100\rangle-|11101011\rangle+|10110001\rangle+|01001110\rangle$$
$$|\psi_7\rangle=|11101000\rangle+|00010111\rangle-|01001101\rangle-|01000010\rangle-|01111110\rangle$$
$$-|10110010\rangle-|10111101\rangle-|10000001\rangle+|11100111\rangle+|11011011\rangle+|11010100\rangle+|00011000\rangle$$
$$+|00100100\rangle+|00101011\rangle-|01110001\rangle-|10001110\rangle$$

dependent error matrices map a code word into a one-dimensional subspace. For instance, if three different errors map code words into a single two-dimensional subspace, this condition will not generally be satisfied.

Given a degenerate quantum error-correcting code of this type that corrects one error, we can list a number of conditions that describe which errors are degenerate. I will call these relations *degeneracy conditions*. As with the stabilizers in Sec. II, each independent condition will reduce the space of possible code words by a factor of 2. Note that I am not requiring that the basis for errors be the X, Y, Z basis I have used in the rest of the paper.[3]

Suppose there are l different degeneracy conditions describing the code. Each one equates two one-qubit errors, so at most $2l$ qubits are affected by the degenerate errors. The errors on the remaining $n-2l$ qubits must produce mutually orthogonal states. There are $3(n-2l)$ possible errors affecting those qubits.

Furthermore, errors on those qubits commute with the degenerate errors, since they act on different qubits, so if $M|\psi_i\rangle=N|\psi_i\rangle$ and E is an error that acts on a qubit unaffected by the degenerate errors,

$$ME|\psi_i\rangle=EM|\psi_i\rangle=EN|\psi_i\rangle=NE|\psi_i\rangle. \qquad \text{(A1)}$$

Thus, the state $E|\psi_i\rangle$ still satisfies the same set of degeneracy conditions. The space of states that satisfy the given set of l degeneracy conditions has dimension at most 2^{n-l}. To fit all the states $E|\psi_i\rangle$ inside it, if $l\leq n/2$, we must have

$$[1+3(n-2l)]2^k\leq 2^{n-l}, \qquad \text{(A2)}$$

or

$$k\leq n-l-\log_2[1+3(n-2l)]=g(l). \qquad \text{(A3)}$$

For $l=0$, this becomes the quantum Hamming bound. Now,

$$\frac{dg}{dl}=-1+\frac{6/\ln 2}{1+3(n-2l)}. \qquad \text{(A4)}$$

Therefore $g(l)$ is decreasing for

$$1+3(n-2l)\geq\frac{6}{\ln 2}, \qquad \text{(A5)}$$

$$l\leq\frac{n}{2}-\left(\frac{1}{\ln 2}-\frac{1}{6}\right). \qquad \text{(A6)}$$

Thus, the quantum Hamming bound holds for $l\leq(n-3)/2$. For $l>(n-3)/2$, we still have $k\leq n-l<(n+3)/2$. This automatically satisfies the quantum Hamming bound for $n\geq 13$ (see Table IV).

For $n<13$, $l>(n-3)/2$, we need a different argument. When $l<n-1$, there must always be at least one degeneracy condition that relates errors on two qubits that are unaffected by any other degeneracy conditions. There are three possible errors on each qubit, and only one pair of them are going to produce the same results, so there are still five different errors, plus the possibility of no error. As above, these errors will remain within the space that satisfies the other $l-1$ degeneracy conditions, so

$$(1+5)2^k\leq 2^{n-(l-1)}, \qquad \text{(A7)}$$

[3]The proof that the dimension of T is 2^{n-l} given in Sec. II only works for the X, Y, Z basis, but for this appendix, I only need the weaker result that the dimension of T is at least halved by any degeneracy condition that constrains a qubit unaffected by any of the other degeneracy conditions. This should be self-evident.

TABLE IV. The maximum k allowed by the quantum Hamming bound for $n \leq 13$.

n	k
5	1
6	1
7	2
8	3
9	4
10	5
11	5
12	6
13	7

or $k \leq n-l+(1-\log_2 6)$ (i.e., $k \leq n-l-2$). When $l>(n-3)/2$, this means $k \leq (n-1)/2$. Applying this condition for $n \leq 12$ restricts violations of the quantum Hamming bound to $n \leq 6$, specifically $n=6$ and $l=2$, and $n=4$ and $l=1$. For these two cases, we can directly apply Eq. (A2) to see that for $n=6$ and $l=2$, $k \leq 1$, in accordance with the quantum Hamming bound, and for $n=4$ and $l=1$, $k=0$.

Finally, for $l=n-1$, there must be at least one qubit that is only affected by a single degeneracy condition. All three errors on this qubit commute with the other $n-2$ degeneracy conditions, so

$$(1+3)2^k \leq 2^{n-(n-2)}. \tag{A8}$$

Therefore $k=0$, and the quantum Hamming bound holds for any degenerate quantum code where linearly independent errors can only map code words into a one-dimensional subspace.

APPENDIX B: PROOF THAT THE NUMBERS FOR Y_i ARE ALL DIFFERENT

The construction of the numbers for X_i and Z_i immediately demonstrates that they are all different. However, it is not as clear that all of the numbers for the Y_i's, which are determined by the numbers for the X_i's and Z_i's, will also be different. The first two bits just enforce the requirement that any Y_i is different from an X or a Z, so I will only consider the last j bits. All references to bit number in this appendix will refer to a position within the last j bits, so bit number "1" is actually bit 3, and bit "l" is actually bit $l+2$.

Consider the pictorial representation of the algorithm to pick the errors' binary numbers given in Table V. The numbers given for X_i are the actual numbers that appear. For Y_i and Z_i, we need to take an XOR with the parity of i (for j even or $i \leq 2^{j-1}$), or an XOR with the reverse of the parity of i (for j odd and $i>2^{j-1}$). We can see that before we apply the XOR, the number for Y_i encodes i in a unique fashion, since if i and i' first differ in the rth bit, then the numbers for Y_i and $Y_{i'}$ will also differ in the rth bit. The only way we could get two of the numbers to be the same would be if the XOR operation reverses one of a pair that would normally have complementary values in all bits.

Does this ever happen? Given a number $f(Y_i)$ for $i \leq n/2$, the number with complementary bits must appear for $i>n/2$, since the first digit does not change until then. The XOR will therefore collapse these two numbers into one whenever the parity of the appropriate i's is the same (for j odd) or different (for j even).

Pick some bit string starting with 0. There will be an $i \leq n/2$ such that Y_i has that number. Which i' will have the complementary bitstring? If we take the binary representation of i, it will begin with a 0 and the binary representation of i' will begin with a 1.[4] The next digit of i can be either 0 or 1, and from Table V we can see i' will have the same value for this digit. The third digits of i and i' will be opposite again. In general, a 0 in the rth digit of i or i' means the two squares relevant to the next digit will read 01, while a 1 in the rth digit will mean the two squares for the next digit will read 10. Thus, if i and i' agree in the rth digit, they will disagree in the next digit, and vice versa. Thus, i and i' agree on even-numbered digits and disagree on odd-numbered digits.

This means the last digit agrees for j even and disagrees for j odd. Therefore, the XOR will not make Y_i the same as $Y_{i'}$—it will either reverse both of them or neither of them. This explains why different rules for odd and even j were necessary.

TABLE V. The first four bits (of the last j) of the numbers for X_i, Y_i, and Z_i. The pth row corresponds to the pth bit and the columns in the pth row correspond to the possible values for the first p bits of i. For Y_i and Z_i, the actual numbers require an additional XOR with the parity or reverse of the parity of i.

<table>
<tr><th colspan="16">X_i:</th></tr>
<tr><td colspan="8">0</td><td colspan="8">1</td></tr>
<tr><td colspan="4">0</td><td colspan="4">1</td><td colspan="4">0</td><td colspan="4">1</td></tr>
<tr><td colspan="2">0</td><td colspan="2">1</td><td colspan="2">0</td><td colspan="2">1</td><td colspan="2">0</td><td colspan="2">1</td><td colspan="2">0</td><td colspan="2">1</td></tr>
<tr><td>0</td><td>1</td><td>0</td><td>1</td><td>0</td><td>1</td><td>0</td><td>1</td><td>0</td><td>1</td><td>0</td><td>1</td><td>0</td><td>1</td><td>0</td><td>1</td></tr>
<tr><th colspan="16">Z_i: Parity XOR</th></tr>
<tr><td colspan="8">0</td><td colspan="8">0</td></tr>
<tr><td colspan="4">0</td><td colspan="4">0</td><td colspan="4">1</td><td colspan="4">1</td></tr>
<tr><td colspan="2">0</td><td colspan="2">0</td><td colspan="2">1</td><td colspan="2">1</td><td colspan="2">0</td><td colspan="2">0</td><td colspan="2">1</td><td colspan="2">1</td></tr>
<tr><td>0</td><td>0</td><td>1</td><td>1</td><td>0</td><td>0</td><td>1</td><td>1</td><td>0</td><td>0</td><td>1</td><td>1</td><td>0</td><td>0</td><td>1</td><td>1</td></tr>
<tr><th colspan="16">Y_i: Parity XOR</th></tr>
<tr><td colspan="8">0</td><td colspan="8">1</td></tr>
<tr><td colspan="4">0</td><td colspan="4">1</td><td colspan="4">1</td><td colspan="4">0</td></tr>
<tr><td colspan="2">0</td><td colspan="2">1</td><td colspan="2">1</td><td colspan="2">0</td><td colspan="2">0</td><td colspan="2">1</td><td colspan="2">1</td><td colspan="2">0</td></tr>
<tr><td>0</td><td>1</td><td>1</td><td>0</td><td>0</td><td>1</td><td>1</td><td>0</td><td>0</td><td>1</td><td>1</td><td>0</td><td>0</td><td>1</td><td>1</td><td>0</td></tr>
</table>

APPENDIX C: PROOF THAT THE GENERATORS OF $\mathcal{H}$ COMMUTE

We can also use Table V to help us understand what the generators $M_1, \ldots, M_a$ of $\mathcal{H}$ look like. M_1 is always the product of all nX_i's, and M_2 is always the product of all the Z_i's. The other generators are a bit more complicated, but still behave systematically. As we advance i, they cycle through the sequence $I \to Z \to X \to Y$, with a change every $2^{j-(r-2)}$ qubits for generator M_r. In addition, the NOT

---

[4] I am ignoring the special case of $i=n/2$, which works on the same principle after the first digit of i.

TABLE VI. Comparisons of M_r and M_s in blocks of size $2^{j-(s-2)}$ when the normal cycle applies and when it is reversed by a NOT.

	$r=s+1$:															
M_r norm	I		I		Z		Z		X		X		Y		Y	
M_s norm	I		Z		X		Y		I		Z		X		Y	
M_r rev	X		X		Y		Y		I		I		Z		Z	
M_s rev	X		Y		I		Z		X		Y		I		Z	
	$r=s+2$:															
M_r norm	I	I	I	I	Z	Z	Z	Z	X	X	X	X	Y	Y	Y	Y
M_s norm	I	Z	X	Y	I	Z	X	Y	I	Z	X	Y	I	Z	X	Y
M_r rev	X	X	X	X	Y	Y	Y	Y	I	I	I	I	Z	Z	Z	Z
M_s rev	X	Y	I	Z	X	Y	I	Z	X	Y	I	Z	X	Y	I	Z

switches $I \leftrightarrow X$ and $Z \leftrightarrow Y$ whenever it applies — odd qubits for even j; odd qubits for the first half and even qubits for the second half for odd j. This immediately implies that every M_r for $r>2$ has equal numbers of X's, Y's, Z's, and I's, namely, 2^{j-2} of each. Since $j \geq 3$, this means there are an even number of Y's, so $M_r^2 = +1$.

Now, do the generators commute? Any time two generators have nontrivial but different operations on a qubit, we get a factor of -1 when we commute them. Therefore, we can determine if M_r and M_s commute by counting the qubits on which they differ and neither is the identity. If this count is even, they commute; if it is odd, they do not.

Since M_1 is all X's, it disagrees with M_r (for $r \geq 3$) whenever M_r has a Y or a Z. M_r has 2^{j-2} of each, so we get 2^{j-1} factors of -1, and $[M_1, M_r]=0$. Similarly, M_2 disagrees with M_r on X's and Y's, producing $2^{j-2}+2^{j-2}$ factors of -1, and $[M_2, M_r]=0$ also. M_1 and M_2 disagree on every qubit, and since there are an even number of qubits, $[M_1, M_2]=0$.

For $r,s \geq 3$, both M_r and M_s follow the pattern described above. I will consider the cases $s=r+1$, $s=r+2$, and $s>r+2$. Table VI compares M_r and M_s on blocks of size $2^{j-(s-2)}$.

In general, half of each block will be normal and half will be reversed by a NOT. Therefore, the number of factors of -1 from commuting M_r and M_s will generally be $2^{j-(s-3)}$ times the total number of nontrivial disagreements for the normal and reversed rows. We also need to consider a few special cases. When $r=3$, the generator never reaches the second half of the cycle, so we need to count up the disagreements only in the first half of the cycle. When $s=a=j+2$, the block size is 1, so the NOT either affects the whole block or it does not affect any of it. In this case, we need to count disagreements only on every other block. For even j, count the normal disagreements on even-numbered blocks and the reversed disagreements on odd-numbered blocks. For odd j, we must count normal disagreements on even-numbered blocks in the first half and odd-numbered blocks in the second half; count reversed disagreements on odd-numbered blocks in the first half and even-numbered blocks in the second half. We must also consider the combined special case of $r=3$, $s=a$.

For $s=r+1$, the general case gives four blocks with normal disagreements and two blocks with reversed disagreements. When $r=3$, there are two blocks with normal disagreements and two blocks with reversed disagreements. When $s=a$, and j is even, there are two blocks with normal disagreements and no blocks with reversed disagreements. When $s=a$ and j is odd, there are also two blocks with normal disagreements and no blocks with reversed disagreements. Because $a \geq 5$, we do not need to consider the combined special case. Thus, whenever $s=r+1$, there are an even number of disagreements and M_r and M_s commute.

For $s=r+2$, the general case gives six blocks with normal disagreements and six blocks with reversed disagreements. For $r=3$, there are two blocks with normal disagreements and four blocks with reversed disagreements. For $s=a$, j even, there are four blocks with normal disagreements and two blocks with reversed disagreements. For $s=a$, j odd, there are two blocks with normal disagreements and two blocks with reversed disagreements. For $r=3$, $s=a$, it does not matter if j is even or odd, since we only consider the first half. In this case, there is one block with a normal disagreement and one block with a reversed disagreement. In all of these cases, the *total* number of disagreements is even, so for $s=r+2$, $[M_r, M_s]=0$.

For $s>r+2$, generator M_s completes 2^{s-r-2} cycles before M_r advances to the next step in the cycle. This means we can just find the number of disagreements by multiplying the number of disagreements for $s=r+2$ by 2^{s-r-2}. We can do this even for the special cases, since the cycle repeats after four steps, which does not change the parity. Thus, there will always be an even number of disagreements, and all of the generators of $\mathcal{H}$ commute.

[1] P. W. Shor, Phys. Rev. A **52**, 2493 (1995).

[2] A. R. Calderbank and P. W. Shor, Phys. Rev. A **54**, 1098 (1996).

[3] A. Steane, Proc. R. Soc. London, Ser. A (to be published).

[4] R. Laflamme, C. Miquel, J. P. Paz, and W. H. Zurek (unpublished).

[5] C. H. Bennett, D. P. DiVincenzo, J. A. Smolin, and W. K. Wootters (unpublished).

[6] E. Knill and R. Laflamme (unpublished).

[7] A. Ekert and C. Macchiavello (unpublished).

[8] P. W. Shor and J. A. Smolin (unpublished).

[9] S. Lloyd (unpublished).

[10] C. H. Bennett, G. Brassard, S. Popescu, B. Schumacher, J. A. Smolin, and W. K. Wootters, Phys. Rev. Lett. **76**, 722 (1996).

[11] A. R. Calderbank, E. M. Rains, P. W. Shor, and N. J. A. Sloane (unpublished).

[12] A. M. Steane (unpublished).

VOLUME 77, NUMBER 15 PHYSICAL REVIEW LETTERS 7 OCTOBER 1996

Fault-Tolerant Error Correction with Efficient Quantum Codes

David P. DiVincenzo[1] and Peter W. Shor[2]
[1]IBM T. J. Watson Research Center, Yorktown Heights, New York 10598
[2]AT&T Research, Murray Hill, New Jersey 07974
(Received 22 May 1996)

We exhibit a simple, systematic procedure for detecting and correcting errors using any of the recently reported quantum error-correcting codes. The procedure is shown explicitly for a code in which one qubit is mapped into five. The quantum networks obtained are fault tolerant, that is, they can function successfully even if errors occur during the error correction. Our construction is derived using a recently introduced group-theoretic framework for unifying all known quantum codes. [S0031-9007(96)01353-1]

PACS numbers: 89.80.+h, 03.65.Bz, 89.70.+c

The past year has witnessed an astonishing rate of progress in the development of error-correction schemes for quantum memory and quantum computation. The initial discovery [1] that a qubit, when suitably encoded in a block of qubits, can withstand a substantial degree of interaction with the environment without degradation of its quantum state has been followed by myriad contributions which have identified many new coding schemes [2–13], considered their application in proposed experimental implementations of quantum computation [14–16], and established the relationship of quantum error-correcting codes to the preservation of quantum entanglement in a noisy environment [17]. The most recent work has unified all the known quantum codes within a group-theoretic framework [18].

Throughout the developments of the past year, there has been a hope that these quantum error-correcting codes would permit quantum computation to be done fault tolerantly. Such an outcome was not guaranteed; in classical computation, the existence of error-correction codes does not by itself ensure that logic can be performed using noisy gates. However, one of us has recently established a complete protocol for performing fault-tolerant quantum computation [19]. The protocol guarantees that, if the loss of fidelity of the quantum state between the operation of one quantum gate and the next, due to both decoherence and inaccuracy in the quantum-gate operation, is p, then the number of steps of quantum computation which can be completed successfully is $O(p^a \exp(b/p^c))$ (for some positive constants a, b, and c), a scaling law which appears very favorable for the ultimate physical implementation of large-scale quantum computation.

This fault-tolerant protocol lays down specific rules for how to use the previously discovered quantum error-correction codes. The class of codes first discovered by Calderbank and Shor [2] and Steane [3] conform to these rules, and can be used fault tolerantly; however, it has not been clear that the more efficient quantum codes which have been discovered more recently (see, e.g., [18]) could be utilized in a fault-tolerant computation. In this note we establish that errors in all known quantum error-correcting codes can be corrected in the necessary fault-tolerant way. We first show explicitly how this is done in one of the simplest efficient quantum codes, one which encodes a single qubit into five [4,17]. This result gives some interesting insights into the relationship between the different presentations of this code which have recently appeared in the literature, and it shows that it is actually necessary to use these different presentations to produce the fault-tolerant implementation of the error-correction procedure. We then show, using the recently developed group-theoretic framework for the quantum codes, that the protocol developed for the five-bit code can be generalized to permit all known codes to be used for error correction in a fault-tolerant way.

We begin with a short review of the five-qubit error-correcting code as presented in [17]. Using this code, an arbitrary qubit $|\xi\rangle = \alpha|0\rangle + \beta|1\rangle$ is represented by the five-qubit state $|\xi\rangle = \alpha|c_0\rangle + \beta|c_1\rangle$, where one choice of the "code words" is the pair of basis states

$$\begin{aligned}|c_0\rangle = |00000\rangle &\\ &+ |11000\rangle + |01100\rangle + |00110\rangle + |00011\rangle + |10001\rangle \\ &- |10100\rangle - |01010\rangle - |00101\rangle - |10010\rangle - |01001\rangle \\ &- |11110\rangle - |01111\rangle - |10111\rangle - |11011\rangle - |11101\rangle\end{aligned} \tag{1}$$

and

$$\begin{aligned}|c_1\rangle = |11111\rangle &\\ &+ |00111\rangle + |10011\rangle + |11001\rangle + |11100\rangle + |01110\rangle \\ &- |01011\rangle - |10101\rangle - |11010\rangle - |01101\rangle - |10110\rangle \\ &- |00001\rangle - |10000\rangle - |01000\rangle - |00100\rangle - |00010\rangle.\end{aligned} \tag{2}$$

When encoded in this way, the qubit can survive an interaction with the environment suffered by any one of the five qubits. For purposes of error correction, it is sufficient to take the error caused by the environment to

 0031-9007/96/77(15)/3260(4)$10.00

be of three different types [5,17]: bit i may suffer a bit-flip error, represented by the operator X_i acting on coded state $|\xi\rangle$; it may suffer a conditional phase-shift error (Z_i), or it may suffer both simultaneously (Y_i). (We use the notation of Refs. [11,18].) The right-hand column of Table I lists the 16 possible error processes P (including the no-error case $P = I$). During error correction, the erroneous state $P|\xi\rangle$ is subjected to some quantum-computation operations (one- and two-bit quantum gates [20]) so that measurements on some of the qubits will reveal the identity of the error process P, without disturbing the superposition of code words. When the error process is determined, the effect of P can be undone, returning the qubit to its undisturbed state $|\xi\rangle$.

It has now been shown by a number of authors [4,14,17] that there exist various quantum circuits which perform the necessary error correction on the five-bit coded state. However, none of them perform this error correction fault tolerantly (unlike the network of Fig. 1 which can operate fault tolerantly). We call a quantum error-correcting network fault tolerant if it can recover from errors *during* the operation of the network. Previous constructions are not fault tolerant because they use two-bit quantum gates involving pairs of qubits within the coded state. If an error occurs on one of these qubits before or during the operation of this two-bit gate, the error will, in general, propagate to both of the qubits, and to yet others if additional two-bit operations are performed. In the five-bit code, two errors are already more than can be recovered from, so such two-bit gates must be avoided. The network of Fig. 1 avoids them by using only two-bit gates which connect the coded bits to ancilla bits a, so that, with small modifications, it can be made perfectly fault tolerant. These modifications are described briefly in [19] and given in detail in [21].

TABLE I. The four measurement outcomes in the fault-tolerant error correction, and the error process P revealed by each.

M_3	M_4	M_0	M_1	P
0	0	0	0	I
0	0	0	1	Z_4
0	0	1	0	X_1
0	0	1	1	Z_3
0	1	0	0	X_3
0	1	0	1	X_0
0	1	1	0	Z_2
0	1	1	1	Y_3
1	0	0	0	Z_0
1	0	0	1	X_2
1	0	1	0	X_4
1	0	1	1	Y_4
1	1	0	0	Z_1
1	1	0	1	Y_0
1	1	1	0	Y_1
1	1	1	1	Y_2

To explain how the network of Fig. 1 works, we note that the code of Eqs. (1) and (2) can be presented in an infinite number of ways, all related by a change of basis of any one of the five qubits. Even if we confine ourselves to bases in which the superpositions all involve equal amplitudes as in Eqs. (1) and (2), the number of alternative presentations is very large. One important class of presentations is symmetric under cyclic permutation of the five qubits, as in the example given above. We will define a particular symmetric presentation, S, as the one in which $|0\rangle$ is coded as $|c_0\rangle + |c_1\rangle$, and $|1\rangle$ is coded as $|c_0\rangle - |c_1\rangle$.

Another class of presentation has been given in the work of Laflamme *et al.* [4]. Their presentation is obtained by starting with presentation S and applying the one-bit rotation $R = 1/\sqrt{2}\begin{pmatrix} 1 & 1 \\ 1 & -1 \end{pmatrix}$ to qubits 0 and 1 (we number the qubits 0–4 as in Fig. 1). In this presentation, the code words are

$$|c_0'\rangle = |00010\rangle + |00101\rangle - |01011\rangle + |01100\rangle + |10001\rangle - |10110\rangle - |11000\rangle - |11111\rangle, \tag{3}$$

and

$$|c_1'\rangle = |00000\rangle - |00111\rangle + |01001\rangle + |01110\rangle + |10011\rangle + |10100\rangle + |11010\rangle - |11101\rangle. \tag{4}$$

We will call this presentation L_3; except for a trivial relabeling of the qubits, this is exactly the one given in [4]. The reason for the subscript is that, since the L_3 presentation is *not* symmetric under cyclic permutation, there are five distinct ones L_{0-4}. The particular label 3 is used for this example because of an important property which this presentation possesses: all the basis states of both the code words in Eqs. (3) and (4) have even parity for the group of four qubits 0, 1, 2, and 4. Thus, a convenient label for this presentation is the qubit which is left out of this parity. Since an error can change this parity, we can learn one bit of information about the error process by collecting up this parity into the ancilla qubit a (done by the first four quantum XOR gates in Fig. 1), and performing measurement M_3 on a.

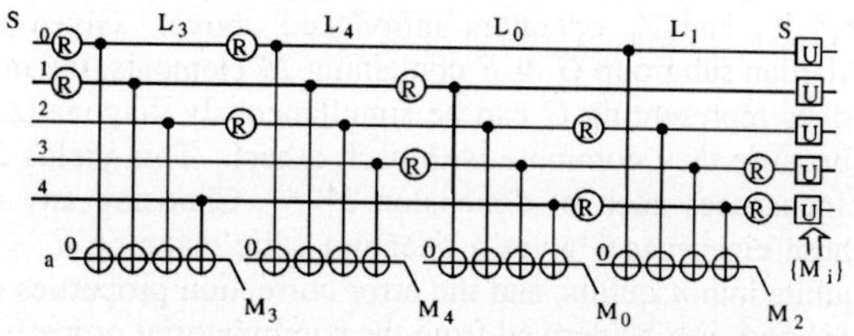

FIG. 1. Quantum network to correct for one-bit errors in the five-bit code in the S presentation. Four different code presentations $L_{3,4,0,1}$ are used in the different stages of error detection. By a simple modification of the ancilla space a, and by appropriate repetitions of the syndrome computation, this error-correction network can be made fault tolerant.

VOLUME 77, NUMBER 15 PHYSICAL REVIEW LETTERS 7 OCTOBER 1996

The remainder of the quantum circuit in Fig. 1 is self-explanatory. By passing in succession into three additional bases, those corresponding to the code presentations L_4, L_0, and L_1, three additional parity bits may be obtained in measurements M_4, M_0, and M_1. (In standard coding theory terminology, the outcome of these four measurements is called the *error syndrome.*) As Table I indicates, these measurements uniquely distinguish the error process P. This error can then be undone by returning the code to the original S basis and selecting the appropriate one-bit operation U.

As presented, this error-correction network is not completely fault tolerant, because an error occurring on one of the a bits can be transmitted back to one of the code qubits through the action of the XOR gates. For instance, if a phase error occurs on the ancilla qubit a between the second and third XOR gates in Fig. 1, the back action of the XOR gates results in two phase errors in the state of the code qubits, rendering them uncorrectable. However, as one of us has recently shown [19], the network may be made completely fault tolerant by replacing the single-bit ancilla a by a set of four qubits, each of which is initialized to a "cat" state $|0000\rangle + |1111\rangle$. If the targets of each of the XOR gates are four different qubits in the cat state, then the parity of the measured state of the four ancilla bits gives the same information as the measurements indicated in Fig. 1. However, the back action that makes the errors on the ancilla a dangerous is avoided. The ancilla errors may still result in a mistake in the measured syndrome; we prevent this from adding errors to the coded state by repetition of the entire network and syndrome measurement, before the one-bit operation U is performed [19]. Once the correct syndrome has been confirmed, the correct U may be applied [21].

The fact that the four measurements $M_{3,4,0,1}$ completely distinguish the error process is no accident; it is guaranteed by the group-theoretic structure of these codes [11,18]. In fact, the procedure devised above can be generalized to give a fault-tolerant error-correction procedure that covers every quantum code which is presently known, all of which are derivable as eigenspaces of Abelian subgroups of a group E [22].

The group E is obtained by taking all products of the X_i, Y_i, and Z_i operators introduced above. Given an Abelian subgroup G of E containing 2^g elements, the matrices representing G can be simultaneously diagonalized (because they commute with each other). This yields 2^g eigenspaces each of dimension 2^{n-g}. Choosing any of these eigenspaces gives a quantum code mapping $n - g$ qubits into n qubits, and the error correction properties of this code can be derived from the combinatorial properties of the subgroup G [11,18]. The subgroup G can be generated by an independent set of g of its elements, which we call generators; again, these generators are products of the X_i, Y_i, and Z_i operators. For instance, one of the generators for the five-bit code in the S presentation is, in the notation of [18], $X(11000)Z(00101)$; a 1 in the ith place in the X list means that X_i is included in the operation, a 1 in the Z list means that Z_i is included, and a 1 in both lists means that Y_i is included.

Each such generator of G gives a prescription for one stage of fault-tolerant error correction, as follows: First, a change of basis involving just one-bit operations is performed, in order to place the generator in the form $X(000,\ldots,0)Z(z_1z_2z_3,\ldots,z_n)$ where $z_i = 0$ or 1 (i.e., so that the generator contains only Z_i factors). The one-bit rotation required for the ith qubit is easily determined: if $X_i = 0$, do nothing; if $X_i = 1$ and $Z_i = 0$, apply R to the ith qubit; and if $X_i = Z_i = 1$, apply R', where $R' = 1/\sqrt{2}\begin{pmatrix} 1 & i \\ i & 1 \end{pmatrix}$. After this change of basis, the nonzero elements of the new Z bit string will be just those for which X or Z were nonzero in the original basis. The next step of the error correction is to collect up and measure the parity of the bits with nonzero entries in the Z string, using the ancilla technique discussed above. Finally, undo the basis transformation. Repeat this procedure for each generator of G.

It is guaranteed that this set of measurements will completely determine the error process P. The measurement on a quantum state corresponding to one of the generator matrices of G gives the eigenvalue of the quantum state with respect to that matrix, reducing the number of eigenspaces which the quantum state might lie in by a factor of 2. Thus, if the measurements are made for every matrix in a generator set for the subgroup G, this guarantees that the complete set of eigenvalues for this state with respect to the subgroup is known. This complete set of eigenvalues places the quantum state uniquely in one of the eigenspaces. The error processes X_i, Y_i, and Z_i permute these eigenspaces [18], so knowing which eigenspace a state belongs to is enough to uniquely determine the unitary transformation U of Fig. 1 which will correct the error. (U is also one of the unitary transformations X_i, Y_i, or Z_i.) The requirement that all the measurements be simultaneously observable can be seen to be the physical justification for the requirement that all the generator matrices commute.

The number of gates this construction gives for error correction of a quantum code can be estimated. Suppose it is applied to a quantum code mapping k qubits into n qubits, correcting t errors. (Many such codes have now been tabulated [12,18].) The syndrome will contain $n - k$ bits, and computing each bit of this syndrome requires at most n XOR gates. Similarly, between 0 and n rotation gates will also be required before and after the computation of each of the bits of the syndrome. Thus, the number of gates required by this technique for an n-qubit code is at most $2n(n - k + 1)$, and the number of ancilla bits needed is no greater than $n(n - k)$. The suitable use of this error-correction network will be fault tolerant: up to t errors can occur during the error correction process itself without irretrievably damaging the state of the k coded qubits.

The class of quantum error-correcting codes given in [2,3] have generators which are either products only of Z's or only of X's. This technique applied to these codes thus reduces to first finding the parity of sets of qubits corresponding to the generators composed of Z's, next applying the basis transformation R to each qubit, then finding the parities corresponding to generators composed of X's, and finally undoing the basis transformation R on each qubit. This is exactly the prescription given by Steane [3]. For this class of codes, the correction procedure for bit-flip (X) errors can be decoupled from the treatment of phase (Z) errors. The bit-flip (X) errors affect the eigenvalues of matrices which are a product of Z's, and vice versa. Each type of error can be thought of classically (in the appropriate basis) and corrected using classical techniques, as is emphasized in Steane [3].

To conclude, we have shown that the group-theoretic structure of all the reported quantum error-correcting codes provides rules for designing very simple quantum networks to detect errors and restore the quantum system to its undisturbed state. These networks are superior to previously reported ones in that they can be implemented in a fault-tolerant way. We note that our result does not provide a complete solution for how to use the most efficient quantum codes in fault-tolerant quantum computation, since this would require a fault-tolerant implementation of multibit gates on the coded qubits [19]. Such fault-tolerant gate implementations are known for the nonoptimal codes of [2,3], but it is not yet clear that they exist for all the codes derived from the group E (however, see [13]). Even without this, though, it is clear that the procedures developed here may ultimately have a variety of applications for quantum memory, quantum communications, and quantum computation.

We would like to thank Rob Calderbank for helpful discussions.

[1] P. W. Shor, Phys. Rev. A **52**, 2493 (1995).

[2] A. R. Calderbank and P. W. Shor, Phys. Rev. A **54**, 1098 (1996).

[3] A. M. Steane, Report No. quant-ph/9601029 [Proc. R. Soc. London A (to be published)]; Phys. Rev. Lett. **77**, 793 (1996).

[4] R. Laflamme, C. Miquel, J.-P. Paz, and W. H. Zurek, Phys. Rev. Lett. **77**, 198 (1996).

[5] A. Ekert and C. Macchiavello, Report No. quant-ph/9602022.

[6] L. Vaidman, L. Goldenberg, and S. Wiesner, Phys. Rev. A **54**, R1745 (1996).

[7] P. W. Shor and J. A. Smolin, Report No. quant-ph/9604006.

[8] S. Lloyd, Report No. quant-ph/9604015.

[9] B. Schumacher, Report No. quant-ph/9604023.

[10] E. Knill and R. Laflamme, Report No. quant-ph/9604034.

[11] D. Gottesman, Phys. Rev. A **54**, 1862 (1996).

[12] A. M. Steane, Report No. quant-ph/9605021.

[13] W. H. Zurek and R. Laflamme, Report No. quant-ph/9605013.

[14] S. L. Braunstein, Report No. quant-ph/9604036.

[15] M. B. Plenio, V. Vedral, and P. L. Knight, Report No. quant-ph/9603022.

[16] T. Pellizzari (private communication).

[17] C. H. Bennett, G. Brassard, S. Popescu, B. Schumacher, J. A. Smolin, and W. K. Wootters, Phys. Rev. Lett. **76**, 722 (1996); C. H. Bennett, D. P. DiVincenzo, J. A. Smolin, and W. K. Wootters, Report No. quant-ph/9604024.

[18] A. R. Calderbank, E. M. Rains, P. W. Shor, and N. J. A. Sloane, Report No. quant-ph/9605005; Report No. quant-ph/9608006.

[19] P. W. Shor, Report No. quant-ph/9605011.

[20] A. Barenco, C. H. Bennett, R. Cleve, D. P. DiVincenzo, N. Margolus, P. W. Shor, T. Sleator, J. A. Smolin, and H. Weinfurter, Phys. Rev. A **52**, 3457 (1995).

[21] M. B. Plenio, V. Vedral, and P. L. Knight, Report No. quant-ph/9608028.

[22] A. R. Calderbank, P. J. Cameron, W. M. Kantor, and J. J. Seidel, Proc. London Math. Soc. (to be published).

Quantum Channels

Quantum Channels

Christopher A. Fuchs
California Institute of Technology

The study of the quantum mechanical channel is the study of the general trace-preserving completely positive linear map from one operator algebra to another. What do these words mean?

To get a bit of the flavor, let us go back to the classical information theoretic notion of a channel. There the idea is that one participant, Alice, would like to send a message (a meaningful message) to another participant Bob who is some distance away. The only means available for carrying this out is to prepare some physical system (the information carrier) one way or the other depending on the message and then send it on its way. The process is complete when Bob finally retrieves the system and a has a look at it. What makes the notion of a channel interesting is all the stuff that can happen in between: along the course of its travel, the carrier might be jostled about in any number of ways. Among these are ones that are beyond Alice and Bob's control and indeed ones that may not be completely predictable by them beforehand. Because of this, one has no recourse but to view a classical channel in the following way:

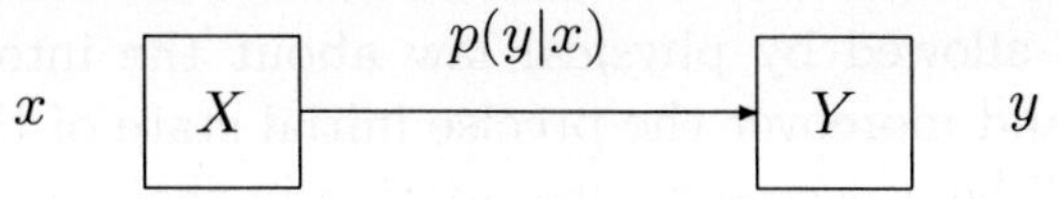

The message is loaded onto the carrier via some alphabet X and ultimately emerges as some new alphabet Y (possibly containing the same characters as X); the unknowns in the channel's action are represented via the transition probabilities $p(y|x)$ between the input and output letters. Of course, one might attempt to give a more detailed account of the message's voyage but such a more detailed account is irrelevant to the information transmission problem: all relevant aspects of the problem are encoded in the transition probabilities $p(y|x)$.

The questions that can be asked about how Alice and Bob can function in spite of the handicap of this unavoidable stochasticity are manifold and make up the subject of classical information theory. For instance, when Alice and Bob are willing to add the redundancy of two, three, or more transmissions, how much more reliability can this give them? What is the largest number of bits per transmission that can be gotten across if one is willing to go the limit of infinite numbers of transmissions? These are just a sampling.

The subject of quantum information theory has much the same motivation as that of the classical theory: Alice needs to send the preparation of a system—perhaps for the purpose of sending a real message or simply for the sheer pleasure of sending a quantum state—and the only way for it to get to Bob is to travel through territories unknown or at least uncontrollable. Those territories unknown or uncontrollable are known as the *quantum channel.* Pictorially, we have this:

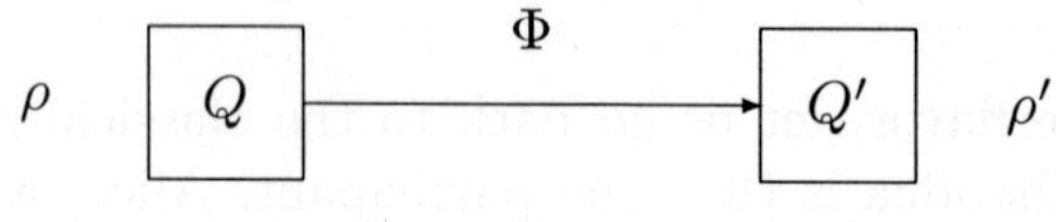

The quantum state ρ is loaded onto the system Q and ultimately emerges as some state ρ' on a system Q' (possibly isomorphic to the original system but not necessarily). The unknowns or uncontrollabilities in the channel's action are represented by way of a mapping $\Phi : \rho \longrightarrow \rho'$.

The question before us in these lectures is to delineate the structure of the mappings Φ. The technical name for this structure is the set of trace-preserving completely positive linear maps. These maps take the place of the simple transition probabilities $p(y|x)$ of the classical case and, in fact, make life much more interesting. Perhaps the most interesting thing about the quantum channel at the outset is this: even if Alice and Bob have the maximum knowledge allowed by physical law about the interaction between the carrier and its environment and moreover the precise initial state of the environment (as depicted below), there may still be a necessary degradation in their signal.

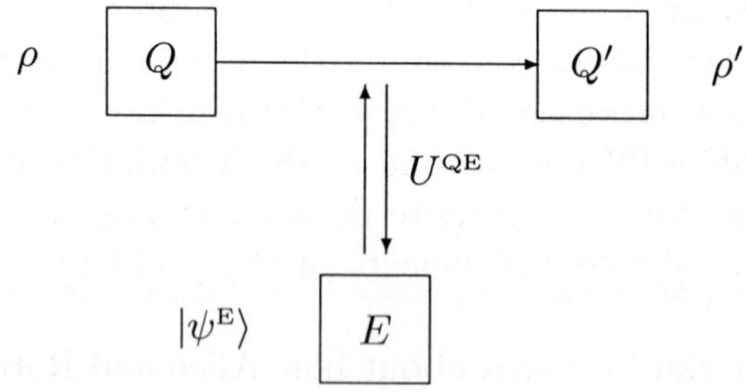

This is because the environment may become entangled with the carrier system. This feature of the quantum channel is precisely what Schrödinger dubbed in 1935 as "not … *one* but rather *the* characteristic trait of quantum mechanics, the one that enforces its entire departure from classical lines of thought."

There are startling nonclassical features in every turn for this new notion a channel. For instance, in the way in which the most reliable transmission of classical information requires that all the separate signals be measured collectively instead of separately. Or in the way in which it is useful to throw away some distinguishability for the signals before even sending them. These lectures should provide a firm foundation for the further exploration of the idiosyncrasies of the quantum channel: from the structure of the set as a whole (convexity features, representation theorems), to relevant quantitative measures of how they degrade signal distinguishability (the quantum no-broadcasting theorem, the monotonicity of the Holevo capacity), to a cluster of open questions that can be asked of them (e.g., the various quantum capacities).

1 Selected Reading

1. K.-E. Hellwig, "General Scheme of Measurement Processes," Int. J. Theor. Phys. **34**, 1467–1479 (1995).

2. M.-D. Choi, "Completely Positive Linear Maps on Complex Matrices," Lin. Alg. Appl. **10**, 285–289 (1975).

3. B. Schumacher, "Sending Entanglement Through Noisy Quantum Channels," Phys. Rev. A **54**, 2614–2617, 2625–2627 (1996).

4. H. Barnum, C. M. Caves, C. A. Fuchs, R. Jozsa, and B. Schumacher, "Noncommuting Mixed States Cannot Be Broadcast," Phys. Rev. Lett. **76**, 2818–2821 (1996).

5. B. Schumacher and M. D. Westmoreland, "Sending Classical Information via Noisy Quantum Channels," Phys. Rev. A **56**, 131–138 (1997).

6. C. A. Fuchs, "Nonorthogonal Quantum States Maximize Classical Information Capacity," Phys. Rev. Lett. **79**, 1163–1166 (1997).

2 Further Reading

1. I. L. Chuang and M. A. Nielsen, "Prescription for Experimental Determination of the Dynamics of a Quantum Black Box," J. Mod. Opt. **44**, 2455–2467 (1997)

2. B. M. Terhal, I. L. Chuang, D. P. DiVincenzo, M. Grassl, and J. A. Smolin, "Simulating Quantum Operations with Mixed Environments," `quant-ph/9806095`.

3. K. Hellwig and K. Kraus, "Pure Operations and Measurements," Comm. Math. Phys. **11**, 214–220 (1969).

4. K. Hellwig and K. Kraus, "Operations and Measurements. II," Comm. Math. Phys. **16**, 142–147 (1970).

5. K. Kraus, "General State Changes in Quantum Theory," Ann. Phys. **64**, 311–335 (1971).

6. K. Kraus, *States, Effects, and Operations: Fundamental Notions of Quantum Theory*, Lecture Notes in Physics, vol. 190, Berlin: Springer-Verlag, 1983.

7. L. J. Landau and R. F. Streater, "On Birkhoff's Theorem for Doubly Stochastic Completely Positive Maps of Matrix Algebras," Lin. Alg. Appl. **193**, 107–127 (1993).

8. J. A. Poluikis and R. D. Hill, "Completely Positive and Hermitian-Preserving Linear Transformations," Lin. Alg. Appl. **35**, 1–10 (1981).

9. C.-K. Li and H. J. Woerdeman, "Special Classes of Positive and Completely Positive Maps," Lin. Alg. Appl. **255**, 247–258 (1997).

10. M. Czachor and M. Kuna, "Complete Positivity of Nonlinear Evolution: A Case Study," Phys. Rev. A **58**, 128–134 (1998).

11. E. H. Lieb, "Convex Trace Functions and the Wigner-Yanase-Dyson Conjecture," Adv. Math. **11**, 267–288 (1973).

12. E. H. Lieb and M. B. Ruskai, "Proof of the Strong Subadditivity of Quantum Mechanical Entropy," J. Math. Phys. **14**, 1938–1941 (1973).

13. G. Lindblad, "Completely Positive Maps and Entropy Inequalities," Comm. Math. Phys. **40**, 147–151 (1975).

14. M. A. Nielsen, *Quantum Information Theory*, Ph.D. thesis, University of New Mexico, 1998.

15. E. H. Lieb and M. B. Ruskai, "Some Operator Inequalities of the Schwarz Type," Adv. Math. **12**, 269–273 (1974).

International Journal of Theoretical Physics, Vol. 34, No. 8, 1995

General Scheme of Measurement Processes

K.-E. Hellwig[1]

Received November 10, 1994

Quantum mechanical operations are motivated and their formal representation is derived from principles of statistics as well as from interaction processes.

1. INTRODUCTION

For almost twenty years the problem of quantum measurement did not attract the interest of a broader physical community. The development of quantum optics, especially the progress with optical waveguides, which opened possibilities for optical communication systems, gave a new impetus to work in this field. Now quantum measurement theory is fundamental to quantum communication and quantum information theory. In the following I will confine myself to describing operations. Operations are the simplest nontrivial quantum measurements. They show the main features and difficulties of the theory of quantum measurements. Operations can be the starting point of more detailed investigations. The main material presented here is taken from my dissertation (Hellwig, 1967, 1969, 1971), some papers together with Kraus (Hellwig and Kraus, 1969, 1970, 1971), later papers by Kraus (1971, 1977), who discovered the complete positivity of the representing maps, as well as his book Kraus (1983), in which this knowledge is collected. Further recommended literature is Pauli (1933), Davies and Lewis (1970), Ludwig (1976), Davies (1976), Gudder (1979), and Busch *et al.* (1991).

2. SOME GENERALITIES

Usually, the result of a measurement is understood as a statement about the presence or absence of some accidental property at an individual physical

[1] Institut für Theoretische Physik, Technische Universität Berlin, Berlin, Germany.

system on which the measurement has been performed. This is well illustrated in classical mechanics. Here the essence of the physical system together with the external conditions is encoded in a symplectic manifold $\mathcal{M}$ and a Hamiltonian function. The accidence of an individual system is represented by a point of $\mathcal{M}$ which is unknown in general. The result of a measurement consists in the statement that this point is contained in a subset $\alpha \subseteq \mathcal{M}$ which may be the preimage of some function on $\mathcal{M}$, the value of which has been measured.

The philosophy of classical individual systems states that any such system at any time is situated at a point $p \in \mathcal{M}$, i.e., exactly the accidentals α containing p are present and any other is absent. For any accidental it is decided whether it is present or absent. On the other hand, it is impossible to prepare the system in such a way that p results with certainty. The result of a real preparation can at most be a probability distribution described by a probability measure on the Borel sets of $\mathcal{M}$ which is absolutely continuous with respect to the Lebesgue measure. Let us consider a preparation procedure by which the distribution on $\mathcal{M}$ is given by the probability measure

$$\mu: \mathfrak{B}(\mathcal{M}) \to [0, 1]$$

where $\mathfrak{B}(\mathcal{M})$ denotes the Borel sets of $\mathcal{M}$.

A measurement answering the question

$$\begin{matrix} m \in \alpha & \text{or} & m \notin \alpha \\ \text{'yes'} & \text{or} & \text{'no'} \end{matrix}$$

where $\alpha \in \mathfrak{B}(\mathcal{M})$ and $\mu(\alpha) \in [0, 1]$, decomposes the ensemble into the statistical mixture of two subensembles according to the classical formula of Bayes:

$$\mu = \mu(\alpha)\,\frac{\mu(\cdot \cap \alpha)}{\mu(\alpha)} + \mu(\mathcal{M}\backslash\alpha)\,\frac{\mu(\cdot \cap (\mathcal{M}\backslash\alpha))}{\mu(\mathcal{M}\backslash\alpha)}$$

The subensembles with the probability measures

$$\hat{\mu}_+ := \frac{\mu(\cdot \cap \alpha)}{\mu(\alpha)}, \qquad \hat{\mu}_- := \frac{\mu(\cdot \cap (\mathcal{M}\backslash\alpha))}{\mu(\mathcal{M}\backslash\alpha)}$$

can be prepared by selecting those individual systems for which the result of the measurement is 'yes' or 'no,' respectively. Bayes' formula expresses the nondisturbance principle of classical measurements in that the set of accidentals present at an individual system is the same before and after the measurement.

The classical philosophy about the accidence of physical systems does not apply to quantum physics. Nevertheless, selection procedures with respect to 'yes'–'no' experiments like the classical one just described can still be performed. They are called quantum mechanical operations. However, the 'nondisturbance principle' and the formula of Bayes do not hold generally. The description of operations leads to generalizations of the Bayes formula which include the necessary quantal state changes.

Instead of the symplectic manifold $\mathcal{M}$ with a distinguished Hamiltonian function we have, as the result of a quantization procedure, a complex Hilbert space $\mathcal{H}$ with a distinguished Hamiltonian operator H. The Boolean algebra formed by the characteristic functions of the Borel sets of $\mathcal{M}$ is in a sense replaced by the lattice formed by the orthogonal projections $\mathcal{P}(\mathcal{H})$, the so-called quantum logic. Recalling that the self-adjoint operators correspond to their spectral measures, quantum logics at first sight seem to extend classical Boolean logics to a more general one, but from the physical point of view this is a restriction because of the incompatibility of position and momentum. What is the accidence of an individual quantum system?

To change the classical philosophy into a suitable one for quantum systems hidden variables have been invented. As shown by the experiments of A. Aspect improving the quantal form of Bell's inequality in Bohm's form of the Einstein–Podolski–Rosen situation, hidden variables cannot be maintained together with the principle of locality. Hence, believing in the absence of action at a distance, one has to forget about hidden variables. But also smaller sets of 'elements of physical reality' in the sense of Einstein, Podolski, and Rosen are ruled out by such experiments.

To understand the following we need not enter into the deep philosophical problems about quantum reality. We only need to speak about:

- Preparation procedures leading to probability measures on $\mathcal{P}(\mathcal{H})$.
- Registration of classically observable effects on macroscopic apparatuses occurring after interaction with a quantum object.
- Selection procedures which render subensembles according to occurred or not occurred effects.

The task is to characterize the generalizations of the classical procedure and of Bayes' formula without using the classical philosophy about the realization of the accidence and the nondisturbance principle.

3. OPERATIONS AND EFFECTS

The fundamental concepts of a statistical theory are:

- A convex structure $\mathfrak{S}$ which represents the set of preparation procedures involved in the theory.

- A set $\tilde{\mathfrak{E}}$ representing the set of 'yes'–'no' registration procedures including a trivial registration procedure which always answers 'yes.'
- A probability law $\tilde{\mu}$: $\tilde{\mathfrak{S}} \times \tilde{\mathfrak{E}} \to [0, 1]$.

A trivial 'yes–no' registration procedure $\tilde{e}$ fulfils $\tilde{\mu}(\cdot, \tilde{e}) = 1$. One forms classes of equivalent preparation procedures and classes of equivalent 'yes'–'no' registration procedures in the obvious way and introduces:

- The convex structure of states $\mathfrak{S}$ consisting of the classes of equivalent preparation procedures.
- The set of effects $\mathfrak{E}$ consisting of the classes of equivalent registration procedures; the class of trivial registration procedures shall be denoted by e.
- The probability law μ: $\mathfrak{S} \times \mathfrak{E} \to [0, 1]$.

For the trivial effect e, $\mu(\cdot, e) = 1$. The advantage of this factorization is that the set of effects is separating the set of states and the set of states is separating the set of effects. By a well-known and simple construction one shows that $\mathfrak{S}$, $\mathfrak{E}$, μ, and e can be identified with objects in a dual pair of partially ordered normed complete vector spaces $\mathcal{B}$, $\mathcal{B}'$, where:

- $\mathfrak{S}$ forms a base of the convex cone $\mathcal{C}^+$ of positive elements of $\mathcal{B}$, the norm of which is the base norm.
- e is the order unit of $\mathcal{B}'$, $\mathfrak{E}$ is its order unit segment, and the norm is the order unit norm.
- μ is just the restriction to $\mathfrak{S} \times \mathfrak{E}$ of the bilinear pairing $\langle \cdot, \cdot \rangle$ of $(\mathcal{B}, \mathcal{B}')$.

Although the following can be formulated in this abstract setting, we will confine ourselves to the Hilbert space model of quantum mechanics. Let $\mathcal{H}$ denote a Hilbert space; then:

- $\mathcal{B}$ is the space $\mathcal{B}_s(\mathcal{H})_\infty$ of Hermitian trace class operators in $\mathcal{H}$ with the trace norm $\|T\| := \mathrm{tr}(|T|)$, $T \in \mathcal{B}_s(\mathcal{H})_\infty$.
- $\mathcal{B}'$ is the space of bounded Hermitian operators $\mathcal{B}_s(\mathcal{H})$ in $\mathcal{H}$ with operator norm.
- $\langle T, A \rangle := \mathrm{tr}(T, A)$, $T \in \mathcal{B}_s(\mathcal{H})_\infty$, $A \in \mathcal{B}_s(\mathcal{H})$.
- $\mathfrak{S}$ is the set of positive elements $W \in \mathcal{B}_s(\mathcal{H})_\infty$ with $\mathrm{tr}(W) = 1$.
- $\mathfrak{E} := \{F \in \mathcal{B}_s(\mathcal{H}) | 0 \leq F \leq 1\}$.
- e is the unit operator 1 of $\mathcal{H}$.

In the following let $W \in \mathfrak{S}$ and $F, G \in \mathfrak{E}$.

It is useful to have a formal scheme of a 'yes'–'no' measurement in mind (Fig. 1), showing preparation and registration apparatuses as black boxes. Let N_+ be the number of results 'yes' and N_- be the number of results 'no' in a series of N experiments. Then for $N \to \infty$

Fig. 1.

$$\frac{N_+}{N} \to \mathrm{tr}(WF), \qquad \frac{N_-}{N} \to \mathrm{tr}(W(1 - F))$$

A nonselective operation is a 'yes'–'no' experiment which does not absorb the objects such that subsequent experiments with them can be performed. The formal scheme of a nonselective operation is shown in Fig. 2, where $W \to \hat{W}$ describes the state change caused by the operation and G is the operator of a subsequent measurement.

In a selective operation the objects which cause the answer 'no' will be absorbed while the objects causing the answer 'yes' become free thereafter and are available for subsequent experiments. By this selection procedure the state change $W \to \hat{W}_+$ is caused. The formal scheme is shown in Fig. 3.

Analogously, a selective operation can be considered in which the objects causing '−' become free thereafter and those causing '+' are absorbed. The state prepared by this procedure is denoted by $\hat{W}_-$.

If the classical nondisturbance principle held, the density operators W, $\hat{W}_+$, $\hat{W}_-$, and F would be related by the formula of Bayes. Since this principle fails to hold, we have to look for more general relations.

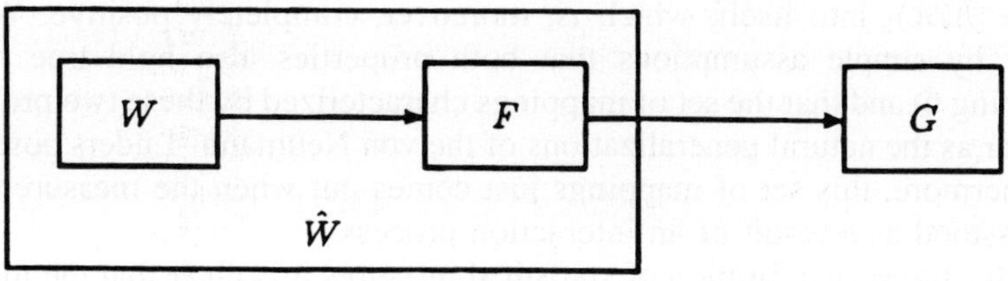

Fig. 2.

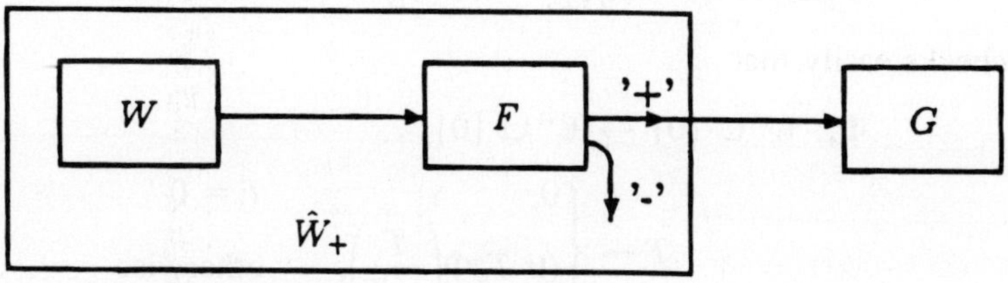

Fig. 3.

Obviously we can state

$$\hat{W} = \begin{cases} \hat{W}_+, & \mathrm{tr}(WF) = 1 \\ \hat{W}_-, & \mathrm{tr}(WF) = 0 \\ \mathrm{tr}(WF)\hat{W}_+ + \mathrm{tr}(W(1 - F))\hat{W}_-, & \text{otherwise} \end{cases}$$

J. von Neumann and G. Lüders (Lüders, 1951) supposed in the case that the effect operator is a projection operator E that

$$\hat{W}_+ = \frac{EWE}{\mathrm{tr}(WF)}, \qquad \hat{W}_- = \frac{(1 - E)W(1 - E)}{\mathrm{tr}(W(1 - F))}$$

This assumption is often called the "minimal disturbance principle." These formulas presuppose that the effect is presented by a projection operator and do not make sense for general effect operators F, $0 \leq F \leq 1$.

Instead of the projection postulate

$$W \mapsto EWE$$

we now introduce a mapping

$$\Phi\colon \mathfrak{S} \to \mathscr{C}^+ \cup \{0\} \subseteq \mathscr{B}$$

$$W \mapsto \begin{cases} \mathrm{tr}(WF)\hat{W}_+, & \mathrm{tr}(WF) \neq 0 \\ 0, & \text{otherwise} \end{cases}$$

which makes sense also in the general case. Observe now that the projection postulate is the restriction to $\mathfrak{S}$ of a complex linear map of the complex space $\mathscr{B}(\mathscr{H})_1$ into itself, which is, moreover, completely positive. We will show by simple assumptions that both properties also hold true for the mapping Φ and that the set of mappings characterized by these two properties appear as the natural generalizations of the von Neumann–Lüders postulates. Furthermore, this set of mappings just comes out when the measurement is understood as a result of an interaction process.

By the very definition of statistical mixtures it is clear that the mapping Φ must be affine, i.e., for $S_1, S_2 \in \mathfrak{S}$ and $0 \leq \lambda \leq 1$ there holds

$$\Phi(\lambda S_1 + (1 - \lambda)S_2) = \lambda\Phi(S_1) + (1 - \lambda)\Phi(S_2)$$

One checks easily that

$$\Phi_+\colon \mathscr{C}^+ \cup \{0\} \to \mathscr{C}^+ \cup \{0\}$$

$$T \mapsto \begin{cases} 0, & T = 0 \\ (\mathrm{tr}\, T)\Phi\left(\dfrac{T}{\mathrm{tr}\, T}\right), & \text{otherwise} \end{cases}$$

extends Φ from $\mathfrak{S}$ to $\mathcal{C}^+ \cup \{0\} \subseteq \mathcal{B}_s(\mathcal{H})_1$, where $\mathcal{C}^+$ is the positive cone of the real space of the symmetric trace class operators, and this extension is homogeneous for positive numbers and additive. Since the positive cone of $\mathcal{B}_1(\mathcal{H})$ is generating, i.e., any $T \in \mathcal{B}_1(\mathcal{H})$ can be written as $T = T_+ - T_-$, T_+, $T_- \in \mathcal{C}^+ \cup \{0\}$, it is almost trivial to check that

$$\Phi_r(T) := \Phi_r(T_+) + \Phi_r(T_-)$$

is well defined and a linear extension of Φ_+ from $\mathcal{C}^+ \cup \{0\}$ to $\mathcal{B}_s(\mathcal{H})_1$, which is positive by construction. By

$$\Phi_c(T_1 + iT_2) := \Phi_r(T_1) + i\Phi_r(T_2)$$

this map extends linearly to the complex space of all trace class operators. Finally, one can show that Φ_c is bounded with respect to the trace norm, i.e.,

$$\|\Phi(T)\|_1 \leq C\|T\|_1, \qquad \|T\|_1 = \mathrm{tr}[(T + T)^{1/2}]$$

and that a bound is given by

$$C = \sup_{W \in \mathfrak{S}} \mathrm{tr}(\Phi(W)) \leq 1$$

By construction there holds for $F \in \mathfrak{F}$ and $W \in \mathfrak{S}$

$$\mathrm{tr}(WF) = \mathrm{tr}(\Phi(W))$$

Since the space of bounded linear operators $\mathcal{B}(\mathcal{H})$ is just the Banach dual of the space $\mathcal{B}(\mathcal{H})_1$ with the trace norm and the extension Φ_c of Φ is bounded, we can introduce uniquely the dual mapping

$$\Phi_c^*: \mathcal{B}(\mathcal{H}) \to \mathcal{B}(\mathcal{H})$$

by the requirement that for any $T \in \mathcal{B}(\mathcal{H})_1$

$$\mathrm{tr}(T\Phi_c^*(X)) = \mathrm{tr}(\Phi_c(T)X), \qquad X \in \mathcal{B}(\mathcal{H})$$

Φ_c^* is a linear, operator-norm bounded, positive map with

$$\|\Phi_c^*(X)\| \leq (\sup_{W \in \mathfrak{S}} \mathrm{tr}(\Phi(W))\|X\|$$

such that the bound is less than one. Obviously there holds

$$F = \Phi_c^*(1)$$

Summing up to this point, we have established, by the purely statistical argument that convex mixtures of preparations can always be prepared, a relation between $W \in \mathfrak{S}$, $\hat{W}_+$, and F, namely, that there is a linear, bounded, positive operator

$$\Phi_c: \mathcal{B}(\mathcal{H})_1 \to \mathcal{B}(\mathcal{H})_1$$

such that

$$\hat{W}_+ = \frac{\Phi_c(W)}{\mathrm{tr}(FW)}$$

Remember that $\mathrm{tr}(FW) = \mathrm{tr}(\Phi_c(W))$, and $F = \Phi_c^*(1)$.

This relation becomes much more specific after establishing the complete positivity of the map Φ_c. If one considers the state change as the result of an interaction with another quantum system, the apparatus on which an observation is described by the von Neumann–Lüders assumption, this property is a formal consequence. But one may also establish it more directly, as we will do now.

Any two quantum systems can be considered as being coupled one to the other such that they are described in the common Hilbert space $\mathcal{H}_1 \otimes \mathcal{H}_2$. The reason for the tensor product structure is the fact that the observables of each component are observables of the composed system and two observables concerning different subsystems have to be commensurable. Consider now our quantum object described in the Hilbert space $\mathcal{H}$ to be coupled with an n-level system described in $\mathbf{C}^n$. By the argument which has been used to establish the tensor product structure $\mathcal{H} \otimes \mathbf{C}^n$ one may assume that operations concerning only the object described in $\mathcal{H}$ must be extended to the composed system in such a manner that at least when correlations are absent the state of the n-level component is not affected.

To that end we realize that the density operators in $\mathcal{H} \otimes \mathbf{C}^n$ are limit points of linear combinations of uncorrelated density operators like $W \otimes V$, $W \in \mathfrak{S}(\mathcal{H})$, $V \in \mathfrak{S}(\mathbf{C}^n)$ and have the shape

$$\begin{pmatrix} W_{11} & W_{12} & \cdots & W_{1n} \\ W_{21} & W_{22} & \cdots & W_{2n} \\ \cdots & \cdots & \cdots & \cdots \\ W_{n1} & W_{n2} & \cdots & W_{nn} \end{pmatrix}, \qquad W_{ij} \in \mathcal{B}(\mathcal{H})_1$$

where

$$\mathrm{tr}((W_{ij})) = \sum_{i=1}^{n} \mathrm{tr}\, W_{ii} = 1$$

Especially, an uncorrelated state has the shape

$$W \otimes ((v_{ij})) = \begin{pmatrix} Wv_{11} & Wv_{12} & \cdots & Wv_{1n} \\ Wv_{21} & Wv_{22} & \cdots & Wv_{2n} \\ \cdots & \cdots & \cdots & \cdots \\ Wv_{n1} & Wv_{n2} & \cdots & Wv_{nn} \end{pmatrix}$$

Hence, an operation Φ on the object Hilbert space $\mathcal{H}$ must obviously be extended to act on uncorrelated states in $\mathcal{H} \otimes \mathbf{C}^n$ in such a way that

$$\Phi_n\colon W \otimes ((v_{ij})) \mapsto \Phi(W) \otimes ((v_{ij}))$$

$$= \begin{pmatrix} \Phi(W)v_{11} & \Phi(W)v_{12} & \cdots & \Phi(W)v_{1n} \\ \Phi(W)v_{21} & \Phi(W)v_{22} & \cdots & \Phi(W)v_{2n} \\ \cdots & \cdots & \cdots & \cdots \\ \Phi(W)v_{n1} & \Phi(W)v_{n2} & \cdots & \Phi(W)v_{nn} \end{pmatrix}$$

Since Φ is linear and bounded, the mappings just defined for uncorrelated states can be extended to a linear and bounded map onto the linear hull of the uncorrelated density operators, and because it is bounded, it can be continuously extended onto the space of density operators in $\mathcal{H} \otimes \mathbf{C}^n$ to give

$$\Phi_n\colon \begin{pmatrix} W_{11} & W_{12} \cdots & W_{1n} \\ W_{21} & W_{22} \cdots & W_{2n} \\ \cdots & \cdots & \cdots \\ W_{n1} & W_{n2} \cdots & W_{nn} \end{pmatrix} \mapsto \begin{pmatrix} \Phi(W_{11}) & \Phi(W_{12}) & \cdots \Phi(W_{2n}) \\ \Phi(W_{21}) & \Phi(W_{22}) & \cdots \Phi(W_{1n}) \\ \cdots & \cdots & \cdots \\ \Phi(W_{n1}) & \Phi(W_{n2}) & \cdots \Phi(W_{nn}) \end{pmatrix}$$

Now the representation of an operation Φ_n has to be a positive map, hence, the operation Φ on the object acting on $\mathcal{H}$ is by definition an n-positive map. Since there is no restriction to the number n of levels, Φ has to be completely positive.

As a consequence of the Stinespring representation theorem for completely positive mappings there is a series $\{A_k\}_{k\in K\subseteq \mathbf{N}}$ of linear operators in $\mathcal{H}$ fulfilling

$$\sum_{k\in K} A_k^+ A_k \leq 1$$

such that

$$\Phi(W) = \sum_{k\in K} A_k W A_k^+$$

In Table I the main formulas for the von Neumann–Lüders operation are compared with those of a general operation.

Up to now, no quantum dynamic principles have been taken into consideration. The complete positivity has been established using only general statistical and quantum kinematic principles.

4. EXPLICIT EXPRESSIONS OF THE A_k FOR A MEASURING PROCESS

The measuring process will be treated as an interaction process between the object and the measuring apparatus. Both systems are quantum systems in principle. That the measuring apparatus has an additional structure as a

Table I.

Operation	von Neumann–Lüders	General
Mathematical representative	$E = E^{+} = E^2 \leq 1$	$\{A_k\}_{k\in K\subseteq \mathbf{N}}$, $\Sigma_{k\in K} A_k^{+} A_k \leq 1$
Probability for the result '+'	$\mathrm{tr}(WE)$, $E \leq 1$	$\mathrm{tr}(WF)$, $F := \Sigma_{k\in K} A_k^{+} A_k \leq 1$
Selection according to the result '+'	$W \mapsto \Phi(W) = EWE$	$W \mapsto \Phi(W) = \Sigma_{k\in K} A_k W A_k^{+}$
Operator of two subsequent measurements	$G \mapsto \Phi^*(G) = EGE$	$G \mapsto \Phi^*(G) = \Sigma_{k\in K} A_k^{+} G A_k$
No subsequent measurement	$E = \Phi^*(1) = E^{+}E = E$	$F = \Phi^*(1) = \Sigma_{k\in K} A_k^{+} A_k$

many-particle system with a macroscopically observable decomposition of the unit operator does not enter into the following computation. Denote by:

- $\mathcal{H}_o$ the Hilbert space of the quantum objects.
- $\mathcal{H}_a$ the Hilbert space of the apparatus.
- $\mathcal{B}(\mathcal{H}_o) \supseteq \mathfrak{S}_o \ni W_o$ the density operator of the ensemble of objects on which the measurements are performed.
- $\mathcal{B}(\mathcal{H}_a) \supseteq \mathfrak{S}_a \ni W_a$ the density operator of the ensemble of measuring apparatuses by which the measurements are performed.
- $\mathcal{B}(\mathcal{H}_o \otimes \mathcal{H}_a) \ni \mathbf{S}$ the unitarian representing the solution of the Schrödinger equation for an interaction of finite duration or a scattering operator.

We remark that unitarity is not an essential requirement. S may contain irreversible dynamics of the measuring apparatus, thus representing a solution of a Schrödinger equation with dissipation. But at least for $\mathcal{B}(\mathcal{H}_o \otimes \mathcal{H}_a) \supseteq \mathfrak{S}_{(o,a)} \ni \mathbf{W}$ the equality

$$\mathrm{tr}\,\mathbf{W} = \mathbf{S}\mathbf{W}\mathbf{S}^{+}$$

must hold. Finally:

- $\mathcal{B}(\mathcal{H}_a) \supseteq \{B_j\}_{j\in J}$ is the representing sequence of the operation to be observed on the measuring apparatus.

The measuring process is then described as follows.

We are considering a series of experiments in which the object and the apparatus are prepared independently from one another. The initial state of the combined system is therefore uncorrelated. Let

$$W_o \otimes W_a$$

denote that state. The interaction (which may include irreversible motion) leads to the correlated state

$$\mathbf{S}(W_o \otimes W_a)\mathbf{S}^+$$

On this state an operation is performed. Whether it concerns a macroscopically observable property of the apparatus only or a general property of the combined system as an operation, it must be described by a sequence of operators $\{\mathbf{B}_j\}_{j\in J}$ with $\Sigma_{j\in J}\, \mathbf{B}_j^+ \mathbf{B}_j \leq 1$. That we write $1 \otimes B_j$ instead of more general $\mathbf{B}_j$ has only aesthetic reasons and no consequences for the later computations. Let $\mathbf{\Phi}$ denote the complete positive map corresponding to this operation, i.e.,

$$\mathbf{\Phi}(\mathbf{S}(W_o \otimes W_a)\mathbf{S}^+) = \sum_{j\in J} (1 \otimes B_j)\mathbf{S}(W_o \otimes W_a)\mathbf{S}^+(1 \otimes B_j^+)$$

Since we want to describe the operation on the object component of the combined system, which means that we are interested only in the results of further measurements concerning the first component described in the Hilbert space $\mathcal{H}_o$, we can use the partial trace formalism. Let tr_a denote the partial trace with respect to the Hilbert space of the apparatus. With

$$\Phi(W_o) := \mathrm{tr}_a(\mathbf{\Phi}(\mathbf{S}(W_o \otimes W_a)\mathbf{S}^+))$$

the representing map of the desired operation is given. The density operator of the ensemble of objects prepared by selection according to the result '+' on the apparatus is given by

$$\hat{W}_+ = \frac{\Phi(W_o)}{\mathrm{tr}(\Phi(W_o))} \in \mathfrak{S}_o$$

To find the corresponding representing sequence of operators we write the map $\mathbf{\Phi}$ corresponding to the operation on the combined system in the form

$$\mathbf{\Phi}(\mathbf{S}(W_o \otimes W_a)\mathbf{S}^+)$$

$$= \mathrm{tr}_a\left(\sum_{j\in J} (1 \otimes B_j)\mathbf{S}(1 \otimes \sqrt{W_a})(W_o \otimes 1)(1 \otimes \sqrt{W_a})\mathbf{S}^+(1 \otimes B_j^+)\right)$$

We get the map Φ corresponding to the operation on the object system by forming the partial trace tr_a. This becomes more explicit if we write the bilinear form

$$\langle\varphi, \Phi(W_o)\psi\rangle = \langle\varphi, \mathrm{tr}_a\, \mathbf{\Phi}(\mathbf{S}(W_o \otimes W_a)\mathbf{S}^+)\psi\rangle$$

$$= \sum_{j\in J}\sum_{\nu=0}^{\infty} \langle\varphi \otimes \phi_\nu, (1 \otimes B_j)\mathbf{S}(1 \otimes \sqrt{W_a})$$

$$\times (W_o \otimes 1)(1 \otimes \sqrt{W_a})\mathbf{S}^+(1 \otimes B_j^+)\psi \otimes \phi_\nu\rangle$$

where $\{\phi_\nu\}_{\nu\in\mathbf{N}}$ is a complete orthonormal system in the Hilbert space $\mathcal{H}_a$ of the apparatus. This expression suggests we introduce the series of operators $\{A_{j\nu\mu}\}_{j\in J,\nu\in\mathbf{N},\mu\in\mathbf{N}}$ by the definition

$$\langle\varphi, A_{j\nu\mu}\psi\rangle := \langle\varphi\otimes\phi_\nu, (1\otimes B_j)\mathbf{S}(1\otimes\sqrt{W_a})\psi\otimes\phi_\mu\rangle$$

The adjoint operators are then defined by the bilinear forms

$$\langle\varphi, A^+_{j\nu\mu}\psi\rangle := \langle\varphi\otimes\phi_\mu, (1\otimes\sqrt{W_a})\mathbf{S}^+(1\otimes B_j^+)\psi\otimes\phi_\nu\rangle$$

Replacing now the unit operator in the expression $(W_o\otimes 1)$ in the equation for $\Phi(W_o)$ by $\Sigma_{\mu=0}^{\infty} \mid\phi_\mu\times\phi_\mu\mid$ we get the equality

$$\langle\varphi, \Phi(W_o)\psi\rangle = \sum_{j\in J}\sum_{\nu=0}^{\infty}\sum_{\mu=0}^{\infty}\langle\varphi, A_{j\nu\mu}W_oA^+_{j\nu\mu}\psi\rangle$$

Since φ and ψ are arbitrary in $\mathcal{H}$, we have proved

$$\Phi(W_o) = \sum_{j\in J}\sum_{\nu=0}^{\infty}\sum_{\mu=0}^{\infty} A_{j\mu\nu}W_oA^+_{j\nu\mu}$$

Hence $\{A_{j\mu\nu}\}_{j\in J,\nu\in\mathbf{N},\mu\in\mathbf{N}}$ is a sequence of operators representing the operation for the object system. Since for $\varphi\in\mathcal{H}_o$, $\|\varphi\| = 1$, we have

$$\operatorname{tr}\Phi(|\varphi\rangle\langle\varphi|) = \operatorname{tr}\sum_{j\in J}\sum_{\nu=0}^{\infty}\sum_{\mu=0}^{\infty} A_{j\nu\mu}|\varphi\rangle\langle\varphi|A^+_{j\nu\mu}$$

$$= \left\langle\varphi, \sum_{j\in J}\sum_{\nu=0}^{\infty}\sum_{\mu=0}^{\infty} A^+_{j\nu\mu}A_{j\nu\mu}|\varphi\right\rangle$$

and, on the other hand,

$$\operatorname{tr}\Phi(|\varphi\rangle\langle\varphi|) = \mathbf{\Phi}(\mathbf{S}(W_o\otimes W_a)\mathbf{S}^+)\leq 1$$

it follows that

$$F := \sum_{j\in J}\sum_{\nu=0}^{\infty}\sum_{\mu=0}^{\infty} A^+_{j\nu\mu}A_{j\nu\mu}\leq 1$$

as it should be.

Considering effects and operations as a result of interaction processes, this form of the representations of operations was derived by K. Kraus and myself at the end of the 1960s by assuming that selections by observations on the apparatus are to be described on a von Neumann–Lüders operation. Later Kraus realized that it is just the Stinespring representation of a complete positive mapping and he gave the more general arguments that this must be fulfilled by the very definition of an operation and the kinematics of coupled systems. Hence, the artificial assumption about macroscopic observations could be dropped.

The following can be proved: Let be given a complete positive linear mapping Φ operating on $\mathcal{B}(\mathcal{H}_o)_1$ and a Hilbert space $\mathcal{H}_a$. There are always triplets

$$(W_a, \{B_j\}_{j\in J\subseteq \mathbf{N}}, \mathbf{S}) \in \mathcal{B}(\mathcal{H}_o)_1 \times (\mathcal{B}'(\mathcal{H}_o))^2 \times \mathcal{U}(\mathcal{H}_o \otimes \mathcal{H}_a)$$

where $(\mathcal{B}'(\mathcal{H}_o))^2$ means that the sum of $B_j^+ B_j$ exists and is a bounded operator, such that Φ arises in the manner just described.

Moreover, one may prove that coexistent effects can be produced together in one and the same interaction process and many other properties fitting well into the philosophy of quantum measurements. The operations described here are the elementary building blocks by which the theory of quantum measurements is formed.

REFERENCES

Busch, P., Lahti, P. J., and Mittelstaed, P. (1991). *The Quantum Theory of Measurements*, Springer, Berlin.

Davies, E. B. (1976). *Quantum Theory of Open Systems*, Academic Press, New York, Chapters 2–4.

Davies, E. B., and Lewis, J. C. T. (1970). An operational approach to quantum probability, *Communications in Mathematical Physics*, **17**, 239–269.

Gudder, S. (1979). *Stochastic Methods in Quantum Mechanics*, Elsevier/North-Holland, Amsterdam, Chapter 4.

Hellwig, K.-E. (1967). Makroskopische Effekte und Quantenmechanische Messung, Dissertation, Universität Marburg.

Hellwig, K.-E. (1969). Coexistent effects in quantum mechanics, *International Journal of Theoretical Physics*, **2**, 147–155.

Hellwig, K.-E. (1971). Measuring process and additive conservation laws, in *Foundation of Quantum Mechanics*, B. d'Espagnat, ed., Academic Press, New York, pp. 338–345.

Hellwig, K.-E., and Kraus, K. (1969). Pure operations and measurements, *Communications in Mathematical Physics*, **11**, 214–220.

Hellwig, K.-E., and Kraus, K. (1970). Operations and measurements II. *Communications in Mathematical Physics*, **16**, 142–147.

Hellwig, K.-E., and Kraus, K. (1971). Formal description of measurements in local quantum field theory, *Physical Review D*, **1**, 566–571.

Kraus, K. (1971). General state changes in quantum theory, *Annals of Physics*, **64**, 311–335.

Kraus, K. (1977). Position observables of the photon, in *The Uncertainty Principle and Foundations of Quantum Mechanics*, W. C. Price and S. S. Chissick, eds., Wiley, London.

Kraus, K. (1983). *States, Effects, and Operations*, Springer, Berlin.

Ludwig, G. (1976). *Einführung in die Grundlagen der Theoretischen Physik*, Vol. 3, Chapter XII, Vieweg, Braunschweig.

Lüders, G. (1951). Über die Zustandsänderung durech den Messprozess, *Annalen der Physik*, **8**, 322–328.

Pauli, W. (1933). Die allgemeinen Prinzipien der Wellemechanik, in *Handbuch der Physik*, Vol. 24, H. Geiger und K. Scheel, eds., Springer, Berlin.

Completely Positive Linear Maps on Complex Matrices

Man-Duen Choi
Department of Mathematics, University of California,
Berkeley, California 94720

Recommended by Chandler Davis

ABSTRACT

A linear map Φ from $\mathfrak{M}_n$ to $\mathfrak{M}_m$ is completely positive iff it admits an expression $\Phi(A)=\Sigma_i V_i^* A V_i$ where V_i are $n\times m$ matrices.

In this paper, we describe the tractable structure of completely positive linear maps between complex matrix algebras. The objective is (pursuing the work of Stinespring [8], Størmer [9], and Arveson [1,2]) to establish that completely positive linear maps, rather than positive linear maps, are the natural generalization of positive linear functionals. The results presented here are 'finite' and 'concrete' in essence. The reader may consult [1, Chapter 1] for general abstract information about the infinite-dimensional case.

Our main theorems reveal that the class of completely positive linear maps is the positive cone of the class of hermitian-preserving maps endowed with a natural ordering. Thus, a thorough structure theory follows immediately (Theorem 5). Finally (Theorem 7), we show that positive linear maps have the same effect as completely positive linear maps on 2×2 symmetric matrices.

For a complex matrix A, A^* denotes the transpose of the complex conjugate of A. We say a square matrix A is *symmetric* iff A equals its transpose, A is *hermitian* iff $A=A^*$, A is *positive* (or *positive semi-definite* for exactness) iff A is hermitian and its eigenvalues are nonnegative. We denote by $\mathfrak{M}_n$ the collection of $n\times n$ complex matrices. The Kronecker delta δ_{jk} equals 1 if $j=k$, and 0 if $j\neq k$; hence $I=(\delta_{jk})\in\mathfrak{M}_n$ is the identity $n\times n$ matrix. $E_{jk}\in\mathfrak{M}_n$ is the $n\times n$ matrix with 1 at the j,k component and zeros elsewhere. $\mathfrak{M}_n(\mathfrak{M}_m)=\mathfrak{M}_m\otimes\mathfrak{M}_n$ is the collection of all $n\times n$ block matrices with $m\times m$ matrices as entries; each element of $\mathfrak{M}_n(\mathfrak{M}_m)$ can also be regarded as an $nm\times nm$ matrix with numerical entries.

A linear map $\Phi:\mathfrak{M}_n\to\mathfrak{M}_m$ is *positive* (resp. *hermitian-preserving*) iff $\Phi(A)$ is positive (resp. hermitian) for all positive (resp. hermitian) A in $\mathfrak{M}_n$. We define $\Phi\otimes 1_p:\mathfrak{M}_p(\mathfrak{M}_n)\to\mathfrak{M}_p(\mathfrak{M}_m)$ by $\Phi\otimes 1_p((A_{jk})_{1\leq j,k\leq p})=(\Phi(A_{jk}))_{1\leq j,k\leq p}$. Φ is *completely positive* iff $\Phi\otimes 1_p$ is positive for all positive integers p. The reader is referred to [4] for the discrimination in a precise way between completely positive linear maps and positive linear maps.

For each $n\times m$ matrix V, it is evident that the map: $\mathfrak{M}_n\to\mathfrak{M}_m$ with $A\to V^*AV$ is completely positive. In the following, we show that the combinations of maps of the above form constitute all completely positive linear maps.

THEOREM 1. *Let $\Phi:\mathfrak{M}_n\to\mathfrak{M}_m$. Then Φ is completely positive iff Φ is of the form $\Phi(A)=\Sigma_i V_i^*AV$ for all A in $\mathfrak{M}_n$ where V_i are $n\times m$ matrices.*

Proof. The 'if' part is obvious. We proceed to prove the converse. Each $1\times nm$ matrix v can be regarded as a $1\times n$ block matrix $(x_1,\ldots,x_n)$ with $1\times m$ matrices x_j as entries; hence we associate with it the $n\times m$ matrix V which has x_j as the j-th row. A simple computation leads to

$$(V^*E_{jk}V)_{1\leq j,k\leq n}=(x_j^*x_k)_{1\leq j,k\leq n}=v^*v.$$

Now suppose $\Phi:\mathfrak{M}_n\to\mathfrak{M}_m$ is completely positive. As $(E_{jk})_{1\leq j,k\leq n}$ is positive, so $(\Phi(E_{jk}))_{1\leq j,k\leq n}\in\mathfrak{M}_n(\mathfrak{M}_m)$ is positive; thus there exist vectors v_i^* (regarded as $nm\times 1$ matrices) such that $(\Phi(E_{jk}))_{jk}=\Sigma_i v_i^*v_i$. Let V_i be the $n\times m$ matrices associated with v_i. Then by the preceding result, $(\Phi(E_{jk}))_{jk}=\Sigma_i(V_i^*E_{jk}V_i)_{jk}$. Therefore, we conclude that $\Phi(A)=\Sigma_i V_i^*AV_i$ for all A. ■

Each linear map $\Phi:\mathfrak{M}_n\to\mathfrak{M}_m$ is determined by its values on E_{jk}. Hence Φ is completely determined by the single element $(\Phi(E_{jk}))_{1\leq j,k\leq n}\in\mathfrak{M}_n(\mathfrak{M}_m)$. The proof of Theorem 1 also provides another characterization of completely positive linear maps:

THEOREM 2. *Let Φ be a linear map from $\mathfrak{M}_n$ to $\mathfrak{M}_m$. Then Φ is completely positive iff $(\Phi(E_{jk}))_{1\leq j,k\leq n}$ is positive.*

REMARK 3. For a linear map $\Phi:\mathfrak{M}_n\to\mathfrak{M}_m$, it is obvious that Φ is hermitian-preserving iff $(\Phi(E_{jk}))_{jk}$ is hermitian. Endowed with the natural ordering induced by $\mathfrak{M}_n(\mathfrak{M}_m)$, the class of hermitian-preserving maps is a partially ordered vector space over the reals, while the class of completely positive linear maps is just the positive cone.

Referring again to the proof of Theorem 1, we deduce another pertinent fact (cf., [7, p. 134, Theorem 2.1] and [5, p. 259, Theorem 2]): $\Phi : \mathfrak{M}_n \to \mathfrak{M}_m$ is hermitian-preserving iff Φ admits an expression $\Phi(A) = \Sigma \epsilon_i V_i^* A V_i$ where $\epsilon_i = \pm 1$, V_i are $n \times m$ matrices. Since there are no such elegant expressions for positive linear maps, we may be convinced that completely positive linear maps, rather than positive linear maps, deserve the adjective 'positive'.

REMARK 4. In the proof of Theorem 1, the expression $(\Phi(E_{jk}))_{jk} = \Sigma_i v_i^* v_i$ is not unique, hence $\{V_i\}$ is not uniquely determined. For some improvement, we may require $\{v_i^*\}$ to be linearly independent, then $\{V_i\}$ must be linearly independent too.

This additional condition on $\{V_i\}_i^\ell$ ensures that $\Phi(A) = \Sigma_i^\ell V_i^* A V_i$ is a 'canonical' expression for Φ, in the following precise sense: Let $\{W_p\}_p^{\ell'}$ be a class of $n \times m$ matrices, then Φ has the expression $\Phi(A) = \Sigma_p^{\ell'} W_p^* A W_p$ iff there exists an isometric $\ell' \times \ell$ matrix $(\mu_{pi})_{pi}$, such that $W_p = \Sigma_i \mu_{pi} V_i$ for all p. Moreover, if $\{W_p\}_p^{\ell'}$ is also a linearly independent set, then $\ell' = \ell$, and $(\mu_{pi})_{pi}$ is unitary.

Proof. The 'if' part follows by direct computation. We proceed to prove the 'only if' part. Denote by w_p, the display of W_p as a $1 \times nm$ matrix. As in the proof of Theorem 1, $\Sigma_p w_p^* w_p = (\Phi(E_{jk}))_{jk} = \Sigma_i v_i^* v_i$. Thus w_p^* belongs to $sp\{v_i^*\}_i$, the linear span of $\{v_i^*\}_i$; namely, there exists $(\mu_{pi})_{pi}$ such that $w_p^* = \Sigma_i \overline{\mu_{pi}}\, v_i^*$. It follows that $W_p = \Sigma_i \mu_{pi} V_i$.

Since $\{v_i^*\}_i$ is a linearly independent set, $\{v_i^* v_j\}_{ij}$ is also a linearly independent set. (In fact, $\{v_i^* v_j\}_{ij}$ is a basis of the linear transformation space on $sp\{v_i^*\}_i$.) From

$$\Sigma_i v_i^* v_i = \Sigma_p w_p^* w_p = \Sigma_{pij} \overline{\mu_{pi}}\, \mu_{pj} v_i^* v_j,$$

we obtain $\Sigma_p \overline{\mu_{pi}}\, \mu_{pj} = \delta_{ij}$. Hence $(\mu_{pi})_{pi}$ is an isometry. In case that $\{W_p\}_p$ is also a linearly independent set, from $sp\{v_i^*\}_i^\ell = sp\{w_p^*\}_p^{\ell'}$, we derive that $\ell = \ell'$ and $(\mu_{pi})_{pi}$ is unitary. ■

For each fixed positive K in $\mathfrak{M}_m$, we write $\mathbf{CP}[\mathfrak{M}_n, \mathfrak{M}_m; K] = \{\Phi : \mathfrak{M}_n \to \mathfrak{M}_m | \Phi$ is completely positive and $\Phi(I) = K\}$. It is evident that $\mathbf{CP}[\mathfrak{M}_n, \mathfrak{M}_m; K]$ is a convex set, hence it is the convex hull of its extreme points. The following theorem gives a thorough description of the structure of completely positive linear maps.

THEOREM 5. *Let* $\Phi : \mathfrak{M}_n \to \mathfrak{M}_m$. *Then* Φ *is extreme in* $\mathbf{CP}[\mathfrak{M}_n, \mathfrak{M}_m; K]$ *iff* Φ *admits an expression* $\Phi(A) = \Sigma_i V_i^* A V_i$ *for all* A *in* $\mathfrak{M}_n$, *where* V_i *are* $n \times m$ *matrices,* $\Sigma_i V_i^* V_i = K$, *and* $\{V_i^* V_j\}_{ij}$ *is a linearly independent set.*

Proof. 'The only if part'

Assume Φ is extreme in $\mathbf{CP}[\mathfrak{M}_n, \mathfrak{M}_m; K]$. Express Φ in canonical form (Remark 4) $\Phi(A) = \Sigma V_i^* A V_i$ with $\{V_i\}$ linearly independent. Now suppose $\Sigma_{ij}\lambda_{ij} V_i^* V_j = 0$, we wish to prove that $(\lambda_{ij})_{ij} = 0$.

We may assume that $(\lambda_{ij})_{ij}$ is a hermitian matrix. (In fact, from $\Sigma\lambda_{ij} V_i^* V_j = 0$ we infer that $\Sigma(\lambda_{ij} \pm \overline{\lambda_{ji}}) V_i^* V_j = 0$. Then, if we prove $(\lambda_{ij} \pm \overline{\lambda_{ji}})_{ij} = 0$, that will yield $(\lambda_{ij})_{ij} = 0$.) By a scalar multiplication, we may further assume $-I \leqslant (\lambda_{ij})_{ij} \leqslant I$.

Define $\Psi_\pm : \mathfrak{M}_n \to \mathfrak{M}_m$ by $\Psi_\pm(A) = \Sigma V_i^* A V_i \pm \Sigma\lambda_{ij} V_i^* A V_j$. Hence $\Psi_\pm(I) = \Sigma V_i^* V_i = K$. Let $I + (\lambda_{ij})_{ij} = (\alpha_{ij})_{ij}^*(\alpha_{ij})_{ij}$ and $W_i = \Sigma_j \alpha_{ij} V_j$. By direct computation, $\Psi_+(A) = \Sigma W_i^* A W_i$; thus Ψ_+ is completely positive. In the same manner, Ψ_- is also completely positive. From $\Phi = \frac{1}{2}(\Psi_+ + \Psi_-)$ and the extremeness of Φ, we obtain $\Phi = \Psi_+$. By Remark 4, $(\alpha_{ij})_{ij}$ is an isometry. Therefore, $I + (\lambda_{ij})_{ij} = I$, i.e., $(\lambda_{ij})_{ij} = 0$ as required. ∎

Proof. 'The if part'.

Assume $\Phi(A) = \Sigma V_i^* A V_i$, $\Sigma V_i^* V_i = K$ and $\{V_i^* V_j\}_{ij}$ is a linearly independent set. (Consequently, $\{V_i\}_i$ is a linearly independent set.) Now suppose $\Phi = \frac{1}{2}(\Psi_1 + \Psi_2)$ with $\Psi_1(A) = \Sigma W_p^* A W_p$, $\Psi_2(A) = \Sigma Z_q^* A Z_q$, and $\Sigma W_p^* W_p = \Sigma Z_q^* Z_q = K$. Since $\Psi(A) = \frac{1}{2}\Sigma W_p^* A W_p + \frac{1}{2}\Sigma Z_q^* A Z_q$, W_p and Z_q can be expressed in terms of V_i (Remark 4). Let $W_p = \Sigma_i \mu_{pi} V_i$ for each p. Then $\Sigma V_i^* V_i = \Sigma W_p^* W_p = \Sigma_{pij} \overline{\mu_{pi}}\, \mu_{pj} V_i^* V_j$, so $\Sigma_p \overline{\mu_{pi}}\, \mu_{pj} = \delta_{ij}$; i.e., $(\mu_{pi})_{pi}$ is an isometry. From Remark 4 again, we conclude that $\Phi = \Psi_1$; therefore Φ is extreme in $\mathbf{CP}[\mathfrak{M}_n, \mathfrak{M}_m; K]$. ∎

REMARK 6. Suppose $\Phi : \mathfrak{M}_n \to \mathfrak{M}_m$ is completely positive. From Remark 4, we can write Φ in the form $\Phi(A) = \Sigma_i^\ell V_i^* A V_i$ where $\{V_i\}_i^\ell$ is a class of linearly independent $n \times m$ matrices; hence $\ell \leqslant nm$. In case Φ is extreme in $\mathbf{CP}[\mathfrak{M}_n, \mathfrak{M}_m; K]$, ℓ can be reduced to $\leqslant m$. In fact, $\{V_i^* V_j\}_{ij}$ is a linearly independent set only if the cardinal number of $\{V_i^* V_j\}_{ij} \leqslant \dim \mathfrak{M}_m$, hence only if $\ell^2 \leqslant m^2$, i.e., $\ell \leqslant m$.

In particular, if $m = 1$, we obtain the well-known fact that Φ is a 'pure state' (an extreme identity-preserving positive functional) on $\mathfrak{M}_n$ iff Φ is a 'vector state' (i.e., there exists a unit vector v such that $\Phi(A) = v^* A v$ for all A).

The general structure of positive linear maps is very complicated (see [9, Chapter 8]). We will treat it for 2×2 matrices only. Following is an 'almost global' property of positive linear maps.

THEOREM 7. *If $\Phi : \mathfrak{M}_2 \to \mathfrak{M}_m$ is positive, then there exist $2 \times m$ matrices V_i such that $\Phi(A) = \Sigma_i V_i^* A V_i$ for all 2×2 symmetric A.*

Proof. We will associate each linear map with a matrix-coefficient quadratic form. First, we call attention to a known result (see [3, Theorem 2] for an elegant proof; the statement also appeared in an earlier paper [6, Appendix III]): Let F be an $\mathfrak{M}_m$-coefficient quadratic form $F(s,t)=B_1s^2+B_2st+B_3t^2$ with real indeterminates s, t. If $F(s,t)$ is positive for all s, t, then there exist $k\times m$ matrices C, D, such that $F(s,t)=(Cs+Dt)^*(Cs+Dt)$. (k is a certain integer.)

Now suppose Φ is positive, then $\Phi\left(\begin{bmatrix} s^2 & st \\ st & t^2 \end{bmatrix}\right)$ is positive for all real s, t; i.e., $F(s,t)=\Phi(E_{11})s^2+\Phi(E_{12}+E_{21})st+\Phi(E_{22})t^2$ is a positive $\mathfrak{M}_m$-coefficient quadratic form. From the preceding paragraph, there exist matrices C, D such that $\Phi(E_{11})=C^*C$, $\Phi(E_{12}+E_{21})=C^*D+D^*C$, and $\Phi(E_{22})=D^*D$. Define $\Psi:\mathfrak{M}_2\to\mathfrak{M}_m$ by

$$\left(\Psi(E_{jk})\right)_{jk}=\begin{bmatrix} C^*C & C^*D \\ D^*C & D^*D \end{bmatrix};$$

then Ψ is completely positive from Theorem 2. Since Φ agrees with Ψ on every symmetric matrix, we obtain the desired expression from Theorem 1. ∎

We remark that Theorem 7 is not valid for higher order matrices. This is just because the quoted result for an $\mathfrak{M}_m$-coefficient quadratic form cannot be generalized to the case of more than 2 real indeterminates, as will be shown in a forthcoming paper.

The author would like to express his thanks to Prof. Chandler Davis for stimulating discussions on related topics. Partial material of this paper has appeared in the author's PhD. Thesis at the University of Toronto.

REFERENCES

1 W. B. Arveson, Subalgebras of C*-algebras, *Acta Math.* **123**, 141–224 (1969).
2 W. B. Arveson, Subalgebras of C*-algebras II, *Acta Math.* **128**, 271–308 (1972).
3 A. P. Calderón, A note on biquadratic forms, *Linear Alg. Appl.* 7, 175–177 (1973).
4 M. D. Choi, Positive linear maps on C*-algebras, *Canad. J. Math.* **24**, 520–529 (1972).
5 R. D. Hill, Linear transformations which preserve hermitian matrices, *Linear Alg. Appl.* 6, 257–262 (1973).
6 T. Koga, Synthesis of finite passive n-ports with prescribed positive real matrices of several variables, *IEEE Trans. Circuit Theory*, **CT-15**, 2–23 (1968).

7 J. dePillis, Linear transformations which preserve hermitian and positive semi-definite operators, *Pacific J. Math.* **23**, 129–137 (1967).

8 W. F. Stinespring, Positive functions on C^*-algebras, *Proc. Amer. Math. Soc.* **6**, 211–216 (1955).

9 E. Størmer, Positive linear maps of operator algebras, *Acta Math.* **110**, 233–278 (1963).

Received 24 October 1973; revised 4 October 1974

PHYSICAL REVIEW A VOLUME 54, NUMBER 4 OCTOBER 1996

Sending entanglement through noisy quantum channels

Benjamin Schumacher*
Theoretical Astrophysics, T-6 M.S. B288, Los Alamos National Laboratory, Los Alamos, New Mexico 87545
(Received 26 April 1996)

This paper addresses some general questions of quantum information theory arising from the transmission of quantum entanglement through (possibly noisy) quantum channels. A pure entangled state is prepared of a pair of systems R and Q, after which Q is subjected to a dynamical evolution given by the superoperator $\mathcal{E}^Q$. Two interesting quantities can be defined for this process: the entanglement fidelity F_e and the entropy exchange S_e. It turns out that neither of these quantities depends in any way on the system R, but only on the initial state and dynamical evolution of Q. F_e and S_e are related to various other fidelities and entropies and are connected by an inequality reminiscent of the Fano inequality of classical information theory. Some insight can be gained from these techniques into the security of quantum cryptographic protocols and the nature of quantum error-correcting codes. [S1050-2947(96)03909-1]

PACS number(s): 03.65.Bz, 05.30.−d, 89.70.+c

I. INTRODUCTION

In recent years, considerable progress has been made toward developing a general quantum theory of information [1], analogous to classical information theory founded by Shannon [2]. Distinctively quantum-mechanical notions of coding [3] and channel fidelity [4] have been developed and the role of entangled states in storing and transferring quantum information has been explored [5]. Recently, the study of noisy quantum channels has yielded important results about quantum error-correcting codes [6] and the purification of noisy entangled states [7].

The aim of this paper is to further clarify our understanding of noisy quantum channels by defining and exploiting notions of fidelity and entropy associated with the quantum transmission process. These quantities are based on an analysis of the transmission of entangled states through the noisy channel, although (as we shall see) the use of entanglement is not essential to their definition. A number of applications of these ideas will be outlined.

Here is the general situation that we will consider. Suppose R and Q are two quantum systems and Q is described by a Hilbert space $\mathcal{H}_Q$ of finite dimension d. Initially the joint system RQ is prepared in a pure entangled state $|\Psi^{RQ}\rangle$. The system R is dynamically isolated and has a zero internal Hamiltonian, while the system Q undergoes some evolution that possibly involves interaction with the environment E. The evolution of Q might, for example, represent a coding, transmission, and decoding process via some quantum channel for the quantum information in Q. The final state of RQ is possibly mixed and is described by the density operator $\rho^{RQ'}$.

The *fidelity* of this process is $F_e=\langle\Psi^{RQ}|\rho^{RQ'}|\Psi^{RQ}\rangle$, which is the probability that the final state $\rho^{RQ'}$ would pass a test checking whether it agreed with the initial state $|\Psi^{RQ}\rangle$. (This imagined test would be a measurement of a joint observable on RQ.) F_e measures how successfully the quantum channel preserves the entanglement of Q with the "reference system" R.

We will demonstrate three important results. First, the fidelity F_e can be defined entirely in terms of the initial state and evolution of the system Q. Furthermore, $F_e\leq\bar{F}$, where $\bar{F}$ is the average fidelity when the channel carries one of an ensemble of pure states of Q described by $\rho^Q=\mathrm{Tr}_R|\Psi^{RQ}\rangle\langle\Psi^{RQ}|$. Thus channels that can convey entanglement faithfully will also convey ensembles of pure states faithfully.

Second, there exists a quantity S_e called *entropy exchange*, also defined in terms of the internal properties of the system Q. This quantity can be viewed as the amount of information that is exchanged with the environment during the interaction of Q and E and it characterizes the amount of "quantum noise" in the evolution of Q.

Finally, we will find an inequality (resembling the Fano inequality of classical information theory) that bounds F_e in terms of the dimension d and the entropy exchange S_e in Q. In other words, the faithfulness of Q's dynamical evolution in preserving entanglement is limited by the amount of information that is exchanged with the environment.

The Appendix uses some ideas from the paper to give a derivation of two representation theorems for trace-preserving, completely positive maps, which are the most general descriptions for quantum dynamical evolutions [8].

Throughout this paper, the systems relevant to a particular vector, operator, or superoperator will be indicated by a superscript. Thus $|\psi^Q\rangle$ is a state vector for the system Q, while A^{RQ} is an operator acting on $\mathcal{H}_{RQ}=\mathcal{H}_R\otimes\mathcal{H}_Q$. (If no superscript is given, the quantum system is supposed to be generic.) A prime denotes that a particular state or density operator arises as a result of some dynamical evolution. A tilde is usually present when a particular state vector or operator is not normalized, so that $\langle\psi|\psi\rangle=1$, but $\langle\tilde{\psi}|\tilde{\psi}\rangle\neq1$ in general.

II. CHANNEL DYNAMICS

A. Completely positive maps

Imagine that the system Q is prepared in an initial state ρ^Q and then subjected to some dynamical process, after

*Permanent address: Department of Physics, Kenyon College, Gambier, OH 43022.

which the state is $\rho^{Q'}$. The dynamical process is described by a map $\mathcal{E}^Q$, so that the evolution is

$$\rho^Q \to \rho^{Q'} = \mathcal{E}^Q(\rho^Q). \tag{1}$$

In the most general case, the map $\mathcal{E}^Q$ must be a trace-preserving, completely positive linear map [8]. In other words, we have the following.

(i) $\mathcal{E}^Q$ must be linear in the density operators. That is, if $\rho^Q = p_1\rho_1^Q + p_2\rho_2^Q$, then

$$\begin{aligned}\mathcal{E}^Q(\rho^{Q'}) &= p_1\rho_1^{Q'} + p_2\rho_2^{Q'}\\ &= p_1(\mathcal{E}^Q(\rho_1^Q)) + p_2(\mathcal{E}^Q(\rho_2^Q)).\end{aligned}$$

A probabilistic mixture of inputs to $\mathcal{E}^Q$ leads to a probabilistic mixture of outputs. This means that $\mathcal{E}^Q$ must be a *superoperator*, that is, a linear operator acting on the space of linear operators (e.g., density operators) on $\mathcal{H}_Q$.

(ii) $\mathcal{E}^Q$ must be trace-preserving, so that $\mathrm{Tr}\rho^{Q'} = \mathrm{Tr}\rho^Q = 1$.

(iii) $\mathcal{E}^Q$ must be positive. This means that if ρ^Q is positive[1] then $\rho^{Q'} = \mathcal{E}^Q(\rho^Q)$ must be positive.

These three conditions mean that the superoperator $\mathcal{E}^Q$ takes normalized density operators to normalized density operators in a reasonable way. The requirement of *complete* positivity is somewhat more subtle.

(iv) $\mathcal{E}^Q$ must be completely positive. That is, suppose we extend the evolution superoperator $\mathcal{E}^Q$ in a trivial way to an evolution superoperator for a compound system RQ, yielding $\mathcal{I}^R \otimes \mathcal{E}^Q$, where $\mathcal{I}^R$ is the identity superoperator on R states. Physically, this means adjoining a system R that has trivial dynamics (no state of R is changed) and which does not interact with Q. $\mathcal{E}^Q$ is completely positive if, for all such trivial extensions, the resulting superoperator $\mathcal{I}^R \otimes \mathcal{E}^Q$ is positive.

A completely positive map is not only a reasonable map from density operators to density operators for Q, but it is *extensible* in a trivial way to a reasonable map from density operators to density operators on any larger system RQ. Since we cannot exclude *a priori* that our system Q is in fact initially entangled with some distant isolated system R, any acceptable $\mathcal{E}^Q$ had better satisfy this condition.

B. Representations of $\mathcal{E}^Q$

Completely positive, trace-preserving linear maps obviously include all unitary evolutions of the state $\rho^{Q'} = U^Q\rho^Q U^{Q\dagger}$. They also include unitary evolutions involving interactions with an external system. Suppose we consider an environment system E that is initially in the pure state $|0^E\rangle$. Then we could have

$$\mathcal{E}^Q(\rho^Q) = \mathrm{Tr}_E U^{QE}(\rho^Q \otimes |0^E\rangle\langle 0^E|)U^{QE\dagger}, \tag{2}$$

[1]We will use the term "positive" to refer generically to operators that are *positive semidefinite*, i.e., those that are Hermitian and have no negative eigenvalues.

where U^{QE} is some arbitrary unitary evolution on the joint system QE. This map is also trace preserving and completely positive.

If we can write a superoperator $\mathcal{E}^Q$ as a unitary evolution on an extended system QE followed by a partial trace over E, we say that we have a "unitary representation" of the superoperator. Such a representation is not unique since many different unitary operators U^{QE} will lead to the same $\mathcal{E}^Q$.

Another useful sort of representation for completely positive maps employs only operators on $\mathcal{H}_Q$. Let A_μ^Q be a collection of such operators indexed by μ. Then the map $\mathcal{E}^Q$ given by

$$\mathcal{E}^Q(\rho^Q) = \sum_\mu A_\mu^Q \rho^Q A_\mu^{Q\dagger} \tag{3}$$

is a completely positive map. If, in addition, the A_μ operators satisfy

$$\sum_\mu A_\mu^{Q\dagger} A_\mu^Q = 1^Q, \tag{4}$$

then the map is also trace preserving. Such a representation for $\mathcal{E}^Q$ in terms of operators A_μ^Q will be called an "operator-sum representation" for $\mathcal{E}^Q$. A single $\mathcal{E}^Q$ will admit many different operator-sum representations.

Some insight into the connection between these representations for $\mathcal{E}^Q$ can be gained by explicitly writing down the partial trace Tr_E from Eq. (2). Suppose that $\rho^Q = |\phi^Q\rangle\langle\phi^Q|$ and let $|\mu^E\rangle$ be a complete orthonormal set of states of E. Then

$$\mathcal{E}^Q(\rho^Q) = \sum_\mu \langle\mu^E|U^{QE}(|\phi^Q\rangle\langle\phi^Q| \otimes |0^E\rangle\langle 0^E|)U^{QE\dagger}|\mu^E\rangle. \tag{5}$$

If we define the operator A_μ^Q by

$$A_\mu^Q|\phi^Q\rangle = \langle\mu^E|U^{QE}(|\phi^Q\rangle \otimes |0^E\rangle), \tag{6}$$

then we recover an expression identical to Eq. (3). Since every input state ρ^Q is a convex combination of pure states, we recover Eq. (3) for arbitrary ρ^Q by linearity.

A pair of important representation theorems [9] state the following.

(i) Every trace-preserving, completely positive linear map $\mathcal{E}^Q$ has a unitary representation, as in Eq. (2).

(ii) Every trace-preserving, completely positive linear map $\mathcal{E}^Q$ has an operator-sum representation, as in Eq. (3).

(By our argument above, the second statement follows from the first.) These statements, particularly the first, motivate us to assert that the trace-preserving, completely positive linear maps is exactly the class of allowed evolutions of a quantum system. Any reasonable evolution should be such a map and every such map could be accomplished by unitary dynamics (i.e., Hamiltonian evolution) on a larger system. A relatively simple proof of both of these representation theorems is found in the Appendix.

From now on we will assume that a particular $\mathcal{E}^Q$ has been specified, giving the evolution of states of the system Q. We will use unitary representations and operator-sum representations as convenient.

III. MIXED STATES AND PURIFICATIONS

A. Entangled states

Given a pure state $|\Psi^{RQ}\rangle$ of a joint system RQ, we can form the reduced state ρ^Q for one of the subsystems Q by means of a partial trace operation

$$\rho^Q = \mathrm{Tr}_R|\Psi^{RQ}\rangle\langle\Psi^{RQ}| = \sum_k \langle k^R|\Psi^{RQ}\rangle\langle\Psi^{RQ}|k^R\rangle, \qquad (7)$$

where $|k^R\rangle$ is an orthonormal basis for $\mathcal{H}_R$. We can define the reduced state ρ^Q given a mixed joint state ρ^{RQ} in the same fashion.

We have made use of a *partial inner product* between states of R and states of a larger system RQ. This is easy to understand. The vector

$$|\xi^Q\rangle = \langle\phi^R|\Psi^{RQ}\rangle \qquad (8)$$

is defined to be the unique vector in $\mathcal{H}_Q$ such that

$$\langle\alpha^Q|\xi^Q\rangle = \langle\phi^R\alpha^Q|\Psi^{RQ}\rangle \qquad (9)$$

for all vectors $|\alpha^Q\rangle$ in $\mathcal{H}_Q$ (where $|\phi^R\alpha^Q\rangle = |\phi^R\rangle\otimes|\alpha^Q\rangle$). We could also write this as

$$\langle\phi^R|\Psi^{RQ}\rangle = \sum_k \langle\phi^R\xi_k^Q|\Psi^{RQ}\rangle|\xi_k^Q\rangle \qquad (10)$$

for some orthonormal basis set $|\xi_k^Q\rangle$ for $\mathcal{H}_Q$.

There are, of course, many different pure entangled states $|\Psi^{RQ}\rangle$ that give rise to a given reduced state ρ^Q. These are generically called *purifications* of ρ^Q. Suppose $|\Psi_1^{RQ}\rangle$ and $|\Psi_2^{RQ}\rangle$ are two such purifications. Then we can write each of them using the Schmidt decomposition

$$|\Psi_1^{RQ}\rangle = \sum_k \sqrt{\lambda_k}|\xi_{1k}^R\rangle\otimes|\lambda_k^Q\rangle, \qquad (11)$$

$$|\Psi_2^{RQ}\rangle = \sum_k \sqrt{\lambda_k}|\xi_{2k}^R\rangle\otimes|\lambda_k^Q\rangle, \qquad (12)$$

where the λ_k and $|\lambda_k^Q\rangle$ are eigenvalues and eigenstates of ρ^Q and the $|\xi_{1k}^R\rangle$ and $|\xi_{2k}^R\rangle$ are two orthonormal sets of states in $\mathcal{H}_R$. Since the two purifications differ only in the choice of orthonormal set in $\mathcal{H}_R$, they are connected by a unitary operator of the form $U^R\otimes 1^Q$. Any purification of ρ^Q can be converted to any other by a unitary rotation acting on the auxilliary "reference" system R.

The Schmidt decomposition also makes clear the fact that, given a pure entangled state

$$|\Psi^{RQ}\rangle = \sum_k \sqrt{\lambda_k}|\xi^R\rangle\otimes|\lambda_k^Q\rangle, \qquad (13)$$

the reduced states $\rho^Q = \mathrm{Tr}_R|\Psi^{RQ}\rangle\langle\Psi^{RQ}|$ and $\rho^R = \mathrm{Tr}_Q|\Psi^{RQ}\rangle\langle\Psi^{RQ}|$ will have exactly the same set of nonzero eigenvalues, namely, the λ_k.

B. Mixed-state fidelity

The notion of purification is used to define the *fidelity* between two density operators ρ_1 and ρ_2. This is

$$F(\rho_1,\rho_2) = \max|\langle 1|2\rangle|^2, \qquad (14)$$

where the maximum is taken over all purifications $|1\rangle$ and $|2\rangle$ of ρ_1 and ρ_2 [4]. The fidelity has several important properties: $0\leq F(\rho_1,\rho_2)\leq 1$, with $F(\rho_1,\rho_2)=1$ if and only if $\rho_1=\rho_2$; $F(\rho_1,\rho_2)=F(\rho_2,\rho_1)$; and if $\rho_1=|\psi_1\rangle\langle\psi_1|$ is a pure state, then

$$F(\rho_1,\rho_2) = \mathrm{Tr}\rho_1\rho_2 = \langle\psi_1|\rho_2|\psi_1\rangle. \qquad (15)$$

This is just the probability that the state ρ_2 would pass a measurement testing whether or not it is the state $|\psi_1\rangle$. The fidelity is a general way of defining the "closeness" of a pair of states.

If we have two states ρ_1^{RQ} and ρ_2^{RQ}, we can form

$$\rho_1^Q = \mathrm{Tr}_R\rho_1^{RQ}, \qquad (16)$$

$$\rho_2^Q = \mathrm{Tr}_R\rho_2^{RQ}. \qquad (17)$$

Then $F(\rho_1^{RQ},\rho_2^{RQ})\leq F(\rho_1^Q,\rho_2^Q)$. This can be seen directly from the definition by noting that every purification of ρ_1^{RQ} is also a purification of ρ_1^Q, and so on.

C. Ensembles of pure states

A mixed state ρ^Q may arise from a statistical ensemble $\mathcal{S}$ of pure states $|\psi_i^Q\rangle$ of Q. In this case we can write

$$\rho^Q = \sum_i p_i|\psi_i^Q\rangle\langle\psi_i^Q|, \qquad (18)$$

where p_i is the probability of the state $|\psi_i^Q\rangle$ in the ensemble $\mathcal{S}$.

If $\rho^Q = \mathrm{Tr}_R|\Psi^{RQ}\rangle\langle\Psi^{RQ}|$ for a pure entangled state $|\Psi^{RQ}\rangle$ of RQ, we can "realize" an ensemble of pure states for ρ^Q by performing a complete measurement on the system R. (This and other characterizations of the ensembles described by ρ^Q are given in [10].) Let $|\epsilon_i^R\rangle$ be the basis for this complete measurement. Each outcome of the R measurement will be associated with a relative state [11] of the system Q. If p_i is the probability of the ith outcome of the R measurement and $|\psi_i^Q\rangle$ is the relative state of Q associated with this outcome, then

$$\sqrt{p_i}|\psi_i^Q\rangle = \langle\epsilon_i^R|\Psi^{RQ}\rangle. \qquad (19)$$

(Note that, in dealing with ensembles of pure states, it is sometimes useful to consider the non-normalized vectors $|\tilde{\psi}_i^Q\rangle = \sqrt{p_i}|\psi_i^Q\rangle$. In other words, we can normalize the component states in $\mathcal{S}$ by their probabilities. The resulting vectors are in themselves a complete description of the ensemble $\mathcal{S}$. See [10] for fuller details.) It follows that

$$\sum_i p_i|\psi_i^Q\rangle\langle\psi_i^Q| = \sum_i \langle\epsilon_i^R|\Psi^{RQ}\rangle\langle\Psi^{RQ}|\epsilon_i^R\rangle$$
$$= \mathrm{Tr}_R|\Psi^{RQ}\rangle\langle\Psi^{RQ}| = \rho^Q, \quad (20)$$

so that the ensemble $\mathcal{S}$ of relative states is a pure state ensemble for ρ^Q. In fact, any pure state ensemble for ρ^Q can be realized in just this way. That is, we can fix a particular purification $|\Psi^{RQ}\rangle$ for ρ^Q and give a prescription for realizing any pure state ensemble for ρ^Q as a relative state ensemble for some complete measurement on R.

Let $\mathcal{S}_1$ be a pure state ensemble for ρ^Q given by probabilities p_i and states $|\psi_i^Q\rangle$ and suppose that $\mathcal{H}_R$ has arbitrarily high dimension, at least as large as the number of distinct pure states in the ensembles we consider. Then we can construct a purification $|\Psi_1^{RQ}\rangle$ by

$$|\Psi_1^{RQ}\rangle = \sum_i \sqrt{p_i}|\alpha_i^R\rangle\otimes|\psi_i^Q\rangle, \quad (21)$$

where the $|\alpha_i^R\rangle$ are a basis for $\mathcal{H}_R$. (Only some of these basis vectors may appear in this superposition.) Clearly, $\rho^Q = \mathrm{Tr}_R|\Psi_1^{RQ}\rangle\langle\Psi_1^{RQ}|$. Similarly, if we have another ensemble $\mathcal{S}_2$ for ρ^Q given by probabilities q_i and states $|\phi_i^Q\rangle$, we can construct a purification

$$|\Psi_2^{RQ}\rangle = \sum_i \sqrt{q_i}|\beta_i^R\rangle\otimes|\phi_i^Q\rangle \quad (22)$$

for some other R basis $|\beta_i^R\rangle$. Since both of these are purifications of the same ρ^Q, there is a unitary operator U^R such that $|\Psi_2^{RQ}\rangle = (U^R\otimes 1^Q)|\Psi_1^{RQ}\rangle$.

We can clearly realize the ensemble $\mathcal{S}_2$ by making a measurement of the $|\beta_i^R\rangle$ basis on the state $|\Psi_2^{RQ}\rangle$ of R; but this is equivalent to making a measurement of the basis $|\gamma_i^R\rangle = U^{R\dagger}|\beta_i^R\rangle$ on the state $|\Psi_1^{RQ}\rangle$:

$$\langle\gamma_i^R|\Psi_1^{RQ}\rangle = (\langle\beta_i^R|U^R)|\Psi_1^{RQ}\rangle$$
$$= \langle\beta_i^R|[(U^R\otimes 1^Q)|\Psi_1^{RQ}\rangle]$$
$$= \langle\beta_i^R|\Psi_2^{RQ}\rangle$$
$$= \sqrt{q_i}|\phi_i^Q\rangle. \quad (23)$$

Thus the ensemble $\mathcal{S}_2$ can be realized by making an R measurement on the purification $|\Psi_1^{RQ}\rangle$. It follows that we could pick a particular purification $|\Psi^{RQ}\rangle$ and obtain *any* pure state ensemble for ρ^Q by a suitable choice of measurement basis for the system R.

We have assumed that $\dim\mathcal{H}_R$ is arbitrarily large so that we can have an arbitrarily large number of basis vectors (since the pure state ensembles may have an arbitrarily large number of components). But this is not really necessary. If we allow positive operator measurements (POMs) [12] on R, then the dimension of $\mathcal{H}_R$ need be no greater than the dimension of $\mathcal{H}_Q$, which is the minimum size necessary to purify all mixed states ρ^Q. The only relevant part of the basis $|\alpha_i^R\rangle$ is the set of subnormalized vectors $|\tilde{\alpha}_i^R\rangle = \Pi|\alpha_i^R\rangle$, where Π is the projection onto the subspace of $\mathcal{H}_R$ that supports $\rho^R = \mathrm{Tr}_Q|\Psi^{RQ}\rangle\langle\Psi^{RQ}|$. Since $\dim\mathcal{H}_Q = d$, this subspace need have only up to d dimensions. The $|\tilde{\alpha}_i^R\rangle\langle\tilde{\alpha}_i^R|$ are elements of a POM on this subspace. We can use this POM on the d-dimensional subspace of $\mathcal{H}_R$ to find a POM for a purification that uses another reference system R_*, with $\dim\mathcal{H}_{R_*} = d$.

D. Entropy

Since entropy will be of central importance for our results, we will review some of the relevant properties of classical and quantum entropy. Suppose the non-negative numbers $p_1, p_2, \ldots$ sum to unity and thus form a probability distribution. The Shannon entropy $H(\vec{p})$ of this probability distribution (represented by the vector $\vec{p}$) is just

$$H(\vec{p}) = -\sum_k p_k \log p_k. \quad (24)$$

We specify the base of our logarithms to be 2 and take $0\log 0 = 0$. If $\vec{p}$ forms the probability for some random variable X, so that $p(x_k) = p_k$ for various values x_k of X, then we will often write this entropy as $H(X)$.

The Shannon entropy $H(X)$ is the fundamental quantity in classical information theory and it represents the average number of binary digits (or *bits*) required to represent the value of X [2]. It can be thought of as a measure of the uncertainty in the value of X expressed by the probability distribution. We can use it to define various information-theoretic quantities, such as the conditional entropy

$$H(X|Y) = \sum_k p(y_k)H(X|y_k) = -\sum_{j,k} p(x_j, y_k)\log p(x_j|y_k) \quad (25)$$

for a joint distribution $p(x_j, y_k)$ over values of two variables X and Y. A very important quantity is the *mutual information* $I(X{:}Y)$ between two random variables X and Y:

$$I(X{:}Y) = H(X) - H(X|Y), \quad (26)$$

which is the average amount that the uncertainty about X decreases when the value of Y is known. If X represents the input of a communications channel and Y represents the output, then $I(X{:}Y)$ represents the amount of information conveyed by the channel. It turns out that $I(X{:}Y) = I(Y{:}X)$.

The quantum-mechanical definition of entropy was first given by von Neumann [13]. Suppose ρ^Q is a density operator representing a mixed state of Q. Then the entropy is

$$S(\rho^Q) = -\mathrm{Tr}\rho^Q\log\rho^Q. \quad (27)$$

If $\lambda_1, \lambda_2, \ldots$ are the eigenvalues of ρ^Q, then $S(\rho^Q) = H(\vec{\lambda})$. The von Neumann entropy also has a signficance for coding similar to the Shannon entropy: it is the average number of two-level quantum systems (or *qubits*) needed to faithfully represent one of the pure states of an ensemble described by ρ^Q [3].

Suppose that systems R and Q are in a pure entangled state $|\Psi^{RQ}\rangle$. Then $S(\rho^{RQ}) = 0$. However, unlike the classical Shannon entropy, it is possible for the von Neumann entropy of the subsystems R and Q to be nonzero even when the entropy of the joint system RQ is zero. We saw above that the density operators ρ^Q and ρ^R have the same nonzero eigenvalues. Thus $S(\rho^R)=S(\rho^Q)$. That is, if a pair of quantum systems are in a pure entangled state, the reduced mixed states will have the same von Neumann entropy.

The von Neumann entropy has a number of important properties (usefully reviewed in [14]). Suppose A and B are quantum systems with joint state ρ^{AB} and reduced states ρ^A and ρ^B. Then

$$S(\rho^{AB}) \leqslant S(\rho^A)+S(\rho^B), \tag{28}$$

$$S(\rho^{AB}) \geqslant S(\rho^A)-S(\rho^B). \tag{29}$$

Equation (28) is the *subadditivity* property of the von Neumann entropy and Eq. (29) is sometimes called the "triangle inequality" for the entropy functional.

Another useful property of the von Neumann entropy relates it to the Shannon entropy of the probability distribution for the measurement outcomes of a complete observable. Let ρ be a mixed state with eigenvalues λ_k, so that

$$\rho=\sum_k \lambda_k |\lambda_k\rangle\langle\lambda_k|. \tag{30}$$

Now imagine that a measurement is performed of some complete ordinary observable, that is, the state is resolved using an orthonormal basis $|a_j\rangle$. The probability p_j that the jth outcome is obtained is thus

$$p_j=\langle a_j|\rho|a_j\rangle=\sum_k \lambda_k\langle a_j|\lambda_k\rangle\langle\lambda_k|a_j\rangle$$

$$=\sum_k M_{jk}\lambda_k. \tag{31}$$

The matrix $V_{jk}=\langle a_j|\lambda_k\rangle$ is unitary, so the matrix $M_{jk}=|V_{jk}|^2$ is doubly stochastic. That is, the rows and columns of V_{jk} are orthonormal vectors, so that the rows and columns of M_{jk} all sum to one:

$$\sum_i M_{ij}=1 \text{ for all j}, \quad \sum_j M_{ij}=1 \text{ for all i}.$$

It is a standard theorem of information theory that the Shannon entropy $H(\vec{q})=-\Sigma_i q_i \log q_i$ cannot decrease if the probabilities q_i are changed via a doubly stochastic matrix [15]. Therefore,

$$H(\vec{p}) \geqslant H(\vec{\lambda})=S(\rho). \tag{32}$$

The von Neumann entropy is thus a lower bound on the Shannon entropy for the outcome of a complete measurement on the system.

IV. ENTANGLEMENT FIDELITY

A. Definition

Suppose that an entangled state $|\Psi^{RQ}\rangle$ is prepared for the joint system RQ and that Q is subjected to a dynamical evolution described by $\mathcal{E}^Q$ (so that the overall evolution is given by $\mathcal{I}^R\otimes\mathcal{E}^Q$). The final state is

$$\rho^{RQ'}=\mathcal{I}^R\otimes\mathcal{E}^Q(|\Psi^{RQ}\rangle\langle\Psi^{RQ}|). \tag{33}$$

The fidelity of this process is

$$F_e=\mathrm{Tr}|\Psi^{RQ}\rangle\langle\Psi^{RQ}|\rho^{RQ'}=\langle\Psi^{RQ}|\rho^{RQ'}|\Psi^{RQ}\rangle. \tag{34}$$

We call F_e the *entanglement fidelity* of the process.

Written in these terms, F_e depends on the initial and final states of the system RQ. We will next show that F_e depends only on the map $\mathcal{E}^Q$ and the initial reduced state ρ^Q obtained by a partial trace

$$\rho^Q=\mathrm{Tr}_R|\Psi^{RQ}\rangle\langle\Psi^{RQ}|. \tag{35}$$

That is, the entanglement fidelity F_e, which is associated with an entangled state including Q, is (rather surprisingly) a property *intrinsic* to the system Q itself.

The superoperator $\mathcal{I}^R\otimes\mathcal{E}^Q$ can be expressed

$$\mathcal{I}^R\otimes\mathcal{E}^Q(\rho^{RQ})=\sum_\mu (1^R\otimes A_\mu^Q)\rho^{RQ}(1^R\otimes A_\mu^Q)^\dagger. \tag{36}$$

Suppose that the initial states $|\Psi_1^{RQ}\rangle$ and $|\Psi_2^{RQ}\rangle$, both purifications of ρ^Q, lead to final states $\rho_1^{RQ'}$ and $\rho_2^{RQ'}$, respectively, under the action of the superoperator $\mathcal{I}^R\otimes\mathcal{E}^Q$ and let U^R be the unitary operator for R such that

$$|\Psi_2^{RQ}\rangle=(U^R\otimes 1^Q)|\Psi_1^{RQ}\rangle. \tag{37}$$

Clearly, $U^R\otimes 1^Q$ commutes with $1^R\otimes A_\mu^Q$ for all μ. Therefore,

$$\begin{aligned}
\rho_2^{RQ'} &= \sum_\mu (1^R\otimes A_\mu^Q)|\Psi_2^{RQ}\rangle\langle\Psi_2^{RQ}|(1^R\otimes A_\mu^Q)^\dagger \\
&= \sum_\mu (1^R\otimes A_\mu^Q)(U^R\otimes 1^Q)|\Psi_1^{RQ}\rangle \\
&\quad \times\langle\Psi_1^{RQ}|(U^R\otimes 1^Q)^\dagger(1^R\otimes A_\mu^Q) \\
&= (U^R\otimes 1^Q)\left(\sum_\mu (1^R\otimes A_\mu^Q)|\Psi_1^{RQ}\rangle\langle\Psi_1^{RQ}|(1^R\otimes A_\mu^Q)^\dagger\right) \\
&\quad \times(U^R\otimes 1^Q)^\dagger\rho_2^{RQ'} \\
&= (U^R\otimes 1^Q)\rho_1^{RQ'}(U^R\otimes 1^Q)^\dagger.
\end{aligned} \tag{38}$$

[Note that Eq. (38) implies that $\rho_1^{RQ'}$ and $\rho_2^{RQ'}$ must have the same eigenvalues. This will be important later in the definition of entropy exchange.] From Eq. (38) it follows that

$$\begin{aligned}
F_{e2} &= \langle\Psi_2^{RQ}|\rho_2^{RQ'}|\Psi_2^{RQ}\rangle \\
&= \langle\Psi_1^{RQ}|(U^R\otimes 1^Q)^\dagger(U^R\otimes 1^Q)\rho_1^{RQ'}(U^R\otimes 1^Q)^\dagger \\
&\quad \times(U^R\otimes 1^Q)|\Psi_1^{RQ}\rangle \\
&= \langle\Psi_1^{RQ}|\rho_1^{RQ'}|\Psi_1^{RQ}\rangle \\
&= F_{e1}.
\end{aligned} \tag{39}$$

Hence the fidelity F_e does not depend on *which* purification for ρ^Q is chosen. It only depends on ρ^Q and the superoperator $\mathcal{E}^Q$.

B. Intrinsic expression for F_e

It is instructive to derive an expression for F_e in terms of things that are intrinsic to the system Q, i.e., an expression that does not refer to R. Suppose we have an operator-sum representation for $\mathcal{E}^Q$, as in Eq. (3). Consider a particular pure entangled state for RQ

$$|\Psi^{RQ}\rangle=\sum_k \sqrt{p_k}|k^R\rangle\otimes|\phi_k^Q\rangle, \tag{40}$$

where the $|k^R\rangle$ are orthonormal states in $\mathcal{H}_R$. (We do not need to require the $|\phi_k^Q\rangle$ to be orthonormal.) This state evolves under $\mathcal{I}^R\otimes\mathcal{E}^Q$ into $\rho^{RQ'}$. The initial state of Q is

$$\rho^Q=\mathrm{Tr}_R|\Psi^{RQ}\rangle\langle\Psi^{RQ}|=\sum_k p_k|\phi_k^Q\rangle\langle\phi_k^Q|. \tag{41}$$

Now, for any operator X^Q acting on $\mathcal{H}_Q$,

$$\begin{aligned}\langle\Psi^{RQ}|(1^R\otimes X^Q)|\Psi^{RQ}\rangle&=\sum_{jk}\sqrt{p_jp_k}\langle j^R|1^R|k^R\rangle\langle\phi_j^Q|X^Q|\phi_k^Q\rangle\\&=\sum_{jk}\sqrt{p_jp_k}\delta_{jk}\langle\phi_j^Q|X^Q|\phi_k^Q\rangle\\&=\sum_k p_k\langle\phi_k^Q|X^Q|\phi_k^Q\rangle=\mathrm{Tr}\rho^QX^Q.\end{aligned} \tag{42}$$

We can now work out the fidelity very easily:

$$\begin{aligned}F_e&=\langle\Psi^{RQ}|\rho^{RQ'}|\Psi^{RQ}\rangle\\&=\sum_\mu\langle\Psi^{RQ}|(1^R\otimes A_\mu^Q)|\Psi^{RQ}\rangle\langle\Psi^{RQ}|(1^R\otimes A_\mu^Q)^\dagger|\Psi^{RQ}\rangle,\end{aligned}$$

$$F_e=\sum_\mu(\mathrm{Tr}\rho^QA_\mu^Q)(\mathrm{Tr}\rho^QA_\mu^{Q\dagger}). \tag{43}$$

Although this is written with respect to a particular operator-sum representation of $\mathcal{E}^Q$ (which is not unique), the value of F_e will clearly be independent of this representation. Equation (43) expresses F_e entirely in terms of the initial state ρ^Q of the system Q and the evolution superoperator $\mathcal{E}^Q$.

C. Relations to other fidelities

It is worth noting what F_e is not. It is not the simple fidelity of the input and output states of Q. This fidelity can be written $F(\rho^Q,\rho^{Q'})$, where $\rho^{Q'}=\mathcal{E}^Q(\rho^Q)$. We can show that $F_e\neq F(\rho^Q,\rho^{Q'})$ in general by considering an operation defined by

$$A_\mu^Q=|\mu^Q\rangle\langle\mu^Q| \tag{44}$$

for some orthonormal basis $|\mu^Q\rangle$. The effect of the operation is to completely destroy any coherences between different elements of the basis. That is, the superposition $\Sigma_\mu c_\mu|\mu^Q\rangle$ would be transformed into the mixed state

$$\rho^{Q'}=\sum_\mu|c_\mu|^2|\mu^Q\rangle\langle\mu^Q|. \tag{45}$$

Now suppose $\rho^Q=\Sigma_\mu\lambda_\mu|\mu^Q\rangle\langle\mu^Q|$. Then $\rho^{Q'}=\rho^Q$ and thus $F(\rho^Q,\rho^{Q'})=1$. However, let $|\Psi^{RQ}\rangle$ be a purification of ρ^Q, for example,

$$|\Psi^{RQ}\rangle=\sum_\mu\sqrt{\lambda_\mu}|\phi_\mu^R\rangle\otimes|\mu^Q\rangle. \tag{46}$$

The action of the superoperator $\mathcal{I}^R\otimes\mathcal{E}^Q$ on this state yields

$$\rho^{RQ'}=\sum_\mu\lambda_\mu|\phi_\mu^R\rangle\langle\phi_\mu^R|\otimes|\mu^Q\rangle\langle\mu^Q|. \tag{47}$$

If more than one of the λ_μ's is nonzero, then $F_e=F(\rho^{RQ},\rho^{RQ'})\neq1$. Thus $F_e\neq F(\rho^Q,\rho^{Q'})$.

However, there is a general relation between F_e and $F(\rho^Q,\rho^{Q'})$,

$$F_e=F(\rho^{RQ},\rho^{RQ'})\leq F(\rho^Q,\rho^{Q'}). \tag{48}$$

The entanglement fidelity F_e is thus a lower bound to the input-output fidelity $F(\rho^Q,\rho^{Q'})$ for states of Q.

F_e and $F(\rho^Q,\rho^{Q'})$ do sometimes agree. Suppose that the initial state ρ^Q is in fact a pure state of Q, so that there is no entanglement between R and Q. Then, letting $\rho^Q=|\psi^Q\rangle\langle\psi^Q|$,

$$\begin{aligned}F(\rho^Q,\rho^{Q'})&=\langle\psi^Q|\rho^{Q'}|\psi^Q\rangle\\&=\sum_\mu\langle\psi^Q|A_\mu^Q|\psi^Q\rangle\langle\psi^Q|A_\mu^{Q\dagger}|\psi^Q\rangle\\&=\sum_\mu(\mathrm{Tr}\rho^QA_\mu^Q)(\mathrm{Tr}\rho^QA_\mu^{Q\dagger})\\&=F_e.\end{aligned} \tag{49}$$

The entanglement fidelity equals the ''input-output'' fidelity when the input state is a pure state.

Now suppose that ρ^Q is a mixed state of Q arising from an ensemble $\mathcal{S}$ in which the pure state $|\psi_i^Q\rangle$ occurs with probability p_i. The average input-output fidelity for this ensemble is

$$\begin{aligned}\bar{F}&=\sum_i p_iF(|\psi_i^Q\rangle\langle\psi_i^Q|,\rho_i^{Q'})\\&=\sum_i p_i\langle\psi_i^Q|\rho_i^{Q'}|\psi_i^Q\rangle,\end{aligned} \tag{50}$$

where $\rho_i^{Q'}=\mathcal{E}^Q(|\psi_i^Q\rangle\langle\psi_i^Q|)$.

It turns out that $\bar{F}\geq F_e$. Some such connection is reasonable physically, since we can realize a pure state ensemble

$\mathcal{S}$ by means of an R measurement on a purification of ρ^Q, and this measurement may be performed either before or after the dynamical evolution given by $\mathcal{S}^Q$. A full proof follows.

Let $|\alpha_i^R\rangle$ be an orthonormal set in $\mathcal{H}_R$ (assumed to have as many dimensions as there are elements in the ensemble $\mathcal{S}$) and let

$$|\Psi^{RQ}\rangle = \sum_i \sqrt{p_i}|\alpha_i^R\rangle \otimes |\psi_i^Q\rangle. \tag{51}$$

$|\Psi^{RQ}\rangle$ is clearly a purification of ρ^Q and the $|\alpha_i^R\rangle$ basis is the basis in $\mathcal{H}_R$ that, when measured, generates the ensemble $\mathcal{S}$ as an ensemble of relative states in Q. That is, $\sqrt{p_i}|\psi_i^Q\rangle = \langle\alpha_i^R|\Psi^{RQ}\rangle$, which we could also write as

$$(|\alpha_i^R\rangle\langle\alpha_i^R|\otimes 1^Q)|\Psi^{RQ}\rangle = \sqrt{p_i}|\alpha_i^R\rangle\otimes|\psi_i^Q\rangle. \tag{52}$$

Now consider the operator Γ^{RQ} given by

$$\begin{aligned}\Gamma^{RQ} &= \sum_j |\alpha_j^R\rangle\langle\alpha_j^R|\otimes|\psi_j^Q\rangle\langle\psi_j^Q| \\ &= \sum_j (1^R\otimes|\psi_j^Q\rangle\langle\psi_j^Q|)(|\alpha_j^R\rangle\langle\alpha_j^R|\otimes 1^Q).\end{aligned} \tag{53}$$

Since Γ^{RQ} is the sum of an orthogonal set of projections, it is itself a projection operator onto some subspace of $\mathcal{H}_R\otimes\mathcal{H}_Q$. $|\Psi^{RQ}\rangle$ itself is in this subspace:

$$\begin{aligned}\Gamma^{RQ}|\Psi^{RQ}\rangle &= \sum_j (1^R\otimes|\psi_j^Q\rangle\langle\psi_j^Q|)(|\alpha_j^R\rangle\langle\alpha_j^R|\otimes 1^Q)|\Psi^{RQ}\rangle \\ &= \sum_j (1^R\otimes|\psi_j^Q\rangle\langle\psi_j^Q|)\sqrt{p_j}|\alpha_j^R\rangle\otimes|\psi_j^Q\rangle \\ &= \sum_j \sqrt{p_j}|\alpha_j^R\rangle\otimes|\psi_j^Q\rangle \\ &= |\Psi^{RQ}\rangle.\end{aligned} \tag{54}$$

Therefore, we have the operator inequality $\Gamma^{RQ} \geqslant |\Psi^{RQ}\rangle\langle\Psi^{RQ}|$. This means that, for any vector $|\chi^{RQ}\rangle$,

$$\langle\chi^{RQ}|\Gamma^{RQ}|\chi^{RQ}\rangle \geqslant \langle\chi^{RQ}|(|\Psi^{RQ}\rangle\langle\Psi^{RQ}|)|\chi^{RQ}\rangle, \tag{55}$$

which in turn implies that, for all positive operators X^{RQ},

$$\mathrm{Tr}\Gamma^{RQ}X^{RQ} \geqslant \mathrm{Tr}|\Psi^{RQ}\rangle\langle\Psi^{RQ}|X^{RQ} = \langle\Psi^{RQ}|X^{RQ}|\Psi^{RQ}\rangle. \tag{56}$$

Let A_μ^Q be the operators in an operator-sum representation of the evolution superoperator $\mathcal{E}^Q$. Then

$$\begin{aligned}&\Gamma^{RQ}(1^R\otimes A_\mu^Q)|\Psi^{RQ}\rangle \\ &= \sum_j (1^R\otimes|\psi_j^Q\rangle\langle\psi_j^Q|)(|\alpha_j^R\rangle\langle\alpha_j^R|\otimes 1^Q)(1^R\otimes A_\mu^Q)|\Psi^{RQ}\rangle \\ &= \sum_j (1^R\otimes|\psi_j^Q\rangle\langle\psi_j^Q|)(1^R\otimes A_\mu^Q)(|\alpha_j^R\rangle\langle\alpha_j^R|\otimes 1^Q)|\Psi^{RQ}\rangle \\ &= \sum_j (1^R\otimes|\psi_j^Q\rangle\langle\psi_j^Q|)(1^R\otimes A_\mu^Q)\sqrt{p_j}|\alpha_j^R\rangle\otimes|\psi_j^Q\rangle \\ &= \sum_j (1^R\otimes|\psi_j^Q\rangle\langle\psi_j^Q|)\sqrt{p_j}|\alpha_j^R\rangle\otimes A_\mu^Q|\psi_j^Q\rangle \\ &= \sum_j \sqrt{p_j}\langle\psi_j^Q|A_\mu^Q|\psi_j^Q\rangle|\alpha_j^R\rangle\otimes|\psi_j^Q\rangle.\end{aligned} \tag{57}$$

If $\rho^{RQ\prime} = \mathcal{I}^R\otimes\mathcal{E}^Q(|\Psi^{RQ}\rangle\langle\Psi^{RQ}|)$, then

$$\begin{aligned}F_e &= \mathrm{Tr}|\Psi^{RQ}\rangle\langle\Psi^{RQ}|\rho^{RQ\prime} \\ &\leqslant \mathrm{Tr}\Gamma^{RQ}\rho^{RQ\prime} \\ &= \mathrm{Tr}\Gamma^{RQ}\rho^{RQ\prime}\Gamma^{RQ} \\ &= \sum_\mu \mathrm{Tr}\Gamma^{RQ}(1^R\otimes A_\mu^Q)|\Psi^{RQ}\rangle\langle\Psi^{RQ}|(1^R\otimes A_\mu^Q)^\dagger\Gamma^{RQ} \\ &= \sum_{j,k}\sum_\mu \sqrt{p_jp_k}\langle\psi_j^Q|A_\mu^Q|\psi_j^Q\rangle\langle\psi_k^Q|A_\mu^{Q\dagger}|\psi_k^Q\rangle\langle\psi_k^Q|\psi_j^Q\rangle \\ &\quad\times\langle\alpha_k^R|\alpha_j^R\rangle \\ &= \sum_k\sum_\mu p_k\langle\psi_k^Q|A_\mu^Q|\psi_k^Q\rangle\langle\psi_k^Q|A_\mu^{Q\dagger}|\psi_k^Q\rangle \\ &= \sum_k p_k\langle\psi_k^Q|\left(\sum_\mu A_\mu^Q|\psi_k^Q\rangle\langle\psi_k^Q|A_\mu^{Q\dagger}\right)|\psi_k^Q\rangle \\ &= \sum_k p_k\langle\psi_k^Q|\rho_k^{Q\prime}|\psi_k^Q\rangle \\ &= \bar{F}.\end{aligned} \tag{58}$$

Thus $\bar{F} \geqslant F_e$, as we wished to show. The average input-output fidelity under the evolution superoperator $\mathcal{E}^Q$ for any ensemble of pure states with density operator ρ^Q is bounded below by the entanglement fidelity F_e.

V. ENTROPY EXCHANGE

A. Definition

As shown in Eq. (38) above, if $|\Psi_1^{RQ}\rangle$ and $|\Psi_2^{RQ}\rangle$ are two purifications of ρ^Q and each is subjected to the same evolution superoperator $\mathcal{I}^R\otimes\mathcal{E}^Q$, the resulting states $\rho_1^{RQ\prime}$ and $\rho_2^{RQ\prime}$ will have exactly the same eigenvalues. Therefore,

$$S(\rho_1^{RQ\prime}) = S(\rho_2^{RQ\prime}), \tag{59}$$

where $S(\rho)$ is the von Neumann entropy of the density operator ρ. In other words, the entropy of the final joint state of

RQ is independent of which purification is chosen. Again, rather surprisingly, we have a quantity that depends only on the initial state ρ^Q and the evolution superoperator $\mathcal{E}^Q$; that is, we have a quantity that is *intrinsic* to Q. For a given ρ^Q and $\mathcal{E}^Q$, we therefore define the *entropy exchange* S_e to be

$$S_e = -\operatorname{Tr}\rho^{RQ'}\log\rho^{RQ'} \tag{60}$$

where $\rho^{RQ'} = \mathcal{I}^R \otimes \mathcal{E}^Q(|\Psi^{RQ}\rangle\langle\Psi^{RQ}|)$ and $|\Psi^{RQ}\rangle$ is some purification of ρ^Q.

Why call S_e the entropy "exchange"? Suppose we have two systems A and B, initially in the state $\rho^{AB} = \rho^A \otimes \rho^B$, which interact according to a unitary evolution operator U^{AB}. The evolution of each system will be describable in terms of a superoperator. That is,

$$\mathcal{E}^A(\rho^A) = \operatorname{Tr}_B U^{AB}(\rho^A \otimes \rho^B)U^{AB\dagger}, \tag{61}$$

$$\mathcal{E}^B(\rho^B) = \operatorname{Tr}_A U^{AB}(\rho^A \otimes \rho^B)U^{AB\dagger}. \tag{62}$$

(In the definition of $\mathcal{E}^A$ we imagine that ρ^B is given, and vice versa.) We can thus calculate the entropy exchanges S_e^A and S_e^B This can be done by including reference systems R_A and R_B to purify the initial state:

$$|\Psi^{ABR_AR_B}\rangle = |\Psi^{AR_A}\rangle \otimes |\Phi^{BR_B}\rangle. \tag{63}$$

Now, since the overall evolution is unitary, the final state $|\Psi^{ABR_AR_B\prime}\rangle$ is also pure. This means that $\rho^{AR_A'}$ and $\rho^{BR_B'}$ have exactly the same nonzero eigenvalues and thus the same entropy. Thus $S_e^A = S_e^B$ In other words, the entropy exchange is a *common* quantity for two initially uncorrelated systems that interact unitarily.

We will now derive an explicit expression for S_e in terms of ρ^Q and $\mathcal{E}^Q$. Suppose we have an operator-sum representation for $\mathcal{E}^Q$ and we define

$$|\tilde{\Phi}_\mu^{RQ'}\rangle = (1^R \otimes A_\mu^Q)|\Psi^{RQ}\rangle. \tag{64}$$

(These are not normalized vectors in general.) Then

$$\rho^{RQ'} = \sum_\mu (1^R \otimes A_\mu^Q)|\Psi^{RQ}\rangle\langle\Psi^{RQ}|(1^R \otimes A_\mu^Q)^\dagger$$

$$= \sum_\mu |\tilde{\Phi}_\mu^{RQ'}\rangle\langle\tilde{\Phi}_\mu^{RQ'}|. \tag{65}$$

Thus the vectors $|\tilde{\Phi}_\mu^{RQ'}\rangle$ give us a pure state ensemble for $\rho^{RQ'}$. We can use these states to construct a purification for $\rho^{RQ'}$. Let us adjoin a system E whose Hilbert space $\mathcal{H}_E$ has at least as many dimensions as the number of A_μ^Q operators. Then the state

$$|\Upsilon^{RQE'}\rangle = \sum_\mu |\tilde{\Phi}_\mu^{RQ'}\rangle \otimes |\mu^E\rangle \tag{66}$$

(where the $|\mu^E\rangle$ are an orthonormal set of E states) will be a purification for $\rho^{RQ'}$.

Since the state $|\Upsilon^{RQE'}\rangle$ is a pure state, the reduced states

$$\rho^{RQ'} = \operatorname{Tr}_E|\Upsilon^{RQE'}\rangle\langle\Upsilon^{RQE'}|, \tag{67}$$

$$\rho^{E'} = \operatorname{Tr}_{RQ}|\Upsilon^{RQE'}\rangle\langle\Upsilon^{RQE'}| \tag{68}$$

will have the same entropy. Therefore, $S_e = S(\rho^{RQ'}) = S(\rho^{E'})$. We can write down the density operator $\rho^{E'}$,

$$\rho^{E'} = \operatorname{Tr}_{RQ}|\Upsilon^{RQE'}\rangle\langle\Upsilon^{RQE'}|$$

$$= \sum_{\mu,\nu} \langle\tilde{\Phi}_\nu^{RQ'}|\tilde{\Phi}_\mu^{RQ'}\rangle|\mu^E\rangle\langle\nu^E|. \tag{69}$$

That is, $\rho^{E'} = \Sigma_{\mu\nu}W_{\mu\nu}|\mu^E\rangle\langle\nu^E|$, where

$$W_{\mu\nu} = \langle\tilde{\Phi}_\nu^{RQ'}|\tilde{\Phi}_\mu^{RQ'}\rangle$$

$$= \operatorname{Tr}|\tilde{\Phi}_\mu^{RQ'}\rangle\langle\tilde{\Phi}_\nu^{RQ'}|$$

$$= \operatorname{Tr}(1^R \otimes A_\mu^Q)|\Psi^{RQ'}\rangle\langle\Psi^{RQ'}|(1^R \otimes A_\nu^Q)^\dagger$$

$$= \operatorname{Tr}_Q A_\mu^Q(\operatorname{Tr}_R|\Psi^{RQ}\rangle\langle\Psi^{RQ}|)A_\nu^{Q\dagger}$$

$$= \operatorname{Tr}_Q A_\mu^Q \rho^Q A_\nu^{Q\dagger}. \tag{70}$$

In other words, we have the following prescription. Let W be a density operator with components (in some orthonormal basis)

$$W_{\mu\nu} = \operatorname{Tr}A_\mu^Q \rho^Q A_\nu^{Q\dagger}. \tag{71}$$

Then

$$S_e = S(W). \tag{72}$$

As explained in the Appendix, any two operator-sum representations for $\mathcal{E}^Q$ are related by a unitary matrix $U_{\mu\nu}$. This simply corresponds to the freedom to write the matrix $W_{\mu\nu}$ with respect to any basis (which obviously does not affect S_e). Let $P_\mu = W_{\mu\mu}$ be the diagonal elements of $W_{\mu\nu}$. These would be the probabilities given the state W for a complete measurement using the basis that yields the matrix elements $W_{\mu\nu}$. Therefore, $H(\vec{P}) \geq S(W)$. But we could, by choosing the unitary matrix that diagonalizes $W_{\mu\nu}$, find a representation such that $H(\vec{P}) = S(W)$. This yields another expression for S_e,

$$S_e = \min\left(-\sum_\mu P_\mu \log P_\mu\right), \tag{73}$$

where $P_\mu = \operatorname{Tr}A_\mu^Q \rho^Q A_\mu^{Q\dagger}$ and the minimum is taken over all operator-sum representations of $\mathcal{E}^Q$.

For a given input state ρ^Q, there is a "diagonal" operator-sum representation, in which $W_{\mu\nu}$ is diagonal. In this representation,

$$\operatorname{Tr}A_\mu^Q \rho^Q A_\nu^{Q\dagger} = 0 \quad \text{for } \mu \neq \nu. \tag{74}$$

If $\rho^Q = d^{-1}1^Q$ (the "maximally mixed" state), then this simply means that the various A_μ^Q operators are orthogonal in the operator inner product $\langle B, C\rangle = \operatorname{Tr}B^\dagger C$. This diagonal representation is minimal, in the sense that no other operator-sum representation includes a smaller number of A_μ^Q operators.

The evolution $\mathcal{E}^Q$ might in fact be due to unitary evolution of a larger system that includes an environment E, with E initially in a pure state and RQ initially in a pure entangled state. In this case the final state of RQE will be also be a pure state. Then $S(\rho^{E'})=S(\rho^{RQ'})=S_e$. In other words, the entropy exchange S_e is just the entropy produced in the environment, if it is initially in a pure state.

Note that the same $\rho^{E'}$ would have been obtained if we ignored the reference system R entirely and simply considered the unitary evolution of QE with an initial state ρ^Q for Q. The entropy produced in the environment does not depend on the dynamically isolated reference system R.

The assumption that the environment is initially in a pure state $|0^E\rangle$ at first seems too restrictive. For example, we may wish to consider environments that are initially in some thermal equilibrium state ρ^E. However, we may imagine that the environment consists of a "near" environment E_n and a "far" environment E_f. The system Q interacts only with the near environment E_n. The initial state of the full environment may be an entangled pure state, but the system Q will "see" a mixed state for E_n.

To summarize, the entropy exchange S_e has the following properties.

(i) S_e is a quantity intrinsic to the system Q and can be defined entirely in terms of the initial state ρ^Q and the superoperator $\mathcal{E}^Q$.

(ii) If the initial state ρ^Q arises because a larger system RQ is in a pure entangled state and if the reference system R has trivial dynamics, then the entropy exchange S_e is the entropy of the final state $\rho^{RQ'}$ of RQ. (It is easy to generalize this to the case when R itself can have arbitrary unitary evolution, i.e., when R is dynamically isolated but may have a nonzero internal Hamiltonian.)

(iii) If the nonunitary evolution of Q arises because Q interacts with an environment E that is initially in a pure state, then S_e is the entropy of the final state $\rho^{E'}$ of the environment.

(iv) If the initial state ρ^Q of the system Q is a pure state, we can adopt a unitary representation for $\mathcal{E}^Q$ in which E is also initially in a pure state. Then $\rho^{Q'}$ and $\rho^{E'}$ have the same eigenvalues. In this case, $S_e=S(\rho^{Q'})$, the entropy produced in the system Q.

B. Relation to other entropies

Once again, it is useful to emphasize what S_e is not. It is not, in general, the increase in the entropy of the system Q; in fact, this entropy may actually decrease, whereas S_e is never negative. It is also not always the entropy increase of the environment if the initial environment state is mixed. The entropy exchange S_e simply characterizes the information exchange between the system Q and the external world during the evolution given by $\mathcal{E}^Q$.

There are, however, inequalities relating S_e to entropy changes in Q and E. First we will relate the entropy exchange to changes in the entropy of Q. Suppose an evolution superoperator $\mathcal{E}^Q$ is given, together with an initial state ρ^Q of Q. We can always find a representation for $\mathcal{E}^Q$ as a unitary evolution on a larger system QE with an initial pure state $|0^E\rangle$ for the environment system. With this representation, the entropy of the joint initial state $S(\rho^{QE})=S(\rho^Q)$. The joint system QE evolves unitarily, so the entropy of the joint state remains unchanged. Thus $S(\rho^{QE'})=S(\rho^Q)$. The entropy exchange in this case is the final entropy of the environment $S(\rho^{E'})$. The triangle inequality [Eq. (29)] yields

$$S(\rho^Q)\geq S(\rho^{Q'})-S(\rho^{E'}),\quad S_e\geq S(\rho^{Q'})-S(\rho^Q). \tag{75}$$

In other words, the entropy exchange is no less than the increase in entropy of the system Q. We can also in this way establish that

$$S_e\leq S(\rho^Q)+S(\rho^{Q'}). \tag{76}$$

Now we relate S_e to the entropy change in the environment. In this case, we are given a particular (possibly mixed) initial state ρ^E for the environment and a particular unitary evolution U^{QE} for the joint system QE. Again, the initial state of Q is ρ^Q, but now we will imagine that this is a partial state of a pure entangled state $|\Psi^{RQ}\rangle$, where R is an isolated reference system. The entropy of the joint system RQE is initially $S(\rho^{RQE})=S(\rho^E)$ and remains unchanged during the unitary evolution of the joint system. By definition, the entropy exchange is just the entropy $S(\rho^{RQ'})$ of the final state of RQ. Thus

$$S(\rho^E)\geq S(\rho^{E'})-S(\rho^{RQ'}),\quad S_e\geq S(\rho^{E'})-S(\rho^E), \tag{77}$$

so that the entropy exchange is no less than the increase in the entropy of the environment. We can also derive

$$S_e\leq S(\rho^E)+S(\rho^{E'}), \tag{78}$$

which, for a large environment, is probably not very useful.

Similar arguments based on the subaddtivity of the entropy functional [Eq. (28)], also demonstrate that S_e is no smaller than the entropy *decrease* in either the system Q or the environment E. To summarize the lower bounds for S_e,

$$S_e\geq|\Delta S^Q|, \tag{79}$$

$$S_e\geq|\Delta S^E|, \tag{80}$$

where ΔS^Q and ΔS^E are the changes in entropy of the system Q and environment E, respectively.

C. Entropy exchange and eavesdropping

There is a simple application of these ideas to quantum cryptography [16]. Suppose Alice prepares the state ρ_k^Q of Q with probability p_k and then conveys the system Q to Bob as part of a quantum cryptographic protocol. (Alternatively, we could imagine that Alice prepares Q in a state entangled with a system R, which she retains, as part of an entanglement-based protocol [17]. But, in such protocols, Alice usually later makes a measurement on R, giving rise to an ensemble of relative states of Q.) Along the way Q may interact with the rest of the world, represented by the environment system E, producing some level of "noise" in Q. The environment, however, may also contain the measuring apparatus of an eavesdropper Eve. We will assume that the environment is initially in a pure state (but see the remark above about the possibility of an entangled state of near and far zones within the environment).

The dynamical evolution of Q is given by the evolution superoperator $\mathcal{E}^Q$. Let $S_{e,k}$ be the entropy exchange in Q for the input state ρ_k^Q which equals the entropy of the final environment state $\rho_k^{E'}$ resulting from the input of ρ_k^Q and let S_e be the entropy exchange associated with the "average" input state $\rho^Q=\Sigma_k p_k \rho_k^Q$ which equals the entropy of the average final environment state $\rho^{E'}$.

The eavesdropper Eve will try to infer the preparation ρ_k^Q by examining the state of her measuring apparatus, that is, by trying to distinguish the various environment states $\rho_k^{E'}$ Denote Alice's preparation, and thus the final environment state produced by that preparation, by the random variable X and the reading on Eve's measuring apparatus by Y. Then a theorem of Kholevo [18] limits the mutual information $I(X:Y)$, which is the amount of information about X that Eve obtains from a knowledge of Y. This limit is

$$I(X:Y)\leq S(\rho^{E'})-\sum_k p_k S(\rho_k^{E'})=S_e-\sum_k p_k S_{e,k} \quad (81)$$

$$\leq S_e. \quad (82)$$

[If the eavesdropper Eve only has access to part of the environment system E, then she will be able to do no better and $I(X:Y)$ will still be bounded in this way.]

Thus the entropy exchange associated with the ensemble of input states and the evolution superoperator $\mathcal{E}^Q$, both of which can be determined, in principle, from repeated use of the channel Q, limits the amount of information that any eavesdropper might obtain about the input. Put another way, any process by which the eavesdropper obtains information about the channel system Q disturbs the system, leaving traces in the evolution superoperator $\mathcal{E}^Q$. The disturbance produced by the eavesdropper (and other interactions with the environment) is characterized by the entropy exchange S_e.

VI. THE QUANTUM FANO INEQUALITY

A. Classical theorem

In classical information theory, there is a simple relation between the noise in a channel and probability of error in that channel [15]. This relation is Fano's inequality. We will derive an analogous quantum relation.

Let X be a classical random variable representing the input of a noisy channel and suppose that X can take on up to N different values. The output of the noisy channel is represented by the random variable Y. The channel itself is represented by the conditional probabilities $p(y_k|x_j)$ of an output value y_k given an input value x_j. These probabilities, together with the input probability distribution $p(x_j)$, characterize the situation. The receiver makes an estimate $\hat{X}$ of the input X based only on the channel output Y. The probability of error P_E is the total likelihood that $\hat{X}\neq X$.

Fano's inequality (in its stronger form) states that

$$h(P_E)+P_E\log(N-1)\geq H(X|Y), \quad (83)$$

where $h(P_E)=-P_E\log P_E-(1-P_E)\log P_E$ and $H(X|Y)$ is the Shannon conditional entropy of X given Y. $H(X|Y)$, the average residual information uncertainty about the input given the output, is a measure of the noise in the channel. $H(X|Y)=0$ for a noiseless channel, in which the input X can be exactly determined by the output Y. Noting that $h(P_E)\leq 1$ (since our logarithms are base 2), we can derive a simpler but slightly weaker form of Fano's inequality,

$$1+P_E\log N>H(X|Y). \quad (84)$$

Fano's inequality is used to prove the "weak converse" of the classical noisy coding theorem, which states that information cannot be sent at a rate greater than the channel capacity with arbitrarily low probability of error [15].

B. Quantum theorem

We now turn to the quantum problem. As before, we suppose that the system RQ is initially in the entangled state $|\Psi^{RQ}\rangle$ and that Q is subjected to an evolution described by $\mathcal{E}^Q$. The reference system R is isolated and has trivial dynamics described by $\mathcal{I}^R$. The dimensions of $\mathcal{H}_Q$ and $\mathcal{H}_R$ are both finite and equal to d. After the evolution, the system is described by a joint state $\rho^{RQ'}$.

Now suppose that we subject the final state $\rho^{RQ'}$ to a measurement of a complete ordinary observable on the system RQ, which is described by a basis of d^2 orthogonal states for RQ. Let the random variable X represent the outcome of this measurement. Then we know [from Eq. (32)] that

$$S_e=S(\rho^{RQ'})\leq H(X). \quad (85)$$

Further suppose that one of these basis vectors is chosen to be the original state $|\Psi^{RQ}\rangle$. Then the fidelity $F_e=\langle\Psi^{RQ}|\rho^{RQ'}|\Psi^{RQ}\rangle$ is just the probability of this outcome. Given this probability, the largest possible value of $H(X)$ would occur when all of the d^2-1 other outcomes have equal probability. Then

$$\max H(X)=-F_e\log F_e-(d^2-1)\frac{1-F_e}{d^2-1}\log\frac{1-F_e}{d^2-1}$$

$$=-F_e\log F_e-(1-F_e)\log(1-F_e)$$

$$+(1-F_e)\log(d^2-1). \quad (86)$$

Therefore we can conclude that

$$h\ F_e \quad 1 \quad F_e \log d^2 \quad 1 \quad S_e. \quad 87$$

This is our quantum version of the Fano inequality, relating the entanglement fidelity F_e with the entropy exchange S_e. Although we have made use of the reference system R in deriving this inequality, both F_e and S_e have meanings that are intrinsic to the system Q.

As before, we can give a slightly weaker form of the inequality:

$$1 \quad 2 \quad 1 \quad F_e \log d \quad S_e. \quad 88$$

It is instructive to compare the form of this equation to that of Eq. (84). The number N of possible input states is analogous the dimension d of $\mathcal{H}$. The probability of error P_E roughly corresponds $1-F_e$, the amount by which the final entangled state fails to correspond to the initial one. The noise term $H(X|Y)$ is replaced by the entropy exchange S_e. Finally, a factor of 2 appears in the error term in the quantum case, which in fact corresponds to replacing N by d^2, the dimension of $\mathcal{H}_Q \otimes \mathcal{H}_R$.

We can strengthen the quantum Fano inequality in a number of ways. First, if the reference system R has a Hilbert space of dimension $d_R < d$, the quantity d^2 can be replaced by the product $d_R d$. The required dimension d_R is in fact just the dimension of the subspace that supports ρ^R and so $d_R \leq d$ even if R is much larger than Q. Since we wish to consider F_e and S_e to be quantities intrinsic to Q, though, we will simply adopt $d_R = d$.

Finally, we note that the fidelity F_e can be lowered by *internal dynamics* of Q as well as by information exchange with the environment. To take this into account, we could allow the final state of the system to be "processed" via any unitary transformation U^Q on Q and define

$$\hat{F}_e = \max_{U^Q} \langle \Psi^{RQ} | (1^R \otimes U^Q) \rho^{RQ'} (1^R \otimes U^Q)^\dagger | \Psi^{RQ} \rangle. \tag{89}$$

($\hat{F}_e$ is also independent of the particular purification for ρ^Q and is thus an quantity intrinsic to Q.) Clearly $\hat{F}_e \geq F_e$. A derivation very similar to the one we have given allows us to replace F_e by $\hat{F}_e$ in Eq. (87), obtaining

$$h(\hat{F}_e) + (1-\hat{F}_e)\log(d^2-1) \geq S_e, \tag{90}$$

$$1 + 2(1-\hat{F}_e)\log d \geq S_e. \tag{91}$$

We could further extend this by allowing Q to be subjected to a second arbitrary completely positive map after $\mathcal{E}^Q$ and obtain a similar relation. However, in this case the relevant entropy exchange $\hat{S}_e$ would be that due to the total evolution, both $\mathcal{E}^Q$, and the subsequent "processing." Since it is possible that $\hat{S}_e < S_e$, we do not obtain a useful general relation. (This is precisely what happens in quantum error-correcting codes, as explained below.)

VII. REMARKS

One possible application of entanglement fidelity and entropy exchange is in the study of nonideal quantum computers [19]. In a typical state of a quantum computer, the different parts of the computer are in a highly entangled state. The elements of the computer's memory must maintain their states in such a fashion that this entanglement is preserved. The considerations in these notes are thus particularly suited to studying the effects of noise and decoherence in this context.

What we have found is that the capability of a system Q to preserve its entanglement with some other system R can be determined from the initial state and the dynamics of Q itself. Destruction or distortion of entanglement, and information exchange with the environment, leave distinct traces in the dynamics of the system itself. We can characterize these by the entanglement fidelity F_e and the entropy exchange S_e.

F_e is properly thought of not as the fidelity of one state with another (though it can be given that interpretation by including a reference system R) but as the fidelity of a *process* given by the input state ρ^Q and the system dynamics $\mathcal{E}^Q$. F_e does not just measure how *well* the state of Q is preserved by $\mathcal{E}^Q$, but also how *coherently*. If the input state is a pure state, these amount to the same thing; but otherwise, F_e is a stronger measure of the amount of disturbance the state experiences.

S_e is also properly thought of not as the entropy of some state but as the entropy associated with the dynamical process given by ρ^Q and $\mathcal{E}^Q$. Information exchange with the environment, even if it does not change the entropy of either the system Q or the environment E, can lead to nonzero entropy exchange S_e. Entropy exchange is therefore a clearer measure of this exchange than the changes in entropy of either system.

The relationship between F_e and S_e amounts to a quantum Fano inequality, connecting the information exchange with the environment to the disturbance of the state. This illustrates very clearly a general principle: In quantum information theory, noise is exactly information exchange with an external system. In a classical system, information can be "leaked" into the environment with arbitrarily little disturbance to the system: the environment can simply make a copy of the information, leaving the original intact within the system. But quantum information cannot be copied. Any departure of information into the environment necessarily yields an irreducible disturbance of the system. (This is the fundamental idea behind quantum cryptography see [20].) The departing information leaves its "footprints" behind in the entropy exchange S_e and associated imperfect entanglement fidelity F_e.

These ideas shed an interesting light on the recently discovered quantum error-correcting codes [6]. In these codes, input quantum states are represented by massively entangled states of a system Q composed of many qubits: $Q = Q_1 \cdots Q_n$. The environment is assumed to act independently on these systems, which in our language corresponds to the requirement that the evolution superoperator for the system Q factorizes:

$$\mathcal{E}^Q = \mathcal{E}^{Q_1} \otimes \cdots \otimes \mathcal{E}^{Q_2}. \tag{92}$$

The resulting state is then subjected to a second process, which typically involves an incomplete measurement on Q followed by a unitary evolution (which depends on the measurement result). Under certain circumstances, the original state of the system may be restored with very high fidelity.

The action of the channel and the subsequent restoration process of the sequence of qubits can be written as a single superoperator for $Q_1 \cdots Q_n$. Since the fidelity of this combined process is high, we can conclude, rather surprisingly, that the total entropy exchange is quite low. At first this seems paradoxical since the individual entropy exchanges of the noise process and the restoration measurement may both be high.

But this is not too difficult to understand. Let E represent the environment system that interacts with the qubits during the noise stage and let M represent the apparatus that performs the restoration process. To begin with, we might imagine that E and M are in pure states. After Q interacts with E (and thus exchanges information), the state of QE becomes entangled. In the second stage, M interacts and exchanges information with Q, and the entanglement of Q with the rest of the world is reduced: it is passed to M. At the end of the process, both Q and the "rest of the world" EM are in near-pure states, but E and M have now become entangled.

Thus the process of quantum error correction can be thought of as a process of passing entanglement (produced by a previous interaction with the environment) to the apparatus, in such a way that the entropy exchange for the total process (noise followed by restoration) on Q is very low. If S_e is very low, then the overall dynamics for Q is nearly unitary, so that the original state of Q can be approximately recovered. It is not yet known under what general circumstances, and to what fidelity, this can be accomplished.

ACKNOWLEDGMENTS

The author is indebted to many people for extensive conversations about the issues discussed in this paper, including H. Barnum, C. H. Bennett, C. M. Caves, I. Chuang, A. Ekert, C. A. Fuchs, E. H. Knill, R. Jozsa, R. Laflamme, J. Smolin, M. D. Westmoreland, W. K. Wootters, and W. H. Zurek. He also wishes to acknowledge the hospitality and support of the Theoretical Astrophysics group (T-6) at Los Alamos National Laboratory.

APPENDIX: REPRESENTATION THEOREMS

1. Index states and relative states

In this appendix we will use some of the ideas from the main text to show that any trace-preserving, completely positve linear map has both an operator-sum representation and a unitary representation. This derivation is somewhat more direct than that found in [9]. We will also suggest a useful characterization of all such representations.

Suppose R and Q are quantum systems with $\dim\mathcal{H}_R=\dim\mathcal{H}_Q=d$ and let $|\alpha_k^R\rangle$ and $|\beta_k^Q\rangle$ be orthonormal basis vectors for $\mathcal{H}_R$ and $\mathcal{H}_Q$. We can write down a maximally entangled pure state of RQ,

$$|\Psi^{RQ}\rangle=\frac{1}{\sqrt{d}}\sum_k |\alpha_k^R\rangle\otimes|\beta_k^Q\rangle. \tag{A1}$$

It will be convenient to consider instead the non-normalized vector

$$|\tilde{\Psi}^{RQ}\rangle=\sqrt{d}|\Psi^{RQ}\rangle=\sum_k |\alpha_k^R\rangle\otimes|\beta_k^Q\rangle. \tag{A2}$$

(Using $|\tilde{\Psi}^{RQ}\rangle$ rather than $|\Psi^{RQ}\rangle$ will eliminate some factors of $\sqrt{d}$ in our expressions.)

For every state $|\zeta^R\rangle$ of R there is a unique state $|\xi^Q\rangle$ such that

$$\frac{1}{\sqrt{d}}|\xi^Q\rangle=\langle\zeta^R|\Psi^{RQ}\rangle \tag{A3}$$

$$|\xi^Q\rangle=\langle\zeta^R|\tilde{\Psi}^{RQ}\rangle. \tag{A4}$$

The relation between $|\zeta^R\rangle$ and $|\xi^Q\rangle$ is a one-to-one correspondence. We call $|\xi^Q\rangle$ the *relative state* in Q to $|\zeta^R\rangle$ and we call $|\zeta^R\rangle$ the *index state* in R that yields $|\xi^Q\rangle$.

Given a state $|\phi^Q\rangle$, let us denote the associated index state in R by $|\phi^{*R}\rangle$. We can give a simple prescription for finding $|\phi^{*R}\rangle$ from $|\phi^Q\rangle$. Suppose

$$|\phi^Q\rangle=\sum_k c_k|\beta_k^Q\rangle. \tag{A5}$$

Then

$$|\phi^{*R}\rangle=\sum_k c_k^*|\alpha_k^R\rangle, \tag{A6}$$

as can be easily seen:

$$\langle\phi^{*R}|\tilde{\Psi}^{RQ}\rangle=\sum_{kl} c_k\langle\alpha_k^R|\alpha_l^R\rangle|\beta_l^Q\rangle$$

$$=\sum_k c_k|\beta_k^Q\rangle=|\phi^Q\rangle. \tag{A7}$$

It is also clear that

$$|\phi^{*R}\rangle\langle\phi^{*R}|\otimes|\phi^Q\rangle\langle\phi^Q|$$
$$=(|\phi^{*R}\rangle\langle\phi^{*R}|\otimes 1^Q)|\tilde{\Psi}^{RQ}\rangle\langle\tilde{\Psi}^{RQ}|(|\phi^{*R}\rangle\langle\phi^{*R}|\otimes 1^Q), \tag{A8}$$

a relation that will be useful later on.

The function that takes $|\phi^Q\rangle$ to $|\phi^{*R}\rangle$ is conjugate linear. If $|\phi^Q\rangle=a_1|\phi_1^Q\rangle+a_2|\phi_2^Q\rangle$, then

$$|\phi^{*R}\rangle=a_1^*|\phi_1^{*R}\rangle+a_2^*|\phi_2^{*R}\rangle, \tag{A9}$$

$$\langle\phi^{*R}|=a_1\langle\phi_1^{*R}|+a_2\langle\phi_2^{*R}|. \tag{A10}$$

2. Operator-sum representations

Let $\mathcal{E}^Q$ be the trace-preserving, completely positive linear map that describes the dynamical evolution of the system Q. Since $\mathcal{E}^Q$ is completely positive, any trivial extension of it is positive; in particular, the superoperator $\mathcal{I}^R\otimes\mathcal{E}^Q$ is positive. Thus the state

$$\rho^{RQ'}=\mathcal{I}^R\otimes\mathcal{E}^Q(|\Psi^{RQ}\rangle\langle\Psi^{RQ}|) \tag{A11}$$

is a positive operator, as is

$$D^{RQ'}=d\rho^{RQ'}=\mathcal{I}^R\otimes\mathcal{E}^Q(|\tilde{\Psi}^{RQ}\rangle\langle\tilde{\Psi}^{RQ}|). \tag{A12}$$

Of course, $\rho^{RQ'}$ has unit trace, so it is a normalized density operator, while $\mathrm{Tr}D^{RQ'}=d$.

The operation of realizing a state of Q via choosing an index state of R commutes with the dynamical operation given by $\mathcal{I}^R\otimes\mathcal{E}^Q$. In other words, if we wish to write down the final state $\rho^{Q'}=\mathcal{E}^Q(\rho^Q)$, where $\rho^Q=|\phi^Q\rangle\langle\phi^Q|$, we can either apply the index state $|\phi^{*R}\rangle$ to $|\tilde{\Psi}^{RQ}\rangle$ and then apply $\mathcal{E}^Q$ or we can apply the extended superoperator $\mathcal{I}^R\otimes\mathcal{E}^Q$ to the joint state and then apply the index state; thus

$$\rho^{Q'}=\langle\phi^{*R}|D^{RQ'}|\phi^{*R}\rangle. \quad \text{(A13)}$$

This makes sense on physical grounds. A measurement of an observable on R involves a completely different system than the dynamical evolution of Q, and the two operations might take place arbitrarily far apart. The time order of the two should irrelevant to the result.

A more formal argument runs as follows. Let Φ^R be the superoperator (i.e., a linear map on operators on $\mathcal{H}_R$) associated with multiplication by $|\phi^{*R}\rangle\langle\phi^{*R}|$ on both sides. That is, if T^R is an operator on $\mathcal{H}_R$, then $\Phi^R(T^R)=|\phi^{*R}\rangle\langle\phi^{*R}|T^R|\phi^{*R}\rangle\langle\phi^{*R}|$. The superoperator $\Phi^R\otimes\mathcal{I}^Q$ (which is just multiplication on both sides by $|\phi^{*R}\rangle\langle\phi^{*R}|\otimes 1^Q$) obviously commutes with the dynamical superoperator $\mathcal{I}^R\otimes\mathcal{E}^Q$. Therefore,

$$\begin{aligned}\Phi^R\otimes\mathcal{I}^Q[D^{RQ'}]&=\Phi^R\otimes\mathcal{I}^Q[\mathcal{I}^R\otimes\mathcal{E}^Q(|\tilde{\Psi}^{RQ}\rangle\langle\tilde{\Psi}^{RQ}|)]\\&=\mathcal{I}^R\otimes\mathcal{E}^Q[\Phi^R\otimes\mathcal{I}^Q(|\tilde{\Psi}^{RQ}\rangle\langle\tilde{\Psi}^{RQ}|)]\\&=\mathcal{I}^R\otimes\mathcal{E}^Q[(|\phi^{*R}\rangle\langle\phi^{*R}|\otimes 1^Q)|\tilde{\Psi}^{RQ}\rangle\\&\quad\times\langle\tilde{\Psi}^{RQ}|(|\phi^{*R}\rangle\langle\phi^{*R}|\otimes 1^Q)]\\&=\mathcal{I}^R\otimes\mathcal{E}^Q(|\phi^{*R}\rangle\langle\phi^{*R}|\otimes|\phi^Q\rangle\langle\phi^Q|)\\&=|\phi^{*R}\rangle\langle\phi^{*R}|\otimes\rho^{Q'}. \end{aligned}\quad \text{(A14)}$$

From this we can see that

$$\rho^{Q'}=\mathcal{E}^Q(|\phi^Q\rangle\langle\phi^Q|)=\langle\phi^{*R}|D^{RQ'}|\phi^{*R}\rangle \quad \text{(A15)}$$

as we wished to show.

The operator $D^{RQ'}$ is positive; thus we can find a set of vectors $|\tilde{\mu}^{RQ'}\rangle$ such that

$$D^{RQ'}=\sum_\mu |\tilde{\mu}^{RQ'}\rangle\langle\tilde{\mu}^{RQ'}|. \quad \text{(A16)}$$

These vectors, for example, might be constructed from the eigenvectors of $D^{RQ'}$, normalized by their eigenvalues; but there are many such decompositions. In fact, it is easy to see that the $|\tilde{\mu}^{RQ'}\rangle$ vectors are simply related to the representation of $\rho^{RQ'}$ by an ensemble of pure states. That is, given such a representation

$$\rho^{RQ'}=\sum_\mu p_\mu|\psi_\mu^{RQ'}\rangle\langle\psi_\mu^{RQ'}|, \quad \text{(A17)}$$

we can simply set $|\tilde{\mu}^{RQ'}\rangle=\sqrt{p_\mu d}|\psi_\mu^{RQ'}\rangle$. It is also clear that there is a decomposition of $D^{RQ'}$ with no more than d^2 vectors $|\tilde{\mu}^{RQ'}\rangle$, since the dimension of the space $\mathcal{H}_R\otimes\mathcal{H}_Q$ is d^2.

Here comes the essential trick. Define the operator A_μ^Q by

$$A_\mu^Q|\phi^Q\rangle=\langle\phi^{*R}|\tilde{\mu}^{RQ'}\rangle \quad \text{(A18)}$$

for each state $|\phi^Q\rangle$ of Q. Because of the conjugate linear relation between $|\phi^Q\rangle$ and $|\phi^{*R}\rangle$, each A_μ^Q thus defined is a perfectly good linear operator on $\mathcal{H}_Q$. Furthermore,

$$\begin{aligned}\sum_\mu A_\mu^Q|\phi^Q\rangle\langle\phi^Q|A_\mu^{Q\dagger}&=\sum_\mu \langle\phi^{*R}|\tilde{\mu}^{RQ'}\rangle\langle\tilde{\mu}^{RQ'}|\phi^{*R}\rangle\\&=\langle\phi^{*R}|D^{RQ'}|\phi^{*R}\rangle\\&=\mathcal{E}^Q(|\phi^Q\rangle\langle\phi^Q|).\end{aligned}\quad \text{(A19)}$$

We have thus derived an operator-sum representation for the completely positive map $\mathcal{E}^Q$ for all pure input states $|\phi^Q\rangle\langle\phi^Q|$. Extending this to mixed state inputs is trivial, of course, since every mixed state is a linear (convex) combination of pure states. We can further see that each completely positive map $\mathcal{E}^Q$ has an operator-sum representation with no more than d^2 terms.

We also find that, for our operator-sum representation for $\mathcal{E}^Q$,

$$\begin{aligned}\sum_\mu \langle\phi^Q|A_\mu^Q A_\mu^{Q\dagger}|\phi^Q\rangle&=\mathrm{Tr}\sum_\mu A_\mu^Q|\phi^Q\rangle\langle\phi^Q|A_\mu^{Q\dagger}\\&=\mathrm{Tr}\rho^{Q'}\\&=1\end{aligned}\quad \text{(A20)}$$

since $\mathcal{E}^Q$ is trace preserving by assumption. Since this is true for all states $|\phi^Q\rangle$, including the eigenstates of the positive operator $\Sigma_\mu A_\mu^{Q\dagger}A_\mu^Q$, we conclude that

$$\sum_\mu A_\mu^{Q\dagger}A_\mu^Q=1^Q. \quad \text{(A21)}$$

3. Unitary representations

Having derived an operator-sum representation for $\mathcal{E}^Q$, it is easy to arrive at a unitary representation. Add an extra quantum system E and write down a purification $|\tilde{\Upsilon}^{RQE'}\rangle$ for $D^{RQ'}$ as

$$|\tilde{\Upsilon}^{RQE'}\rangle=\sum_\mu |\tilde{\mu}^{RQ'}\rangle\otimes|\epsilon_\mu^E\rangle \quad \text{(A22)}$$

for an orthonormal set of vectors $|\epsilon_\mu^E\rangle$ in $\mathcal{H}_E$. (Again, finding a purification for $D^{RQ'}$ is equivalent to finding a purification for $\rho^{RQ'}$, but it is slightly easier to work with the non-normalized states.) We note that we require no more than d^2 dimensions in $\mathcal{H}_E$ to construct this purification since there are decompositions of $D^{RQ'}$ with no more than d^2 vectors $|\tilde{\mu}^{RQ'}\rangle$. Fix some state $|0^E\rangle$ of E. We can define an operator U^{QE} on a subspace of $\mathcal{H}_Q\otimes\mathcal{H}_E$ by

$$\begin{aligned}U^{QE}(|\phi^Q\rangle\otimes|0^E\rangle)&=\langle\phi^{*R}|\tilde{\Upsilon}^{RQE'}\rangle=\sum_\mu \langle\phi^{*R}|\tilde{\mu}^{RQ'}\rangle\otimes|\epsilon_\mu^E\rangle\\&=\sum_\mu A_\mu^Q|\phi^Q\rangle\otimes|\epsilon_\mu^E\rangle=|\Phi^{QE'}\rangle\end{aligned}\quad \text{(A23)}$$

for all $|\phi^Q\rangle$ in $\mathcal{H}_Q$. Once again, the conjugate linear relation of index state and relative state guarantees that this is a linear operator. Furthermore, given two states $|\phi_1^Q\rangle$ and $|\phi_2^Q\rangle$,

$$\langle\Phi_1^{QE'}|\Phi_2^{QE'}\rangle=\langle\tilde{\Upsilon}^{RQE'}|\phi_1^{*R}\rangle\langle\phi_2^{*R}|\tilde{\Upsilon}^{RQE'}\rangle$$

$$=\sum_{\mu,\nu}\langle\phi_1^Q|A_\mu^{Q\dagger}A_\nu^Q|\phi_2^Q\rangle\langle\epsilon_\mu^Q|\epsilon_\nu^E\rangle$$

$$=\sum_\mu\langle\phi_1^Q|A_\mu^{Q\dagger}A_\mu^Q|\phi_2^Q\rangle$$

$$=\langle\phi_1^Q|\phi_2^Q\rangle. \qquad (A24)$$

The operator U^{QE} preserves inner products on this subspace of states; it can therefore be extended to a unitary operator on the entire space $\mathcal{H}_Q\otimes\mathcal{H}_E$.

Thus we have a unitary representation for $\mathcal{E}^Q$,

$$\mathrm{Tr}_E U^{QE}(|\phi^Q\rangle\langle\phi^Q|\otimes|0^E\rangle\langle 0^E|)U^{QE\dagger}$$

$$=\mathrm{Tr}_E\sum_{\mu,\nu}(A_\mu^Q|\phi^Q\rangle\langle\phi^Q|A_\nu^{Q\dagger})\otimes|\epsilon_\mu^E\rangle\langle\epsilon_\nu^E|$$

$$=\sum_{\mu,\nu}(A_\mu^Q|\phi^Q\rangle\langle\phi^Q|A_\nu^{Q\dagger})\langle\epsilon_\nu^E|\epsilon_\mu^E\rangle$$

$$=\sum_\mu A_\mu^Q|\phi^Q\rangle\langle\phi^Q|A_\nu^{Q\dagger}=\mathcal{E}^Q(|\phi^Q\rangle\langle\phi^Q|). \qquad (A25)$$

Once again, we can extend this unitary representation to mixed state inputs since these are linear (convex) combinations of pure states.

4. Remarks

In the above arguments, we arrived at an operator-sum representation for $\mathcal{E}^Q$ by a decomposition of $D^{RQ'}$, that is, by a pure state ensemble for $\rho^{RQ'}$. It is also easy to see that every operator-sum representation for $\mathcal{E}^Q$, when extended and applied to $|\Psi^{RQ}\rangle$, will yield such a decomposition. [Simply define $|\widetilde{\mu}^{RQ'}\rangle=(1^R\otimes A_\mu^Q)|\Psi^{RQ}\rangle$.] Thus the operator-sum representations for $\mathcal{E}^Q$ are in a one-to-one correspondence with the pure state ensembles for $\rho^{RQ'}$.

Similarly, we obtained a unitary representation for $\mathcal{E}^Q$ by finding a purification for $D^{RQ'}$ or equivalently for $\rho^{RQ'}$. But every unitary representation will be associated with such a purification because the initial total state $|\Psi^{RQ}\rangle\otimes|0^E\rangle$ of RQE will evolve unitarily to a pure state, from which the state $\rho^{RQ'}$ is obtained by a partial trace over E. Now, any such purification of $\rho^{RQ'}$ can be obtained from any other by means of a unitary transformation that acts on $\mathcal{H}_E$, which corresponds to an internal rotation of the environment system E that acts *after* the interaction of Q and E.

The nonuniqueness of the operator-sum representation and the unitary representations are related since every pure state ensemble for $\rho^{RQ'}$ can be realized by fixing a purification $|\Upsilon^{RQE'}\rangle$ and choosing a complete ordinary measurement for E (i.e., an orthonormal basis for $\mathcal{H}_E$). Equivalently, we might fix a measurement basis for $\mathcal{H}_E$ and a particular purification. A change of representation in each case will be associated with a unitary matrix corresponding to a rotation in $\mathcal{H}_E$. That is, suppose that for all ρ^Q,

$$\mathcal{E}^Q(\rho^Q)=\sum_\mu A_\mu^Q\rho^Q A_\mu^{Q\dagger}=\sum_\nu B_\nu^Q\rho^Q B_\nu^{Q\dagger}, \qquad (A26)$$

so that the A_μ^Q and the B_ν^Q operators both form operator-sum representations for $\mathcal{E}^Q$. Then there is a unitary matrix $U_{\mu\nu}$ so that

$$A_\mu^Q=\sum_\nu U_{\mu\nu}B_\nu^Q. \qquad (A27)$$

(Note that we may have to extend one operator-sum representation by a finite number of zero operators so that the two representations have the same number of operators.) The matrix $U_{\mu\nu}$ is in fact the matrix that relates two different bases in E, corresponding to two purifications related, in the sense outlined above, to the two operator-sum representations.

[1] C. H. Bennett, Physics Today **48**, 24 (1995).
[2] C. E. Shannon, Bell Syst. Tech. J. **27**, 379 (1948).
[3] R. Jozsa and B. Schumacher, J. Mod. Opt. **41**, 2343 (1994); B. Schumacher Phys. Rev. A **51**, 2738 (1995); H. Barnum, C. A. Fuchs, R. Jozsa, and B. Schumacher (unpublished).
[4] R. Jozsa, J. Mod. Opt. **41**, 2315 (1995).
[5] C. H. Bennett and S. J. Wiesner, Phys. Rev. Lett. **68**, 3121 (1992); C. H. Bennett, G. Brassard, C. Crépeau, R. Jozsa, A. Peres, and W. K. Wootters, *ibid.* **69**, 2881 (1992).
[6] P. W. Shor, Phys. Rev. A **52**, 2493 (1995); A. R. Calderbank and P. W. Shor, Phys. Rev. A **54**, 1098 (1996); A. Steane, Phys. Rev. Lett. **77**, 793 (1996); R. Laflamme, C. Miquel, J. P. Paz, and W. H. Zurek, *ibid.* **77**, 198 (1996).
[7] C. H. Bennett, G. Brassard, S. Popescu, B. Schumacher, J. A. Smolin, and W. K. Wootters, Phys. Rev. Lett. **76**, 722 (1996).
[8] W. F. Stinespring, Proc. Am. Math. Soc. **6**, 211 (1955); K. Kraus, Ann. of Phys. (N.Y.) **64**, 311 (1971).
[9] K. Hellwig and K. Kraus, Commun. Math. Phys. **16**, 142 (1970); M.-D. Choi, Linear Algebra Appl. **10**, 285 (1975); K. Kraus, *States, Effects, and Operations: Fundamental Notions of Quantum Theory* (Springer-Verlag, Berlin, 1983).
[10] L. P. Hughston, R. Jozsa, and W. K. Wooters, Phys. Lett. A **183**, 14 (1993).
[11] H. Everett III, Rev. Mod. Phys. **29**, 454 (1957).
[12] C. W. Helstrom, *Quantum Detection and Estimation Theory* (Academic, New York, 1976), A. Peres, Found. Phys. **20**, 1441 (1990).
[13] J. von Neumann, *Mathematical Foundations of Quantum Mechanics*, translated by E. T. Beyer (Princeton University Press, Princeton, 1955).
[14] A. Wehrl, Rev. Mod. Phys. **50**, 221 (1978).
[15] T. M. Cover and J. A. Thomas, *Elements of Information*

Theory (Wiley, New York, 1991).
[16] C. H. Bennett and G. Brassard, *Proceedings of the IEEE Conference on Computers, Systems, and Signal Processing, Bangalore, 1984* (IEEE, New York, 1984), p. 175; C. H. Bennett, F. Bessette, G. Brassard, L. Salvail, and J. Smolin, J. Crypt. **5**, 3 (1992).
[17] A. K. Ekert, Phys. Rev. Lett. **67**, 661 (1991).
[18] A. S. Kholevo, Prob. Peredachi Info. **9**, 3 (1973) Prob. Info. Transm. (USSR) **9**, 177 (1973); C. Caves and C. Fuchs, Phys. Rev. Lett. **73**, 3047 (1994); B. Schumacher, M. D. Westmoreland, and W. K. Wootters, Phys. Rev. Lett. **76**, 3452 (1996).
[19] S. Lloyd, Sci. Am. **273**, 140 (1995); A. Ekert and R. Jozsa, Rev. Mod. Phys. (to be published).
[20] C. A. Fuchs, Ph.D. thesis, The University of New Mexico, Albuquerque, NM (1996); C. A. Fuchs and A. Peres, Phys. Rev. A **53**, 2308 (1996).

VOLUME 76, NUMBER 15 PHYSICAL REVIEW LETTERS 8 APRIL 1996

Noncommuting Mixed States Cannot Be Broadcast

Howard Barnum, Carlton M. Caves, Christopher A. Fuchs, Richard Jozsa,* and Benjamin Schumacher†
Center for Advanced Studies, Department of Physics and Astronomy, University of New Mexico, Albuquerque, New Mexico 87131-1156
(Received 8 November 1995)

We show that, given a general mixed state for a quantum system, there are no physical means for *broadcasting* that state onto two separate quantum systems, even when the state need only be reproduced marginally on the separate systems. This result extends the standard no-cloning theorem for pure states.

PACS numbers: 89.70.+c, 02.50.–r, 03.65.Bz

The fledgling field of quantum information theory [1] draws attention to fundamental questions about what is physically possible and what is not. An example is the theorem [2,3] that there are no physical means by which an *unknown pure* quantum state can be reproduced or copied—a result summarized by the phrase "quantum states cannot be cloned." In this paper we formulate and prove an impossibility theorem that extends the pure-state no-cloning theorem to (invertible) mixed quantum states. The theorem answers the question: Are there any physical means for *broadcasting* an unknown quantum state onto two separate quantum systems? By broadcasting we mean that the marginal density operator of each of the separate systems is the same as the state to be broadcast.

The pure-state "no-cloning" theorem [2,3] prohibits broadcasting pure states, for the only way to broadcast a pure state $|\psi\rangle$ is to put the two systems in the product state $|\psi\rangle \otimes |\psi\rangle$, i.e., to clone $|\psi\rangle$. Things are more complicated when the states are mixed. A mixed-state no-cloning theorem is not sufficient to demonstrate no broadcasting, for there are many conceivable ways to broadcast a mixed state ρ without the joint state being in the product form $\rho \otimes \rho$, the mixed-state analog of cloning; the systems might be correlated or entangled in such a way as to give the right marginal density operators. For instance, if the density operator has the spectral decomposition $\rho = \sum_b \lambda_b |b\rangle\langle b|$, a potential broadcasting state is the highly correlated joint state $\tilde{\rho} = \sum_b \lambda_b |b\rangle|b\rangle\langle b|\langle b|$, which, though not of the product form $\rho \otimes \rho$, reproduces the correct marginal density operators.

The general problem, posed formally, is this. A quantum system AB is composed of two parts, A and B, each having an N-dimensional Hilbert space. System A is secretly prepared in one state from a set $\mathcal{A} = \{\rho_0, \rho_1\}$ of two quantum states. System B, slated to receive the unknown state, is in a standard quantum state Σ. The initial state of the composite system AB is the product state $\rho_s \otimes \Sigma$, where $s = 0$ or 1 specifies which state is to be broadcast. We ask whether there is any physical process $\mathcal{E}$, consistent with the laws of quantum theory, that leads to an evolution of the form $\rho_s \otimes \Sigma \rightarrow \mathcal{E}(\rho_s \otimes \Sigma) = \tilde{\rho}_s$, where $\tilde{\rho}_s$ is *any* state on the N^2-dimensional Hilbert space AB such that

$$\mathrm{tr}_A(\tilde{\rho}_s) = \rho_s \quad \text{and} \quad \mathrm{tr}_B(\tilde{\rho}_s) = \rho_s . \qquad (1)$$

Here tr_A and tr_B denote partial traces over A and B. If there is an $\mathcal{E}$ that satisfies Eq. (1) for both ρ_0 and ρ_1, then the set $\mathcal{A}$ can be *broadcast*. A special case of broadcasting is the evolution specified by $\mathcal{E}(\rho_s \otimes \Sigma) = \rho_s \otimes \rho_s$; we reserve the word *cloning* for this strong form of broadcasting.

The most general action $\mathcal{E}$ on AB consistent with quantum theory is to allow AB to interact unitarily with an auxiliary quantum system C in some standard state and thereafter to ignore the auxiliary system [4]; that is,

$$\mathcal{E}(\rho_s \otimes \Sigma) = \mathrm{tr}_C[U(\rho_s \otimes \Sigma \otimes \Upsilon)U^\dagger], \qquad (2)$$

for some auxiliary system C, some standard state Υ on C, and some unitary operator U on ABC. We show that such an evolution can lead to broadcasting if and only if ρ_0 and ρ_1 commute. This result strikes close to the heart of the difference between the classical and quantum theories, because it provides another physical distinction between *commuting* and *noncommuting* states. We further show that $\mathcal{A}$ is clonable if and only if ρ_0 and ρ_1 are identical or orthogonal ($\rho_0\rho_1 = 0$).

To see that the set $\mathcal{A}$ can be broadcast when the states commute, we do not need to attach an auxiliary system. Since orthogonal pure states can be cloned, broadcasting can be obtained by cloning the simultaneous eigenstates of ρ_0 and ρ_1. Let $|b\rangle$, $b = 1, \ldots, N$, be an orthonormal basis for A in which both ρ_0 and ρ_1 are diagonal, and let their spectral decompositions be $\rho_s = \sum_b \lambda_{sb}|b\rangle\langle b|$. Consider any unitary operator U on AB consistent with $U|b\rangle|1\rangle = |b\rangle|b\rangle$. If we choose $\Sigma = |1\rangle\langle 1|$ and let

$$\tilde{\rho}_s = U(\rho_s \otimes \Sigma)U^\dagger = \sum_b \lambda_{sb}|b\rangle|b\rangle\langle b|\langle b| , \qquad (3)$$

we immediately have that $\tilde{\rho}_0$ and $\tilde{\rho}_1$ satisfy Eq. (1).

The converse of this statement—that if $\mathcal{A}$ can be broadcast, ρ_0 and ρ_1 commute—is more difficult to prove. Our proof is couched in terms of the concept of *fidelity* between two density operators. The fidelity $F(\rho_0, \rho_1)$ is defined by

$$F(\rho_0, \rho_1) = \mathrm{tr}\sqrt{\rho_0^{1/2}\rho_1\rho_0^{1/2}} , \qquad (4)$$

where for any positive operator O, i.e., any Hermitian operator with *non-negative* eigenvalues, $O^{1/2}$ denotes its

 0031-9007/96/76(15)/2818(4)$10.00

unique positive square root. (Note that Ref. [5] defines fidelity to be the square of the present quantity.) Fidelity is an analog of the modulus of the inner product for pure states [5,6] and can be interpreted as a measure of distinguishability for quantum states: it ranges between 0 and 1, reaching 0 if and only if the states are orthogonal and reaching 1 if and only if $\rho_0 = \rho_1$. It is invariant under the interchange $0 \rightarrow 1$ and under the transformation $\rho_0 \rightarrow U\rho_0 U^\dagger$, $\rho_1 \rightarrow U\rho_1 U^\dagger$ for any unitary operator U [5,7]. Also, from the properties of the direct product, one has that $F(\rho_0 \otimes \sigma_0, \rho_1 \otimes \sigma_1) = F(\rho_0, \rho_1) F(\sigma_0, \sigma_1)$.

Another reason $F(\rho_0, \rho_1)$ defines a good notion of distinguishability [8] is that it equals the minimal overlap between the probability distributions $p_0(b) = \mathrm{tr}(\rho_0 E_b)$ and $p_1(b) = \mathrm{tr}(\rho_1 E_b)$ generated by a generalized measurement or *positive operator-valued measure* (POVM) $\{E_b\}$ [4]. That is [7],

$$F(\rho_0, \rho_1) = \min_{\{E_b\}} \sum_b \sqrt{\mathrm{tr}(\rho_0 E_b)} \sqrt{\mathrm{tr}(\rho_1 E_b)}, \tag{5}$$

where the minimum is taken over all sets of positive operators $\{E_b\}$ such that $\sum_b E_b = \mathbb{1}$. This representation of fidelity has the advantage of being defined operationally in terms of measurements. We call a POVM that achieves the minimum in Eq. (5) an *optimal* POVM.

One way to see the equivalence of Eqs. (5) and (4) is through the Schwarz inequality for the operator inner product $\mathrm{tr}(AB^\dagger)$: $\mathrm{tr}(AA^\dagger)\,\mathrm{tr}(BB^\dagger) \geq |\mathrm{tr}(AB^\dagger)|^2$, with equality if and only if $A = \alpha B$ for some constant α. Going through this exercise is useful because it leads directly to the proof of the no-broadcasting theorem. Let $\{E_b\}$ be any POVM and let U be any unitary operator. Using the cyclic property of the trace and the Schwarz inequality, we have that

$$\sum_b \sqrt{\mathrm{tr}(\rho_0 E_b)} \sqrt{\mathrm{tr}(\rho_1 E_b)} = \sum_b \sqrt{\mathrm{tr}(U\rho_0^{1/2} E_b \rho_0^{1/2} U^\dagger)} \sqrt{\mathrm{tr}(\rho_1^{1/2} E_b \rho_1^{1/2})} \geq \sum_b |\mathrm{tr}(U\rho_0^{1/2} E_b^{1/2} E_b^{1/2} \rho_1^{1/2})| \tag{6}$$

$$\geq \left| \sum_b \mathrm{tr}(U\rho_0^{1/2} E_b \rho_1^{1/2}) \right| = |\,\mathrm{tr}(U\rho_0^{1/2} \rho_1^{1/2})|. \tag{7}$$

We can use the freedom in U to make the inequality as tight as possible. To do this, we recall [5,9] that $\max |\mathrm{tr}(VO)| = \mathrm{tr}\sqrt{O^\dagger O}$, where O is any operator and the maximum is taken over all unitary operators V. The maximum is achieved only by those V such that $VO = e^{i\phi}\sqrt{O^\dagger O}$, ϕ being arbitrary; that there exists at least one such V is ensured by the operator polar decomposition theorem [9]. Therefore, by choosing

$$U\rho_0^{1/2}\rho_1^{1/2} = \sqrt{\rho_1^{1/2}\rho_0\rho_1^{1/2}}, \tag{8}$$

we get that $\sum_b \sqrt{\mathrm{tr}(\rho_0 E_b)}\sqrt{\mathrm{tr}(\rho_1 E_b)} \geq F(\rho_0, \rho_1)$.

Consulting the conditions for equality in steps (6) and (7), we find that a POVM is optimal if and only if

$$U\rho_0^{1/2} E_b^{1/2} = \mu_b \rho_1^{1/2} E_b^{1/2} \tag{9}$$

and the terms in the sum (7) have a common phase. By absorbing this phase into U by virtue of its phase freedom, this second condition becomes

$$\mathrm{tr}(U\rho_0^{1/2} E_b \rho_1^{1/2}) = \mu_b\, \mathrm{tr}(\rho_1 E_b) \geq 0 \Leftrightarrow \mu_b \geq 0. \tag{10}$$

When ρ_1 is invertible, Eq. (9) becomes

$$ME_b^{1/2} = \mu_b E_b^{1/2}, \tag{11}$$

where

$$M = \rho_1^{-1/2} U \rho_0^{1/2} = \rho_1^{-1/2} \sqrt{\rho_1^{1/2}\rho_0\rho_1^{1/2}}\, \rho_1^{-1/2} \tag{12}$$

is a positive operator. Therefore one way to satisfy Eq. (9) with $\mu_b \geq 0$ is to take $E_b = |b\rangle\langle b|$, where the vectors $|b\rangle$ are an orthonormal eigenbasis for M, with μ_b the eigenvalue of $|b\rangle$. When ρ_1 is noninvertible, there are still optimal POVMs. One can choose the first E_b to be the projector onto the null space of ρ_1. In the support of ρ_1 (the orthocomplement of its null space), ρ_1 is invertible, so we may construct the analog of M restricted to the support and choose the remaining E_b's to project onto its eigenvectors. Note that if both ρ_0 and ρ_1 are invertible, M is invertible.

We begin the proof of the no-broadcasting theorem by using Eq. (5) to show that fidelity cannot decrease under the operation of partial trace; this gives rise to an elementary constraint on all potential broadcasting processes $\mathcal{E}$. Suppose Eq. (1) is satisfied for the process $\mathcal{E}$ of Eq. (2), and let $\{E_b\}$ denote an optimal POVM for distinguishing ρ_0 and ρ_1. Then, for each s, $\mathrm{tr}[\tilde{\rho}_s(E_b \otimes \mathbb{1})] = \mathrm{tr}_A[\mathrm{tr}_B(\tilde{\rho}_s)E_b] = \mathrm{tr}_A(\rho_s E_b)$; it follows that

$$\begin{aligned} F_A(\rho_0, \rho_1) &\equiv \sum_b \sqrt{\mathrm{tr}[\tilde{\rho}_0(E_b \otimes \mathbb{1})]} \sqrt{\mathrm{tr}[\tilde{\rho}_1(E_b \otimes \mathbb{1})]} \\ &\geq \min_{\{\tilde{E}_c\}} \sum_c \sqrt{\mathrm{tr}(\tilde{\rho}_0 \tilde{E}_c)} \sqrt{\mathrm{tr}(\tilde{\rho}_1 \tilde{E}_c)} \\ &= F(\tilde{\rho}_0, \tilde{\rho}_1). \end{aligned} \tag{13}$$

Here $F_A(\rho_0, \rho_1)$ denotes the fidelity $F(\rho_0, \rho_1)$; the subscript A emphasizes that $F_A(\rho_0, \rho_1)$ stands for the particular representation on the first line. The inequality in Eq. (13) comes from the fact that $\{E_b \otimes \mathbb{1}\}$ might not be an optimal POVM for distinguishing $\tilde{\rho}_0$ and $\tilde{\rho}_1$; this demonstrates the said partial-trace property. Similarly

$$\begin{aligned} F_B(\rho_0, \rho_1) &\equiv \sum_b \sqrt{\mathrm{tr}[\tilde{\rho}_0(\mathbb{1} \otimes E_b)]} \sqrt{\mathrm{tr}[\tilde{\rho}_1(\mathbb{1} \otimes E_b)]} \\ &\geq F(\tilde{\rho}_0, \tilde{\rho}_1), \end{aligned} \tag{14}$$

where the subscript B emphasizes that $F_B(\rho_0, \rho_1)$ stands for the representation on the first line.

On the other hand, we can just as easily derive an inequality that is opposite to Eqs. (13) and (14). By the direct product formula and the invariance of fidelity under unitary transformations,

$$F(\rho_0, \rho_1) = F(\rho_0 \otimes \Sigma \otimes \Upsilon, \rho_1 \otimes \Sigma \otimes \Upsilon) = F(U(\rho_0 \otimes \Sigma \otimes \Upsilon)U^\dagger, U(\rho_1 \otimes \Sigma \otimes \Upsilon)U^\dagger). \tag{15}$$

Therefore, by the partial-trace property,

$$F(\rho_0, \rho_1) \le F(\mathrm{tr}_C[U(\rho_0 \otimes \Sigma \otimes \Upsilon)U^\dagger], \mathrm{tr}_C[U(\rho_1 \otimes \Sigma \otimes \Upsilon)U^\dagger]), \tag{16}$$

or, more succinctly,

$$F(\rho_0, \rho_1) \le F(\mathcal{E}(\rho_0 \otimes \Sigma), \mathcal{E}(\rho_1 \otimes \Sigma)) = F(\tilde\rho_0, \tilde\rho_1). \tag{17}$$

The elementary constraint now follows, for the only way to maintain Eqs. (13), (14), *and* (17) is with strict equality. In other words, we have that if the set $\mathcal{A}$ can be broadcast, then there are density operators $\tilde\rho_0$ and $\tilde\rho_1$ on AB satisfying Eq. (1) *and*

$$F_A(\rho_0, \rho_1) = F(\tilde\rho_0, \tilde\rho_1) = F_B(\rho_0, \rho_1). \tag{18}$$

Let us pause at this point to consider the restricted question of cloning. If $\mathcal{A}$ is to be clonable, there must exist a process $\mathcal{E}$ such that $\tilde\rho_s = \rho_s \otimes \rho_s$ for $s = 0, 1$. But then, by Eq. (18), we must have

$$F(\rho_0, \rho_1) = F(\rho_0 \otimes \rho_0, \rho_1 \otimes \rho_1) = F(\rho_0, \rho_1)^2, \tag{19}$$

which means that $F(\rho_0, \rho_1) = 1$ or 0; i.e., ρ_0 and ρ_1 are identical or orthogonal. There can be no cloning for density operators with nontrivial fidelity. The converse, that orthogonal and identical density operators can be cloned, follows, in the first case, from the fact that they can be distinguished by measurement and, in the second case, because they need not be distinguished at all.

Like the pure-state no-cloning theorem [2,3], this no-cloning result for mixed states is a consistency requirement for the axiom that quantum measurements cannot distinguish nonorthogonal states with perfect reliability. If nonorthogonal quantum states could be cloned, there would exist a measurement procedure for distinguishing those states with arbitrarily high reliability: one could make measurements on enough copies of the quantum state to make the probability of a correct inference of its identity arbitrarily high. That this consistency requirement, as expressed in Eq. (18), should also exclude more general kinds of broadcasting is not immediately obvious. Nevertheless, this is the content of our claim that Eq. (18) generally cannot be satisfied; any broadcasting process can be viewed as creating distinguishability *ex nihilo* with respect to measurements on the larger Hilbert space AB. Only for commuting density operators does broadcasting not create any extra distinguishability.

We now show that Eq. (18) implies that ρ_0 and ρ_1 commute. We assume that ρ_0 and ρ_1 are invertible. We proceed by studying the conditions necessary for the representations $F_A(\rho_0, \rho_1)$ and $F_B(\rho_0, \rho_1)$ in Eqs. (13) and (14) to equal $F(\tilde\rho_0, \tilde\rho_1)$. Recall that the optimal POVM $\{E_b\}$ for distinguishing ρ_0 and ρ_1 can be chosen so that the POVM elements $E_b = |b\rangle\langle b|$ are a complete set of orthogonal one-dimensional projectors onto orthonormal eigenstates of M. Then, repeating the steps leading from Eqs. (7) to (10), one finds that the necessary conditions for equality in Eq. (18) are that each $E_b \otimes \mathbb{1} = (E_b \otimes \mathbb{1})^{1/2}$ and each $\mathbb{1} \otimes E_b = (\mathbb{1} \otimes E_b)^{1/2}$ satisfy

$$\tilde U \tilde\rho_0^{1/2}(\mathbb{1} \otimes E_b) = \alpha_b \tilde\rho_1^{1/2}(\mathbb{1} \otimes E_b), \tag{20}$$

$$\tilde V \tilde\rho_0^{1/2}(E_b \otimes \mathbb{1}) = \beta_b \tilde\rho_1^{1/2}(E_b \otimes \mathbb{1}), \tag{21}$$

where α_b and β_b are non-negative numbers and $\tilde U$ and $\tilde V$ are unitary operators satisfying

$$\tilde U \tilde\rho_0^{1/2} \tilde\rho_1^{1/2} = \tilde V \tilde\rho_0^{1/2} \tilde\rho_1^{1/2} = \sqrt{\tilde\rho_1^{1/2} \tilde\rho_0 \tilde\rho_1^{1/2}}. \tag{22}$$

Although ρ_0 and ρ_1 are assumed invertible, one cannot demand that $\tilde\rho_0$ and $\tilde\rho_1$ be invertible—a glance at Eq. (3) shows that to be too restrictive. This means that $\tilde U$ and $\tilde V$ need not be the same. Also we cannot assume that there is any relation between α_b and β_b.

The remainder of the proof consists in showing that Eqs. (20)–(22), which are necessary (though perhaps not sufficient) for broadcasting, are nevertheless restrictive enough to imply that ρ_0 and ρ_1 commute. The first step is to sum over b in Eqs. (20) and (21). Defining the positive operators

$$G = \sum_b \alpha_b |b\rangle\langle b| \quad \text{and} \quad H = \sum_b \beta_b |b\rangle\langle b|, \tag{23}$$

we obtain

$$\tilde U \tilde\rho_0^{1/2} = \tilde\rho_1^{1/2}(\mathbb{1} \otimes G) \quad \text{and} \quad \tilde V \tilde\rho_0^{1/2} = \tilde\rho_1^{1/2}(H \otimes \mathbb{1}). \tag{24}$$

The next step is to demonstrate that G and H are invertible and, in fact, equal to each other. Multiplying the two equations in Eq. (24) from the left by $\tilde\rho_0^{1/2}\tilde U^\dagger$ and

$\tilde{\rho}_0^{1/2}\tilde{V}^\dagger$, respectively, and tracing the first over A and the second over B, we get

$$\rho_0 = \mathrm{tr}_A(\tilde{\rho}_0^{1/2}\tilde{U}^\dagger\tilde{\rho}_1^{1/2})G \quad \text{and} \quad \rho_0 = \mathrm{tr}_B(\tilde{\rho}_0^{1/2}\tilde{V}^\dagger\tilde{\rho}_1^{1/2})H\,. \tag{25}$$

Since, by assumption, ρ_0 is invertible, it follows that G and H are invertible. Returning to Eq. (24), multiplying both parts from the left by $\tilde{\rho}_1^{1/2}$, and tracing over A and B, respectively, we obtain

$$\mathrm{tr}_A(\tilde{\rho}_1^{1/2}\tilde{U}\tilde{\rho}_0^{1/2}) = \rho_1 G \quad \text{and} \quad \mathrm{tr}_B(\tilde{\rho}_1^{1/2}\tilde{V}\tilde{\rho}_0^{1/2}) = \rho_1 H\,. \tag{26}$$

Conjugating the two parts of Eq. (26) and inserting the results into the two parts of Eq. (25) yields

$$\rho_0 = G\rho_1 G \quad \text{and} \quad \rho_0 = H\rho_1 H\,. \tag{27}$$

This shows that $G = H$, because these equations have a unique positive solution, namely, the operator M of Eq. (12). This can be seen by multiplying Eq. (27) from the left and right by $\rho_1^{1/2}$ to get $\rho_1^{1/2}\rho_0\rho_1^{1/2} = (\rho_1^{1/2}G\rho_1^{1/2})^2$. The positive operator $\rho_1^{1/2}G\rho_1^{1/2}$ is thus the unique positive square root of $\rho_1^{1/2}\rho_0\rho_1^{1/2}$.

Knowing that $G = H = M$, we return to Eq. (24). The two parts, taken together, imply that

$$\tilde{V}^\dagger\tilde{U}\tilde{\rho}_0^{1/2} = \tilde{\rho}_0^{1/2}(M^{-1}\otimes M)\,. \tag{28}$$

If $|b\rangle$ and $|c\rangle$ are eigenvectors of M, with eigenvalues μ_b and μ_c, Eq. (28) implies that

$$\tilde{V}^\dagger\tilde{U}(\tilde{\rho}_0^{1/2}|b\rangle|c\rangle) = \frac{\mu_c}{\mu_b}(\tilde{\rho}_0^{1/2}|b\rangle|c\rangle)\,. \tag{29}$$

This means that $\tilde{\rho}_0^{1/2}|b\rangle|c\rangle$ is zero or it is an eigenvector of the unitary operator $\tilde{V}^\dagger\tilde{U}$. In the latter case, since the eigenvalues of a unitary operator have modulus 1, it must be true that $\mu_b = \mu_c$. Hence we can conclude that

$$\tilde{\rho}_0^{1/2}|b\rangle|c\rangle = 0 \quad \text{when} \quad \mu_b \neq \mu_c\,. \tag{30}$$

This is enough to show that M and ρ_0 commute and hence $[\rho_0, \rho_1] = 0$. Consider the matrix element

$$\begin{aligned}\langle b'|(M\rho_0 - \rho_0 M)|b\rangle &= (\mu_{b'} - \mu_b)\langle b'|\rho_0|b\rangle \\ &= (\mu_{b'} - \mu_b)\sum_c \langle b'|\langle c|\tilde{\rho}_0|c\rangle|b\rangle\,.\end{aligned} \tag{31}$$

If $\mu_b = \mu_{b'}$, this is automatically zero. If, on the other hand, $\mu_b \neq \mu_{b'}$, then the sum over c must vanish by Eq. (30). It follows that ρ_0 and M commute. Hence, using Eq. (27),

$$\rho_1\rho_0 = M^{-1}\rho_0 M^{-1}\rho_0 = \rho_0 M^{-1}\rho_0 M^{-1} = \rho_0\rho_1\,. \tag{32}$$

This completes the proof that noncommuting quantum states cannot be broadcast.

Note that, by the same method as above, $\tilde{\rho}_1^{1/2}|b\rangle|c\rangle = 0$ when $\mu_b \neq \mu_c$. This condition, along with Eq. (30), determines the conceivable broadcasting states, in which the correlations between the systems A and B range from purely classical to purely quantum. For example, since ρ_0 and ρ_1 commute, the states of Eq. (3) satisfy these conditions, but so do the perfectly entangled pure states $\sum_b \sqrt{\lambda_{sb}}\,|b\rangle|b\rangle$. Not all such broadcasting states can be realized by a physical process $\mathcal{E}$, but sufficient conditions for realizability are not known.

In closing, we mention an application of this result. In some versions of quantum cryptography [10], the legitimate users of a communication channel encode the bits 0 and 1 into nonorthogonal pure states. This is done to ensure that any eavesdropping is detectable, since eavesdropping necessarily disturbs the states sent to the legitimate receiver [11]. If the channel is noisy, however, causing the bits to evolve to noncommuting mixed states, the detectability of eavesdropping is no longer a given. The result presented here shows that there are no means available for an eavesdropper to obtain the signal, noise and all, intended for the legitimate receiver without in some way changing the states sent to the receiver.

We thank Richard Hughes for useful discussions. This work was supported in part by the Office of Naval Research (Grant No. N00014-93-1-0116).

---

*Permanent address: School of Mathematics and Statistics, University of Plymouth, Drake Circus, Plymouth, Devon PL4 8AA, England.

†Permanent address: Department of Physics, Kenyon College, Gambier, OH 43022.

[1] C. H. Bennett, Phys. Today **48**, No. 10, 24 (1995).
[2] W. K. Wootters and W. H. Zurek, Nature (London) **299**, 802 (1982).
[3] D. Dieks, Phys. Lett. **92A**, 271 (1982).
[4] K. Kraus, *States, Effects, and Operations: Fundamental Notions of Quantum Theory* (Springer, Berlin, 1983).
[5] R. Jozsa, J. Mod. Opt. **41**, 2315 (1994).
[6] A. Uhlmann, Rep. Math. Phys. **9**, 273 (1976).
[7] C. A. Fuchs and C. M. Caves, Open Sys. Inf. Dyn. **3**, 1 (1995).
[8] W. K. Wootters, Phys. Rev. D **23**, 357 (1981).
[9] R. Schatten, *Norm Ideals of Completely Continuous Operators* (Springer, Berlin, 1960).
[10] C. H. Bennett, Phys. Rev. Lett. **68**, 3121 (1992).
[11] C. H. Bennett, G. Brassard, and N. D. Mermin, Phys. Rev. Lett. **68**, 557 (1992).

PHYSICAL REVIEW A VOLUME 56, NUMBER 1 JULY 1997

Sending classical information via noisy quantum channels

Benjamin Schumacher[1] and Michael D. Westmoreland[2]
[1]*Department of Physics, Kenyon College, Gambier, Ohio 43022*
[2]*Department of Mathematical Sciences, Denison University, Granville, Ohio 43023*
(Received 12 February 1997)

This paper extends previous results about the classical information capacity of a noiseless quantum-mechanical communication channel to situations in which the final signal states are mixed states, that is, to channels with noise. [S1050-2947(97)02007-6]

PACS number(s): 03.65.Bz, 42.50.Dv, 89.70.+c

I. INTRODUCTION

Suppose Alice wishes to convey classical information to Bob by using a quantum system Q as a communication channel. Alice prepares the channel in one of various quantum states W_x with *a priori* probabilities p_x. Bob makes a measurement on the system Q, and from its result he tries to infer which state Alice prepared. A theorem stated by Gordon [1] and Levitin [2], proved by Kholevo [3], gives an upper bound to the amount of information that Bob can obtain about Alice's signal. If $W=\Sigma_x p_x W_x$ is the density operator describing the ensemble of Alice's signals, then the mutual information $H(X:Y)$ between Alice's input X and Bob's output Y is bounded by

$$H(X:Y)\leq H(W)-\sum_x p_x H(W_x), \qquad (1)$$

where $H(W)=-\mathrm{Tr}W\log_2 W$, the von Neumann entropy of the density operator W. The upper bound in Eq. (1) is in general a weak one, in that Bob may not be able to choose an observable that gives him an amount of information near the upper bound [4].

Recently, Hausladen *et al.* [5] showed that, if Alice's signal states W_x are pure states, then it is possible to approach the Kholevo bound $H(W)$ for an appropriate choice of Alice's code and Bob's decoding observable. This is done by (i) employing long strings of signals to send many independent messages together, (ii) "pruning" the set of strings used as codewords so that the codewords are sufficiently distinguishable, and (iii) choosing a suitable decoding observable that acts on entire strings of signals. For large enough L, codewords of L "letters" may be used to transmit up to $LH(W)$ bits of information [thus $H(W)$ bits per letter] with arbitrarily low probability of error.

This naturally suggests a generalization, which was presented in [5] as a conjecture. Suppose that Alice employs signal states W_x that are *mixed* states. Then can Alice and Bob find a choice of code and decoding observable so that the general Kholevo bound [Eq. (1)] can be approached arbitrarily closely? In this paper, we show that the answer to this question is "yes." That is, we prove the following result.

Theorem. Suppose we have letter states W_x with *a priori* probabilities p_x and let

$$\chi=H(W)-\sum_x p_x H(W_x).$$

Fix $\epsilon,\delta>0$. Then for sufficiently large L, there exist a code (whose codewords are strings of L letters) and a decoding observable such that the information carried per letter is at least $\chi-\delta$ and the probability of error $P_E<\epsilon$.

As in [5], we employ an average over randomly generated codes to establish the existence of a satisfactory code. (If the average probability of error is small for an ensemble of codes, the ensemble must contain specific codes with small probability of error.) We also use a similar prescription for Bob's decoding observable. The chief refinement in the proof presented here is the enforcement of stronger "typicality" conditions on various quantities associated with the channel.

The mixed states W_x may be thought of as the outputs of a *noisy* quantum channel. Thus our main result will enable us to draw conclusions about the classical information capacity of a noisy quantum channel.

Our main result is the same as that given recently in independent work by Holevo [6]. Holevo's proof, like ours, follows the general strategy of [5], though there are substantial differences of detail.

II. SETTING IT UP

We will assume that we have an alphabet of mixed states W_x, each of which has an *a priori* probability p_x. The average density matrix is $W=\Sigma_x p_x W_x$. We wish to show that, if we use long strings of these letters (suitably pruning the set of codewords to improve distinguishability) and an appropriate decoding observable, we can send reliably an amount of information up to

$$\chi=H(W)-\sum_x p_x H(W_x) \qquad (2)$$

per letter.

We will be considering strings of L letters. In what follows we will assume that the index a refers to a whole string of letters: $a=x_1\cdots x_L$. $P_a=p_{x_1}\cdots p_{x_L}$ is the *a priori* probability of the sequence a and $\rho_a=W_{x_1}\otimes\cdots\otimes W_{x_L}$ is the state associated with the string. The average state is

$$\rho = \sum_a P_a \rho_a = \underbrace{W \otimes \cdots \otimes W}_{L \text{ times}}. \tag{3}$$

Consider the state ρ_a. This has a complete orthogonal set of eigenstates, which we will denote $|s_{ak}\rangle$ (where k ranges over the dimension of the space), and a corresponding set of eigenvalues $p_{k|a}$. As the notation suggests, we may think of the $p_{k|a}$'s as "conditional probabilities" for k given a, and this motivates us to form the "joint probability" distribution $P_{ak}=P_a p_{k|a}$. Of course, $\rho=\Sigma_{ak}P_{ak}|s_{ak}\rangle\langle s_{ak}|$. It will be convenient to refer to the index k identifying the string eigenstate as the *syndrome* of the codeword ρ_a.

When we construct our decoding observable, we will be trying not only to distinguish the string a, but also (a seemingly harder task) to determine the syndrome k as well. An error will occur if either the codeword or the syndrome is incorrectly identified. For a given codeword ρ_a, the various $|s_{ak}\rangle$'s are orthogonal and hence perfectly distinguishable from one another, so this will not really be more difficult than identifying the codeword only.

III. TYPICALITY

Let $\epsilon,\delta>0$. Then we can find a length L large enough to enforce the following typicality conditions on strings of length L.

(i) There exists a typical subspace [8,9] for the states. That is, there is a subspace Λ spanned by eigenstates of ρ such that, if Π is the projection onto Λ, $\mathrm{Tr}\rho\Pi>1-\epsilon$. Further, if we denote by $|\lambda_n\rangle$ the eigenstate of ρ with eigenvalue λ_n,

$$2^{-L[H(W)+\delta]}<\lambda_n<2^{-L[H(W)-\delta]} \tag{4}$$

for all $|\lambda_n\rangle\in\Lambda$.

One key property of the typical subspace is that

$$\mathrm{Tr}\rho^2\Pi<2^{-L[H(W)-2\delta]}. \tag{5}$$

This property was used by Hausladen *et al.* [5] to bound the probability of error, and it will play that role again.

(ii) There exist a typical set of strings (relative to the distribution P_a) and a typical set of string-syndrome pairs (relative to the joint distribution P_{ak}). Let $H(A)$ be the Shannon entropy associated with the string distribution P_a and let $H(A,K)$ be the Shannon entropy associated with the joint distribution P_{ak}. Notice that $H(A)=LH(X)$, where $H(X)$ is the Shannon entropy of the letter distribution. Also,

$$H(A,K)=H(A)+\sum_a P_a H(\rho_a)$$

$$=L\left(H(X)+\sum_x p_x H(W_x)\right), \tag{6}$$

where the x sum is over the letters. Typicality means the following.

(a) For a typical string a,

$$2^{-L[H(X)+\delta]}<P_a<2^{-L[H(X)-\delta]}. \tag{7}$$

Furthermore, the sum of the P_a's for the typical strings is greater than $1-\epsilon$.

(b) For a typical string-syndrome pair ak,

$$2\exp\left[-L\left(H(X)+\sum_x p_x H(W_x)+\delta\right)\right]<P_{ak}$$

$$<2\exp\left[-L\left(H(X)+\sum_x p_x H(W_x)-\delta\right)\right], \tag{8}$$

where $2\exp[x]$ means 2^x. Furthermore, the sum of P_{ak} over the typical string-syndrome pairs is also greater than $1-\epsilon$.

For each string a, we define a set of *relatively typical* syndromes as follows: k is relatively typical to a if a is a typical string and ak is a typical string-syndrome pair. (Note that atypical strings have no relatively typical syndromes.) If k is relatively typical to a, then

$$2\exp\left[-L\left(\sum_x p_x H(W_x)+2\delta\right)\right]<p_{k|a}$$

$$<2\exp\left[-L\left(\sum_x p_x H(W_x)-2\delta\right)\right] \tag{9}$$

since $p_{k|a}=P_{ak}/P_a$. We can take advantage of the definition of χ above to write this as

$$2^{-L[H(W)-\chi+2\delta]}<p_{k|a}<2^{-L[H(W)-\chi-2\delta]}. \tag{10}$$

We adopt the following notations for sums: $\Sigma_{k|a}$ means sum over k for a given value of a, $\Sigma'_{k|a}$ means sum restricted to relatively typical k's only (note that this sum may have no terms), and Σ'_{ak} means $\Sigma_a\Sigma'_{k|a}$. If we restrict sums to relatively typical syndromes only, we do not lose much weight in the ensemble. That is, consider the pairs ak in which k is relatively typical. This excludes all atypical a's (a set of total probability less than ϵ) and all atypical pairs ak (also of probability less than ϵ). It follows that

$$\sum_{ak}{}' P_{ak}=\sum_a P_a \sum_{k|a}{}' p_{k|a}>1-2\epsilon. \tag{11}$$

The total ensemble $\rho=\Sigma_{ak}P_{ak}|s_{ak}\rangle\langle s_{ak}|$. If we restrict the ensemble to string-syndrome pairs in which the syndrome is relatively typical, then we get a subnormalized density operator $\tilde{\rho}$ for which

$$\mathrm{Tr}\tilde{\rho}=\mathrm{Tr}\left(\sum_{ak}{}' P_{ak}|s_{ak}\rangle\langle s_{ak}|\right)>1-2\epsilon. \tag{12}$$

We also note that $\tilde{\rho}\leqslant\rho$ under the usual partial ordering of positive operators. (That is, $\langle\psi|\tilde{\rho}|\psi\rangle\leqslant\langle\psi|\rho|\psi\rangle$ for all $|\psi\rangle$.)

IV. CODING AND DECODING

Now we discuss our code and our decoding procedure. The code will consist of N codewords (each a string of length L), which we will use with equal frequency. Codewords in our code will be indexed by a greek index such as α. Thus the latin characters $a,b,\ldots$ index the whole set of strings, while the greek characters $\alpha,\beta,\ldots$ index the codewords in our code. Greek indices thus take on N possible values.

The decoding procedure will be a variation of the "pretty good measurement" used in [5]. We will attempt to identify not only the codeword but also the syndrome. Our decoding observable will be a "positive operator measurement" (POM), described by a set of positive operators summing to unity. For each codeword-syndrome pair αk we will have a (possibly subnormalized) vector $|\tilde{\mu}_{\alpha k}\rangle$ such that $|\tilde{\mu}_{\alpha k}\rangle\langle\tilde{\mu}_{\alpha k}|$ is an element of our decoding POM. The probability of error is thus

$$\begin{aligned} P_E &= 1-\frac{1}{N}\sum_{\alpha k} p_{k|\alpha}|\langle\tilde{\mu}_{\alpha k}|s_{\alpha k}\rangle|^2 \\ &= \frac{1}{N}\sum_{\alpha}\sum_{k\alpha} p_{k|\alpha}(1-|\langle\tilde{\mu}_{\alpha k}|s_{\alpha k}\rangle|^2) \\ &\leqslant 2\left(1-\frac{1}{N}\sum_{\alpha k} p_{k|\alpha}\left|\langle\tilde{\mu}_{\alpha k}|s_{\alpha k}\rangle\right|\right). \end{aligned} \tag{13}$$

We next describe how to specify the decoding observable. If k is not relatively typical to α, we let $|\tilde{\mu}_{\alpha k}\rangle=0$. For the rest, we construct the operator

$$\Upsilon=\sum_{\alpha k}{}' \Pi|s_{\alpha k}\rangle\langle s_{\alpha k}|\Pi, \tag{14}$$

where Π is the projection onto the typical subspace Λ for the *a priori* ensemble, as described above. We define

$$|\tilde{\mu}_{\alpha k}\rangle=\Upsilon^{-1/2}|s_{\alpha k}\rangle \tag{15}$$

for k relatively typical to α. (Since Υ is not generally fully invertible, this $\Upsilon^{-1/2}$ is the pseudoinverse of $\Upsilon^{1/2}$, supported only on the support of Υ.) It follows that

$$\sum_{\alpha,k}|\tilde{\mu}_{\alpha k}\rangle\langle\tilde{\mu}_{\alpha k}|=\sum_{\alpha k}{}'|\tilde{\mu}_{\alpha k}\rangle\langle\tilde{\mu}_{\alpha k}|=1 \tag{16}$$

on the range space of Υ (which is a subspace of Λ). We can add an element of the POM (labeled "error") on the orthogonal space, if necessary, to give overall normalization.

Since Υ (and thus $\Upsilon^{-1/2}$) is positive, the inner product $\langle\tilde{\mu}_{\alpha k}|s_{\alpha k}\rangle$ is real and non-negative. We do not need the modulus signs in our bound for the probability of error. Furthermore, our construction of the $|\tilde{\mu}_{\alpha k}\rangle$'s means that the only contributions come from those terms in which k is relatively typical to α. Thus we can write

$$P_E\leqslant 2\left(1-\frac{1}{N}\sum_{\alpha k}{}' p_{k|\alpha}\langle\tilde{\mu}_{\alpha k}|s_{\alpha k}\rangle\right). \tag{17}$$

As was mentioned in [5], this definition has some nice properties connected with a matrix of inner products of $|s_{\alpha k}\rangle$. For k relatively typical to α and l relatively typical to β, we define

$$\mathcal{S}_{\alpha k,\beta l}=\langle s_{\alpha k}|\Pi|s_{\beta l}\rangle. \tag{18}$$

The matrix $\mathcal{S}$ is a positive square matrix. It turns out that

$$\langle\tilde{\mu}_{\alpha k}|s_{\alpha k}\rangle=(\sqrt{\mathcal{S}})_{\alpha k,\alpha k}. \tag{19}$$

(This could be used as an implicit definition of the $|\tilde{\mu}_{\alpha k}\rangle$'s.) We will employ the same inequality that was used in [5]: for $x\geqslant 0$, $\sqrt{x}\geqslant\frac{3}{2}x-\frac{1}{2}x^2$. This means that

$$(\sqrt{\mathcal{S}})_{\alpha k,\alpha k}\geqslant\frac{3}{2}\mathcal{S}_{\alpha k,\alpha k}-\frac{1}{2}\sum_{\beta,l}{}'\mathcal{S}_{\alpha k,\beta l}\mathcal{S}_{\beta l,\alpha k}. \tag{20}$$

This gives us a bound for the probability of error

$$P_E\leq 2-\underbrace{\frac{3}{N}\sum_{\alpha}\sum_{k|\alpha}{}' p_{k|\alpha}\langle s_{\alpha k}|\Pi|s_{\alpha k}\rangle}_{\#1} + \underbrace{\frac{1}{N}\sum_{\alpha\beta}\sum_{k|\alpha}{}'\sum_{l|\beta}{}' p_{k|\alpha}\langle s_{\alpha k}|\Pi|s_{\beta l}\rangle\langle s_{\beta l}|\Pi|s_{\alpha k}\rangle}_{\#2} \tag{21}$$

We will deal with the terms labeled # 1 and # 2 separately.

V. RANDOM CODES

Now we will average the probability of error P_E over random codes. These codes are constructed by choosing the N codewords independently according to the *a priori* string distribution P_a. This will have the effect of turning averages over the codewords in the code into averages over the *a priori* string ensemble.

Denote the random code average by $\langle\ \rangle_c$. Consider term #1 above,

$$\begin{aligned} \langle\#1\rangle_c &= \left\langle\frac{3}{N}\sum_{\alpha}\sum_{k|\alpha}{}' p_{k|\alpha}\langle s_{\alpha k}|\Pi|s_{\alpha k}\rangle\right\rangle_c \\ &= 3\left(\sum_a P_a\sum_{k|a}{}' p_{k|a}\mathrm{Tr}\Pi|s_{ak}\rangle\langle s_{ak}|\Pi\right) \\ &= 3(\mathrm{Tr}\Pi\tilde{\rho}\Pi). \end{aligned} \tag{22}$$

(Notice that the average over random codes transformed the sum over the codewords Σ_α into N times the average over the string ensemble described by P_a since each codeword is chosen independently according to P_a.) Now let $\Delta=\rho-\tilde{\rho}$, a positive operator since $\tilde{\rho}\leqslant\rho$. Then

$$\begin{aligned} \langle\#1\rangle_c &= 3(\mathrm{Tr}\Pi\rho\Pi-\mathrm{Tr}\Pi\Delta\Pi)>3[(1-\epsilon)-\mathrm{Tr}\Delta] \\ &>3(1-3\epsilon) \end{aligned} \tag{23}$$

since $\mathrm{Tr}\Delta<2\epsilon$.

Next, examine term #2. The double sum over α and β may be split into two parts: a part in which $\alpha=\beta$ and a part in which $\alpha\neq\beta$. The advantage in this is that, if $\alpha\neq\beta$, the codewords are chosen independently in a random code:

$$\#2 = \underbrace{\frac{1}{N}\sum_{\alpha}\sum_{k|\alpha}{}'\sum_{l|\alpha}{}' p_{k|\alpha}\langle s_{\alpha k}|\Pi|s_{\alpha l}\rangle\langle s_{\alpha l}|\Pi|s_{\alpha k}\rangle}_{\#\,2\mathrm{a}} + \underbrace{\frac{1}{N}\sum_{\alpha,\beta\neq\alpha}\sum_{k|\alpha}{}'\sum_{l|\beta}{}' p_{k|\alpha}\langle s_{\alpha k}|\Pi|s_{\beta l}\rangle\langle s_{\beta l}|\Pi|s_{\alpha k}\rangle}_{\#\,2\mathrm{b}}. \tag{24}$$

We consider term #2a:

$$\begin{aligned}\#2a &= \frac{1}{N}\sum_{\alpha}\sum_{k|\alpha}{}'\sum_{l|\alpha}{}' p_{k|\alpha}\langle s_{\alpha k}|\Pi|s_{\alpha l}\rangle\langle s_{\alpha l}|\Pi|s_{\alpha k}\rangle \\ &\leq \frac{1}{N}\sum_{\alpha}\sum_{k|\alpha}{}'\sum_{l|\alpha} p_{k|\alpha}\langle s_{\alpha k}|\Pi|s_{\alpha l}\rangle\langle s_{\alpha l}|\Pi|s_{\alpha k}\rangle \\ &= \frac{1}{N}\sum_{\alpha}\sum_{k|\alpha}{}' p_{k|\alpha}\left\langle s_{\alpha k}\middle|\Pi\left(\sum_{l|a}|s_{\alpha l}\rangle\langle s_{\alpha l}|\right)\Pi\middle|s_{\alpha k}\right\rangle \\ &= \frac{1}{N}\sum_{\alpha}\sum_{k|\alpha}{}' p_{k|\alpha}\langle s_{\alpha k}|\Pi|s_{\alpha k}\rangle \end{aligned} \tag{25}$$

since for any α, the $|s_{\alpha l}\rangle$ form a complete set. But $\langle s_{\alpha k}|\Pi|s_{\alpha k}\rangle \leq 1$, so

$$\#2a \leq \frac{1}{N}\sum_{\alpha}\sum_{k|\alpha}{}' p_{k|\alpha} \leq \frac{1}{N}\sum_{\alpha}\sum_{k\alpha} p_{k|\alpha} = 1. \tag{26}$$

Therefore, of course, $\langle \#2a\rangle_c \leq 1$.

Now we consider the much more interesting term #2b:

$$\#2b = \frac{1}{N}\sum_{\alpha,\beta\neq\alpha}\sum_{k|\alpha}\sum_{l|\beta} p_{k|\alpha}\mathrm{Tr}\Pi|s_{\alpha k}\rangle\langle s_{\alpha k}|\Pi|s_{\beta l}\rangle\langle s_{\beta l}|\Pi. \tag{27}$$

The only terms that appear in this sum are terms in which l is typical relative to β. But for such codeword-syndrome pairs, we have a uniform lower bound on $p_{l|\beta}$, which allows us to say that, for all l and β that appear in our sum,

$$1 < p_{l|\beta}2^{L[H(W)-\chi+2\delta]}. \tag{28}$$

Therefore,

$$\#2b \leq 2^{L[H(W)-\chi+2\delta]}\frac{1}{N}\sum_{\alpha,\beta\neq\alpha}\sum_{k|\alpha}{}'\sum_{l|\beta}{}' \times p_{k|\alpha}p_{l|\beta}\mathrm{Tr}\Pi|s_{\alpha k}\rangle\langle s_{\alpha k}|\Pi|s_{\beta l}\rangle\langle s_{\beta l}|\Pi. \tag{29}$$

Taking the average of #2b over random codes,

$$\begin{aligned}\langle \#2b\rangle_c &= 2^{L[H(W)-\chi+2\delta]}\frac{N(N-1)}{N} \\ &\quad\times\left(\sum_a P_a\sum_b P_b\sum_{k|a}{}' p_{k|a}\sum_{l|b}{}' p_{l|b} \times\mathrm{Tr}\Pi|s_{ak}\rangle\langle s_{ak}|\Pi|s_{bl}\rangle\langle s_{bl}|\Pi\right) \\ &\leq N2^{L[H(W)-\chi+2\delta]}\mathrm{Tr}\Pi\tilde{\rho}\Pi\tilde{\rho}\Pi.\end{aligned} \tag{30}$$

(Notice again that each term in the sums over the codewords has been replaced by the appropriate string-ensemble average.) We note that if A, B, and C are positive operators with $B\leq A$, $\mathrm{Tr}BC\leq\mathrm{Tr}AC$. Thus

$$\begin{aligned}\mathrm{Tr}\Pi\tilde{\rho}\Pi\tilde{\rho}\Pi &= \mathrm{Tr}\Pi\tilde{\rho}\Pi\tilde{\rho} \leq \mathrm{Tr}\Pi\tilde{\rho}\Pi\rho = \mathrm{Tr}\tilde{\rho}\Pi\rho\Pi \leq \mathrm{Tr}\rho\Pi\rho\Pi \\ &= \mathrm{Tr}\rho^2\Pi,\end{aligned} \tag{31}$$

where the last line uses the fact that ρ and Π commute:

$$\begin{aligned}\langle \#2b\rangle_c &\leq N2^{L[H(W)-\chi+2\delta]}\mathrm{Tr}\rho^2\Pi \\ &< N2^{L[H(W)-\chi+2\delta]}2^{-L[H(W)-2\delta]} = N2^{-L(\chi-4\delta)}.\end{aligned} \tag{32}$$

Combining these results, we can find an upper bound for the probability of error averaged over all random codes:

$$\begin{aligned}\langle P_E\rangle_c &\leq 2-\langle\#1\rangle_c+\langle\#2a\rangle_c+\langle\#2b\rangle_c \\ &< 2-3(1-3\epsilon)+1+N2^{-L(\chi-4\delta)} = 9\epsilon+N2^{-L(\chi-4\delta)}.\end{aligned} \tag{33}$$

For L sufficiently large, we can choose N nearly as big as $2^{L\chi}$ and still have the probability of error small.

If the *average* probability of error is below this bound, then Alice and Bob will be able to find some particular code for which

$$P_E \leq 9\epsilon + N2^{-L(\chi-4\delta)}. \tag{34}$$

If L is very large, Alice can use up to $N=2^{L(\chi-5\delta)}$ codewords and still have $P_E\leq 10\epsilon$. In this case, Alice encodes $\chi-5\delta$ bits per letter. This proves our main theorem.

We have shown the existence of a satisfactory code without actually constructing it. Consequently, we do not know much about the structure of the code. In particular, we have not guaranteed in our proof that the letter states occur in the codewords with frequencies that closely match their *a priori* probabilities p_x. (This is something that we might wish to require since the distribution p_x might be chosen to optimize some resource, such as the energy required per letter.) It turns out, however, that we can satisfy such a requirement. Since we generate the codewords in our ensemble of codes by using the *a priori* probabilities, the law of large numbers implies that the letter frequencies will match the *a priori* distribution within any specified tolerance for a set of "typical codes." The set of typical codes includes almost the entire weight of the code ensemble and thus many of the particular codes with low probability of error. See [5] for the details of this argument applied to the pure state case.

VI. FIXED-ALPHABET CAPACITY

We have shown that it is possible to send information at any rate up to χ bits per letter with arbitrarily low probability of error. The capacity of a channel is defined as the maximum information per letter that may be sent through the channel with P_E arbitrarily small. Thus χ provides a lower bound to the capacity of the quantum channel.

Classical information theory together with Kholevo's theorem also allows us to use χ to establish an *upper bound* for the capacity of the channel. Suppose X represents Alice's input and Y represents Bob's decoding measurement outcome. Then the Fano inequality [7] states that

$$-P_E\log_2 P_E-(1-P_E)\log_2(1-P_E)+P_E\log_2(N_X-1) \geq H(X|Y), \tag{35}$$

where P_E is the probability of error and N_X is the number of possible values of X. $H(X|Y)$ is the conditional Shannon entropy of X given Y, that is, the entropy of the conditional distribution $p(x|y)$, averaged over the various values of y [13]. It is related to the mutual information $H(X:Y)$ by

$$H(X|Y)=H(X)-H(X:Y). \tag{36}$$

In the channel, Alice uses some signal states ρ_a with probabilities P_a. Kholevo's theorem places an upper bound on the mutual information $H(X:Y)$:

$$H(X:Y)\leq H(\rho)-\sum_a P_a H(\rho_a).$$

(Note that if the channel used by Alice and Bob consists of L letters used independently, then the Kholevo bound is just $L\chi$, where χ is the Kholevo bound for a single letter.) If the Alice's input X has an entropy $H(X)$ that exceeds $H(\rho)-\Sigma_a P_a H(\rho_a)$, then $H(X|Y)>0$ and it will not be possible to make the probability of error P_E arbitrarily small.

Suppose we fix an alphabet $\Gamma=\{W_x\}$ of letter states W_x and require that Alice use codewords a that are length-L strings of these letter states: $a=x_1\cdots x_L$. Then the probability distribution P_a yields marginal probability distributions $p(x_1),\ldots,p(x_L)$ and average density operators $W_1,\ldots,W_L$ for the L different letters. It follows that

$$\begin{aligned}
H(\rho)-\sum_a P_a H(\rho_a) &= H(\rho)-\sum_a P_a(H(W_{x_1})+\cdots+H(W_{x_L})) \\
&= H(\rho)-\left(\sum_{x_1} p(x_1)H(W_{x_1})+\cdots + \sum_{x_L} p(x_L)H(W_{x_L})\right) \\
&\leq \left(H(W_1)-\sum_{x_1} p(x_1)H(W_{x_1})\right)+\cdots + \left(H(W_L)-\sum_{x_L} p(x_L)H(W_{x_L})\right),
\end{aligned} \tag{37}$$

where we have used the subadditivity of the entropy $H(\rho)$. We might write this as

$$\chi^{(L)}\leq \chi_1+\cdots+\chi_L, \tag{38}$$

where $\chi^{(L)}$ represents the Kholevo bound for the ensemble of codewords of length L and $\chi_1,\ldots,\chi_L$ represent Kholevo bounds for the individual letter ensembles.

We define the *fixed-alphabet capacity* C_Γ to be

$$C_\Gamma=\sup_{p(x)}\chi, \tag{39}$$

where $p(x)$ is the probability distribution over the letter states in Γ and χ is the single-letter Kholevo bound. This quantity represents the maximum information rate per letter that Alice can send to Bob with arbitrarily low probability of error.

This claim follows directly from our results so far. Suppose Alice uses codewords of length L. Then $\chi^{(L)}\leq LC_\Gamma$; by the above argument, if Alice attempts to send more than LC_Γ bits using these codewords then the probability of error will not be arbitrarily small. Conversely, we can choose the letter probabilities so that χ is as close as required to C_Γ, and we have previously shown that a suitable choice of code and decoding observable can convey up to χ bits per letter with arbitrarily low P_E. Thus the capacity C_Γ cannot be exceeded, but can be approached arbitrarily closely.

VII. NOISY CHANNELS

The mixed states W_x used in our alphabet are the states available to Bob for decoding. They may in fact not be the original states of the channel Q chosen by Alice. In the interval between Alice's encoding and Bob's decoding, the system Q may have undergone unitary internal evolution (which Bob can correct by a suitable choice of "rotated" decoding observable) and interaction with the external environment (which Bob cannot in general correct).

The most general description of the evolution of a quantum system Q interacting with an environment is provided by a trace-preserving completely positive linear map on the set of density operators of Q [11]. Such a map is described by a superoperator $\mathcal{E}$:

$$\rho\to\rho'=\mathcal{E}(\rho), \tag{40}$$

where ρ is the initial state of the system and ρ' is the final state. The superoperator $\mathcal{E}$ acts linearly, so that a convex combination of input states yields a convex combination of output states. This description clearly includes unitary evolution of Q as a special case, but it also can account for interaction with the environment.

A noisy quantum channel is defined by a superoperator $\mathcal{E}$ that describes the evolution of each letter as it is transmitted from Alice to Bob. We assume that the channel is memoryless, i.e., that the evolution of each letter is independent.

This means, among other things, that a product state of several input letters will evolve into a product state output.

Alice's basic problem is to use input states w_x so that the output states $W_x=\mathcal{E}(w_x)$ can be distinguished by Bob. If Alice has a fixed alphabet $\{w_x\}$ of input states, then the maximum achievable information rate per letter is still given by our fixed-alphabet capacity C_Γ, where Γ is the alphabet of *output* states.

Now suppose that Alice is allowed to choose her input states in order to maximize the information conveyed to Bob over the noisy quantum channel, subject to the constraint that Alice must transmit codewords that are represented by product states of the letters. This *almost* reduces to the fixed-alphabet problem, where the fixed alphabet Γ now includes all of the possible output states of the channel. The maximum over probability distributions is now a maximum over all input ensembles of states chosen by Alice.

We say that this problem *almost* reduces to the fixed alphabet problem in that the argument that χ is an upper bound of the capacity must be modified in this case. Recall from Sec. VI that we applied the classical Fano inequality to show that if Alice attempts to send information at a rate exceeding χ, then the probability of error cannot be made arbitrarily small. If we attempt to use the same argument in the present case, then the Fano inequality does not help us for at least two reasons. First, the number of possible input states N_X is unbounded. Second, we do not have a characterization of $H(X|Y)$ that allows us to compare it with N_X. Thus we will modify the Fano inequality to understand the behavior of the probability of error in the present case.

We first note that the probability of "getting it right"

$$1-P_E=\frac{1}{N}\sum_{\alpha k} p_{k|\alpha}|\langle\tilde{\mu}_{\alpha k}|s_{\alpha k}\rangle|^2 \tag{41}$$

is linear in the elements of the POM. Thus the probability of error P_E is a convex function on the elements of the POM. We may modify the proof of a result of Davies (Theorem 3 of [14]) to show that the convex function P_E is minimized by a POM having no more than d^2 elements, where d is the dimension of the support of the POM. Thus the probability of error is minimized by a decision scheme in which at most d^2 of the inputs are identified by the decision scheme. Let us denote the output of such a scheme by $Y_{\min}$. Fano's inequality gives us that

$$-P_E\log_2 P_E-(1-P_E)\log_2(1-P_E)+P_E\log_2(d^2-1)$$
$$\geqslant H(X|Y_{\min}). \tag{42}$$

Note that

$$H(X|Y_{\min})=H(X)-H(X:Y_{\min}\min) \tag{43}$$

$$\geqslant H(X)-\chi, \tag{44}$$

so that we conclude

$$-P_E\log_2 P_E-(1-P_E)\log_2(1-P_E)+P_E\log_2(d^2-1)$$
$$\geqslant H(X)-\chi. \tag{45}$$

Note that this is a relation between the minimum probability of error and a quantity $[H(X)-\chi]$ that does not depend on the particular decision scheme. We see that if Alice attempts to send information at a rate $H(X)$ in excess of χ, then the probability of error cannot be made arbitrarily small.

We now turn to a demonstration that this rate can be achieved. Alice wishes to choose a set of input states w_x (together with input probabilities p_x) so that χ is maximized for the output states W_x. We next show that Alice can do no better than choose the input states w_x to be pure. Let a set of (possibly mixed) input states w_x be given along with their *a priori* probabilities and let

$$W=\sum_x p_x W_x=\sum_x p_x\mathcal{E}(w_x) \tag{46}$$

be the average output state. Then

$$\chi=H(W)-\sum_x p_x H(\mathcal{E}(w_x)). \tag{47}$$

Construct a new set of pure state inputs by resolving each mixed state input into a convex combination of pure states:

$$w_x=\lambda_{xk}|\psi_{xk}\rangle\langle\psi_{xk}|. \tag{48}$$

We will use the state $|\psi_{xk}\rangle$ with probability $p_{xk}=p_x\lambda_{xk}$. By linearity,

$$W_x=\mathcal{E}(w_x)=\sum_k \lambda_{xk}\mathcal{E}(|\psi_{xk}\rangle\langle\psi_{xk}|), \tag{49}$$

so that the average output state is still W, as before. By the convexity of the von Neumann entropy,

$$H(W_x)\geqslant\sum_k \lambda_{xk}H(\mathcal{E}(|\psi_{xk}\rangle\langle\psi_{xk}|)). \tag{50}$$

It follows that

$$\chi'=H(W)-\sum_{xk} p_{xk}H(\mathcal{E}(|\psi_{xk}\rangle\langle\psi_{xk}|))$$

$$\geqslant H(W)-\sum_x p_x H(\mathcal{E}(w_x))=\chi. \tag{51}$$

In other words, for any ensemble of mixed input states, we can find an ensemble of pure input states whose output states have a χ at least as great. The optimal inputs for the noisy quantum channel are pure states.

To sum up, if Alice is required to use product states to represent her codewords, then the capacity $C^{(1)}$ of the noisy quantum channel is

$$C^{(1)}=\max\chi, \tag{52}$$

where χ is the Kholevo bound for the output states of the channel and the maximum is taken over all ensembles of pure state inputs. Alice can reliably transmit information to

Bob at any rate below $C^{(1)}$. We will refer to $C^{(1)}$ as the product state capacity. The superscript (1) reminds us that Alice is required to use the multiple available copies of the channel *one at a time*, coding her messages into product states.

The product state capacity $C^{(1)}$ is a function only of the superoperator $\mathcal{E}$ describing the dynamical evolution of a single channel. To emphasize this, we will calculate $C^{(1)}$ in the simple case of a one-quantum-bit (one-qubit) depolarizing channel. A two-level system, or qubit, is sent through the channel. With probability P, the state of the qubit is left intact; with probability $1-P$, the state is completely randomized, so that the output state of the qubit is a completely mixed density operator. For any pure state input $w_x=|\psi_x\rangle\langle\psi_x|$, the output state is

$$W_x=\mathcal{E}(w_x)=P|\psi_x\rangle\langle\psi_x|+\frac{1-P}{2}I, \tag{53}$$

where I is the identity operator. Any such state has eigenvalues $\frac{1}{2}(1+P)$ and $\frac{1}{2}(1-P)$ and thus an entropy

$$\begin{aligned}H(W_x)&=-\frac{1+P}{2}\log_2\frac{1+P}{2}-\frac{1-P}{2}\log_2\frac{1-P}{2}\\&=1-\frac{1}{2}[(1+P)\log_2(1+P)+(1-P)\log_2(1-P)].\end{aligned} \tag{54}$$

To calculate the capacity, we maximize the output χ over all ensembles of pure state inputs. But the entropy of each output state will be the same, so we only need to maximize the entropy of the average output state W. This is easily seen to be 1, so that

$$C^{(1)}=\frac{1}{2}[(1+P)\log_2(1+P)+(1-P)\log_2(1-P)]. \tag{55}$$

If $P=0$, then the product state capacity is (reassuringly) zero; but for any $P>0$, the product state capacity $C^{(1)}>0$, with $C^{(1)}=1$ bit for $P=1$.

However, Alice can do more than we have so far allowed her to do. It might conceivably be to her advantage to use *entangled* states to represent her codewords. The output states will in general be entangled states. (This will present no additional difficulties for Bob; even to distinguish product states, we have allowed Bob to use a collective decoding observable for strings of L letters.)

In this case, it is no longer true that the Kholevo bound $\chi^{(L)}$ for the output codewords satisfies

$$\chi^{(L)}\leqslant\chi_1+\cdots+\chi_L, \tag{56}$$

where the χ_k denote the Kholevo bounds for the individual letters. That is, χ is not necessarily subadditive for systems that may be entangled.

Suppose that Alice is permitted to prepare entangled states of L copies of the channel. Then we can treat these L copies as a single "extended" channel, which Alice can prepare in any state. Our main theorem applied to this extended channel means that for any $\chi^{(L)}$ of the output states, Alice can reliably send up to $\chi^{(L)}/L$ bits of information per letter to Bob. Thus we define

$$C^{(L)}=\frac{1}{L}\max\chi^{(L)}, \tag{57}$$

where the maximum is taken over all input ensembles, including entangled states, for the L elementary channels. (By our previous arguments, it suffices to consider only ensembles of pure input states.) $C^{(L)}$ is the capacity if Alice is allowed to use the channels in entangled blocks of length L. Since product states are allowed, it is clear that $C^{(L)}\geqslant C^{(1)}$. The asymptotic capacity will be

$$C=\lim_{L\to\infty}C^{(L)}. \tag{58}$$

This will be the ultimate information capacity of the noisy quantum channel. (Similar considerations are discussed in [10].)

Like $C^{(1)}$, C will be a function only of the dynamical superoperator $\mathcal{E}$. No examples are known where $C>C^{(1)}$ (though the example in [12] is suggestive). Thus it is not known whether or not $C=C^{(1)}$.

ACKNOWLEDGMENTS

The authors wish to thank L. Levitin, R. Jozsa, and W. Wootters for many helpful discussions and suggestions about this and related work. The authors also wish to thank the Institute for Scientific Interchange in Turin, Italy for the opportunities afforded by a workshop on quantum computation. M.D.W. was supported by the Robert C. Good Foundation at Denison University.

[1] J. P. Gordon, in *Quantum Electronics and Coherent Light*, Proceedings of the International School of Physics "Enrico Fermi," Course XXXI, edited by P. A. Miles (Academic, New York, 1964), pp. 156–181.

[2] L. B. Levitin, *Information, Complexity, and Control in Quantum Physics*, edited by A. Blaquière, S. Diner, and G. Lochak (Springer, Vienna, 1987), pp. 111–115.

[3] A. S. Kholevo, Probl. Peredachi Inf. **9**, 177 (1973).

[4] C. A. Fuchs and C. M. Caves, Phys. Rev. Lett. **73**, 3047 (1994).

[5] P. Hausladen, R. Josza, B. Schumacher, M. Westmoreland, and W. K. Wootters, Phys. Rev. A **54**, 1869 (1996).

[6] A. S. Kholevo, IEEE Trans. Inf. Theory (to be published).

[7] T. M. Cover and J. A. Thomas, *Elements of Information Theory* (Wiley, New York, 1991).

[8] B. Schumacher, Phys. Rev. A **51**, 2738 (1995).

[9] B. Schumacher and R. Jozsa, J. Mod. Opt. **41**, 2343 (1994).
[10] A. S. Kholevo, Probl. Peredachi Inf. **15**, 3 (1979).
[11] K. Hellwig and K. Kraus, Commun. Math. Phys. **16**, 142 (1970); M.-D. Choi, Linear Algebr. Appl. **10**, 285 (1975); K. Kraus, *States Effects and Operations: Fundamental Notions of Quantum Theory* (Springer-Verlag, Berlin, 1983).
[12] C. H. Bennett, C. A. Fuchs, and J. Smolin (unpublished).
[13] C. E. Shannon, Bell Syst. Tech. J. **27**, 379 (1948); **27**, 623 (1948).
[14] E. B. Davies, IEEE Trans. Inf. Theory **IT-24**, 596 (1978).

VOLUME 79, NUMBER 6 PHYSICAL REVIEW LETTERS 11 AUGUST 1997

Nonorthogonal Quantum States Maximize Classical Information Capacity

Christopher A. Fuchs
Norman Bridge Laboratory of Physics 12-33, California Institute of Technology, Pasadena, California 91125
(Received 24 March 1997)

I demonstrate that, rather unexpectedly, there exist noisy quantum channels for which the optimal classical information transmission rate is achieved only by signaling alphabets consisting of nonorthogonal quantum states. [S0031-9007(97)03754-X]

PACS numbers: 89.70.+c, 02.50.−r, 03.65.Bz

Within the framework of classical information theory, there is a tacit but basic assumption that a communication channel's possible inputs correspond to a set of mutually exclusive properties for the information carriers. In the brief instant after a signal leaves the sender's hand, but before it enters a noisy channel, an independent observer or wire tap should be able—in principle, at least—to read out the signal with complete reliability. Anything less than complete reliability in this readout represents an extra source of noise over and above that which is supplied by the channel. This is a situation that both the sender and receiver work to avoid.

When quantum systems are used as information carriers, one's natural inclination is that the same basic assumption should hold. For instance, one might think that encoding distinct signals in nonorthogonal quantum states must be less than optimal for information transfer. This is because the readout possibilities for times intermediate to the signal's generation and its entrance into the channel are excluded automatically: it is a matter of physical law that nonorthogonal quantum states cannot be distinguished with perfect reliability [1] and any attempt to do so (even imperfectly) imparts a disturbance to them [2]. These are the principles that encourage the use of nonorthogonal signals for cryptographic purposes [3]; however, just because of this, one would not expect them to play a role in questions to do with reliable, public communication.

In what follows, I present an example that dispels this prejudice: signals encoded in *nonorthogonal* quantum states are sometimes required to achieve the highest information transfer rate that a channel can yield. In particular, I present a noisy quantum mechanical channel for which the channel capacity expression recently derived by Holevo [4] and Schumacher and Westmoreland [5] is only achieved by signals consisting of nonorthogonal states.

In order to state the result, I first review the standard notion of a quantum discrete memoryless channel (QDMC). For such a channel, the information carriers are quantum systems with a finite dimensional Hilbert space $\mathcal{H}_d$, d denotes the dimension. The action of the channel is assumed to be due to interactions between the carrier and an independent environment outside the sender's and receiver's control. Thus, the channel's action on the carrier's quantum state ρ—most generally, a density operator—can be represented as an evolution of the form $\rho \rightarrow \Phi(\rho) = \mathrm{tr}_{\mathrm{E}}(U(\rho \otimes \tau)U^\dagger)$, where τ denotes the standard state of the environment, U is some unitary operator, and tr_{E} denotes a partial trace over the environmental degrees of freedom. A convenient theorem of Kraus [6] is that a mapping Φ holds the form above if and only if it can also be represented as

$$\rho \rightarrow \Phi(\rho) = \sum_i A_i \rho A_i^\dagger \tag{1}$$

for some set of (possibly non-Hermitian) operators A_i satisfying $\sum_i A_i^\dagger A_i = \mathbb{1}$ ($\mathbb{1}$ = the identity operator). The channel is memoryless when the evolution for arbitrary states σ (including entangled ones) on $\mathcal{H}_d^{\otimes n}$ is

$$\Phi^{\otimes n}(\sigma) = \sum_{i_1 \cdots i_n} (A_{i_1} \otimes \cdots \otimes A_{i_n})\sigma(A_{i_1}^\dagger \otimes \cdots \otimes A_{i_n}^\dagger),$$

for each n. That is to say, the noise acts independently on each information carrier sent down the channel.

Let us now consider using a QDMC for the purpose of transmitting classical information. What we imagine here is a sender encoding various messages u, $u = 1, \ldots, \mathcal{M}$, into an equal number of pure state preparations (i.e., one-dimensional projectors) Π_u on $\mathcal{H}_d^{\otimes n}$. Along the way to the intended receiver, the states evolve according to the rule above, generally emerging as mixed states $\rho_u = \Phi^{\otimes n}(\Pi_u)$. Finally, the receiver performs some measurement—mathematically, a positive operator-valued measure (POVM) [6]—$\{E_u\}$, with one outcome for each message u. The game here is that the measurement outcome is used to represent the receiver's best guess of the quantum state ρ_u—and consequently the message u—appearing at the output of the channel.

Note that the formulation so far is completely general in its usage of the QDMC. In particular, the quantum states used to encode the messages may be massively entangled across the n transmissions [7]. Moreover, the POVM $\{E_u\}$ may be a collective quantum measurement over the whole Hilbert space $\mathcal{H}_d^{\otimes n}$, and need not factorize into measurements on the individual carriers [8]. For the considerations here, however, I restrict attention to senders using encodings based on a *finite alphabet*. A sender is said to make use of a finite alphabet when his signals are restricted to be product states on $\mathcal{H}_d^{\otimes n}$, all of which are drawn from some fixed finite set $\mathcal{X} = \{\Pi_\ell\}$, $\ell = 1, \ldots, m$, of pure states on $\mathcal{H}_d$. That is to say, the sender is now imagined to encode messages $u = (\ell_1, \ldots, \ell_n)$ into quantum states of the form $\Pi_u = \Pi_{\ell_1} \otimes \cdots \otimes \Pi_{\ell_n}$. Such an encoding, taken as a whole, is called a *code*.

With this, we can turn to the issue of reliable transmission of information through the channel. A $(\lfloor 2^{nR} \rfloor, n, \lambda_n)$ code, $0 \leq R \leq 1$, is a set of $\lfloor 2^{nR} \rfloor$ code words Π_u (each of length n) such that the maximum probability of error in guessing a message is λ_n, i.e., $\lambda_n = \max_u (1 - \mathrm{tr}\,(\rho_u E_u))$. The number R appearing in this definition is known as the *rate* of information transfer of the code. A rate R is said to be achievable if there exists a sequence of $(\lfloor 2^{nR} \rfloor, n, \lambda_n)$ codes with $\lambda_n \to 0$ as $n \to \infty$. The *capacity* C of the QDMC is the supremum of all achievable rates, where the supremum is taken explicitly over all alphabets used for coding, all codes making use of that alphabet, and all possible POVMs used for decoding at the receiver. Our main concern here is in finding the optimal alphabet for the encoding, the issue being whether the optimal alphabet must consist of orthogonal states or not.

A method of calculating the capacity has been known for some time when the POVM elements E_u are, like the code words in this scenario, restricted to be tensor product operators on $\mathcal{H}_d^{\otimes n}$ [9]. This restriction is equivalent to saying that collective measurements on code words are excluded from the game; each information carrier is measured individually. The restricted capacity C_1 is given by the supremum *accessible information* $I_1(\mathcal{E})$ [1] over all signal ensembles $\mathcal{E} = \{p_i; \Pi_i\}$, $p_i \geq 0$, $\sum_i p_i = 1$; i.e.,

$$C_1 = \sup_{\mathcal{E}} I_1(\mathcal{E}), \tag{2}$$

where

$$I_1(\mathcal{E}) = \sup_{\{E_b\}} [H(\mathrm{tr}(\rho E_b)) - \sum_i p_i H(\mathrm{tr}(\rho_i E_b))], \tag{3}$$

$\rho_i = \Phi(\Pi_i)$ are the output states, $\rho = \sum_i p_i \rho_i$, and $H(\mathrm{tr}(\tau E_b)) = -\sum_b \mathrm{tr}(\tau E_b) \log \mathrm{tr}(\tau E_b)$ is the Shannon entropy for the probability distribution $\mathrm{tr}(\sigma E_b)$ derived from a POVM $\{E_b\}$. (All logarithms throughout are calculated base 2.)

Expression (2) coincides with the standard classical capacity theorem of Shannon [10] for a discrete memoryless channel: it is just that in the quantum case extra care must be taken to optimize both the input alphabet and the output observable—neither is given *a priori*. Note that the supremization in Eq. (3) is over *all* POVMs on $\mathcal{H}_d$: for this expression there is no restriction that the number of POVM elements be the same as the number of states in the alphabet $\mathcal{X}$. However, convexity arguments can be used to show that Eqs. (2) and (3) are achievable by ensembles and POVMs *each* with no more than d^2 elements [11,12].

Recently, an elegant expression for the capacity C has been derived [4,5], which dispenses with an explicit optimization over the receiver's measurement. The theorem is that

$$C = \sup_{\mathcal{E}} I(\mathcal{E}), \tag{4}$$

where

$$I(\mathcal{E}) = S(\rho) - \sum_i p_i S(\rho_i), \tag{5}$$

ρ_i and ρ are defined as above, and $S(\tau) = -\mathrm{tr}(\tau \log \tau)$ is the von Neumann entropy of a density operator τ. Here again, convexity arguments [12,13] give that the supremum can be achieved by signal ensembles consisting of no more than d^2 terms.

It is important to note that, depending upon the channel, C can be strictly greater than C_1. This is a result of the fact that collective measurements generally afford more power for distinguishing product states than do product measurements [8,14,15]. Moreover, this point is doubly significant for the task at hand because collective measurements also appear to be the key for eliciting the optimality of nonorthogonal inputs.

With the Theorems (2) and (4) for the capacities C_1 and C in hand, the last remark can be made precise. The question is this. Do there exist channels for which Eq. (4) is achieved *only* by an ensemble of nonorthogonal states? I will answer this in the affirmative by explicitly constructing an example of a channel on $\mathcal{H}_2$ that requires, at the very least, a nonorthogonal binary alphabet to achieve capacity. That is to say, I shall exhibit a particular Φ and a particular ensemble $\mathcal{E}_p = \{\frac{1}{2}, \frac{1}{2}; \Pi_0, \Pi_1\}$ with $\mathrm{tr}(\Pi_0 \Pi_1) \neq 0$ for which $C \geq I(\mathcal{E}_p) > \sup_{\mathcal{E}_\perp} I(\mathcal{E}_\perp)$. The rightmost supremization in this is taken exclusively over ensembles of orthogonal states.

As stated earlier, this situation is somewhat surprising. Indeed it can be shown for general Φ and $\mathcal{H}_d$ that when the issue is of distinguishing two outputs in an optimal way—rather than optimizing information rate—*and* there are no restrictions on the inputs or the POVMs, then orthogonal inputs are always sufficient [7]. Moreover, when $d = 2$ and the input alphabet is binary, the capacity C_1 is always achievable by an orthogonal alphabet: this will be demonstrated later in the paper. For the present, I turn to the particular example.

The "splaying" channel acting on density operators of $\mathcal{H}_2$ is described simply enough by means of a Kraus representation as in Eq. (1). The A_i used to define it are

$$A_x = \sqrt{\frac{2}{3}}\, |x\rangle\langle x|, \qquad A_y = \sqrt{\frac{2}{3}}\, |y\rangle\langle +|,$$

$$A_{\bar{y}} = \sqrt{\frac{2}{3}}\, |\bar{y}\rangle\langle -|,$$

where, fixing an orthonormal basis $\{|x\rangle, |\bar{x}\rangle\}$ on $\mathcal{H}_2$,

$$|y\rangle = \frac{1}{\sqrt{2}}(|x\rangle + |\bar{x}\rangle), \qquad |\bar{y}\rangle = \frac{1}{\sqrt{2}}(|x\rangle - |\bar{x}\rangle), \tag{6}$$

$$|+\rangle = \frac{1}{2}|x\rangle + \frac{\sqrt{3}}{2}|\bar{x}\rangle, \qquad |-\rangle = \frac{1}{2}|x\rangle - \frac{\sqrt{3}}{2}|\bar{x}\rangle. \tag{7}$$

VOLUME 79, NUMBER 6 PHYSICAL REVIEW LETTERS 11 AUGUST 1997

The action of this channel can be thought of in more graphic terms as follows. Let us make a switch to Bloch-sphere notation for all operators. The channel, personified as Eve, begins by performing the symmetric three-outcome "trine" POVM as the quantum states make their way from sender to receiver. That is, the positive operators in her POVM are given by

$$E_i = \frac{1}{3}(\mathbb{1} + \boldsymbol{n}_i \cdot \boldsymbol{\sigma}), \tag{8}$$

where $\boldsymbol{\sigma}$ is the vector of Pauli matrices, $\boldsymbol{n}_x = (1, 0, 0)$, and $\boldsymbol{n}_\pm = (-1/2, \pm\sqrt{3}/2, 0)$. The three vectors here are 120° apart and confined to the x-y plane; as must be the case for all POVMs, $E_x + E_+ + E_- = I$. Upon receiving outcome i, Eve forwards a quantum state η_i to Bob according to the rule

$$\eta_x = \frac{1}{2}(\mathbb{1} + \boldsymbol{x} \cdot \boldsymbol{\sigma}) \quad \text{and} \quad \eta_\pm = \frac{1}{2}(\mathbb{1} \pm \boldsymbol{y} \cdot \boldsymbol{\sigma}), \tag{9}$$

where $\boldsymbol{x} = (1, 0, 0)$ and $\boldsymbol{y} = (0, 1, 0)$. The key idea is that if E_x is detected, the state corresponding to the outcome is forwarded to the receiver; however, if E_+ or E_- are detected, orthogonal or "splayed" versions of the outcomes are sent.

If the sender transmits a general pure state

$$\Pi_{\alpha\beta} = \frac{1}{2}(\mathbb{1} + \boldsymbol{s}_{\alpha\beta} \cdot \boldsymbol{\sigma}), \tag{10}$$

where $\boldsymbol{s}_{\alpha\beta} = (\cos\alpha \sin\beta, \sin\alpha \sin\beta, \cos\beta)$, for $\alpha \in [0, 2\pi)$ and $\beta \in [0, \pi)$, the upshot of Eve's interference—as far as the sender and receiver are concerned—is the evolution $\Pi_{\alpha\beta} \to \Phi(\Pi_{\alpha\beta})$ where

$$\Phi(\Pi_{\alpha\beta}) = \sum_i \mathrm{tr}(\Pi_{\alpha\beta} E_i)\eta_i = \frac{1}{2}(\mathbb{1} + \boldsymbol{t}_{\alpha\beta} \cdot \boldsymbol{\sigma}) \tag{11}$$

and $\boldsymbol{t}_{\alpha\beta} = \frac{1}{3}(1 + \cos\alpha \sin\beta, \sqrt{3}\sin\alpha \sin\beta, 0)$. This follows since $\mathrm{tr}(\Pi_{\alpha\beta} E_x) = (1 + \cos\alpha \sin\beta)/3$ and also $\mathrm{tr}(\Pi_{\alpha\beta} E_\pm) = (2 - \cos\alpha \sin\beta \pm \sqrt{3}\sin\alpha \sin\beta)/6$.

With Eq. (11), one can readily calculate Eq. (5) for an arbitrary ensemble of *orthogonal* input states. Suppose the state in Eq. (10) and one orthogonal to it (i.e., with Bloch vector $-\boldsymbol{s}_{\alpha\beta}$) are sent through the channel with prior probabilities t and $1 - t$, respectively. Calling the result of Eq. (5) $I(\alpha, \beta, t)$, this gives

$$\begin{aligned} I(\alpha, \beta, t) = {} & \phi\{[1 + (2t - 1)\cos\alpha \sin\beta]^2 + 3[(2t - 1)\sin\alpha \sin\beta]^2\} - t\phi[(1 + \cos\alpha \sin\beta)^2 + 3(\sin\alpha \sin\beta)^2] \\ & - (1 - t)\phi[(1 - \cos\alpha \sin\beta)^2 + 3(\sin\alpha \sin\beta)^2], \end{aligned} \tag{12}$$

where $\phi(x) = -h(\sqrt{x}/3)$ and

$$2h(z) = (1 + z)\log(1 + z) + (1 - z)\log(1 - z). \tag{13}$$

One can easily check that Eq. (12) is maximized when $\alpha = \beta = \pi/2$ and $t = 1/2$, yielding a value of

$$C_{\mathrm{ortho}} = \frac{1}{6}\log\left(\frac{3125}{1024}\right) \approx 0.268\,273 \text{ bits}. \tag{14}$$

Now consider the following ensemble of inputs. Let Π_α be a state given by Eq. (10) but with $\beta = \pi/2$, and let $\overline{\Pi}_\alpha = \Pi_{-\alpha}$. Assume each of these occurs with prior probability 1/2. Thus, the two signaling states in this ensemble are (generally) nonorthogonal, but restricted to the plane of the POVM elements and reflecting their symmetry. Again, one readily calculates Eq. (5) to get

$$\begin{aligned} I(\alpha) = {} & \phi((1 + \cos\alpha)^2) \\ & - \phi((1 + \cos\alpha)^2 + 3\sin^2\alpha). \end{aligned}$$

The analytic maximization of this quantity depends upon the solution of a transcendental equation. Therefore, the maximization requires some numerical work: it turns out to be attained when $\alpha = 1.521\,808 \neq \pi/2$, roughly 87.2°. The value of the maximum is

$$C_{\mathrm{nono}} \approx 0.268\,932 \text{ bits}. \tag{15}$$

This completes the demonstration that a QDMC's classical information capacity need not be achievable by orthogonal states. The difference in this particular example is not large, but it is enough to prove the principle.

Heuristically, what is going on with the splaying channel is that, from a "God's eye view," the output η_x acts like an erasure flag, signifying the disappearance of a bit. As the angle α is reduced, the probability of a flagged erasure increases, and so the information rate decreases. As α is made larger, the transmission probability for distinguishable bits (i.e., η_+ and η_-) increases, but there is an accompanying increased probability that a bit will have flipped. The angle α in Eq. (15) represents the optimal tradeoff between these tensions, as quantified by the capacity formula for C in Eq. (4)—in other words, when the full power of collective quantum measurements is made available at the receiver.

The last point appears to be crucial for understanding the origin of this effect. When each qubit is measured individually, the optimal tradeoff between the tensions is quantified by the capacity C_1 given in Eq. (2). In that case, it should be noted that the erasure flag's contribution to the tensions effectively disappears; with respect to individual measurements, the erasure flag always manifests itself as a probability for a bit flip error. This is seen easily with an example. If the ensemble $\{\Pi_\alpha, \overline{\Pi}_\alpha\}$ (equal prior probabilities) is used, but no collective measurements, then it turns out that there is enough symmetry in the problem that Eq. (3) can be calculated explicitly. When two equiprobable states with equal-length Bloch vectors $\boldsymbol{a}$ and $\boldsymbol{b}$ are to be distinguished, the optimal measurement is specified by the unit vectors parallel and antiparallel to $\boldsymbol{d} = \boldsymbol{a} - \boldsymbol{b}$, and the accessible information is given by $I_1(\boldsymbol{a},\boldsymbol{b}) = -\phi(9(\boldsymbol{a}\cdot\boldsymbol{d})/2)$ [1,16]. For the case at hand, we obtain $I_1(\alpha) = \phi(3\sin^2\alpha)$, which is achieved by a measurement basis consisting of the projectors η_+ and η_-. As far as this measurement is concerned, an erasure-flag output has equal probabilities for leading to correct and incorrect identifications by the receiver. In particular, $I_1(\alpha)$ has a maximum of 0.255 992 bits at $\alpha = \pi/2$, i.e., for an ensemble of orthogonal input states.

Indeed, it is a generic property of channels on $\mathcal{H}_2$ with *binary* input alphabets that the maximum achievable rate with respect to individual measurements can be attained by an orthogonal alphabet. Furthermore, since a standard orthogonal projection-valued measurement always suffices for achieving capacity here [2,16], this remains true even without optimizing the ensemble prior probabilities or the measurement observable. This fact arises in the following manner.

Suppose a fixed measurement is given by the Bloch vectors $\boldsymbol{n}$ and $-\boldsymbol{n}$, and the binary signal alphabet (yet to be optimized) is associated with fixed prior probabilities $1 - t$ and t ($t \neq 0, 1$) for its letters. Let $\boldsymbol{a}$ and $\boldsymbol{b}$ denote the respective Bloch vectors associated with the signal alphabet, and let $\boldsymbol{c} = (1 - t)\boldsymbol{a} + t\boldsymbol{b}$. The effect of the channel on these Bloch vectors is to transform them according to some affine transformation [12]: $\boldsymbol{a} \rightarrow \boldsymbol{a}' = M\boldsymbol{a} + \boldsymbol{e}$, $\boldsymbol{b} \rightarrow \boldsymbol{b}' = M\boldsymbol{b} + \boldsymbol{e}$, etc., where M is a real 3×3 matrix and $\boldsymbol{e}$ is a fixed vector within the Bloch sphere. With these notations, the mutual information J between input and output for this ensemble is

$$\begin{aligned} J &= -h(\boldsymbol{c}'\cdot\boldsymbol{n}) + (1-t)h(\boldsymbol{a}'\cdot\boldsymbol{n}) + th(\boldsymbol{b}'\cdot\boldsymbol{n}) \\ &= -h(\boldsymbol{c}\cdot\tilde{\boldsymbol{n}} + w) + (1-t)h(\boldsymbol{a}\cdot\tilde{\boldsymbol{n}} + w) \\ &\quad + th(\boldsymbol{b}\cdot\tilde{\boldsymbol{n}} + w), \end{aligned}$$

where $\tilde{\boldsymbol{n}} = M^{\mathrm{T}}\boldsymbol{n}$ and $w = \boldsymbol{e}\cdot\boldsymbol{n}$. A necessary condition on any candidates $\boldsymbol{a}$ and $\boldsymbol{b}$ for optimizing this mutual information is that it be invariant to first order with respect to small variations about these vectors. Taking into account the constraint that the inputs be pure states, this leads to the following two variational equations:

$$\mathbf{0} = \log\left[\frac{(1 - w - \boldsymbol{c}\cdot\tilde{\boldsymbol{n}})(1 + w + \boldsymbol{a}\cdot\tilde{\boldsymbol{n}})}{(1 + w + \boldsymbol{c}\cdot\tilde{\boldsymbol{n}})(1 - w - \boldsymbol{a}\cdot\tilde{\boldsymbol{n}})}\right]\boldsymbol{n}_a, \quad (16)$$

$$\mathbf{0} = \log\left[\frac{(1 - w - \boldsymbol{c}\cdot\tilde{\boldsymbol{n}})(1 + w + \boldsymbol{b}\cdot\tilde{\boldsymbol{n}})}{(1 + w + \boldsymbol{c}\cdot\tilde{\boldsymbol{n}})(1 - w - \boldsymbol{b}\cdot\tilde{\boldsymbol{n}})}\right]\boldsymbol{n}_b, \quad (17)$$

where $\mathbf{0}$ is the zero vector, $\boldsymbol{n}_a = \tilde{\boldsymbol{n}} - (\tilde{\boldsymbol{n}}\cdot\boldsymbol{a})\boldsymbol{a}$, and $\boldsymbol{n}_b = \tilde{\boldsymbol{n}} - (\tilde{\boldsymbol{n}}\cdot\boldsymbol{b})\boldsymbol{b}$. It is easy to check that the only solutions to these occur when either $\boldsymbol{a} = \boldsymbol{b}$, $\tilde{\boldsymbol{n}}\cdot\boldsymbol{a} = \tilde{\boldsymbol{n}}\cdot\boldsymbol{b}$, or $\boldsymbol{a} = -\boldsymbol{b}$. In the first two cases the mutual information vanishes; in the last case, it is maximal. This proves the point.

In summary, I have shown that, contrary to some intuition, there exist noisy quantum channels for which nonorthogonal input states lead to the largest reliable information transfer rate. In the particular example here, and indeed for all possible channels on $\mathcal{H}_2$, collective measurements appear to play a crucial role in bringing about the effect *whenever it exists.* However, it remains an open question whether *collective measurements following product-state inputs* is the one and only ingredient required for bringing about the optimality of nonorthogonal inputs: for instance, it is not known whether there exists a channel on $\mathcal{H}_d$, $d \geq 3$, for which the capacity C_1 is only attained for a nonorthogonal input alphabet.

The particular example exhibited here was somewhat contrived, being built explicitly to show the desired effect. However, since the completion of this work, several "real world" channels have been discovered (through numerical simulation) to require nonorthogonal inputs to achieve capacity. In fact, the effect appears to be generic for channels of a certain dissipative character—the standard amplitude damping channel being one such example. An extended discussion of these channels will appear elsewhere [17].

I thank H. Barnum, C. Bennett, H. Mabuchi, P. Shor, J. Smolin, and A. Uhlmann for helpful discussions. This work was supported by a Lee A. DuBridge Fellowship and by DARPA through the Quantum Information and Computing (QUIC) Institute administered by ARO.

[1] C. A. Fuchs, Ph.D. thesis, University of New Mexico, 1996. LANL archive Report No. quant-ph/9601020.
[2] C. A. Fuchs and A. Peres, Phys. Rev. A **53**, 2038 (1996).
[3] C. H. Bennett, G. Brassard, and N. D. Mermin, Phys. Rev. Lett. **68**, 557 (1992).
[4] A. S. Holevo, LANL archive Report No. quant-ph/9611023.
[5] B. Schumacher and M. D. Westmoreland, Phys. Rev. A **56**, 131 (1997).
[6] K. Kraus, *States, Effects, and Operations: Fundamental Notions of Quantum Theory* (Springer, Berlin, 1983).
[7] C. H. Bennett, C. A. Fuchs, and J. A. Smolin, in *Quantum Communication, Computing and Measurement,* edited by O. Hirota, A. S. Holevo, and C. M. Caves (Plenum, New York, 1997).
[8] A. S. Kholevo, Probl. Inf. Transm. **15**, 247 (1979).
[9] A. S. Kholevo, Probl. Inf. Transm. **9**, 177 (1973).
[10] C. E. Shannon, Bell Syst. Tech. J. **27**, 379 (1948); **27**, 623 (1948).
[11] E. B. Davies, IEEE Trans. Inf. Theory **IT-24**, 569 (1978).
[12] A. Fujiwara and H. Nagaoka (to be published).
[13] A. Uhlmann, LANL archive Report No. quant-ph/9701014.
[14] A. Peres and W. K. Wootters, Phys. Rev. Lett. **66**, 1119 (1991).
[15] P. Hausladen *et al.,* Phys. Rev. A **54**, 1869 (1996).
[16] L. B. Levitin, in *Workshop on Physics and Computation: PhysComp '92* (IEEE Computer Society Press, Los Alamitos, CA, 1993).
[17] C. A. Fuchs, P. W. Shor, J. A. Smolin, and B. Terhal (to be published).

Entanglement Purification and Long-Distance Quantum Communication

Long-distance quantum communication

Hans J. Briegel
Ludwig-Maximilians University of Munich

Quantum communication exploits the quantum properties of its information carriers for communication purposes such as the distribution of secure cryptographic keys in quantum cryptography [1] (see the contribution of Gisin to this volume) and the communication between distant quantum computers in a network [2]. A central problem of quantum communication is how to faithfully transmit unknown quantum states through a noisy quantum channel [3]. While information is sent through such a channel (for example an optical fiber), the carriers of the information interact with the channel, which gives rise to the phenomenon of decoherence; an initially pure quantum state becomes a mixed state when it leaves the channel. For quantum communication purposes, it is however necessary that the transmitted qubits retain their genuine quantum properties, for example in form of an entanglement with qubits on the other side of the channel.

There are two well-established methods to deal with the problem of noisy channels (see the contributions of DiVincenzo and of Fuchs to this volume). The theory of quantum error correction has mainly been developed to make quantum computation possible despite the effects of decoherence and imperfect apparatus. Since data transmission – like data storage – can be regarded as a special case of a computational process, clearly quantum error correction can also be used for quantum communication through noisy channels. An alternative approach, which has been developed roughly in parallel with the theory of quantum error correction, is the purification of mixed entangled states [4, 5, 6, 7]. This approach allows for the creation of maximally entangled states of particles at different locations, even if the channel that connects those locations is noisy. The entangled particles can then be used for faithful teleportation [4, 9] or for secure key distribution in entanglement-based quantum cryptography [6, 8]. Entanglement purification is a specific and efficient tool for quantum communication; by exploiting classical communication between the parties, it allows highly efficient two-way protocols which cannot be realized with quantum error correction procedures. A detailed quantitative connection between the efficiency of quantum error correcting codes and of entanglement purification protocols has been established by Bennett *et al.* in Ref. [5].

Long-distance quantum communication describes a situation where the length of the channel connecting the parties is typically much longer than its coherence length. In such a situation, the accumulation of errors during the process of repeated quantum error correction becomes a serious problem, if the fidelites of the gates and measurements used in the process are less than 100%. A general solution for this problem is provided by the theory of fault-tolerant quantum computing [10], which implies that any computation can be performed with a certain overhead in memory space and time, if the error probability for each gate operation can be made sufficiently small. Insofar as transmission over arbitrary long distances corresponds to a computation of arbitrary length in quantum computing, this implies that quantum communication can be achieved over arbitrary distances with a similar overhead [11]. An explicit scheme for data transmission and storage has been discussed by Knill and Laflamme [12], using the method of concatenated quantum coding. Their method requires to encode each qubit into an entangled state of a certain number of qubits that scales polynomially with the length of the channel, and to apply error correction on this state repeatedly during the transmission process. Although the theory of fault-tolerant quantum computation solves, in principle, the problem of quantum data transmission over noisy channels, its requirements for the precision with which the operations need to be carried out, are extremely high [10, 12].

It seems therefore natural to ask, how the more efficient methods of entanglement purification may be utilized for long-distance communication. There are, in fact, two different questions:

1. What is, in analogy with fault-tolerant quantum error correction, the role of imperfect apparatus in entanglement purification protocols? And how tolerant are these protocols with respect to noise introduced by local operations such as quantum gates and measurements that are used by the protocols themselves?

2. Can we design a quantum repeater based on entanglement purification that exploits the higher efficiency of these protocols?

These problems are treated in Refs. [13, 14, 15].

In the lectures, we will first study the applicability and the efficiency of entanglement purification protocols in the situation of imperfect local operations. The general conditions under which standard purification protocols can be used in the presence of errors have been studied by Giedke *et al.* [15]. This includes, in particular, thresholds and lower bounds for the attainable fidelities. For a generic class of stochastic errors, one can give explicit representations of the imperfect operations in terms of completely positive maps and/or POVMs [13]. In terms of these maps, we will derive recursion formulas for the fidelities, which generalize some of the formulas given in Refs. [4] and [6]. From these results, we estimate that, for a generic class of errors, standard entanglement purifcation protocols work even for error probabilities of the order of a few per cent. Equipped with these results, we will then be ready to treat the problem of long distance communication. We will develop a purification procedure that merges the ideas of entanglement purification and entanglement swapping [9, 16] into a single (meta-)protocol. This procedure, which we call *nested entanglement purification*, is the analog of *concatenated quantum coding* in the theory of quantum error correction. It allows quantum communication via noisy channels of arbitrary length. Since it explicitly exploits two-way classical communication [5], this procedure turns out to be much more efficient for quantum communication than concatenated quantum coding [12]. Specifically, we will find that a quantum repeater based on entanglement purification tolerates errors on the per-cent level; it requires a polynomial overhead in time and an overhead in local resources that grows only logarithmically with the length of the channel.

We will briefly discuss a possible quantum optical implementation of such a repeater using photonic channels [17] (see also the contribution of Mabuchi to this volume). Finally, we will discuss some implications of these results for the problem of secure quantum key distribution over long distances [18].

The selected papers reprinted in this volume are those of Refs. [13, 4, 6, 17]. The scheme of a quantum repeater based on entanglement purification is introduced in [13]. A central ingredient of this work is the consideration of imperfect local operations for quantum communication. Since the paper uses some of the results of earlier work by Bennett *et al.* and Deutsch *et al.* in Refs. [4] and [6], respectively, those papers are reprinted together with [13] to underline this connection. Specifically, the influence of imperfect apparatus on entanglement purification has been studied with the protocols of [4, 6], as an example. The paper by van Enk *et al.* [17] discusses a specific physical system in which most of these ideas could be implemented.

Part of this research is supported through a grant of the Schwerpunktsprogramm "Quanten-Informationsverarbeitung" der Deutschen Forschungsgemeinschaft and by the European Community under the TMR network ERB-FMRX-CT96-0087.

References

[1] C. H. Bennett, G. Brassard, and A. K. Ekert, *Scientific American*, Oct. 1992, p. 50.

[2] J. I. Cirac, P. Zoller, H. Mabuchi, and H. J. Kimble, *Phys. Rev. Lett.* **78**, 3221 (1997).

[3] B. Schumacher, *Phys. Rev. A* **54**, 2614 (1996).

[4] C. H. Bennett, G. Brassard, S. Popescu, B. Schumacher, J. A. Smolin, and W. K. Wootters, *Phys. Rev. Lett.* **76**, 722 (1996).

[5] C. H. Bennett, D. P. DiVincenzo, J. A. Smolin, and W. K. Wootters, *Phys. Rev. A* **54** 3825 (1996).

[6] D. Deutsch, A. Ekert, R. Josza, C. Macchiavello, S. Popescu, and A. Sanpera, *Phys. Rev. Lett.* **77**, 2818 (1996); see also C. Macchiavello, *Phys. Lett. A* **246**, 385 (1998).

[7] N. Gisin, *Phys. Lett. A***210**, 151 (1996).

[8] A. Ekert, *Phys. Rev. Lett.* **67**, 661 (1991).

[9] C. H. Bennett, G. Brassard, G. Crepeau, R. Josza, A. Peres, and W. Wootters, *Phys. Rev. Lett.* **70**, 1895 (1993).

[10] J. Preskill, *Proc. Roy. Soc. Lond. A* **454**, 385 (1998); e-print quant-ph/9705031.

[11] In the language of computer science, the *asymptotic complexity* of the transmission process must be the same as for computations.

[12] E. Knill and R. Laflamme, e-print quant-ph/9608012.

[13] H.-J. Briegel, W. Dür, J. I. Cirac, and P. Zoller, *Phys. Rev. Lett.* **81**, 5932 (1998).

[14] W. Dür, H.-J. Briegel, J. I. Cirac, and P. Zoller, *Phys. Rev. A* **59**, 169 (1999), *Phys. Rev. A* **60**, 725 (1999).

[15] G. Giedke, H.-J. Briegel, J. I. Cirac, and P. Zoller, *Phys. Rev. A* **59**, 2641 (1999).

[16] M. Zukowski, A. Zeilinger, M. A. Horne, and A. Ekert, *Phys. Rev. Lett.* **71**, 4287 (1993). The first experimental demonstration of entanglement swapping is reported in J. W. Pan, D. Bouwmeester, H. Weinfurter, and A. Zeilinger, *ibid.* **80**, 3891 (1998).

[17] S. J. van Enk, J. I. Cirac, and P. Zoller, *Science* **279**, 205 (1998).

[18] D. Mayers and A. C. C. Yao, e-print quant-ph/9802025; H.-K. Lo and H. F. Chau, e-print quant-ph/9803006.

VOLUME 81, NUMBER 26 PHYSICAL REVIEW LETTERS 28 DECEMBER 1998

Quantum Repeaters: The Role of Imperfect Local Operations in Quantum Communication

H.-J. Briegel,[1,2,*] W. Dür,[1] J. I. Cirac,[1,2] and P. Zoller[1]
[1]*Institut für Theoretische Physik, Universität Innsbruck, Technikerstrasse 25, A-6020 Innsbruck, Austria*
[2]*Departamento de Fisica, Universidad de Castilla-La Mancha, 13071 Ciudad Real, Spain*
(Received 20 March 1998)

In quantum communication via noisy channels, the error probability scales exponentially with the length of the channel. We present a scheme of a quantum repeater that overcomes this limitation. The central idea is to connect a string of (imperfect) entangled pairs of particles by using a novel nested purification protocol, thereby creating a single distant pair of high fidelity. Our scheme tolerates general errors on the percent level, it works with a polynomial overhead in time and a logarithmic overhead in the number of particles that need to be controlled locally. [S0031-9007(98)08063-6]

PACS numbers: 03.67.Hk, 03.65.Bz, 42.50.–p

Quantum communication deals with the transmission and exchange of quantum information between distant nodes of a network. Remarkable experimental progress has been reported recently, for example, on secret key distribution for quantum cryptography [1,2], teleportation of the polarization state of a single photon [3,4], and the creation of entanglement between different atoms [5]. On the other hand, first steps towards the implementation of quantum logical operations, which are the building blocks of quantum computing, have been demonstrated [6]. In view of this progress, it is not farfetched to expect the creation of small quantum networks in the near future. Such networks will involve nodes, where qubits are stored and locally manipulated, and which are connected by quantum channels over which communication takes place by sending qubits. This will open the possibility for more complex activities such as multiparty communication and distributed quantum computing [7].

The bottleneck for communication between distant nodes is the scaling of the error probability with the length of the channel connecting the nodes. For channels such as an optical fiber, the probability for both absorption and depolarization of a photon (i.e., the qubit) grows exponentially with the length l of the fiber. This has two effects: (i) to transmit a photon without absorption, the number of trials scales exponentially with l; (ii) even when a photon arrives, the fidelity of the transmitted state decreases exponentially with l. One may think that this last problem can be circumvented by standard purification schemes [8–10]. However, purification schemes require a certain minimum fidelity $F_{\min}$ to operate, which cannot be achieved as l increases. Furthermore, in any realistic situation, the operations that are part of the purification protocol are themselves imperfect, and this defines a maximum attainable fidelity $F_{\max}$ and limits the efficiency of the scheme. For this reason, it is not obvious, first, what the allowed error tolerances of local operations are for entanglement purification to be applicable at all and, second, how the resources that are needed for purification grow with the length of the channel. In the experiments, the distance between the nodes is presently limited by (a few times) the absorption length of the fiber [11].

The theory of fault-tolerant quantum computing [12] implies that any computation can be performed with polylogarithmic cost in time and space [13], if the error probability for each gate operation can be made sufficiently small. A special case of a computation is the transmission of information, for which these fault-tolerant methods must therefore have the same (or a better) asymptotic complexity. An explicit scheme for quantum transmission has been discussed by Knill and Laflamme, using concatenated quantum codes [14]. Their method requires one to encode a single qubit into an entangled state of a polynomially large number of qubits, and to operate on this code repeatedly during the transmission process. The tolerable error probabilities for transmission are less than 10^{-2}, whereas for local operations they are less than 5×10^{-5}. This seems to be outside the range of any practical implementation in the near future. A crucial figure for any experiment will be the number of particles that can be manipulated locally in a coherent fashion, together with the precision with which such local manipulations can be realized.

In this Letter, we present a model of a *quantum repeater* that allows the creation of an entangled (EPR) pair of particles over arbitrary large distances with a tolerability of errors in the percent region. Once an EPR pair is created, it can be employed to teleport any quantum information [15,16]. Our solution of this problem comprises three novel elements: (i) entanglement *purification with imperfect means,* including analytic results for the range and the working conditions of standard protocols; (ii) a method for creation of entanglement between particles at distant nodes that uses auxiliary particles at intermediate "connection points" and a *nested purification protocol;* (iii) a scheme for which the time needed for entanglement creation scales polynomially whereas the required material resources per connection point grow only *logarithmically* with the distance. Since our model is based on two-way classical communication, it is qualitatively different from

 0031-9007/98/81(26)/5932(4)$15.00

quantum error correction. By exploiting this property we will obtain a higher efficiency and significantly more favorable error tolerances.

In classical communication, the problem of exponential attenuation can be overcome by using repeaters at certain points in the channel, which amplify the signal and restore it to its original shape. Guided by these ideas, for quantum communication, we divide the channel into N segments with connection points (i.e., auxiliary nodes) in between. We then create N elementary EPR pairs of fidelity F_1 between the nodes A and C_1, C_1 and $C_2, \ldots, C_{N-1}$ and B, as in Fig. 1(a). The number N is chosen such that $F_{\min} < F_1 \lesssim F_{\max}$. Subsequently, we connect these pairs by making Bell measurements at the nodes C_i and classically communicating the results between the nodes as in the schemes for teleportation [15] and entanglement swapping [15,17]. Unfortunately, with every connection, the fidelity F' of the resulting pair will decrease: on the one hand, the connection process involves imperfect operations which introduce noise; on the other hand, even for perfect connections, the fidelity decreases. Both effects lead to an exponential decrease of the fidelity F_N with N of the final pair shared between A and B. Eventually, the value of F_N drops below $F_{\min}$, and therefore it will not be possible to increase the fidelity by purification. The number of pairs $L \ll N$ that may be connected by this method seems therefore to be restricted by the condition $F_L > F_{\min}$.

Our proposal, the *nested purification protocol,* combines the methods of entanglement swapping and purification into a single (meta) protocol that circumvents this restriction. For simplicity, assume that $N = L^n$ for some integer n. On the first level, we simultaneously connect the pairs (initial fidelity F_1) at all of the checkpoints except at $C_L, C_{2L}, \ldots, C_{N-L}$. As a result, we have N/L pairs of length L and fidelity F_L between A and C_L, C_L and C_{2L}, and so on. To purify these pairs, we need a certain number M of copies that we construct in parallel fashion. We then use these copies on the segments A and C_L, C_L and C_{2L}, etc., to purify and obtain one pair of fidelity $\geq F_1$ on each segment. This last condition determines the (average) number of copies M that we need, which will depend

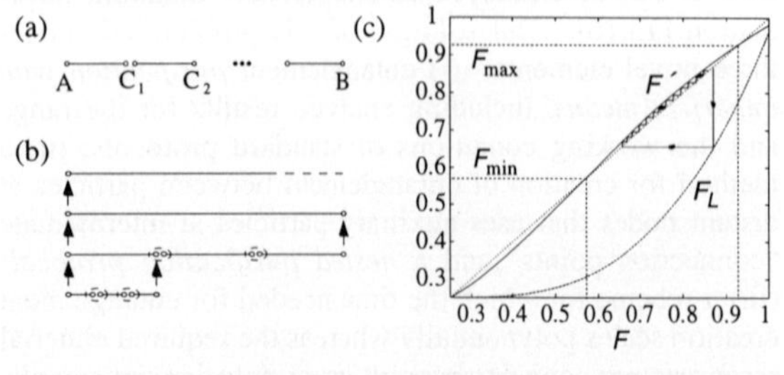

FIG. 1. (a) Connection of a sequence of N EPR pairs; (b) nested purification with repeated creation of auxiliary pairs; (c) "purification loop" for connecting and purifying EPR pairs. Parameters are $L = 3$, $\eta = p_1 = 1$, and $p_2 = 0.97$.

on the initial fidelity, the degradation of the fidelity under connections, and the efficiency of the purification protocol. The total number of elementary pairs involved in constructing one of the more distant pairs of length L is LM. On the second level, we connect L of these more distant pairs at every checkpoint C_{kL} ($k = 1, 2, \ldots$) except at $C_{L^2}, C_{2L^2}, \ldots, C_{N-L^2}$. As a result, we have N/L^2 pairs of length L^2 between A and C_{L^2}, C_{L^2} and C_{2L^2}, and so on, of fidelity $\geq F_L$. Again, we need M parallel copies of these long pairs to repurify up to the fidelity $\geq F_1$. The total number of elementary pairs involved in constructing one pair of length L^2 is thus $(LM)^2$. We iterate the procedure to higher and higher levels, until we reach the nth level. As a result, we have obtained a final pair between A and B of length N and fidelity $\geq F_1$. In this way, the total number R of elementary pairs will be $(LM)^n$. We can reexpress this result in the form

$$R = N^{\log_L M + 1}, \tag{1}$$

which shows that the resources grow polynomially with the distance N. A similar formula was obtained in [14] for the overhead required in propagating the concatenated quantum code. Note that R depends only on L and M. In order to evaluate M, we need to know the specific form of the error mechanisms involved in the purification and connections, which in turn depend on the specific physical implementation of the quantum network. In general, we have only limited knowledge of these details. In order to estimate M, we will choose a generic error model for imperfect operations and measurements.

We define *imperfect operations* on states of one or more qubits by the following maps:

$$\rho \to O_1 \rho = p_1 O_1^{\text{ideal}} \rho + \frac{1 - p_1}{2} \operatorname{tr}_1\{\rho\} \otimes I_1 , \tag{2}$$

$$\rho \to O_{12} \rho = p_2 O_{12}^{\text{ideal}} \rho + \frac{1 - p_2}{4} \operatorname{tr}_{12}\{\rho\} \otimes I_{12} , \tag{3}$$

the first of which describes an imperfect one-qubit operation on particle 1, and the second an imperfect two-qubit operation on particles 1 and 2. In these expressions, O^{ideal} is the *ideal* operation, and I_1 and I_{12} denote unit operators on the subspace where the ideal operation acts. The quantities p_1 and p_2 measure the *reliability* of the operations. The expressions (2) and (3) describe a situation where we have no knowledge about the result of an error occurring during some operation ("depolarization"), except that it happens with a certain probability $(1 - p_j)$. Any sequence of two one-qubit operations on the same qubit is equivalent to a single one-qubit operation, and is therefore described by a single parameter p_1. Similarly, a sequence of a two- and a one-qubit operation counts as a single two-qubit operation and is thus described by p_2. An *imperfect measurement* on a single qubit in the computational basis is described by a POVM (positive-operator-valued measure) corresponding to

$$\begin{aligned} P_0^\eta &= \eta |0\rangle\langle 0| + (1 - \eta) |1\rangle\langle 1| , \\ P_1^\eta &= \eta |1\rangle\langle 1| + (1 - \eta) |0\rangle\langle 0| . \end{aligned} \tag{4}$$

VOLUME 81, NUMBER 26 PHYSICAL REVIEW LETTERS 28 DECEMBER 1998

The parameter η is a measure for the quality of the projection onto the basis states. For example, for the state $\rho = |0\rangle\langle 0|$ the measuring apparatus will give the wrong result ("1") with probability $1 - \eta \geq 0$. A detailed discussion of this and more general models for imperfect operations will be given elsewhere [18]. With these error models, we have a toolbox to analyze all of the processes involved in the connection and purification procedures. For example, the Bell measurement required in the connection can be decomposed into a controlled-NOT (CNOT) operation, effecting, e.g., $|0\rangle|0\rangle \pm |1\rangle|1\rangle \to (|0\rangle \pm |1\rangle)|0\rangle$, followed by two single-qubit measurements.

The basic elements of the nested purification protocol are (i) pair connections and (ii) purification. In the following we analyze these elements using the error models introduced above. Assume now that all of the pairs in Fig. 1(a) are in Werner states (see [8]). These states can be produced using depolarization (as in Ref. [8]) after each connection and purification process. This depolarization works even in the presence of errors which we take into account. Connecting L neighboring pairs as explained earlier, one obtains a new "L pair" with fidelity

$$F_L = \frac{1}{4} + \frac{3}{4}\left(\frac{p_1^2 p_2(4\eta^2 - 1)}{3}\right)^{L-1}\left(\frac{4F - 1}{3}\right)^L. \quad (5)$$

This formula describes an exponential decrease of the resulting fidelity, unless both the elementary pairs and all of the operations involved in the connection process are perfect. There are several possibilities to do the purification, and we first analyze the scheme introduced by Bennett *et al.* [8] in the case of imperfect gate and measurement operations. In short, the scheme takes two adjacent L pairs of fidelity F, performs local (1-bit and 2-bit) operations on the particles at the same ends of the pairs, and obtains with a certain probability p_{succ} a new pair of fidelity

$$F' = \frac{[F^2 + (\frac{1-F}{3})^2][\eta^2 + (1-\eta)^2] + [F(\frac{1-F}{3}) + (\frac{1-F}{3})^2][2\eta(1-\eta)] + (\frac{1-p_2^2}{8p_2^2})}{[F^2 + \frac{2}{3}F(1-F) + \frac{5}{9}(1-F)^2][\eta^2 + (1-\eta)^2] + [F(\frac{1-F}{3}) + (\frac{1-F}{3})^2][8\eta(1-\eta)] + 4(\frac{1-p_2^2}{8p_2^2})}. \quad (6)$$

The value of p_{succ} is given by p_2^2 times the denominator of this expression. For perfect operations, $\eta = 1$ and $p_2 = 1$, (6) reduces to the formula given in Ref. [8].

Figure 1(c) shows the curves for connection (5) and purification (6) for a certain set of parameters. The purification curve has three intersection points with the diagonal, which are the real fixpoints of the map (6). In addition to the trivial point at $F = 1/4$, there are two nontrivial fixpoints. The upper point, $F_{\max} < 1$, is an attractor and gives the maximum value of the fidelity beyond which no pair can be purified. Note also the existence of the minimum value $F_{\min} > 1/2$. Together, they define the interval within which purification is possible. The limiting situation $F_{\max} = F_{\min}$ defines the threshold for the applicability of the purification protocol. For all pairs (p_2, η) for which there is only one real fixpoint (at $F = 1/4$), the imperfections of the local operations introduce more noise than one gains from the purification, so the scheme breaks down. For example, for $\eta = 1$ the threshold is at $p_2 \simeq 0.95$; that is, the CNOT gate must work with a reliability of 95%, at least. Please note that the fixpoints and the threshold condition can all be given analytically from (6). The connection curve, which looks like a simple power in Fig. 1(c), stays below the diagonal for all values of F between 1/4 and 1. The offset of this curve at $F = 1$ from the ideal value $F' = 1$ quantifies the amount of noise that is introduced through imperfect operations in the connection process.

With the above results, we can now analyze the nested purification protocol. Let us consider a given level k in this protocol, where we have N/L^{k-1} pairs of fidelity F each. The two-step process connection-purification can now be visualized as follows [see Fig. 1(c)]. Starting from F, the fidelity F_L after connecting L pairs can be read off from the curve below the diagonal. Reflecting this value back to the diagonal line, as indicated by the arrows in Fig. 1(c), sets the starting value for the purification curve. If F_L lies within the purification interval, then iterated application of (6) leads back to the initial value F (staircase). Once the initial value F is reobtained, we have N/L^k pairs and we can start with the level $k + 1$. In summary, each level in the protocol corresponds to one cycle in Fig. 1(c). Note that if, in the loop, $F_L \leq F_{\min}$ then purification is not possible. Being polynomial in F, the lower curve gets steeper and steeper near $F = 1$ for higher values of L. From this, one sees that for a given starting fidelity F, there is a maximum number of pairs one can connect before purification becomes impossible.

For the resources we obtain $M = \prod_m^{m_{\max}} 2/p_{\text{succ}}^{(m)}$, where $p_{\text{succ}}^{(m)}$ is the probability of obtaining the required outcome (00 or 11) in the measurement at the mth purification step. The total number of steps, $m_{\max}$, is the same as in the staircase of Fig. 1(c).

In Fig. 2(a), M is plotted against the working fidelity F. Because of the discrete nature of the purification process, the fidelity of the repurified pairs need not be exactly the same on each nesting level. The working fidelity is thus defined as the fidelity maintained *on average* when going through different nesting levels. The error parameters for this plot are $\eta = p_1 = p_2 = 0.995$. One can see that there exists an optimum working fidelity of about 0.94 which requires a minimum number of about 15 resources.

A purification protocol that converges faster and therefore involves less parallel channels was proposed by Deutsch *et al.* [9]. We have employed this protocol,

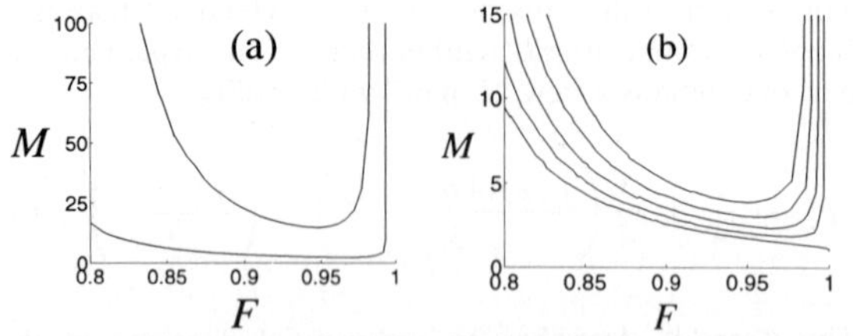

FIG. 2. M (see text) versus working fidelity F. (a) Realization of the repeater with the aid of the purification schemes of Refs. [8] (upper curve) and [9] (lower curve). The error probabilities of all operations are 0.5% (error parameters 0.995), and $L = 2$. (b) Lower curve in (a) for different error probabilities. From bottom to top: 0%, 0.25%, 0.5%, 0.75%, 1%.

using imperfect operations (2)–(4). As is demonstrated in Fig. 2(a), M can be reduced by a factor of the order of 10. Since this number has to be taken to the nth power, this reduces the number of total resources that are required at each connection point by many orders of magnitude. In Fig. 2(b), M is plotted versus the working fidelity for different error parameters. One can see that for errors in the one percent region, a working fidelity can be maintained with, on average, five L pairs on each nesting level. We note that the procedure also works for error probabilities up to about 3%, but the number of purification resources gets larger.

In the remainder of this paper, we propose a method for which the resources grow only logarithmically with the distance, whereas the total *time* needed for building the pair scales polynomially. Imagine that we purify a pair not with the help of M copies, but instead with one auxiliary pair of constant fidelity π_0 that is repeatedly created at each purification step. The purification with the help of such a pair leads to a maximum achievable fidelity $F_{\max}(\pi_0)$ that depends on the value of π_0 and, more generally, on the state of the auxiliary pair. This purification method is different from the standard schemes [8,9], and the purification limit $F_{\max}$ is usually smaller than for the destillation method. In the context of the repeater protocol, it is therefore not *a priori* clear whether the fidelity that is lost by the connection process can be regained with this method.

When connecting L pairs of fidelity F as in Fig. 1(b), we obtain a resulting L pair of fidelity $\pi_0 \equiv F_L$. In the first step, this pair is swapped to two auxiliary particles at the ends of the L pair, as indicated by the arrows in Fig. 1(b). In the next step, an L-pair of fidelity π_0 is again created by using the same string of particles as before, which is now used to purify the pair stored between the auxiliary particles. This procedure can be iterated and thus the stored pair be purified back to the fidelity F *given that the nesting condition* $F_{\max}(F_L) > F$ *is satisfied.* If this is the case, then the same procedure can be applied at higher levels, thereby purifying correlations between more and more distant particles as indicated in Fig. 1(b). Here, the dependence of the fixpoint on the form of the auxiliary pair becomes quite important: When we use our method together with the recurrence protocol of Ref. [8], which is based on Werner states, the fixpoint $F(\pi_0)$ is too small and the nesting condition cannot be satisfied for any $L \geq 2$. On the other hand, the nesting condition can be satisfied if we adopt a similar sequence of local operations as in Ref. [9], which does not involve a depolarization to Werner states.

Using this method, the vertical axes in Fig. 2 are essentially translated into temporal resources [18]. On the other hand, the number of particles at each node [see Fig. 1(b)] increases by unity with every additional nesting level, and thus depends only logarithmically on the distance between the initial and the final node. In the context of a quantum optical implementation [19], for example, this would correspond to the number of ions that need to be controlled in a cavity at each node [20]. Note, however, that this method requires perfect memory during the process. In this particular implementation, the storage decoherence time is orders of magnitude longer than the estimated duration of the process [20].

This work was supported in part by the Austrian Science Foundation, and by the TMR network ERB-FMRX-CT96-0087.

*On leave from Institut für Theoretische Physik, Universität München, Theresienstrasse 37, D-80333 München, Germany.

[1] W. Tittel *et al.*, Phys. Rev. Lett. **81**, 3563 (1998).
[2] W. T. Buttler *et al.*, Phys. Rev. Lett. **81**, 3283 (1998).
[3] D. Bouwmeester *et al.*, Nature (London) **390**, 575 (1997).
[4] D. Boschi *et al.*, Phys. Rev. Lett. **80**, 1121 (1998).
[5] E. Hagley *et al.*, Phys. Rev. Lett. **79**, 1 (1997).
[6] C. Monroe *et al.*, Phys. Rev. Lett. **75**, 4714 (1995); Q. A. Turchette *et al.*, *ibid.* **75**, 4710 (1995).
[7] L. K. Grover, quant-ph/9704012.
[8] C. H. Bennett *et al.*, Phys. Rev. Lett. **76**, 722 (1996).
[9] D. Deutsch *et al.*, Phys. Rev. Lett. **77**, 2818 (1996).
[10] N. Gisin, Phys. Lett. A **210**, 151 (1996).
[11] For optical fibers, this length is typically 10 km (see [1]).
[12] P. Shor, quant-ph/9605011; A. M. Steane, Phys. Rev. Lett. **78**, 2252 (1997).
[13] E. Knill, R. Laflamme, and W. Zurek, Science **279**, 342 (1998); D. Aharonov and M. Ben-Or, quant-ph/9611025; A. Yu. Kitaev, Russ. Math. Surv. **52**, 1191 (1997).
[14] E. Knill and R. Laflamme, quant/ph-9608012.
[15] C. H. Bennett *et al.*, Phys. Rev. Lett. **70**, 1895 (1993).
[16] C. H. Bennett *et al.*, Phys. Rev. A **54**, 3824 (1996).
[17] M. Zukowski *et al.*, Phys. Rev. Lett. **71**, 4287 (1993).
[18] W. Dür *et al.*, quant-ph/9808065; G. Giedke *et al.*, quant-ph/9809043.
[19] S. J. van Enk *et al.*, Phys. Rev. Lett. **78**, 4293 (1997); Science **279**, 205 (1998).
[20] For a distance of 1280 km and a local node every 10 km [11], this amounts to, e.g., $7 = \log_2(128)$ particles per node and an estimated purification time of less than 1 s [18].

VOLUME 76, NUMBER 5 PHYSICAL REVIEW LETTERS 29 JANUARY 1996

Purification of Noisy Entanglement and Faithful Teleportation via Noisy Channels

Charles H. Bennett,[1,*] Gilles Brassard,[2,†] Sandu Popescu,[3,‡] Benjamin Schumacher,[4,§]
John A. Smolin,[5,‖] and William K. Wootters[6,¶]
[1]*IBM Research Division, Yorktown Heights, New York 10598*
[2]*Département IRO, Université de Montréal, C.P. 6128, Succursale centre-ville, Montréal, Québec, Canada H3C 3J7*
[3]*Physics Department, Tel Aviv University, Tel Aviv, Israel*
[4]*Physics Department, Kenyon College, Gambier, Ohio 43022*
[5]*Physics Department, University of California at Los Angeles, Los Angeles, California 90024*
[6]*Physics Department, Williams College, Williamstown, Massachusetts 01267*
(Received 24 April 1995)

Two separated observers, by applying local operations to a supply of not-too-impure entangled states (e.g., singlets shared through a noisy channel), can prepare a smaller number of entangled pairs of arbitrarily high purity (e.g., near-perfect singlets). These can then be used to faithfully teleport unknown quantum states from one observer to the other, thereby achieving faithful transmission of quantum information through a noisy channel. We give upper and lower bounds on the yield $D(M)$ of pure singlets ($|\Psi^-\rangle$) distillable from mixed states M, showing $D(M) > 0$ if $\langle\Psi^-|M|\Psi^-\rangle > \frac{1}{2}$.

PACS numbers: 03.65.Bz, 42.50.Dv, 89.70.+c

The techniques of quantum teleportation [1] and quantum data compression [2,3] exemplify a new goal of quantum information theory, namely, to understand the kind and quantity of channel resources needed for the transmission of intact quantum states, rather than classical information, from a sender to a receiver. In this approach, the quantum source S is viewed as an ensemble of pure states ψ_i, typically not all orthogonal, emitted with known probabilities p_i. Transmission of quantum information through a channel is considered successful if the channel outputs closely approximate the inputs as quantum states. Because nonorthogonal states, in principle, cannot be observed without disturbing them, their faithful transmission requires that the entire transmission processes be carried out by a physical apparatus that functions obliviously, that is, without knowing or learning which ψ_i are passing through.

Just as classical data compression techniques allow data from a classical source to be faithfully transmitted using a number of bits per signal asymptotically approaching the source's Shannon entropy, $-\sum_i p_i \log_2 p_i$, quantum data compression [2,3] allows quantum data to be transmitted, with asymptotically perfect fidelity, using a number of 2-state quantum systems or *qubits* (e.g., spin-$\frac{1}{2}$ particles) asymptotically approaching the source's von Neumann entropy

$$S(\rho) = -\mathrm{Tr}\rho \log_2\rho, \quad \text{where } \rho = \sum_i p_i|\psi_i\rangle\langle\psi_i|. \quad (1)$$

Quantum teleportation achieves the goal of faithful transmission in a different way, by substituting classical communication and prior entanglement for a direct quantum channel. Using teleportation, an arbitrary unknown qubit can be faithfully transmitted via a pair of maximally entangled qubits (e.g., two spin-$\frac{1}{2}$ particles in a pure singlet state) previously shared between sender and receiver, and a 2-bit classical message from the sender to the receiver.

Both quantum data compression and teleportation require a noiseless quantum channel—in the former case for the direct quantum transmission and in the latter for sharing the entangled particles—yet available channels are typically noisy. Since quantum information cannot be cloned [4], it would perhaps appear impossible to use redundancy in the usual way to correct errors. Nevertheless, quantum error-correcting codes have recently been discovered [5] which operate in a subtler way, essentially by embedding the quantum information to be protected in a subspace so oriented in a larger Hilbert space as to leak little or no information to the environment, within a given noise model. We describe another approach in which the noisy channel is not used to transmit the source states directly, but rather to share entangled pairs (e.g., singlets) for use in teleportation. But before they can be used to teleport reliably, the entangled pairs must be purified—converted to almost perfectly entangled states from the mixed entangled states that result from transmission through the noisy channel. We show below how the two observers can accomplish this purification, by performing local unitary operations and measurements on the shared entangled pairs, coordinating their actions through classical messages, and sacrificing some of the entangled pairs to increase the purity of the remaining ones. Once this is done, the resulting almost perfectly pure, almost perfectly entangled pairs can be used, in conjunction with classical messages, to teleport the unknown quantum states ψ_i from sender to receiver with high fidelity. The overall result is to simulate a noiseless quantum channel by a noisy one, supplemented by local actions and classical communication.

Let M be a general mixed state of two spin-$\frac{1}{2}$ particles, from which we wish to distill some pure entanglement. The state M could result, for example,

 0031-9007/96/76(5)/722(4)$06.00

when one or both members of an initially pure singlet state $\Psi^- = (\uparrow\downarrow - \downarrow\uparrow)/\sqrt{2}$ are transmitted through a noisy channel to two separated observers, whom we shall call Alice and Bob. The purity of M can be conveniently expressed by its fidelity [2]

$$F = \langle\Psi^-|M|\Psi^-\rangle \tag{2}$$

relative to a perfect singlet. Though nonlocally defined, the purity F can be computed from the probability $P_{\|}$ of obtaining parallel outcomes if the two spins are measured locally along the same random axis: One finds that $F = 1 - 3P_{\|}/2$.

The recovery of entanglement from M is best understood in the special case that M is already a pure state of the two particles, $M = |\Upsilon\rangle\langle\Upsilon|$ for some Υ. The quantity of entanglement, $E(\Upsilon)$, in such a pure state is naturally defined by the von Neumann entropy of the reduced density matrix of either particle considered separately:

$$E(\Upsilon) = S(\rho_A) = S(\rho_B), \tag{3}$$

where $\rho_A = \mathrm{Tr}_B(|\Upsilon\rangle\langle\Upsilon|)$, and similarly for ρ_B. For pure states, this entanglement can be efficiently concentrated into singlets by the methods of [6], which use local operations and classical communication to transform n input states Υ into m singlets with a yield m/n approaching $E(\Upsilon)$ as $n \to \infty$. Conversely, given n shared singlets, local actions and classical communication suffice to prepare m arbitrarily good copies of Υ with a yield m/n approaching $1/E(\Upsilon)$ as $n \to \infty$.

Returning now to the problem of obtaining singlets from mixed states, the first step in our purification protocol is to have Alice and Bob perform a *random bilateral rotation* on each shared pair, choosing a random SU(2) rotation independently for each pair and applying it locally to both members of the pair (the same result could also be achieved by choosing from a finite set of rotations $\{B_x, B_y, B_z, I\}$ defined below). This transforms the initial general two-spin mixed state M into a rotationally symmetric mixture,

$$W_F = F|\Psi^-\rangle\langle\Psi^-| + \frac{1-F}{3}|\Psi^+\rangle\langle\Psi^+| + \frac{1-F}{3}|\Phi^+\rangle\langle\Phi^+| + \frac{1-F}{3}|\Phi^-\rangle\langle\Phi^-|, \tag{4}$$

of the singlet state Ψ^- and the three triplet states $\Psi^+ = (\uparrow\downarrow + \downarrow\uparrow)/\sqrt{2}$ and $\Phi^\pm = (\uparrow\uparrow \pm \downarrow\downarrow)/\sqrt{2}$. Because of the singlet's invariance under bilateral rotations, the symmetrized state W_F, which we shall call a Werner state [7] of purity F, has the same F as the initial mixed state M from which it was derived.

At this point, it should be recalled that two mixed states having the same density matrix are physically indistinguishable, even though they may have had different preparations. Therefore, subsequent steps in the purification can be carried out without regard to any properties of the original mixed state M, or of the noisy channel(s) that may have generated it, except for the purity F.

Mixtures of the four states $\Psi^\pm$ and $\Phi^\pm$—known as the four Bell states—are particularly easy to analyze, because the Bell states transform simply under several kinds of local unitary operations. Besides the random bilateral rotation already described, several other local operations will be used in entanglement purification.

(i) Unilateral Pauli rotations (that is, rotations by π rad about the x, y, or z axis) of *one* particle in an entangled pair. These operations map the Bell states onto one another in a 1:1 pairwise fashion, leaving no state unchanged; thus σ_x maps $\Psi^\pm \leftrightarrow \Phi^\pm$, σ_z maps $\Psi^\pm \leftrightarrow \Psi^\mp$ and $\Phi^\pm \leftrightarrow \Phi^\mp$, while σ_y maps $\Psi^\pm \leftrightarrow \Phi^\mp$. We ignore overall phase changes because they do not affect our arguments.

(ii) Bilateral $\pi/2$ rotations B_x, B_y, and B_z of *both* particles in a pair about the x, y, or z axis, respectively. Each of these operations leaves the singlet state and a different one of the triplets invariant, interchanging the other two triplets, with B_x mapping $\Phi^+ \leftrightarrow \Psi^+$, B_y mapping $\Phi^- \leftrightarrow \Psi^+$, and B_z mapping $\Phi^+ \leftrightarrow \Phi^-$. Again we omit phases.

(iii) The quantum-XOR or controlled-NOT operation [8] performed bilaterally by both observers on corresponding members of two shared pairs. The *unilateral* quantum XOR is an operation on two qubits held by the same observer which conditionally flips the second or "target" spin if the first or "source" spin is up, and does nothing otherwise. As a unitary operator it is expressed

$$U_{\mathrm{XOR}} = |\uparrow_S\uparrow_T\rangle\langle\uparrow_S\downarrow_T| + |\uparrow_S\downarrow_T\rangle\langle\uparrow_S\uparrow_T| + |\downarrow_S\downarrow_T\rangle\langle\downarrow_S\downarrow_T| + |\downarrow_S\uparrow_T\rangle\langle\downarrow_S\uparrow_T|. \tag{5}$$

The bilateral XOR (henceforth, BXOR) operates in a similar fashion on corresponding members of two pairs shared between Alice and Bob: If Alice holds spins 1 and 3, and Bob holds spins 2 and 4, a BXOR, with spins 1 and 2 as source and spins 3 and 4 as target, would conditionally flip spin 3 if and only if spin 1 was up, while conditionally flipping spin 4 if and only if spin 2 was up. A BXOR on two Φ^+ states leaves them both invariant. The results of applying BXOR to other combinations of Bell states is shown below, omitting phases.

Before		After (n.c. = no change)	
Source	Target	Source	Target
$\Phi^\pm$	Φ^+	n.c.	n.c.
$\Psi^\pm$	Φ^+	n.c.	Ψ^+
$\Psi^\pm$	Ψ^+	n.c.	Φ^+
$\Phi^\pm$	Ψ^+	n.c.	n.c.
$\Phi^\pm$	Φ^-	$\Phi^\mp$	n.c.
$\Psi^\pm$	Φ^-	$\Psi^\mp$	Ψ^-
$\Psi^\pm$	Ψ^-	$\Psi^\mp$	Φ^-
$\Phi^\pm$	Ψ^-	$\Phi^\mp$	n.c.

(6)

(iv) Besides these unitary operations, Alice and Bob perform one kind of measurement: measuring both spins in a given pair along the z spin axis. This reliably distinguishes Ψ states from Φ states, but cannot distinguish + from − states. Of course, after the measurement has been performed, the measured pair is no longer entangled.

We now show that, given two Werner pairs of fidelity $F > \frac{1}{2}$, Alice and Bob can use local operations and two-way classical communication to obtain, with probability greater than $\frac{1}{4}$, one Werner pair of fidelity $F' > F$, where the F' satisfies the recurrence relation

$$F' = \frac{F^2 + \frac{1}{9}(1-F)^2}{F^2 + \frac{2}{3}F(1-F) + \frac{5}{9}(1-F)^2}. \tag{7}$$

To achieve this, the following protocol is used.

(A1) A unilateral σ_y rotation is performed on each of the two pairs, converting them from mostly Ψ^- Werner states to the analogous mostly Φ^+ states, i.e., states with a large component $F > \frac{1}{2}$ of Φ^+ and equal components of the other three Bell states.

(A2) A BXOR is performed on the two impure Φ^+ states, after which the target pair is locally measured along the z axis. If the target pair's z spins come out parallel, as they would if both inputs were true Φ^+ states, the unmeasured source pair is kept; otherwise, it is discarded.

(A3) If the source pair has been kept, it is converted back to a mostly Ψ^- state by a unilateral σ_y rotation, then made rotationally symmetric by a random bilateral rotation (cf. [9]).

Because $F'(F)$ is continuous and exceeds F over the entire range $\frac{1}{2} < F < 1$, iteration of the above protocol can distill Werner states of arbitrarily high purity $F_{\text{out}} < 1$ from a supply of input mixed states M of any purity $F_{\text{in}} > \frac{1}{2}$. The yield (purified output pairs per impure input pair) is rather poor and tends to zero in the limit $F_{\text{out}} \to 1$; but by BXORing a variable number $k(F) \approx 1/\sqrt{1-F}$ of source pairs, rather than 1, into each target pair before measuring it, the yield can be increased and made to approach a positive limit as $F_{\text{out}} \to 1$. [For this choice of k, to lowest order in $1-F$, the iteration formula for purity agrees with Eq. (7): $F'(F) = 1 - \frac{2}{3}(1-F)$. The expected fraction of the pairs discarded at each step, also to lowest order, is $\frac{2}{3}(1-F)^{\frac{1}{2}}$. One thus obtains a nonzero yield as $F_{\text{out}} \to 1$.] We do not give the asymptotic yield from this method, because a higher yield can be obtained by combining it with another method to be described below, which uses a supply of previously purified Φ^+ pairs in the manner of a breeder reactor, consuming some in order to produce more than the number consumed.

The basic step (a "BXOR test") used in this method consists of bilaterally XORing a subset of the impure pairs, used as sources, into one of the pure Φ^+ states, used as a target, followed by measurement of the target. Consulting table (6) above, we see that each Ψ^+ or Ψ^- source pair toggles the target between Φ^+ and Ψ^+, without affecting the source. Thus a BXOR test, like a parity check on classical data, tells Alice and Bob whether there are an even or odd number of Ψ states in the tested subset. By performing a number of BXOR tests, on different subsets of the original impure pairs, all the Ψ states can be found and corrected to Φ states. A similar procedure is then used to find all the Φ^- states and correct them to the desired Φ^+. The full protocol is described below.

(B1) Alice and Bob start with n impure pairs each described by the same Bell-diagonal density matrix W with $S(W) < 1$, and $n[S(W) + \delta]$ prepurified Φ^+ states, prepared, for example, by the variable blocksize recurrence method described above. Here δ is a positive constant that can be allowed to approach 0 in the limit of large n.

(B2) Using the prepurified Φ^+ pairs as targets, Alice and Bob perform BXOR tests on sufficiently many random subsets of the impure pairs to locate all Ψ states, with high probability, without distinguishing Ψ^+ from Ψ^-. Once found, the $\Psi^\pm$ are converted, respectively, to $\Phi^\pm$ by applying a unilateral σ_x rotation to each of them. The impure pairs now consist of only Φ^+ and Φ^- states.

(B3) Next Alice and Bob do a bilateral B_y to convert the Φ^- states into Ψ^+, while leaving the Φ^+ states invariant. This done, they perform BXOR tests on sufficiently many more random subsets to find all the new Ψ^+ states with high probability. Once found, these states are corrected to the desired Φ^+ form by unilateral σ_x rotations.

The number of BXOR test per impure pair required to find all the errors, with arbitrarily small chance of failure, approaches the entropy of the impure pairs, $S(W) = -\text{Tr}W\log_2 W$, in the limit of large n. This follows from the following facts: (i) For any two distinct n-bit strings the probability that they agree on the parities of r independent random subsets of their bits is $\leq 2^{-r}$ [10].

(ii) The probability distribution P_X over n-bit strings x, where x represents the original sequence of Φ/Ψ values of the impure pairs, receives almost all its weight from a set of "typical" strings containing $N_1 = 2^{H(X)+O(\sqrt{n})}$ members, where $H(X)$ is the Shannon entropy of P_X. Similarly, the conditional distribution $P_{Y|X=x}$ of n-bit strings y, representing the ± values of a sequence of impure pairs whose Φ/Ψ sequence is x, receives almost all its weight from a set of typical strings containing $N_2 = 2^{H(Y|X=x)+O(\sqrt{n})}$ members.

Let r_1 BXOR tests be performed in the first round, whose goal is to find x uniquely. The expected number of "false positives"—strings in the typical set, other than the correct x, which agree with it on r_1 subset parities—is $\leq N_1 2^{-r_1}$. Thus the chance of a false positive becomes negligible when $r_1 > \log_2 N_1$. Similarly, the chance of a false positive in the second round after r_2 BXOR tests is negligible when $r_2 > \log_2 N_2$. Combining these results, and recalling that $\log_2(N_1 N_2) = nS(W) + O(\sqrt{n})$, we obtain the desired result, viz., that asymptotically $S(W)$ BXOR tests per impure pair suffice to find all the errors.

The breeding method has a yield $1 - S(W)$, producing more pure pairs than consumed if the mixed state's von Neumann entropy, $S(W)$, is less than 1. For Werner states, the yield

$$1 - S(W_F) = 1 + F\log_2 F + (1 - F)\log_2 \frac{1 - F}{3} \tag{8}$$

is positive for $F > 0.8107$.

The use of prepurified pairs as targets simplifies analysis of the protocol by avoiding backaction of the targets on the sources, but is not strictly necessary. Even without the prepurified pairs, using only impure Bell-diagonal states W as input, it is possible [11] to design a sequence of BXOR's and local rotations that eliminate approximately half the candidates for x or y at each step, achieving the same asymptotic yield $1 - S(W)$ as the breeding method. This nonbreeding protocol requires only one-way classical communication, allowing it to be used to protect quantum information from errors during storage (cf. [5,11]) as well as during transmission.

We do not yet know the optimal asymptotic yield $D(M)$ of purified singlets distillable from general mixed states M, nor even from Werner states. Figure 1 compares the yields of several purification methods for Werner states W_F with an upper bound $E(W_F)$ given by

$$E(W_F) = \begin{cases} H_2(\frac{1}{2} + \sqrt{F(1 - F)}), & \text{if } F > 1/2, \\ 0, & \text{if } F \leq 1/2. \end{cases} \tag{9}$$

Here $H_2(x) = -x\log_2 x - (1 - x)\log_2(1 - x)$ is the dyadic Shannon entropy. This upper bound is based on the fact that W_F, for $F > \frac{1}{2}$, can be expressed as an equal mixture of eight pure states

$$\sqrt{F}\,|\Psi^-\rangle + \sqrt{\frac{1 - F}{3}}\,(\pm|\Psi^+\rangle \pm |\Phi^-\rangle \pm i|\Phi^+\rangle), \tag{10}$$

each having entropy of entanglement equal to the right side of Eq. (9), while for $F \leq \frac{1}{2}$, W_F can be expressed as a mixture of unentangled product states ↑↑, ↓↓, ↑↓, and

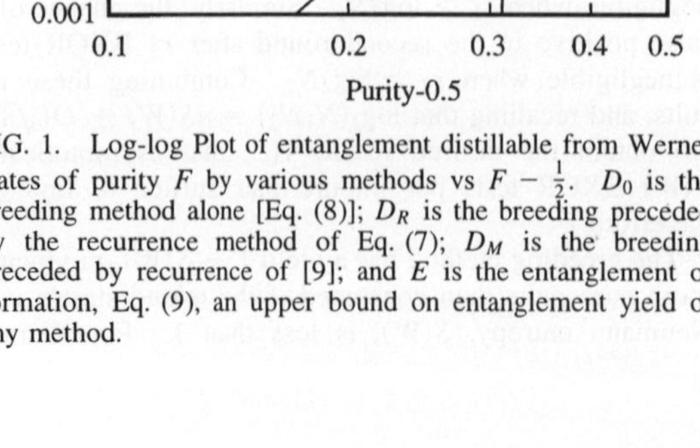

FIG. 1. Log-log Plot of entanglement distillable from Werner states of purity F by various methods vs $F - \frac{1}{2}$. D_0 is the breeding method alone [Eq. (8)]; D_R is the breeding preceded by the recurrence method of Eq. (7); D_M is the breeding preceded by recurrence of [9]; and E is the entanglement of formation, Eq. (9), an upper bound on entanglement yield of any method.

↓↑. In fact [11], these are the least entangled ensembles realizing W_F; therefore, $E(W_F)$ may be viewed as the Werner state's "entanglement of formation"—the asymptotic number of singlets required to prepare one W_F by local actions. Because expected entanglement cannot be increased by local actions and classical communication [11], a mixed state's distillable entanglement $D(M)$ cannot exceed its entanglement of formation $E(M)$.

We have seen that $F = \frac{1}{2}$ is a threshold below which Werner states can be made from unentangled ingredients, and above which they can be used as a starting material to make pure singlets. This further grounds (cf. also [12]) for regarding all Werner states with $F > \frac{1}{2}$ as nonlocal even though only those with $F > (2 + 3\sqrt{2})/8 \approx 0.78$ violate the Clauser-Horne-Shimony-Holt [13] inequality. Distillable entanglement and entanglement of formation are two alternative extensions of the definition of entanglement from pure to mixed states, but for most mixed states M, we do not know the value of either quantity, nor do we know an M for which they probably differ.

We thank David DiVincenzo for extensive and valuable advice, and Chiara Macchiavello and the Oxford quantum information group for sharing their unpublished results. Technion (Haifa), the Institute for Scientific Research (Torino), ELSAG-Bailey (Genoa), and IBM Research sponsored workshops greatly facilitating our work.

*Electronic address: bennetc@watson.ibm.com
†Electronic address: brassard@iro.umontreal.ca
‡Electronic address: spopescu@ccsg.tau.ac.il
§Electronic address: schumacb@kenyon.edu
‖Electronic address: smolin@vesta.physics.ucla.edu
¶Electronic address: wwootters@williams.edu

[1] C. H. Bennett, G. Brassard, C. Crépeau, R. Jozsa, A. Peres, and W. K. Wootters, Phys. Rev. Lett. **70**, 1895 (1993).
[2] B Schumacher, Phys. Rev. A **51**, 2738 (1995).
[3] R. Jozsa and B. Schumacher, J. Mod. Opt. **41**, 2343 (1994).
[4] W. K. Wootters and W. H. Zurek, Nature (London) **299**, 802 (1982).
[5] P. Shor, Phys. Rev. A **52**, R2493 (1995).
[6] C. H. Bennett, H. Bernstein, S. Popescu, and B. Schumacher, "Concentrating Partial Entanglement by Local Operations" (to be published).
[7] R. F. Werner, Phys. Rev. A **40**, 4277 (1989).
[8] D. Deutsch, Proc. R. Soc. London A **425**, 73 (1989).
[9] C. Macchiavello (private communication) found an improved recurrence using B_x, σ_y in place of our step A3.
[10] C. H. Bennett, G. Brassard, C. Crépeau, and U. M. Maurer, "Generalized Privacy Amplification," IEEE Trans. Inf. Theory (to be published).
[11] C. H. Bennett, D. DiVincenzo, J. A. Smolin, and W. K. Wootters, "Mixed State Entanglement and Quantum Error Correcting Codes" (to be published).
[12] S. Popescu, Phys. Rev. Lett. **72**, 797 (1994).
[13] J. F. Clauser, M. A. Horne, A. Shimony, and R. A. Holt, Phys. Rev. Lett. **23**, 880 (1980).

VOLUME 77, NUMBER 13 PHYSICAL REVIEW LETTERS 23 SEPTEMBER 1996

Quantum Privacy Amplification and the Security of Quantum Cryptography over Noisy Channels

David Deutsch,[1] Artur Ekert,[1] Richard Jozsa,[2] Chiara Macchiavello,[1] Sandu Popescu,[3] and Anna Sanpera[1]
[1]*Clarendon Laboratory, Department of Physics, University of Oxford, Parks Road, Oxford OX1 3PU, United Kingdom*
[2]*School of Mathematics and Statistics, University of Plymouth, Plymouth, Devon PL4 8AA, United Kingdom*
[3]*Department of Electrical, Computer and Systems Engineering, Boston University, Boston, Massachusetts 02215*
(Received 26 April 1996)

Existing quantum cryptographic schemes are not, as they stand, operable in the presence of noise on the quantum communication channel. Although they become operable if they are supplemented by classical privacy-amplification techniques, the resulting schemes are difficult to analyze and have not been proved secure. We introduce the concept of quantum privacy amplification and a cryptographic scheme incorporating it which is provably secure over a noisy channel. The scheme uses an "entanglement purification" procedure which, because it requires only a few quantum controlled-not and single-qubit operations, could be implemented using technology that is currently being developed. [S0031-9007(96)01288-4]

PACS numbers: 89.70.+c, 02.50.–r, 03.65.Bz, 89.80.+h

Quantum cryptography [1–3] allows two parties (traditionally known as Alice and Bob) to establish a secure random cryptographic key if, first, they have access to a quantum communication channel, and second, they can exchange classical public messages which can be monitored but not altered by an eavesdropper (Eve). Using such a key, a secure message of equal length can be transmitted over the classical channel. However, the security of quantum cryptography has so far been proved only for the idealized case where the quantum channel, in the absence of eavesdropping, is *noiseless*. That is because, under existing protocols, Alice and Bob detect eavesdropping by performing certain quantum measurements on transmitted batches of qubits and then using statistical tests to determine, with any desired degree of confidence, that the transmitted qubits are not entangled with any third system such as Eve. The problem is that there is in principle no way of distinguishing entanglement with an eavesdropper (caused by her measurements) from entanglement with the environment caused by innocent *noise*, some of which is presumably always present.

This implies that all existing protocols are, strictly speaking, inoperable in the presence of noise, since they require the transmission of messages to be suspended whenever an eavesdropper (or, therefore, noise) is detected. Conversely, if we want a protocol that is secure in the presence of noise, we must find one that allows secure transmission to continue even in the presence of eavesdroppers. To this end, one might consider modifying the existing protocols by reducing the statistical confidence level at which Alice and Bob accept a batch of qubits. Instead of the astronomically high level envisaged in the idealized protocol, they would set the level so that they would accept most batches that had encountered a given level of noise. They would then have to assume that some of the information in the batch was known to an eavesdropper. It seems reasonable that classical privacy amplification [4] could then be used to distill, from large numbers of such qubits, a key in whose security one could have an astronomically high level of confidence [5]. However, no such scheme has yet been proved to be secure. Existing proofs of the security of classical privacy amplification apply only to classical communication channels and classical eavesdroppers. They do not cover the new eavesdropping strategies that become possible in the quantum case: for instance, causing a quantum ancilla to interact with the encrypted message, storing the ancilla and later performing a measurement on it that is chosen according to the data that Alice and Bob exchange publicly.

In this paper we present a protocol that is secure in the presence of noise and an eavesdropper. It uses entanglement-based quantum cryptography [2], but with a new element, an "entanglement purification" procedure. This allows Alice and Bob to generate a pair of qubits in a state that is close to a pure, maximally entangled state, and whose entanglement with any outside system is arbitrarily low. They can generate this from any supply of pairs of qubits in mixed states with nonzero entanglement, even if an eavesdropper has had access to those qubits (see also [6,7]).

Our procedure—a *quantum privacy amplification* algorithm—(abbreviated as QPA algorithm) can be performed by Alice and Bob at distant locations by a sequence of local operations which are agreed upon by communication over a public channel. It is related to the procedure described in [8], but is more efficient.

In the idealized theory of entanglement-based quantum cryptography, Alice and Bob have a supply of qubit pairs,

 0031-9007/96/77(13)/2818(4)$10.00

each pair being in the pure, maximally entangled state $|\phi^+\rangle$, where

$$
\begin{aligned}
|\phi^\pm\rangle &= \tfrac{1}{\sqrt{2}}(|00\rangle \pm |11\rangle), \\
|\psi^\pm\rangle &= \tfrac{1}{\sqrt{2}}(|01\rangle \pm |10\rangle).
\end{aligned} \tag{1}
$$

These are the so-called "Bell states" which form a convenient basis for the state space of a qubit pair. Alice and Bob each have one qubit from each pair. In the presence of noise, each pair would in general have become entangled with other pairs and with the environment, and would be described by a density operator on the space spanned by (1).

Note that any two qubits that are jointly in a pure state cannot be entangled with any third physical object. Therefore any algorithm that delivers qubit pairs in pure states must also have eliminated the entanglement between any of those pairs and any other system. Our scheme is based on an iterative quantum algorithm which, if performed with perfect accuracy, starting with a collection of qubit pairs in mixed states, would discard some of them and leave the remaining ones in states converging to $|\phi^+\rangle\langle\phi^+|$.

Our first departure from existing quantum cryptographic schemes is to assume that Eve *does* interact with all the qubits that are transmitted or received by either Alice or Bob. Indeed we analyze the scenario that is most favorable for eavesdropping, namely where Eve herself is allowed to prepare all the qubit pairs that Alice and Bob will subsequently use for cryptography. Any realistic situation would also involve environmental noise that is not under Eve's control, but this may be treated as a special case in which Eve is not using the full information available to her.

Suppose, then, that Eve has prepared two qubit pairs in some manner of her own choosing and sends one qubit from each pair to both Alice and Bob. Let the density operators of the two pairs be $\hat{\rho}$ and $\hat{\rho}'$, respectively. Alice performs a unitary operation

$$|0\rangle \rightarrow \tfrac{1}{\sqrt{2}}(|0\rangle - i|1\rangle), \tag{2}$$

$$|1\rangle \rightarrow \tfrac{1}{\sqrt{2}}(|1\rangle - i|0\rangle) \tag{3}$$

on each of her two qubits; Bob performs the inverse operation

$$|0\rangle \rightarrow \tfrac{1}{\sqrt{2}}(|0\rangle + i|1\rangle), \tag{4}$$

$$|1\rangle \rightarrow \tfrac{1}{\sqrt{2}}(|1\rangle + i|0\rangle) \tag{5}$$

on his. If the qubits are spin-$\frac{1}{2}$ particles and the computation basis is that of the eigenstates of the z components of their spins, then the two operations correspond, respectively, to rotations by $\pi/2$ and $-\pi/2$ about the x axis.

Then Alice and Bob each perform two instances of the quantum controlled-not operation

$$\overset{\text{control}}{|a\rangle}\ \overset{\text{target}}{|b\rangle} \rightarrow \overset{\text{control}}{|a\rangle}\ \overset{\text{target}}{|a \oplus b\rangle} \quad (a,b) \in \{0,1\}, \tag{6}$$

where one pair ($\hat{\rho}$) comprises the two control qubits and the other one ($\hat{\rho}'$) the two target qubits [9]. Alice and Bob then measure the target qubits in the computational basis (e.g., they measure the z components of the targets' spins). If the outcomes coincide (e.g., both spins up or both spins down) they keep the control pair for the next round and discard the target pair. If the outcomes do not coincide, both pairs are discarded.

To see the effect of this procedure, consider the special case in which each pair is in state $\hat{\rho}$ and the joint state of the two pairs is the simple product $\hat{\rho} \otimes \hat{\rho}$. This case will suffice for our applications. We express the density operator $\hat{\rho}$ in the Bell basis $\{|\phi^+\rangle, |\psi^-\rangle, |\psi^+\rangle, |\phi^-\rangle\}$ and denote by $\{A, B, C, D\}$ the diagonal elements in that basis. Note that the first diagonal element $A = \langle\phi^+|\hat{\rho}|\phi^+\rangle$, which we call the "fidelity," is the probability that the qubit would pass a test for being in the state $|\phi^+\rangle$. Thus we wish to drive the fidelity to 1 (which implies that the other three diagonal elements go to 0). Now, in the case where the control qubits are retained, their density operator $\hat{\rho}^{\sim}$ will have diagonal elements $\{\tilde{A}, \tilde{B}, \tilde{C}, \tilde{D}\}$ which depend on average *only* on the diagonal elements of $\hat{\rho}$ (the average is taken over the two different coincident outcomes, e.g., both spins up and both spins out):

$$
\begin{aligned}
\tilde{A} &= \frac{A^2+B^2}{N}, \\
\tilde{B} &= \frac{2CD}{N}, \\
\tilde{C} &= \frac{C^2+D^2}{N}, \\
\tilde{D} &= \frac{2AB}{N},
\end{aligned} \tag{7}
$$

where $N = (A + B)^2 + (C + D)^2$ is the probability that Alice and Bob obtain coinciding outcomes in the measurements on the target pair. That is, if the procedure is carried out many times on an ensemble of such pairs of pairs, then $\tilde{A}$, $\tilde{B}$, $\tilde{C}$, and $\tilde{D}$ give the average diagonal entries of the surviving pairs. Note that if the average $\tilde{A}$ is driven to 1 then each of the surviving pairs must individually approach the pure state $|\phi^+\rangle\langle\phi^+|$.

In passing, we note that if the two input pairs have *different* states $\hat{\rho}$ and $\hat{\rho}'$ with diagonal elements $\{A, B, C, D\}$ and $\{A', B', C', D'\}$, respectively, then the retained control pairs will, on average, have diagonal elements given by

$$
\begin{aligned}
\tilde{A} &= \frac{AA'+BB'}{N}, \\
\tilde{B} &= \frac{C'D+CD'}{N}, \\
\tilde{C} &= \frac{CC'+DD'}{N}, \\
\tilde{D} &= \frac{AB'+A'B}{N},
\end{aligned} \tag{8}
$$

where $N = (A + B)(A' + B') + (C + D)(C' + D')$, which generalizes (7).

Suppose that Eve has provided L pairs of qubits, with density operators $\hat{\rho}_1, \hat{\rho}_2, \dots, \hat{\rho}_L$. This is *not* to say that their overall density operator is necessarily of the product form

$$\hat{\rho}_1 \otimes \hat{\rho}_2 \otimes \cdots \otimes \hat{\rho}_L \tag{9}$$

for Eve may have prepared them in an entangled state. However, let us consider first the case in which the pairs are not entangled with each other, i.e., the overall state *is* of the form (9) above. Alice and Bob know nothing about the state preparation, they are simply presented with an ensemble of L pairs of qubits from which they can (if they wish) estimate the average density operator $\hat{\rho}_{\text{ave}}$:

$$\hat{\rho}_{\text{ave}} = \frac{1}{L}(\hat{\rho}_1 + \hat{\rho}_2 + \cdots + \hat{\rho}_L), \tag{10}$$

which characterizes the ensemble of pairs.

Alice and Bob now select pairs at random from the ensemble of provided pairs and apply the QPA procedure to pairs of these selected pairs. Thus we may set $\hat{\rho} = \hat{\rho}_{\text{ave}}$ in (7) and we are in effect studying the properties of the map

$$\begin{pmatrix} A \\ B \\ C \\ D \end{pmatrix} \rightarrow \begin{pmatrix} \tilde{A} \\ \tilde{B} \\ \tilde{C} \\ \tilde{D} \end{pmatrix} = \frac{1}{N} \begin{pmatrix} A^2 + B^2 \\ 2CD \\ C^2 + D^2 \\ 2AB \end{pmatrix}. \tag{11}$$

$\{\tilde{A}, \tilde{B}, \tilde{C}, \tilde{D}\}$ in (11) gives the average diagonal entries for the states of the surviving pairs, i.e., the diagonal entries of the average density operator of the ensemble of surviving pairs. Therefore the repeated application of the QPA procedure—generating successive ensembles of surviving pairs—corresponds to iteration of the map in (11).

Several interesting properties of this map can be easily verified. For example, if at any stage the fidelity A exceeds $\frac{1}{2}$, then after one more iteration, it still exceeds $\frac{1}{2}$. Although A does not necessarily increase monotonically, our target point, $A = 1, B = C = D = 0$, is a fixed point of the map and is the only fixed point in the region $A > \frac{1}{2}$. It is a local attractor. We have been unable to obtain a proof that it is also a global attractor in the region $A > \frac{1}{2}$, but we have verified this by computer simulation. In other words, if we begin with pairs whose average fidelity exceeds $\frac{1}{2}$, but which are otherwise in an arbitrary state (unentangled with each other), then the states of pairs surviving after successive iterations always converge to the unit-fidelity pure state $|\phi^+\rangle$. Since this is a pure state, none of the surviving pairs is, in the limit, entangled with any other system.

To illustrate the behavior of the iteration in Fig. 1 we plot the fidelity as a function of the initial fidelity and the number of iterations, in cases where $A > \frac{1}{2}$ and $B = C = D$ initially.

The above analysis applies to the case in which Eve does not entangle the pairs with each other [c.f. Eq. (9)].

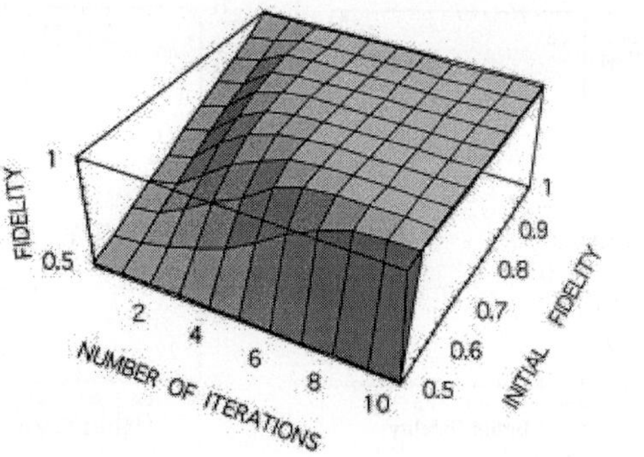

FIG. 1. Average fidelity as a function of the initial fidelity and the number of iterations.

However, if Eve provides pairs which *are* entangled with each other, then Eq. (11) no longer holds, and the QPA iterations may or may not converge to the pure state $|\phi^+\rangle\langle\phi^+|$. Nevertheless it is *never* of advantage to Eve to entangle pairs with each other: Eve knows that Alice and Bob will apply the QPA procedure to the distributed pairs. In the course of the QPA iterations Alice and Bob will periodically check the average fidelity of the surviving pairs, which is achieved by purely local operations and classical communication between them. Thus they determine whether they have achieved an acceptably high fidelity. If Eve provides pairs which are entangle with each other then the QPA procedure may not converge. In this case the protocol will force Alice and Bob to discard the entire transmission, and Eve is merely in effect blocking the quantum channel. (This would also be the case if, for example, she distributed pairs unentangled with each other, but having $A < \frac{1}{2}$.) On the other hand, if Eve provides pairs which *do* converge to $|\phi^+\rangle\langle\phi^+|$ (at an acceptable rate, i.e., at least the rate corresponding to the starting values of A, B, C, and D, which can be measured before starting the QPA procedure), then the QPA procedure is effective in excluding Eve despite the initial entanglement between the pairs. Thus Eve never benefits from providing pairs which are entangled with each other, and hence the above analysis suffices to prove the security of the protocol.

The QPA procedure is rather wasteful in terms of discarded particles—at least on half of the particles (the ones used as targets) are lost at every iteration. The efficiency of the procedure (i.e., the ratio of the number of surviving pairs to the number of initial pairs) depends on the final fidelity required and on the initial state. As an example, in Fig. 2(a) we plot the efficiency as a function of the initial fidelity A (taking $B = C = D$), for purification to fidelity 0.99, and in Fig. 2(b) we show the number of iterations used. The efficiency of our scheme compares very favorably with the entanglement purification scheme as described in [8], and it can be

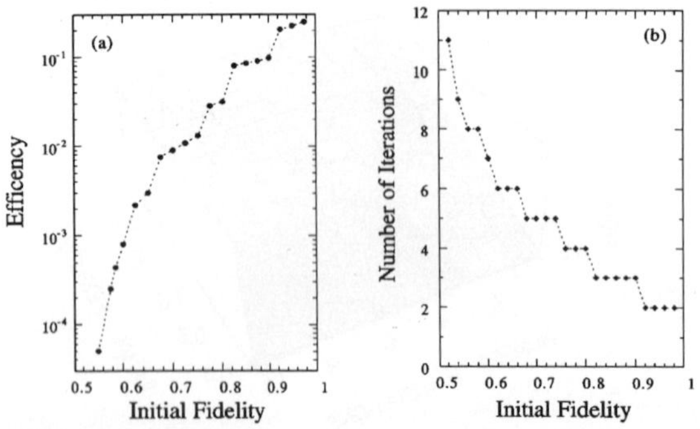

FIG. 2. States with $B = C = D$ are purified up to a fidelity of 0.99. (a) The efficiency of the purification as a function of the initial fidelity A. (b) The number of iterations used in the QPA procedure as a function of the initial fidelity.

directly applied to purify states which are not necessarily of the Werner form [10].

Even though the efficiency of our procedure may be low in many cases, it nevertheless establishes that there *exist* unconditionally secure quantum key distribution protocols. This is in contrast to recent claims [11] that quantum bit commitment protocols can never be unconditionally secure.

The QPA procedure is capable of purifying a collection of pairs in any state $\hat{\rho}$ of the product form (9), whose average fidelity with respect to at least one maximally entangled state (i.e., a Bell state or a state obtained from a Bell state via local unitary operations) is greater than $\frac{1}{2}$ (because any state of that type can be transformed into $|\phi^+\rangle$ via local unitary operations [12]). If we denote by $\mathcal{B}$ a class of pure, maximally entangled states (the generalized Bell states) then the condition that the state $\hat{\rho}$ can be purified using the QPA procedure is

$$\max_{\phi \in \mathcal{B}} \langle \phi | \hat{\rho} | \phi \rangle > \tfrac{1}{2}. \quad (12)$$

Note that this condition is not equivalent to the Horodecki condition [13] characterizing mixed states which can violate a generalized Bell inequality (CHSH inequality [14]). Indeed there exist mixed states which satisfy *both* our condition (12) *and* the CHSH inequalities. Thus the QPA algorithm reveals a more complete characterization of nonlocality than that given by Bell's theorem (c.f. also [6,7,15–17]). We hope to elaborate this in a forthcoming paper.

The practical implementation of the QPA procedure would require efficient quantum controlled-not gates operating directly on information carriers. Perhaps the most promising implementation of gates of this type (in the QPA context) is the one proposed by Turchette *et al.* [18]. It operates on polarized photons and allows the polarization of the target photon to be rotated depending on the polarization of the control photon. Although the current efficiency of the device is quite low, recent experimental progress in this field raises hopes for a successful QPA experiment in the not too distant future.

This research was supported in part by Elsag-Bailey PLC. We would like to thank A. Barenco and W.K. Wootters for stimulating discussions. A.E. and R.J. are sponsored by The Royal Society, London. C.M. is sponsored by the European Union HCM Programme. A.S. is sponsored by U.K. Engineering and Physical Sciences Research Council. A.E., R.J., and S.P. acknowledge Rabezzana Grignolino d'Asti.

[1] C.H. Bennett and G. Brassard, in *Proceedings of the IEEE International Conference on Computers, Systems and Signal Processing, Bangalore, India, 1984* (IEEE, New York, 1984), p. 175.
[2] A.K. Ekert, Phys. Rev. Lett. **67**, 661 (1991).
[3] C.H. Bennett, Phys. Rev. Lett. **68**, 3121 (1992).
[4] C.H. Bennett, G. Brassard, and J.-M. Robert, SIAM J. Comput. **17**, 210 (1988); C.H. Bennett, G. Brassard, C. Crépeau, and U.M. Maurer, IEEE Trans. Inf. Theory **41**, 1915 (1995).
[5] H.K. Lo and H.F. Chau, Los Alamos Report No. quant-ph/9511025; D. Mayers, *Lecture Notes in Computer Science* (Springer-Verlag, Berlin, 1995), Vol. 963, pp. 124–135.
[6] M. Horodecki, P. Horodecki, and R. Horodecki, Los Alamos Report No. quant-ph/9605038.
[7] M. Horodecki, P. Horodecki, and R. Horodecki, Los Alamos Report No. quant-ph/9607009.
[8] C.H. Bennett, G. Brassard, S. Popescu, B. Schumacher, J. Smolin, and W.K. Wootters, Phys. Rev. Lett. **76**, 722 (1996).
[9] A. Barenco, D. Deutsch, A. Ekert, and R. Jozsa, Phys. Rev. Lett. **74**, 4083 (1995).
[10] R.F. Werner, Phys. Rev. A **40**, 4277 (1989).
[11] H.K. Lo and H.F. Chau, Los Alamos Report No. quant-ph/9603004; D. Mayers, Los Alamos Report No. quant-ph/9603015.
[12] C.H. Bennett and S.J. Wiesner, Phys. Rev. Lett. **69**, 2881 (1992).
[13] R. Horodecki, P. Horodecki, and M. Horodecki, Phys. Lett. A **200**, 340 (1995).
[14] J. Clauser, M. Horne, A. Shimony, and R. Holt, Phys. Rev. Lett. **23**, 880 (1969).
[15] S. Popescu, Phys. Rev. Lett. **72**, 797 (1994).
[16] S. Popescu, Phys. Rev. Lett. **74**, 2619 (1995).
[17] N. Gisin, Phys. Lett. A **210**, 151 (1996).
[18] Q.A. Turchette, C.J. Hood, W. Lange, H. Mabuchi, and H.J. Kimble, Phys. Rev. Lett. **75**, 4710 (1995).

Photonic Channels for Quantum Communication

S. J. van Enk, J. I. Cirac, P. Zoller

A general photonic channel for quantum communication is defined. By means of local quantum computing with a few auxiliary atoms, this channel can be reduced to one with effectively less noise. A scheme based on quantum interference is proposed that iteratively improves the fidelity of distant entangled particles.

Security for communication of sensitive data over public channels such as the Internet is indispensable nowadays. Quantum mechanics offers the possibility of storing, processing, and distributing information in a proven secure way by exploiting the fragility of quantum states and the fact that they cannot be cloned (*1*). In practice, many obstacles stand in the way of implementing a reliable quantum network. Although remarkable progress has recently been made experimentally in the context of quantum cryptography and computation (*2*), the presence of errors during the transmission and processing of quantum information remains as the main obstacle. In principle, these problems could be circumvented with ingenious schemes for purifying states (*3*) and correcting errors (*4*), because they allow the transmission of intact quantum states even in the presence of errors. These "standard" methods require a large (in principle, infinite) number of extra quantum bits (qubits) to store intermediate information. However, in the first generations of experiments on quantum networks, one expects to be able to store and manip-

Institut für Theoretische Physik, Universität Innsbruck, Technikerstrasse 25, A-6020 Innsbruck, Austria.

ulate only a few qubits in each location. Thus, new methods are needed to overcome the presence of errors during quantum communication in small physical systems.

Recently we proposed a physical implementation based on cavity quantum electrodynamics (QED) (5) to accomplish ideal transmission over a noisy channel where the dominant error is due to photon absorption (6). We modeled photon absorption by a Markovian process and showed how this property can be exploited to convey intact quantum information within a quantum network composed of small physical systems. Although this assumption is restrictive, it shows that one has to reconsider the definition of quantum channels, which leads to nonstandard methods of purification and error correction.

In this work we define a general channel for communication via photons and show how to transmit quantum information via that channel. This channel is not based on a particular physical model, does not use the Markov property, and includes all possible errors during transmission. Moreover, in contrast to usual definitions of noisy quantum channels [such as the depolarizing channel or the erasure channel (7)], we do not describe the action of the channel only in terms of classical probabilities but allow for quantum interference effects. In fact, these quantum interferences allow one, under certain conditions, to transmit quantum states over channels that have so much noise in terms of classical probabilities that one would be led to believe no quantum information could be transmitted at all. The scheme we propose is based on "channel reduction," which consists of combining local operations and measurements with multiple uses of the channel to reduce the description to a simplified effective channel. Using this effective channel and exploiting quantum interference effects, we show how to create a perfect distant maximally entangled state [Einstein-Podolsky-Rosen (EPR) pair] utilizing only three qubits at each location. With teleportation, one can then send any unknown quantum state securely without distortions (8).

To define the photonic channel, we denote by $|0\rangle$ and $|1\rangle$ the states of the qubit that a sender, traditionally called Alice, transmits to the receiver Bob, and by $|E\rangle$ the initial state of the environment. The action of the most general channel leaving the qubit inside its original two-dimensional (2D) Hilbert space is

$$|0\rangle|E\rangle \mapsto (|0\rangle T_{00} + |1\rangle T_{01})|E\rangle \quad (1a)$$

$$|1\rangle|E\rangle \mapsto (|0\rangle T_{10} + |1\rangle T_{11})|E\rangle \quad (1b)$$

where the operators T act on the environment. In analogy with the definition of classical channels, one typically characterizes a channel by the probability that a qubit is transmitted without distortion, as well as the probability of occurrence of certain specific errors. For example, the depolarizing channel (7) assumes that with probability F the qubit is transmitted intact and with probabilities $(1 - F)/3$ it suffers a sign flip, a spin flip, or both, which are represented by the Pauli spin operators $\sigma_{z,x,y}$ acting on the qubit. One usually has in mind a situation where the states of the qubit correspond to two orthogonal polarizations of a photon. Errors are changes of polarization, of the relative phase, or both. However, this description of the channel does not take into account the possibility of photon absorption or photon emission. In fact, for realistic channels, photon absorption is the dominant error, whereas the creation of photons in a particular given mode at optical frequency can be safely neglected. With this in mind, the best choice for encoding information in photons is to assign the state $|0\rangle$ to sending no photon, with the simple idea that, if one sends no photon, it cannot be absorbed. The state $|1\rangle$ is chosen as one of the polarization states. Therefore, this channel acts as follows

$$|0\rangle \mapsto |0\rangle T_0 \quad (2a)$$

$$|1\rangle \mapsto |1\rangle T_1 + |0\rangle T_a \quad (2b)$$

where we have omitted the initial state of the environment. The operator T_a describes the disappearance of a photon of the chosen polarization, due either to photon absorption or to a polarization change. We emphasize that this formulation of encoding and transmission (Eq. 2) incorporates more physical processes (that is, is more general) and yet is simpler than the one using two polarizations (see Eq. 1).

We must include in the description of the channel the fact that the photon is created by matter. In general, we can assume that the photon is produced by making an atom change its internal state. We wish to describe this process in the most general fashion. We consider two atoms A and B belonging to Alice and Bob, respectively. We denote by $|0\rangle$ and $|1\rangle$ two internal levels of the atoms, and by $|x\rangle$ any other level that may be involved in the process. As in (5), we consider a transmission process in which the sending atom will produce a photon only if it is in the state $|1\rangle$. Under ideal conditions, this photon will be absorbed by the receiving atom, which will be transferred from the state $|0\rangle \mapsto |1\rangle$. In reality, there will be errors involving both atoms and photons. For the photons, all possible errors are described by Eq. 2. For the atoms, we only require that if the sending atom is in the state $|0\rangle$, then no photon is produced; and if no photon reaches the receiving atom, which is in the state $|0\rangle$, it does not change state. Any other error can take place; for example, transfer of the atom to any other state $|x\rangle$. In order to keep the atoms in the 2D Hilbert space after the transmission, we optically pump the sending atom to the state $|0\rangle$; and in the receiving atom, we pump any state $|x \neq 0, 1\rangle$ to the state $|0\rangle$ (9). The states of the atoms undergo the following process:

$$|0\rangle_A|0\rangle_B \mapsto |0\rangle_A|0\rangle_B T_0 \quad (3a)$$

$$|1\rangle_A|0\rangle_B \mapsto |0\rangle_A(|1\rangle_B T_1 + |0\rangle_B T_a) \quad (3b)$$

T_0, T_1, and T_a contain spontaneous emission errors, photon absorption, and transitions to other states, followed by repumping to $|0\rangle$; all the physics is in these formulas. A possible way of implementing the process described by Eq. 3 is to use the scheme of (5). In a quantum network, there might be other atoms entangled with A and B. We emphasize that the above definition also applies to this situation. In the following, we will call a channel defined by Eq. 3 the photonic channel. The goal is thus to establish a perfect EPR pair, using the photonic channel. It is instructive to consider the channel as defined in Eq. 3, using its classical definition. There are nonzero probabilities of errors described by the operators σ_z and $\sigma_- = (\sigma_x - i\sigma_y)/2$. Straightforward application of the standard purification scheme to a situation with a finite number of atoms is not possible.

The possibility of (error-free) local quantum computing allows us to reduce the photonic channel (Eq. 3) to a channel without the absorptive term T_a. We will first present an outline of the key idea and then describe the process in detail. Let us assume that Alice has an initial arbitrary state in atom A (which could be entangled with other atoms in the network). Bob has atom B initially in state $|0\rangle_B$. In addition, Alice and Bob need two and one auxiliary atoms in state $|0\rangle$, respectively. Alice performs local operations with her particles and makes several transmissions to Bob using the photonic channel. Bob performs local operations and measurements. For a positive outcome of the measurement (see below), the mapping between the initial and the final state is given by

$$|0\rangle_A|0\rangle_B \mapsto |0\rangle_A|0\rangle_B S_0 \quad (4a)$$

$$|1\rangle_A|0\rangle_B \mapsto |1\rangle_A|1\rangle_B S_1 \quad (4b)$$

where the operators S act on the environment (see below for the specific form), and the auxiliary atoms end up in $|0\rangle$. For the opposite outcome, we recover the initial state of all atoms perfectly. By repeating the above scheme until a positive outcome is obtained, one accomplishes the mapping of Eq. 4 with certainty. The above protocol defines an "effective channel" that is absorption free: By comparing Eq. 3 with Eq. 4 we see that the effective channel acts like the photonic channel but without an absorption term (T_a). In the following, we will call channel reduction a protocol that combines local quantum computing with several transmissions to obtain an effective, less noisy channel.

The proof of the channel reduction involves two layers of protocols, which we describe here. (i) Alice applies a controlled-NOT operation to atom A and an auxiliary atom, and then uses the photonic channel to transmit the state of this atom to B. The mapping of this protocol will be

$$|0\rangle_A|0\rangle_B \mapsto |0\rangle_A|0\rangle_B T_0 \quad (5a)$$

$$|1\rangle_A|0\rangle_B \mapsto |1\rangle_A|1\rangle_B T_1 + |1\rangle_A|0\rangle_B T_a \quad (5b)$$

where the state of the auxiliary atom factorizes out. Equation 5 corresponds to an effective channel, which will be used in the following. (ii) We apply a Hadamard transformation to atom A, followed by a controlled-NOT with the auxiliary atom A_1 (which acts as a backup). Then we transmit the qubit A to B (at time t) according to Eq. 5, apply the operation NOT to atom A, transmit the qubit A to B_1 (at time t') according to Eq. 5, and apply a NOT operation to atom A again. Now a measurement is performed on atoms B and B_1 to check whether they are in the state $|0\rangle_B|0\rangle_{B_1}$: (a) If the outcome is "no," we perform the unitary transformation $|0\rangle_B|1\rangle_{B_1} \mapsto |0\rangle_B|0\rangle_{B_1}$, and $|1\rangle_B|0\rangle_{B_1} \mapsto |1\rangle_B|0\rangle_{B_1}$, and measure the state of A_2. If the outcome is $|0\rangle_{A_1}$, then we have Eq. 4 with $S_0 = T_1(t')T_0(t)$ and $S_1 = T_0(t')T_1(t)$, and similarly if $|1\rangle_{A_1}$. Here $T_{0,1}(t)$ and $T_{0,1}(t')$ denote the environment operators acting at time t and t' (first and second transmission, respectively). (b) If the outcome is "yes," we measure the state of A and then swap the state of A_1 into A. If the outcome was $|0\rangle_A$, then one has

$$|0\rangle_A|0\rangle_B \mapsto |0\rangle_A|0\rangle_B S_a \quad (6a)$$

$$|1\rangle_A|0\rangle_B \mapsto |1\rangle_A|0\rangle_B S_a \quad (6b)$$

with $S_a = T_a(t')T_0(t)$, and similarly if it is $|1\rangle_A$. This mapping is the identity, because the environment operator factors out. Therefore, we can repeat this protocol until we obtain a "no" in the measurement.

We define a stationary channel as the one fulfilling

$$T_1(t')T_0(t) = T_0(t')T_1(t) \quad (7)$$

when acting on the environment. In particular, this is true if the Markovian property considered in our previous work (6) holds. In that work, photon absorption was modeled with a Markovian master equation, and the other errors were assumed to be systematic (that is, the same in two subsequent transmissions). In the stationary limit (Eq. 7), the channel in Eq. 4 allows for ideal transmission in a single try. In contrast, we are interested in the general case where the stationarity property does not apply. In particular, this will be the case where there are additional random errors and when a Markovian description of decoherence is questionable. In the following, we will show how to establish distant EPR pairs using the channel in Eq. 4.

Alice and Bob repeatedly perform the process described below. In the Nth intermediate stage, the state of particles A and B is a superposition of two Bell states ("right" and "wrong") $|R\rangle_{AB} = |0\rangle_A|0\rangle_B + |1\rangle_A|1\rangle_B$ and $|W\rangle_{AB} = |0\rangle_A|1\rangle_B + |1\rangle_A|0\rangle_B$ (we omit normalization factors of $1/\sqrt{2}$)

$$|\Psi^{(N)}\rangle = |R\rangle_{AB}|E_R^{(N)}\rangle + |W\rangle_{AB}|E_W^{(N)}\rangle \quad (8)$$

where $|E_{R,W}^{(N)}\rangle$ are unnormalized states of the environment. In order to characterize the quality of the state in Eq. 8, we define its fidelity as $F_N = \||E_R^{(N)}\rangle\|^2$. The goal is to increase the fidelity so that for large N the state of the system will tend to $|R\rangle$.

Initially, Alice prepares her qubit A in the state $|+\rangle_A$ and Bob prepares his qubit B in $|0\rangle_B$. They use the channel in Eq. 4, and then both of them apply the local Hadamard operation $|0\rangle \mapsto |+\rangle$, and $|1\rangle \mapsto |-\rangle$, where we have defined $|\pm\rangle = |0\rangle \pm |1\rangle$. They obtain Eq. 8 with $|E^1_{R,W}\rangle = 1/2(S_0 \pm S_1)|E\rangle$, where $|E\rangle$ is the initial state of the environment. They repeatedly perform the following process using two auxiliary atoms A_2 and B_2.

1) The auxiliary qubit A_2 is locally entangled with the qubit A according to the transformation $|0\rangle_A|0\rangle_{A_2} \mapsto |0\rangle_A|+\rangle_{A_2}$, and $|1\rangle_A|0\rangle_{A_2} \mapsto |1\rangle_A|-\rangle_{A_2}$.

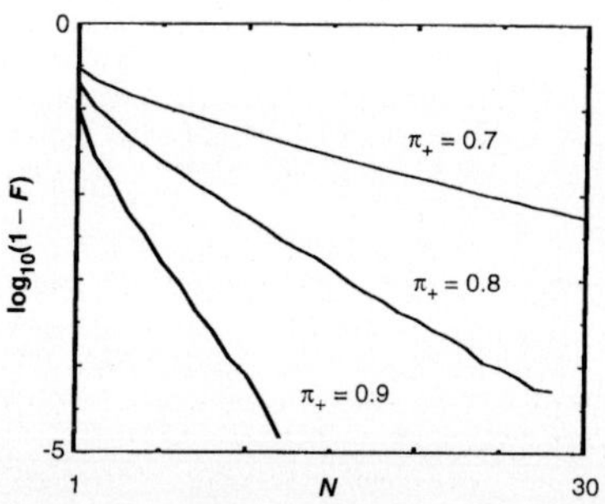

Fig. 1. Plot of the logarithm of the mean value of $1 - F_N$ as function of the number of steps N for $\pi_+ = 0.9$, 0.8, and 0.7.

2) The qubit A_2 is transmitted to the auxiliary qubit B_2 according to the effective channel of Eq. 4. Then the qubit A_2 is measured in the $|\pm\rangle_{A_2}$ basis. If the result is $|-\rangle_{A_2}$, one applies the unitary operation $|1\rangle_{B_2} \mapsto -|1\rangle_{B_2}$. Then one applies the transformation $|1\rangle_B|1\rangle_{B_2} \mapsto -|1\rangle_B|1\rangle_{B_2}$. The state after the transmission will be

$$|\Phi^{(N)}\rangle = |R\rangle_{AB}(|0\rangle_{B_2}S_0 + |1\rangle_{B_2}S_1)|E_R^{(N)}\rangle + |W\rangle_{AB}(|0\rangle_{B_2}S_0 - |1\rangle_{B_2}S_1)|E_W^{(N)}\rangle \quad (9)$$

3) The auxiliary qubit B_2 is measured. If the outcome is $|\pm\rangle_{B_2}$, the state becomes Eq. 8 with

$$|E_{R,W}^{(N+1)}\rangle = \frac{1}{2}(S_0 \pm S_1)|E_{R,W}^{(N)}\rangle \quad (10a)$$

$$|E_{R,W}^{(N+1)}\rangle = \frac{1}{2}(S_0 \mp S_1)|E_{R,W}^{(N)}\rangle \quad (10b)$$

respectively. We will denote the probability of these outcomes by $P_\pm$.

We analyze how the fidelity changes after each step, for which we need to evaluate $P_\pm$. To this end we define

$$\pi_\pm \equiv \frac{\left\|\frac{1}{2}(S_0 \pm S_1)|E\rangle\right\|^2}{\||E\rangle\|^2} \quad (11)$$

This parameter gives the probability that, starting from a perfect EPR pair, after one step we obtain the outcome $|\pm\rangle_{B_2}$. To calculate $\pi_\pm$, one needs to know the specific form of the operators and states at all times. We will estimate the change in the fidelity by assuming that $\pi_\pm$ does not depend on $|E\rangle$. Using the definition of Eq. 11, we have $P_\pm = \pi_\pm F_N + \pi_\mp(1 - F_N)$. Then, depending on the outcome of the measurement $|\pm\rangle_{B_2}$ the new fidelity is

$$F_{N+1} = \frac{\pi_\pm F_N}{\pi_\pm F_N + \pi_\mp(1 - F_N)} \quad (12)$$

respectively. For $\pi_+ > \pi_-$, the outcome $|+\rangle_{B_2}$ increases the fidelity and occurs with a higher probability. Because the decrease in fidelity after a $|-\rangle_{B_2}$ measurement is compensated for by a subsequent $|+\rangle_{B_2}$ measurement, the protocol consists of a random walk along a set of particular values of F, where it is more likely to go up than to go down, thus achieving $F \uparrow 1$ asymptotically. The process depends only on the value of π_+, which characterizes the effective channel; a good channel has a $\pi_+ \simeq 1$ (for the stationary channel, $\pi_+ = 1$).

We have simulated the improvement of the fidelity for several values of the probability π_+. In Fig. 1, we have plotted the logarithm of the mean value of $1 - F_N$ as a function of the number of steps N for π_+

$= 0.9$, 0.8, and 0.7. For example, we obtain $F = 1 - 10^{-5}$ after $N = 12$ steps for $\pi_+ = 0.9$. For general channels the fidelity approaches $F \sim 1 - e^{-cN}$, with N the number of steps. We emphasize that for $\pi_+ = 1$ we have a stationary channel. In this case, we obtain $F_1 = 1$ in a single step.

For a "good" standard channel (see Eq. 3), T_0 is close to T_1. As a consequence of our reduction procedure, this implies that $S_0 \simeq S_1$ and therefore $\pi_+ \simeq 1$, and $\pi_- \ll 1$. We emphasize that the reverse statement is not true; namely, one can have $S_0 \simeq S_1$ but a "bad" standard channel. Consider, for example, a very simple toy model in which the environment is a qubit in the initial state $|0\rangle$, and with $T_0 = \mathbb{1}$ and $T_1 = \sigma_x$. With the classical definition of a channel, one can easily show that this channel cannot produce entanglement; suppose that Alice has an entangled state of two qubits A and A_2, $|\alpha\rangle_A|0\rangle_{A_2} + |\beta\rangle_A|1\rangle_{A_2}$, and sends the second qubit to the qubit B of Bob via such a channel. The state after this transmission will be a mixed state $|\alpha\rangle_A\langle\alpha||0\rangle_B\langle 0| + \langle\beta\rangle_A\langle\beta||1\rangle_B\langle 1|$ and therefore is not entangled. However, for this channel $S_0 = S_1 = \sigma_x$, and therefore one can establish an EPR pair with the procedure introduced above. By twice using the channel as we proposed, the state of the environment after both transmission factorizes out, and therefore entanglement can be produced. When $S_0 \simeq S_1$, then $\pi_- \propto \|(S_0 - S_1)|E\rangle\|^2 \approx 0$, which is due to quantum interference between the first and second transmission, using the reduction scheme of Eq. 4.

We have defined a photonic channel where $|0\rangle$ is assigned to sending no photon and $|1\rangle$ is assigned to sending one photon. Using local quantum computing with three and two auxiliary atoms in the first and second node, we have reduced it to an absorption-free channel. We have proposed a scheme based on this channel that iteratively improves the fidelity of distant EPR pairs, using quantum interference between two transmissions. For a stationary channel, one obtains a pure EPR pair in a single step. For a general channel, the fidelity approaches 1 exponentially with the number of steps.

REFERENCES AND NOTES

1. W. K. Wootters and W. H. Zurek, *Nature* **299**, 802 (1982).
2. C. H. Bennett, *Phys. Today* **48** (no. 1), 24 (1995).
3. C. H. Bennett *et al.*, *Phys. Rev. Lett.* **76**, 722 (1996); A. Ekert and C. Macchiavello, *ibid.* **77**, 2585 (1996).
4. P. W. Shor, *Phys. Rev. A* **52**, 2493 (1995); A. M. Steane, *Phys. Rev. Lett.* **77**, 793 (1996); E. Knill and R. Laflamme, *Phys. Rev. A* **55**, 900 (1997); J. I. Cirac, T. Pellizzari, P. Zoller, *Science* **273**, 1207 (1996).
5. J. I. Cirac, P. Zoller, H. Mabuchi, H. J. Kimble, *Phys. Rev. Lett.* **78**, 3221 (1997).
6. S. J. van Enk, J. I. Cirac, P. Zoller, *Phys. Rev. Lett.* **78**, 4293 (1997).
7. See, for example, B. Schumacher, *Phys. Rev. A* **45**, 2614 (1996); C. H. Bennett, D. P. DiVincenzo, J. A. Smolin, *Phys. Rev. Lett.* **78**, 3217 (1997).
8. C. H. Bennett *et al.*, *Phys. Rev. Lett.* **70**, 1895 (1993).
9. One can also use repumping techniques to supply fresh ancillas in error correction and purification. However, one still needs an infinite number to correct all errors.

10. This work was supported in part by the TMR network ERB-FMRX-CT96-0087 and by the Austrian Science Foundation.

2 July 1997; accepted 12 November 1997

Quantum Key Distribution

Quantum key distribution

Grgoire Ribordy, Nicolas Gisin and Hugo Zbinden
Universit de Genve, Group of Applied Physics

The expansion of telecommunications during the past two decades has induced the need for the development of new cryptographic techniques offering high security as well as easy key management. The so-called public key cryptosystems, first introduced by Diffie and Hellman [1] in 1975, were the first attempt to solve the key distribution problem. They make use of mutually inverse transformations for encrypting and decrypting. The encryption algorithm and key are made public - hence the name of public key cryptosystems - and the safety relies on the high complexity of the inverse transformation, unless the decryption - or private - key can be used. These ciphers form the class of asymmetric cryptosystems, of which RSA [2], well known to all internet surfers, is the most widely used. However, they suffer from a major flaw, namely the fact that their security relies on unproven assumptions about complexity theory. They could indeed be broken by a fast algorithm. Although such an algorithm has already been introduced for quantum computers, it is not sure whether a similar one could also be found for classical computers. The fact that this discovery would immediately jeopardize the secrecy of most electronic communications explains the strong interest in other encryption techniques. One could of course use symmetric ciphers. This class of cryptosystems includes Vernam's "one time pad", which is the only one offering proven absolute secrecy. In this case however, the problem of secrecy is shifted to the distribution of the secret key safely between the emitter and the receiver. A spy could indeed intercept the key material and copy it. Quantum key distribution [3, 4] - often referred to as quantum cryptography - prevents this from happening. It complements symmetric ciphers, to simplify their implementation. This promising application has triggered a very strong interest in the industry as well as in the public. Its security is based on the principles of physics, and not on any unproven assumptions about complexity theory.

Quantum key distribution was first proposed in 1984 by Bennett and Brassard [5], respectively of IBM and University of Montreal, and allows two remote parties to generate a secret random key, used to carry out secure communications, without meeting or resorting to the services of a courier. It is based on the fact that, according to quantum physics, performing a measurement on an unknown state will in most cases disturb it. This property can be exploited to reveal an eavesdropper. Indeed, if the sequence of transmitted bits, encoded in non-orthogonal states, does not contain any errors, it can be inferred that no unlegitimate party tried to listen in. In addition to the first proposal by Bennett and Brassard, other key distribution protocols have been introduced [6,7].

A long way has been covered by the quantum optics community, since the first demonstration of quantum key distribution over 30 cm in air by Bennett and his coworkers in

1989 [8]. Several groups have developed and tested key distribution schemes over distances ranging between 1 and 50 kilometers. They have all used photons as information carrying quanta, because of the existence of efficient channels to transport them - namely optical fibers - and of the availability and rapid improvement of photonic components, induced by the development of optical telecommunications.

The research activity in this field focuses on two main issues. The first center of interest is the development of practical quantum key distribution schemes for real applications. The values of the bits can indeed be encoded into various properties of the photons. The most common choices are phase and polarization. It is essential to select a property that is robust to decoherence. In addition, the system should feature good stability and simple adjustment. The first demonstrations were carried out with polarization encoding [9,10,11]. This choice is however not optimal. If the polarization mode dispersion of the fiber link becomes too large, the photons are depolarized, making key distribution impossible. Besides, this birefringence varies with time, causing random changes in the output polarization state of photons transmitted through a fiber. Although these fluctuations are slow enough to allow active polarization tracking, such a compensation would be very unpractical in a real system. Although it is more robust to decoherence, choosing the phase of the photons to carry bits values [12,13] is the source of other difficulties. As such a scheme is based on single photon interference, the photons travel successively through two interferometers separated by the fiber link. Their path differences must be set equal within less than a wavelength and stabilized. Polarization tracking then becomes superfluous, but it is replaced by interferometer adjustment. Moreover this scheme still requires polarization control in the interferometers in order to ensure high visibility. A third type of systems, using an alternate interferometer [14,15,16], combines the advantages of both previous schemes. Polarization independence is ensured by the use of Faraday mirrors, while path length adjustment is achieved by time-multiplexing. This system represents one of the most serious candidate for the application of quantum cryptography in the real world. There exists others proposals for quantum key distribution [17,18,19,20], but the three systems discussed above underwent the most thorough experimental tests. It should be noted that the most critical components in all of these systems are the photon counting detectors. Their poor performance sets a limit to the distance and the key generation rate achievable. As one must use avalanche photodiodes to detect single photons at the telecommunications wavelength [21,22], these devices could only be improved by semiconductor components manufacturer, which lack the interest.

The second focus of the research activity in this field is theoretical. On the one hand, the study of eavesdropping strategies is essential, to quantify the minimum perturbation induced by a clever eavesdropper [23,24,25,26,27]. On the other hand, the development of protocols is becoming important, with the improvement of experimental systems towards prototypes. In real applications, there are always errors in the key distributed, due to detectors noise and other experimental factors. To guarantee secrecy, it is essential to attribute these errors to a spy and to deal with them. One must indeed remove them through the application of an error correction algorithm, before reducing the information that may have leaked to the eavesdropper with a technique called "privacy amplification" [28,29]. Both of these operations result in a shortening of the key. In order to preserve as much of the key material as possible, it is necessary to come up with optimal algorithms. This issue may yield valuable crossfertilization between classical and quantum information theory [30].

In conclusion, quantum cryptography has come a long way since the first experiments, about ten years ago. It has now reached the point where the development and test of a prototype would be feasible. The study of dedicated protocols is hence of major importance. In spite of the progresses, there is still room for interesting experiments on alternate schemes. As a result, quantum key distribution may well be the first direct application of quantum physics in everyday life.

1. W.Diffie, and M. Hellman, "New directions in cryptography", IEEE Trans. Inf. Theory IT-22, pp. 644-654 (1976)

2. R. Rivest, A. Shamir, and L. Adleman, "On digital signatures and public key cryptosystems", MIT Lab. For Comp. Science, Technical Report, MIT/LCS/TR-212 (January 1979)

3 see below H. Zbinden, N. Gisin, B. Huttner, A. Muller, and W. Tittel, "Practical aspects of quantum key distribution", J. Cryptology 11, pp. 1-14 (1998)

4 H. Zbinden, H. Bechmann- Pasquinucci, N. Gisin, G. Ribordy, "Quantum Cryptography", Appl. Phys. B 67, pp. 743-748 (1998)

5 C. H. Bennett and G. Brassard, "Quantum cryptography: public key distribution and coin tossing", Proc. Internat. Conf. Computer Systems and Signal Processing, Bangalore, pp. 175-179 (1984)

6 C. Bennett, "Quantum cryptography using any two nonorthogonal states", Phys. Rev. Lett., 68 (21), pp. 3121- 3124 (1992)

7 A. K. Ekert, "Quantum cryptography based on Bell's theorem", Phys. Rev. Lett., 67 (6), pp. 661-663 (1991)

8 C. H. Bennett, F. Bessette, G. Brassard, L. Salvail, and J. Smolin, "Experimental quantum cryptography", J. Cryptology 5, pp. 3-28 (1992)

9 A. Muller, J. Breguet, and N. Gisin, "Experimental demonstration of quantum cryptography using polarized photons in optical fibre over more than 1 km", Europhys. Lett., 23 (6), pp. 383-388 (1993)

10 J. D. Franson, and B. C. Jacobs, "Operational system for quantum cryptography", Electron. Lett., 31 (3), pp. 232-234 (1995)

11 P. D. Townsend, "Experimental investigation of the performance limits for first telecommunication window quantum cryptography systems", IEEE Photon. Technol. Lett., 10 (7), pp. 1048-1050 (1998)

12 C. Marand, and P. D. Townsend, "Quantum key distribution over distances as long as 30 km", Opt. Lett., 20 (16), pp. 1695-1697 (1995)

13 R. J. Hughes, G. G. Luther, G. L. Morgan, C. G. Peterson, and C. Simmons, "Quantum cryptography over underground optical fibers", Lect. Notes in Comp. Sci., 1109, pp. 329-342 (1996)

14 A. Muller, T. Herzog, B. Huttner, W. Tittel, H. Zbiden, and N. Gisin, "Plug and Play systems for quantum cryptography", Appl. Phys. Lett. 70 (7), pp. 793-795 (1997)

15 H. Zbinden, J. D. Gautier, N. Gisin, B. Huttner, A. Muller, and W. Tittel, "Interferometry with Faraday mirrors fo rquantum cryptography", Electron. Lett., 33 (7), pp. 586-588 (1997)

16 G. Ribordy, J.-D. Gautier, N. Gisin, O. Guinnard, and H. Zbiden, "Automated 'plug & play' quantum key distribution", Electron. Lett., 34 (22), pp. 2116-2117 (1998)

17 A. Ekert, J. Rarity, P. Tapster, and M. Palma, "Pratcical quantum cryptography based on two-photon interferometry", Phys. Rev. Lett., 69 (9), pp. 1293-1295 (1992)

18 B. C. Jacobs, and J. D. Franson, "Quantum cryptography in free space", Opt. Lett., 21 (22), pp. 1854-1856 (1996)

19 W. T. Butler, R. J. Hughes, P. G. Kwiat, S. K. Lamoreaux, G. G. Luther, G. L. Morgan, J. E. Nordholt, C. G. Peterson, and C. M. Simmons, "Practical free-space quantum key distribution over 1 km", Phys. Rev. Lett., 81 (15), pp. 3283-3286 (1998)

20 J.-M- Mrolla, Y. Mazurenko, J.-P. Goedgebuer, H. Porte, W. T. Rhodes, "A phase-modulation transmission system for quantum cryptography", Opt. Lett., 24 (2), pp.

21 P. Owens, J. Rarity, P. Tapster, D. Knight, and P. D. Townsend, "Photon counting with passively quenched germanium avalanche photodiodes", Appl. Opt., 33 (30), pp. 6895-6901 (1994)

22 G. Ribordy, J. D. Gautier, H. Zbinden, and N. Gisin, "Performance of InGaAs/InP avalanche photodiodes as gated-mode photon counters", Appl. Opt., 37 (12), pp. 2272-2277 (1998)

23 I. Cirac and N. Gisin, "Coherent eavesdropping strategies for the four state quantum cryptography protocol", Phys. Lett. A, 229, pp. 1-7 (1997)

24 C. Fuchs, N. Gisin, R. Griffiths, C.-S. Niu, and A. Peres, "Optimal eavesdropping in quantum cryptography. I. Information bound and optimal strategy", Phys. Rev. A, 56 (2), pp. 1163-1172 (1997)

25 B. Slutsky, R. Rao, P.-C. Sun, and Y. Fainman, "Security of quantum cryptography against individual attacks", Phys. Rev. A, 57 (4), pp. 2383-2398 (1998)

26 E. Biham, M. Boyer, G. Brassard, J. van de Graaf, T. Mor, "Security of Quantum Key Distribution Against All Collective Attacks", quant-ph/9801022

27 D. Mayers, "Unconditional security in Quantum Cryptography", quant-ph/9802025

28 N. Lutkenhaus, "Estimates for practical quantum cryptography", quant-ph/9806008

29 C. Bennett, G. Brassard, C. Crepeau, U. Maurer, "Generalized privacy amplification", IEEE Trans. Inf. Theory, 41 (6), pp. 1915-1923 (1994)

30 N. Gisin and S. Wolf, "Quantum cryptography on noisy channels: quantum versus classical key-agreement protocols", quant-ph/9902048

QUANTUM INFORMATION

Telephones and faxes are not perfectly secure, but send a secret message made up of quantum bits, and you can know for sure if it was read before reaching its intended target

Quantum cryptography

Wolfgang Tittel, Grégoire Ribordy and Nicolas Gisin

WE LIVE in a quantum world – something that physicists have considered with amazement for more than seventy years. But we now realize that quantum physics is more than a radical departure from classical physics. It also offers many new possibilities for information processing.

Quantum cryptography is the most mature prospect of this fascinating new field. It is based on the fundamental postulate of quantum physics that "every measurement perturbs a system". Imagine sending a message carried by single quantum states, such as linearly polarized photons oriented at various angles. If the bits are not altered during transmission, you can be sure that no eavesdropper has measured the values of those bits. In other words, quantum cryptography turns an apparent limitation – namely that a measurement perturbs the system – into a potentially useful process, in which the perturbation uncovers the presence of an eavesdropper.

This idea of turning quantum conundrums into potentially useful processes is a characteristic of the whole field of "quantum information processing". For example, the famous Einstein–Podolsky–Rosen paradox has lead to novel techniques such as "dense coding" and "quantum teleportation" (see "Fundamentals of quantum information" by Zeilinger on page 35). "Quantum entanglement", meanwhile, could make it possible to build quantum computers that could factorize large integers exponentially faster than the best-known algorithm for classical computers (see "Quantum computation" by Deutsch and Ekert on page 47).

1 Carry on spying

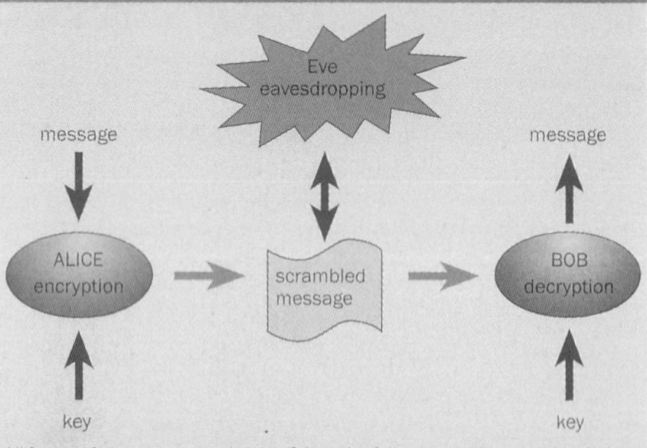

All forms of secret communication follow the following basic principles. Using an algorithm of some sort, the sender, Alice, combines a message with a key to create a scrambled message that she sends to Bob. He decrypts the scrambled message using his key to reveal the real message. Eve is an unauthorized eavesdropper. Quantum versions of this set-up enable Alice and Bob to exchange the key with absolute security and to find out whether or not Eve read their message.

Standard crypto-systems

Cryptography is the art of hiding information in a string of bits that are meaningless to any unauthorized party. To achieve this goal, an algorithm is used to combine a message with some additional information – known as the "key" – to produce a cryptogram. This technique is known as "encryption" (figure 1). The person who encrypts and transmits the message is traditionally known as Alice, while the person who receives it is called Bob. Eve is the unauthorized, malevolent eavesdropper. For a crypto-system to be secure, it should be impossible to unlock the cryptogram without Bob's key. In practice, this demand is often softened so that the system is just extremely difficult to crack. The idea is that the message should remain protected as long as the information it contains is valuable.

Crypto-systems come in two main classes – depending on whether the key is shared in secret or in public. The "one-time pad" system, which was proposed by Gilbert Vernam at AT&T in 1935, involves sharing a secret key and is the only crypto-system that provides proven, perfect secrecy. In this scheme, Alice encrypts a message using a randomly generated key and then simply adds each bit of the message to the corresponding bit of the key (figure 2). The scrambled text is then sent to Bob, who decrypts the message by subtracting the same key. Because the bits of the scrambled text are as random as those of the key, they do not contain any information.

Although perfectly secure, the problem with this system is that it is essential for Alice and Bob to share a common secret key, which must be at least as long as the message itself. They can also only use the key for a single encryption – hence the name "one-time pad". (If they used the key more than once, Eve could record all of the scrambled messages and start to build up a picture of the key.) Furthermore, the key has to be transmitted by some trusted means, such as a courier, or through a personal meeting between Alice and Bob. This procedure can be complex and expensive, and may even amount to a loophole in the system. (It is interesting to note that if Eve wanted to crack the one-time pad by trying out all possible keys one by one, she would obtain a message for each key and would then have to search through all of them. But she would have absolutely no way of knowing which was the right one!)

The other class of crypto-systems shares a public key. The first "public key crypto-systems" were proposed in 1976 by Whitfield Diffie and Martin Hellman, who were then at Stanford University in the US. These systems are based on so-called one-way functions, in which it is easy to compute the function $f(x)$ given the variable x, but difficult to go in the opposite direction and compute x from $f(x)$. In this context, the word "difficult" means that the time to do a task grows exponentially with the number of bits in the input. Factoring large integers is a candidate for such a one-way function. For example, it only takes a few seconds to work out that 107×53 is 5671, but it takes much longer to find the prime factors of 5671.

However, some of these one-way functions have a "trapdoor", which means that there is in fact an easy way of doing the computation in the difficult direction, provided that you have some additional information. For example, if you were told that 107 was one of the prime factors of 5671, the calculation would be relatively easy.

2 Perfect security

Alice		
Message		10110101
Add key	+	01101001
Scrambled text	=	11011100
	↓	
Transmit		
	↓	
Bob		
Scrambled text		11011100
Subtract key	−	01101001
Message	=	10110101

The "one-time pad" crypto-system allows messages to be sent with perfect security. Alice chooses a random number for the key and encrypts her message by adding the key to her message. She then transmits the scrambled message to Bob, who decrypts it by subtracting the key to reveal the real message. In this example, all of the calculations are performed in modulo 2 (i.e. 1+1=0+0=0; 1+0=0+1=1; 1−1=0−0=0; and 1−0=0−1=1). The problem with this system is that both the sender and recipient have to share the key.

For Alice to transmit a message with a public-key crypto-system, Bob first chooses a private key. He uses this key to compute a public key, which he discloses publicly. Alice then uses this public key to encrypt her message. She transmits the encrypted message to Bob, who decrypts it with his private key. The encryption–decryption process can be described mathematically as a one-way function with a trapdoor – namely, the private key. One therefore only needs to know this key to obtain the original message. In other words, if Bob knows what the trapdoor is, he can do the reverse calculation and reveal the message from the encrypted text.

Public-key crypto-systems are convenient and they have become very popular over the last 20 years. The security of the Internet, for example, is partially based on such systems. The most common example is the RSA crypto-system, which was developed by Ronald Rivest, Adi Shamir and Leonard Adleman of the Massachusetts Institute of Technology in 1977. Its secrecy is actually based on the fact that (as far as we know) the time needed to calculate the prime factors of an integer – and hence to work out the private key – increases exponentially with the number of input bits.

However, this system suffers from two potential major flaws. First, nobody knows for sure if factorization is actually as difficult as we currently think. Of course, one could easily improve the safety of the RSA by choosing a longer key, but if an algorithm were found that could factorize numbers quickly, it would immediately annihilate the security of the RSA system. Although such an algorithm has not yet been discovered – or if it has, it has not been published! – there is no guarantee that such an algorithm does not exist.

The second drawback to the RSA system is that problems that are difficult for a classical computer could become easy for a quantum computer (see box). With the recent developments in the theory of quantum computation, there are reasons to believe that it will eventually become possible to build these machines. If either of these possibilities were fulfilled, the RSA system would become obsolete. Meanwhile, other public-key crypto-systems also rely on unproven assumptions for their security, which could themselves be weakened or suppressed by theoretical or practical advances. One would then have no choice but to turn to secret-key crypto-systems.

Quantum cryptography on paper

The principles of cryptography that we have so far described have all been entirely general. Vernam's system, however, requires Bob and Alice to share a secret key, and it is here that quantum physics enters the scene. Quantum cryptography allows two physically separated parties to create a random secret key without resorting to the services of a courier. It also allows them to verify that the key has not been intercepted. ("Quantum key distribution" is therefore really a better name for quantum cryptography.) When used with Vernam's one-time pad scheme, the key allows the message to be transmitted with proven and absolute security. Quantum cryptography is not therefore a totally new crypto-system. But it does allow a key to be securely distributed and is consequently a natural complement to Vernam's cipher.

To understand how quantum cryptography works, consider the "BB84" communication protocol, which was introduced in 1984 by Charles Bennett of IBM in Yorktown Heights, US, and Gilles Brassard from the University of Montreal in Canada (figure 3). Alice and Bob are connected by a quantum channel and a classical public channel. If single photons are being used to carry the information, the quantum channel is usually an optical fibre. The public channel, however, can be any communication link, such as a phone line or an Internet connection. In practice, the public link is usually also an optical fibre, with both channels differing only in the intensity of the light pulses that code the bits: one photon per bit for the quantum channel, hundreds of photons per bit for the classical public channel. So how does it work?

First, Alice has four polarizers, which can transmit single photons that are linearly polarized either vertically, horizontally, at +45° or at −45°. She sends a series of photons down the quantum channel, having chosen at random one of the polarization states for each photon. She also records her choice.

Second, Bob has two analysers. One analyser allows him to distinguish between horizontally and vertically polarized photons. The other allows him to distinguish between photons polarized at +45° and −45°. Bob selects one analyser at random, and uses it to record each photon. He writes down which analyser he used and what it recorded. Note that every time Bob uses an analyser that is not compatible with Alice's choice of polarization, he will not be able to get any information about the state of the photon. For example, if Alice sent a vertically polarized photon and Bob chose the analyser designed to detect photons at ±45°, there is a 50% chance that he will find the photon in either the +45° channel or the −45° channel. And even if he finds out later that he chose the wrong analyser, he will have no way of finding out which polarization state Alice sent.

Third, after exchanging enough photons, Bob announces on the public channel the sequence of analysers he used, but

3 Encryption with polarized light

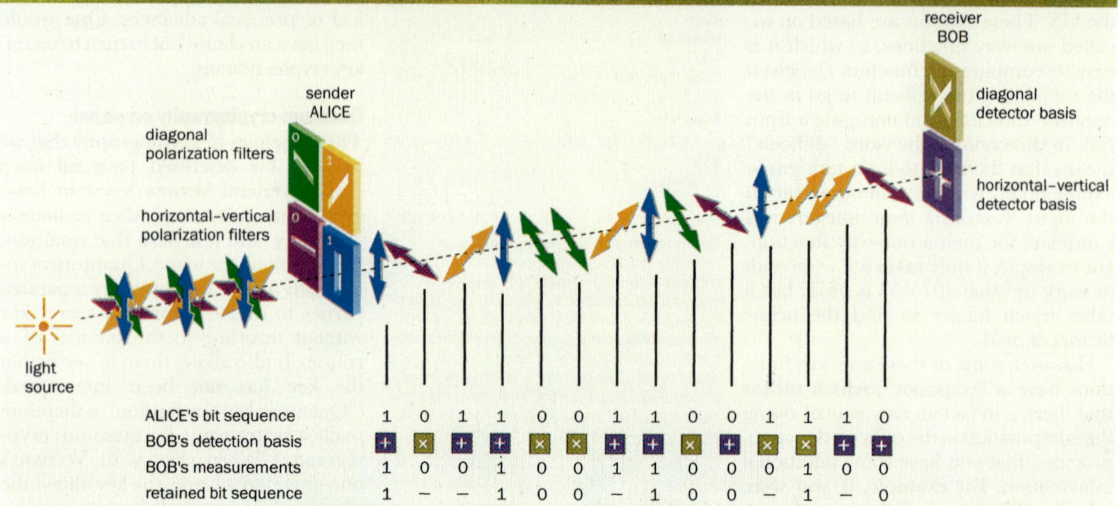

Quantum cryptography relies on creating keys that can be used with absolute secrecy to encrypt and decrypt a message. In the "BB84" communication protocol, Alice has four filters that can linearly polarize photons either vertically, horizontally, at +45° or at –45°. For each photon she sends down an optical fibre, she chooses one of these filters at random (row 1). Bob has an analyser that can distinguish between horizontally and vertically polarized photons, and another one that can distinguish between those polarized at ±45°. Every time he expects a photon to arrive, he chooses one analyser at random (row 2). He records whether or not he detects a signal, which analyser he used and which detector registered the count (row 3). He then tells Alice which photons he detected, the sequence of analysers he used, but not the results he obtained. Alice looks at her data and tells Bob when his analyser was compatible with the polarization of the photon she sent. If the analyser was incompatible, or if Bob did not detect the photon, the bit is discarded. For the bits that remain (row 4), Alice and Bob know for sure that they have the same values and the retained bits can now be used to generate a secret key.

not the results that he obtained.

Fourth, Alice compares this sequence with the list of bits that she originally sent, and tells Bob on the public channel on which occasions his analyser was compatible with the photon's polarization. She does not, however, tell him which polarization states she sent. If Bob used an analyser that was not compatible with Alice's photon, the bit is simply discarded. For the bits that remain, Alice and Bob know that they have the same values – provided that an eavesdropper did not perturb the transmission. They can now use these bits to generate a key, and send encrypted messages to one another.

To assess the secrecy of their communication, Alice and Bob select a random part of their key, and compare it over the public channel. Obviously, the disclosed bits cannot then be used for encryption any more. If their key had been intercepted by an eavesdropper, the correlation between the values of their bits will have been reduced. For example, if Eve has the same equipment as Bob and cuts the fibre and measures the signal, she will always get a random bit whenever she chooses the wrong analyser, i.e. in 50% of cases. But having intercepted the signal, Eve still has to send a photon to Bob, to cover her tracks. Therefore in half of the cases in which Alice's and Bob's analysers match, Eve will have sent a photon that is incorrectly polarized. However, in half of these cases, the photon will accidentally leave Bob's analyser through the correct channel – in which case, Eve's presence goes undetected. The point is that if Eve had been listening in, one in four of Alice's and Bob's bit values would disagree. In other words, her eavesdropping strategy could be easily detected.

There are other eavesdropping strategies that produce a lower disagreement rate. But since all measurements perturb either the vertical–horizontal polarization states or the diagonal states, or all four states, all eavesdropping strategies perturb the system to some extent. Hence, if Alice and Bob do not notice any discrepancies in the subset of their keys, they can be sure that their key has not been intercepted by Eve. They can then use their key with total confidence to encrypt a message.

Quantum cryptography in the real world

So how do people achieve quantum cryptography in practice? Photons are the best candidates for carrying the different quantum states. They are relatively easy to produce and can be transmitted using existing optical fibres. Over the last 25 years, the attenuation of light at a wavelength of 1300 nm has been reduced from several decibels per metre of fibre to just 0.35 decibels per kilometre. This means that photons can travel up to 10 km in a fibre before half of them are absorbed, which is sufficient to perform quantum cryptography in local networks. (Amplifiers cannot be used to transmit the photons further, because quantum states cannot be copied.) Although most quantum-key distribution prototypes use optical fibres, there are some projects aiming to establish quantum communication from a satellite down to earth or to another satellite.

As always happens in physics, however, there is a gap between theory and experiment. In practice, there will always be some errors in the transmission, usually up to a few per cent. The number of errors that are transmitted as a fraction of the total number of detected bits is called the quantum-bit error rate and is one of the parameters that characterizes how well a quantum-cryptography system works.

Uncorrelated bits may originate from several experimental imperfections. For example, Alice has to ensure that she creates photons that are in exactly the states she chose. If, for instance, a vertical photon is incorrectly polarized at an angle of 84°, there is a 1% possibility that Bob will find it in the channel for horizontally polarized photons. A similar problem arises for Bob. If his polarizer cannot distinguish perfectly between two orthogonal states, he will detect photons in the wrong channel from time to time. Another difficulty is ensuring that the encoded bits are maintained during transmission. A vertically polarized photon, for example, should still be vertically polarized by the time it reaches Bob. But due to the birefringence of the fibre, the polarization states received by Bob will, in general, be different from those sent by Alice.

Even worse, changes to the mechanical or thermal environment can produce fluctuations on a time-scale of seconds or minutes, which means that the alignment of the two analysers has to be continuously monitored. This is possible in principle, but is not very convenient. In fact, the number of transmission errors – and hence the quantum-bit error rate – is dominated by the noise of the detector. In other words, most errors are not due to photons that have been incorrectly detected. The errors arise when a photon fails to reach a detector as expected and the wrong detector registers a dark count instead. Unfortunately, at the wavelengths where the fibre losses are low (i.e. 1310 nm), relatively noisy, low-efficiency home-made single-photon detectors have to be used.

To overcome these problems, Alice and Bob have to apply a classical error-correction algorithm to their data so that they can reduce the errors below an error rate of 10^{-9} – the industry standard for digital telecommunications. And since they cannot be sure if the presence of uncorrelated bits was due to the poor performance of their set-up or to an eavesdropper, they have to assume the worst-case scenario – namely that all of the errors were caused by Eve. To reduce the amount of information that Eve may have obtained, Alice and Bob therefore use a procedure known as "privacy amplification", in which several bits are combined into one. This procedure ensures that the combined bits only correlate if Alice and Bob's initial bits are the same. But Eve ends up with a totally different series of bits, because she only knows a fraction of the initial bits. The problem with privacy amplification is that it shortens the key length a lot and it is only possible up to certain error, which means that Alice and Bob have to be careful to introduce as few errors as possible when they initially send their quantum bits.

How fast are computers?

The continuous dialogue between basic quantum physics and fascinating potential applications leads to one basic question: are quantum computers really faster than classical ones? The consequences of solving this question will be dramatic whatever the answer. If quantum computers are indeed much faster, it would obviously be worth investing money in this field, although the very concept of information would then have to be changed. Instead of being part of mathematics, information would become part of physics! On the other hand, if classical computers can be as fast as quantum ones, then presumably the best classical algorithms have not yet been found. This finding could destroy all of the major security systems, which our IT-dependent society relies so heavily upon.

One of the fathers of quantum computing, David Deutsch of Oxford University, has recently argued that physics is more fundamental than mathematics, because answers to mathematical questions (like working out to which class of complexity a mathematical problem belongs) depend on physics. This claim has come as a shattering blow to mathematicians, who in an attempt to keep their science as the root of all others, are now trying to prove that classical computers are actually as efficient as quantum computers.

It is amusing to follow these debates that have been provoked by quantum physics, but it is important to realize that progress on these fundamental issues could happen soon, since some excellent theorists have recently joined the field. (Security managers, however, might be having nightmares!) But whatever the outcome of these debates, quantum cryptography and other applications of quantum communication are already proving that quantum mechanics can do useful things that are impossible with classical physics.

Cryptography experiments

Quantum cryptography moved from the realms of theory to experiment in 1989, when researchers at IBM built the first prototype that could securely distribute a key. They coded their message using polarized photons (figure 3), and managed to send it over a distance of 30 cm in air. Since then, the improvements have been immense, and several groups have shown that quantum cryptography works outside the lab as well. (We will only consider those systems that use 1310 nm photons, which could one day be used over long distances.) At Geneva University in 1995, the authors also demonstrated the feasibility of the polarization-encoding scheme with installed Swisscom fibres, and BT (formerly British Telecom) followed in 1997 with a similar system.

Another set-up, which encodes the message using the photons' phase rather than their polarization, was developed in 1993 by Paul Townsend and colleagues at BT. In this scheme, both Alice and Bob use identical unbalanced "Mach–Zehnder interferometers", in which one arm is longer than the other (figure 4). They are used to produce and detect photons with a particular phase shift. This scheme is also being used by Richard Hughes and his group at the Los Alamos National Laboratory in New Mexico.

Pulses that go down the short arm in Alice's interferometer and then the long arm in Bob's interferometer interfere with pulses that take the long first and then the short one. When Alice sends her message, she randomly applies phase shifts of 0, $\pi/2$, π or $3\pi/2$ to her photons. Bob, however, only has the option of applying a phase shift of $\pi/2$ or none at all. If Bob applies no phase shift, he can work out whether Alice's photon has a phase shift of 0 or π. On the other hand, if Bob applies a phase shift of $\pi/2$, he can distinguish between Alice's choice of $\pi/2$ and $3\pi/2$. After the message has been sent, Alice and Bob compare their settings using the public channel. If they chose compatible settings, Bob knows which phase Alice applied. A secret key can therefore be established by interpreting phase shifts of 0 and $\pi/2$ as "1", and π and $3\pi/2$ as "0". Incompatible measurements are disregarded.

As with polarization encoding, this scheme has to be actively controlled. For example, the arms in the two interferometers have to be adjusted so that the differences in the path length are the same. These differences also have to be kept stable. Another problem is that the two pulses at Bob's interferometer interfere perfectly only if they are in the same polarization state, which means that the scheme also requires an active polarization control.

In collaboration with Swisscom, we have recently proposed and tested a new type of interferometer that is self-balanced

4 Phase encryption

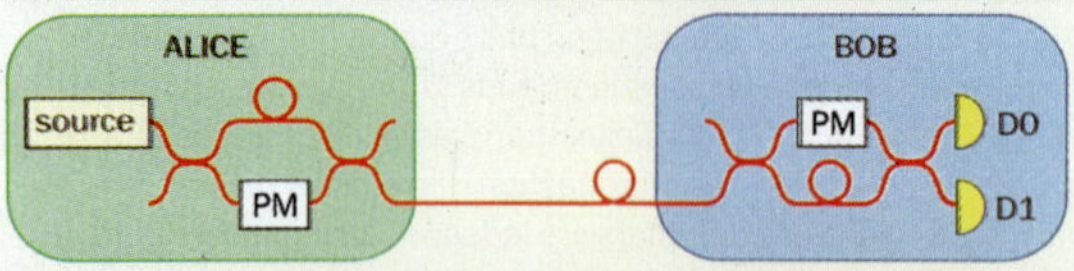

This set-up has been used by researchers at British Telecom and Los Alamos to encode and send a message using a photon's phase rather than polarization. Alice and Bob both have identical unbalanced Mach–Zehnder interferometers, each of which consists of one short arm containing a phase modulator (PM) and one long arm (denoted by the circle). Light entering Alice's interferometer is split in two and passes down the separate arms, before recombining where the arms join up. Alice's phase modulator is used to add a phase shift of either 0, $\pi/2$, π or $3\pi/2$. Bob can either apply a phase shift of 0 or $\pi/2$. Depending on whether the photon is detected by detector D0 or D1, this allows him to distinguish between Alice's phase shift of 0 or π, or between a phase shift of $\pi/2$ or $3\pi/2$.

and in which all birefringence fluctuations are automatically compensated. This set-up uses "time-multiplexed interferometry" – in other words, the pulses that interfere travel along precisely the same paths, but at different times. The advantage is that thermal drifts do not have to be controlled. Moreover, any fluctuations in the polarization of the interfering pulses are wiped out using "Faraday mirrors" at the end of the fibres – instruments that reflect light and transform the state of polarization to the orthogonal one.

There are also prototypes that work at other wavelengths. However, due to higher losses in the fibres, these systems cannot be used to transmit quantum bits any further than a few kilometres. For example, James Franson from Johns Hopkins University in Baltimore demonstrated polarization encoding in 1995 using 830 nm photons. Last year, BT tested a similar system, working at a frequency of 1.2 MHz, which is the highest transmission rate for quantum-key distribution to have so far been achieved.

Cryptography on noisy channels

Although quantum cryptography on noiseless channels has proved to be perfectly secure, noisy channels are much more difficult to handle. The problem with noisy channels is that if Eve intercepts and reads a message, she could then pass on a partially garbled message and get away with it. And if the quantum-bit error rate on her message is lower than the level of noisè, Alice and Bob would never suspect anything.

Before they send any messages, Alice and Bob therefore have to evaluate how much information Eve could possibly obtain. They assume that Eve has unlimited technology, and that her eavesdropping strategy is only restricted by the laws of physics. Once Alice and Bob establish an upper limit on the amount of information that Eve knows, they can – provided this limit is not too high – use error-correction and privacy-amplification algorithms to reduce the information that she can get her hands on. Although this approach will produce a final key that is shorter than the raw data, Eve's information about the final key will then be arbitrarily small.

The drawback is that the complete solution to this problem is not yet known. However, if one assumes that Eve can only interact one by one with the quantum bits that Alice sent to Bob, it turns out that Eve will never know as much as Bob, provided that the quantum-bit error rate is less than 15%. Remarkably, this result establishes a connection with the famous "Bell inequality" – an inequality that is satisfied by all local hidden variable theories, but not by quantum mechanics. Eve's information is lower than Bob's if and only if Bob's results cannot be explained by any local hidden variable theory! This point nicely illustrates the fascinating dual nature of quantum information theory. It deals on the one hand with practical issues, such as the security of crypto-systems and fundamental questions about quantum physics – like non-locality – on the other.

The future starts here

Several groups have now shown that quantum cryptography is possible outside the laboratory. The error rates in sending quantum bits are now low enough to guarantee that the key can be securely distributed. Although the systems still suffer from low transmission rates – and messages can only be sent over a few tens of kilometres – they could, even today, provide a means of securely transmitting messages if the public-key systems that are used on the Internet were suddenly cracked. But, above all, quantum cryptography is fun. Not only does it naturally complement standard crypto-systems, it is also an excellent example of the interplay between fundamental and applied research.

Further reading

C A Fuchs 1997 Optimal eavesdropping in quantum cryptography. 1. Information bound and optimal strategy *Phys. Rev.* **A56** 1163–1172

R J Hughes 1995 Quantum cryptography *Contemp. Phys.* **36** 149–163

N D Mermin 1981 Bringing home the atomic world: quantum mysteries for anybody *Am. J. Phys.* **49** 940–943

S J D Phoenix and P D Townsend 1995 Quantum cryptography: how to beat the code breakers using quantum mechanics *Contemp. Phys.* **36** 165–195

J G Rarity June 1994 Dreams of a quiet light *Physics World* July pp46–51

T Spiller 1996 Q information processing: cryptography, computation and teleportation *Proc. IEEE* **84** 1719–46

Wolfgang Tittel, **Grégoire Ribordy** and **Nicolas Gisin** are in GAP-Optique, Université de Genève, 20 rue de l'Ecole de Medicine, CH-1211 Genève, Switzerland

J. Cryptology (1998) 11: 1–208

Journal of
CRYPTOLOGY

Practical Aspects of Quantum Cryptographic Key Distribution

H. Zbinden, N. Gisin, B. Huttner, A. Muller, and W. Tittel
Group of Applied Physics, University of Geneva,
CH-1211 Geneva 4, Switzerland

Communicated by Gilles Brassard

Received 11 April 1997 and revised 21 July 1997

Abstract. Performance of various experimental realizations of quantum cryptographic protocols using polarization or phase coding are compared, including a new self-balanced interferometric setup using Faraday mirrors. The importance of detector noise is illustrated and means of reducing it are presented. Maximal distances and bit rates achievable with present day technologies are evaluated. Practical eavesdropping strategies taking advantages of the optical fiber that could open a gate into the transmitter's and receiver's offices are discussed.

Key words. Quantum cryptography, Key distribution, Interferometry, Single photon counting, Optical fibers.

1. Introduction

The use of quantum mechanical properties in cryptography has been proposed by Wiesner [24] and developed by Bennett and Brassard in 1984 [3]. In cryptography safety can be guaranteed by a common secret key, known only to the two users, Alice and Bob. Quantum cryptography (QC) provides means to establish such a key at a distance and to check its confidentiality. It is based on the fact that any measurement of incompatible quantities on a quantum system will inevitably modify the state of this system. Therefore an eavesdropper, Eve, might get information out of a quantum channel by performing measurements, but the legitimate users will detect her and hence not use the key. For convenience the quantum system is in practice a single photon (or a weak pulse) propagating through an optical fiber, and the key can be encoded either by its polarization or by its phase. A variation of the general principle based on entangled photon pairs was proposed by Ekert [8]. The first experimental demonstration of quantum cryptography was performed in 1989 over 30 cm in air with polarized photons [2]. Since then, several groups presented realizations of both the polarization [10], [21] and the phase coding scheme in optical fibers over distances of up to 30 km [19], [14].

Three parameters describe the performance of experimental quantum cryptography systems: the transmission distance, the bit rate, and the error rate. The losses in optical fibers are typically around 2 dB/km at 800 nm, 0.35 dB/km in the 1300 nm telecom window, and 0.2 dB/km in the 1550 nm telecom window. Hence, at 1300 nm the bit rate is reducted by a factor of ten after about 30 km. At this wavelength germanium avalanche photo diodes (Ge APD) have to be used instead of commercial silicon photon counting modules. This means lower detection efficiencies, hence lower bit rates and higher dark count rates, hence higher error rates. Actually the noise of the available photon counters in the near infrared is one major problem of experimental QC that finally limits the transmission distance. Note that incompatible modes of a quantum channel cannot be amplified without noise (no cloning theorem [25]). On the one hand this is essential for the security of QC, on the other hand this limits the possible transmission distance.

Another experimental problem is that most QC systems need continuous alignment of the setup. In polarization-based QC systems, the polarization has to be maintained stable over tens of kilometers, in order to keep the polarizers at the emitter's and at the receiver's ends aligned. In fact, depending on the environment the output polarization can fluctuate randomly on time scales of hours to seconds. Therefore, these systems have to compensate actively changes of the outcoming polarization. These fluctuations are generally slow enough that automatic tracking would be feasible [21]. Interferometric QC systems are usually based on two unbalanced Mach-Zehnder interferometers, one at each end. Since two interfering pulses do not follow the same path within the two interferometers, the difference in arm lengths must be kept stable to a fraction of a wavelength for both interferometers, in order to obtain high visibility. Consequently, every few seconds, one interferometer has to be adjusted to the other by a piezoelectric transducer to compensate for thermal drifts [19].

In this article we show first that the performance of Ge APDs can be considerably improved using fast active biasing electronics. Next, we introduce an interferometric system with Faraday mirrors [20]. This phase coding setup needs no alignment of the interferometer nor polarization control, and therefore considerably facilitates the experiment. Moreover, it features excellent fringe visibility. Thirdly, we present the realization of a secret key over 23 km of installed telecom fiber. The performance of this setup is compared with polarization and phase coding setups presented before. Finally, the susceptibility of the different setups to different eavesdropping strategies is briefly discussed.

2. QC and Sources of Errors

We recall the principle of QC (based on the four-states protocol BB84 [3]) using the example of a polarization coding setup shown in Fig. 1. Experimental setups published before were based on one laser followed by a polarization rotator. The present scheme proposes using four lasers with polarizers oriented at 0°, 90°, 45°, and 135°.[1] The lasers fire at random at a rate ν. Their polarization states are adjusted to compensate for the transformation in the following fiber link with a total loss L. Bob randomly selects one

[1] The use of four laser may have experimental advantages. However, one has to make sure that Eve cannot find out which laser has fired due to differences in spectrum or timing.

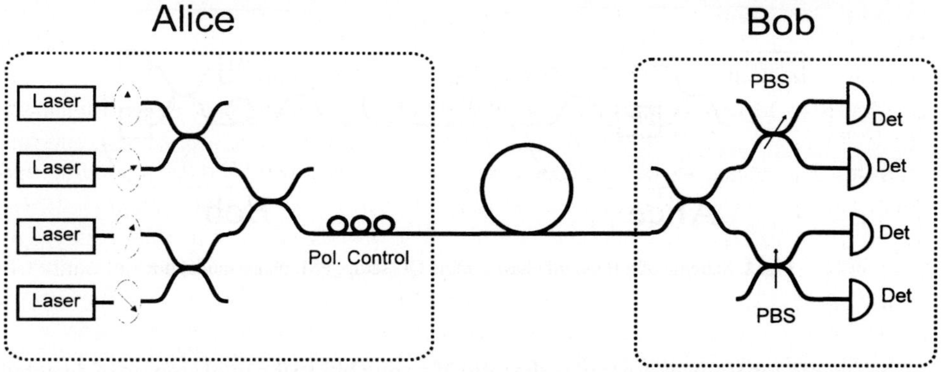

Fig. 1. Scheme of a polarization coding QC setup. PBS: polarization beam splitter.

of the two analyzers oriented at 45° (in this setup this is automatically done by a passive coupler). To prevent the simplest eavesdropping strategy, that is just splitting the pulse in two and measuring the polarization of one-half, at most 1 photon per pulse must be used. In practice the laser pulses are attenuated to an average number of photons per pulse well below 1 ($\mu = 0.1$, say), to limit the probability of obtaining more than 1 photon per pulse. The photons are then detected with a photon counter and acquisition electronics collect the data. After the measurement Alice and Bob publicly compare the chosen bases (0°/90° or 45°/135°) of emission and detection, without revealing the polarization states transmitted and measured. Incompatible measurements are disregarded. With the other results a secret key can be established by interpreting 0° and 45° as bit 1, and 90° and 135° as bit 0. If, for example, Eve uses a simple intercept–resend strategy, i.e., would just measure the polarization of every photon, she would introduce an error of 25% which can be easily detected by Alice and Bob by simply comparing a sample of their key.

For comparison, the standard phase coding setup is shown in Fig. 2. There are two unbalanced Mach-Zehnder interferometers, one at Alice's and one at Bob's side. Pulses taking the short path in Alice's and the long one in Bob's interferometers will interfere with pulses taking the long path in Alice's and the short one in Bob's interferometers. In one arm, Alice randomly applies phase shifts of 0, π, or $\pi/2$, and $3\pi/2$; Bob chooses a base by applying a phase shift of 0 or $\pi/2$. If compatible bases have been chosen, i.e., the phase difference is 0 or π, the outcome is deterministic. Hence a secret key can be established by interpreting 0 and $\pi/2$ as bit 1, and π and $3\pi/2$ as bit 0. Again incompatible measurements are disregarded.

For every QC scheme the same simple equation for the raw data rate R, i.e., the number of exchanged bits per second before any error correction, can be applied:

$$R = q\mu\nu(1 - L)\eta, \tag{1}$$

where ν is the pulse rate of the laser, μ is the average number of photon per pulse, L is the losses in the fiber lnk, η is the quantum efficiency of the detector, and q is a systematic factor smaller than (or equal to) $\frac{1}{2}$ depending on the chosen implementation. For example, in the case of the polarization scheme of Fig. 1, q equals the maximum value $\frac{1}{2}$ due to the random selection of the polarizer basis.

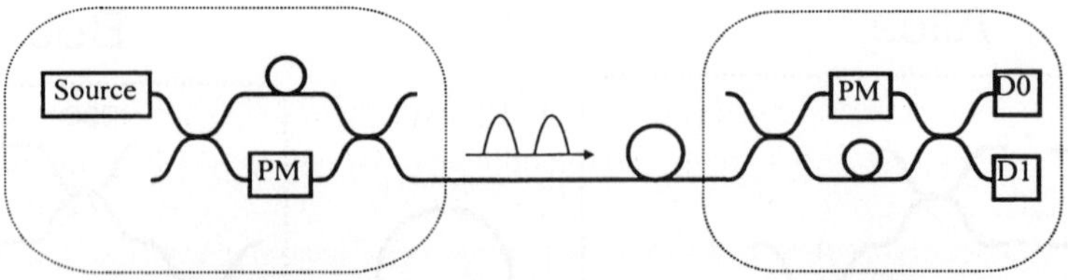

Fig. 2. Scheme of a standard phase coding QC setup. PM: phase modulator.

The error is generally expressed as the ratio of wrong bits to the total amount of detected bits (or the ratio of probability of obtaining a false detection to the total probability of detection). We call this quantity the quantum bit error rate (QBER)[2]:

$$\mathrm{QBER} = \frac{p_{\mathrm{dark}} + p_{\mathrm{opt}} \cdot p_{\mathrm{phot}}}{2 \cdot p_{\mathrm{dark}} + p_{\mathrm{phot}}} = \frac{n_{\mathrm{dark}} \cdot \Delta\tau + p_{\mathrm{opt}} \cdot \mu(1-L)\eta}{2 \cdot n_{\mathrm{dark}} \cdot \Delta\tau + \mu(1-L)\eta}$$
$$\cong \frac{n_{\mathrm{dark}} \cdot \Delta\tau}{\mu(1-L)\eta} + p_{\mathrm{opt}} \equiv \mathrm{QBER}_{\mathrm{det}} + \mathrm{QBER}_{\mathrm{opt}}, \quad (2)$$

where p_{dark}, p_{phot}, and p_{opt} are the probabilities of obtaining a dark count, of detecting a photon, and the probability that a photon went to an erroneous detector, respectively. n_{dark} is the dark count rate of the detector and $\Delta\tau$ is the detection time window. This formula applies for a setup with two detectors. Since a dark count will with a 50% chance not lead to an error, but just to an additional count, there is a factor two in the denominator, but not in the numerator. Of course, we do not consider dark counts when incompatible bases are used. Hence, the factor q of (1) does not appear in the denominator.

The QBER consists of two parts: The first part $\mathrm{QBER}_{\mathrm{det}}$ is due to the dark count rate of the detector, this part is proportional to $\Delta\tau$. Hence a good detector must not only be efficient and have a small dark count rate, it should also have a small time jitter, at least smaller than the pulse length of the laser diodes. The second part is what we call $\mathrm{QBER}_{\mathrm{opt}}$, that is, the fraction of photons p_{opt} whose polarization or phase is erroneously determined, i.e., the fraction of photons who end up in the wrong detector. This is mainly due to depolarization and to poor polarization alignments or due to the limited visibility of the interferometers. For example, for our first long distance experiment below Lake Geneva using polarization coding [21] we computed a $\mathrm{QBER}_{\mathrm{det}}$ of 3% and a $\mathrm{QBER}_{\mathrm{opt}}$ of 0.5%, which fitted to the measured total QBER of 3.4%. We discuss the first source of errors and have a closer look at the photon counters used.

[2] Physicists often call this quantity the bit error rate (BER). In telecommunications BER is commonly used for the total error in a transmission and is in the order of 10^{-9}. In QC the BER is in the order of 1%. Of course, this does not correspond to the final error in the message, since error correction will be applied. However, to prevent any confusion of Telecom specialists we renounce the expression BER and call it QBER. Note that in theoretical papers about eavesdropping the QBER introduced by Eve is often called the disturbance (D).

3. The Performance of Photon Counters

The photons are detected by liquid nitrogen (LN_2) cooled Ge avalanche photodiodes (NEC NDL5131) working in the passively quenched Geiger mode [22]. In this mode the diodes are driven above breakdown, i.e., the bias voltage is so high that one electron hole pair created by an absorbed photon will be able to produce an avalanche of thousands of carriers. The avalanche only stops when the current created through the resistance in series to the diode lowers the applied voltage below the breakdown value. The noise in such detectors is due to carriers generated in the detector volume by other causes than an impinging photon (dark counts). These carriers can be created thermally or by band-to-band tunneling processes, or they can be emitted from trapping levels that were populated in previous avalanches (after pulsing). The quantum efficiency and the dark count rate n_{dark} both increase with increasing bias voltage U_{bias}. To obtain a low QBER a tradeoff between high efficiency and low noise has to be found. In the early experiment mentioned above [21] we worked at $\eta = 0.2\%$ with $n_{\text{dark}} = 700$ Hz and we obtained 3% QBER_{det} (following (2), for $\mu = 0.1$ and $L = 0.9$). For $\eta = 10\%$ we would have expected more than 20% QBER_{det}.

For LN_2 cooled Ge diodes the thermal contribution can be neglected and the dark counts mainly consist of tunneled electrons and afterpulses, the latter being more important if the total charge through the device is large [17]. The afterpulse rate is decreasing almost exponentially with a time constant ($1/e$) of about 200 ns. This fact opens the door to a further reduction of the dark count rate: If the diode is biased only immediately before a photon is expected, no spontaneous avalanches can occur before the detection and consequently no afterpulses will fall into the detection time interval. So we developed the following electronic circuit. The bias voltage of a diode is the sum of a DC part well below the threshold and a 2 ns long almost rectangular pulse of 7.5 V amplitude that pushes the diode about 1.4 V over the threshold at the time when the photon is expected. This allows us to increase considerably the efficiency without excessively increasing the noise. Moreover, the time jitter is reduced to a value below 100 ps. The short bias pulse induces a parasite signal. A discriminator in combination with a temporal coincidence window allows us to recover the true avalanche signal from this parasitic signal. A time-to-amplitude converter followed by a window-discriminator of 300 ps width, allows us to reduce the noise level further. Thanks to this technique we get 7 and 22 dark counts per 1 million pulses ($p_{\text{dark}} = 22 \cdot 10^{-6}$ and $7 \cdot 10^{-6}$) for detection efficiencies of 10% and 20%, respectively. This corresponds to a QBER_{det} of $0.72 \pm 0.13\%$ at 10% efficiency.

Recent progress in photon counting with InGaAs APDs could allow us to replace the LN_2 cooled Ge detectors [18]. A QC experiment has been performed with InGaAs detectors [14]. Performances similar to that of Ge APDs seem to be possible. Moreover, these diodes would open the second telecom window at 1550 nm.

We compare Ge detector specifications to those of commercial silicon single photon counting modules at 800 nm. These modules have about 50% efficiency with extremely low dark count rates of down to 10 Hz. The QBER_{det} and R for the different wavelengths with corresponding detector performances are summarized in Table 1 for different fiber lengths.

Note that the wavelength of 800 nm is a good choice only for distances shorter than 5 km, taking advantage of the efficient and commercially available Si detectors. The

Table 1. Quantum bit error rates (QBER) and raw data rates R for different wavelengths and detector performances for two different fiber lengths with $\nu = 10$ MHz, $\mu = 0.1$ (or as indicated), and $q = 0.5$.*

	5 km		20 km		$QBER_{max} = 15\%$	
	$QBER_{det}$ (%)	R (kHz)	$QBER_{det}$ (%)	R (kHz)	L_{max} (km)	R (Hz)
800 nm, $\eta = 0.5$, $p_{dark} = 10^{-8}$	0.0022	25	0.2	0.025	29	0.3
1300 nm, $\eta = 0.1$, $p_{dark} = 7 \cdot 10^{-6}$	0.11	33	0.35	10	67	233
1300 nm, $\eta = 0.2$, $p_{dark} = 21 \cdot 10^{-6}$	0.16	67	0.53	20	62	700
1300 nm, $\eta = 0.2$, $p_{dark} = 21 \cdot 10^{-6}$ $\mu = 1, \nu = 1$ MHz†	0.016	67	0.053	20	90	70
1550 nm, $\eta = 0.1$, $p_{dark} = 10^{-5}$	0.13	40	0.25	20	109	333
1550 nm, $\eta = 0.1$, $p_{dark} = 10^{-5}$ $\mu = 1, \nu = 1$ MHz†	0.013	40	0.025	20	159	33

* The transmission losses are assumed to be 2 dB/km at 800 nm, 0.35 dB/km at 1300 nm, and 0.2 dB/km at 1550 nm. At 1550 nm, the estimated performance of InGaAs detector according to first results [14], [18].
† 1 MHz single photon production rate.

disadvantage is that fibers and modulators are generally conceived for the longer telecom wavelengths. Consequently, when Peltier cooled InGaAs counters with the expected performance are available, the telecom wavelengths will clearly be preferable, especially at 1550 nm for long distance QC. According to recent calculations QC could be performed securely with QBER up to 15% [11]. In the last column of Table 1, the maximum length of the link leading to this QBER is calculated. The limit for 1550 is around 110 km, a limit, however, that depends strongly on the performance of the detector, and its development in future. The given QBERs and L_{max} could be improved using single photon states ($\mu = 1$) [4].[3] The attainable raw data rates would be in the same order of magnitude, supposing that both a 1 MHz single photon production rate and a 10 MHz pulse rate for weak pulses are feasible.

Of course, the raw bit rates obtained will be reduced further, due to error correction and privacy amplification depending on the corresponding QBER. So the above-mentioned tradeoff between efficiency and noise of the detector depends not only on the transmission length, but also on the error correction algorithms.

With present day detector performances the QBER limit for transmission lengths

[3] The single photon source can be a two photon source (based on parametric down-conversion) where one photon serves as a trigger for the presence of its twin. The wavelength of the trigger photon is chosen in the detection range of high efficiency and low noise Si detectors. However, these are not really single photon states, because the two photon distribution is chaotic. Taking into account our time resolution the photon number can be considered as Poisson distributed as for attenuated laser pulses. For 1 MHz production rate the probability of having a second photon in a 1 ns time window pulse is 0.05%, equal to that of a laser pulse with $\mu = 0.001$.

below 20 km is set by $QBER_{opt}$ that is in the order of 0.5%. Therefore, we have a closer look at the sources of this part of the QBER.

4. Polarization Control

The fiber optic implementation of the polarimetric scheme faces three difficulties:

The first one is a topological problem related to the transport of a vector along a curve. Since the path taken by the light in the optical fiber is three-dimensional, its polarization rotates by an angle related to Berry's phase [5]. This effect does not limit the distance or the quality of the transmission if the fiber link is stable. It is clear from that consideration that an aerial cable or cable sustaining strong vibrational perturbations are not suited.

The second difficulty arises from the intrinsic birefringence of optical fibers. Changes in mechanical stress that can cause birefringence will change the state of polarization at the output of the fiber. However, these changes are usually quite slow in the order of tens of minutes depending on the mechanical and thermal stability of the environment [12]. Another effect of the birefringence is polarization mode dispersion (PMD) [13]. An optical cable behaves as a concatenation of pieces of birefringent fibers. The result of this is a spread of the pulses growing with the square root of length for long distances. This evolution is the same as a random walk. To prevent depolarization of the light pulses, lasers with a coherence time larger than the polarization mode delay must be used. This is not a real limitation since typical PMDs are between 0.1 ps/km^2 and 1 ps/km^2 and semiconductor lasers with 1 ns coherence time are available.

A third potential problem are polarization dependent losses in optical components that could arise in Passive Optical Networks (PONs). In this case the relation between the polarization state at the input and the output of the optical link is no longer unitary [16].

As for the topological effects, polarization instabilities are due to mechanical stresses and temperature variations. This requires the optical fiber to be kept as stable as possible. However, an active polarization controller is necessary to align Alice's and Bob's polarizers and keep them aligned, compensating temporal evaluation. The error rate p_{opt} can be determined simply by aligning at the receiver a polarization analyzer on the outgoing state of polarization and measure the ratio of the intensities of the two arms. In our experiments, both in the laboratory over 26 km and in the field over 23 km, we obtained a separation of the polarization of 23 dB that corresponds to an error fraction p_{opt} of 0.5%. The stability of the polarization alignment in the field experiment was excellent most of the time, and measurements could be performed for an hour without realigning the system. However, from time to time there were quite fast polarization instabilities of 2π within a few seconds. In such moments we could not of course compensate the fluctuations with our manual polarization controller. An automated polarization controller with a response time of some tens of milliseconds should be able to guarantee an uninterrupted operation.

One might think that one could spare the polarization controller by using the phase coding scheme of Fig. 2. In fact, to prevent that only every fourth photon chooses interfering paths (to increase q from $\frac{1}{8}$ to $\frac{1}{2}$ in (1)), a polarizing beamsplitter (PBS) is used at the receiver's end. Consequently, the phase coding scheme requires polarization controllers, too [23]. Ignoring the delay loops (which are actually no longer necessary

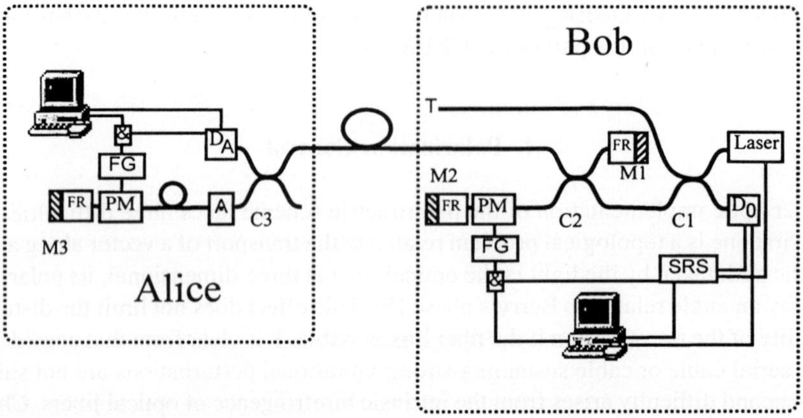

Fig. 3. Experimental setup of an interferometric QC system with Faraday mirrors. C1, C2, and C3: fiber optic couplers; M1, M2, and M3: Faraday mirrors (ordinary mirrors in combination with Faraday rotators, FR); PM: phase modulator; A: Attenuator; D_0: photon counter; D_A: photodiode; T: optional trigger output; SRS: delay generator; FG: function generator; &: and-gate.

using PBSs) the two Mach-Zehnder interferometers with the phase shifters can simply be regarded as polarization modulators. The interferometric setup is finally equivalent to the polarization code scheme. It has just the additional inconveniences that in each Mach-Zehnder interferometer polarization has to be controlled to improve the fringe visibility and the path length differences have to be balanced every few seconds [19]. The fringe visibility obtained in phase coding is 0.99 [19], corresponding to a polarization separation of 20 dB and leading to $\mathrm{QBER}_{\mathrm{opt}} = 1\%$.

To summarize, polarization separation of 23 dB over 23 km can be achieved, leading to $\mathrm{QBER}_{\mathrm{opt}} = 0.5\%$. For a practical system, however, the main drawback is the need for active polarization controllers to compensate for fluctuations due to thermal and mechanical disturbances of the fiber. In the next section we present a novel QC setup that at the same time needs no alignment and reduces $\mathrm{QBER}_{\mathrm{opt}}$ further.

5. QC Using Faraday Mirrors

5.1. *An Interferometer with Faraday Mirrors*

Let us have a closer look at the QC scheme depicted in Fig. 3 [20], [26], disregarding the Faraday rotators (FR) for the moment, their crucial effect will be explained later. In principle Bob has a very unbalanced Michelson interferometer (beamsplitter C2) with one long arm going all the way to Alice. The laser pulse impinging on C2 is split in two pulses P1 and P2. P2 propagates through the short arm first (mirror M2 then M1) and then travels to Alice and back, whereas P1 propagates first to Alice and next passes through the short arm. As both pulses run exactly the same path length, they interfere maximally at C2 (disregarding polarization for the time being). To encode their bits, Alice acts with her phase modulator (PM) only on P2 (phase shift φ_a), whereas Bob lets pulse P2 pass unaltered and modulates the phase of P1 (phase shift φ_b). If no phase shifts are applied or if the difference $\varphi_a - \varphi_b = 0$, then the interference will be constructive. On the contrary, when $\varphi_a - \varphi_a = \pi$ the interference will be destructive and no light will be detected by detector D_0. Since the interfering pulses travel the same path, the interferometer is automatically aligned. The visibility of the fringes is also independent of the splitting ratio of C2.

However, visibility depends also strongly on the polarization states of the interfering pulses. Let $\mathcal{M}_{in}$ be the vector representing the polarization state of the incoming laser pulse at C2 on the Poincaré sphere, then the polarization states of the interfering pulses P1 and P2 are

$$\mathcal{M}1_{out} = R_2 R_1 R_3 \mathcal{M}_{in} \quad \text{and} \quad \mathcal{M}2_{out} = R_3 R_1 R_2 \mathcal{M}_{in},$$

where R_i is the matrix describing the polarization rotation in a round trip path to mirror M_i. Because rotation operators do not commute, these two operations are in general not identical, hence the two outcoming polarizations are not parallel. This is where the Faraday mirrors (FM) enter the game. An FM is composed of a 45° Faraday rotator and a mirror. A light pulse injected in any arbitrary polarization into a fiber terminated by an FM will come back exactly orthogonally polarized, regardless of the polarization transformations in the fiber due to induced birefringence.[4] Hence a round trip path in any fiber terminated with an FM will lead to a polarization transformation $R = -1$. This is true since there are no significant mechanical or thermal variations during the time of flight of the photons [21], which is 300 μs for a 30 km link. However, this applies only if there is no Faraday rotation inside the fiber. In fact, although the Verdet constant of a standard optical fiber is low, Faraday rotation due to the geomagnetic field may not be completely neglected for optical fibers of several tens of kilometers,[5] hence $R_3 \neq -1$. However, with $R_1 = R_2 = -1$ we obtain $\mathcal{M}1_{out} = R_3\mathcal{M}_{in} = \mathcal{M}2_{out}$. To quantify the performance of our interferometer, we measure the ratio of the count rates for constructive and destructive interference. In practice, we change the attenuation (A) at Alice to obtain the same count rate with and without phaseshift. When we apply a phaseshift at Bob's piezo-optic modulator we obtain an attenuation of 30 ± 1 dB, while when we apply the phaseshift at Alice's $LiNbO_3$ integrated optic phase modulator the extinction is 27 ± 1 dB. Obviously the integrated phase shifter is slightly less precise. These values were reproducible within the given errors over weeks. An extinction of 30 dB corresponds to a classical fringe visibility $V = (I_{max} - I_{min})/(I_{max} + I_{min})$ of 99.8%. The measured values of 30 dB and 27 dB result in a $QBER_{opt}$ of 0.1% and 0.2%, respectively. The average, decisive for key creation, is therefore 0.15%. Replacing one Faraday mirror by an ordinary mirror, the extinction is strongly fluctuating and can be reduced to 20 dB. If two Faraday mirrors are removed, essentially no interference is visible.

[4] This description of Faraday mirrors requires that after a reflection one switches from a right-handed to a left-handed reference frame, or vice versa. This is no problem as long as the interfering paths each undergo the same (the same parity of) numbers of reflections.

[5] The horizontal component of the geomagnetic field $H = B/\mu_0$ is 17 A/m in Geneva, the Verdet constant is ca. $0.6 \cdot 10^{-4}$ °/A at 1300 nm. Therefore the polarization is turned by about twice 1° per km displacement in the north–south direction. However, polarization mode coupling strongly reduces this effect.

5.2. *Key Creation*

For the key exchange we used the two-states protocol B92 [1], because our driving electronics for the phase modulators could not be used for the four-states protocol [3]. In principle, our setup could be quite easily adapted to the latter protocol by inserting another coupler and detector. We tested that using also $\pi/2$ and $3\pi/2$ phase shifts the same excellent performances of the interferometer are obtained. Alice and Bob choose at random 0 or π phase shifts, defined as bit values 0 and 1. Since very weak pulses are used, in most cases no photon will be detected in D_0. If a detection, i.e., constructive interference occurred, Alice and Bob know that they applied the same phase shift, and they register the same bit value. In our interferometric setup the pulses leaving Bob carry no phase information. The information is in the phase difference of the two pulses P1 and P2 leaving Alice. The attenuator (A) is set such that the weaker pulse P2 that already passed through Bob's delay line has 0.05 photons on average when leaving Alice. The information that Eve could obtain depends on the number of photons in the weaker pulse. Therefore, to measure the phase difference, she must attenuate P1 to the intensity of P2 in order to obtain complete interference. She actually performs the same measurement as Bob does. More generally, such a kind of measurement can be called a Loss Induced Generalized Measurement [16]. Consequently, 0.05 photons in the weaker pulse is equivalent to an average number of $\mu = 0.1$ for the pulse pair. Of course, this reasoning applies also for the standard time multiplexed interferometer setup (Fig. 2), where the two pulses may also have different intensities.

5.3. *Experimental Realization*

The heart of our experiment is a delay generator (SRS 535) at Bob (see Fig. 3). It beats at 1 kHz and triggers the laser, Bob's phase modulator (PM), the actively biased photon counter (D_0), and Bob's computer. The 1300 nm DFB-laser (Fujitsu, driven by an Avtech pulser) delivers 300 ps pulses. The phase modulator is a fiber wrapped around piezoelectric-tube. It is driven by a sinus function from a function generator (SRS DG 345). The modulation frequency of the piezo of about 10 kHz is high enough since the time delay between the two pulses is about 230 μs. Only if the computer gives a logical 1 to the and-gate at the external trigger input of the function generator is a phase applied. The optical fiber is a 22.8 km long telecom link between Geneva and Nyon, Switzerland, featuring 8.6 dB loss. The pulse P1 detected at Alice by D_A (Newport AD-300/AC) triggers Alice's phase modulator and Alice's computer. At Alice the delay between the two pulses is smaller, hence a 1 GHz $LiNbO_3$ waveguide phase modulator is used. Again this modulator is driven by a function generator, in case Alice's computer supplies a logical 1. Back at Bob's, the interfering photon directly runs to the detector D_0 via the 10 dB coupler C1 to limit the losses. The photon counter electronics are precisely triggered to coincide with the arrival of the photon at Bob and the biasing of diode. The adjustment must be precise within 100 ps, which can be easily obtained with the delay generator. Every detection is registered by Bob and assigned to the number of the pulse after the beginning of the measurement. Alice and Bob disposed of 100 files of 65,536 bits of random numbers. These numbers have been generated by an apparatus based on thermal noise of an electrical resistor [7].

Table 2. Results of the key distribution with a QC setup with Faraday mirrors.

Photons per pulse μ	Measured QBER (%)	$QBER_{det}$ (%)	$QBER_{opt}$ (%)	Length of key (bit)	Bit rate (Hz)
0.2	0.5 ± 0.1	0.40 ± 0.07	0.15 ± 0.03	2,980	0.9
0.1	1.35 ± 0.08	0.81 ± 0.14	0.15 ± 0.03	20,142	0.5

5.4. *Results and Discussion*

After having registered the results of the measurement, Alice and Bob compare their random lists in order to determine the QBER. The results are summarized in Table 2.

To our knowledge, these QBER rates are the lowest ever obtained for the corresponding numbers of photons per pulse and over a distance of more than 20 km. The measurement for $\mu = 0.1$ lasted more than 11 hours and no realignment was performed, hence the stability of the setup was extraordinary. However, we notice that the measured QBER is higher than the sum of the detector and interferometer noise. We believe that the increase in the error rate is not due to any fluctuations of the interferometer, but rather due to an increasing $QBER_{det}$ in the course of the measurement. Variations in the photon counter, its electronics and timing, which proved to be quite delicate, might be the reason for this increase. We also tried to trigger the photon counter by the strong laser pulse at the trigger output (T) running down another fiber to Alice and back, in order to obtain less time jitter and to be less sensitive to changes in the optical path length due to temperature variations. We gave up this optical trigger signal, because it did not improve the results for short time measurements (tens of minutes) significantly enough to justify the need of an additional fiber. Under difficult environmental conditions with large temperature fluctuations, however, the use of an auxiliary fiber for timing improvements might be appropriate, or periodical readjustments of the detector timing could be envisaged.

The obtained bit rates are quite low, in agreement with the expected values following (1). This is simply due to the low pulse rate and could be increased by replacing the piezoelectric modulator and adapting the computer steering. We have noticed that the noise of our detector increases if a relatively strong light pulse is impinging before the detection window. This might cause a problem going to higher frequencies, since in our setup we have to deal with different parasite pulses.

It is noteworthy that the timing of Alice's apparatus can be preadjusted in the laboratory and will not change, even if the apparatus is plugged into another fiber to communicate with a third party. The timing of Bob's apparatus, especially of his photon counter, has to be adjusted once for every link, this could be done using an Optical Time Domain Reflectometer (OTDR).

6. Practical Eavesdropping

We have seen that the simplest attack of Eve can be prevented using weak pulses with, e.g., $\mu = 0.1$. More elaborated strategies are analyzed in [2], [11], [15], [6], and [9]. However, in practice, Eve could follow another strategy: She could chop the fiber and try to measure actively the phase or polarization settings applied by Alice. Eve could mount an interferometer similar to Bob's one and measure with intense light pulses the phase shifts applied by Alice. Then she can apply the same phase shifts to the pulses received from Bob and send them back to him, as if she was Alice. However, as Alice attenuates the incoming pulses by more than 40 dB down to the 0.1 photon level before sending them back, Eve is forced to send intense pulses to Alice, which can be detected by the detector D_A, inserted for this purpose. However, by assumption, Eve has perfect technology at her disposal. Therefore, she could for example try to sense Alice's phase with a very short pulse beyond the bandwidth of D_A. Alice, in return, could prevent such an intrusion with a narrow line filter. Probably any kind of intrusion could be prevented with the appropriate means, but security would no longer be guaranteed solely by the fundamental laws of quantum mechanics. In fact, all other QC schemes face the same problem. In the standard phase scheme the position of the phase shifter could be sensed interferometrically using small reflections at Alice's or Bob's ends. Hypothetically, Eve might find an optical technique to find out which laser fired or which detector clicked in the polarization scheme proposed above. In these setups optical isolators could be introduced in contrast to the Faraday mirror setup. We cannot discuss all possible strategies of Eve and the technical means to fight them. A general assumption implicit in all discussions of QC security is that Alice's and Bob's offices are absolutely safe. This is a reasonable assumption, necessary also for all other cryptosystems. However, as illustrated by the above discussion, care should be given to the fact that the fiber-obtic quantum channel provides a potential entrance gate for malevolent intruders.

In yet another eavesdropping strategy, that applies to the two-states system only [1], Eve interrupts the transmission and measures as many pulses as possible. She sends to Bob only the pulses for which she obtained the phase or the polarization. To prevent this, Bob has to introduce another detector to monitor the stronger pulse P1 to make sure that Eve cannot suppress this pulse. If Eve suppresses only the weak one, because she did not get the phase information, the strong pulse alone will introduce 50% error on detector D_0. To render the power of P1 measurable by a conventional detector, the losses of Bob's delay line could be increased and the attenuation applied at Alice's side reduced by the same amount. The attenuation at Alice applies also to pulses needed by Eve to spy on Alice's phase, following the strategy mentioned above. With the laser power and the detectors at our disposal, it is not possible to monitor P1 at Bob's and P2 at Alice's (hence Eve's spying pulse) at the same time (also with appropriate choice of the splitting ratio of the couplers C2 and C3, presently 3 dB couplers). So, the present implementation of the B92 protocol is insecure, and the BB84 protocol should be applied.

In the four-states protocol BB84 [3] the eavesdropping strategy mentioned in the previous paragraph fails because Eve would introduce errors when she chooses the wrong basis. However, suppose that Eve has a lossless line and a way to sense how many photons are in the pulse. For $\mu = 0.1$ there is about a 6% chance of having two photons in a nonempty pulse. In these cases Eve could let one photon pass and store the other until Alice and Bob publicly communicate their bases and get full information on this bit. Eve would then send only these pulses to Bob, and block the others. Bob would not notice Eve's presence, since he expects considerable losses in his line. Therefore, Eve could obtain 100% of the information if the line had, e.g., just 6% transmission. In conclusion, as a function of μ and the losses in the line Eve could win a considerable fraction of the information. Again this could be prevented by measuring the intensity of a stronger pulse, to force Eve to send a pulse every time [15].

QC with correlated photon pairs would have the advantage that, since in this case real single photon states are used, all strategies dealing with the fraction of pulses containing more than one photon must fail. Unfortunately, the self-aligning setup with Faraday mirrors is not suited for such a photon source, due to the high losses in a complete round trip.

In practice, a tradeoff has to be found between the complexity, hence the price, and the absolute security of the setup. We mention in this context that since the interferometer in the Faraday mirror setup is not stabilized, the absolute phase difference between the pulses P1 and P2 will randomly fluctuate, rendering Eve's job very hard. This contrasts with the standard phase coding setup, where the intense pulses sent by Alice to adjust Bob's interferometer can also be used by Eve to adjust hers.

7. Conclusions

We have discussed the experimental advantages and drawbacks of different QC setups. We have seen that one major problem is the availability of good photon counters. It is essentially the noise of these detectors, in combination with the losses in the optical fiber, that limits the maximum distance of a QC link. This maximum distance would be about 100 km working at 1550 nm in combination with InGaAs photon counters. The other problem of standard polarization and phase coding setups is the need for continuous alignment. We introduced and demonstrated an interferometric QC setup using Faraday mirrors which requires no continuous alignment. It features impressive stability and a fringe visibility of 99.8%. Using this new QC setup, we produced a secret key of 20 kbit length with a QBER of 1.35% for 0.1 photon per pulse.

Acknowledgments

We would like to thank the Swiss Telecom for financial support and for placing at our disposal the Nyon–Geneva optical fiber link. We appreciate stimulating discussions with our colleagues within the TMR network on the physics of quantum information.

References

[1] C. H. Bennett, Quantum cryptography using any two non-orthogonal states, *Phys. Rev. Lett.* 68, 3121 (1992).

[2] C. H. Bennett, F. Bessette, G. Brassard, L. Salvail, and J. Smolin, Experimental quantum cryptography, *J. Cryptology* 5, 3–28 (1992).

[3] C. H. Bennet and G. Brassard, Quantum cryptography: public key distribution and coin tossing, *Proc. Internat. Conf. Computer Systems and Signal Processing*, Bangalore, 1984, pp. 175–179.

[4] C. H. Bennett, G. Brassard, and N. D. Mermin, Quantum cryptography without Bell's theorem, *Phys. Rev. Lett.* 68 (5), 557–559 (1992).

[5] R. Y. Chiao and Y. S. Wu, *Phys. Rev. Lett.* 57 (8), 933–936 (1986).

[6] J. Cirac and N. Gisin, Coherent eavesdropping strategies for the four state quantum cryptography protocol, *Phys. Lett. A*, in press.

[7] F. Devillard, Etude d'un générateur de bruit et de ses applications, Travail de diplôme, Ecole d'ingénieurs de Genéve, 1996.

[8] A. K. Ekert, Quantum cryptography based on Bell's theorem, *Phys. Rev. Lett.* 67, 661–661, 1991.

[9] A. K. Ekert, B. Huttner, G. M. Palma, and A. Peres, Eavesdropping on quantum-cryptographical systems, *Phys. Rev. A* 50 (2), 1047–1056 (1994).

[10] J. D. Franson and B. C. Jacobs, Operational system for quantum cryptography, *Electron. Lett.* 31 (3), 232–234 (1995).

[11] C. A. Fuchs, N. Gisin, R. B. Griffiths, C.-S. Niu, and A. Peres, Optimal eavesdropping in quantum cryptography, submitted to *Phys. Rev. A*.

[12] N. Gisin, R. Passy, J. C. Bishoff, and B. Perny, Experimental investigations of the statistical properties of polarization mode dispersion in single mode fibers, *IEEE Phot. Technol. Lett.* 5, 819 (1993).

[13] N. Gisin, J. P. Pellaux, and J. P. Von Der Weid, Polarization mode dispersion of short and long single mode fibers, *IEEE J. Lightwave Technol.* 9, 821–827 (1991).

[14] R. J. Hughes, G. G. Luther, G. L. Morgan, C. G. Peterson, and C. Simmons, Quantum cryptography over underground optical fibers, *Proc. Crypto.* '96, Lecture Notes in Computer Science, vol. 1109, p. 329 (1996).

[15] B. Huttner, N. Imoto, N. Gisin, and T. Mor, Quantum cryptography with coherent states, *Phys. Rev. A* 51 (3), 1863–1869 (1995).

[16] B. Huttner, A. Muller, J. D. Gautier, H. Zbinden, and N. Gisin, Unambiguous quantum measurements of non-orthogonal states, *Phys. Rev. A* 54 (5), 3783–3789 (1996).

[17] A. Lacaita, P. A. Francese, F. Zappa, and S. Cova, Single-photon detection beyond 1 μm: performance of commercially available germanium photodiodes, *Appl. Opt.* 33 (30), 6902–6918 (1996).

[18] A. Lacaita, F. Zappa, S. Cova, and P. Lovati, Single-photon detection beyond 1 μm: performance of commercially available InGaAs/InP detectors, *Appl. Opt.* 35 (16), 2986–2996 (1996).

[19] Ch. Marand and P. D. Townsend, Quantum key distribution over distances as long as 30 km, *Opt. Lett.* 20 (16), 1695–1697 (1995).

[20] A. Muller, T. Herzog, B. Huttner, W. Tittel, H. Zbinden, and N. Gisin, "Plug and play" systems for quantum cryptography, *Appl. Phys. Lett.* 70 (7), 793–795 (1997). US patent pending.

[21] A. Muller, H. Zbinden, and N. Gisin, Quantum cryptography over 23 km in installed under-lake telecom fibre, *Europhys. Lett.* 33 (5), 335–339 (1996).

[22] P. C. M. Owens, J. G. Rarity, P. R. Tapster, D. Knight, and P. D. Townsend, Photon counting with passively quenched germanium avalanche diodes, *Appl. Opt.* 33 (30), 6895–6901 (1994).

[23] P. D. Townsend, J. G. Rarity, and P. R. Tapster, Enhanced single photon fringe visibility in a 10 km long prototype quantum cryptography channel, *Electron. Lett.* 29 (14), 1291–1293 (1993).

[24] S. Wiesner, Conjugate coding, *Sigact News*, 77–88, 1983.

[25] W. K. Wootters and W. H. Zurek, A single quantum cannot be cloned, *Nature*, **299**, 802 (1982).

[26] H. Zbinden, J. D. Gautier, N. Gisin, B. Huttner, A. Muller, and W. Tittel, Interferometry with Faraday mirrors for quantum cryptography, *Electron. Lett.* 33 (7), 586–588 (1997).

Applicable Algebra in Engineering, Communication & Computing
Manuscript-Nr. AAECC #387

Quantum Key Distribution: from Principles to Practicalities

Dagmar Bruß[1]* and Norbert Lütkenhaus[2]

[1]ISI, Villa Gualino, Viale Settimio Severo 65, I-10133 Torino, Italy
[2]Helsinki Institute of Physics, PL 9, FIN-00014 Helsingin yliopisto, Finland

Received: 15.10.1998; Revised: 20.01.199/08.06.1999

Summary. We review the main protocols for key distribution based on principles of quantum mechanics, describing the general underlying ideas, discussing implementation requirements and pointing out directions of current experiments. The issue of security is addressed both from a principal and real-life point of view.

1. Principles

The desire and necessity to transmit secret messages from one person to another is probably as old as the capability of human beings to communicate. Cryptography is the art to encode a text in such a way that a spy (or eavesdropper) can get as little information as possible about it, and only the authorized receiver can decode it perfectly. The methods to perform this task have been improved over thousands of years. An important class of today's schemes are public-key cryptosystems [14], in which mutually inverse transformations are used for encoding and decoding. The instruction for encoding is made public, and safety relies on the high complexity of the inverse transformation (factorization of large prime numbers). In principle this system could be broken, though, by faster algorithms (see Shor's algorithm in quantum computation).

The only crypto-system that has been proven to be safe is using a random key which is only known to the sender and the receiver. The recipe for the sender is to translate the text with a look-up table into a sequence of 0's and 1's, e.g., A$\rightarrow$ 00001, B $\rightarrow$ 00011, etc., (this translation alone is fairly easy to decipher by an enemy) and then to add modulo 2 the random key (a random sequence of 0's and 1's), which needs to be of the same length as the message.

* Present affiliation: Inst. für Theoret. Physik, Universität Hannover, Appelstr. 2, D-30167 Hannover, Germany

The result is that letters which were the same in the original message are encoded into completely uncorrelated strings. Only the receiver can decode the message by adding again the secret key. This method is only safe, though, if the key is used just once, otherwise consecutive messages reveal information about the messages.[1] Therefore this type of protocol is also labeled with the key word "one-time pad", because in the second world war the key would be written on a sheet torn from a pad.

Unfortunately, the problem of secrecy is hereby only shifted to the problem of distributing the key in a safe way to the receiver. In principle, a spy can get hold of the key, copy it and send it on to the receiver. This is the point where quantum physics enters the stage: if the key distribution makes use of quantum states (this is possible in different ways which will be explained in detail in the following) the spy cannot measure them without disturbing them. Thus principles of quantum mechanics can help to make the key distribution safe. Often this young research area is, slightly misleading, also referred to as quantum cryptography (for an introduction see [6]).

In the modern communication society there is widespread need of secure transmission of secret information (e.g. credit card numbers, passwords). Therefore, a practical realization of these ideas is certainly very desirable, and some experimental results have indeed already been achieved. We will summarize the occurring problems and solutions for some of them and point out the open questions.

Let us list the main ingredients of quantum mechanics which allow for different protocols of secure key distribution - these will be explained in more detail in the following chapter:

- *non-orthogonal states cannot be distinguished perfectly*
  A quantum mechanical two-state system cannot only be in the state $|0\rangle$ or $|1\rangle$, but more generally in a linear superposition $|\psi\rangle = \alpha|0\rangle + \beta|1\rangle$ with complex coefficients α and β satisfying $|\alpha|^2 + |\beta|^2 = 1$. Due to the laws of quantum mechanics, it is impossible to distinguish reliably between
  $$\begin{aligned} |\psi_1\rangle &= \alpha_1|0\rangle + \beta_1|1\rangle \quad \text{and} \\ |\psi_2\rangle &= \alpha_2|0\rangle + \beta_2|1\rangle \end{aligned} \tag{1}$$
  unless the state overlap is $\langle\psi_1|\psi_2\rangle = 0$, i.e. the states are orthogonal.
- *no-cloning theorem*
  It is impossible, due to linearity and unitarity of quantum mechanics, to create perfect copies of an unknown quantum state [33]. Thus a spy is not able to produce perfect copies of a quantum state in transit in order to measure it, while sending on the original.
- *entanglement (quantum correlation)*
  Two or more quantum systems can be correlated or entangled. An entangled state cannot be written as a direct product of the subsystems. The singlet of two spin-$\frac{1}{2}$-systems is an example of a maximally entangled state:
  $$|\psi^-\rangle = \frac{1}{\sqrt{2}}(|01\rangle - |10\rangle) \quad . \tag{2}$$

[1] By adding two messages encoded with the same key one obtains the sum of the two original messages. This narrows down the possible combinations and reveals a considerable amount of information to an eavesdropper.

(The four maximally entangled states of two spin-$\frac{1}{2}$-systems are called Bell states.) If a measurement is done on one of these two quantum systems (in any basis), the result will be 0 or 1 with equal probability. The state of the other system is anti-correlated, i.e. if the first system collapsed into 0, the second collapses into 1 and vice versa. Without any measurement, though, none of the two systems *is* in a fixed state.

– *causality and superposition*
Causality is *not* an ingredient of non-relativistic quantum mechanics. Nevertheless it is mentioned in this list of principles because together with the superposition principle it can be used for secure key distribution: if the two terms of which a superposition consists are sent with a time delay relative to each other, such that they are not causally connected, the eavesdropper cannot spy on them.

2. Concrete Protocols

In this chapter we explain different approaches to the task of establishing a common secret key between two parties. The sender of the key is usually called Alice and the receiver Bob. Here we will assume that no enemy (usually called Eve) is present. In chapter 3 we will then discuss how a spy can gain some information on the key.

We can distinguish the following main three classes of protocols.

1) *BB84 class:*
In 1984 Bennett and Brassard suggested a quantum cryptographic protocol that relies on the use of non-orthogonal states [4]. It is often referred to as BB84. There have been several ideas for variations of this protocol which will for this review be included in the BB84-class.

– *BB84:*
In the BB84 protocol [4] Alice sends randomly one of the four quantum states

$$\begin{aligned} |\,0\rangle & \quad , \\ |\,1\rangle & \quad , \\ |\,\bar{0}\rangle &= \tfrac{1}{\sqrt{2}}(|\,0\rangle + |\,1\rangle) \ , \\ |\,\bar{1}\rangle &= \tfrac{1}{\sqrt{2}}(|\,0\rangle - |\,1\rangle) \ , \end{aligned} \tag{3}$$

with equal probability. Here the states $|\,0\rangle$ and $|\,\bar{0}\rangle$ represent bit value '0', the states $|\,1\rangle$ and $|\,\bar{1}\rangle$ stand for bit value '1'. The first two states in equation (3) correspond to a spin-$\frac{1}{2}$-particle being polarized in positive or negative z-direction, the last two to polarization in positive or negative x-direction. This can be graphically visualized as in figure 1. (All figures in connection with the protocols show directions corresponding to polarization vectors of spin-$\frac{1}{2}$-particles.)

The states in eq. (3) can also be represented by linearly polarized photons: the first two states then correspond to vertically and horizontally polarized photons, the last two to polarization angles 45° and 135° with respect to the vertical axis.

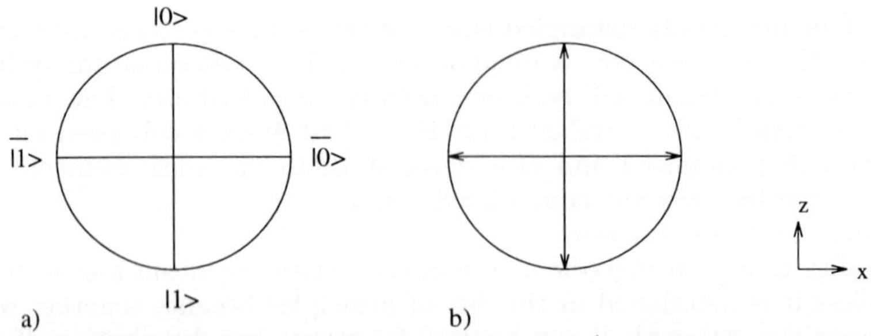

Fig. 1. Directions corresponding to polarization of a spin-$\frac{1}{2}$-particle for the BB84 protocol: a) ensemble of states Alice sends, b) Bob's directions of measurement. Note that orthogonal states point in opposite directions, see e.g. $|0\rangle$ and $|1\rangle$, which point in $+z$ and $-z$ direction, respectively.

When Bob receives a state from Alice, he chooses randomly either the x- or the z-basis for making a measurement. His result will always be either $|0\rangle$ or $|1\rangle$. But only in the cases where he picked the "right" basis, i.e. the one which Alice used, is his result correlated with the bit Alice sent. If, e.g., Alice sent $|\bar{0}\rangle$, but Bob measures along the z-direction, he will find either $|0\rangle$ or $|1\rangle$ with equal probability. After Alice sent and Bob measured the necessary number of states, Alice phones Bob (or uses some other "classical" channel) and tells him when she used which basis. They throw away the cases in which they used different bases, and thus have established a secret key. This key is called the *sifted key*.

– *B92:*
In this protocol by Bennett [2] Alice chooses between two non-orthogonal states to be sent to Bob. It was shown that in principle *any* two non-orthogonal states of a quantum system can be used for quantum key distribution. Let $|u_0\rangle$ and $|u_1\rangle$ be the two non-orthogonal states which represent the bit values 0 and 1, see figure 2.
Bob makes a measurement with a set of so-called POVM's (positive operator valued measurements), which gives as result either "$|u_0\rangle$" or "$|u_1\rangle$" or "I don't know" (see, e.g., [30]). For example, if Alice sends $|u_0\rangle$, Bob will either find $|u_0\rangle$ or an inconclusive result, but never $|u_1\rangle$. They can then use the public channel to discard inconclusive results, thus arriving at a correlated string of bits.
In practice the two non-orthogonal states can be realized by two low-intensity coherent states (note that two different coherent states are never exactly orthogonal, and for low intensities they become significantly non-orthogonal). An additional strong reference pulse is used in order to enhance security of the protocol (see section 4.1).

– *4+2 protocol:*
The protocol described in [22] combines ideas from BB84 and B92: as in BB84 Alice chooses between two different bases (so the number of possible states to send is 4), and as in B92 the two states within a basis, representing bit '0' and '1', are non-orthogonal. As in B92, a strong reference pulse is used.

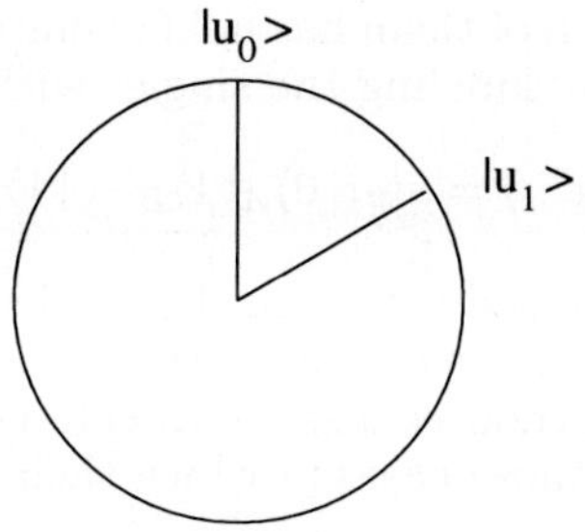

Fig. 2. B92 protocol: two non-orthogonal states.

Thus, this protocol corresponds to realizing BB84 with coherent states and a strong reference pulse.

- *Six state protocol:*
  In the six state protocol [10, 1] Alice enlarges her ensemble of quantum states she sends across to Bob, using in addition to the four states in BB84 the states

$$|\bar{\bar{0}}\rangle = \frac{1}{\sqrt{2}}(|0\rangle + i|1\rangle) \text{ and}$$
$$|\bar{\bar{1}}\rangle = \frac{1}{\sqrt{2}}(|0\rangle - i|1\rangle) , \quad (4)$$

  which describe a spin-$\frac{1}{2}$-particle polarized in positive or negative y-direction. (In the case of photons, these states represent circular polarization.) The six states are shown in figure 3.
  Thus, Alice sends a state randomly polarized in positive or negative x-, y-, or z-direction to Bob, who measures randomly in the $x-, y-$ or z-basis. As in BB84 they communicate over a public channel and keep only those cases in which their basis was the same.

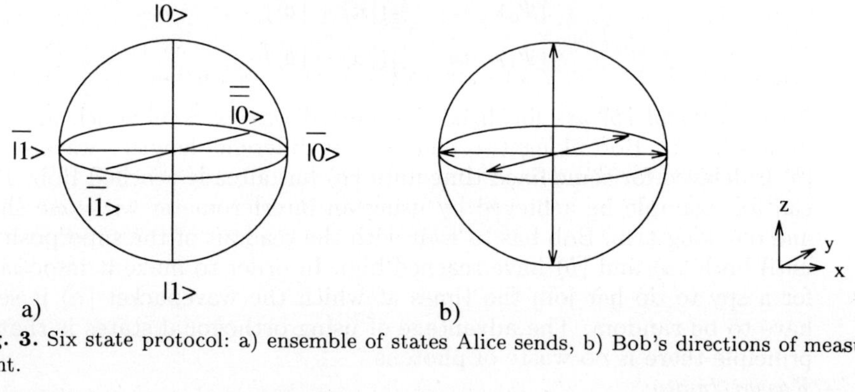

Fig. 3. Six state protocol: a) ensemble of states Alice sends, b) Bob's directions of measurement.

2) *Ekert scheme:*
   In the key distribution scheme designed by Ekert [15] Alice and Bob are sharing a number of maximally entangled states consisting of two two-state

systems, such that each of them has hold of one of the two correlated systems. Let us indicate this by labeling the singlet with indices A and B:

$$|\psi^-\rangle = \frac{1}{\sqrt{2}}(|0\rangle_A|1\rangle_B - |1\rangle_A|0\rangle_B) \quad . \tag{5}$$

They store their entangled states until they decide to establish the key, then Alice chooses randomly one of the three measurement directions indicated in figure 4 whereas Bob chooses a set of directions rotated by 45°.
They use again just those cases in which their measurement directions were

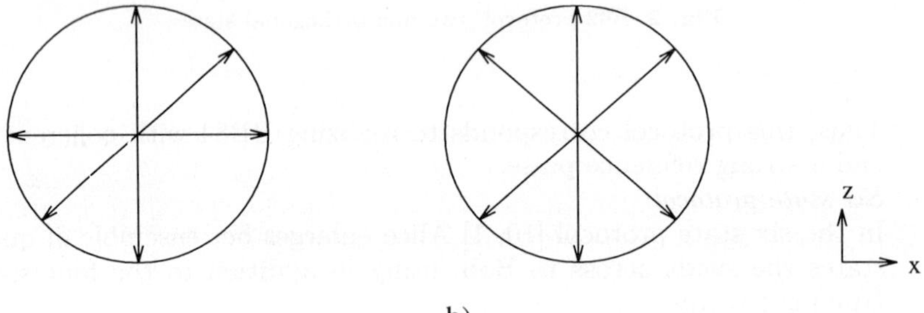

a) b)

Fig. 4. Ekert protocol: a) Alice's directions of measurement, b) Bob's directions of measurement.

the same. Only then their results are correlated. The runs where they used different directions can be used to test the Bell inequality and thus find out whether anybody has interfered with their systems.

3) *Goldenberg/Vaidman class:*
The idea of this class of protocols is to use a superposition of states, which arrive at different times at Bob's site.
   - *Goldenberg/Vaidman:*
   The scheme described in [19] uses two orthogonal states, $|\Psi_0\rangle$ and $|\Psi_1\rangle$, to represent bits '0' and '1', given by

$$\begin{aligned} |\Psi_0\rangle &= \frac{1}{\sqrt{2}}(|a\rangle + |b\rangle) \quad , \\ |\Psi_1\rangle &= \frac{1}{\sqrt{2}}(|a\rangle - |b\rangle) \quad , \end{aligned} \tag{6}$$

   where $|a\rangle$ and $|b\rangle$ are localized normalized wavepackets which are sent from Alice to Bob along two channels of different 'length': wavepacket $|b\rangle$ is delayed for some fixed time until $|a\rangle$ has already reached Bob. This can for example be achieved by using an interferometer with one short and one long arm. Bob has to wait with the readout of the superposition until both $|a\rangle$ and $|b\rangle$ have reached him. In order to make it impossible for a spy to do her job, the times at which the wavepacket $|a\rangle$ is sent, have to be random. The advantage of using orthogonal states is that in principle there is no waste of photons.
   - *Koashi/Imoto:*
   The authors of [23] show how to circumvent the necessity of random timing by making the interferometer asymmetric, i.e. by using beamsplitters that do not have equal transmittivity and reflectivity. This means that the amplitudes in eq. (6) change to

$$|\Psi_0\rangle = -i\sqrt{R}|a\rangle + \sqrt{T}|b\rangle \; ,$$
$$|\Psi_1\rangle = \sqrt{T}|a\rangle - i\sqrt{R}|b\rangle \; . \quad (7)$$

The different amplitudes for $|a\rangle$ deprive Eve of the possibility (given she knows the sending times) to use the simple strategy to send Bob a dummy $|a\rangle$ and later, after learning the phase, to send him $\pm|b\rangle$.

3. Security

Due to the principles of quantum mechanics described above, it is impossible for the spy Eve to gain *perfect* knowledge of the quantum state sent from Alice to Bob. Nevertheless, she can acquire *some* knowledge. Without interaction of a spy, each two-level quantum system carries 1 bit of information (commonly called qubit) from Alice to Bob. When Eve gets hold of part of this information, she cannot prevent causing a disturbance to the state arriving at Bob's side, and thus introducing a non-zero error rate. In principle, Bob can find out about this error rate and thus about the existence of a spy by communicating with Alice.

The source for Eve's knowledge are measurements performed on the signals (quantum states). The simplest eavesdropping attack for Eve would be to measure each signal just as Bob would do, and then to resend a signal to Bob which corresponds to the measurement result.

However, quantum mechanics allows more general measurements than these simple projection measurements. Eve can bring an auxiliary quantum system (a probe) in contact with the signal so that they interact, and then perform a projection measurement on the auxiliary system to draw some information about the signal from it. All measurements, including the simple projection measurements, can be described in this fashion [20, 30]. Another opportunity arises for Eve: she might delay the measurement of the auxiliary system until she learns more about the signal during public discussion. An example for useful information is the signal set from which a signal has been drawn. More involved strategies within quantum mechanics correlate measurements of several signals, thereby attacking the key as a whole rather than the individual components. This scenario is referred to as *coherent* eavesdropping. A simpler class is that of *collective* eavesdropping where to each signal an individual probe is attached just as in the individual attack. These probes, however, now can be read out together in a coherent process.

As mentioned above, in the ideal case we are always able to identify an eavesdropping activity by the occurrence of errors in the transmission. In a real world this becomes a tricky issue. We will always have some detector noise, misalignments of detectors and so on. It should be pointed out that we cannot even in principle distinguish errors due to noise from errors due to eavesdropping activity. We therefore assume that all errors are due to eavesdropping. An other issue, not discussed here, is that of statistics. Eavesdroppers can be lucky: they create errors only on average, so in any specific realization the error rate might be zero (with probability exponentially small in the key length, of course). We are guided by the idea that a small error rate, for example 1 %, implies that an eavesdropper was not very active, while a big error rate is the signature of a serious eavesdropping attempt. But what is the meaning of "small" and "big"?

From an information theoretic point of view, the natural measure of "knowledge" about some signal is the Shannon information. It is measured in bits and can be defined for any two parties, the sender of the signal and the observer (receiver). In general terms, the knowledge of the observer consists of obtained measurement results and any additional gathered knowledge, like the announced basis of signals in the BB84 protocol. All this knowledge will be denoted by M.

From the receiver's point of view there will be an a-priori $p(x)$ and an a-posteriori $p(x|M)$ probability distribution for the signal x. The knowledge M will turn up with probability $q(M)$. The Shannon information can now be defined as the *expected* change in entropy of the two probability distributions. It is therefore given by

$$I = -\sum_x p(x) \log_2 p(x) + \sum_M q(M) \sum_x p(x|M) \log_2 p(x|M) \ . \tag{8}$$

For a binary channel with equal a-priori probabilities for the two signals the Shannon information can be expressed in terms of the error probability e with which the signals are received. It is given (in bits per signal) by

$$I[e] = 1 + e \log_2 e + (1-e) \log_2 (1-e) \ . \tag{9}$$

This is the Shannon information, per element of the sifted key, between Alice and Bob, I_{AB}, with the observed error rate e of the channel. On the other hand we will use the information I_E, generally given by equation (8), which Eve obtains on the key where M then represents her measurement results and all the information exchange between Alice and Bob over the public channel.

Another proposed measure of Eve's knowledge is the probability that the eavesdropper guesses the correct key given her knowledge about it.

A fundamental difference between classical cryptography and the use of a one-time pad together with quantum key distribution is that the former one is vulnerable to technological improvements (faster computers and algorithms) and therefore has to be designed to keep the secret secure against improvements which occur during the whole period of time in which the secrecy is required. Quantum key distribution, on the other hand, needs to be designed to be secure only against technology available at the time (and location) of the quantum part of key distribution. Therefore it makes sense to give the estimates of the Shannon information for various scenarios. They differ by the technology available to Eve. Examples for potential improvement of Eve's knowledge are the ability to perform delayed measurements (needs physical storage of auxiliary quantum systems), the availability of quantum channels superior to those used by Alice and Bob (for example in form of an optical fibre which is less lossy and noisy), and the ability to perform coherent eavesdropping attacks (needs ability to manipulate and store coherently several quantum systems).

Let us now quote some results on maximal information leakage to the eavesdropper. They are valid under the assumption of ideal BB84 signal states, for example single photons.

It has been shown that the simple intercept-resend strategy leads for the BB84 protocol to an average error rate of 25 % while it yields at best 0.5 bit of information per signal [16, 21]. The optimal probability of a correct guess would be 75% in that case.

Bounds on the obtainable Shannon information for eavesdropping on single bits can be found in the literature for different protocols. Fuchs et al. give bounds for the BB84 [17] and the B92 protocol [18]. A bound for the six state protocol was obtained in [10]. These bounds are illustrated in figure 5. Note the trade-off between Eve's information gain and the disturbance she causes: more information for Eve means higher error rate for Bob. For reasonably low error rates Eve's maximal information is smallest in the six-state protocol, as it uses the biggest ensemble of input states.

Bounds for the Shannon information in more general attacks are studied in [12] for BB84 and [1] for the six-state protocol.

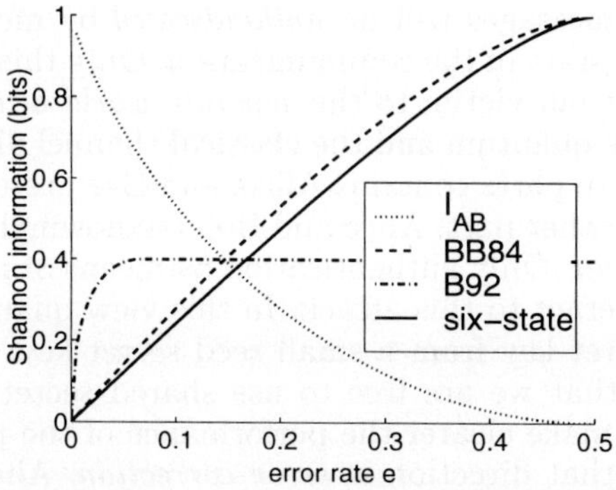

Fig. 5. Maximal mutual information I_E on the sifted key shared between Alice and Eve as function of Bob's error rate e for the protocols BB84 [17] and the six-state protocol [10] together with the the mutual information between Alice and Bob given by the curve I_{AB}. The graph for the B92 protocol [18] with state overlap of $1/\sqrt{2}$ displays I_E for the raw key and not for the sifted key.

The important result from these estimates is that even for small error rates the eavesdropper might be in possession of information about the key at a level deemed dangerous for secure communication. For example, at an observed error rate of 1% we find that an eavesdropper might have gained up to 0.024 bit of Shannon information per bit of key even for the six-state protocol. This is far too high to allow the direct use of the obtained key for encryption. Instead, one uses the tool of privacy amplification [5] (see following section) to extract a short secure key from the long insecure key.

One of the advantages of the Ekert scheme is that by storing the states at both ends of the transmission line and coherent manipulation on each side between the accumulated states the performance of the key distribution could be enhanced. This technique is called *quantum privacy amplification* [13] and effectively gives a new, shorter key with lower error rate.

4. Elements for realistic implementations

In the previous section we have seen that the Shannon information available to an eavesdropper about the *sifted key* (that is the key directly after the key exchange) is too high to allow secret communication directly. Fortunately, it is possible to process this key with help of a purely classical protocol in order to distill a new, shorter key from the sifted key which exponentially approximates a secret key. We present the procedure here in a form which is valid only if Eve's activity is restricted to attacks on individual signals (as opposed to coherent or collective attacks). However, the steps executed in a quantum key distribution apparatus are the same in the general case, only the reasoning behind them changes, as indicated below.

More details about the full protocol to deal with restricted attacks in a realistic scenario can be found in [26]. Here we will concentrate only on the main points. An important point for practical realization is that in a realistic protocol no ideal public channel exists which can be overheard but not changed by an eavesdropper. This property of a channel can only be approximated by using an open channel where messages will be *authenticated* by means of a small secret key shared before the start of the communication. Only this method ensures that Alice and Bob do not fall victim to the *separate world attack.* In this attack an eavesdropper cuts the quantum and the classical channel dividing the world into two parts. One of these parts contains Alice, and Eve pretends to her to be Bob and vice versa in the other part. Alice and Bob unknowingly never communicate directly with each other. Only authentication by means of previous shared secret knowledge can counteract to this attack. In this view quantum key distribution will grow a large secret key from a small seed secret key. A by-product of this changed scenario is that we are free to use shared secret bits in intermediate states to enhance or make clearer the performance of the protocol.

The first step in that direction is *error correction.* Alice and Bob exchange redundant information over the public channel to reconcile their versions of the key. Obviously, the amount of exchanged redundant information has to be kept as small as possible, since the information flow to Eve has to be taken account of. (One possibility is to encode it using part of the initially shared secret key.) What is the minimum amount of exchanged redundant bits? A correctly received binary string of length n_{sif} carries exactly n_{sif} bits of Shannon information. On the other hand, if Bob received this key with an error rate e then he is in possession of $n_{sif} I_{AB}$ bits only. He, therefore, has to get hold of the difference of $n_{sif} - n_{sif} I_{AB}[e]$ bits of information. Since the public channel can be made error free[2] the information per signal sent there is the ideal 1 bit, so for each bit of information missing, Alice has to send on average one signal. Therefore the minimum amount n_{min} of bits to be exchanged is given by the Shannon bound,

$$n_{min} = -n_{sif} \left(e \log_2 e + (1-e) \log_2 (1-e) \right) . \quad (10)$$

The best known practical protocol is that of Brassard and Salvail [8]. It uses an interactive information exchange between the two sides. The requirements for a good error correction protocol are to be as close as possible to the minimum number of exchanged bits given by the Shannon bound and a success rate of correction as high as possible. In contrast to a standard problem in error correction the channel used for transmission of the redundant bits can be assumed to be error free, which allows for improved, specialized error correction schemes.

Starting from the reconciled key, Alice and Bob now use *privacy amplification* [5] to establish a secret key. The idea behind privacy amplification is to hash

[2] Any information sent through the public channel can be put into code words, using any error correction scheme, to protect it against errors. This encoding into codewords does not change the amount of Shannon information contained, and one codeword can be regarded as one signal.

the reconciled key of length n_{rec} into a shorter key of length n_{fin} using random hashing. An example for hashing is to taky bits of random subsets of the reconciled key to form the new key. In general, we shorten the reconciled key by the fraction τ_1 and then by additional n_S bits to a final key length of $n_{fin} = (1 - \tau_1)n_{rec} - n_S$. As shown by Bennett et al. [5] Eve's Shannon information on the final key is bounded by

$$I_E^{final} \leq \log_2(2^{-n_S} + 1) \approx \frac{2^{-n_S}}{\ln 2} \quad . \tag{11}$$

A consequence is that I_{final} can be made exponentially small by means of the number of security bits n_S.

The central quantity in this context is the collision probability P_{coll}, and the fraction τ_1 is given by $\tau_1 = 1 + \frac{1}{n_{rec}} \log P_{coll}$. Here P_{coll} is a measure of the a posteriori probability distribution P_{post} of the reconciled key conditioned on all information available to the eavesdropper. It is defined by the relation $P_{coll} = \sum_x (P_{post})^2$ where the sum is taken over all reconciled keys. For security against eavesdropping strategies attacking individual signals only it is essential to find an upper bound on the collision probability. Bounds for P_{coll} and expressions for τ_1 for the BB84 protocol are given in [25, 32, 26], for the B92 protocol in [32] and for the six-state protocol in [1].

With these results it is possible to calculate the optimal rate at which one can extract secure bits from the sifted key. We assume error correction at the Shannon bound of equation (10) and encryption of the redundant bits. Then the balance between new secure bits being created and old secure bits being used up gives an average creation rate per bit of the sifted key of

$$R_{corr} = I_{AB}[e] - \tau_1[e] \tag{12}$$

if we use error correction, and

$$R_{del} = I_{AB}[e] - \tau_1[e](1 - e) - e \tag{13}$$

if we discard errors from the key. To obtain the creation rate of secure bits as a fraction of the sent quantum signals we have to multiply R_{corr} and R_{del} by a factor 1/2 for the B92 and the BB84 protocol, and by 1/3 for the six-state protocol. A direct comparison for the resulting rates in case of discarded errors is made in figure 6. The results show that the restriction to eavesdropping attacks on individual signals allows secure quantum key distribution with existing experiments. The tolerable error rates, leading to positive rates, are 4%, 10.5%, and 12% for the three protocols respectively. The six-state protocol gives the lowest gain for error rates below $\approx 0.65\%$ while it becomes superior to the BB84 protocol for error rates bigger than approximately 8%. Though tolerable error rates are achievable with present day experiments, some work still has to be done to cope with the signal states which are not the ideal one-photon states (see section 4.1).

For more general strategies than those measuring individual signals the presented way of error correction, estimation of collision probability, and privacy amplification is no longer valid since in that case Eve might make use of the knowledge of the particular hashing function (choice of random subsets for parity bits) to optimize her measurements. Instead, one has to directly estimate the

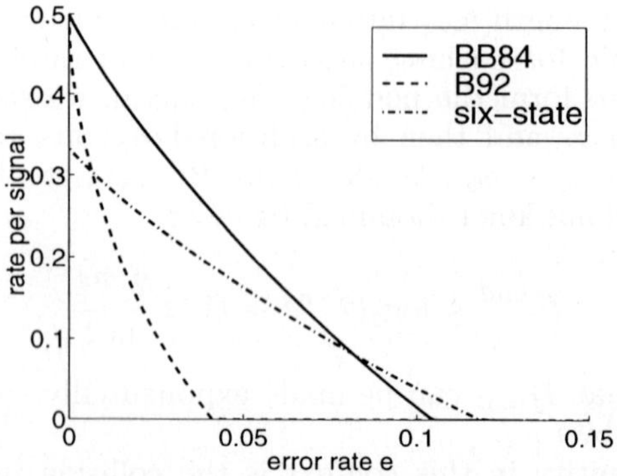

Fig. 6. The rate $\frac{1}{2}R_{del}$ for the B92 protocol with overlap $1/\sqrt{2}$ between the two signal states has been calculated using the results by Slutsky et al. [32]. The rate $\frac{1}{2}R_{del}$ for the BB84 protocol is obtained with the estimates from [32, 26] and the estimates leading to the rate $\frac{1}{3}R_{del}$ for the six-state protocol are taken from [1].

Shannon information on the final key. This has been done for a wide class of *collective attacks* in [7], while bounds in the most general case are obtained in [29, 24]. The proof given in [29] leads to a maximal tolerated error rate of circa 7 %. The proof of [24] uses the Ekert scheme in connection with [13] to tolerate higher error rates, as mentioned in the discussion of the Ekert scheme, but it needs local operations operating with an error rate below the threshold set for fault tolerant quantum computing.

It is important to note that the key generated by quantum key distribution is different from the key assumed in the one-time pad or as seed for the Data Encryption Standard. These keys are assumed to be absolutely secure and certainly shared between Alice and Bob. The key established in quantum key distribution does not carry those absolute attributes. It is not absolute secure. Instead, we can make a statement about it of the following form: With probability $1-\alpha$ an eavesdropper has less Shannon information than a tolerated value I_E^{tol} on that key (secrecy) and it is shared between Alice and Bob with probability $1-\beta$. The two probabilities α and β can be made arbitrary small (on cost of the key rate) as long as the initial error rate is below the cut-off rate mentioned above for the different scenarios. To our knowledge, this subtle difference between the key properties assumed in applications and the key properties resulting from quantum key distribution has not been explored sufficiently yet. Especially, it would be interesting to explore what values for I_E^{tol}, α and β are required for applications.

4.1. Problems and practicalities

All current implementations of quantum key distribution make use of quantum optical methods. In this context we will discuss realization issues important for the security aspect without going into technical details. The problem of realizing quantum cryptography consists of three parts: realization of the signal states, transportation of the signals to Bob and an efficient measurement of the signals.

The simplest choice of signal states, from the theoretical point of view, are single photons with the polarization as carrier of the signal. However, at present we do not have a source which would give us single photons on demand. Instead, one uses weak laser pulses. On average, each pulse contains typically 0.1 photons. The photon number distribution is such that most pulses contain no photon, around 10% contain one photon and 1% contain more than one photon. The pulses containing more than one photon endanger security of transmission, since an eavesdropper could split off one photon and extract the full information about the signal later on without causing any disturbance of the channel. This has to be taken account of when calculating the amount by which the key is shortened during privacy amplification. The transmission is totally insecure if the number of received signals is smaller than the number of multiple-photon signals sent.

One of the big problems in quantum key distribution is loss of signals in the fibre. It has been shown that strong loss in the transmission going together with multi-photon components of the signal states renders key distribution in all key distribution schemes insecure unless a strong reference pulse is used [22]. This strong reference pulse is an original part of the B92 protocol [2]. It fights the problem that the eavesdropper has means to suppress a signal without causing errors by sending a vacuum state to Bob. A strong reference pulse, however, makes sure that no such state exists.

To keep the error rate low, the set-up should be stable under influence of the environment. In the case of polarization based cryptography the main error source is cross talk between the two polarization modes and a random (classical) rotation of the polarization along the propagation direction of the fibre. Here the proposals of the BB84 or B92 type are easier to implement than the time separated ideas of Goldenberg/Vaidman and Koashi/Imoto. In the first group the signal travels from Alice to Bob and is influenced by the environment as an entity, while in the second group we have two parts of a signal interacting with two different environments. We therefore cannot expect the error rates of the second group to be as good as the 1% error rates of the first group. This is the reason why no experimental realization of the second group has been tackled yet.

For the detection schemes we find that it poses a problem to lower the amplitude of coherent states below a certain point in order to improve the single-photon approximation. Bob's detectors will give false alarm (dark counts) with a fixed probability proportional to the time the detector is gated. Using weaker pulses will increase the number of dark counts with respect to the real counts, which effectively increases the error rate because a dark count will give a random measurement result.

One of the advantages of the Ekert scheme is that is allows to use quantum privacy amplification, thereby giving a new raw key with lower error rate than the original key. This allows to go below the cut-off rate for the tolerated error rate even with a noisy channel. However, the necessary storing and manipulation devices are not available at present.

4.2. Experiments

Quantum key distribution was implemented for the first time by Bennett et al. in a demonstration set-up [3]. The transfer of the signals took place over 32 cm of free air with (incoherent) faint pulses. Experimental demonstrations of the BB84 protocol near to commercial realizations are reported by the group at British Telecom by Marand and Townsend [28]. Over a distance up to 30 km they achieved an error rate of 1.5–4 % with an average photon number per pulse of 0.1–0.2 photons.

Several experiments have been done implementing an approximate B92 protocol. In these experiments the strong reference pulse of the original scheme is omitted, thereby using the idea of two non-orthogonal pulses only. It is known that this omission renders the scheme more insecure. Best results regarding low error rate are achieved here by the group in Geneva. They achieve error rates of about 0.5–1.35 % over distances of 23 km with an average photon number of 0.1–0.2 [34]. An initial problem of their scheme to give a low key rate only has now been resolved. Other schemes in free-space key distribution and over fibre are reported by the Los Alamos group, going over 40 km in fibre and 1 km in free space [11]. Variations of the Ekert scheme have been implemented by Rarity et al. [31].

5. Open questions and summary

From the point of fundamental physics the most interesting question is to show security against the most general coherent eavesdropping attack on single photon signals. This has been achieved by Mayer [29] and by Lo and Chau [24]. From the practical point of view these proofs are not relevant yet since they do not deal with realistic situations. For this one would need the use of efficient error correction methods, the ability to cope with large losses and with realistic error rates and, finally, the extension to realistic signals like dim coherent states or photons from parametric downconversion.

For practical purposes it makes sense to restrict eavesdropping strategies to attacks on individual signals. For this scenario workable schemes for single-photon states have been presented in [32, 26]. The extension to realistic signal states has been achieved recently [27].

The experimental groups will have to look for set-ups improving the rate at which the key is generated. It is essential to keep in mind that it is not the aim to minimize the error rate but to maximize the key rate!

The questions directed to the audience dealing with the classical part of quantum key distribution are: a) what is a good goal for the security of the final key? b) How good does it have to be? (In terms of I_E^{tol}, α and β as introduced in section 4.) c) What is the optimal reconciliation protocol in these circumstances?

In summary, quantum key distribution is a truly interdisciplinary topic in quantum information. It brings together cryptologists, classical information scientists, and experimental and theoretical physicists. At present, there are physical systems which already produce sifted keys at a reasonable rate with a low error rate. Although the implementation is not ideal, theoretical work should soon show in which scenario it is possible to extract secure keys from that. To optimize procedures more work in error correction etc. is needed. After realizing the nature of security of the final key, we need more input about the specific requirements for applications - as quantum key distribution has already passed the first threshold towards implementation.

Acknowledgments

The authors took benefit from the 1998 quantum information workshops at ISI (Italy) and Benasque Center for Physics (Spain) and wish to thank their organizers and Elsag–Bailey for support. DB acknowledges support by the European TMR Research Network ERP-4061PL95-1412 and by Deutsche Forschungsgemeinschaft under grant SFB 407, and NL by the Academy of Finland.

References

1. H. Bechmann-Pasquinucci and N. Gisin: Incoherent and Coherent Eavesdropping in the 6-state Protocol of Quantum Cryptography. quant-ph/9807041 (1998).
2. C. Bennett: Quantum cryptography using any two nonorthogonal states. Phys. Rev. Lett. **68**, 3121 (1992).
3. C. H. Bennett, F. Bessette, G. Brassard, and L. Savail: Experimental quantum cryptography. J. Crypt. **5**, 3–28 (1992).
4. C. H. Bennett and G. Brassard: Quantum cryptography: Public-key distribution and coin tossing. In: *Proceedings of the IEEE International Conference on Computers, Systems, and Signal Processing, Bangalore, India* (IEEE, New York, 1984), pp. 175-179.
5. C. H. Bennett, G. Brassard, C. Crépeau, and U. M. Maurer: Generalized privacy amplification. IEEE Trans. Inf. Theo. **41**, 1915 (1995).
6. C. H. Bennett, G. Brassard, and A. Ekert: Quantum Cryptography. Scient. American, 50 (Oct. 1992).
7. E. Biham, M. Boyer, G. Brassard, J. van de Graaf, and T. Mor: Security of quantum key distribution against all collective attacks. quant-ph/9801022, (1998).
8. G. Brassard and L. Salvail: Secret-key reconciliation by public discussion. In *Proceedings of Eurocrypt '93, held in Lofthus, Norway, 1993*, (1993).
9. H. J. Briegel, W. Dür, J. I. Cirac, P. Zoller: Quantum Repeaters: The Role of Imperfect Local Operations in Quantum Communication. Phys. Rev. Lett. **81**, 5932 (1998).
10. D. Bruß: Optimal eavesdropping in quantum cryptography with six states. Phys. Rev. Lett. **81**, 3018 (1998).
11. W. T. Buttler, R. J. Hughes, P. G. Kwiat, G. G. Luther, G. L. Morgan, J. E. Nordholt, C. G. Peterson, and C. M. Simmons: Free-space quantum-key distribution. Phys. Rev. A **57**, 2379–2382 (1998).
12. I. Cirac, and N. Gisin: Coherent eavesdropping strategies for the 4-state quantum cryptography protocol. Phys. Lett. A **229**, 1 (1997).
13. D. Deutsch, A. Ekert, R. Josza, C. Macchiavello, S. Popescu, and A. Sanpera: Quantum privacy amplification and the security of quantum cryptography over noisy channels. Phys. Rev. Lett. **77**, 2818–2821 (1996).
14. W. Diffie and M. Hellman, IEEE Trans. Inf. Theory IT-22, 644 (1977);
R. Rivest, A. Shamir, and L. Adleman, "*On Digital Signatures and Public-Key Cryptosystems*", MIT Lab. for Comp. Science, Technical report, MIT/LCS/TR-212 (Jan. 1979).
15. A. Ekert: Quantum cryptography based on Bell's theorem. Phys. Rev. Lett. **67**, 661 (1991).
16. A. K. Ekert, B. Huttner, G. M. N. Palma and A. Peres: Eavesdropping on quantum-cryptographical systems. Phys. Rev. A **50**, 1047–1056 (1994).
17. C. A. Fuchs, N. Gisin, R. B. Griffiths, C.-S. Niu, and A. Peres: Optimal Eavesdropping in Quantum Cryptography I. Phys. Rev. A **56**, 1163 (1997).
18. C. A. Fuchs and A. Peres: Quantum State Disturbance vs. Information Gain: Uncertainty Relations for Quantum Information. Phys. Rev. A **53**, 2038–2045 (1996).
19. L. Goldenberg and L. Vaidman: Quantum Cryptography based on Orthogonal States. Phys. Rev. Lett. **75**, 1239 (1995);
A. Peres: Quantum Cryptography with Orthogonal States? Phys. Rev. Lett. **77**, 3264 (1996);
L. Goldenberg and L. Vaidman: Reply to Comment: Quantum Cryptography based on Orthogonal States. Phys. Rev. Lett. **77**, 3265 (1996).

20. C. W. Helstrom, *Quantum detection and estimation theory* (Academic Press, New York, 1976).
21. B. Huttner and A. K. Ekert: Information gain in quantum eavesdropping. J. Mod. Opt. **41,** 2455–2466 (1994).
22. B. Huttner, N. Imoto, N. Gisin, and T. Mor: Quantum Cryptography with Coherent States. Phys. Rev. A **51**, 1863–1869 (1995).
23. M. Koashi and N. Imoto: Quantum Cryptography Based on Split Transmission of One-Bit Information in Two Steps. Phys. Rev. Lett. **79**, 2383 (1997).
24. H. K. Lo and H. F. Chau: Quantum computers render quantum key distribution unconditionally secure over arbitrarily long distance. quant-ph/9803006, (1998).
25. N. Lütkenhaus: Security against eavesdropping in quantum cryptography. Phys. Rev. A **54,** 97 (1996).
26. N. Lütkenhaus: Estimates for practical quantum cryptography Phys. Rev. A **59,** 3301 (1999).
27. N. Lütkenhaus, Security of quantum cryptography with realistic sources. Acta Physica Slovaca **49,** 549 (1999).
28. C. Marand and P. T. Townsend: Quantum key distribution over distances as long as 30 km. Opt. Lett. **20,** 1695–1697 (1995).
29. D. Mayers: Unconditional security in quantum cryptography. quant-ph/9802025v4, (1998).
30. A. Peres, "Quantum Theory: Concepts and Methods", Kluwer (1995).
31. J. G. Rarity, P. C. M. Owens, and P. R. Tapster: Quantum random-number generation and key sharing. J. Mod. Opt. **41,** 2435–2444 (1994).
32. B. Slutsky, R. Rao, P. C. Sun, and Y. Fainman: Security of quantum cryptography against individual attacks. Phys. Rev. A **57,** 2383–2398 (1998).
33. W.K. Wootters and W.H. Zurek: A single quantum cannot be cloned. Nature **299**, 802 (1982).
34. H. Zbinden, N. Gisin, B. Huttner, A. Muller, and W. Tittel: Practical aspects of quantum cryptographic key distribution. (submitted to the Journal of Cryptology), (1998).

This article was processed by the author using the LaTeX style file *cljour1* from Springer-Verlag.

Cavity Quantum Electrodynamics

Cavity Quantum Electrodynamics

Hideo Mabuchi
California Institute of Technology

1 Introduction

Of all the physical interactions one might consider as a basis for quantum information processing, the electromagnetic interaction between photons and bound electrons offers superlative operational versatility coupled with distinct practical advantages. For example, recent theoretical proposals have described realistic strategies for implementing networks of quantum gates [1, 2], for robust transmission of quantum information over noisy channels [3], for quantum state synthesis [4], and for deterministic generation of single-photon wavepackets [5]. The pursuit of such goals in the laboratory is facilitated by the advanced state of optical and photonic technologies, as well as the fact that radiative interactions may be included among the most well-studied aspects of quantum physics. The formidable difficulty that must be overcome in such endeavors is the need to induce strong coupling between one atom (or molecule, quantum dot, etc.) and a single photon, while minimizing the decohering effects of radiative decay and thermal motion.

In the field of cavity quantum electrodynamics (cavity QED), one employs highly reflecting boundaries to confine an individual photon within a sufficiently small volume of space, and for a sufficiently long period of time, to saturate its interaction with a given atom. The physical scenario is depicted in Figure 1 – for simplicity, we consider a two-level atom located within a single-mode, high finesse electromagnetic resonator. The dynamics of this system are determined by three fundamental rates: the vacuum Rabi frequency g, the atomic dipole-decay rate γ, and the cavity field decay rate κ. The rate g, which essentially characterizes the strength of the coherent atom–cavity coupling, is given by $g = d \cdot E_1/2\hbar$, where d is the electric-dipole transition matrix element for the atom and E_1 is the electric field per photon in the cavity mode. If an atom, prepared initially in its excited state, is placed into an empty cavity, the basic quantum dynamical evolution will consist of an oscillatory exchange of energy between atom and cavity (via emission and re-absorption) at the rate $2g$. However, this coherent process competes with the dissipative processes of atomic spontaneous emission into non-cavity modes (at rate 2γ) and loss of the photon from the cavity mode (at rate 2κ). The regime of *strong coupling* may loosely be defined as that in which $g \gg (\gamma, \kappa)$ and the dwell time of an individual atom within the cavity mode volume is much longer than the inverse of any of these three rates [6, 7].

One of the definitive achievements in experimental cavity QED during this decade has been the unambiguous demonstration of strong coupling in both optical [8] and microwave [9] regimes. Current efforts have turned towards the *application* of strong coupling to

=3in cavityqed.eps

Figure 1: Dynamics of the cavity QED system are determined by three fundamental rates (see text).

tasks in quantum information processing, quantum measurement, and related fields. The utility of strong coupling for quantum information processing may be illustrated through the introduction of two dimensionless parameters, the critical photon number m_0 and critical atom number N_0:

$$m_0 \sim \frac{\gamma^2}{2g^2}, \qquad N_0 \sim \frac{2\gamma\kappa}{g^2}. \tag{1}$$

Note that in the strong coupling regime, both m_0 and N_0 are 1. For a single intracavity atom, the critical photon number m_0 roughly measures the number of photons that must be circulating in the cavity mode before the atomic response becomes saturated, and therefore nonlinear. Hence with strong coupling, one should be able to perform nonlinear optics with only one photon per mode. In optical incarnations of quantum logic, nonlinearity at the single-photon level is precisely what is required to implement universal quantum gates. The critical atom number N_0 provides a complimentary measure of the number of atoms that must be placed inside a cavity in order to substantially alter its optical properties. For example, when $N_0 \ll 1$ the absence or presence of just one atom inside a cavity can switch the cavity from being transmissive to reflective for incident light. Hence, one may contemplate the possibility of making a "quantum switch," in which an intracavity atom may be prepared in a superposition of one state that couples to the cavity mode and a second "dark" state that does not [10]. The expectation is that the cavity as a whole should thereby be placed in a superposition of transmitting and reflecting "states," in the sense that one should be able to use such a device to prepare entanglements between the atomic internal state and the outgoing propagation direction of light that impinges upon the cavity.

Before moving on to a discussion of the reprints, one further distinctive feature of optical cavity QED should be pointed out. In cases where cavity decay (as opposed to atomic spontaneous emission) constitutes the dominant channel for loss of photons from the atom-cavity system, the "environment" responsible for quantum decoherence is easily accessible to measurements that can be made with both high bandwidth and high efficiency. For example when $\kappa \gg \gamma$, most of the decoherence suffered by the atom-cavity system is simply caused by photons leaking out of the cavity. Such photons are radiated into a small solid angle and may therefore be easily re-focused onto a high quantum-efficiency photodetector. In the parlance of decoherence theory, this would correspond to measuring the state of the environment with which the atom-cavity system becomes entangled. As a result of this capability, one can devise specialized schemes for quantum error correction with less overhead than is required in scenarios where the environment is not accessible to direct measurement. In addition, one expects that cavity QED should provide an important experimental testbed for the investigation of outstanding issues in decoherence theory.

2 Selected reprints

Current research in the field of cavity QED, broadly defined, investigates the interaction of optical and microwave photons with atoms, molecules, excitons, and quasi-electrons bound in quantum dots and quantum wells. The reprints contained in this section, however, are drawn exclusively from the literature on cavity QED with alkali atoms. These papers were chosen on the basis of their significance for the field of quantum information, but the reader should be aware that all such work draws from and builds upon a very rich foundation of research in cavity QED that has been pursued—for the intrinsic interest of the subject—since the 1970's.

Measurement of Conditional Phase Shifts for Quantum Logic

As discussed in the introduction, nonlinear optical responses can be obtained in cavity QED (under the condition of strong coupling) with only one photon per mode. This 1995 paper by Turchette *et al* presents the first experimental measurement of such single-photon nonlinearities, and describes a quantum logic scheme that would utilize the observed effect to construct a universal quantum gate.

Real-Time Cavity QED with Single Atoms

From the perspective of quantum information science, this 1998 paper by Hood *et al* provides a preview of the techniques that will be used to construct the next generation of cavity-QED devices. Extremely strong coupling is achieved through careful cavity design ($m_0 \sim 2 \times 10^{-4}$, $N_0 \sim 10^{-2}$), and laser cooling methods are used (building upon earlier work by Mabuchi *et al* [8]) to minimize the thermal velocities of intracavity atoms. The next stage in the evolution of optical cavity QED will be to trap and localize these laser-cooled atoms within the cavity.

Quantum Memory with a Single Photon in a Cavity

Whereas the first two papers described work conducted in the optical regime, this and the following paper present the state of the art in quantum information processing in microwave cavity QED. This 1997 work by Maitre *et al* directly demonstrates the experimental capability to map the quantum state of a two-level atom (*i.e.*, one qubit) to that of a microwave cavity, and vice-versa. In addition, a measurement is presented of the "holding time" of the microwave cavity as a quantum memory device.

Observing the Progressive Decoherence of the "Meter" in a Quantum Measurement

This 1996 paper by Brune *et al* demonstrates the power of cavity QED as an experimental paradigm for fundamental investigations of decoherence. "Schrödinger-Cat" states of the microwave cavity field are prepared, and their decoherence rates measured as a function of phase-space separation between the superposed components.

Inversion of Quantum Jumps in Quantum Optical Systems under Continuous Observation

This 1996 paper by Mabuchi and Zoller describes a theoretical scheme for quantum error correction in a cavity-QED context. The details of this scheme are tied to the specific decoherence mechanisms of cavity QED, and direct measurements of the cavity output channels are utilized as described above.

References

[1] T. Pellizzari, S. A. Gardiner, J.-I. Cirac, and P. Zoller, "Decoherence, Continuous Observation, and Quantum Computing – a Cavity QED Model," Phys. Rev. Lett. **75**, 3788-3791 (1995).

[2] P. Domokos, J. M. Raimond, M. Brune, and S. Haroche, "Simple Cavity-QED 2-Bit Universal Quantum Logic Gate – the Principle and Expected Performances," Phys. Rev. A **52**, 3554-3559 (1995).

[3] S. J. van Enk, J.-I. Cirac, and P. Zoller, "Ideal Quantum Communication over Noisy Channels: a Quantum Optical Implementation," Phys. Rev. Lett. **78**, 4293-4296 (1997).

[4] A. S. Parkins, P. Marte, P. Zoller, and H. J. Kimble, "Synthesis of Arbitrary Quantum States via Adiabatic Transfer of Zeeman Coherence," Phys. Rev. Lett. **71**, 3095-3098 (1993).

[5] C. K. Law and H. J. Kimble, "Deterministic generation of a bit-stream of single-photon pulses," J. Mod. Opt. **44** 2067-2074 (1997).

[6] H. J. Kimble, "Strong interactions of single atoms and photons in cavity QED," Phys. Scripta **T76**, 127-137 (1998).

[7] S. Haroche, "Tests of quantum mechanics with single atoms in high Q cavities," Hyperfine Interact. **114**, 87-101 (1998).

[8] H. Mabuchi, Q. A. Turchette, M. S. Chapman, and H. J. Kimble, "Real-time detection of individual atoms falling through a high finesse optical cavity," Opt. Lett. **21**, 1393-1395 (1996).

[9] M. Brune, F. Schmidt-Kaler, A. Maali, J. Dreyer, E. Hagley, J. M. Raimond, and S. Haroche, "Quantum Rabi Oscillation: a Direct Test of Field Quantization in a Cavity," Phys. Rev. Lett. **76**, 1800-1803 (1996).

[10] L. Davidovich, A. Maali, M. Brune, J. M. Raimond, and S. Haroche, "Quantum Switches and Nonlocal Microwave Fields," Phys. Rev. Lett. **71**, 2360-2363 (1993).

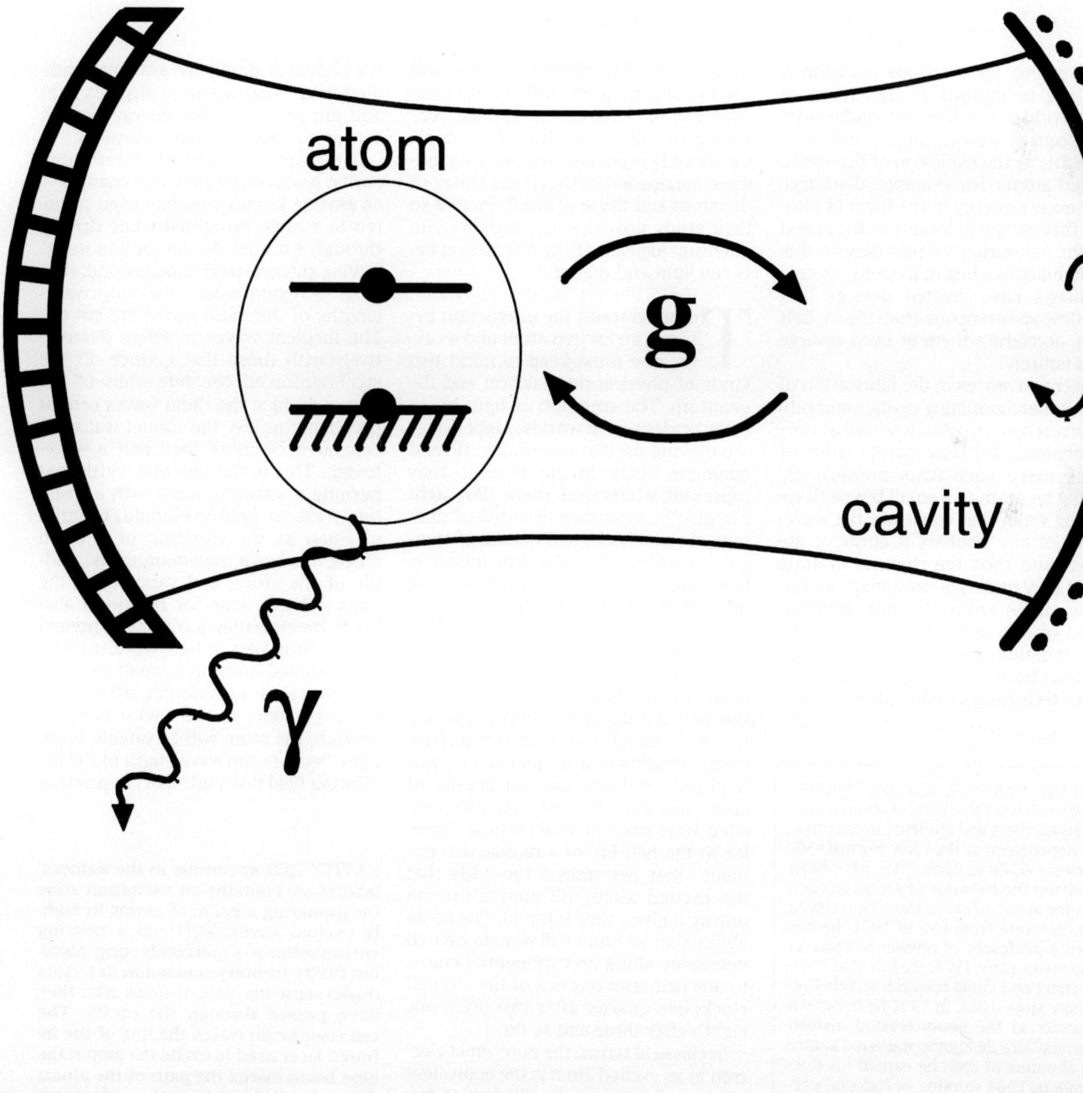
atom
g
cavity
γ

Cavity Quantum Electrodynamics

Atoms and photons in small cavities behave completely unlike those in free space. Their quirks illustrate some of the principles of quantum physics and make possible the development of new sensors

by Serge Haroche and Jean-Michel Raimond

Fleeting, spontaneous transitions are ubiquitous in the quantum world. Once they are under way, they seem as uncontrollable and as irreversible as the explosion of fireworks. Excited atoms, for example, discharge their excess energy in the form of photons that escape to infinity at the speed of light. Yet during the past decade, this inevitability has begun to yield. Atomic physicists have created devices that can slow spontaneous transitions, halt them, accelerate them or even reverse them entirely.

Recent advances in the fabrication of small superconducting cavities and other microscopic structures as well as novel techniques for laser manipulation of atoms make such feats possible. By placing an atom in a small box with reflecting walls that constrain the wavelength of any photons it emits or absorbs—and thus the changes in state that it may undergo—investigators can cause single atoms to emit photons ahead of schedule, stay in an excited state indefinitely or block the passage of a laser beam. With further refinement of this technology, cavity quantum electrodynamic (QED) phenomena may find use in the generation and precise measurement of electromagnetic fields consisting of only a handful of photons. Cavity QED processes engender an intimate correlation between the states of the atom and those of the field, and so their study provides new insights into quantum aspects of the interaction between light and matter.

SERGE HAROCHE and JEAN-MICHEL RAIMOND work in a team of about a dozen researchers and students in the physics department of the École Normale Supérieure (ENS) in Paris. They have been studying the behavior of atoms in cavities for about 10 years. Haroche received his doctorate from ENS in 1971; he has been a professor of physics at Paris VI University since 1975. He has also been teaching and doing research at Yale University since 1984. In 1991 he became a member of the newly created Institut Universitaire de France. Raimond is also an alumnus of ENS; he earned his doctorate in 1984 working in Haroche's research group and is also a professor of physics at Paris VI University.

To understand the interaction between an excited atom and a cavity, one must keep in mind two kinds of physics: the classical and the quantum. The emission of light by an atom bridges both worlds. Light waves are moving oscillations of electric and magnetic fields. In this respect, they represent a classical event. But light can also be described in terms of photons, discretely emitted quanta of energy. Sometimes the classical model is best, and sometimes the quantum one offers more understanding.

When an electron in an atom jumps from a high energy level to a lower one, the atom emits a photon that carries away the difference in energy between the two levels. This photon typically has a wavelength of a micron or less, corresponding to a frequency of a few hundred terahertz and an energy of about one electron volt. Any given excited state has a natural lifetime—similar to the half-life of a radioactive element—that determines the odds that the excited atom will emit a photon during a given time interval. The probability that an atom will remain excited decreases along an exponential curve: to one half after one tick of the internal clock, one quarter after two ticks, one eighth after three and so on.

In classical terms, the outermost electron in an excited atom is the equivalent of a small antenna, oscillating at frequencies corresponding to the energy of transitions to less excited states, and the photon is simply the antenna's radiated field. When an atom absorbs light and jumps to a higher energy level, it acts as a receiving antenna instead.

If the antenna is inside a reflecting cavity, however, its behavior changes—as anyone knows who has tried to listen to a radio broadcast while driving through a tunnel. As the car and its receiving antenna pass underground, they enter a region where the long wavelengths of the radio waves are cut off. The incident waves interfere destructively with those that bounce off the steel-reinforced concrete walls of the tunnel. In fact, the radio waves cannot propagate unless the tunnel walls are separated by more than half a wavelength. This is the minimal width that permits a standing wave with at least one crest, or field maximum, to build up—just as the vibration of a violin string reaches a maximum at the middle of the string and vanishes at the ends. What is true for reception also holds for emission: a confined antenna cannot broadcast at long wavelengths.

An excited atom in a small cavity is precisely such an antenna, albeit a microscopic one. If the cavity is small enough, the atom will be unable to radiate because the wavelength of the oscillating field it would "like" to produce

CAVITY QED apparatus in the authors' laboratory contains an excitation zone for preparing a beam of atoms in highly excited states (*left*) and a housing surrounding a superconducting niobium cavity (*center*). Ionization detectors (*right*) sense the state of atoms after they have passed through the cavity. The red laser beam traces the line of the infrared laser used to excite the atoms; the blue beam marks the path of the atoms themselves. When in use, the entire apparatus is enclosed in a liquid-helium cryostat that cools it to less than one kelvin.

cannot fit within the boundaries. As long as the atom cannot emit a photon, it must remain in the same energy level; the excited state acquires an infinite lifetime.

In 1985 research groups at the University of Washington and at the Massachusetts Institute of Technology demonstrated suppressed emission. The group in Seattle inhibited the radiation of a single electron inside an electromagnetic trap, whereas the M.I.T. group studied excited atoms confined between two metallic plates about a quarter of a millimeter apart. The atoms remained in the same state without radiating as long as they were between the plates.

Millimeter-scale structures are much too wide to alter the behavior of conventionally excited atoms emitting micron or submicron radiation; consequently, the M.I.T. experimenters had to work with atoms in special states known as Rydberg states. An atom in a Rydberg state has almost enough energy to lose an electron completely. Because this outermost electron is bound only weakly, it can assume any of a great number of closely spaced energy levels, and the photons it emits while jumping from one to another have wavelengths ranging from a fraction of a millimeter to a few centimeters. Rydberg atoms are prepared by irradiating ground-state atoms with laser light of appropriate wavelengths and are widely used in cavity QED experiments.

The suppression of spontaneous emission at an optical frequency requires much smaller cavities. In 1986 one of us (Haroche), along with other physicists at Yale University, made a micron-wide structure by stacking two optically flat mirrors separated by extremely thin metallic spacers. The workers sent atoms through this passage, thereby preventing them from radiating for as long as 13 times the normal excited-state lifetime. Researchers at the University of Rome used similar micron-wide gaps to inhibit emission by excited dye molecules.

The experiments performed on atoms between two flat mirrors have an interesting twist. Such a structure, with no sidewalls, constrains the wavelength only of photons whose polarization is parallel to the mirrors. As a result, emission is inhibited only if the atomic dipole antenna oscillates along the

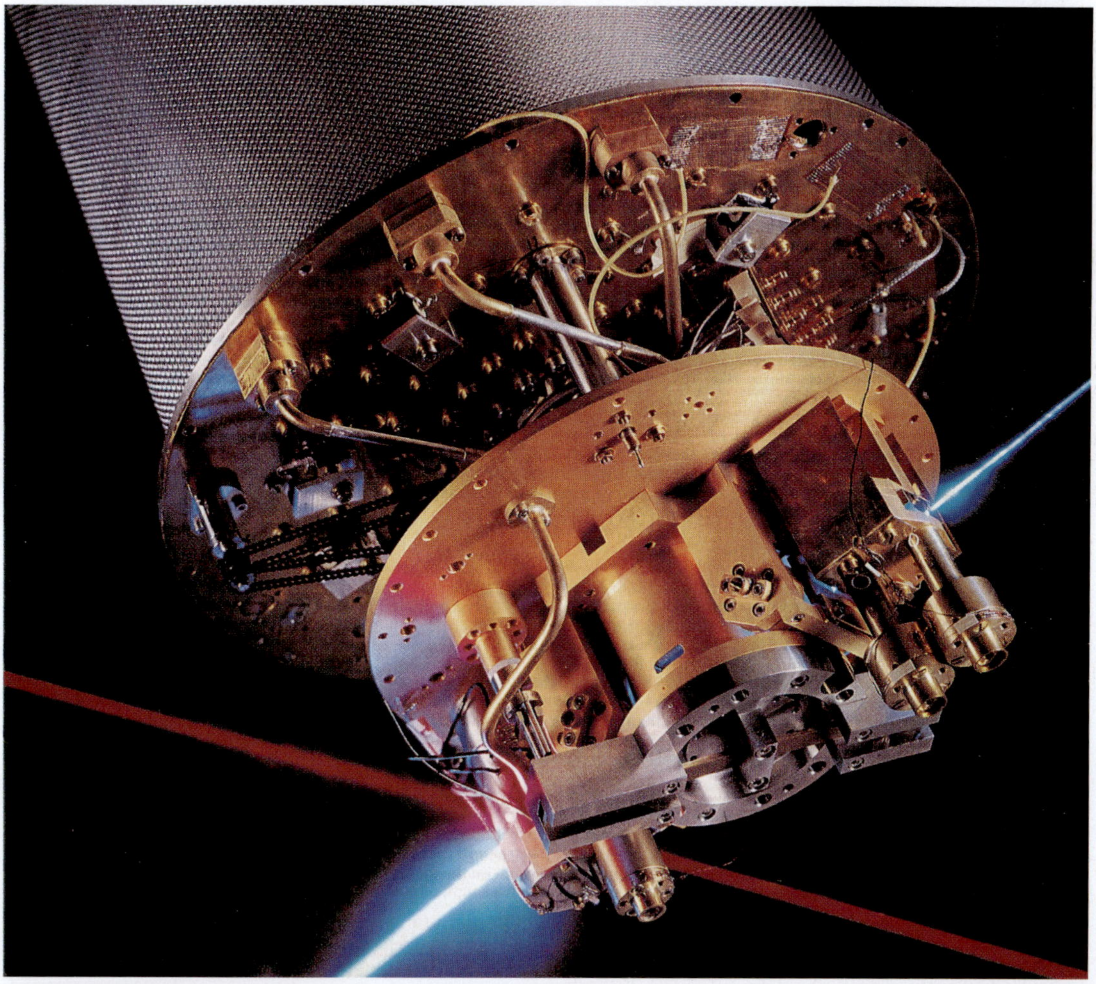

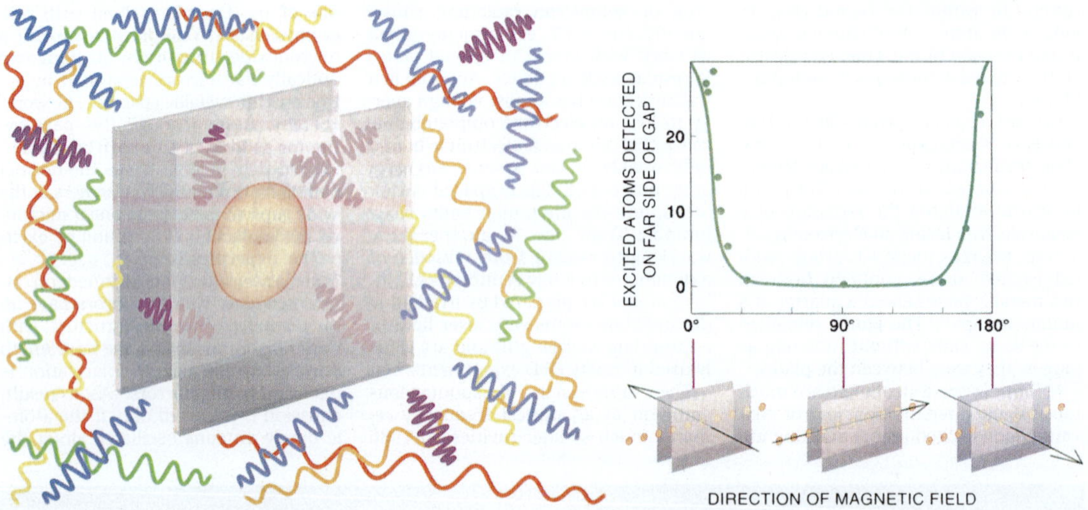

EXCITED ATOM between two mirrors (*left*) cannot emit a photon. The atom is sensitive to long-wavelength vacuum fluctuations whose polarization is parallel to the mirrors, but the narrow cavity prevents such fluctuations. Atoms passing through a micron-wide gap between mirrors have remained in the excited state for 13 natural lifetimes. Subjecting the atoms to a magnetic field causes their dipole axes to precess and changes the transmission of excited atoms through the gap (*right*). When the field is parallel to the mirrors, the atom rotates out of the plane of the mirrors and can quickly lose its excitation.

plane of the mirrors. (It was essential, for example, to prepare the excited atoms with this dipole orientation in the M.I.T. and Yale spontaneous-emission inhibition experiments.) The Yale researchers demonstrated these polarization-dependent effects by rotating the atomic dipole between the mirrors with the help of a magnetic field. When the dipole orientation was tilted with respect to the mirrors' plane, the excited-state lifetime dropped substantially.

Suppressed emission also takes place in solid-state cavities—tiny regions of semiconductor bounded by layers of disparate substances. Solid-state physicists routinely produce structures of submicron dimensions by means of molecular-beam epitaxy, in which materials are built up one atomic layer at a time. Devices built to take advantage of cavity QED phenomena could engender a new generation of light emitters [see "Microlasers," by Jack L. Jewell, James P. Harbison and Axel Scherer; SCIENTIFIC AMERICAN, November 1991].

These experiments indicate a counterintuitive phenomenon that might be called "no-photon interference." In short, the cavity prevents an atom from emitting a photon because that photon would have interfered destructively with itself had it ever existed. But this begs a philosophical question: How can the photon "know," even before being emitted, whether the cavity is the right or wrong size?

Part of the answer lies in yet another odd result of quantum mechanics. A cavity with no photon is in its lowest-energy state, the so-called ground state, but it is not really empty. The Heisenberg uncertainty principle sets a lower limit on the product of the electric and magnetic fields inside the cavity (or anywhere else for that matter) and thus prevents them from simultaneously vanishing. This so-called vacuum field exhibits intrinsic fluctuations at all frequencies, from long radio waves down to visible, ultraviolet and gamma radiation, and is a crucial concept in theoretical physics. Indeed, spontaneous emission of a photon by an excited atom is in a sense induced by vacuum fluctuations.

The no-photon interference effect arises because the fluctuations of the vacuum field, like the oscillations of more actual electromagnetic waves, are constrained by the cavity walls. In a small box, boundary conditions forbid long wavelengths—there can be no vacuum fluctuations at low frequencies. An excited atom that would ordinarily emit a low-frequency photon cannot do so, because there are no vacuum fluctuations to stimulate its emission by oscillating in phase with it.

Small cavities suppress atomic transitions; slightly larger ones, however, can enhance them. When the size of a cavity surrounding an excited atom is increased to the point where it matches the wavelength of the photon that the atom would naturally emit, vacuum-field fluctuations at that wavelength flood the cavity and become stronger than they would be in free space. This state of affairs encourages emission; the lifetime of the excited state becomes much shorter than it would naturally be. We observed this emission enhancement with Rydberg atoms at the École Normale Supérieure (ENS) in Paris in one of the first cavity QED experiments, in 1983.

If the resonant cavity has absorbing walls or allows photons to escape, the emission is not essentially different from spontaneous radiation in free space—it just proceeds much faster. If the cavity walls are very good reflectors and the cavity is closed, however, novel effects occur. These effects, which depend on intimate long-term interactions between the excited atom and the cavity, are the basis for a series of new devices that can make sensitive measurements of quantum phenomena.

Instead of simply emitting a photon and going on its way, an excited atom in such a resonant cavity oscillates back and forth between its excited and unexcited states. The emitted photon remains in the box in the vicinity of the atom and is promptly reabsorbed. The atom-cavity system oscillates between two states, one consisting of an excited atom and no photon, and the other of a de-excited atom and a photon trapped in the cavity. The frequency of this oscillation depends on the transition energy, on the size of the atomic dipole and on the size of the cavity.

This atom-photon exchange has a deep analogue in classical physics. If two identical pendulums are coupled by a weak spring and one of them is set in motion, the other will soon start swinging while the first gradually comes to rest. At this point, the first pendulum starts swinging again, commencing an ideally endless exchange of energy. A state in which one pendulum is excited and the other is at rest is clearly not stationary, because energy moves continuously from one pendulum to the other. The system does have two steady states, however: one in which the pendulums swing in phase with each other, and the other in which they swing alternatively toward and away from each other. The system's oscillation in each of these "eigenmodes" differs because of the additional force imposed by the coupling—the pendulums oscillate slightly slower in phase and slightly faster out of phase. Furthermore, the magnitude of the frequency difference between the two eigenmodes is precisely equal to the rate at which the two pendulums exchange their energy in the nonstationary states.

Researchers at the California Institute of Technology recently observed this "mode splitting" in an atom-cavity system. They transmitted a weak laser beam through a cavity made of two spherical mirrors while a beam of cesium atoms also crossed the cavity. The atomic beam was so tenuous that there was at most one atom at a time in the cavity. Although the cavity was not closed, the rate at which it exchanged photons with each atom exceeded the rate at which the atoms emitted photons that escaped the cavity; consequently, the physics was fundamentally the same as that in a closed resonator.

The spacing between the mirrors was an integral multiple of the wavelength of the transition between the first excited state of cesium and its ground state. Experimenters varied the wavelength (and hence frequency) of the laser and recorded its transmission across the cavity. When the cavity was empty, the transmission reached a sharp maximum at the resonant frequency of the cavity. When the resonator contained one atom on average, however, a symmetrical double peak appeared; its valley matched the position of the previous single peak. The frequency splitting, about six megahertz, marked the rate of energy exchange between the atom and a single photon in the cavity.

This apparatus is extremely sensitive: when the laser is tuned to the cavity's resonant frequency, the passage of a single atom lowers transmission significantly. This phenomenon can be used to count atoms in the same way one currently counts cars or people intercepting an infrared light in front of a photodetector.

Although simple in principle, such an experiment is technically demanding. The cavity must be as small as possible because the frequency splitting is proportional to the vacuum-field amplitude, which is inversely proportional to the square root of the box's volume. At the same time, the mirrors must be very good reflectors so that the photon remains trapped for at least as long as it takes the atom and cavity to exchange a photon. The group at Caltech used mirrors that were coated to achieve 99.996 percent reflectivity, separated by about a millimeter. In such a trap, a photon could bounce back and forth about 100,000 times over the course of a quarter of a microsecond before being transmitted through the mirrors.

Experimenters have been able to achieve even longer storage times—as great as several hundred milliseconds—by means of superconducting niobium cavities cooled to temperatures of about one kelvin or less. These cavities are ideal for trapping the photons emitted by Rydberg atoms, which typically range in wavelength from a few millimeters to a few centimeters (corresponding to frequencies between 10 and 100 gigahertz). In a recent experiment in our laboratory at ENS, we excited rubidium atoms with lasers and sent them across a superconducting cylindrical cavity tuned to a transition connecting the excited state to another Rydberg level 68 gigahertz higher in energy. We observed a mode splitting of about 100 kilohertz when the cavity contained two or three atoms at the same time.

There is a striking similarity between the single atom-cavity system and a laser or a maser. Either device, which emits photons in the optical and microwave domain, respec-

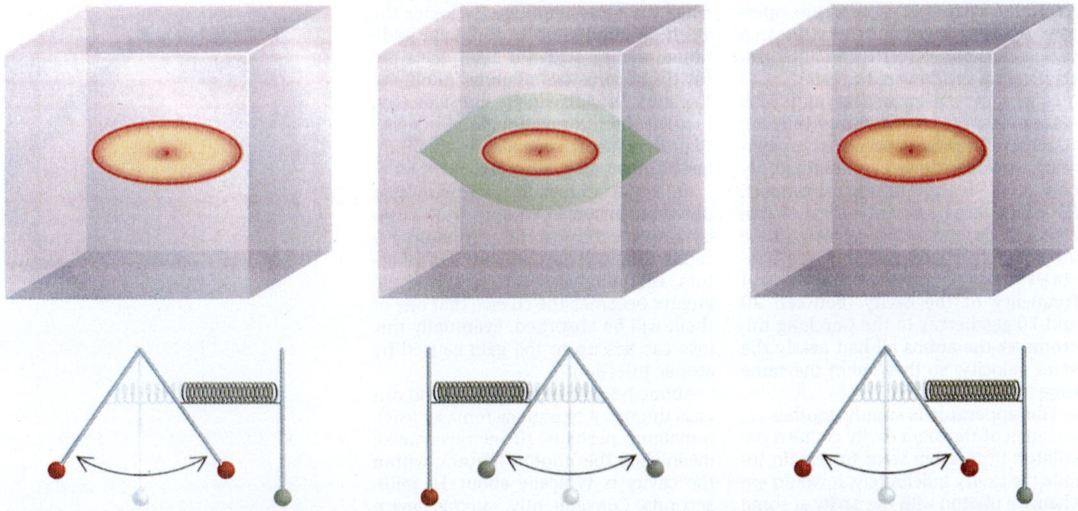

ATOM IN A CAVITY with highly reflective walls can be modeled by two weakly coupled pendulums. The system oscillates between two states. In one, the atom is excited, but there is no photon in the cavity (*left* and *right*). In the other, the atom is de-excited, and the cavity contains a photon (*center*). The atom and the cavity continually exchange energy.

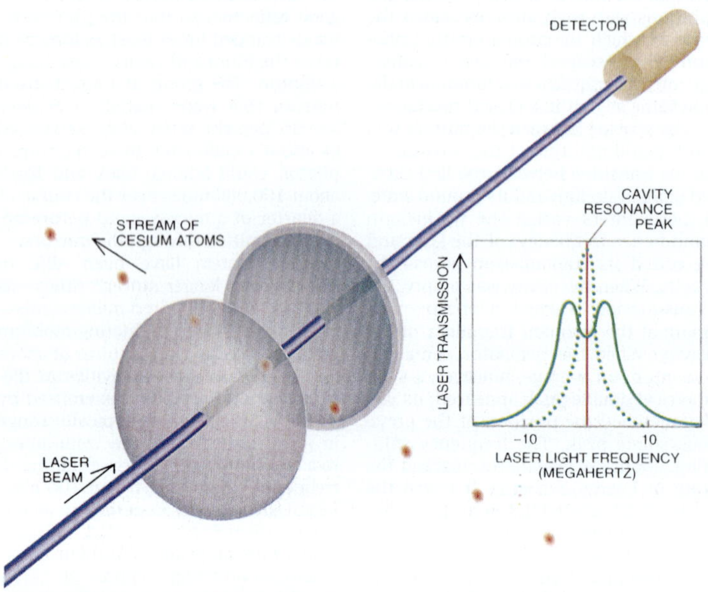

LASER BEAM TRANSMISSION through a cavity made of two closely spaced spherical mirrors is altered by the passage of individual atoms. When the cavity is empty, transmission peaks at a frequency set by the cavity dimensions (*dotted curve*). When an atom resonant with the cavity enters, however, the atom and cavity form a coupled-oscillator system. Transmission peaks at two separate frequencies corresponding to the "eigenmodes" of the atom-cavity system. The distance between the peaks marks the frequency at which the atom and cavity exchange energy.

tively, consists of a tuned cavity and an atomic medium that can undergo transitions whose wavelength matches the length of the cavity. When energy is supplied to the medium, the radiation field inside the cavity builds up to a point where all the excited atoms undergo stimulated emission and give out their photons in phase. A maser usually contains a very large number of atoms, collectively coupled to the radiation field in a large, resonating structure. In contrast, the cavity QED experiments operate on only a single atom at a time in a very small box. Nevertheless, the principles of operation are the same.

Indeed, in 1984 physicists at the Max Planck Institute for Quantum Optics in Garching, Germany, succeeded in operating a "micromaser" containing only one atom. To start up the micromaser, Rydberg atoms are sent one at a time through a superconducting cavity. These atoms are prepared in a state whose favored transition matches the resonant frequency of the cavity (between 20 and 70 gigahertz). In the Garching micromaser the atoms all had nearly the same velocity, so they spent the same time inside the cavity.

This apparatus is simply another realization of the atom-cavity coupled oscillator; if an atom were to remain inside the cavity indefinitely, it would exchange a photon with the cavity at some characteristic rate. Instead, depending on the atom's speed, there is some fixed chance that an atom will exit unchanged and a complementary chance that it will leave a photon behind.

If the cavity remains empty after the first atom, the next one faces an identical chance of exiting the cavity in the same state in which it entered. Eventually, however, an atom deposits a photon; then the next atom in line encounters sharply altered odds that it will emit energy. The rate at which atom and field exchange energy depends on the number of photons already present—the more photons, the faster the atom is stimulated to exchange additional energy with the field. Soon the cavity contains two photons, modifying the odds for subsequent emission even further, then three and so on at a rate that depends at each step on the number of previously deposited photons.

In fact, of course, the photon number does not increase without limit as atoms keep crossing the resonator. Because the walls are not perfect reflectors, the more photons there are, the greater becomes the chance that one of them will be absorbed. Eventually this loss catches up to the gain caused by atomic injection.

About 100,000 atoms per second can pass through a typical micromaser (each remaining perhaps 10 microseconds); meanwhile the photon lifetime within the cavity is typically about 10 milliseconds. Consequently, such a device running in steady state contains about 1,000 microwave photons. Each of them carries an energy of about 0.0001 electron volt; thus, the total radiation stored in the cavity does not exceed one tenth of one electron volt. This amount is much smaller than the electronic excitation energy stored in a single Rydberg atom, which is on the order of four electron volts.

Although it would be difficult to measure such a tiny field directly, the atoms passing through the resonator provide a very simple, elegant way to monitor the maser. The transition rate from one Rydberg state to the other depends on the photon number in the cavity, and experimenters need only measure the fraction of atoms leaving the maser in each state. The populations of the two levels can be determined by ionizing the atoms in two small detectors, each consisting of plates with an electric field across them. The first detector operates at a low field to ionize atoms in the higher-energy state; the second operates at a slightly higher field to ionize atoms in the lower-lying state (those that have left a photon behind in the cavity).

With its tiny radiation output and its drastic operational requirements, the micromaser is certainly not a machine that could be taken off a shelf and switched on by pushing a knob. It is nevertheless an ideal system to illustrate and test some of the principles of quantum physics. The buildup of photons in the cavity, for example, is a probabilistic quantum phenomenon—each atom in effect rolls a die to determine whether it will emit a photon—and measurements of micromaser operation match theoretical predictions.

An intriguing variation of the micromaser is the two-photon maser source. Such a device was operated for the first time five years ago by our group at ENS. Atoms pass through a cavity tuned to half the frequency of a transition between two Rydberg levels. Under the influence of the cavity radiation, each atom is stimulated to emit a pair of identical photons, each bringing half the energy required for the atomic transition. The maser field builds up as a result of the emission of successive photon pairs.

The presence of an intermediate energy level near the midpoint between the initial and the final levels of the transition helps the two-photon process along. Loosely speaking, an atom goes from its initial level to its final one via a "virtual" transition during which it jumps down to the middle level while emitting the first photon; it then jumps down again while emitting the second photon. The intermediate step is virtual because the energy of the emitted photons, whose frequency is set by the cavity, does not match the energy differences between the intermediate level and either of its neighbors. How can such a paradoxical situation exist? The Heisenberg uncertainty principle permits the atom briefly to borrow enough energy to emit a photon whose energy exceeds the difference between the top level and the middle one, provided that this loan is paid back during the emission of the second photon.

Like all such quantum transactions, the term of the energy loan is very short. Its maximum duration is inversely proportional to the amount of borrowed energy. For a mismatch of a few billionths of an electron volt, the loan typically lasts a few nanoseconds. Because larger loans are increasingly unlikely, the probability of the two-photon process is inversely proportional to this mismatch.

The micromaser cavity makes two-photon operation possible in two ways. It inhibits single-photon transitions that are not resonant with the cavity, and it strongly enhances the emission of photon pairs. Without the cavity, Rydberg atoms in the upper level would radiate a single photon and jump down to the intermediate level. This process would deplete the upper level before two-photon emission could build up.

Although the basic principle of a two-photon micromaser is the same as that of its simple one-photon cousin, the way in which it starts up and operates differs significantly. A strong fluctuation, corresponding to the unlikely emission of several photon pairs in close succession, is required to trigger the system; as a result, the field builds up only after a period of "lethargy." Once this fluctuation has occurred, the field in the cavity is relatively strong and stimulates emission by subsequent atoms, causing the device to reach full power (about 10^{-18} watt) rapidly. A two-photon laser system recently developed by a group at Oregon State University operates along a different scheme but displays essentially the same metastable behavior.

The success of micromasers and other similar devices has prompted cavity QED researchers to conceive new experiments, some of which would have been dismissed as pure science fiction only a few years ago. Perhaps the most remarkable of these as yet hypothetical experiments are those that deal with the forces experienced by an atom in a cavity containing only a vacuum or a small field made of a few photons.

The first thought experiment starts with a single atom and an empty cavity tuned to a transition between two of the atom's states. This coupled-oscillator system has two nonstationary states: one corresponds to an excited atom in an empty cavity, the other to a de-excited atom with one photon. The system also has two stationary states, obtained by addition or subtraction of the nonstationary ones—addition of the nonstationary states corresponds to the in-phase oscillation mode of the two-pendulum model, and subtraction of the states corresponds to the

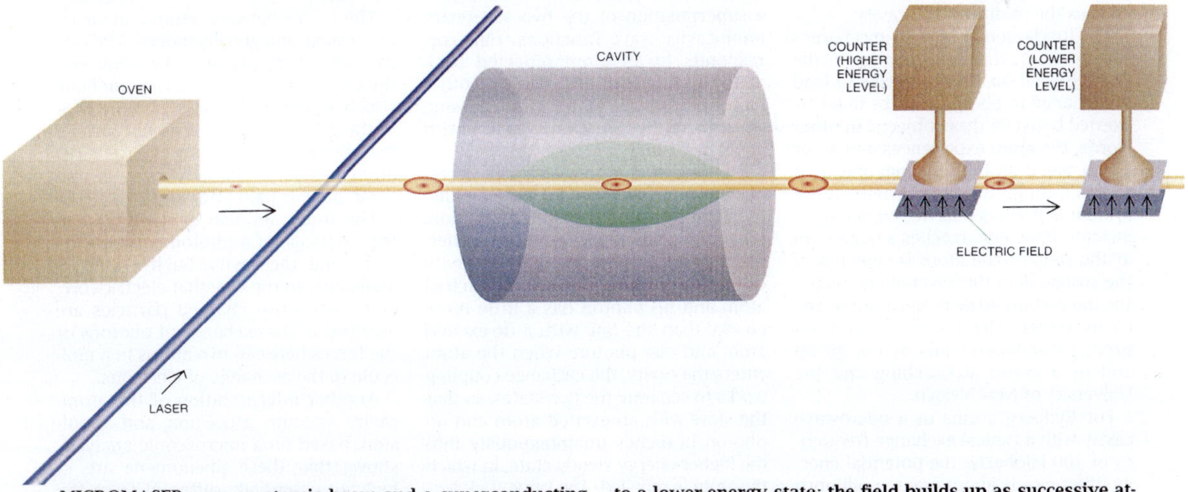

MICROMASER uses an atomic beam and a superconducting cavity to produce coherent microwave radiation. A laser beam (*left*) strikes atoms coming out of an oven and excites them into high-energy Rydberg states. The atoms pass one at a time through a cavity tuned to the frequency of a transition to a lower-energy state; the field builds up as successive atoms interact with the cavity and deposit photons in it. The micromaser field can be inferred from the readings of counters that monitor the number of atoms leaving the cavity in either the higher- or lower-energy state.

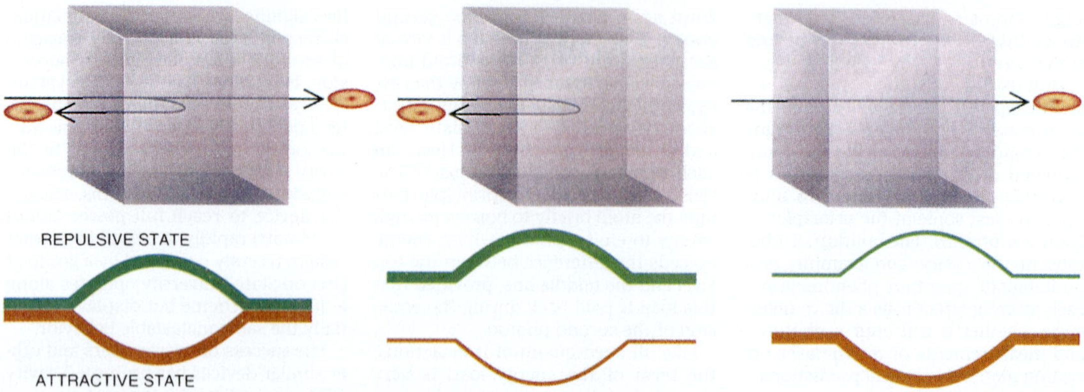

EMPTY CAVITY can repel or attract slow-moving, excited atoms. The strength of the coupling between an atom and a tuned cavity typically vanishes at the walls and reaches a maximum in the center. (Curves at the bottom show the energy of the atom-cavity system as a function of the atom's position within the cavity.) The change in energy results in a force on atoms moving through the cavity. If the cavity wavelength matches the atomic transition exactly, this force can be either attractive or repulsive (*left*). If the atomic transition has a slightly higher frequency than the resonant frequency of the cavity, the force will be repulsive (*center*); if the transition has a lower frequency, the force will be attractive (*right*).

out-of-phase mode. These stationary states differ in energy by a factor equal to Planck's constant, *h*, times the exchange frequency between the atom and the cavity.

This exchange frequency is proportional to the amplitude of the cavity's resonant vacuum field. Typically this field vanishes at the walls and near the ports by which the atom enters and leaves the cavity. It reaches a maximum at the cavity center. As a result, the atom-cavity coupling (and thus the energy difference between the system's two stationary states) is zero when the atom enters and leaves the cavity and goes to a maximum when the atom reaches the middle of the cavity.

The fundamental laws of mechanics say, however, that for a change in the relative position of two objects to lead to a change in energy, a force must be exerted between these objects. In other words, the atom experiences a push or a pull, albeit an infinitesimal one, as it moves through the empty cavity. If the system is prepared in the higher-energy state, its energy reaches a maximum at the center—the atom is repelled. If the system is in the lower-energy state, the interaction attracts the atom to the cavity center. These forces have been predicted independently by our group and by a group at Garching and the University of New Mexico.

For Rydberg atoms in a microwave cavity with a typical exchange frequency of 100 kilohertz, the potential energy difference is about one ten-billionth of an electron volt. This corresponds to a temperature of a few microkelvins and to the kinetic energy of an atom moving with a velocity of a few centimeters per second. If the speed of the incoming atom is less than this critical value, the potential barrier caused by the atom-cavity interaction will reflect the atom back, or, conversely, the potential well will be deep enough to trap it near the cavity center. Atoms in such slow motion can now be produced by laser cooling [see "Laser Trapping of Neutral Particles," by Steven Chu; SCIENTIFIC AMERICAN, February 1992]; these tiny forces may yet be observed.

If a very slow moving, excited atom is sent into a resonant, empty cavity, these forces result in a kind of atomic beam splitter. The nonstationary initial state of the system consists of the sum of the repelling and attractive states—a superposition of the two stationary atom-cavity wave functions. Half corresponds to an atom reflected back at the cavity entrance, and the other half corresponds to an atom passing through; either outcome occurs with equal probability.

To prepare a pure attractive or repelling state, one should detune the cavity slightly from the atomic transition. When the transition is a bit more energetic than the photon that the cavity can sustain, the state with an excited atom and no photon has a little more energy than the one with a de-excited atom and one photon. When the atom enters the cavity, the exchange coupling works to separate the two states, so that the state with an excited atom and no photon branches unambiguously into the higher-energy steady state, in which the atom is repelled. The same trick just as easily makes an attractive state if the cavity photon energy is slightly higher than the atomic transition.

This evolution of the atom-cavity system relies on the so-called adiabatic theorem, which says that if a quantum system's rate of change is slow enough, the system will continuously follow the state it is initially prepared in, provided the energy of that state does not coincide at any time with that of another state. This adiabaticity criterion is certainly met for the very slow atoms considered here.

These atom-cavity forces persist as long as the atom remains in its Rydberg state and the photon is not absorbed by the cavity walls. This state of affairs can typically last up to a fraction of a second, long enough for the atom to travel through the centimeter-size cavity.

The forces between atom and cavity are strange and ghostly indeed. The cavity is initially empty, and so in some way the force comes from the vacuum field, which suggests that it is obtained for nothing. Of course, that is not strictly true, because if the cavity is empty, the atom has to be initially excited, and some price is paid after all.

The force can also be attributed to the exchange of a photon between the atom and the cavity. Such a view is analogous to the way that electric forces between two charged particles are ascribed to the exchange of photons or the forces between two atoms in a molecule to the exchange of electrons.

Another interpretation of the atom-cavity vacuum attraction and repulsion, based on a microscopic analysis, shows that these phenomena are in fact not essentially different from the electrostatic forces whose demonstration was a society game in the 18th-century French court. If one charges a needle and brings small pieces of paper into its vicinity, the pieces stick to the metal. The strong electric field at the tip polarizes the pieces, pulling their electrons onto one side and leaving a net positive charge on the other, essentially making small electric dipoles. The attraction between the needle and the charges on the near side of the paper exceeds the repulsion between the needle and those on the far side, creating a net attractive force.

The atom and the cavity contain the same ingredients, albeit at a quantum level. The vacuum field bounded by the cavity walls polarizes the Rydberg atom, and the spatial variations of the field produce a net force. The atomic dipole and the vacuum field are oscillating quantities, however, and their respective oscillations must maintain a constant relative phase if a net force is to continue for any length of time. As it turns out, the photon exchange process does in fact lock the atomic dipole and the vacuum fluctuations.

The tiny force experienced by the atom is enhanced by adding photons to the cavity. The atom-cavity exchange frequency increases with the field intensity, so that each photon adds a discrete quantum of height to the potential barrier in the repelling state and a discrete quantum of depth to the potential well in the attractive state. As a result, it should be possible to infer the number of photons inside the cavity by measuring the time an atom with a known velocity takes to cross it or, equivalently, by detecting the atom's position downstream of the cavity at a given time.

One could inject perhaps a dozen or so photons into a cavity and then launch through it, one by one, Rydberg atoms whose velocity is fixed at about a meter per second. The kinetic energy of these atoms would be greater than the atom-cavity potential energy, and they would pass through the cavity after experiencing a slight positive or negative delay, depending on the sign of the atom-cavity detuning. To detect the atom's position after it has passed through the cavity, researchers could fire an array of field ionization detectors simultaneously some time after the launch of each atom. A spatial resolution of a few microns should be good enough to count the number of photons in the cavity.

Before measurement, of course, the photon number is not merely a classically unknown quantity. It also usually contains an inherent quantum uncertainty. The cavity generally contains a field whose description is a quantum wave function assigning a complex amplitude to each possible number of photons. The probability that the cavity stores a given number of photons is the squared modulus of the corresponding complex amplitude.

The laws of quantum mechanics say that the firing of the detector that registers an atom's position after it has crossed the cavity collapses the ambiguous photon-number wave function to a single value. Any subsequent atom used to measure this number will register the same value. If the experiment is repeated from scratch many times, with the same initial field in the cavity, the statistical distribution of photons will be revealed by the ensemble of individual measurements. In any given run, however, the photon number will remain constant, once pinned down.

This method for measuring the number of photons in the cavity realizes the remarkable feat of observation known as quantum nondemolition. Not only does the technique determine perfectly the number of photons in the cavity, but it also leaves that number unchanged for further readings.

Although this characteristic seems to be merely what one would ask of any measurement, it is impossible to attain by conventional means. The ordinary way to measure this field is to couple the cavity to some kind of photodetector, transforming the photons into electrons and counting them. The absorption of photons is also a quantum event, ruled by chance; thus, the detector adds its own noise to the measured intensity. Furthermore, each measurement requires absorbing photons; thus, the field irreversibly loses energy. Repeating such a procedure therefore results in a different, lower reading each time. In the nondemolition experiment, in contrast, the slightly nonresonant atoms interact with the cavity field without permanently exchanging energy.

Quantum optics groups around the world have discussed various versions of quantum nondemolition experiments for several years, and recently they have begun reducing theory to practice. Direct measurement of an atom's delay is conceptually simple but not very sensitive. More promising variants are based on interference effects involving atoms passing through the cavity—like photons, atoms can behave like waves. They can even interfere with themselves. The so-called de Broglie wavelength of an atom is inversely proportional to velocity; a rubidium atom traveling 100 meters per second, for example, has a wavelength of 0.45 angstrom.

If an atom is slowed while traversing the cavity, its phase will be shifted by an angle proportional to the delay. A delay that holds an atom back by a mere 0.22 angstrom, or one half of a de Broglie wavelength, will replace a crest of the matter wave by a trough. This shift can readily be detected by atomic interferometry.

If one prepares the atom itself in a superposition of two states, one of which is delayed by the cavity while the other is unaffected, then the atomic wave packet itself will be split into two parts. As these two parts interfere with each other, the resulting signal yields a measurement of the phase shift of the matter wave and hence of the photon number in the cavity. Precisely this experiment is now under way at our laboratory in Paris, using Rydberg atoms that are coupled to a superconducting cavity in an apparatus known as a Ramsey interferometer.

Such an apparatus has many potential uses. Because the passing atoms can monitor the number of photons in a cavity without perturbing it, one can witness the natural death of photons in real time. If a photon disappears in the cavity walls, that disappearance would register immediately in the atomic interference pattern. Such experiments should provide more tests of quantum theory and may open the way to a new generation of sensors in the optical and microwave domains.

FURTHER READING

RADIATIVE PROPERTIES OF RYDBERG STATES IN RESONANT CAVITIES. S. Haroche and J.-M. Raimond in *Advances in Atomic and Molecular Physics*, Vol. 20, pages 350–409; 1985.

THE SINGLE ATOM MASER AND THE QUANTUM ELECTRODYNAMICS IN A CAVITY. H. Walther in *Physica Scripta*, Vol. T23, pages 165–169; 1988.

CAVITY QUANTUM ELECTRODYNAMICS. S. Haroche and D. Kleppner in *Physics Today*, Vol. 42, No. 1, pages 24–30; January 1989.

CAVITY QUANTUM ELECTRODYNAMICS. E. A. Hinds in *Advances in Atomic, Molecular, and Optical Physics*, Vol. 28, pages 237–289; 1991.

CAVITY QUANTUM OPTICS AND THE QUANTUM MEASUREMENT PROCESS. P. Meystre in *Progress in Optics*, Vol. 30. Edited by E. Wolf. Elsevier Science Publishers, 1992.

CAVITY QUANTUM ELECTRODYNAMICS. S. Haroche in *Fundamental Systems in Quantum Optics*. Proceedings of Les Houches Summer School, Session LIII. Edited by J. Dalibard, J.-M. Raimond and J. Zinn-Justin. North-Holland, 1992.

VOLUME 75, NUMBER 25 PHYSICAL REVIEW LETTERS 18 DECEMBER 1995

Measurement of Conditional Phase Shifts for Quantum Logic

Q. A. Turchette,* C. J. Hood, W. Lange, H. Mabuchi, and H. J. Kimble

Norman Bridge Laboratory of Physics 12-33, California Institute of Technology, Pasadena, California 91125

(Received 12 June 1995)

Measurements of the birefringence of a single atom strongly coupled to a high-finesse optical resonator are reported, with nonlinear phase shifts observed for an intracavity photon number much less than one. A proposal to utilize the measured conditional phase shifts for implementing quantum logic via a quantum-phase gate (QPG) is considered. Within the context of a simple model for the field transformation, the parameters of the "truth table" for the QPG are determined.

PACS numbers: 89.80.+h, 32.80.–t, 33.55.Ad, 42.65.Pc

Although the theory of quantum computation dates back more than a decade to the seminal works of Feynman and Deutsch [1], there has recently been an explosion of new activity driven in large measure by Shor's quantum algorithm [2] for efficient factorization. While most attention has been directed toward theoretical issues, several strategies have also been proposed for laboratory investigations [3]. However, the demands on experimental systems for building quantum computational networks [4] are quite severe, requiring strong coupling between quantum carriers of information ("qubits") in an environment with minimal dissipation. Hence, experimental progress has lagged behind the remarkable theoretical developments in quantum information theory.

Within this context, we present a significant experimental step toward realizing quantum logic with individual photons as qubits. Moreover, our work bears import for related experimental challenges such as quantum nondemolition (QND) measurement and quantum cryptography. Specifically, we report the demonstration of conditional dynamics *at the single-photon level* between two frequency-distinct fields in an optical resonator. Our measurements utilize the circular birefringence of an atom strongly coupled to the resonator to rotate the linear polarization of a transmitted probe beam. The phase shift between circular polarization states $\sigma_\pm$ is conditioned upon the intensity of a pump beam via a Kerr-type nonlinearity, with conditional phase shifts $\Delta \sim 16°$ per intracavity photon extracted from our data. To explore further the prospects for quantum logic based on these capabilities, we have experimentally investigated a candidate quantum-phase gate (QPG) and, within the context of a simple model, have extracted relevant phase shifts for the "truth table" of the QPG. In our proposed implementation, "flying qubits" are single-photon pulses propagating in two frequency-offset channels, with internal states specified by $\sigma_\pm$ polarization.

It should be noted at the outset that necessary and sufficient testing procedures have not yet been established for providing *direct experimental verification* that a given "black box" laboratory system can perform quantum logic transformations with sufficient fidelity to implement Deutsch's quantum Turing machine [1]. In particular, it is not known what level of dissipation (if any) can be tolerated in experimental systems before the advantages of unitary information processing are lost. However, any laboratory quantum gate must exhibit coherence and demonstrably produce entanglement between qubits. The practical application of such criteria requires the formulation of new measurement strategies, which we consider explicitly for our experiment.

Our efforts here focus on the implementation of quantum logic by exploiting the extremely large optical nonlinearities realizable in cavity quantum electrodynamics (CQED) [5,6]. In CQED systems, individual photons circulating in a high-finesse resonator can interact strongly via their mutual coupling to a single intracavity atom. The critical parameters that characterize our apparatus are g, the dipole coupling rate of atom to cavity; κ, the cavity-field damping rate; and γ, the transverse atomic decay rate to noncavity modes. The current work is performed with parameters such that $\kappa > g^2/\kappa > \gamma$. In this *bad cavity regime* the atom's coherent coupling to the cavity mode (at rate g^2/κ) dominates incoherent emission into free space (at rate γ), making it possible to couple strongly a single atom to the cavity mode in a manner that allows for efficient transfer of electromagnetic fields from input to output channels (at rate κ), thus creating an effectively *one-dimensional atom* [6]. The atom-cavity system may therefore be viewed as a quantum-optical device (a *nonlinear one-atom wave plate*), which is exploited for processing field states.

Conditional dynamics in our system originate from the nonlinear optical response of a cesium atom coupled to the cavity field. For the particular optical frequencies used, the relevant atomic states form a three-level system shown in the inset of Fig. 1. The transitions couple to cavity modes with orthogonal circular polarizations $\sigma_\pm$ with rates $g_\pm$, where the σ_+ transition corresponds to $(6S_{1/2}, F = 4, m = 4) \to (6P_{3/2}, F' = 5, m' = 5)$ and the σ_- transition connects $(m = 4)$ to $(m' = 3)$. Since $g_- = g_+/\sqrt{45}$, we set $g_- = 0$ to simplify the current discussion (this is not an essential approximation). As shown in Fig. 1, the ground state $(F = 4, m = 4)$ is prepared by optical pumping of an atomic beam of cesium just before it enters the high finesse ($\mathcal{F} = 18\,000$)

 0031-9007/95/75(25)/4710(4)$06.00

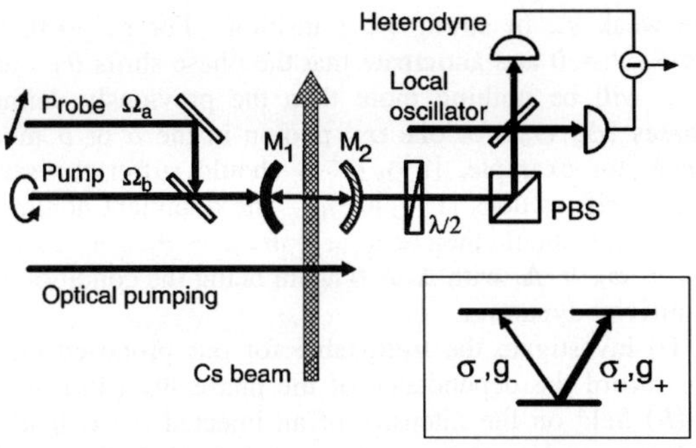

FIG. 1. Schematic of the experimental apparatus.

cavity. The cavity length and Gaussian waist are 56 μm and 35 μm. The mirrors (M_1, M_2) have transmission coefficients $(1.1 \times 10^{-6}, 3.5 \times 10^{-4})$. Together with the atomic lifetime $\tau = 32$ ns and transit time $T_0 = 7\tau$, these parameters lead to the set of rates $(g_+, \kappa, \gamma, T_0^{-1})/2\pi = (20, 75, 2.5, 0.7)$ MHz. Hence, the intracavity saturation photon number $m_0 = 4\gamma^2/3g_+^2 = 0.02$ photons, the critical atom number $N_0 = 2\kappa\gamma/g_+^2 = 0.94$ atoms, and the one-photon tipping angle $2g_+T_0 = 15\pi$.

To characterize photon-photon interactions inside our atom-cavity device, we investigate the transmission of monochromatic coherent-state pump and probe beams, which are independently tunable in frequency, power, and polarization [6] (see Fig. 1). After passing through the cavity, these beams are analyzed for polarization state with a rotatable half-wave plate, a polarizer, and balanced heterodyne detectors.

Turning now to our measurements, we present in Fig. 2 the weak-field response (average intracavity photon number $\ll m_0$) of the atom-cavity system for the case of coincident atomic (ω_A) and cavity (ω_C) resonances. For these scans the average intracavity atom number is $\overline{N} = 1.0 \pm 0.1$ atom, as determined by fits to the data as discussed in Refs. [6,7]. The inset data in Fig. 2 give the ratio T_a of transmitted power with atoms present to that without as a function of the detuning Ω_a of the probe, which is σ_+ polarized to interact with the strong g_+ transition. The main data of Fig. 2 represent the phase of the transmission function and are taken by injecting a linearly polarized probe beam, with the σ_+ component of this beam attaining a phase shift due to the composite atom-cavity system, while the σ_- component only receives a phase shift corresponding to an empty cavity (in the approximation $g_- \rightarrow 0$). The differential phase ϕ_a between the $\sigma_\pm$ components combines with changes in amplitude to produce an elliptically polarized output beam with its major axis rotated by $\phi_a/2$ relative to the linearly polarized input, so that ϕ_a can be determined by analysis of the polarization state of the output beam.

To utilize these phase shifts for conditional dynamics, we next consider measurements of nonlinear dispersion. We fix the detuning Ω_a of the weak linearly polarized probe beam ($m_a \approx 10^{-4}$ photons) at a position on the dispersion curve of Fig. 2 corresponding to relatively low intracavity loss as determined from T_a. As a controlling field, we inject a σ_+-polarized pump beam at detuning Ω_b. Figure 3 displays the variation of the phase Φ_a of the probe beam for a wide range of pump powers, with Φ_a measured by polarization interferometry as discussed above. In the limit $m_b \rightarrow 0$, $\Phi_a \rightarrow \phi_a$, which is the phase shift for the probe field alone. Note that the pump-probe coupling is manifest for $m_b \approx 0.1$, with a 30% reduction of $|\Phi_a|$ as m_b goes from 0.1 to 0.3 photon. The Fig. 3 inset shows the corresponding nonlinearity of T_a for single frequency resonant excitation.

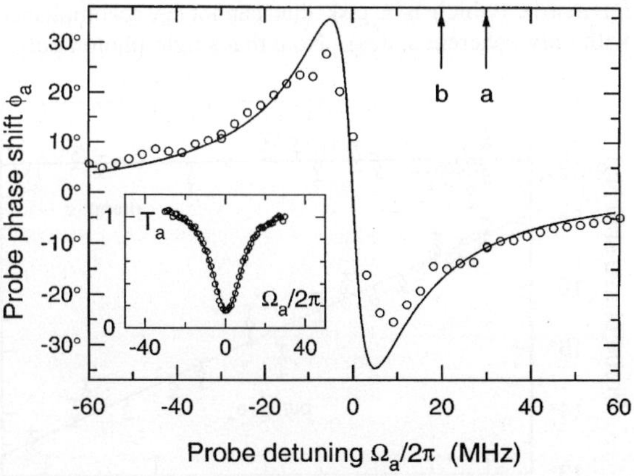

FIG. 2. Measured weak-field response of the atom-cavity system for $\overline{N} = 1.0 \pm 0.1$ atom. Full curves represent the theoretical model from Ref. [6]. The inset shows the squared modulus of the normalized probe transmission T_a and the main axes show probe phase shift ϕ_a. Ω_a denotes detuning from the resonance frequency $\omega_A = \omega_C$.

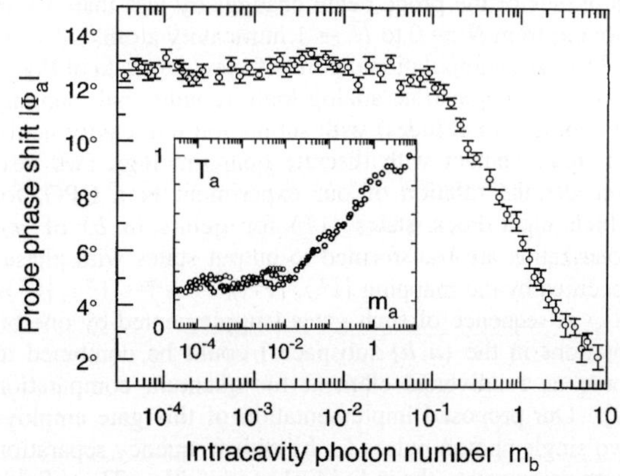

FIG. 3. Probe phase shift Φ_a vs m_b for an injected σ_+ pump, for $\overline{N} = 0.9$ atom and pump (probe) detuning of +20 (+30) MHz from atomic resonance as shown in Fig. 2. Error bars indicate uncertainties in least squares fits used for evaluating the phase shifts. The inset shows transmission T_a vs m_a for a resonant probe without pump, with $\overline{N} = 0.6$.

These measurements represent the realization of a nonlinear optical susceptibility at the single-photon level and unambiguously demonstrate the conditional dynamics necessary for implementing quantum logic. To quantify further the interaction strength involved, we note that the pump and probe input fields are prepared as uncorrelated coherent states with small amplitudes $|\alpha|^2, |\beta|^2 \ll 1$. Hence, their composite state can be expanded in the form $|\psi\rangle \propto [|0\rangle_a + \alpha|1\rangle_a] \otimes [|0\rangle_b + \beta|1\rangle_b]$. Our *ansatz* for the transformation of field states is

$$|j\rangle_a|k\rangle_b \mapsto e^{i\mu_{jk}}|j\rangle_a|k\rangle_b, \quad j,k = \{0,1\}, \tag{1}$$

which amounts to the physically motivated assumption that Fock states asymptotically connect to the dressed states of the atom-cavity system and hence are the appropriate eigenstates of the transformation. For $\mu_{00} + \mu_{11} \neq \mu_{01} + \mu_{10}$, this unitary transformation exhibits conditional dynamics suitable for quantum logic. Setting $\mu_{00} = 0$, $\mu_{10} = \phi_a$, $\mu_{01} = \phi_b$, and defining a parameter Δ by $\mu_{11} = \phi_a + \phi_b + \Delta$, we find the output state

$$|\psi_{\text{out}}\rangle = |\overline{\alpha}\rangle_a|\overline{\beta}\rangle_b + \overline{\alpha}\overline{\beta}[e^{i\Delta} - 1]|1\rangle_a|1\rangle_b, \tag{2}$$

where $\overline{\alpha} \equiv \alpha e^{i\phi_a}$ and $\overline{\beta} \equiv \beta e^{i\phi_b}$. This state is clearly entangled for $\Delta \neq 0$. To connect this model to our observations, we examine the reduced density operator for the a field alone and find that in the limit $\beta \to 0$, Eq. (2) leads to $\Phi_a \approx \phi_a - 2m_b \sin(\Delta/2)$. Therefore Δ may be determined directly from measurements of the initial slope $\partial\Phi_a/\partial m_b$ in a plot of the phase Φ_a of the probe field versus pump intensity m_b.

Note that although the effects of dissipation are neglected in Eqs. (1) and (2), they could be incorporated via a density matrix corresponding to $|\psi_{\text{out}}\rangle$. However, we shall temporarily set aside such considerations since we are operating with large detunings from atomic resonance in order to approximate purely dispersive interactions. For example, for the measurements of Fig. 3 the amplitude of the probe beam changes by less than 3% in moving from $N = 0$ to $\overline{N} = 1$ intracavity atom.

From the computational point of view, the data of Fig. 3 explicitly demonstrate analog logic (conditional mapping of complex amplitudes) with subphoton intracavity fields. To make contact with discrete quantum logic, we next consider the relation of our experiment to a QPG, for which input Fock states $|1^{\pm}\rangle$ for qubits (a,b) of $\sigma_{\pm}$ polarization are transformed to output states with phases specified by the mapping $|1^{\pm}\rangle_a|1^{\pm}\rangle_b \mapsto e^{i\theta_{\pm\pm}}|1^{\pm}\rangle_a|1^{\pm}\rangle_b$ [8]. A sequence of such gates [supplemented by one bit rotations in the (a,b) subspaces] could be combined to serve as a universal element for quantum computation [9]. Our proposed implementation of this gate employs two single-photon pulses (a,b) with frequency separation large compared to the individual bandwidths. These fields would be incident on the cavity mirror M_2 of Fig. 1, interact with the atom-cavity system, and then reflect with high efficiency [10]. The basis states $|1^{\pm}\rangle_a|1^{\pm}\rangle_b$ of the truth table for the QPG are associated with $\sigma_{\pm}$ polarizations for the (a,b) fields, which couple to either the weak g_- or strong g_+ transition. For $g_- \to 0$, we set $\theta_{--} = 0$ and anticipate that the phase shifts θ_{+-} and θ_{-+} will be nothing more than the previously defined phases (ϕ_a, ϕ_b) for one σ_+ photon in the a or b mode since, for example, $|1^+\rangle_a|1^-\rangle_b$ should suffer the same phase shift as does $|1^+\rangle_a|0^-\rangle_b$. The dominant nonlinear phase shift should then be $\theta_{++} = \theta_{+-} + \theta_{-+} + \Delta_{++} \simeq \phi_a + \phi_b + \Delta$, with $\Delta \neq 0$ again being the condition for nontrivial dynamics.

To investigate the truth table for our proposed QPG, we record the dependence of the phase Φ_a (Φ_b) of the a (b) field on the intensity of an injected b (a) field of either $\sigma_{\pm}$ polarization, as shown in Fig. 4. Following the discussion of Eq. (2), we extract one-photon phase shifts from initial linear slopes. The straight-line fits shown in Fig. 4 yield $\Delta_{++} \approx \Delta = (16 \pm 3)°$ and $\theta_{+-} - \phi_a = (0.3 \pm 2)° \approx 0$ as anticipated. With the roles of the (a,b) modes interchanged, we can likewise find that $\theta_{-+} \approx \phi_b$. Hence, subject to the validity of our model (1), the experimentally determined parameters for our QPG read

$$\begin{array}{llll} |1^-\rangle_a|1^-\rangle_b & \to & & |1^-\rangle_a|1^-\rangle_b\,, \\ |1^+\rangle_a|1^-\rangle_b & \to & e^{i\phi_a} & |1^+\rangle_a|1^-\rangle_b\,, \\ |1^-\rangle_a|1^+\rangle_b & \to & e^{i\phi_b} & |1^-\rangle_a|1^+\rangle_b\,, \\ |1^+\rangle_a|1^+\rangle_b & \to & e^{i(\phi_a+\phi_b+\Delta)} & |1^+\rangle_a|1^+\rangle_b\,, \end{array} \tag{3}$$

where for data as in Fig. 4, $\phi_a \approx (17.5 \pm 1)°$, $\phi_b \approx (12.5 \pm 1)°$, and $\Delta \approx (16 \pm 3)°$.

We believe that this demonstration of *polarization-conditional* phase shifts holds great promise for the implementation of quantum logic with "flying qubits" encoded by the polarization of single-photon pulses. Given the ability to generate a $|1\rangle_a|1\rangle_b$ state of arbitrary polarization, it would then be straightforward to derive states of mutually orthogonal polarization to span the four-dimensional qubit Hilbert space, and hence to measure directly the diagonal elements of the SU(4) transfer matrix (which is a task that cannot be accomplished with only coherent states). Note that single-photon pulses

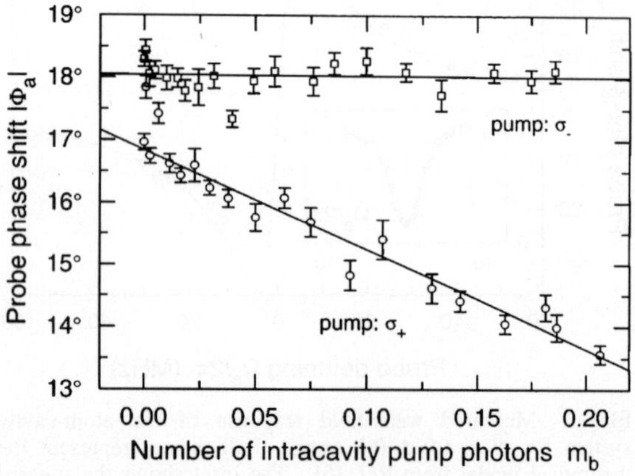

FIG. 4. Dependence of probe phase shift on intensity for two orthogonal polarizations of the pump beam. Pump (probe) detuning is +30 (+20) MHz and $\overline{N} = 0.9$ atom.

could be generated for this purpose by a variety of techniques and that the optical response of our system to pulses with duration long compared to the inverse cavity damping time $1/\kappa$ should closely reproduce the steady-state behavior investigated here [10]. Furthermore, operation in a regime of strong coupling with $g > \kappa > \gamma$ [7] affords the possibility of yet larger conditional phase shifts for our quantum-phase gate in cavity QED [10].

We wish to stress that the parameter Δ has model-independent significance as the strength of the dispersive nonlinear interaction between intracavity fields, quoted in degrees per unit of stored energy. Its large measured value represents a unique achievement within the field of nonlinear optics. Our *ansatz* (1), on the other hand, may be viewed with some skepticism, for although our assumptions seem reasonable, we have not explicitly verified the full transformation (2). We are thus led to consider the question of how to evaluate *operationally* the potential of our system for performing quantum logic, without relying on any particular theoretical model of the appropriate state transformation. From the example provided by Shor's algorithm, it seems reasonable to adopt the observation of coherence and the production of entanglement as *necessary* conditions for calling a candidate device a quantum gate. With these conditions in mind, we briefly consider strategies for evaluating our laboratory system.

Let us first consider damping of coherences in the output fields by writing their joint density matrix in the generalized form $\rho_{jk}d_{jk}$. Here ρ_{jk} represents a pure-state density matrix in a basis $\{j,k\} = \{0_{a,b}, 1_{a,b}\}$ for Eqs. (1) and (2) and $\{j,k\} = \{1^-_{a,b}, 1^+_{a,b}\}$ for Eq. (3), and the parameters d_{jk} provide a phenomenological characterization of decoherence. Physical considerations require that $\mathrm{Tr}[\rho_{jk}d_{jk}] = 1$, but dissipative processes could in principle cause complete dephasing of the output density matrix ($d_{j\neq k} \to 0$). Fortunately, with optical fields there exists a straightforward procedure for establishing that this is not the case—heterodyne detection such as implemented in the current work provides signals that are proportional to off-diagonal matrix elements $\rho_{jk}d_{jk}$.

As regards the second criterion, we note that the output state (2) clearly shows entanglement between the pump and probe fields for $\Delta \neq 0$. Hence there must exist a Clauser-Horne-Shimony-Holt (CHSH) inequality [11] violated by correlation measurements on $|\psi_{\mathrm{out}}\rangle$. Following, e.g., the method of Gisin and Peres [12] we could explicitly formulate the optimal correlation measurement for our particular gate in terms of $\overline{\alpha}$, $\overline{\beta}$, and Δ. Unfortunately the violation must necessarily be of order $|\overline{\alpha}\overline{\beta}(1 - \cos\Delta)|^2 \ll 1$ and therefore quite difficult to detect experimentally. In order to quantify the degree of entanglement that could be generated in our current apparatus we consider the input state $(|1^-\rangle_a + |1^+\rangle_a) \otimes (|1^-\rangle_b + |1^+\rangle_b)/2$, for which the sum of expectation values in the appropriate CHSH inequality is $2\sqrt{1 + \sin^2(\Delta/2)}$. Note that 2 corresponds to the classical upper limit, while the measured conditional phase shift $\Delta \approx 16°$ per photon would generate a value of 2.02 [13]. Although we do not know of any rigorous procedure to compute a "transfer matrix" analogous to (3) for compactly specifying the mapping of input to output states *in the presence of finite dissipation,* the correlation functions appearing in any relevant CHSH inequality can be calculated for arbitrary input fields using Heisenberg equations of motion and the quantum regression theorem. Thus the dependence of entanglement production on d_{jk} could be investigated in quantitative detail [14].

We acknowledge the contributions of R. J. Thompson, S. Lloyd, A. Ekert, and J. Preskill. H.M. holds an NDSEG fellowship. W.L. is supported by DFG. This work is supported by the National Science Foundation (Grant No. PHY-9014547) and the U.S. Office of Naval Research (Contract No. N00014-90-J-1058).

*Electronic address: quat@cco.caltech.edu

[1] R.P. Feynman, Found. Phys. **16**, 507 (1986); D. Deutsch, Proc. R. Soc. London A **400**, 97 (1985).

[2] P.W. Shor, in *Proceedings of the 35th Annual Symposium on FOCS,* edited by S. Goldwasser (IEEE Computer Society Press, New York, 1994); see also A. Ekert, in *Proceedings of the ICAP '94, Boulder,* edited by D. Wineland, C. Wieman, and S. Smith, AIP Conf. Proc. 323 (AIP, New York, 1995), p. 450.

[3] S. Lloyd, Science **261**, 1569 (1993); A. Barenco *et al.*, Phys. Rev. Lett. **74**, 4083 (1995); T. Sleator and H. Weinfurter, *ibid.* **74**, 4087 (1995); J.I. Cirac and P. Zoller, *ibid.* **74**, 4091 (1995).

[4] D. Deutsch, Proc. R. Soc. London A **425**, 73 (1989).

[5] See, for example, *Cavity Quantum Electrodynamics,* edited by P.R. Berman (Academic, San Diego, 1994).

[6] Q.A. Turchette, R.J. Thompson, and H.J. Kimble, Appl. Phys. B **60**, S1 (1995).

[7] R.J. Thompson, G. Rempe, and H.J. Kimble, Phys. Rev. Lett. **68**, 1132 (1992); see also H.J. Kimble, in Ref. [5].

[8] This proposal for a quantum-phase gate has been developed in collaboration with S. Lloyd.

[9] S. Lloyd, Phys. Rev. Lett. **75**, 346 (1995); A. Barenco, D. Deutsch, and A. Ekert, Proc. R. Soc. London A **449**, 669 (1995).

[10] W. Lange *et al.* (to be published). Note that for the parameters of our apparatus, the reflection output channel (via M_2) dominates over transmission (via M_1) and mirror absorption and scattering losses (see [6]). Atomic spontaneous emission is also small because $\kappa > g^2/\kappa > \gamma$ and fields are detuned from resonance.

[11] J.F. Clauser *et al.*, Phys. Rev. Lett. **23**, 880 (1969).

[12] N. Gisin and A. Peres, Phys. Lett. A **162**, 15 (1992).

[13] CHSH violation actually represents a conservative measure of entanglement in mixed states, as compared to alternate measures utilized in quantum communication theory [C.H. Bennett (private communication)].

[14] H. Mabuchi *et al.* (to be published). For example, for the parametrization $\rho_{jk}d_{jk}$ we find that violations of the CHSH inequality require both self-coherence and mutual coherence for the (a,b) qubits and scale linearly with the damping of coherences in the joint density matrix.

VOLUME 80, NUMBER 19 PHYSICAL REVIEW LETTERS 11 MAY 1998

Real-Time Cavity QED with Single Atoms

C. J. Hood, M. S. Chapman, * T. W. Lynn, and H. J. Kimble
Norman Bridge Laboratory of Physics 12-33, California Institute of Technology, Pasadena, California 91125
(Received 18 November 1997)

The combination of cold atoms and large coherent coupling enables investigations in a new regime in cavity QED with single-atom trajectories monitored in real time with high signal-to-noise ratio. The underlying "vacuum-Rabi" splitting is clearly reflected in the frequency dependence of atomic transit signals recorded atom by atom, with evidence for mechanical light forces for intracavity photon number <1. The nonlinear optical response of one atom in a cavity is observed to be in accord with the one-atom quantum theory but at variance with semiclassical predictions. [S0031-9007(98)06037-2]

PACS numbers: 42.50.Ct, 32.80.–t, 42.50.Vk

An important trend in modern physics has been the increasing ability to isolate and manipulate the dynamical processes of individual quantum systems, with interactions studied quantum by quantum. In optical physics, examples include cavity QED with single atoms and photons [1] and trapped ions cooled to the motional zero point [2], while in condensed matter physics, an example is the Coulomb blockade with discrete electron energies [3]. An essential ingredient in these endeavors is that the components of a complex quantum system should interact in a controlled fashion with minimal decoherence. More quantitatively, if the off-diagonal elements of the system's interaction Hamiltonian are characterized by $\langle H_{\rm int}\rangle \sim \hbar g$, where g is the rate of coherent, reversible evolution, then a necessary requirement is to achieve *strong coupling* for which $g > \beta \equiv \max[\Gamma, T^{-1}]$ with T as the interaction time and Γ as the set of decoherence rates for the system.

Although there are many facets to investigations of such open quantum systems, our primary motivation has been to exploit strong coupling in cavity QED to enable research in quantum measurement and more generally, in the emerging field of quantum information dynamics [4]. Several experiments in cavity QED have investigated the nonperturbative interaction of an atom with the electromagnetic field at the level of a single photon; for this system $2g_0$ is the single-photon Rabi frequency and $\Gamma \equiv \{\gamma_\perp, \kappa\}$, with $\gamma_\perp$ as the atomic dipole decay rate and κ as the rate of decay of the cavity field [5–8]. However, without exception these experiments have employed atomic beams in settings for which the information per atomic transit (of duration T) is $I \equiv \frac{g_0^2 T}{\beta} \sim 1$, so that measurements over an ensemble of atoms are required. For example, the passage of a Rydberg atom through a microwave cavity and its subsequent measurement provides a single bit of information [5,7].

By contrast, an exciting recent development in cavity QED has been the ability to observe single-atom trajectories in *real time* with $I \gg 1$ [9]. In this method the transmitted power of a probe beam is monitored as cold atoms fall between the mirrors of a high-finesse optical resonator, with the probe transmission significantly altered by the position-dependent interaction between atom and cavity field [10,11].

Similarly enabled by the use of cold atoms, the research reported in this Letter exploits the largest coupling g_0 achieved to date to explore a new regime in cavity QED, for which single-atom trajectories directly reveal the nature of the underlying one-atom master equation. More specifically, for atoms taken one by one, we map the frequency response of the atom-cavity system, and thereby directly determine g_0 from the vacuum-Rabi splitting. For probe excitation near the coincident atom-cavity resonance, the nonlinear saturation behavior of the atom-cavity system is found to be in accord with the single-atom master equation but at variance with semiclassical theory. However, for probe detunings $\Delta \sim \pm g_0$, we observe a marked asymmetry in the vacuum-Rabi spectrum; few trajectories achieve optimal coupling with a blue detuned probe, an effect which we attribute to light forces even for photon numbers <1. Notably, this is the first experiment for which the interaction energy $\hbar g_0$ is greater than the atomic kinetic energy.

Our apparatus is shown schematically in Fig. 1. The Fabry-Perot cavity consists of two superpolished spherical mirrors of radius of curvature 10 cm, forming a cavity of length 10.1 μm and finesse $\mathcal{F} = 1.8 \times 10^5$. In this cavity $(g_0, \kappa, \gamma_\perp, T^{-1})/2\pi = (120, 40, 2.6, 0.002)$ MHz, where the atom-field coupling coefficient g_0 is determined by the cavity geometry (and the known transition dipole moment [12]), κ is the measured linewidth of the TEM_{00} mode of the cavity, $\gamma_\perp$ is the dipole decay

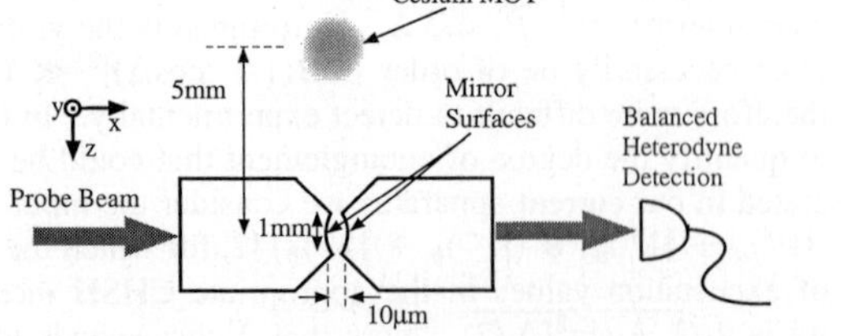

FIG. 1. Schematic of the experimental apparatus.

rate for the Cs $(6S_{1/2}, F = 4, m_F = 4) \rightarrow (6P_{3/2}, F = 5, m_F = 5)$ transition ($\lambda = 852.36$ nm) [12], and typical transit times for atoms through the cavity mode (waist $w_0 \simeq 15$ μm) are $T \simeq 75$ μs. These rates correspond to critical photon and atom numbers ($m_0 \equiv \gamma_\perp^2/2g_0^2, N_0 \equiv 2\kappa\gamma_\perp/g_0^2$) $= (2.3 \times 10^{-4}, 0.015)$, and to optical information per atomic transit $I \sim 5.4 \times 10^4 \pi$ [4]. The probe transmission (typical power 10 pW) is measured using balanced heterodyne detection with overall efficiency 40%. The length of the cavity is actively stabilized by chopping an auxiliary locking beam [6].

Our experimental procedure consists of loading the magneto-optical trap (MOT) for 0.5 s, performing sub-Doppler cooling to 20 μK and then dropping the atoms, all the while monitoring transmission of a circularly polarized probe beam with fixed detuning $\Delta \equiv \omega_P - \omega_{AC}$ (where $\omega_{\text{atom}} = \omega_{\text{cavity}} \equiv \omega_{AC}$) [13]. As an atom falls into the cavity, it encounters a spatially dependent coupling coefficient $g(\vec{r}) \simeq g_0 \cos(2\pi x/\lambda) \exp[-(y^2 + z^2)/w_0^2] \equiv g_0 \psi(\vec{r})$. Hence as $g(\vec{r})$ increases from zero, the otherwise coincident atomic and cavity resonances map to a nondegenerate superposition of dressed states for the atom-cavity system, so the probe spectrum evolves from a simple Lorentzian to a "vacuum-Rabi" spectrum with two peaks at $\omega_{AC} \pm g(\vec{r})$, as illustrated in Fig. 2. Displayed is a series of theoretical transmission spectra $\overline{m}(\Delta) \equiv |\langle a \rangle|^2$ associated with the mean intracavity amplitude $\langle a \rangle$ calculated from the steady-state solution of the master equation for a single atom (of infinite mass) placed at sites $\vec{r}_l$ with coupling $g(\vec{r}_l)$ [11]. Of course, the spectrum at $g(\vec{r}) = 0$ is simply the response of the cavity with no atom present, hereafter denoted by $\overline{n}(\Delta) = \overline{n}_0/[1 + (\Delta/\kappa)^2]$, while that at $g(\vec{r}_l) = g_0$ corresponds to an optimally coupled atom.

Although most atoms never reach a region of optimal coupling, some do enter in the desired $m_F = 4$ sublevel and fall through antinodes of the field; these encounter an increasing $g(\vec{r})$ which sweeps the vacuum-Rabi sidebands outward in frequency to a maximum of $\pm g_0/2\pi = \pm 120$ MHz. The bold traces in Fig. 2 illustrate the corresponding evolution of $\overline{m}$ for three probe detunings Δ relevant to our observations. The process reverses as the atom leaves the cavity.

Turning to our measurements, we present in Fig. 3(a)–3(c) examples of the time-dependent transmission $T(t) \equiv \overline{m}(t)/\overline{n}$ of the atom-cavity system at the probe detunings of Fig. 2. With $\omega_P \approx \omega_{AC}$ [Fig. 3(a)] we observe first decreasing probe transmission [due to increasing $g(\vec{r})$ as the atom enters the mode volume], then a minimum in transmission [when $g(\vec{r}) \approx g_0$], and finally transmission increasing to its original value (as the atom exits the cavity). $T_{\min} \approx 10^{-2}$ is regularly observed for single atom transits. Conversely, for $\Delta/2\pi = -120$ MHz $\approx g_0/2\pi$ [Fig. 3(c)], the transmission increases as the atom enters the cavity mode, peaking at $T_{\max} \approx 3.5$ when $g(\vec{r}) \approx g_0$, and then falling as the atom exits. Finally, an intermediate regime $\Delta/2\pi = -40$ MHz $\approx \frac{1}{3} g_0/2\pi$ exhibits more complicated behavior [Fig. 3(b)]. Here, as the atom enters the cavity, the transmission first increases as the lower Rabi peak sweeps past ω_P, then decreases to a minimum when $g(\vec{r}) \approx g_0$, and finally passes through a second maximum as $g(\vec{r})$ decreases with the atom's departure [14].

To confirm the qualitative characteristics of the vacuum-Rabi spectrum during a single atom transit we simultaneously record the transmission of two probe beams, as in Figs. 3(d),3(e). For probes with detunings $\Delta_{1,2} \approx \mp g_0$, the cavity transmission increases simultaneously for each

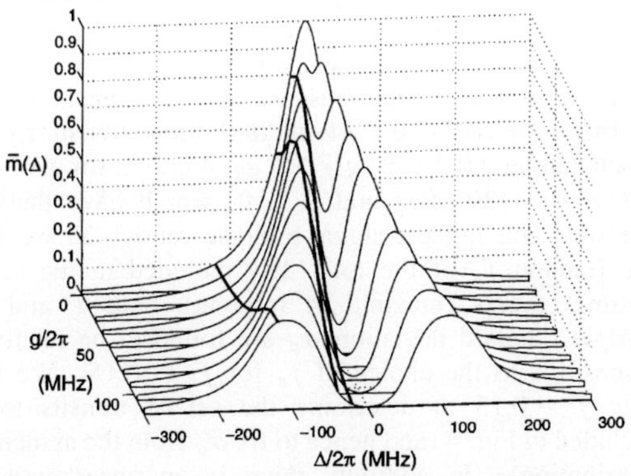

FIG. 2. $\overline{m}(\Delta) \equiv |\langle a \rangle|^2$ as a function of probe detuning Δ for atomic positions $\vec{r}_l$ such that $g(\vec{r}_l) = \{0, g_0/9, \ldots, g_0\}$, with probe intensity fixed at $\overline{n}_0 = 1$. For an atom transiting the cavity, this *position dependent coupling* yields a *time dependent transmission*, indicated by the bold curves for fixed probe detunings $\Delta/2\pi = \{-20, -40, -120\}$ MHz.

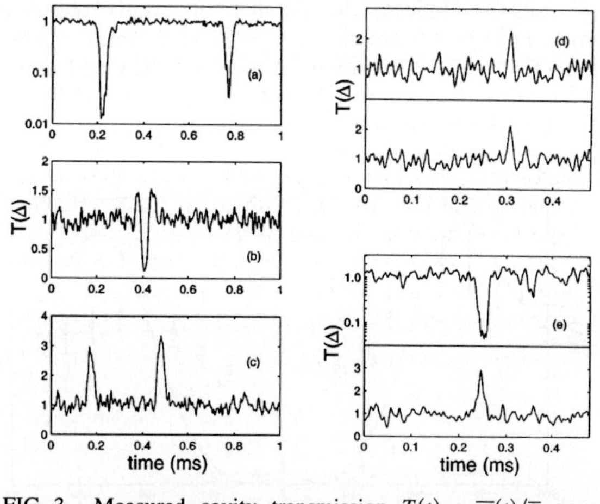

FIG. 3. Measured cavity transmission $T(t) = \overline{m}(t)/\overline{n}$ as a function of time for individual atom transits. Traces (a)–(c) are for $\Delta/2\pi = \{-20, -40, -120\}$ MHz with $\overline{n}_0 = 0.7, 0.6, 1.0$. (d) $\Delta_{1,2}/2\pi = \{-100, +100\}$ MHz with $\overline{n}_{01,02} = 0.38, 0.22$. (e) $\Delta_{1,2}/2\pi = \{-20, -100\}$ MHz with $\overline{n}_{01,02} = \{0.05, 0.3\}$. All traces are acquired with 100 kHz resolution bandwidth and digitized at 500 kHz sampling rate.

probe during the atom transit [Fig. 3(d)]. For one probe near resonance ($\Delta_1 \approx 0$) and the other red-detuned ($\Delta_2 \approx -g_0$), there is a reduction in the transmission at Δ_1, and an increase in the transmission at Δ_2 [Fig. 3(e)]. Note that the signal-to-noise for these traces is less than that for single-probe measurements due to saturation, reflecting a limitation *in principle* to the rate at which information can be extracted from this quantum system.

We next map the frequency response of the atom-cavity system over a range of detunings $\frac{|\Delta|}{2\pi} \leq 200$ MHz (Fig. 4). Clearly evidenced is a double-peaked structure reminiscent of the "vacuum-Rabi" splitting, with peaks near $\pm g_0/2\pi$, as was first observed in Ref. [15]. In contrast to previous work with atomic beams, here atoms are observed one by one with negligible effect from background atoms in the tails of the cavity-mode function [16] (such "spectator" atoms contribute in aggregate an effective atom number $\overline{N}_e < 0.04$).

At each value of Δ, a series of about 50 trap drops is made, yielding up to 800 single-atom events, from which the maximum and/or minimum relative transmissions, shown in Fig. 4, are determined. Note that at small (large) detunings only decreases (increases) in transmission are observed [cf. Figs. 3(a),3(c)], whereas for detunings 40 MHz $\leq |\Delta|/2\pi \leq 60$ MHz both increases and decreases are observed [cf. Fig. 3(b)], hence both a maximum *and* a minimum transmission are shown. Again, the transit signals are normalized to the transmission of the empty cavity at each frequency to give $T(\Delta)$, with $\overline{n}_0$ varying from $\approx$0.6 photons near resonance to $\approx$1.4 photons at $\Delta/2\pi = \pm 200$ MHz.

One of the most striking features of the data in Fig. 4 is the asymmetry of the spectrum between red and blue probe detunings, both in the magnitude and abundance of transits. Indeed, the number of events observed with $T(\Delta) \approx 2.5$ around $\Delta \approx +g_0$ is 5 times smaller than for $T(\Delta) \approx 3.3$ around $\Delta \approx -g_0$. Residual atom-cavity detunings are insufficient to explain the observed asymmetry (the cavity lock results in $\omega_{\text{atom}} \simeq \omega_{\text{cavity}}$

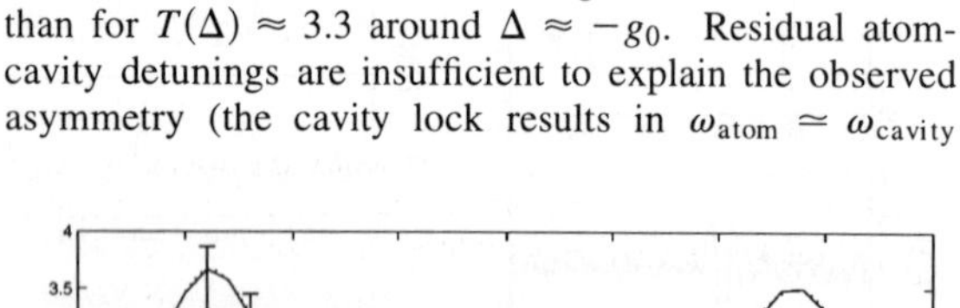

FIG. 4. Maximum (○) and minimum (□) normalized transmission $T(\Delta)$ versus detuning Δ measured via single atom events. The solid curve gives $T(\Delta)$ for an atom with $g(\vec{r}) = g_0$ (the vacuum-Rabi spectrum), while the dashed line is the maximum transmission for any coupling $g(\vec{r}) \leq g_0$.

with systematic offsets below ± 2 MHz and peak-to-peak excursions less than ± 5 MHz). We attribute this asymmetry to mechanical light forces from the probe beam affecting the atom's trajectory. As analyzed in Ref. [17], weak excitation by a coherent probe tuned to $\Delta_\pm \approx \pm g_0$ gives rise to a pseudopotential (for times $\gg \kappa^{-1} \sim 4$ nsec), with depth $\pm \hbar g_0 p_\pm$, where $p_\pm \propto \overline{m}(\Delta_\pm)$ is the probability of occupation of the upper (lower) dressed state. Since $\hbar g_0/k_B \approx 7$ mK, such light forces can be significant even for $\overline{m} \sim 0.5$ photons. We thus expect significant channeling of atomic trajectories into regions of high light intensity and strong coupling for a red-detuned probe ($\Delta < 0$). Conversely, a blue-detuned probe ($\Delta > 0$) creates a potential barrier and prevents an atom from reaching areas of optimal coupling. Apart from its relevance to the spectrum of Fig. 4, this phenomenon suggests the possibility of trapping single atoms in the cavity mode with single photons.

For comparison with theory, the solid curve in Fig. 4 gives $T(\Delta)$ obtained from the steady-state solution of the master equation for a single stationary atom with $g(\vec{r}) = g_0$. Because the largest increases in transmission for $|\Delta| \gtrsim g_0$ and similarly the deepest downgoing transits near $\Delta = 0$ correspond to atoms with maximal coupling g_0, these data points track the solid curve well. However, for intermediate detunings $40 \lesssim |\Delta|/2\pi \lesssim 100$ MHz, the maximum observed transmission corresponds to a *smaller* value of coupling, $g(\vec{r}) \simeq |\Delta| < g_0$, and so these points are not expected to fall on the solid curve. We can, however, determine the maximum expected transmission at each Δ by considering all couplings $g(\vec{r}) \leq g_0$, with the result plotted as the dashed curve in Fig. 4. Agreement between this *ideal one-atom* theory and experiment is evident for $\Delta < 0$, providing direct confirmation of the quoted value for g_0.

Note that because $g(\vec{r})$ for most atoms never reaches g_0 as they transit the cavity, we record a continuous distribution of transit sizes at each Δ, from which the maximum and minimum values of $T(\Delta)$ of Fig. 4 and the associated uncertainties are determined as follows. First note that in the absence of mechanical forces, a fraction $f_t(\Delta) \sim 0.1$ of all *detectable* transit signals reach coupling $0.9 g_0 \leq g(\vec{r}) \leq g_0$. Further, for data with $\Delta/2\pi = -120$ MHz and $\Delta \sim 0$ (which have the best statistics and highest signal-to-noise ratios), as we vary the fraction f_e of the total data included in the set of optimal events (maxima or minima), both T and the sample standard deviation σ_d are found to be relatively insensitive to the choice of f_e for $f_e \lesssim 0.15$. We thus take $f = 0.15$ to determine the set of transits to be included in Fig. 4 (and hence to fix σ_d from the associated distribution). In addition, there is an uncertainty σ_q arising from the noise of the detected probe beam itself, which is estimated by an appropriate scaling of the noise for "no-atom" data bracketing a given transit signal. The quantity $\sigma = \sqrt{\sigma_d^2 + \sigma_q^2}$ is shown in Fig. 4 to estimate the error in T at each Δ. For all our data,

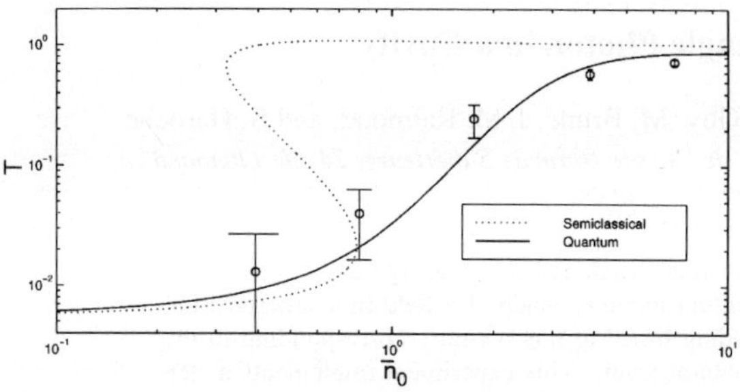

FIG. 5. Transmission T versus probe photon number $\overline{n}_0$ for maximally coupled atom transits for fixed $\Delta/2\pi = -20$ MHz. The solid line results from the quantum master equation for one atom with $g(\vec{r}) = g_0$, while the dashed line is the semiclassical bistability state equation.

the absolute uncertainty in the quoted photon numbers is $\simeq \pm 30\%$.

In a final series of measurements shown in Fig. 5, we explore the nonlinear saturation behavior of the atom-cavity system. We vary $\overline{n}_0$ with the probe beam fixed at detuning $\Delta/2\pi = -20$ MHz. At each $\overline{n}_0$ we again digitize the cavity transmission for a large number of transits, with a set of "optimal" single-atom events determining the value of T and its uncertainty σ. The solid curve of Fig. 5 is from the steady-state solution of the master equation for a single (stationary) atom with $g(\vec{r}) = g_0$, with reasonable agreement between the data and this ideal quantum model. By contrast, the dashed line is the semiclassical transmission function [18] evaluated for the parameters of our experiment, and exhibits bistable behavior. Shifts from the semiclassical bistability curve have also been predicted for other regimes of cavity parameters [19].

In conclusion, by exploiting laser cooled atoms in cavity QED, a unique optical system has been realized which approximates the ideal situation of one atom strongly coupled to a cavity, with $\hbar g$ larger than even the atomic kinetic energy. The system's characteristics have been explored atom by atom, leading to measurements of the "vacuum-Rabi" splitting and of the nonlinear transmission for probe photon number ~ 1. Because $I \gg 1$, the system offers considerable opportunity for long interaction times and controlled quantum dynamics, as in our current efforts to generate a bit stream containing $m \sim 10^4$ photons with a single falling atom [20] as well as to trap one atom in the quantized cavity field. Although the atomic center-of-mass (CM) motion has here been treated classically, this work sets the stage for investigations of quantum dynamics involving the quantized CM and the internal atomic dipole + cavity field degrees of freedom [21,22], including trapping by way of the "well-dressed" states for single quanta [23].

We gratefully acknowledge the contributions of D. Bass, H. Mabuchi, and Q. Turchette. T.W.L. acknowledges support from the NSF. This work has been supported by DARPA via the QUIC Institute administered by ARO, by the NSF, and by the ONR.

---

*Permanent address: School of Physics, Georgia Institute of Technology, Atlanta, GA 30332-0430.

[1] *Cavity Quantum Electrodynamics,* edited by P. Berman (Academic Press, San Diego, 1994).
[2] C. Monroe *et al.*, Phys. Rev. Lett. **75**, 4011 (1995).
[3] D.C. Ralph, C.T. Black, and M. Tinkham, Phys. Rev. Lett. **78**, 4087 (1997).
[4] H.J. Kimble, Philos. Trans. R. Soc. London A **355**, 2327 (1997).
[5] G. Rempe, F. Schmidt-Kaler, and H. Walther, Phys. Rev. Lett. **64**, 2783 (1990).
[6] G. Rempe *et al.*, Phys. Rev. Lett. **23**, 1727 (1991).
[7] M. Brune *et al.*, Phys. Rev. Lett. **76**, 1800 (1996).
[8] J.J. Childs *et al.*, Phys. Rev. Lett. **77**, 2901 (1996).
[9] H. Mabuchi *et al.*, Opt. Lett. **21**, 1393 (1996).
[10] G. Rempe, Appl. Phys. B **60**, 233 (1995).
[11] A.C. Doherty *et al.*, Phys. Rev. A **56**, 833 (1997).
[12] C.E. Tanner *et al.*, Nucl. Instrum. Methods Phys. Res., Sect. B **99**, 117 (1995).
[13] The circularly polarized probe beam drives the cycling transition $m_F = 4 \rightarrow m_F = 5$ and provides optical pumping to this sublevel as an atom enters the cavity mode. The magnetic field is zeroed at the site of the MOT to allow sub-Doppler cooling and from the coil geometry, should also be well zeroed at the cavity mode. Explicit spatial quantization obtained by switching a magnetic field "on" along the cavity axis as the atom transits produced no significant impact on the transit signals. Note that transit signals from atoms that are incorrectly optically pumped and driven on transitions other than $mF = 4 \rightarrow m_F = 5$ are eliminated by our selection of optimal events.
[14] The transit signals of Fig. 3 are smoothly varying without the rapid oscillations recorded in Ref. [9], which were tentatively attributed to motion along the standing wave. Here we suspect that the tenfold increase in g leads to mechanical forces which inhibit this motion.
[15] R.J. Thompson, G. Rempe, and H.J. Kimble, Phys. Rev. Lett. **68**, 1132 (1992).
[16] Q. Turchette *et al.*, Phys. Rev. A (to be published).
[17] A.S. Parkins (unpublished); P. Horak *et al.*, Phys. Rev. Lett. (to be published).
[18] L.A. Lugiato, *Progress in Optics,* edited by E. Wolf (Elsevier Science Publishers, Amsterdam, 1984), Vol. XXI.
[19] C.M. Savage and H.J. Carmichael, IEEE J. Quantum Electron. **24**, 1495 (1988).
[20] C.K. Law and H.J. Kimble, Quantum Semiclass. Opt. **44**, 2067 (1997).
[21] R. Quadt, M. Collett, and D.F. Walls, Phys. Rev. Lett. **74**, 351 (1995).
[22] A.C. Doherty *et al.*, Phys. Rev. A (to be published).
[23] D.W. Vernooy and H.J. Kimble, Phys. Rev. A **56**, 4207 (1997).

VOLUME 79, NUMBER 4 PHYSICAL REVIEW LETTERS 28 JULY 1997

Quantum Memory with a Single Photon in a Cavity

X. Maître, E. Hagley, G. Nogues, C. Wunderlich, P. Goy, M. Brune, J. M. Raimond, and S. Haroche
Laboratoire Kastler Brossel, Département de Physique de l'Ecole Normale Supérieure, 24 rue Lhomond, F-75231 Paris Cedex 05, France*
(Received 31 March 1997)

The quantum information carried by a two-level atom was transferred to a high-Q cavity and, after a delay, to another atom. We realized in this way a quantum memory made of a field in a superposition of 0 and 1 photon Fock states. We measured the "holding time" of this memory corresponding to the decay of the field intensity or amplitude at the single photon level. This experiment implements a step essential for quantum information processing operations. [S0031-9007(97)03701-0]

PACS numbers: 89.70.+c, 03.65.-w, 32.80.-t, 42.50.-p

The manipulation of simple quantum systems interacting in a well-controlled environment is a very active field in quantum optics, with strong connections to the theory of quantum information [1]. Atoms and photons can be viewed as carriers of "quantum bits" (or qubits) storing and processing information in a nonclassical way. The interaction between two qubit carriers can model the operation of a quantum gate in which the evolution of one qubit is conditioned by the state of the other [2,3]. Combining a few qubits and gates could lead to the realization of simple quantum networks in which an "engineered entanglement" between the interacting qubits carriers could be achieved. Even if practical applications to large scale quantum computing are likely to remain inaccessible [4], fundamental tests of quantum theory could be performed, such as demonstrations of new quantum nonlocal effects [5], decoherence studies, etc.

Several quantum optics systems are investigated in this context, including trapped ions [6,7], combinations of photon pairs [8], or atoms in cavities [9]. In the latter case, atoms cross one at a time a high-Q cavity. The qubits are carried either by the atom, schematized as a two-level system, or by the quantum field in the cavity, which is in a superposition of 0 and 1 photon states. The interaction between the atom and the cavity field mode provides the conditional dynamics required for the operation of a quantum gate, as has been demonstrated recently in microwave [10] and in optical cavity QED experiments [11].

To implement quantum logic, the information should be transferable between qubit carriers and preserved between gate operations. This involves the existence of a quantum memory whose holding time is limited by the carrier relaxation processes. We report here the realization of a quantum memory in a cavity QED experiment. We have transferred a qubit from an atomic carrier to a field one, then to another atom. The initial atom was either in one of its two energy eigenstates, or in a superposition of them. The mediating field was prepared either in a 0 or 1 photon number state (Fock state) or in a superposition of the two. These are highly nonclassical states of radiation. By varying the delay between the two transfer processes, we have measured the qubit holding time of the cavity. We have directly determined in this way the lifetime of a single photon and of a superposition of 0 and 1 photon.

The principle of the quantum information transfer relies on the Rabi precession at frequency $\Omega/2\pi$ of an atom between two energy eigenstates e and g in the cavity vacuum $|0\rangle$ [12]. If the atom starts in the upper level e and the effective resonant atom-cavity interaction time t [12] is such that $\Omega t = \pi$, the combined system evolves from the $|e,0\rangle$ into the $|g,1\rangle$ state: the atomic excitation is transferred to the field. If the atom is initially in level g, the system starts in the $|g,0\rangle$ state and no evolution occurs. If the atom is initially in a superposition $\alpha|e\rangle + \beta|g\rangle$, the linearity of quantum mechanics implies that the combined system evolves into the state $(\alpha|1\rangle + \beta|0\rangle)|g\rangle$. The interaction has transferred the quantum superposition from the atom to the field, leaving the former in g. This information can then be transferred to a second atom initially in g and crossing the cavity after a delay, in a process reverse of the one experienced by the first atom.

The main elements of our setup, schematized in Fig. 1, have been described elsewhere [12,13]. Rubidium atoms

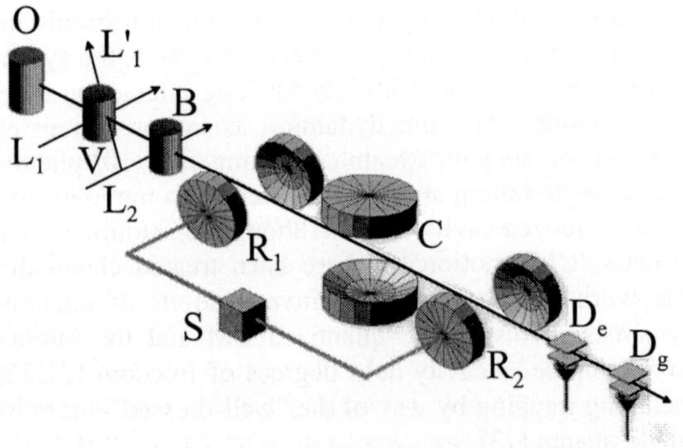

FIG. 1. Sketch of the experimental setup.

effusing from an oven O and velocity selected in zone V, are prepared in box B in the circular Rydberg state with principal quantum number 51 (level e) or 50 (level g) [14]. The atoms then cross a low-Q cavity R_1 in which a classical microwave pulse resonant with the transition at 51.1 GHz between e and g can be applied to prepare a controlled superposition of these two states.

The atoms then pass through a high-Q superconducting cavity C in which the Rabi precession in vacuum produces the quantum information transfer. The cavity, made of two niobium mirrors in a Fabry-Perot configuration (mirror separation 2.7 cm), sustains two orthogonally polarized TEM_{900} modes M_1 and M_2 with a spacing of 70 kHz. The vacuum Rabi frequency of the Rydberg atom at cavity center is $\Omega/2\pi = 48$ kHz for both modes [12]. The field energy damping times, measured by standard microwave techniques, are $T_r = 112\ \mu$s and 84 μs for M_1 and M_2, respectively. Both modes are close to resonance with the $e \rightarrow g$ transition. Either of them can be tuned in exact resonance by Stark shifting the atomic transition with the help of a time-varying electric field $F(t)$ applied across the gap between the cavity mirrors. When a mode is not exactly resonant, it has no effect on the evolution of the atomic populations in C. By proper adjustment of $F(t)$, one can induce an exact π pulse of the atom interacting either with M_1 or M_2. After leaving C, each atom crosses a second auxiliary cavity R_2, identical to R_1, which can mix again e and g. Finally the atoms are detected by state-selective field ionization in detectors D_e and D_g for levels e and g, respectively (detection efficiency: 35%). The combination of R_2 and $D_{e,g}$ analyzes either the atomic energy (no pulse applied in R_2) or the quantum coherence between levels e and g (pulse applied in R_2). The distances between the exit of B and the centers of R_1, C, and R_2 are 5.4, 9.95, and 14.5 cm, respectively. The zone from B to D is cooled to 0.6 K by a ^{3}He-^{4}He refrigerator to avoid blackbody radiation (0.02 thermal photon on the average in C).

The control of the atomic velocity and of the atomic timing across the setup are essential. The velocity selection involves the optical depumping of the $F = 3$ ground hyperfine sublevel of rubidium with a diode laser L_1, followed by a Doppler selective repumping of this level with the help of a laser beam L_1' oriented at an angle with the atomic beam. By tuning the frequency of L_1', a velocity profile centered at 400 m/s with a ± 30 m/s width is selected in the Maxwellian distribution of the atomic beam. L_1' is pulsed with a 2 μs duration. The circular state preparation in box B is a pulsed process starting from the $F = 3$ hyperfine level which involves a stepwise excitation (lasers L_2) and radio frequency transitions. It prepares within a time window of 2 μs a pulse of velocity selected atoms in e or g. The circularization process cuts a very thin slice of ± 0.4 m/s in the already selected atomic velocity profile. This velocity selection procedure is checked by time-of-flight measurements. The position of each atom can thus be determined at any time between preparation and detection with a precision better than 1 mm. This allows us to fire microwave pulses in R_1 and R_2 and the Stark-switching field in C exactly when the atom reaches the corresponding position with the possibility of exposing successive atoms to different interactions. The intensity of the lasers L_2 is reduced so that about 0.3 atom on the average is prepared in each pulse, and the probability to have more than one atom is small.

A quantum information transfer sequence consists in sending from B a pair of atomic pulses with variable velocities separated by an adjustable delay. In 1% of the sequences, one atom is detected in each pulse (useful events). The atomic interactions with C are separated by a known delay T which is adjusted between 30 and 400 μs. The state of the two atoms are detected by D_e and D_g. The sequence is repeated every 1.75 ms, and statistics are accumulated to reconstruct the joint probabilities P_{ee}, P_{eg}, P_{ge}, and P_{gg} that the pair of atoms is found in any configuration of quantum states.

In a first experiment, we prepare a single photon Fock state and exchange energy between the two atoms of each pair. No state mixing pulses are applied in R_1 or R_2. The first atom is prepared in e, the second in g. Both are coupled to the same C mode (either M_1 or M_2) and undergo a π pulse. Ideally, if the pulses were perfect and the cavity Q infinite, the first atom would emit exactly one photon which would be picked up with unit probability by the second atom. As a result, the conditional probability to detect the second atom in e provided the first one is detected in g, $\Pi_{ge} = P_{ge}/(P_{ge} + P_{gg})$, should be exactly one. When cavity relaxation is taken into account, Π_{ge} is expected to decay exponentially with the time constant T_r.

Figure 2 shows the measured Π_{ge} probability as a function of the delay T between the atoms in units of T_r. Each point averages 7000 useful events. Data corresponding to the two cavity modes have been merged. The experimental points fit to an exponential curve displaying the decay of a single photon in the cavity with the expected rate $1/T_r$. The maximum probability extrapolated to zero delay is 74%. Several experimental imperfections explain this reduced value. The vacuum Rabi pulse in C cannot transfer more than 94% of the atoms, due to coupling dispersions related to the atomic position spread in the cavity mode. When an atom is detected, there is also a 20% probability to have a second atom in the pulse which may be undetected. Finally, an atom in g is erroneously counted by D_e in 13% of the cases (and an atom in e by D_g in 10% of the cases). This last point explains the 13% background at long times in Fig. 2. Taking all these effects into account, we get a maximum conditional probability at $T = 0$ of 70%, in good agreement with the observed value.

In a second experiment, we perform a transfer of coherence between the two atoms. The first one is prepared in

VOLUME 79, NUMBER 4 PHYSICAL REVIEW LETTERS 28 JULY 1997

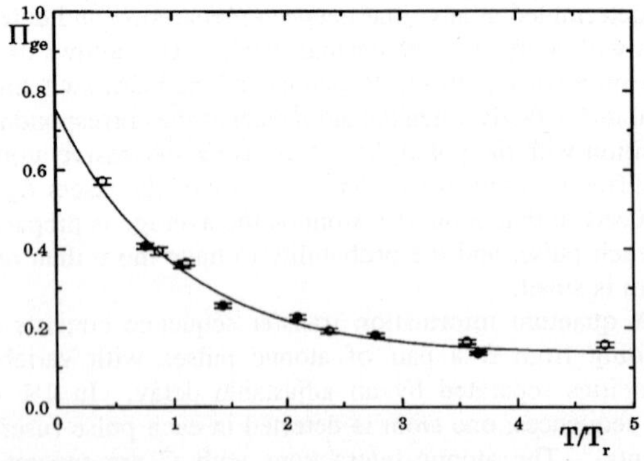

FIG. 2. Decay of a one photon Fock state in the cavity: conditional probability $\Pi_{ge}(T)$ versus the delay T between the two atoms expressed in units of cavity mode damping times T_r. Solid and open circles correspond, respectively, to a photon stored in mode M_1 ($T_r = 112\ \mu$s) or M_2 ($T_r = 84\ \mu$s). The line is an exponential fit with unit time constant and a 13% offset accounting for atomic energy detection errors.

e, undergoes a $\pi/2$ pulse in R_1, and is thus injected in C in a superposition $(|e\rangle + |g\rangle)/\sqrt{2}$. A π pulse in C transfers this coherence to the field (superposition of 0 and 1 photon states) and the first atom is finally detected in g. The second atom, prepared in g, experiences no pulse in R_1 and a π pulse in C. It enters thus R_2 in a coherent superposition of e and g. A $\pi/2$ pulse applied in R_2 analyzes the transferred quantum coherence. The conditional probability Π_{ge}, measured as a function of the common frequency ν of the microwave fields applied to the cavities R_1 (first atom) and R_2 (second atom), exhibits fringes which reveal the transfer of coherence. The signal is shown in Figs. 3(a), 3(b), and 3(c) for various values of the delay T between the two atoms. Each scan corresponds to 9000 useful events.

These recordings are reminiscent of Ramsey separated oscillatory field signals [15], the fringe period corresponding to the inverse of the time delay T' between the two interactions in R_1 and R_2. Here, however, the separated fields are applied to two different atoms. From the selected atomic velocities and the R_1 to C and C to R_2 distances, we get $T' = T + 216\ \mu$s. As a test of the consistency of our results, we have checked that the probability of detecting the second atom in e or g is independent of ν when the first atom is not sent in the apparatus.

Alternatively, one may see this experiment as the preparation of a nonclassical field in C, a superposition state with equal weights of 0 and 1 photon. Such a state, like a coherent one, has a nonzero expectation value of the electric field. It is different, however, from a coherent state, since it does not have a Poisson photon number distribution.

The fringe amplitude in Fig. 3 shrinks when the delay T is increased, measuring the decay of the coherence

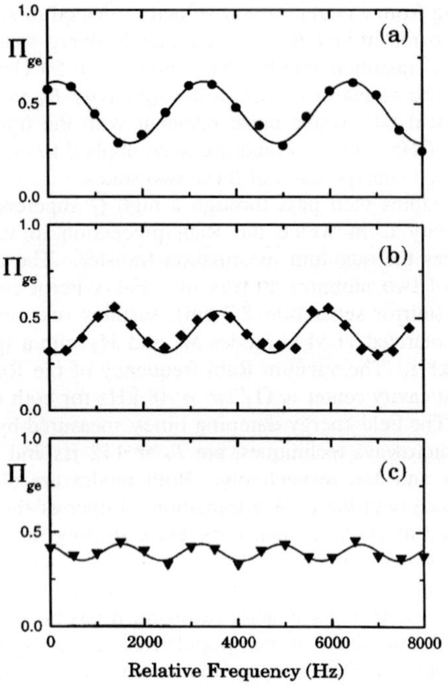

FIG. 3. Transfer of coherence between two atoms: conditional probability $\Pi_{ge}(\nu)$ versus the frequency ν of the microwave pulses applied to the first atom in R_1 and to the second in R_2. The delays $T' = T + 216\ \mu$s between the two microwave pulses in R_1 and R_2 are 301, 436, and 581 μs, respectively from (a) to (c). Cavity mode M_1 is used.

stored in the cavity field. Figure 4 shows this decay as a function of T/T_r. The experimental points fit now to an exponential with a characteristic time $2T_r$. The coherence between the 0 and 1 photon states lives twice

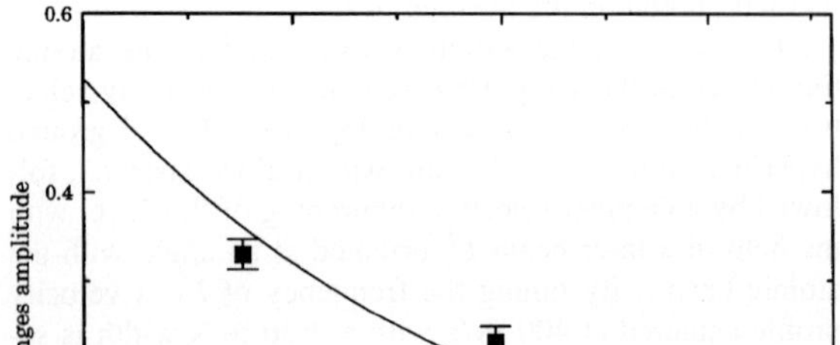

FIG. 4. Decay of the cavity field coherence: amplitudes of the $\Pi_{ge}(\nu)$ fringes of Fig. 3 versus the delay T expressed in units of $T_r = 112\ \mu$s. Solid line: exponential curve with a time constant of 2.

as long as a single photon. We are, in fact, measuring the average decay rate of a 1 photon state (rate $1/T_r$) and of the vacuum (rate 0). We can also remark that this experiment measures the field amplitude in C, whereas the previous one was measuring the field intensity. The maximum contrast value extrapolated to $T = 0$ (52%) derives from the single atom Ramsey fringes contrast (65%) by taking into account the various experimental imperfections discussed above.

This experiment shows that it is possible, via resonant atom-field interaction, to prepare and measure in a cavity a single-photon microwave quantum field which can serve as a mediator to transfer quantum information between two atoms. We thus realize a quantum memory which will be useful for further quantum information processing experiments. The blueprint for the realization of a cavity QED quantum gate [3] entangling a control and a target atomic qubit requires a transfer of the control qubit to the cavity field. This field is then coupled dispersively to the target atomic qubit and conditions its evolution, before being finally transferred back to a third atom, leaving the cavity empty. The exchange of information demonstrated in the present work plays an essential role in this program.

Combining a few gates to perform simple quantum logic operations is very challenging. This requires in particular a much better control of decoherence processes. With the improvements of the cavity modes quality factor under way in our laboratory, holding times 10 to 100 times longer than in this demonstration experiment could be obtained, opening the way to entanglement studies involving several atoms.

*Laboratoire de l'Université Pierre et Marie Curie et de l'ENS, associé au CNRS (URA18).

[1] D. P. DiVincenzo, Science **270**, 255 (1995); A. Ekert and R. Josza, Rev. Mod. Phys. **68**, 3733 (1997).
[2] A. Barenco, D. Deutsch, A. Ekert, and R. Josza, Phys. Rev. Lett. **74**, 4083 (1995); T. Sleator and H. Weinfurter, Phys. Rev. Lett. **74**, 4087 (1995); J. I. Cirac and P. Zoller, Phys. Rev. Lett. **74**, 4091 (1995).
[3] P. Domokos *et al.*, Phys. Rev. A **52**, 3554 (1995).
[4] R. Landauer, Phys. Lett. A **217**, 188 (1996); W. Unruh, Phys. Rev. A **51**, 992 (1995); M. Plenio and P. L. Knight, Phys. Rev. A **53**, 2986 (1996); S. Haroche and J. M. Raimond, Physics Today, Aug. 1996, p. 51.
[5] D. M. Greenberger, M. A. Horne, and A. Zeilinger, Am. J. Phys. **58**, 1131 (1990); S. Haroche, in *Fundamental Problems in Quantum Theory*, edited by D. Greenberger and A. Zeilinger, Ann. N.Y. Acad. Sci. **755**, 73 (1995).
[6] J. I. Cirac, S. Parkins, R. Blatt, and P. Zoller, Adv. At. Mol. Phys. **37**, 238 (1996).
[7] C. Monroe *et al.*, Phys. Rev. Lett. **75**, 4714 (1995).
[8] T. J. Herzog, P. G. Kwiat, H. Weinfurter, and A. Zeilinger, Phys. Rev. Lett. **75**, 3034 (1995).
[9] *Cavity Quantum Electrodynamics*, edited by P. Berman (Academic Press, New York, 1994).
[10] M. Brune *et al.*, Phys. Rev. Lett. **72**, 3339 (1994).
[11] Q. A. Turchette *et al.*, Phys. Rev. Lett. **75**, 4710 (1995).
[12] M. Brune *et al.*, Phys. Rev. Lett. **76**, 1800 (1996).
[13] M. Brune *et al.*, Phys. Rev. Lett. **77**, 4887 (1996).
[14] R. G. Hulet and D. Kleppner, Phys. Rev. Lett. **51**,1430 (1983); P. Nussenzveig *et al.*, Phys. Rev. A **48**, 3991 (1993).
[15] N. F. Ramsey, *Molecular Beams* (Oxford University Press, New York, 1985).

VOLUME 77, NUMBER 24 PHYSICAL REVIEW LETTERS 9 DECEMBER 1996

Observing the Progressive Decoherence of the "Meter" in a Quantum Measurement

M. Brune, E. Hagley, J. Dreyer, X. Maître, A. Maali, C. Wunderlich, J. M. Raimond, and S. Haroche
Laboratoire Kastler Brossel, Département de Physique de l'Ecole Normale Supérieure, 24 Rue Lhomond, F-75231 Paris Cedex 05, France*
(Received 10 September 1996)

A mesoscopic superposition of quantum states involving radiation fields with classically distinct phases was created and its progressive decoherence observed. The experiment involved Rydberg atoms interacting one at a time with a few photon coherent field trapped in a high Q microwave cavity. The mesoscopic superposition was the equivalent of an "atom + measuring apparatus" system in which the "meter" was pointing simultaneously towards two different directions—a "Schrödinger cat." The decoherence phenomenon transforming this superposition into a statistical mixture was observed while it unfolded, providing a direct insight into a process at the heart of quantum measurement. [S0031-9007(96)01848-0]

PACS numbers: 32.80.-t, 03.65.-w, 42.50.-p

The transition between the microscopic and macroscopic worlds is a fundamental issue in quantum measurement theory [1]. In an ideal model of measurement, the coupling between a macroscopic apparatus ("meter") and a microscopic system ("atom") results in their entanglement and produces a quantum superposition state of the "meter + atom" system. Such a superposition is however never observed. Schrödinger has illustrated vividly this problem, replacing the meter by a "cat" [2] and considering the dramatic superposition of dead and alive animal "states." Although such a striking image can only be a metaphor, quantum superpositions involving "meter states" are often called "Schrödinger cats." Following von Neumann [3], it is postulated that an irreversible reduction process takes the quantum superposition into a statistical mixture in a "preferred" basis, corresponding to the eigenvalues of the observable measured by the meter. From then on, the information contents in the system can be described classically. The nature of this reduction has been much debated, with recent theories stressing the role of quantum decoherence [4,5]. According to these approaches, the meter coordinate is always coupled to a large reservoir of microscopic variables inducing a fast dissipation of macroscopic coherences.

The simplest model of a quantum measurement involves a two-level atom (e, g) coupled to a quantum oscillator (meter or cat). An oscillator in a coherent state [6] is indeed defined by a c number α, represented by a vector in phase space ($|\alpha| = \sqrt{n}$ where n is the mean number of oscillator quanta). Quantum fluctuations make the tip of this vector uncertain, with a circular gaussian distribution of radius unity [Fig. 1(a)]. Consider the ideal measurement where the "atom-meter" interaction entangles the phase of the oscillator $(\pm\phi)$ to the state of the atom, leading to the combined state

$$|\Psi\rangle = \frac{1}{\sqrt{2}}(|e, \alpha e^{i\phi}\rangle + |g, \alpha e^{-i\phi}\rangle). \qquad (1)$$

When the "distance" $D = 2\sqrt{n}\sin\phi$ between the meter states is larger than 1, a Schrödinger cat is obtained [Fig. 1(b)].

Decoherence is modeled by coupling the oscillator to a reservoir, which damps its energy in a characteristic time T_r. When $D \gg 1$, decoherence is found to occur within a time scale $2T_r/D^2$ [7,8]. This result illustrates the basic feature of the quantum to classical transition [4]. Mesoscopic superpositions made of a few quanta are expected to decohere in a finite time interval shorter than T_r, while macroscopic ones ($n \gg 1$) decohere instantaneously and cannot be observed in practice.

Recently, a Schrödinger cat of a material oscillator was generated by preparing a single trapped ion in a superposition of two spatially separated wave packets entangled with internal states of the ion [9]. Quantum decoherence was however not studied. Various schemes have been proposed to prepare Schrödinger cats of a field oscillator [10]. Some of these proposals involve a dispersive coupling between a single atom and a field in free space [11] or in a cavity [8]. Implementing this last scheme, we report here the generation of a Schrödinger cat like state of radiation in a cavity and the first dynamical observation of quantum decoherence in a measurement process.

The mesoscopic state is generated by sending a rubidium atom, prepared in a superposition of two circular Rydberg states e and g [12], across a high Q microwave

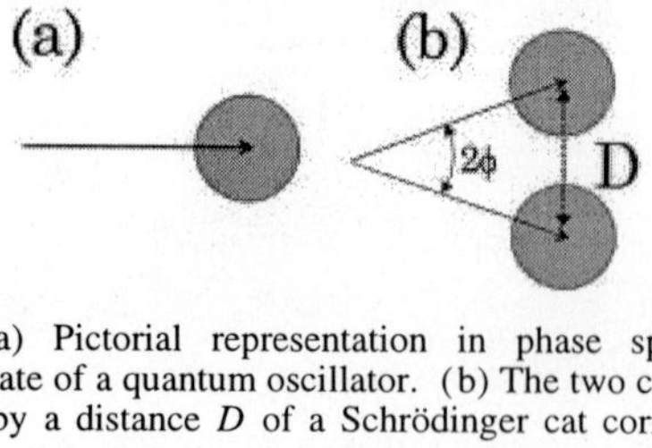

FIG. 1. (a) Pictorial representation in phase space of a coherent state of a quantum oscillator. (b) The two components separated by a distance D of a Schrödinger cat corresponding to Eq. (1).

cavity C storing a small coherent field $|\alpha\rangle$. The coupling between the atom and the cavity is measured by the "Rabi frequency" Ω [13]. The $e \rightarrow g$ atomic transition and the cavity frequencies are slightly off resonance (detuning δ), so that the atom and the field cannot exchange energy but only undergo $1/\delta$ dispersive frequency shifts (single atom index effect). The atom-field coupling during time t produces an atomic-level dependent dephasing of the field and generates an entangled state given (for $\Omega/\delta \ll 1$) by Eq. (1) with $\phi = \Omega^2 t/\delta$ [13].

The states e and g are circular Rydberg levels with principal quantum numbers 51 and 50 (transition frequency $\nu_0 = 51.099$ GHz). They have a long radiative lifetime (30 ms) and a very strong coupling to radiation. The cavity C is a Fabry-Pérot resonator with its axis normal to the atomic trajectory. It is made of two superconducting niobium mirrors (mirror distance 2.7 cm; mode waist 6 mm). $\Omega/2\pi$ is 24 kHz [14]. The cavity Q factor is 5.1×10^7 (photon lifetime $T_r = 160$ μs). The cavity is tuned by adjusting the mirror separation, thus varying $\delta/2\pi$ between 70 and 800 kHz. The effective interaction time t is set to 19 μs by selecting atoms with a velocity of 400 m/s. For $\delta = 100$ kHz, ϕ is 0.69 radian which is an unusually large single atom index effect.

The setup is sketched in Fig. 2. It is cooled to 0.6 K by a He^4-He^3 cryostat making thermal radiation negligible (mean blackbody photon number in C: 0.05). All Rb atoms effusing from the oven O are pumped out of the $F = 3$ ground hyperfine level by a diode laser L_1 and optically repumped into this level by a diode laser beam L_1' oriented at 58° relative to the atomic beam. With a proper tuning of L_1' in the Doppler profile, only atoms at 400 ± 6 m/s are prepared in $F = 3$. The atoms are then excited into the circular state e in box B [15]. This pulsed process involves laser excitation from $F = 3$ (lasers L_2) and prepares, on the average, 0.5 atom within a 2 μs time window, every 1.5 ms.

Each circular atom is prepared in a superposition of e and g by a resonant microwave $\pi/2$ pulse in a low Q cavity R_1. It then crosses C in which a small coherent field with an average photon number n varying from 0 to 10 is injected by a pulsed source S (see below how n is measured). The field, which evolves freely while each atom crosses C, relaxes to vacuum before being regenerated for the next atom ($T_r \ll 1.5$ ms). We make sure that the field is coherent by checking that n is proportional to the square of the injection pulse duration. After leaving C, each atom undergoes another $\pi/2$ pulse in a cavity R_2 identical to R_1. R_1 and R_2 are fed by a cw source S' whose frequency ν is swept across ν_0. The atoms are finally counted in e and g by two field ionization detectors (D_e, D_g; detection efficiency $40 \pm 15\%$). With 50 000 events recorded in 10 min, the probability $P_g^{(1)}(\nu)$ to find the atom in g as a function of ν is reconstructed.

Figure 3(a) shows the signal obtained when C is empty ($\delta/2\pi = 712$ kHz). $P_g^{(1)}(\nu)$ exhibits Ramsey fringes [16] typical of atoms subjected to two pulses separated by a time interval $T = 230$ μs. The fringes result from a quantum interference. The $e \rightarrow g$ transition can occur either in R_1 or in R_2 (atom crossing C in g or e). These two "paths" are indistinguishable, leading, in the final transition rate, to an interference term between the corresponding probability amplitudes. The phase difference between these amplitudes is $2\pi(\nu - \nu_0)T$ so that $P_g^{(1)}(\nu)$ oscillates with the period $1/T = 4.2$ kHz. The fringe contrast,

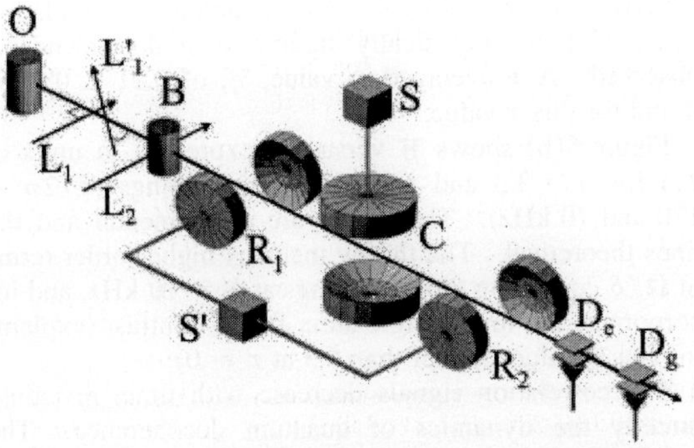

FIG. 2. Sketch of the experimental setup.

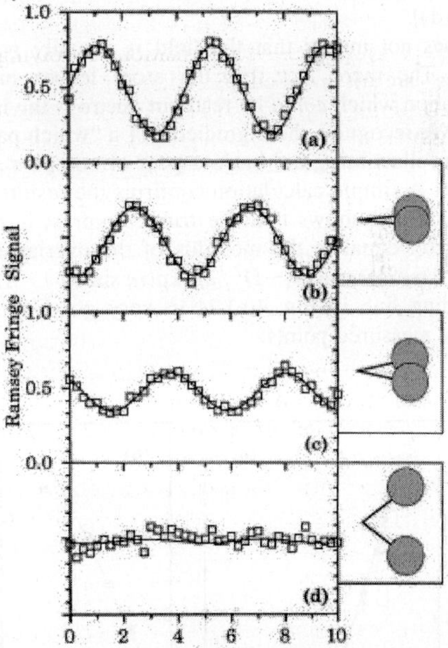

FIG. 3. $P_g^{(1)}(\nu)$ signal exhibiting Ramsey fringes: (a) C empty, $\delta/2\pi = 712$ kHz; (b)–(d) C stores a coherent field with $|\alpha| = \sqrt{9.5} = 3.1$, $\delta/2\pi = 712, 347$, and 104 kHz, respectively. Points are experimental and curves are sinusoidal fits. Insets show the phase space representation of the field components left in C.

ideally 100%, is reduced to 55 ± 5% by various effects (static and microwave fields inhomogeneities between R_1 and R_2 over the 0.7 mm atomic beam diameter, finite atomic lifetime, atom count noise).

Figures 3(b) to 3(d) show the fringes for $\delta/2\pi =$ 712, 347, and 104 kHz when there is a field in C ($n =$ 9.5; $|\alpha| = 3.1$). Two features are striking: when δ is reduced, the contrast of the fringes decreases and their phase is shifted. The fringe contrast and shift are plotted versus ϕ in Figs. 4(a) and 4(b) (points are experimental and lines are from theory; see below).

The fringe contrast reduction demonstrates the separation of the field state into two components and provides a measurement of D^2. When an atom leaves C, the system is prepared in the entangled state of Eq. (1), so that the field phase "points" towards *e and g at the same time*. The insets in Fig. 3 show the phase space representation of the two field components. When δ is large (ϕ small), the measurement of the field phase would give no information on the state of the atom in C (large overlap of the field components). The two "paths" (atom crosses C in e or g) are still indistinguishable and the contrast remains large [Fig. 3(b)]. When δ is reduced (ϕ increased), the field contains more information about the atomic state in C. The two paths thus become partially distinguishable and the fringe contrast decreases [Fig. 3(c)]. It vanishes when the field components do not overlap at all [Fig. 3(d)].

It does not matter that the field is actually not measured. The mere fact that the atom leaves in C an information which *could* be read out destroys the interference. We recognize the ingredients of a "which path" experiment illustrating the basic aspect of complementarity [17,18]. A simple calculation confirms the results of this discussion and shows that the fringe contrast is reduced by a factor equal to the modulus of the overlap integral $\langle \alpha e^{-i\phi} \mid \alpha e^{i\phi} \rangle = \exp(-D^2/2) \exp(in \sin 2\phi)$. The corresponding line in Fig. 4(a) is in very good agreement with the measured points.

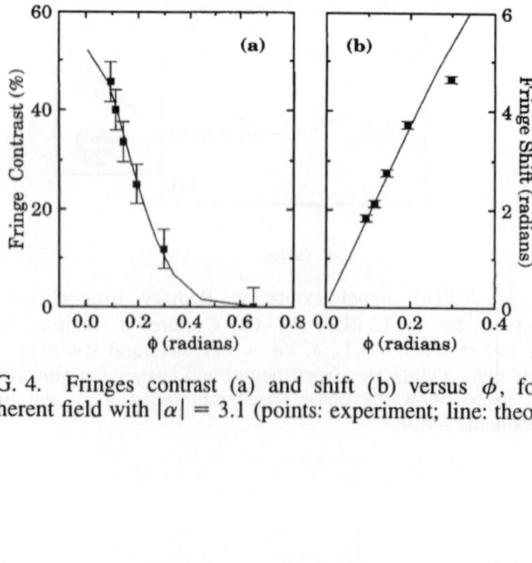

FIG. 4. Fringes contrast (a) and shift (b) versus ϕ, for a coherent field with $|\alpha| = 3.1$ (points: experiment; line: theory).

The same analysis shows that the phase of the fringes [Fig. 4(b)] is shifted by an angle equal to the phase of $\langle \alpha e^{-i\phi} \mid \alpha e^{i\phi} \rangle$, $n \sin 2\phi$. The fringe phase shift is proportional to n [19], which is determined from this set of data. The line on Fig. 4(b) corresponds to the best fitted value $n = 9.5 \pm 0.2$.

The coherence between the two components of the state and its quantum decoherence were revealed by a subsequent two-atom correlation experiment, whose principle follows closely a proposal described in [20]. A first atom creates a superposition state involving two field components. A second "probe" atom crosses C with the same velocity after a short delay τ and dephases again the field by an angle $\pm\phi$. The two field components turn into three, with phases $\pm 2\phi$ and zero.

The "zero-phase" component is obtained via two different paths since the atoms may have crossed C either in the (e, g) or in the (g, e) configurations. The probe atom "undoes" the phase splitting produced by the first one and recombines partially the state components. Since the atomic states are mixed after C in R_2, the (e, g) and (g, e) paths are indistinguishable. As a result, there is an interference term in the joint probabilities $P_{ee}^{(2)}$, $P_{eg}^{(2)}$, $P_{ge}^{(2)}$, and $P_{gg}^{(2)}$ of detecting any of the four possible two-atom configurations. These probabilities can be calculated analytically, provided a few simplifying assumptions are made [21]. The difference between conditional probabilities $\eta = [P_{ee}^{(2)}/(P_{ee}^{(2)} + P_{eg}^{(2)})] - [P_{ge}^{(2)}/(P_{ge}^{(2)} + P_{gg}^{(2)})]$ is independent of ν, except around $\phi = 0$ and $\phi = \pi/2$. Equal to 0.5 at short times τ when the quantum coherence is fully preserved, η is shown to decay to 0 when the "first atom + field" system has evolved into a fully incoherent statistical mixture.

To measure η, the Rydberg state preparation pulse is replaced by a pair of pulses separated by τ, varied from 30 to 250 μs. The sequence is, as before, repeated every 1.5 ms and statistics on double detection events are accumulated. Because of the low atom flux, the atom pair rate is 10 times smaller than the single atom count rate. For each delay τ, 15 000 coincidences are detected. Figure 5(a) shows η versus ν for $n = 3.3$, $\delta/2\pi =$ 70 kHz, and $\tau = 40$ μs. As predicted, a correlation signal with no statistically significant ν dependence is observed. A ν-averaged η value, $\overline{\eta}$, of 0.11 ± 0.01 is found for this τ value.

Figure 5(b) shows $\overline{\eta}$ versus τ (expressed in units of T_r) for $n = 3.3$ and two different detunings ($\delta/2\pi =$ 170 and 70 kHz). The points are experimental and the lines theoretical. The theory includes higher order terms in Ω/δ correcting the dephasings at $\delta = 70$ kHz, and incorporates the finite single atom fringe contrast (explaining the $\overline{\eta}$ value smaller than 0.5 at $\tau = 0$).

The correlation signals decrease with time, revealing directly the dynamics of quantum decoherence. The agreement with the simple analytical model is excellent.

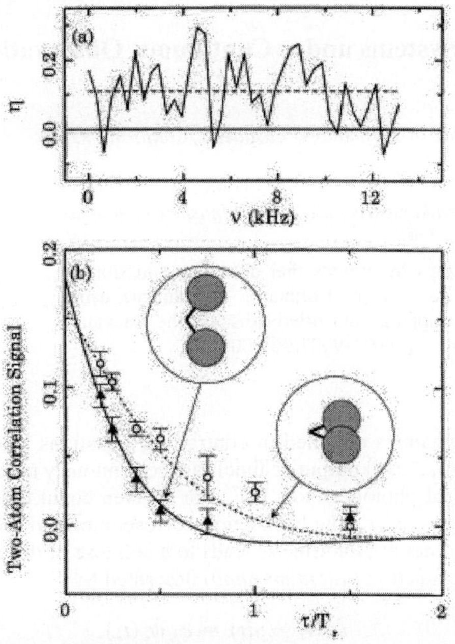

FIG. 5. (a) Two-atom correlation signal η versus ν for $n = 3.3$, $\delta/2\pi = 70$ kHz, and $\tau = 40$ μs. (b) ν-averaged η values versus τ/T_r for $\delta/2\pi = 170$ kHz (circles) and $\delta/2\pi = 70$ kHz (triangles). Dashed and solid lines are theoretical. Insets: pictorial representations of corresponding field components separated by 2ϕ.

Most strikingly, we observe that decoherence proceeds at a faster rate when the distance between the two state components is increased. An effective decoherence time of $0.24T_r$, much shorter than the photon decay time, is found for $\delta = 70$ kHz. A similar agreement with theory is obtained when comparing for the same $\delta/2\pi$ value (70 kHz) the correlations signals corresponding to different n values (5.1 and 3.3). We thus demonstrate the basic features of the decoherence theory on this simple model, namely, the fast evolution in a measurement process of the "atom + meter" state towards a statistical mixture and the increasing difficulty to maintain quantum coherence when the distance between the components of the mesoscopic superposition is increased. Using higher Q cavities, we intend to increase n further and to study decoherence processes occurring even faster on the scale of T_r. We can now continuously vary, from microscopic to macroscopic, the size of the meter in an ideal measurement process, allowing us to explore the elusive boundary between the quantum and classical worlds.

We thank P. Goy for assistance with microwave technology and AB Millimetre for the loan of equipment. This work was supported in part by EEC (Grant No. ERBCHRXCT930114).

*Laboratoire de l'Université Pierre et Marie Curie et de l'ENS, associé au CNRS (URA18).

[1] J.A. Wheeler and W.H. Zurek, *Quantum Theory of Measurement* (Princeton Univ. Press, Princeton, NJ, 1983).
[2] E. Schrödinger, Naturwissenschaften **23**, 807, 823, 844 (1935); reprinted in English in [1].
[3] J. von Neumann, in *Matematische Grundlagen der Quantenmechanik* (Springer, Berlin, 1932); reprinted in English in [1].
[4] W.H. Zurek, Phys. Today **44**, No. 10, 36 (1991).
[5] W.H. Zurek, Phys. Rev. D **24**, 1516 (1981); **26**, 1862 (1982); A.O. Caldeira and A.J. Leggett, Physica (Amsterdam) **121A**, 587 (1983); E. Joos and H.D. Zeh, Z. Phys. B **59**, 223 (1985); R. Omnès, The Interpretation of Quantum Mechanics (Princeton University Press, Princeton, NJ, 1994).
[6] R.J. Glauber, Phys. Rev. **131**, 2766 (1963).
[7] D.F. Waals and G.J. Milburn, Phys. Rev. A **31**, 2403 (1985).
[8] M. Brune *et al.*, Phys. Rev. A **45**, 5193 (1992).
[9] C. Monroe *et al.*, Science **272**, 1131 (1996).
[10] B. Yurke and D. Stoler, Phys. Rev. Lett. **57**, 13 (1986); B. Yurke *et al.*, Phys. Rev. A **42**, 1703 (1990); G. Milburn, *ibid.* **33**, 674 (1986); V. Buzek *et al.*, Phys. Rev. A **45**, 8190 (1992).
[11] C.N. Savage *et al.*, Opt. Lett. **15**, 628 (1990).
[12] R.G. Hulet and D. Kleppner, Phys. Rev. Lett. **51**, 1430 (1983).
[13] S. Haroche and J.M. Raimond, in *Cavity Quantum Electrodynamics*, edited by P. Berman (Academic Press, New York, 1994), p. 123.
[14] M. Brune *et al.*, Phys. Rev. Lett. **76**, 1800 (1996).
[15] P. Nussenzveig *et al.*, Phys. Rev. A **48**, 3991 (1993).
[16] N.F. Ramsey, *Molecular Beams* (Oxford Univ. Press, New York, 1985).
[17] M.O. Scully *et al.*, Nature (London) **351**, 111 (1991); S. Haroche *et al.*, Appl. Phys. B **54**, 355 (1992).
[18] T. Pfau *et al.*, Phys. Rev. Lett. **73**, 1223 (1994); M.S. Chapman *et al.*, Phys. Rev. Lett. **75**, 3783 (1995).
[19] M. Brune *et al.*, Phys. Rev. Lett. **72**, 3339 (1994).
[20] L. Davidovich *et al.*, Phys. Rev. A **53**, 1295 (1996).
[21] The calculation— to be published—generalizes to arbitrary ϕ values the derivation of [20], which was restricted to $\phi = \pi/2$. It describes independently atom-field interaction and relaxation and neglects fluctuations of atomic velocity as well as of atom number (possibility of two atoms in each preparation pulse).

VOLUME 76, NUMBER 17 PHYSICAL REVIEW LETTERS 22 APRIL 1996

Inversion of Quantum Jumps in Quantum Optical Systems under Continuous Observation

H. Mabuchi
Norman Bridge Laboratory of Physics 12-33, California Institute of Technology, Pasadena, California 91125

P. Zoller
Institut für Theoretische Physik, Universität Innsbruck, Technikerstrasse 25, A-6020 Innsbruck, Austria
(Received 11 January 1996)

We formulate conditions for invertibility of quantum jumps in systems that decay by emission of quanta into a continuously monitored reservoir. We propose proof-of-principle experiments using techniques from cavity quantum electrodynamics and ion trapping, and briefly discuss the relevance of such methods for error correction in quantum computation. [S0031-9007(96)00057-9]

PACS numbers: 42.50.Lc, 42.50.–p, 89.70.+c

Many current investigations of fundamental quantum phenomena would benefit greatly from the implementation of methods to stabilize quantum states against noise and dissipation [1]. For example, the realizability of quantum computers [2] seems to depend critically on development of robust techniques for preserving the coherence of quantum memory elements. In this Letter we shall describe a scheme for *inversion of quantum jumps* which, under ideal experimental conditions, makes possible the complete preservation of quantum coherences *within a subspace of initial states* for specially constructed systems in quantum optics. In the context of quantum computation, our scheme provides a means for dissipation-free storage of quantum bits (qubits).

Decoherence and decay of a quantum optical system may be viewed as the result of weak coupling between the system of interest and a reservoir of electromagnetic field modes whose correlation time is much shorter than the time scale set by system dynamics [1,3]. Under the assumption of vanishing correlation time (Markov approximation), one typically traces over reservoir states in the global equations of motion to derive a master equation that describes evolution of the reduced density operator $\hat{\rho}$ for the system alone. The master equation for $j = \{1, \ldots, d\}$ decay channels is ($\hbar = 1$)

$$\frac{d}{dt}\hat{\rho} = -i[\hat{H}, \hat{\rho}] + \sum_{j=1}^{d}\left(\hat{c}_j\hat{\rho}\hat{c}_j^\dagger - \frac{1}{2}\hat{c}_j^\dagger\hat{c}_j\hat{\rho} - \hat{\rho}\frac{1}{2}\hat{c}_j^\dagger\hat{c}_j\right), \quad (1)$$

where $\hat{H}$ is the system Hamiltonian and $\{\hat{c}_j\}$ are the system operators that appear in the system-reservoir coupling. Such a master equation will generally map pure states of the system into statistical mixtures, reflecting the decoherence which results from loss of information into the unobserved reservoir modes.

Indeed, by tracing over the reservoir state to derive (1), one implicitly and essentially assumes that no measurements are ever performed on the reservoir. Much recent work in quantum optics has investigated the *a posteriori* dynamics obtained in contrasting situations where all $j = \{1, \ldots, d\}$ output channels are continuously monitored by ideal photodetectors [4]. For a given count trajectory $j_1, t_1, \ldots, j_n, t_n$, the backaction corresponding to observation of count j_r at time t_r leads to a collapse of the system wave function *(quantum jump)* described by

$$\tilde{\psi}_c(t_r + dt) = \hat{c}_{j_r}\tilde{\psi}_c(t_r). \quad (2)$$

Here $\hat{c}_j$ denotes the system *jump operator* corresponding to counts in channel j, while $\hat{H}_{\rm eff} = \hat{H} - i\frac{1}{2}\sum_j \hat{c}_j^\dagger\hat{c}_j$ is an effective non-Hermitian Hamiltonian. Between counts, the system wave function obeys a Schrödinger equation

$$\tilde{\psi}_c(t) = e^{-i\hat{H}_{\rm eff}(t-t_r)}\tilde{\psi}_c(t_r). \quad (3)$$

This quantum-jump picture of dissipative dynamics underlies the recently developed "quantum trajectories" method for Monte Carlo integration of quantum optical master equations [5]. Starting from a known initial (pure) state, count trajectories $j_1, t_1, \ldots, j_n, t_n$ may be generated by taking the probability density for a jump to occur at time t to be $\|\hat{c}_j\psi_c(t)\|^2$. Using the *a posteriori* evolution rules described above, the system wave function at time t is then given by the normalized state vector $\psi_c(t) = \tilde{\psi}_c(t)/\|\tilde{\psi}_c(t)\|$. For a physical system in which the count trajectories are not actually detected and recorded, one can average over Monte Carlo wave functions to recover a system density operator $\hat{\rho} = \langle|\psi_c\rangle\langle\psi_c|\rangle$ which obeys the quantum master equation (1). However, for a laboratory setup which *actually* incorporates complete and continuous photodetection, the individual trajectories and *a posteriori* dynamics may be interpreted (with some caution) as reflecting the "real" dynamics *of single quantum realizations.* This principle motivates our scheme for the preservation of quantum coherence in dissipative systems—if discrete quantum jumps constitute the entire noncoherent component of Markovian dynamics, one can indeed hope to suppress decoherence by performing appropriate operations to invert the operation $\hat{c}_{j_r}$ every time a count of type j_r is recorded. In general, however, a quantum jump

 0031-9007/96/76(17)/3108(4)$10.00

(2) will destroy superpositions so that information is irreversibly lost. Hence $\hat{c}_j$ will not necessarily be invertible on the entire system Hilbert space $\mathcal{H}$.

Let us therefore formulate conditions under which a quantum jump (2) can be inverted for a system initial state $\tilde{\psi}_c(t)$ which is *known* to lie within a certain *subspace* $\mathcal{H}_s \subset \mathcal{H}$ of the system Hilbert space. We are particularly interested in the case where a detected quantum jump $\tilde{\psi}_c \in \mathcal{H}_s \to \tilde{\psi}_c' = \hat{c}_j \tilde{\psi}_c \in \mathcal{H}_s^{(j)}$ can be inverted using feedback [6] described by a unitary time evolution operator $\hat{U}_j$, so that $\tilde{\psi}_c(t_r + dt) = \hat{U}_j \hat{c}_{j_r} \tilde{\psi}_c(t_r) \propto \tilde{\psi}_c(t_r)$. Thus, as a first condition (A), we require

$$\hat{c}_j = k_j \hat{U}_j^{-1}|_{\mathcal{H}_s \to \mathcal{H}_s^{(j)}} \quad (k_j \in C); \tag{4}$$

together with the inverse relation $\hat{c}_j^\dagger = k_j^* \hat{U}_j|_{\mathcal{H}_s^{(j)} \to \mathcal{H}_s}$, i.e., for the mapping $\hat{c}_j$: $\mathcal{H}_s \to \mathcal{H}_s^{(j)}$ there exists a unitary extension $\hat{U}_j$ to the whole Hilbert space [7] which can be generated by an appropriate feedback Hamiltonian. The feedback is assumed to be instantaneous on the time scale of the system dynamics. Equation (4) implies $\hat{c}_j^\dagger \hat{c}_j = |k_j|^2 \mathbb{1}|_{\mathcal{H}_s \to \mathcal{H}_s}$. If we add the requirement (B) that the system Hamiltonian $\hat{H}$ leaves the subspace of interest $\mathcal{H}_s$ invariant, the system dynamics between two quantum jumps is governed by

$$\begin{aligned}\tilde{\psi}_c(t) &= e^{-i\hat{H}_{\text{eff}} t} \tilde{\psi} \\ &= e^{-1/2 \sum_j |k_j|^2 t} e^{-i\hat{H} t} \tilde{\psi} \quad [\tilde{\psi}_c(t) \in \mathcal{H}_s]\end{aligned}$$

so that the damping terms factor out and thus do not distort the system dynamics between jumps. Furthermore, if each decay is detected and is followed by a feedback $\hat{U}_j$ to "undo" the effect of the quantum jump, we have essentially eliminated the effects of decoherence on system states in the subspace $\mathcal{H}_s$:

$$\begin{aligned}\psi_c(t) &= e^{-i\hat{H}_{\text{eff}}(t-t_n)} \hat{U}_{j_n} \hat{c}_{j_n} \cdots \hat{U}_{j_1} \hat{c}_{j_1} e^{-i\hat{H}_{\text{eff}} t_1} \psi / \| \cdots \| \\ &= e^{-i\hat{H} t} \psi,\end{aligned} \tag{5}$$

where $\| \cdots \|$ denotes normalization of the state.

For the derivation of Eq. (5) to be valid, it is essential that the system dynamics conform to the model of a quantum Markov process [1]. The underlying physical assumption is a separation of time scales where the correlation time τ_c of the environment is much shorter than all time scales characterizing the system evolution, including, in particular, the system decay time [3]. This separation admits the treatment of system dynamics with "coarse-grained" time resolution, and it is only on coarse timescales ($\gg \tau_c$) that the system wave function appears to evolve according to a non-Hermitian Hamiltonian (3) with stochastic, "instantaneous" quantum jumps [Eq. (2)]. Likewise, it is only on coarse time scales that coherence can be preserved in the system according to Eq. (5), while the state of the environment will not (and need not) be restored at all in the present scheme. We wish to further stress that quantum optical systems are *known* to realize quantum Markovian models to an excellent approximation, so we fully expect our conclusions drawn from this assumption to be directly applicable to *realistic* experimental systems.

Significantly, the type of jump-inversion procedure described above seems to be realizable with familiar experimental techniques in several systems of current interest in quantum optics. Our first example utilizes recent ideas from the field of cavity quantum electrodynamics (CQED) [8]. Consider the apparatus shown in Fig. 1, in which the output modes of two identical single-sided Fabry-Pérot resonators are mixed by a 50/50 beam splitter before impinging upon photon-counting detectors. We assume that the high-reflector (HR) mirror of each resonator is perfect, and that the output couplers (OC) have no scattering or absorption losses but have some small transmissivity $t > 0$. The beam splitter is likewise assumed to be lossless, and we treat the photodetectors as having unit quantum efficiency. Note that we are not invoking any sort of Zeno effect, so that the time resolution of the detectors is taken as being very short compared to the cavity decay times but long compared to the optical time scale $\sim 1/\omega_0$ (ω_0 being the optical frequency of the resonator modes). We assume a separation of time scales in which all operations described below can be performed in a time much less than the cavity decay times, which we assume to be equal.

Let $\hat{a}$ and $\hat{b}$ be the annihilation operators for the optical modes of cavities **a** and **b**, respectively. The master equation for the resonator modes may be written

$$\dot{\hat{\rho}} = -i(\hat{H}_{\text{eff}} \hat{\rho} - \hat{\rho} \hat{H}_{\text{eff}}^\dagger) + \Gamma(\hat{a} \hat{\rho} \hat{a}^\dagger + \hat{b} \hat{\rho} \hat{b}^\dagger), \tag{6}$$

with $\hat{H}_{\text{eff}} = (\omega_0 - i\frac{1}{2}\Gamma)(\hat{a}^\dagger \hat{a} + \hat{b}^\dagger \hat{b})$ the effective Hamiltonian. We identify $\hat{a}$ and $\hat{b}$ as the jump operators

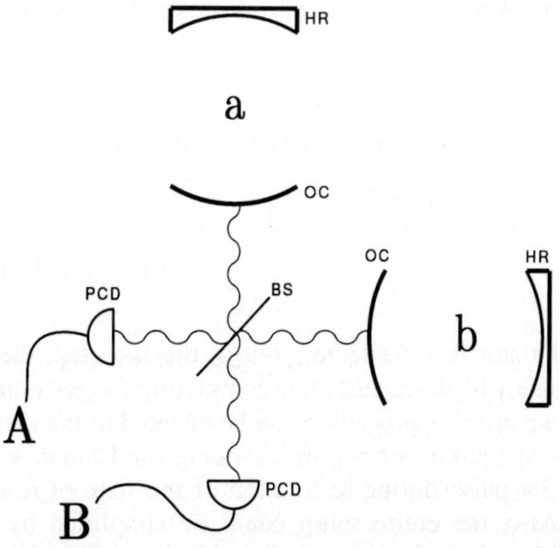

FIG. 1. Schematic of a cavity-QED gedanken experiment. Components are labeled (see text) HR, high-reflector mirrors; OC, output-coupler mirrors; BS, 50/50 beam splitter; and PCD, photon-counting detectors.

associated with the detection of dissipative events in cavities **a** and **b**. In order to account for the mixed-output measurement scheme of Fig. 1, we must make a basis transformation by defining

$$\hat{A} = \frac{1}{\sqrt{2}}(\hat{a} + \hat{b}), \qquad \hat{B} = \frac{1}{\sqrt{2}}(\hat{a} - \hat{b}), \qquad (7)$$

which we interpret as jump operators corresponding to the registration of photons by detectors **A** and **B**. In terms of the new jump operators, Eq. (6) becomes

$$\dot{\hat{\rho}} = -i(\hat{H}_{\rm eff}\hat{\rho} - \hat{\rho}\hat{H}_{\rm eff}^{\dagger}) + \Gamma(\hat{A}\hat{\rho}\hat{A}^{\dagger} + \hat{B}\hat{\rho}\hat{B}^{\dagger}), \quad (8)$$

where $\hat{H}_{\rm eff} = (\omega_0 - i\frac{1}{2}\Gamma)(\hat{A}^{\dagger}\hat{A} + \hat{B}^{\dagger}\hat{B})$ remains invariant under the change of basis.

In this example we consider stabilization of the subspace spanned by the Fock states $|0\rangle_L \equiv |2_a 0_b\rangle$ and $|1\rangle_L \equiv |0_a 2_b\rangle$, representing a logical zero and one, respectively. Let the initial state of the two-cavity system be given by

$$|\psi\rangle = c_0|2_a 0_b\rangle + c_1|0_a 2_b\rangle \equiv c_0|0\rangle_L + c_1|1\rangle_L. \quad (9)$$

We first note that states of this form are stationary under the time evolution (3), since $|2_a 0_b\rangle$ and $|0_a 2_b\rangle$ are degenerate eigenstates of $\hat{H}_{\rm eff}$. Therefore the superposition (9) remains unchanged during periods of time in which no photons are detected. When photodetection events do occur, the postjump state $|\psi_c\rangle$ will be either

$$\begin{aligned} \hat{A}|\psi\rangle &= c_0|1_a 0_b\rangle + c_1|0_a 1_b\rangle, \\ \hat{B}|\psi\rangle &= c_0|1_a 0_b\rangle - c_1|0_a 1_b\rangle. \end{aligned} \quad (10)$$

As both coefficients (c_0, c_1) survive in either case, and remain attached to orthogonal state vectors, the original state $|\psi\rangle$ may, in principle, be fully restored by the application of the appropriate feedback operator $\hat{U}_A$ or $\hat{U}_B$. Note that one knows which of these to apply based upon which detector registered the photon. The inverse jump operators correspond to the doubling of the photon number in both resonators $(0 \to 0, 1 \to 2)$, with or without a phase change of π in resonator **b**. Since we must employ only coherent processes for the doubling operation, the states of resonators **a** and **b** may be independently manipulated without compromising the entanglement between them. This allows us to consider the simplified task of "independently" doubling the photon number in each resonator.

We now proceed to give an explicit example of a process to achieve this photon-number doubling. Our proposal employs adiabatic state-mapping techniques described in [9], by which one can "swap" the state of a resonator field with the internal Zeeman state of an atom. Consider an atom having an angular momentum $J \to J - 1$ transition $(J > 1)$ with frequency ω_0, prepared in the $|g_{m_J=-J}\rangle$ ground state as depicted in Fig. 2. If we wish to invert an $\hat{A}$-type jump, the combined state of the atom plus resonator fields will initially be

$$|\Psi\rangle = |g_{-J}\rangle\hat{A}|\psi\rangle = |g_{-J}\rangle(c_0|1_a 0_b\rangle + c_1|0_a 1_b\rangle). \quad (11)$$

After performing adiabatic state mapping [assuming the resonator mode has σ_+ polarization, see Fig. 2(a)],

$$|\Psi\rangle \to (c_0|g_{-J+1}\rangle|0_b\rangle + c_1|g_{-J}\rangle|1_b\rangle)|0_a\rangle. \quad (12)$$

We can now effect the photon-number doubling for resonator **a** by applying a Raman π pulse to the atom, with π- and σ_--polarized lasers having frequency $\omega_0 - \delta$. The detuning δ should be chosen large enough to eliminate any possibility of populating the excited atomic state. After the π pulse, we have [Fig. 2(b)]

$$|\Psi\rangle \to (c_0|g_{-J+2}\rangle|0_b\rangle + c_1|g_{-J}\rangle|1_b\rangle)|0_a\rangle. \quad (13)$$

Note that polarization selection rules prevent the $|-J\rangle$ atomic state from coupling to the specified Raman fields. With a final "reverse" application of the state-mapping procedure [Fig. 2(c)], the total state of the system becomes

$$|\Psi\rangle \to |g_{-J}\rangle(c_0|2_a 0_b\rangle + c_1|0_a 1_b\rangle). \quad (14)$$

Thus the photon-number doubling has been accomplished for the first resonator. An analogous procedure for resonator **b** completes the process, with the sequence of intermediate states given by

$$\begin{aligned} |g_{-J}\rangle(c_0|2_a 0_b\rangle + c_1|0_a 1_b\rangle) &\to (c_0|g_{-J}\rangle|2_a\rangle + c_1|g_{-J+1}\rangle|0_a\rangle)|0_b\rangle \to (c_0|g_{-J}\rangle|2_a\rangle + c_1|g_{-J+2}\rangle|0_a\rangle)|0_b\rangle \\ &\to |g_{-J}\rangle(c_0|2_a 0_b\rangle + c_1|0_a 2_b\rangle) \equiv |g_{-J}\rangle|\psi\rangle. \end{aligned} \quad (15)$$

As the atomic state factors out in the last step, the atom can safely be discarded after completing the restoration.

Note that this procedure can be adapted to the inversion of $\hat{B}$-type jumps simply by changing the Raman π pulse to a 3π pulse during restoration of the state of resonator **b**. Also, the entire setup could be simplified by using two optical modes of opposite circular polarization in a single Fabry-Pérot cavity. The appropriate observation basis would then be photon counting with discrimination of the *linear* polarization of the leaking photons.

Our second example could be implemented using trapped ions [10]. Consider an ion having a $J_g = \frac{1}{2} \to J_e = \frac{3}{2}$ optical transition, with the initial state

$$|\psi\rangle = c_0|e_{-3/2}\rangle + c_1|e_{3/2}\rangle \equiv c_0|0\rangle_L + c_1|1\rangle_L. \quad (16)$$

The decay channels for this initial state are ordinary spontaneous emission, with $|e_{-3/2}\rangle \mapsto |g_{-1/2}\rangle$ producing a σ_--polarized photon and $|e_{3/2}\rangle \mapsto |g_{1/2}\rangle$ producing a σ_+-polarized photon. Although polarization-preserving imaging of the entire dipole emission pattern would be

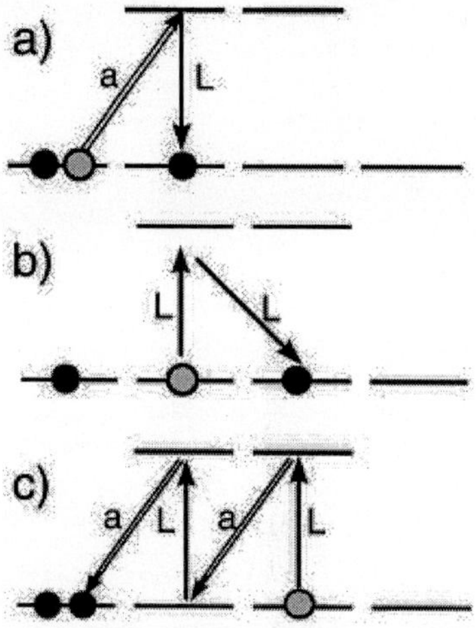

FIG. 2. Level diagram of a $J_g = 3/2 \rightarrow J_e = 1/2$ Zeeman atom used for photon-number doubling: (a) adiabatic state mapping via a Raman process according to Eq. (12), (b) doubling (13), and (c) mapping the Zeeman superposition back to cavity field (14) (σ_+-cavity mode a, laser L).

experimentally difficult, let us imagine for the moment that the decay photons can be detected with perfect efficiency after the circular-polarization modes are mixed by a linearly polarizing beam splitter. The jump operators for the system are then

$$\hat{x}, \hat{y} = \frac{1}{\sqrt{2}} (|g_{-1/2}\rangle\langle e_{-3/2}| \pm i|g_{1/2}\rangle\langle e_{3/2}|), \qquad (17)$$

where the operator $\hat{x}$ is associated with the detection of an x-polarized photon and $\hat{y}$ with the detection of a y-polarized photon. The associated reset operations can be achieved by simple π pulsing on the $\pm\frac{1}{2} \rightarrow \mp\frac{3}{2}$ transitions, as long as measures are taken to avoid the unwanted but degenerate transitions $\pm\frac{1}{2} \rightarrow \mp\frac{1}{2}$. This degeneracy could be lifted by selectively shifting the $|e_{\pm 1/2}\rangle$ states, for example, by applying π-polarized ac Stark fields on a transition to an auxiliary atomic level with $J_{e'} = 1/2$. In this scenario, the coherent restoration of superposition (16) could be verified by Ramsey-interferometry techniques. A proof-of-principle demonstration could be performed even with low photodetection efficiency by selecting the subensemble of events in which the decay photon is successfully detected [11].

The problem of storing and manipulating entangled atomic and photon states has lately attracted considerable attention within the context of recent proposals for implementing quantum computation and quantum cryptography [2]. In a quantum computer, *quantum registers* are defined as product states of L (logical) qubits, and the general state is an entangle state of these product states. We note that state restoration by inversion of quantum jumps is also possible in such a composite system. If the subsystems (the individual qubits) are coupled to independent reservoirs, a detected decay of one of the qubits can be restored by single bit operation. Finally, we remark that the present scheme complements a recent proposal by Shor [12] on quantum error correction via redundant coding. In contrast to the Shor proposal the present scheme involves no overhead of stored and manipulated qubits, but, on the other hand, incorporates a specific quantum optical model for damping (which must be reliably known to apply to the system in question). Whereas Shor's protocol may be viewed as having quite general applicability, our scheme benefits from its context of well-established models for dissipation in concrete physical systems. In addition, we have shown recently that the methods proposed in the present paper can be extended to provide an error correction procedure for quantum *gates* [13].

H. M. is supported by a National Defense Science and Engineering Graduate Fellowship. This work was supported in part by the Austrian Science Foundation. Discussions with A. Barenco, T. Beth, R. Blatt, J. I. Cirac, and H. J. Kimble are acknowledged.

[1] C. W. Gardiner, *Quantum Noise* (Springer, Berlin, 1991).
[2] For an overview, see A. Ekert, in *Proceedings of the 14th ICAP,* edited by D. Wineland *et al.* (AIP Press, New York, 1995), p. 450.
[3] In a rotating frame where the optical frequencies have been eliminated.
[4] A. Barchielli and V. P. Belavkin, J. Phys. A **24**, 1495 (1991), and references therein.
[5] H. J. Carmichael, in *An Open Systems Approach to Quantum Optics* (Springer, Berlin, 1993); J. Dalibard *et al.,* Phys. Rev. Lett. **68**, 580 (1992); C. W. Gardiner, A. S. Parkins, and P. Zoller, Phys. Rev. A **46**, 4363 (1992).
[6] An operational definition of feedback in a quantum trajectory picture was first given by H. M. Wiseman and G. J. Milburn, Phys. Rev. Lett. **70**, 548 (1993); H. M. Wiseman, Ph.D. thesis, University of Queensland, 1994.
[7] Equation (4) implies that an orthonormal basis (ONB) of $\mathcal{H}$ is mapped by $\hat{c}_j$ to an ONB of $\mathcal{H}_s^{(j)}$.
[8] See, for example, *Cavity Quantum Electrodynamics,* edited by P. R. Berman (Academic, San Diego, 1994).
[9] A. S. Parkins *et al.,* Phys. Rev. Lett. **71**, 3095 (1993).
[10] D. J. Wineland *et al.,* Phys. Rev. A **50**, 67 (1994); R. Blatt, in *Proceedings of the 14th ICAP* (Ref. [2]).
[11] This example illustrates clearly the relation and differences to the *quantum eraser* as discussed in T. J. Herzog *et al.,* Phys. Rev. Lett. **75**, 3034 (1995); P. G. Kwiat *et al.,* Phys. Rev. A **49**, 61 (1994); P. G. Kwiat *et al.,* Phys. Rev. A **45**, 7729 (1992). In the quantum eraser, *interference* is restored by selection of an appropriate subensemble "without path information," while in our scheme we completely restore the original *state.*
[12] P. W. Shor, Phys. Rev. A **52**, R2493 (1995).
[13] J. I. Cirac, T. Pellizzari, and P. Zoller (unpublished); for recent experiments on quantum gates, see Q. A. Turchette *et al.,* Phys. Rev. Lett. **75**, 4710 (1995); C. Monroe *et al.,* Phys. Rev. Lett. **75**, 4714 (1995).

Quantum Computation with Ion Traps

Quantum Computation with Ion Traps

Rainer Blatt
Institut für Experimentalphysik, Universität Innsbruck
Wolfgang Lange
Max-Planck-Institut für Quantenoptik

The experimental implementation of a quantum computer requires coherent control of the dynamics of a physical system and at the same time isolation from the decohering influence of the environment. Crystals of laser-cooled ions in a radio-frequency trap almost ideally meet these requirements [1, 8, 9, 10, 11, 12]. For quantum information processing, a linear trap geometry [2, 7] is most suitable. As in a quadrupole massfilter, ions are strongly confined in the radial direction by the effective potential of a transverse rf-field. Axial confinement is provided by an additional static potential well in the longitudinal direction. For smaller numbers of ions, other trap geometries are also possible [13, 14].

An important application of such a trap is ultra-high precision spectroscopy [15, 16, 17, 18]. In order to eliminate Doppler-broadening of the transition frequencies, laser-cooling of the ions' motion is applied [19, 20, 21]. At sufficiently low temperatures, the particles take fixed positions in a crystal-like structure [22, 23]. If the radial confinement is stronger than the axial one, the ions become arranged in a linear chain along the trap axis at distances determined by the equilibrium of their mutual Coulomb repulsion and the static axial potential [3, 4]. For realistic trap parameters, the spacing is on the order of tens of microns. This is large enough to allow individual addressing with laser beams.

Quantum information may be stored in individual ions using two internal (electronic) states [24] to realize one qubit. To sustain coherence in the course of an extended computation, the radiative lifetime of both levels should be sufficiently long. Possible choices are the ground state and a metastable state, two metastable states, two hyperfine components of the ground state, or two Zeeman substates of the ground state, all of which are stable against electric dipole decay. Suitable candidates include $^9Be^+$, $^{25}Mg^+$ and $^{40}Ca^+$.

In order to process quantum information in a well-defined way, the electronic quantum state of each single ion in the string has to be carefully prepared and modified. This is achieved with optical techniques, manipulating arbitrary ions in the chain by addressing them individually with a laser beam. For example, to coherently modify the contents of a single ionic quantum register (single-qubit rotations), laser induced Rabi-cycling between the qubit states is applied. Depending on the level scheme used, either two-photon Raman transitions or single-photon transitions are employed.

To read out the result at the end of a calculation, the state of the qubit-register must be determined. With ion traps, this can be achieved with nearly 100% detection efficiency by exciting the ions on a fast transition, coupled to only one of the qubit basis states, and detecting the emitted fluorescent light [25, 26, 27, 28]. Thus the presence or absence of scattered light from a given ion indicates which basis state is occupied.

The operations described so far manipulate single qubits independently from each other. In order to perform non-trivial quantum computations, logic gates between two ions in the chain must be implemented so that the state of a given ion can condition the dynamic evolution of the state of another ion. Due to their large separation, any direct interaction which depends on the internal states of neighboring ions is too weak to provide a useful coupling. Thus a mediator is needed to exchange quantum information between different qubits (quantum data bus).

According to a proposal by Cirac and Zoller [1], the required coupling between ions may be provided by the quantized vibration of the ion chain in the external potential. In their scheme, the strong Coulomb repulsion provides the necessary interaction between ions at different sites.

In an N-ion linear crystal, there are N normal modes of vibration along the axis. Each vibrational mode corresponds to a highly correlated harmonic oscillation of the ions around their equilibrium positions [4, 29] with a characteristic eigenfrequency. Any normal mode can be selectively excited or de-excited through laser interaction with a single ion in the chain by driving an internal atomic transition on a vibrational sideband at the desired eigenfrequency.

Cirac and Zoller [1] have suggested the center-of-mass (COM) mode, in which all the ions oscillate in phase along the trap axis, to transfer quantum information between the ions. However, higher order modes may also be used which may be less sensitive to external perturbations [30].

The logical coupling between two widely separated ions in the Cirac-Zoller scheme via a vibrational mode proceeds in three steps. Initially, the electronic state of the first ion is mapped to the state of the vibrational mode by means of a laser-induced sideband transition in this ion. A well-defined processing of quantum information requires that the oscillator associated with a vibrational mode, is initially in its quantum mechanical ground state (see below).

In the second step, the actual conditional dynamics is achieved by addressing the second ion with a laser, which changes its internal state depending on the state of the vibrational mode. A CNOT-gate for a single ion, controlled by its quantized vibrational state, has been demonstrated in a recent experiment [5]. Alternative schemes have been suggested as well [31]. Finally, the initial step is reversed and the quantum information stored in the vibrational mode is mapped back to the state of the first ion. In this process the vibrational mode is reset to the ground state, and thus prepared for another gate sequence.

One of the main technological challenges in ion trap quantum computation is reaching the required quantum mechanical ground state of vibration. A special cooling technique called "resolved sideband cooling" [32, 33] must be applied for this purpose. Since the line width of electric dipole transitions is much larger than the vibrational sideband-splitting, either electric dipole forbidden lines or Raman transitions are usually employed. The method has been successfully demonstrated for single trapped ions [34, 35] and, recently, for a string of two ions [30].

The degree of quantum control that can be achieved in a coupled system of internal and vibrational degrees of freedom was demonstrated in experiments on the generation of non-classical states of motion of a single trapped ion [36, 37, 38]. Recently, the attention has turned to systems of several ions with well controlled interactions between them, in order to realize extended quantum registers. As a first result, the preparation of an entangled state of two ions was accomplished [6].

The Cirac-Zoller scheme is not the only possibility of achieving dynamics which are conditioned on the internal states of different ions. A logic gate between two ions can also be realized by using two-photon transitions, addressing both of the ions simultaneously and only virtually exciting the vibrational degrees of freedom [39, 40]. In yet another proposal, strong coupling to an optical cavity mode is employed to entangle the internal states of the ions [41]. For two ions, their state-dependent recoil may be used to achieve a splitting of their spatial wave function, resulting in entanglement of position and internal state of the ions. Local laser excitation is then sufficient to generate a conditional evolution [42]. What all of these schemes have in common is that gates are realized as a well-defined series of laser pulses addressing different ions in the string.

In a realistic ion trap quantum computer, the phenomenon of decoherence limits the size of the quantum register and the length of calculations that may be implemented. Early experiments on quantum control of internal and external degrees of freedom of the ions have shown that decoherence is an important issue [38]. For ion traps, two types of decoherence have to be considered [9, 10]: decoherence of the internal levels and decoherence of the vibrational motion For hyperfine- and metastable states, radiative decay rates are negligible [44], so that the coherence time of internal superposition states is limited by uncontrolled magnetic field fluctuations and collisions with background gas in the vacuum chamber.

Another restriction is the decoherence of the vibrational states of the ion string [46, 47]. For a single $^{198}Hg^{+}$- ion, a transition out of the zero-point vibrational level occurred in 0.15 s [34], while in the case of $^{9}Be^{+}$ a lifetime of 1 ms [35, 37] was measured. At present, these processes put an upper limit to the number of operations that may be performed with a quantum computer before coherence is lost. Additional problems compromising the fidelity of a quantum calculation are inaccurate settings of the system parameters [48, 49]. In future realizations, errors must be taken care of by the implementation of error correcting codes and protocols [45, 50, 43].

From a long-term perspective, ion traps offer the benefit that there is no fundamental limit to the number of ions in a chain and thus to the number of qubits that can be stored in the system. Thus ion traps that are scaled up to provide long qubit registers, are the most promising candidates for realizing complex networks of quantum gates as well as schemes for quantum error correction.

There are prospects for combining the computational power of ion traps with the fiber optical technology already employed for long distance quantum communication. The quantum interface between the trapped ions and the photonic channels may be provided by a high finesse optical cavity [51, 52, 53].

In this way, ion traps are an important tool for storing, manipulating and distributing quantum information and provide unique opportunities for studying highly entangled quantum systems [18, 39, 54, 55].

References

[1] J. I. Cirac and P. Zoller, *Quantum computations with cold trapped ions*, Phys. Rev. Lett. **74**, 4091 (1995).

[2] J. D. Prestage, G. J. Dick, and L. Maleki, *New ion trap for frequency standard applications*, J. Appl. Phys. **66**, 1013 (1989).

[3] H. C. Nägerl, W. Bechter, J. Eschner, F. Schmidt-Kaler, and R. Blatt, *Ion strings for quantum gates*, Appl. Phys. B-Lasers Opt. **66**, 603 (1998).

[4] D. F. V. James, *Quantum dynamics of cold trapped ions with application to quantum computation*, Appl. Phys. B-Lasers Opt. **66**, 181 (1998).

[5] C. Monroe, D. M. Meekhof, B. E. King, W. M. Itano, and D. J. Wineland, *Demonstration of a fundamental quantum logic gate*, Phys. Rev. Lett. **75**, 4714 (1995).

[6] Q. A. Turchette, C. S. Wood, B. E. King, C. J. Myatt, D. Leibfried, W. M. Itano, C. Monroe, and D. J. Wineland, *Deterministic entanglement of two trapped ions*, Phys. Rev. Lett. **81**, 3631 (1998).

[7] P. K. Gosh, *Ion Traps* (Clarendon, Oxford, 1995).

[8] A. Steane, *The ion trap quantum information processor*, Appl. Phys. B-Lasers Opt. **64**, 623 (1997).

[9] D. J. Wineland, C. Monroe, W. M. Itano, D. Leibfried, B. E. King, and D. Meekhof, *Experimental issues in coherent quantum-state manipulation of trapped atomic ions*, J. Res. Natl. Inst. Stand. Technol. **103**, 259 (1998), LANL e-print quant-ph/9710025.

[10] D. J. Wineland, C. Monroe, W. M. Itano, B. E. King, D. Leibfried, D. M. Meekhof, C. Myatt, and C. Wood, *Experimental primer on the trapped ion quantum computer*, Fortschritte Phys.-Prog. Phys. **46**, 363 (1998).

[11] V. Vedral and M. B. Plenio, *Basics of quantum computation*, Prog. Quantum Electron. **22**, 1 (1998).

[12] R. J. Hughes, D. F. V. James, J. J. Gomez, M. S. Gulley, M. H. Holzscheiter, P. G. Kwiat, S. K. Lamoreaux, C. G. Peterson, V. D. Sandberg, M. M. Schauer, C. M. Simmons, C. E. Thorburn, D. Tupa, P. Z. Wang, and A. G. White, *The Los Alamos trapped ion quantum computer experiment*, Fortschritte Phys.-Prog. Phys. **46**, 329 (1998).

[13] S. R. Jefferts, C. Monroe, E. W. Bell, and D. J. Wineland, *Coaxial-resonator-driven rf (Paul) trap for strong confinement*, Phys. Rev. A **51**, 3112 (1995).

[14] R. G. DeVoe, *Elliptical ion traps and trap arrays for quantum computation*, Phys. Rev. A **58**, 910 (1998).

[15] D. J. Wineland, W. M. Itano, and R. S. Vandyck, *High-resolution spectroscopy of stored ions*, Adv. At. Mol. Phys. **19**, 135 (1983).

[16] D. J. Wineland, W. M. Itano, J. C. Bergquist, and R. G. Hulet, *Laser-cooling limits and single-ion spectroscopy*, Phys. Rev. A **36**, 2220 (1987).

[17] R. Blatt, in *Atomic Physics*, edited by D. J. Wineland, C. E. Wieman, and S. J. Smith (AIP, New York, 1995), Vol. 14, p. 219.

[18] J. J. Bollinger, W. M. Itano, D. J. Wineland, and D. J. Heinzen, *Optimal frequency measurements with maximally correlated states*, Phys. Rev. A **54**, R4649 (1996).

[19] D. J. Wineland, R. E. Drullinger, and F. L. Walls, *Radiation-pressure cooling of bound resonant absorbers*, Phys. Rev. Lett. **40**, 1639 (1978).

[20] D. J. Wineland and W. M. Itano, *Laser cooling of atoms*, Phys. Rev. A **20**, 1521 (1979).

[21] S. Stenholm, *The semiclassical theory of laser cooling*, Rev. Mod. Phys. **58**, 699 (1986).

[22] I. Waki, S. Kassner, G. Birkl, and H. Walther, *Observation of ordered structures of laser-cooled ions in a quadrupole storage ring*, Phys. Rev. Lett. **68**, 2007 (1992).

[23] G. Birkl, S. Kassner, and H. Walther, *Multiple-shell structures of laser-cooled* $^{24}Mg^{+}$ *ions in a quadrupole storage ring*, Nature **357**, 310 (1992).

[24] W. M. Itano, J. C. Bergquist, J. J. Bollinger, J. M. Gilligan, D. J. Heinzen, F. L. Moore, M. G. Raizen, and D. J. Wineland, *Quantum projection noise - population fluctuations in 2-level systems*, Phys. Rev. A **47**, 3554 (1993).

[25] W. Nagourney, J. Sandberg, and H. Dehmelt, *Shelved optical electron amplifier - observation of quantum jumps*, Phys. Rev. Lett. **56**, 2797 (1986).

[26] J. C. Bergquist, R. G. Hulet, W. M. Itano, and D. J. Wineland, *Observation of quantum jumps in a single atom*, Phys. Rev. Lett. **57**, 1699 (1986).

[27] T. Sauter, W. Neuhauser, R. Blatt, and P. E. Toschek, *Observation of quantum jumps*, Phys. Rev. Lett. **57**, 1696 (1986).

[28] D. Stevens, J. Brochard, and A. M. Steane, *Simple experimental methods for trapped-ion quantum processors*, Phys. Rev. A **58**, 2750 (1998).

[29] H. C. Nägerl, D. Leibfried, F. Schmidt-Kaler, J. Eschner, and R. Blatt, *Coherent excitation of normal modes in a string of Ca+ ions*, Opt. Express **3**, 89 (1998).

[30] B. E. King, C. S. Wood, C. J. Myatt, Q. A. Turchette, D. Leibfried, W. M. Itano, C. Monroe, and D. J. Wineland, *Cooling the collective motion of trapped ions to initialize a quantum register*, Phys. Rev. Lett. **81**, 1525 (1998).

[31] C. Monroe, D. Leibfried, B. E. King, D. M. Meekhof, W. M. Itano, and D. J. Wineland, *Simplified quantum logic with trapped ions*, Phys. Rev. A **55**, R2489 (1997).

[32] D. Wineland and H. Dehmelt, *Proposed 1014 delta upsilon less than upsilon laser fluorescence spectroscopy on Tl+ mono-ion oscillator III*, Bull. Amer. Phys. Soc. **20**, 637 (1975).

[33] G. Morigi, J. I. Cirac, M. Lewenstein, and P. Zoller, *Ground-state laser cooling beyond the Lamb-Dicke limit*, Europhys. Lett. **39**, 13 (1997).

[34] F. Diedrich, J. C. Bergquist, W. M. Itano, and D. J. Wineland, *Laser cooling to the zero-point energy of motion*, Phys. Rev. Lett. **62**, 403 (1989).

[35] C. Monroe, D. M. Meekhof, B. E. King, S. R. Jefferts, W. M. Itano, D. J. Wineland, and P. Gould, *Resolved-side-band raman cooling of a bound atom to the 3D zero-point energy*, Phys. Rev. Lett. **75**, 4011 (1995).

[36] J. I. Cirac, A. S. Parkins, R. Blatt, and P. Zoller, *Nonclassical states of motion in ion traps*, Adv. At. Molec. Opt. Physics **37**, 237 (1996).

[37] D. M. Meekhof, C. Monroe, B. E. King, W. M. Itano, and D. J. Wineland, *Generation of nonclassical motional states of a trapped atom*, Phys. Rev. Lett. **76**, 1796 (1996).

[38] C. Monroe, D. M. Meekhof, B. E. King, and D. J. Wineland, *A "Schrödinger cat" superposition state of an atom*, Science **272**, 1131 (1996).

[39] K. Mølmer and A. Sørensen, *Multiparticle entanglement of hot trapped ions*, Phys. Rev. Lett. **82**, 1835 (1999).

[40] A. Sørensen and K. Mølmer, *Quantum computation with ions in thermal motion*, Phys. Rev. Lett. **82**, 1971 (1999).

[41] T. Pellizzari, S. A. Gardiner, J. I. Cirac, and P. Zoller, *Decoherence, continuous observation, and quantum computing - a cavity QED model*, Phys. Rev. Lett. **75**, 3788 (1995).

[42] J. F. Poyatos, J. I. Cirac, and P. Zoller, *Quantum gates with "hot" trapped ions*, Phys. Rev. Lett. **81**, 1322 (1998).

[43] M. B. Plenio, V. Vedral, and P. L. Knight, *Quantum error correction in the presence of spontaneous emission*, Phys. Rev. A **55**, 67 (1997).

[44] D. J. Wineland, J. J. Bollinger, W. M. Itano, and D. J. Heinzen, *Squeezed atomic states and projection noise in spectroscopy*, Phys. Rev. A **50**, 67 (1994).

[45] J. I. Cirac, T. Pellizzari, and P. Zoller, *Enforcing coherent evolution in dissipative quantum dynamics*, Science **273**, 1207 (1996).

[46] A. Garg, *Decoherence in ion trap quantum computers*, Phys. Rev. Lett. **77**, 964 (1996).

[47] D. F. V. James, *Theory of heating of the quantum ground state of trapped ions*, Phys. Rev. Lett. **81**, 317 (1998).

[48] R. J. Hughes, D. F. V. James, E. H. Knill, R. Laflamme, and A. G. Petschek, *Decoherence bounds on quantum computation with trapped ions*, Phys. Rev. Lett. **77**, 3240 (1996).

[49] S. Schneider and G. J. Milburn, *Decoherence in ion traps due to laser intensity and phase fluctuations*, Phys. Rev. A **57**, 3748 (1998).

[50] C. Dhelon and G. J. Milburn, *Correcting the effects of spontaneous emission on cold-trapped ions*, Phys. Rev. A **56**, 640 (1997).

[51] J. I. Cirac, P. Zoller, H. J. Kimble, and H. Mabuchi, *Quantum state transfer and entanglement distribution among distant nodes in a quantum network*, Phys. Rev. Lett. **78**, 3221 (1997).

[52] S. J. van Enk, J. I. Cirac, and P. Zoller, *Ideal quantum communication over noisy channels: A quantum optical implementation*, Phys. Rev. Lett. **78**, 4293 (1997).

[53] T. Pellizzari, *Quantum networking with optical fibres*, Phys. Rev. Lett. **79**, 5242 (1997).

[54] J. Steinbach and C. C. Gerry, *Efficient scheme for the deterministic maximal entanglement of N trapped ions*, Phys. Rev. Lett. **81**, 5528 (1998).

[55] D. J. Wineland, C. Monroe, W. M. Itano, B. E. King, D. Leibfried, C. Myatt, and C. Wood, *Trapped-ion quantum simulator*, Phys. Scr. **T76**, 147 (1998).

Electromagnetic traps for charged and neutral particles*

Wolfgang Paul

Physikalisches Institut, Universität Bonn, Bonn, Germany

Experimental physics is the art of observing the structure of matter and of detecting the dynamic processes within it. But in order to understand the extremely complicated behavior of natural processes as an interplay of a few constituents governed by as few as possible fundamental forces and laws, one has to measure the properties of the relevant constituents and their interaction as precisely as possible. And as all processes in nature are interwoven, one must separate and study them individually. It is the skill of the experimentalist to carry out clear experiments in order to get answers to his questions undisturbed by undesired effects, and it is his ingenuity to improve the art of measuring to ever higher precision. There are many examples in physics showing that higher precision revealed new phenomena, inspired new ideas, or confirmed or dethroned well-established theories. On the other hand, new experimental techniques conceived to answer special questions in one field of physics became very fruitful in other fields, too, be it in chemistry, biology, or engineering. In awarding the Nobel prize to my colleagues Norman Ramsey, Hans Dehmelt, and me for new experimental methods, the Swedish Academy indicates her appreciation for the aphorism the Göttingen physicist Georg Christoph Lichtenberg wrote two hundred years ago in his notebook "one has to do something new in order to see something new." On the same page Lichtenberg said: "I think it is a sad situation in all our chemistry that we are unable to suspend the constituents of matter free."

Today the subject of my lecture will be the suspension of such constituents of matter or, in other words, about traps for free charged and neutral particles without material walls. Such traps permit the observation of isolated particles, even of a single one, over a long period of time and therefore according to Heisenberg's uncertainty principle enable us to measure their properties with extremely high accuracy.

In particular, the possibility to observe individual trapped particles opens up a new dimension in atomic measurements. Until a few years ago all measurements were performed on an ensemble of particles. Therefore the measured value—for example, the transition probability between two eigenstates of an atom—is a value averaged over many particles. Tacitly one assumes that all atoms show exactly the same statistical behavior if one attributes the result to the single atom. On a trapped single atom, however, one can observe its interaction with a radiation field and its own statistical behavior alone.

The idea of building traps grew out of molecular-beam physics, mass spectrometry, and particle accelerator physics I was involved in during the first decade of my career as a physicist more than 30 years ago. In these years (1950–55) we had learned that plane electric and magnetic multipole fields are able to focus particles in two dimensions acting on the magnetic or electric dipole moment of the particles. Lenses for atomic and molecular beams (Friedburg and Paul, 1951; Bennewitz and Paul, 1954, 1955) were conceived and realized, improving considerably the molecular-beam method for spectroscopy or for state selection. The lenses found application as well to the ammonia as to the hydrogen maser (Townes, 1983).

The question "What happens if one injects charged particles, ions or electrons, in such multipole fields" led to the development of the linear quadrupole mass spectrometer. It employs not only the focusing and defocusing forces of a high-frequency electric quadrupole field acting on ions, but also exploits the stability properties of their equations of motion in analogy to the principle of strong focusing for accelerators which had just been conceived.

If one extends the rules of two-dimensional focusing to three dimensions, one possesses all ingredients for particle traps.

As already mentioned the physics or the particle dynamics in such focusing devices is very closely related to that of accelerators or storage rings for nuclear or particle physics. In fact, multipole fields were used in molecular-beam physics first. But the two fields have complementary goals; the storage of particles, even of a single one, of extremely low energy down to the micro-electron-volt region on the one side and of as many as possible of extremely high energy on the other. Today we will deal with the low-energy part.

At first I will talk about the physics of dynamic stabilization of ions in two- and three-dimensional radio-frequency quadrupole fields, the quadrupole mass spectrometer, and the ion trap. In a second part I shall report on trapping of neutral particles with emphasis on an experiment with magnetically stored neutrons.

As in most cases in physics, especially in experimental physics, the achievements are not the achievements of a single person, even if he contributed in posing the problems and some basic ideas in solving them. All the exper-

*This lecture was delivered 8 December 1989, on the occasion of the presentation of the 1989 Nobel Prize in Physics.

iments I am awarded for were done together with research students or young colleagues in mutual inspiration. In particular, I have to mention H. Friedburg and H. G. Bennewitz, C. H. Schlier and P. Toschek in the field of molecular-beam physics, and in conceiving and realizing the linear quadrupole spectrometer and the rf ion trap H. Steinwedel, O. Osberghaus, and especially the late Erhard Fischer. Later H. P. Reinhard, U. Zahn, and F. V. Busch played an important role in developing this field.

What are the principles of focusing and trapping particles? Particles are elastically bound to an axis or a coordinate in space if a binding force acts on them which increases linearly with their distance r

$$F=-cr\ ;$$

in other words, if they move in a parabolic potential

$$\Phi\sim(\alpha x^2+\beta y^2+\gamma z^2)\ .$$

The tools appropriate to generate such fields of force to bind charged particles or neutrals with a dipole moment are electric or magnetic multipole fields. In such configurations the field strength, or the potential, respectively, increases according to a power law and shows the desired symmetry. Generally if m is the number of "poles" or the order of symmetry the potential is given by

$$\Phi\sim r^{m/2}\cos\left[\frac{m}{2}\varphi\right]\ .$$

For a quadrupole $m=4$, it gives $\Phi\sim r^2\cos2\varphi$; and for a sextupole $m=6$, one gets $\Phi\sim r^3\cos3\varphi$ corresponding to a field strength increasing with r and r^2, respectively.

Trapping of charged particles in two- and three-dimensional quadrupole fields

In the electric quadrupole field the potential is quadratic in the Cartesian coordinates,

$$\Phi=\frac{\Phi_0}{2r_0^2}(\alpha x^2+\beta y^2+\gamma z^2)\ . \tag{1}$$

The Laplace condition $\Delta\Phi=0$ imposes the condition $\alpha+\beta+\gamma=0$. There are two simple ways to satisfy this condition.

(a) $\alpha=1=-\gamma$, $\beta=0$ results in the two-dimensional field

$$\Phi=\frac{\Phi_0}{2r_0^2}(x^2-z^2)\ , \tag{2}$$

(b) $\alpha=\beta=1$, $\gamma=-2$ generates the three-dimensional configuration, in cylindrical coordinates

$$\Phi=\frac{\phi_0(r^2-2z^2)}{r_0^2+2z_0^2}\ \text{with}\ 2z_0^2=r_0^2\ . \tag{3}$$

The two-dimensional quadrupole or the mass filter

Configuration (a) is generated by four hyperbolically shaped electrodes linearly extended in the y-direction as is shown in Fig. 1. The potential on the electrodes is $\pm\Phi_0/2$ if one applies the voltage Φ_0 between the electrode pairs. The field strength is given by

$$E_x=-\frac{\Phi_0}{r_0^2}x,\ E_z=\frac{\Phi_0}{r_0^2}z,\ E_y=0\ .$$

If one injects ions in the y direction, it is obvious that for a constant voltage Φ_0 the ions will perform harmonic oscillations in the x-y plane; but due to the opposite sign in the field E_z, their amplitude in the z direction will increase exponentially. The particles are defocused and will be lost by hitting the electrodes.

This behavior can be avoided if the applied voltage is periodic. Due to the periodic change of the sign of the electric force, one gets focusing and defocusing in both the x and z directions alternating in time. If the applied voltage is given by a dc voltage U plus an rf voltage V with the driving frequency ω

$$\Phi_0=U+V\cos\omega t\ ,$$

the equations of motion are

$$\begin{aligned}\ddot{x}+\frac{e}{mr_0^2}(U+V\cos\omega t)x&=0\ ,\\ \ddot{z}-\frac{e}{mr_0^2}(U+V\cos\omega t)z&=0\ .\end{aligned} \tag{4}$$

At first sight one expects that the time-dependent term of the force cancels out in the time average. But this would be true only in a homogeneous field. In a periodic inhomogeneous field, like the quadrupole field, there is a small average force left, which is always in the direction of the lower field, in our case toward the center. Therefore, certain conditions exist that enable the ions to traverse the quadrupole field without hitting the electrodes; i.e., their motion around the y axis is stable with limited amplitudes in x and z directions. We learned these rules from the theory of the Mathieu equations, as this type of differential equation is called.

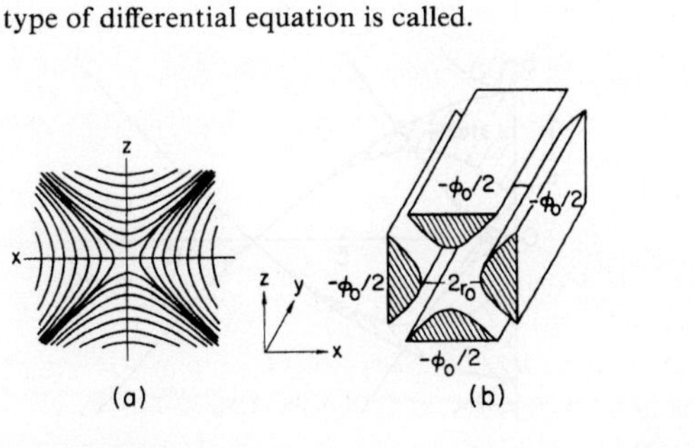

FIG. 1. (a) Equipotential lines for a plane quadrupole field. (b) The electrode structure for the mass filter.

In dimensionless parameters these equations are written

$$\frac{d^2x}{d\tau^2}+(a+2q\cos 2\tau)x=0\ ,$$
$$\frac{d^2z}{d\tau^2}-(a+2q\cos 2\tau)z=0\ . \quad (5)$$

By comparison with Eq. (4) one gets

$$a=\frac{4eU}{mr_0^2\omega^2},\quad q=\frac{2eV}{mr_0^2\omega^2},\quad \tau=\frac{\omega t}{2}\ . \quad (6)$$

The Mathieu equation has two types of solution.

(1) Stable motion: The particles oscillate in the x- z plane with limited amplitudes. They pass the quadrupole field in the y direction without hitting the electrodes.

(2) Unstable motion: The amplitudes grow exponentially in x,z, or in both directions. The particles will be lost.

Whether stability exists depends only on the parameters a and q and not on the initial parameters of the ion motion, e.g., their velocity. Therefore, in a a-q map there are regions of stability and instability (Fig. 2). Only the overlapping region for x and z stability is of interest for our problem. The most relevant region $0<a,q<1$ is plotted in Fig. 3. The motion is stable in x and z only within the triangle.

For fixed values r_0, ω, U, and V, all ions with the same m/e have the same operating point in the stability diagram. Since a/q is equal to $2U/V$ and does not depend on m, all masses lie along the operating line $a/q=$const. On the q axis ($a=0$, no dc voltage) one has stability from $0<q<q_{\max}=0.92$ with the consequence that all masses between $\infty>m>m_{\min}$ have stable orbits. In this case the quadrupole field works on as a high-pass mass filter. The mass range Δm becomes narrower with increasing dc voltage U, i.e., with a steeper operating line and approaches $\Delta m=0$, if the line goes through the tip of the stability region. The bandwidth in this case is given only by the fluctuation of the field parameters. If one changes U and V simultaneously and proportionally in such a way that a/q remains constant, one brings the ions of the various masses successively in the stability region scanning through the mass spectrum in this way. Thus the quadrupole works as a mass spectrometer (Paul and Steinwedel, 1953a, 1953b; Paul and Raether, 1955).

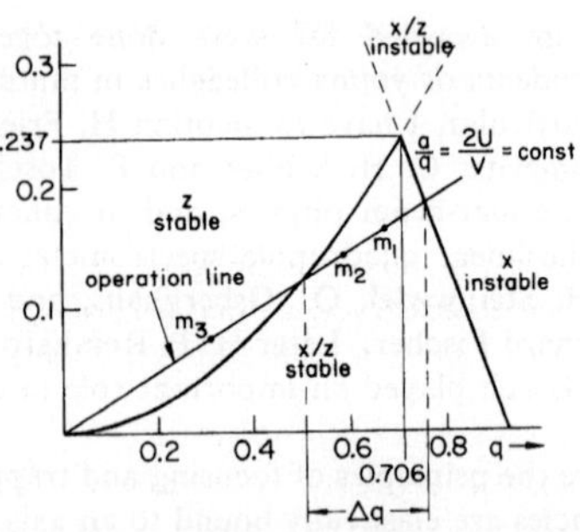

FIG. 3. The lowest region for simultaneous stability in x and z directions. All ion masses lie on the operation line. $m_2>m_1$.

A schematic view of such a mass spectrometer is given in Fig. 4. In Figs. 5(a) and 5(b) the first mass spectra obtained in 1954 are shown (Paul and Raether, 1955). Clearly one sees the influence of the dc voltage U on the resolving power.

In quite a number of theses the performance and application of such instruments was investigated at Bonn University (Paul, Reinhardt, and von Zahn, 1958; von Busch and Paul, 1961; von Zahn, 1962). We studied the influence of geometrical and electrical imperfections giving rise to higher multipole terms in the field. A very long instrument ($l=6$ m) for high-precision mass measurements was built achieving an accuracy of 2×10^{-7} in determining mass ratios at a resolving power $m/\Delta m=16\,000$. Very small ones were used in rockets to measure atomic abundances in the high atmosphere. In another experiment we succeeded in separating isotopes in amounts of milligrams using a resonance method to shake single masses out of an intense ion beam guided in the quadrupole.

In recent decades the rf quadrupole, whether as mass spectrometer or beam guide due to its versatility and technical simplicity, has found broad applications in many fields of science and technology. It became a kind

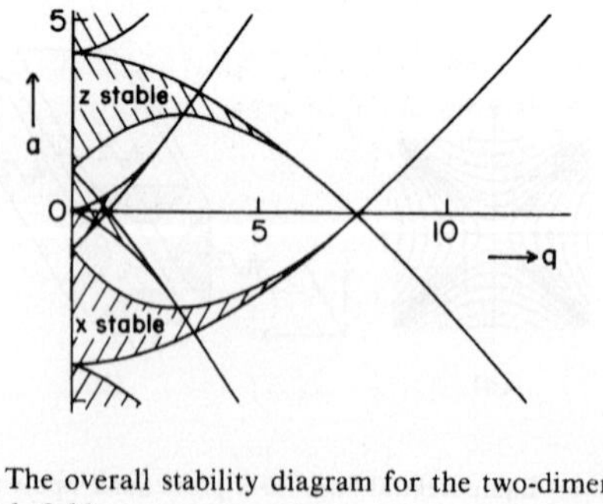

FIG. 2. The overall stability diagram for the two-dimensional quadrupole field.

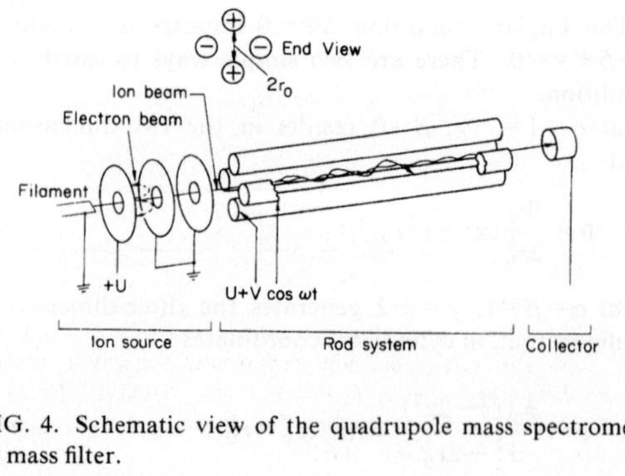

FIG. 4. Schematic view of the quadrupole mass spectrometer or mass filter.

of standard instrument and its properties were treated extensively in the literature (Dawson, 1976).

The ion trap

Already at the very beginning of our thinking about the dynamic stabilization of ions we were aware of the possibility of using it for trapping ions in a three-dimensional field. We called such a device "Ionenkäfig." Nowadays the word "ion trap" is preferred (Berkling, 1956; Paul, Osberghaus, and Fischer, 1958; Fischer, 1959).

The potential configuration in the ion trap has been given in Eq. (3). This configuration is generated by an hyperbolically shaped ring and two hyperbolic rotationally symmetric caps as it is shown schematically in Fig. 6(a). Figure 6(b) gives the view of the first realized trap in 1954.

If one brings ions into the trap, which is easily achieved by ionizing inside a low-pressure gas by electrons passing through the volume, they perform the same forced motions as in the two-dimensional case. The only difference is that the field in z direction is stronger by a factor 2. Again a periodic field is needed for the stabilization of the orbits. If the voltage $\Phi_0 = U + V\cos\omega t$ is applied between the caps and the ring electrode, the equations of motion are represented by the same Mathieu function of Eq. (5). The relevant parameters for the r motion correspond to those in the x direction in the plane field case. Only the z parameters are changed by a factor 2.

Accordingly, the region of stability in the a-q map for the trap has a different shape, as is shown in Fig. 7. Again the mass range of the storable ions (i.e., ions in the stable region) can be chosen by the slope of the operation line $a/q = 2U/V$. Starting with operating parameters in the tip of the stable region, one can trap ions of a single mass number. By lowering the dc voltage one brings the ions near the q axis where their motions are much more stable.

For many applications it is necessary to know the frequency spectrum of the oscillating ions. From mathematics we learn that the motion of the ions can be described as a slow (secular) oscillation with the fundamental frequencies $\omega_{r,z} = \beta_{r,z}\omega/2$ modulated with a micromotion, a much faster oscillation of the driving frequency ω, if one neglects higher harmonics. The frequency determining factor β is a function only of the Mathieu parameters a and q and therefore mass dependent. Its

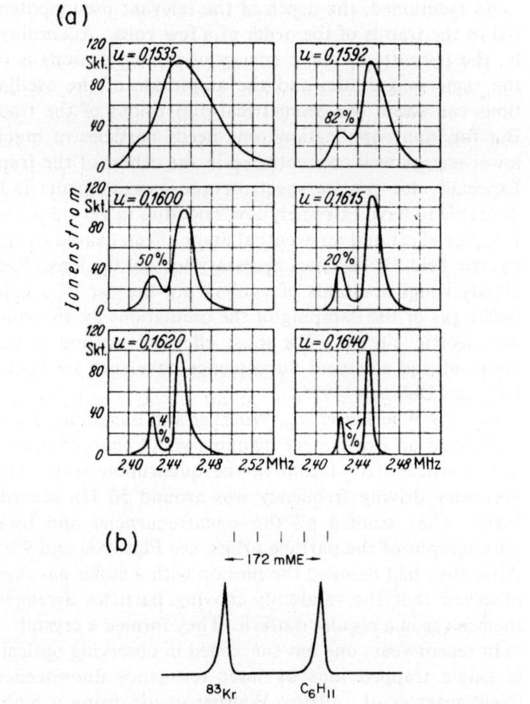

FIG. 5. (a) Very first mass spectrum of rubidium. Mass scanning was achieved by periodic variation of the driving frequency ν. Parameter: $u = U/V$, at $u = 0.164$ ^{85}Rb and ^{87}Rb are fully resolved. (b) Mass doublet ^{83}Kr-C_6H_{11} Resolving power $m/\delta m = 6500$ (von Zahn, 1962).

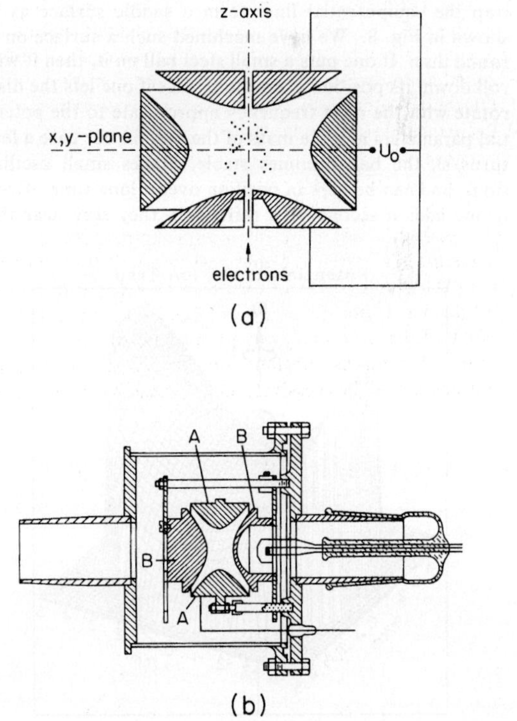

FIG. 6. (a) Schematic view of the ion trap. (b) Cross section of the first trap (1955).

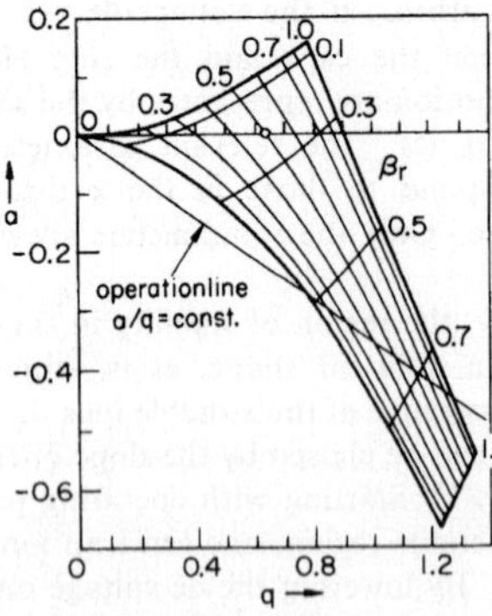

FIG. 7. The lowest region for stability in the ion trap. On the lines inside the stability region β_z and β_r are constant.

value varies between 0 and 1; lines of equal β are drawn in Fig. 7.

Due to the stronger field the frequency ω_z of the secular motion becomes twice ω_r. The ratio ω/ω_z is a criterion for the stability. Ratios of 10:1 are easily achieved and therefore the displacement by the micromotion averages out over a period of the secular motion.

The dynamic stabilization in the trap can easily be demonstrated in a mechanical analogue device. In the trap the equipotential lines form a saddle surface as is shown in Fig. 8. We have machined such a surface on a round disc. If one puts a small steel ball on it, then it will roll down; its position is unstable. But if one lets the disk rotate with the right frequency appropriate to the potential parameters and the mass of the ball (in our case a few turns/s), the ball becomes stable, makes small oscillations, and can be kept in position over a long time. Even if one adds a second or a third ball, they stay near the

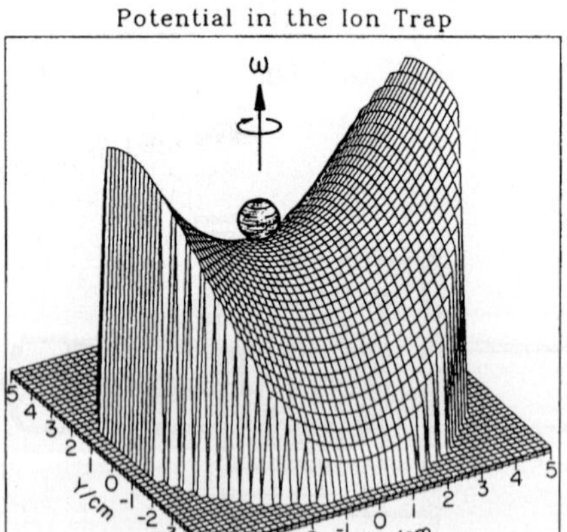

FIG. 8. Mechanical analogue model for the ion trap with steel-ball as "particle."

center of the disc. The only condition is that the related Mathieu parameter q be in the permitted range. I brought the device with me. It is made out of Plexiglas, which allows demonstration of the particle motions with the overhead projector.

This behavior gives us a hint of the physics of the dynamic stabilization. The ions oscillating in the r and z directions to first approximation harmonically, behave as if they are moving in a pseudopotential well quadratic in the coordinates. From their frequencies ω_r and ω_z we can calculate the depth of this well for both directions. It is related to the amplitude V of the driving voltage and to the parameters a and q. Without any dc voltage the depth is given by $D_z=(q/8)V$; in the r direction it is half of this. As in practice V amounts to a few hundred volts; the potential depth is of the order of 10 volts. The width of the well is given by the geometric dimensions. The resulting configuration of the pseudopotential (Dehmelt, 1967) is therefore given by

$$\Phi=D\frac{r^2+4z^2}{r_0^2+2z_0^2}$$

Cooling process

As mentioned, the depth of the relevant pseudopotential in the trap is of the order of a few volts. Accordingly, the permitted kinetic energy of the stored ions is of the same magnitude, and the amplitude of the oscillations can reach the geometrical dimensions of the trap. But for many applications one needs particles of much lower energy well concentrated in the center of the trap. Especially for precise spectroscopic measurements it is desirable to have extremely low velocities to get rid of the Doppler effect and an eventual Stark effect, caused by the electric field. It becomes necessary to cool the ions. Relatively rough methods of cooling are the use of a cold buffer gas or the damping of the oscillations by an external electric circuit. The most effective method is the laser-induced sideband fluorescence developed by Wineland and Dehmelt (1975).

In 1959 Wuerker *et al.* (Wuerker and Langmuir, 1959) performed an experiment trapping small charged aluminum particles ($\phi \sim 1$ μm) in the quadrupole trap. The necessary driving frequency was around 50 Hz accordingly. They studied all the eigenfrequencies and took photographs of the particle orbits; see Figs. 9(a) and 9(b). After they had damped the motion with a buffer gas they observed that the randomly moving particles arranged themselves in a regular pattern. They formed a crystal

In recent years one has succeeded in observing optically single trapped ions by laser resonance fluorescence (Neuhauser *et al.*, 1980). Walther *et al.*, using a high-resolution image intensifier, observed the pseudocrystallization of ions in the trap after cooling the ions with laser light. The ions are moving to such positions where the repulsive Coulomb force is compensated by the focusing forces in the trap and the energy of the ensemble has a minimum. Figure 10(a) and 10(b) show such a pattern with seven ions. Their distance is of the order of a few micrometers. These observations opened a new field of research (Dietrich *et al.*, 1987).

The ion trap as mass spectrometer

As mentioned, the ions perform oscillations in the trap with frequencies ω_r and ω_z which at fixed field parameters are determined by the mass of the ion. This enables a mass selective detection of the stored ions. If one connects the cap electrodes with an active rf circuit with the eigenfrequency Ω, in the case of resonance $\Omega = \omega_z$ the amplitude of the oscillations increases linearly with time. The ions hit the cap or leave the field through a bore hole and can easily be detected by an electron multiplier device. By modulating the ion frequency determining voltage V in a sawtooth mode, one brings the ions of the various masses one after the other into resonance, scanning the mass spectrum. Figure 11 shows the first spectrum of this kind achieved by Rettinghaus (1967).

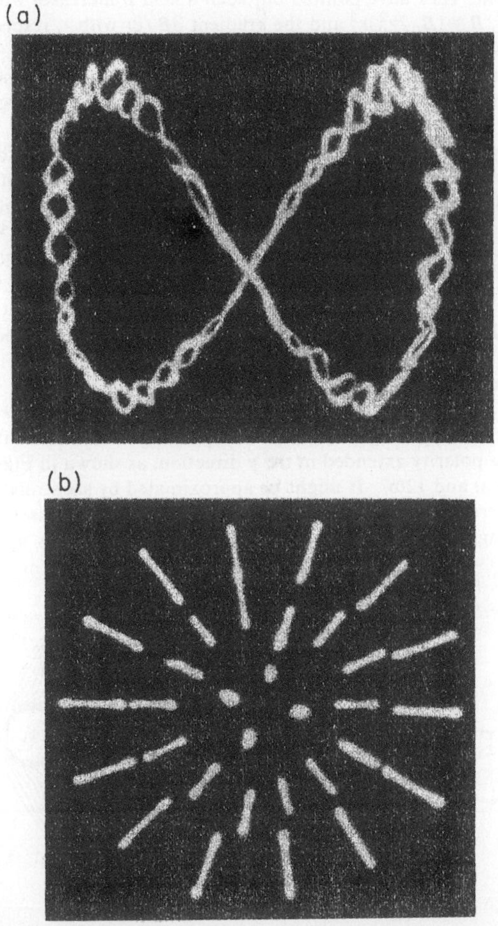

FIG. 9. (a) Photomicrograph of a Lissajous orbit in the r-z plane of a single charged particle of aluminum powder. The micromotion is visible. (b) Pattern of "condensed" Al particles (Wuerker and Langmuir, 1959).

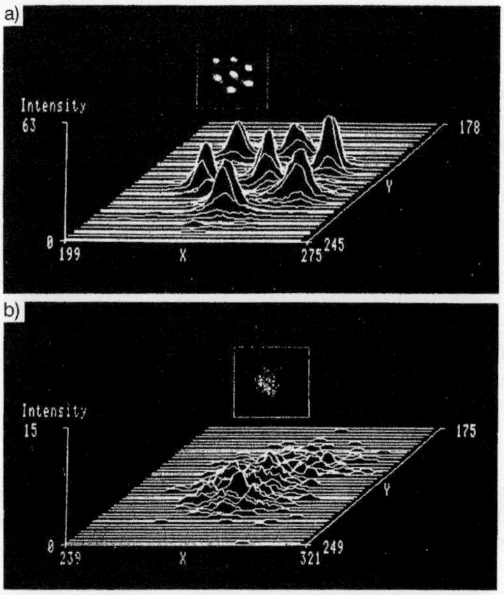

FIG. 10. (a) Pseudocrystal of seven magnesium ions. Particle distance 23 μm. (b) The same trapped particles at "higher temperature." The crystal has melted (Diedrich *et al.*, 1988).

The same effect with a faster increase of the amplitude is achieved if one inserts a small band of instability into the stability diagram. It can be generated by superimposing on the driving voltage $V \cos\omega t$ a small additional rf voltage, e.g., with frequency $\omega/2$, or by adding a higher multipole term to the potential configuration (Paul and Steinwedel, 1953b; von Busch and Paul, 1961a).

In summary the ion trap works as ion source and mass spectrometer at the same time. It became the most sensitive mass analyzer available, as only a few ions are necessary for detection. Its theory and performance are reviewed in detail by R. E. March (March and Hughes, 1989).

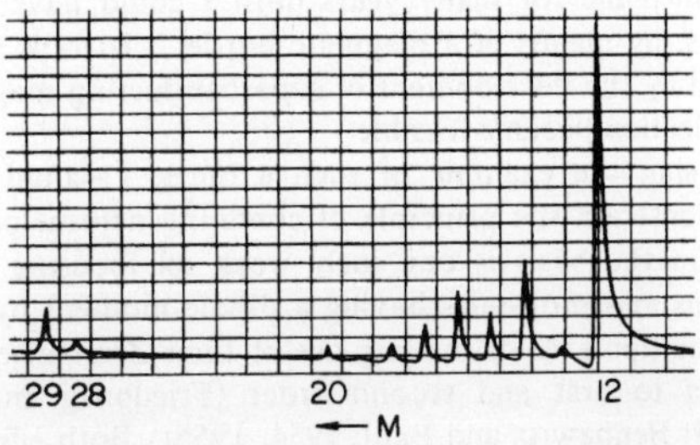

FIG. 11. First mass spectrum achieved with the ion trap. Gas: air at 2×10^{-9} Torr (Rettinghaus, 1967).

The Penning trap

If one applies to the quadrupole trap only a dc voltage in such a polarity that the ions perform stable oscillations in the z direction with the frequency $\omega_z^2=2eU/mr_0^2$, the ions are unstable in the x-y plane, since the field is directed outwards. Applying a magnetic field in the axial direction, the z motion remains unchanged but the ions perform a cyclotron motion ω in the x-y plane. It is generated by the Lorentz force F_L directed towards the center. This force is partially compensated by the radial electric force $F_r=eU\cdot r/r_0^2$. As long as the magnetic force is much larger than the electric one, stability exists in the x-y plane as well. No rf field is needed. The resulting rotation frequency calculates to

$$\omega=\omega_c-\frac{\omega_z^2}{2\omega} .$$

It is slightly smaller than the undisturbed cyclotron frequency eB/m. The difference is due to the magnetron frequency

$$\omega_M=\frac{\omega_z^2}{2\omega} ,$$

which is independent of the particle mass.

The Penning trap (Penning, 1936), as this device is called, is of advantage if magnetic properties of particles have to be measured, as, for example, Zeeman transitions in spectroscopic experiments, or cyclotron frequencies for a very precise comparison of masses as are performed, e.g., by G. Werth. The most spectacular application the trap has found in the experiments of G. Gräff (Gräff *et al.*, 1969) and H. Dehmelt for measuring the anomalous magnetic moment of the electron. It was brought by Dehmelt (van Dyck *et al.*, 1977) to an admirable precision by observing only a single electron stored for many months.

Traps for neutral particles

In the last examination I had to pass as a young man I was asked if it would be possible to confine neutrons in a bottle in order to prove if they are radioactive. This question, at that time only to be answered with "no," pursued me for many years until I could have replied: "Yes, by means of a magnetic bottle." It took 30 years until by the development of superconducting magnets its realization became feasible.

Using the example of such a bottle I would like to demonstrate the principle of confining neutral particles. Again the basis is our early work on focusing neutral atoms and molecules having a dipole moment by means of multipole fields making use of their Zeeman or Stark effect to first and second order (Friedburg and Paul, 1951; Bennewitz and Paul, 1954, 1955). Both effects can be used for trapping. Until now only magnetic traps were realized for atoms and neutrons. Particularly, B. Martin, U. Trinks, and K. J. Kügler contributed to their development with great enthusiasm.

The principle of magnetic bottles

The potential energy U of a particle with a permanent magnetic moment μ in a magnetic field is given by $U=-\mu B$. If the field is inhomogeneous, it corresponds to a force $F=\text{grad}(\mu B)$. In the case of the neutron with its spin $\hbar/2$, only two spin directions relative to the field are permitted. Therefore, its magnetic moment can be oriented only parallel or antiparallel to B. In the parallel position the particles are drawn into the field, and in the opposite orientation they are repelled. This permits their confinement to a volume with magnetic walls.

The appropriate field configuration to bind the particles harmonically is in this case a magnetic sextupole field. As I have pointed out such a field B increases with r^2, $B=(B_0/r_0^2)r^2$ and the gradient $\partial B/\partial r$ with r, respectively.

In such a field neutrons with orientation $\mu\uparrow\uparrow B$ satisfy the confining condition as their potential energy $U=+\mu B\sim r^2$ and the restoring force $\mu\,\text{grad}B=-cr$ is always oriented towards the center. They oscillate in the field with the frequency $\omega^2=2\mu B_0/mr_0^2$. Particles with $\mu\uparrow\downarrow B$ are defocused and leave the field. This is valid only as long as the spin orientation is conserved. Of course, in the sextupole the direction of the magnetic field changes with the azimuth, but as long as the particle motion is not too fast the spin follows the field direction adiabatically conserving the magnetic quantum state. This behavior permits the use of a magnetic field constant in time, in contrast to the charged particle in an ion trap.

An ideal linear sextupole in the x-z plane is generated by six hyperbolically shaped magnetic poles of alternating polarity extended in the y direction, as shown in Figs. 12(a) and 12(b). It might be approximated by six straight current leads with alternating current directions arranged in a hexagon. Such a configuration works as a

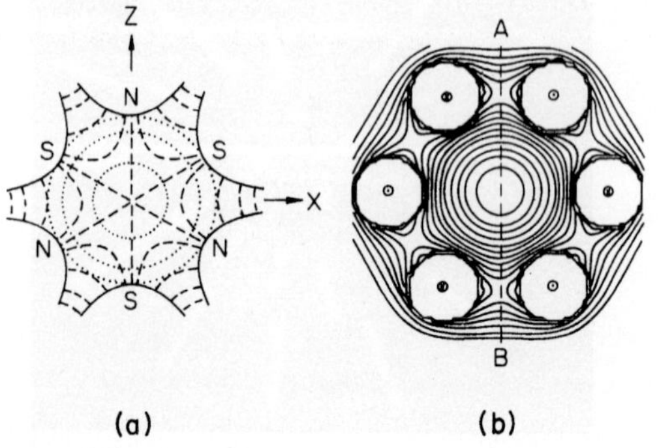

FIG. 12. (a) Ideal sextupole field. Dashed: magnetic field lines; dotted: lines of equal magnetic potential, $B=$const. (b) Linear sextupole made of six straight current leads with alternating current direction.

lense for particles moving along the y axis.

There are two possibilities to achieve a *closed storage volume:* a sextupole sphere and a sextupole torus. We have realized and studied both.

The spherically symmetric field is generated by three ring currents in an arrangement shown in Fig. 13. The field B increases in all directions with r^2 and has its maximum value B_0 at the radius r_0 of the sphere. Using superconducting current leads we achieved $B_0=3$ T in a sphere with a radius of 5 cm. But due to the low magnetic moment of the neutron $\mu=6\times10^{-8}$ eV/T the potential depth μB_0 is only 1.8×10^{-7} eV and hence the highest velocity of storable neutrons is only $v_{max}=6$ m/s. Due to their stronger moment for Na atoms these values are 2.2×10^{-4} eV and 37 m/s, respectively.

The main problem with such a closed configuration is the filling process, especially the cooling inside. However, in 1975 in a test experiment we succeeded in observing a storage time of 3 s for sodium atoms evaporated inside the bottle with its helium-cooled walls (Martin, 1975). But the breakthrough in confining atoms was achieved by W. D. Phillip and H. J. Metcalf using the modern technique of laser cooling (Migdal *et al.*, 1985).

The problem of storing neutrons becomes easier if one uses a linear sextupole field bent to a closed torus with a radius R as is shown in Fig. 14. The magnetic field in the torus volume is unchanged $B=(B_0/r_0^2)r^2$ and has no component in azimuthal direction. The neutrons move in a circular orbit with radius R_S if the centrifugal force is compensated by the magnetic force

$$F_c=\frac{m_{v_\varphi^2}}{R_S}=\mu\frac{\partial B}{\partial r}\bigg|_{R_S} .$$

In such a ring the permitted neutron energy is limited by

$$E_{max}=\mu B_0\left[\frac{R}{r_0}+1\right] .$$

It is increased by a factor (R/r_0+1) compared to the case of the sextupole sphere. As the neutrons have not

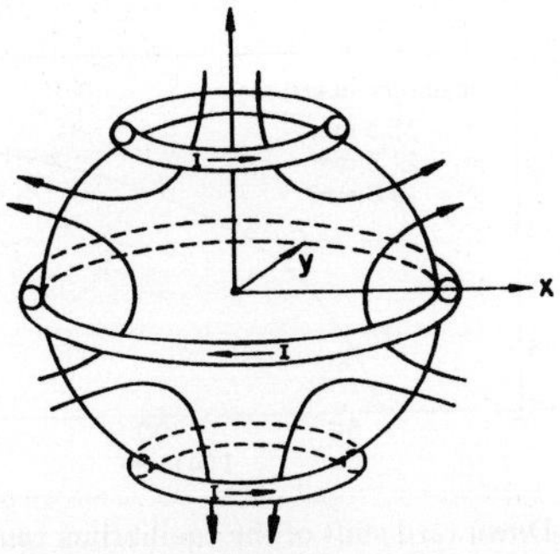

FIG. 13. Sextupole sphere.

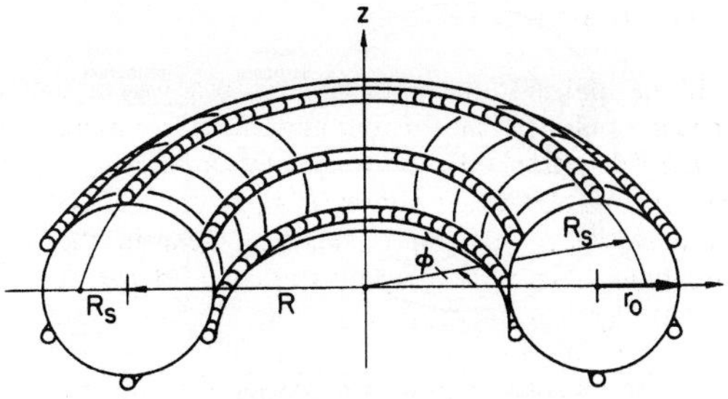

FIG. 14. Sextupole torus. R_s orbit of circulating neutrons.

only an azimuthal velocity but also components in r and z directions, they are oscillating around the circular orbit.

But this toroidal configuration has not only the advantage of accepting higher neutron velocities, it also permits an easy injection of the neutrons in the ring from the inside. The neutrons are not only moving in the magnetic potential well but they also experience the centrifugal barrier. Accordingly, one can lower the magnetic wall on the inside by omitting the two inward current leads. The resulting superposition of the magnetic and the centrifugal potential still provides a potential well with its minimum at the beam orbit. But there is no barrier for the inflected neutrons.

It is obvious that the toroidal trap in principle works analogous to the storage rings for high-energy charged particles. In many respects the same problems of instabilities of the particle orbits by resonance phenomena exist causing the loss of the particles. But also new problems arise, like, e.g., undesired spin flips or the influence of the gravitational force. In accelerator physics one has learned to overcome such problems by shaping the magnetic field by employing the proper multipole components.

This technique is also appropriate in case of the neutron storage ring. The use of the magnetic force μ grad$B$ instead of the Lorentz force being proportional to B just requires multipole terms of one order higher. Quadrupoles for focusing have to be replaced by sextupoles and, e.g., octupoles for stabilization of the orbits by decapoles.

In the seventies we have designed and constructed such a magnetic storage ring with a diameter of the orbits of 1 m. The achieved usable field of 3.5 T permits the confinement of neutrons in the velocity range of 5–20 m/s corresponding to a kinetic energy up to 2×10^{-6} eV. The neutrons are injected tangentially into the ring by a neutron guide with totally reflecting walls. The inflector can be moved mechanically into the storage volume and shortly afterwards be withdrawn.

The experimental setup is shown in Fig. 15. A detailed description of the storage ring, its theory, and performance is given in (Kügler *et al.*, 1985).

In 1978 in a first experiment we have tested the instrument at the Grenoble high-flux reactor. We could ob-

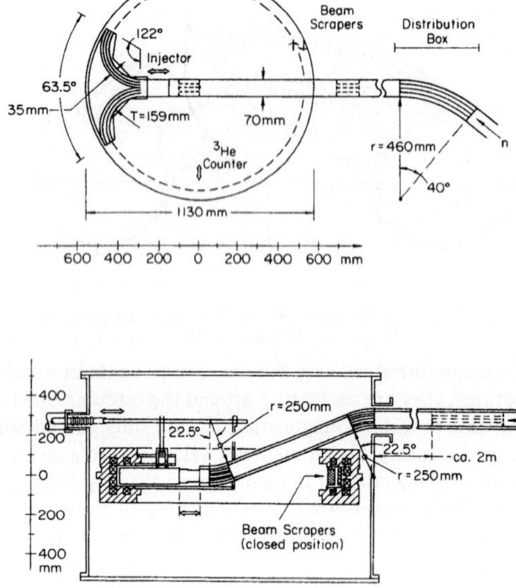

FIG. 15. Schematic top and side view of the neutron storage ring experiment.

serve neutrons stored up to 20 min after injection by moving a neutron counter through the confined beam after a preset time. As by the detection process the neutrons are lost, one has to refill the ring starting a new measurement. But due to the relatively low flux of neutrons in the acceptable velocity range, their number was too low to make relevant measurements with it.

In a recent experiment Paul *et al.*, 1989 at a new neutron beam with a flux improved by a factor 40 we could observe neutrons up to 90 min, i.e., roughly 6 times the decay time of the neutron due to radioactive decay. Figure 16 shows the measured profile of the neutron beam circulating inside the magnetic gap. Measuring carefully the number of stored neutrons as a function of time we could determine the lifetime to $\tau = 877 \pm 10$ s (Fig. 17).

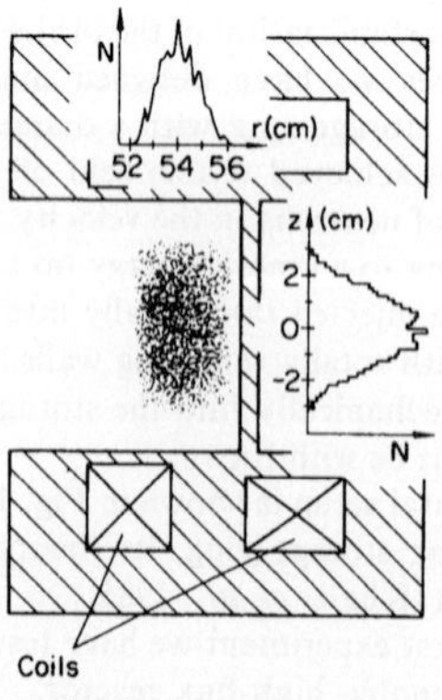

FIG. 16. Beam profile of the stored neutrons inside the magnet gap 400 s after injection.

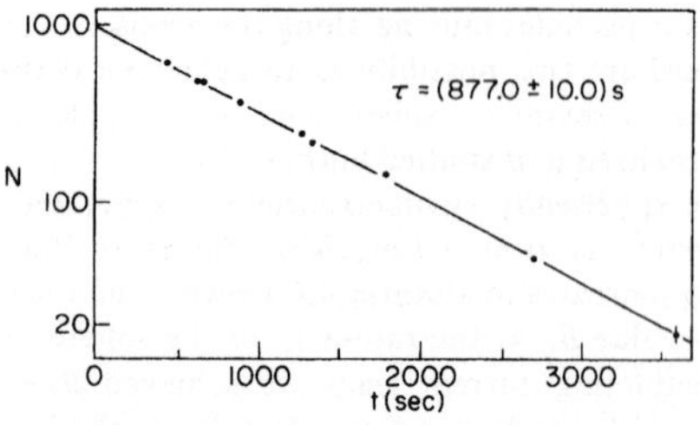

FIG. 17. Logarithmic decrease of the number of stored neutrons with time.

The analysis of our measurements lets us conclude that the intrinsic storage time of the ring for neutrons is at least one day. It shows that we had understood the relevant problems in its design.

The storage ring as a balance

This very reproducible performance permitted another interesting experiment. As I explained, the neutrons are elastically bound to the symmetry plane of the magnetic field. Due to the low magnetic moment the restoring force is of the order of the gravitational force. Hence it follows that the weight of the neutron stretches the magnetic spring the particle is hanging on; the equilibrium center of the oscillating neutrons is shifted downwards.

The shift z_0 is given by the balance $mg = \mu \, \mathrm{grad} B$. One needs a gradient $\partial B / \partial z = 173$ G/cm for compensating the weight. As the gradient in the ring in first approximation increases with z and is proportional to the magnetic current I, one calculates the shift z_0 to

$$z_0 = \mathrm{const} \times mg / I \ .$$

It amounts in our case to $z_0 = 1.2$ mm at the highest magnet current $I = 200$ A and 4.8 mm at 50 Å, accordingly.

By moving a thin neutron counter through the storage volume we could measure the profile of the circulating neutron beam and its position in the magnet. Driving al-

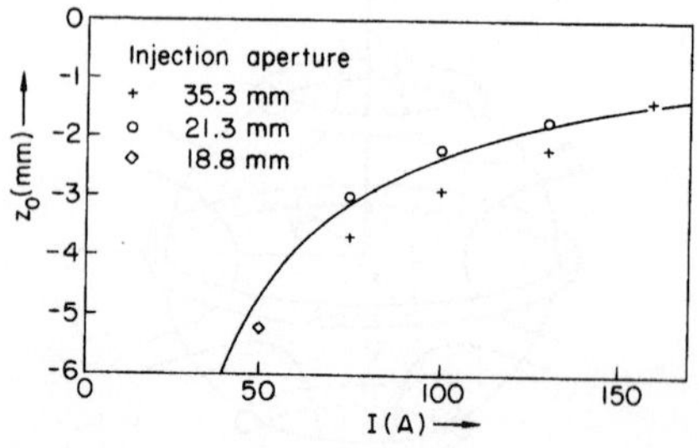

FIG. 18. Downward shift of the equilibrium center of the neutron orbits due to the weight of the neutron as function of the magnetic current.

ternately the counter downwards and upwards in many measuring runs we determined z_0 as a function of the magnet current.

The result is shown in Fig. 18. The measured data taken with different experimental parameters are following the predicted line. A detailed analysis gives for the gravitational mass of the neutron the value

$$m_g = 1.63 \pm 0.06 \times 10^{-24}\ \text{g}\ .$$

It agrees within 4% with the well-known inertial mass.

Thus the magnetic storage ring represents a balance with a sensitivity of 10^{-25} g. It is only achieved because the much higher electric forces play no role at all.

I am convinced that the magnetic bottles developed in our laboratory as described will be useful and fruitful instruments for many other experiments in the future as the Ion Trap has already proved.

REFERENCES

Bennewitz, H. G., and W. Paul, 1954, Z. Phys. **139**, 489.

Bennewitz, H. G., and W. Paul, 1955, Z. Phys. **141**, 6.

Berkling, K., 1956, thesis (Bonn).

Dawson, P. H., 1976, *Quadrupole Mass Spectrometry and its Application* (Elsevier, Amsterdam).

Dehmelt, H., 1967, in *Advances in Atomic and Molecular Physics,* Vol. 3, edited by D. R. Bates and I. Estermann (Academic, New York).

Diedrich, F., E. Chen, J. W. Quint, and H. Walther, 1987, Phys. Rev. Lett. **59**, 2931.

Diedrich, F. E. Peik, M. Chen, and H. Walther, 1988, Physik Blätter **44**, 12.

Fischer, E., 1959, Z. Phys. **156**, 1.

Friedburg, H., and W. Paul, 1951, Naturwissenschaften **38**, 159.

Gräff, G., E. Klempt, and G. Werth, 1969, Z. Phys. **222**, 201.

Kügler, K. J., W. Paul, and U. Trinks, 1985, Nucl. Instrum. Methods A **228**, 240.

March, R. E., and R. J. Hughes, 1989, *Quadrupole Storage Mass Spectrometry* (Wiley, New York).

Martin, B., 1975, thesis (Bonn University).

Migdal A. L., J. Prodan, W. D. Phillips, Th. H. Bergmann, and H. J. Metcalf, 1985, Phys. Rev. Lett. **54**, 2596.

Neuhauser, W., M. Hohenstett, P. Toschek, and H. Dehmelt, 1980, Phys. Rev. A **22**, 1137.

Paul, W., F. Anton, L. Paul, S. Paul, and W. Mampe, 1989, Z. Phys. C **45**, 25.

Paul, W., O. Osberghaus, and E. Fischer, 1958, Forsch. Berichte des Wirtschaftsministeriums Nordrhein-Westfalen Nr. 415.

Paul, W., and M. Raether, 1955, Z. Phys. **140**, 262.

Paul, W., H. P. Reinhardt, and U. von Zahn, 1958, Z. Phys. **152**, 143.

Paul, W., and H. Steinwedel, 1953, Z. Naturforsch. Teil A **8**, 448.

Paul, W., and H. Steinwedel, 1953, German Patent No. 944 900; U.S. Patent 2939958.

Penning, F. M., 1936, Physica **3**, 873.

Rettinghaus, G., 1967, Z. Angew. Phys. **22**, 321.

Townes, C. H., 1983, Proc. Nat. Acad. Sci. **80**, 7679.

van Dyck, R. S., P. B. Schwinberg, H. G. Dehmelt, 1977, Phys. Rev. Lett. **38**, 310.

von Busch, F., and W. Paul, 1961, Z. Phys. **164**, 580.

von Busch, F., and W. Paul, 1961, Z. Phys. **164**, 581.

von Zahn, U., 1962, Z. Phys. **168**, 129.

Wineland, D. J., and H. Dehmelt, 1975, Bull. Am. Phys. Soc. **20**, 637.

Wuerker, R. F., and R. V. Langmuir, 1959, Appl. Phys. **30**, 342.

Quantum Computations with Cold Trapped Ions

J. I. Cirac and P. Zoller*

Institut für Theoretische Physik, Universiät Innsbruck, Technikerstrasse 25, A-6020 Innsbruck, Austria
(Received 30 November 1994)

A quantum computer can be implemented with cold ions confined in a linear trap and interacting with laser beams. Quantum gates involving any pair, triplet, or subset of ions can be realized by coupling the ions through the collective quantized motion. In this system decoherence is negligible, and the measurement (readout of the quantum register) can be carried out with a high efficiency.

PACS numbers: 89.80.+h, 03.65.Bz, 12.20.Fv, 32.80.Pj

A quantum computer (QC) obeys the laws of quantum mechanics, and its unique feature is that it can follow a superposition of computation paths simultaneously and produce a final state depending on the interference of these paths [1]. Recent results in quantum complexity theory and the development of algorithms indicate that quantum computers can solve some problems efficiently which are considered intractable on classical Turing machines. An example is the factorization of large composite numbers into primes [2], a problem which is the basis of the security of many classical key cryptosystems.

The task of designing a QC is equivalent to finding a physical implementation of quantum gates between quantum bits (or qubits), where a qubit refers to a two-state system $\{|0\rangle, |1\rangle\}$ [3]. It has been shown [4] that any operation can be decomposed into controlled-NOT gates between two qubits and rotations on a single qubit, where a controlled-NOT is defined by $\hat{C}_{12} : |\epsilon_1\rangle|\epsilon_2\rangle \rightarrow |\epsilon_1\rangle|\epsilon_1 \oplus \epsilon_2\rangle$ with $\epsilon_{1,2} = 0, 1$, and $\oplus$ denoting addition modulo 2. While there exist promising proposals to demonstrate the basic principle of gates in cavity QED [4], the experimental steps necessary to realize even a controlled-NOT gate indicate that extended networks would be exceedingly difficult to build. On the other hand, a number of interactions have been proposed for the construction of quantum computers [1,5], but so far no explicit physical system has been shown to serve as a realistic model. The main obstacle for a practical realization is the existence of decoherence processes due to the interaction of the system (the QC) with the environment [6].

In this Letter we show that a set of N cold ions interacting with laser light and moving in a linear trap [7] provides a realistic physical system to implement a quantum computer. The distinctive features of this system are (i) it allows the implementation of n-bit quantum gates between any set of (not necessarily neighboring) ions, (ii) decoherence can be made negligible during the whole computation, and (iii) the final readout can be performed with unit efficiency.

The basic elements of the computer (i.e., the qubits) are the ions themselves. The two states of the nth qubit are identified with two of the internal states of the corresponding ion; for example, a ground state $|g\rangle_n \equiv |0\rangle_n$ and an excited state $|e\rangle_n \equiv |1\rangle_n$. The state of the QC is a macroscopic superposition

$$|\psi\rangle = \sum_{x=0}^{2^N-1} c_x|x\rangle \equiv \sum_{\underline{x}=\{0,1\}^N} c_{\underline{x}}|\underline{x}\rangle$$

of quantum registers $|\underline{x}\rangle = |x_{N-1}\rangle_{N-1}\cdots|x_0\rangle_0$ with $x = \sum_{n=0}^{N-1} x_n 2^n$ the binary decomposition of x. In this system independent manipulation of each individual qubit is accomplished by directing different laser beams to each of the ions. The quantum controlled-NOT, and, more generally, the (controlled)^n-NOT gate between n arbitrary ions in the trap can be implemented by exciting the collective quantized motion of the ions with lasers [8]. The coupling of the motion of the ions is provided by the Coulomb repulsion which is much stronger than any other interaction for typical separations between the ions of a few optical wavelengths. Decoherence in an ion trap is due to spontaneous decay of the internal atomic states and damping of the motion of the ion. Application of stored ions in ultrahigh precision spectroscopy and time and frequency standards [9,10] shows that this decoherence time can be extremely long, much longer than the time required to perform many operations in a QC. Spontaneous emission is suppressed using metastable transitions [10]. Collisions with background atoms can be avoided at sufficiently low pressures for very long times, and other couplings that affect the moving charges can be made sufficiently small [9]. Furthermore, the final readout of the quantum register (state measurement of the individual qubits) at the end of the computation can be accomplished using the quantum jumps technique with unit efficiency [11].

The situation we have in mind is depicted in Fig. 1. N ions are confined in a linear trap, and interact with different laser beams [Fig. 1(a)] in standing wave configurations [12]. The confinement of the motion along X, Y, and Z directions can be described by an (anisotropic) harmonic potential of frequencies $\nu_x \ll \nu_y, \nu_z$. Nonharmonic traps can also be used leading to very similar results. The ions have been previously laser cooled in all three dimensions so that they undergo very small oscillations around the equilibrium position. In this case, the

VOLUME 74, NUMBER 20 PHYSICAL REVIEW LETTERS 15 MAY 1995

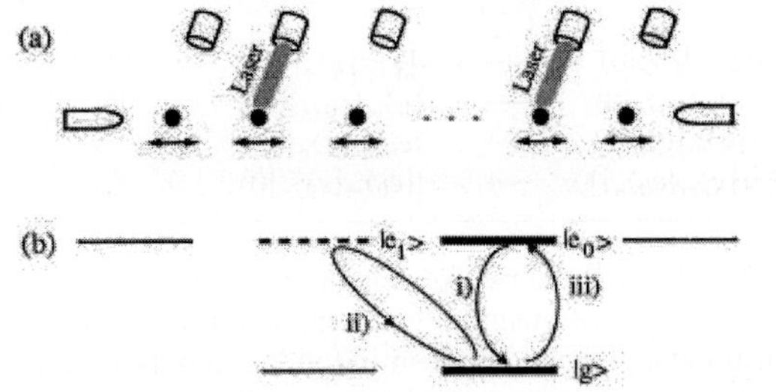

FIG. 1. (a) N ions in a linear trap interacting with N different laser beams; (b) atomic level scheme.

motion of the ions is described in terms of normal modes. Furthermore, we will assume that sideband cooling has left all the normal modes in their corresponding (quantum) ground states [13]. For this to be possible, one has to assume that the Lamb-Dicke limit (LDL) holds for all the modes [10]. This implies that their frequency is larger than the photon recoil frequency corresponding to the transition used for laser cooling. For example, for the $S_{1/2} \rightarrow D_{5/2}$ dipole-forbidden transition of a barium ion, this requires $\nu_{x,y,z} \gg 3$ kHz; in typical situations $\nu_{y,z} \gg \nu_x \sim 2\pi \times 50$ kHz [7]. The minimum frequency is that of the center-of-mass (CM) mode moving in the X direction, and coincides with ν_x. The next frequency is $\sqrt{3}\nu_x$, and all the others are larger. A remarkable feature of this system is that the frequency spacing is independent of the number of ions N in the trap.

Figure 1(b) shows a typical level scheme for an alkaline earth ion, corresponding to an electric dipole-forbidden transition [10]. The two-level system that we choose as the qubit is marked with thicker lines ($|g\rangle$ and $|e_0\rangle$). The other levels do not disturb the computation process. On the contrary, some of them are needed for implementing quantum gates, as we will show below.

When a laser beam acts on one of the ions, it induces transitions between its (internal) ground and excited levels and can change the state of the collective normal modes. However, in the LDL and for sufficiently low intensities, the laser beam will only cause transitions that modify the state of one of the modes. For example, with a laser frequency so that the detuning equals minus the CM mode frequency ($\delta_n = -\nu_x$), one excites the CM mode exclusively. This is so since the frequencies of the different normal modes are well separated in the excitation spectrum. This fact allows one to control the interactions between the ions through the CM motion, by selecting appropriately the frequency of the lasers.

Let $\hat{H}_0$ be the Hamiltonian for the system in the absence of any laser field. Now, consider that the laser acting on the nth ion is turned on. Obviously, this laser will leave the internal state of all the other ions unaffected. The laser frequency is chosen such that $\delta_n = -\nu_x$ and the equilibrium position of the ion coincides with the node of the laser standing wave [12]. The Hamiltonian describing this situation in an interaction picture defined by the operator $\exp(-i\hat{H}_0 t)$ is ($\hbar = 1$)

$$\hat{H}_{n,q} = \frac{\eta}{\sqrt{N}} \frac{\Omega}{2} [|e_q\rangle_n \langle g| a e^{-i\phi} + |g\rangle_n \langle e_q| a^\dagger e^{i\phi}]. \quad (1)$$

Here $a^\dagger$ and a are the creation and annihilation operators of CM phonons, respectively, Ω is the Rabi frequency, ϕ is the laser phase, and $\eta = [\hbar k_\theta^2/(2M\nu_x)]^{1/2}$ is the LDL parameter [$k_\theta = k\cos(\theta)$, with k the laser wave vector and θ the angle between the X axis and the direction of propagation of the laser]. The subscript $q = 0, 1$ refers to the transition excited by the laser, which depends on the laser polarization [see Fig. 1(b)]. Equation (1) can be derived as a generalization of the single ion Hamiltonian for the case of a linear trap [14]. Physically, the factor $\sqrt{N}$ appears since the effective mass of the CM motion is NM, and the amplitude of the mode scales like $1/\sqrt{NM}$ (Mössbauer effect). A careful analysis shows that the model Hamiltonian (1) is valid for $(\Omega/2\nu_x)^2\eta^2 \ll 1$. Note that in the LDL $\eta \ll 1$. On the other hand, corrections to this Hamiltonian can be made arbitrarily small for sufficiently low laser intensities.

If the laser beam is on for a certain time $t = k\pi/(\Omega\eta/\sqrt{N})$ (i.e., using a $k\pi$ pulse), the evolution of the system will be described by the unitary operator

$$\hat{U}_n^{k,q}(\phi) = \exp\left[-ik\frac{\pi}{2}(|e_q\rangle_n \langle g| a e^{-i\phi} + \text{H.c.})\right]. \quad (2)$$

It is easy to prove that this transformation keeps the state $|g\rangle_n|0\rangle$ unaltered, whereas

$$|g\rangle_n|1\rangle \longrightarrow \cos(k\pi/2)|g\rangle_n|1\rangle - ie^{i\phi}\sin(k\pi/2)|e_q\rangle_n|0\rangle,$$
$$|e\rangle_n|0\rangle \longrightarrow \cos(k\pi/2)|e_q\rangle_n|0\rangle - ie^{-i\phi}\sin(k\pi/2)|g\rangle_n|1\rangle,$$

where $|0\rangle$ ($|1\rangle$) denotes a state of the CM mode with no (one) phonon.

Let us now show how a two-bit gate can be performed using this interaction. We consider the following three-step process [see Fig. 1(b)]. (i) A π laser pulse with polarization $q = 0$ and $\phi = 0$ excites the mth ion. The evolution corresponding to this step is given by $\hat{U}_m^{1,0} \equiv \hat{U}_m^{1,0}(0)$. (ii) The laser directed on the nth ion is then turned on for a time of a 2π pulse with polarization $q = 1$ and $\phi = 0$. The corresponding evolution operator $\hat{U}_n^{2,1}$ changes the sign of the state $|g\rangle_n|1\rangle$ (without affecting the others) via a rotation through the auxiliary state $|e_1\rangle_n|0\rangle$. (iii) Same as (i). Thus the unitary operation for the whole process is $\hat{U}_{m,n} \equiv \hat{U}_m^{1,0}\hat{U}_n^{2,1}\hat{U}_m^{1,0}$ which is represented diagrammatically as follows:

VOLUME 74, NUMBER 20 PHYSICAL REVIEW LETTERS 15 MAY 1995

$$\begin{array}{ccccccc}
 & \hat U_m^{1,0} & & \hat U_n^{2,1} & & \hat U_m^{1,0} & \\
|g\rangle_m|g\rangle_n|0\rangle & \longrightarrow & |g\rangle_m|g\rangle_n|0\rangle & \longrightarrow & |g\rangle_m|g\rangle_n|0\rangle & \longrightarrow & |g\rangle_m|g\rangle_n|0\rangle\,, \\
|g\rangle_m|e_0\rangle_n|0\rangle & \longrightarrow & |g\rangle_m|e_0\rangle_n|0\rangle & \longrightarrow & |g\rangle_m|e_0\rangle_n|0\rangle & \longrightarrow & |g\rangle_m|e_0\rangle_n|0\rangle\,, \\
|e_0\rangle_m|g\rangle_n|0\rangle & \longrightarrow & -i|g\rangle_m|g\rangle_n|1\rangle & \longrightarrow & i|g\rangle_m|g\rangle_n|1\rangle & \longrightarrow & |e_0\rangle_m|g\rangle_n|0\rangle\,, \\
|e_0\rangle_m|e_0\rangle_n|0\rangle & \longrightarrow & -i|g\rangle_m|e_0\rangle_n|1\rangle & \longrightarrow & -i|g\rangle_m|e_0\rangle_n|1\rangle & \longrightarrow & -|e_0\rangle_m|e_0\rangle_n|0\rangle\,.
\end{array} \quad (3)$$

The effect of this interaction is to change the sign of the state only when both ions are initially excited. Note that the state of the CM mode is restored to the vacuum state $|0\rangle$ after the process. Equation (3) is equivalent to a controlled-NOT gate. To show this, let us denote by $|\pm\rangle = (|g\rangle \pm |e_0\rangle)/\sqrt{2}$. Then, this procedure can be summarized as $|g\rangle_m|\pm\rangle_n \rightarrow |g\rangle_m|\pm\rangle_n$ and $|e_0\rangle_m|\pm\rangle_n \rightarrow |e_0\rangle_m|\mp\rangle_n$. With an appropriate individual (one-bit) rotation applied to the nth ion this procedure then yields the controlled-NOT. These individual rotations acting on a single ion (without modifying the CM motion) can be performed using a laser frequency on resonance with the internal transition ($\delta_n = 0$), polarization $q = 0$, and with the equilibrium position of the ion coinciding with the antinode of the laser standing wave. In this case, the Hamiltonian is

$$\hat H_n = (\Omega/2)[|e_0\rangle_n\langle g|e^{-i\phi} + |g\rangle_n\langle e_0|e^{i\phi}]\,. \quad (4)$$

For an interaction time $t = k\pi/\Omega$ (i.e., using a $k\pi$ pulse), this process is described by the following unitary evolution operator:

$$\hat V_n^k(\phi) = \exp\left[-ik\frac{\pi}{2}(|e_0\rangle_n\langle g|e^{-i\phi} + \text{H.c.})\right], \quad (5)$$

so that

$$|g\rangle_n \longrightarrow \cos(k\pi/2)|g\rangle_n - ie^{i\phi}\sin(k\pi/2)|e_0\rangle_n\,,$$
$$|e_0\rangle_n \longrightarrow \cos(k\pi/2)|e_0\rangle_n - ie^{-i\phi}\sin(k\pi/2)|g\rangle_n\,.$$

Thus the complete controlled-NOT gate for the states $|\epsilon_m\rangle|\epsilon_n\rangle$ ($\epsilon_{m,n} = g, e_0$) is given by $\hat C_{mn} = \hat V_n^{1/2}(\frac{\pi}{2})\hat U_{m,n}\hat V_n^{1/2}(-\frac{\pi}{2})$ [15].

Nonlocal three-bit gates can be implemented in a similar way between ions n, m, and l. The process takes place in five steps: (j) Same as (i); (jj) same as (ii), but with a π pulse; (jjj) same as step (ii) but with ion l; (jv) same as (jj); (v) same as (j). The corresponding unitary operation for this process is $\hat U_m^{1,0}\hat U_n^{1,1}\hat U_l^{2,1}\hat U_n^{1,1}\hat U_m^{1,0}$. This procedure only changes the sign of the state if all three ions were initially excited. One can easily generalize this procedure to the case of many ions. For example, a (control)^p-NOT gate acting on ions $n_1, n_2, \ldots, n_q$ corresponds to the unitary evolution

$$\hat V_p^{\frac{1}{2}}\left(\frac{\pi}{2}\right)\hat U_{n_1}^{1,0}\left[\prod_{j=2}^{p-1}\hat U_{n_j}^{1,1}\right]\hat U_{n_p}^{2,1}\left[\prod_{j=p-1}^{2}\hat U_{n_j}^{1,1}\right]\hat U_{n_1}^{1,0}\hat V_p^{\frac{1}{2}}\left(-\frac{\pi}{2}\right).$$

Using similar ideas with different laser phases and interaction times one can implement other n-bit gates [8].

In summary, the two key elements behind the above implementation of quantum gates are as follows. First, nonlocal entanglement between individual qubits is achieved by transferring the internal atomic coherence to and from the CM motion shared by all the ions ($\hat U_n^{k=1,q=0}$). Second, as an intermediate step we "hide atomic amplitudes" corresponding to the qubits in a third internal atomic level $|e_1\rangle$ ($\hat U_n^{k=1,q=1}$), and induce 2π rotations via this state to selectively change the sign of atomic amplitudes ($\hat U_n^{k=2,q=1}$). We note that no population is left in these auxiliary atomic and CM levels after the complete gate operation. Any population left in these states is an indication of an imprecise realization. This could be used to implement an error detection scheme by probing the population of these intermediate states, for example, with a laser inducing fluorescence after each gate operation [16].

The core of Shor's factorization scheme [2] is the high efficiency of a QC to find the period r of a given function by doing a discrete Fourier transform (FT) on a periodic state vector of the form $|\Psi\rangle \propto \sum_l |lr + k\rangle$. Here k is an integer number and $l = 0, \ldots, [(2^N - k)/r]$ with $[\ldots]$ the integer part. The FT is defined by the operation

$$\hat{FT}|x\rangle = 1/\sqrt{2^N}\sum_y e^{2\pi ixy/2^N}|y\rangle$$

on the quantum registers. This FT can be decomposed into a sequence of one- and two-bit operations [17,18]. The probability to measure the state $|y\rangle$ of the quantum register is then $P_y = |\langle y|\hat{FT}|\Psi\rangle|^2$. Shor has shown that this measurement gives with high probability an outcome that allows one to calculate r.

To show the capabilities of an ion trap as a QC, and to analyze how experimental uncertainties may affect the final results [6], we have simulated the above scheme on a (digital) computer. Figure 2 shows a comparison

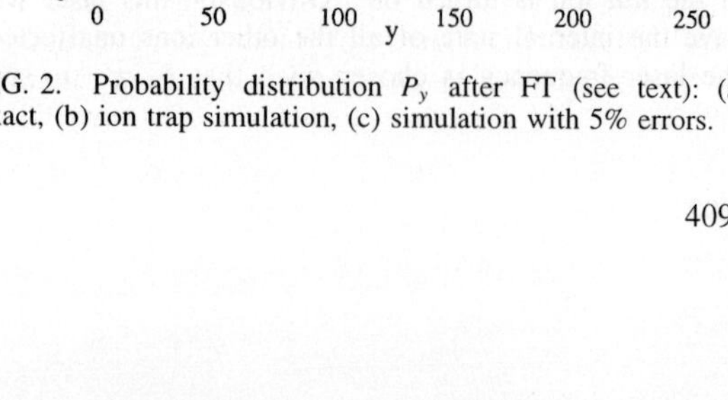

FIG. 2. Probability distribution P_y after FT (see text): (a) exact, (b) ion trap simulation, (c) simulation with 5% errors.

between the exact results [Fig. 2(a)] for P_y and the ion trap simulation [Figs. 2(b) and 2(c)] for a state with $k = 4$, $r = 7$, and eight ions. The existence of peaks in this spectrum (separated by $\simeq 2^N/r = 256/7$) allows one to determine the period r. Similar to Ref. [17] one can show that this probability distribution P_y can be obtained from the physical process corresponding to the sequence of operations $\hat{V}_0\hat{W}_0\hat{W}_1\hat{V}_1\cdots\hat{W}_{N-2}\hat{V}_{N-1}$. Here $\hat{W}_n = \hat{W}_n^{N-1}\hat{W}_n^{N-2}\ldots\hat{W}_n^{n+1}$ is a sequence of two-bit operations $\hat{W}_n^m = \hat{U}_m^{1,0}[\pi(1-2^{(n-m)})]\hat{U}_n^{1,1}[\pi(1-2^{(n-m)})]\hat{U}_n^{1,1}\hat{U}_m^{1,0}$ $(n < m)$, and $\hat{V}_n \equiv \hat{V}_n^{1/2}(-\pi/2)$ is a one-ion rotation [see (2) and (5)]. The specific form of the pulse sequence can be directly deduced from the definition of the operators W, and requires two- and one-bit gates between the ions. The simulation has been performed with the full Hamiltonian (to all orders in the Lamb-Dicke expansion) for $N = 8$ Ba^+ ions in a trap with $\nu_x = 2\pi \times 50$ kHz. The Rabi frequencies have been chosen as follows: $\Omega = 2\pi \times 1.5$ kHz for resonant excitations (at the antinode) and $\Omega = 2\pi \times 15$ kHz for off-resonant excitations (at the node). The rest of the parameters correspond to those of the Ba^+ ions. As shown in Fig. 2(b), with these realistic parameters, the result is nearly indistinguishable from the exact one. From our numerical simulations we could see that this result can even be improved by increasing the trap frequency (or decreasing the Rabi frequencies), in agreement with a perturbation theory analysis for the terms neglected in (1) and (4). Note that the total time required for the whole operation is about 35 ms, much smaller than the decoherence time due to spontaneous emission (the lifetime of the metastable state of Ba^+ is about 45 s, so that the decoherence time is $\simeq 6$ s). To analyze how experimental uncertainties affect the final results we have carried out numerical simulations assuming a 5% error in *all* the interaction times involved in the operation, 1 kHz of error in *all* the laser detunings, and a 5% $\pi/2$ error in all the angles in the problem (situation of the standing waves with respect to the position of the ions, and phases of the lasers). Figure 2(c) shows that even with all these errors the peaks in the distribution are still maintained, and the system of ions is remarkably robust to perform quantum computations.

Apart from one- and two-bit operations, (5) and (2), one can also prepare the most general entangled state of N ions [9,19]. For example, the maximal entangled state

$$|\Psi\rangle = 1/\sqrt{2}(|-\rangle_{N-1}\cdots|-\rangle_1 \times |-\rangle_0 - |+\rangle_{N-1}\cdots|+\rangle_1|+\rangle_0)$$

can be obtained starting from $|g\rangle_{N-1}|g\rangle_{N-2}\cdots|g\rangle_0$ (as obtained after sideband cooling), by using the operations $\hat{V}_0\hat{U}_{N-1,0}\cdots\hat{U}_{2,0}\hat{U}_{1,0}\hat{V}_{N-1}\cdots\hat{V}_1\hat{V}_0$ [18].

In summary, linear ion traps are well suited to implement a QC. This is due to the negligible decoherence in these systems [9], as well as the possibility to manipulate the internal and CM degrees of freedom with external fields, and to perform efficient state measurements. We have shown how to implement n-bit gates between n arbitrary ions, and have illustrated the performance of such a system with a numerical simulation. We believe that the present system provides a realistic implementation of a QC which can be built with present or planned technology.

We thank R. Blatt, A. Ekert, M. Lewenstein, and D. Wineland for helpful comments. This work was supported by the Austrian Science Foundation.

*Permanent address: Departamento de Fisica, Universidad de Castilla-La Mancha, 13071 Ciudad Real, Spain.

[1] For a review, see A. Ekert, in Proc. ICAP '94. edited by S. Smith, C. Wieman, and D. Wineland (to be published).
[2] P. W. Shor, in *Proceedings of the 35th Annual Symposium on the Foundations of Computer Science, Los Alamitos, CA* (IEEE Computer Society Press, New York, 1994), p. 124.
[3] D. Deutsch, Proc. R. Soc. London A **425**, 73 (1989).
[4] T. Sleator and H. Weinfurter, Phys. Rev. Lett. **74**, 4087 (1995).
[5] K. Obermayer, W. G. Teich, and G. Mahler, Phys. Rev. B **37**, 8096 (1988); S. Lloyd, Science **261**, 1569 (1993); D. P. Di Vincenzo, Phys. Rev. A **50**, 1015 (1995).
[6] R. Landauer, Proc. R. Soc. London A (to be published); W. G. Unruh (to be publishd).
[7] M. G. Raizen *et al.*, Phys. Rev. A **45**, 6493 (1992); H. Walther, Adv. At. Mol. Opt. Phys. **32**, 379 (1994).
[8] Although a (controlled)^n-NOT can be decomposed into a finite number of controlled-NOT gates plus one-bit rotations, this may require many operations. [H. Weinfurter (private communication)] Thus a direct implementation of the (controlled)^n-NOT gate may be interesting from a practical point of view.
[9] D. J. Wineland *et al.*, Phys. Rev. A **50**, 67 (1994); **46**, R6797 (1992).
[10] R. Blatt, in Proc. ICAP '94, Ref. [1].
[11] W. Nagourney *et al.*, Phys. Rev. Lett. **56**, 2797 (1986); J. C. Bergquist *et al.*, *ibid.* **56**, 1699 (1986); Th. Sauter *et al.*, *ibid.* **56**, 1696 (1986).
[12] A similar scheme can be used with traveling wave configurations. However, the standing wave minimizes the effects of unwanted transitions; see Ref. [14].
[13] F. Diedrich *et al.*, Phys. Rev. Lett. **62**, 403 (1989); here only the CM has to be cooled to the ground state.
[14] J. I. Cirac *et al.*, Phys. Rev. Lett. **70**, 762 (1993).
[15] The two-bit gate (3) (instead of the controlled-NOT) together with single bit rotations are sufficient to generate arbitrary unitary operations.
[16] Nonobservation of fluorescence corresponds to a projection of the state vector on $|g\rangle$, $|e_0\rangle$. This might be the basis of a partial error correction scheme.
[17] D. Coppersmith, IBM Research Report No. RC19642, 1994.
[18] FT and the preparation of general entangled states could be performed more efficiently using general n-bit gates (instead of a sequence of two-bit gates).
[19] D. M. Greenberger *et al.*, Am. J. Phys. **58**, 1131 (1990); see also N. D. Mermin, *ibid.* **58**, 8 (1990).

New ion trap for frequency standard applications

J. D. Prestage, G. J. Dick, and L. Maleki
California Institute of Technology, Jet Propulsion Laboratory, 4800 Oak Grove Drive, Pasadena, California 91109

(Received 3 February 1989; accepted for publication 10 April 1989)

We have designed a novel linear ion trap which permits storage of a large number of ions with reduced susceptibility to the second-order Doppler effect caused by the rf confining fields. This new trap should store about 20 times the number of ions as a conventional rf trap with no corresponding increase in second-order Doppler shift from the confining field. In addition the sensitivity of this shift to trapping parameters, i.e., rf voltage, rf frequency, and trap size, is greatly reduced.

INTRODUCTION

There has been much recent activity directed toward the development of trapped ion frequency standards, in part because ions confined in an electromagnetic trap are subjected to very small perturbations of their atomic energy levels. The inherent immunity to environmental changes that is afforded by suitably chosen ions suspended in dc or rf quadrupole traps has led to the development of frequency standards with very good long term stability.[1] Indeed, the trapped $^{199}Hg^+$ ion clock of Ref. 2 is the most stable clock yet developed for averaging times $\geqslant 10^6$ s. However, certain applications such as millisecond pulsar timing[3] and low frequency gravity wave detection across the solar system[4] require stabilities beyond that of present day standards.

While the basic performance of the ion frequency source depends fundamentally on the number of ions in the trap, the largest source of frequency offset stems from the motion of the atoms caused by the trapping fields via the second-order Doppler or relativistic time dilation effect.[5] Moreover, instability in certain trapping parameters, e.g., trap field strength, temperature, and the actual number of trapped particles will influence the frequency shift and lead to frequency instabilities. Since this offset also depends strongly on the number of ions, a trade-off situation results, where fewer ions are trapped in order to reduce the (relatively) large frequency offset which would otherwise result.

We have designed and constructed a hybrid rf/dc linear ion trap which should allow an increase in the stored ion number with no corresponding increase in second-order Doppler instabilities. The 20 times larger ion storage capacity should improve clock performance substantially. Alternatively, the Doppler shift from the trapping fields may be reduced by a factor of 10 below comparably loaded hyperbolic traps.

SECOND-ORDER DOPPLER SHIFT FOR IONS IN A rf TRAP

Trapping forces in a rf ion trap are due to time-varying electric fields which increase in every direction from the trap's center. A single particle at rest in such a trap at its very center (where, in an ideal trap, these fields are zero) would have no velocity and thus no second-order Doppler shift.

A very different condition holds for many particles in such a trap. In this case, electrostatic repulsion tends to keep the ions away from each other and from the center of the trap. As the number of ions increases, the size of the cloud also increases, pushing the ions into regions of larger and larger rf fields. The resultant velocity gives rise to a downward shift of atomic transition frequencies with increasing ion number.

Calculation of the second order Doppler shift requires a detailed knowledge of the ionic distribution density which results from the balance between trapping and (repulsive) Coulomb forces. A method has been developed in which an average over one cycle of the rf field reduces its effect to that of a pseudopotential acting on the charge of the particle.[6] The effect is subsequently further reduced to that of a pseudocharge distribution which produces the equivalent effective potential. Ionic distribution density is then calculated by considering the response of charges to this resultant "background" pseudocharge.

This method has been previously applied[7] to the trap shown in Fig. 1. Calculation shows a spatially uniform pseudocharge giving rise to a spherical ion cloud, also with uniform density. The resulting average frequency shift can be expressed in terms of the total number of trapped ions, together with a trap strength parameter ω, ion mass m, and charge q.

In the following sections we perform a similar calculation for a cylindrical geometry, a case not previously exam-

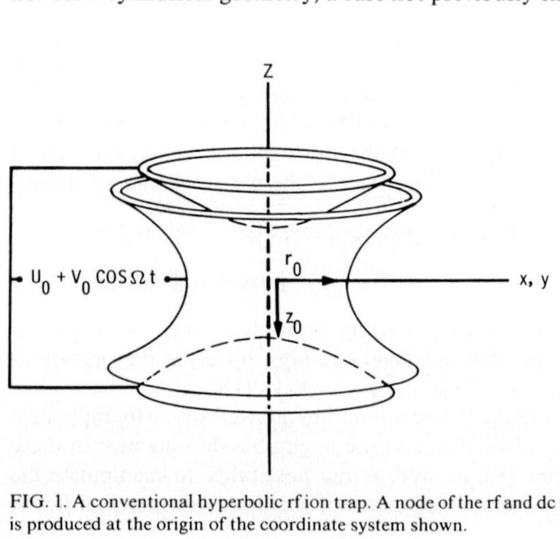

FIG. 1. A conventional hyperbolic rf ion trap. A node of the rf and dc fields is produced at the origin of the coordinate system shown.

 0021-8979/89/151013-05$02.40

ined. The cloud forms a cylinder of uniform density,[8] in a manner analogous to that of the spherical trap. Comparison between the consequences of the two geometries shows a very different story. While physically similar in overall size, the linear trap can hold many more ions than the spherical one with no increase in the second-order Doppler shift, or conversely, the shift can be greatly reduced. Furthermore, its dependence on trap parameters is qualitatively different, allowing miniaturization of the transverse trap dimensions without penalty in performance.

CALCULATION FOR A SPHERICAL TRAP

Figure 1 shows a conventional rf ion trap along with the applied voltages. Trapping forces are generated by the driven motion of the ions (at frequency Ω) in an inhomogeneous rf electric field created by hyperbolic trap electrodes.[6] The motion in each of three directions for a single ion in a rf trap is characterized by two frequencies, the fast driving frequency Ω and a slower secular frequency ω which is related to the harmonic force binding the particle to the trap center. An exact solution to the equations of motion shows that frequencies $k\Omega \pm \omega$, $k = 2,3,...$ are also present. However, in the limit $\omega/\Omega \ll 1$ (which is the primary condition for stability of the ion orbits) the ω and $\Omega \pm \omega$ frequencies dominate and the kinetic energy (KE) of a particle, averaged over one cycle of Ω, separates into the kinetic energy of the secular motion and the kinetic energy of the driven motion. The average kinetic energy is transferred from the secular to the driven motion and back while the sum remains constant just as a harmonic oscillator transfers energy from kinetic to potential and back.

We consider two cases. A hot ion cloud, or one containing a very few ions where interactions between ions are negligible, shows a second-order Doppler shift given by

$$\left(\frac{\Delta f}{f}\right)_{\text{hot}} = -\frac{1}{2}\frac{\langle v^2\rangle}{c^2} = -\frac{\langle \text{total KE}\rangle}{mc^2} \quad (1)$$

$$= -\frac{\langle \text{secular KE} + \text{driven KE}\rangle}{mc^2} \quad (2)$$

$$= -2(\langle \text{secular KE}\rangle/mc^2) \quad (3)$$

$$= -3k_B T/mc^2, \quad (4)$$

where m is the ionic mass, T the temperature, and $\langle\ \rangle$ indicates a time average over one cycle of Ω. We have also averaged over one cycle of ω to equate the secular and driven KE. This is analogous to a simple harmonic oscillator where the average KE is equal to the average potential energy. The consequence is a frequency shift that is twice as large as that due to thermal motion alone.

Of greater interest is the case where many ions are contained in a trap and interactions between ions dominate. In this cold cloud model[7] displacements of individual ions from the trap center are primarily due to electrostatic repulsion between the ions, and random thermal velocities associated with temperature can be assumed to be small compared to driven motion due to the trap fields.

The electric potential inside the trap of Fig. 1 is

$$\phi(\rho,z) = \{[U_0 + V_0\cos(\Omega t)]/\epsilon^2\}(\rho^2 - 2z^2), \quad (5)$$

where $\epsilon^2 = r_0^2 + 2z_0^2$ describes the trap size, and U_0 and V_0 represent the amplitudes of dc and ac trap voltages, respectively.

The trapping force generated by the rf field alone can be described by an electric pseudopotential[6]:

$$\psi(\rho,z) \equiv q[\mathbf{E}_0(\rho,z)]^2/4m\Omega^2, \quad (6)$$

where q is the ionic charge and $\mathbf{E}_0$ is the peak local rf field. This becomes

$$\psi(\rho,z) = (qV_0^2/m\Omega^2\epsilon^4)/(\rho^2 + 4z^2) \quad (7)$$

for the effect of the rf part of Eq. (5). Adding the dc potential from Eq. (5) gives the total potential energy for an ion in the trap of Fig. 1:

$$\phi(\rho,z) = \tfrac{1}{2}(m\omega_\rho^2\rho^2 + m\omega_z^2 z^2), \quad (8)$$

where

$$\omega_\rho^2 = 2q^2V_0^2/m^2\Omega^2\epsilon^4 + 2qU_0/m\epsilon^2 \quad (9)$$

and

$$\omega_z^2 = 8q^2V_0^2/m^2\Omega^2\epsilon^4 - 4qU_0/m\epsilon^2 \quad (10)$$

describe secular frequencies for radial and longitudinal ion motion.

The pseudopotential can be further analyzed in terms of an effective pseudocharge by applying Poisson's equation to Eq. (7) or (8). The result of this calculation is a uniform background charge density throughout the trap region which is given by

$$Q_b = -[\epsilon_0 m(2\omega_\rho^2 + \omega_z^2)]/q. \quad (11)$$

An easy solution for the charge configuration can be found if we assume that the dc and rf voltages are adjusted to make the trapping forces spherical, i.e., $\omega_\rho = \omega_z = \omega$. In this case the ion cloud is also spherical and trapped positive ions exactly neutralize the negative background of charge, matching its density out to a radius where the supply of ions is depleted. Ion density is given by

$$n_0 = 3\epsilon_0 m\omega^2/q^2, \quad (12)$$

and the total number of ions by

$$N = n_0(4\pi/3)R_{\text{sph}}^3, \quad (13)$$

where R_{sph} is the radius of the sphere of trapped ions.

The oscillating electric field which generates the trapping force grows linearly with distance from the trap center. The corresponding amplitude of any ion's driven oscillation is proportional to the strength of the driving field, i.e., also increasing linearly with the distance from the trap center. The average square velocity of the driven motion for an ion at position (ρ,z) is

$$\langle v^2\rangle = \tfrac{1}{2}\omega^2(\rho^2 + 4z^2). \quad (14)$$

For a given trapping strength, reflected in force constant ω^2, the density is fixed by Eq. (12) while the radius of the spherical cloud is determined once the ion number N has been specified. The second-order Doppler shift due to the micromotion is found by taking a spatial average of Eq. (1) over the spherical ion cloud. Using Eq. (14) for the spatial dependence of the micromotion:

$$(\Delta f/f)_{\text{sph}} = -\tfrac{1}{2}(\overline{\langle v^2\rangle}/c^2) \quad (15)$$

$$= -\tfrac{3}{10}(\omega^2 R_{\rm sph}^2/c^2) \quad (16)$$

$$= -(3/10c^2)(N\omega q^2/4\pi\epsilon_0 m)^{2/3}. \quad (17)$$

For typical operating conditions,[7] $N = 2\times10^6$ and $\omega = (2\pi)$ 50 kHz, $\Delta f/f = 2\times10^{-12}$. This second-order Doppler shift is about 10 times larger than the shift for free ^{199}Hg ions at room temperature, $\Delta f/f = 3k_B T/2mc^2 = 2\times10^{-13}$.

If the temperature is not too high, its effect on the ion cloud is to broaden the sharp edge at its outside radius. In this case the plasma density falls off in a distance characterized by the Debye length[8]:

$$\lambda_D = \sqrt{k_B T\epsilon_0/n_0 q^2}. \quad (18)$$

The cold cloud model should be useful provided the ion cloud size is large compared to the Debye length. This ratio is given by

$$\left(\frac{\lambda_D}{R_{\rm sph}}\right)^2 = \frac{1}{30}\frac{(\Delta f/f)_{\rm hot}}{(\Delta f/f)_{\rm sph}}. \quad (19)$$

This indicates a relatively small fractional Debye length throughout the regime of interest. For the typical conditions indicated above, the Debye length is about 1/5 mm in comparison to a spherical cloud diameter of 2.5 mm.

CALCULATION FOR A LINEAR TRAP

For increased signal to noise in the measured atomic resonance used in frequency standard applications, it is desirable to have as many trapped ions as possible. However, as we have just seen, larger ion clouds have larger second-order Doppler shifts. This frequency offset must be stabilized to a high degree in order to prevent degradation of long term performance.

To reduce this susceptibility to second-order Doppler shift we now propose a hybrid rf/dc ion trap which replaces the single field node of the hyperbolic rap with a line of nodes. The rf electrode structure producing this line of nodes of the rf field is shown in Fig. 2. Ions are trapped in the radial direction by the same type of rf trapping forces used in the previously discussed hyperbolic trap and we follow a similar analysis in terms of an equivalent pseudopotential and background pseudocharge.

Ions are prevented from escaping along the axis of the trap by dc biased "endcap" needle electrodes mounted on each end as shown in Fig. 3. These electrodes approximate the electrostatic effect of the missing parts of an infinitely long ion cloud. Their diameter is the same as the ion cloud to be trapped and is small compared to the trap diameter so that the rf trapping field is perturbed only slightly. Because these endcaps reach well inside the rf electrodes, any end effect of the rf fields on the ion cloud should be small. Unlike conventional rf traps this linear trap will hold positive or negative ions, but not both simultaneously.

Near the central axis of the trap we assume a quadrupolar rf electric potential:

$$\phi = [V_0(x^2 - y^2)\cos(\Omega t)]/2R^2, \quad (20)$$

from which, as in the previous section, we derive a corresponding pseudopotential:

$$\psi = (qV_0^2/4m\Omega^2 R^4)(x^2 + y^2), \quad (21)$$

for a total ionic potential given by

$$\phi = q\psi = m(\omega^2/2)\rho^2; \quad (22)$$

where, for the cylindrical electrodes of Fig. 2, R is an approximate distance from the trap center to an electrode's surface, and

$$\omega^2 = q^2 V_0^2/2m^2\Omega^2 R^4. \quad (23)$$

Here ω is the characteristic frequency for transverse or radial motion in the trap. Longitudinal motion is described in terms of motion at thermal velocities between the trap ends.

Application of Poisson's equation shows Eq. (21) to be equivalent to a uniform background pseudocharge with density:

$$Q_b = -(2\epsilon_0 m\omega^2/q). \quad (24)$$

Solving for the charge configuration for an infinitely long trap follows a nearly identical process to that of the preceding section since, from Gauss's law, cylindrical or spherical surfaces of charge induce no fields in their interior. Thus we find a uniform cylinder of ions just canceling the

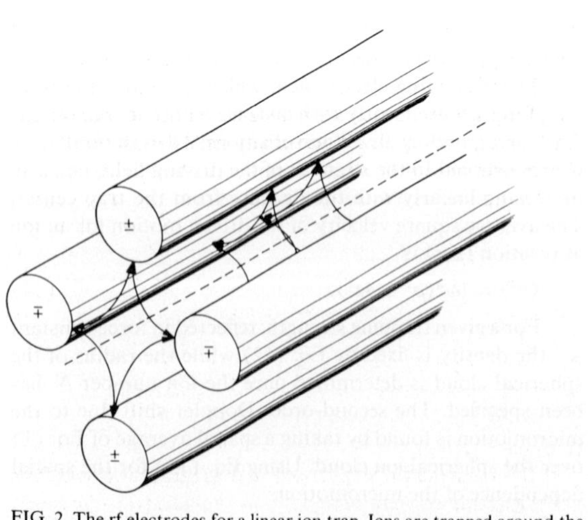

FIG. 2. The rf electrodes for a linear ion trap. Ions are trapped around the line of nodes of the rf field with reduced susceptibility to second-order Doppler frequency shift.

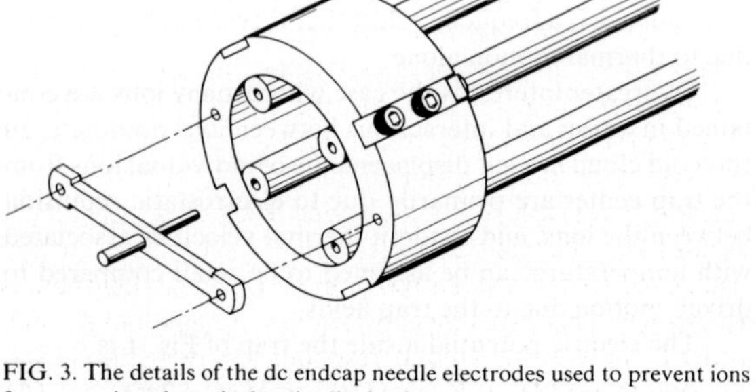

FIG. 3. The details of the dc endcap needle electrodes used to prevent ions from escaping along the longitudinal axis.

background pseudocharge out to a radius R_c with density:

$$n_0 = (2\epsilon_0 m\omega^2/q^2), \tag{25}$$

with ion number per unit length of

$$N/L = n_0 \pi R_c^2. \tag{26}$$

The motion induced by the rf trapping field is purely transverse and is given by

$$\langle v^2 \rangle = \omega^2 \rho^2. \tag{27}$$

As before we average this quantity over the ion cloud to find the second-order Doppler shift:

$$\left(\frac{\Delta f}{f}\right)_{\text{lin}} = -\frac{1}{2}\frac{\overline{\langle v^2 \rangle}}{c^2} = -\frac{\omega^2 R_c^2}{4c^2}. \tag{28}$$

We assume for simplicity a cylindrical ion cloud of radius R_c and length L. Equation (28) can be written in terms of total ion number N, and trap length L, as

$$\left(\frac{\Delta f}{f}\right)_{\text{lin}} = -\left(\frac{q^2}{8\pi\epsilon_0 mc^2}\right)\frac{N}{L}. \tag{29}$$

In conrast to the spherical case, this expression contains no dependence on trap parameters except for the linear ion density N/L. This is also true for the relative Debye length:

$$\left(\frac{\lambda_D}{R_c}\right)^2 = \frac{1}{24}\frac{(\Delta f/f)_{\text{hot}}}{(\Delta f/f)_{\text{lin}}}, \tag{30}$$

which must be small to insure the validity of our "cold cloud" model.

From this it is seen that the transverse dimension R of the trap may be reduced without penalty of performance, providing that operational parameters are appropriately scaled. This requires ω and Ω to vary as R^{-1}, and the applied voltage V_0 to be held constant.

COMPARISON

We can compare the second-order Doppler shift for the two traps assuming both hold the same number of ions by

$$\left(\frac{\Delta f}{f}\right)_{\text{lin}} = \frac{5}{3}\frac{R_{\text{sph}}}{L}\left(\frac{\Delta f}{f}\right)_{\text{sph}}. \tag{31}$$

As more ions are added to the linear trap their average second-order Doppler shift will increase. It will equal that of the spherical ion cloud in the hyperbolic trap when

$$N_{\text{lin}} = \frac{3}{5}(L/R_{\text{sph}})N_{\text{sph}}. \tag{32}$$

A linear trap can thus store $\frac{3}{5}(L/R_{\text{sph}})$ times the ion number as a conventional rf trap with no increase in average second-order Doppler shift. For the trap we have designed, L is 75 mm. Taking $R_{\text{sph}} = 2.5$ mm for 2×10^6 $^{199}Hg^+$ ions in a spherical trap with similar overall size, we find that the linear trap capacity is about 18 times larger. Furthermore, it seems likely that the transverse dimension of the linear trap can be reduced to a value 100 or more times smaller than its length while maintaining constant ion number and second-order Doppler shift. This corresponds to a reduction in volume of 10 000 times.

DESIGN OF A LINEAR ION TRAP

We have designed a linear trap consisting of four molybdenum rods equally spaced on an approximately 1 cm radius. Axial confinement is accomplished by means of OFHC copper pins with dc bias which are located at each end and which are about 75 mm apart. The proximity of the four rods also aids axial confinement by localizing the coulomb interaction to an axial region with a length approximately equal to the trap's transverse dimension R. We calculate that a peak electrode potential of $V_0 = 180$ V at $\Omega = 2\pi\,500$ kHz is required to obtain a secular frequency $\omega = 2\pi\,50$ kHz.

The input optical system which performs state selection and also determines which hyperfine state the ions are in has been modified from the previous system.[9] The present system illuminates about 1/3 of the 75-mm-long cylindrical ion cloud. An ion's room temperature thermal motion along the axis of the trap will give an average round trip time of 1.4 ms, a value which is much smaller than our optical pumping and interrogation times. Thus, during the time of the optical pulse all ions will be illuminated, and pumping and interrogation completed. The only change is that a somewhat longer optical pulse is required.

In order to operate within the Lambe-Dicke regime[10] the 40.5-GHz microwave resonance radiation will be propagated perpendicular to the line of ions. The ions should then all experience phase variations of this radiation which are less than π so that the first-order Doppler absorption in sidebands induced by an ions motion will not degrade the 40.5-GHz fundamental.

The optical axis of the fluorescence collection system is perpendicular to the axis of the input optical system as in the previous system. There is one difference, however. In the hyperbolic trap the collection has in its field of view the ion cloud and the semitransparent mesh of both endcap trap electrodes. This mesh can scatter stray light into the collection system which will degrade the signal-to-noise ratio in the clock resonance. This linear trap has no trap electrodes, mesh or otherwise, in its field of view and, consequently, should have less detected stray light, allowing further performance improvement over the spherical trap.

CONCLUSIONS

Trapped ion frequency standards eliminate containing walls and their associated perturbations of the atomic transition frequencies by using electromagnetic fields alone to confine the particles. For any given trap, however, there exists a tradeoff between the number of ions in the trap and a frequency shift due to second-order Doppler effects. This tradeoff directly affects performance of the standard since the frequency shift is typically very much larger than the ultimate stability required and since the statistical limit to performance is directly related to ion number. We have calculated this performance tradeoff for a rf trap with cylindrical geometry, a case not previously considered for a trapped ion frequency source.

By replacing the single node in the rf trapping field for a spherical trap by a line of nodes, a cylindrical trap effectively increases effective volume without increasing overall size. Furthermore, this performance is found to be independent of its transverse dimensions, as long as the driving frequency is scaled appropriately, with the driving voltage unchanged. More specifically, for the same frequency shift, we find that a linear trap with length L can hold as many ions as a spherical trap with diameter $6L/5$. In addition to the practical advantage of greatly reduced overall volume, a fundamental advantage is also allowed since operation within the Lambe-Dicke regime places a limit on the size of the ion cloud, a requirement which may be met for a cylindrical trap by irradiating the microwave atomic transition in a direction perpendicular to the trap's longitudinal axis.

We have designed a trapped ion frequency source in which a cylindrical trap is implemented with a combination of rf and dc electric fields. With similar overall size and improved optical performance, this trap has 15 to 20 times the ion storage volume as conventional rf traps with no increase in second-order Doppler shift.

ACKNOWLEDGMENTS

We wish to thank Dave Seidel for assisting in the design of the linear trap described here and G. R. Janik for helpful comments. This work represents the results of one phase of research carried out at the Jet Propulsion Laboratory, California Institute of Technology, under contract sponsored by the National Aeronautics and Space Administration.

[1]D. W. Allan, in *Proceedings of the 19th Annual Precise Time and Time Interval Applications and Planning Meeting*, edited by R. L. Sydnor (U.S. Naval Observatory, Washington, DC, 1987), p. 375.

[2]L. S. Cutler, R. P. Giffard, P. J. Wheeler, and G. M. R. Winkler, Proceedings of the 41st Annual Symposium Frequency Control, IEEE Cat. No. 87CH2427-3, 12 (1987).

[3]L. A. Rawley, J. H. Taylor, M. M. Davis, and D. W. Allan, Science **238**, 761 (1987).

[4]J. W. Armstrong, F. B. Estabrook, and H. D. Wahlquist, Astrophys. J. **318**, 536 (1987).

[5]rf and Penning traps are both subject to second-order Doppler problems related to the trapping fields. In this paper we confine our attention specifically to the case of rf traps. We simply note that, for the same number of stored ions, magnetron rotation of the ion cloud in a Penning trap leads to a second-order Doppler shift of comparable magnitude to that for a rf trap. A Penning trap based clock is described in: J. J. Bollinger, J. D. Prestage, W. M. Itano, and D. J. Wineland, Phys. Rev. Lett. **54**, 1000 (1985).

[6]H. G. Dehmelt, Adv. At. Mol. Phys. **3**, 53 (1967).

[7]L. S. Cutler, R. P. Giffard, and M. D. McGuire, Appl. Phys. B **36**, 137 (1985).

[8]S. S. Prasad and T. M. O'Neil, Phys. Fluids **22**, 278 (1979).

[9]J. D. Prestage, G. J. Dick, and L. Maleki, in *Proceedings of the 19th Annual Precise Time and Time Interval Applications and Planning Meeting*, edited by R. L. Sydnor (U.S. Naval Observatory, Washington, DC, 1987), p. 285.

[10]R. H. Dicke, Phys. Rev. **89**, 472 (1953).

Appl. Phys. B 66, 603–608 (1998)

Applied Physics B
Lasers
and Optics

Ion strings for quantum gates

H.C. Nägerl, W. Bechter, J. Eschner, F. Schmidt-Kaler, R. Blatt

Institut für Experimentalphysik, Universität Innsbruck, Technikerstraße 25, A-6020 Innsbruck, Austria

Received: 6 August 1997/Revised version: 21 October 1997

Abstract. Crystal structures of calcium ions have been prepared in a linear Paul trap. The trapped ions are laser-cooled by simultaneous resonant excitation near 397 nm and 866 nm. Images of the fluorescing ions are obtained with a CCD camera and show the individual ions spatially resolved. Complex crystal structures of more than 60 ions have been observed whereas smaller crystals with up to 10 ions arrange in a linear string. Measured distances between the ions in strings of different lengths are in good agreement with expected values obtained from a harmonic trap potential. The application of a calcium ion string for quantum computation is discussed.

PACS: 32.80.Pj, 42.50.Vk

Ion traps have been shown to provide an ideal environment for isolated quantum systems such as a single, trapped and laser-cooled atoms. Ion storage has long been applied to ultra-high precision spectroscopy and the development of frequency standards [1]. More recently, single trapped ions have been used to demonstrate and test some of the intriguing features of quantum mechanics [2, 3]. In particular, both the internal electronic state and the motional state of a trapped ion can be modified using laser light. Decoherence of internal superposition states is nearly negligible even for very long interaction times. To explore these properties, several schemes have been proposed for the preparation of non-classical states of motion in a trap [4]. Their experimental realization [2] promises further improvement for the precision of spectroscopic measurements [5, 6]. With almost perfect control of the quantum state of a single ion, attention has turned to systems of few ions with well-controlled interactions between them. Manipulations of their overall quantum state include the preparation of entangled states that have no classical counterpart. The possibility of entangling massive particles opens up many prospects for new experiments including measurements with Bell states and GHZ states [7] which would allow for new tests of quantum mechanics. Moreover, entanglement of particles will allow one to study quantum measurements such as the investigation of decoherence processes [3, 8] in more detail.

A very exciting proposal is the application of linear ion traps and the collective quantum motion of a trapped string of ions for the realization of a quantum gate [9]. Quantum gates are the basic building blocks of a quantum computer and their operation relies fundamentally on the entanglement of internal degrees of freedom of the ions (electronic excitation) and their collective motion (vibrational excitation). A quantum computer works with quantum registers made up of quantum bits (qbits) which can be manipulated analogously to classical bits using gate operations. The quantum-mechanical analogue of a classical XOR gate, the so-called controlled-NOT gate operation, can be realized using a linear string of ions and a well-defined series of laser pulses to address different ions. It has been shown that a controlled-NOT gate operating on two ions is a realization of a universal quantum gate, so that in principle universal computations can be carried out using just the two-ion quantum gate and one-bit rotations [10]. The realization of these gate operations based on a string of ions would therefore be of fundamental interest and furthermore all basic algorithms could be tested using just a string of trapped ions.

Many species of ions have been used for ion trapping, and strings in linear traps have been experimentally demonstrated with Be^+ ions [11] and Mg^+ ions [12]. However, it was shown that $^{40}Ca^+$ is one of the most promising candidates for the realization of quantum gates, because of its mass, level structure, and transition frequencies and widths [13]. Furthermore, quantum gates require certain characteristics of the linear trap that have not been met in earlier experiments, in particular optical access from many directions.

In this paper, we report the first trapping of strings of $^{40}Ca^+$ ions in a linear trap, as a significant step towards the realization of a quantum gate. The novel trap that we use and describe in detail, has been designed especially for storing small numbers (up to a few tens) of $^{40}Ca^+$ ions with the aim of producing linear strings and using them as a quantum register. The paper is organized as follows. In Sect. 1 we present the requirements for an application of ion strings to quantum computation in the particular case of $^{40}Ca^+$. We also highlight the advantages of the specific choice of this ion. In Sect. 2 the experimental setup is described and in

Sect. 3 experimental results are presented and discussed. In the concluding section our next experimental steps towards the operation of a quantum gate are described.

1 $^{40}Ca^+$ for quantum computation

Quantum mechanical information is delicate and must be stored in systems which are essentially free of environmental perturbations. Any interactions with an environment, such as collisions with walls or surrounding atoms of background gas, have an effect similar to a measurement process in that they tend to alter or destroy the quantum-mechanical information by inducing decoherence. Furthermore, the necessity to manipulate quantum states in order to implement algorithms requires that the quantum information can be stored for times that are long enough to allow for coherent interaction of the ions with external fields. These requirements are all met by the ion storage technique and therefore that is why stored ions have been proposed for future use in quantum computation. In that spirit, a single laser-cooled Be^+ ion has been used to demonstrate the manipulations necessary for the implementation of a controlled-NOT gate [14]. The single qbit quantum information (the so-called target bit) was stored in two hyperfine ground states of the Be^+ while the other qbit (the so-called control bit) was encoded in the quantized vibration. The gate operations were realized with optical Raman transitions. Thus the potential of trapped ions for quantum bits has been clearly demonstrated. However, since in that experiment the control and target bit are internal and external states of the same single ion, it cannot be scaled up to realize a larger quantum register. Instead, several ions with controlled interaction between them are required. For that purpose, a promising choice would be to use a string of $^{40}Ca^+$ ions. Each ion in the linear string represents a qbit with the quantum information stored in a metastable optical transition. Gate operations involve one additional bit (the so-called bus bit) for which the common center-of-mass vibration of the ion string would be used.

The relevant energy levels of $^{40}Ca^+$ are shown in Fig. 1. All transitions can be driven by diode lasers, frequency-doubled diode lasers, or Ti:sapphire lasers. Optical cooling and detection of resonance fluorescence is achieved by simultaneous application of laser light at 397 nm and 866 nm. The $D_{3/2}$ and $D_{5/2}$ levels have lifetimes of about 1 s [15, 16] and together with the $S_{1/2}$ ground state they can be used to store quantum information. Performing a quantum computation with a string of trapped $^{40}Ca^+$ ions prepared in the vibrational ground state will require a number of laser pulses on the $S_{1/2}$ to $D_{5/2}$ transition to be applied coherently to individual ions in accord with the algorithm given by the computational problem. Determining which of the ions are left in the $S_{1/2}$ ground state will conclude the computational cycle and yield the output of the quantum computer. A necessary ingredient is therefore the ability to measure state populations with a 100% detection efficiency. This is routinely done with trapped ions using the electron shelving technique [17–19] and can be realized with $^{40}Ca^+$ by exciting the ion to the $D_{5/2}$ state. Subsequent probing on the $S_{1/2}$ to $P_{1/2}$ transition results in fluorescence being either generated or not, indicating with certainty whether the ground state is populated or not.

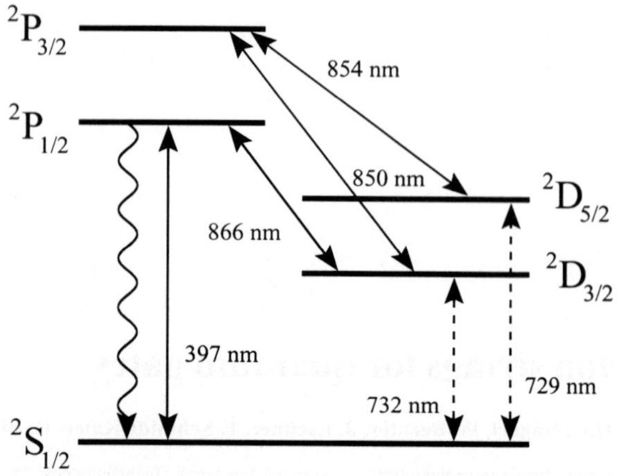

Fig. 1. Level scheme of $^{40}Ca^+$

This operation of quantum gates, as proposed by Cirac and Zoller [9], also requires the trapped string of ions to be cooled to the ground state of their collective vibrational motion. This cannot be achieved with cooling on the allowed transitions only but requires additional cooling techniques such as Raman cooling or sideband cooling with coupled transitions. A discussion of these and their experimental demonstration with *single* ions is given in [20–22]. With the transitions available in $^{40}Ca^+$, both advanced cooling techniques are possible. In particular, simultaneous excitation of the low-energy vibrational sideband of the $S_{1/2}$ to $D_{5/2}$ transition at 729 nm, and the $D_{5/2}$ to $P_{3/2}$ transition near 854 nm, would provide efficient sideband cooling to the motional ground state of the ion string.

It is clear that coherent manipulation of the $^{40}Ca^+$ ions requires a highly stabilized laser for the S–D transition near 729 nm. Decoherence will set in on a time scale proportional to the inverse laser bandwidth and limit the number of coherent manipulations that are possible. With the present laser system, already several gate operations could be performed. Decoherence of the qbits during their manipulation could be further suppressed by using ground state Zeeman coherences which would be controlled via radio-frequency techniques and Raman excitations [13, 14]. Another possibility is to store the quantum information in superpositions of the two metastable D states. At the expense of employing an additional laser source near 850 nm, phaselocked to the 854 nm laser with, for example, a comb generator [23], many coherent manipulations then become possible using optical transitions.

2 Trap design, laser sources, and fluorescence detection

Linear traps for ion clouds or for a few laser-cooled ions have been investigated by many groups [11, 12, 24–26]. For computational purposes the linear trap should be optimized to hold linear strings of ions and the motion should be as harmonic as possible to allow for optimal cooling. In addition, optical access to the trap should be very good to ensure optimal imaging and application of the manipulating beams which address individual ions. This in turn requires that the mean spacing between the ions should be such that a laser beam near 729 nm can be easily focused on any chosen ion without exciting adjacent ones.

Figure 2 shows the realized trap, which is mounted inside the UHV system. It consists of 4 stainless steel rods with a diameter of 0.6 mm at a center distance of 2 mm, diagonally connected to generate the quadrupole rf field for dynamic confinement in the x–y plane perpendicular to the trap z axis. Two ring electrodes with dc potentials serve as the axial endcaps for longitudinal static confinement. The endcap rings have an inner diameter of about 4 mm and the spacing between them is 10 mm. The rf drive frequency ($\Omega/2\pi = 18$ MHz) is amplified, resonantly enhanced by a helical resonator (loaded $Q \approx 250$) and coupled to one pair of rod electrodes with the other pair grounded. The alternating rf potential yields a trap (quasi-)potential with secular frequencies of up to $\omega_r = \omega_{x,y} \approx 1.2$ MHz. For longitudinal confinement, dc voltages between 20 V and 400 V are applied to the rings resulting in an axial vibrational frequency ω_z of between 20 kHz and 400 kHz. At standard operating conditions, ω_z is about 180 kHz. Numerical calculations reveal that the axial static trap potential is a very good approximation to a harmonic potential. According to these calculations, within a distance in the z direction of about 50 μm from the trap centre, contributions of higher-order potentials are as small as 2×10^{-6}. For the frequencies $(\omega_r, \omega_z) = (1.2, 0.2)$MHz we obtain the Lamb–Dicke parameters $\eta_{r,z} = \sqrt{\hbar k^2/2m\omega_{r,z}} = (0.09, 0.22)$ for the transition near $\lambda = 729$ nm and $(\eta_r, \eta_z) = (0.16, 0.4)$ for the $\lambda = 397$ nm cooling transition. Here, $k = \sin\alpha 2\pi/\lambda$ is the effective wavevector of the light beam (applied at an angle $\alpha = 45°$ with respect to the r and z axes) and $m = 40$ amu denotes the mass of the Ca^+ ion. The rms size $\langle q \rangle = \sqrt{\hbar/2m\omega_q}$ of the ground state wavepacket corresponds to $(\langle r \rangle, \langle z \rangle) = (10, 25)$ nm. Within the Doppler cooling limit $E_D = \hbar\Gamma/2$, the mean vibrational quantum numbers of $(\langle n_r \rangle, \langle n_z \rangle) \approx (8, 50)$ are reached.

In order to achieve optical cooling in all dimensions, the ions are illuminated by laser light from the $(x, y, z) = (0, 1, -1)$ and the $(x, y, z) = (-1, 0, -1)$ direction. The emitted fluorescence light at 397 nm is collected simultaneously in opposite directions (along x) and recorded with a photomultiplier and a CCD camera, respectively. For the CCD camera, we use an $f/1$ optics with 40-fold magnification and an overall photon detection efficiency of 10^{-3}. Ca^+ ions are produced from a weak atomic beam by impact ionization from an electron beam focused to the trap centre. The magnetic field is controlled by Helmholtz coils in all dimensions.

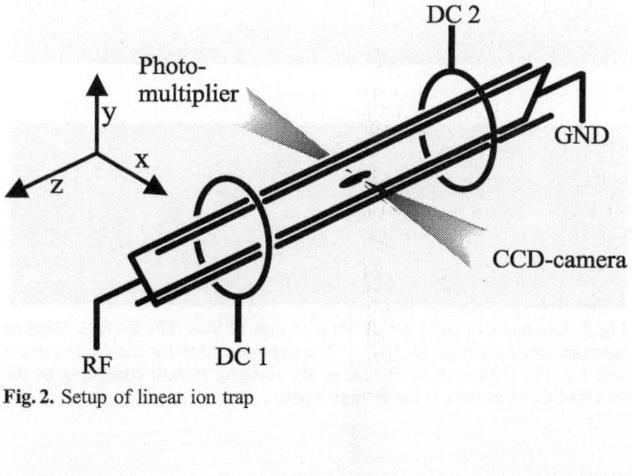

Fig. 2. Setup of linear ion trap

For excitation and manipulation of the trapped ions, solid-state and diode laser sources near 397 nm, 866 nm, 729 nm, and 854 nm are used. In order to generate the light at 397 nm, a Ti:sapphire ring laser is radio-frequency stabilized to an external cavity resulting in an rms linewidth of 250 kHz and a long-term drift stability of a few MHz/h. The output of up to 1.5 W at 793 nm is frequency doubled using an LBO crystal inside an external enhancement resonator. Typically, 25–30 mW of light near 397 nm is coupled with a fiber to the ion trap located on a different optical table. The fiber output of 5 mW is spatially filtered and allows for stable optical adjustment. About 15 mW of light at 866 nm is generated with an external-grating cavity diode laser in a Littrow arrangement. This laser is locked to a temperature-stabilized cavity yielding an rms linewidth of about 30 kHz. Thus, a low drift rate (few MHz/h) and a sufficient short-term stability of the diode laser is achieved. About 50 mW of light near 729 nm is generated with a second Ti:sapphire laser and is also fiber-coupled to the ion trap. This laser has been radio-frequency stabilized to an external cavity and provides a bandwidth near 10 kHz. Another diode laser with an external grating cavity produces 15 mW of light at 854 nm with a free-running bandwidth of 1 MHz. Stable and uninterrupted locking of all lasers has been accomplished for more than 5 h. The wavelengths of the lasers can be monitored with wavemeters, and for the laser sources at 397 nm, 866 nm, and 854 nm we obtain optogalvanic signals from a hollow cathode lamp. Laser frequencies and intensities are computer-controlled by acousto-optical modulators.

3 Experimental results

A typical experimental run starts with the preparation of a large cloud of ions. Ions are generated with the oven and electron gun operating, and the alignment of the laser beams with respect to the cloud is optimized. The laser frequencies are set to optimal cooling. Optionally, background gas cooling can be used by simply switching off the ion getter pump. The oven and electron gun are then switched off, and the getter pump is switched on again. The following procedure depends on the ordered structures to be produced. For large crystal structures, it is preferable to block the laser at 397 nm for several minutes. Radio-frequency heating of the extended cloud [27,28] then results in a reduction in the cloud size, and when the laser is switched on again with the detuning chosen correctly, a large crystal ionic structure is usually obtained. We observed structures with more than 60 ions. An example is shown in Fig. 3 where the ions order themselves in opposing pairs of perpendicular orientation. Note the bright central spots in Fig. 3, which are due to two ions contributing to the fluorescence light in this particular line of sight.

Crystallization of the cloud can also be observed in a spectroscopic measurement [12]. This is shown in Fig. 4 where the fluorescent light, detected with the photomultiplier tube, is recorded as a function of the detuning near 397 nm. The detuning is scanned into resonance from the

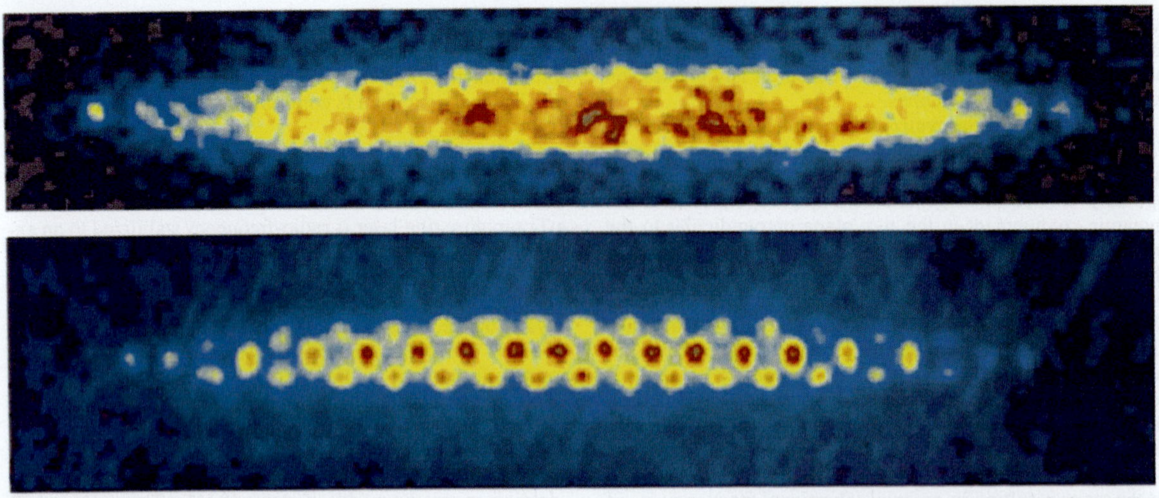

Fig. 3. *Upper part*: large hot ion cloud consisting of 62 ions, laser detuning not optimized for cooling (cf. Fig. 4a). *Lower part*: same sample for optimized optical cooling (cf. Fig. 4b), ion crystal consists of 62 ions. Note that the bright central spots represent the overlapping light of two ions. The total length of this ion crystal is 180 μm at $\omega_z = 195 \pm 5\,\text{kHz}$

low-frequency side (from left to right in Fig. 4). Far below resonance (see (a) in Fig. 4) the fluorescence intensity increases as observed for a hot ion cloud (see upper part of Fig. 3). Laser cooling is not efficient because of the large detuning. At a certain detuning closer to resonance, crystallization shows up as a sharp feature in the spectrum which is the result of the sudden change from the Doppler-broadened cloud to a crystal. The fluorescence response of the crystal is similar to the excitation spectrum of a single laser-cooled ion. The sharp feature at the right of Fig. 4 indicates again the change to a cloud-like behavior, i.e. the melting of the ion crystal. The decrease of fluorescence

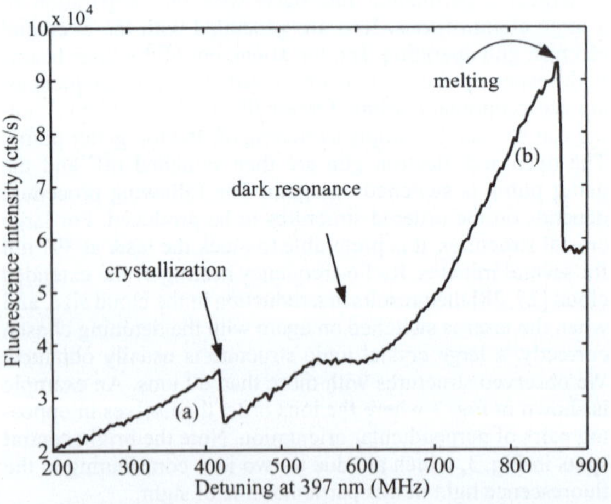

Fig. 4. Crystallization and melting of an ion cloud is observed when the cooling laser is tuned across the atomic resonance. The line shape to the left and to the right of the sharp features corresponds to a Doppler-broadened (cloud-like) spectrum, whereas the central part corresponds to a crystalline state. The upper part of Fig. 3 was taken at a detuning marked with (a), the lower part of Fig. 3 was taken at the detuning marked with (b)

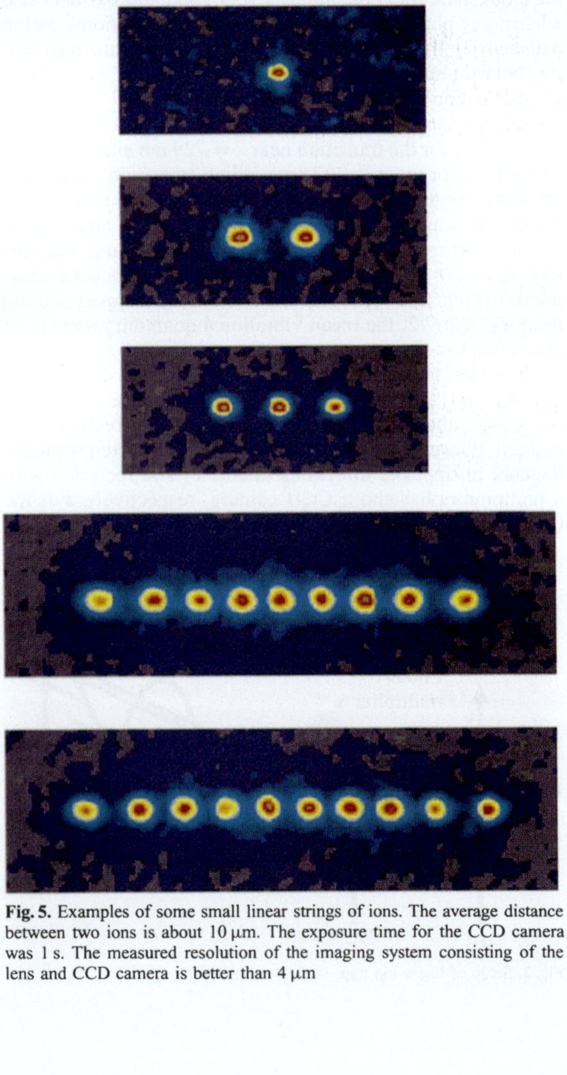

Fig. 5. Examples of some small linear strings of ions. The average distance between two ions is about 10 μm. The exposure time for the CCD camera was 1 s. The measured resolution of the imaging system consisting of the lens and CCD camera is better than 4 μm

light in the central part of the spectrum indicates a dark resonance in Ca^+ [29] which is broadened by the cooling laser.

For the trap parameters indicated above and the current trap design, ion numbers larger than 10 result in three-dimensional ordered structures. Smaller ion numbers and subsequently linear strings can be prepared by using heating procedures to reduce the ion number further. This is achieved either by applying blue detuned laser radiation on the 397 nm transition (laser heating) or by application of a radio frequency tuned to the secular vibration frequency of the ions. Since laser heating acts only on the ^{40}Ca isotope, the probability of other Ca^+ isotopes staying trapped is increased. These ions do not interact with the laser light and show up as non-fluorescing lattice sites in the crystal.

For single ions or for small strings of up to 5 ions it is easier experimentally to load them directly. This is achieved with the lasers set to optimal cooling, the oven operating continuously and the electron gun being pulsed for a few seconds. Images of the fluorescing ions are then readily obtained with the CCD camera with a spatial resolution of 4 μm. Several examples are shown in Fig. 5. The images were taken with an exposure time of 1 s and a spatial average is taken over 5 pixels which corresponds to about 2.5 μm. Note that the background light is subtracted to obtain the images.

For the realization of a two-bit quantum gate, individual ions have to be addressed by a laser beam. In a harmonic linear trap a single parameter, the trap frequency ω_z, suffices to describe the distances between the ions of a string, which allows one to calculate the expected positions of the ions. For up to 4 ions analytical results can be given, and for larger strings numerical results are readily derived [13]. Figure 6 shows a comparison of the experimental results with numerical calculations based on a value of ω_z = 181 kHz. Note that the deviations of the experimental values from the calculated positions are smaller than the optical resolution of the imaging system. These systematic deviations arise since most of the data were obtained during different experimental runs which required renewed loading of the trap. This procedure (calcium atomic beam and electron impact ionization) causes stray surface potentials on the electrodes owing to patch effects, which in turn modify the trap frequency ω_z slightly. In fact, each measurement in Fig. 6 (i.e. each string) could be individually fitted and an individual trap frequency could be assigned resulting in even better agreement with theory. However, the data presented in Fig. 6 prove that even a single value of the longitudinal trap frequency suffices to describe the ion positions well within the resolution of our optical system and, in particular, accurately enough to allow the steering of the addressing laser. Furthermore, the average distances in the order of 10–20 μm should be sufficient to focus radiation near 729 nm onto the individual ions.

The trap frequency ω_z was also measured directly. For this, a frequency signal was applied to the axially confining rings and tuned across the frequency range around ω_z. On resonance, the collective axial motion is excited and the ions heat up, which results in blurred pictures on the CCD camera. The observed values of ω_z are in good agreement with those calculated from the ion distances. Furthermore, from the well-resolved resonances we expect to be able to selectively excite the center-of-mass vibration in both side-

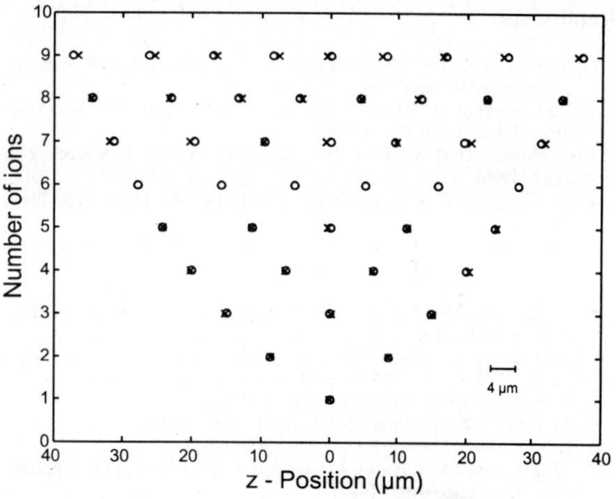

Fig. 6. Measured positions (x) of ions in strings of indicated number (1–9), compared with values (o) from analytic and numeric calculations using an axial frequency of ω_z = 181 kHz. No experimental data were taken for 6 ions. The deviations are within the 4-μm resolution of the imaging system indicated by the horizontal bar

band cooling and gate operation, without coupling into other modes of vibration.

4 Conclusions and outlook

We have built a linear ion trap optimized for trapping a small number of $^{40}Ca^+$ ions and performing quantum gate operations between them. We have shown how strings of Ca^+ ions in this trap can be laser cooled and we observed them crystallize in a linear string. Using a CCD camera, we have been able to image the individual ions with high spatial resolution. The distances measured between the ions agree with analytically and numerically calculated values based on a harmonic trap potential. Also, the directly measured frequency of axial vibration is consistent with the value of ω_z obtained from the ion positions. Addressing individual ions will be achieved using an acousto-optic deflector driven by appropriate radio frequencies which can be derived from the distance measurements above.

The agreement between the expected and measured trap characteristics is encouraging for a future application of the $^{40}Ca^+$ strings as a quantum register. With additional cooling schemes it will be possible to reach the vibrational ground state [20–22] and then the ion strings can be applied to perform a quantum gate operation. Currently, the laser near 729 nm is being stabilized to an ultrahigh-finesse resonator to provide the frequency stability which is ultimately necessary for multiple gate operations.

Acknowledgements. This work is supported by the Fonds zur Förderung der wissenschaftlichen Forschung (FWF) under contract number P11467-PHY and in parts by the TMR networks "Quantum Information" (ERB-FMRX-CT96-0087), and "Quantum Structures" (ERB-FMRX-CT96-0077).

References

1. R. Blatt: In Atomic Physics **14**, ed. by D.J. Wineland, C.E. Wieman, S.J. Smith (AIP, New York 1995), pp. 219
2. D.M. Meekhof, C. Monroe, B.E. King, W.M. Itano, D.J. Wineland: Phys. Rev. Lett. **76** 1796 (1996)
3. C. Monroe, D.M. Meekhof, B.E. King, D.J. Wineland: Science **272**, 1131 (1996)
4. J.I. Cirac, A.S. Parkins, R. Blatt, P. Zoller: Adv. At. Molec. Opt. Phys. **37**, p. 238 (1996)
5. W.M. Itano, J.C. Bergquist, J.J. Bollinger, J.M. Gilligan, D.J. Heinzen, F.L. Moore, M.G. Raizen, D.J. Wineland: Phys. Rev. A **47**, 3554 (1993)
6. J.J. Bollinger, W.M. Itano, D.J. Wineland, D.J. Heinzen: Phys. Rev. A **54**, R4649 (1996)
7. D.M. Greenberger, M.A. Horne, A. Shimony, A. Zeilinger: Am. J. Phys. **58**, 1131 (1990); N.D. Mermin: Phys. Today (June, 9 1990)
8. W.H. Zurek: Phys. Today (October, 36 1991)
9. J.I. Cirac, P. Zoller: Phys. Rev. Lett. **74**, 4091 (1995)
10. D.P. DiVincenzo: Phys. Rev. A **51**, 1015 (1995)
11. M.G. Raizen, J.M. Gilligan, J.C. Bergquist, W.M. Itano, D.J. Wineland: Phys. Rev. A **45**, 6493 (1992)
12. I. Waki, S. Kassner, G. Birkl, H. Walther: Phys. Rev. Lett. **68**, 2007 (1992)
13. A. Steane: Appl. Phys. B **64**, 623 (1997)
14. C. Monroe, D.M. Meekhof, B.E. King, W.M. Itano, D.J. Wineland: Phys. Rev. Lett. **75**, 4714 (1995)
15. T. Gudjons, B. Hilbert, P. Seibert, G. Werth: Europhys. Lett. **33**, 595 (1996)
16. M. Knoop, M. Vedel, F. Vedel: Phys. Rev. A **52**, 3763 (1995)
17. W. Nagourney, J. Sandberg, H. Dehmelt: Phys. Rev. Lett. **56**, 2797 (1986)
18. Th. Sauter, W. Neuhauser, R. Blatt, P.E. Toschek: Phys. Rev. Lett. **57**, 1696 (1986)
19. J.C. Bergquist, R. Hulet, W.M. Itano, D.J. Wineland: Phys. Rev. Lett. **57**, 1699 (1986)
20. C. Monroe, D.M. Meekhof, B.E. King, S.R. Jeffers, W.M. Itano, D.J. Wineland, P. Gould: Phys. Rev. Lett. **74**, 4011 (1995)
21. F. Dietrich, J.C. Bergquist, W.M. Itano, D.J. Wineland: Phys. Rev. Lett. **62**, 403 (1989)
22. I. Marzoli, J.I. Cirac, R. Blatt, P. Zoller: Phys. Rev. A **49**, 2771 (1994)
23. M. Kourogi, B. Widiystomoko, Y. Takeuchi, M. Ohtsu: IEEE J. Quantum Electron. **31**, 2120 (1995)
24. M.E. Poitzsch, J.C. Bergquist, W.M. Itano, D.J. Wineland: Rev. Sci. Instrum. **67**, 129 (1996)
25. P.T.H. Fisk, M.J. Sellars, M.A. Lawn, C. Coles: Appl. Phys. B **60**, 519 (1995)
26. J.D. Prestage, G.J. Dick, L. Maleki: J. Appl. Phys. **66**, 1013 (1989)
27. H.G. Dehmelt: Adv. At. Mol. Opt. Phys. **3**, 53 (1967); *ibid.* **5**, 109 (1969)
28. R. Blümel, C. Kappler, W. Quint, H. Walther: Phys. Rev. A **40**, 808 (1989)
29. I. Siemers, M. Schubert, R. Blatt, W. Neuhauser, P.E. Toschek: Europhys. Lett. **18**, 139 (1992)

Appl. Phys. B 66, 181–190 (1998)

Applied Physics B
Lasers
and Optics

Quantum dynamics of cold trapped ions with application to quantum computation

D.F.V. James

Theoretical Division (T-4), Los Alamos National Laboratory, Los Alamos, NM 87545, USA
(Fax: +505/665-3909, E-mail: dfvj@t4.lanl.gov)

Received: 10 March 1997/Revised version: 10 July 1997

Abstract. The theory of interactions between lasers and cold trapped ions as it pertains to the design of Cirac–Zoller quantum computers is discussed. The mean positions of the trapped ions, the eigenvalues and eigenmodes of the ions' oscillations, the magnitude of the Rabi frequencies for both allowed and forbidden internal transitions of the ions, and the validity criterion for the required Hamiltonian are calculated. Energy level data for a variety of ion species are also presented.

PACS: 32.80.Qk; 42.50.Vk; 89.80.+h

A quantum computer is a device in which data can be stored in a network of quantum mechanical two-level systems, such as spin-1/2 particles or two-level atoms. The quantum mechanical nature of such systems allows the possibility of a powerful new feature to be incorporated into data processing, namely, the capability of performing logical operations upon quantum mechanical superpositions of numbers. Thus in a conventional digital computer each data register is, throughout any computation, always in a definite state "1" or "0"; however in a quantum computer, if such a device can be realized, each data register (or "qubit") will be in an undetermined quantum superposition of two states, $|1\rangle$ and $|0\rangle$. Calculations would then be performed by external interactions with the various two-level systems that constitute the device, in such a way that conditional gate operations involving two or more different qubits can be realized. The final result would be obtained by measurement of the quantum mechanical probability amplitudes at the conclusion of the calculation. Much of the recent interest in practical quantum computing has been stimulated by the discovery of a quantum algorithm that allows the determination of the prime factors of large composite numbers efficiently [1] and of coding schemes that, provided operations on the qubits can be performed within a certain threshold degree of accuracy, will allow arbitarily complicated quantum computations to be performed reliably *regardless of operational error* [2].

So far, the most promising hardware proposed for implementation of such a device seems to be the cold-trapped ion system devised by Cirac and Zoller [3]. Their design, which is shown schematically in Fig. 1, consists of a string of ions stored in a linear radiofrequency trap and cooled sufficiently so that their motion, which is coupled together due to the Coulomb force between them, is quantum mechanical in nature. Each qubit would be formed by two internal levels of each ion, a laser being used to perform manipulations of the quantum mechanical probability amplitudes of the states, conditional two-qubit logic gates being realized with the aid of the excitation or de-excitation of quanta of the ions' collective motion. For a more detailed description of the concept of cold-trapped ion quantum computation, the reader is referred to the article by Steane [4].

There are two distinct possibilities for the choice of the internal levels of the ion: first, the two states could be the ground state and a metastable excited state of the ion (or more precisely, sublevels of these states) and second, the two states could be two nearly degenerate sublevels of the ground state. In the first case, a single laser would suffice to perform the required operations; in the second, two lasers would be required to perform Raman transitions between the states, via a third level. Both of these schemes have advantages: the first, which I will refer to as the "single photon" scheme, has the great advantage of conceptual and experimental simplicity; the second, the "Raman scheme", offers the advantages of a very low

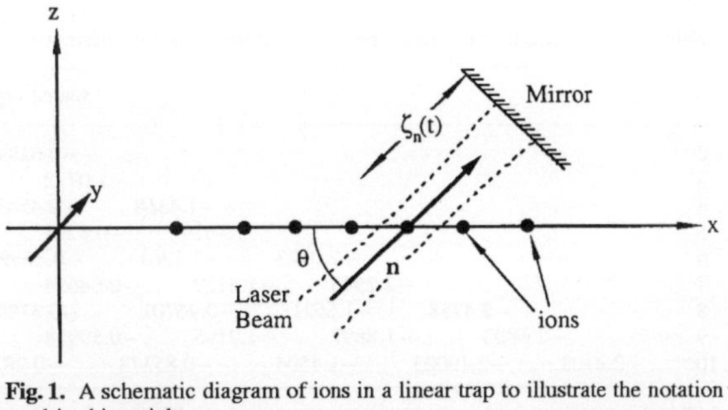

Fig. 1. A schematic diagram of ions in a linear trap to illustrate the notation used in this article

rate for spontaneous decay between the two nearly degenerate states and resilience against fluctuations of the phase of the laser. This later scheme was recently used by the group headed by Dr. D.J. Wineland at the National Institute of Science and Technology at Boulder, Colorado to realize a quantum logic gate using a *single* trapped Beryllium ion [5].

In this article, I will discuss the theory of laser interactions with cold trapped ions as it pertains to the design of a Cirac–Zoller quantum computer. I will concentrate on the "single photon scheme" as originally proposed by those authors, although much of the analysis is also relevant to the "Raman scheme". Fuller accounts of aspects of this are available in the literature: see, for example, [4, 6, 7]; however the derivation of several results are presented here for the first time. I will also present relevant data gleaned from various sources on some species of ion suitable for use in a quantum computation.

1 Equilibrium positions of ions in a linear trap

Let us consider a chain of N ions in a trap. The ions are assumed to be strongly bound in the y and z directions but weakly bound in an harmonic potential in the x direction. The position of the mth ion, where the ions are numbered from left to right, will be denoted $x_m(t)$. The motion of each ion will be influenced by an overall harmonic potential due to the trap electrodes and by the Coulomb force exerted by all of the other ions. We will assume that the binding potential in the y and z directions is sufficiently strong that motion along these axes can be neglected. However, motion of the ions transverse to the trap axis can be important in some circumstances: Garg [8] has pointed out that such motion can be a source of decoherence; furthermore if a large number of ions are stored in the trap, the transverse vibrations can become unstable, and the ions will adopt a zigzag configuration [9]. Hence the potential energy of the ion chain is given by the following expression:

$$V=\sum_{m=1}^{N}\frac{1}{2}M\nu^2 x_m(t)^2+\sum_{\substack{n,m=1\\ m\neq n}}^{N}\frac{Z^2e^2}{8\pi\epsilon_0}\frac{1}{|x_n(t)-x_m(t)|}\,, \tag{1}$$

where M is the mass of each ion, e is the electron charge, Z is the degree of ionization of the ions, ϵ_0 is the permitivity of free space, and ν is the trap frequency, which characterizes the strength of the trapping potential in the axial direction. Note that this is an unconventional use of the symbol ν, which often denotes frequency rather than angular frequency; following Cirac and Zoller, I will use ω to denote the angular frequencies of the laser or the transitions between internal states of the ions, and ν to denote angular frequencies associated with the motion of the ions.

Assume that the ions are sufficiently cold that the position of the mth ion can be approximated by the formula

$$x_m(t)\approx x_m^{(0)}+q_m(t) \tag{2}$$

where $x_m^{(0)}$ is the equilibrium position of the ion, and $q_m(t)$ is a small displacement. The equilibrium positions will be determined by the following equation:

$$\left[\frac{\partial V}{\partial x_m}\right]_{x_m=x_m^{(0)}}=0\,. \tag{3}$$

If we define the length scale ℓ by the formula

$$\ell^3=\frac{Z^2e^2}{4\pi\epsilon_0 M\nu^2} \tag{4}$$

and the dimensionless equilibrium position $u_m=x_m^{(0)}/\ell$, then (3) may be rewritten as the following set of N coupled algebraic equations for the values of u_m:

$$u_m-\sum_{n=1}^{m-1}\frac{1}{(u_m-u_n)^2}+\sum_{n=m+1}^{N}\frac{1}{(u_m-u_n)^2}=0 \qquad (m=1,2,\ldots N)\,. \tag{5}$$

For $N=2$ and $N=3$, these equations may be solved analytically:

$$N=2:\quad u_1=-(1/2)^{2/3},\quad u_2=(1/2)^{2/3}, \tag{6}$$

$$N=3:\quad u_1=-(5/4)^{1/3},\quad u_2=0,\quad u_3=(5/4)^{1/3}\,. \tag{7}$$

For larger values of N it is necessary to solve for the values of u_m numerically. The numerical values of the solutions to these equations for 2 to 10 ions is given in Table 1. Determining the solutions for larger numbers of ions is a straightforward but time consuming task.

By inspection, the minimum value of the spacing between two adjacent ions occurs at the center of the ion chain. Compiling the numerical data for the minimum value of the separation for different numbers of trapped ions, we find that it

Table 1. Scaled equilibrium positions of the trapped ions for different total numbers of ions [a]

N	Scaled equilibrium positions									
2	−0.62996	0.62996								
3	−1.0772	0	1.0772							
4	−1.4368	−0.45438	0.45438	1.4368						
5	−1.7429	−0.8221	0	0.8221	1.7429					
6	−2.0123	−1.1361	−0.36992	0.36992	1.1361	2.0123				
7	−2.2545	−1.4129	−0.68694	0	0.68694	1.4129	2.2545			
8	−2.4758	−1.6621	−0.96701	−0.31802	0.31802	0.96701	1.6621	2.4758		
9	−2.6803	−1.8897	−1.2195	−0.59958	0	0.59958	1.2195	1.8897	2.6803	
10	−2.8708	−2.10003	−1.4504	−0.85378	−0.2821	0.2821	0.85378	1.4504	2.10003	2.8708

[a] This data was obtained by numerical solutions of (5). The length scale is given by (4)

obeys the following relation:

$$u_{\min}(N) \approx \frac{2.018}{N^{0.559}}\,. \tag{8}$$

This relation is illustrated in Figure 2. Thus the minimum inter-ion spacing for different numbers of ions is given by the following formula:

$$x_{\min}(N) = \left(\frac{Z^2e^2}{4\pi\epsilon_0 M\nu^2}\right)^{1/3} \frac{2.018}{N^{0.559}}\,. \tag{9}$$

This relationship is important in determining the capabilities of cold-trapped ion quantum computers [10].

2 Quantum fluctuations of the ions

This section discusses the equations of motion that describe the displacements of the ions from their equilibrium positions. Because of the Coulomb interactions between the ions, the displacements of different ions will be coupled together. The Lagrangian describing the motion is then

$$L = \frac{M}{2}\sum_{m=1}^{N}(\dot{q}_m)^2 - \frac{1}{2}\sum_{n,m=1}^{N} q_n q_m \left[\frac{\partial^2 V}{\partial x_n \partial x_m}\right]_0, \tag{10}$$

where the subscript 0 denotes that the double partial derivative is evaluated at $q_n = q_m = 0$, and we have neglected terms $O[q_n^3]$. The partial derivatives may be calculated explicitly to give the following expression:

$$L = \frac{M}{2}\left[\sum_{m=1}^{N}(\dot{q}_m)^2 - \nu^2 \sum_{n,m=1}^{N} A_{nm} q_n q_m\right], \tag{11}$$

where

$$A_{nm} = \begin{cases} 1+2\sum\limits_{\substack{p=1\\p\neq m}}^{N} \dfrac{1}{|u_m-u_p|^3} & \text{if } n=m\,,\\ \dfrac{-2}{|u_m-u_n|^3} & \text{if } n\neq m\,. \end{cases} \tag{12}$$

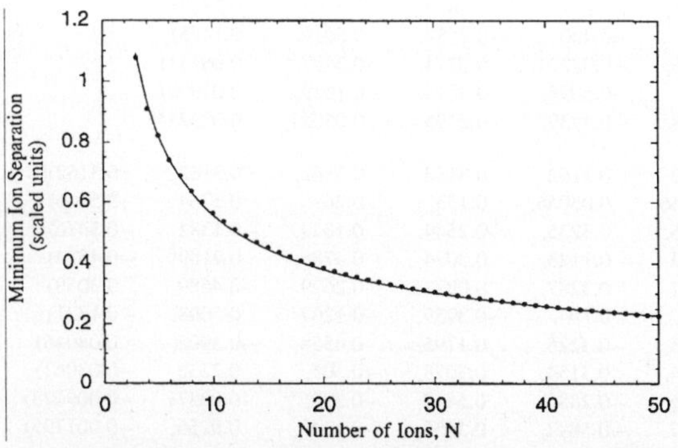

Fig. 2. The relationship between the number of trapped ions N and the minimum separation. The curve is given by (8) while the points come from the numerical solutions of the algebraic equations (5)

Since the matrix A_{nm} is real, symmetric, and non-negative definite, its eigenvalues must be non-negative. The eigenvectors $b_m^{(p)}$ $(p = 1, 2, \cdots N)$ are therefore defined by the following formula:

$$\sum_{n=1}^{N} A_{nm} b_n^{(p)} = \mu_p b_m^{(p)}\ (p = 1, \ldots, N)\,, \tag{13}$$

where $\mu_p \geq 0$. The eigenvectors are assumed to be numbered in order of increasing eigenvalue and to be properly normalized so that

$$\sum_{p=1}^{N} b_n^{(p)} b_m^{(p)} = \delta_{nm} \tag{14}$$

$$\sum_{n=1}^{N} b_n^{(p)} b_n^{(q)} = \delta_{pq}\,. \tag{15}$$

The first eigenvector (i.e., the eigenvector with the smallest eigenvalue) can be shown to be

$$b^{(1)} = \frac{1}{\sqrt{N}}\{1, 1, \cdots, 1\}, \quad \mu_1 = 1\,. \tag{16}$$

The next eigenvector can be shown to be

$$b^{(2)} = \frac{1}{\left(\sum_{m=1}^{N} u_m^2\right)^{1/2}}\{u_1, u_2, \cdots, u_N\}, \quad \mu_2 = 3\,. \tag{17}$$

Higher eigenvectors must, in general, be determined numerically; (15) and (16) imply that

$$\sum_{m=1}^{N} b_m^{(p)} = 0 \quad \text{if } p \neq 1\,. \tag{18}$$

For $N = 2$ and $N = 3$, the eigenvectors and eigenvalues may be determined algebraically:

$$N=2:\quad b^{(1)} = \frac{1}{\sqrt{2}}(1, 1), \quad \mu_1 = 1\,,$$
$$b^{(2)} = \frac{1}{\sqrt{2}}(-1, 1), \quad \mu_2 = 3\,, \tag{19}$$

$$N=3:\quad b^{(1)} = \frac{1}{\sqrt{3}}(1, 1, 1)\,, \quad \mu_1 = 1\,,$$
$$b^{(2)} = \frac{1}{\sqrt{2}}(-1, 0, 1)\,, \quad \mu_2 = 3\,,$$
$$b^{(3)} = \frac{1}{\sqrt{6}}(1, -2, 1)\,, \quad \mu_3 = 29/5\,. \tag{20}$$

For larger values of N, the eigenvalues and eigenvectors must be determined numerically; their numerical values for 2 to 10 ions are given in Table 2.

The *normal modes* of the ion motion are defined by the formula

$$Q_p(t) = \sum_{m=1}^{N} b_m^{(p)} q_m(t)\,. \tag{21}$$

Table 2. Numerically determined eigenvalues and eigenvectors of the matrix A_{nm} defined by (12), for 2 to 10 ions [a]

	Eigenvalue	Eigenvector									
N=2	1	(0.7071,	0.7071)								
	3	(−0.7071,	0.7071)								
N=3	1	(0.5774,	0.5774,	0.5774)							
	3	(−0.7071,	0,	0.7071)							
	5.8	(0.4082,	−0.8165,	0.4082)							
N=4	1	(0.5,	0.5,	0.5,	0.5)						
	3	(−0.6742,	−0.2132,	0.2132,	0.6742)						
	5.81	(0.5,	−0.5,	−0.5,	0.5)						
	9.308	(−0.2132,	0.6742,	−0.6742,	0.2132)						
N=5	1	(0.4472,	0.4472,	0.4472,	0.4472,	0.4472)					
	3	(−0.6395,	−0.3017,	0,	0.3017,	0.6395)					
	5.818	(0.5377,	−0.2805,	−0.5143,	−0.2805,	0.5377)					
	9.332	(−0.3017,	0.6395,	0,	−0.6395,	0.3017)					
	13.47	(0.1045,	−0.4704,	0.7318,	−0.4704,	0.1045)					
N=6	1	(0.4082,	0.4082,	0.4082,	0.4082,	0.4082,	0.4082)				
	3	(−0.608,	−0.3433,	−0.1118,	0.1118,	0.3433,	0.608)				
	5.824	(−0.5531,	0.1332,	0.4199,	0.4199,	0.1332,	−0.5531)				
	9.352	(0.3577,	−0.5431,	−0.2778,	0.2778,	0.5431,	−0.3577)				
	13.51	(0.1655,	−0.5618,	0.3963,	0.3963,	−0.5618,	0.1655)				
	18.27	(−0.04902,	0.2954,	−0.6406,	0.6406,	−0.2954,	0.04902)				
N=7	1	(0.378,	0.378,	0.378,	0.378,	0.378,	0.378,	0.378)			
	3	(−0.5801,	−0.3636,	−0.1768,	0,	0.1768,	0.3636,	0.5801)			
	5.829	(−0.5579,	0.031,	0.3213,	0.4111,	0.3213,	0.031,	−0.5579)			
	9.369	(−0.3952,	0.445,	0.3818,	0,	−0.3818,	−0.445,	0.3952)			
	13.55	(−0.213,	0.5714,	−0.1199,	−0.4769,	−0.1199,	0.5714,	−0.213)			
	18.32	(0.08508,	−0.4121,	0.5683,	0,	−0.5683,	0.4121,	−0.08508)			
	23.66	(0.02222,	−0.1723,	0.4894,	−0.6787,	0.4894,	−0.1723,	0.02222)			
N=8	1	(0.3536,	0.3536,	0.3536,	0.3536,	0.3536,	0.3536,	0.3536,	0.3536)		
	3	(−0.5556,	−0.373,	−0.217,	−0.07137,	0.07137,	0.217,	0.373,	0.5556)		
	5.834	(−0.5571,	−0.0425,	0.2362,	0.3634,	0.3634,	0.2362,	−0.0425,	−0.5571)		
	9.383	(0.4212,	−0.3577,	−0.4093,	−0.1647,	0.1647,	0.4093,	0.3577,	−0.4212)		
	13.58	(−0.2508,	0.5479,	0.0669,	−0.364,	−0.364,	0.0669,	0.5479,	−0.2508)		
	18.37	(0.1176,	−0.4732,	0.4123,	0.3039,	−0.3039,	−0.4123,	0.4732,	−0.1176)		
	23.73	(−0.04169,	0.2703,	−0.561,	0.3324,	0.3324,	−0.561,	0.2703,	−0.04169)		
	29.63	(−0.009806,	0.09504,	−0.3398,	0.6127,	−0.6127,	0.3398,	−0.09504,	0.009806)		
N=9	1	(0.3333,	0.3333,	0.3333,	0.3333,	0.3333,	0.3333,	0.3333,	0.3333,	0.3333)	
	3	(−0.5339,	−0.3764,	−0.2429,	−0.1194,	0,	0.1194,	0.2429,	0.3764,	0.5339)	
	5.838	(−0.5532,	−0.09692,	0.1658,	0.3078,	0.3531,	0.3078,	0.1658,	−0.09692,	−0.5532)	
	9.396	(−0.4394,	0.2828,	0.4019,	0.2558,	0,	−0.2558,	−0.4019,	−0.2828,	0.4394)	
	13.6	(0.2812,	−0.5108,	−0.1873,	0.2228,	0.3881,	0.2228,	−0.1873,	−0.5108,	0.2812)	
	18.41	(0.1465,	−0.5015,	0.2582,	0.4005,	0,	−0.4005,	−0.2582,	0.5015,	−0.1465)	
	23.79	(0.06133,	−0.3407,	0.5274,	−0.02271,	−0.4505,	−0.02271,	0.5274,	−0.3407,	0.06133)	
	29.71	(−0.01969,	0.1639,	−0.4614,	0.5098,	0,	−0.5098,	0.4614,	−0.1639,	0.01969)	
	36.16	(−0.004234,	0.05021,	−0.2195,	0.4939,	−0.6408,	0.4939,	−0.2195,	0.05021,	−0.004234)	
N=10	1	(0.3162,	0.3162,	0.3162,	0.3162,	0.3162,	0.3162,	0.3162,	0.3162,	0.3162,	0.3162)
	3	(−0.5146,	−0.3764,	−0.26,	−0.153,	−0.05056,	0.05056,	0.153,	0.26,	0.3764,	0.5146)
	5.841	(−0.5476,	−0.1382,	0.1079,	0.2544,	0.3235,	0.3235,	0.2544,	0.1079,	−0.1382	−0.5476)
	9.408	(0.4524,	−0.2189,	−0.3786,	−0.3024,	−0.1123,	0.1123,	0.3024,	0.3786,	0.2189	−0.4524)
	13.63	(0.3059,	−0.4689,	−0.2629,	0.09726,	0.3287,	0.3287,	0.09726,	−0.2629,	−0.4689	0.3059)
	18.45	(0.1721,	−0.5098,	0.1267,	0.3959,	0.194,	−0.194,	−0.3959,	−0.1267,	0.5098,	−0.1721)
	23.85	(0.08046,	−0.3902,	0.4528,	0.1795,	−0.3225,	−0.3225,	0.1795,	0.4528,	−0.3902,	0.08046)
	29.79	(0.03062,	−0.2232,	0.505,	−0.3078,	−0.3154,	0.3154,	0.3078,	−0.505,	0.2232,	−0.03062)
	36.26	(−0.009023,	0.09371,	−0.338,	0.5419,	−0.2886,	−0.2886,	0.5419,	−0.338,	0.09371	−0.009023)
	43.24	(0.001795,	−0.0256,	0.134,	−0.3656,	0.5897,	−0.5897,	0.3656,	−0.134,	0.0256,	−0.001795)

[a] The eigenvectors are normalized as defined by (15)

The first mode $Q_1(t)$ corresponds to all of the ions oscillating back and forth as if they were rigidly clamped together; this is referred to as the *center of mass* mode. The second mode $Q_2(t)$ corresponds to each ion oscillating with an amplitude proportional to its equilibrium distance form the trap center; This is called the *breathing mode*. The Lagrangian for the ion oscillations (11) may be rewritten in terms of these normal modes as follows:

$$L = \frac{M}{2}\sum_{p=1}^{N}\left[\dot{Q}_p^2 - \nu_p^2 Q_p^2\right], \tag{22}$$

where the angular frequency of the pth mode is defined by

$$\nu_p = \sqrt{\mu_p}\,\nu . \tag{23}$$

This expression implies that the modes Q_p are uncoupled. Thus the canonical momentum conjugate to Q_p is $P_p = M\dot{Q}_p$ and one can immediately write the Hamiltonian as

$$\hat{H} = \frac{1}{2M}\sum_{p=1}^{N} P_p^2 + \frac{M}{2}\sum_{p=1}^{N}\nu_p^2 Q_p^2 . \tag{24}$$

The quantum motion of the ions can now be considered by introducing the operators[1]

$$Q_p \to \hat{Q}_p = i\sqrt{\frac{\hbar}{2M\nu_p}}\left(\hat{a}_p - \hat{a}_p^\dagger\right), \tag{25}$$

$$P_p \to \hat{P}_p = \sqrt{\frac{\hbar M\nu_p}{2}}\left(\hat{a}_p + \hat{a}_p^\dagger\right). \tag{26}$$

where $\hat{Q}_p$ and $\hat{P}_p$ obey the canonical commutation relation $[\hat{Q}_p, \hat{P}_p] = i\hbar\delta_{pq}$ and the creation and annihilation operators $\hat{a}_p^\dagger$ and $\hat{a}_p$ obey the usual commutation relation $[\hat{a}_p, \hat{a}_q^\dagger] = \delta_{pq}$.

Using this notation, the interaction picture operator for the displacement of the mth ion from its equilibrium position is given by the formula:

$$\begin{aligned}\hat{q}_m(t) &= \sum_{p=1}^{N} \boldsymbol{b}_m^{(p)}\hat{Q}_p(t) \\ &= i\sqrt{\frac{\hbar}{2M\nu N}}\sum_{p=1}^{N} s_m^{(p)}\left(\hat{a}_p e^{-i\nu_p t} - \hat{a}_p^\dagger e^{-i\nu_p t}\right), \end{aligned}\tag{27}$$

where the coupling constant is defined by

$$s_m^{(p)} = \frac{\sqrt{N}\boldsymbol{b}_m^{(p)}}{\mu_p^{1/4}} . \tag{28}$$

For the center of mass mode,

$$s_m^{(1)} = 1 \quad \nu_1 = \nu , \tag{29}$$

[1] There is some arbitariness in the definition of the operators $\hat{P}_p$ and $\hat{Q}_p$, which is related to the arbitrariness of the phase of the Fock states. I have used the definitions given by Kittel ([11], p. 16), which differs from that given in other texts on quantum mechanics (see, for example, [12] p. 183 or [13] p. 36).

and for the breathing mode

$$s_m^{(2)} = \frac{\sqrt{N}}{\sqrt[4]{3}}\frac{1}{\left(\sum_{m=1}^{N} u_m^2\right)^{1/2}} u_m \quad \nu_2 = \sqrt{3}\,\nu . \tag{30}$$

The Lagrangian equation (10) was derived from a Taylor expansion of the potential function about the equilibrium positions of the ions, terms $O[q_n^3]$ being neglected. The ratio of the strengths of the neglected terms to the strength of the quadratic terms, which are included, is, for low phonon numbers, of the order of $(\hbar V/8Mc^2N^3\alpha^2)^{1/6}$, where α is the fine structure constant. Clearly this dimensionless quantity must be small if the approximation we have made is to be valid; for example, if we consider a single Ca II ion in a trap with axial frequency $\nu = (2\pi) \times 500\,\mathrm{KHz}$, it has the value 2.2×10^{-3}. The neglected terms will however important because they give a coupling between different phonon modes which may be a source of decoherence.

3 Laser-ion interactions

I will now consider the interaction of a laser field with the trapped ions. The theory must take into acount both the internal and vibrational degrees of freedom of the ions. I will consider two types of transition between internal ionic levels: the familiar electric-dipole allowed (E1) transitions and dipole forbidden electric quadrupole (E2) transitions. Electric quadrupole transitions have been considered in detail by Freedhoff [14, 15]. The reason for considering forbidden transitions is that they have very long decay lifetimes; spontaneous emission will destroy the coherence of a quantum computer, and therefore is a major limitation on the capabilities of such devices [10, 16]. Magnetic dipole (M1) transitions, which also have long lifetimes, tend only to occur between sub-levels of a configuration and will therefore require, when using the single photon scheme, long wavelength lasers in order to excite them. As it is necessary to resolve individual ions in the trap using the laser, the use of long wavelengths will seriously degrade performance. Transitions between sub-levels of a configuration are however possible using the Raman scheme. More highly forbidden transitions are also a possibility for use in a quantum computer. In particular, there is an octupole allowed (E3) transition of the ion Yb II at 467 nm with a theoretical lifetime of 1.325×10^8 sec [17], which has recently been observed at the National Physical Laboratory at Teddington, England [18]. However, such weak transitions can only be excited by either very long laser pulses or by very powerful lasers. Since it is impossible to maintain the phase stability of a laser indefinitely, very long duration pulses (i.e., more than ~ 100 msec) are not practicable. Very high laser power can cause a break-down of the two-level approximation, as highly detuned dipole transitions can become excited. Thus it appears that such *very* long lived states may not in fact give any particular advantages for quantum computing.

The interaction picture Hamiltonians for electric dipole (E_1) and electric quadrupole (E_2) transitions of the mth ion, located at x_m are

$$\hat{H}_I^{(E1)} = ie \sum_{MN} \omega_{MN} |N\rangle\langle M| \langle N|\hat{r}_i|M\rangle A_i(x_m, t) e^{i\omega_{MN} t}, \quad (31)$$

$$\hat{H}_I^{(E2)} = \frac{ie}{2} \sum_{MN} \omega_{MN} |N\rangle\langle M| \langle N|\hat{r}_i \hat{r}_j|M\rangle \partial_i A_j(x_m, t) e^{i\omega_{MN} t}, \quad (32)$$

where $A_j(x, t)$ is the jth component of the vector potential of the laser field, ∂_i denotes differentiation along the ith direction and summation over repeated indices $(i, j = x, y, z)$ is implied; $\hat{r}_i$ is the ith component of the position operator for the valence electron of the ion; $\{|N\rangle\}$ is the set of all eigenstates of the unperturbed ion and the transition frequency is $\omega_{MN} = \omega_M - \omega_N$ where the energy of the Nth state is $\hbar\omega_N$.

For a laser beam in a standing wave configuration (see Fig. 1), propagating along a direction specified by the unit vector $\boldsymbol{n}$, the vector potential and its derivative are given by the formulas

$$A_i(x_m, t) = -\epsilon_i \frac{E}{i\omega} \sin\left[k\hat{\xi}_m(t)\right] e^{i\omega t} + \text{c.c.}, \quad (33)$$

$$\partial_i A_j(x_m, t) = -n_i \epsilon_j \frac{E}{c} \cos\left[k\hat{\xi}_m(t)\right] e^{i\omega t} + \text{c.c.} \quad (34)$$

In (34), I have approximated the laser beam as a plane wave, ϵ being the polarization vector, E is the amplitude of the electric field, ω is the laser frequency and $k = \omega/c$ is the wavenumber. The operator $\hat{\xi}_m(t)$ is the distance between the mth ion and the plane mirror used to form the standing wave.

If we restrict our consideration to just two states, $|1\rangle$ and $|2\rangle$, and make the rotating wave equation, the interaction Hamiltonians may be rewritten as follows:

$$\hat{H}_I^{(E1)} = \hbar\Omega_0^{(E1)} \sin\left[k\hat{\xi}_m(t)\right] e^{i(t\Delta - \phi)} |1\rangle\langle 2| + \text{h.a.}, \quad (35)$$

$$\hat{H}_I^{(E2)} = i\hbar\Omega_0^{(E2)} \cos\left[k\hat{\xi}_m(t)\right] e^{i(t\Delta - \phi)} |1\rangle\langle 2| + \text{h.a.}, \quad (36)$$

where the detuning is $\Delta = \omega - \omega_{21}$ and the Rabi frequencies are given by

$$\Omega_0^{(E1)} = \left| \frac{eE}{\hbar} \langle 1|\hat{r}_i|2\rangle \epsilon_i \right|, \quad (37)$$

$$\Omega_0^{(E2)} = \left| \frac{eE\omega_{21}}{2\hbar c} \langle 1|\hat{r}_i \hat{r}_j|2\rangle \epsilon_i n_j \right|. \quad (38)$$

If the standing wave of the laser is so contrived that the equilibrium position of the mth ion is located at a *node*, i.e., the electric field strength is zero, then

$$\hat{\xi}_m(t) = l\lambda + \cos\theta \hat{q}_m(t) \quad (39)$$

where l is some integer, λ is the wavelength, and θ is the angle between the laser beam and the trap axis and we have assumed that the fluctuations of the ions transverse to the trap axis are negligible. In this case the two Hamiltonians become

$$\hat{H}_I^{(E1)} = \hbar\Omega_0^{(E1)} k \cos\theta \hat{q}_m(t) e^{i(t\Delta - \phi + l\pi)} |1\rangle\langle 2| + \text{h.a.}, \quad (40)$$

$$\hat{H}_I^{(E2)} = \hbar\Omega_0^{(E2)} e^{i[t\Delta - \phi - (l+1/2)\pi]} |1\rangle\langle 2| + \text{h.a.}, \quad (41)$$

where we have neglected terms involving $\hat{q}_m(t)^2$. It is convenient to write the displacement of the ion in terms of the creation and annihilation operators of the phonon modes, viz.,

$$k\cos\theta \hat{q}_m(t) = i\frac{\eta}{\sqrt{N}} \sum_{p=1}^{N} s_m^{(p)} \left(\hat{a}_p e^{-i\nu_p t} - \hat{a}_p^\dagger e^{i\nu_p t}\right), \quad (42)$$

where $\eta = \sqrt{\hbar k^2 \cos^2\theta / 2M\nu}$ is called the Lamb–Dicke parameter.

Similarly if the standing wave is arranged so that the ion is at an *antinode*, i.e.,

$$\hat{\xi}_m(t) = \frac{(2l-1)\lambda}{2} + \cos\theta \hat{q}_m(t), \quad (43)$$

then the Hamiltonians are

$$\hat{H}_I^{(E1)} = \hbar\Omega_0^{(E1)} e^{i(t\Delta - \phi + l\pi)} |1\rangle\langle 2| + \text{h.a.}, \quad (44)$$

$$\hat{H}_I^{(E2)} = \hbar\Omega_0^{(E2)} k\cos\theta \hat{q}_m(t) e^{i[t\Delta - \phi - (l+1/2)\pi]} |1\rangle\langle 2| + \text{h.a.} \quad (45)$$

Thus we have two basic types of Hamiltonian:

$$\hat{H}_V = \hbar\Omega_0 e^{i(t\Delta - \phi_u)} |1\rangle\langle 2| + \text{h.a.}, \quad (46)$$

$$\hat{H}_U = \hbar\Omega_0 k\cos\theta \hat{q}_m(t) e^{i(t\Delta - \phi_u)} |1\rangle\langle 2| + \text{h.a.}, \quad (47)$$

where Ω_0 stands for either $\Omega_0^{(E1)}$ or $\Omega_0^{(E2)}$.

By changing the node to an antinode, by moving the reflecting mirror, for example, we can switch from one type of Hamiltonian to the other. In the first case, the laser beam will only interact with internal degrees of freedom of the ion, while in the second case the collective motion of the ions will be affected as well.

4 Evaluation of the Rabi frequencies

We can relate the matrix elements appearing in the definitions of the Rabi frequencies to the Einstein A coefficients for the transitions. In order to do this we will rewrite the matrix elements in terms of the Racah tensors:

$$\langle 1|\hat{r}_i|2\rangle \epsilon_i = \sum_{q=-1}^{1} \langle 1|rC_q^{(1)}|2\rangle c_i^{(q)} \epsilon_i, \quad (48)$$

$$\langle 1|\hat{r}_i \hat{r}_j|2\rangle \epsilon_i n_j = \sum_{q=-2}^{2} \langle 1|r^2 C_q^{(2)}|2\rangle c_{ij}^{(q)} \epsilon_i n_j, \quad (49)$$

where we have used the fact that $\boldsymbol{\epsilon} \cdot \boldsymbol{n} = 0$. The vectors $\boldsymbol{c}^{(q)}$ and the second rank tensors $c_{ij}^{(q)}$ may be calculated quite easily; explicit expressions are given in the appendix. If we assume LS coupling, the states $|1\rangle$ and $|2\rangle$ are specified by the angular momentum quantum numbers; thus we will use the notation $|1\rangle = |jm_j\rangle$ and $|2\rangle = |j'm_j'\rangle$, where j is the total angular momentum quantum number and m_j is the magnetic quantum number of the lower state and j' is the total angular momentum quantum number and m_j' the magnetic quantum number of the upper state. Using the Wigner–Eckart theorem ([19],

Section 11.4), the matrix elements may be rewritten as

$$\langle 1|\hat{r}_i|2\rangle\epsilon_i = \langle 1||rC^{(1)}||2\rangle \sum_{q=-1}^{1} \begin{pmatrix} j & 1 & j' \\ -m_j & q & m'_j \end{pmatrix} c_i^{(q)}\epsilon_i\,, \quad (50)$$

$$\langle 1|\hat{r}_i\hat{r}_j|2\rangle\epsilon_i n_j = \langle 1||r^2C^{(2)}||2\rangle \sum_{q=-2}^{2} \begin{pmatrix} j & 2 & j' \\ -m & q & m' \end{pmatrix} c_{ij}^{(q)}\epsilon_i n_j\,, \quad (51)$$

the terms containing six numbers in brackets being Wigner $3-j$ symbols ([19], Section 5.1), and $\langle 1||r^qC^{(q)}||2\rangle$ being the reduced matrix element. The Einstein A coefficients for the two levels are given by the expressions:

$$\overline{A}_{12}^{(E1)} = \frac{4c\alpha k_{12}^3}{3} \sum_{q=-1}^{1} \left|\langle 1|rC_q^{(1)}|2\rangle\right|^2 \quad (52)$$

$$\overline{A}_{12}^{(E2)} = \frac{c\alpha k_{12}^5}{15} \sum_{q=-2}^{2} \left|\langle 1|r^2C_q^{(2)}|2\rangle\right|^2 . \quad (53)$$

Using the Wigner–Eckart theorem again, these expressions reduce to the following:

$$\overline{A}_{12}^{(E1)} = \frac{4c\alpha k_{12}^3}{3} \left|\langle 1||rC^{(1)}||2\rangle\right|^2 \sum_{q=-1}^{1} \begin{pmatrix} j & 1 & j' \\ -m_j & q & m'_j \end{pmatrix}^2 \quad (54)$$

$$\overline{A}_{12}^{(E2)} = \frac{c\alpha k_{12}^5}{15} \left|\langle 1||r^2C^{(2)}||2\rangle\right|^2 \sum_{q=-2}^{2} \begin{pmatrix} j & 2 & j' \\ -m_j & q & m'_j \end{pmatrix}^2 . \quad (55)$$

These coefficients are the rates for spontaneous decay from the upper level $|1\rangle$ to the lower level $|2\rangle$. A simpler expression for the total rate of spontaneous decay of $|2\rangle$ to all of the sublevels of the lower state may be found by summing these rates over all values of m_j:

$$A_{12}^{(E1)} \equiv \sum_{m=-j}^{j} \overline{A}_{12}^{(E1)} = \frac{4c\alpha k_{12}^3}{3(2j'+1)} \left|\langle 1||rC^{(1)}||2\rangle\right|^2 , \quad (56)$$

$$A_{12}^{(E2)} \equiv \sum_{m=-j}^{j} \overline{A}_{12}^{(E2)} = \frac{c\alpha k_{12}^5}{15(2j'+1)} \left|\langle 1||r^2C^{(2)}||2\rangle\right|^2 . \quad (57)$$

These decay rates, which are the same for all of the sublevels of the upper level, are the quantities usually quoted in data tables. Using (37), (38), (50), (51), (56) and (57), we then obtain the following formula for the Rabi frequencies:

$$\Omega_0 = \frac{e|E|}{\hbar\sqrt{c\alpha}} \sqrt{\frac{A_{12}}{k_{12}^3}}\,\sigma\,, \quad (58)$$

where

$$\sigma^{(E1)} = \sqrt{\frac{3(2j'+1)}{4}} \left| \sum_{q=-1}^{1} \begin{pmatrix} j & 1 & j' \\ -m_j & q & m'_j \end{pmatrix} c_i^{(q)}\epsilon_i^{(q)} \right| , \quad (59)$$

$$\sigma^{(E2)} = \sqrt{\frac{15(2j'+1)}{4}} \left| \sum_{q=-2}^{2} \begin{pmatrix} j & 2 & j' \\ -m_j & q & m'_j \end{pmatrix} c_{ij}^{(q)}\epsilon_i^{(q)} n_j \right| . \quad (60)$$

The values of these quantities will be dependent on the choice of states of ions used for the upper and lower levels, and upon the polarization and direction of the laser beam. As a specific example, we will assume that the ions are in a weak magnetic field, which serves to define the z-direction of quantization. Furthermore, we will assume that the lower level $|1\rangle$ is the $m_j = -1/2$ sublevel of a $^2S_{1/2}$ ground state, the nucleus having spin zero. The upper level for the dipole transition is a sublevel of a $^2P_{1/2}$ state, while for the quadrupole transition it is a sublevel of a $^2D_{3/2}$ state:

$$\Omega_0^{(E1)} = \frac{e|E|}{\hbar} \sqrt{\frac{A_{12}^{(E1)}}{4c\alpha k_{12}^3}}\,, \quad (61)$$

$$\Omega_0^{(E2)} = \frac{e|E|}{\hbar} \sqrt{\frac{A_{12}^{(E2)}}{2c\alpha k_{12}^3}}\,. \quad (62)$$

5 Validity of Cirac and Zoller's Hamiltonian

Equations (42) and (47) give the following expression for the Hamiltonian for the case when the laser standing wave is so configured that it can excite the vibration modes of the ions:

$$\hat{H}_U = \frac{i\eta\hbar\Omega_0}{\sqrt{N}} \sum_{p=1}^{N} s_m^{(p)} \left(\hat{a}_p e^{-i\nu_p t} - \hat{a}_p^\dagger e^{i\nu_p t}\right) e^{i(t\Delta-\phi_u)}|1\rangle\langle 2| + \text{h.a.} \quad (63)$$

In their paper [3], Cirac and Zoller assumed that the laser can interact with only the center of mass mode of the ions' fluctuations. This interaction forms a vitally important element of their proposed method for implementing a quantum controlled not logic gate. Thus they used a Hamiltonian of the following form [cf. [3], Eq. (1)]:

$$\hat{H}_U^{(CZ)} = \frac{i\eta\hbar\Omega_0}{\sqrt{N}} \left(\hat{a}_1 e^{-i\nu_1 t} - \hat{a}_1^\dagger e^{i\nu_1 t}\right) e^{i(t\Delta-\phi_u)}|1\rangle\langle 2| + \text{h.a.} \quad (64)$$

This is an approximate form of (63), in which all of the other "extraneous" phonon modes have been neglected. We will now investigate under what circumstances these modes may be ignored.

We will assume that the wavefunction for a single ion interacting with the laser beam may be written as follows:

$$|\Psi(t)\rangle = \alpha_0(t)|1\rangle|vac\rangle + b_0(t)|2\rangle|vac\rangle + \sum_{p=1}^{N} \alpha_p(t)|1\rangle|1_p\rangle + \sum_{p=1}^{N} b_p(t)|2\rangle|1_p\rangle\,, \quad (65)$$

where $|1\rangle$ and $|2\rangle$ are the energy eigenstates of the mth ion's internal degrees of freedom, $|1_p\rangle$ is the state of the ions' collective vibration in which the pth mode has been excited by one quantum, and $|vac\rangle$ is the vibrational ground state. To avoid ambiguity, the ket for the ion's internal state appears first, the ket for the vibrational state second.

The equation of motion for this wavefunction is

$$i\hbar\frac{\partial}{\partial t}|\Psi(t)\rangle = \hat{H}_U|\Psi(t)\rangle\,. \quad (66)$$

By using (63), and assuming that one cannot excite states with two phonons, one obtains the following equations:

$$\dot{\alpha}_0 = \frac{\Omega_0 \eta}{\sqrt{N}} \sum_{p=1}^{N} s_m^{(p)} \beta_p(t), \tag{67}$$

$$\dot{\beta}_0 = \frac{\Omega_0 \eta}{\sqrt{N}} \sum_{p=1}^{N} s_m^{(p)} \alpha_p(t), \tag{68}$$

$$\dot{\alpha}_p = -i(\nu_p - \nu_1)\alpha_p - \frac{\Omega_0 \eta}{\sqrt{N}} s_m^{(p)} \beta_0(t), \tag{69}$$

$$\dot{\beta}_p = -i(\nu_p + \nu_1)\beta_p - \frac{\Omega_0 \eta}{\sqrt{N}} s_m^{(p)} \alpha_0(t). \tag{70}$$

We have assumed that $\Delta = -\nu_1$, so that the laser is tuned to the specific sideband resonance required to perform Cirac and Zoller's universal gate operation ([3], Eq. 3), namely, the two level transition $|1\rangle_n|1_1\rangle \leftrightarrow |2\rangle_n|vac\rangle$.

Since $|\alpha_0(t)|, |\beta_0(t)| \leq 1$, we can consider the following upper limits on the amplitudes of the states which include excitation of "extraneous" phonon modes (i.e., phonon modes other than the center of mass mode):

$$|\alpha_p(t)| \leq |A_p(t)|, \quad |\beta_p(t)| \leq |B_p(t)|, \tag{71}$$

where

$$\dot{A}_0 + i(\nu_p - \nu_1)A_p = -\frac{\Omega_0 \eta}{\sqrt{N}} s_m^{(p)}, \tag{72}$$

$$\dot{B}_0 + i(\nu_p + \nu_1)B_p = -\frac{\Omega_0 \eta}{\sqrt{N}} s_m^{(p)}. \tag{73}$$

Solving these equations one finds that

$$|A_p(t)| \leq \frac{2\Omega_0 \eta}{\sqrt{N}(\nu_p - \nu_1)} |s_m^{(p)}|, \tag{74}$$

$$|B_p(t)| \leq \frac{2\Omega_0 \eta}{\sqrt{N}(\nu_p + \nu_1)} |s_m^{(p)}|. \tag{75}$$

Thus the total probability that "extraneous" modes are excited has the following upper limit:

$$P_{\text{ext}} = \sum_{p=2}^{N} |\alpha_p(t)|^2 + |\beta_p(t)|^2 \leq 2\left(\frac{2\Omega_0 \eta}{\sqrt{N}\nu}\right)^2 \left[\sum_{p=2}^{N} \frac{\mu_p + 1}{(\mu_p - 1)^2} \left(s_m^{(p)}\right)^2\right], \tag{76}$$

where we have used the definition of the mode frequencies (23) and the fact that the eigenvalue for the center of mass mode is $\mu_1 = 1$. This quantity will be different for each ion in the string; taking its average value, we find

$$\overline{P}_{\text{ext}} \equiv \frac{1}{N} \sum_{m=1}^{N} P_{\text{ext}} \leq 2\left(\frac{2\Omega_0 \eta}{\sqrt{N}\nu}\right)^2 \Sigma(N), \tag{77}$$

where we have used the definition of the coupling constants (28) and the orthonormality of the eigenvectors (15).

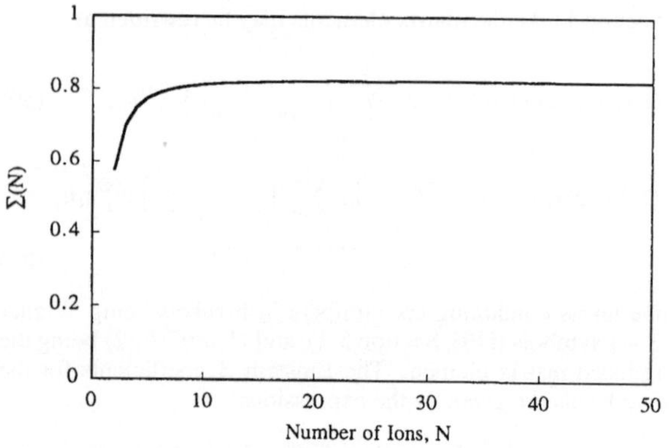

Fig. 3. The function $\Sigma(N)$ defined by (78)

The function $\Sigma(N)$ is defined by the formula

$$\Sigma(N) = \sum_{p=2}^{N} \frac{\mu_p + 1}{(\mu_p - 1)^2 \sqrt{\mu_p}}. \tag{78}$$

This must be evaluated numerically by solving for the eigenvalues of the trap normal modes for different numbers of trapped ions N. The results are shown in Fig. 3. The function varies slowly with the value of N, and, for $N \geq 10$, we can, to a good approximation, replace it by a constant $\Sigma(N) \approx 0.82$. Thus we obtain the following upper limit on the total probability of the "extraneous" phonon modes becoming excited:

$$\overline{P}_{\text{ext}} \leq \left(\frac{2.6\Omega_0 \eta}{\sqrt{N}\nu}\right)^2. \tag{79}$$

Thus we obtain the following sufficiency condition for the validity of Cirac and Zoller's Hamiltonian (64):

$$\left(\frac{2.6\Omega_0 \eta}{\sqrt{N}\nu}\right)^2 \ll 1. \tag{80}$$

6 Conclusion

In the preceding sections, we have reviewed the theoretical basis for cold-trapped ion quantum computation. How these various laser–ion interaction effects may be combined to perform fundamental quantum logic gates is described in the seminal work of Cirac and Zoller [3]. By using the formulas given here one can determine, for example, the laser field strength required or the separation between ions in the trap. Such things are of great importance in the engineering of practical devices.

Finally there is the question of what type of ion to use. Figure 4 shows the energy levels of four suitable species of ion. These have been chosen based on two criteria: that the lowest excited state has a forbidden transition to the ground state, and their popularity among published ion trapping experiments. It is not intended that this is an exhaustive list of suitable ions, but rather it is to show the properties of typical ions.

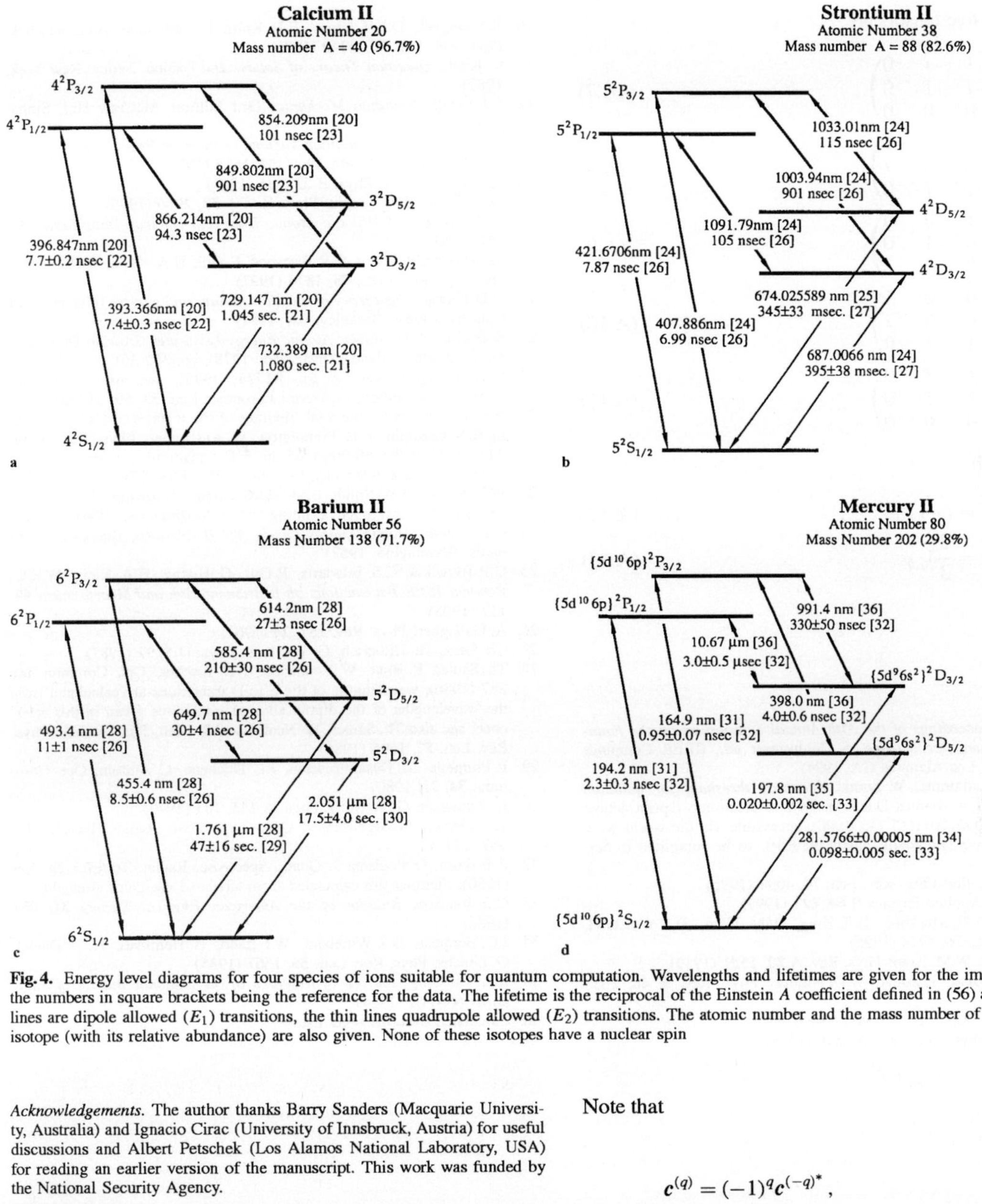

Fig. 4. Energy level diagrams for four species of ions suitable for quantum computation. Wavelengths and lifetimes are given for the important transitions, the numbers in square brackets being the reference for the data. The lifetime is the reciprocal of the Einstein *A* coefficient defined in (56) and (57). The thick lines are dipole allowed (E_1) transitions, the thin lines quadrupole allowed (E_2) transitions. The atomic number and the mass number of the most abundant isotope (with its relative abundance) are also given. None of these isotopes have a nuclear spin

Acknowledgements. The author thanks Barry Sanders (Macquarie University, Australia) and Ignacio Cirac (University of Innsbruck, Austria) for useful discussions and Albert Petschek (Los Alamos National Laboratory, USA) for reading an earlier version of the manuscript. This work was funded by the National Security Agency.

Appendix

The vectors $c_i^{(q)}$ are usual normalized spherical basis vectors:

$$c^{(1)} = -\frac{1}{\sqrt{2}}(1, -i, 0)\,, \tag{A.1}$$

$$c^{(0)} = (0, 0, 1)\,, \tag{A.2}$$

$$c^{(-1)} = \frac{1}{\sqrt{2}}(1, i, 0)\,. \tag{A.3}$$

Note that

$$c^{(q)} = (-1)^q c^{(-q)^*}\,, \tag{A.4}$$

$$c^{(q)} \cdot \left\{c^{(q')}\right\}^* = \delta_{q,q'}\,. \tag{A.5}$$

The second rank tensors $c_{ij}^{(q)}$ are given by the formula

$$c_{ij}^{(q)} = \sqrt{\frac{10}{3}}(-1)^q \sum_{m_1,m_2=-1}^{1} \begin{pmatrix} 1 & 1 & 2 \\ m_1 & m_2 & -q \end{pmatrix} c_i^{(m_1)} c_j^{(m_2)}\,. \tag{A.6}$$

190

Explicity these five tensors are:

$$c_{ij}^{(2)} = \frac{1}{\sqrt{6}} \begin{pmatrix} 1 & -i & 0 \\ -i & -1 & 0 \\ 0 & 0 & 0 \end{pmatrix}, \tag{A.7}$$

$$c_{ij}^{(1)} = \frac{1}{\sqrt{6}} \begin{pmatrix} 0 & 0 & -1 \\ 0 & 0 & i \\ -1 & i & 0 \end{pmatrix}, \tag{A.8}$$

$$c_{ij}^{(0)} = \frac{1}{3} \begin{pmatrix} -1 & 0 & 0 \\ 0 & -1 & 0 \\ 0 & 0 & 2 \end{pmatrix}, \tag{A.9}$$

$$c_{ij}^{(-1)} = \frac{1}{\sqrt{6}} \begin{pmatrix} 0 & 0 & 1 \\ 0 & 0 & i \\ 1 & i & 0 \end{pmatrix}, \tag{A.10}$$

$$c_{ij}^{(-2)} = \frac{1}{\sqrt{6}} \begin{pmatrix} 1 & i & 0 \\ i & -1 & 0 \\ 0 & 0 & 0 \end{pmatrix}. \tag{A.11}$$

Note that

$$c_{ij}^{(q)} = (-1)^q c_{ij}^{(-q)*}, \tag{A.12}$$

$$\sum_{ij} c_{ij}^{(q)} \left\{c_{ij}^{(q')}\right\}^* = \frac{2}{3}\delta_{q,q'}. \tag{A.13}$$

References

1. P.W. Shor: *Proceedings of the 35th Annual Symposium on the Foundations of Computer Science*, S. Goldwasser ed., (IEEE Computer Society Press, Los Alamitos CA, 1994)
2. E. Knill, R. Laflamme, W. Zurek: *Accuracy threshold for quantum computation*, Los Alamos Quantum Physics electronic reprint achive paper number 9610011(15 Oct 1996), accessible via the world wide web at http://xxx.lanl.gov/list/quant-ph/9610; to be submitted to *Science*, 1997
3. J.I. Cirac, P. Zoller: Phys. Rev. Lett. **74**, 4094 (1995)
4. A.M. Steane: Applied Physics B **64**, 623 (1997)
5. C. Monroe, D.M. Meekhof, B.E. King, W.M. Itano, D.J. Wineland: Phys. Rev. Lett. **75**, 4714 (1995)
6. D.J. Wineland, W.M. Itano: Phys. Rev. A **20**, 1521 (1979)
7. J.I. Cirac, R. Blatt, P. Zoller, W.D. Phillips: Phys. Rev. A **46**, 2668 (1992)
8. A. Garg: Phys. Rev. Lett. **77**, 964 (1996)
9. J.P. Schiffer: Phys. Rev. Lett. **70**, 818 (1993)
10. R.J. Hughes, D.F.V. James, E.H. Knill, R. Laflamme, A.G. Petschek: Phys. Rev. Lett. **77**, 3240 (1996)
11. C. Kittel: *Quantum Theory of Solids* (2nd edition, Wiley, New York, 1987)
12. L.I. Schiff: *Quantum Mechanics* (3rd Edition, McCraw Hill, Singapore, 1968)
13. P.W. Milonni: *The Quantum Vacuum* (Academic Press, Boston, 1994)
14. H.S. Freedhoff: J. Chem. Phys. **54**, 1618 (1971)
15. H.S. Freedhoff: J. Phys. B **22**, 435 (1989)
16. M.B. Plenio, P.L. Knight: Phys. Rev. A **53**, 2986 (1995)
17. B.C. Fawcett, M. Wilson: *Atomic Data and Nuclear Data Tables* **47**, 241 (1991)
18. M. Roberts, P. Taylor, G.P. Barwood, P. Gill, H.A. Klein, W.R.C. Rowley: Phys. Rev. Lett., **78**, 1876 (1997)
19. R.D. Cowan: *The theory of atomic structure and spectra* (University of California Press, Berkeley, CA, 1981)
20. S. Bashkin, J.O. Stoner: *Atomic Energy-Level and Grotrian Diagrams, Vol II* (North Holland, Amsterdam, 1978), pp. 360-361
21. S. Liaw: Phys. Rev. A **51**, R1723 (1995); see also T. Gudjons, B. Hilbert, P. Seibert, G. Werth: Europhys. Lett. **33**, 595 (1996)
22. Accurate values for the total lifetimes of the P states of Ca^+ are given in R.N. Gosselin, E.H. Pinnington, W. Ansbacher: Phys. Rev. A **38**, 4887 (1988); the lifetimes of the S-P transitions can be calculated using this data and the data from the NBS tables [23]
23. W.L. Wiese, M.W. Smith, B.M. Miles: *Atomic Transition Probabilities, Vol II* (U.S. Government Printing Office, Washington, 1969), p. 251
24. C.E. Moore: *Atomic Energy Levels, Vol II* (National Bureau of Standards, Washington, 1952)
25. G.P. Barwood, C.S. Edwards, P. Gill, G. Huang, H.A. Klein, W.R.C. Rowley: *IEEE Transactions on Instrumentation and Measurement* **44**, 117 (1995)
26. A. Gallagher: Phys. Rev. **157**, 24 (1967)
27. Ch. Gerz, Th. Hilberath, G. Werth: Z. Phys. D **5**, 97 (1987)
28. Th. Sauter, R. Blatt, W. Neuhauser, P.E. Toschek: Opt. Commun. **60**, 287 (1986); wavelengths of the S to D transitions are calculated from the wavelengths of the dipole allowed transitions given in this reference; see also Th. Sauter, W. Neuhauser, R. Blatt, P.E. Toschek: Phys. Rev. Lett. **57**, 1696 (1986)
29. F. Plumelle, M. Desaintfuscien, J.L. Duchene, C. Audoin: Opt. Commun. **34**, 71(1980)
30. R. Schneider, G. Werth: Z.Phys. A **293**, 103 (1979)
31. T. Andersen, G. Sørensen: J. Quant. Spectrosc. Radiat. Transfer **13**, 369 (1973)
32. P. Eriksen, O. Poulsen: J. Quant. Spectrosc. Radiat. Transfer **23**, 599 (1980); lifetimes are calculated from tabulated oscillator strengths
33. C.E. Johnson: *Bulletin of the American Physical Society* **31**, 957 (1986)
34. J.C. Berquist, D.J. Wineland, W.I. Itano, H. Hemmati, H.-U. Daniel, G. Leuchs: Phys. Rev. Lett. **55**, 1567 (1985)
35. R.H. Garstang: *Journal of Research of the National Bureau of Standards* **68A**, 61 (1964)
36. Calculated from the other data

VOLUME 75, NUMBER 25 PHYSICAL REVIEW LETTERS 18 DECEMBER 1995

Demonstration of a Fundamental Quantum Logic Gate

C. Monroe, D. M. Meekhof, B. E. King, W. M. Itano, and D. J. Wineland
National Institute of Standards and Technology, Boulder, Colorado 80303
(Received 14 July 1995)

We demonstrate the operation of a two-bit "controlled-NOT" quantum logic gate, which, in conjunction with simple single-bit operations, forms a universal quantum logic gate for quantum computation. The two quantum bits are stored in the internal and external degrees of freedom of a single trapped atom, which is first laser cooled to the zero-point energy. Decoherence effects are identified for the operation, and the possibility of extending the system to more qubits appears promising.

PACS numbers: 89.80.+h, 03.65.–w, 32.80.Pj

We report the first demonstration of a fundamental quantum logic gate that operates on prepared quantum states. Following the scheme proposed by Cirac and Zoller [1], we demonstrate a controlled-NOT gate on a pair of quantum bits (qubits). The two qubits comprise two internal (hyperfine) states and two external (quantized motional harmonic oscillator) states of a single trapped atom. Although this minimal system consists of only two qubits, it illustrates the basic operations necessary for, and the problems associated with, constructing a large scale quantum computer.

The distinctive feature of a quantum computer is its ability to store and process superpositions of numbers [2]. This potential for parallel computing has led to the discovery that certain problems are more efficiently solved on a quantum computer than on a classical computer [3]. The most dramatic example is an algorithm presented by Shor [4] showing that a quantum computer should be able to factor large numbers very efficiently. This appears to be of considerable interest, since the security of many data encryption schemes [5] relies on the inability of classical computers to factor large numbers.

A quantum computer hosts a register of qubits, each of which behaves as quantum mechanical two-level systems and can store arbitrary superposition states of 0 and 1. It has been shown that any computation on a register of qubits can be broken up into a series of two-bit operations [6], for example, a series of two-bit "controlled-NOT" (CN) quantum logic gates, accompanied by simple rotations on single qubits [7,8]. The CN gate transforms the state of two qubits ϵ_1 and ϵ_2 from $|\epsilon_1\rangle|\epsilon_2\rangle$ to $|\epsilon_1\rangle|\epsilon_1 \oplus \epsilon_2\rangle$, where the $\oplus$ operation is addition modulo 2. Reminiscent of the classical exclusive-OR (XOR) gate, the CN gate represents a computation at the most fundamental level: the "target" qubit $|\epsilon_2\rangle$ is flipped depending on the state of the "control" qubit $|\epsilon_1\rangle$.

Experimental realization of a quantum computer requires isolated quantum systems that act as the qubits, and the presence of controlled unitary interactions between the qubits that allow construction of the CN gate. As pointed out by many authors [6,9,10], if the qubits are not sufficiently isolated from outside influences, decoherences can destroy the quantum interferences that form the computation. Several proposed experimental schemes for quantum computers and CN gates involving a dipole-dipole interaction between quantum dots or atomic nuclei [6,7,11,12] may suffer from decoherence efforts. The light shifts on atoms located inside electromagnetic cavities have been shown to be large enough [13,14] that one could construct a quantum gate where a single photon prepared in the cavity acts as the control qubit [7,15] for the atomic state. However, extension to large quantum registers may be difficult. Cirac and Zoller [1] have proposed a very attractive quantum computer architecture based on laser-cooled trapped ions in which the qubits are associated with internal states of the ions, and information is transferred between qubits through a shared motional degree of freedom. The highlights of their proposal are that (i) decoherence can be small, (ii) extension to large registers is relatively straightforward, and (iii) the qubit readout can have nearly unit efficiency.

In our implementation of a quantum CN logic gate, the target qubit $|S\rangle$ is spanned by two $^2S_{1/2}$ hyperfine ground states of a single $^9\mathrm{Be}^+$ ion (the $|f = 2, m_F = 2\rangle$ and $|F = 1, m_F = 1\rangle$ states, abbreviated by the equivalent spin-1/2 states $|\downarrow\rangle$ and $|\uparrow\rangle$) separated in frequency by $\omega_0/2\pi \simeq 1.250$ GHz. The control qubit $|n\rangle$ is spanned by the first two quantized harmonic oscillator states of the trapped in ($|0\rangle$ and $|1\rangle$), separated in frequency by the vibrational frequency $\omega_x/2\pi \simeq 11$ MHz of the harmonically trapped ion. Figure 1 displays the relevant $^9\mathrm{Be}^+$ energy levels. Manipulation between the four basis eigenstates spanning the two-qubit register ($|n\rangle|S\rangle = |0\rangle|\downarrow\rangle, |0\rangle|\uparrow\rangle, |1\rangle|\downarrow\rangle, |1\rangle|\uparrow\rangle$) is achieved by applying a pair of off-resonant laser beams to the ion, which drives stimulated Raman transitions between basis states. When the difference frequency δ of the beams is set near $\delta \simeq \omega_0$ (the carrier), transitions are coherently driven between internal states $|S\rangle$ while preserving $|n\rangle$. Likewise, for $\delta \simeq \omega_0 - \omega_x$ (the red sideband), transitions are coherently driven between $|1\rangle|\downarrow\rangle$ and $|0\rangle|\uparrow\rangle$, and for $\delta \simeq \omega_0 + \omega_x$ (the blue sideband), transitions are coherently driven between $|0\rangle|\downarrow\rangle$ and $|1\rangle|\uparrow\rangle$. Note that when δ is tuned to either sideband, the stimulated Raman transitions entangle $|S\rangle$ with $|n\rangle$, a crucial part of the trapped-ion quantum CN gate.

We realize the controlled-NOT gate by sequentially applying three pulses of the Raman beams to the ion

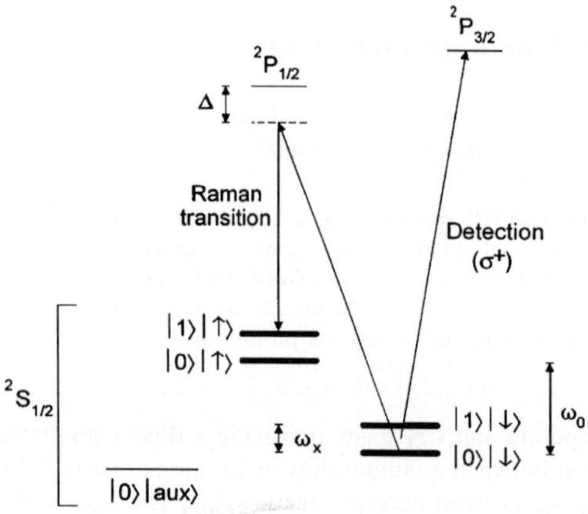

FIG. 1. $^9\mathrm{Be}^+$ energy levels. The levels indicated with thick lines form the basis of the quantum register: internal levels are $|S\rangle = |\downarrow\rangle$ and $|\uparrow\rangle$ ($^2S_{1/2}|F = 2, m_F = 2\rangle$ and $^2S_{1/2}|F = 1, m_F = 1\rangle$ levels, respectively, separated by $\omega_0/2\pi \simeq 1.250$ GHz), and $|\mathrm{aux}\rangle = {}^2S_{1/2}|F = 2, m_F = 0\rangle$ (separated from $|\downarrow\rangle$ by $\simeq$2.5 MHz); external vibrational levels are $|n\rangle = |0\rangle$ and $|1\rangle$ (separated by $\omega_x/2\pi \simeq 11.2$ MHz). Stimulated Raman transitions between $^2S_{1/2}$ hyperfine states are driven through the virtual $^2P_{1/2}$ level ($\Delta \simeq 50$ GHz) with a pair of $\simeq$313 nm laser beams. Measurement of S is accomplished by driving the cycling $|\downarrow\rangle \rightarrow {}^2P_{3/2}|F = 3, m_F = 3\rangle$ transition with σ^+-polarized light and detecting the resulting ion fluorescence.

according to the following format:

(a) A $\pi/2$ pulse is applied on the carrier transition. The effect is described by the operator $V^{1/2}(\pi/2)$ in the notation of Ref. [1].

(b) A 2π pulse is applied on the blue sideband transition between $|\uparrow\rangle$ and an auxiliary atomic level $|\mathrm{aux}\rangle$ (see Fig. 1). (1)

(c) A $\pi/2$ pulse is applied on the carrier transition, with a π phase shift relative to (a), leading to the operator $V^{1/2}(-\pi/2)$ of Ref. [1].

The $\pi/2$ pulses in steps (a) and (c) cause the spin $|S\rangle$ to undergo $+1/4$ and $-1/4$ of a complete Rabi cycle, respectively, while leaving $|n\rangle$ unchanged. The auxiliary transition in step (b) simply reverses the sign of any component of the register in the $|1\rangle|\uparrow\rangle$ state by inducing a complete Rabi cycle from $|1\rangle|\uparrow\rangle \rightarrow |0\rangle|\mathrm{aux}\rangle \rightarrow -|1\rangle|\uparrow\rangle$. The auxiliary level $|\mathrm{aux}\rangle$ is the $^2S_{1/2}\ |F = 2, m_F = 0\rangle$ ground state, split from the $|\downarrow\rangle$ state by virtue of a Zeeman shift of $\simeq$2.5 MHz resulting from a 0.18 mT applied magnetic field (see Fig. 1). Any component of the quantum register in the $|n\rangle = |0\rangle$ state is unaffected by the blue sideband transition of step (b), and the effects of the two Ramsey $\pi/2$ pulses cancel. On the other hand, any component of the quantum register in the $|1\rangle|\uparrow\rangle$ state acquires a sign change in step (b), and the two Ramsey pulses add constructively, effectively "flipping" the target qubit by π radians. The truth table of the CN operation is as follows:

$$
\begin{array}{ccc}
\text{Input state} & \rightarrow & \text{Output state} \\
|0\rangle|\downarrow\rangle & \rightarrow & |0\rangle|\downarrow\rangle \\
|0\rangle|\uparrow\rangle & \rightarrow & |0\rangle|\uparrow\rangle \\
|1\rangle|\downarrow\rangle & \rightarrow & |1\rangle|\uparrow\rangle \\
|1\rangle|\uparrow\rangle & \rightarrow & |1\rangle|\downarrow\rangle .
\end{array}
\qquad (2)
$$

The experiment apparatus is described elsewhere [16,17]. A single $^9\mathrm{Be}^+$ ion is stored in a coaxial-resonator rf-ion trap [17], which provides pseudopotential oscillation frequencies of $(\omega_x, \omega_y, \omega_z)/2\pi \simeq (11.2, 18.2, 29.8)$ MHz along the principal axes of the trap. We cool the ion so that the $n_x = 0$ vibrational ground state is occupied $\simeq$95% of the time by employing resolved-sideband stimulated Raman cooling in the x dimension, exactly as in Ref. [16]. The two Raman beams each contain $\simeq$1 mW of power at $\simeq$313 nm and are detuned $\simeq$50 GHz red of the $^2P_{1/2}$ excited state. The Raman beams are applied to the ion in directions such that their wave-vector difference $\boldsymbol{\delta k}$ points nearly along the x axis of the trap; thus the Raman transitions are highly insensitive to motion in the other two dimensions. The Lamb-Dicke parameter is $\eta_x = \delta k\, x_0 \simeq 0.2$, where $x_0 \simeq 7$ nm is the spread of the $n_x = 0$ wave function. The carrier ($|n\rangle|\downarrow\rangle \rightarrow |n\rangle|\uparrow\rangle$) Rabi frequency is $\Omega_0 2\pi \simeq 140$ kHz, the red ($|1\rangle|\downarrow\rangle \rightarrow |0\rangle|\uparrow\rangle$) and blue ($|0\rangle|\downarrow\rangle \rightarrow |1\rangle|\uparrow\rangle$) sideband Rabi frequencies are $\eta_x\Omega_0/2\pi \simeq 30$ kHz, and the auxiliary transition ($|1\rangle|\uparrow\rangle \rightarrow |0\rangle|\downarrow\rangle$) Rabi frequency is $\eta_x\Omega_{\mathrm{aux}}/2\pi \simeq 12$ kHz. The difference frequency of the Raman beams is tunable from 1200 to 1300 MHz with the use of a double pass acousto-optic modulator (AOM), and the Raman pulse durations are controlled with additional switching AOMs. Since the Raman beams are generated from a single laser and an AOM, broadening of the Raman transitions due to a finite laser linewidth is negligible [18].

Following Raman cooling to the $|0\rangle|\downarrow\rangle$ state, but before application of the CN operation, we apply appropriately tuned and timed Raman pulses to the ion, which can prepare an arbitrary state of the two-qubit register. For instance, to prepare a $|1\rangle|\downarrow\rangle$ eigenstate, we apply a π pulse on the blue sideband followed by a π pulse on the carrier ($|0\rangle|\downarrow\rangle \rightarrow |1\rangle|\uparrow\rangle \rightarrow |1\rangle|\downarrow\rangle$). We perform two measurements to detect the population of the register after an arbitrary sequence of operations. First, we measure the probability $P\{S = \downarrow\}$ that the target qubit $|S\rangle$ is in the $|\downarrow\rangle$ state by collecting the ion fluorescence when σ^+-polarized laser radiation is applied resonant with the cycling $|\downarrow\rangle \rightarrow {}^2P_{3/2}|F = 3, m_F = 3\rangle$ transition (radiative linewidth $\gamma/2\pi \simeq 19.4$ MHz at $\lambda \simeq 313$ nm; see Fig. 1). Since this radiation does not appreciably couple to the $|\uparrow\rangle$ state (relative excitation probability: $\simeq 5 \times 10^{-5}$), the fluorescence reading is proportional to $P\{S = \downarrow\}$. For $S = \downarrow$, we collect on average $\simeq$1 photon per measurement cycle [16]. Once S is measured, we

VOLUME 75, NUMBER 25 PHYSICAL REVIEW LETTERS 18 DECEMBER 1995

perform a second independent measurement that provides the probability $P\{n = 1\}$ that the control bit $|n\rangle$ is in the $|1\rangle$ state: After the same operation sequence is repeated, an appropriate Raman pulse is added just prior to the detection of S. This "detection preparation" pulse maps $|n\rangle$ onto $|S\rangle$. For instance, if we first measure S to be $\downarrow$, we repeat the experiment with the addition of a "π pulse" on the red sideband. Subsequent detection of S resulting in the presence (absence) of fluorescence indicates that $n = 0$ (1). Likewise, if we first measure S to be $\uparrow$, we repeat the experiment with the addition of a "π pulse" on the *blue* sideband. Subsequent detection of S resulting in the presence (absence) of fluorescence indicates that $n = 1$ (0).

In the above measurement scheme, we do not obtain phase information about the quantum state of the register and therefore do not measure the complete transformation matrix associated with the CN operation. The phase information could be obtained by performing additional operations (similar to those described above) prior to the measurement of S. Here, we demonstrate the key features of the CN gate by (i) verifying that the populations of the register follow the truth table given in (2), and (ii) demonstrating the conditional quantum dynamics associated with the CN operation.

To verify the CN truth table, we separately prepare each of the four eigenstates spanning the register ($|n\rangle|S\rangle = |0\rangle|\downarrow\rangle, |0\rangle|\uparrow\rangle, |1\rangle|\downarrow\rangle, |1\rangle|\uparrow\rangle$), then apply the CN operation given in (1). We measure the resulting register population as described above after operation of the CN gate, as shown in Fig. 2. When the control qubit is prepared in the $|n\rangle = |0\rangle$ state, the measurements show that the gate preserves S with high probability, whereas when the initial control qubit is prepared in the $|n\rangle = |1\rangle$ state, the CN gate flips the value of S with high probability. In contrast, the gate preserves the population n of the control qubit $|n\rangle$ with high probability, verifying that the register populations follow the CN truth table expressed in (2). The fact that the measured probabilities are not exactly zero or one is primarily due to imperfect laser-cooling, imperfect state preparation and detection preparation, and decoherence effects.

To illustrate the conditional dynamics of a quantum logic gate, we desire to perform a unitary transformation on one physical system conditioned upon the quantum state of another subsystem [19]. To see this in the present experiment, it is useful to view steps (a) and (c) of the CN operations given in (1) as Ramsey radiation pulses [20], which drive the $|n\rangle|\downarrow\rangle \to |n\rangle|\uparrow\rangle$ transition—with the addition of the perturbation (b) inserted between the pulses. If we now vary the frequency of the Ramsey pulses, we obtain the typical sinusoidal Ramsey interference pattern indicative of the coherent evolution between states $|S\rangle = |\downarrow\rangle$ and $|\uparrow\rangle$. However, the final population S depends on the status of the control qubit $|n\rangle$. This is illustrated in Fig. 3 where we plot the measured probability $P\{S = \downarrow\}$ as a function of detuning of the Ramsey pulses. For initial state $|0\rangle|\downarrow\rangle$, we obtain the normal Ramsey spectrum since step (b) is inactive. For initial state $|1\rangle|\downarrow\rangle$, the Ramsey spectrum is inverted indicating the conditional control (by quantum bit $|n\rangle$) of the dynamics of the Ramsey pulses. Appropriate Ramsey curves are also obtained for initial states $|0\rangle|\uparrow\rangle$ and $|1\rangle|\uparrow\rangle$.

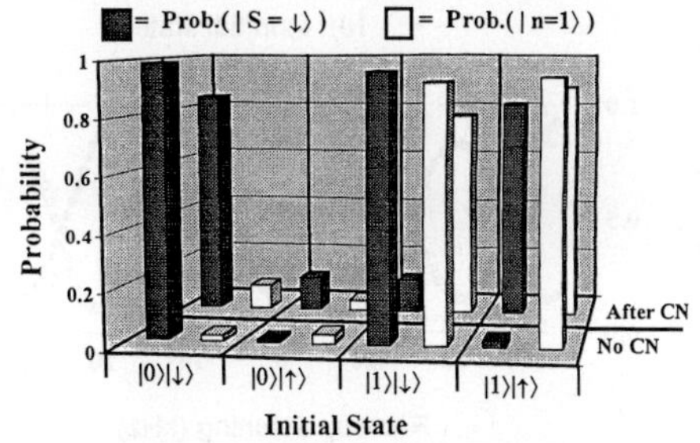

FIG. 2. Controlled-NOT (CN) truth table measurements for eigenstates. The two horizontal rows give measured final values of n and S with and without operation of the CN gate, expressed in terms of the probabilities $P\{n = 1\}$ and $P\{S = \downarrow\}$. The measurements are grouped according to the initial prepared eigenstate of the quantum register ($|0\rangle|\downarrow\rangle$, $|0\rangle|\uparrow\rangle$, $|1\rangle|\downarrow\rangle$, or $|1\rangle|\uparrow\rangle$). Even without CN operations, the probabilities are not exactly 0 or 1 due to imperfect laser-cooling, state preparation and detection preparation, and decoherence effects. However, with high probability, the CN operation preserves the value of the control qubit n, and flips the value of the target qubit S only if $n = 1$.

The switching speed of the CN gate is approximately 20 kHz, limited mainly by the auxiliary 2π pulse in step (b) given in (1). This rate could be increased with more Raman beam laser power, although a fundamental limit in switching speed appears to be the frequency separation of the control qubit vibrational energy levels, which can be as high as 50 MHz in our experiment [17].

We measure a decoherence rate of a few kHz in the experiment, adequate for a single CN gate operating at a speed of $\simeq$20 kHz, but certainly not acceptable for a more extended computation. We identify several sources responsible for decoherence, including instabilities in the laser beam power and the relative position of the ion with respect to the beams, fluctuating external magnetic fields (which can modulate the qubit phases), and instabilities in the rf-ion trap drive frequency and voltage amplitude. Substantial reduction of these sources of decoherence can be expected. Other sources of decoherence that may become important in the future include external heating and dissipation of the ion motion [16,21], and spontaneous emission caused by off-resonant transitions. We note that decoherence rates of under 0.001 Hz have been achieved for internal-state ion qubits [22].

The single-ion quantum register in the experiment comprises only two qubits and is therefore not useful for computation. However, if the techniques described here are applied to a collection of many ions cooled to the $n = 0$ state of collective motion, it should be possible to imple-

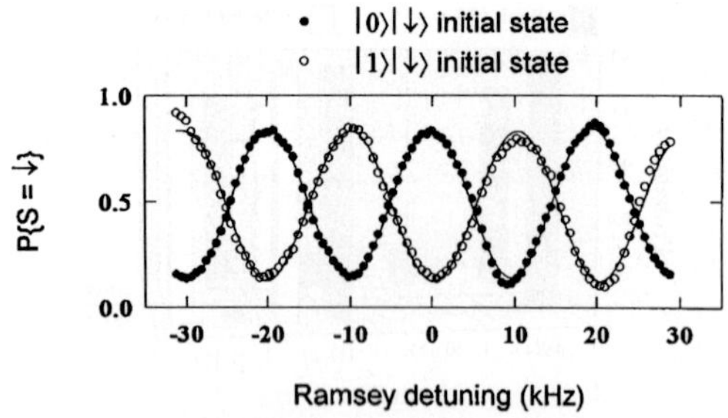

FIG. 3. Ramsey spectra of the controlled-NOT (CN) gate. The detuning of the Ramsey $\pi/2$ pulses of the CN gate [steps (a) and (c)] is swept, and S is measured, expressed in terms of the probability $P\{S = \downarrow\}$. The solid points correspond to initial preparation in the $|n\rangle|S\rangle = |0\rangle|\downarrow\rangle$ state, and the hollow points correspond to preparation in the $|n\rangle|S\rangle = |1\rangle|\downarrow\rangle$ state. The resulting patterns are shifted in phase by π rad. This flipping of $|S\rangle$ depending on the state of the control qubit indicates the conditional dynamics of the gate. Similar curves are obtained when the $|n\rangle|S\rangle = |0\rangle|\uparrow\rangle$ and $|1\rangle|\uparrow\rangle$ states are prepared. The lines are fits by a sinusoid, and the width of the Ramsey fringes is consistent with the $\simeq 50$ μsec duration of the CN operation.

ment computations on larger quantum registers. For example, the CN gate between two ions (m and n) might be realized by mapping the internal state of the mth ion onto the collective vibrational state of all ions, applying the single-ion CN operation demonstrated in this work to the nth ion, then returning the vibrational state back to the internal state of the mth ion. (This mapping may be achieved by simply driving a π pulse on the red of blue sideband of the mth ion.) This approach is equivalent to the scheme proposed by Cirac and Zoller [1,23]. An arbitrary computation may then be broken into a number of such operations on different pairs of ions, accompanied by single qubit rotations on each ion (carrier transitions) [6–8].

We are currently devoting effort into the multiplexing of the register to many ions. Several technical issues remain to be explored in this scaling, including laser-cooling efficiency, the coupling of internal vibrational modes due to trap imperfections, and the unique addressing of each ion with laser beams. Although we can trap and cool a few ions in the current apparatus, other geometries such as the linear rf-ion trap [24] or an array of ion traps each confining a single ion [25] might be considered.

This work is supported by the U.S. Office of Naval Research and the U.S. Army Research Office. We acknowledge useful contributions from J.C. Bergquist and J.J. Bollinger. We thank Robert Peterson, Dana Berkeland, and Chris Myatt for helpful suggestions on the manuscript.

[1] J.I. Cirac and P. Zoller, Phys. Rev. Lett. **74**, 4091 (1995).
[2] R.P. Feynman, Int. J. Theor. Phys. **21**, 467 (1982); Opt. News **11**, 11 (1985).
[3] D. Deutsch, Proc. R. Soc. London A **425**, 73 (1989); D. Deutsch and R. Jozsa, Proc. Soc. London A **439**, 554 (1992).
[4] P. Shor, in Proceedings of the 35th Annual Symposium on the Foundations of Computer Science (IEEE Computer Society Press, New York, 1994), p. 124.
[5] R.L. Rivest, A. Shamir, and L. Adelman, Comm. ACM, **28**, 120 (1978).
[6] D.P. DiVincenzo, Phys. Rev. A **51**, 1015 (1995).
[7] A. Barenco, D. Deutsch, A. Ekert, and R. Jozsa, Phys. Rev. Lett. **74**, 4083 (1995).
[8] A. Barenco, C.H. Bennett, R. Cleve, D.P. DiVincenzo, N. Margolus, P. Shor, T. Sleator, J. Smolin, and H. Weinfurter, Phys. Rev. A **52**, 3457 (1995).
[9] R. Landauer, in Proceedings of the Drexel-4 Symposium on Quantum Nonintegrability-Quantum Classical Correspondence, edited by D.H. Feng and B-L. Hu (International Press, Boston, to be published).
[10] W.G. Unruh, Phys. Rev. A **51**, 992 (1995).
[11] K. Obermeyer, W.G. Teich, and G. Mahler, Phys. Rev. B **37**, 8111 (1988).
[12] S. Lloyd, Science **261**, 1569 (1993).
[13] M. Brune, P. Nussenzveig, F. Schmidt-Kaler, F. Bernardot, A. Maali, J.M. Raimond, and S. Haroche, Phys. Rev. Lett. **72**, 3339 (1994); M. Brune, F. Schmidt-Kaler, J. Deyer, A. Maali, and J.M. Raimond, *"Laser Spectroscopy XII,"* edited by M. Inguscio (World Scientific, Singapore, to be published).
[14] H.J. Kimble, in *"Laser Spectroscopy XIII"* Ref. [13].
[15] T. Sleator and H. Weinfurter, Phys. Rev. Lett. **74**, 4087 (1995).
[16] C. Monroe, D.M. Meekhof, B.E. King, S.R. Jefferts, W.M. Itano, D.J. Wineland, and P. Gould, Phys. Rev. Lett. **75**, 4011 (1995).
[17] S.R. Jefferts, C. Monroe, E.W. Bell, and D.J. Wineland, Phys. Rev. A **51**, 3112 (1995).
[18] J.E. Thomas, P.R. Hemmer, S. Ezekiel, C.C. Leiby, R.H. Picard, and C.R. Willis, Phys. Rev. Lett. **48**, 867 (1982).
[19] A. Ekert and R. Jozsa, Rev. Mod. Phys. (to be published).
[20] N.F. Ramsey, *Molecular Beams* (Oxford University Press, London, 1956).
[21] F. Diedrich, J.C. Bergquist, W.M. Itano, and D.J. Wineland, Phys. Rev. Lett. **62**, 403 (1989).
[22] J.J. Bollinger, D.J. Heinzen, W.M. Itano, S.L. Gilbert, and D.J. Wineland, IEEE Trans. Instrum. Meas. **40**, 126 (1991).
[23] The controlled-NOT operator proposed in Ref. [1] is $V_n^{1/2}(\pi/2)U_m^{1,0}U_n^{2,1}U_m^{1,0}V_n^{1/2}(-\pi/2)$, adopting their notation. This is equivalent to the controlled-NOT operator proposed here between ions m and n, $U_m^{1,0}V_n^{1/2}(\pi/2)U_n^{2,1}V_n^{1/2}(-\pi/2)U_m^{1,0}$, since V_n and U_m commute.
[24] M.G. Raizen, J.M. Gilligan, J.C. Bergquist, W.M. Itano, and D.J. Wineland, Phys. Rev. A **45**, 6493 (1992).
[25] F. Major, J. Phys. (Paris) Lett. **38**, L221 (1977); R. Brewer, R.G. DeVoe, and R. Kallenbach, Phys. Rev. A **46**, R6781 (1992).

VOLUME 81, NUMBER 17 PHYSICAL REVIEW LETTERS 26 OCTOBER 1998

Deterministic Entanglement of Two Trapped Ions

Q. A. Turchette,* C. S. Wood, B. E. King, C. J. Myatt, D. Leibfried,† W. M. Itano, C. Monroe, and D. J. Wineland
Time and Frequency Division, National Institute of Standards and Technology, Boulder, Colorado 80303
(Received 26 May 1998)

We have prepared the internal states of two trapped ions in both the Bell-like singlet and triplet entangled states. In contrast to all other experiments with entangled states of either massive particles or photons, we do this in a deterministic fashion, producing entangled states *on demand* without selection. The deterministic production of entangled states is a crucial prerequisite for large-scale quantum computation. [S0031-9007(98)07411-0]

PACS numbers: 42.50.Ct, 03.65.Bz, 03.67.Lx, 32.80.Pj

Since the seminal discussions of Einstein, Podolsky, and Rosen, two-particle quantum entanglement has been used to magnify and confirm the peculiarities of quantum mechanics [1]. More recently, quantum entanglement has been shown to be not purely of pedagogical interest, but also relevant to computation [2], information transfer [3], cryptography [4], and spectroscopy [5,6]. Quantum computation (QC) exploits the inherent parallelism of quantum superposition and entanglement to perform certain tasks more efficiently than can be achieved classically [7].

Relatively few physical systems are able to approach the severe requirements of QC: Controllable coherent interaction between the quantum information carriers (quantum bits or qubits), isolation from the environment, and high-efficiency interrogation of individual qubits. Cirac and Zoller have proposed a *scalable* scheme utilizing trapped ions for QC [8]. In it, the qubits are two internal states of an ion; entanglement and computation are achieved by quantum logic operations on pairs of ions involving shared quantized motion. Previously, trapped-ion quantum logic operations were demonstrated between a single ion's motion and its spin [9]. In this Letter, we use conditional quantum logic transformations to entangle and manipulate the qubits of two trapped ions.

Previous experiments have studied entangled states of photons [10,11] and of massive particles [12–14]. These experiments rely on *random processes,* either in creation of the entanglement in photon cascades [10], photon down-conversion [11], and proton scattering [12], or in the selection of appropriate atom pairs from a larger sample of trials in cavity QED [13]. Recent results in NMR of bulk samples have shown entanglement of particle spins [14,15], but because pseudopure states are selected through averaging over a thermal distribution, the signal is exponentially degraded as the number of qubits is increased. In the preceding experiments the efficiency of state generation will exponentially decrease with the system size (both particles and operations). This is because the preceding processes are *selectable* but not *deterministic* generators of entanglement. We mean deterministic as defined in Ref. [16] which in the present context is "the property that if the [entanglement] source is switched on, then with a high degree of certainty [the desired quantum state of all of a given set of particles is generated] at a known, user-specified time." Deterministic entanglement coupled with the ability to store entangled states for future use is crucial for the realization of large-scale quantum computation. Ion-trap QC has no fundamental scaling limits; moreover, even the simple two-ion manipulations described here can, in principle, be incorporated into large-scale computing by coupling two-ion subsystems via cavities [17], or by using accumulators [6].

In this Letter, we describe the deterministic generation of a state which under ideal conditions is given by

$$|\psi_e(\phi)\rangle = \tfrac{3}{5}\,|\downarrow\uparrow\rangle - e^{i\phi}\,\tfrac{4}{5}\,|\uparrow\downarrow\rangle, \quad (1)$$

where $|\downarrow\rangle$ and $|\uparrow\rangle$ refer to internal electronic states of each ion (in the usual spin-1/2 analogy) and ϕ is a controllable phase factor. For $\phi = 0$ or π, $|\psi_e(\phi)\rangle$ is a good approximation to the usual Bell singlet (−) or triplet (+) state $|\psi_B^{\mp}\rangle = [|\downarrow\uparrow\rangle \mp |\uparrow\downarrow\rangle]/\sqrt{2}$ since $|\langle\psi_B^-|\psi_e(0)\rangle|^2 = |\langle\psi_B^+|\psi_e(\pi)\rangle|^2 = 0.98$ and $E[\psi_e(\phi)] = 0.94$ where E is the *entanglement* defined in [18]. We also describe a novel means of differentially addressing each ion to generate the entanglement and a state-sensitive detection process to characterize it, leading to a measured fidelity of our experimentally generated state described by density matrix $\rho^{\pm}$ of $\langle\psi_e(\pi,0)|\rho^{\pm}|\psi_e(\pi,0)\rangle \approx \langle\psi_B^{\mp}|\rho^{\pm}|\psi_B^{\mp}\rangle \approx 0.70$.

The apparatus is described in Ref. [19]. We confine $^9Be^+$ ions in an elliptical rf Paul trap (major axis $\approx 525\ \mu$m, aspect ratio 3:2) with a potential applied between ring and end caps of $V_0 \cos \Omega_T t + U_0$ with $\Omega_T/2\pi \approx 238$ MHz, $V_0 \approx 520$ V. The trap is typically operated over the range $12 < U_0 < 17$ V leading to secular frequencies of $(\omega_x, \omega_y, \omega_z)/2\pi = (7.3, 16, 12.6)$ to $(8.2, 17.2, 10.1)$ MHz. The ion-ion spacing (along $\hat{x}$) is $l \approx 2\ \mu$m.

The relevant level structure of $^9Be^+$ is shown in Fig. 1a. The qubit states are the $2s\ ^2S_{1/2}\ |F = 2, m_F = 2\rangle \equiv |\downarrow\rangle$ and $2s\ ^2S_{1/2}\ |F = 1, m_F = 1\rangle \equiv |\uparrow\rangle$ states. Laser beams D1 and D2 provide Doppler precooling and beam D3 prevents optical pumping to the $|F = 2, m_F = 1\rangle$ state. The cycling $|\downarrow\rangle \rightarrow 2p\ ^2P_{3/2}$

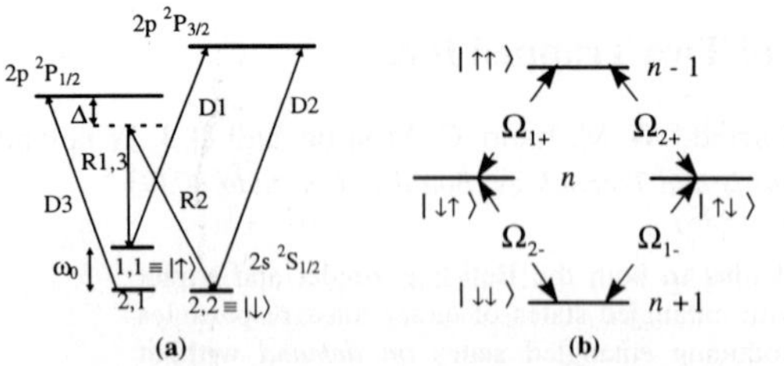

FIG. 1. (a) Relevant $^9Be^+$ energy levels. All optical transitions are near $\lambda = 313$ nm, $\Delta/2\pi = 40$ GHz, and $\omega_0/2\pi = 1.25$ GHz. R1–R3: Raman beams. D1–D3: Doppler cooling, optical pumping, and detection beams. (b) The internal basis qubit states of two spins shown with the vibrational levels connected on the red motional sideband. The labeled atomic states are as in (a); n is the motional-state quantum number (note that the motional mode frequency $\omega_{\rm str} \ll \omega_0$). $\Omega_{i\pm}$ are the Rabi frequencies connecting the states indicated.

$|F = 3, m_F = 3\rangle$ transition driven by the σ^+-polarized D2 laser beam allows us to differentiate $|\uparrow\rangle$ from $|\downarrow\rangle$ in a single ion with $\approx$90% detection efficiency by observing the fluorescence.

Transitions $|\downarrow\rangle|n\rangle \leftrightarrow |\uparrow\rangle|n'\rangle$ (where n, n' are vibrational quantum numbers) are driven by stimulated Raman processes from pairs of laser beams in one of two geometries. Additionally, two types of transitions are driven: the "carrier" with $n' = n$, and the red motional sideband (rsb) with $n' = n - 1$ [20]. With reference to Fig. 1a, the pair of Raman beams R1⊥R2 has difference wave vector $\delta\vec{k} \parallel \hat{x}$ and is used for sideband cooling (to prepare $|\downarrow\downarrow\rangle|0\rangle$), driving the $\hat{x}$ rsb, and to drive the "$\hat{x}$ carrier." Beam pair R2 ∥ R3 with $\delta\vec{k} \approx 0$ drives the "copropagating carrier" and is insensitive to motion.

Two trapped ions aligned along $\hat{x}$ have two modes of motion along $\hat{x}$: the center-of-mass (c.m.) mode (at ω_x) and the stretch mode (at $\omega_{\rm str} = \sqrt{3}\,\omega_x$) in which the two ions move in opposite directions. We sideband cool both of these modes to near the ground state, but use the stretch mode on transitions which involve the motion since it is colder (99% probability of $|n = 0\rangle$) than the c.m. and heats at a significantly reduced rate [19]. Figure 1b shows the relevant states coupled on the rsb with Rabi frequencies (in the Lamb-Dicke limit)

$$\Omega_{i+} = \sqrt{n}\,\eta'\Omega_i; \qquad \Omega_{i-} = \sqrt{n+1}\,\eta'\Omega_i, \quad (2)$$

where $\eta' = \eta/\sqrt{2\sqrt{3}}$ is the stretch-mode two-ion Lamb-Dicke parameter (with single-ion $\eta \approx 0.23$ for $\omega_x/2\pi \approx 8$ MHz) and Ω_i is the carrier Rabi frequency of ion i [9]. On the carrier the time evolution is simply that of independent Rabi oscillations with Rabi frequencies Ω_i. On the copropagating carrier, $\Omega_1 = \Omega_2 \equiv \Omega_c$.

In the Cirac-Zoller scheme, each of an array of tightly focused laser beams illuminates one and only one ion for individual state preparation. Here, each ion is equally illuminated, and we pursue an alternative technique to attain $\Omega_1 \neq \Omega_2$. Differential Rabi frequencies can be used conveniently for individual addressing on the $\hat{x}$ carrier: for example, if $\Omega_1 = 2\Omega_2$, then ion 1 can be driven for a time $\Omega_1 t = \pi$ (2π pulse, no spin flip) while ion 2 is driven for a π pulse resulting in a spin flip.

For differential addressing, we control the ion micromotion. To a good approximation, we can write [21]

$$\Omega_i = \Omega_c J_0(|\delta\vec{k}|\xi_i), \quad (3)$$

where J_0 is the zero-order Bessel function and ξ_i is the amplitude of micromotion at Ω_T (along $\hat{x}$) associated with ion i, proportional to the ion's mean $\hat{x}$ displacement from trap center. The Bessel function arises because the micromotion effectively smears out the position of an ion, thereby suppressing the laser-atom interaction [21]. The micromotion is controlled by applying a static electric field to push the ions [22] along $\hat{x}$, moving ion 2 (ion 1) away from (toward) the rf null position, inducing a smaller (larger) Rabi frequency. The range of Rabi frequencies explored experimentally is shown in Fig. 2a.

We determine $\Omega_{1,2}$ by observing the Rabi oscillations of the ions (between $|\downarrow\rangle$ and $|\uparrow\rangle$) driven on the $\hat{x}$ carrier. An example with $\Omega_1 = 2\Omega_2$ is shown in Fig. 2b. We

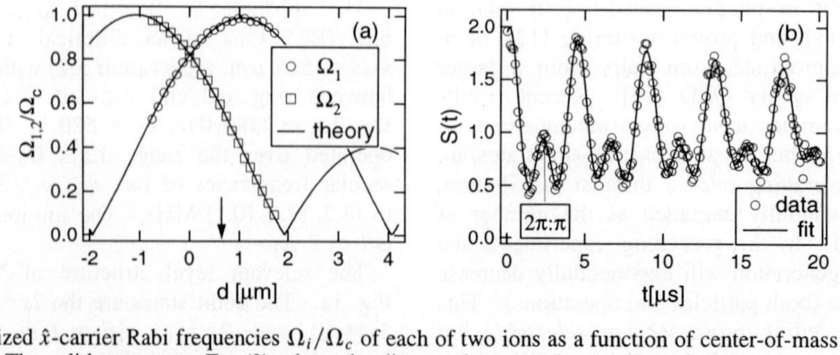

FIG. 2. (a) Normalized $\hat{x}$-carrier Rabi frequencies Ω_i/Ω_c of each of two ions as a function of center-of-mass displacement d from the rf-null position. The solid curves are Eq. (3) where the distance between the maxima of the two curves sets the scale of the ordinate, based on the known ion-ion spacing of $l \approx 2.2$ μm at $\omega_x/2\pi = 8.8$ MHz. (b) Example of Rabi oscillations starting from the initial state $|\downarrow\downarrow\rangle|n = 0\rangle$ with $\Omega_1 = 2\Omega_2$. A fit to Eq. (4) determines that $\Omega_1/2\pi = 2\Omega_2/2\pi \approx 225$ kHz, $\gamma/2\pi \approx 6$ kHz, and $\alpha \approx -0.05$. The arrow in (a) indicates the conditions of (b).

detect a fluorescence signal $S(t) = 2P_{\downarrow\downarrow} + (1 + \alpha)P_{\downarrow\uparrow} + (1 - \alpha)P_{\uparrow\downarrow}$ where $P_{kl} = |\langle\psi(t)|kl\rangle|^2$, $k, l \in \{\uparrow, \downarrow\}$, $\psi(t)$ is the state at time t and $|\alpha| \ll 1$ describes a small differential detection efficiency due to the induced differential micromotion. Driving on the $\hat{x}$ carrier for time t starting from $|\downarrow\downarrow\rangle|0\rangle$, $S(t)$ can be described by

$$S(t) = 1 + (1/2)(1 + \alpha)\cos(2\Omega_1 t)e^{-\gamma t} + (1/2)(1 - \alpha)\cos(2\Omega_2 t)e^{-(\Omega_2/\Omega_1)\gamma t}, \quad (4)$$

where γ allows for decay of the signal [20]. The local maximum at $t = 2.4\ \mu$s on Fig. 2b is the $2\pi{:}\pi$ point at which ion 1 has undergone a 2π pulse while ion 2 has undergone a π pulse resulting in $|\downarrow\downarrow\rangle|0\rangle \rightarrow |\downarrow\uparrow\rangle|0\rangle$. Driving a $\pi{:}\pi$ pulse on the copropagating carrier transforms $|\downarrow\uparrow\rangle|0\rangle$ to $|\uparrow\downarrow\rangle|0\rangle$ and $|\downarrow\downarrow\rangle|0\rangle$ to $|\uparrow\uparrow\rangle|0\rangle$, completing the generation of all four internal basis states of Fig. 1b.

Now consider the levels coupled by the first rsb [20] shown in Fig. 1b. If we start in the state $|\psi(0)\rangle = |\downarrow\uparrow\rangle|0\rangle$ and drive on the (stretch mode) rsb for time t, the Schrödinger equation can be integrated to yield

$$|\psi(t)\rangle = -\frac{i\Omega_{2-}}{G}\sin(Gt)|\downarrow\downarrow\rangle|1\rangle + \left[\frac{\Omega_{2-}^2}{G^2}(\cos Gt - 1) + 1\right]|\downarrow\uparrow\rangle|0\rangle + e^{i\phi}\left[\frac{\Omega_{2-}\Omega_{1-}}{G^2}(\cos Gt - 1)\right]|\uparrow\downarrow\rangle|0\rangle, \quad (5)$$

where $G = (\Omega_{2-}^2 + \Omega_{1-}^2)^{1/2}$ and Ω_{i-} is from Eq. (2) with $n = 0$. The phase factor $\phi = \delta\vec{k} \cdot \langle\vec{x}_1 - \vec{x}_2\rangle$ depends on the spatial separation of ions and the arises because each ion sees different laser phases. The ion-ion spacing varies by $\delta l \approx 100$ nm over the range of U_0 cited previously ($\phi = 0$ for $U_0 = 16.3$ V and $\phi = \pi$ for $U_0 = 12.6$ V, with $d\phi/dU_0$ in good agreement with theory). For $Gt = \pi$ and $\Omega_1 = 2\Omega_2$, the final state is $\psi_e(\phi)$ from Eq. (1). Note that $\Omega_1 = (\sqrt{2} + 1)\Omega_2$ would generate the Bell states (but we would not have access to the initial state $|\downarrow\uparrow\rangle$, since Ω_i are fixed throughout an experiment).

We now describe our two-ion state-detection procedure. We first prepare a two-ion basis state $|kl\rangle$, apply the detection beam D2 for a time $\tau_d \approx 500\ \mu$s, and record the number of photons m detected in time τ_d. We repeat this sequence for $N \approx 10^4$ trials and build a histogram of the photons collected (Fig. 3). To determine the population of an unknown state, we fit its histogram to a weighted sum of the four basis histograms with a simple linear least-squares procedure.

We observe that the $|\uparrow\uparrow\rangle$ count distribution (Fig. 3a) is not the expected single peak at $m = 0$, but includes contributions at $m = 1$ and $m = 2$ due to background counts. The signal in bins $m > 2$ (which accounts for $\sim$10% of the area) is due to a depumping process in which

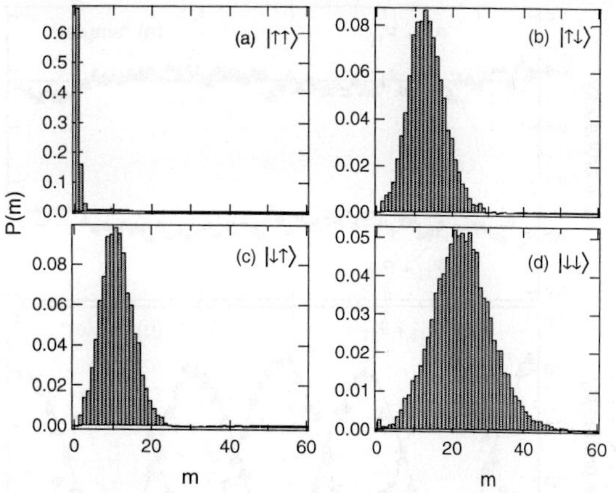

FIG. 3. Photon-number distributions for the four basis qubit states. Plotted in each graph is the probability of occurrence $P(m)$ of m photons detected in 500 μs vs m, taken over $\sim 10^4$ trials. Note the different scales for each graph.

D2 off-resonantly drives an ion out of $|\uparrow\rangle$, ultimately trapping it in the cycling transition. We approximately double the depumping time by applying two additional Raman "shelving" pulses ($|\uparrow\rangle \rightarrow {}^2S_{1/2}|F = 2, m_F = 0\rangle \rightarrow {}^2S_{1/2}|F = 1, m_F = -1\rangle$; $|\downarrow\rangle$ unaffected) after every state preparation. This results in an average difference of 10–15 detected photons between an initial $|\downarrow\rangle$ and $|\uparrow\rangle$ state, as shown in Fig. 3. The distributions associated with $|\downarrow\uparrow\rangle$, $|\uparrow\downarrow\rangle$, and $|\downarrow\downarrow\rangle$ are non-Poissonian due to detection laser intensity and frequency fluctuations, the depumping described previously and $|\downarrow\rangle \rightarrow |\uparrow\rangle$ transitions from imperfect polarization of D2.

One may ask: What is our overall two-ion state-detection efficiency on a *per experiment* basis? To address this issue, we distinguish three cases: (1) $|\uparrow\uparrow\rangle$, (2) $|\uparrow\downarrow\rangle$ or $|\downarrow\uparrow\rangle$, and (3) $|\downarrow\downarrow\rangle$. Now define case 1 to be true when $m \leq 3$, case 2 when $3 < m < 17$, and case 3 when $m \geq 17$. This gives an optimal 80% probability that the correct case is diagnosed.

We have generated states described by density operators $\rho^{\pm}$ in which the populations (diagonals of $\rho^{\pm}$) are measured to be $P_{\downarrow\uparrow} \approx P_{\uparrow\downarrow} \approx 0.4$, $P_{\downarrow\downarrow} \approx 0.15$, and $P_{\uparrow\uparrow} \approx 0.05$. To establish coherence, consider first the Bell singlet state ψ_B^- which has $P_{\downarrow\uparrow} = P_{\uparrow\downarrow} = 1/2$. Since ψ_B^- has total spin $J = 0$, any J-preserving transformation, such as an equal rotation on both spins, must leave this state unchanged, whereas such a rotation on a mixed state with populations $P_{\downarrow\uparrow} = P_{\uparrow\downarrow} = 1/2$ and no coherences will evolve quite differently. We perform a rotation on both spins through an angle θ by driving on the copropagating carrier for a time t such that $\theta = \Omega_c t$. Figure 4a shows the time evolution of an experimental state which approximates the singlet Bell state. Contrast this with the approximate "triplet" state shown in Fig. 4b. The data show that $\rho^{\pm}$ is decomposed as $\rho^{\pm} = C|\psi_B^{\pm}\rangle\langle\psi_B^{\pm}| + (1 - C)\rho_m$

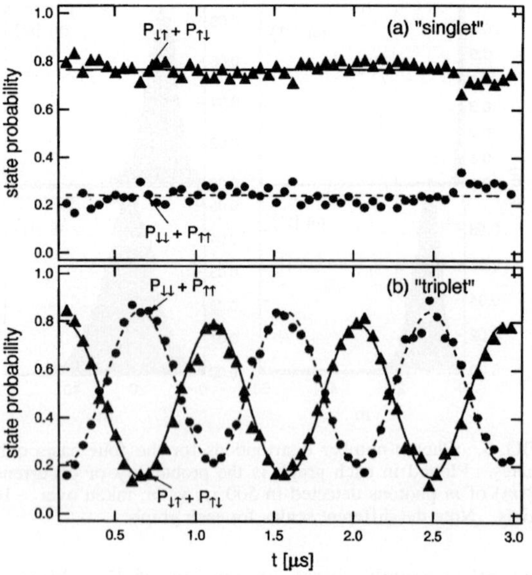

FIG. 4. Probabilities $P_{\downarrow\uparrow} + P_{\uparrow\downarrow}$ and $P_{\downarrow\downarrow} + P_{\uparrow\uparrow}$ as a function of time t driving on the copropagating carrier, starting from (a) the "singlet" $\psi_e(0)$ and (b) the "triplet" $\psi_e(\pi)$ entangled states. The equivalent rotation angle is $2\Omega_c t$ ($\Omega_c/2\pi \approx 200$ kHz for these data). The solid and dashed lines in (a) and (b) are sinusoidal fits to the data, from which the contrast is extracted.

in which ρ_m has no coherences which contribute to the measured signal (off-diagonal elements connecting $|\uparrow\downarrow\rangle$ with $|\downarrow\uparrow\rangle$ and $|\uparrow\uparrow\rangle$ with $|\downarrow\downarrow\rangle$), and $C = 0.6$ is the contrast of the curves in Fig. 4. This leads to a fidelity of $\langle\psi_B^{\pm}|\rho^{\pm}|\psi_B^{\pm}\rangle = (P_{\downarrow\uparrow} + P_{\uparrow\downarrow} + C)/2 \approx 0.7$.

The nonunit fidelity of our states arises from Raman laser intensity noise and a second-order (in η) effect on Ω_i due to excitation of the c.m. mode [19]. These effects can be seen in Fig. 2b as a decay envelope on the data [modeled by γ of Eq. (4)] and cause a 10% loss of fidelity in initial state preparation [23].

The micromotion-induced selection of Rabi frequencies as here demonstrated is sufficient to implement two-ion universal quantum logic with individual addressing [8]. To start, we arrange the trap strength and static electric field in such a way that $|\delta\vec{k}|\xi_1 = 0$ and $|\delta\vec{k}|\xi_2 = a_0$, where $J_0(a_0) = 0$. To isolate ion 1, note that by Eq. (3) $\Omega_1 = \Omega_c J_0(0) = \Omega_c$ and $\Omega_2 = \Omega_c J_0(a_0) = 0$. To isolate ion 2, we add $\Omega_T/2\pi = \pm 238$ MHz to the difference frequency of the Raman beams. This drives the first sideband of the rf micromotion so that the J_0 of Eq. (3) is replaced by J_1, resulting in $\Omega_1 = \Omega_c J_1(0) = 0$ and $\Omega_2 = \Omega_c J_1(a_0) \neq 0$.

In conclusion, we have taken a first step which is crucial for quantum computations with trapped ions. We have *engineered* entangled states deterministically; that is, there is no inherent probabilistic nature to our quantum entangling source. We have developed a two-ion state-sensitive detection technique which allows us to measure the diagonal elements of the density matrix $\rho^{\pm}$ of our states, and have performed transformations which directly measure the relevant off-diagonal coherences of $\rho^{\pm}$.

We acknowledge support from the U.S. National Security Agency, Office of Naval Research, and Army Research Office. We thank Eric Cornell, Tom Heavner, David Kielpinski, and Matt Young for critical readings of the manuscript.

*Electronic address: quentint@boulder.nist.gov.
†Present address: Univ. Innsbruck, Innsbruck, Austria.

[1] A. Einstein, B. Podolsky, and N. Rosen, Phys. Rev. **47**, 777 (1935); J. S. Bell, *Speakable and Unspeakable in Quantum Mechanics* (Cambridge University Press, Cambridge, England, 1987).
[2] P. W. Shor, SIAM J. Comput. **26**, 1484 (1997); L. K. Grover, Phys. Rev. Lett. **79**, 325 (1997).
[3] A. Barenco and A. Ekert, J. Mod. Opt. **42**, 1253 (1995); C. H. Bennett, Phys. Today **48**, No. 10, 24 (1995).
[4] A. Ekert, Phys. Rev. Lett. **67**, 661 (1991); C. H. Bennett, Sci. Am. **267**, No. 4, 50 (1992).
[5] J. J. Bollinger *et al.*, Phys. Rev. A **54**, R4649 (1996); S. F. Huelga *et al.*, Phys. Rev. Lett. **79**, 3865 (1997).
[6] D. J. Wineland *et al.*, J. Res. Natl. Inst. Stand. Technol. **103**, 259 (1998).
[7] A. Ekert and R. Jozsa, Rev. Mod. Phys. **68**, 733 (1996); A. Steane, Rep. Prog. Phys. **61**, 117 (1998).
[8] J. I. Cirac and P. Zoller, Phys. Rev. Lett. **74**, 4091 (1995).
[9] C. Monroe *et al.*, Phys. Rev. Lett. **75**, 4714 (1995).
[10] S. J. Freedman and J. F. Clauser, Phys. Rev. Lett. **28**, 938 (1972); E. S. Fry and R. C. Thompson, Phys. Rev. Lett. **37**, 465 (1976); A. Aspect, P. Grangier, and G. Roger, Phys. Rev. Lett. **49**, 91 (1982).
[11] Z. Y. Ou and L. Mandel, Phys. Rev. Lett. **61**, 50 (1988); Y. H. Shih and C. O. Alley, Phys. Rev. Lett. **68**, 3663 (1992); Z. Y. Ou *et al.*, Phys. Rev. Lett. **68**, 3663 (1992); P. Kwiat *et al.*, Phys. Rev. Lett. **75**, 4337 (1995); W. Tittel *et al.*, Europhys. Lett. **40**, 595 (1997); D. Bouwmeester *et al.*, Nature (London) **390**, 575 (1997).
[12] M. Lamehi-Rachti and W. Mittig, Phys. Rev. D **14**, 2543 (1976).
[13] E. Hagley *et al.*, Phys. Rev. Lett. **79**, 1 (1997).
[14] R. Laflamme *et al.*, quant-ph/9709025 (unpublished).
[15] I. L. Chuang, N. Gershenfeld, and M. Kubinec, Phys. Rev. Lett. **80**, 3408 (1998); I. L. Chuang *et al.*, Nature (London) **393**, 143 (1998); D. G. Cory, M. D. Price, and T. F. Havel, Physica (Amsterdam) **120D**, 82 (1998); D. G. Cory *et al.*, Phys. Rev. Lett. **81**, 2152 (1998).
[16] C. K. Law and H. J. Kimble, J. Mod. Opt. **44**, 2067 (1997).
[17] J. I. Cirac *et al.*, Phys. Rev. Lett. **78**, 3221 (1997).
[18] C. H. Bennett *et al.*, Phys. Rev. A **53**, 2046 (1996).
[19] B. E. King *et al.*, Phys. Rev. Lett. **81**, 1525 (1998).
[20] D. M. Meekhof *et al.*, Phys. Rev. Lett. **76**, 1796 (1996); **77**, 2346 (1996).
[21] D. J. Wineland and W. M. Itano, Phys. Rev. A **20**, 1521 (1979).
[22] S. R. Jefferts *et al.*, Phys. Rev. A **51**, 3112 (1995).
[23] The reference histograms for $|\downarrow\uparrow\rangle$ and $|\uparrow\downarrow\rangle$ (Figs. 3b and 3c) have had this 10% contamination from $|\uparrow\uparrow\rangle$ and $|\downarrow\downarrow\rangle$ removed.

Josephson Junctions and Quantum Computation

Josephson Junctions and Quantum Computation

Rosario Fazio[a], Yuriy Makhlin[b], and Gerd Schön[b]
(a) Universita' di Catania & INFM
(b) Institut für Theoretische Festkörperphysik, Universität Karlsruhe

In a superconducting tunnel junction (which consists of two electrodes separated by an insulating barrier) a supercurrent can flow at zero voltage. The supercurrent depends on the phase difference ϕ of the order parameter on the two sides of the junction. It depends on the coupling energy $\Delta E = -E_J \cos\phi$ through the relation $I = (2e/\hbar)\partial_\phi \Delta E$. If the junction is biased at a voltage V then the supercurrent will oscillate with a frequency $\nu = 2eV/\hbar$. This is the *Josephson effect* [1, 2]. The essence of superconductivity is the macroscopic phase coherence and the Josephson current is one of the most basic implications.

Recent progress in micro-fabrication technology made it possible to fabricate in a controlled way metallic tunnel junctions with capacitances in the range of $C = 10^{-15}$F. In this case the charging energy associated with a single-electron charge, $E_{ch} = e^2/2C$ is of the order of $10^{-4}e$V, which corresponds to a temperature scale $\approx$ 1K. This implies that electron transport in the sub-Kelvin regime is strongly affected by charging effects [3], the so called *Coulomb Blockade.* When the small junctions are made of superconducting material a variety of new phenomena appear since the phase difference of the superconducting condensates and the charge Q at the junction are conjugate variables. Consequently there is an uncertainty relation

$$\Delta Q \, \Delta\phi \geq e \quad .$$

The charging energy of a tunnel junction depends on the electron number and the applied voltage. The simplest example is the single-electron box, which consists of a small island, coupled via a tunnel junction with capacitance C_J to an electrode and via a capacitor C_G to a gate voltage source V_G. For $V_G = 0$ the lowest energy state of the system is charge neutral, i.e. there are $n = 0$ excess electrons on the island. If the gate voltage is turned on, polarization charges build up at the capacitors until the number of excess electrons on the island can change due to tunneling across the junction in *discrete* steps to $n = \pm 1, \pm 2, ...$ Elementary considerations show that the charging energy depends quadratically on the gate voltage

$$E_{ch}(n, Q_G) = \frac{(2ne - Q_G)^2}{2C} \,. \tag{1}$$

Here $C = C_J + C_G$ is the total capacitance of the island. The effect of the voltage source is contained in the "gate charge" defined as $Q_G = C_G V_G$. In the regime $\Delta > E_C \gg k_B T$ (with Δ being the superconducting gap) quasi-particle tunneling can be ignored and the dynamics of an ideal Josephson junction is governed by the Hamiltonian

$$\mathcal{H} = -4E_{ch}\left(-i\frac{\partial}{\partial\phi} - \frac{Q_G}{2e}\right)^2 - E_J\cos\phi \tag{2}$$

where $Q/2e = -i\partial/\partial\phi$. In this case, at low voltages quasi-particle tunneling is suppressed, and the island charge can change only by Cooper-pair tunneling in units of $2e$. The ratio E_J/E_{ch} characterizes the properties of the junction. If $E_J/E_{ch} \gg 1$ then the junction behaves classically with a well defined critical current, i.e. phase fluctuations are very small. In the opposite limit the charge becomes localized and the Josephson coupling provides a mechanism to coherently tunnel between two different charge states. The tunneling is strong near points of degeneracy. For instance for $Q_G \approx e$ the states with $n = 0$ and $n = 2$ are nearly degenerates, and one can restrict the attention to these two charge states. The coherent tunneling between both is described by the two state system Hamiltonian

$$H = \begin{pmatrix} E_{ch}(0) & -E_J/2 \\ -E_J/2 & E_{ch}(2) \end{pmatrix} \tag{3}$$

which is the building block for the use of Josephson junctions in quantum computation. By now many properties related to the coherent tunneling of Cooper pairs have been considered theoretically and found in experiments, the interested reader may found additional material in Refs.[8, 10]. Very recently the quantum nature of a quantum Josephson junction has been probed in time domain by Nakamura et *al.* [11].

The possibility to realize an artificial two-state system (when the gate voltage is close to degeneracy point) and the macroscopic quantum coherence due to the superconductivity make Josephson junctions very good candidates to implement a solid state qubit [12, 13, 14, 15].

The first reprint [R1] of this section gives an introduction to mesoscopic superconductivity. The reprints [R2, R3] discuss the implementation of the Josephson qubit and the design of the read-out process.

References

[1] B.D. Josephson, Phys. Lett. **1**, 251 (1962).

[2] M. Tinkham, *Introduction to Superconductivity*, McGraw Hill (New York), 1996.

[3] see the articles in *Single Charge Tunneling*, H. Grabert and M.H. Devoret Eds., NATO ASI Series B: Physics Vol. 294; D. V. Averin and K. K. Likharev, in *Mesoscopic Phenomena in Solids*, B. L. Altshuler, P. A. Lee and R. A. Webb, eds., p. 173 (Elsevier, Amsterdam, 1991).

[4] A. Maassen van den Brink, G. Schön, and L.J. Geerligs, Phys. Rev. Lett. **67**, 3030 (1991).

[5] M.T. Tuominen, J.M. Hergenrother, T.S. Tighe, and M. Tinkham, Phys. Rev. Lett. **69**, 1997 (1992).

[6] K.A. Matveev, M. Gisselfält, L. I. Glazman, M. Jonson and R. I. Shekter, Phys. Rev. Lett. **70**, 2940 (1993).

[7] M. Matters, W.J. Elion and J.E. Mooij, Phys. Rev. Lett. **75**, 721 (1995).

[8] V. Bouchiat, D. Vion, P. Joyez, D. Esteve, and M.H. Devoret, Physica Scripta **T76**, 165 (1998).

[9] J. Siewert and G. Schön, Phys. Rev. B **54**, 7421 (1996).

[10] Y. Nakamura, C.D. Chen, and J.S. Tsai, Phys. Rev. Lett. **79**, 2328 (1997).

[11] Y. Nakamura, Yu. A. Pashkin, J. S. Tsai to be published in Nature, cond-mat/9904003.

[12] A. Shnirman, G. Schön, and Z. Hermon, Phys. Rev. Lett. **79**, 2371 (1997).

[13] D.V. Averin, Solid State Commun. **105**, 659 (1998).

[14] J.E. Mooij, T.P. Orlando, L. Tian, C.H. van der Wal, L. Levitov, S. Lloyd, and J.J. Mazo, to be published

[15] L.B. Ioffe, V.D. Geshkenbein, M.V. Feigel'man, A.L. Fauchere, and G. Blatter, cond-mat/9809116.

[REPRINTS]

[R1] R. Fazio and G. Schön, in *Mesoscopic Electron Transport*, NATO ASI Series E - Vol. 345, pp. 407-446, Kluwer (1997).

[R2] G. Schön, A. Shnirman, and Yu. Makhlin in: *Exploring the Quantum - Classical Frontier: Recent Advances in Macroscopic and Mesoscopic Quantum Phenomena*, Eds. J.R. Friedman and S. Han, Nova Science Publishers, Commack, NY cond-mat/9811029

[R3] Y. Makhlin, G. Schön, and A. Shnirman, Nature 398, 305 (1999).

MESOSCOPIC EFFECTS IN SUPERCONDUCTIVITY *

Rosario Fazio[(1)] and Gerd Schön[(2)]
[(1)] *Istituto di Fisica, Università di Catania viale A. Doria 6, 95129 Catania, Italy*
[(2)] *Institut für Theoretische Festkörperphysik Universtität Karlsruhe, 76128 Karlsruhe, Germany*

I. INTRODUCTION

Several chapters in this book elaborate on the concepts of mesoscopic physics. This includes phase-coherent quantum transport combined with concepts from the macroscopic world such as reservoirs and dissipation, as well as single-electron effects. Mesoscopic physics is displayed in the electronic transport properties of small systems with spatial dimensions in the range of a few nanometers to micrometers, at low temperatures typically below 1 K. The progress in nano-fabrication allowed the controlled fabrication of these structures and led to an increased interest in this physics.

Characteristic for superconductivity is the macroscopic phase coherence of the order parameter and the supercurrent flow, as well as the modifications of quasiparticle properties by the energy gap. Superconductivity adds new degrees of freedom and makes the description of mesoscopic electron transport richer. On the other hand, typical superconducting properties are influenced by mesoscopic effects, e.g. by charging effects, and the question arises whether superconductivity persists in ultrasmall systems (see e.g. the chapter by Ralph *et al.* in this volume).

In this chapter we will investigate mesoscopic superconducting systems and heterostructures of normal metals and superconductors. We will first discuss in Section 2 in a few illustrative examples the single-electron and charging effects in superconducting tunnel junction systems. We show how, on the one hand, the superconducting gap influences single-electron tunneling and how, on the other hand, charging effects influence Andreev reflection processes. Cooper pairs can tunnel coherently; associated with their quantum dynamics is the 'macroscopic quantum tunneling' of the phase in low capacitance junctions (see e.g. the chapter of Devoret and Grabert in this volume). In junctions with even lower capacitance quantum mechanical superposition of charge states play a role. The combination of coherent Cooper pair tunneling, Andreev reflection, and quasiparticle tunneling leads to richly structured dissipative I-V characteristics.

We then turn in Section 3 to the properties of superconductor-normal metal heterostructures. The key words here are proximity effect and, again, Andreev reflection. It has been known for a long time that the proximity effect and the conversion between normal and supercurrents modifies the system properties over a finite length near the interfaces. Recent experiments could spatially resolve these properties either by probes placed close enough to the interface, or in samples with small spatial dimensions L, such that the Thouless energy D/L^2 becomes comparable to the temperatures in the experiment. We present some examples and several theoretical approaches to these physical questions.

II. CHARGING EFFECTS IN LOW-CAPACITANCE SUPERCONDUCTING JUNCTION SYSTEMS

Modern technology has made it possible to fabricate, in a controlled way, metallic tunnel junctions with capacitances in the range of $C = 10^{-15}$F and below. In this case the charging energy associated with a single-electron charge, $E_{\rm C} \equiv e^2/2C$, is of the order of 10^{-4}eV or larger, which corresponds to a temperature scale $E_{\rm C}/k_{\rm B} \geq 1$K. This implies that electron transport in the sub-Kelvin regime is strongly affected by charging effects (see the introductory chapter and Refs. [1,2]).

If part of the system is superconducting further interesting effects are found: at subgap voltages single-electron tunneling (SET) is suppressed. This makes higher-order processes such as Andreev reflection in normal-superconductor (NS) junctions a dominant transport process. Here we discuss how this process is affected by the charging effects.

The charging energy allows the control of the electron number of small islands. Adding one electron to a small superconducting island necessarily puts it into an excited state with an energy exceeding the gap. Only when a second electron is added, can both recombine to form a Cooper pair. If this happens in a coherent way, the energy of the excitation created in the first tunneling process can be regained in the second. This leads to "parity effects", which

*published in *Mesoscopic Electron Transport* vol. 345 of NATO-ASI series E, pag. 407, edited by L.L. Sohn, L.P. Kouwenhoven, and G. Schön distinguish between states with even or odd electron number in the superconducting island. As an example we analyze the I-V characteristics of a NSN SET transistor with a superconducting island between normal conducting leads.

Also the coherent tunneling of Cooper pairs is influenced by charging effects. The charge and the phase difference in a Josephson junction, although macroscopic degrees of freedom, are quantum mechanical conjugate variables. The eigenstates in general are superpositions of different charge states. We discuss the consequences for the dissipative I-V characteristics of superconducting SET transistors. In an NSS transistor the Andreev reflection process in the NS junction can be used to probe the eigenstates emerging from coherent Cooper pair tunneling in the SS junction. In SSS transistors we analyze the combination of coherent Cooper pair tunneling and quasiparticle tunneling, which leads to richly structured I-V characteristic.

Reviews of single-electron effects in normal metal systems can be found in the book *Single Charge Tunneling* [2]. Tinkham's *Introduction to Superconductivity* [3] includes some topics of the present chapter. Recent work is presented in the proceedings of the workshop *Mesoscopic Superconductivity* [4] and reviews by Bruder [5] and Schön [6].

A. The charging energy

The charging energy of systems of tunnel junctions depends on the electron number in various parts of the system and the applied voltages. An important example is the single-electron transistor shown in Fig. 1. An island is coupled via two tunnel junctions to a transport voltage source, $V = V_{\rm L} - V_{\rm R}$, such that a current can flow. The island is, furthermore, coupled capacitively to a gate voltage $V_{\rm G}$. The charging energy of the system depends on the integer number of excess electrons $n = \pm 1, \pm 2, ...$ on the island and the continuously varied voltages. Elementary electrostatics [2] yields the "charging energy"

$$E_{\rm ch}(n, Q_{\rm G}) = \frac{(ne - Q_{\rm G})^2}{2C} \,. \tag{1}$$

Here $C = C_{\rm L} + C_{\rm R} + C_{\rm G}$ is the total capacitance of the island. The effect of the voltage sources is contained in the "gate charge" $Q_{\rm G} = C_{\rm G} V_{\rm G} + C_{\rm L} V_{\rm L} + C_{\rm R} V_{\rm R}$. Similar expressions hold for the "single-electron box", an even simpler system which consists of one junction only and a gate capacitance.

In a tunneling process, which changes the number of excess electrons in the island from n to $n+1$, the charging energy changes. Tunneling in the left junction is possible at low temperatures only if the energy in the left lead, $eV_{\rm L}$, is high enough to compensate for the increase in charging energy $eV_{\rm L} > E_{\rm ch}(n+1, Q_{\rm G}) - E_{\rm ch}(n, Q_{\rm G})$. Similarly, tunneling from the island (transition from $n+1$ to n) to the right lead is possible at low T only if $E_{\rm ch}(n+1, Q_{\rm G}) - E_{\rm ch}(n, Q_{\rm G}) > eV_{\rm R}$. Both conditions have to be satisfied simultaneously for a current to flow through the transistor. If this is not the case the current is exponentially suppressed, which is denoted as "Coulomb blockade". Varying the gate voltage produces the Coulomb oscillations, i.e. an e-periodic dependence of the conductance on $Q_{\rm G}$.

Many properties of the SET transistor and its extensions can be understood by considering only the energy of the different charge configurations. A further understanding of the I-V characteristic requires the knowledge of the tunneling rates of the electrons, which will be next topic.

B. Single-Electron Tunneling Rates

The SET transistor, shown in Fig. 1, is described by the Hamiltonian

$$H = H_{\rm L} + H_{\rm I} + H_{\rm R} + H_{\rm ch} + H_{\rm t} \,. \tag{2}$$

Here, $H_{\rm L} = \sum_{k,\sigma} \epsilon_k c^\dagger_{k\sigma} c_{k\sigma}$ describes noninteracting electrons with wave vector k in the left lead, with similar expressions for the island (with states denoted by q) and the right lead. The Coulomb interaction, $H_{\rm ch} = (ne - Q_{\rm G})^2/(2C)$, is assumed to depend only on the total (net) charge $n = \sum_{k,\sigma} c^\dagger_{k\sigma} c_{k\sigma} - n_+$ on the island (electronic and ionic background), as discussed above. Charge transfer processes are described by a tunneling Hamiltonian, for instance tunneling in the left junction by

$$H_{\rm t,L} = \sum_{k,q,\sigma} T_{kq} c^\dagger_{k\sigma} c_{q\sigma} + \text{h.c.} \,. \tag{3}$$

We determine the transition rates by Golden-rule arguments. An electron tunnels from one of the states k in the left lead into one of the available states q in the island, thereby changing the electron number from n to $n+1$, with rate

$$\gamma_{\rm LI}(n) = \frac{1}{e^2 R_{\rm t,L}} \int_{-\infty}^{\infty} {\rm d}E \int_{-\infty}^{\infty} {\rm d}E' \, \mathcal{N}_{\rm L}(E)\mathcal{N}_{\rm I}(E') \times f_{\rm L}(E)[1 - f_{\rm I}(E')]\delta(E' - E + \delta E_{\rm ch}) \, . \tag{4}$$

The crucial point is that the conservation of energy, expressed by the δ-function, includes apart from the energies of the electron states $\epsilon_{k/q}$ also the charging energy. The latter depends on the change of the electron number and the applied voltages. In the process considered it changes by $\delta E_{\rm ch} = E_{\rm ch}(n+1, Q_{\rm G}) - E_{\rm ch}(n, Q_{\rm G}) - eV_{\rm L}$. We further introduced the normal state tunnel conductance of the junction $R_{\rm t,L}^{-1} = (4\pi e^2/\hbar) N_{\rm I}(0)\Omega_{\rm I} N_{\rm L}(0)\Omega_{\rm L}|T|^2$. At this stage, the tunnel matrix elements T_{kq} can be considered as constants; $N_{\rm I/L}(0)$ and $\Omega_{\rm I/L}$ denote the normal densities of states and volumes of the island and lead. If the electrodes are superconducting we have to account for the reduced densities of states. In ideal systems they take the BCS form $\mathcal{N}_{\rm I/L}(E) = \Theta(|E| - \Delta_{\rm I/L})\,|E|/\sqrt{E^2 - \Delta_{\rm I/L}^2}$.

Usually the distribution functions $f_{\rm I/L}$ can be chosen to be Fermi functions. If both electrodes are normal conducting the integrals over the electron states in Eq. (4) can be performed, resulting in the "single-electron tunneling" (SET) rate [1]

$$\gamma_{\rm LI}(n, Q_g) = \frac{1}{e^2 R_{\rm t,L}} \frac{\delta E_{\rm ch}}{\exp[\delta E_{\rm ch}/k_{\rm B}T] - 1} \, . \tag{5}$$

At low temperatures, $k_{\rm B}T \ll |\delta E_{\rm ch}|$, a tunneling process which would increase the charging energy is suppressed, $\gamma \to 0$. This phenomenon is called "Coulomb blockade" of electron tunneling.

If one or both electrodes are superconducting the rate still can be expressed in a transparent way

$$\gamma_{\rm LI}(n, Q_g) = \frac{1}{e} I_{\rm qp}\left(\frac{\delta E_{\rm ch}}{e}\right) \frac{1}{\exp[\delta E_{\rm ch}/k_{\rm B}T] - 1} \, . \tag{6}$$

The function $I_{\rm qp}(V)$ is the well-known quasiparticle tunneling characteristic (see e.g. Ref. [3]), which is suppressed at voltages below the superconducting gap(s). Charging effects reduce the quasiparticle tunneling further. At zero temperature the rate is nonzero only if the gain in charging energy compensates the energy needed to create the excitations $\delta E_{\rm ch} + \Delta_{\rm I} + \Delta_{\rm L} \le 0$.

The rates describe the stochastic time evolution of the charge of the junction system. For the theoretical analysis Monte Carlo schemes or – in small systems – a master equation approach can be used. Examples of the resulting I-V characteristics of normal metal junctions are presented in the introductory chapter of this volume. Characteristic is the e-periodic dependence of the current and conductance on the applied gate charge $Q_{\rm G}$. Examples of superconducting junction systems will be presented below.

C. Two-Electron Tunneling, Andreev Reflection

In the regime where quasiparticle tunneling is suppressed by the superconducting gap higher-order processes involving multi-electron tunneling play a role. Cooper-pair tunneling is such a process, and will be discussed later. If only one of the electrodes is superconducting there still exists a 2-electron tunneling process, denoted as Andreev reflection[1]. In this process an electron approaching from the normal side with energy below the gap is reflected as a hole, while a Cooper pair propagates into the superconductor.

We will determine now the rate of Andreev tunneling taking into account charging effects, as discussed in Ref. [7]. For this purpose we consider a SET transistor with a superconducting island and normal leads (NSN). The tunneling Hamiltonian is rewritten in terms of the Bogoliubov creation and annihilation operators for the excitations in the superconducting island

$$H_{\rm t,L} = \sum_{k,q,\sigma} T_{kq}[u_{q,\sigma}\gamma_{q,\sigma}^{\dagger} + v_{q,\sigma}^{*}\gamma_{-q,-\sigma}]c_{k,\sigma} + {\rm h.c.} \, . \tag{7}$$

---

[1] Andreev considered a normal metal and a superconductor in good metallic contact, but his physical picture can be generalized to tunnel junctions.

Here, $u_{q,\sigma}$ and $v_{q,\sigma}$ are the standard BCS coherence factors with magnitudes $\sqrt{(1 \pm \epsilon_q/E_q)/2}$, and $E_q = \sqrt{\epsilon_q^2 + \Delta^2}$ is the energy of the quasiparticles.

Andreev reflection is a second-order coherent process. In the first part of the transition one electron is transferred from an initial state, e.g. $k\uparrow$ of the normal lead, into an intermediate excited state $q\uparrow$ of the superconducting island. In the second part of the coherent transition an electron tunnels from $k'\downarrow$ into the partner state $-q\downarrow$ of the first electron, such that both form a Cooper pair. The final state contains two excitations in the normal lead and an extra Cooper pair in the superconducting island. The amplitude for this process, to which we add the amplitude of the process in reverse order, is given by [7]

$$A_{kk'} = \sum_q T_{kq} T_{k'-q} u_q v_q \left(\frac{1}{\delta E_{\mathrm{ch},1} + E_q - \epsilon_k} + \frac{1}{\delta E_{\mathrm{ch},1} + E_q - \epsilon_{k'}} \right) . \tag{8}$$

Here spin indices have been suppressed and the relation $v_{q,\uparrow} = v_{q,\downarrow}^*$ has been used. The change in the charging energy $\delta E_{\mathrm{ch},1} \equiv E_{\mathrm{ch}}(n+1, Q_{\mathrm{G}}) - E_{\mathrm{ch}}(n, Q_{\mathrm{G}}) - eV$ corresponds to the virtual intermediate state where *one* electron has tunneled from the lead (at voltage V) to the island. Finally, the rate for the Andreev reflection process is

$$\gamma_{\mathrm{LI}}^A = \frac{2\pi}{\hbar} \sum_{k,k'} |A_{kk'}|^2 \, f_{\mathrm{L}}(\epsilon_k) f_{\mathrm{L}}(\epsilon'_k) \delta(\epsilon_k + \epsilon'_k + \delta E_{\mathrm{ch},2}) . \tag{9}$$

Here, the change in the charging energy $\delta E_{\mathrm{ch},2} = E_{\mathrm{ch}}(n+2, Q_{\mathrm{G}}) - E_{\mathrm{ch}}(n, Q_{\mathrm{G}}) - 2eV$ corresponds to the real final state where *two* electron charges have been added to the superconducting island.

If we approximate the product of tunneling matrix elements by its average the q-summation in (8) can be performed, with the result $A_{kk'} = \pi N_{\mathrm{I}}(0) a\left(\Delta/\delta E_{\mathrm{ch},1}\right) \langle T_{kq} T_{k'-q} \rangle_q$, where $a(x) \equiv \frac{4}{\pi} \frac{x}{\sqrt{x^2-1}} \arctan \sqrt{\frac{x-1}{x+1}}$. Andreev reflection is most important if the gap Δ is larger than the relevant energy differences $|\delta E_{\mathrm{ch},1}|$. In this limit the function a reduces to $a(x \gg 1) \approx 1$. Henceforth, we disregard this weak energy dependence: As such the integrations in (9) can be performed, resulting in

$$\gamma_{\mathrm{LI}}^A(n, Q_{\mathrm{G}}) = \frac{G^{\mathrm{A}}}{4e^2} \frac{\delta E_{\mathrm{ch},2}}{\exp(\delta E_{\mathrm{ch},2}/k_{\mathrm{B}}T) - 1} . \tag{10}$$

Note that the functional dependence of this rate coincides with that for single-electron tunneling in a normal junction, Eq. (5). Hence Andreev reflection is subject to Coulomb blockade like normal-state single-electron tunneling [8] with the exception that:

(i) The charge transferred in an Andreev reflection is $2e$, and the charging energy changes accordingly.

(ii) The effective Andreev conductance is of second-order in the tunneling conductance

$$G^{\mathrm{A}} = \frac{1}{4} \frac{R_{\mathrm{K}}}{N_{\mathrm{ch}} R_{\mathrm{t}}^2} . \tag{11}$$

(iii) We have to account for the number of independent parallel channels for both the normal state conductance, $1/R_{\mathrm{t}} = N_{\mathrm{ch}}/R_{\mathrm{t},0}$, and the Andreev conductance, $G^{\mathrm{A}} \propto N_{\mathrm{ch}} R_{\mathrm{K}}/R_{\mathrm{t},0}^2$. (Note that in Eq. (11) we express the latter by $1/R_{\mathrm{t}}$. Hence the factor N_{ch} appears in the denominator.) In the tunneling Hamiltonian approach N_{ch} is expressed by the correlations between the matrix elements [7]

$$\frac{1}{N_{\mathrm{ch}}} = \frac{\langle |\langle T_{kq} T_{k'-q} \rangle_q|^2 \rangle_{k,k'}}{\left(\langle |T_{kq}|^2 \rangle_{kq}\right)^2} . \tag{12}$$

A more detailed analysis [9] shows that the second-order Andreev process is sensitive to spatial correlations in the normal metal, which can be expressed by the Cooperon propagator. For the moment we consider N_{ch} as a fit parameter; a comparison of the Andreev and the normal state conductance shows that even in small junctions it is much larger than one.

As an example we show in in Fig. 2 the *I-V* characteristic of a NNS SET transistor. The structure observed there with two characteristic scales arises due to a combination of single-electron tunneling with rate (6) and Andreev reflection with rate (10).

D. Parity effects

1. The Superconducting Electron Box

In a normal-metal electron box, sweeping the applied gate voltage increases the electron number on the island in unit steps, and the voltage of the island shows a periodic saw-tooth behavior. The periodicity in the gate charge Q_G is e. If the island is superconducting and the gap Δ smaller than the charging energy E_C, the charge and the voltage show at low temperatures a characteristic long-short cyclic, $2e$-periodic dependence on Q_G. This effect arises because single-electron tunneling from the ground state, where all electrons near the Fermi surface of the superconducting island are paired, leads to a state where one extra electron – the "odd" one – is in an excited state [11]. In a small island, as long as charging effects prevent further tunneling, the odd electron does not find another excitation for recombination. Hence the energy of this state stays (at least metastable) above that of the equivalent normal system by the gap energy. Only at larger gate voltages can another electron enter the island, and the system can relax to the ground state. This scenario repeats with periodicity $2e$ in Q_G, as displayed in Fig. 3.

At low temperatures the even-odd asymmetry has been observed in the electron box [12] as well as in the I-V characteristics of superconducting SET transistors [13–15]. However, at higher temperatures, above a cross-over value $T_{cr} \ll \Delta$, the e-periodic behavior typical for normal-metal systems is recovered. We can explain this cross-over as well as the structure in the I-V characteristics by analyzing the rate of tunneling of electrons between the lead and the island, paying particular attention to the fate of the "odd" electron [16,17].

We first consider an electron box with a superconducting island and a normal lead. If the distribution functions of lead and island are Fermi functions, the rate of tunneling is given by Eq. (6). At low temperature the rate γ_{LI} is finite only at voltages where the gain in charging energy (i.e. $\delta E_{ch} < 0$) exceeds the energy of the excitations ($\epsilon_k \geq 0, E_p \geq \Delta$) created in the lead and island, i.e. for $\delta E_{ch} + \Delta < 0$. It is exponentially suppressed otherwise. The assumption of equilibrium Fermi distributions is sufficient when we start from the even state. For definiteness let us assume that we started from $n = 0$ at gate voltage $0 \leq Q_G \leq e$. As such, the relevant change in charging energy is $\delta E_{ch} = E_{ch}(1, Q_G) - E_{ch}(0, Q_G)$ and the rate of tunneling from an even to an odd state is given by eq. (6)

$$\gamma^{eo} = \gamma_{LI}(n = 0, Q_G) . \tag{13}$$

In the odd state the quasiparticle distribution differs from an equilibrium Fermi function. There is extra charge in the normal component. After thermalization the excitations in the island can be described by a Fermi function, $f_{\delta\mu}(\epsilon) = [e^{(\epsilon-\delta\mu)/k_BT}+1]^{-1}$, but with a shifted chemical potential $\mu_N = \mu_S + \delta\mu$ relative to the condensate. The shift in chemical potential is fixed by the constraint of *one* excess electron charge $1 = N_I(0)\Omega_I \int_{-\infty}^{\infty} dE\mathcal{N}_I(E)[f_{\delta\mu}(E) - f_0(E)]$. This reduces at low temperatures to

$$\delta\mu = \Delta - k_BT \ln N_{eff}(T), \tag{14}$$

where

$$N_{eff}(T) = N_I(0)\Omega_I\sqrt{2\pi\Delta k_BT} \tag{15}$$

is the number of states in the island available for quasiparticles near the gap [13]. Parity effects are observable as long as the shift of the chemical potential is relevant $\delta\mu > k_BT$. This is the case for temperatures below the cross-over temperature

$$k_BT_{cr} = \Delta/\ln N_{eff}(T_{cr}) . \tag{16}$$

The tunneling rate back from the odd state (here $n = 1$) to the even state ($n = 0$), $\gamma^{oe} = \gamma_{IL,\delta\mu}(n = 1, Q_G)$, is given by (4) with the island distribution function replaced by $f_{\delta\mu}(\epsilon)$. For $\exp(-\Delta/k_BT) \ll 1$ the ratio of the rates of the two transitions is

$$\gamma^{oe}/\gamma^{eo} = e^{[E_{ch}(\text{odd})+\delta\mu-E_{ch}(\text{even})]/k_BT} = e^{\delta F/k_BT} . \tag{17}$$

In other words, they obey a detailed balance relation, depending on a "free energy" difference, which, in addition to the charging energy, contains the shift of the chemical potential $\delta\mu$. This free energy difference coincides with that introduced in Ref. [13].

For the following discussion it is useful to decompose the rate as

$$\gamma^{\rm oe} = \gamma_{\rm IL}(1, Q_{\rm G}) + \delta\gamma(Q_{\rm G})\,, \tag{18}$$

where $\gamma_{\rm IL}$ is given by the equilibrium form, equivalent to (6), and

$$\delta\gamma(Q_{\rm G}) = \frac{1}{e^2 R_{\rm t}} \int_{-\infty}^{\infty} d\epsilon_k \int_{-\infty}^{\infty} {\rm d}E\mathcal{N}_{\rm I}(E) \times [f_{\delta\mu}(E) - f_0(E)][1 - f_0(\epsilon_k)]\delta(\epsilon_k - E - \delta E_{\rm ch}) \tag{19}$$

describes the rate of tunneling of the odd, excited electron [16]. In the important range of parameters $\Delta + \delta E_{\rm ch} > k_{\rm B}T$ this "odd-electron tunneling rate" reduces to

$$\delta\gamma(Q_{\rm G}) = \frac{1}{2e^2 R_{\rm t} N_{\rm I}(0)\Omega_{\rm I}}\,, \tag{20}$$

whereas it is exponentially suppressed otherwise. It contains a small prefactor $1/N_{\rm I}(0)\Omega_{\rm I}$ as compared to $\gamma_{\rm IL}$. On the other hand, in the considered range of gate voltages – since the energy of the excitation in the island is regained in the tunneling process – $\delta\gamma$ is not exponentially suppressed. Hence it may be larger than $\gamma_{\rm IL}$.

In the range $0 \le Q_{\rm G} \le e$ tunneling connects the island states $n = 0$ and $n = 1$. The range $e \le Q_{\rm G} \le 2e$ can be treated analogously. The tunneling now connects the states $n = 1$ and $n = 2$. In this case, except for the single-electron tunneling processes which create further excitations with rate (6), one electron can tunnel into one specific state $(-k, -\sigma)$, the partner state of the excitation (k, σ) which is already present. Both condense immediately; the state with two excitations only exists virtually. The latter process occurs again with rate $\delta\gamma(Q_{\rm G})$. The symmetry implies $\gamma^{\rm eo/oe}(Q_{\rm G}) = \gamma^{\rm eo/oe}(2e - Q_{\rm G})$. Since the properties of the system are $2e$-periodic in $Q_{\rm G}$, we have provided a complete description for all gate voltage.

The sequential tunneling of charges between the island and the lead is described by a master equation for the occupation probabilities of the even and odd states $P_{\rm e}(Q_{\rm G})$ and $P_{\rm o}(Q_{\rm G})$,

$$\frac{dP_{\rm e}(Q_{\rm G})}{dt} = -\gamma^{\rm eo}(Q_{\rm G})P_{\rm e}(Q_{\rm G}) + \gamma^{\rm oe}(Q_{\rm G})P_{\rm o}(Q_{\rm G}) \tag{21}$$

with $P_{\rm e}(Q_{\rm G}) + P_{\rm o}(Q_{\rm G}) = 1$. With $\gamma_\Sigma(Q_{\rm G}) = \gamma^{\rm oe}(Q_{\rm G}) + \gamma^{\rm eo}(Q_{\rm G})$ the equilibrium solution follows to be $P_{\rm e(o)}(Q_{\rm G}) = \gamma^{\rm oe(eo)}(Q_{\rm G})/\gamma_\Sigma(Q_{\rm G})$. For $\gamma^{\rm oe} \gg \gamma^{\rm eo}$ we have $P_{\rm e}(Q_{\rm G}) \approx 1$, i.e. the system occupies the even state, while for $\gamma^{\rm eo} \gg \gamma^{\rm oe}$ the island is in the odd state.

The cross-over value $Q_{\rm cr}$ of the gate charge, where the system switches between the even and the odd state, is determined by the condition $P_{\rm e} \approx P_{\rm o}$, i.e. $\gamma^{\rm oe}(Q_{\rm cr}) \approx \gamma^{\rm eo}(Q_{\rm cr})$. At low temperatures this condition coincides with the condition that the energy is minimal, see Fig. 3. At finite, but low temperature we find

$$Q_{\rm cr}(T) = \frac{e}{2} + \frac{C}{e}[\Delta - k_{\rm B}T \ln N_{\rm eff}(T)]\,, \tag{22}$$

where $N_{\rm eff}(T)$ was introduced in (15). This means the short plateaus in Fig. 3 get longer until, above $T_{\rm cr}$, we have $Q_{\rm cr} = e/2$, and the e-periodic behavior known from normal systems is recovered.

2. I-V Characteristics of NSN Transistors

The analysis presented above can be extended such that we can derive the I-V characteristics of SET transistors with a superconducting island. We first consider an NSN transistor with an energy gap smaller than the charging energy scale $\Delta < E_{\rm C}$. In this system the important processes are single-electron tunneling processes in the left and right junction, causing transitions between even and odd states, with rates $\gamma_{\rm L}^{\rm eo/oe}$ and $\gamma_{\rm R}^{\rm eo/oe}$ which are obvious generalizations of Eq. (13) and (18). At low T it is sufficient to consider only one even and one odd state of the island. The solution of the corresponding master equation yields the single-electron tunneling current

$$I = e(\gamma_{\rm L}^{\rm eo}P_{\rm e} - \gamma_{\rm L}^{\rm oe}P_{\rm o}) = e\frac{\gamma_{\rm L}^{\rm eo}\gamma_{\rm R}^{\rm oe} - \gamma_{\rm R}^{\rm eo}\gamma_{\rm L}^{\rm oe}}{\gamma_{\rm L}^{\rm eo} + \gamma_{\rm R}^{\rm eo} + \gamma_{\rm L}^{\rm oe} + \gamma_{\rm R}^{\rm oe}}\,. \tag{23}$$

At high temperatures, $T > T_{\rm cr}$, this current (23) shows the Coulomb oscillations known from normal systems with parabola-shaped maxima at the points $Q_{\rm G} = e/2 + ne$ with integer n. At low temperatures, $T < T_{\rm cr}$, the current is limited by the odd-electron tunneling rate γ in one of the junctions. In the window $Q_{\rm cr}(T) < Q_{\rm G} < \frac{e}{2} + \Delta C/e + Q_{\rm cr}/2 < e$ it is

$$I_{\text{plateau}} = e\delta\gamma = \frac{1}{2eR_t N_I(0)\Omega_I} , \tag{24}$$

while it is exponentially small outside. A second current plateau exists in the window $e < 3e/2 - \Delta C/e - Q_{cr}/2 < Q_G < 2e - Q_{cr}$. Both plateaus create a double structure which repeats 2e-periodically. For $\Delta + eV/2 > E_C$ the two plateaus merge to form a 2e-periodic single plateau structure. The resulting I-V characteristic is visualized in Fig. 4.

In NSN transistors with a larger superconducting gap $\Delta > E_C$ the odd states have a large energy. Hence a mechanism which transfers two electrons between the normal metal and the superconductor becomes important. Andreev reflection with rate (10) provides such a mechanism [7]. The master equation description can be generalized to include also this process. At low temperatures a set of parabolic current peaks is found centered around the degeneracy points $Q_G = \pm e, \pm 3e, \ldots$ [7]

$$I^A(\delta Q_G, V) = G^A \left(V - 4\frac{\delta Q_G^2}{VC^2} \right) \Theta \left(V - 4\frac{\delta Q_G^2}{VC^2} \right) . \tag{25}$$

Here δQ_G is $\delta Q_G = Q_G - e$ for Q_G close to e, and similar near the other degeneracy points.

At larger transport voltages, single-electron tunneling sets in, even if $\Delta > E_C$, and Andreev reflection gets "poisoned" [7]. This occurs for

$$V \geq V_{\text{poison}} = \frac{2}{e} \left(E_C - \frac{eQ_G}{C} + \Delta \right) . \tag{26}$$

The rate for this transition, from the even to the odd state, is of the order of $\gamma^{eo} \sim (V - V_{\text{poison}})/eR_t$. It puts the system into an excited state, making it energetically favorable that a second electron tunnels into the partner state of the excitation created in the first process. The rate for the second process is given by $\delta\gamma$, which in the considered range of parameters takes the value given in Eq. (20). Typically the rate for the second transition, from odd to even, is smaller than that of the first processes and, hence, creates the bottleneck in the sequence of SET processes. The same inequality also implies that above V_{poison} the system is most likely in the odd state, $P_o/P_e = \gamma^{eo}/\delta\gamma \gg 1$. Hence the current produced by the cycle is given by Eq. (24).

Fig. 5 shows the current-voltage characteristic of a NSN transistor with $\Delta > E_C$. At small transport voltages the 2e-periodic peaks due to Andreev reflection dominate; they get poisoned above a threshold voltage. The peaks at larger transport voltages arise from a combination of single-electron tunneling and Andreev reflection. The shape and size of the even-even Andreev peaks and some of the single-electron tunneling features at higher transport voltages agree well with those observed in the experiments of Hergenrother et al. [15].

E. Cooper pair tunneling

1. Macroscopic Quantum Effects

In "classical" Josephson junctions, Cooper pairs can tunnel free of dissipation between the superconducting electrodes. The coupling is described by the Josephson energy $-E_J \cos\varphi$, which depends on φ, the phase difference across the barrier. The energy scale $E_J = \hbar I_{cr}/2e$ is related to the critical current of the junction, which in turn can be expressed by the tunneling resistance of the junction and the energy gap of the superconductor, $I_{cr}(T=0) = \pi\Delta/(2eR_t)$.

Charging effects introduce quantum dynamics: The phase difference and the charge on the electrodes, Q, are quantum mechanical conjugate variables. An ideal Josephson junction is governed by the Hamiltonian

$$H_0 = \frac{Q^2}{2C} - E_J \cos\varphi , \quad Q = \frac{\hbar}{\mathrm{i}} \frac{\partial}{\partial(\hbar\varphi/2e)} . \tag{27}$$

(For simplicity, we first describe a single junction; generalizations are presented below.) An important question, addressed in Refs. [18–21], is how to account for dissipation due to the flow of normal currents and/or quasiparticle tunneling. The so-called "macroscopic quantum effects" like macroscopic quantum tunneling of the phase, or quantum coherent oscillations are derived from the Hamiltonian (27). Macroscopic quantum tunneling has been observed in tunnel junctions with small capacitances of the order of 10^{-12} F. These values are still orders of magnitude to large for single-electron effects to play a role.

2. Superposition of Charge States

We now turn to mesoscopic Josephson junctions or junction systems, where the number of electrons or Cooper pairs in small islands is the relevant degree of freedom. The charging energy has been discussed above. The Josephson coupling describes the transfer of Cooper-pair charges forward or backward, and can be written in a basis of charge states as

$$\langle n|E_{\rm J}\cos\varphi|n'\rangle = \frac{E_{\rm J}}{2}(\delta_{n',n+2} + \delta_{n',n-2}) \ . \tag{28}$$

Below, we will consider situations where Cooper pairs tunnel coherently, which shows features known from the phenomenon of resonant tunneling. Coherent Cooper pair tunneling is non-dissipative and strongest near points of degeneracy. First we will show that these quantum fluctuations broaden the steps in the expectation value of the charge on the island of a superconducting electron box. Then we will discuss how coherent Cooper-pair tunneling can be probed by Andreev reflection and observed in the dissipative I-V characteristic of a NSS transistor [22]. Finally we describe how the combination of coherent Cooper-pair tunneling and dissipative quasiparticle tunneling leads to a dissipative I-V characteristic of SSS transistors [23,24,13,25,26]. Further examples of coherent tunneling of Cooper pairs can be found in the literature, e.g. the gate-voltage dependence of the critical current of SSS or SNS transistors [27,28].

We first consider an electron box with superconducting island and lead with large energy gap at low temperatures, $\Delta > E_{\rm C} \gg k_{\rm B}T$. In this case, at low voltages, quasiparticle tunneling is suppressed, and the island charge can change only by Cooper-pair tunneling in units of $2e$ as described by Eq. (28). The tunneling is strong near points of degeneracy. For instance for $Q_{\rm G} \approx e$ the charging energies of the states with $n = 0$ and $n = 2$ are comparable, and we can restrict our attention to these two charge states. The coherent tunneling between both is described by the 2×2 Hamiltonian

$$H = \begin{pmatrix} E_{\rm ch}(0) & -E_{\rm J}/2 \\ -E_{\rm J}/2 & E_{\rm ch}(2) \end{pmatrix} \ . \tag{29}$$

This Hamiltonian is easily diagonalized: the eigenstates and energies are

$$\begin{aligned} \psi_0 &= \alpha|0\rangle + \beta|2\rangle \quad , \quad \psi_1 = \beta|0\rangle - \alpha|2\rangle \ , \\ \alpha^2 &= \frac{1}{2}\Big[1 + \frac{\delta E_{\rm ch}}{\sqrt{\delta E_{\rm ch}^2 + E_{\rm J}^2}}\Big] = 1 - \beta^2 \ , \\ E_{0/1} &= \frac{1}{2}\left[E_{\rm ch}(0) + E_{\rm ch}(2) \mp \sqrt{\delta E_{\rm ch}^2 + E_{\rm J}^2}\right] \ . \end{aligned} \tag{30}$$

Here we have introduced the difference in charging energy $\delta E_{\rm ch} \equiv E_{\rm ch}(2) - E_{\rm ch}(0) = 4E_{\rm C}\,(Q_{\rm G}/e - 1)$. The coefficient α approaches unity if the charging energy of the state $|2\rangle$ lies far above that of $|0\rangle$, i.e. for $\delta E_{\rm ch} > 0$, and vanishes in the opposite limit, while β has the complementary behavior.

The expectation value of the charge on the island in the ground state is given by

$$\langle\psi_0|n|\psi_0\rangle = 2\beta^2 \ . \tag{31}$$

It changes near $Q_{\rm G} = e$ from 0 to 2 over a width of order $\delta Q_{\rm G} \approx E_{\rm J}/E_{\rm C}$. This has recently been observed experimentally [29].

3. NSS Transistors

Next we consider a NSS transistor. In this system the coherent tunneling of Cooper pairs in the Josephson (SS) junction can be probed by the dissipative current due Andreev reflection across the NS junction [22]. We restrict ourselves to low temperatures, $k_{\rm B}T \ll E_{\rm J}$. In order to describe coherent Cooper-pair tunneling in a situation with nonzero transport voltage we have to account in the Hamiltonian for the work done by the voltage sources during the transitions. We, therefore, keep track also of the number of electrons $N_{\rm L}$ and $N_{\rm R}$ in the left and right electrode. A basis set of states is denoted by $|N_{\rm L}, n, N_{\rm R}\rangle$, and the corresponding charging energy (for symmetric bias $V_{\rm L} = -V_{\rm R} = V/2$) is

$$H_{\rm ch}(N_{\rm L}, n, N_{\rm R}) = (ne - Q_{\rm G})^2/2C - (N_{\rm R} - N_{\rm L})eV/2\,. \tag{32}$$

In a situation where only two charge states get appreciably mixed the eigenstates and energies of the corresponding 2×2 Hamiltonian are

$$\psi_0 = \alpha|0,0,0\rangle + \beta|0,2,-2\rangle \quad , \quad \psi_1 = \beta|0,0,0\rangle - \alpha|0,2,-2\rangle\,,$$
$$E_{0/1} = \frac{1}{2}\left[E_{\rm ch}(0,0,0) + E_{\rm ch}(0,2,-2) \mp \sqrt{\delta E_{\rm ch}^2 + E_{\rm J}^2}\right]\,. \tag{33}$$

The coefficients coincide with those of the box discussed above, except for the obvious change of notation, and $\delta E_{\rm ch} = E_{\rm ch}(0,2,-2) - E_{\rm ch}(0,0,0)$.

In the low-bias regime, the dominant mechanism of transport in the NS junction of the transistor is Andreev reflection. Starting from a state $|0,0,0\rangle$ we are led in such a process to the state $|-2,2,0\rangle$. The Josephson coupling mixes this state with the state $|-2,0,2\rangle$. Hence we have to consider a second set of eigenstates

$$\psi_0' = \alpha|-2,0,2\rangle + \beta|-2,2,0\rangle \quad , \quad \psi_1' = \beta|-2,0,2\rangle - \alpha|-2,2,0\rangle\,. \tag{34}$$

The coefficients α and β are the same as for the other pair, but the corresponding energies are shifted $E_{0/1}' = E_{0/1} - 2eV$.

Andreev reflection causes transitions between the two set of eigenstates $\psi_0 \to \psi_0'$. The rate for this process can be derived along the lines described in an earlier. Compared to Eq. (8) a modification arises since the charge transfer operators pick from the initial state the component with zero charge on the island, which has amplitude α, and select from the final state the component with two extra charges, which has amplitude β. Hence the amplitude for a Andreev reflection process between the states ψ_0 and ψ_0' with two electrons tunneling from the states $k,\uparrow$ and $k',\downarrow$ of the normal electrode is

$$A_{k,k'}(\psi_0 \to \psi_0') = \alpha\beta \sum_q T_{kq}T_{k'-q}u_q v_q \left(\frac{1}{E_0 - E_{kq}} + \frac{1}{E_0 - E_{k'q}}\right)\,. \tag{35}$$

The energy of the virtual intermediate state $|-1_k, 1_q, 0\rangle$, with one electron added to the island and two excited quasiparticles with energies ϵ_k and $E_q = \sqrt{\epsilon_q^2 + \Delta^2}$ in the normal and superconducting electrode, is given by $E_{kq} = E_{\rm ch}(-1,1,0) - \epsilon_k + E_q$.

The summation in Eq. (35) can be performed, and the rate for the Andreev reflection process is obtained by the Golden rule. After summation over the initial states k and k' one finds for $E_0' - E_0 = -2eV \le 0$

$$\gamma(\psi_0 \to \psi_0') = (\alpha\beta)^2\, a_0^2\, \frac{G^{\rm A}}{4e^2}\, 2eV\,. \tag{36}$$

The rate is proportional to the product

$$\alpha^2\beta^2 = \frac{1}{4}\frac{E_{\rm J}^2}{(\delta E_{\rm ch})^2 + E_{\rm J}^2}\,, \tag{37}$$

which displays a typical resonance structure. Here $G^{\rm A}$ is the Andreev conductance (11),and the function $a_0 = a\,(\Delta/[E_{\rm ch}(-1,1,0) - E_0])$ has been defined below Eq. (9). We further assumed that the energy $\Delta + E_{\rm ch}(-1,1,0)$ of the intermediate state lies above E_0. If $\Delta \gg E_{\rm C}$ the function a_0 reduces to $a_0 \approx 1$.

Andreev reflection processes can also lead to transitions between the other states introduced above, with rates

$$\gamma(\psi_0 \to \psi_1') = \alpha^4\, a_0^2\, \frac{G^{\rm A}}{4e^2}\,[E_0 + 2eV - E_1]\,\Theta[E_0 + 2eV - E_1)]\,,$$
$$\gamma(\psi_1 \to \psi_0') = \beta^4\, a_1^2\, \frac{G^{\rm A}}{4e^2}\,[E_1 + 2eV - E_0]\,,$$
$$\gamma(\psi_1 \to \psi_1') = (\alpha\beta)^2\, a_1^2\, \frac{G^{\rm A}}{4e^2}\, 2eV\,. \tag{38}$$

The function a_1 is defined similar as a_0, but the energy of the initial state E_0 is replaced by E_1.

Below the threshold voltage $V < V_{\rm th} = (E_1 - E_0)/2e$ the only transition at low temperatures is Andreev reflection between the states ψ_0 and ψ_0'. The resulting current, $I = 2e\gamma(\psi_0 \to \psi_0')$, shows a pronounced resonant structure due to the overlap of the functions α and β. At higher voltages Andreev reflection can take the transistor to the excited state ψ'_1. A master equation yields the probabilities for the ground and excited states

$$P_0 = \frac{\gamma(\psi_1 \to \psi'_0)}{\gamma(\psi_0 \to \psi'_1) + \gamma(\psi_1 \to \psi'_0)} \ , \ P_1 = 1 - P_0 \neq 0 \ \text{for} \ V > V_{\text{th}} \ . \tag{39}$$

The current then is

$$\frac{I}{2e} = [\gamma(\psi_0 \to \psi'_0) + \gamma(\psi_0 \to \psi'_1)]\, P_0 + [\gamma(\psi_1 \to \psi'_1) + \gamma(\psi_1 \to \psi'_0)]\, P_1 \ . \tag{40}$$

A plot of the current-voltage-characteristic, as a function of the gate and bias voltage is shown in Fig. 6 for the case where the superconducting gap is larger than the charging energy $\Delta \geq E_0$.

4. SSS Transistors

Next we consider the case of an SSS-SET transistor with superconducting electrodes and island below the crossover temperature T_{cr} where parity effects can be observed. The charging energy and coherent Cooper pair tunneling in this system are described by the model Hamiltonian [24]

$$H_0 = \sum_{n,\bar{n}} \left(\left[\frac{(ne - Q_G)^2}{2C} - \frac{1}{2} e\bar{n}V \right] |n,\bar{n}\rangle\langle n,\bar{n}| - \frac{E_J}{2} \sum_{\pm}\sum_{\pm} |n \pm 2, \bar{n} \pm 2\rangle\langle n,\bar{n}| \right) \ . \tag{41}$$

Here we shortened the notation as compared to the previous subsection Eq. (32) and introduced $\bar{n} \equiv (N_R - N_L)$, the number of electrons which have tunneled through the transistor. The eigenstates of H_0 are linear combinations of different charge states

$$|\Psi_\alpha\rangle = \sum_{n,\bar{n}} a^{\alpha}_{n,\bar{n}} |n,\bar{n}\rangle \ , \tag{42}$$

and the energies are E_α.

Quasiparticle tunneling can cause transitions between different eigenstates $|\Psi_\alpha\rangle$. It is accounted for by

$$\begin{aligned} H_t^{\text{qp}} \ &= \sum_{k\in L, q\in I} T^{(L)}_{kq} |n+1, \bar{n}+1\rangle\langle n,\bar{n}| c^\dagger_q c_k + \text{h.c.} \\ &+ \sum_{q\in I, k'\in R} T^{(R)}_{qk'} |n-1, \bar{n}+1\rangle\langle n,\bar{n}| c^\dagger_{k'} c_q + \text{h.c.} \ . \end{aligned} \tag{43}$$

If the junction resistances are large compared to the quantum resistance $R_{t,L/R} > R_K = h/e^2$ the transition rates can be calculated by the Golden rule. A quasiparticle tunneling process in the left junction gives rise to a transition with rate

$$\gamma^{(L)}_{\alpha\to\beta} = \sum_{n,\bar{n},\pm} \left(\frac{I^{(L)}_{\text{qp}}(\varepsilon_{\alpha\beta})/e}{1 - \exp(-\varepsilon_{\alpha\beta}/k_B T)} + \gamma \right) |\langle \Psi_\beta | n \pm 1, \bar{n} \pm 1\rangle\langle n,\bar{n}|\Psi_\alpha\rangle|^2 \ . \tag{44}$$

Here $I^{(L)}_{\text{qp}}$ is the I-V characteristic for quasiparticle tunneling in the left junction [3], and $\varepsilon_{\alpha\beta} = E_\alpha - E_\beta$ is the energy difference between initial and final state. We describe parity effects by including the escape rate $\delta\gamma$ of an odd quasiparticle in the island. It is given by an expression similar to Eq. (20), modified by the density of state in the superconducting electrode. It is

$$\delta\gamma \simeq \frac{1}{2e^2 R_t N_I(0)} \frac{\varepsilon_{\alpha\beta} + \Delta}{\sqrt{(\varepsilon_{\alpha\beta} + \Delta)^2 - \Delta^2}} \theta(\varepsilon_{\alpha\beta}) \tag{45}$$

if n is odd and vanishes in the even state.

In order to determine the dc-current we follow the procedure described in Ref. [26] and first determine the eigenstates of H_0, either in an expansion in the Josephson coupling or numerically taking into account a sufficient number of charge states. This procedure converges for not too large Josephson coupling energies, $E_J < E_C$. Given the eigenstates

$|\Psi_\alpha\rangle$ we calculate the rates in Eq. (44), which then enter a master equation $\partial_t P_\alpha = \sum_{\beta\neq\alpha}(P_\beta\gamma_{\beta\to\alpha} - P_\alpha\gamma_{\alpha\to\beta})$ for the probabilities P_α to find the system in the α-th eigenstate. The stationary solution $\partial_t P_\alpha = 0$ is sufficient to evaluate the dc-current

$$I = \frac{e}{2}\sum_{\alpha,\beta\neq\alpha} P_\alpha\gamma_{\alpha\to\beta}\left(\langle\Psi_\beta|\bar{n}|\Psi_\beta\rangle - \langle\Psi_\alpha|\bar{n}|\Psi_\alpha\rangle\right) \ . \tag{46}$$

The combination of coherent Cooper pair tunneling and single-electron tunneling leads to a dissipative I-V characteristic. Results are shown in Fig. 7 with parameters corresponding to those in Ref. [13]. We note that the I-V characteristic is $2e$-periodic and observe a rich structure deep in the subgap region. For transport voltages $eV \gtrsim 2.5E_{\mathrm{C}}$ the $2e$-periodic features disappear and the current becomes e-periodic in Q_{G} again. This is not surprising since on a current scale $I \gg e\delta\gamma$ the unpaired quasiparticle in the island looses its importance.

For the parameters chosen at low transport voltages only few (two or three) states $|\Psi_\alpha\rangle$ are noticeably populated. Therefore, we can calculate the eigenstates of H_0, i.e. the coefficients $a^\alpha_{n,\bar{n}}$ in Eq. (42), by expanding in E_J. Away from certain resonant situations, the α-th eigenstate has only one coefficient $a^\alpha_{n,\bar{n}}$ of order unity, whereas all other coefficients are considerably smaller. To fix ideas, let us consider the state $|\Psi_0\rangle$ in the range of gate charges $Q_{\mathrm{G}} \in [0, e/2]$. In this eigenstate the most likely charge state is $|n = 0, \bar{n} = 0\rangle$, i.e. $a^0_{0,0} \approx 1$. Due to coherent tunneling of one Cooper pair, there is a non-zero amplitude $a^0_{\pm 2,\pm 2} \propto E_{\mathrm{J}}/E_{\mathrm{C}}$ for the system to be in the charge states $|n = \pm 2, \bar{n} = \pm 2\rangle$. Higher order Cooper pair tunneling leads to a population of higher charge states with smaller amplitude. Off resonance the system is in the charge state $|2,6\rangle$ with amplitude $a^0_{2,6} \propto (E_{\mathrm{J}}/E_{\mathrm{C}})^3$.

At resonance these amplitudes are much larger. For instance along the line

$$3eV = 4E_{\mathrm{C}}(1 - Q_{\mathrm{G}}) \tag{47}$$

the charge states $|0,0\rangle$ and $|2,6\rangle$ have the same energy, and a three-Cooper-pair tunneling process is in resonance. As a result the amplitude is drastically increased $a^0_{2,6} \propto (E_{\mathrm{J}}/E_{\mathrm{C}})$.

A transition from $|\Psi_0\rangle$ to another eigenstate can occur if it is energetically favorable and the matrix element Eq. (44) is nonzero. When analyzing the energies we find that the process

$$|\Psi_0\rangle \approx |0,0\rangle \longrightarrow |\Psi_1\rangle \approx |1,7\rangle \qquad \text{(process a)}$$

is possible. Off resonance the rate of process a) is of the order $\gamma_{(\mathrm{a})} \propto (E_{\mathrm{J}}/E_{\mathrm{C}})^6$. However, in a narrow strip of width is proportional to E_{J} around the resonance line (47) we find

$$\gamma_{(\mathrm{a})} \propto \left(\frac{E_{\mathrm{J}}/2}{4E_{\mathrm{C}}(1 - Q_{\mathrm{G}}) + eV}\right)^2 \propto \left(\frac{E_{\mathrm{J}}}{E_{\mathrm{C}}}\right)^2 \ . \tag{48}$$

This process leads to the most significant resonance in the I-V characteristic. We are, thus, led to the conclusion that the dominant transport process in the subgap region is tunneling of a quasiparticle accompanied by simultaneous tunneling of 3 Cooper pairs. This combination provides enough energy to overcome the quasiparticle tunneling gap 2Δ. The importance of this type of transport mechanism was first noted by Fulton *et al.* [23].

So far we have studied the conditions for the system to leave the initial state. However, a dc charge transport through the system requires cycles, after which the island returns to a state equivalent to the initial one. The simplest version is a two-step cycle of subsequent transitions of the same type in the left and right junction. Such cycles dominate in NNN or NSN transistors at low bias voltages. The cycle which leads to the pronounced feature in Fig. 7, at $3eV = 4E_{\mathrm{C}}(1 - Q_{\mathrm{G}})$, arise due to two-step cycles as well, but the second step is different from the first one. The transition completing the cycle which starts with process a) is

$$|\Psi_1\rangle \approx |1e,7\rangle \longrightarrow |\Psi_2\rangle \approx |0,12\rangle \qquad \text{(process b)} \ .$$

This means a quasiparticle transfer is accompanied by 2 Cooper-pair tunneling processes. The latter process is not in resonance and, therefore, the rate is $\gamma_{(\mathrm{b})} \propto (E_{\mathrm{J}}/E_{\mathrm{C}})^4$. Whereas off resonance the process a) is the bottleneck for the current, at resonance the process b) has the smaller rate. This explains the value of the current at the resonance.

For further discussions of the structures manifest in Fig. 7, including extensions such as the influence of fluctuations of the electromagnetic environment, as well as a comparison with experiments on SSS transistor [13,28] we refer to Ref. [26].

F. Extensions

In the examples discussed above, the charging energy is the dominant energy, while tunneling could be described in low order perturbation theory or – in the case of coherent Cooper pair tunneling – by diagonalization of a simple Hamiltonian. The expansion parameter is the dimensionless tunneling conductance $R_K/(4\pi^2 R_t)$, where $R_K = h/e^2 = 25.8k\Omega$ is the quantum of the resistance. In situations where this parameter is not small a more general approach is required. H. Schoeller describes in his Chapter of this volume a diagrammatic expansion to account for strong tunneling through quantum dots [30,31]. Strong tunneling in normal metal junctions has been studied by several authors [32–35].

A formulation in terms of path integrals displays in a transparent way the interplay of charging effects and tunneling phenomena [31,34]. Here we would like to draw attention to the equivalent path-integral description of superconducting junction systems, presented in Refs. [5,6,36]. In these articles applications to selected problems have been discussed, such as (i) the influence of charging effects on the Josephson current through a SNS system, where earlier results of Bauernschmitt *et al.* [37] have been reproduced, (ii) the influence of charging effects on Andreev reflection, and the proximity effect, which extends earlier results of Aslamazov *et al.* [38].

III. HYBRID NORMAL-METAL/SUPERCONDUCTOR STRUCTURES

A. Review

In the last few years new experiments revived the interest in equilibrium and non-equilibrium properties of superconductor-normal metal (SN) hybrid structures. Two key words in this context are: *proximity effect* and *Andreev reflection*. The hybrid structures can be grouped in two classes depending on the transparency of the interface between superconductor and normal metal. If they are separated by an insulating barrier with low transparency the process of two-electron tunneling is the relevant transport mechanism at low bias. If they are in good metallic contact nearly all particles are transmitted; here the dominant process is Andreev reflection. Various excellent reviews [39,40,5] discuss many aspects of SN structures. Our aim here is to introduce the basic concepts and theoretical techniques, and to review some of the current literature. Some examples are discussed explicitly to demonstrate the physics involved.

When a superconductor is put in contact with a normal metal, Cooper pairs can leak across the interface. As a result there exists a non-vanishing pair amplitude in the normal metal, defined by

$$F(\vec{r}) = \langle \psi_\uparrow(\vec{r})\psi_\downarrow(\vec{r}) \rangle \,, \tag{49}$$

where $\psi_\sigma(\vec{r})$ is the annihilation operator for an electron with spin σ. The pair amplitude is a two-particle property, related to the probability of finding two time-reversed electrons at position $\vec{r}$. The decay of $F(\vec{r})$ away from the interface depends strongly on properties – diffusive vs. ballistic, noninteracting vs. interacting – of the normal metal [41]. At finite temperature it decays in the normal metal exponentially on a scale ξ_T given by

$$\xi_T = \frac{\hbar v_{\rm F}}{2\pi T} \quad \text{or} \quad \sqrt{\frac{\hbar D}{2\pi T}} \,, \tag{50}$$

depending on whether the metal is in the clean or diffusive limit. Here D is the diffusion constant. (Henceforth, we use units where $\hbar = k_{\rm B} = 1$.) At zero temperature, if interaction effects can be be disregarded, the decay follows a power law, $F(\vec{r}) \propto 1/r$. The appearance of the pair amplitude on the normal side of the interface is accompanied by a depression of the order parameter on the superconducting side.

A nonvanishing pair amplitude implies the coherence of two electrons in the normal metal induced by the coupling to the superconductor. It does not necessarily lead to a gap in the spectrum, $\Delta(\vec{r}) = \lambda F(\vec{r})$, since both are related by the interaction strength λ, which may vanish in normal metals in the absence of an attractive or repulsive interaction.

The proximity effect is intimately related to the microscopic mechanism which governs the transport through SN interfaces. At voltages and temperatures below the superconducting gap single particle tunneling is exponentially suppressed. The dominant process is then Andreev reflection [42], where an incoming quasi-electron from N is reflected at the interface as a quasi-hole, as a result of which a Cooper pair is injected into the superconductor. The reflected hole has a momentum which is opposite (to order $|k - k_{\rm F}|/k_{\rm F}$) to the one of the incident electron. The small difference in the momentum implies that the particle and the hole maintain their phase coherence up to distance of the order of $L_\epsilon \sim \sqrt{D/\epsilon}$ where ϵ is the energy of the particle relative to the Fermi energy. If ϵ is the thermal energy, this length coincides with the correlation length given in Eq. (50) [43]. This demonstrates that the proximity effect and Andreev reflection, though seemingly different concepts, are closely related. Also in the presence of a tunnel barrier at the NS boundary the dominant mechanisms of transport is the transfer of two electrons across the barrier. We call also this process Andreev tunneling, although the momentum perpendicular to the interface is not conserved. Andreev processes are also responsible for the Josephson effect in S-N-S sandwiches [38,44]. If the thickness of the normal region is comparable to or less than its coherence length $\xi_{T,\mathrm{N}}$, phase coherence can be maintained and a supercurrent can flow through it, depending on the phase difference of the two superconductors.

Although many properties of hybrid SN system have already been studied in the past, the interest in proximity devices has been renewed recently. The reason is that it became possible to study mesoscopic hybrid systems with dimensions smaller than ξ_T. In this case the particle and the hole preserve their phase coherence across the entire conductor. Another relevant length scale, the phase-coherence length L_ϕ of single electrons in a normal metal, might well be larger than ξ_T.

In mesoscopic proximity systems the interplay between phase-coherent electron propagation in N and macroscopic phase coherence in S gives rise to interesting new physics [4]. For instance, Andreev reflection in mesoscopic N-S tunnel junctions is strongly influenced by electronic interference. The transport through NS-QUIDS (Normal Metal-Superconductor QUantum Interference DeviceS) was studied theoretically by Hekking and Nazarov [9] and experimentally by the Saclay group [45], showing the existence of a modulated current as a function of the flux piercing the device. Nakano and Takayanagi [46] considered a different type of interferometer where the phase difference is created by a current which passes through the superconductor. Petrashov *et al.* [47] and Courtois *et al.* [48] performed a series of experiments on interference effects in transport through mesoscopic samples containing superconducting arms. Proximity systems with clean N-S interfaces show a remarkable non-monotonic temperature dependence [49,43]. In these systems the presence of the superconductor renders the diffusion constant of the normal metal effectively energy dependent.

Since electrons from the normal metal can enter the superconductor and then return to the normal metal not only the *off-diagonal* properties of the metal are modified, but also the single particle properties (*diagonal* in the Nambu space). Very recently, the local electron density of states (DOS) of a normal metal in contact with a superconductor has been studied at mesoscopic distances from the N-S interface [50,51]. Close to the Fermi energy, a suppression of the DOS below its normal value has been observed.

Due to the development of superconductor-semiconductor (S-Sc) integration technology, it is now possible to observe the transport of Cooper pairs through S-Sc mesoscopic interfaces as well [4]. Examples are the supercurrent through a two-dimensional electron gas (2DEG) with Nb contacts (S-Sc-S junction) [52,53] or through quantum point contacts [54,55]. The critical current was predicted to be quantized in units of $e\Delta/\hbar$ analogously to the quantization of the normal state conductance in ordinary quantum point contacts. Another example is the excess low-voltage conductance due to Andreev scattering in Nb-InGaAs (S-Sc) junctions [56].

Electron-electron interactions in the normal metal modify the proximity effect, both qualitatively and quantitatively [41]. If the interactions between the electrons are repulsive, the induced pair amplitude in N decays faster than in the noninteracting case. This is because interactions scatter the two electrons out of their initial, time-reversed state. If the interaction is attractive, e.g. if N becomes superconducting at a lower transition temperature, $T_{\mathrm{cN}} < T < T_{\mathrm{cS}}$, the pair amplitude decays slower because of the presence of superconducting correlations, and ξ_T diverges at T_{cN}. A perturbative treatment of the interactions [57] shows that an additional contribution to the supercurrent arises. Its sign depends on the nature of the interactions in the slab (attractive or repulsive), and its phase-dependence has period π (in contrasts to 2π in the non-interacting case). In the tunneling regime, if the dimensions of the normal metal and its electric capacitance are small, the phenomenological capacity model described in section 2 can be used. In this case the critical current of an S-N-S system depends strongly on charging effects and can be tuned by a gate voltage applied to the island [37].

In low-dimensional semiconductor nano-structures with low electron concentration the Coulomb interactions cannot be treated as a weak perturbation. Rather a non-perturbative, microscopic treatment of interactions is required. For 1D systems, e.g. in a 2DEG gated to form a quantum wire, this can be done in the framework of the Luttinger liquid (LL) model [58]. Hybrid superconductor - Luttinger liquid (S-LL) systems are interesting since they enable one to study how the Coulomb interaction influences the phase-coherent propagation of *two* electrons through a 1D normal region. The Josephson current through a S-LL-S device has been evaluated in Refs. [59,60]. Due to the interactions the Andreev current in a junction between a superconductor and a chiral Luttinger liquid depends in a non-linear way of the voltage [61]. Recently also the single particle properties (DOS) of a LL connected to a superconductor have been studied, where the combined effect of interaction and Andreev tunneling leads to a behavior compared which differs qualitatively from that of an isolated LL [62].

Various theoretical approaches have been employed to study SN heterostructures. One school generalizes the scattering approaches of Landauer to include Andreev tunneling. In this case the Bogoliubov-de Gennes equations are used to construct the scattering matrix (the interested reader is invited to read the reviews on the topic [39,40]). Another school uses quasiclassical methods starting from the Eilenberger equations (or the Usadel equations for dirty metals) with the inclusion of the appropriate boundary conditions for the Green's functions at the interface. The two complementary methods provide a framework to tackle various problems involving hybrid structures. The quasiclassical methods have been useful to extract analytic results in the diffusive limits. On the other hand the scattering approach is more appropriate in multi-terminal geometries or in the regimes where neither the ballistic nor the diffusive limit are appropriate. In this case numerical solutions have been worked out. The next sections are devoted to a summary of the two approaches. In the final part of this chapter we discuss the influence of interactions on the proximity effect in superconductor - Luttinger liquid systems.

B. Scattering theory

Transport properties of mesoscopic systems have been described successfully within the scattering (Landauer) formalism [63]. The conductance is related to the transmission, and the transport theory is reduced to an analysis of the properties of the scattering matrix. This approach has been generalized to systems containing SN interfaces by Lambert [64] and Beenakker [65].

We consider an n-terminal geometry where the n-th reservoir is superconducting (the case of two or more superconducting reservoirs can be described in the same fashion). Each lead contains N incoming and outgoing modes (for simplicity we assume here that N is the same for each reservoir). In the scattering approach, one needs to evaluate the S-matrix, defined as

$$\underline{O}_\alpha = \hat{S}_{\alpha\beta} \underline{I}_\beta \ . \tag{51}$$

Here $\alpha \equiv (a,p)$ and $\beta \equiv (b,q)$, $a,b = 1,...n$ refer to the reservoirs while $q,p = 1,...N$ refer to the channel indices. The superconducting reservoir is characterized by the pair $(n, l = 1,...N)$ and it will be denoted by the index s. In Eq. (51) $\underline{O}$ and $\underline{I}$ are the amplitudes of outgoing and incoming channels, respectively. The underline denotes the two components in particle - hole space (in the absence of the superconductor the S-matrix is block-diagonal in this space).

The aim of this section is to express the scattering matrix $\hat{S}_{\alpha\beta}$ as a function of the scattering in the mesoscopic region and the scattering which takes place at the SN-interface. In order to pursue this scheme, it is conceptually simpler to separate the scattering in the normal region, which is determined by the geometry and disorder in the the mesoscopic conductor, from the scattering at the SN interface, where the Andreev processes occur. For this purpose it is assumed that a normal region, free of disorder, lies between the scattering region and the SN boundary, as illustrated in Fig. 8. This ballistic region can be thought of as arbitrarily small, and its properties do not appear in the physical results.

If the superconductor were not present the scattering matrix is determined exclusively by the geometry of the mesoscopic region

$$\underline{O}_\alpha = \hat{S}^{(0)}_{\alpha\beta} \underline{I}_\beta \ . \tag{52}$$

On the other hand, the scattering at the SN interface is described by the 2×2 matrix

$$\begin{pmatrix} \underline{I}_{\sigma^\star} \\ \underline{O}_s \end{pmatrix} = \begin{pmatrix} \hat{S}^{(A)}_{\sigma^\star\sigma^\star} & \hat{S}^{(A)}_{\sigma^\star s} \\ \hat{S}^{(A)}_{s\sigma^\star} & \hat{S}^{(A)}_{ss} \end{pmatrix} \begin{pmatrix} \underline{O}_{\sigma^\star} \\ \underline{I}_s \end{pmatrix} , \tag{53}$$

where the index $\sigma^\star \equiv (n, r = 1,..N)$ refers to the intermediate ballistic region, separating the scattering region from the superconducting reservoir (see Fig.8). The components of the Andreev scattering matrix are constructed by solving the Bogoliubov - de Gennes equations [66,67]. Note also that the outgoing states in the previous equation are $\underline{I}_{\sigma^\star}$ and $\underline{O}_S$, since the incoming wave in the reservoir n (as defined in Eq. (51)) is outgoing with respect to the SN interface.

Using Eq. (52) and Eq. (53) it is possible is to eliminate $\underline{I}_{\sigma^\star}$ and $\underline{O}_{\sigma^\star}$, which allows us to express $\hat{S}_{\alpha\beta}$ in terms of $\hat{S}^{(0)}_{\alpha\beta}$ and $\hat{S}^{(A)}_{\alpha\beta}$

$$
\begin{aligned}
\hat{S}_{\alpha\beta} &= \hat{S}^{(0)}_{\alpha\beta} + \hat{S}^{(0)}_{\alpha\sigma^\star}\left[\mathbf{1} - \hat{S}^{(A)}_{\sigma^\star\sigma^\star}\hat{S}^{(0)}_{\sigma^\star\sigma^\star}\right]^{-1}\hat{S}^{(A)}_{\sigma^\star\sigma^\star}\hat{S}^{(0)}_{\sigma^\star\beta} & (a,b\neq n)\\
\hat{S}_{\alpha s} &= \hat{S}^{(0)}_{\alpha\sigma^\star}\left[\mathbf{1} - \hat{S}^{(A)}_{\sigma^\star\sigma^\star}\hat{S}^{(0)}_{\sigma^\star\sigma^\star}\right]^{-1}\hat{S}^{(A)}_{\sigma^\star s} & (a\neq n)\\
\hat{S}_{s\beta} &= \hat{S}^{(A)}_{s\sigma^\star}\hat{S}^{(0)}_{\sigma^\star\beta} + \hat{S}^{(A)}_{s\sigma^\star}\hat{S}^{(0)}_{\sigma^\star\sigma^\star}\left[\mathbf{1} - \hat{S}^{(A)}_{\sigma^\star\sigma^\star}\hat{S}^{(0)}_{\sigma^\star\sigma^\star}\right]^{-1}\hat{S}^{(A)}_{\sigma^\star\sigma^\star}\hat{S}^{(0)}_{\sigma\star\beta} & (b\neq n)\\
\hat{S}_{ss} &= \hat{S}^{(0)}_{ss} + \hat{S}^{(A)}_{s\sigma^\star}\hat{S}^{0}_{\sigma^\star\sigma^\star}\left[\mathbf{1} - \hat{S}^{(A)}_{\sigma^\star\sigma^\star}\hat{S}^{(0)}_{\sigma^\star\sigma^\star}\right]^{-1}\hat{S}^{(A)}_{\sigma^\star s}
\end{aligned}
\tag{54}
$$

The previous expressions for the S-matrix have a simple physical meaning: By expanding the denominators one can identify each term of the series as a sequence of scattering processes (reflections and transmissions) at the various reservoirs.

The final step is to express physical quantities in the scattering formalism. Let us first consider the current operator in the (normal) lead α

$$
\hat{I}_\alpha = \frac{e}{2mi}\mathrm{Tr}\int dy_\alpha\, \underline{\Psi}_\alpha(\vec{r}_\alpha)\sigma_z\nabla\underline{\Psi}^\dagger_\alpha(\vec{r}) - \text{h.c.}\,. \tag{55}
$$

Here a trace is performed in Nambu and spin space, and the matrix σ_z accounts for the different signs of the current in the electron and hole channels. The integration is over the transverse coordinate y_α in lead α. Using scattering states [68,69] as a (more convenient) basis, with destruction and creation operators a and $a^\dagger$ of incoming scattering states, the current can be expressed as

$$
\hat{I}_\alpha(t) = e\sum_{\beta\gamma}\int dEdE'\,\underline{a}^\dagger_\beta(E)\hat{\sigma}_z\left[\delta_{\alpha\beta}\delta_{\alpha\gamma} - \hat{S}^\dagger_{\alpha\beta}\hat{S}_{\alpha\gamma}\right]\underline{a}_\gamma(E')e^{-i(E-E')t}. \tag{56}
$$

Since the reservoirs are in thermal equilibrium, the occupation probabilities of the scattering states are given by Fermi distributions. Combining Eq. (56) with the expressions Eq. (54) one arrives at the desired result for the transport properties in terms of the geometric properties of the scattering region and the Andreev scattering at the boundary with the superconducting lead.

In a two terminal geometry, where $\alpha \equiv (1, q = 1...N)$ is the index for the normal contact and $s \equiv (2, p = 1...N)$ for the superconductor, the average current is

$$
I_{\mathrm{NS}} = 2\pi e\int dE\,[f(E) - f(E+eV)]\left\{1 - |S^{(\mathrm{ee})}_{\alpha\alpha}|^2 + |S^{(\mathrm{he})}_{\alpha\alpha}|^2\right\}\,. \tag{57}
$$

A trace over the channels is implied. The current depends on the reflection coefficients. Note that the normal reflection (ee) and Andreev reflection (he) enter with opposite sign. If there is no potential barrier at the NS interface, at energies below the gap there is no normal reflection but only Andreev reflection. The linear conductance in this regime has been obtained by Beenakker [65]

$$
G_{\mathrm{NS}} = \frac{4e^2}{h}\sum_{q=1,N}\frac{T_q^2}{(2-T_q)^2}\,. \tag{58}
$$

This result is the multi-terminal generalization of the formula obtained by Blonder, Tinkham and Klapwijk [66] and by Shelankov [67]. The amplitudes T_q are the eigenvalues of the geometrical transmission matrix $(S^{(0)}_{12})$, i.e. the same coefficients which enter in the multichannel Landauer formula $G = (2e^2/h)\sum_q T_q$, and the index q runs over the transverse channels in the normal lead. Various applications of the previous expression and extensions can be found in Ref. [39]. The general formula for the current beyond the Andreev approximation and at finite voltage has been discussed in Refs. [70,71], extensions to d-wave superconductors have been considered in Ref. [72]. The use of the scattering approach is not limited to the study of the average current. Eq. (56) also allows the evaluation of the current noise (see the chapter by de Jong and Beenakker in this volume).

C. Quasiclassical approach

1. Equilibrium Theory

A complementary approach, developed to study hybrid structures, employs Green's functions

$$\mathcal{G}(\vec{r},\vec{r'},t) = -\mathrm{i}\langle T\psi(\vec{r},t)\psi^\dagger(\vec{r'},0)\rangle, \quad \mathcal{F}(\vec{r},\vec{r'},t) = -\mathrm{i}\langle T\psi(\vec{r},t)\psi(\vec{r'},0)\rangle \ . \tag{59}$$

Despite their apparent simplicity, the Gor'kov equations governing the dynamics of $\mathcal{G}$ and $\mathcal{F}$ are almost impossible to handle in inhomogeneous situations. On the other hand, the information contained in these equations is redundant, since usually only properties close to the Fermi energy are interesting. It is possible in these cases to reduce the Gor'kov equations to transport-like equations which are much easier to study. These are the Eilenberger [73] and Usadel [74] equations for clean and dirty systems, respectively.

The main steps are as follows. It is convenient to introduce the center of mass $\vec{R} = (\vec{r}+\vec{r'})/2$ and relative coordinates $\vec{\rho} = \vec{r}-\vec{r'}$ and to consider the Fourier transform of the Green's functions with respect to the latter (since we are dealing with time-independent situations we use the energy representation). The Green's function show strong oscillations as a function of the relative coordinate on the scale of the Fermi wavelength λ_F. If one is interested only in variations on scales much larger than λ_F it is sufficient to consider the quasiclassical Green's functions obtained by integrating $\mathcal{G},\mathcal{F}$ over $\xi_p = p^2/2m - \mu$,

$$\left\{ \begin{matrix} g \\ f \end{matrix} \right\} (E,\vec{R},\vec{v}_\mathrm{F}) = \frac{\mathrm{i}}{\pi} \int d\xi_p \int d^3\rho \left\{ \begin{matrix} \mathcal{G} \\ \mathcal{F} \end{matrix} \right\} (E,\vec{R},\vec{\rho}) \mathrm{e}^{-\mathrm{i}\vec{p}\cdot\vec{\rho}} \ . \tag{60}$$

Further simplifications are possible if the system is dirty and the dependence on the direction of $\vec{v}_\mathrm{F}$ is weak (the system is nearly isotropic). In this case g and f can be expanded in spherical harmonics, $g(E,\vec{R},\vec{v}_\mathrm{F}) = G(E,\vec{R}) + \vec{v}_\mathrm{F}\cdot\vec{G}(E,\vec{R})$ and $f(E,\vec{R},\vec{v}_\mathrm{F}) = F(E,\vec{R}) + \vec{v}_\mathrm{F}\cdot\vec{F}(E,\vec{R})$. An expansion yields the Usadel equation

$$-\mathrm{i}EF - \Delta G = \frac{D}{2}\left[G\widehat{\nabla}^2 F - F\nabla^2 G\right] \ . \tag{61}$$

Inelastic interactions can be accounted for by the shift $-\mathrm{i}E \to -\mathrm{i}E + 1/(2\tau_\mathrm{in})$, by the inelastic scattering rate. Pair-breaking effects add a further term $(1/\tau_\mathrm{s})GF$ on the left hand side. The magnetic field is introduced through the gauge invariant derivative, $\widehat{\nabla} \to \nabla - 2ie\vec{A}$, acting on F.

The diagonal and off-diagonal component satisfy a normalization condition, $G^2 + F^2 = 1$, which is automatically guaranteed if we choose a parameterization $F = \sin\theta$ and $G = \cos\theta$.

The formalism is completed by the self-consistency equation for the gap

$$\Delta(\vec{r}) \ln\frac{T}{T_c} = 2\pi T \sum_{\omega_\mu > 0} \left[F(\mathrm{i}\omega_\mu,\vec{r}) - \frac{\Delta(\vec{r})}{\omega_\mu} \right] \ . \tag{62}$$

We further have to specify the boundary conditions at the SN interface (which we assume to be located in the $x=0$ plane). In the absence of an extra boundary potential these read [75]

$$F(E,0_\mathrm{S}) = F(E,0_\mathrm{N}) \qquad \sigma_\mathrm{S}\frac{\mathrm{d}}{\mathrm{d}x}F(E,0_\mathrm{S}) = \sigma_\mathrm{N}\frac{\mathrm{d}}{\mathrm{d}x}F(E,0_\mathrm{N}) \ . \tag{63}$$

Hence, the parameter which describes the properties of the interface is the ratio of the coherence lengths over the ratio of the conductivities in the two materials $\Gamma = \sigma_\mathrm{N}\xi_{T\mathrm{S}}/\sigma_\mathrm{S}\xi_{T\mathrm{N}}$.

This semiclassical approach has recently been applied to study the local density of states (DOS) in hybrid structures [51]. Earlier theoretical treatments of this problem can be found in Refs. [76,77]. Experimentally the DOS is studied by attaching several tunnel junctions at certain distances from the interface and measuring the *I*-*V* characteristics [50].

The local DOS is defined through the retarded Green's function $\mathcal{G}^\mathrm{R}(x,x';t) \equiv -\mathrm{i}\langle\{\psi(x,t),\psi(x',0)^\dagger\}\rangle\theta(t)$ as

$$N(x,E) = -\frac{1}{\pi}\,\mathrm{Im}\int_{-\infty}^{\infty} \mathrm{d}t\, e^{\mathrm{i}Et}\,\mathcal{G}^\mathrm{R}(x,x;t) = N(0)\,\mathrm{Re}\,G(E) \ . \tag{64}$$

If the metal is a Fermi liquid the DOS is almost featureless $\sim N(0)$ while in the superconductor it behaves like $N(E) = N(0)E/\sqrt{E^2-\Delta^2}$. This raises the question how the DOS behaves close to an NS interface to interpolate between these two very different limits.

Due to the proximity effect, the DOS indeed acquires nontrivial structure. Results are shown in Fig. 9 (a) for the normal side of the interface. It shows a subgap structure (a bump) and a depression close to the Fermi energy. These features tend to disappear when one moves away from the interface. In the absence of pair breaking the DOS vanishes at the Fermi level. On the superconducting side the singularity at Δ is suppressed and a finite DOS appears also at low energies, as shown in Fig. 9 (b).

If the dimensions of the normal metal are finite (a slab of thickness L), a true gap E_g appears in the DOS of the normal metal. This mini-gap scales with the length and is related to the Thouless energy D/L^2. A fit to the numerical results is

$$E_g \sim (\xi + L)^{-2} ,$$

where $\xi = \sqrt{D/2\Delta}$, implying that the effective diffusion length is $\sim \xi + L$.

2. Nonequilibrium Situations

To describe systems with a finite applied voltage, the formalism of nonequilibrium superconductivity [78–80] should be used. It is based on the real-time Keldysh technique [80,81], which involves matrix Green's functions

$$\check{G} = \begin{pmatrix} \hat{G}^R & \hat{G}^K \\ 0 & \hat{G}^A \end{pmatrix} \tag{65}$$

having retarded, advanced and Keldysh (R,A,K) components. Each of these are 2×2 matrices in Nambu space typical for superconductivity

$$\hat{G} = \begin{pmatrix} G & F \\ F^\dagger & -G \end{pmatrix} , \tag{66}$$

whose entries are quasiclassical Green's functions, which in the dirty limit satisfy the Usadel equation (61). The boundary conditions for Keldysh Green's functions at NS-interfaces have been derived by Zaitsev [82]. Applications to diffusive NS heterostructures have been discussed by Volkov *et al.* [83].

As an example, and application of the Keldysh technique, we consider the transport through a diffusive wire of length L connected to a normal and a superconducting reservoir via metallic contacts. One of the striking effects in the transport of this system is the *non-monotonic* temperature dependence when the temperature is of the order of the Thouless energy $E_L = D/L^2$. The resistance of this system initially decreases when the temperature is lowered but approaches again the normal state resistance at $T = 0$. This effect was analyzed theoretically in Refs. [49,84,85] and experimentally in Ref. [43].

The differential conductance, normalized to its value if the superconductor was not present, can be expressed as

$$G_N = \frac{1}{2T} \int_0^\infty dE \frac{D(E)}{\cosh^2(E/2T)} \tag{67}$$

where $D(E)$ is the effective energy-dependent transparency to be determined microscopically from the quasiclassical equations. It is the presence of the electric field combined with the proximity effect which renders this situation a nonequilibrium one.

At temperatures much lower than the Thouless energy E_L the conductance increases quadratically with temperature

$$G_N = 1 + A \left(\frac{T}{E_L} \right)^2 , \tag{68}$$

where A is a constant. Although the low-temperature conductance coincides with the normal state value, the wire is influenced by the superconductor, as can be seen in the local density of states [85]. At higher temperatures $T > E_L$ the conductance decreases with rising temperature

$$G_N = 1 + B \sqrt{\frac{E_L}{T}} , \tag{69}$$

where B is a constant. In this regime the coherence length ξ_T in the normal wire is shorter than L. Hence the resistance of the structure is determined by the portion of the wire, $\sim L - \xi_T$, which is still normal [85].

In the non-equilibrium situation considered here, it is essential to analyze the penetration of the electrical field in the wire. This problem was considered in Ref. [86] for the infinite wire case and in Ref. [85] for the case where the wire is attached to two reservoirs. At high temperatures, the field is essentially constant, but at low temperatures it has a non-monotonic behavior. This in turn is responsible for the non-monotonic temperature dependence of the conductance. The electric field in the wire is plotted in Fig. 10.

In the presence of tunnel barriers with resistance larger than the Drude resistance of the wire, the electric field is confined to the barrier. In this case the nonequilibrium effects related to the electric field in the wire, and responsible for the non-monotonic temperature dependence disappear.

Recent experiments [47,48] have also studied more complex NS structures where supercurrents can flow between at least two superconducting reservoirs [47] or are induced by a magnetic flux through a loop within the structure [48]. In the picture of Andreev reflections, interference between quasiparticles acquiring a superconducting phase during the reflection process occurs. As the Usadel equations (61) describe the modulation of the Green's functions by possible gradients of the superconducting phase, the influence of these currents on the system conductance (67) can be easily calculated within the quasiclassical approach. These effects remain pronounced even at higher temperatures, when the coherence length ξ_{TN} is smaller than the geometrical lengths of the system. In this case one could expect that superconducting correlations are destroyed before interference occurs, hence the effect should be absent and supercurrents are exponentially small. However, low energy channels ($E \ll E_L$) can still interfere and contribute to the conductance with a relative weight of E_L/T. Their contribution remains pronounced. This is characteristic for linear response quantities, in contrast to thermodynamic ones like the supercurrent.

D. Superconductor-Luttinger liquid systems

As discussed already in the first part of this chapter, electron-electron interaction plays an important role in systems of reduced dimensionality. The interplay of proximity effect and charging was discussed in [36]. Another class of systems in which interaction is of fundamental important is that of quantum wires. In this case the capacitance model cannot applied any longer, instead a paradigm model for interacting one-dimensional electron systems is the Luttinger model (see the chapter by Fisher and Glazman in this volume for an introduction) In this last section we briefly review some properties of hybrid systems of superconductors and a Luttinger liquid. In 1D, interactions have drastic consequences. For instance, there are no fermionic quasiparticle excitations, and the transport properties cannot be described in terms of the conventional Fermi-liquid approach. Instead the low-energy excitations of the system are independent long-wavelength oscillations of the charge (ρ) and spin density (σ), which propagate with different velocities. For a quantum wire with an arbitrarily small barrier this leads to a complete suppression of transport at low energies [87–89].

Hybrid S-LL have been studied in the two extremes of tunneling junction and of perfectly transparent interfaces. In the first case the tunneling Hamiltonian is used [59], while in the second case a new type of bosonization developed in Ref. [60] is employed. In this section we consider only the tunneling density of states in a LL with a highly transparent S-LL interface [62].

The Hamiltonian of a LL can be written in bosonized form as

$$\hat{H}_L = \frac{1}{2}\sum_j v_j \int dx \left[\frac{g_j}{2}(\nabla\theta_j)^2 + \frac{2}{g_j}(\nabla\phi_j)^2\right] , \tag{70}$$

where $j = \rho, \sigma$, and $v_j = (2/g_j)v_F$ are the renormalized interaction-dependent Fermi velocities for charge and spin density excitations. For repulsive, spin-independent interactions we have $g_\rho < 2$ and $g_\sigma = 2$. The Fermi field operators are decomposed in right- and left-moving Fermion operators $\psi_{+,s}$ and $\psi_{-,s}$, respectively, $\psi_s = e^{ik_F x}\psi_{+,s} + e^{-ik_F x}\psi_{-,s}$, where k_F is the Fermi wave vector. The fields $\psi_{\pm,s}$ in turn can be expressed through Boson operators

$$\psi^\dagger_{\pm,s} = \sqrt{\rho_0}e^{i\sqrt{\pi}[\mp\phi_s(x)+\theta_s(x)]}, \tag{71}$$

where $\theta_s = \frac{1}{\sqrt{2}}(\theta_\rho + s\theta_\sigma)$ and $\phi_s = \frac{1}{\sqrt{2}}(\phi_\rho + s\phi_\sigma)$. The density of electrons per spin in the LL is $\rho_0 = k_F/2\pi$. The fields θ, ϕ can be decomposed in a normal mode expansion which incorporates the boundary conditions at the S-LL interfaces. For a LL coupled to two superconductors at a distance L, Maslov *et al.* [60] obtained the result

$$\theta_\rho(x) = \sqrt{\frac{\pi}{2}}(J+\chi)\frac{x}{L} + i\sqrt{\frac{2}{g_\rho}}\sum_{q>0}\gamma_q \sin(qx)(\hat{b}^\dagger_{\rho,q} - \hat{b}_{\rho,q}); \tag{72}$$

$$\theta_\sigma(x) = \frac{1}{\sqrt{\pi}}\theta^{(0)}_\sigma + \sqrt{\frac{2}{g_\sigma}}\sum_{q>0}\gamma_q \cos(qx)(\hat{b}^\dagger_{\sigma,q} + \hat{b}_{\sigma,q}); \tag{73}$$

$$\phi_\sigma(x) = \sqrt{\frac{\pi}{2}} M \frac{x}{L} + i\sqrt{\frac{g_\sigma}{2}} \sum_{q>0} \gamma_q \sin(qx)(\hat{b}^\dagger_{\sigma,q} - \hat{b}_{\sigma,q}); \tag{74}$$

$$\phi_\rho(x) = \frac{1}{\sqrt{\pi}} \phi_\rho^{(0)} + \sqrt{\frac{g_\rho}{2}} \sum_{q>0} \gamma_q \cos(qx)(\hat{b}^\dagger_{\rho,q} + \hat{b}_{\rho,q}) \; . \tag{75}$$

Here, $\hat{b}^{(\dagger)}_{j,q}$ are Bose operators and $\gamma_q = \exp\{-q\alpha/2\pi\}/\sqrt{qL}$ where α is a short range cut-off. The expansion (72) – (75) is valid at energies smaller than the superconducting gap Δ. The phase difference between the two superconductors is χ; J and M describe topological excitations satisfying the constraint $J + M = \text{odd}$. Finally, $\theta_\sigma^{(0)}$ and $\phi_\rho^{(0)}$ are canonically conjugate to M, J. The local density of states (per spin) of the LL measured at a distance x from the superconducting contact is related to the retarded one-electron Green's function of the LL by Eq. (64).

As an example we discuss the space and frequency dependent DOS of a LL contacted at $x = 0$ with a superconductor. This corresponds to the limit $L \to \infty$ in the mode expansion given by Eqs. (72) – (75). In this case only the non-zero modes ($q > 0$) contribute to the local DOS. The correlation function $\langle \psi_s^\dagger(x,t)\psi_s(x,0)\rangle$ can be evaluated using the boson representation Eq. (71), with the result

$$\langle \psi_s^\dagger(x,0)\psi_s(x,t)\rangle = 2\rho_0 \prod_{j=\rho,\sigma} \left(\frac{\alpha^2 + (2x)^2}{\alpha^2}\right)^{\gamma_j} \left(\frac{\alpha^2}{(\alpha - iv_j t)^2}\right)^{\eta_j} \left[\frac{\alpha^2}{(\alpha - i(2x + v_j t))(\alpha + i(2x - v_j t))}\right]^{\gamma_j} , \tag{76}$$

at distance x from the LL-S interface. Here $\gamma_j = (g_j/16 - 1/(4g_j))$ and $\eta_j = (g_j/16 + 1/(4g_j))$. At small energies $\omega \ll \Delta$, the DOS behaves as

$$N_{S-LL}(\omega) \sim \omega^{g_\rho/4 - 1/2}. \tag{77}$$

The exponent of the DOS is negative ($g_\rho < 2$), which implies a *strong enhancement* at low energies whereas in the absence of the superconductor the DOS of the LL *vanishes* at the Fermi energy

$$N(\omega) \sim \omega^{(g_\rho + 4/g_\rho - 4)/8} \; . \tag{78}$$

Thus the presence of the superconductor changes the properties of the Luttinger liquid in a qualitative way. Although we consider a clean S-LL interface, backscattering is induced by the superconducting gap, which reflects low-energy electrons either directly or via (multiple) Andreev processes. The enhanced DOS as a function of frequency, Eq. (77), is schematically drawn in Fig. 11; for comparison we also show the vanishing DOS in absence of the superconductor, Eq. (78).

On the other hand, at low energies ω the enhancement of the DOS persists over large distances $x(\omega) \sim v_\rho/\omega$ from the interface. On the other hand, the induced pair amplitude in the LL, which is characteristic of the presence of the superconductor, decays as a power [60] of the distance x. This profound difference in the space dependence demonstrates that the DOS provides different information compared to the proximity effect. The reason why the DOS does not approach the well-known behavior of an Luttinger liquid far from the superconducting contact is in part related to the fact that we are considering a clean wire. In this case the states in the LL are extended and the DOS enhancement does not depend on x.

ACKNOWLEDGMENTS

We would like to thank our colleagues with whom we had been working on the problems reviewed in this article, W. Belzig, C. Bruder, G. Falci, F. W. J. Hekking, A. Odintsov, J. Siewert, L.L. Sohn, F. K. Wilhelm, and A. D. Zaikin. The work has been supported by the 'Sonderforschungsbereich' 195 of the 'Deutsche Forschungsgemeinschaft'. Also the support by the A.v.Humboldt award of the Academy of Finland (G.S.) is gratefully acknowledged.

---

[1] D. V. Averin and K. K. Likharev, in *Mesoscopic Phenomena in Solids*, B. L. Altshuler, P. A. Lee and R. A. Webb, eds., p. 173 (Elsevier, Amsterdam, 1991).

[2] *Single Charge Tunneling*, NATO ASI Series, Vol. B 294, eds. H. Grabert and M. H. Devoret, (New York, Plenum Press 1992).
[3] M. Tinkham, *Introduction to Superconductivity*, 2nd Edition, McGraw Hill (1996).
[4] *Mesoscopic Superconductivity*, Proceedings of the NATO ARW, F. W. J. Hekking, G. Schön, and D. V. Averin, eds., Physica B **203** (1994).
[5] C. Bruder, in *Superconductivity Review* **1**, 261 (1996).
[6] G. Schön, in *Quantum Transport and Dissipation*, VCH Publishers, Chapter 4, to be published.
[7] F. W. J. Hekking, L. I. Glazman, K. A. Matveev, and R. I. Shekhter, Phys. Rev. Lett. **70**, 4138 (1993).
[8] F. Guinea and G. Schön, Physica B **152**, 165 (1988).
[9] F. W. J. Hekking and Yu. V. Nazarov, Phys. Rev. Lett. **71**, 1625 (1993); Phys. Rev. B **49**, 6847 (1994).
[10] W. Tichy, *Diplomthesis, University Karlsruhe* (1996).
[11] D. V. Averin and Yu. V. Nazarov, Phys. Rev. Lett. **69**, 1993 (1992).
[12] P. Lafarge, P. Joyez, D. Esteve, C. Urbina, and M. H. Devoret, Phys. Rev. Lett. **70**, 994 (1993).
[13] M. T. Tuominen, J. M. Hergenrother, T. S. Tighe, and M. Tinkham, Phys. Rev. Lett. **69**, 1997 (1992); Phys. Rev. B **47**, 11599 (1993).
[14] T. M. Eiles, J. M. Martinis, and M. H. Devoret, Phys. Rev. Lett. **70**, 1862 (1993).
[15] J. M. Hergenrother, M. T. Tuominen, and M. Tinkham, Phys. Rev. Lett. **72**, 1742 (1994); J. M. Hergenrother et al., page 327 in Ref. 4.
[16] G. Schön and A. D. Zaikin, Europhys. Lett. **26**, 695 (1994).
[17] G. Schön, J. Siewert, and A. D. Zaikin, p. 340 in Ref. [4], and in *Quantum Dynamics of Submicron Structures*, NATO ASI Series B **291**, eds. H. Cerdeira, B. Kramer, and G. Schön, p. 489 (1995).
[18] A. O. Caldeira and A. J. Leggett, Ann. Phys. (NY) **149**, 374 (1983).
[19] U. Weiss, *Quantum Dissipative Systems*, Series in Modern Condensed Matter Physics, Vol. 2 (World Scientific, 1993).
[20] U. Eckern, G. Schön, and V. Ambegaokar, Phys. Rev. B **30**, 6419 (1984).
[21] G. Schön and A. D. Zaikin, Phys. Rep. **198**, 237 (1990).
[22] F. W. J. Hekking, L. I. Glazman, and G. Schön, Phys. Rev. B **51**, 15312 (1995).
[23] T.A. Fulton, P.L. Gammel, D.J. Bishop, L.N. Dunkleberger, and G.J. Dolan, Phys. Rev. Lett. **63**, 1307 (1989).
[24] A. Maassen v.d. Brink, G. Schön, and L. J. Geerligs, Phys. Rev. Lett. **67**, 3030 (1991); A. Maassen v.d. Brink et al., Z. Phys. B **85**, 459 (1991).
[25] D. B. Haviland, Y. Harada, P. Delsing, C. D. Chen, and T. Claeson, Phys. Rev. Lett. **73**, 1541 (1994).
[26] J. Siewert and G. Schön, Phys. Rev. B **54**, 7421 (1996).
[27] K.A. Matveev, M. Gisselfält, L. I. Glazman, M. Jonson, and R. I. Shekter, Phys. Rev. Lett. **70**, 2940 (1993); K. A. Matveev, L. I. Glazman, and R. I. Shekter, Mod. Phys. Lett. B **8**, 15 (1994).
[28] P. Joyez, Ph.D. Thesis, Université Paris 6 (1995)
[29] M. H. Devoret, private communication.
[30] H. Schoeller and G. Schön, Phys. Rev. B **50**, 18436 (1994).
[31] J. König, H. Schoeller, G. Schön, Europhys. Letters **31**, 31 (1995); J. König et al. in *Quantum Dynamics of Submicron Structures*, NATO ASI Series B **291**, eds. H. Cerdeira, B. Kramer, and G. Schön, p. 221 (1995).
[32] L. I. Glazman and K. A. Matveev, Sov. Phys. JETP. **71**, 1031 (1990); K. A. Matveev, Sov. Phys. JETP **72**, 892 (1991).
[33] S. V. Panyukov and A. D. Zaikin, Phys. Rev. Lett. **67**, 3168 (1991); D. S. Golubev and A. D. Zaikin, Phys. Rev. B **50**, 8736 (1994).
[34] G. Falci, G. Schön, and G. T. Zimanyi, Phys. Rev. Lett. **74**, 3257 (1995); and p. 409 in [4].
[35] H. Grabert, Phys. Rev. B **50**, 17364 (1994); X. Wang and H. Grabert, Phys. Rev. B **53**, 12621 (1996).
[36] C. Bruder, R. Fazio, and G. Schön, Phys. Rev. B **50**, 12766 (1994); and p. 240 in [4].
[37] R. Bauernschmitt, J. Siewert, A. A. Odintsov, and Yu. V. Nazarov, Phys. Rev. B **49**, 4076 (1994).
[38] L. G. Aslamazov, A. I. Larkin, and Yu. N. Ovchinnikov, Zh. Eksp.Teor. Fiz. **55**, 323 (1968) [Sov. Phys. JETP **28**, 171 (1969)].
[39] C.W.J. Beenakker in *Mesoscopic Quantum Physics*, edited by E. Akkermans, G. Montambaux, and J.-L Pichard (North Holland, Amsterdam) 1995.
[40] C.J. Lambert and R. Raimondi, J. Phys. Cond. Matt. (to be published).
[41] G. Deutscher and P. G. de Gennes, in *Superconductivity*, edited by R. D. Parks (Marcel Dekker, New York, 1965), Vol.II, p. 1005.
[42] A. F. Andreev, Zh. Eksp. Teor. Fiz. **46**, 1823 (1964) [Sov. Phys. JETP **19**, 1228 (1964)].
[43] P. Charlat, H. Courtois, Ph. Gandit, D. Mailly, A.F. Volkov, and B. Pannetier, Phys. Rev. Lett., to be published.
[44] I.O. Kulik, Zh. Eksp. Teor. Fiz. **57**, 1745 (1969) [Sov. Phys. JETP **30**, 944 (1970)].
[45] H. Pothier, S. Guéron, D. Esteve, and M. H. Devoret, in Ref [4]; Phys. Rev. Lett. **73**, 2488 (1994).
[46] H. Nakano and H. Takayanagi, Sol. St. Comm. **80**, 997 (1991).
[47] V.T. Petrashov, V.N. Antonov, and M. Persson, Physica Scripta **42**, 136 (1992); V.T. Petrashov, V.N. Antonov, P. Delsing, and, T. Claeson, Phys. Rev. Lett. **70**, 347 (1993); Phys. Rev. Lett. **74**, 5268 (1995).
[48] H. Courtois, Ph. Gandit, D. Mailly, and B. Pannetier, Phys. Rev. Lett. **76**, 130 (1996).
[49] Yu. V. Nazarov and T. H. Stoof, Phys. Rev. Lett. **76**, 823 (1996); Phys. Rev. B **53**, 14496 (1996).

[50] S. Guéron, H. Pothier, N. O. Birge, D. Esteve, and M. Devoret, Phys. Rev. Lett. **77**, 3025 (1996) .
[51] W. Belzig, C. Bruder, and G. Schön, Phys. Rev. B **54**, 9443 (1996).
[52] J. Nitta, T. Azaki, H. Takayanagi, and K. Arai, Phys. Rev. B **46**, 14286 (1992).
[53] A. Dimoulas, J. P. Heida, B. J. van Wees, T. M. Klapwijk, W. v.d. Graaf, and G. Borghs, Phys. Rev. Lett. **74**, 602 (1995).
[54] C. W. J. Beenakker and H. van Houten, Phys. Rev. Lett. **66**, 3056 (1991).
[55] A. Furusaki, H. Takayanagi, and M. Tsukada, Phys. Rev. Lett. **67**, 132 (1991).
[56] A. Kastalsky, A. W. Kleinsasser, L. H. Greene, R. Bhat, F. P. Milliken, and J. P. Harbison, Phys. Rev. Lett. **67**, 3026 (1991).
[57] B. L. Altshuler, D. E. Khmelnitskii, and B. Z. Spivak, Solid State Comm. **48**, 841 (1983).
[58] J. Voit, Rep. Prog. Phys. **58**, 977 (1995).
[59] R. Fazio, F.W.J. Hekking,, and A.A. Odintsov, Phys. Rev. Lett. **74**, 1843 (1995).
[60] D.L. Maslov, M. Stone, P.M. Goldbart, and D. Loss, Phys. Rev. B **53**, 1548 (1996).
[61] M.P.A. Fisher, Phys. Rev. B **49**, 14550 (1994).
[62] C. Winkelholz, R. Fazio, F.W.J. Hekking, and G. Schön, Phys. Rev. Lett. **77**, 3200 (1996).
[63] see e.g. the textbook by S. Datta, *Mesoscopic Electron Transport*, Cambridge University Press (1995).
[64] C.J. Lambert, J. Phys. C **3**, 6579 (1991).
[65] C.W.J. Beenakker, Phys. Rev. B **46**, 12481 (1992).
[66] G.E. Blonder, M. Tinkham, and T.M. Klapwijk, Phys. Rev. B **25**, 4515 (1982).
[67] A.L. Shelankov Sov. Phys. Solid St. bf 26, 981 (1984).
[68] M. Büttiker, Phys. Rev. B **46**, 12485 (1992).
[69] M.P. Anantram and S. Datta, Phys. Rev. B **53**, 16390 (1996).
[70] G.B. Lesovik, A.L. Fauchère, and G. Blatter, Phys. Rev. B **55**, 3146 (1997).
[71] J. Sánchez-Canizares and F. Sols, Phys. Rev. B **55**, 531 (1997).
[72] P. Cook, R. Raimondi and C.J. Lambert, Phys. Rev. B **54**, 9491 (1996).
[73] G. Eilenberger, Z. Phys. **214**, 195 (1968).
[74] K. Usadel, Phys. Rev. Lett. **25**, 507 (1970).
[75] M.Y. Kupriyanov and V.F. Lukichev, Zh. Eksp.Teor. Fiz. **94**, 139 (1988) [Sov. Phys. JETP **67**, 1163 (1988)].
[76] W.L. McMillan, Phys. Rev. **175**, 537 (1968).
[77] A.A. Golubov and M.Y. Kupriyanov, J. Low. Temp. Phys. **70**, 83 (1988).
[78] A.I. Larkin and Yu.N. Ovchinnikov in *Nonequilibrium Superconductivity*, eds. D.N. Langenberg and A.I. Larkin, (Elsevier, Amsterdam, 1985).
[79] A. Schmid and G. Schön, J. Low. Temp. Phys. **20**, 207 (1975).
[80] A. Schmid in *Nonequilibrium Superconductivity, Phonons, and Kapitza Boundaries*, NATO ASI Series B 65, ed. K.E. Gray, (Plenum, New York 1981)
[81] J. Rammer and H. Smith, Rev. Mod. Phys. **58**, 323 (1986).
[82] A.V. Zaitsev, Sov. Phys. JETP **59**, 1015 (1984); M.Yu. Kuprianov and V.F. Lukichev, Sov. Phys. JETP **94**, 139 (1988).
[83] A.F. Volkov, A.V. Zaitsev, and T.M. Klapwijk, Physica C **59**, 21 (1993).
[84] A.F. Volkov and C.J. Lambert, J. Cond. Mat., **8**, L45 (1996).
[85] A.A. Golubov, F.K. Wilhelm, and A.D. Zaikin, to be published in Phys. Rev. B.
[86] F. Zhou, B. Spivak, and A. Zyuzin, Phys. Rev. B 52,4467 (1995)
[87] C.L. Kane and M.P.A. Fisher, Phys. Rev. Lett. **68**, 1220 (1992).
[88] K.A. Matveev and L.I. Glazman, Phys. Rev. Lett. **70**, 990 (1993).
[89] A. Furusaki and N. Nagaosa, Phys. Rev. B 47,4631 (1993).

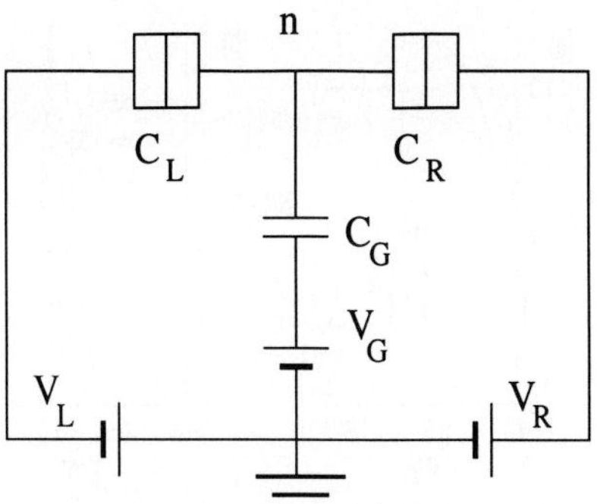

FIG. 1. The SET transistor.

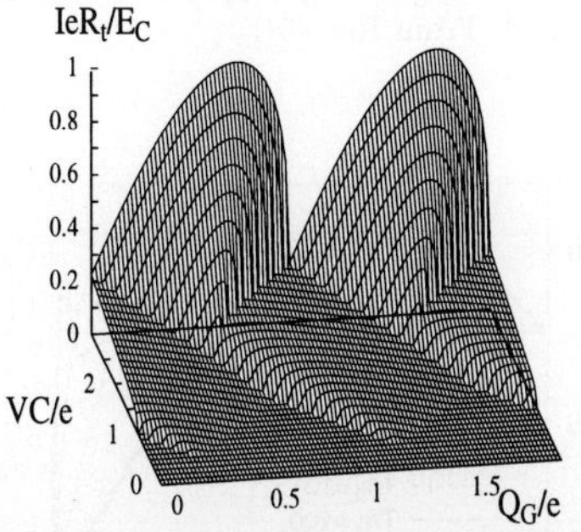

FIG. 2. I-V characteristic of an NNS transistor. Both junctions have the same normal-state conductance. The ratio of Andreev and normal-state conductance is $G^{\mathrm{A}} R_{\mathrm{t}} = 0.02$, and $\Delta = 4E_{\mathrm{C}}$. From Ref. [10].

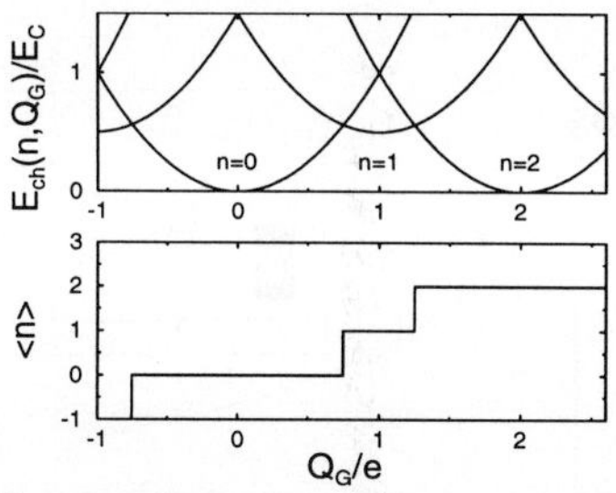

FIG. 3. The charging energy of a superconducting single-electron box as a function of the gate voltage shows a difference between even and odd numbers n of electron charges on the island. Accordingly the average island charge $\langle n \rangle$ is found in a broader range of gate voltages in the even state than in the odd state.

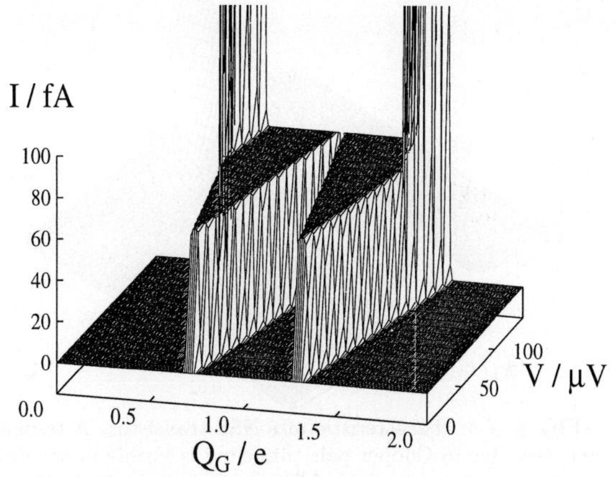

FIG. 4. The current $I(Q_{\mathrm{G}}, V)$ through a NSN transistor with $\Delta < E_{\mathrm{C}}$. From Ref. [17].

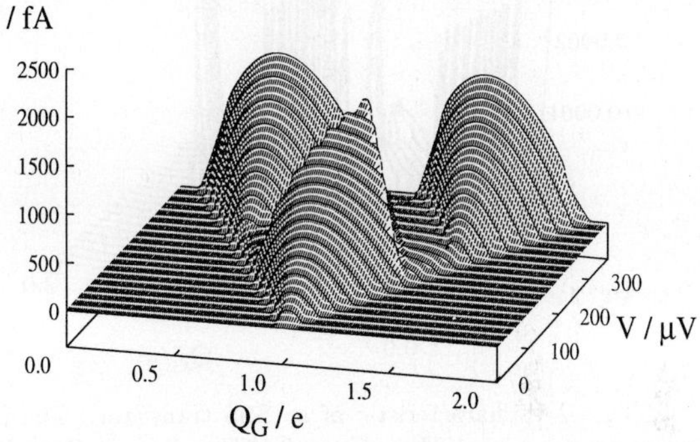

FIG. 5. The current $I(Q_{\mathrm{G}}, V)$ through a NSN transistor with $\Delta > E_{\mathrm{C}}$. The parameters correspond to those of the experiments [15], $E_{\mathrm{C}} = 100\mu\mathrm{eV}$, $\Delta = 245\mu\mathrm{eV}$, $R_{\mathrm{tl/R}} = 43k\Omega, 1/G^{\mathrm{A}} \approx 1.2(2.4)10^8\,\Omega$ for the left (right) junction. From Ref. [17].

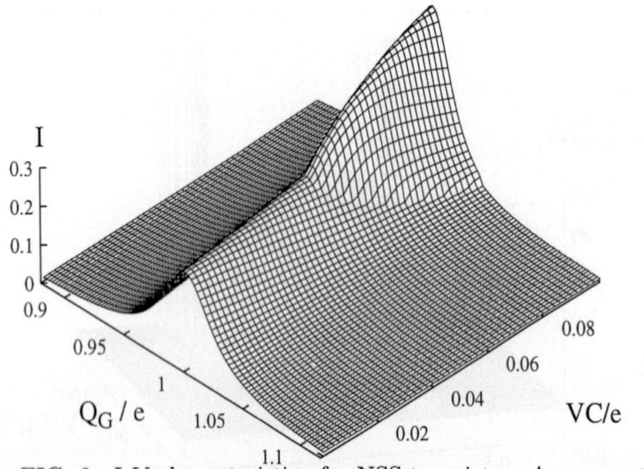

FIG. 6. *I-V* characteristic of a NSS transistor. A resonant structure due to Cooper pair tunneling is visible in the dissipative current due to Andreev reflection. From Ref. [22].

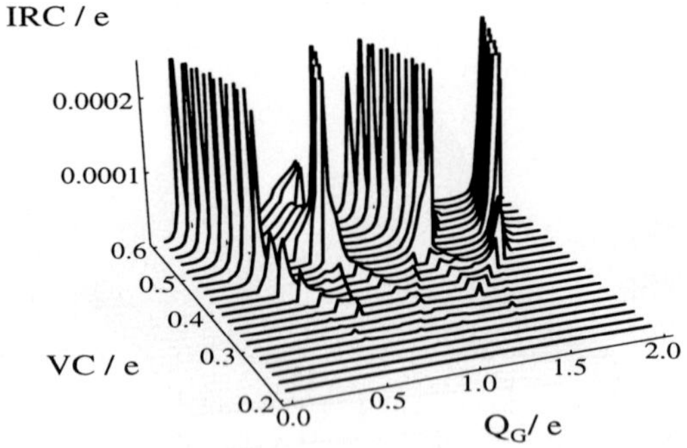

FIG. 7. *I-V* characteristic of an SSS transistor. The parameters are $\Delta = 1.3E_C$, $E_J = 0.17E_C$, $R_{l/r} = R \approx R_K$, $\gamma = 2.5 \cdot 10^{-5}(RC)^{-1}$. From Ref. [26].

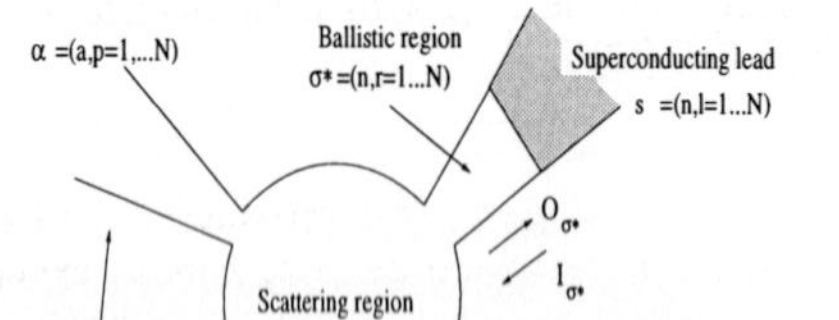

FIG. 8. Sketch of the scattering region including the small ballistic region in contact with the superconducting reservoir.

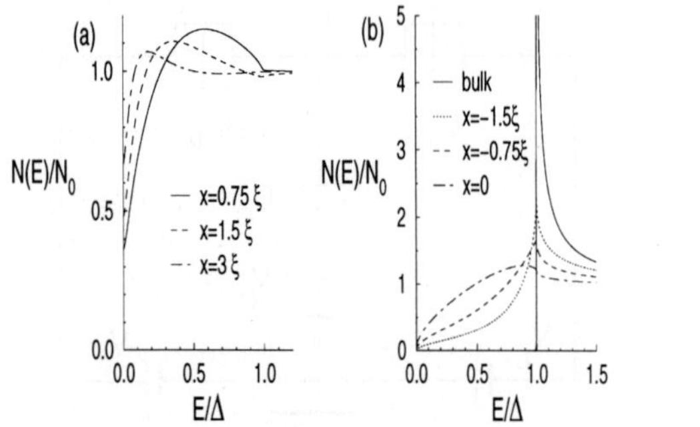

FIG. 9. The density of states on the normal (a) and superconducting side (b) of a NS heterostructure at different distances from the interface. The length scale is $\xi = \sqrt{D/2\Delta}$. A nonzero pair breaking strength $1/\tau_s = 0.03\Delta$ and $\Gamma = 1$ have been assumed. From Ref. [51].

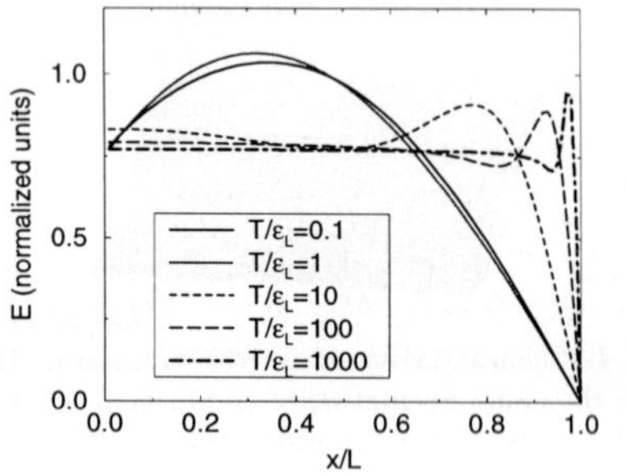

FIG. 10. Electric field in the normal bridge between a bulk superconductor ($x \geq L$) and a bulk normal metal ($x \leq 0$). From Ref. [85].

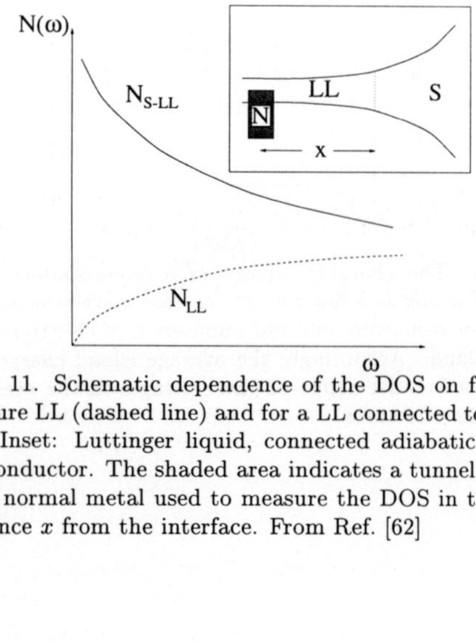

FIG. 11. Schematic dependence of the DOS on frequency for a pure LL (dashed line) and for a LL connected to S (solid line). Inset: Luttinger liquid, connected adiabatically to a superconductor. The shaded area indicates a tunnel junction with a normal metal used to measure the DOS in the LL at a distance x from the interface. From Ref. [62]

VOLUME 79, NUMBER 12 PHYSICAL REVIEW LETTERS 22 SEPTEMBER 1997

Quantum Manipulations of Small Josephson Junctions

Alexander Shnirman,[1,2] Gerd Schön,[1] and Ziv Hermon[1]
[1]*Institut für Theoretische Festkörperphysik, Universität Karlsruhe, D-76128 Karlsruhe, Germany*
[2]*School of Physics and Astronomy, Tel Aviv University, 69978 Tel Aviv, Israel*
(Received 6 June 1997)

Low-capacitance Josephson junction arrays in the parameter range where single charges can be controlled are suggested as possible physical realizations of the elements which have been considered in the context of quantum computers. We discuss single and multiple quantum-bit systems. The systems are controlled by applied gate voltages, which also allow the necessary manipulation of the quantum states. We estimate that the phase-coherence time is sufficiently long for experimental demonstration of the principles of quantum computation. [S0031-9007(97)04084-2]

PACS numbers: 85.25.Cp, 03.65.Sq, 73.23.–b

The issue of quantum computation has attracted much attention recently [1]. Quantum algorithms can perform certain types of calculations much faster than classical computers [2]. The basic concepts of quantum computation are quantum operations (gates) on quantum bits (qubits) and registers (arrays of qubits). A qubit can be a two-level system which can be prepared in arbitrary superpositions of its two eigenstates, usually denoted as $|0\rangle$ and $|1\rangle$. Quantum computation requires "quantum state engineering," i.e., the controlled preparation and manipulation of these quantum states. For quantum registers, "entangled" many-qubits states (like the EPR state of two spins) have to be constructed as well. This necessitates a coupling between different qubits. A serious limitation is the requirement that the phase coherence time is sufficiently long to allow the coherent quantum manipulations. Several physical systems have been proposed as qubits; the most advanced so far appears to be a chain with trapped ions [3,4].

In this Letter we propose an alternative system, composed of low-capacitance Josephson junctions. The coherent tunneling of Cooper pairs mixes different charge states. By controlling the gate voltages we can control the strength of the mixing. The physics of coherent Cooper-pair tunneling in this system has been established before [5–7]. The algorithms of quantum computation introduce new, well-defined rules. Their realization in experiments creates a new challenge. We consider first an ideal one-bit system, and describe the possible ways of constructing quantum states. Then we focus on a two-bit system, where we propose a controllable coupling and discuss the construction of two-bit states. Finally, we include the coupling to a realistic external electrodynamic environment which limits the phase coherence time.

The ideal system which we propose as a qubit is shown in Fig. 1(a) (with $R = 0$ and $L = 0$). It consists of two small superconducting grains connected by a tunnel junction with capacitance C_J and Josephson coupling energy E_J. An ideal voltage source is connected to the system via two external capacitors, C. We assume that Δ is the largest energy in the problem. At low temperatures quasiparticle tunneling is suppressed. It is further well established, from the study of parity effects [6,8,9], that below a crossover temperature, T^*, the superconducting state is either totally paired (when the number of electrons is even) or it has exactly one quasiparticle (when the number of electrons is odd). The crossover temperature is $T^* \approx \Delta / \ln N_{\text{eff}}$, where N_{eff} is the number of electrons in the system near the Fermi energy. Typical values for aluminum are in the range of 100–200 mK. In the following we require that the total number of electrons in both grains is even. This condition is naturally satisfied for 50% of the qubits. If only one of the islands has an unpaired excitation it can escape to the normal parts of the system—if such a channel is provided—since the gap energy Δ is gained in such a process [9].

Possible quantum states of the system are then characterized by the numbers of extra Cooper pairs on the up and down islands, n_u and n_d. Because of the external capacitors C, the total number $N \equiv n_u + n_d$ is fixed. Hence the set of basis states is parametrized by the number of Cooper pairs on one island or the difference $n \equiv (n_u - n_d)/2$. The Hamiltonian of this system is

$$H = \frac{(n - CV/2)^2}{C + 2C_J} - E_J \cos\Theta \,, \qquad (1)$$

where Θ is the conjugate to the variable n. To shorten notations we use units where $2e = 1$, $\hbar = 1$, except where

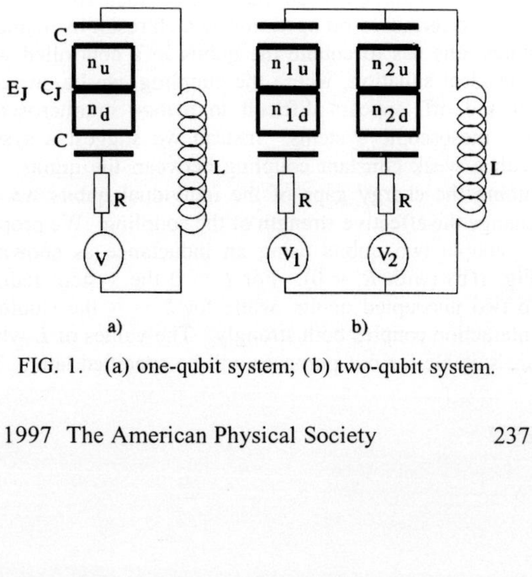

FIG. 1. (a) one-qubit system; (b) two-qubit system.

0031-9007/97/79(12)/2371(4)$10.00

it helps to keep the results transparent. We consider systems where the charging energy of the internal capacitor $E_{C_J} = (2e)^2/2C_J$ is much larger than E_J. In this regime, for most values of the external voltage V, the energies of the states are dominated by the charging part of (1). However, for those values of V where the charging energies of two neighboring states $|n\rangle$ and $|n+1\rangle$ are nearly degenerate, the Josephson coupling becomes relevant. The eigenstates are now superpositions of $|n\rangle$ and $|n+1\rangle$ with a minimum energy gap E_J between them.

We concentrate on a voltage interval where only two adjacent charge states play a role. Then it is convenient to rewrite (1) in a spin-$\frac{1}{2}$ language:

$$H = \frac{CV}{2(C + 2C_J)}\sigma_z + \frac{E_J}{2}\sigma_x, \tag{2}$$

where $|\uparrow\rangle \equiv |n\rangle$ and $|\downarrow\rangle \equiv |n+1\rangle$. Using this language we propose a few one-bit operations. If one chooses the operating point (i.e., the voltage) sufficiently far away from the degeneracy, the eigenstates are just $|\downarrow\rangle$ and $|\uparrow\rangle$. Then, switching the system suddenly to the degeneracy point for a time Δt and suddenly back, we can perform one of the basic one-bit operations—a spin flip:

$$U_{\text{flip}}(\Delta t) = \begin{pmatrix} \cos(E_J\Delta t/2) & i\sin(E_J\Delta t/2) \\ i\sin(E_J\Delta t/2) & \cos(E_J\Delta t/2) \end{pmatrix}. \tag{3}$$

We got rid of time-dependent phases by working in the interaction picture, where the zero-order Hamiltonian is the one at the operation point. To estimate the time width Δt of the voltage pulse needed for a total spin flip (the operation time), we note that a typical experimental value of E_J is of order 1 K. It cannot be chosen much smaller, since the condition $k_B T \ll E_J$ must be satisfied. Therefore the operation time is very short: $\Delta t \approx 10^{-10}$ s.

An alternative way to perform a coherent spin flip is probably easier to realize: The system is pushed adiabatically to the degeneracy point, and an ac voltage with frequency $E_J/\hbar$ is applied. The process is analogous to the paramagnetic resonance (here the constant magnetic field component is in the x direction, while the oscillating one is in the z direction). The time width of the ac pulse needed for the total spin flip depends on its amplitude; therefore it can be chosen much longer than 10^{-10} s.

To perform two-bit operations which result in entangled states, one has to couple the qubits in a controlled way. The ideal situation, where the coupling can be switched on and off, appears difficult to realize in microscopic and mesoscopic systems. Instead we suggest a system with a weak constant coupling between the qubits. By tuning the energy gaps of the individual qubits we can change the effective strength of the coupling. We propose to couple two qubits using an inductance as shown in Fig. 1(b) (with $R = 0$). For $L = 0$ the system reduces to two uncoupled qubits, while for $L = \infty$ the Coulomb interaction couples both strongly. The values of L which are suitable for our purposes will be specified later. The Hamiltonian describing this system is

$$H = \sum_{i=1,2}\left[\frac{(n_i - V_i C_t)^2}{2C_J} - E_J\cos\Theta_i\right] + \frac{q^2}{2(2C_t)} + \frac{\phi^2}{2L} - \frac{(n_1+n_2)q}{2C_J} - \frac{C_t}{4C_J^2}(n_1 - n_2)^2. \tag{4}$$

Here q denotes the total charge on the external capacitors of both qubits, ϕ is its conjugate variable, and $C_t^{-1} = C_J^{-1} + 2C^{-1}$. The (q, ϕ) oscillator produces an effective mean-field coupling between the qubits for frequencies smaller than $\omega_{LC}^{(2)} = 1/\sqrt{2C_t L}$. In order to have this coupling in a wide enough voltage range around the degeneracy point, we demand

$$A \equiv \frac{\hbar\omega_{LC}^{(2)}}{E_J} \gg 1. \tag{5}$$

To obtain the mean-field coupling of the qubits we eliminate the variables q and ϕ. For this purpose we first perform a canonical transformation $\tilde{q} = q - \frac{(n_1+n_2)C_t}{C_J}$, $\tilde{\Theta}_i = \Theta_i + \frac{C_t}{C_J}\phi$ (ϕ and n_i unchanged), which leads to the new Hamiltonian (we omit the tildes):

$$H = \sum_{i=1,2}\left[\frac{(n_i - CV_i/2)^2}{C + 2C_J} - E_J\cos\left(\Theta_i - \frac{C_t}{C_J}\phi\right)\right] + \frac{q^2}{2(2C_t)} + \frac{\phi^2}{2L}. \tag{6}$$

We assume that the fluctuations of ϕ are weak

$$(C_t/C_J)\sqrt{\langle\phi^2\rangle} \ll 2\pi. \tag{7}$$

Otherwise the Josephson tunneling terms in the Hamiltonian (6) are washed out. (Below we will show this in a more rigorous way.) Assuming (7), we expand the $E_J\cos(\ldots)$ terms of (6) in powers of ϕ and neglect powers higher than linear. Then we can trace out the variables q and ϕ. As a result we obtain an effective Hamiltonian, consisting of two one-bit Hamiltonians (1) and a coupling term: $H_{\text{coup}} = E_L[\sin\Theta_1 + \sin\Theta_2]^2$, where

$$E_L = 2\pi^2\frac{C_t^2}{C_J^2}\frac{E_J^2 L}{\Phi_0^2}, \tag{8}$$

and $\Phi_0 \equiv h/2e$ is the flux quantum. In the spin-$\frac{1}{2}$ language we get

$$H_{\text{coup}} = -(E_L/4)(\sigma_y^{(1)} + \sigma_y^{(2)})^2. \tag{9}$$

This term provides the required weak coupling if it is small, i.e., if $E_L \ll E_{C_J}$.

The mixed term in (9) is important in certain situations. If the voltages V_1 and V_2 are such that both qubits are out of degeneracy, to a good approximation, the eigenstates of the two-bit system without coupling are $|\downarrow\downarrow\rangle$, $|\downarrow\uparrow\rangle$, $|\uparrow\downarrow\rangle$, and $|\uparrow\uparrow\rangle$. In a general situation, these states are separated by energies which are larger or much larger than E_J or E_L. Therefore, the effect of the coupling is small.

If, however, a pair of these states is degenerate, the coupling may lift the degeneracy, changing the eigenstates drastically. For example, if $V_1 = -V_2$, the states $|\downarrow\uparrow\rangle$ and $|\uparrow\downarrow\rangle$ are degenerate. In this case the correct eigenstates are $\frac{1}{\sqrt{2}}(|\downarrow\uparrow\rangle + |\uparrow\downarrow\rangle)$ and $\frac{1}{\sqrt{2}}(|\downarrow\uparrow\rangle - |\uparrow\downarrow\rangle)$ with the energy splitting E_L between them.

Now we propose a way to perform two-bit operations which result in entangled states. For this we choose the operating points for the qubits at different voltages, switch suddenly the voltages to be equal for a time Δt, and switch suddenly back. The result is a "generalized" spin flip, which may be described in the basis $\{|\downarrow\downarrow\rangle, |\downarrow\uparrow\rangle, |\uparrow\downarrow\rangle, |\uparrow\uparrow\rangle\}$ by a matrix:

$$U^{(2)}_{\rm flip}(\Delta t) = \begin{pmatrix} 1 & 0 & 0 & 0 \\ 0 & \cos(\frac{E_L\Delta t}{2}) & i\sin(\frac{E_L\Delta t}{2}) & 0 \\ 0 & i\sin(\frac{E_L\Delta t}{2}) & \cos(\frac{E_L\Delta t}{2}) & 0 \\ 0 & 0 & 0 & 1 \end{pmatrix}. \tag{10}$$

Instead of applying very short voltage pulses, one can push the system adiabatically to the degeneracy point ($V_1 = -V_2$), and apply an ac voltage pulse in the symmetric channel $V_1 + V_2 = A\exp(iE_L t)$.

The idealized picture outlined above has to be extended to account for possible dissipation mechanisms which cause decoherence and energy relaxation. In this Letter we focus on the effect of Ohmic dissipation in the circuit, which originates mostly from the voltage sources (the quasiparticle tunneling is strongly suppressed at $T \leq T^*$ [8,9]). We also consider the effect of LC resonances in the circuit. The system is shown in Fig. 1(a), including the inductance L explicitly, since the LC oscillatory mode plays an important role in the two-bit system. The Hamiltonian of the system is

$$H = \frac{(n - VC_t)^2}{2C_J} - E_J\cos\Theta + \frac{q^2}{2C_t} + \frac{\phi^2}{2L} - \frac{nq}{C_J} + \sum_j \left[\frac{p_j^2}{2m_j} + \frac{m_j\omega_j^2}{2}\left(x_j - \frac{\lambda_j}{m_j\omega_j^2}q\right)^2\right], \tag{11}$$

with $\frac{\pi}{2}\sum_j \frac{\lambda_j^2}{m_j\omega_j}\delta(\omega - \omega_j) = R\,\omega$.

First, we estimate the energy relaxation time, τ_r, due to the Ohmic dissipation. We assume that the system is prepared away from the degeneracy point in one of its eigenstates ($|n\rangle$ or $|n+1\rangle$). To apply the standard golden rule results for the transition rate, we perform two consecutive canonical transformations:

$$\tilde{\phi} = \phi + \sum_j \frac{\lambda_j}{m_j\omega_j^2}p_j, \qquad \tilde{x}_j = x_j - \frac{\lambda_j}{m_j\omega_j^2}q, \tag{12}$$

(q and p_j unchanged), and $\bar{q} = \tilde{q} - \frac{nC_t}{C_J}$, $\bar{\Theta} = \Theta + \frac{C_t}{C_J}\tilde{\phi}$, ($\tilde{\phi}$ and n unchanged). Then, the part of the Hamiltonian connecting the states $|n\rangle$ and $|n+1\rangle$ is $H_t = \frac{E_J}{2}\exp(i\bar{\Theta})\exp(-i\frac{C_t}{C_J}\tilde{\phi}) + \text{H.c.}$, and the transition rate from $|n\rangle$ to $|n+1\rangle$ is given by [10,11]

$$\Gamma(\Delta E) = \frac{\pi}{2\hbar}E_J^2 P(\Delta E), \tag{13}$$

$$P(\Delta E) = \frac{1}{2\pi\hbar}\int_{-\infty}^{\infty} dt \exp\left[4\frac{C_t^2}{C_J^2}K(t) + \frac{i}{\hbar}\Delta E t\right], \tag{14}$$

$$K(t) \equiv \langle[\tilde{\phi}(t) - \tilde{\phi}(0)]\tilde{\phi}(0)\rangle = 2\int_0^{\infty}\frac{d\omega}{\omega}\frac{\text{Re}\,Z_t(\omega)}{R_K} \times \left[\coth\left(\frac{\hbar\omega}{2k_BT}\right)[\cos(\omega t) - 1] - i\sin(\omega t)\right]. \tag{15}$$

Here $Z_t^{-1} = i\omega C_t + (R + i\omega L)^{-1}$ and ΔE is the energy gap between the two states. The qualitative behavior of the system is controlled by the dimensionless conductance $g = R_K/4R$ ($R_K \equiv h/e^2$ is the quantum resistance). In our system the controlling parameter is renormalized. From (14) one can observe that $\tilde{g} = (C_J^2/C_t^2)g$ is the relevant parameter. Thus, choosing the external capacitances, C, smaller than the internal one, C_J, we can reduce the effect of the dissipation. Physically, this means that the fluctuations produced by the resistor are screened by the small capacitors, and have little effect on the junction.

To be more concrete, we exploit the asymptotic formula for $P(\Delta E)$ [11]

$$P(\Delta E) = \frac{\exp(-2\gamma/\tilde{g})}{\Gamma(2/\tilde{g})}\frac{1}{\Delta E}\left[\frac{\pi}{\tilde{g}}\frac{\Delta E}{E_{C_t}}\right]^{2/\tilde{g}}, \tag{16}$$

where $\Gamma(\dots)$ is the gamma function and $E_{C_t} = (2e)^2/2C_t$. For large values of $\tilde{g}$ we obtain

$$\tau_r \equiv \frac{1}{\Gamma(\Delta E)} \approx \tau_{\rm op}\frac{\tilde{g}}{2\pi^2}\frac{\Delta E}{E_J}, \tag{17}$$

where $\tau_{\rm op} \approx h/E_J$ is the operation time [see (3)].

At the degeneracy point the system is equivalent to the two-level model with a weak Ohmic dissipation, which has been studied extensively [12]. It is well known that when $\tilde{g} \gg 1$ coherent oscillations take place. These oscillations make the spin-flip operation (3) possible. The decay time of the coherent oscillations is given by

$$\tau_d \approx \frac{\tilde{g}}{2\pi^2}\frac{h}{E_J} = \frac{\tilde{g}}{2\pi^2}\tau_{\rm op}, \tag{18}$$

and the energy gap E_J is slightly renormalized: $E_J \to E_J(E_J/\hbar\omega_c)^{1/(\tilde{g}-1)}$. The physical cutoff ω_c is usually a system-dependent property. For a pure Ohmic dissipation caused by a metallic resistor it may be as high as the Drude frequency. However, when additional capacitances and inductances are present in the circuit, the cutoff is lowered to the characteristic LC frequencies.

As indicated above, the LC phase fluctuations can wash out the Josephson coupling. To see this, we begin with the Hamiltonian (11) and trace out the bath variables and the oscillatory mode variables—ϕ, q. The partition function reads

$$Z = \sum_{n_0} \int_{n_0}^{n_0} Dn\, D\Theta \exp\left\{ \int_0^\beta d\tau \left[i\Theta\dot{n} - \frac{(n - VC_t)^2}{2C_J} + E_J \cos\Theta \right] - \int_0^\beta \int_0^\beta d\tau\, d\tau' \frac{1}{2} G(\tau - \tau') n(\tau) n(\tau') \right\}, \tag{19}$$

where $G(\omega_n) = -(C_t/C_J^2)(1 + C_t L \omega_n^2 + C_t R|\omega_n|)^{-1}$. Below we show that in the relevant parameters' range the following inequality holds:

$$1/(C_t R) \gg 1/\sqrt{LC_t} \gg R/L. \tag{20}$$

Therefore, the natural cutoff for $G(\omega_n)$ is $\omega_c = \omega_{LC}^{(1)} = 1/\sqrt{LC_t}$. We approximate $G(\omega_n) \approx -\frac{C_t}{C_J^2}(1 - C_t L\omega_n^2 - C_t R|\omega_n|)$ for $\omega_n < \omega_c$, and $G(\omega_n) = 0$ otherwise. We focus on the inductive (second) term of $G(\omega_n)$, and apply the standard charge representation technique [13]. Expanding $\exp[\int_0^\beta d\tau\, E_J \cos(\Theta)]$ in powers of E_J and integrating over Θ term by term, one obtains a path integral over integer charge paths with instantaneous "jumps" between the different values of n. Each jump contributes a multiplicative factor of $E_J/2\hbar$ to the weight of the path. The inductive term contributes another multiplicative factor for each jump, so that E_J is renormalized as $E_J \to E_J \exp(-\frac{2LC_t^2\omega_c}{\pi R_K C_J^2})$. One can immediately observe that the condition that E_J is not renormalized to zero coincides with the small fluctuations condition (7). We emphasize that the phase fluctuations which may wash out the Josephson coupling are related to the "weakly fluctuating" phase ϕ, rather than to the "strongly fluctuating" $\tilde{\phi}$ [see (12)]. Thus the effects of the inductance and the dissipation are well separated in this regime. One arrives at another way of viewing this separation by noting that the LC phase fluctuations are fast; therefore they effectively wash out the slower processes (like Josephson tunneling). These are the fast small fluctuations of ϕ that are responsible for the two-bit coupling (9). On the other hand, the phase fluctuations caused by the resistor are large only at low frequencies. In [14] we have extended the present arguments and showed also that the two-bit coupling is stable under the influence of the dissipation.

Several conditions have been assumed in this Letter in order to obtain a controlled manipulation of qubits. Here we repeat these conditions and discuss the appropriate range of parameters. We start with $E_J \approx 1$ K as a suitable experimental condition. To satisfy $E_J \ll E_{C_J}$ we take $C_J \approx 10^{-16}$ F, which is an experimentally accessible value. As we would like A to be large (5), it seems that L and C_t should be as small as possible. However, the two-bit coupling energy, E_L (8), should be larger than the temperature of the experiment. Assuming a reasonable working temperature of 20 mK, we demand $E_L \approx 0.1$ K. From (5) and (8) we get $C_t = E_L C_J^2 A^2/e^2$. To have a wide enough operation voltage interval we take $A \approx 10$, and obtain $C_t \approx 10^{-17}$–10^{-16} F and $L \approx 10^{-8}$–10^{-7} H. Thus the renormalization of g is of the order of 10, and $\tau_r/\tau_{op} \approx 10^2$–$10^3$ (assuming the realistic value $R \approx 100\Omega$) (17). Finally we observe that in this range of parameters the inequalities (7) and (20) are always satisfied. We conclude that the quantum manipulations we have discussed in this Letter can be tested experimentally using the currently available lithographic and cryogenic techniques. Application of the Josephson junction system as an element of a quantum computer is a more subtle issue, demanding either the fabrication of junctions with $C_J < 10^{-16}$ F, or a further reduction of the working temperature.

We thank T. Beth, J. E. Mooij, A. Zaikin, and P. Zoller for stimulating discussions. This work is supported by the Graduiertenkolleg "Kollektive Phänomene im Festkörper," by the SFB 195 of the DFG, and by the German Israeli Foundation (Contract No. G-464-247.07/95).

[1] A. Barenco, Contemp. Phys. **37**, 375 (1996); D. P. DiVincenzo, e-print cond-mat/9612126.
[2] P. W. Shor, in *Proceedings of the 35th Annual Symposium on the Foundations of Computer Science,* edited by S. Goldwasser (IEEE Computer Society Press, Los Alamos, CA, 1994), p. 124.
[3] J. I. Cirac and P. Zoller, Phys. Rev. Lett. **74**, 4091 (1995).
[4] D. Loss and D. P. DiVincenzo, e-print cond-mat/9701055.
[5] A. Maassen v.d. Brink, G. Schön, and L. J. Geerligs, Phys. Rev. Lett. **67**, 3030 (1991); A. Maassen v.d. Brink *et al.*, Z. Phys. B **85**, 459 (1991).
[6] M. T. Tuominen, J. M. Hergenrother, T. S. Tighe, and M. Tinkham, Phys. Rev. Lett. **69**, 1997 (1992).
[7] J. Siewert and G. Schön, Phys. Rev. B **54**, 7421 (1996).
[8] M. Tinkham, *Introduction to Superconductivity* (McGraw-Hill, New York, 1996), 2nd ed.
[9] P. Lafarge *et al.*, Phys. Rev. Lett. **70**, 994 (1993).
[10] S. V. Panyukov and A. D. Zaikin, J. Low. Temp. Phys. **73**, 1 (1988); A. A. Odintsov, Sov. Phys. JETP **67**, 1265 (1988).
[11] M. H. Devoret *et al.*, Phys. Rev. Lett. **64**, 1824 (1990).
[12] A. J. Leggett *et al.*, Rev. Mod. Phys. **59**, 1 (1987).
[13] G. Schön and A. D. Zaikin, Phys. Rep. **198**, 237 (1990).
[14] A. Shnirman, G. Schön, and Z. Hermon, e-print cond-mat/9706016.

Josephson-junction qubits with controlled couplings

Yuriy Makhlin*†, Gerd Schön* & Alexander Shnirman‡

* *Institut für Theoretische Festkörperphysik, Universität Karlsruhe, D-76128 Karlsruhe, Germany*
† *Landau Institute for Theoretical Physics, Kosygin Street 2, 117940 Moscow, Russia*
‡ *Department of Physics, University of Illinois at Urbana-Champaign, Urbana, Illinois 61801, USA*

Quantum computers, if available, could perform certain tasks much more efficiently than classical computers by exploiting different physical principles[1–3]. A quantum computer would be comprised of coupled, two-state quantum systems or qubits, whose coherent time evolution must be controlled in a computation. Experimentally, trapped ions[4,5], nuclear magnetic resonance[6–8] in molecules, and quantum optical systems[9] have been investigated for embodying quantum computation. But solid-state implementations[10–14] would be more practical, particularly nanometre-scale electronic devices: these could be easily embedded in electronic circuitry and scaled up to provide the large numbers of qubits required for useful computations. Here we present a proposal for solid-state qubits that utilizes controllable, low-capacitance Josephson junctions. The design exploits coherent tunnelling of Cooper pairs in the superconducting state, while employing the control mechanisms of single-charge devices: single- and two-bit operations can be controlled by gate voltages. The advantages of using tunable Josephson couplings include the simplification of the operation and the reduction of errors associated with permanent couplings.

Two versions of Josephson-junction qubits are shown in Fig. 1. The simpler one (Fig. 1a), proposed earlier[10], consists of a superconducting electron box, that is, a low-capacitance island coupled via a Josephson tunnel junction to a lead. The Coulomb interaction (charging energy) restricts the number, n, of Cooper-pair charges, $Q = 2ne$ (where e is the charge on an electron), on the island. If biased near a degeneracy point the system constitutes a qubit with two states differing by one Cooper-pair charge. Quantum logic operations can be performed by switching the gate voltage. Before describing the systems in detail we will first present an ideal model. This puts in perspective the possibilities and drawbacks of the simple design, as well as the advantages of the new design with Josephson coupling controlled by a superconducting quantum interference device (SQUID: Fig. 1b). These are, first, during idle periods between operations the energy splitting between logical states is tuned to zero, thus avoiding an undesired phase evolution. With this drawback of most proposals overcome, the requirement on the precision of time control is substantially reduced; second, the 2-bit couplings can be switched on and off avoiding errors associated with permanent couplings.

To realize a quantum computer we search for a system with the following "ideal" model hamiltonian:

$$\hat{H} = -\sum_{i=1}^{N}[H_z^i(t)\hat{\sigma}_z^i + H_x^i(t)\hat{\sigma}_x^i] + \sum_{i\neq j} J^{ij}(t)\hat{\sigma}_+^i\hat{\sigma}_-^j \qquad (1)$$

A spin notation is used for the qubits with Pauli matrices $\hat{\sigma}_z$, $\hat{\sigma}_x$, $\hat{\sigma}_\pm = \frac{1}{2}(\hat{\sigma}_x \pm i\hat{\sigma}_y)$. Ideally, each energy $H_z^i(t)$, $H_x^i(t)$ and the (real symmetric) couplings $J^{ij}(t)$ can be switched separately for controlled times between zero and finite values. We assume that H_z^i is the largest energy, suggesting the choice of basis states $|\uparrow_i\rangle$ and $|\downarrow_i\rangle$ aligned along the z-axis. Residual inelastic interactions (which destroy the coherence), and the measurement device (when turned on) should be accounted for by extra terms $\hat{H}_{res}$ and $\hat{H}_{meas}(t)$, respectively.

Quantum computation requires four elementary steps.
(1) The system has to be prepared in a well defined initial state. For this we turn on at low temperature all $H_z^i \gg k_B T$, while $H_x^i = J^{ij} = 0$. After sufficient time the residual interaction $\hat{H}_{res}$ relaxes all spins to the ground state, $|\uparrow\uparrow\ldots\rangle$. Then $H_z^i(t)$ is set back to zero.
(2) Single-bit operations (gates) have to be performed. They are controlled by turning on one of the fields. If H_x^i is switched on, the spin i evolves according to the unitary transformation $U_{1b}^i(\tau) = \exp(iH_x^i\tau\hat{\sigma}_x^i/\hbar)$. Depending on the time span τ, a π- or $\pi/2$-rotation is performed, producing a spin flip or an equal-weight superposition of spin states. Switching on one H_z^i produces another needed operation: a phase shift between $|\uparrow_i\rangle$ and $|\downarrow_i\rangle$. Back in the idle state, where $\hat{H} = 0$, the relative phase shift of the states does not evolve further.
(3) A two-bit operation on qubits i and j is achieved by turning on the corresponding J^{ij}. In the basis $|\uparrow_i\downarrow_j\rangle$, $|\downarrow_i\uparrow_j\rangle$, the result is described by $U_{2b}^{ij}(\tau) = \begin{pmatrix} \cos\alpha & i\sin\alpha \\ i\sin\alpha & \cos\alpha \end{pmatrix}$, with $\alpha = J^{ij}\tau/\hbar$, while the states $|\uparrow_i\uparrow_j\rangle$,$|\downarrow_i\downarrow_j\rangle$ are not affected. For $\alpha = \pi/2$ the result is a spin-swap operation, while $\alpha = \pi/4$ yields a 'square-root swap'. The latter transforms the state $|\uparrow_i\downarrow_j\rangle$ into the entangled state $(|\uparrow_i\downarrow_j\rangle + i|\downarrow_i\uparrow_j\rangle)/\sqrt{2}$. The combination with single-bit operations allows us to perform the 'controlled-not' gate; in fact, they provide a universal set, sufficient for all logic gates of quantum computations[15].
(4) The final state has to be read out, which constitutes a quantum measurement process[16].

Searching for nanometre-scale electronic realizations of qubits, one might consider normal-metal single-electron devices. But they are ruled out, because in the normal state different tunnelling processes are incoherent. Ultrasmall quantum dots with discrete levels or spin degrees of freedom in nanostructured circuits[13,14] are candidates, but are difficult to fabricate in a controlled way. More

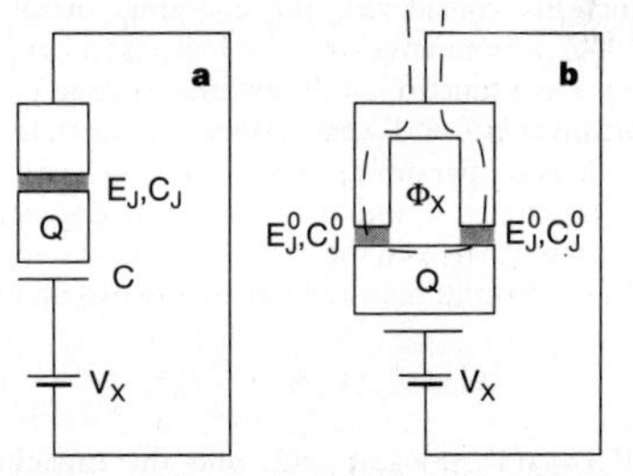

Figure 1 Josephson junction qubits. **a**, A simple realization of a qubit is provided by the superconducting electron box. A superconducting metallic island is coupled by a Josephson tunnel barrier (with capacitance C_J and Josephson coupling energy E_J; grey area) to a superconducting lead and through a gate capacitor C to a voltage source. The important degree of freedom is the Cooper-pair charge $Q = 2ne$ on the island. **b**, The improved design of the qubit. The island is coupled to the circuit via two Josephson junctions with parameters C_J^0 and E_J^0. This d.c.-SQUID can be tuned by the external flux Φ_x which is controlled by the current through the inductor loop (dashed line). If the self-inductance L_Φ of the SQUID is low, $\Phi_0^2/L_\Phi \gg 4\pi^2 E_J^0$, e^2/C_J^0, fluctuations of the flux from Φ_x are weak. Furthermore, if the frequency of flux oscillations is high, $\hbar\omega_\Phi = \hbar(L_\Phi C_J^0/2)^{-1/2} \gg E_J^0$, E_{ch}, $k_B T$, the Φ-degree of freedom is in the ground state. In this case, the set-up allows switching the effective Josephson coupling to zero. ($E_J = 0$ requires the Josephson energies of two junctions in the loop to be equal. This has been reached with a precision of 1% in quantum tunnelling experiments[17]. Even with this precision, taking into account[10] the finite value of E_J one can perform a large number of logical gates. On the other hand, by replacing one junction in **b** by another SQUID, one can tune the Josephson couplings to be equal.) The effective junction capacitance is $C_J = 2C_J^0$.

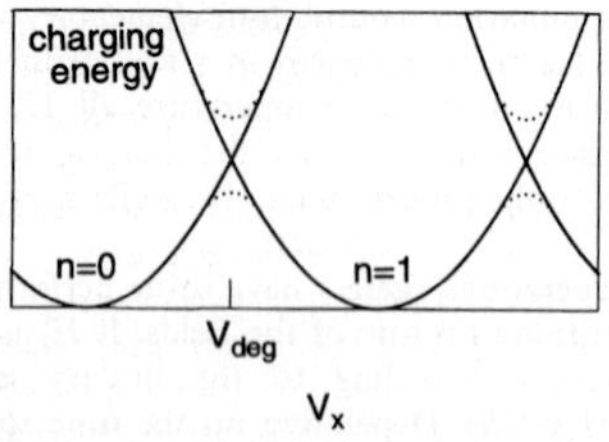

Figure 2 Spectrum of a superconducting electron box. The charging energy, $(Q - CV_x)^2/(2(C + C_J))$, of the superconducting electron box is shown (solid lines) as a function of the applied gate voltage V_x for different numbers n of extra Cooper pairs on the island. Near degeneracy points, the weaker Josephson coupling energy mixes the charge states and modifies the energy of the eigenstates (dotted line). In this regime, the system effectively reduces to a 2-state quantum system.

promising are systems built from Josephson junctions, where the coherence of the superconducting state can be exploited. Quantum extension of elements based on single-flux logic have been considered (ref. 17, and J. E. Mooij, personal communication). Encouraged by successful experiments that demonstrated the superposition of charge states[12,18,19], we suggest here the use of superconducting electron boxes with low-capacitance Josephson junctions as qubits.

In the system of Fig. 1a Cooper pairs tunnel coherently, while Coulomb blockade effects allow the control of the charge. The relevant conjugate variables are the phase difference γ across the junction and the charge $Q = 2ne$ on the island. If quasiparticle tunnelling is suppressed by the superconducting gap and only 'even-parity' states are involved[20], the circuit dynamics is governed by the hamiltonian:

$$\hat{H} = \frac{(Q - CV_x)^2}{2(C + C_J)} - E_J \cos\gamma; \quad Q = \frac{\hbar}{i} \frac{\partial}{\partial(\hbar\gamma/2e)} \tag{2}$$

For the junctions considered, the charging energy with scale $E_C \equiv e^2/2(C + C_J)$ dominates over the Josephson coupling E_J. It is plotted in Fig. 2 as a function of the external voltage V_x for different n. In equilibrium at $k_B T \ll E_C$, the system is in the state corresponding to the lowest parabola. But, near the voltages $V_{deg} = (2n + 1)e/C$, the states n and $n + 1$ are near-degenerate, and E_J mixes them strongly. Here, in the basis of charge states $|\uparrow\rangle = |n\rangle$ and $|\downarrow\rangle = |n + 1\rangle$, the hamiltonian reduces to a two-state model

$$\hat{H} = E_{ch}(V_x)\hat{\sigma}_z - \frac{1}{2}E_J\hat{\sigma}_x \tag{3}$$

where $E_{ch}(V_x) = e(V_x - V_{deg})C_{qb}/C_J$, and the capacitance of the qubit in the circuit is $C_{qb}^{-1} = C_J^{-1} + C^{-1}$.

On the way towards the model of equation (1), we achieved a tunable $H_z^i(t)$; but the Josephson coupling is fixed, $H_x^i(t) = E_J/2$. Still, single-bit operations can be performed by controlling the bias voltage V_x (ref. 10). Furthermore, when the qubits are connected in parallel with an inductor (as in Fig. 3), the common LC-oscillator mode provides a two-bit coupling with weak, but constant $J^{ij} \approx (C^2/C_J^2)(E_J^2 L/\Phi_0^2)$, where $\Phi_0 = h/2e$. This coupling provides a two-bit gate if two qubits, i and j, are brought into resonance by biasing them with the same gate voltage $V_{xi} = V_{xj}$. Out of resonance, the two-bit coupling provides only a weak perturbation.

The external voltage source is part of a dissipative circuit with effective resistance R_V. Its Johnson–Nyquist voltage fluctuations destroy the phase coherence. The dephasing rate varies slightly during manipulations[21,22]. At the degeneracy point, the decoherence time is:

$$\tau_V = \frac{1}{4\pi}\frac{R_K}{R_V}\left(\frac{C_J}{C_{qb}}\right)^2 \frac{\hbar}{E_J} \tanh\left(\frac{E_J}{2k_B T}\right) \tag{4}$$

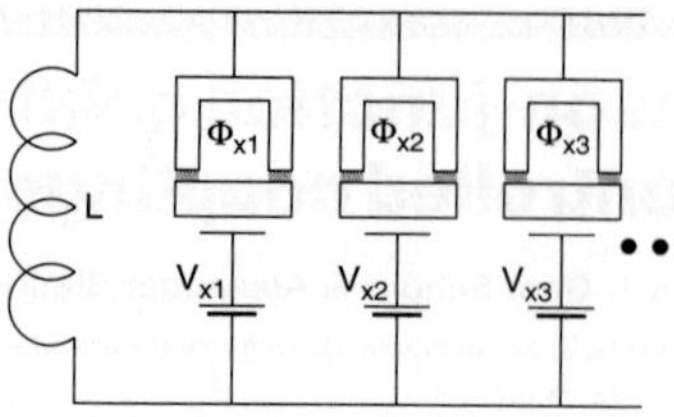

Figure 3 Design of a quantum computer. The coupling of the qubits is provided by the LC-oscillator mode in circuit shown. If the frequency of the LC-mode in the resulting circuit is large, $\hbar\omega_{LC} = \hbar(NC_{qb}L)^{-1/2} \gg E_J, E_{ch}, k_B T$, the fast oscillations produce an effective coupling of the qubits. We note that the system can be scaled to large numbers of qubits. In the idle state all effective Josephson couplings are tuned to zero, and the voltages are chosen such that the charge states are degenerate. Single-bit operations are performed by changing the gate voltage or flux of one qubit at a time. Two-bit operations between any two qubits are triggered by turning on the corresponding two Josephson couplings. The two lowest states of the qubit are separated from higher states, which exist in the real system, by the energies E_C, $\hbar\omega_{LC}$, $\hbar\omega_\Phi$. These should be larger than the energy scales of the qubit, $E_J, E_{ch}, k_B T$. If, in addition, switching processes of V_x and Φ_x are slow on the corresponding timescales, the requirements presented above also ensure that the higher states are not excited. Alternatively, instead of sudden switching, one can change the biases adiabatically. Another advantage of the presented design is that the result of a single-bit operation depends only on the time integral of energies $E_J(t)$ and $E_{ch}(t)$ over the operation period, but not on the profile of their time dependences.

Here R_V is compared to the quantum resistance $R_K = h/e^2 \approx 26\,\mathrm{k\Omega}$. A small gate capacitance $C \approx C_{qb} \ll C_J$ helps further decoupling of the qubit from the environment. Both can be optimized to yield a phase coherence time that is long compared to typical operation times $\hbar/E_J$.

A problem with the simple design is that the eigenstates of the hamiltonian shown in equation (3) are non-degenerate at all V_x. Therefore, the relative phase of two logical states evolves even during idle periods. We can still store quantum information in the qubit, as becomes apparent after a transformation to the interaction representation. But this introduces an explicit time dependence in the operators, with the result that the unitary transformations not only depend on the time span τ of the operations but also on the time t_0 when they start. Hence the time elapsed since the beginning of the computation, multiplied by the energy spacing between the logical states should be controlled with high accuracy. A second problem of the simple design is the non-vanishing two-bit coupling, even out of resonance. It introduces an error in the computation. The design discussed below overcomes both these problems.

A crucial step towards the ideal model (equation (1)) is to tune the Josephson coupling. This is achieved in the design of Fig. 1b, where each Josephson junction is replaced by a d.c.-SQUID (see, for example, ref. 20). The SQUID is biased by an external flux Φ_x, coupled into the system through an inductor loop. If the loop self-inductance L_Φ is low the SQUID-controlled qubit is described by a hamiltonian of the form of equation (2), but with potential energy $2E_J^0 \cos(\pi\Phi_x/\Phi_0)\cos\gamma$. Hence, the effective Josephson coupling is tunable by the external flux Φ_x between $2E_J^0$ and zero:

$$E_J(\Phi_x) = 2E_J^0 \cos(\pi\Phi_x/\Phi_0) \tag{5}$$

The SQUID-controlled qubit is described by the first two terms of the model hamiltonian shown in equation (1), with z- and x-components controlled independently by the gate voltage and the flux. In the idle state we keep $V_x = V_{deg}$ and $\Phi_x = \Phi_0/2$, so that the hamiltonian $\hat{H} = 0$. Changing one of them generates z- or x-rotations, respectively, that is, the elementary one-qubit operations.

With the improved design there is no need to control the total operation time t_0, while the time dependence of the voltage and flux can be optimized such that the time span of the manipulations τ is long enough to simplify time control and short enough to speed up the computation.

Also, the circuit of the current source, with resistance R_I, which couples the flux Φ_x to the SQUID by the mutual inductance M, introduces fluctuations and may destroy the coherence of the qubit dynamics. At the degeneracy point, the decoherence time is[21,22] $\tau_I = (1/\pi^3)(R_I/R_K)[\Phi_0^2/(E_J^0 M)]^2(\hbar/k_B T)$. This dephasing is slow if the current source is coupled weakly to the qubit (small M) and its resistance is high.

The control of the Josephson energies $E_J(\Phi_{xi})$ provides the possibility of coupling each selected pair of qubits, while keeping all the other ones uncoupled, bringing us close to the ideal model of equation (1). The simplest implementation of the coupling is to connect all N qubits in parallel with each other, and with inductor L (Fig. 3). Fast oscillations in the resulting LC-circuit produce an effective coupling of the qubits

$$\hat{H}_{\rm int} = -\sum_{i<j} \frac{E_J(\Phi_{xi})E_J(\Phi_{xj})}{E_L}\hat{\sigma}_y^i\hat{\sigma}_y^j \qquad (6)$$

where $E_L = [\Phi_0^2/(\pi^2 L)](C_J/C_{\rm qb})^2$. The coupling shown in equation (6) can be understood as the magnetic energy of the inductor which is biased by a current composed of contributions from all qubits, $I^i \propto E_J^i\hat{\sigma}_y^i$.

With this design we can perform all gate operations. In the idle state the interaction hamiltonian of equation (6) is zero as all the Josephson couplings are turned off. The same is true during a one-qubit operation, as long as we perform one such operation at a time that is, only one $E_J^i \neq 0$. To perform a two-qubit operation with any given pair of qubits, say 1 and 2, E_J^1 and E_J^2 are switched on simultaneously, yielding the total hamiltonian $\hat{H} = -(E_J^1/2)\hat{\sigma}_x^1 - (E_J^2/2)\sigma_x^2 - (E_J^1E_J^2/E_L)\hat{\sigma}_y^1\hat{\sigma}_y^2$. Although not identical to equation (1), these two-bit gates, in combination with the single-bit operations discussed above, also provide a complete set of gates required for quantum computation.

To demonstrate that the constraints on the set of system parameters can be met by available technology, we suggest a suitable set. We choose junctions with capacitance $C_J = 300$ aF, corresponding to a charging energy (in temperature units) $E_C \approx 3$ K, and a smaller gate capacitance $C = 30$ aF to reduce the coupling to the environment (even lower C are available and improve the performance further). The superconducting gap has to be slightly larger, $\Delta > E_C$. Thus at a working temperature of the order of $T = 50$ mK, the initial thermalization is assured. We further choose $E_J^0 = 50$ mK; so the timescale of one-qubit operations is $\tau_{\rm op} = \hbar/E_J \approx 70$ ps. Fluctuations associated with the gate voltages (equation (4)), with resistance $R_V \approx 50\,\Omega$, limit the coherence time to $\tau_V/\tau_{\rm op} \approx 4{,}000$ operations. With the parameters of the flux-circuit $L_\Phi = 0.1$ nH, $M = 1$ nH and $R_I = 10^2$–$10^6\,\Omega$, current fluctuations have a weak dephasing effect. To assure fast two-bit operations, we choose the energy scale E_L to be of the order of $10E_J$, which is achieved for $L \approx 3\,\mu$H. With these parameters, the number of qubits in the circuit can be chosen in the range of 10–50, of course at the expense of shorter coherence times $\tau_{V,I}/N$.

Some further remarks are in order.

(1) After the gate operations, the resulting quantum state has to be read out. This can be achieved by coupling a normal-state single-electron transistor capacitively to a qubit. The important aspect is that during computation the transistor is kept in a zero-current state and adds only to the total capacitance. When the transport voltage is turned on, the phase coherence of the qubit is destroyed, and the dissipative current in the transistor, which depends on the state of the qubit, can be read out. This quantum measurement process has been described explicitly in ref. 16 by an analysis of the time-evolution of the density matrix of the coupled system.

(2) Inaccuracy in the control of fluxes, voltages and the time-span of operations leads to diffusion of the actual quantum state from the one that exists in the absence of errors[23]. A random error of order ϵ per gate limits the number of operations to a value which is of order ϵ^{-2}. For the circuit parameters above, $\epsilon = 1\%$ would lead to smaller effects than those produced by environment.

(3) Many powerful quantum algorithms make use of parallel operations on different qubits. Although this is not possible with the present system, it may be achievable by a more advanced design, making use of further tunable SQUIDs decoupling different parts of the circuit. Such modifications, as well as the further progress of nanotechnology, should provide longer coherence times and allow scaling to larger numbers of qubits. □

Received 16 October 1998; accepted 12 January 1999.

1. Lloyd, S. A potentially realizable quantum computer. *Science* **261,** 1569–1571 (1993).
2. Bennett, C. H. Quantum information and computation. *Phys. Today* **48**(10), 24–30 (1995).
3. DiVincenzo, D. P. Quantum computation. *Science* **269,** 255–261 (1995).
4. Cirac, J. I. & Zoller, P. Quantum computations with cold trapped ions. *Phys. Rev. Lett.* **74,** 4091–4094 (1995).
5. King, B. E. *et al.* Demonstration of a fundamental quantum logic gate. *Phys. Rev. Lett.* **81,** 1525–1528 (1998).
6. Chuang, I. L., Gershenfeld, N. A. & Kubinec, M. Experimental implementation of fast quantum searching. *Phys. Rev. Lett.* **80,** 3408–3411 (1998).
7. Cory, D. G. *et al.* Experimental error correction. *Phys. Rev. Lett.* **81,** 2152–2155 (1998).
8. Jones, J. A., Mosca, M. & Hansen, R. H. Implementation of a quantum search algorithm on a quantum computer. *Nature* **393,** 344–346 (1998).
9. Turchette, Q. A., Hood, C. J., Lange, W., Mabuchi, H. & Kimble, W. Measurement of conditional phase shifts for quantum logic. *Phys. Rev. Lett.* **75,** 4710–4713 (1995).
10. Shnirman, A., Schön, G. & Hermon, Z. Quantum manipulations of small Josephson junctions. *Phys. Rev. Lett.* **79,** 2371–2374 (1997).
11. Averin, D. V. Adiabatic quantum computation with Cooper pairs. *Solid State Commun.* **105,** 659–664 (1998).
12. Bouchiat, V., Vion, D., Joyez, P., Esteve, D. & Devoret, M. H. Quantum coherence with a single Cooper pair. *Phys. Scripta T* **76,** 165–170 (1998).
13. Kane, B. E. A silicon-based nuclear spin quantum computer. *Nature* **393,** 133–137 (1998).
14. Loss, D. & DiVincenzo, D. P. Quantum computation with quantum dots. *Phys. Rev. A* **57,** 120–126 (1998).
15. Barenco, A. *et al.* Elementary gates for quantum computation. *Phys. Rev. A* **52,** 3457–3467 (1995).
16. Shnirman, A. & Schön, G. Quantum measurements performed with a single-electron transistor. *Phys. Rev. B* **57,** 15400–15407 (1998).
17. Rouse, R., Han, S. & Lukens, J. E. Observation of resonant tunneling between macroscopically distinct quantum levels. *Phys. Rev. Lett.* **75,** 1614–1617 (1995).
18. Maassen v.d. Brink, A., Schön, G. & Geerligs, L. J. Combined single-electron and coherent-Cooper-pair tunneling in voltage-biased Josephson junctions. *Phys. Rev. Lett.* **67,** 3030–3033 (1991).
19. Nakamura, Y., Chen, C. D. & Tsai, J. S. Spectroscopy of energy level spitting between two macroscopic quantum states of charge coherently superposed by Josephson coupling. *Phys. Rev. Lett.* **79,** 2328–2331 (1997).
20. Tinkham, M. *Introduction to Superconductivity* (McGraw-Hill, New York, 1996).
21. Leggett, A. J. *et al.* Dynamics of the dissipative two-state system. *Rev. Mod. Phys.* **59,** 1–85 (1987).
22. Weiss, U. *Quantum Dissipative Systems* (World Scientific, Singapore, 1993).
23. Miquel, C., Paz, J. P. & Zurek, W. H. Quantum computation with phase drift errors. *Phys. Rev. Lett.* **78,** 3971–3974 (1997).

Acknowledgements. We thank T. Beth, M. Devoret, D. P. DiVincenzo, E. Knill, K. K. Likharev and J. E. Mooij for discussions.

Correspondence and requests for materials should be addressed to Y.M. (e-mail: makhlin@tfp.physik.uni-karlsruhe.de).

Quantum Computing in Optical Lattices

Quantum computing in optical lattices

Hans J. Briegel
Ludwig-Maximilians University of Munich.

The potential power of a quantum computer is based on its ability of processing entangled data. Quantum algorithms make use of this ability, together with the possibility of interference of computational paths, which may enhance the efficiency of certain computations compared with classical algorithms (see the contribution of Ekert to this volume). Generating and controlling entanglement in real physical systems, on the other hand, requires precise control of the Hamiltonian interactions and a high degree of coherence. Achieving these conditions in the laboratory is extremely demanding and therefore only a few systems, including trapped ions, cavity QED, and NMR, have been investigated experimentally as candidates for implementing quantum logic.

Recently, *ultracold controlled collisions* have been identified as a possibility of entangling neutral atoms [4]. Controlled atomic collisions can be achieved by manipulating microscopic potentials that are capable of storing individual atoms. Examples are given by magnetic microtraps and by optical lattices. In such systems, it is possible to vary the shape the trapping potentials depending on the internal state of the atoms. Atoms in certain internal states can thereby made to interact via s-wave scattering. For sufficiently low temperatures, this interaction is coherent and can be used to implement a quantum gate. Given the impressive experimental progress that has been made in the fields of neutral atom trapping and cooling [1], and in the studies of Bose-Einstein condensation (BEC) of ultracold gases [2, 3], this proposal of using controlled atomic collisions for quantum logic opens new scenarios for quantum computing and for the experimental study of quantum information [5].

In the lectures, we will mainly concentrate on one particular scenario, that is optical lattices and their potential use for quantum computing (see also [6]). Optical lattices combine two important features. First, they provide a variety of "turns and knobs" which allow for a high degree of control on the internal and external state of the trapped atoms [7]. Second, they offer a massive parallelism not available in other systems. Together, as we shall see, these features make optical lattices an ideal prototype model of a quantum computer including elements of *parallel processing* of the type we expect to see in future systems based e.g. on nanostructures (see also the contributions of Schön and Fazio, and of Loss to this volume).

In the discussion of these ideas, one has to distinguish short-term goals from long-term perspectives. Although there has been remarkable experimental progress with cooling and manipulating atoms in highly-detuned optical lattices recently [8], the implementation of quantum information concepts and of quantum computing in this system will require further experimental steps. These include the creation of regular filling structures [9] – which has not been achieved in the laboratory yet – and, like in ion-traps, the possibility of addressing single atoms individually. We will pay close attention to these requirements and make suggestions for different "generations" of experiments, ranging from basic entanglement studies in present-day experimental set-ups (i.e. with random occupation of the lattice sites and without any control of the individual atomic positions), to efficient quantum-error-correction schemes and implementations for fault-tolerant computing [10] in more advances set-ups. Some of these ideas have been summarizes in a recent review article [5] by the Innsbruck group and myself, which is reprinted in this volume and may serve as a guide accompanying the lectures. It includes further references to relevant work from other fields of Atomic Physics and of Quantum Computing.

Part of this work is supported through a grant of the Schwerpunktsprogramm "Quanten-Informationsverarbeitung" der Deutschen Forschungsgemeinschaft, and by the European Community under the TMR network ERB-FMRX-CT96-0087.

References

[1] CHU, S., 1998, *Rev. Mod. Phys.* **70**, 686; COHEN-TANNOUDJI, C., *ibid.*, 707; PHILLIPS, W. D., *ibid.*, 721. For an earlier review see, for example, *Laser manipulation of Atoms and Ions,*1992, edited by E. ARIMONDO, W. D. PHILLIPS, and F. STRUMIA,(North Holland, Amsterdam). A very recent review on optical dipol trapping is given by GRIMM, R., WEIDENMÜLLER, M., and OVCHINNIKOV, Y. B., 1999, e-print physics/9902072.

[2] ANDERSON, M. H., ENSHER, J. R., MATTHEWS, M. R., WIEMAN, C. E., and CORNELL, E. A., 1995, *Science*, **269**, 198; BRADLEY, C. C., SACKETT, C. A., TOLLETT, J. J., and HULET, R. G., 1995, *Phys. Rev. Lett.* **75**, 1687; DAVIES, K. B., MEWES, M.-O., ANDREWS, M. R., VAN DRUTEN, N. J., DURFEE, D. S., KURN, D. M., and KETTERLE, W., 1995, *Phys. Rev. Lett.* **75**, 3969. See the BEC homepage at `http://amo.phy.gasou.edu/bec.html`

[3] HALL, D. S., MATTHEWS, M. R., WIEMAN, C. E., and CORNELL, E. A., 1998, *Phys. Rev. Lett.***81**, 1539; STAMPER-KURN, D. M., ANDREWS, M. R., CHIKKATUR, A. P., INOUYE, S., MIESNER, H.-J., STENGER, J., and KETTERLE, W., 1998, *Phys. Rev. Lett.***80**, 2027.

[4] JAKSCH, D., BRIEGEL, H.-J., CIRAC, J. I., GARDINER, C. W., and ZOLLER, P., 1999, *Phys. Rev. Lett.* **82,** 1975.

[5] BRIEGEL, H.-J., CALARCO, T., JAKSCH, D., CIRAC, J. I., and ZOLLER, P., *Quantum computing with neutral atoms*, 1999, submitted for publication in the Special Issue on Quantum Computing of the Journal of Modern Optics.

[6] BRENNEN, G.K., CAVES, C.M., JESSEN, P.S., and DEUTSCH, I.H., 1999, *Phys. Rev. Lett.* **82**, 1060.

[7] DEUTSCH I. H. and JESSEN, P.S., Phys. Rev. A **57**, 1972 (1998).

[8] DePue, M. T., McCormick, C., Winoto, S.L., Oliver, S. and Weiss, D.W., 1999, *Phys. Rev. Lett.* **82** 2262; Mennerat-Robilliard, C., Lukas, D., Guibal, S., Tabosa, J., Jurczak, C. Courtois, J.-Y., and Grynberg, G., 1999, *Phys. Rev. Lett.* **82** 85; Hamann, S. E., Haycock, D. L., Klose, G., Pax, P. H., Deutsch, I. H., and Jessen, P. S., 1998, *Phys. Rev. Lett.* **80**, 4149; Friebel, S., D'Andrea, C., Walz, J., Weitz, M., and Hänsch, T. W., 1998, *Phys. Rev. A* **57**, R20.

[9] Jaksch, D., Bruder, C., Cirac, J. I., Gardiner, C. W., and Zoller, P, 1998 *Phys. Rev. Lett.* **81**, 3108.

[10] Preskill, J., 1997, e-print quant-ph/9705031.

Quantum computing with neutral atoms

H.-J. Briegel,[1] T. Calarco,[2,3] D. Jaksch,[2] J.I. Cirac,[2] and P. Zoller[2]
[1] *Sektion Physik, Ludwig-Maximilians-Universität München, D-80333 München, Germany*
[2] *Institut für Theoretische Physik, Universität Innsbruck, Technikerstrasse 25, A-6020 Innsbruck, Austria*
[3] *ECT*, European Centre for Theoretical Studies in Nuclear Physics and Related Areas*
Villa Tambosi, Strada delle Tabarelle 286, Villazzano (Trento), Italy 38050

We develop a method to entangle neutral atoms using cold controlled collisions. We analyze this method in two particular set-ups: optical lattices and magnetic micro-traps. Both offer the possibility of performing certain multi-particle operations in parallel. Using this fact, we show how to implement efficient quantum error correction and schemes for fault-tolerant computing.

I. INTRODUCTION

Entanglement is one of the most intriguing features of Quantum Mechanics. However, there are very few physical systems in which entanglement can be systematically studied in a controlled way. Those systems include ion-traps [1–8], cavity QED [9–16], photons [17–25], and molecules in the context of NMR [26–29] (see [30] however). Very recently, we have identified a new way of entangling particles by using *cold controlled collisions* with which one could study experimentally basic issues of quantum information theory [31]. Given the impressive experimental advances made so far in the fields of neutral atom trapping and cooling [32–35], and in the studies of Bose Einstein condensation (BEC) of ultracold gases [36–41], that proposal opens a new perspective to several experimental groups who so far have concentrated their efforts in other fields of Atomic Physics.

In the present paper, we build upon the work in [31] and explore the idea of using atomic controlled cold collisions for entangling neutral atoms in optical lattices (see also [42]) and in arrays of magnetic micro-traps. We show how to perform two-qubit gate operations with those systems obtaining very high fidelities. We propose a variety of experiments to entangle particles using state-of-the-art technology. We also concentrate on the unique possibilities that these set-ups offer to perform multi-particle entanglement operations in parallel [43–46]. Using such parallelism, we show how to implement efficient error correction [47–54] and fault-tolerant quantum computation schemes [55–62].

The paper is organized as follows. In Sec. II we discuss the use of ultracold collisions as a mechanism for entangling neutral atoms. Such collisions can be brought about by either moving the potentials in certain spatial directions or by modifying the shape of the trapping potentials. In Sec. III we describe two systems in which such operations can be implemented. These are optical lattices [42,63–66] and magnetic microtraps [67–71] both of which have been studied experimentally in detail in the past. In Sec. IV we describe a class of multi-particle entanglement operations that can be realized in these systems (we concentrate here on optical lattices). The usefulness of such operations for quantum computing depends on certain conditions that need to be satisfied in an experiment. Among these conditions, the filling problem, i.e. how to fill the potentials with regular patterns of atoms, is most outstanding. We discuss these matters and show that even under present-day experimental conditions, very interesting entanglement studies could be performed. Section V summarizes the main results and discusses their relevance for future research.

II. ENTANGLEMENT OF ATOMS VIA COLD CONTROLLED COLLISIONS

In this Section, we consider two bosonic neutral atoms with two internal states trapped by conservative potentials and cooled to the motional ground states. Initially these two particles are sufficiently far apart so that they do not interact with each other. We then assume the shape of the potentials to be varied in a way that depends on the internal state of the atoms so that the two particles come close to each other if they are in certain internal states. As we will show, this can be done e.g. by moving the center position of the trapping potentials state selectively, or by switching off a potential barrier between the two atoms for one of the two internal states. In both cases the particles will interact via s-wave scattering with each other in a coherent way when they are close to each other. After the interaction has taken place the particles are restored to their initial position. In this way one can implement conditional dynamics and realize a fundamental two-qubit gate.

Note that we are dealing with bosons. Therefore, we have to use symmetrized wave functions for describing the two particles. It will turn out that if the center positions of the trapping potentials are moved state selectively, particles in the same internal state will always be so far apart that their wave functions never overlap. Thus, we will not care about the symmetrization in this case. On the other hand, if the potential barrier is switched off for one internal state, particles in the same internal state will come close to each other and symmetrizing the wave function is essential.

A. Hamiltonian

Here we deal with the interaction Hamiltonian of two neutral atoms 1 and 2 with internal states $|a\rangle_{1,2}$ and $|b\rangle_{1,2}$ trapped by conservative potentials $V^{\beta_i}(\mathbf{x}_i,t)$ whose functional dependence on the coordinate $\mathbf{x}_i$, with $i=1,2$ the particle index, depends on the internal state of the particle $\beta_{1,2}=a,b$. Initially, the two particles are in the ground state of the trapping potentials and the centers of the two potential wells are sufficiently far apart so that the particles do not interact. Then the form of the potential wells is changed such that there is some overlap of the wave functions of the two atoms, and the particles will interact with each other. This interaction between the atoms in two given internal states β_1 and β_2 can be described by a contact potential

$$u^{\beta_1\beta_2}(\mathbf{x}_1-\mathbf{x}_2)=\frac{4\pi a_s^{\beta_1\beta_2}\hbar^2}{m}\delta^3(\mathbf{x}_1-\mathbf{x}_2), \quad (1)$$

where $a_s^{\beta_1\beta_2}$ is the s-wave scattering length for the corresponding internal states describing elastic collisions and m is the mass of the particles. This zero energy s-wave scattering approximation will be valid as long as we assume that $v_{\rm osc}$, the rms velocity of the atoms in the vibrational ground state, approximately given by $v_{\rm osc}\approx a_0\omega$, is sufficiently small [72]. Here a_0 is the size of the ground state of the trap potential, and ω is the first excitation frequency. Thus we can describe the evolution of the system by the Hamiltonian

$$H=\sum_{\beta_1,\beta_2}H^{\beta_1\beta_2}\otimes|\beta_1\rangle_1\langle\beta_1|\otimes|\beta_2\rangle_2\langle\beta_2|, \quad (2)$$

where

$$H^{\beta_1\beta_2}=\sum_{i=1,2}\left[\frac{(\mathbf{p}_i)^2}{2m}+V^{\beta_i}(\mathbf{x}_i,t)\right]+u^{\beta_1\beta_2}(\mathbf{x}_1-\mathbf{x}_2). \quad (3)$$

Here $\mathbf{p}_i$ is the momentum operator.

1. Interaction in perturbation theory

We want to treat the interaction term in the Hamiltonian Eq. (3) perturbatively. For particles in two different internal states $\beta_1\neq\beta_2$ we find the correction to the energy due to the interaction as

$$\Delta E^{\beta_1\beta_2}(t)=\frac{4\pi a_s^{ab}\hbar^2}{m}\int d\mathbf{x}\prod_i\left|\psi_i^{\beta_i}(\mathbf{x},t)\right|^2, \quad (4)$$

where $\psi_i^{\beta_i}(\mathbf{x},t)$ is the normalized one-particle wave function of particle i in internal state β_i in the time dependent potential $V^{\beta_i}(\mathbf{x},t)$. If the particles are in the same internal state $\beta_1=\beta_2=\beta$, we have to account for the Bose statistics i.e. use the properly normalized symmetrized two-particle wave function for calculating the energy shift. We therefore find

$$\Delta E^{\beta\beta}(t)=\frac{8\pi a_s^{\beta\beta}\hbar^2}{m(1+|\alpha|^2)}\int dx\prod_i\left|\psi_i^{\beta}(\mathbf{x},t)\right|^2, \quad (5)$$

where

$$\alpha=\int dx\left(\psi_1^{\beta}(\mathbf{x},t)\right)^*\psi_2^{\beta}(\mathbf{x},t). \quad (6)$$

For general β_1, β_2 we find the phase accumulated due to the interaction in the time interval $[-\tau,\tau]$ by

$$\phi^{\beta_1\beta_2}=\frac{1}{\hbar}\int_{-\tau}^{\tau}dt\,\Delta E^{\beta_1\beta_2}(t). \quad (7)$$

B. Moving potentials

One way of controlling the interaction between the particles is to move the center position of the potentials $V^{\beta_i}(\mathbf{x}_i,t)=V\left(\mathbf{x}_i-\bar{\mathbf{x}}_i^{\beta_i}(t)\right)$ towards each other in a state-dependent way while leaving the shape of the potential unchanged. By moving the potential we get two kinds of phase shifts. A kinetic phase which is a single-particle phase due to the kinetic energy of the particles and an interaction phase due to coherent interactions between two atoms. First we will define these two phases for general trapping potentials and afterwards specialize them to moving harmonic potentials. Finally, we will show how conditional dynamics can be realized.

1. Kinetic phase

First we want to consider a single atom in internal state $|\beta\rangle$ trapped in the instantaneous ground state ψ_0 of a moving potential well $V(\mathbf{x}-\bar{\mathbf{x}}^\beta(t))$. The center position of the potential is moved along a trajectory $\bar{\mathbf{x}}^\beta(t)$. Ideally, we want the atom to remain in the ground state of its trapping potential and to preserve its internal state during the motion. This corresponds to the transformation from $t=-\tau$ to $t=\tau$

$$\psi_0[\mathbf{x}-\bar{\mathbf{x}}^\beta(-\tau)]\to e^{-i\phi^\beta}\psi_0[\mathbf{x}-\bar{\mathbf{x}}^\beta(\tau)], \quad (8)$$

where the atom remains in the ground state of the trapping potential and preserves its internal state. Transformation (8) can be realized in the *adiabatic limit* [73], where we move the potentials so that the atoms remain in the instantaneous motional ground state. Adiabaticity requires $|\dot{\bar{\mathbf{x}}}^\beta(t)|\ll\hbar\omega/2mv_{\rm osc}$, for all times t. Here ω is the smallest excitation frequency in the potential and $v_{\rm osc}^2=\langle\psi_0(0)|q^2|\psi_0(0)\rangle/m^2$. q is the momentum operator in the direction the optical lattice is moved. The phase

ϕ^β can be easily calculated in the limit $|\ddot{\mathbf{x}}^\beta(t)| \ll v_{\rm osc}/\tau$. We find the *kinetic phase*

$$\phi^\beta = \frac{m}{2\hbar}\int_{-\tau}^{\tau} dt\,\left(\dot{\mathbf{x}}^\beta(t)\right)^2. \tag{9}$$

2. Interaction phase

Let us now consider two particles $i = 1, 2$ in different internal states $|\beta_i\rangle_i$ trapped in the ground states of two moving potentials. Initially, at time $t = -\tau$, these wells are centered at positions $\bar{\mathbf{x}}_i$, sufficiently far apart (distance $d = \bar{\mathbf{x}}_1 - \bar{\mathbf{x}}_2$) so that the particles do not interact. The positions of the potentials are moved along trajectories $\bar{\mathbf{x}}_i^{\beta_i}(t)$ so that the wave packets of the atoms overlap for certain time, until finally they are restored to the initial position at the time $t = \tau$. We assume that: (i) $|\dot{\bar{\mathbf{x}}}_i^{\beta_i}(t)| \ll v_{\rm osc}$ (adiabatic condition) so that the particles remain in the ground states of the moving trapping potentials; (ii) The interaction can be treated perturbatively, where $|\Delta E^{\beta_1\beta_2}(t)| \ll \hbar\omega$ so that no sloshing motion is excited. In that case, we realize the transformation

$$\begin{aligned}\psi_0(\mathbf{x}_1 - \bar{\mathbf{x}}_1)\psi_0(\mathbf{x}_2 - \bar{\mathbf{x}}_2) \to\\ e^{-i\phi}\psi_0(\mathbf{x}_1 - \bar{\mathbf{x}}_1)\psi_0(\mathbf{x}_2 - \bar{\mathbf{x}}_2),\end{aligned} \tag{10}$$

where $\phi = \phi^{\beta_1} + \phi^{\beta_2} + \phi^{\beta_1\beta_2}$ with the *collisional phase* $\phi^{\beta_1\beta_2}$ defined in Eq. (7).

3. Moving harmonic potentials

Here we specialize to harmonic trapping potentials. The wave function $\psi_i^{\beta_i}(\mathbf{x}, t)$ of a particle in a moving harmonic potential can be found analytically. In the Appendix A we show that when we start to move the harmonic potential at time $-\tau$ with the particle in its motional ground state and stop to move the potential at time τ, the condition for the particle to end up in the motional ground state at τ is given by

$$\left|\int_{-\tau}^{\tau} \dot{\bar{\mathbf{x}}}_i^{\beta_i}(t')e^{i\omega t'}dt'\right| \ll a_0. \tag{11}$$

This condition is weaker than the condition $|\dot{\bar{\mathbf{x}}}_i^{\beta_i}(t)| \ll v_{\rm osc}$ for adiabaticity, and means that the particle need not be in the instantaneous ground state of the moving potential at all times, but only at the final time. The kinetic phases can be found exactly (cf. Eq. (A7)). If $|\Delta E^{\beta_1\beta_2}(t)| \ll \hbar\omega$ is satisfied, the interaction phase can be found by Eq. (7) since the $\psi_i^{\beta_i}(\mathbf{x}, t)$ are known. It is also possible to generalize these results to the case in which the trap frequency changes with time [74].

4. Implementation of conditional dynamics

Let us now assume that we can design the potentials such that atoms in the internal state $|\beta_i\rangle_i$ experience a potential $V^{\beta_i}(\mathbf{x}_i, t) = V(\mathbf{x}_i - \bar{\mathbf{x}}_i^{\beta_i}(t))$ which is initially ($t = -\tau$) centered at position $\bar{\mathbf{x}}_i$. We assume that we can move the centers of the potentials as follows: $\bar{\mathbf{x}}_i^{\beta_i}(t) = \bar{\mathbf{x}}_i + \delta\mathbf{x}^{\beta_i}(t)$. As shown in Fig. 1 the trajectories $\delta\mathbf{x}^{\beta_i}(t)$ are chosen in such a way that $\delta\mathbf{x}^{\beta_i}(-\tau) = \delta\mathbf{x}^{\beta_i}(\tau) = 0$ and the first atom collides with the second one only if they are in states $|a\rangle$ and $|b\rangle$, respectively ($|\bar{\mathbf{x}}_1^b(t) - \bar{\mathbf{x}}_2^a(t)| \gg a_0\ \forall t$). This choice is motivated by the physical implementation considered in Sec. III A. The fact that $\bar{\mathbf{x}}_i$ does not depend on the internal atomic state and the shape of the two potentials is the same at times $\pm\tau$ allows one to easily change the internal state at times $t = \pm\tau$ by applying laser pulses. If the conditions stated above are fulfilled, depending on the initial internal atomic states we have:

$$\begin{aligned}|a\rangle_1|a\rangle_2 &\to e^{-i2\phi^a}|a\rangle_1|a\rangle_2,\\ |a\rangle_1|b\rangle_2 &\to e^{-i(\phi^a+\phi^b+\phi^{ab})}|a\rangle_1|b\rangle_2,\\ |b\rangle_1|a\rangle_2 &\to e^{-i(\phi^a+\phi^b)}|b\rangle_1|a\rangle_2,\\ |b\rangle_1|b\rangle_2 &\to e^{-i2\phi^b}|b\rangle_1|b\rangle_2,\end{aligned} \tag{12}$$

where the motional states remain unchanged. The kinetic phases ϕ^β and the collisional phase ϕ^{ab} can be calculated as stated above. We emphasize that the ϕ^β are (trivial) one-particle phases that, as long as they are known, can always be incorporated in the definition of the states $|a\rangle$ and $|b\rangle$. This realizes a fundamental two-qubit quantum gate for certain values of ϕ^{ab}, e.g. $\phi^{ab} = \pi$.

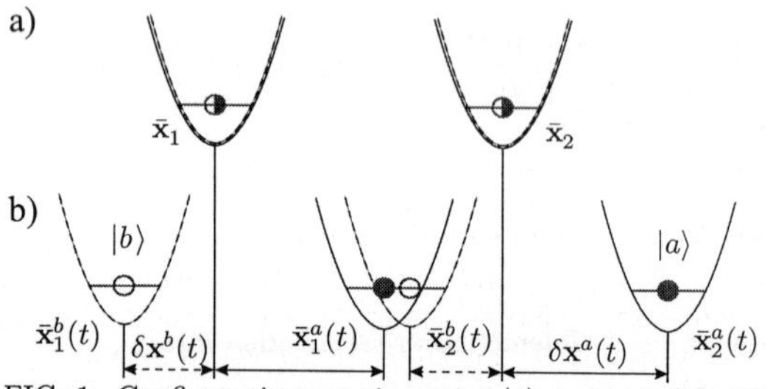

FIG. 1. Configurations at times $\pm\tau$ (a) and at t (b). The solid (dashed) curves show the potentials for particles in the internal state $|a\rangle$ ($|b\rangle$). Center positions $\bar{\mathbf{x}}_i^{\beta_i}(t)$ and displacements $\delta\mathbf{x}^{\beta_i}(t)$ as defined in the text.

C. Switching potentials

The interaction between the particles can be controlled also in another way, for example by changing with time the shape of the potentials depending on the particles' internal states. Different regimes for the time-dependence of the potential are possible. The two limits of extremely slow (adiabatic) or extremely fast (sudden) potential changes are both interesting and lead to peculiar schemes. Here we will analyze the latter case. We consider two atoms initially trapped in two displaced wells. At a certain time the barrier between the wells is suddenly removed in a selective way for atoms in state $|b\rangle$, whereas it remains unchanged for atoms in state $|a\rangle$. The atoms are allowed to oscillate for some time, and then the barrier is raised again suddenly such as to trap them back at the original positions. During the process they will acquire both a kinematic phase due to the oscillations within their respective wells, and an interaction phase due to the collision. We will calculate such quantities and look for the optimal switching time required in order to maximize the fidelity for a quantum gate relying on this scheme, which we will estimate quantitatively for the relevant physical example in Sec. III B.

1. Kinematic phase

Let us first consider the time-independent problem of an atom subject to a three-dimensional potential whose functional form along x depends on the internal atomic state $\beta = a, b$:

$$V^{\beta}(\mathbf{x}) = v^{\beta}(x) + v_{\perp}(y) + v_{\perp}(z). \tag{13}$$

Here the v's are single-well trapping potentials, and v^a, v^b are centered around $x = x_0$, $x = 0$, respectively. We assume that the atom is initially prepared in the motional state $|\Psi_+\rangle$, where

$$\langle \mathbf{x}|\Psi_+\rangle \equiv \Psi_+(\mathbf{x}) = \psi_+(x)\psi_{\perp}(y)\psi_{\perp}(z) \tag{14}$$

and ψ_+, $\psi_{\perp}$ are the ground-state wave functions of v^a, $v_{\perp}$ with eigenvalues E^a, $E_{\perp}$ respectively. Thus $\Psi_+(\mathbf{x})$ is peaked around the position $\mathbf{x}_0 \equiv (x_0, 0, 0)$, coinciding with the center of $V^a(\mathbf{x})$ but displaced from the one of $V^b(\mathbf{x})$. Therefore, if the atom is in internal state $|a\rangle$, its motional state after a time t will be unchanged up to a phase $\phi^a = (E^a + 2E_{\perp})t/\hbar$. If it is instead in state $|b\rangle$, it will start oscillating within the well, thus picking up a different phase ϕ^b due to the kinematical evolution, and possibly coming back at the initial position after some time.

2. Interaction phase

We now consider two atoms 1 and 2 initially (at $t = 0$) prepared in the motional states $|\Psi_+\rangle$ and $|\Psi_-\rangle$, the latter being defined as in Eq. (14) but with $\psi_-(x) \equiv \psi_+(-x)$ replacing $\psi_+(x)$. We assume that the particles are subject to the potentials $\sum_{\varsigma=+,-}\Theta(\varsigma x)V^{\beta_i}(\varsigma \mathbf{x}_i)$, where Θ denotes the step function. If any one of them is in state $|b\rangle$, for $t > 0$ it will start oscillating within the well, eventually interacting with the other one. If $v_{\perp}$ is much steeper than v^{β}, then the probability of transversal excitations can be neglected, *i.e.* each atom remains in the ground state along y and z. By integrating over these variables, the problem is then reduced to a one-dimensional two-particle Schrödinger equation, with

$$\mathcal{H}^{\beta_1\beta_2} = \sum_{i=1}^{2}\left[\frac{(p_i)^2}{2m} + w^{\beta_i}(x_i, t)\right] + u_x^{\beta_1\beta_2}(x_1 - x_2) \tag{15}$$

replacing the Hamiltonian (3) in Eq. (2). Here $w^{\beta}(x,t)$ is a combination of the $v^{\beta}(x)$ whose form changes with time, and $u_x^{\beta_1\beta_2}$ is an effective interaction potential taking into account the integration over y and z, and therefore depending on the shape of $v_{\perp}$. We shall study the dynamics at $t \geq 0$ for different values of (β_1, β_2) separately. If $\beta_1 = \beta_2 \equiv \beta$ the total initial normalized state, symmetric under particle interchange, is

$$|\psi^{\beta\beta}(0)\rangle \approx \frac{|\psi_+\rangle_1|\psi_-\rangle_2 + |\psi_-\rangle_1|\psi_+\rangle_2}{\sqrt{2}} \otimes |\beta\rangle_1 \otimes |\beta\rangle_2, \tag{16}$$

where the initial overlap $\langle\psi_-|\psi_+\rangle \ll 1$ has been neglected in computing the normalization. If both particles are in state $|a\rangle$, no interaction takes place and thus the collisional phase $\phi^{aa} = 0$. Therefore, we shall now consider in more detail the situation in which both particles are in state $|b\rangle$ and thus move within the well v^b. In the absence of interaction, after an oscillation period T they would come back exactly to the initial state. Due to the interaction, two effects arise: an additional phase, which is accumulated by the wave function as the number of undergone oscillations increases; and a slight decrease in their frequency, because the atoms acquire a small delay in their motion inside the trap as they come out from a collision. These effects have to be evaluated in detail, since they influence the attainable fidelity for a quantum gate based on this scheme. For symmetry reasons, the relative coordinate motion decouples from the center of mass motion, which is not affected by the interaction and can be solved analytically. For an explicit calculation, it is now needed to specify the form of the potentials in Eq. (13).

3. Switching harmonic potentials

In order to perform the calculations analytically, the potentials in Eq. (13) are chosen to be harmonic:

$$v^a(x) = \frac{m\omega_0^2}{2}(x - x_0)^2, \tag{17a}$$

$$v^b(x) = \frac{m\omega^2}{2}x^2, \tag{17b}$$

$$v_{\perp}(y) = \frac{m\omega_{\perp}^2}{2}y^2, \tag{17c}$$

where $\omega_\perp \gg \omega_0 > \omega$. Our scheme for gate operation is as follows: initially the two particles are separately stored in two displaced harmonic wells at $\pm x_0$ as described above, *i.e.* with the potential (Fig. 2a)

$$w^a(x, t<0) = \sum_{\varsigma=+,-} \Theta(\varsigma x) v^a(\varsigma x), \tag{18a}$$

$$w^b(x, t<0) = w^a(x, t<0) \tag{18b}$$

in the one-dimensional Hamiltonian Eq. (15). At $t = 0$ the potential undergoes a sudden change, namely the barrier between the two wells is selectively switched off for state $|b\rangle$ only (Fig. 2b):

$$w^a(x, 0<t<\tau) = w^a(x, t<0); \tag{19a}$$

$$w^b(x, 0<t<\tau) = v^b(x). \tag{19b}$$

Then at $t = \tau$, the potential barrier is suddenly restored: $w^{a,b}(x, t>\tau) = w^{a,b}(x, t<0)$. The time evolution at $0 < t < \tau$ is characterized by oscillations with periodicity $T \equiv 2\pi/\omega$. The projection of the evolved CM wave function on the initial one

$$\left|\langle \psi^{bb}_{\rm CM}(t)|\, \psi^{bb}_{\rm CM}(0)\rangle\right|^2 = \left[1 + \frac{(\omega_0^2 - \omega^2)^2}{4\omega_0^2\omega^2} \sin^2(\omega t)\right]^{-\frac{1}{2}} \tag{20}$$

has instead a period of $T/2$, because of the parity of the spatial wave function. The time-dependent energy shift (4) due to the interaction turns out to be

$$\Delta E^{bb}(t) = a_s^{bb} \hbar\omega_\perp \sqrt{\frac{8m\Omega(t)}{\pi\hbar}}\, e^{-\frac{2m\omega_0}{\hbar} x_0^2 \left[1 - \sin^2(\omega t)\frac{\omega_0 \Omega(t)}{\omega^2}\right]}, \tag{21}$$

where $\Omega(t) = \omega^2\omega_0/[\omega^2\cos^2(\omega t) + \omega_0^2 \sin^2(\omega t)]$. Hence the interaction-induced phase shift (7) accumulated after each oscillation period T is (evaluating the integral in a saddle-point approximation)

$$\phi_T^{bb} = \int_0^T \frac{\Delta E^{bb}(t)}{\hbar} dt$$

$$\approx 8 a_s^{bb} \sqrt{\frac{m\omega_0}{\hbar}} \frac{\omega_y \omega_z}{\omega_0^2 + \omega^2(4x_0^2\, m\omega_0/\hbar - 1)}. \tag{22}$$

If the particles are in different internal states, the center of mass does not decouple from the relative motion. No analytical solution is found in this case, and one must resort to numerical techniques to evaluate the collisional phase ϕ^{ab}.

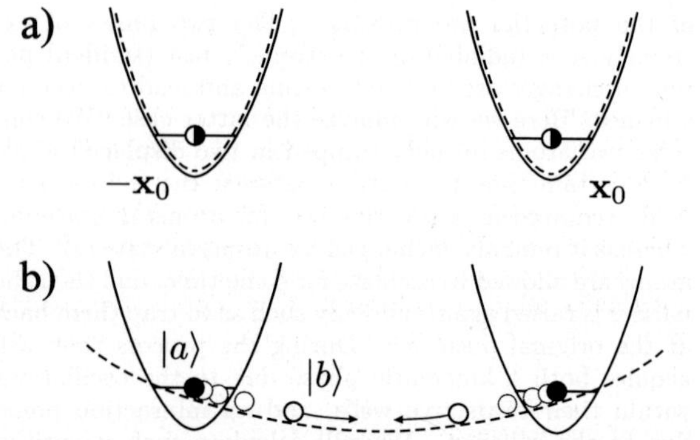

FIG. 2. Configuration at times $t < 0$, $t > \tau$ (a) and at $0 \le t \le \tau$ (b). The solid (dashed) curves show the potentials for particles in the internal state $|a\rangle$ ($|b\rangle$).

4. Implementation of conditional dynamics

If at time τ the atoms have come back to their initial spatial distribution, corresponding to a symmetrized product of the ground states of the two wells, then after the barrier is raised they will remain trapped around the original position. The only change in the overall state will be a phase $\phi_\tau^{\beta_1} + \phi_\tau^{\beta_2} + \phi_\tau^{\beta_1\beta_2}$ as discussed in Sect. II B 2. Therefore, the gate operation time τ has to be chosen in such a way as to maximize the overlap $|\langle \psi^{\beta_1\beta_2}(\tau)|\psi^{\beta_1\beta_2}(0)\rangle|^2$ for all β_1, β_2. If the modifications in the atomic motion due to interaction are not too strong, this condition will be satisfied to a good approximation after an integer number n of oscillations. Thus, for $\tau \approx nT$, the following mapping is realized:

$$\begin{aligned}
|a\rangle_1|a\rangle_2 &\to e^{-i(2\phi_\tau^a + \phi_\tau^{aa})}|a\rangle_1|a\rangle_2,\\
|a\rangle_1|b\rangle_2 &\to e^{-i(\phi_\tau^a + \phi_\tau^b + \phi_\tau^{ab})}|a\rangle_1|b\rangle_2,\\
|b\rangle_1|a\rangle_2 &\to e^{-i(\phi_\tau^b + \phi_\tau^a + \phi_\tau^{ba})}|b\rangle_1|a\rangle_2,\\
|b\rangle_1|b\rangle_2 &\to e^{-i(2\phi_\tau^b + \phi_\tau^{bb})}|b\rangle_1|b\rangle_2,
\end{aligned} \tag{23}$$

where $\phi_\tau^{aa} = 0$ as discussed in Sect. II C 2. If we apply a further single-bit rotation $|0\rangle\langle 0|e^{-i\phi_\tau^a} + |1\rangle\langle 1|e^{-i(\phi_\tau^b + \phi_\tau^{ab})}$ (where the logical states are defined as $|0\rangle \equiv |a\rangle$ and $|1\rangle \equiv |b\rangle$) and take into account that for symmetry reasons $\phi_\tau^{ab} = \phi_\tau^{ba}$, the mapping Eq. (23) realizes the fundamental phase gate

$$\begin{aligned}
|0\rangle|0\rangle &\to |0\rangle|0\rangle,\\
|0\rangle|1\rangle &\to |0\rangle|1\rangle,\\
|1\rangle|0\rangle &\to |1\rangle|0\rangle,\\
|1\rangle|1\rangle &\to e^{-in(\phi_T^{bb} - 2\phi_T^{ab})}|1\rangle|1\rangle,
\end{aligned} \tag{24}$$

where the phase difference $\phi_T^{bb} - 2\phi_T^{ab}$ has to be adjusted to $\pm\pi$ by a proper choice of the trap parameters.

III. PHYSICAL REALIZATIONS

A physical implementation of the scenarios described in Sec. II requires an interaction which produces internal-state-dependent conservative trap potentials and the possibility of manipulating these potentials independently. The choice of the internal atomic states $|a\rangle$ and $|b\rangle$ has to be such that they are elastic (i.e. the internal states do not change after the collision). To achieve entanglement operations with high fidelity, one has to be able to load or cool the atoms to the ground states of the trapping potentials. Finally, for the application of parallel quantum computing one needs periodic structures (e.g. optical lattices), together with the ability to control the positions of the atoms and to fill the lattice sites selectively.

A. Two-qubit gates in optical lattices

In this Section we want to discuss how a number of difficulties can be overcome that one encounters when trying to use optical lattices for quantum computing. We will first show how one can achieve a filling factor of 1 with particles in the ground states (lowest band) of the lattice. This can be achieved by using an ultracold very dense sample of weakly interacting atoms, namely a Bose-Einstein condensate, and slowly turning on an optical potential. The repulsive interaction between the particles increases as the optical potential is made deeper. At the same time the hopping rate at which particles move from one site to the next decreases. If the optical lattice is turned on on a time scale much slower than the hopping rate and the temperature kT can be kept much smaller than the interaction energy between two particles in one site, one can achieve a filling of the optical lattice with exactly one particle per lattice site. [75] Finally, we note that a filling factor of one out of two lattice sites has been achieved in very recent optical lattice experiments. [65]

We will also discuss how the lattice potentials can be moved in a state-selective way for implementing the two-qubit gate [31]. For alkali atoms with a nuclear spin equal to 3/2 we show how atoms in different hyperfine levels can be moved into different directions. It is clear that other difficulties like e.g. addressing single qubits exist, but they will not be discussed here since their experimental solution is not specific to the present implementation.

1. Hamiltonian for a Bose-Einstein condensate in an optical lattice

We assume a Bose-Einstein condensate of atoms in internal state $|a\rangle$ to be loaded into an optical lattice potential $V_T(\mathbf{x}) + V_0(\mathbf{x})$, where

$$V_0(\mathbf{x}) = V_{x0}\sin^2(kx) + V_{y0}\sin^2(ky) + V_{z0}\sin^2(kz) \quad (25)$$

is a periodic optical lattice potential and $V_T(\mathbf{x})$ is a superlattice potential slowly varying in space compared to $V_0(\mathbf{x})$. k is the wave number of the lasers producing the lattice potential. The Hamiltonian reads [75]

$$H = \int d^3x\psi^\dagger(\mathbf{x})\left(-\frac{\hbar^2}{2m}\nabla^2 + V_0(\mathbf{x}) + V_T(\mathbf{x}) - \mu\right)\psi(\mathbf{x}) + \frac{1}{2}\frac{4\pi a_s^{aa}\hbar^2}{m}\int d^3x\psi^\dagger(\mathbf{x})\psi^\dagger(\mathbf{x})\psi(\mathbf{x})\psi(\mathbf{x}), \quad (26)$$

where $\psi(\mathbf{x})$ is the bosonic field operator and μ is the chemical potential i.e. a Lagrangian multiplier to fix the number of particles. Expanding the field operators in the Wannier basis while keeping only the dominant terms [75] Eq. (26) reduces to the Bose-Hubbard Hamiltonian

$$H = -J\sum_{<i,j>} b_i^\dagger b_j + \sum_i (\epsilon_i - \mu)\,\hat{n}_i + \frac{1}{2}U\sum_i \hat{n}_i(\hat{n}_i - 1), \quad (27)$$

where the operators $\hat{n}_i = b_i^\dagger b_i$ count the number of bosonic atoms at lattice site i; the annihilation and creation operators b_i and $b_i^\dagger$ obey the canonical commutation relations $[b_i, b_j^\dagger] = \delta_{ij}$. J is the tunneling matrix element and U describes the (repulsive) interaction between particles at the same lattice site. $\epsilon_i = V_T(x_i)$ is the value of the slowly varying superlattice potential at site i. The ratio U/J is controlled by the depth of the optical lattice potential V_{j0}. Increasing V_{j0} (via the intensity of the trapping lasers) reduces the tunneling matrix element J and increases the repulsive interaction between the atoms U [75].

2. Loading the lattice

In order to perform gate operations in optical lattices we have to be able to selectively fill the lattice sites with exactly one particle. This can be achieved by making use of the phase transition from a superfluid BEC phase to a Mott insulator (MI) phase at low temperatures, which can be induced by increasing the ratio of the onsite interaction U to the tunneling matrix element J predicted by the Bose-Hubbard model [76,77]. In the MI phase the density ρ_i (occupation number per site) is pinned at integer $n = 0, 1, 2, \ldots$ corresponding to a commensurate filling of the lattice, and thus represents an *optical crystal* with diagonal long range order with period imposed by the laser light. Particle number fluctuations are thereby drastically reduced and thus the number of particles per lattice site is fixed. The number of particles per lattice site depends on the chemical potential μ in the isotropic case $\epsilon_i = 0$ [76]. In the non-isotropic case we may view $\mu - \epsilon_i$ as a local chemical potential. Therefore ρ_i can be controlled by the superlattice potential $V_T(\mathbf{x})$.

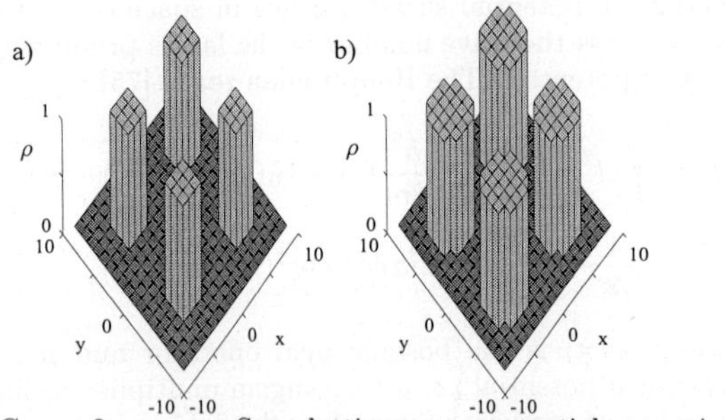

FIG. 3. Superlattice potential in 2D with $V_T(x,y) = 40J\ (\sin^2(\pi x/9a) + \sin^2(\pi y/9a))$ with a the spacing between the lattice sites. The particle density $\rho(x,y)$ for four superlattice wells is shown. Parameters: a) $U = 30J$ and $\mu = 15J$, b) $U = 50J$ and $\mu = 27J$.

Using a Gutzwiller ansatz [75,78,79] for the wave function we have performed a mean field calculation to demonstrate how, by a proper choice of the potential $V_T(\mathbf{x})$, one can fill certain blocks of the optical lattice with exactly one particle at temperature $T = 0$. Figure 3 shows the result of this mean field calculation, a MI phase where the lattice sites are either filled with 0 or 1 particles. The number fluctuations are almost equal to zero and thus not shown in this plot. To achieve a MI phase at finite temperature $T \neq 0$ one has to fulfill the requirement $kT \ll U$ where the interaction strength U gives the order of magnitude of the first excitation energy in a MI phase. One also has to ensure that particles do not move from a filled site with energy ϵ_i to an adjacent empty site with energy ϵ_j i.e. the temperature has to be much smaller than the energy difference between these two sites $kT \ll \epsilon_j - \epsilon_i$. In Sec. IV, we will need periodic fillings of optical lattices as shown in Fig. 3 to implement efficient multi-particle entanglement operations and for parallel quantum computing.

3. Moving the lattice potentials state selectively

We consider the example of alkali atoms with a nuclear spin equal to 3/2 (^{87}Rb, ^{23}Na) trapped by standing waves in three dimensions and thus confined by a potential of the shape as given in Eq. (25). The internal states of interest are hyperfine levels corresponding to the ground state $S_{1/2}$ as shown in Fig. 5b. Along the z axis, the standing waves are in the lin∠lin configuration (two linearly polarized counter-propagating traveling waves with the electric fields $\vec{E}_1$ and $\vec{E}_2$ forming an angle 2θ [80]) as shown in Fig. 4.

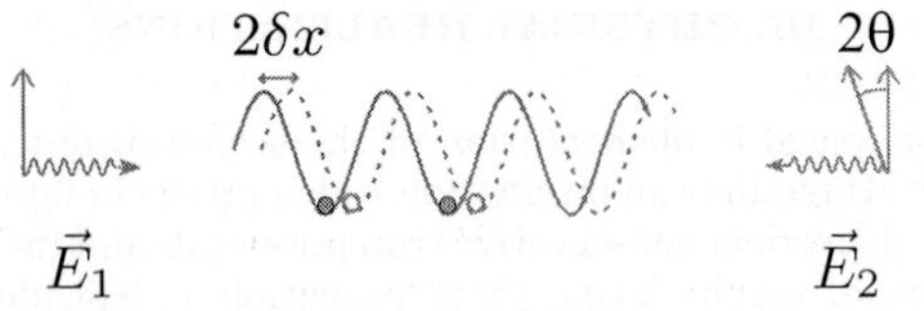

FIG. 4. Laser configuration along the z-axis.

The total electric field is a superposition of right and left circularly polarized standing waves ($\sigma^\pm$) which can be shifted with respect to each other by changing θ,

$$\vec{E}^+(z,t) = E_0 e^{-i\nu t}\left[\vec{\epsilon}_+ \sin(kz+\theta) + \vec{\epsilon}_- \sin(kz-\theta)\right], \quad (28)$$

where $\vec{\epsilon}_\pm$ denote unit right and left circular polarization vectors, $k = \nu/c$ is the laser wave vector and E_0 the amplitude. The lasers are tuned between the $P_{1/2}$ and $P_{3/2}$ levels so that the dynamical polarizabilities $\alpha_{\pm\mp}$ of the two fine structure $S_{1/2}$ states corresponding to $m_s = \pm 1/2$ due to the laser polarization $\sigma^\mp$ vanish ($\alpha_{+-} = \alpha_{-+} = 0$), whereas the polarizabilities $\alpha_{\pm\pm}$ due to $\sigma^\pm$ are identical ($\alpha_{++} = \alpha_{--} \equiv \alpha$). This configuration is shown in Fig. 5a and can be achieved by tuning the lasers between the $P_{3/2}$ and $P_{1/2}$ fine state levels so that the ac-Stark shifts of these two levels cancel each other. The optical potentials for these two states are $V_{m_s=\pm 1/2}(z,\theta) = \alpha|E_0|^2 \sin^2(kz \pm \theta)$.

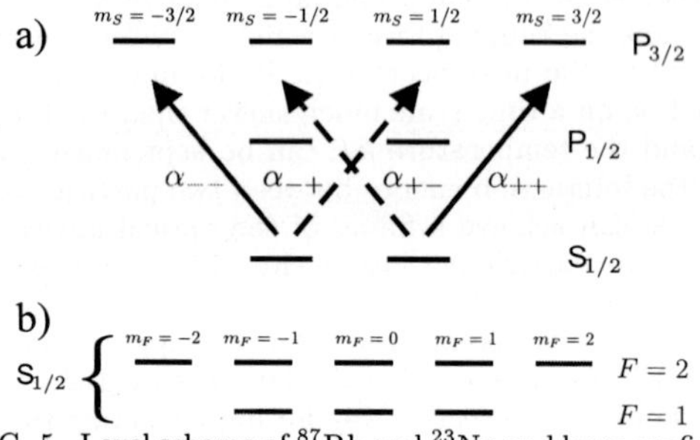

FIG. 5. Level scheme of ^{87}Rb and ^{23}Na and laser configuration. (a) Fine structure energy levels and laser configuration. The detuning is chosen such that the polarizabilities α_{+-} and α_{-+} vanish. (b) Hyperfine level structure.

We choose for the states $|a\rangle$ and $|b\rangle$ the hyperfine structure states $|a\rangle \equiv |F=1, m_f=1\rangle$ and $|b\rangle \equiv |F=2, m_f=2\rangle$. Due to angular momentum conservation, these states are stable under collisions (for the dominant central electronic interaction [81,82]). The potentials "seen" by the atoms in these internal states are

$$V^a(z,\theta) = \left[V_{m_s=1/2}(z,\theta) + 3V_{m_s=-1/2}(z,\theta)\right]/4 \quad (29a)$$

$$V^b(z,\theta) = V_{m_s=1/2}(z,\theta). \quad (29b)$$

If one stores atoms in these potentials and they are deep enough, there is no tunneling to neighboring wells and we can approximate them by harmonic potentials. By varying the angle θ from $\pi/2$ to 0, the potentials V^b and V^a move in opposite directions until they completely overlap. Then, going back to $\theta = \pi/2$ the potentials return to their original positions. The shape of the potential V^a changes as it moves. By choosing $\theta(t) = \pi\left(1 - \left(1+\exp(-(\tau_i/\tau_r)^2)\right) / \left(1+\exp((t^2-\tau_i^2)/\tau_r^2)\right)\right)/2$ with $\tau_r = 25/\omega$ and $\tau_i = 25/\omega$, the frequencies and displacements of the harmonic potentials approximating (29) are exactly those plotted in Fig. 6a.

4. Gate fidelity

We use the minimum fidelity F [83] to characterize the quality of the gate. F is defined as

$$F = \min_{\varphi} \langle\tilde{\varphi}|\mathrm{tr}_{\mathrm{ext}}\left(\mathcal{U}|\varphi\rangle\langle\varphi| \otimes \rho_{\mathrm{ext}}\mathcal{U}^\dagger\right)|\tilde{\varphi}\rangle, \tag{30}$$

where $|\varphi\rangle$ is an arbitrary internal state of both atoms, $|\tilde{\varphi}\rangle$ is the state resulting from $|\varphi\rangle$ using the mapping (12). The trace is taken over motional states, $\mathcal{U}$ is the evolution operator for the internal states coupled to the external motion (including the collision), and ρ_{ext} is the density operator corresponding to both atoms being at a temperature T at time $t = -\tau$ [31]. In Fig. 6b the fidelity F is plotted as a function of the temperature T for the displacements and trap frequencies shown in Fig. 6a. This figure shows that one can achieve very high fidelities in realistic situations.

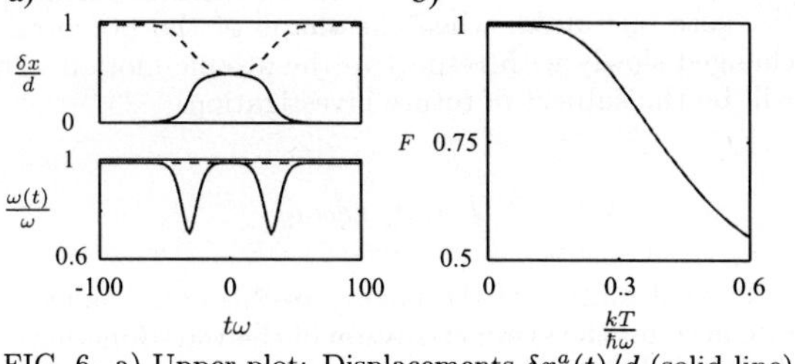

FIG. 6. a) Upper plot: Displacements $\delta x^a(t)/d$ (solid line) and $1 + \delta x^b(t)/d$ (dashed line). Lower plot: Trap frequencies $\omega^a(t)/\omega$ (solid line) and $\omega^b(t)/\omega$ (dashed line). b) Fidelity F against temperature $kT/\hbar\omega$ for ^{87}Rb with $a_s = 5.1$nm. Here $\omega = 2\pi \times 100$kHz and $d = 390$nm.

B. Two-qubit gates in magnetic microtraps

We now consider the implementation of a switching potential by means of electromagnetic trapping forces. We first discuss the possibility of obtaining the desired state dependence by assuming some improvements on devices which are now experimentally available [84–86]. Then we compute the performance of a quantum gate for realistic trapping parameters.

1. Microscopic electromagnetic trapping potential

The interaction between the magnetic dipole moment of an atom in some hyperfine state $|F, m_F\rangle$ and an external static magnetic field $\vec{B}$ entails an energy $U_{\mathrm{magn}} \approx g_F \mu_B m_F |\vec{B}|$ depending on the atomic internal state via the quantum number m_F (here μ_B is the Bohr magneton and g_F is the Landé factor). On the other hand, the Stark shift induced on an atom by an electric field $\vec{E}$ gives a state-independent energy $U_{\mathrm{el}} \approx \frac{1}{2}\alpha|\vec{E}|^2$, where α is the atomic polarizability. The interplay between these two effects can be exploited in order to obtain a trapping potential whose shape depends on the atomic internal state. As an example, we consider an atomic mirror with an external magnetic field [84–86], providing confinement along two directions with trapping frequencies which can range from a few tens of kHz up to some MHz. Microscopic electrodes can be plugged on the mirror's surface [87], thus allowing for the design of a potential with the characteristics described in Sect. II C.

2. Loading and moving atoms within the trap

Several schemes of loading atoms into the trap have been envisaged (see for example [84–86]). Most of them rely on an intermediate stage where atoms can be trapped and cooled without coming in contact with the magnetic mirror. This pre-loading trap can be either initially displaced from the surface, or close to it but based on a different trapping mechanism (for instance an evanescent wave mirror, where different internal states can be trapped by gravity [88] before the atoms can be put in the right states for magnetic trapping), to be replaced by the electromagnetic microtrap with a gradual switch-on of the electric and bias magnetic fields in the final stage of loading [87]. This could also allow for implementing a controlled filling of the trap sites, in a similar way to that already discussed in Sec. III A. A further feature to be implemented in view of performing more complex algorithms is the arrangement of several gate potentials in a periodic pattern, and the possibility of transporting atoms within this structure. An example would be given by two adjacent rows of potential minima, shiftable with respect to each other, where atoms could be loaded. A system like the one suggested in Sect. III B 1 could allow in principle to obtain such a configuration, since the magnetic field minima can be shifted parallel to the surface by rotating the bias magnetic field. In this way it should be possible to move some atoms, while holding others in place by means of additional local electric fields [84–86]. Provided that atoms can be addressed individually, which is needed even for performing a one-bit quantum gate, a procedure for implementing a simple quantum algorithm could be the following: perform a gate between two suitably chosen atoms, being close to each other but belonging to different rows, then mutually displace the rows and select another pair of atoms, including one of those coming out from the previous gate. Repeat until the algorithm has been operated, applying the required one-qubit rotations in between the above steps and possibly performing some of them in parallel.

3. Switching the trap potentials state selectively

We choose for the states $|a\rangle$ and $|b\rangle$ the same hyperfine structure states of ^{87}Rb considered in the previous Section, which are low-magnetic field seekers. If both particles are in state $|a\rangle$, there is no interaction-induced phase shift, as already discussed in Sec. II C 2. The results for both particles in state $|b\rangle$ are shown in Fig. 7.

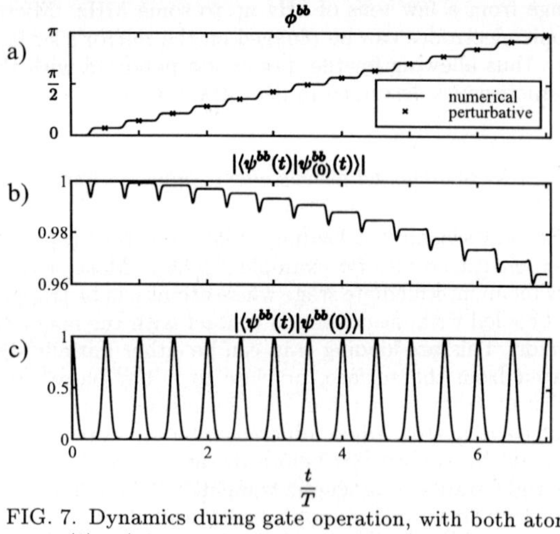

FIG. 7. Dynamics during gate operation, with both atoms in state $|b\rangle$: a) interaction-induced phase shift - the circles refer to the perturbative calculation (22); b) projection of the evolved state on the corresponding state evolved without interaction; c) projection of the evolved state on the initial one. We choose $\omega \approx 23.4$ kHz and $\omega_y = \omega_z = 150$ kHz, corresponding to ground-state widths $a_x \approx 50$ nm, $a_y = a_z \approx 28$ nm, with the initial wells having frequency $\omega_0 = 2\omega$ and displaced by $x_0 = 3\sqrt{2}a_x$. We take for the scattering length the known value for ^{87}Rb, *i.e.* $a_s^{bb} \approx a_s^{ab} = 5.1$ nm. Time is in units of the oscillation period T.

The time dependence of ϕ^{bb} is step-like (Fig. 7a): the collisional phase is incremented at times $t_k \equiv (2k-1)T/4$, when the atoms meet at the center of the well, and remains constant at intermediate times, when they separate again. The influence of the interaction on the atomic motion can be seen from Fig. 7b, depicting the overlap between the evolved interacting two-atom state $|\psi^{bb}(t)\rangle$ and the corresponding state $|\psi^{bb}_{(0)}(t)\rangle$ computed without taking into account the interaction. The curve has local minima at times t_k, signalling that a collision is taking place, and shows a global decrease corresponding to a slight delay of the interacting motion with respect to the non-interacting one. As it can be seen from Fig. 7c, this effect is not dramatic: the oscillation period in the presence of interaction is increased just by $\delta T \approx 2 \times 10^{-3}T$ (with the parameters used here), and the harmonic potential ensures that the system comes periodically back to its initial state. After 7 oscillations we get a phase shift due to the interaction of π, whereas the perturbative formula (22) gives $7\phi_T^{bb} \approx 0.98\pi$. Therefore we choose $\tau = 7(T + \delta T) \approx 0.15$ms: the overlap between the initial and the evolved wave function at that time is $|\langle\psi^{bb}(\tau)|\psi^{bb}(0)\rangle|^2 \approx 0.996$. The behavior turns out to be quite different [89] when the atoms are in different internal states: the phase shift increases more rapidly, but after a few oscillations the system does no longer come back to the initial state. This has a simple explanation. The two atoms collide as soon as the one being in state $|b\rangle$, moving within the potential $m\omega^2 x^2/2$, reaches its turning point, where the other atom is trapped. The interaction time is therefore longer than if both atoms were in state $|b\rangle$. Indeed, in that case they meet at the trap center, with their maximal velocity. This explains why the system picks up a bigger phase shift per oscillation period in the present case. On the other hand, the collision excites the motion of the atom in state $|a\rangle$ within its own well, and therefore the initial state is no longer recovered. This problem can be avoided if the potential minimum for state $|a\rangle$ is displaced along the transverse direction from the one for state $|b\rangle$ by means of an additional electrostatic field [84–86], so that the atoms interact if and only if they are both in state $|b\rangle$. This problem would not exist in an adiabatic scheme for the gate operation, when the shape of the potential is changed slowly with respect to the atomic motion. This will be the subject of future investigation.

4. Gate fidelity

The calculation of the fidelity in this case has to take into account the symmetrization of the wave function under particle interchange, expressed by an operator S to be explicitly inserted into Eq. (30):

$$F = \min_{\varphi} \left\{ \mathrm{tr}_{\mathrm{ext}} \left[\langle\tilde{\varphi}|\mathcal{U}S\left(|\varphi\rangle\langle\varphi| \otimes \rho_{\mathrm{ext}}\right) S^\dagger \mathcal{U}^\dagger |\tilde{\varphi}\rangle \right] \right\}, \quad (31)$$

With the parameters quoted above, we obtain $F > 0.98$. In order to reach such a fidelity, timing has to be quite precise, with a resolution of the order of $10^{-3}T$ corresponding to tens of ns in this case.

IV. PARALLEL QUANTUM COMPUTING

In this Section, we will discuss how quantum gates based on controlled collisions can be exploited for quantum computing. It is clear that, with the realization of a universal two-bit gate, any quantum computation can be performed, just as it is the case with other implementations. On the other hand, manipulations such as moving and switching potentials offer a great deal of parallelism [43,44] not available in other systems.

We will focus our attention on implementations in optical lattices. Some of the ideas could readily be translated into arrays of magnetic microtraps, if the distances between the individual potential wells could be made much shorter than present-day state-of-the-art of nanofabrication. In such a situation, adiabatic variants of the switching operations (see comment at the beginning of Sec. II C) can be used to create multi-particle entangled clusters of neighboring atoms, similar as with moving potentials. Details of this analysis will be presented somewhere else [89].

One may ask, what can be done in optical lattices that cannot be done in other implementations? The answer to this question depends on a number of experimental conditions such as the possibility of creating regular filling structures and, like in ion-traps, on the possibility of addressing single atoms individually. In the following, we will first (Sec. IV A) give an example of what can be done with controlled lattice movements in conventional set-ups i.e. with random filling of the lattice sites and without any control of the position of individual atoms. We will see that this already allows one to perform interesting spectroscopic studies of the degree of entanglement between the atoms thus created. Next (Sec. IV B), we will describe what can be done if one achieves a regular occupation of the lattice sites and can address the atoms individually. Under such circumstances, an efficient implementation of quantum error correction and of a quantum memory (concatenated Shor code) is possible. Furthermore, fault tolerant versions of certain quantum gates and of quantum error correction can be implemented straightforwardly, as will be sketched in (Sec. IV C). Finally, in Sec. IV D, we describe how auxiliary atomic levels can be used to realize highly selective entanglement operations, where individually selected atoms are swept across the lattice to create GHZ states [90] of a large number of particles. Together with IV B and IV C, this scheme has all the ingredients that are necessary for an efficient realization of fault-tolerant quantum computing.

A. Multi-particle entanglement operations

The two-qubit gates described in Sec. III correspond abstractly to an atom interferometer as shown in Fig. 8. The interferometer has two inputs which are the two atoms trapped at neighboring potential wells. By shifting the potentials back and forth as described in Secs. II B, only one combination of paths of the two particles overlaps and leads to a phase shift, namely the paths corresponding to state $|a\rangle_1$ for the left particle and $|b\rangle_2$ for the right particle. To emphasize the role of the internal states as *logical* states, we shall henceforth use the notation $|0\rangle \equiv |a\rangle$ and $|1\rangle \equiv |b\rangle$ and neglect the kinetic phases ϕ^a, ϕ^b as they appear in (12). Furthermore, we drop the atomic index as long as there is no danger of confusion.

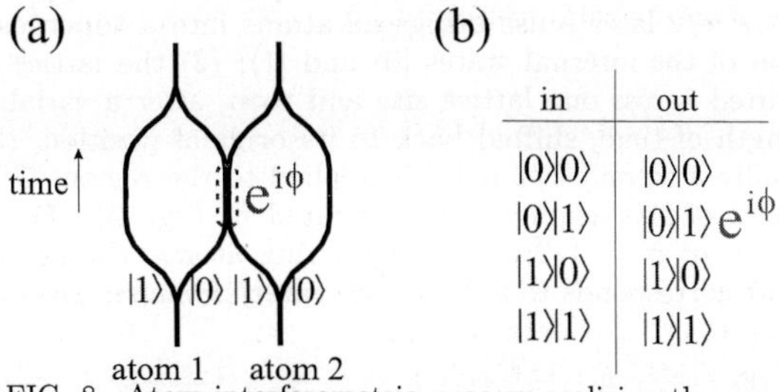

FIG. 8. Atom-interferometric process realizing the quantum gate. (a) Two-particle interferometer; (b) Truth table.

The logical truth table corresponding to the interferometric process is shown in Fig. 8. [A similar identification of logical states can be made in magnetic traps as is pointed out in Sec. II C 4. The labelling of the paths for the left particle in the interferometer has to be interchanged in this case.] For $\phi = \phi^{01} = \pi$ this realizes a phase gate [1]. The phase gate and the set of all one-bit unitary transformations, which can be realized by Raman laser pulses on the internal states $|0\rangle$ and $|1\rangle$, define a *universal set of quantum gates.* [91–94]

An important difference between optical lattices and other implementations is given by the global effect of the lattice manipulations. To illustrate this point, consider first a two dimensional lattice as in Fig. 9 with random occupation of the sites and a filling factor $\eta \ll 1$, where η is defined as the average number of atoms per lattice site. Let us assume that the loading of the lattice can be accomplished in such a way that there are no multiply occupied lattice sites, i.e. that each lattice site is occupied by no more than a single atom.

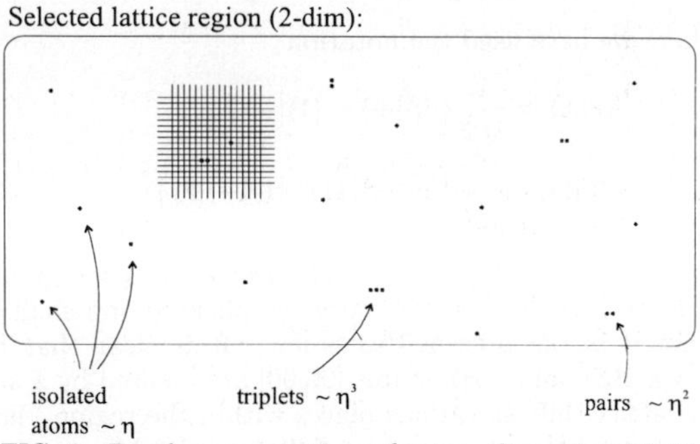

FIG. 9. Random occupation of a two dimensional lattice with single atoms.

Then, in any region of the lattice, one will find isolated atoms, pairs of neighboring atoms, triplets, and so forth, with a relative frequency proportional to η, η^2, η^3, respectively. Consider now the following *Ramsey experiment* [95] where initially all atoms are prepared in the internal state $|0\rangle$ and in the motional ground state of their individual potential wells. In some selected region of the lattice, the following sequence of operations is applied: (1) a $\pi/2$ laser pulse brings all atoms into a superposition of the internal states $|0\rangle$ and $|1\rangle$; (2) the lattice is shifted across one lattice site and then, after a variable length of time, shifted back to its original position, (3) finally a second $\pi/2$ pulse is applied to the region. The effect of this sequence is illustrated in Fig. 10. For a group of $N = 1, 2, 3, \ldots$ neighboring atoms, the lattice shift corresponds to a *N-particle interferometric process.*

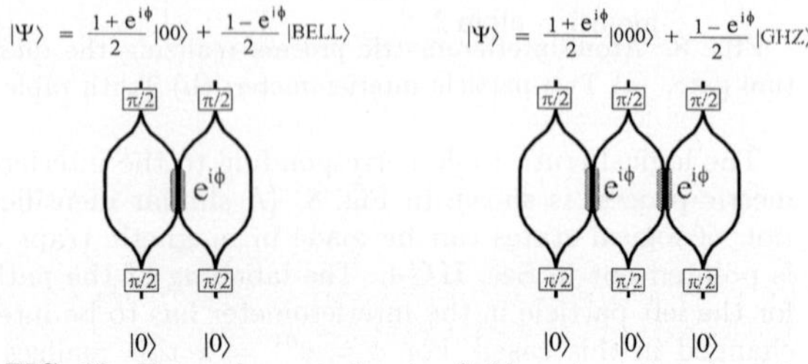

FIG. 10. Entanglement of pairs (left) and triplets (right) of neighboring atoms by a single lattice shift.

Specifically, one obtains the following transformations. For isolated atoms:

$$|0\rangle \longrightarrow |0\rangle; \tag{32}$$

for pairs of neighboring atoms:

$$|00\rangle \longrightarrow \frac{1+e^{i\phi}}{2}|00\rangle + \frac{1-e^{i\phi}}{2}|\text{BELL}\rangle\,; \tag{33}$$

and for triplets of neighboring atoms:

$$|000\rangle \longrightarrow \frac{1+e^{i\phi}}{2}|000\rangle + \frac{1-e^{i\phi}}{2}|\text{GHZ}\rangle\,; \tag{34}$$

where we have used the notation

$$|\text{BELL}\rangle = \frac{1}{\sqrt{2}}\left\{|0\rangle|+\rangle - |1\rangle|-\rangle\right\}\,,$$

$$|\text{GHZ}\rangle = \frac{1}{\sqrt{2}}\left\{|0\rangle|+\rangle|1\rangle - |1\rangle|-\rangle|0\rangle\right\}\,, \tag{35}$$

and $|\pm\rangle = (|0\rangle \pm |1\rangle)/\sqrt{2}$. The expressions for groups of more particles become more complicated and shall be ignored in the present discussion. It is clear that for $\phi = \pi$ Bell- and GHZ states [96,90] are created by a *single* lattice shift at various places within the region. This corresponds to an ensemble of 2-bit and 3-bit quantum gates, respectively, acting simultaneously at different lattice sites. To analyze the states (33) and (34) spectroscopically one could measure the state of the atoms in a final step of the above Ramsey sequence e.g. by a fluorescence measurement. It is clear that by such a measurement the entangled states will be destroyed. On the other hand, by repeating this sequence many times with different samples, one can measure the fluorescence signal as function of the phase ϕ (interaction time). Under ideal circumstances, all isolated atoms will remain in the dark state $|0\rangle$ while all fluorescence signals come from Bell ($\sim \eta^2$) or GHZ ($\sim \eta^3$) states [97]. To check that entangled states, rather than mixtures, are created in the process, the experiment is performed with different interaction times, e.g. times corresponding to $\phi = \pi$ and $\phi = 2\pi$. For entangled states as in (33) and (34) all fluorescence signals will vanish at $\phi = 2\pi$, while this will not be the case if the states created by the atomic collisions are mixtures of classical many-particle states. More generally, by measuring the ***visibility*** of the fluorescence signal one may study the *fidelity of the entanglement* created in the process, and its dependence on certain noise sources such as a finite temperature of the atoms. This way, the curve plotted in Fig. 6b) could be tested experimentally.

B. Quantum error correction

To employ these entanglement operations for quantum computing, one has to have precise control over the number and the location of atoms that are involved in the collisional process. In addition to the ability of addressing single atoms, one therefore has to achieve a certain *ordered occupation of the lattice sites.* As described in Sec. III.A., Fig. 3, this can be by achieved by controlling the intensity of the trapping laser at sufficiently low temperatures. This way optical crystals with periodic patterns of atoms can be created as indicated in Figs. 11 and 15 [98]

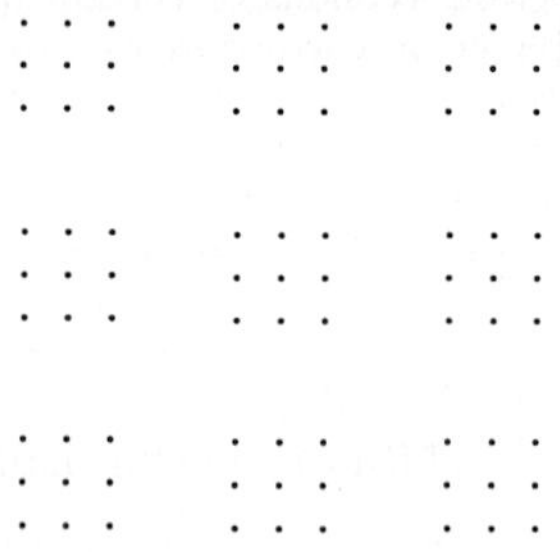

FIG. 11. Ordered arrangement of atoms in an optical lattice (see also Fig. 3).

Under such circumstances the parallelism of the lattice manipulations can be exploited advantageously. On one side, similar logical operations can be performed simultaneously at different locations on the lattice. On the other side, as we have seen in Fig. 10, a single lattice shift can entangle whole groups of atoms. Two types of such entanglement operations are shown in Fig. 12. One involves only the logical states $|0\rangle$ and $|1\rangle$, while the second uses a third atomic level as a "transport state" (see Sec. IV D), into which any atom must first be activated, before it can participate in an entanglement operation. In the following, we will first discuss applications of the shift operation as in Fig. 12(a). Later, in Sec. IV D we will consider a more flexible ("sweep") operation shown in Fig. 12(b).

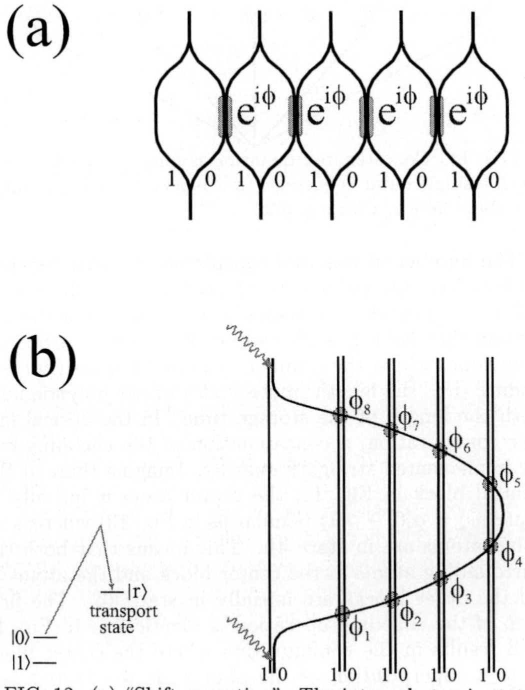

FIG. 12. (a) "Shift operation": The internal atomic states $|0\rangle$ and $|1\rangle$ couple to different lattice potentials that are moved against each other as explained in Sec. III A 3. This corresponds to a multi-particle interferometer where the same phase shift is acquired whenever two paths temporary overlap. By simple lattice manipulations, therefore, entire groups of atoms become entangled. (b) "Sweep operation": For more selective entanglement operations, a third atomic level $|r\rangle$ is used [99]. In this scheme, only atoms in the level $|r\rangle$ are moved, whereas the states $|0\rangle$ and $|1\rangle$ are kept in the same potential. At the beginning of an entanglement operation, the atoms are first excited from one of the states $|0\rangle$ or $|1\rangle$ to the state $|r\rangle$ before the lattice is moved. This scheme is much more selective in the sense that those atoms which shall participate in a gate operation are first activated, before they couple to the moving lattice, and the collisional phases ϕ_j can be varied for each interaction individually.

An application of shift operations as described in Fig. 12(a) concerns the realization of a *quantum memory*, where a qubit $\alpha|0\rangle + \beta|1\rangle$ (with unknown coefficients α and β) is encoded in the quantum state of a larger block of atoms and stabilized against decoherence with the help of quantum error correction [47–54]. A particular quantum code that is able to protect a qubit against general 1-bit errors (spin flip and phase flip) has been proposed by Shor [47]. It is a 9-bit code where the codewords

$$\begin{aligned}|0_S\rangle &= 2^{-3/2}(|000\rangle + |111\rangle)(|000\rangle + |111\rangle)(|000\rangle + |111\rangle)\\ |1_S\rangle &= 2^{-3/2}(|000\rangle - |111\rangle)(|000\rangle - |111\rangle)(|000\rangle - |111\rangle)\end{aligned} \quad (36)$$

consist of products of certain GHZ states. Abstractly speaking, the encoding operation consists of a mapping (embedding) of the qubit's 2-dimensional Hilbert space $\mathcal{H}$ into a 2^9-dimensional Hilbert space of the form

$$\mathcal{H} \ni \alpha|0\rangle + \beta|1\rangle \mapsto \alpha|0_S\rangle + \beta|1_S\rangle \in \mathcal{H}_E \subset \mathcal{H}^{\otimes 9}. \quad (37)$$

To stabilize the encoded information against decoherence, the code must be measured and corrected on a time scale $\tau \ll 1/9\gamma$ where γ is the rate of decoherence for a single qubit. This is possible since all errors that may occur on any one of the qubits of the codewords (36) map the code into a family of 2-dimensional subspaces of $\mathcal{H}^{\otimes 9}$ which are all orthogonal on $\mathcal{H}_E$ [47].

The Shor code (36) can be implemented efficiently in a two dimensional lattice configuration [100] as in Fig. 11, by using the shift operation of Fig. 12(a). To see this, imagine that the qubit/atom whose state is to be encoded is surrounded by neighboring atoms as in Fig. 13. The idea is of course to encode the central qubit in the whole block of 3×3 qubits. Initially the central atom is in the unknown state $|\psi\rangle = \alpha|0\rangle + \beta|1\rangle$ while all neighboring atoms are in state $|0\rangle$. As is shown in Appendix B, the initial state is transformed into the Shor code by a simple sequence of horizontal and vertical lattice shifts combined with certain 1-bit rotations, as indicated in Fig. 13(a). By this process, the information contained in ψ is so to speak de-localized over the whole block of 9 atoms. To check whether an error has occurred on one of the qubits, the block is first decoded by the inverse transformation [50], which involves the same sequence of lattice shifts as the encoding. Subsequently, one measures which of the neighboring atoms are in the state $|1\rangle$.

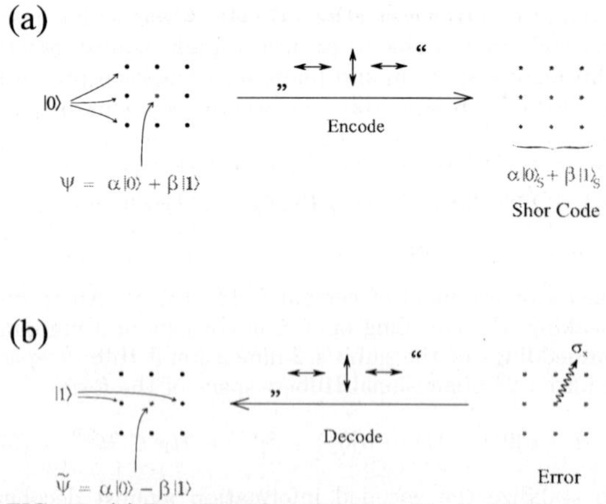

FIG. 13. (a) Encoding of a qubit into a block of 3×3 atoms; (b) Decoding and syndrome measurement.

In the language of quantum error correction, the surrounding atoms of the central qubit in Fig. 13(b) are the carriers of the *error syndrome* [50], meaning that their state gives information regarding what type of error occurred and, more importantly, which unitary 1-bit rotation has to be applied to the central qubit to restore it to the original state. In a fluorescence measurement, this information corresponds to a specific pattern of bright and dark atoms surrounding the central qubit. For example, in Fig. 13(b) a spin flip has occurred in the central qubit. This means that the state of the *block* is transformed into $\alpha\sigma_x^{\rm ctr}|0_{\rm S}\rangle + \beta\sigma_x^{\rm ctr}|1_{\rm S}\rangle$ where $\sigma_x^{\rm ctr}$ is the corresponding Pauli spin operator associated with the central atom. By the decoding operation of Fig. 13(b), this state is mapped into a product state of 9 qubits in which the neighboring atoms in the central row are both in the (bright) state $|1\rangle$, while the other surrounding atoms are in the (dark) state $|0\rangle$. The central atom is in a state equivalent to $|\psi\rangle$ up to a unitary transformation. Similar fluorescence patterns are obtained if a phase error occurs in the central atom or an arbitrary 1-bit error on any of the atoms in the block. A complete table of the error syndrome is given in the Appendix.

The essential point is that the measurement on the surrounding atoms does not reveal nor destroy the state of the central atoms (the coefficients α and β remain unknown throughout the process). If the sequence of operations "decode-correct-encode" is repeated sufficiently often within the decoherence time $1/9\gamma$ of the block, the state ψ may be protected over arbitrarily long times, in principle. Here one assumes, of course, that the decoding and encoding operations themselves are *free of errors*. In our situation this means that all phases acquired in the atomic collisions can be perfectly controlled. Since these operations will always bear some imperfection/imprecision, the probability that an error is introduced by an imperfect operation increases/accumulates with repeated applications of these operations. The general solution to this quantum-memory problem was given by Knill and Laflamme [57] and by others [58–60], and requires a *concatenation of encoding operations* as shown schematically in Fig. 14.

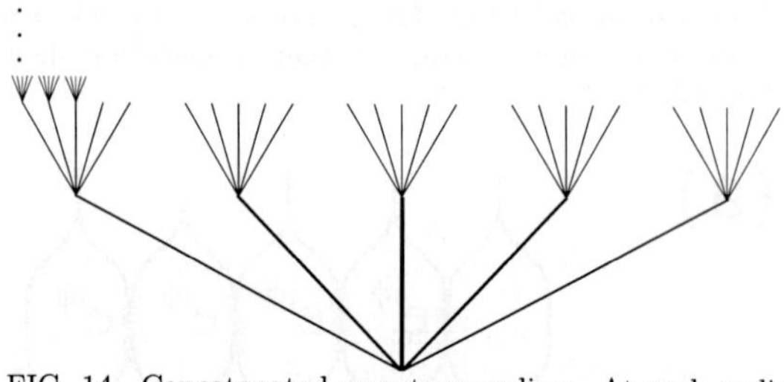

FIG. 14. Concatenated quantum coding. At each coding level, a single qubit is encoded in a block of a larger number of qubits, here 5. (See e.g. [62]).

The number of required concatenation steps depends on how long the qubit is to be stored. It can be shown [57] that, given the precision of the operations is above a certain threshold, a qubit can be stored for an arbitrary long time, where the number of qubits required for encoding (i.e. the length of the code) grows polynomially with the length of the storage time. In the optical lattice configuration, a concatenation of the encoding can be implemented straightforwardly. Imagine that, in the central block in Fig. 11, the center atom is initially in state $|\psi\rangle = \alpha|0\rangle + \beta|1\rangle$ (similar as in Fig. 13) whereas all other atoms are in state $|0\rangle$. This means that both the surrounding atoms in the center block and the atoms of all the other blocks are initially in state $|0\rangle$. The first step of the encoding operation is identical as in Fig. 13 and results in the configuration where the center block is in a superposition of the Shor code words $|0_{\rm S}\rangle$ and $|1_{\rm S}\rangle$, whereas the surrounding blocks remain in state $|0\rangle$. In the second step, the same operation is repeated on a larger scale, i.e. the lattice is shifted across a larger distance such as to make the *blocks* temporarily overlap while the 1-bit operations of the first step are now repeated on corresponding atoms of the outer blocks. As a result, the information $|\psi\rangle$ originally carried by the center atom is now delocalized over $9 \times 9 = 81$ atoms! This scheme may be iterated as indicated in Fig. 15.

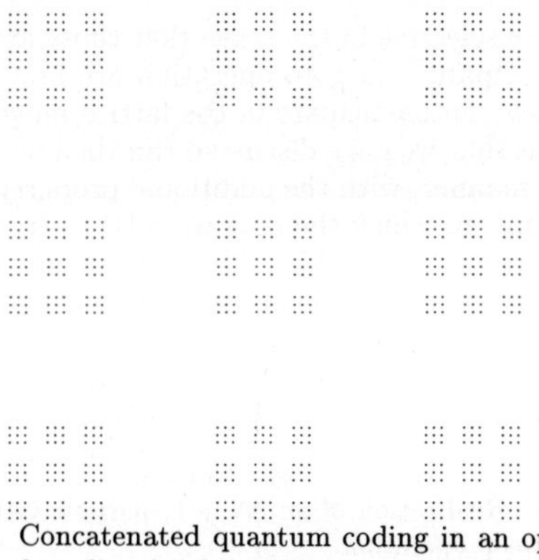

FIG. 15. Concatenated quantum coding in an optical lattice. At each coding level, two-dimensional lattice displacements with an increasing periodicity are applied. The nested character of Fig. 14 is here reflected by a self-similar filling pattern of the lattice.

When in the second (and higher-order) encoding step the blocks are brought to overlap, one has to make sure that only phases between corresponding atoms of the different blocks are accumulated. The most elegant way to achieve this would be with the aid of a technique where the 0 and 1 states are displaced vertically before the atoms are moved. This could be implemented in a three-dimensional lattice configuration [101]. The shift operation is then really a "lift & shift" operation. The collisional interaction is then only switched on by varying the vertical displacement, after the blocks have been moved horizontally. If such a lifting technique can not be implemented, e.g. in a truly two-dimensional configuration, then during the horizontal motion there will be also collisions between non-corresponding atoms, for example the atoms in the right column of one of the blocks with atoms in the left column of a neighboring block. To avoid these unwanted phase shifts, it is possible to vary the velocity of the lattice movement in such a way that during unwanted collisions a phase of $e^{2\pi i}$ is acquired. This method is clearly more susceptible to decoherence. On the other hand, our numerical studies have shown [31], that by an appropriate choice of the displacement function $\theta(t)$ in Fig. 4, the phase of a single collision can, in principle, be controlled with a very high precision (with fidelity ≥ 0.9997) and the probability for exciting phonons remains correspondingly small [102].

It does not seem impossible that $\theta(t)$ could be controlled precisely enough to meet the threshold of fault-tolerant computation [62], but we have not yet made detailed numerical investigations for this situation. In summary, the method of *concatenated coding* can be implemented in optical lattices by repeated sequences of lattice displacements on *self-similar filling structures.*

C. Fault-tolerant computing

In a quantum computer, we do not only wish to store quantum information, but also to process it in a quantum algorithm. To prevent an accumulation of errors during the calculation due to imperfect gate operations, one needs to use fault-tolerant quantum gates that act on the encoded information. Furthermore, errors should be corrected fault tolerantly, that is, without decoding the information (and therefore exposing the qubit to decoherence). The general theory of fault-tolerant computation has been developed by several researchers [62]. In optical lattices, many of such fault-tolerant operations have a geometrically intuitive implementation. For example, if two qubits are encoded in blocks of 9 atoms each, as in Fig. 13, a controlled-NOT operation can be implemented by moving one block on top of the other so that each pair of corresponding atoms from the two blocks share a single potential well and acquire a phase shift $e^{i\pi}$. [This is a straightforward generalization of the situation in Fig. 8].

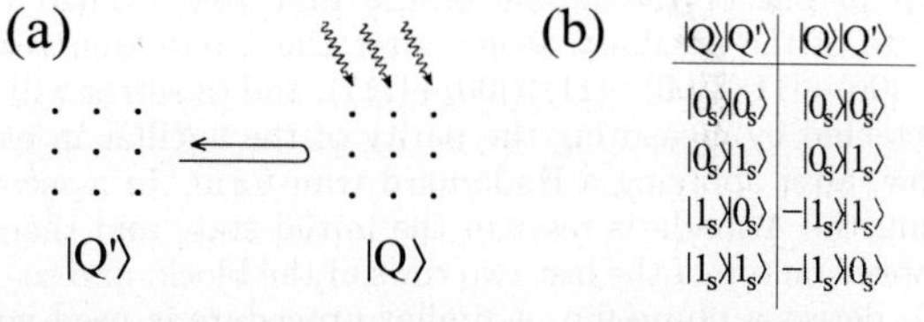

FIG. 16. Implementation of a fault-tolerant CNOT gate.

When a $\pi/2$ pulse is applied on one of the blocks before and after the blocks are shifted, a fault-tolerant realization of the CNOT gate, with a truth table as in Fig. 16 is realized. The minus sign may be eliminated by applying a $3\pi/2$ pulse instead of the second $\pi/2$ pulse. Similarly, one can find a simple fault-tolerant realization of the NOT gate, while for example the Hadamard transform is more involved and requires a measurement with auxiliary qubits. Whether or not one can find similarly *efficient* implementations for a complete set of fault tolerant gates, is still under investigation.

To check whether an error has occurred during a gate operation, one has to measure whether the blocks are still in a superposition of the correct codewords. For the Shor code, this can be done in the following way [47], see Fig. 17: To detect a spin-flip, one has to measure the parity of the first two atoms in any row and compare it to the parity of the last two atoms of every row. To do this one would use an "Armada" of 3×2 ancillas in the state $(|00\rangle + |11\rangle)(|00\rangle + |11\rangle)(|00\rangle + |11\rangle)$, which approaches the block from the left in Fig. 17 by moving the lattice horizontally.

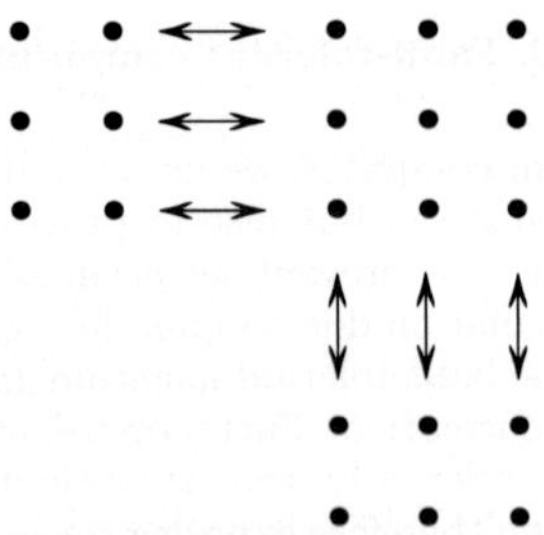

FIG. 17. Implementation of fault-tolerant error correction.

To measure the parities, the Armada is moved on top of the first two columns of the data block so that the atoms interact pairwise with atoms of the data block and acquire a phase shift of $e^{i\pi}$. To satisfy the criteria for fault tolerance, we need to avoid collisions while the ancillas are moved on top of the code, and thus need a "lift & shift" implementation of the operation, as mentioned earlier. Suppose there was a spin-flip in one of the atoms of the first row. Then the state of the ancillary atoms after the interaction reads $(-|00\rangle+|11\rangle)(|00\rangle+|11\rangle)(|00\rangle+|11\rangle)$, and the error will be detected by measuring the parity of the ancillas in each row, after applying a Hadamard transform. In a second run, the Armada is reset in the initial state and then is moved on top of the last two rows of the block, and so on. To detect a phase-flip, a similar procedure is used with an Armada of 2×3 atoms that approaches the block in Fig. 17 from *below* by moving the lattice vertically. Since these ancillas should measure any change of sign in any of the GHZ states that make up the codewords (36), they have to be prepared in the state $|000000\rangle + |111111\rangle$. A phase flip can then be detected as previously, where now a Hadamard transform has to be applied to the block first, before the "attack" starts from below. In the specific implementation using optical lattices, one could also think about other schemes using only a single row of ancilla atoms on each side of the data block in Fig. 17 as realized in Fig. 3b).

D. Selectivity and "sweep operations"

The examples discussed so far make use of the parallelism of the lattice shift to implement certain multi-particle entanglement (or gate) operations efficiently. On the other hand, the shift operation as described in Fig. 12(a) is too rigid, when certain operations should apply to a selected group of atoms only. This problem can in principle be solved by using a third atomic level $|r\rangle$ as indicated in Fig. 12(b). In this scheme, the level $|r\rangle$ couples dominantly to a transport lattice [99], while the "logical states" $|0\rangle$ and $|1\rangle$ are kept in the same potential. At the beginning of an entanglement operation, the atoms are first excited from one of the states $|0\rangle$ or $|1\rangle$ to the state $|r\rangle$, before the lattice is moved. This scheme is much more selective in the sense that those atoms which shall participate in a gate operation are first activated, before they can participate in the lattice movement. All operations that we have discussed can then be realized in the same manner, with the additional property that only those atoms, to which the operation $|1\rangle \to |\mathrm{r}\rangle$ is applied, will participate. With this additional feature, it is clear, that *universal computations* can be implemented.

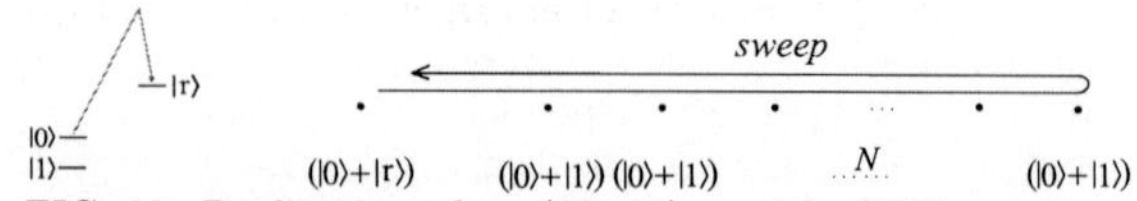

FIG. 18. Realization of an $(N+1)$-particle GHZ state by a single sweep operation.

Another merit of this scheme is that one can realize more flexible entanglement operations. Consider, for example, a 1-dimensional situation as in Fig. 18 with a string of N atoms initially prepared in the product state $(|0\rangle + |1\rangle)^{\otimes N}$ and a selected additional atom (left) in the state $(|0\rangle + |\mathrm{r}\rangle)$. By moving the transport lattice, the selected atom is swept across the N lattice sites. During that motion, it interacts with each of the N atoms thereby transforming the state of each atom into $e^{i\phi^0}|0\rangle + e^{i\phi^1}|1\rangle$, with a differential phase $\phi = \phi^1 - \phi^0$. The resulting total state is of the form

$$\begin{aligned} &|0\rangle(|0\rangle + |1\rangle)(|0\rangle + |1\rangle)\cdots(|0\rangle + |1\rangle) \\ +&|\mathrm{r}\rangle e^{iN\phi^0}(|0\rangle + e^{i\phi}|1\rangle)(|0\rangle + e^{i\phi}|1\rangle)\cdots(|0\rangle + e^{i\phi}|1\rangle). \end{aligned} \quad (38)$$

As long as the collisional phases are different ($\phi \neq 0$) for the two logical states, ϕ can by varied with the speed of this sweep operation. For $\phi = \pi$ one obtains a $N+1$-particle GHZ state (see Fig. 18) which can easily be brought to the standard form

$$|\psi\rangle = \frac{1}{\sqrt{2}}\left(|0\rangle|0\rangle|0\rangle\cdots|0\rangle + |1\rangle|1\rangle|1\rangle\cdots|1\rangle\right). \quad (39)$$

Note that for the creation of this state only a single sweep operation is required!

This scheme can be generalized in several directions. By varying the speed by which the lattice is moved during the sweep operation, the phases can be controlled individually for each atom of the string as indicated in Fig. 12, allowing for more complex entanglement operations. As a final example consider a configuration as in Fig. 19(b), with a "source register" consisting of a string of m atoms in the state $|a\rangle = |a_1\ a_2\ a_3\ \cdots\ a_m\rangle$, $a_j \in \{0,1\}$ and a "target register" of m further atoms in the state $(|0\rangle + |1\rangle)^{\otimes m}$, similar as in Fig. 18.

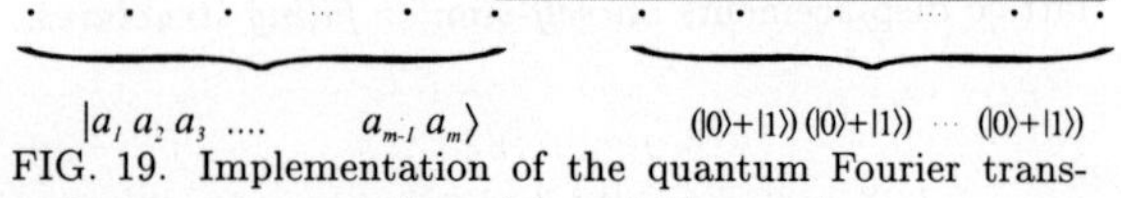

FIG. 19. Implementation of the quantum Fourier transform by a sweep operation with variable speed.

The state vector $|a\rangle = |a_1\ a_2\ a_3\ \cdots\ a_m\rangle$ should be interpreted as a binary representation of the number $a = a_1 2^{m-1} + a_2 2^{m-2} + \ldots + a_m 2^0$. Consider now the following operation where the source register is first activated to couple to the transport lattice, meaning that each of the atoms 1 to m that is in state $|1\rangle$ is excited to state $|r\rangle$. Next, the lattice is moved to the right so that atoms of the source and the target register interact; this motion continues with variable speed until the source register completely overlaps with the target register. It is helpful to mentally decompose this operation into discrete steps. In the first step, the transport lattice is shifted one lattice site to the right such that the mth atom of the source register interacts with the first atom of the target register. One can tune the interaction time such that a certain phase shift is acquired during this interaction, namely $\phi = 2\pi/2^m$. In the next step, the transport lattice is moved one lattice site further to the right such that now the m th atom of the source register interacts with the *second* atom of the target register, while at the same time the $m-1$ th atom of the source register interacts with the first atom of the target register. In this step, the interaction time is made double as long as in the first step, so that $\phi = 2\pi/2^{m-1}$, and so on. After the lattice has been moved across m sites in this vein, the total state of the source and the target register is given by

$$|a_1\ a_2\ a_3\ \cdots\ a_m\rangle \otimes (|0\rangle + e^{2\pi i 0.a_1 a_2 \ldots a_m}|1\rangle) \\ (|0\rangle + e^{2\pi i 0.a_2 \ldots a_m}|1\rangle) \\ \cdots\ (|0\rangle + e^{2\pi i 0.a_m}|1\rangle)\,. \tag{40}$$

Finally, the lattice is shifted back to the original position without changing the phases any more (modulo 2π, see earlier remark, or the process can be made symmetric such that only half the phase values are accumulated during the motion to the right while the second halves of the phase values are accumulated when the lattice is brought back to its original position.) The overall effect of this sweep operation can be summarized in the form

$$|a\rangle|\ \ \rangle \longrightarrow e^{i\Phi(a)}|a\rangle|\mathcal{F}(a)\rangle \tag{41}$$

wherein $|a\rangle$ and $|\ \ \rangle$ denote the initial state of the source and the target register, and $|\mathcal{F}(a)\rangle = \frac{1}{\sqrt{2^m}}\sum_{y=0}^{2^m-1} e^{2\pi i a y/2^m}|y\rangle$ is the quantum Fourier transform of $|a\rangle$ [103]. The additional phase factor $e^{i\Phi(a)}$ accounts for a possible phase shift arising from collisions among different atoms of the source, if no vertical displacement of the transport lattice is possible.

As described, this method gives a very immediate way of implementing the quantum Fourier transform. Note that for a superposition of different input states the source and the target register become entangled. To apply the method in the Shor algorithm [104,105], for example, additional steps have to be taken. A detailed discussion of this method, together with possible applications, will be presented somewhere else [106]. This final example demonstrates a remarkable flexibility of the entanglement operations that are possible in optical lattices and similar systems, offering new perspectives for efficient implementations of quantum algorithms.

V. FINAL REMARKS

It is clear that, at the present time, most of the experimental requirements have yet to be realized, before one can implement quantum computing. There are, however, recent achievements in cooling and trapping of atoms in optical lattices and in magnetic microtraps which make it seem possible that some of these elements could be implemented in the laboratory in the near future. There are short-term and long-term perspectives. Essential for all quantum information experiments is a successful *cooling* of the atoms to the ground state of a three dimensional lattice. Numerical calculations [31] using realistic parameters give $kT < 0.2\hbar\omega$ as a critical value. Under these circumstances, one could perform interesting Ramsey-type spectroscopic studies of the fidelity of multi-particle entanglement as discussed earlier. To do this, neither single-atom addressability is required nor are regular filling structures. When the latter requirements can be realized, on the other hand, coding experiments can be done and a quantum memory be implemented. Finally, if one can find three-level schemes with different scattering phases for the logical states, universal computations can be performed. The parallelism of the lattice could then be exploited for efficient implementations of fault-tolerant quantum computing.

We have discussed multi-particle entanglement schemes mainly in the context of optical lattice implementations. Some of these ideas could readily be adopted in implementations with magnetic microtraps if one uses adiabatic schemes. A basic requirement for this is the possibility of creating quantum dots that are spatially sufficiently close to each other. These ideas will be discussed somewhere else [89].

ACKNOWLEDGMENTS

We thank E. Hinds, J. Schmiedmayer, M. Weitz and T. W. Hänsch for many useful discussions. We also thank David DiVincenzo and Andrew Steane for helpful discussions on fault-tolerant quantum computing during the Benasque Workshop 1998. H.-J.B. likes to thank Manny Knill for a helpful discussion on the problem of concatenated coding. One of us (T. C.) thanks M. Traini and S. Stringari for the kind hospitality at the Physics Department of Trento University, and the ECT* for partial support during the completion of this work. This work was supported in part by the Österreichischer Fonds zur Förderung der wissenschaftlichen Forschung, the European Community under the TMR network ERB-FMRX-CT96-0087, the Institute for Quantum Information

GmbH, and by the Schwerpunktsprogramm "Quanten-Informationsverarbeitung" der Deutschen Forschungsgemeinschaft.

APPENDIX A: ONE PARTICLE IN A MOVING HARMONIC POTENTIAL

1. Hamiltonian

The center of the potential with frequencies ω_x, ω_y and ω_z is assumed to be given by $\bar{\mathbf{x}}(t) = (\bar{x}(t), 0, 0)$ and the Hamiltonian reads

$$H = H_x + H_y + H_z, \tag{A1}$$

where $H_z = \hbar\omega_z(a_z^\dagger a_z + 1/2)$, $H_y = \hbar\omega_y(a_y^\dagger a_y + 1/2)$, and

$$H_x = \hbar\omega_x\left(a_x^\dagger a_x + \frac{1}{2} + (a_x^\dagger + a_x)\frac{\bar{x}(t)}{\sqrt{2}} + \frac{\bar{x}(t)^2}{2}\right). \tag{A2}$$

The a's are bosonic destruction operators and $\bar{x}(t)$ is given in harmonic oscillator units. We will concentrate on the x-direction leave out the subscript x and normalize energies to $\hbar\omega$.

2. Exact solution

The Schrödinger equation for the Hamiltonian Eq. A2 can be solved exactly [107,73]. To do so we define

$$H_0 = a^\dagger a + \frac{1}{2}, \tag{A3}$$

$$K(t, -\tau) = \frac{1}{\sqrt{2}}\int_{-\tau}^{t} ds\, \bar{x}(s)e^{i(s+\tau)} \tag{A4}$$

and

$$\beta(t, -\tau) = i\int_{-\tau}^{t} ds\left(K(s, -\tau)\{\partial_s K^*(s, -\tau)\} + \frac{i\bar{x}(s)^2}{2}\right). \tag{A5}$$

If initially at time $-\tau$ the system is in the state $|\Psi(-\tau)\rangle = |0\rangle$, where $|n\rangle$ is the n-th harmonic oscillator eigenstate we get

$$|\Psi(t)\rangle = e^{i\beta(t,-\tau)}\sum_n \frac{(iK(t, -\tau)e^{-i(t+\tau)})^n}{\sqrt{n!}}|n\rangle. \tag{A6}$$

The kinetic phase ϕ is thus given by the phase of the overlap of $|\Psi(t)\rangle$ with the instantaneous ground state $D(\bar{\mathbf{x}}(t))|0\rangle$, where $D(\gamma) = \exp(\gamma a^\dagger - \gamma^* a)$ denotes the displacement operator

$$\phi = -\arg\left(\langle 0|D(\bar{\mathbf{x}}(t))^\dagger|\Psi(t)\rangle\right). \tag{A7}$$

The interaction phase can be found by Eq. (7) with the known $|\Psi(t)\rangle$.

3. Corrections to the adiabatic approximation

We assume $\bar{x}(t)$ to be an analytic function of t and that $\bar{x}(t) \gg \partial_t\bar{x}(t) \gg \partial_t^2\bar{x}(t) \gg \ldots \gg \partial_t^n\bar{x}(t)$. By expanding in orders of the time derivatives we can write for $K(t', -\tau)$

$$K(t', -\tau) = \frac{1}{\sqrt{2}}\left(i^{N+1}\int_{-\tau}^{t'} ds\{\partial_s^{N+1}\bar{x}(s)\}e^{i(s+\tau)} - \sum_{n=0}^{N} i^{n+1}\{\partial_s^n\bar{x}(s)\}e^{i(s+\tau)}\Big|_{s=-\tau}^{s=t'}\right), \tag{A8}$$

where N is a positive integer. Note that if we may neglect all terms of order greater than $\partial_t\bar{x}(t)$ and start in a coherent state $|\Psi(-\tau)\rangle = D(\bar{x}(-\tau) + i\partial_t\bar{x}(t)|_{t=-\tau})|0\rangle$ the state will always be a coherent state with $\langle x(t)\rangle = \bar{x}(t)$ and $\langle p(t)\rangle = \partial_t\bar{x}(t)$.

Now we assume for simplicity that $\bar{x}(\tau) = \bar{x}(-\tau) = 0$, $(\partial_t\bar{x}(t))|_{t=-\tau} = (\partial_t\bar{x}(t))|_{t=\tau} = 0$ and $(\partial_t^n\bar{x}(t))|_{t=-\tau} = (-1)^n(\partial_t^n\bar{x}(t))|_{t=\tau}$ for $n > 1$. The system is assumed to be in the state $|\Psi(-\tau)\rangle = |0\rangle$, initially. We keep all the terms to fourth order in the derivatives (in the integrand) and find

$$K(t, -\tau) = \frac{1}{\sqrt{2}}\Big(\{i(\partial_t^2 - \partial_t^4)\bar{x}(t)\}\left(e^{i(t+\tau)} - 1\right) - \{\partial_t^3\bar{x}(t)\}(e^{i(t+\tau)} + 1)\Big), \tag{A9}$$

and

$$\beta(t, -\tau) = \frac{1}{2}\int_{-\tau}^{t} ds\left(\{\partial_s\bar{x}(s)\}^2 + \{\partial_s^2\bar{x}(s)\}^2\right) + \frac{1}{2}\{\partial_s^2\bar{x}(t)\}^2(1 - e^{-i(t+\tau)}). \tag{A10}$$

If we choose $(t + \tau) = 2n\pi$ with integer n the largest correction to the approximation to the kinetic phase discussed in Sec. II B 1 is of third order. Also the amplitude of the first excited state is of third order as can be seen from Eq. (A9).

APPENDIX B: QUANTUM ERROR CORRECTION AND THE IMPLEMENTATION OF SHOR'S CODE

Consider a one-dimensional configuration with a string of n atoms, where $x_0, x_1, x_2, \ldots, x_n \in \{0, 1\}$ label the internal state of the atoms at position $0, 1, 2, \ldots, n$ of the lattice. An elementary lattice-shift operation as given in Fig. 12(a) is then described as

$$LX: \quad |x_0, x_1, x_2, \ldots, x_n\rangle \longmapsto e^{-i\sum_j (x_j+1 \bmod 2)x_{j+1}\varphi_{j+1}}|x_0, x_1, x_2, \ldots, x_n\rangle \tag{B1}$$

where the phase φ_{j+1} in the exponent depends on the interaction time and the interaction strength between two atoms at the lattice site $j+1$, and the addition is performed modulo 2. Note that two neighboring atoms at the sites j and $j+1$ contribute to the exponent if and only if $x_j = 0$ and $x_{j+1} = 1$. The variables x_j can only take on the values 0 and 1. In all examples we discuss here, $\varphi_j = \varphi =$ constant and is the same for all lattice sites.

The operation (B1) defines a generalized phase gate that acts on a group of n neighboring atoms via shifting the lattice across one lattice site. It can easily be seen that, for example when combined with $\pi/2$-pulses as in Fig. 10, LX produces the entangled states (33) and (34) for $n=2$ and $n=3$.

In two dimensional lattices, the logical variables x_{kl} are labeled by two indices, where k is the horizontal index and l the vertical index. The phase gates corresponding to horizontal and vertical lattice shifts are then defined as

$$LX|\{x_{kl}\}\rangle = e^{-i\sum_j (x_{jl}+1 \bmod 2) x_{j+1,l}\varphi_{j+1,l}}|\{x_{kl}\}\rangle$$
$$LY|\{x_{kl}\}\rangle = e^{-i\sum_j (x_{kj}+1 \bmod 2) x_{k,j+1}\varphi_{k,j+1}}|\{x_{kl}\}\rangle \quad \text{(B2)}$$

as an obvious generalization of (B1).

It is clear that the operations can be further generalized to lattice shifts across an arbitrary number of lattice sites and along arbitrary directions. There are interesting topological questions in this general situation. For the present discussion, however, the gates LX, LY as defined in (B2) are sufficient and we will set $\varphi_{k,l} = \pi$. Apart from these gates, we will only need single-particle operations, in particular the Pauli-operators $\sigma_{x,j}, \sigma_{y,j}, \sigma_{z,j}$ and the Hadamard transformation ($\pi/2$ pulse) $H_j = (\sigma_{z,j} + \sigma_{x,j})/\sqrt{2}$ applied to an atom with index j.

Consider now a configuration of 3×3 atoms as in Fig. 13, where the central atom is in the unknown state $|\psi\rangle = \alpha|0\rangle + \beta|1\rangle$ while all surrounding atoms are initially in the state $|0\rangle$. Let us first look at the special case when $|\psi\rangle = |0\rangle$, that is, the central atom is in the state $|0\rangle$ as well. If we apply a $\pi/2$ pulse to each atom of the block and then the operation LX, we obtain a tensor product of three GHZ states where each row of the block is in the same state $(0+1)0(0-1)-(0-1)1(0+1)$. (For notational brevity, we suppress the bracket notation in the following and identify $0 \equiv |0\rangle$ and $1 \equiv |1\rangle$). This state can be transformed to the form $000-111$ by applying H_1 to the first atom and $H_3\sigma_{z,3}$ to the third atom of each row. The operation LX, supplemented by one-qubit rotations, produces thus one of the code words in (36).

To realize a quantum memory, an *unknown* state $\psi = \alpha 0 + \beta 1$ of the central atom is to be encoded into an entangled 9-bit state as in (37). Let us write the initial (unencoded) state of the block in the form

$$|\mathtt{bare}\rangle = 0_1 0_2 0_3\ 0_4(\alpha 0_5 + \beta 1_5)0_6\ 0_7 0_8 0_9 \quad \text{(B3)}$$

where the first, second, and third triplet refers to the upper, center, and lower row of the block in Fig. 13. To encode ψ into a corresponding superposition of both codewords, lattice movements in both horizontal and in vertical direction are required. In detail, the *encoding operation* is given by

$$ENC = H_{46} \circ LX \circ H_{456} \circ LY \circ \sigma_{x;369} H_{134679} \circ LX \circ H^{s} \quad \text{(B4)}$$

whose essential part is a sequence of three lattice movements, `horizontal - vertical - horizontal`, with certain 1-bit unitary transformations in between. In the notation used here, H^s denotes a Hadamard transform applied to each of the 8 syndrome atoms, whereas $H_{ijk\ldots}$ and $\sigma_{z;ijk\ldots}$ are single-qubit rotations applied to the selected atoms $i, j, k, \ldots$, only. Applied to the state (B3), ENC produces

$$\begin{aligned} ENC\,|\mathtt{bare}\rangle &= \alpha(000-111)(001-110)(000+111) \\ &\quad \beta(000+111)(100+011)(000-111) \\ &\equiv \alpha 0_{\mathrm{L}} + \beta 1_{\mathrm{L}}\,. \end{aligned} \quad \text{(B5)}$$

The codewords 0_{L} and 1_{L} are equivalent to the Shor code (36), as we shall see presently.

The *decoding operation* is given by the inverse of (B4),

$$DEC = H^{s} \circ LX \circ H_{134679}\sigma_{x;369} \circ LY \circ H_{456} \circ LX \circ H_{46} \quad \text{(B6)}$$

involving the same lattice movements, but the 1-bit operations carried out in reverse order. To see explicitly how one can correct an error occurring on one of the qubits $j = 1, 2, \ldots 9$, we apply the error operators $\sigma_{x,j}$, $\sigma_{y,j}$, or $\sigma_{z,j}$ to the encoded state $\alpha 0_{\mathrm{L}} + \beta 1_{\mathrm{L}}$. Then we apply the decoding operation DEC and measure the state of the syndrome atoms. The code 0_{L} and 1_{L} is error correcting if every possible error is mapped into a syndrome state different from $0_1 0_2 0_3\ 0_4 0_6\ 0_7 0_8 0_9$, and for each syndrome we can tell which unitary transformation has to be applied to the central qubit to restore it to its original state [it is not necessary that all errors are mapped to mutually orthogonal subspaces [47]]. The following table gives for each error the corresponding syndrome and the state of the central qubit:

error	syndrome	central qubit
none	000 00 000	$\alpha 0 + \beta\ 1$
$\sigma_{x,1}$	110 00 000	$\alpha 0 - \beta\ 1$
$\sigma_{x,2}$	101 00 000	$\alpha 0 - \beta\ 1$
$\sigma_{x,3}$	011 00 000	$\alpha 0 - \beta\ 1$
$\sigma_{x,4}$	010 10 010	$\alpha 0 + \beta\ 1$
$\sigma_{x,5}$	000 11 000	$\alpha 0 - \beta\ 1$
$\sigma_{x,6}$	010 01 010	$\alpha 0 + \beta\ 1$
$\sigma_{x,7}$	000 00 110	$\alpha 0 - \beta\ 1$
$\sigma_{x,8}$	000 00 101	$\alpha 0 - \beta\ 1$
$\sigma_{x,9}$	000 00 011	$\alpha 0 - \beta\ 1$
$\sigma_{y,1}$	100 00 000	$\alpha 0 - \beta\ 1$
$\sigma_{y,2}$	111 00 000	$\alpha 0 - \beta\ 1$
$\sigma_{y,3}$	001 00 000	$\alpha 0 - \beta\ 1$
$\sigma_{y,4}$	000 01 000	$\alpha 1 - \beta\ 0$
$\sigma_{y,5}$	010 00 010	$\alpha 1 - \beta\ 0$
$\sigma_{y,6}$	000 10 000	$\alpha 1 - \beta\ 0$
$\sigma_{y,7}$	000 00 100	$\alpha 0 - \beta\ 1$
$\sigma_{y,8}$	000 00 111	$\alpha 0 - \beta\ 1$
$\sigma_{y,9}$	000 00 001	$\alpha 0 - \beta\ 1$
$\sigma_{z,1}$	010 00 000	$\alpha 0 + \beta\ 1$
$\sigma_{z,2}$	010 00 000	$\alpha 0 + \beta\ 1$
$\sigma_{z,3}$	010 00 000	$\alpha 0 + \beta\ 1$
$\sigma_{z,4}$	010 11 010	$\alpha 1 - \beta\ 0$
$\sigma_{z,5}$	010 11 010	$\alpha 1 + \beta\ 0$
$\sigma_{z,6}$	010 11 010	$\alpha 1 - \beta\ 0$
$\sigma_{z,7}$	000 00 010	$\alpha 0 + \beta\ 1$
$\sigma_{z,8}$	000 00 010	$\alpha 0 + \beta\ 1$
$\sigma_{z,9}$	000 00 010	$\alpha 0 + \beta\ 1$

In Fig. 13, the syndrome atoms visually encircle the unknown qubit that is to be protected. If any of the 9 qubit suffers a spin flip, a phase flip, or both, the error can be detected by measuring the state of the syndrome atoms after the decoding operation has been applied to the group. This could be done by a fluorescence measurement where atoms in state 1 and 0 correspond to "bright" and "dark", respectively. For example, according to above table, the pattern

$$\begin{matrix} 0 & 0 & 0 \\ 1 & \psi' & 1 \\ 0 & 0 & 0 \end{matrix}$$

tells us that a spin flip has occurred in the central atom, whereas

$$\begin{matrix} 0 & 0 & 0 \\ 0 & \psi' & 0 \\ 1 & 1 & 0 \end{matrix}$$

reveals a spin flip in the left atom of the lower row, and

$$\begin{matrix} 0 & 1 & 0 \\ 1 & \psi' & 1 \\ 0 & 1 & 0 \end{matrix}$$

corresponds to a phase flip in any of the atoms of the central row. In any case, the state ψ' of the central qubit after the detection of an error is related to the initial state ψ via a (known) unitary operation U: $\psi' = U\psi$, which can be obtained from the third column of the syndrome table given above.

The fact that the encoding operation involves only 3 lattice movements provides a specific example of a "parallelization of a quantum circuit" [43,44]. We have not proven that 3 is really the minimum number of entanglement operations needed; there might be still faster sequences. The original Shor code can be recovered from this code by applying an additional vertical lattice shift, LY, and certain 1-bit rotations.

[1] CIRAC, J. I., and ZOLLER, P., 1995, *Phys. Rev. Lett.* **74**, 4091.

[2] MONROE, C., MEEKHOF, D. M., KING, B. E., ITANO, W. M., and WINELAND, D. J., 1995 *Phys. Rev. Lett.* **75**, 4714.

[3] TURCHETTE, Q. A., WOOD, C. S., KING, B. E., MYATT, C. J., LEIBFRIED, D., ITANO, W. M., MONROE, C., and WINELAND, D. J., 1998, *Phys. Rev. Lett.* **81** 3631;

[4] KING, B.E., MYATT, C.J., TURCHETTE, Q.A., LEIBFRIED, D., ITANO, W.M., MONROE, C., and WINELAND, D.J., 1998, *Phys. Rev. Lett.* **81** 1525;

[5] MONROE, C., LEIBFRIED, D., KING, B.E., MEEKHOF, D.M., ITANO, W.M., and WINELAND, D.J., 1997, *Phys. Rev. A* **55**, R2489.

[6] STEANE, A., 1997, *Appl. Phys. B* **64**, 623.

[7] STEVENS, D., BROCHARD, J., and STEANE, A., 1998, *Phys. Rev. A* **64**, 623.

[8] JAMES, D.F.V., GULLEY, M.S., HOLZSCHEITER, M.H., HUGHES, R.J., KWIAT, P.G., LAMOREAUX, S.K., PETERSON, C.G., SANDBERG, V.D., SCHAUER, M.M., SIMMONS, C.M., TUPA, D., WANG, P.Z., and WHITE, A.G., quant-ph/9807071.

[9] TURCHETTE, Q. A., HOOD, C. J., LANGE, W., MABUCHI, H., and KIMBLE, H. J., 1995, *Phys. Rev. Lett.* **75**, 4710.

[10] MAÎTRE, X., HAGLEY, E., NOGUES, G., WUNDERLICH C., GOY, P., BRUNE, M., RAIMOND, J. M., and HAROCHE, S., 1997, *Phys. Rev. Lett.* **79**, 769.

[11] HAGLEY, E., MATRE, X., NOGUES, G., WUNDERLICH, C., BRUNE, M., RAIMOND, J. M., and S. HAROCHE, S., 1997, *Phys. Rev. Lett.* **79**, 1.

[12] PELLIZZARI, T., GARDINER, S.A., CIRAC, J.I., and ZOLLER, P., 1995, *Phys. Rev. Lett.* **75**, 3788.

[13] CIRAC, J. I., ZOLLER, P., MABUCHI, H., and KIMBLE, H. J., 1997, *Phys. Rev. Lett.* **78**, 3221.

[14] VAN ENK, S. J., CIRAC, J. I., and ZOLLER, P., 1998, *Science* **279,** 205.

[15] LAW, C.K. and KIMBLE, H.J., 1997 *J. Mod. Opt.* **44**, 2067.

[16] GHERI, K.M. SAAVEDRA, C., TÖRMÄ, P., CIRAC, J.I., and ZOLLER, P., 1998, *Phys. Rev. A* **58**, R2627.

[17] ASPECT, A., GRANGIER, P., and ROGER, G., 1981, *Phys. Rev. Lett.* **47**, 460.

[18] ASPECT, A., GRANGIER, P., and ROGER, G., 1982, *Phys. Rev. Lett.* **49**, 91.

[19] ASPECT, A., DALIBARD, A., and ROGER, G., 1982, *Phys. Rev. Lett.* **49**, 1804.

[20] KWIAT, P.G., MATTLE, K., WEINFURTER, H., ZEILINGER, A., SERGIENKO, A.V., and SHIH, Y.H., 1995, *Phys. Rev. Lett.* **75**, 4337; BRENDEL, J., GISIN, N., TITTEL, W., and ZBINDEN, H., quant-ph/9809034; KWIAT, P.G., WAKS, E., WHITE, A.G., APPELBAUM, I., and EBERHARD, P.H., quant-ph/9810003; CABRILLO, C. CIRAC, J.I. GARCIA-FERNANDEZ, P., and ZOLLER, P., 1999, *Phys. Rev. A* **59**, 1025.

[21] BOUWMEESTER, D., PAN, J. I., MATTLE, K., EIBL, M., WEINFURTER, H., and ZEILINGER, A., 1997, *Nature* **390**, 575.

[22] BOSCHI, D., BRANCA, S., DE MARTINI, F., HARDY, L., and POPESCU, S., 1998, *Phys. Rev. Lett.* **80**, 1121.

[23] FURUSAWA, A., SORENSEN, J. L., BRAUNSTEIN, S. L., FUCHS, C. A., KIMBLE, H. J., and POLZIK, E. S., 1998, *Science* **282**, 706.

[24] PAN, J. W., BOUWMEESTER, D., WEINFURTER, H., and ZEILINGER, A., 1998, *Phys. Rev. Lett.* **80**, 3891.

[25] BOUMEESTER, D., PAN, J.-W., DANIELL, M., WEINFURTER, H., and ZEILINGER, A., 1999, *Phys. Rev. Lett.* **82**, 1345.

[26] CORY, D. G., FAHMY, A. F., and HAVEL, T. F., 1997, *Proc. Natl. Acad. Sci. USA* **94**, 1634.

[27] GERSHENFELD, N. A., and CHUANG, I. L., 1997, *Science* **275,** 350.

[28] CORY, D. G., MASS, W., PRICE, M., KNILL, E., LAFLAMME, R., ZUREK, W. H., HAVEL, T. F., and SOMAROO, S. S., 1998, *Phys. Rev. Lett.* **81,** 2152.

[29] JONES, J. A., MOSCA, M., and HANSEN, R. H., 1998, *Nature,* **393**, 344.

[30] BRAUNSTEIN, S.L., CAVES, C.M., JOZSA, R., LINDEN, N., POPESCU, S., and SCHACK R., quant-ph/9811018.

[31] JAKSCH, D., BRIEGEL, H.-J., CIRAC, J. I., GARDINER, C. W., and ZOLLER, P., 1999, *Phys. Rev. Lett.* **82,** 1975.

[32] *Laser manipulation of Atoms and Ions,*1992, edited by E. ARIMONDO, W. D. PHILLIPS, and F. STRUMIA, (North Holland, Amsterdam); see also *Special issue on Laser cooling and trapping of atoms,* JOSA B **6** No. 11 (1989); *ibid.* **2** No. 11 (1985).

[33] CHU, S., 1998, *Rev. Mod. Phys.* **70**, 686.

[34] COHEN-TANNOUDJI, C., 1998, *Rev. Mod. Phys.* **70**, 707.

[35] PHILLIPS, W. D., 1998, *Rev. Mod. Phys.* **70**, 721.

[36] ANDERSON, M. H., ENSHER, J. R., MATTHEWS, M. R., WIEMAN, C. E., and CORNELL, E. A., 1995, *Science,* **269**, 198.

[37] BRADLEY, C. C., SACKETT, C. A., TOLLETT, J. J., and HULET, R. G., 1995, *Phys. Rev. Lett.* **75**, 1687.

[38] DAVIES, K. B., MEWES, M.-O., ANDREWS, M. R., VAN DRUTEN, N. J., DURFEE, D. S., KURN, D. M. and KETTERLE, W., 1995, *Phys. Rev. Lett.* **75**, 3969.

[39] HALL, D. S., MATTHEWS, M. R., WIEMAN, C. E., and CORNELL, E. A., 1998, *Phys. Rev. Lett.* **81**, 1543; *ibid.* **81**, 4532.

[40] HALL, D. S., MATTHEWS, M. R., ENSHER, J. R., WIEMAN, C. E., and CORNELL, E. A., 1998, *Phys. Rev. Lett.* **81**, 1539; *ibid.* **81**, 4531.

[41] STAMPER-KURN, D. M., ANDREWS, M. R., CHIKKATUR, A. P., INOUYE, S., MIESNER, H.-J., STENGER, J., and KETTERLE, W., 1998, *Phys. Rev. Lett.***80**, 2027.

[42] BRENNEN, G.K., CAVES, C.M., JESSEN, P.S., and DEUTSCH, I.H., 1999, *Phys. Rev. Lett.* **82**, 1060.

[43] MOORE, C., and NILSSON, M., 1998, Los Alamos preprint server quant-ph/9804034.

[44] MOORE, C., and NILSSON, M., 1998, Los Alamos preprint server quant-ph/9808027.

[45] FREEDMAN, M. H.,and MEYER, D. A., Los Alamos preprint server quant-ph/9810055.

[46] KITAEV, A. YU., and BRAVYI, S. B., Los Alamos preprint server quant-ph/9811052.

[47] SHOR, P. W., 1995, *Phys. Rev. A* **52**, R2493.

[48] STEANE, A. M., 1996, *Phys. Rev. Lett.* **77**, 793; *ibid.*, 1996, *Proc. Roy. Soc. Lond. A* **452**, 2551.

[49] CALDERBANK, A. R., and SHOR, P. W., 1996, *Phys. Rev. A* **54**, 1098.

[50] LAFLAMME, R., MIQUEL, C., PAZ, J.-P., and ZUREK, W. H., 1996 *Phys. Rev. Lett.* **77**, 198.

[51] EKERT, A., and MACCHIAVELLO, C., 1996, *Phys. Rev. Lett.* **77**, 2585.

[52] BENNETT, C. H., DIVINCENZO, D. P., SMOLIN, J. A., and WOOTTERS, W. K., 1996, *Phys. Rev. A* **54**, 3825.

[53] KNILL, E., and LAFLAMME, R., 1997, *Phys. Rev. A* **55**, 900.

[54] For a review on the subject of quantum error correction (and on quantum computing in general) see, for example, STEANE, A.M., 1998, *Rept. Prog. Phys.* **61**, 117-173.

[55] SHOR, P. W., 1996, in *Proc. 37th Symposium on the Foundations of Computer Science*, (IEEE Computer Society Press), p. 56; quant-ph/9605011.

[56] STEANE, A. M., 1997, *Phys. Rev. Lett.* **78**, 2252.

[57] KNILL, E., and LAFLAMME, R., 1996, quant-ph/9608012.

[58] AHARONOV, D., and BEN-OR, M., 1996, quant-ph/9611025.

[59] KITAEV, A. YU. 1997, *Russ. Math. Surv.* **52**, 1191; *ibid.*, quant-ph/9707021.

[60] KNILL, E., LAFLAMME, R., and ZUREK, W., 1997, *Science* **279**, 342.

[61] GOTTESMAN, D., 1998, *Phys. Rev. A* **57**, 127.

[62] For a review on the subject of fault-tolerant computing and for further references see, for example, PRESKILL, J., 1997, quant-ph/9705031.

[63] HAMANN, S.E., HAYCOCK, D.L., KLOSE, G., PAX, P.H., DEUTSCH, I.H., and JESSEN, P.S., 1998, *Physical*

Review Letters **80** 4149.

[64] MENNERAT-ROBILLIARD, C., LUCAS, D., GUIBAL, S., TABOSA, J., JURCZAK, C., COURTOIS, J.-Y., and GRYNBERG, G., 1999, *Physical Review Letters* **82** 851.

[65] DEPUE, M. T. , MCCORMICK, C. , WINOTO, S.L. , OLIVER, S. and WEISS, D.W., 1999, *Physical Review Letters* **82** 2262.

[66] FRIEBEL, S., D'ANDREA, C., WALZ, J., WEITZ, M., and HÄNSCH, T.W., 1998, *Phys. Rev. A* **57**, R20.

[67] HINDS, E. A., BOSHIER, M. G., and HUGHES, I. G., 1998, *Phys. Rev. Lett.* **80**, 645.

[68] WEINSTEIN, J.D., and LIBBRECHT, K.G., 1995, *Phys. Rev. A* **52**, 4004.

[69] VULETIC, V., HÄNSCH, T. W., and ZIMMERMANN, C., 1996, *Europhys. Lett.* **36**, 349.

[70] VULETIC, V., FISCHER, T., PRAEGER, M., HÄNSCH, T. W., and ZIMMERMANN, C., 1998, *Phys. Rev. Lett.* **80**, 1634.

[71] DENSCHLAG, J., UMSHAUS, G., and SCHMIEDMAYER, J., 1998, *Phys. Rev. Lett.* **81**, 737; SCHMIEDMAYER, J., 1995, *Phys. Rev. A* **52**, R13.

[72] The justification parallels the arguments presented in the context of the pseudo-potential approximation in BEC see, e.g., H. T. C. STOOF, M. BIJLSMA, and M. HOUBIERS, *J. Res. Natl. Inst. Stand. Technol.* **101**, 443 (1996).

[73] see, e.g., GALINDO, A., PASCUAL, P., *Quantum Mechanics II* (Springer Berlin 1991).

[74] KAGAN, YU., SURKOV, E.L., and SHLYAPNIKOV, G.V., 1996, *Phys. Rev. A* **54**, R1753.

[75] JAKSCH, D., BRUDER, C., CIRAC, J. I., GARDINER, C. W., and ZOLLER, P, 1998 *Phys. Rev. Lett.* **81**, 3108.

[76] FISHER, M.P.A., WEICHMAN, P.B., GRINSTEIN, G., and FISHER, D.S., 1989, *Phys. Rev. B* **40**, 546.

[77] BRUDER, C., FAZIO, R., and SCHÖN, G., 1993, *Phys. Rev. B* **47**, 342.

[78] KRAUTH, W., CAFFAREL, M., and BOUCHARD, J.-P., 1992, *Phys. Rev. B* **45**, 3137.

[79] SHESHADRI, K., KRISHNAMURTHY, H.R., PANDIT, R., and RAMAKRISHNAN, T. V., 1993, *Europhys. Lett.* **22**, 257.

[80] FINKELSTEIN, V., BERMAN, P.R., and GUO, J., 1992, *Phys. Rev. A*, **45**, 1829.

[81] WEINER, J., BAGNATO, V.S., ZILIO, S., and JULIENNE, P.S., 1999, *Rev. Mod. Phys.* **71**, 1.

[82] TIESINGA, E., VERHAAR, B.J., and STOOF, H.T.C., 1993, *Phys. Rev. A*, **47**, 4114.

[83] SCHUMACHER, B., 1996, *Phys. Rev. A* **54**, 2614.

[84] ROACH, T.M., ABELE, H., BOSHIER, M.G., GROSSMAN, H.L., ZETIE, K.P., and HINDS, E.A., 1995, *Phys. Rev. Lett.* **75**, 629.

[85] HINDS, E.A., BOSHIER, M.G., and HUGHES, I.G., 1998. *Phys. Rev. Lett.* **80**, 645;

[86] HINDS, E.A., 1999, to appear in *Proc. R. Soc. London.*

[87] SCHMIEDMAYER, J., 1998, *Eur. Phys. J. D* **4**, 57.

[88] OVCHINNIKOV, YU.B., MANEK, I., and GRIMM, R., 1997, *Phys. Rev. Lett.* **79**, 2225.

[89] CALARCO, T. *et al.*, unpublished.

[90] GREENBERGER, D. M., HORNE, M., and ZEILINGER, A., 1989, in *Bell's Theorem, Quantum Theory, and Conceptions of the Universe,* M. Kafatos, editor, Kluwer, Dordrecht, The Netherlands, p. 69.

[91] BARENCO, A., BENNETT, C.H., CLEVE, R., DIVINCENZO, D.P., MARGOLUS, N., SHOR, P., SLEATOR, T., SMOLIN, J.A., and WEINFURTER, H., 1995, *Phys. Rev. A* **52**, 3457; and references cited therein.

[92] DEUTSCH, D., BARENCO, A., and EKERT, A., 1995, quant-ph/9505018.

[93] BENNETT, C. H., 1995, *Phys. Today* **48**, 24.

[94] DIVINCENZO, D. P., 1995, *Phys. Rev. A* **51**, 1015.

[95] N. F. Ramsey, Rev. Mod. Phys. **62**, 541 (1990).

[96] BELL, J.S., 1964, *Physics* **1**, 195.

[97] For small filling factors, the signals coming from different groups of atoms could even be spatially resolved.

[98] Other proposals for selectively filling optical lattice sites exist (Martin Weitz, private communication), using infrared superlattices to load the (logical) lattice with a certain, well defined spacing between the atoms.

[99] In ^{87}Rb and ^{23}Na one could use $|a\rangle \equiv |F = 1, m_F = -1\rangle$, $|b\rangle \equiv |F = 2, m_F = 2\rangle$ and $|c\rangle \equiv |F = 1, m_F = 1\rangle$.

[100] Freedman and Meyer [45] have independently identified Shor's code as a two-dimensional topological quantum code (see also Kitaevs work [46,59]). The connection between their results and our lattice implementation remains to be clarified.

[101] Here we assume that one is able to selectively load a two-dimensional "sheet" in a three-dimensional lattice.

[102] It remains the practical question, how well the polarization $\theta(t)$ can be controlled in the experiment.

[103] CLEVE, R., EKERT, A., MACCHIAVELLO, C., and MOSCA, M., 1998, *Proc. Roy. Soc. Lond.* **454**, 339.

[104] SHOR, P. W., 1995, quant-ph/9508027; *ibid.*, 1997, *SIAM J. Computing* **26**, 1484.

[105] EKERT, A., and JOSZA, R., 1996, *Rev. Mod. Phys.* **68**, 733.

[106] BRIEGEL, H.-J. *et al.*, unpublished.

[107] GARDINER, C.W., *Quantum Noise* (Springer Berlin 1991)

Quantum Computation and Quantum Communication with Electrons

Quantum Computing and Quantum Communication with Electrons

Daniel Loss
Department of Physics and Astronomy, University of Basel,

1 Coupled quantum dots as quantum gates

Semiconductor quantum dots, sometimes referred to as artificial atoms, are small devices in which charge carriers are confined in all three dimensions[1]. The confinement is usually achieved by electrical gating and/or etching techniques applied e.g. to a two-dimensional electron gas (2DEG). Since the dimensions of quantum dots are on the order of the Fermi wavelength, their electronic spectrum consists of discrete energy levels which have been studied in great detail in conductance [1, 2] and spectroscopy measurements[1, 3, 4]. In GaAs heterostructures the number of electrons in the dots can be changed one by one starting from zero[5]. Typical laboratory magnetic fields ($B \approx 1\,\mathrm{T}$) correspond to magnetic lengths on the order of $l_B \approx 10\,\mathrm{nm}$, being much larger than the Bohr radius of real atoms but of the same size as artificial atoms. As a consequence, the dot spectrum depends strongly on the applied magnetic field[1, 2, 3]. In coupled quantum dots which can be considered to some extent as artificial molecules, Coulomb blockade effects[6] and magnetization[7] have been observed as well as the formation of a delocalized "molecular state" [8].

Motivated by the rapid down-scaling of integrated circuits, there has been continued interest in classical logic devices made of electrostatically coupled quantum dots[9]. More recently, the discovery of new principles of computation based on quantum mechanics[10] has led to the idea of using coupled quantum dots for quantum computation[11]; many other proposed implementations have been explored, involving NMR[12, 13, 14], trapped ions[15], cavity QED[16], and Josephson junctions[17]. Solid-state devices open up the possibility of fabricating large integrated networks which would be required for realistic applications of quantum computers. A basic feature of the quantum-dot scenario[11, 30, 31] is to consider the *electron spin* **S** as the qubit (the qubit being the basic unit of information in the quantum computer). This stands in contrast to alternative proposals also based on quantum dots[18, 19, 20, 21], in which it is the *charge* (orbital) degrees of freedom out of which a qubit is formed and represented in terms of a pseudospin-1/2. However, there are two immediate advantages of real spin over pseudospin: First, the qubit represented by a real spin-1/2 is always a well defined qubit; the two-dimensional Hilbert space is the entire space available, thus there are no extra dimensions into which the qubit state could "leak" [22]. Second, during a quantum computation phase coherence of the qubits must be preserved. It is thus an essential advantage of real spins that their dephasing times in GaAs can be on the order of microseconds[23], whereas for charge degrees of freedom dephasing times are typically much less, on the order of nanoseconds[24, 25].

In addition to a well defined qubit, we also need a controllable or deterministic "source of entanglement", i.e. a mechanism by which two specified qubits at a time can be entangled[26] so as to produce the fundamental quantum XOR (or controlled-NOT) gate operation, represented by a unitary operator $U_{\rm XOR}$[27]. This can be achieved by temporarily coupling two spins[11]. As we will show in detail below, due to the Coulomb interaction and the Pauli exclusion principle the ground state of two coupled electrons is a spin singlet, i.e. a highly entangled spin state.

This physical picture translates into an exchange coupling $J(t)$ between the two spins $\mathbf{S}_1$ and $\mathbf{S}_2$ described by a Heisenberg Hamiltonian

$$H_{\rm s}(t) = J(t)\ \mathbf{S}_1 \cdot \mathbf{S}_2. \tag{1}$$

If the exchange coupling is pulsed such that $\int dt J(t)/\hbar = J_0\tau_s/\hbar = \pi \pmod{2\pi}$, the associated unitary time evolution $U(t) = T\exp(i\int_0^t H_{\rm s}(\tau)d\tau/\hbar)$ corresponds to the "swap" operator $U_{\rm sw}$ which simply exchanges the quantum states of qubit 1 and 2[11]. Furthermore, the quantum XOR can be obtained[11] by applying the sequence

$$\exp(i(\pi/2)S_1^z)\exp(-i(\pi/2)S_2^z)U_{\rm sw}^{1/2}\exp(i\pi S_1^z)U_{\rm sw}^{1/2} \equiv U_{\rm XOR}$$

i.e. a combination of "square-root of swap" $U_{\rm sw}^{1/2}$ and single-qubit rotations $\exp(i\pi S_1^z)$, etc. Since $U_{\rm XOR}$ (combined with single-qubit rotations) is proven to be a universal quantum gate[18, 26], it can therefore be used to assemble any quantum algorithm. Thus, the study of a quantum XOR gate is essentially reduced to the study of the *exchange mechanism* and how the exchange coupling $J(t)$ can be controlled experimentally. We wish to emphasize that the switchable coupling mechanism described in the following need not be confined to quantum dots: the same principle first introduced in [11] can be applied to other systems, e.g. coupled atoms in a Bravais lattice, or overlapping shallow donors in semiconductors (such as P in Si, as subsequently discussed in Ref.[28]), and so on. The main reason to concentrate here on quantum dots is that these systems are at the center of many ongoing experimental investigations in mesoscopic physics, and thus there seems to be reasonable hope that these systems can be made into quantum gates functioning along the lines proposed by us.

In view of this motivation we will discuss in these notes the spin dynamics of two laterally coupled quantum dots containing a single electron each. We show that the exchange coupling $J(B, E, a)$ can be controlled by a magnetic field B (leading to wave function compression), or by an electric field E (leading to level detuning), or by varying the barrier height or equivalently the inter-dot distance $2a$ (leading to a suppression of tunneling between the dots). The dependence on these parameters is of direct practical interest, since it opens the door to tailoring the exchange $J(t)$ for the specific purpose of creating quantum gates.

We further present calculations of the static and dynamical magnetization responses in the presence of perpendicular and parallel magnetic fields, and show that they give experimentally accessible information about the exchange J. The analytical analysis is based on an adaptation of Heitler-London and Hund-Mulliken variational techniques[29] to parabolically confined coupled quantum dots. In particular, we present an extension of the Hubbard approximation induced by the long-range Coulomb interaction. We find a striking dependence of the Hubbard parameters on the magnetic field and inter-dot distance which is of relevance also for atomic-scale Hubbard physics in the presence of long-range Coulomb interactions. Finally, we discuss the effects of dephasing induced by nuclear spins in GaAs and show that dephasing can be strongly reduced by dynamically polarizing the nuclear spins and/or by magnetic fields. The remaining weak dephasing effects can then be described in terms of a generalized Master equation[11] obtained in a weak coupling expansion with subsequent Markovian approximation. The effect of the environmnet causing the dephasing is described generically in terms of a Caldeira-Leggett model that couples to the magnetic moments of the spins. We will discuss some properties of the superoperator formalism[32, 33, 34, 11] needed to derive the equation and to discuss some important properties such as complete positivity of the linear map.

Very recently we have described an electron-spin switching mechanism based on cavity-QED[35]. Such a mechanism has the potential advantage that it allows us to connect two distant qubits directly (the nearest neighbor coupling scheme above allows only indirect coupling of distant qubits via swapping operations).

2 Quantum Communication with electrons

In the second part of these notes we would like to address the following question[36]: is it possible to use mobile electrons, prepared in a definite (entangled) spin state, for the purpose of quantum communication? Such a question, for instance, is of central importance in a solid state quantum computer where one wishes to exchange quantum information between distant parts of a quantum network. The question is of course also of broader interest: if we could use electrons for creating entangled states, in particular so-called EPR pairs, and if we could move them around separately while preserving their spin entanglement, then we would be able to implement, for instance, tests of Bell's inequality; thereby, we could obtain tests of non-locality—one of the most striking concepts of quantum mechanics—for the first time with electrons. So far, such tests have been done on photons [37, 38], It is quite amusing to note here that the Gedanken experiment which has been formulated by Einstein, Podolsky, and Rosen[39], and which underlies the Bell inequalities, makes use of point particles and not of massless particles such as photons. Thus, there can be no doubt that it would be highly desirable to extend tests of non-locality also to quantities which have a rest mass such as electrons in particular.

One basic ingredient for quantum communication are entangled pairs of qubits which are shared by two parties. There are three separate requirements involved here which must be satisfied. First of all we need *mobile qubits* which can be transported from position A to position B. Second, we need a source of entanglement for such qubits which can be operated in a controllable way, and third, it must be possible to transport each of the qubits separately in a phase-coherent manner such that the entanglement between the two qubits of interest is not destroyed in the process of transporting them to their desired locations.

Our choice of representing the qubit in terms of the spin of a mobile electron satisfies the first requirement trivially whereas qubits defined as pseudospins are typically not mobile. The second requirement, to have a source of entanglement, can be satisfied by using the quantum gate mechanism based on coupled quantum dots (see above). The third requirement regarding phase coherent transport requires a detailed discussion of many-body transport physics in mesoscopic structures (e.g. 2DEGs in GaAs heterjunctions). We will present results of a tunneling Hamiltonian approach to analyze a double dot system attached to two in- and two outgoing leads where the noise correlation[40, 41] in a non-equilibrium set-up provide a tool for detecting entanglement between two electron–states[36]. In particular, we will show that the noise power spectrum of the current cross-correlations decays as $1/\omega$ (up to logarithmic corrections) for frequencies larger than the inverse life-time of the quasiparticle states[42]. We will discuss the conditions under which such predictions could be tested experimentally e.g. in GaAs heterostructures.

References

[1] L. Jacak, P. Hawrylak, and A. Wójs, *Quantum Dots* (Springer, Berlin, 1997).

[2] L. P. Kouwenhoven, C. M. Marcus, P. L. McEuen, S. Tarucha, R. M. Westervelt, and N. S. Wingreen, Proceedings of the Advanced Study Institute on *Mesoscopic Electron Transport*, edited by L. L. Sohn, L. P. Kouwenhoven, G. Schön (Kluwer, 1997).

[3] R. C. Ashoori, Nature **379**, 413 (1996).

[4] R. J. Luyken, A. Lorke, M. Haslinger, B. T. Miller, M. Fricke, J. P. Kotthaus, G. Medeiros-Ribiero, and P. M. Petroff, preprint.

[5] S. Tarucha, D. G. Austing, T. Honda, R. J. van der Hage, and L. P. Kouwenhoven, Phys. Rev. Lett. **77**, 3613 (1996); L. P. Kouwenhoven, T. H. Oosterkamp, M. W. S. Danoesastro, M. Eto, D. G. Austing, T. Honda, and S. Tarucha, Science **278**, 1788 (1997).

[6] F. R. Waugh, M. J. Berry, D. J. Mar, R. M. Westervelt, K. L. Chapman, and A. C. Gossard, Phys. Rev. Lett. **75**, 705 (1995); C. Livermore, C. H. Crouch, R. M. Westervelt, K. L. Chapman, and A. C. Gossard, Science **274**, 1332 (1996).

[7] T. H. Oosterkamp, S. F. Godijn, M. J. Uilenreef, Y. V. Nazarov, N. C. van der Vaart, and L. P. Kouwenhoven, Phys. Rev. Lett. **80**, 4951 (1998).

[8] R. H. Blick, D. Pfannkuche, R. J. Haug, K. v. Klitzing, and K. Eberl, Phys. Rev. Lett. **80**, 4032 (1998); R. H. Blick, D. W. van der Weide, R. J. Haug, K. Eberl, Phys. Rev. Lett. **81**, 689 (1998).

[9] K. Nomoto, R. Ugaijn, T. Suzuki, and I. Hase, J. Appl. Phys. **79**, 291 (1996); A. O. Orlov, I. Amlani, G. H. Bernstein, C. S. Lent, and G. L. Snider, Science **277**, 928 (1997).

[10] D. Deutsch, Proc. R. Soc. Lond. A **400**, 97 (1985).

[11] D. Loss and D. P. DiVincenzo, Phys. Rev. A **57**, 120 (1998).

[12] I. L. Chuang, N. A. Gershenfeld, and M. Kubinec, Phys. Rev. Lett. **80**, 3408 (1998).

[13] D. Cory, A. Fahmy, and T. Havel, Proc. Nat. Acad. Sci. U.S.A. **94**, 1634 (1997).

[14] J. A. Jones, M. Mosca, R. H. Hansen, Nature **393**, 344 (1998).

[15] J. I. Cirac and P. Zoller, Phys. Rev. Lett. **74**, 4091 (1995); C. Monroe, D. M. Meekhof, B. E. King, W. M. Itano, and D. J. Wineland, Phys. Rev. Lett. **75**, 4714 (1995).

[16] Q. A. Turchette, C. J. Hood, W. Lange, H. Mabuchi, and H. J. Kimble, Phys. Rev. Lett. **75**, 4710 (1995).

[17] D. V. Averin, Solid State Commun. **105**, 659 (1998); A. Shnirman, G. Schön, and Z. Hermon, Phys. Rev. Lett. **79**, 2371 (1997).

[18] A. Barenco, D. Deutsch, A. Ekert, and R. Josza, Phys. Rev. Lett. **74**, 4083 (1995).

[19] R. Landauer, Science **272**, 1914 (1996).

[20] J. A. Brum and P. Hawrylak, Superlattices and Microstructures **22**, 431 (1997).

[21] P. Zanardi and F. Rossi, quant-ph/9804016.

[22] Such leakage can happen e.g. in the switching process of single qubits or by coupling two qubits together etc. and can easily lead to uncontrollable errors. This concern is especially relevant in quantum dots where the energy level spacing is (nearly) uniform (in contrast to real atoms) so that the levels defining the qubit are of similar scale as the separation to neighboring energy levels.

[23] J. M. Kikkawa and D. D. Awschalom, Phys. Rev. Lett. **80**, 4313 (1998).

[24] A. G. Huibers, M. Switkes, C. M. Marcus, K. Campman, and A. C. Gossard, Phys. Rev. Lett. **81**, 200 (1998).

[25] The dephasing times of Refs.[23, 24] are both measured in GaAs semiconductors which involve many electrons. It would be highly desirable to get direct experimental information about dephasing times in isolated quantum dots of low filling as considered here.

[26] D. P. DiVincenzo, Phys. Rev. A **51**, 1015 (1995).

[27] A. Barenco, C. H. Bennett, R. Cleve, D. P. DiVincenzo, N. Margolus, P. Shor, T. Sleator, J. A. Smolin, and H. Weinfurter, Phys. Rev. A **52**, 3457 (1995).

[28] B. E. Kane, Nature **393**, 133 (1998).

[29] D.C. Mattis, *The Theory of Magnetism*, Vol. I, Sec. 4.5, Springer Series in Solid-State Sciences **17** (Springer, New York, 1988).

[30] D. P. DiVincenzo and D. Loss, Superlattices and Microstructures **23**, 419 (1998).

[31] G. Burkard, D. Loss, and D.P. DiVincenzo. Phys. Rev. B **59** (1999) 2070-2078. See condmat/9808026.

[32] E. Fick and G. Sauermann, *The Quantum Statistics of Dynamic Processes*, Springer Series in Solid-State Sciences 86 (Springer-Verlag, Berlin, 1990).

[33] E. B. Davies, *Quantum Theory of Open Systems*, Academic Press, New York (1976).

[34] M. Celio and D. Loss, Physica **A 150**, 769 (1989).

[35] A. Imamoglu, A. Small, G. Burkard, D. Awschalom, D. DiVincenzo, D. Loss, and M. Sherwin. Preprint.

[36] D.P. DiVincenzo and D. Loss. To appear in special issue of J. Mag. Magn. Matl. (vol. 200), "Magnetism beyond 2000". See cond-mat/9901137.

[37] A. Aspect, J. Dalibard, and G. Roger, Phys. Rev. Lett. **49**, 1804 (1982).

[38] W. Tittel, J. Brendel, H. Zbinden, and N. Gisin, Phys. Rev. Lett. **81**, 3563 (1998).

[39] A. Einstein, B. Podolsky, and N. Rosen, Phys. Rev.**47**, 777 (1935).

[40] R. Hanbury Brown and R. Q. Twiss, Nature (London) **177**, 27 (1956). M. Büttiker, Phys. Rev. B**46**, 12485 (1992). R. C. Liu, B. Odom, Y. Yamamoto, and S. Tarucha, Nature **391**, 263 (1998).

[41] E. Sukhorukov and D. Loss. Phys. Rev. Lett. **80** (1998) 4959-4962 . See cond-mat/9802050. E. Sukhorukov and D. Loss. To appear in Phys. Rev. B (May 15, 1999). See cond-mat/9809239.

[42] D. Loss, E. Sukhorukov, and G. Burkard. Preprint.

PHYSICAL REVIEW A VOLUME 57, NUMBER 1 JANUARY 1998

Quantum computation with quantum dots

Daniel Loss[1,2,*] and David P. DiVincenzo[1,3,†]

[1]*Institute for Theoretical Physics, University of California, Santa Barbara, Santa Barbara, California 93106-4030*

[2]*Department of Physics and Astronomy, University of Basel, Klingelbergstrasse 82, 4056 Basel, Switzerland*

[3]*IBM Research Division, T.J. Watson Research Center, P.O. Box 218, Yorktown Heights, New York 10598*

(Received 9 January 1997; revised manuscript received 22 July 1997)

We propose an implementation of a universal set of one- and two-quantum-bit gates for quantum computation using the spin states of coupled single-electron quantum dots. Desired operations are effected by the gating of the tunneling barrier between neighboring dots. Several measures of the gate quality are computed within a recently derived spin master equation incorporating decoherence caused by a prototypical magnetic environment. Dot-array experiments that would provide an initial demonstration of the desired nonequilibrium spin dynamics are proposed. [S1050-2947(98)04501-6]

PACS number(s): 03.67.Lx, 89.70.+c, 75.10.Jm, 89.80.+h

I. INTRODUCTION

The work of the past several years has greatly clarified both the theoretical potential and the experimental challenges of quantum computation [1]. In a quantum computer the state of each bit is permitted to be any quantum-mechanical state of a *qubit* (quantum bit, or two-level quantum system). Computation proceeds by a succession of "two-qubit quantum gates" [2], coherent interactions involving specific pairs of qubits, by analogy to the realization of ordinary digital computation as a succession of Boolean logic gates. It is now understood that the time evolution of an arbitrary quantum state is intrinsically more powerful computationally than the evolution of a digital logic state (the quantum computation can be viewed as a coherent superposition of digital computations proceeding in parallel).

Shor has shown [3] how this parallelism may be exploited to develop polynomial-time quantum algorithms for computational problems, such as prime factoring, which have previously been viewed as intractable. This has sparked investigations into the feasibility of the actual physical implementation of quantum computation. Achieving the conditions for quantum computation is extremely demanding, requiring precision control of Hamiltonian operations on well-defined two-level quantum systems and a very high degree of quantum coherence [4]. In ion-trap systems [5] and cavity quantum electrodynamic experiments [6], quantum computation at the level of an individual two-qubit gate has been demonstrated; however, it is unclear whether such atomic-physics implementations could ever be scaled up to do truly large-scale quantum computation, and some have speculated that solid-state physics, the scientific mainstay of digital computation, would ultimately provide a suitable arena for quantum computation as well. The initial realization of the model that we introduce here would correspond to only a modest step towards the realization of quantum computing, but it would at the same time be a very ambitious advance in the study of controlled nonequilibrium spin dynamics of magnetic nanosystems and could point the way towards more extensive studies to explore the large-scale quantum dynamics envisioned for a quantum computer.

II. QUANTUM-DOT IMPLEMENTATION OF TWO-QUBIT GATES

In this paper we develop a detailed scenario for how quantum computation may be achieved in a coupled quantum-dot system [7]. In our model the qubit is realized as the spin of the excess electron on a single-electron quantum dot; see Fig. 1. We introduce here a mechanism for two-qubit quantum-gate operation that operates by a purely elec-

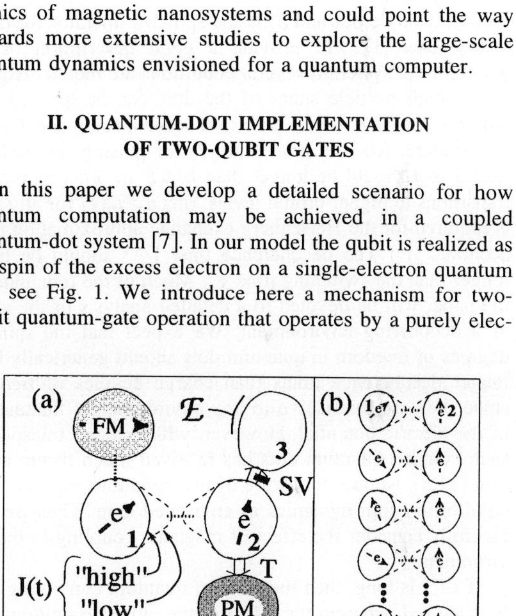

FIG. 1. (a) Schematic top view of two coupled quantum dots labeled 1 and 2, each containing one excess electron (e) with spin 1/2. The tunnel barrier between the dots can be raised or lowered by setting a gate voltage "high" (solid equipotential contour) or "low" (dashed equipotential contour). In the low state virtual tunneling (dotted line) produces a time-dependent Heisenberg exchange $J(t)$. Hopping to an auxiliary ferromagnetic dot (FM) provides one method of performing single-qubit operations. Tunneling (T) to the paramagnetic dot (PM) can be used as a POV read out with 75% reliability; spin-dependent tunneling (through "spin valve" SV) into dot 3 can lead to spin measurement via an electrometer $\mathcal{E}$. (b) Proposed experimental setup for initial test of swap-gate operation in an array of many noninteracting quantum-dot pairs. The left column of dots is initially unpolarized, while the right one is polarized; this state can be reversed by a swap operation [see Eq. (31)].

*Electronic address: loss@ubaclu.unibas.ch

†Electronic address: divince@watson.ibm.com

trical gating of the tunneling barrier between neighboring quantum dots rather than by spectroscopic manipulation as in other models. Controlled gating of the tunneling barrier between neighboring single-electron quantum dots in patterned two-dimensional electron-gas structures has already been achieved experimentally using a split-gate technique [8]. If the barrier potential is "high," tunneling is forbidden between dots and the qubit states are held stably without evolution in time (t). If the barrier is pulsed to a "low" voltage, the usual physics of the Hubbard model [9] says that the spins will be subject to a transient Heisenberg coupling,

$$H_s(t)=J(t)\vec{S}_1\cdot\vec{S}_2, \tag{1}$$

where $J(t)=4t_0^2(t)/u$ is the time-dependent exchange constant [10] that is produced by the turning on and off of the tunneling matrix element $t_0(t)$. Here u is the charging energy of a single dot and $\vec{S}_i$ is the spin-1/2 operator for dot i.

Equation (1) will provide a good description of the quantum-dot system if several conditions are met. (i) Higher-lying single-particle states of the dots can be ignored; this requires $\Delta E \gg kT$, where ΔE is the level spacing and T is the temperature. (ii) The time scale τ_s for pulsing the gate potential low should be longer than $\hbar/\Delta E$ in order to prevent transitions to higher orbital levels. (iii) $u>t_0(t)$ for all t; this is required for the Heisenberg exchange approximation to be accurate. (iv) The decoherence time Γ^{-1} should be much longer than the switching time τ_s. Much of the remainder of the paper will be devoted to a detailed analysis of the effect of a decohering environment. We expect that the spin-1/2 degrees of freedom in quantum dots should generically have longer decoherence times than charge degrees of freedom since they are insensitive to any environmental fluctuations of the electric potential. However, while charge transport in such coupled quantum dots has received much recent attention [11,8], we are not aware of investigations on their non-equilibrium spin dynamics as envisaged here. Thus we will carefully consider the effect of magnetic coupling to the environment.

If Γ^{-1} is long, then the ideal of quantum computing may be achieved, wherein the effect of the pulsed Hamiltonian is to apply a particular unitary time evolution operator $U_s(t) = T\exp\{-i\int_0^t H_s(t')dt'\}$ to the initial state of the two spins: $|\Psi(t)\rangle = U_s|\Psi(0)\rangle$. The pulsed Heisenberg coupling leads to a special form for U_s: For a specific duration τ_s of the spin-spin coupling such that $\int dt J(t)=J_0\tau_s=\pi(\mathrm{mod}2\pi)$ [12], $U_s(J_0\tau_s=\pi)=U_{sw}$ is the "swap" operator: If $|ij\rangle$ labels the basis states of two spins in the S_z basis with $i,j=0,1$, then $U_{sw}|ij\rangle=|ji\rangle$. Because it conserves the total angular momentum of the system, U_{sw} is not by itself sufficient to perform useful quantum computations, but if the interaction is pulsed on for just half the duration, the resulting square root of the swap operator is very useful as a fundamental quantum gate: For instance, a quantum XOR gate is obtained by a simple sequence of operations

$$U_{XOR}=e^{i(\pi/2)S_1^z}e^{-i(\pi/2)S_2^z}U_{sw}^{1/2}e^{i\pi S_1^z}U_{sw}^{1/2}, \tag{2}$$

where $e^{i\pi S_1^z}$, etc., are single-qubit operations only, which can be realized, e.g., by applying local magnetic fields (see Sec. III B) [13]. It has been established that XOR along with single-qubit operations may be assembled to do any quantum computation [2]. Note that the XOR of Eq. (2) is given in the basis where it has the form of a conditional phase-shift operation; the standard XOR is obtained by a simple basis change for qubit 2 [2].

III. MASTER EQUATION

We will now consider in detail the *nonideal* action of the swap operation when the two spins are coupled to a magnetic environment. A master equation model is obtained that explicitly accounts for the action of the environment during switching, to our knowledge, the first treatment of this effect. We use a Caldeira-Leggett–type model in which a set of harmonic oscillators are coupled linearly to the system spins by $H_{int}=\lambda\Sigma_{i=1,2}\vec{S}_i\cdot\vec{b}_i$. Here $b_i^j=\Sigma_\alpha g_\alpha^{ij}(a_{\alpha,ij}+a_{\alpha,ij}^\dagger)$ is a fluctuating quantum field whose free motion is governed by the harmonic-oscillator Hamiltonian $H_B=\Sigma\omega_\alpha^{ij}a_{\alpha,ij}^\dagger a_{\alpha,ij}$, where $a_{\alpha,ij}^\dagger$ ($a_{\alpha,ij}$) are bosonic creation (annihilation) operators (with $j=x,y,z$) and ω_α^{ij} are the corresponding frequencies with spectral distribution function $J_{ij}(\omega)=\pi\Sigma_\alpha(g_\alpha^{ij})^2\delta(\omega-\omega_\alpha)$ [14]. The system and environment are initially uncorrelated with the latter in thermal equilibrium described by the canonical density matrix ρ_B with temperature T. We assume for simplicity that the environment acts isotropically and is equal and independent on both dots. We do not consider this to be a microscopically accurate model for these as-yet-unconstructed quantum-dot systems, but rather as a generic phenomenological description of the environment of a spin, which will permit us to explore the complete time dependence of the gate action on the single coupling constant λ and the controlled parameters of $H_s(t)$ [15].

A. Swap gate

The quantity of interest is the system density matrix $\rho(t)=\mathrm{Tr}_B\bar{\rho}(t)$, which we obtain by tracing out the environment degrees of freedom. The full density matrix $\bar{\rho}$ itself obeys the von Neumann equation

$$\dot{\bar{\rho}}(t)=-i[H,\bar{\rho}]\equiv -i\mathcal{L}\bar{\rho}, \tag{3}$$

where

$$\mathcal{L}=\mathcal{L}_s(t)+\mathcal{L}_{int}+\mathcal{L}_B \tag{4}$$

denotes the Liouvillian [16] corresponding to the full Hamiltonian

$$H=H_s(t)+H_{int}+H_B. \tag{5}$$

Our goal is to find the linear map (superoperator) $\mathcal{V}(t)$ that connects the input state of the gate $\rho_0=\rho(t=0)$ with the output state $\rho(t)$ after time $t>\tau_s$ has elapsed, $\rho(t)=\mathcal{V}(t)\rho_0$. $\mathcal{V}(t)$ must satisfy three physical conditions: (i) trace preservation $\mathrm{Tr}_s\,\mathcal{V}\rho=1$, where Tr_s denotes the system trace; (ii) Hermiticity preservation $(\mathcal{V}\rho)^\dagger=\mathcal{V}\rho$; and (iii) complete positivity, $(\mathcal{V}\otimes 1_B)\bar{\rho}\geq 0$. Using the Zwanzig master equation approach [16], we sketch the derivation for $\mathcal{V}$ in the Born and

Markov approximations, which respects these three conditions. The situation we analyze here is unusual in that H_s is explicitly time dependent and changes abruptly in time. It is this fact that requires a separate treatment for times $t \leq \tau_s$ and $t > \tau_s$. To implement this time scale separation and to preserve positivity it is best to start from the exact master equation in pure integral form

$$\rho(t) = \mathcal{U}_s(t,0)\rho_0 - \int_0^t d\sigma \int_0^\sigma d\tau\, \mathcal{U}_s(t,\sigma)\mathcal{M}(\sigma,\tau)\rho(\tau), \tag{6}$$

where

$$\mathcal{U}_i(t,t') = T\exp\left\{-i\int_{t'}^t d\tau \mathcal{L}_i(\tau)\right\}, \tag{7}$$

where $i = s$, B, int, or q. Here q indicates the projected Liouvillian

$$\mathcal{L}_q = (1-P)\mathcal{L} = (1-\rho_B \mathrm{Tr}_B)\mathcal{L}. \tag{8}$$

Also, the "memory kernel" is

$$\mathcal{M}(\sigma,\tau) = \mathrm{Tr}_B \mathcal{L}_{int}\mathcal{U}_q(\sigma,\tau)\mathcal{L}_{int}\rho_B. \tag{9}$$

We solve Eq. (6) in the Born approximation and for $t \gg \tau_s$. To this end the time integrals are split up into three parts: (i) $0 \leq \tau \leq \sigma \leq \tau_s < t$, (ii) $0 \leq \tau \leq \tau_s \leq \sigma < t$, and (iii) $0 \leq \tau_s \leq \tau \leq \sigma < t$. Keeping only leading terms in τ_s, we retain the contribution from interval (ii) as it is proportional to τ_s, whereas we can drop interval (i), which leads to higher-order terms. However, note that terms containing $J_0\tau_s$ must be kept to all orders [12]. Interval (iii) is independent of τ_s.

Rewriting the expressions and performing a Born approximation (i.e., keeping only lowest-order terms in λ^2) with subsequent Markov approximation we find, for $t \gg \tau_s$,

$$\mathcal{V}(t) = e^{-(t-\tau_s)\mathcal{K}_3}\mathcal{U}_s(\tau_s)\,(1-\mathcal{K}_2), \tag{10}$$

where $\mathcal{U}_i(\tau_s) = \mathcal{U}_i(\tau_s,0)$, $\mathcal{K}_2$ describes the effect of the environment during the switching,

$$\mathcal{K}_2 = \mathcal{U}_s^\dagger(\tau_s)\int_0^{\tau_s} d\tau \int_0^\infty dt\ \mathrm{Tr}_B \mathcal{L}_{int}\mathcal{U}_s(\tau)\mathcal{U}_B(t) \times \mathcal{L}_{int}\rho_B\mathcal{U}_s(\tau_s-\tau), \tag{11}$$

while

$$\mathcal{K}_3 = \int_0^\infty dt\ \mathrm{Tr}_B \mathcal{L}_{int}\mathcal{U}_B(t)\mathcal{L}_{int}\rho_B \tag{12}$$

is independent of H_s. We also note that $\mathcal{U}_s(1-\mathcal{K}_2)$ has a simple interpretation as being the "transient contribution" that changes the initial value ρ_0 at $t=0$ to $\mathcal{U}_s(\tau_s)(1-\mathcal{K}_2)\rho_0$ at $t=\tau_s$. We show in the Appendix that, to leading order, our superoperator $\mathcal{V}$ indeed satisfies all three conditions stated above, in particular complete positivity. Such a proof for spins with an explicit time-dependent and direct interaction (1) is not simply related to the case of a master equation for noninteracting spins (and without explicit time dependence) considered in the literature (see, for example, [17,16]). We also note that the above Born and Markov approximations could also be introduced in the master equation in the more usual differential-integral representation. However, it is well known from studies in noninteracting spin problems [18] that in this case the resulting propagator is in general *no longer* completely positive.

Next, we evaluate the above superoperators more explicitly, obtaining

$$\mathcal{K}_2\rho = (\Gamma + i\Delta)\sum_i \int_0^{\tau_s} d\tau [\vec{S}_i(\tau_s), \vec{S}_i(\tau)\rho] + \mathrm{H.c.}, \tag{13}$$

$$\mathcal{K}_3\rho = \Gamma\left(3\rho - 2\sum_i \vec{S}_i\rho\cdot\vec{S}_i\right), \tag{14}$$

where in the commutator in Eq. (13) a dot product is understood between the vector parts of the two factors, and where Γ, Δ are real and given by

$$\Gamma = \frac{\lambda^2}{\pi}\int_0^\infty dt \int_0^\infty d\omega\ J(\omega)\cos(\omega t)\coth\left(\frac{\omega}{2k_BT}\right), \tag{15}$$

$$\Delta = \frac{\lambda^2}{\pi}\int_0^\infty dt \int_0^\infty d\omega\ J(\omega)\sin(\omega t). \tag{16}$$

In our model, the transverse and longitudinal relaxation or decoherence rates of the system spins are the same and given by Γ. For instance, for Ohmic damping with $J(\omega) = \eta\omega$, we get $\Gamma = \lambda^2\eta k_BT$ and $\Delta = \lambda^2\eta\omega_c/\pi$, with ω_c some high-frequency cutoff. Requiring for consistency that $\Gamma\tau_s, \Delta\tau_s \ll 1$, we find that $\mathcal{K}_2$ is in fact a small correction. However, we emphasize again that, to our knowledge, this is the first time that any analysis of this $\mathcal{K}_2$ term, describing the action of the environment during the finite time that the system Hamiltonian is switched on, has been given.

For further evaluation of $\mathcal{V}$ we adopt a matrix representation, defined by

$$\mathcal{V}_{ab|cd} = (e_{ab}, \mathcal{V}e_{cd}) \equiv \mathrm{Tr}_s e_{ab}^\dagger \mathcal{V}e_{cd}, \tag{17}$$

where $\{e_{ab}|a,b=1,\dots,4\}$ is an orthonormal basis, i.e., $(e_{ab}, e_{cd}) = \delta_{ac}\delta_{bd}$. In this notation we then have

$$\rho(t)_{ab} = \sum_{c,d}\mathcal{V}_{ab|cd}(\rho_0)_{cd}, \tag{18}$$

with $\mathcal{V}$ being a 16×16 matrix.

Note that $\mathcal{K}_{2,3}$ and $\mathcal{U}_s$ are not simultaneously diagonal. However, since $\mathcal{K}_3(1,\vec{S}_i) = 2\Gamma(0,\vec{S}_i)$ we see that $\exp\{-(t-\tau_s)\mathcal{K}_3\}$ is diagonal in the "polarization basis" $\{e_{ab}^p = e_a^1 e_b^2; e_{1,\dots,4}^i = (1/\sqrt{2}, \sqrt{2}S_i^x, \sqrt{2}S_i^y, \sqrt{2}S_i^z), i=1,2\}$, while $\mathcal{L}_s$ and thus $\mathcal{U}_s$ are diagonal in the "multiplet basis" $\{e_{\alpha\beta}^m = |\alpha\rangle\langle\beta|, \alpha,\beta=1,\dots,4; |1\rangle = (|01\rangle - |10\rangle)/\sqrt{2}, |2\rangle = (|01\rangle + |10\rangle)/\sqrt{2}, |3\rangle = |00\rangle, |4\rangle = |11\rangle\}$, with

$$\mathcal{U}_s(t)_{\alpha\beta|\alpha'\beta'} = \delta_{\alpha\alpha'}\delta_{\beta\beta'}e^{-it(E_\alpha^m - E_\beta^m)}, \tag{19}$$

where $E_1^m = -3J_0/4$ and $E_{2,3,4}^m = J_0/4$ are the singlet and triplet eigenvalues. Finally, $\mathcal{K}_2$ is most easily evaluated also in the multiplet basis; after some calculation we find that $\mathcal{K}_2 = \mathcal{K}_2^d - \mathcal{K}_2^{nd}$, with

$$(\mathcal{K}_2^d)_{\alpha\beta|\gamma\delta} = \sum_{i,\alpha'} [\delta_{\alpha\gamma}\langle\delta|\vec{S}_i|\alpha'\rangle\cdot\langle\alpha'|\vec{S}_i|\beta\rangle k^*_{\alpha'\alpha'|\delta\beta} + \delta_{\beta\delta}\langle\alpha|\vec{S}_i|\alpha'\rangle\cdot\langle\alpha'|\vec{S}_i|\gamma\rangle k_{\alpha'\alpha'|\gamma\alpha}], \quad (20)$$

$$(\mathcal{K}_2^{nd})_{\alpha\beta|\gamma\delta} = \sum_i \langle\alpha|\vec{S}_i|\gamma\rangle\cdot\langle\delta|\vec{S}_i|\beta\rangle[k_{\alpha\beta|\gamma\delta} + (k_{\beta\alpha|\delta\gamma})^*]. \quad (21)$$

Here

$$k_{\alpha\beta|\gamma\delta} = (\Gamma + i\Delta)e^{i(E_\delta^m - E_\beta^m)\tau_s}\int_0^{\tau_s} d\tau\, e^{i(E_\alpha^m - E_\gamma^m)\tau}$$
$$= \frac{1}{2\omega_{\alpha\gamma}}[\Gamma c_{\delta\beta} - \Delta s_{\delta\beta} + i(\Gamma s_{\delta\beta} + \Delta c_{\delta\beta})] \times [s_{\alpha\gamma} + i(1 - c_{\alpha\gamma})],$$

$$c_{ij} = \cos(\tau_s\omega_{ij}), \quad s_{ij} = \sin(\tau_s\omega_{ij}), \quad \omega_{ij} = E_i^m - E_j^m.$$

Using the above matrix notation, we can write explicitly

$$\mathcal{V}_{\alpha\beta|\gamma\delta} = \sum_{a,b,\alpha',\beta'} (C_{ab|\alpha\beta})^*(e^{-(t-\tau_s)\mathcal{K}_3})_{ab|ab} \times C_{ab|\alpha'\beta'}e^{-i\tau_s(E_{\alpha'}^m - E_{\beta'}^m)}(1 - \mathcal{K}_2)_{\alpha'\beta'|\gamma\delta}, \quad (22)$$

where $C_{ab|\alpha\beta} = (e_{ab}^p, e_{\alpha\beta}^m)$ is the unitary basis change between the polarization and the multiplet basis.

B. One-bit gates

We now repeat the preceding analysis for single-qubit rotations such as $e^{i(\pi/2)S_i^z}$ as required in Eq. (2). Such rotations can be achieved if a magnetic field $\vec{H}_i$ could be pulsed exclusively onto spin i, perhaps by a scanning-probe tip. An alternative way, which would become attractive if further advances are made in the synthesis of nanostructures in magnetic semiconductors [19], is to use, as indicated in Fig. 1(a), an auxiliary dot (FM) made of an insulating, ferromagnetically ordered material that can be connected to dot 1 (or dot 2) by the same kind of electrical gating as discussed above [8]. If the the barrier between dot 1 and dot FM were lowered so that the electron's wave function overlaps with the magnetized region for a fixed time τ_s, the Hamiltonian for the qubit on dot 1 will contain a Zeeman term during that time. For all earlier and later times the magnetic field seen by the qubit should be zero; any stray magnetic field from the dot FM at neighboring dots 1, 2, etc., could be made small by making FM part of a closure domain or closed magnetic circuit.

In either case, the spin is rotated and the corresponding Hamiltonian is given by

$$\int_0^{\tau_s} dt H_s^H = \sum_{i=1}^2 \omega_i\tau_s S_i^z, \quad (23)$$

with $\omega_i = g\mu_B H_i^z$, where we assume that the H field acting on spin i is along the z axis. The calculation proceeds along the same line as the one described above: Just as in Eq. (10), the expression obtained for the superoperator is

$$\mathcal{V}_H(t) = e^{-(t-\tau_s)\mathcal{K}_3}\mathcal{U}_s^H(\tau_s)(1 - \mathcal{K}_2^H). \quad (24)$$

$\mathcal{K}_3$ is exactly the same as before, Eq. (14). $\mathcal{U}_s^H(\tau_s)$ is again given by Eq. (7) with the modification that the Liouvillian [see Eq. (4)] corresponding to the magnetic-field Hamiltonian of Eq. (23) is used rather than that for the exchange Hamiltonian H_s [Eqs. (5) and (1)]. The explicit matrix representation is

$$(\mathcal{U}_s^H(\tau_s))_{rs|r's'} = \delta_{rr'}\delta_{ss'}\exp\left(-i\sum_{i=1}^2 (E_r^i - E_s^i)\tau_s\right). \quad (25)$$

Here we are employing another basis, the S_z basis for the two spins $\{e_{rs}^z = |r\rangle\langle s|,\ r,s = 1,2,3,4; |s\rangle = |00\rangle, |01\rangle, |10\rangle, |11\rangle\}$. The energies are

$$\{E_r^1\} = \{E_{1,2,3,4}^1\} = \frac{\omega_1}{2}\{1,1,-1,-1\},$$
$$\{E_r^2\} = \{E_{1,2,3,4}^2\} = \frac{\omega_2}{2}\{1,-1,1,-1\}. \quad (26)$$

The $\mathcal{K}_2^H$ calculation also proceeds as before [see Eq. (13)] using the new Hamiltonian; the result is $\mathcal{K}_2^H = \mathcal{K}_2^{H,d} - \mathcal{K}_2^{H,nd}$, with

$$(\mathcal{K}_2^{H,d})_{rs|tu} = \sum_{i,r'} [\delta_{rt}\langle u|\vec{S}_i|r'\rangle\cdot\langle r'|\vec{S}_i|s\rangle(k_{r'r'|us}^i)^* + \delta_{su}\langle r|\vec{S}_i|r'\rangle\cdot\langle r'|\vec{S}_i|t\rangle k_{r'r'|tr}^i], \quad (27)$$

$$(\mathcal{K}_2^{H,nd})_{rs|tu} = \sum_i \langle r|\vec{S}_i|t\rangle\cdot\langle u|\vec{S}_i|s\rangle[k_{rs|tu}^i + (k_{sr|ut}^i)^*]. \quad (28)$$

Here

$$k_{rs|tu}^i = (\Gamma + i\Delta)e^{i(E_u^i - E_s^i)\tau_s}\int_0^{\tau_s} d\tau e^{i(E_r^i - E_t^i)\tau}$$
$$= \frac{1}{2\omega_{rt}^i}[\Gamma c_{us}^i - \Delta s_{us}^i + i(\Gamma s_{us}^i + \Delta c_{us}^i)] \times [s_{rt}^i + i(1 - c_{rt}^i)], \quad (29)$$

$$c_{ij}^k = \cos(\tau_s\omega_{ij}^k), \quad s_{ij}^k = \sin(\tau_s\omega_{ij}^k), \quad \omega_{ij}^k = E_i^k - E_j^k.$$

The E^k's are from Eq. (26). Finally, the explicit matrix form for $\mathcal{V}^H$ may be written

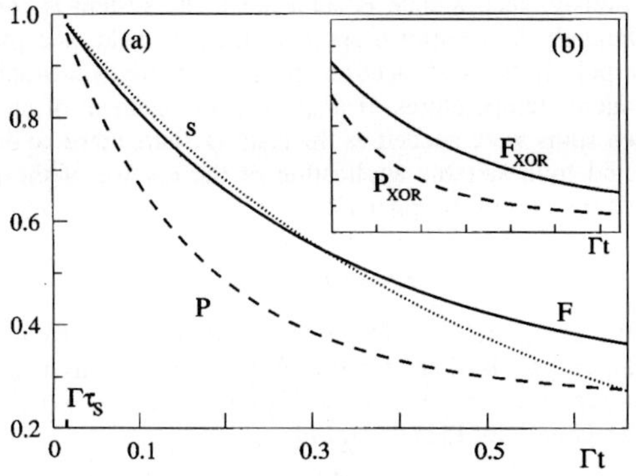

FIG. 2. (a) Swap polarization $s \equiv 2\langle S_1^z(t)\rangle$ [see Eq. (31)], gate fidelity F, and gate purity P vs Γt for "swap" using parameters $J_0\tau_s=\pi$, $\Gamma\tau_s=0.017$, and $\Delta\tau_s=-0.0145$. (b) Same for XOR obtained using the four operations in Eq. (2) (the final two single-spin operations done simultaneously). The same parameters and scales as in (a) are used; the pulse-to-pulse time is taken to be $3\tau_s$. Γt is measured from the end of the fourth pulse.

$$\mathcal{V}^H_{ab|a'b'} = \sum_{r,s,r',s'} (e^{-(t-\tau_s)\mathcal{K}_3})_{ab|ab}(D_{rs|ab})^* \times \exp\left(-i\sum_{i=1}^{2} \tau_s(E_r^i - E_s^i)\right)(1-\mathcal{K}_2^H)_{rs|r's'} \times D_{r's'|a'b'}, \tag{30}$$

where $D_{rs|ab}=(e^z_{rs}, e^p_{ab})$ is now the unitary basis change between the S_z basis and the polarization basis.

C. Numerical study for swap gate and XOR gate

Having diagonalized the problem, we can now calculate any system observable; the required matrix calculations are involved and complete evaluation is done with MATHEMATICA. We will consider three parameters (s, F, and P in Fig. 2) relevant for characterizing the gate operation. We first perform this analysis for the swap operation introduced above.

The swap operation would provide a useful experimental test for the gate functionality: Let us assume that at $t=0$ spin 2 is (nearly) polarized, say, along the z axis, while spin 1 is (nearly) unpolarized, i.e., $\rho_0=(1+2S_2^z)/4$. This can be achieved, e.g., by selective optical excitation or by an applied magnetic field with a strong spatial gradient. Next we apply a swap operation by pulsing the exchange coupling such that $J_0\tau_s=\pi$ and observe the resulting polarization of spin 1 described by

$$\langle S_1^z(t)\rangle = \frac{1}{2}\mathcal{V}(t)_{41|14}, \tag{31}$$

where $\mathcal{V}$ is evaluated in the polarization basis. After time τ_s spin 1 is almost fully polarized (whereas spin 2 is now unpolarized) and, due to the environment, decays exponentially with rate of order Γ. To make the signal (31) easily measurable by conventional magnetometry, we can envisage a setup consisting of a large array of identical, noninteracting pairs of dots as indicated in Fig. 1(b).

To further characterize the gate performance we follow Ref. [20] and calculate the gate fidelity $F=\overline{\langle\psi_0|\mathcal{U}^\dagger(\tau_s)\rho(t)|\psi_0\rangle}$ and the gate purity $P=\overline{Tr_s[\rho(t)]^2}$, where the overbar means an average over all initial system states $|\psi_0\rangle$. Expressing $\mathcal{V}$ in the multiplet basis and using trace and Hermiticity preservation we find

$$F(t)=\frac{1}{6}+\frac{1}{24}\mathrm{Re}\left[\sum_{\alpha}\mathcal{V}_{\alpha\alpha|\alpha\alpha}+\sum_{\alpha,\beta}\mathcal{V}_{\alpha\beta|\alpha\beta}e^{i\tau_s(E^m_\alpha-E^m_\beta)}\right], \tag{32}$$

$$P(t)=\frac{1}{24}\sum_{i,k,k'}\left[|\mathcal{V}_{kk'|ii}|^2+\sum_j(\mathcal{V}_{kk'|ii}\mathcal{V}^*_{kk'|jj}+|\mathcal{V}_{kk'|ij}|^2)\right] \tag{33}$$

[in fact, the expression for $P(t)$ holds in any basis]. Evaluations of these functions for specific parameter values are shown in Fig. 2. For the parameters shown, the effect of the environment during the switching, i.e., $\mathcal{K}_2$ in Eq. (10), is on the order of a few percent.

The dimensionless parameters used here would, for example, correspond to the following actual physical parameters: If an exchange constant $J_0=80$ μeV$\approx$1 K were achievable, then pulse durations of $\tau_s\approx 25$ ps and decoherence times of $\Gamma^{-1}\approx 1.4$ ns would be needed; such parameters, and perhaps much better, are apparently achievable in solid-state spin systems [19].

As a final application, we calculate the full XOR by applying the corresponding superoperators in the sequence associated with the one on the right-hand side of Eq. (2). We use the same dimensionless parameters as above, and as before we then calculate the gate fidelity and the gate purity. Some representative results of this calculation are plotted in the inset of Fig. 2(b). To attain the $\pi/2$ single-bit rotations of Eq. (2) in a τ_s of 25 ps would require a magnetic field H $\approx$0.6 T, which would be readily available in the solid state.

IV. DISCUSSION

As a final remark about the decoherence problem, we note that the parameters that we have chosen in the presentation of our numerical work, which we consider to be realistic for known nanoscale semiconductor materials, of course fall far, far lower than the 0.999 99 levels that are presently considered desirable by quantum-computation theorists [1]; still, the achievement of even much lesser quality quantum gate operation would be a tremendous advance in the controlled, nonequilibrium time evolution of solid-state spin systems and could point the way to the devices that could ultimately be used in a quantum computer. Considering the situation more broadly, we are quite aware that our proposal for quantum-dot quantum computation relies on simultaneous further advances in the experimental techniques of semiconductor nanofabrication, magnetic semiconductor synthesis, single electronics, and perhaps in scanning-probe techniques. Still, we also feel strongly that such proposals should be developed seriously, and taken seriously, at present since we believe that many aspects of the present proposal are testable in the not-too-distant future. This is particularly so for the demonstration of the swap action on an array of dot pairs. Such a demonstration would be of clear interest not only for quantum computation, but would also represent a technique for exploring the nonequilibrium dynamics of spins in quantum dots.

To make the quantum-dot idea a complete proposal for quantum computation, we need to touch on several other important features of quantum-computer operation. As our guideline we follow the five requirements laid out by one of us [4]: (i) identification of well-defined qubits, (ii) reliable state preparation, (iii) low decoherence, (iv) accurate quantum gate operations, and (v) strong quantum measurements. Items (i), (iii), and (iv) have been very thoroughly considered above. We would now like to propose several possible means by which requirements (ii) and (v), for state preparation (read in) and quantum measurement (read out), may be satisfied.

One scheme for qubit measurement that we suggest involves a switchable tunneling [T in Fig. 1(a)] into a supercooled paramagnetic dot (PM). When the measurement is to be performed, the electron tunnels (this will be real tunneling, not the virtual tunneling used for the swap gate above) into PM, nucleating from the metastable phase a ferromagnetic domain whose magnetization direction could be measured by conventional means. The orientation (θ,ϕ) of this magnetization vector is a "pointer" that measures the spin direction; it is a generalized measurement in which the measurement outcomes form a continuous set rather than having two discrete values. Such a case is covered by the general formalism of positive-operator-valued (POV) measurements [21]. If there is no magnetic anisotropy in dot PM, then symmetry dictates that the positive measurement operators would be projectors into the overcomplete set of spin-1/2 coherent states

$$|\theta,\phi\rangle=\cos\frac{\theta}{2}|0\rangle+e^{i\phi}\sin\frac{\theta}{2}|1\rangle. \tag{34}$$

A 75%-reliable measurement of spin up and spin down is obtained if the magnetization direction (θ,ϕ) in the upper hemisphere is interpreted as up and in the lower hemisphere as down; this is so simply because

$$\frac{1}{2\pi}\int_U d\Omega|\langle 0|\theta,\phi\rangle|^2=\frac{3}{4}. \tag{35}$$

Here U denotes integration over the upper hemisphere and 2π is the normalization constant for the coherent states.

Another approach which would potentially give a 100% reliable measurement requires a spin-dependent, switchable "spin valve" tunnel barrier (SV) of the type mentioned, e.g., in Ref. [22]. When the measurement is to be performed, SV is switched so that only an up-spin electron passes into semiconductor dot 3. Then the presence of an electron on 3, measured by electrometer $\mathcal{E}$, would provide a measurement that the spin had been up. It is well known now how to create nanoscale single-electron electrometers with exquisite sensitivity (down to 10^{-8} of one electron) [23].

We need only discuss the state-preparation problem briefly. For many applications in quantum computing, only a simple initial state, such as all spins up, needs to be created. Obviously, such a state is achieved if the system is cooled sufficiently in a uniform applied magnetic field; acceptable spin polarizations of electron spins are readily achievable at cryogenic temperatures. If a specific arrangement of up and down spins were needed as the starting state, these could be created by a suitable application of the reverse of the spin valve measurement apparatus.

ACKNOWLEDGMENTS

We are grateful to D. D. Awschalom, H.-B. Braun, T. Brun, and G. Burkard for useful discussions. This research was supported in part by the National Science Foundation under Grant No. PHY94-07194.

APPENDIX: COMPLETE POSITIVITY OF TIME-EVOLUTION SUPEROPERATOR $\mathcal{V}$

Here we sketch the proof that the superoperator $\mathcal{V}$ in Eq. (10) is completely positive. We analyze the $\mathcal{K}_3$ term first. We write

$$e^{-\tau\mathcal{K}_3}=\lim_{N\to\infty}\left(1-\frac{\tau}{N}\mathcal{K}_3\right)^N. \tag{A1}$$

It is sufficient to prove that the infinitesimal operator is completely positive. It is straightforward to show, using Eq. (14), that

$$\left(1-\frac{\tau}{N}\mathcal{K}_3\right)\rho=Z_3^\dagger\cdot\rho Z_3+O((\tau/N)^2). \tag{A2}$$

Here Z_3 is the seven-component vector operator

$$Z_3=\left(1-\frac{\tau}{2N}\sum_{k=1}^{6}B_k^\dagger B_k,\sqrt{\frac{\tau}{N}}\mathbf{B}\right), \tag{A3}$$

where

$$\mathbf{B}=(B_1,\ldots,B_6)=\sqrt{2\Gamma}(\vec{S}_1,\vec{S}_2). \tag{A4}$$

Note that for this case $B_k^\dagger=B_k$ and $\Sigma_{k=1}^6 B_k^\dagger B_k=3\Gamma$.

We recall that it is easy to prove that any superoperator $\mathcal{S}$ of the form

$$\mathcal{S}\rho=Z^\dagger\rho Z \tag{A5}$$

as in the first term of Eq. (A2) is completely positive. Indeed, considering its action on any state vector of the system plus environment ϕ and taking a positive ρ we get

$$(\phi,\mathcal{S}\rho\phi)=(\phi,Z^\dagger\rho Z\phi)=(Z\phi,\rho Z\phi)\geqslant 0\quad\forall\phi. \tag{A6}$$

Next we consider the $1-\mathcal{K}_2$ term of Eq. (10). Starting from Eq. (13), we put this term in a form corresponding to the completely positive form (A5). We find

$$(1-\mathcal{K}_2)\rho=Z_2^\dagger\cdot\rho Z_2+O(\lambda^4,\tau_s^2,(\lambda^2\tau_s)^2), \tag{A7}$$

with Z_2 being the vector operator

$$Z_2=(1+Y^\dagger\cdot X^\dagger,X-Y^\dagger), \tag{A8}$$

with

$$X=-(\Gamma+i\Delta)(\vec{S}_1(\tau_s),\vec{S}_2(\tau_s)), \tag{A9}$$

$$Y=\int_0^{\tau_s} d\tau(\vec{S}_1(\tau),\vec{S}_2(\tau)). \tag{A10}$$

So, from the same arguments as above, Eq. (A7) establishes that $1-\mathcal{K}_2$ is completely positive up to the order of accuracy discussed in the text.

Finally, we note that the other two general conditions for a physical superoperator also follow immediately: Trace preservation of $\mathcal{V}$ follows from the fact that a Liouvillian $\mathcal{L}$ appears to the left in the basic equations for $\mathcal{K}_2$, Eq. (11), and $\mathcal{K}_3$, Eq. (12). Trace preservation is also reflected in the fact that $Z_2 \cdot Z_2^\dagger = 1$ and $Z_3 \cdot Z_3^\dagger = 1$ to leading order. The form (A5) also obviously preserves Hermiticity of the density operator; this is also clear from the forms of Eqs. (13) and (14).

[1] S. Lloyd, Science **261**, 1589 (1993); C. H. Bennett, Phys. Today **48(10)**, 24 (1995); D. P. DiVincenzo, Science **269**, 255 (1995); A. Barenco, Contemp. Phys. **37**, 375 (1996).

[2] A. Barenco *et al.*, Phys. Rev. A **52**, 3457 (1995).

[3] P. Shor, *Proceedings of the 35th Annual Symposium on the Foundations of Computer Science* (IEEE Press, Los Alamitos, 1994), p. 124.

[4] D. P. DiVincenzo, Report No. cond-mat/9612126; in *Mesoscopic Electron Transport*, Vol. 345 of *NATO Advanced Study Institute, Series E: Applied Sciences*, edited by L. Sohn, L. Kouwenhoven, and G. Schoen (Kluwer, Dordrecht, 1997).

[5] J.-I. Cirac and P. Zoller, Phys. Rev. Lett. **74**, 4091 (1995); J.-I. Cirac, T. Pellizzari, and P. Zoller, Science **273**, 1207 (1996); C. Monroe *et al.*, *ibid.* **75**, 4714 (1995).

[6] Q. A. Turchette *et al.*, Phys. Rev. Lett. **75**, 4710 (1995).

[7] There has been some earlier speculation on how coupled quantum wells might be used in quantum-scale information processing: see R. Landauer, Science **272**, 1914 (1996); A. Barenco *et al.*, Phys. Rev. Lett. **74**, 4083 (1995).

[8] C. Livermore *et al.*, Science **274**, 1332 (1996); F. R. Waugh *et al.*, Phys. Rev. B **53**, 1413 (1996); Phys. Rev. Lett. **75**, 705 (1995).

[9] N. W. Ashcroft and N. D. Mermin, *Solid State Physics* (Saunders, Philadelphia, 1976), Chap. 32.

[10] We can also envisage a superexchange mechanism to obtain a Heisenberg interaction by using three aligned quantum dots where the middle one has a higher-energy level (by the amount ϵ) such that the electron spins of the outer two dots are also Heisenberg coupled, but now with the exchange coupling being $J=4t_0^4(1/\epsilon^2 u+1/2\epsilon^3)$.

[11] L. I. Glazman and K. A. Matveev, Zh. Éksp. Teor. Fiz. **98**, 1834 (1990) [Sov. Phys. JETP **71**, 1031 (1990)]; C. A. Stafford and S. Das Sarma, Phys. Rev. Lett. **72**, 3590 (1994).

[12] We assume for simplicity that the shape of the applied pulse is roughly rectangular with $J_0\tau_s$ constant.

[13] We note that explicitly $U_{XOR}=\frac{1}{2}+S_1^z+S_2^z-2S_1^zS_2^z$, with the corresponding XOR Hamiltonian $\int_0^t dt' H_{XOR} =\pi[1-2S_1^z-2S_2^z+4S_1^zS_2^z]/4$. An alternative way to achieve the XOR operation is given by $U_{XOR} =e^{i\pi S_1^z}U_{sw}^{-1/2}e^{-i(\pi/2)S_1^z}U_{sw}e^{i(\pi/2)S_1^z}U_{sw}^{1/2}$. This form has the potential advantage that the single-qubit operations involve only spin 1.

[14] A simple discussion of the consequences of decoherence models of this type may be found in I. L. Chuang, R. Laflamme, P. Shor, and W. H. Zurek, Science **270**, 1633 (1995).

[15] For a microscopic discussion of dissipation in quantum dots concerning the *charge* degrees of freedom see, e.g., H. Schoeller and G. Schön, Phys. Rev. B **50**, 18 436 (1994); Physica B **203**, 423 (1994).

[16] E. Fick, G. Sauermann, and W. D. Brewer, *Quantum Statistics of Dynamic Processes*, Springer Series in Solid-State Sciences, edited by H. K. V. Lotsch, M. Cardona, P. Fulde, K. v. Klitzing, and H.-J. Queisser, Vol. 86 (Springer-Verlag, Berlin, 1990).

[17] E. B. Davies, *Quantum Theory of Open Systems* (Academic, New York, 1976).

[18] M. Celio and D. Loss, Physica A **150**, 769 (1989).

[19] S. A. Crooker *et al.*, Phys. Rev. Lett. **77**, 2814 (1996).

[20] J. F. Poyatos, J.-I. Cirac, and P. Zoller, Phys. Rev. Lett. **78**, 390 (1997); e-print quant-ph/9611013.

[21] A. Peres, *Quantum Theory: Concepts and Methods* (Kluwer, Dordrecht, 1993).

[22] G. Prinz, Phys. Today **45(4)**, 58 (1995).

[23] M. Devoret, D. Estève, and Ch. Urbina, Nature (London) **360**, 547 (1992).

PHYSICAL REVIEW B VOLUME 59, NUMBER 3 15 JANUARY 1999-I

Coupled quantum dots as quantum gates

Guido Burkard* and Daniel Loss†
Department of Physics and Astronomy, University of Basel, Klingelbergstrasse 82, CH-4056 Basel, Switzerland

David P. DiVincenzo‡
IBM Research Division, Thomas J. Watson Research Center, P.O. Box 218, Yorktown Heights, New York 10598
(Received 3 August 1998)

We consider a quantum-gate mechanism based on electron spins in coupled semiconductor quantum dots. Such gates provide a general source of spin entanglement and can be used for quantum computers. We determine the exchange coupling J in the effective Heisenberg model as a function of magnetic (B) and electric fields, and of the interdot distance a within the Heitler-London approximation of molecular physics. This result is refined by using sp hybridization, and by the Hund-Mulliken molecular-orbit approach, which leads to an extended Hubbard description for the two-dot system that shows a remarkable dependence on B and a due to the long-range Coulomb interaction. We find that the exchange J changes sign at a finite field (leading to a pronounced jump in the magnetization) and then decays exponentially. The magnetization and the spin susceptibilities of the coupled dots are calculated. We show that the dephasing due to nuclear spins in GaAs can be strongly suppressed by dynamical nuclear-spin polarization and/or by magnetic fields.
[S0163-1829(99)01003-6]

I. INTRODUCTION

Semiconductor quantum dots, sometimes referred to as artificial atoms, are small devices in which charge carriers are confined in all three dimensions.[1] The confinement is usually achieved by electrical gating and/or etching techniques applied, e.g., to a two-dimensional electron gas (2DEG). Since the dimensions of quantum dots are on the order of the Fermi wavelength, their electronic spectrum consists of discrete energy levels that have been studied in great detail in conductance[1,2] and spectroscopy measurements.[1,3,4] In GaAs heterostructures the number of electrons in the dots can be changed one-by-one starting from zero.[5] Typical laboratory magnetic fields ($B \approx 1$ T) correspond to magnetic lengths on the order of $l_B \approx 10$ nm, being much larger than the Bohr radius of real atoms but of the same size as artificial atoms. As a consequence, the dot spectrum depends strongly on the applied magnetic field.[1–3] In coupled quantum dots, which can be considered to some extent as artificial molecules, Coulomb blockade effects[6] and magnetization[7] have been observed, as well as the formation of a delocalized "molecular state." [8]

Motivated by the rapid down scaling of integrated circuits, there has been continued interest in classical logic devices made of electrostatically coupled quantum dots.[9] More recently, the discovery of new principles of computation based on quantum mechanics[10] has led to the idea of using coupled quantum dots for quantum computation;[11] many other proposed implementations have been explored, involving NMR,[12–14] trapped ions,[15] cavity QED,[16] and Josephson junctions.[17] Solid-state devices open up the possibility of fabricating large integrated networks that would be required for realistic applications of quantum computers. A basic feature of the quantum-dot scenario[11] is to consider the *electron spin* **S** as the qubit (the qubit being the basic unit of information in the quantum computer). This stands in contrast to alternative proposals also based on quantum dots,[18–21] in which it is the *charge* (orbital) degrees of freedom out of which a qubit is formed and represented in terms of a pseudospin-1/2. However, there are two immediate advantages of real spin over pseudospin: First, the qubit represented by a real spin-1/2 is always a well-defined qubit; the two-dimensional Hilbert space is the entire space available, thus there are no extra dimensions into which the qubit state could "leak." [22] Second, during a quantum computation, phase coherence of the qubits must be preserved. It is thus an essential advantage of real spins that their dephasing times in GaAs can be on the order of microseconds,[23] whereas for charge degrees of freedom dephasing times are typically much less, on the order of nanoseconds.[24,25]

In addition to a well-defined qubit, we also need a controllable "source of entanglement," i.e., a mechanism by which two specified qubits at a time can be entangled[26] so as to produce the fundamental quantum XOR [or controlled-NOT] gate operation, represented by a unitary operator $U_{\rm XOR}$.[27] This can be achieved by temporarily coupling two spins.[11] As we will show in detail below, due to the Coulomb interaction and the Pauli exclusion principle the ground state of two coupled electrons is a spin singlet, i.e., a highly entangled spin state. This physical picture translates into an exchange coupling $J(t)$ between the two spins $\mathbf{S}_1$ and $\mathbf{S}_2$ described by a Heisenberg Hamiltonian

$$H_s(t) = J(t)\mathbf{S}_1 \cdot \mathbf{S}_2 . \tag{1}$$

If the exchange coupling is pulsed such that $\int dt\, J(t)/\hbar = J_0 \tau_s/\hbar = \pi \pmod{2\pi}$, the associated unitary time evolution $U(t) = T\exp[i\int_0^t H_s(\tau)d\tau/\hbar]$ corresponds to the "swap" operator $U_{\rm sw}$, which simply exchanges the quantum states of qubit 1 and 2.[11] Furthermore, the quantum XOR can be obtained[11] by applying the sequence $\exp[i(\pi/$

$2)S_1^z]\exp[-i(\pi/2)S_2^z]U_{sw}^{1/2}\exp(i\pi S_1^z)U_{sw}^{1/2} \equiv U_{XOR}$, i.e., a combination of "square-root of swap" $U_{sw}^{1/2}$ and single-qubit rotations $\exp(i\pi S_1^z)$, etc. Since U_{XOR} (combined with single-qubit rotations) is proven to be a universal quantum gate,[18,26] it can, therefore, be used to assemble any quantum algorithm. Thus, the study of a quantum XOR gate is essentially reduced to the study of the *exchange mechanism* and how the exchange coupling $J(t)$ can be controlled experimentally. We wish to emphasize that the switchable coupling mechanism described in the following need not be confined to quantum dots: the same principle can be applied to other systems, e.g., coupled atoms in a Bravais lattice, overlapping shallow donors in semiconductors such as P in Si,[28] and so on. The main reason to concentrate here on quantum dots is that these systems are at the center of many ongoing experimental investigations in mesoscopic physics, and thus there seems to be reasonable hope that these systems can be made into quantum gates functioning along the lines proposed here.

In view of this motivation we study in the following the spin dynamics of two laterally coupled quantum dots containing a single electron each. We show that the exchange coupling $J(B,E,a)$ can be controlled by a magnetic field B (leading to wave-function compression), or by an electric field E (leading to level detuning), or by varying the barrier height or equivalently the interdot distance $2a$ (leading to a suppression of tunneling between the dots). The dependence on these parameters is of direct practical interest, since it opens the door to tailoring the exchange $J(t)$ for the specific purpose of creating quantum gates. We further calculate the static and dynamical magnetization responses in the presence of perpendicular and parallel magnetic fields, and show that they give experimentally accessible information about the exchange J. Our analysis is based on an adaptation of Heitler-London and Hund-Mulliken variational techniques[29] to parabolically confined coupled quantum dots. In particular, we present an extension of the Hubbard approximation induced by the long-range Coulomb interaction. We find a striking dependence of the Hubbard parameters on the magnetic field and interdot distance, which is of relevance also for atomic-scale Hubbard physics in the presence of long-range Coulomb interactions. Finally, we discuss the effects of dephasing induced by nuclear spins in GaAs and show that dephasing can be strongly reduced by dynamically polarizing the nuclear spins and/or by magnetic fields.

The paper is organized as follows. In Sec. II we introduce the model for the quantum gate in terms of coupled dots. In Sec. III we calculate the exchange coupling first in the Heitler-London and then in the Hund-Mulliken approach. There we also discuss the Hubbard limit and the new features arising from the long-range nature of the Coulomb interactions. In Sec. IV we consider the effects of imperfections leading to dephasing and gate errors; in particular, we consider dephasing resulting from nuclear spins in GaAs. Implications for experiments on magnetization and spin susceptibilities are presented in Sec. V, and Sec. VI contains some concluding remarks on the networks of gates with some suggestions for single-qubit gates operated by local magnetic fields. Finally, we mention that a preliminary account of some of the results presented here has been given in Ref. 30.

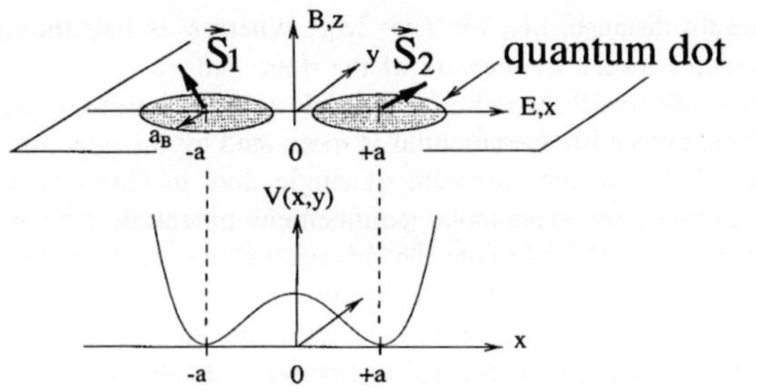

FIG. 1. Two coupled quantum dots with one valence electron per dot. Each electron is confined to the xy plane. The spins of the electrons in dots 1 and 2 are denoted by $\mathbf{S}_1$ and $\mathbf{S}_2$. The magnetic field B is perpendicular to the plane, i.e., along the z axis, and the electric field E is in plane and along the x axis. The quartic potential is given in Eq. (3) and is used to model the coupling of two harmonic wells centered at $(\pm a,0,0)$. The exchange coupling J between the spins is a function of B, E, and the interdot distance $2a$.

II. MODEL FOR THE QUANTUM GATE

We consider a system of two laterally coupled quantum dots containing one (conduction band) electron each (see Fig. 1). It is essential that the electrons are allowed to tunnel between the dots, and that the total wave function of the coupled system must be antisymmetric. It is this fact that introduces correlations between the spins via the charge (orbital) degrees of freedom. For definiteness we shall use in the following the parameter values recently determined for single GaAs heterostructure quantum dots[5] that are formed in a 2DEG; this choice is not crucial for the following analysis but it allows us to illustrate our analytical results with realistic numbers. The Hamiltonian for the coupled system is then given by

$$H=\sum_{i=1,2} h_i+C+H_Z=H_{\rm orb}+H_Z,$$

$$h_i=\frac{1}{2m}\left(\mathbf{p}_i-\frac{e}{c}\mathbf{A}(\mathbf{r}_i)\right)^2+ex_iE+V(\mathbf{r}_i), \qquad (2)$$

$$C=\frac{e^2}{\kappa|\mathbf{r}_1-\mathbf{r}_2|}.$$

The single-particle Hamiltonian h_i describes the electron dynamics confined to the xy-plane. The electrons have an effective mass m ($m=0.067m_e$ in GaAs) and carry a spin-1/2 $\mathbf{S}_i$. The dielectric constant in GaAs is $\kappa=13.1$. We allow for a magnetic field $\mathbf{B}=(0,0,B)$ applied along the z axis, which couples to the electron charge via the vector potential $\mathbf{A}(\mathbf{r})=(B/2)(-y,x,0)$. We also allow for an electric field E applied in plane along the x direction, i.e., along the line connecting the centers of the dots. The coupling of the dots (which includes tunneling) is modeled by a quartic potential,

$$V(x,y)=\frac{m\omega_0^2}{2}\left[\frac{1}{4a^2}(x^2-a^2)^2+y^2\right], \qquad (3)$$

which separates (for x around $\pm a$) into two harmonic wells of frequency ω_0, one for each dot, in the limit of large interdot distance, i.e., for $2a \gg 2a_B$, where a is half the distance between the centers of the dots, and $a_B = \sqrt{\hbar/m\omega_0}$ is the effective Bohr radius of a single isolated harmonic well. This choice for the potential is motivated by the experimental fact[5] that the spectrum of single dots in GaAs is well described by a parabolic confinement potential, e.g., with $\hbar\omega_0 = 3$ meV.[5] We note that increasing (decreasing) the interdot distance is physically equivalent to raising (lowering) the interdot barrier, which can be achieved experimentally by, e.g., applying a gate voltage between the dots.[6] Thus, the effect of such gate voltages is described in our model simply by a change of the interdot distance $2a$. We also note that it is only for simplicity that we choose the two dots to be exactly identical, and no qualitative changes will occur in the following analysis if the dots are only approximately equal and approximately of parabolic shape.

The (bare) Coulomb interaction between the two electrons is described by C. The screening length λ in almost depleted regions like few-electron quantum dots can be expected to be much larger than the bulk 2DEG screening length (which is about 40 nm in GaAs). Therefore, λ is large compared to the size of the coupled system, $\lambda \gg 2a \approx 40$ nm for small dots, and we will consider the limit of unscreened Coulomb interaction ($\lambda/a \gg 1$) throughout this paper.

The magnetic field B also couples to the electron spins via the Zeeman term $H_Z = g\mu_B \Sigma_i \mathbf{B}_i \cdot \mathbf{S}_i$, where g is the effective g factor ($g \approx -0.44$ for GaAs), and μ_B the Bohr magneton. The ratio between the Zeeman splitting and the relevant orbital energies is small for all B values of interest here; indeed, $g\mu_B B/\hbar\omega_0 \lesssim 0.03$, for $B \ll B_0 = (\hbar\omega_0/\mu_B)(m/m_e) \approx 3.5$ T, and $g\mu_B B/\hbar\omega_L \lesssim 0.03$, for $B \gg B_0$, where $\omega_L = eB/2mc$ is the Larmor frequency, and where we used $\hbar\omega_0 = 3$ meV. Thus, we can safely ignore the Zeeman splitting when we discuss the orbital degrees of freedom and include it later into the effective spin Hamiltonian. Also, in the few-electron system we are dealing with, spin-orbit effects can be completely neglected since $H_{so}/\hbar\omega_0 \approx 10^{-7}$, where $H_{so} = (\omega_0^2/2mc^2)\mathbf{L}\cdot\mathbf{S}$ is the spin-orbit coupling of an electron in a parabolic confinement potential.[30] This has the important implication that dephasing effects induced, e.g., by potential or charge fluctuations in the surroundings of the isolated dots can couple only to the charge of the electron so that they have very small influence on the phase coherence of the isolated spin itself (for dephasing induced by coupling the dots see Sec. IV). It is for this reason that it is preferable to consider dots containing electrons instead of holes, since holes will typically have a sizable spin-orbit interaction.[1]

Finally, we assume a low-temperature description where $kT \ll \hbar\omega_0$, so that we can restrict ourselves to the two lowest orbital eigenstates of H_{orb}, one of which is symmetric (spin singlet) and the other one antisymmetric (spin triplet). In this reduced (four-dimensional) Hilbert space, H_{orb} can be replaced by the effective Heisenberg spin Hamiltonian (1), $H_s = J\mathbf{S}_1 \cdot \mathbf{S}_2$, where the exchange energy $J = \epsilon_t - \epsilon_s$ is the difference between the triplet and singlet energy that we wish to calculate. The above model cannot be solved in an analytically closed form. However, the analogy between atoms and quantum dots (artificial atoms) provides us with a powerful set of variational methods from molecular physics for finding ϵ_t and ϵ_s. Note that the typical energy scale $\hbar\omega_0 \approx$ meV in our quantum dot is about a thousand times smaller than the energies (Ry$\approx$eV) in a hydrogen atom, whereas the quantum dot is larger by about the same factor. This is important because their size makes quantum dots much more susceptible to magnetic fields than atoms. In analogy to atomic physics, we call the size of the electron orbitals in a quantum dot the Bohr radius, although it is determined by the confining potential rather than by the Coulomb attraction to a positively charged nucleus. For harmonic confinement $a_B = \sqrt{\hbar/m\omega_0}$ is about 20 nm for $\hbar\omega_0 = 3$ meV.

III. EXCHANGE ENERGY

A. Heitler-London approach

We consider first the Heitler-London approximation, and then refine this approach by including hybridization as well as double occupancy in a Hund-Mulliken approach, which will finally lead us to an extension of the Hubbard description. We will see, however, that the qualitative features of J as a function of the control parameters are already captured by the simplest Heitler-London approximation for the artificial hydrogen molecule described by Eq. (2). In this approximation, one starts from single-dot ground-state orbital wave functions $\varphi(\mathbf{r})$ and combines them into the (anti)symmetric two-particle orbital state vector

$$|\Psi_\pm\rangle = \frac{|12\rangle \pm |21\rangle}{\sqrt{2(1\pm S^2)}}, \tag{4}$$

the positive (negative) sign corresponding to the spin singlet (triplet) state, and $S = \int d^2r\, \varphi^*_{+a}(\mathbf{r})\varphi_{-a}(\mathbf{r}) = \langle 2|1\rangle$ denoting the overlap of the right and left orbitals. A nonvanishing overlap implies that the electrons tunnel between the dots (see also Sec. III B). Here, $\varphi_{-a}(\mathbf{r}) = \langle \mathbf{r}|1\rangle$ and $\varphi_{+a}(\mathbf{r}) = \langle \mathbf{r}|2\rangle$ denote the one-particle orbitals centered at $\mathbf{r} = (\mp a, 0)$, and $|ij\rangle = |i\rangle|j\rangle$ are two-particle product states. The exchange energy is then obtained through $J = \epsilon_t - \epsilon_s = \langle\Psi_-|H_{orb}|\Psi_-\rangle - \langle\Psi_+|H_{orb}|\Psi_+\rangle$. The single-dot orbitals for harmonic confinement in two dimensions in a perpendicular magnetic field are the Fock-Darwin states,[31] which are the usual harmonic oscillator states, magnetically compressed by a factor $b = \omega/\omega_0 = \sqrt{1+\omega_L^2/\omega_0^2}$, where $\omega_L = eB/2mc$ denotes the Larmor frequency. The ground state (energy $\hbar\omega = b\hbar\omega_0$) centered at the origin is

$$\varphi(x,y) = \sqrt{\frac{m\omega}{\pi\hbar}}\, e^{-m\omega(x^2+y^2)/2\hbar}. \tag{5}$$

Shifting the single particle orbitals to $(\pm a, 0)$ in the presence of a magnetic field we obtain $\varphi_{\pm a}(x,y) = \exp(\pm iya/2l_B^2)\varphi(x \mp a, y)$. The phase factor involving the magnetic length $l_B = \sqrt{\hbar c/eB}$ is due to the gauge transformation $\mathbf{A}_{\pm a} = B(-y, x\mp a, 0)/2 \rightarrow \mathbf{A} = B(-y,x,0)/2$. The matrix elements of H_{orb} needed to calculate J are found by adding and subtracting the harmonic potential centered at $x = -(+)a$ for electron 1(2) in H_{orb}, which then takes the form $H_{orb} = h^0_{-a}(\mathbf{r}_1) + h^0_{+a}(\mathbf{r}_2) + W + C$, where $h^0_{\pm a}(\mathbf{r}_i) = [\mathbf{p}_i - e\mathbf{A}(\mathbf{r}_i)/c]^2/2m + m\omega_0^2[(x_i \mp a)^2 + y_i^2]/2$ is the Fock-

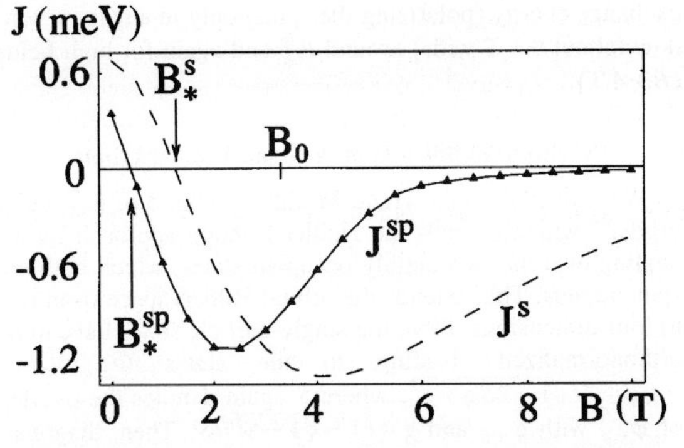

FIG. 2. Exchange energy J in units of meV plotted against the magnetic field B (in units of Tesla), as obtained from the s-wave Heitler-London approximation (dashed line), Eq. (7), and the result from the improved sp-hybridized Heitler-London approximation (triangles), which is obtained numerically as explained in the text. Note that the qualitative behavior of the two curves is similar, i.e., they both have zeroes, the s-wave approximation at B_*^s, and the sp-hybridized approximation at B_*^{sp}, and also both curves vanish exponentially for large fields. $B_0=(\hbar\omega_0/\mu_B)(m/m_e)$ denotes the crossover field to magnetically dominated confining ($B \gg B_0$). The curves are given for a confinement energy $\hbar\omega_0=3$ meV (implying for the Coulomb parameter $c=2.42$), and interdot distance $a=0.7a_B$.

Darwin Hamiltonian shifted to $(\pm a,0)$, and $W(x_1x_2)=\Sigma_{i=1,2}V(x_i,y_i)-m\omega_0^2[(x_1+a)^2+(x_2-a)^2+y_1^2+y_2^2]/2$. We obtain

$$J=\frac{2S^2}{1-S^4}\left(\langle 12|C+W|12\rangle-\frac{\mathrm{Re}\langle 12|C+W|21\rangle}{S^2}\right), \quad (6)$$

where the overlap becomes $S=\exp(-m\omega a^2/\hbar - a^2\hbar/4l_B^4 m\omega)$. Evaluation of the matrix elements of C and W yields (see also Ref. 30)

$$J=\frac{\hbar\omega_0}{\sinh[2d^2(2b-1/b)]}\Bigg[c\sqrt{b}\{e^{-bd^2}\mathrm{I}_0(bd^2) - e^{d^2(b-1/b)}\mathrm{I}_0[d^2(b-1/b)]\}+\frac{3}{4b}(1+bd^2)\Bigg], \quad (7)$$

where we introduce the dimensionless distance $d=a/a_B$, and I_0 is the zeroth-order Bessel function. The first and second terms in Eq. (7) are due to the Coulomb interaction C, where the exchange term enters with a minus sign. The parameter $c=\sqrt{\pi/2}(e^2/\kappa a_B)/\hbar\omega_0$ (≈ 2.4, for $\hbar\omega_0=3$ meV) is the ratio between Coulomb and confining energy. The last term comes from the confinement potential W. The result $J(B)$ is plotted in Fig. 2 (dashed line). Note that typically $|J/\hbar\omega_0|\lesssim 0.2$. Also, we see that $J>0$ for $B=0$, which must be the case for a two-particle system that is time-reversal invariant.[29] The most remarkable feature of $J(B)$, however, is the change of sign from positive to negative at $B=B_*^s$, which occurs over a wide range of parameters c and a. This singlet-triplet crossing occurs at about $B_*^s=1.3$ T for $\hbar\omega_0=3$ meV ($c=2.42$) and $d=0.7$. The transition from antiferromagnetic ($J>0$) to ferromagnetic ($J<0$) spin-spin coupling with increasing magnetic field is caused by the long-

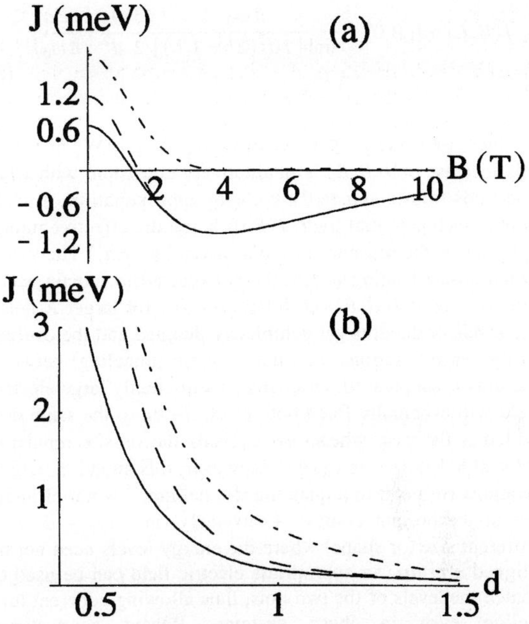

FIG. 3. The exchange coupling J obtained from Hund-Mulliken (full line), Eq. (11), and from the extended Hubbard approximation (dashed line), Eq. (12). For comparison, we also plot the usual Hubbard approximation where the long-range interaction term V is omitted, i.e., $J=4t_H^2/U_H$ (dashed-dotted line). In (a), J is plotted as a function of the magnetic field B at the fixed interdot distance ($d=a/a_B=0.7$), and for $c=2.42$, in (b) as a function of the interdot distance $d=a/a_B$ at zero field ($B=0$), and again $c=2.42$. For these parameter values, the s wave Heitler-London J, Eq. (7), and the Hund-Mulliken J (full line) are almost identical.

range Coulomb interaction, in particular by the negative exchange term, the second term in Eq. (7). As $B \gg B_0$ (≈ 3.5 T for $\hbar\omega_0=3$ meV), the magnetic field compresses the orbits by a factor $b\approx B/B_0 \gg 1$ and thereby reduces the overlap of the wave functions, $S^2=\exp[-2d^2(2b-1/b)]$, exponentially strongly. Similarly, the overlap decays exponentially for large interdot distances $d \gg 1$. Note however, that this exponential suppression is partly compensated by the exponentially growing exchange term $\langle 12|C|21\rangle/S^2 \propto \exp[2d^2(b-1/b)]$. As a result, the exchange coupling J decays exponentially as $\exp(-2d^2b)$ for large b or d, as shown in Fig. 3(b) for $B=0$ ($b=1$). Thus, the exchange coupling J can be tuned through zero and then suppressed to zero by a magnetic field in a very efficient way. We note that our Heitler-London approximation breaks down explicitly (i.e., J becomes negative even when $B=0$) for certain interdot distances when c exceeds 2.8. Finally, a similar singlet-triplet crossing as a function of the magnetic field has been found in *single* dots with two electrons.[32]

The exchange energy J also depends on the applied electric field E. The additional term $e(x_1+x_2)E$ in the potential merely shifts the one-particle orbitals by $\Delta x=eE/m\omega_0^2$, raising the energy of both the singlet and triplet states. Since the singlet energy turns out to be less affected by this shift than the triplet, the exchange energy J increases with increasing E,

$$J(B,E)=J(B,0)+\frac{\hbar\omega_0}{\sinh[2d^2(2b-1/b)]}\frac{3}{2}\frac{1}{d^2}\left(\frac{eEa}{\hbar\omega_0}\right)^2, \tag{8}$$

the increase being proportional to $m\omega_0^2(\Delta x)^2$. (We note that this increase of $J(B,E)$ is qualitatively consistent with what one finds from a standard two-level approximation of a 1D double-well potential [with $J(B,0)$ being the effective tunnel splitting] in the presence of a bias given by eEa.) The variational ansatz leading to Eq. (8) is expected to remain accurate as long as $J(B,E)-J(B,0)\lesssim J(B,0)$; for larger E fields the levels of the dots get completely detuned and the overlap of the wave functions (i.e., the coherent tunneling) between the dots is suppressed. Of course, a sufficiently large electric field will eventually force both electrons on to the same dot, which is the case when eEa exceeds the on-site repulsion $U[\gg J(B,E=0)$, see below]. However, this situation, which would correspond to a quantum-dot helium,[33] is not of interest in the present context. Conversely, in case of dots of different size (or shape) where the energy levels need not be aligned *a priori*, an appropriate electric field can be used to match the levels of the two dots, thus allowing coherent tunneling even in those systems. Recent conductance measurements[8] on coupled dots of different size (containing several electrons) with electrostatic tuning have revealed clear evidence for a delocalized molecular state.

A shortcoming of the simple approximation described above is that solely ground-state single-particle orbitals were taken into account and mixing with excited one-particle states due to interaction is neglected. This approximation is self-consistent if $J\ll\Delta\epsilon$, where $\Delta\epsilon$ denotes the single-particle level separation between the ground state and the first excited state. We find $|J/\Delta\epsilon|<0.25$ at low fields $B\leqslant 1.75$ T, therefore, $J(B)$ is at least qualitatively correct in this regime. At higher fields $|J/\Delta\epsilon|\approx 1$, indicating substantial mixing with higher orbitals. An improved Heitler-London variational ansatz is obtained by introducing sp-hybridized single-dot orbitals (in analogy to molecular physics), i.e., $\phi=\varphi_s+\alpha\varphi_{px}+i\beta\varphi_{py}$, where $\varphi_s=\varphi$ is the s orbital introduced above, $\varphi_{pq}=\sqrt{(2/\pi)}m\omega q\exp(-m\omega r^2/2\hbar)/\hbar$, $q=x,y$, are the lowest two Fock-Darwin excited states (at zero field) with angular momentum $|\ell|=1$, and α and β are real variational parameters to be determined by minimization of the singlet and triplet energies $\epsilon_{s,t}(\alpha,\beta)$, which is done numerically. The φ_{pq} are chosen to be real; they are, however, not eigenstates of the single-particle Hamiltonian, which are $\varphi_{px}\pm i\varphi_{py}$ (with eigenenergy $2\hbar\omega\pm\hbar\omega_L$). Note that while $\epsilon_{s,t}$ decrease only by $\approx 1\%$ due to hybridization, the relative variation of $J=\epsilon_t-\epsilon_s$ can still be substantial. Nevertheless, the resulting exchange energy J^{sp} (Fig. 2) is only quantitatively different from the pure s-wave result $J\equiv J^s$, Eq. (7). At low fields, $J^{sp}<J^s$ and the change of sign occurs already at about $B_*^{sp}\simeq 0.4$ T$<B_*^s$. At high fields, J^{sp} shows a much more pronounced decay as a function of B.

Being a completely orbital effect, the exchange interaction between spins of course competes with the Zeeman coupling H_Z of the spins to the magnetic field. In our case, however, the Zeeman energy H_Z is small and exceeds the exchange energy (polarizing the spins) only in a narrow window (about 0.1 T wide) around B_*^{sp} and again for high fields ($B>4$ T).

B. Hund-Mulliken approach and Hubbard limit

We turn now to the Hund-Mulliken method of molecular orbits,[29] which extends the Heitler-London approach by including also the two doubly occupied states, which both are spin singlets. This extends the orbital Hilbert space from two to four dimensions. First, the single-particle states have to be orthonormalized, leading to the states $\Phi_{\pm a}=(\varphi_{\pm a}-g\varphi_{\mp a})/\sqrt{1-2Sg+g^2}$, where S again denotes the overlap of φ_{-a} with φ_{+a} and $g=(1-\sqrt{1-S^2})/S$. Then, diagonalization of

$$H_{orb}=2\epsilon+\begin{pmatrix} U & X & -\sqrt{2}t_H & 0\\ X & U & -\sqrt{2}t_H & 0\\ -\sqrt{2}t_H & -\sqrt{2}t_H & V_+ & 0\\ 0 & 0 & 0 & V_-\end{pmatrix}, \tag{9}$$

in the space spanned by $\Psi_{\pm a}^d(\mathbf{r}_1,\mathbf{r}_2)=\Phi_{\pm a}(\mathbf{r}_1)\Phi_{\pm a}(\mathbf{r}_2)$, $\Psi_{\pm}^s(\mathbf{r}_1,\mathbf{r}_2)=[\Phi_{+a}(\mathbf{r}_1)\Phi_{-a}(\mathbf{r}_2)\pm\Phi_{-a}(\mathbf{r}_1)\Phi_{+a}(\mathbf{r}_2)]/\sqrt{2}$ yields the eigenvalues $\epsilon_{s\pm}=2\epsilon+U_H/2+V_+\pm\sqrt{U_H^2/4+4t_H^2}$, $\epsilon_{s0}=2\epsilon+U_H-2X+V_+$ (singlet), and $\epsilon_t=2\epsilon+V_-$ (triplet), where the quantities

$$\begin{aligned}
\epsilon&=\langle\Phi_{\pm a}|h_{\pm a}^0|\Phi_{\pm a}\rangle,\\
t_H=t-w&=\langle\Phi_{\pm a}|h_{\pm}^0|\Phi_{\mp a}\rangle-\langle\Psi_+^s|C|\Psi_{\pm a}^d\rangle/\sqrt{2},\\
V=V_--V_+&=\langle\Psi_-^s|C|\Psi_-^s\rangle-\langle\Psi_+^s|C|\Psi_+^s\rangle,\\
X&=\langle\Psi_{\pm a}^d|C|\Psi_{\mp a}^d\rangle,\\
U_H=U-V_++X&=\langle\Psi_{\pm a}^d|C|\Psi_{\pm a}^d\rangle-\langle\Psi_+^s|C|\Psi_+^s\rangle\\
&\quad+\langle\Psi_{\pm a}^d|C|\Psi_{\mp a}^d\rangle,
\end{aligned} \tag{10}$$

all depend on the magnetic field B. The exchange energy is the gap between the lowest singlet and the triplet state

$$J=\epsilon_t-\epsilon_{s-}=V-\frac{U_H}{2}+\frac{1}{2}\sqrt{U_H^2+16t_H^2}. \tag{11}$$

In the standard Hubbard approach for short-range Coulomb interactions (and without the B field)[29] J reduces to $-U/2+\sqrt{U^2+16t^2}/2$, where t denotes the hopping-matrix element, and U the on-site repulsion [cf. Eq. (10)]. Thus, t_H and U_H are the extended hopping-matrix element and the on-site repulsion, respectively, renormalized by long-range Coulomb interactions. The remaining two singlet energies ϵ_{s+} and ϵ_{s0} are separated from ϵ_t and ϵ_{s-} by a gap of order U_H and are, therefore, neglected for the study of low-energy properties. The evaluation of the matrix elements is straightforward but lengthy, and we give the results in the Appendix. Typically, the "Hubbard ratio" t_H/U_H is less than 1, e.g., if $d=0.7$, $\hbar\omega_0=3$ meV, and $B=0$, we obtain $t_H/U_H=0.34$, and it decreases with increasing B. Therefore, we are in an extended Hubbard limit, where J takes the form

$$J=\frac{4t_{\rm H}^2}{U_{\rm H}}+V. \tag{12}$$

The first term has the form of the standard Hubbard approximation[35] (invoked previously[11]) but with $t_{\rm H}$ and $U_{\rm H}$ being renormalized by long-range Coulomb interactions. The second term V is new and accounts for the difference in Coulomb energy between the singly occupied singlet and triplet states $\Psi_{\pm}^{\rm s}$. It is precisely this V that makes J negative for high magnetic fields, whereas $t_{\rm H}^2/U_{\rm H}>0$ for all values of B [see Fig. 3(a)]. Thus, the usual Hubbard approximation (i.e., without V) would not give reliable results, neither for the B dependence [Fig. 3(a)] nor for the dependence on the interdot distance a [Fig. 3(b)].[36] Since only the singlet space has been enlarged, it is clear that we obtain a lower singlet energy $\epsilon_{\rm s}$ than that from the s-wave Heitler-London calculation, but the same triplet energy $\epsilon_{\rm t}$, and, therefore, $J=\epsilon_{\rm t}-\epsilon_{\rm s}$ exceeds the s-wave Heitler-London result [Eq. (7)]. However, the on-site Coulomb repulsion $U\propto c$ strongly suppresses the doubly occupied states $\Psi_{\pm a}^{\rm d}$ and already for the value of $c=2.4$ (corresponding to $\hbar\omega_0=3$ meV) we obtain almost perfect agreement with the s-wave Heitler-London result (Fig. 2). For large fields, i.e., $B\gg B_0$, the suppression becomes even stronger ($U\propto\sqrt{B}$) because the electron orbits become compressed with increasing B and two electrons on the same dot are confined to a smaller area leading to an increased Coulomb energy.

IV. DEPHASING AND QUANTUM-GATE ERRORS

We allow now for imperfections and discuss first the dephasing resulting from coupling to the environment, and then address briefly the issue of errors during the quantum-gate operation. We have already pointed out that dephasing in the charge sector will have little effect on the (uncoupled) spins due to the smallness of the spin-orbit interaction. Similarly, the dipolar interaction between the qubit spin and the surrounding spins is also minute, and it can be estimated as $(g\mu_{\rm B})^2/a_{\rm B}^3\approx 10^{-9}$ meV. Although both couplings are extremely small, they will eventually lead to dephasing for sufficiently long times. We have described such weak-coupling dephasing in terms of a reduced master equation elsewhere,[11] and we refer the interested reader to this work. Since this type of dephasing is small it can be eliminated by error correction schemes.[37]

Next, we consider the dephasing due to nuclear spins in GaAs semiconductors, where both Ga and As possess a nuclear spin $I=3/2$. There is a sizable hyperfine coupling between the electron-spin ($s=1/2$) and all the nuclear spins in the quantum dot, which might easily lead to a flip of the electron spin and thus cause an error in the quantum computation. We shall now estimate this effect and show that it can be substantially reduced by spin polarization or by a field. We consider an electron spin $\mathbf{S}$ in contact with N nuclear spins $\mathbf{I}^{(i)}$ in the presence of a magnetic field $B\| z$. The corresponding Hamiltonian is given by $H=A\mathbf{S}\cdot\mathbf{I}+b_zS_z+\tilde{b}_zI_z=H_0+V$, where

$$H_0=AS_zI_z+b_zS_z+\tilde{b}_zI_z,\quad V=A(S_+I_-+S_-I_+)/2. \tag{13}$$

Here, A is a hyperfine coupling, $\mathbf{I}=\Sigma_{i=1}^N\mathbf{I}^{(i)}$ is the total nuclear spin, and $b_z=g\mu_{\rm B}B_z$, $\tilde{b}_z=g_N\mu_NB_z$ (g_N and μ_N denote the nuclear g factor and magneton). Consider the initial eigenstate $|i\rangle$ of H_0, which we will consider to be one basis vector for the qubit, where the electron spin is up (in the S_z basis), and the nuclear spins are in a product state of $I_z^{(i)}$ eigenstates with total $I_z=pNI$ $(-1\leqslant p\leqslant 1)$, i.e., in a state with polarization p along the z axis; here, $p=\pm 1$ means that the nuclear spins are fully polarized in the positive (negative) z direction, and $p=0$ means no polarization. Due to the hyperfine coupling the electron spin can flip (i.e., dephase) with the entire system going into a final state $|k\rangle$, which is again a product state but now with the electron-spin down, and, due to conservation of total spin, the z component $I_z^{(k)}$ of one and only one nuclear spin having increased by $2s=1$. All final states $|k\rangle$ are degenerate and again eigenstates of H_0 with eigenenergy E_f. We will consider this process now within the time-dependent perturbation theory and up to second order in V. The energy difference between initial and final states amounts to $E_i-E_f\approx 2s[A(pIN+s)+b_z]$, where we have used that $b_z\gg\tilde{b}_z$. For the reversed process with an electron-spin flip from down to up but with the same initial polarization for the nuclear spins the energy difference is $\approx -2s[A(pIN-s)+b_z]$. The total transition probability to leave the initial state $|i\rangle$ after time t has elapsed is then

$$P_i(t)=\left(\frac{2\sin[(E_f-E_i)t/2\hbar]}{E_f-E_i}\right)^2\sum_{k(\neq i)}|\langle k|V|i\rangle|^2. \tag{14}$$

We interpret this total transition probability $P_i(t)$ as the degree of decoherence caused by spin-flip processes over time t. Now, $|\langle k|V|i\rangle|^2=A^2[I(I+1)-I_z^{(k)}(I_z^{(k)}+1)]/4$. Assuming some distribution of the nuclear spins we can replace this matrix element by its average value (denoted by brackets) where $\sqrt{\langle(I_z^{(k)})^2\rangle}$ describes then the variance of the mean value $\langle I_z^{(k)}\rangle=pI$. For example, a Poissonian distribution gives $|\langle k|V|i\rangle|^2\approx A^2[I(I+1)-pI(pI+1)]/4$, in which case the matrix element vanishes for full polarization parallel to the electron spin (i.e., $p=1$), as required by conservation of total spin. $P_i(t)$ is strongly suppressed for final states for which $t_0\equiv 2\pi\hbar/|E_i-E_f|\ll t$, which simply reflects conservation of energy. In particular, for a substantial nuclear polarization, i.e., $p^2N\gg 1$, $P_i(t)$ oscillates in time but with the vanishingly small amplitude $1/p^2N$ (for $B=0$). We can estimate N to be on the order of the number of atoms per quantum dot, which is about 10^5. Such a situation with $p^2N\gg 1$ can be established by dynamically spin polarizing the nuclear spins (Overhauser effect), e.g., via optical pumping[38] or via spin-polarized currents at the edge of a 2DEG.[39] This gives rise to an effective nuclear field $B_{\rm n}=ApNI/g\mu_{\rm B}$, which is reported to be as large as $B_{\rm n}^*=4$ T in GaAs (corresponding to $p=0.85$),[39] and which has a lifetime on the order of minutes.[38] Alternatively, for unpolarized nuclei with $p=0$ but a field B in the Tesla range, the amplitude of $P_i(t)$ vanishes as $(AIN/g\mu_{\rm B}B)^2/N\approx(B_{\rm n}^*/B)^2/N\ll 1$. For B or $B_{\rm n}=1$ T the oscillation frequency $1/t_0$ of $P_i(t)$ is about 10 GHz. Thus, spin-flip processes and hence, dephasing due to nuclear spins can be strongly suppressed, either by dynamically polarizing the nuclear spins and/or by applying a magnetic field B. The remaining dephasing effects (described again by a weak-coupling master equation[11]) should then be small enough to be eliminated by error correction.

We now address the imperfections of the quantum-gate operation. For this we note first that, for the purpose of quantum computing, the qubits must be coupled only for the short time of switching τ_s, while most of the time there is to be no coupling between the dots. We estimate now how small we can choose τ_s. For this we consider a scenario where J (initially zero) is adiabatically switched on and off again during the time τ_s, e.g., by an electrical gate by which we lower and then raise again the barrier $V(t)$ between the dots (alternatively, we can vary B, a, or E). A typical frequency scale during switching is given by the exchange energy (which results in the coherent tunneling between the dots) averaged over the time interval of switching $\bar{J}=(1/\tau_s)\int_0^{\tau_s}dtJ(t)$. Adiabaticity then requires that many coherent oscillations (characterized approximately by $\bar{J}$) have to take place in the double-well system while the control parameter $v=V$, B, a, or E is being changed, i.e., $1/\tau_s \approx |\dot{v}/v| \ll \bar{J}/\hbar$. If this criterion is met, we can use our equilibrium analysis to calculate $J(v)$ and then simply replace $J(v)$ by $J(v(t))$ in case of a time-dependent control parameter $v(t)$.[41] Note that this is compatible with the requirement needed for the XOR operation $J\tau_s/\hbar=n\pi$, n odd, if we choose $n\gg 1$. Our method of calculating J is self-consistent if $J\ll\Delta\epsilon$, where $\Delta\epsilon$ denotes the single-particle level spacing. The combination of both inequalities yields $1/\tau_s\ll\bar{J}/\hbar\ll\Delta\epsilon/\hbar$, i.e., no higher-lying levels can be excited during the switching. Finally, since typically $J\approx 0.2$ meV we see that τ_s should not be smaller than about 50 ps. Now, during the time τ_s spin and charge couple and thus, dephasing in the charge sector described by τ_ϕ^c can induce dephasing of spin via an uncontrolled fluctuation δJ of the exchange coupling. However, this effect is again small, and it can be estimated to be on the order of $\tau_s/\tau_\phi^c\sim 10^{-2}$, since even for large dots τ_ϕ^c is reported to be on the order of nanoseconds.[24] This seems to be a rather conservative estimate and one can expect the spin dephasing to be considerably smaller since not every charge-dephasing event will affect the spin. Finally, weak dephasing of the effective spin Hamiltonian during switching has been described elsewhere[11] in terms of a weak-coupling master equation, which accounts explicitly for decoherence of the spins during the switching process. Based on this analysis,[11] the probability for a gate error per gate operation [described by $\mathcal{K}_2$ in Eq. (13) of Ref. 11] is estimated to be approximately $\tau_s/\tau_\phi^c\sim 10^{-2}$ or better (see above).

V. EXPERIMENTAL IMPLICATIONS

Coherent coupling between the states of neighboring dots is the keystone of our proposal for the quantum-gate operation, and experimental probes of this coupling will be very interesting to explore. The effect of the dot-dot coupling manifests itself in the level structure, which could be measured noninvasively with spectroscopic methods.[3,4] An alternative way is to measure the static magnetization in response to a magnetic field B, which is applied along the z axis. This equilibrium magnetization is given by $M=g\mu_B \,\mathrm{Tr}(S_1^z+S_2^z)e^{-(H_s+H_Z)/kT}$, where H_s is given in Eq. (1), and $H_Z=g\mu_B\Sigma_i \mathbf{B}_i\cdot\mathbf{S}_i$ is the Zeeman term. It is straightforward to

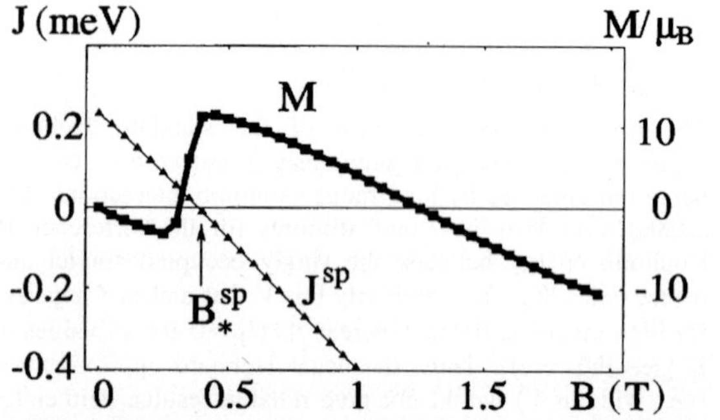

FIG. 4. The equilibrium magnetization M (box-shaped symbols) in units of Bohr magnetons μ_B as a function of magnetic field. M is obtained numerically from the sp-hybridized Heitler-London approximation. Note that the magnetization exhibits a jump at the field value B_*^{sp} for which the exchange J^{sp} (triangle symbols) changes sign. At the left- and right-hand side of the jump the negative slope of $M(B)$ indicates orbital diamagnetism. The temperature for this plot is $T=0.2$ K, while as before $\hbar\omega_0=3$ meV and $a=0.7a_B$.

evaluate M, and in Fig. 4 we plot M as a function of B for a typical temperature $T=0.2$ K. The exchange $J^{sp}(B)$ is also shown in Fig. 4. Both $J^{sp}(B)$ and M are the results of the sp-hybridized Heitler-London approximation. We note that the equilibrium magnetization $M(B)$ is strongly dominated by the orbital response (via the exchange J); we find a diamagnetic response (negative slope of M) for $B<B_*^{sp}$, which is followed by a pronounced jump in the magnetization at the field B_*^{sp} followed again by a diamagnetic response. Experimental observation of this jump would give evidence for the existence of the predicted singlet-triplet level-crossing at B_*^{sp}, and such measurements would allow one to ''map out'' J around the point where it can be tuned to zero, e.g., by also varying the barrier between the dots. The magnetic moment produced by the orbital motion of the electrons in one pair of coupled quantum dots at the peak ($B=B_*^{sp}$) is around 10 μ_B (see Fig. 4). This signal could be further amplified by using an ensemble of pairs of coupled quantum dots.

A further way to get experimental information about the exchange coupling would be to measure the spin response to an ac magnetic field (in the linear-response regime), described by the dynamical spin susceptibilities $\chi_{mn}^{pq}(\omega)=(i/\hbar)\int_0^\infty dt\,\exp(i\omega t)\langle[S_m^p(t),S_n^q(0)]\rangle$, where $m,n=1,2$, and $p,q=x,y,z$. Being interested in the spin response only, we assume this ac field to be applied in plane so that there is no orbital response (for a sufficiently weak field with no subband mixing). We see then that all the transverse spin susceptibilities $\chi_{mn}^{p\neq q,q}$ vanish, and we are left with the longitudinal ones only, where $\chi_{mn}^{xx}=\chi_{mn}^{yy}=\chi_{mn}^{zz}\equiv\chi_{mn}$ due to the rotational symmetry of H_s. It is sufficient to consider the dissipative part $\chi''_{mn}(\omega)=\mathrm{Im}\,\chi_{mn}(\omega)$ for which we obtain $\chi''_{11}=\chi''_{22}=-\chi''_{12}=-\chi''_{21}=-(\pi/4)f(J,B)[\delta(\hbar\omega+J)-\delta(\hbar\omega-J)]$, where $f(J,B)=(e^{J/kT}-1)/[1+e^{J/kT}+2\cosh(g\mu_B B/kT)]$. Also, due to conservation of total spin, the total response $\chi_{1j}+\chi_{2j}$ as well as the response to a spatially uniform field $\chi_{i1}+\chi_{i2}$ vanish. Thus, to observe the spin susceptibilities calculated here one needs to apply the fields locally or to measure the spin of a dot separately; both cases could be realized, e.g., by atomic- or magnetic-force microscopes (see also below, where we briefly discuss local fields produced by field gradients).

VI. CONCLUDING REMARKS

We end with a few comments on a network of coupled quantum dots in the presence of fields (see also Ref. 11). In a setup with only one quantum gate (i.e., two quantum dots) the gate operation can be performed using uniform magnetic fields (besides electric gates), while in a quantum computer with many gates, which have to be controlled individually, local magnetic fields are indispensable, especially for the single-qubit gates.[11,42] However, we emphasize that it is not necessary that every single quantum dot in a network is directly addressable with a local magnetic field. Indeed, using "swap" operations $U_{\rm sw}$, any qubit state can be transported to a region where the single-qubit gate operation is performed, and then back to its original location, without disturbing this or other qubits. In one possible mode of operation a constant field B_*, defined by $J(B_*)=0$, is applied, while smaller time-dependent *local* fields then control the gate operations. We can envision local fields being achieved by a large number of techniques: with neighboring magnetic dots,[11] closure domains, a grid of current-carrying wires below the dots, tips of magnetic- or atomic-force microscopes, or by bringing the qubit into contact (by shifting the dot via electrical gating) with a region containing magnetic moments or nuclear spins with different hyperfine coupling (e.g., AlGaAs instead of GaAs), and others. A related possibility would be to use magnetic field gradients. Single-qubit switching times of the order of $\tau_s\approx 20$ ps require a field of 1 T, and for an interdot distance $2a\approx 30$ nm, we would need gradients of about 1 T/30 nm, which could be produced with commercial disk reading/writing heads. (The operation of several XOR gates via magnetic fields also requires gradients of similar magnitude.) Alternatively, one could use an ac magnetic field $B_{\rm ac}$ and apply electron-spin resonance (ESR) techniques to rotate spins with a single-qubit switching time (at resonance) $\tau_s\approx \pi\hbar/B_{\rm ac}$. To address the dots of an array individually with ESR, a magnetic field gradient is needed, which can be estimated as follows. Assuming a relative ESR linewidth of 1% and again $2a=30$ nm we find about $B_{\rm ac}\times 10^4\ {\rm cm}^{-1}$. Field gradients in excitation sequences for NMR up to 2×10^4 G/cm have been generated,[40] which allows for $B_{\rm ac}\approx 1$ G. The resulting switching times, however, are rather long, on the order of 100 ns, and larger field gradients would be desirable. Finally, such ESR techniques could be employed to obtain information about the effective exchange values J: the exchange coupling between the spins leads to a shift in the spin-resonance frequency, which we found to be of the order of $J/\hbar$ by numerical analysis.[34]

To conclude, we have calculated the exchange energy $J(B,E,a)$ between spins of coupled quantum dots (containing one electron each) as a function of magnetic and electric fields and interdot distance using the Heitler-London, hybridized Heitler-London, and Hund-Mulliken variational approach. We have shown that $J(B,E,a)$ changes sign (reflecting a singlet-triplet crossing) with increasing B field before it vanishes exponentially. Besides being of fundamental interest, this dependence opens up the possibility to use coupled quantum dots as quantum gate devices, which can be operated by magnetic fields and/or electric gates (between the dots) to produce entanglement of qubits.

ACKNOWLEDGMENTS

We would like to thank J. Kyriakidis, S. Shtrikman, and E. Sukhorukov for useful discussions. This paper has been supported in part by the Swiss National Science Foundation.

APPENDIX: HUND-MULLIKEN MATRIX ELEMENTS

Here, we list the explicit expressions for the matrix elements defined in Eqs. (9) and (10) as a function of the dimensionless interdot distance $d=a/a_{\rm B}$ and the magnetic compression factor $b=\sqrt{1+\omega_{\rm L}^2/\omega_0^2}$ where $\omega_{\rm L}=eB/2mc$. The single-particle matrix elements are given by

$$\epsilon=\frac{3}{32}\frac{1}{b^2d^2}+\frac{3}{8}\frac{S^2}{1-S^2}\left(\frac{1}{b}+d^2\right)+b, \tag{A1}$$

$$t=\frac{3}{8}\frac{S}{1-S^2}\left(\frac{1}{b}+d^2\right), \tag{A2}$$

where we used $S=\exp[-d^2(2b-1/b)]$. The (two-particle) Coulomb matrix elements can be expressed as

$$V_+=N^4[4g^2(1+S^2)F_1+(1+g^2)^2F_2+4g^2F_3-16g^2F_4], \tag{A3}$$

$$V_-=N^4(1-g^2)^2(F_2-S^2F_3), \tag{A4}$$

$$U=N^4[(1+g^4+2g^2S^2)F_1+2g^2F_2+2g^2S^2F_3-8g^2F_4], \tag{A5}$$

$$X=N^4\{[(1+g^4)S^2+2g^2]F_1+2g^2F_2+2g^2S^2F_3-8g^2F_4\}, \tag{A6}$$

$$w=N^4[-g(1+g^2)(1+S^2)F_1-g(1+g^2)F_2 -g(1+g^2)S^2F_3+(1+6g^2+g^4)SF_4], \tag{A7}$$

with $N=1/\sqrt{1-2Sg+g^2}$ and $g=(1-\sqrt{1-S^2})/S$. Here, we make use of the functions

$$F_1=c\sqrt{b}, \tag{A8}$$

$$F_2=c\sqrt{b}e^{-bd^2}\,{\rm I}_0(bd^2), \tag{A9}$$

$$F_3=c\sqrt{b}e^{d^2(b-1/b)}{\rm I}_0[d^2(b-1/b)], \tag{A10}$$

$$F_4=c\sqrt{b}e^{-d^2/4b}\sum_{k=-\infty}^{\infty}(-1)^k{\rm I}_{2k}\left(\frac{d^2}{4}(2b-1/b)\right)\times{\rm I}_{2k}\left(i\frac{d^2}{2}\sqrt{b^2-1}\right), \tag{A11}$$

where ${\rm I}_n$ denotes the Bessel function of the nth order. For our purposes, we can neglect terms with $|k|>1$ in the sum in F_4, since for $\hbar\omega_0=3$ meV, $B<30$ T, and $d=0.7$ the relative error introduced by doing so is less than 1%.

*Electronic address: burkard@ubaclu.unibas.ch
†Electronic address: loss@ubaclu.unibas.ch
‡Electronic address: divince@watson.ibm.com

[1] L. Jacak, P. Hawrylak, and A. Wójs, *Quantum Dots* (Springer, Berlin, 1997).

[2] L. P. Kouwenhoven, C. M. Marcus, P. L. McEuen, S. Tarucha, R. M. Westervelt, and N. S. Wingreen, in *Mesoscopic Electron Transport*, Proceedings of the Advanced Study Institute, edited by L. L. Sohn, L. P. Kouwenhoven, and G. Schön (Kluwer, Dordrecht, 1997).

[3] R. C. Ashoori, Nature (London) **379**, 413 (1996).

[4] R. J. Luyken, A. Lorke, M. Haslinger, B. T. Miller, M. Fricke, J. P. Kotthaus, G. Medeiros-Ribiero, and P. M. Petroff (unpublished).

[5] S. Tarucha, D. G. Austing, T. Honda, R. J. van der Hage, and L. P. Kouwenhoven, Phys. Rev. Lett. **77**, 3613 (1996); L. P. Kouwenhoven, T. H. Oosterkamp, M. W. S. Danoesastro, M. Eto, D. G. Austing, T. Honda, and S. Tarucha, Science **278**, 1788 (1997).

[6] F. R. Waugh, M. J. Berry, D. J. Mar, R. M. Westervelt, K. L. Campman, and A. C. Gossard, Phys. Rev. Lett. **75**, 705 (1995); C. Livermore, C. H. Crouch, R. M. Westervelt, K. L. Campman, and A. C. Gossard, Science **274**, 1332 (1996).

[7] T. H. Oosterkamp, S. F. Godijn, M. J. Uilenreef, Y. V. Nazarov, N. C. van der Vaart, and L. P. Kouwenhoven, Phys. Rev. Lett. **80**, 4951 (1998).

[8] R. H. Blick, D. Pfannkuche, R. J. Haug, K. v. Klitzing, and K. Eberl, Phys. Rev. Lett. **80**, 4032 (1998); R. H. Blick, D. W. van der Weide, R. J. Haug, and K. Eberl, *ibid.* **81**, 689 (1998).

[9] K. Nomoto, R. Ugajin, T. Suzuki, and I. Hase, J. Appl. Phys. **79**, 291 (1996); A. O. Orlov, I. Amlani, G. H. Bernstein, C. S. Lent, and G. L. Snider, Science **277**, 928 (1997).

[10] D. Deutsch, Proc. R. Soc. London, Ser. A **400**, 97 (1985).

[11] D. Loss and D. P. DiVincenzo, Phys. Rev. A **57**, 120 (1998).

[12] I. L. Chuang, N. A. Gershenfeld, and M. Kubinec, Phys. Rev. Lett. **80**, 3408 (1998).

[13] D. Cory, A. Fahmy, and T. Havel, Proc. Natl. Acad. Sci. USA **94**, 1634 (1997).

[14] J. A. Jones, M. Mosca, and R. H. Hansen, Nature (London) **393**, 344 (1998).

[15] J. I. Cirac and P. Zoller, Phys. Rev. Lett. **74**, 4091 (1995); C. Monroe, D. M. Meekhof, B. E. King, W. M. Itano, and D. J. Wineland, *ibid.* **75**, 4714 (1995).

[16] Q. A. Turchette, C. J. Hood, W. Lange, H. Mabuchi, and H. J. Kimble, Phys. Rev. Lett. **75**, 4710 (1995).

[17] D. V. Averin, Solid State Commun. **105**, 659 (1998); A. Shnirman, G. Schön, and Z. Hermon, Phys. Rev. Lett. **79**, 2371 (1997).

[18] A. Barenco, D. Deutsch, A. Ekert, and R. Jozsa, Phys. Rev. Lett. **74**, 4083 (1995).

[19] R. Landauer, Science **272**, 1914 (1996).

[20] J. A. Brum and P. Hawrylak, Superlattices Microstruct. **22**, 431 (1997).

[21] P. Zanardi and F. Rossi, quant-ph/9804016 (unpublished).

[22] Such leakage can happen, e.g., in the switching process of single qubits or by coupling two qubits together, etc., and can easily lead to uncontrollable errors. This concern is especially relevant in quantum dots where the energy-level spacing is (nearly) uniform (in contrast to real atoms) so that the levels defining the qubit are of similar scale as the separation to neighboring energy levels.

[23] J. M. Kikkawa and D. D. Awschalom, Phys. Rev. Lett. **80**, 4313 (1998).

[24] A. G. Huibers, M. Switkes, C. M. Marcus, K. Campman, and A. C. Gossard, Phys. Rev. Lett. **81**, 200 (1998).

[25] The dephasing times of Refs. 23 and 24 are both measured in GaAs semiconductors, which involve many electrons. It would be highly desirable to get direct experimental information about dephasing times in isolated quantum dots of low filling as considered here.

[26] D. P. DiVincenzo, Phys. Rev. A **51**, 1015 (1995).

[27] A. Barenco, C. H. Bennett, R. Cleve, D. P. DiVincenzo, N. Margolus, P. Shor, T. Sleator, J. A. Smolin, and H. Weinfurter, Phys. Rev. A **52**, 3457 (1995).

[28] B. E. Kane, Nature (London) **393**, 133 (1998).

[29] D. C. Mattis, in *The Theory of Magnetism*, Springer Series in Solid-State Sciences No. 17 (Springer, New York, 1988), Vol. I, Sec. 4.5.

[30] D. P. DiVincenzo and D. Loss, Superlattices Microstruct. **23**, 419 (1998).

[31] V. Fock, Z. Phys. **47**, 446 (1928); C. Darwin, Proc. Cambridge Philos. Soc. **27**, 86 (1930).

[32] M. Wagner, U. Merkt, and A. V. Chaplik, Phys. Rev. B **45**, 1951 (1992).

[33] D. Pfannkuche, V. Gudmundsson, and P. A. Maksym, Phys. Rev. B **47**, 2244 (1993).

[34] G. Burkard, D. Loss, and D. P. DiVincenzo (unpublished).

[35] See, e.g., E. Fradkin, *Field Theories of Condensed Matter Systems* (Addison-Wesley, Reading, MA, 1991).

[36] We note that the significant changes due to Coulomb long-range interactions are valid down to the scale of real atoms. Since atomic orbitals and the harmonic orbitals used here behave similarly (for $B=0$), we expect to find qualitatively similar results for real molecules (as found here for coupled dots) especially regarding the effect of Coulomb long-range interactions on t_H, U_H, J and their dependence on the interatomic distance a.

[37] J. Preskill, quant-ph/9712048 (unpublished).

[38] M. Dobers, K. v. Klitzing, J. Schneider, G. Weimann, and K. Ploog, Phys. Rev. Lett. **61**, 1650 (1988).

[39] D. C. Dixon, K. R. Wald, P. L. McEuen, and M. R. Melloch, Phys. Rev. B **56**, 4743 (1997).

[40] W. Zhang and D. G. Cory, Phys. Rev. Lett. **80**, 1324 (1998).

[41] If during the change of $v(t)$ the total spin remains conserved, no transitions between the instantaneous singlet and triplet eigenstates can be induced during the switching. Thus, the singlet and triplet states evolve independently of each other, and the condition on adiabatic switching involves $\Delta\epsilon$ (instead of J), i.e., we only need to require that $1/\tau_s \approx |\dot{v}/v| \ll \overline{\Delta\epsilon}/\hbar$, which would be less restrictive. Also, only $\int_0^{\tau_s} dt J(t)$ and not $J(t)$ itself is needed for the gate operation. Therefore, the adiabaticity criterion given in the text, while being sufficient, need not be really necessary. However, the complete analysis of the time-dependent problem in terms of variational wave functions is beyond the scope of the present paper and will be addressed elsewhere.

[42] We note that it is sufficient to have single-qubit rotations about any two orthogonal axes. A preferable choice here are two orthogonal in-plane axes because magnetic fields $\mathbf{B}_\parallel$ parallel to the 2DEG do not affect the exchange coupling $J(B_\perp)$ (assuming that we can exclude subband mixing induced by a sufficiently strong $B_\parallel$).

Quantum Computers and Quantum Coherence

David P. DiVincenzo
IBM Research Division, Thomas J. Watson Research Center, P. O. Box 218, Yorktown Heights, NY 10598, USA

Daniel Loss
Department of Physics and Astronomy, University of Basel, Klingelbergstrasse 82, 4056 Basel, Switzerland

(February 8, 1999)

If the states of spins in solids can be created, manipulated, and measured at the single-quantum level, an entirely new form of information processing, quantum computing, will be possible. We first give an overview of quantum information processing, showing that the famous Shor speedup of integer factoring is just one of a host of important applications for qubits, including cryptography, counterfeit protection, channel capacity enhancement, distributed computing, and others. We review our proposed spin-quantum dot architecture for a quantum computer, and we indicate a variety of first generation materials, optical, and electrical measurements which should be considered. We analyze the efficiency of a two-dot device as a transmitter of quantum information via the propagation of qubit carriers (i.e. electrons) in a Fermi sea.

I. INFORMATION PROCESSING AND QUANTUM MECHANICS

While we will spend much of this chapter considering fairly specifically the application of quantum magnetic systems to quantum computing, we want to first review more broadly the potential "quantum revolution" that is brewing in the area of information science. It is amusing for a physicist to note that quantum mechanics is now being taught as part of the standard curriculum in a growing number of graduate computer science programs! Why would computer scientists find it necessary to take up such an esoteric study from a different field? The problems which have interested them have nothing to do with the quantum world, and this is not being changed by this quantum revolution. Computer scientists have a wide range of tasks which they are interested in accomplishing successfully, safely, and/or efficiently [1]:

1. Given data X, compute $f(X)$ in the fewest number of steps. (*computational complexity* [2])
2. Given two parties holding data X and Y, compute $f(X,Y)$ with the least communication. (*communication complexity* [3])
3. Given two parties holding data X and Y, compute $f(X,Y)$ in such a way that the two learn no more about each other's data than they know from the function value itself. (*discreet function evaluation* ([1], Chap. 5.8))
4. Transmit data X reliably from one party to a second as quickly as possible. (*channel capacity* [4])
5. Protect data X from duplication. (*counterfeit protection*)
6. Transmit data X from one party to a second in such a way that the data cannot be read by any third party. (*key distribution/cryptography* ([1], Chap. 6))
7. Transmit data X from one party to a second in such a way that the receiver can be assured that the data was not corrupted during passage through the channel. (*authentication* [1])
8. Transmit data X from one party to a second in such a way that another party can later confirm that the second party did not alter X, and can confirm that it was produced by the first party. (*digital signature* [1])
9. Divide data X among n parties in such a way that no $n-1$ of them can reconstruct data X, but all n working together can. (*secret sharing* [5])
10. Determine and execute optimal strategies in games. (*game theory; economics* [6])

In this information age, our society's well being increasingly depends on being able to perform these and similar tasks well. Quantum mechanics is never mentioned in this list; nor should it be, since all of these tasks involve the possession and transmission of data in palpable, macroscopic form. "Quantum data" is not useful for members of our very macroscopic society; the inputs and outputs of these tasks must be in classical form. (One might question this assumption in some radically altered definition of "society".)

But what we have increasingly realized is that the *tools* employed to accomplish these tasks can well be quantum mechanical. In addition to "classical" processing primitives involved in completing tasks (place a bit in memory, compute the AND of two bits, launch a bit into a communication channel), we can employ a host of quantum processing primitives: prepare a qubit (two-level quantum system) in a particular pure state; launch a qubit (e.g., a photon or an electron, see Sec. III) into a communications channel; transform the state of two qubits according to the action of some two-body Hamiltonian.

The remarkable fact is that it is known how to achieve improvements in many (but by no means all) of the tasks mentioned above by employing these quantum primitives. We will review here briefly the "quantum state of the art" for our list of tasks:

1. *computational complexity:* Shor's famous work [7] showed that some very important computations, for example prime factorization, have only polynomial complexity if quantum primitives are used, while this computation can (probably) not be done in polynomial time if only classical primitives are used. It is worth reviewing the general way in which the classical specification of the problem is converted into an application of quantum primitives: the data X (the number to be factored) is converted into a time-dependent two-body Hamiltonian function which is applied to a set of qubits prepared in a standard quantum state (e.g., all zeros). Then the answer $f(X)$ (the set of prime factors) is obtained by the results of a quantum measurement performed on each of the qubits. It may be necessary to repeat Shor's procedure several times to obtain a factor.

   It should be noted that there are some other computations for which it has been proved that *no* improvement in computational complexity is achieved by using quantum primitives [8]. For instance, the nth iterate of a function provided as a look-up table takes n references to the table even if quantum primitives can be used [9]. Work continues to explore the cases in which quantum speed-ups are and are not possible [10].

2. *communication complexity:* In this work the advantages gained by communicating using qubits rather than bits have been explored [13]. There are some strong positive results in this area. Quantum communication is provably more efficient for the problem of two-party appointment scheduling: two persons have to compare their appointment books to choose a day to have lunch out of N possible days. For classical bit transmission $O(N)$ bits of communication are required in general. But it has been proved (it is an application of the "Grover" algorithm [12]) that no more than $O(\sqrt{N}\log N)$ quantum bits of transmission are needed to complete this task with high reliability [11]. There is a related task in which the quantum speedup is even more dramatic, in the area of "sampling complexity": two parties must both pick a subset of cardinality $\sqrt{N}$ from a common set of size N in such a way that their subsets are disjoint. Classically, $O(N)$ bits of communication are required to assure disjointness, but just $O(\log N)$ of quantum bit transmission suffices [14]. Such dramatic provable speedups are apparently also possible even in a case where two parties share a string of random bits [15].

3. *discreet function evaluation*: This is an example of a category of task for which there is believed to be *no* quantum solution. This is true, at least, for the principal technique which computer scientists have used to analyze this task [1], which involves reducing it to a procedure called *bit commitment*, in which one party records a bit value of her choosing, locks the record in a safe and sends it to a second party (the "commit" phase); then at a later time of her choosing, she sends the key to the other party (the "opening" phase). Since safes can be x-rayed and locks picked, this protocol is not secure. It has been proved that bit commitment is never secure in a quantum world [16]; using entanglement, the sender can change the value of the bit between the commit phase and the opening phase.

4. *channel capacity:* Here the results are tantalizing, but not conclusive. The problem is this: given a classical bit channel and a quantum bit channel with the same levels of noise (the same probability that the bit will pass through the channel unaffected, roughly speaking), are fewer uses of the qubit channel needed to send a given classical message reliably than of the bit channel? No case has been found in which this "classical capacity of a quantum channel" exceeds the Shannon capacity for the classical channel, although the work of Fuchs *et al.* give indications that it may be possible [17]. Actually, there is one scenario in which the quantum capacity is definitely greater: if the sender and receiver have shared a prior supply of maximally entangled quantum states, which themselves carry no classical information, the quantum capacity can be boosted by the technique of superdense coding [18] by a factor of two or sometimes more (at least up to a factor of three for qubit transmission) [19].

5. *counterfeit protection:* There are really no strong classical techniques for protecting against counterfeiting. The first application of quantum primitives ever conceived, "quantum money" was devised in 1970 by S. Wiesner [20]. It is a beautifully straightforward application of the simple rules of quantum state preparation and measurement. The bank embeds qubits into its banknotes; each qubit is in a pure quantum state, but the states are drawn from a non-orthogonal ensemble (e.g., $|0\rangle$, $|1\rangle$, $|0\rangle+|1\rangle$, and $|0\rangle-|1\rangle$, or in spin language, in the eigenstates of either σ_z or σ_x). A record of the state preparation is kept at the bank, and the bill is sent into circulation. When the note returns to the bank, the bank can use its record to measure each qubit in a "non-demolition" [21] fashion, that is, in the

appropriate σ_z or σ_x basis so that the state is undisturbed and the measurement outcome is deterministic. If all measurements agree with the stored record, the bank can be assured that no attempt has been made by a counterfeiter to read the state of the qubits to duplicate them. This application has not received much attention lately, but perhaps its day will come with the further advance of quantum technology, when qubits can be stored (or error-corrected) over very long times.

6. *key distribution:* The most well-known success of quantum protocols is in "quantum cryptography [22]." The security of quantum transmission of random data (the key) begins with the same trick that is introduced in quantum money, sending one of a set of non-orthogonal quantum states that an eavesdropper cannot reliably distinguish, and that are in fact disturbed if the eavesdropper attempts to learn any information about them. The construction of a secure key from this primitive involves a lot more work, but Mayers has given a proof [23] that the a protocol naturally obtained from the one proposed by Bennett and Brassard in 1984 [22] is unconditionally secure. Another protocol in which state transmission is augmented by local quantum computation is considerably easier to prove secure [24].

7. *authentication:* Wegman and Carter [25] introduced a provably secure authentication technique that assumes that the sender and receiver possess a secret key; therefore, a secure key exchange using quantum primitives leads directly to a way of doing secure authentication. In today's world there is another way to perform authentication: authentication is implied by digital signatures, which are routinely used in present-day cryptography, but —

8. *digital signatures:* The existence of quantum protocols has negated the ability to do digital signatures. First, no quantum protocol can apparently be introduced which can take the place of digital signatures used in public-key cryptography, in which a sender, by appending to the end of a message an encrypted version of that message, produces unalterable evidence that this message originated from him: anyone can later decrypt the "signature" using the sender's public key and compare it with the putative message [1]. Second, the "proof" that this protocol is secure relies on the security of public-key cryptography, which is jeopardized by the ability to factor large numbers by quantum computation. Perhaps some entirely different quantum reasoning will again permit the accomplishment of this information processing task.

9. *secret sharing:* Only a little work has been done on this [26], but it appears that there will be a variety of ways of using multipartite states to split up a secret in such a way that it can only be reconstructed by the cooperative quantum operations of several parties. Buzek *et al.* have shown ways in which this problem can be approached using entangled states; it is perhaps more surprising that it is possible to use unentangled quantum states to perform this task. This arises from the recent discovery that there exist ensembles of multiparty orthogonal product states which can nevertheless not be distinguished by any local operations of those parties, even if they are allowed any amount of classical communication. Only a joint quantum measurement can distinguish them reliably. The detailed application of this discovery to a secure secret sharing protocol has only just begun.

10. *games:* This is a rather ill-defined area at the moment, but one with apparent promise. Meyer and Eisert *et al.* [27] have shown that if the players in a game can perform quantum mechanical manipulations in the game (e.g., moving a chess piece into a superposition of positions by a unitary operation), they can gain some advantages. It seems that some changes will have to take place in our society before some of these game results become applicable — can we have a quantum stock market? A quantum economy?

A final comment about this survey: while in some sense it covers everything that goes on in the research on quantum improvements of information processing tasks, in another way it misses a lot of what workers in this field really think about. Between the bottom level of quantum or classical primitives like data transmission and qubit measurement, and the top level of tasks to be accomplished, lies a whole realm of macros and subprocedures which use the primitives and provide tools for accomplishing the end tasks. We are very familiar with these in classical computing (fetch program instructions, invoke a floating-point multiplier, launch a packet onto an ethernet), but there is a whole host of quantum macros which have no classical analog and which are crucial for facilitating the quantum implementations of many tasks.

An important example of these is quantum error correction and fault-tolerant quantum computation [28], which put together the primitives of state preparation, measurement, and manipulation in such a way that the effective unitary evolution of a quantum computation is carried out reliably despite the intervention of noise ("quantum decoherence"). Another operation which one might consider as a quantum macro is the sharing of a *quantum* secret, recently discussed in [29]. Reliable qubit communication depends on other noise-suppression quantum macros; the most effective approach to this problem involves entanglement purification [30] (in which a large supply of partially entangled mixed quantum states is manipulated locally to produce a smaller supply of pure, maximally entangled quantum states). Another crucial subprocedure in this noise-suppression macro is the celebrated "quantum teleportation [31]." Much of the recent "Star Trek" discussion of teleportation misses the point that it has a well-defined, scientifically valid role as an an enabler of high-level quantum processing for anything involving the transmission of quantum information (e.g., distributed computing, key distribution).

So, quantum information processing isn't just factoring! Quantum factoring alone is interesting and important; seeing the whole picture, though, indicates that we may be just at the beginning of something *really* big.

II. QUANTUM INFORMATION PROCESSING AND MAGNETIC PHYSICS

Specially-crafted magnetic materials and magnetoelectronic structures, we believe, are good candidates for providing some of the important primitive quantum tools for performing many of the tasks itemized above, as we will detail shortly. We will concentrate in this section on those applications which require the creation and manipulation of "fixed" qubits, which include the applications of quantum computing, counterfeit protection, and secret sharing, and pieces of the others, such as the encoding and decoding required in channel transmission. In Sec. III we will discuss a particular scenario based on mobile electrons [32] whose spins provide the "mobile" qubits needed in the other applications; some proposals are now being considered in which the coupling by solid-state optical cavities to photons [33] could provide the tools for the remainder of our tasks as well.

The magnetic structures that we envision are promising because the qubit is naturally defined (in terms of a localized single spin). This localized spin has the potential for being relatively well isolated from its environment – that is, for having low decoherence rates – and it can be manipulated by electrical, magnetic and/or spectroscopic tools and can be measured using advanced magnetometric or electronic techniques.

Of course, the magnetoelectronic structures that we propose are not the only possible approach to the realization of quantum information processing: efforts spanning many of the active areas of experimental quantum physics have led to successful demonstrations of quantum logic gates, and of operating systems for quantum cryptography, superdense coding, and quantum teleportation.

We can only give a brief mention of all the different quantum logic gate demonstrations that have been reported: In 1995, there was the demonstration of the two-qubit controlled-NOT reported using ion trap spectroscopy by the NIST group [34]. Since this demonstration, progress towards realizing the idea of the linear-ion trap quantum computer has been proceeding steadily; this group has recently demonstrated the deterministic creation of entanglement between two ions [35]. In the area of cavity-quantum-electrodynamics, the vacuum cavity version of the solid state microcavity scheme mentioned above was first investigated in 1995 by the Cal Tech group [36], and many proposals have been made for how to use this device in a quantum communication network.

The processing of photons in fiber-optic experiments has also received a lot of attention. Full-scale quantum cryptography demonstrations have now been achieved in many different laboratories [37]. In addition, several other quantum information processing protocols have been realized in such systems: superdense coding has been achieved in systems where photon EPR pairs are created by parametric down conversion, and incomplete Bell measurements are performed using linear optical elements [38]. More recently, teleportation of photon polarization states has been achieved [39,40]. Now it has also become possible to teleport a "continuous" Hilbert space, the quadrature field coordinates of a coherent state of light [41].

Finally, it should be mentioned that there is another condensed matter implementation of quantum gates that has received a lot of experimenatal attention lately, one involving bulk NMR (nuclear magnetic resonance). Following on the original theoretical idea for using NMR for quantum gates [42], the idea was put into practical, realizable form in 1997 [43]. Since then, there has been a plethora of experimental investigations of 2 and 3 spin systems, including demonstrations of the Deutsch-Jozsa and Grover quantum algorithms [44] and of simple quantum error correction techniques [45]. There has even been a realization of intramolecular quantum teleportation [46].

A. Proposed Device Structure

Rather than giving a general discussion of the criteria which a magnetoelectronic device proposal must satisfy in order to be a good candidate for a quantum computer (which we have done previously; see Refs. [47–49]), we will simply proceed to describe the specific model that we have introduced [48–50]. From the discussion here it should be clear what are the critical requirements for this proposal to succeed.

Fig. 1 sketches the model that we have introduced in Ref. [48]. It is a quantum-dot array [51,52], produced in this version of the model by lateral confinement. The figure indicates a two-dimensional layer, for example a quantum well produced in a GaAs heterostructure, above which an array of electrodes is placed. As in the experiments of various groups [53–57,61], voltages on these electrodes can be used to deplete selectively regions of the two-dimensional electron gas below them, leaving isolated regions (the quantum dots shown dotted in the figure) in which electrons can be confined.

The qubit in this scheme is provided by the *electron*

spin of each quantum dot. In order that this qubit be well defined, the electron number must be controlled and constant throughout the operation of the device. This is assured by exploiting the well understood Coulomb blockade effect in these dots. We imagine that a transport (e.g., [55]) or capacitance [53] measurement is performed on every dot in the array separately, and the gate voltages (the gates are shown shaded in the figure) adjusted so that the energy of the N electron state is much lower than that of the $N-1$ or $N+1$ electron state. N will remain fixed throughout the quantum computation operations: this computer has no moving parts, not even the electrons move (at least not much). In order to use the electron spin as a quantum number, it is very likely essential that N be an odd number (if N is even, it would typically be the case that the total spin of the dot would be zero, so that no nearly-degenerate levels would be available to represent the qubit). If N is odd, the spin is at least $1/2$. In fact, $s=1/2$ exactly is the ideal situation for representing a qubit. $s=1/2$ is assured if $N=1$, that is, if there is only one excess electron confined to the dot. For this reason partly, but mainly because of other considerations about the many-body physics of the dot, such as that discussed in Sec. III, we will consider only the $N=1$ case. The $N>1$ case may be usable for quantum computation, but it will require more analysis than we have performed up until now.

$N=1$ is not easy to achieve experimentally. N in the range of a few tens has become relatively routine in the experiments cited, but in the very small-N regime it becomes difficult for electrons to tunnel in and out of the dot, and the quantum-dot potential can become disorder dominated. These are not severe difficulties in principle, but we acknowledge that it is a demanding requirement from the perspective of present-day experiments, and we are committed to studying the effect of using larger electron numbers on our proposed device operation.

B. Decoherence

Among the most crucial requirements for the implementation of quantum logic devices is a high degree of quantum coherence. Coherence is lost when a qubit interacts with other quantum degrees of freedom in its environment and becomes entangled with them. Predicting the coherence time of the electron spin states of the device described above is very difficult, as the possible couplings to all the other quantum degrees of freedom of the system must be considered. We are encouraged, however, by the general fact, observed in many experimental situations in condensed matter physics, that spin degrees of freedom have longer coherence times than charge degrees of freedom (ones for which the different electron states are associated with different orbital wavefunctions), simply due to the weaker couplings of spin states than orbital states to the environment.

This observation does not lead to any simple result about what the available decoherence times in our structure will be. Experiments of spin coherence times have been performed on somewhat related structures [58,59], with the result that a very wide range of decoherence times can be seen for the spins of electrons in semiconductor heterostructures and bulk doped semiconductors. In structures which are intentionally doped with magnetic ions (Mn), the coherence times are seen to be very small, on the order of picoseconds. But times ranging over six orders of magnitude, approaching microseconds in some structures, have now been seen depending on the details of the semiconductor structure. As we will discuss in the next section, microsecond decoherence times would be acceptable for beginning experiments on quantum gate operations, while times of milliseconds would be adequate for even large-scale quantum computing applications (because of the abilities offered by quantum error correction [28]).

We have considered in general the likely mechanisms of decoherence in structures such as Fig. 1, which should be useful in guiding designs of experiments which seek to lengthen the decoherence times. Our estimates [50] indicate that decoherence due to spin-orbit coupling should be negligible for conduction band electrons in GaAs (although not for holes); still, more detailed work needs to be done to quantify this effect. A potentially important mechanism for decoherence in these structures is the coupling to other spin states in the environment. As demonstrated in the Mn-doping experiments [58], this effect will be greatly influenced by the materials preparation of the devices, and can be a very strong pathway to decoherence. Thus, in our work [48] we have studied in detail models in which the qubits are coupled to a bath of other spins. The significance of the effect is entirely determined by the strengths of the coupling constants between the system and bath. The decohering effect of this bath can be enhanced during quantum gate operations [49], that is, when spins in neighboring quantum dots are coupled (see next section).

In addition to other electronic spins, there are unquestionably nuclear spins in the environment as well whose decohering effect must be considered. In GaAs in particular, 100% of the nuclei possess non-zero spin. We have studied the effect of these spins recently [50], and our calculations indicate that these spins can be a serious source of decoherence if the applied magnetic fields are low and the nuclear spins are in their thermal equilibrium state. However, it is relatively easy to modify these conditions, either by dynamically spin-polarizing the nuclear spins e.g. by known optical techniques, and/or by arranging that the operation of the devices is performed with a non-zero applied magnetic field. Actually, the presence of significantly spin-polarized nuclei may actually be very useful for performing gate operations on these qubits (see next section) [50].

Another "trivial" but practically important source of spin decoherence arises from uncertainties in the applied

Hamiltonians to be discussed in the next section. For example, in schemes in which the gate action involves the application of a uniform magnetic field, inhomogeneities in this field will result in inaccuracies in the gate operation. This decoherence effect is analogous to the broadening effect on absorption lines which is well known in traditional spin spectroscopies, where various "refocusing" and "spin-echo" techniques have been devised to ameliorate them. Such techniques may have to be developed and adapted to assure reliable quantum gate operation, but this problem has not been addressed systematically in any detail.

One might think that if fluctuating magnetic fields are a severe problem for quantum-dot quantum bits, then perhaps there would be some value in reconsidering the use of electron orbital states, which, after all, would be insensitive to such magnetic field effects. We are pessimistic on this account, not only because the decoherence times for orbital states are short for myriad other reasons, but also because there is reason to believe that some of the important decoherence mechanisms due to Fermi-sea effects will be non-Markovian. Markovian, or memoryless, decoherence is actually greatly desired over non-Markovian decoherence in quantum computation, as all the powerful techniques introduced in quantum error correction assume a memoryless error scenario [28]. No one has demonstrated that a qubit system with even very weak non-Markovian decoherence would be useful for quantum information processing.

C. Quantum gates

Another crucial requirement for quantum computing, and for many of the other quantum approaches to information processing tasks outlined in the first section, is that it must be possible to apply time-dependent one- and two-body Hamiltonians to the qubits according to the specifications of some program [60].

The structure of Fig. 1 can have many mechanisms for applying such "quantum gates" to the spin qubits. First, the structure has a set of gates which can control the position of the electron's wavefunction within the two-dimensional electron gas, simply by varying the confining voltages on these gates. If two of these electrons in neighboring dots are pushed close together, the overlap of the orbital wavefunctions will, via the Pauli principle, produce an effective two-spin interaction between the two spin qubits. The Hamiltonian produced is that of an exchange interaction which is isotropic in spin space

$$H(t) = J(t)\mathbf{S}_1 \cdot \mathbf{S}_2. \tag{1}$$

Here the time dependence $J(t)$ is regulated by the time variation of the tunneling matrix element Γ of an electron from one dot to the other. According to perturbation theory, $J(t)$ is

$$J(t) \propto \frac{\Gamma(t)^2}{U}. \tag{2}$$

Here U is the Coulomb blockade energy, the charging energy required to add a second electron to one of the dots.

In Ref. [50] we give a more refined and detailed analysis of this switchable spin interaction, in particular we show that the long range part of the Coulomb interaction (if it is not screened) will produce an additional term in (2) of opposite sign that leads to a sign reversal of J for sufficiently large external magnetic fields as a result of competition between long-range Coulomb repulsion and magnetic wave function compression. By working at this magnetic field (where J vanishes) the exchange interaction can be pulsed on, even without changing the tunneling barrier between the dots, either by an application of a local magnetic field, or by exploiting a Stark electric field (which will also make the exchange interaction nonzero). See [50] for further information. We finally note that the exchange energy J can be understood as the level splitting induced by the formation of a molecular state between the two quantum dots [50]. The observation of such a molecular state in a double dot system containing several electrons has indeed been reported recently [61,62].

The exchange interaction of the form Eq. (1) is sufficient for the most general quantum computation, if it is supplemented by a suite of one-body time-dependent interactions (one-bit gates). This is discussed in [48–50], where it is shown that Eq. (1) will produce a quantum gate known as a "square-root of swap" (in which the exchange interaction is turned on for half the time required for it to produce a complete interchange ("swap") of the quantum states of the two qubits). We show [48] that two square roots of swap, in conjuction with a set of one-qubit gates, will produce a quantum XOR (also known as a controlled-NOT) gate, which is known to be employable for any arbitrary quantum computation [63].

The speed at which these switchings are done will be an important parameter; the rule is, the faster the better, consistent with doing the prescribed manipulations with rather high accuracy (error correction theory says that the relative accuracy to be striven for is on the order of 10^{-4}). The fundamental physics says that the switching on and off of the tunneling could be done much faster than a nanosecond [50]–only at much, much shorter time scales will such fundamental limitations as adiabaticity enter the picture. It is necessary that the switching time be smaller than the decoherence time; again, error correction theory says that ultimately, it is desirable that the switching time be smaller than the decoherence time by about 10^{-4}. We think that 10^{-1} will be quite satisfactory for the initial round of measurements. We think that initially, the experimentalist should simply be guided by what is doable. Since high-frequency signals are difficult to transmit into quantum dot structures in the Coulomb blockade regime at 4K or so, we might suggest that one should shoot for switching times in the neighborhood of 10^{-7} sec. A simple calculation indicates that only modest control-voltage excursions are needed to do square-root-of-swap in this time.

We note that the switching of the gates via an external control field $v(t)$ should be performed adiabatically [50], i.e. $|\dot{v}/v| \ll \delta\epsilon/\hbar$, where $\delta\epsilon$ is a characteristic energy scale of the problem. In the present case $\delta\epsilon$ should be taken to be on the order of the orbital energy-level separation. This adiabaticity requirement excludes e.g. switching pulses of rectangular shape, in which case many excitations into higher levels will occur. An adiabatic pulse shape of amplitude v_0 is e.g. given by $v(t) = v_0 \mathrm{sech}(t/\Delta t)$, where $\Delta t = \tau_s/\alpha$ gives the width of the curve and α is chosen such that $v(t = \tau_s)/v_0$ becomes vanishingly small. In this case we have $|\dot{v}/v| = \frac{1}{\Delta t}|\tanh(t/\Delta t)| \leq 1/\Delta t = \alpha/\tau_s$, and thus for adiabaticity we need to choose τ_s such that $\alpha/\tau_s \ll \delta\epsilon/\hbar$. Note that the Fourier transform, $v(\omega) = \Delta t v_0 \pi \mathrm{sech}(\pi\omega\Delta t)$, has the same shape as $v(t)$ but with a width $2/\pi\Delta t$, and we see that $v(\omega)$ decays exponentially in frequency ω, whereas it decays only as $1/\omega$ for a rectangular pulse. [We could, of course, also use a Gaussian pulse shape, however, in this case we would get $|\dot{v}/v| \propto t$ and some cutting of the long-time tails is required in order to satisfy adiabaticity for all times.] It is worth emphasizing, however, that for our quantum gate action the pulse shape is not relevant, the only parameter which counts is the integrated pulse shape, $\int_0^{\tau_s} dt P(v(t))$, where P stands for the exchange J or the magnetic field B which is switched. This stands in contrast to spectroscopic mechanisms based on resonance conditions where more details of the shape of the pulse are relevant. Also, even if adiabaticity is not well satisfied in our switching, not much will happen as long as spin-orbit coupling remains small since typically only charge degrees of freedom will be excited in a non-adiabatic process and not the spins representing the qubits.

The use of an inhomogeneous magnetic field (or an inhomogeneous g-factor) for gating mentioned above for two-bit gates is obligatory, in some form, for the accomplishment of the desired one-qubit gates. That is, every one-body Hamiltonian needed for quantum computing can be written in a standard Zeeman form

$$g\mu_B \mathbf{B}(t) \cdot \mathbf{S}. \tag{3}$$

It is necessary that the field $\mathbf{B}(t)$ (or the effective field) be applicable separately to each qubit (or at least that the effect on neighboring qubits be smaller and known), and that it can be applied along at least two different axes.

There are many ways that we can conceive of applying these local magnetic fields or local Zeeman interactions. If the switching time scale is to be the same as above (10^{-7} sec.), then field strengths of only a few Gauss are necessary, and this could be accomplished by a mechanism as simple as winding a small wire coil or by placing magnetic dots above/below each quantum dot, or by placing the dots between a grid of current-carrying wires as in RAM devices [64]. Other methods of obtaining very localized fields, such as moving magnetic bubbles in a garnet film, using a magnetic-disk writing head, or a magnetic force microscope tip, can be considered.

Although strict localization of the applied field is not necessary, it does make life considerably easier, and there are several ideas which would make this field effectively much more localized. If the nuclear spins of the dot and the material surrounding the dot can be polarized as discussed above, then the electron spin (but not the orbital motion) experiences an effective internal magnetic field, the "Overhauser field", which can be on the order of several Tesla in GaAs [50]. If the Overhauser field is different in the dot and in the confining layers above and below it, then the field as seen by the confined electron can be varied by purely electric gating, that is, by pushing the electron more or less into the insulating barriers. In our original work [48] we introduced another variant of this idea, in which the confining materials possess a *real* magnetization due to a ferromagnetic moment. Such ferromagnetic insulating materials are not so common, but are not unheard of either (the garnets, the ferrites, and the Eu-chalcogenides are some examples); unfortunately, there is little experience in matching these materials epitaxially to the common dot materials such as GaAs, but first promising progress in this direction has been made recently, see [65]. We also would like to emphasize here that our set-up permits the performance of swaps of qubit states in such a way that we can easily move a spin state (not the electron spin itself) of a given quantum dot via a chain of adjacent quantum dots to a desired location in the network where we have localized magnetic fields available, act with the field on the qubit and then swap the qubit back to its original location. This is possible since the swapping operation does not involve single-qubit rotations and since we can swap two states even without knowing their particular state. But either the Overhauser field idea or the magnetic insulator idea can be extended to solve the very important problem of quantum measurement, to be discussed momentarily.

A brief word about error correction, which we have alluded to many times already: error correction provides a way of using redundancy and repeated quantum measurement during the course of computation, which detects and diagnoses the occurrence of decoherence, and undoes its effects. It uses exactly the same gates which we have just introduced, along with qubit measurements to be described shortly. The conventional analysis of quantum error correction [28] assumes that two-qubit gates can be performed between any two qubits. In our computational model, gate operations can only be performed between neighboring qubits. This is not a serious modification, the basic procedures of quantum error correction still work in this case [66]. The more crucial requirement for error correction to work is that two-qubit and one-qubit gates can be performed on many different qubits simultaneously as it is possible in our proposal. There are other popular quantum register designs, for example the well-known linear ion trap model of Cirac and Zoller [67], for which error correction is not possible because gate operations cannot be done in parallel.

Finally, the concept of error correction promises to be important by itself. Indeed, in many areas of mesoscopic physics it would be highly desirable to maintain phase coherence *indefinitely*, a goal which we believe could be achieved with error correction schemes.

D. Quantum measurements

The final requirement which must be addressed for performing quantum information processing with the quantum-dot structure is the need to read out data reliably, which translates into the necessity of doing spin measurements at the single-spin level. It must be possible to address each individual spin in the structure (or at least some subset of the spins) and perform an "up/down" measurement on them. Solid-state magnetometry at the single-Bohr-magneton level has of course proved to be very difficult, as other contributions to this volume will discuss. We forsee, though, that using some of the capabilities of quantum computing, the very difficult single-spin measurement can be turned into a more manageable electrical (i.e., charge) measurement along the lines first proposed by us in Ref. [48].

We have recently reviewed in detail the possib ilities in this area, we will just give an outline here, the interested reader is referred to [68]. The basic idea of turning the spin measurement into a charge measurement [48] (see also [69]) is this: we use the kind of magnetic (either ferromagnetic or nuclear-spin-polarized) barriers mentioned above as tunnel barriers, say in the form of a thin barrier separating two quantum dots or a quantum dot and a single-electron transistor. The tunneling barrier can be made strongly spin dependent (this is the well-known "spin-filter" effect); thus, at the time of measurement, the tunneling of a spin-up electron can be made very probable, while the tunneling of a spin-down electron remains very improbable. Thus, the job of measuring spin is converted into the job of measuring whether an electron has tunneled or not. But this is a feasible (and indeed, almost routine) electrometry measurement–many labs have demonstrated the feasibility of single-electron-charge magnetometry, either with single-electron transistors, quantum point contacts, and other mesoscopic electronic structures.

Another promising idea for single-spin measurement involves near-field optical probing of the spin state. We have not analyzed this approach in any detail, but it deserves future experimental and theoretical attention.

E. Test experiments

It is clear that the above concept, which we have developed over the last three years, has proved far too demanding to be undertaken all at once. It requires a combination of developments, in materials and device fabrication, in precision, high frequency electrical control, in hitherto unexplored, complex, nanoscale architectures, which are far beyond the scope of one generation of experimental investigation.

Therefore, it is very important to pull apart our quantum-dot quantum computer into small pieces, setting feasible shorter-term goals for the demonstration of particular capabilities. We only intend to give a brief idea here of the kind of near-term work which might be done; indeed, it seems that the possible ways of dividing our proposal into smaller, manageable chunks are almost infinite, and finding the most promising ones can only result from a detailed dialog between the theorist and experimentalist. But here is a selection of ideas which we now now might be promising for the next few years:

There is a clear need to demonstrate the controlled fabrication of spin quantum dots. As mentioned above, a desirable goal would be to routinely obtain dots with just *one* excess electron. More theory must be done to see whether using dots with an odd number of excess electrons would be acceptable. One-electron dots have been achieved [53], but not in geometries in which dots could potentially be coupled. Loading by transport in the Coulomb blockade regime would be the obvious way, but doping or optical techniques should also be considered.

If an array of such dots can be obtained, then characterization of the qubit energy levels, g-factors, and especially decoherence times would be the next thing to study. In fact, an initial version of this type of experiment has now been reported [59], which demonstrates that time-resolved optical probes of these systems are extremely promising for these kinds of initial characterizations. Further application of pulsed-spectroscopy techniques should yield further information about the controllability of such qubits (at least at the one-qubit gate level).

Another distinct line of investigation would involve demonstration of two-qubit gate capabilities. We have suggested [48–50] that gated double-dot structures that have been fabricated and studied in GaAs 2DEGs [55,56] could be the starting point of such studies; it will also be desirable to see if other types of dots, say in pillar structures or ones created by chemical nucleation, can be integrated into devices in which their coupling is subject to electrical or magnetic control. We envision experiments in which arrays of these dots can be subjected to identical preparations and probings. It may be that an experiment as straightforward as the measurement of the a.c. magnetic susceptibility of such a dot array as a function of a control voltage [48,49] will be sufficient to demonstrate the basic physics of quantum-mechanical exchange coupling between neighboring spins.

The magnetoelectronic techniques that we have suggested for other gate operations and for single-spin quantum measurement involved additional and quite different experimental challenges. The basic materials issues of the integration of semiconducting and magnetic materials are not yet well enough developed to even propose a likely system to study at this time, although it is promising to note that there is now active research focussed on just this area, finding good matches between magnets and semiconductors which will show clean, reproducible interface properties. If, for example, it proves possible to grow EuS or EuO on GaAs, then an experiment can immediately be considered in which the basic spin filtering phenomenon of carriers in the semiconductor conduction band is looked for. This experiment would be very informative even in a traditional bulk tunneling geometry; there would be no need to even consider integrating these with quantum dot structures at first. Tunneling through Overhauser-polarized barrier materials may be less demanding from the materials science point of view, but will require integration of optical (for nuclear spin polarization) and electrical expertise. A later generation of experiment could consider integrating the spin-filter into a simple point-contact (say of the Ralls type) so that a combined spin-filter/Coulomb blockade effect could be demonstrated. This already takes us quite far into speculative territory.

We would like finally to briefly comment about questions that we have been asked about whether the many experiments on the *charge* degree of freedom in quantum dots could be directed towards the achievement of orbital-level qubits and quantum gates. While there may be a worthwhile approach in this direction, we are pessimistic about its ultimate chance of success compared with the spin approach, even though spin effects are at this time much less well developed in quantum-dot research. We say this based on the fact that orbital (i.e. charge) degrees of freedom of a dot will be much harder to make coherent than the spin of a dot, just based on the typically stronger coupling of charge (compared to magnetic moment) to the environment. A typical Fermi-sea charge environment also has a different, and possibly even worse, problem as already pointed out before: Fermionic baths are very non-Markovian, having power-law decays of correlations. Almost all the well-developed theory of quantum error correction applies only to Markovian baths [28], and it is very unclear whether any useful quantum computation can be done in the presence of a non-Markovian environment (however, see [70]). These considerations have been enough to justify, in our minds, a continued focus on the eventual possibilities of spin quantum dots only.

III. QUANTUM COMMUNICATION WITH ELECTRONS

In this section we would like to address the following question: is it possible to use mobile electrons, prepared in a definite (entangled) spin state, for the purpose of quantum communication? Such a question, for instance, is of central importance in a solid state quantum computer where one wishes to exchange quantum information between distant parts of a quantum network. The question is of course also of broader interest: if we could use electrons for creating entangled states, in particular so-called EPR pairs, and if we could move them around separately while preserving their spin entanglement, then we would be able to implement, for instance, tests of Bell's inequality; thereby, we could obtain tests of non-locality—one of the most striking concepts of quantum mechanics—for the first time with electrons. So far, all such tests have been done on photons [71], most recently by Gisin's group [72] who demonstrated in a remarkable experiment that photons propagating in optical fibers remain in an entangled state over more than 10 km's. It is quite amusing to note here that the Gedanken experiment which has been formulated by Einstein, Podolsky, and Rosen [73], and which underlies the Bell inequalities, makes use of point particles and not of massless particles such as photons. Thus, there can be no doubt that it would be highly desirable to extend tests of non-locality also to quantities which have a rest mass such as electrons in particular.

Now, as we have discussed before, one basic ingredient for quantum communication are entangled pairs of qubits which are shared by two parties. There are three separate requirements involved here which must be satisfied. First of all we need *mobile qubits* which can be transported from position A to position B. Second, we need a source of entanglement for such qubits which can be operated in a controllable way, and third, it must be possible to transport each of the qubits separately in a phase-coherent manner such that the entanglement between the two qubits of interest is not destroyed in the process of transporting them to their desired locations.

Now, our choice of representing the qubit in terms of the spin of a mobile electron satisfies the first requirement trivially (note that qubits defined as pseudospins are typically not mobile). The second requirement, to have a source of entanglement, can be satisfied by using the quantum gate mechanism based on coupled quantum dots [48–50] as we have described it in the preceeding sections.

To assess the third requirement, transport of entangled qubits, we need to be more specific of how we actually envisage such transport. One realistic scenario is to attach leads to the quantum dots into which the electrons can be injected (e.g. by lowering the gate barriers between dot and lead). From an experimental point of view it is best to make leads and dots out of the same material.

For instance, if the dots are formed in a two-dimensional electron gas (2DEG) such as GaAs heterostructures it is not difficult to connect them to leads formed also in the 2DEG by electrostatic confinement or some etching techniques [61,55]. In a first step we inject an electron into quantum dot 1 and another one into quantum dot 2. In a second step, we perform a quantum gate operation to produce an entangled state out of the two electrons, say a singlet state, $|\psi_{kk'}\rangle = \frac{1}{\sqrt{2}}(|k,k'\rangle + |k',k\rangle)\chi_s$, where $\chi_s = (|\uparrow\rangle_1|\downarrow\rangle_2 - |\downarrow\rangle_1|\uparrow\rangle_2)/\sqrt{2}$ is the two spinor describing a spin-singlet state. The orbital part of the state, characterized by the quantum numbers k, k', is symmetric whereas the spin singlet is antisymmetric. As a measure of correlations we consider transition amplitudes between an initial and a final state. We begin with the simplest case given by the wave function overlap of $|\psi_{kk'}\rangle$ with $|\psi_{qq'}\rangle$,

$$\langle\psi_{qq'}|\psi_{kk'}\rangle = \delta_{qk}\delta_{q'k'} + \delta_{qk'}\delta_{q'k} \,. \tag{4}$$

Thus, if e.g. $q = k$, and $q' = k'$, the overlap assumes its maximum value one, simply reflecting maximum correlation between the two states. If we prepare the two electrons in a triplet state instead of a singlet we will find a minus sign instead of the plus sign in Eq. (4). This means that this sign simply reflects the symmetry of the orbital part of the wave function, and thus the overlap (4) distinguishes only triplet from singlet states but not necessarily entangled from unentangled states. Indeed, the triplet states with $m_z = \pm 1$ are not entangled, whereas the triplet state with $m_z = 0$ as well as the singlet state are entangled. Since the (anti-)symmetry of the orbital part of the wave function leads to (anti-)bunching behavior in the noise spectrum [74], we can in principle distinguish singlet from triplet states. The triplet states themselves can be further distinguished by measuring the z-component of the total spin, S_z, which could be achieved e.g. by making use of spin filters in the leads and/or leads that are connected up to other quantum dots into which the electrons can tunnel and then be detected via SET measurements [32]. In this way it is possible (in principle) to distinguish all four spin states, in particular also to distinguish between entangled and unentangled states (provided we deal with these four particular states only–otherwise the expectation value of S_z does not distinguish between entangled and unentangled states in general).

Next we generalize this concept of the overlap to a dynamical situation as well as to the leads which contain many interacting electrons besides the two entangled electrons of interest. Again, we use a similar overlap as a measure of how much weight remains in the final state $|\psi_{qq'}, \psi_0, t\rangle$ when we start from some given initial state $|\psi_{kk'}, \psi_0\rangle$, where ψ_0 denotes the fermionic ground state of the electrons in the leads, which is simply given by a filled Fermi sea. For further discussion it is now convenient to make use of the standard second quantization formalism in terms of fermionic creation ($a^\dagger_{k\sigma}$) and annihilation ($a_{k\sigma}$) operators, where $\sigma = \pm 1$ denotes spin $\uparrow$ ($\downarrow$) in the S_z-basis. The (normalized) initial state, choosing a singlet, can then be written as

$$|\psi_{kk'}, \psi_0\rangle = \frac{1}{\sqrt{2}}(a^\dagger_{k\uparrow}a^\dagger_{k'\downarrow} - a^\dagger_{k\downarrow}a^\dagger_{k'\uparrow})\,|\psi_0\rangle \,, \tag{5}$$

and similarly for the final state, again chosen to be a singlet state. The overlap (4) now becomes a singlet-singlet correlation function which we denote by $G^s(q', q, t; k, k')$, $t \geq 0$, and which is explicitly given by

$$G^s(q', q, t; k, k') = \frac{1}{2}\sum_{\sigma=\pm 1}[G(q', -\sigma; q, \sigma; t; k, \sigma; k', -\sigma) - G(q', -\sigma; q, \sigma; t; k, -\sigma; k', \sigma)] \,, \tag{6}$$

where

$$G(q', -\sigma; q, \sigma; t; k, \sigma; k', -\sigma) = -\langle T\, a_{q'\sigma}(t)\, a_{q-\sigma}(t)\, a^\dagger_{k-\sigma}\, a^\dagger_{k'\sigma}\rangle \tag{7}$$

is a standard 2-particle Green's function, and $k = (\mathbf{k}, k_l)$, where $k_l = \pm 1$ refers to lead 1 (2). Here, T is the time-ordering operator and $\langle ... \rangle$ the zero-temperature or ground state expectation value. We assume a time- and spin-independent Hamiltonian, $H = H_0 + \sum_{i<j} V_{ij}$, where H_0 describes the free motion of the N electrons, and V_{ij} is the bare Coulomb interaction between electrons i and j (extensions to more complicated situations including spin interactions will be considered elsewhere). This four-point correlation function is of the type $G(12; 1'2')$ and it provides a measure of how much overlap (or transition amplitude) is left after time t between an initial and final singlet state of two electrons which have been injected into a Fermi sea (leads) of $N-2$ interacting electrons, and which propagate during time t in the leads before they are taken out again. We emphasize that after injection the two electrons of interest are, of course, no longer distinguishable from the electrons of the leads, and consequently the two electrons taken out of the leads will, in general, not be the same as the ones injected.

It is now a non-trivial many-body problem to find an explicit value for $G(12; 1'2')$. On the other hand, we can expect some simplification: without spin-dependent forces we know that the total spin must be conserved even if the two electrons strongly interact with the rest (and among themselves) via Coulomb interaction. It is thus not unreasonable to expect that we still find some spin correlations, in particular entanglement, between initial and final states. But how much is it? And why and how do we loose some of the correlations, etc.? These questions are of fundamental interest, and we can find answers to them by evaluating $G(12; 1'2')$ explicitly with the help of standard many-body techniques [75,32]. Omitting most of the details [32] here we briefly state the main results. First we note that the four-point

Green's function considerably simplifies for the realistic situation where there is no Coulomb interaction between the electrons in lead 1 and the electrons in lead 2. As a result the 2-particle vertex part vanishes and we get $G(12;1'2') = G(11')G(22') - G(12')G(21')$, i.e. the Hartree-Fock approximation is exact and the problem is reduced to the evaluation of single-particle Green's functions $G_1(\mathbf{k},t)$, $G_2(\mathbf{k}',t)$ pertaining to lead 1 and 2, resp. (these leads are still interacting many-body systems though). In particular, we now find

$$G^{s/t}(q',q,t;k,k') = -\{G_1(\mathbf{q},t)\,G_2(\mathbf{q}',t)\,\delta_{qk}\delta_{q'k'} \pm G_1(\mathbf{q}',t)\,G_2(\mathbf{q},t)\,\delta_{qk'}\delta_{q'k}\}\,, \quad (8)$$

where the upper (lower) sign refers to the spin singlet (triplet), and where we have chosen $k_l = 1$. For the special case $t = 0$, $N = 2$, and no interactions, we have $G_j = -i$, and thus G^s reduces to the rhs of Eq. (4). For the general case, we evaluate the (time-ordered) single-particle Green's functions G_j close to the Fermi surface and get the standard result [75]

$$G_j(\mathbf{q},t) \approx -iz_q\Theta(\epsilon_q - \epsilon_F)e^{-i\epsilon_q t - \Gamma_q t}\,, \quad (9)$$

where $\epsilon_q = q^2/2m$ is the quasiparticle energy (of our additional electron), ϵ_F is the Fermi energy, and $1/\Gamma_q$ is the quasiparticle lifetime. In a 2DEG, $\Gamma_q \propto (\epsilon_q - \epsilon_F)^2 \log(\epsilon_q - \epsilon_F)$ [76] within the random phase approximation (RPA), which accounts for screening and which is obtained by summing all polarization diagrams [75]. Thus, the lifetime becomes infinite when the energy of the added electron approaches ϵ_F. Eq. (9) is valid for $0 \le t \lesssim 1/\Gamma_q$, in which case the incoherent part of the Green's function is negligible. Now, we come to the most important quantity in the present context, the renormalization factor or quasiparticle weight, $z_F = z_{q_F}$, evaluated at the Fermi surface; it is defined by

$$z_F = \frac{1}{1 - \frac{\partial}{\partial\omega}Re\Sigma(q_F,\omega=0)}\,, \quad (10)$$

where $\Sigma(q,\omega)$ is the irreducible self-energy occurring in the Dyson equation. The quasiparticle weight, $0 \le z_q \le 1$, describes the weight of the bare electron in the quasiparticle state $\mathbf{q}$, i.e. when we add an electron with energy $\epsilon_q \ge \epsilon_F$ to the system, some weight (given by $1 - z_q$) of the original state $\mathbf{q}$ will be distributed among all the electrons due to the Coulomb interaction. This rearrangement of the Fermi system due to interactions happens very quickly, at a speed given approximately by the plasmon velocity, which exceeds the Fermi velocity (typically 10^5 m/s in GaAs). Restricting ourselves now to momenta close to the Fermi surface and to identical leads (i.e. $G_1 = G_2$) we then have

$$|G^{s/t}(q',q,t;k,k')| = z_F^2\,|\,\delta_{qk}\delta_{q'k'} \pm \delta_{qk'}\delta_{q'k}| \quad (11)$$

for all times satisfying $0 < t \lesssim 1/\Gamma_q$. Thus we see that it is the quasiparticle weight squared, z_F^2, which is the measure of our spin correlation function G^s we were looking for. It is thus interesting to evaluate z_F explicitly. This is indeed possible, again within RPA, and we find after some calculation [32]

$$z_F = 1 - r_s(\frac{1}{2} + \frac{1}{\pi})\,, \quad (12)$$

in leading order of the interaction parameter $r_s = 1/q_F a_B$, where $a_B = \epsilon_0\hbar^2/me^2$ is the Bohr radius. In particular, in a GaAs 2DEG we have $a_B = 10.3$ nm, and $r_s = 0.614$, and thus we obtain from (12) the value $z_F = 0.665$. We note that a more accurate numerical evaluation of the exact RPA self-energy yields $z_F = 0.691155$ [32], again for GaAs. [For 3D metallic leads with say $r_s = 2$ (e.g. $r_s^{Cu} = 2.67$) the loss of correlation is somewhat less strong, since then the quasiparticle weight becomes $z_F = 0.77$ [77].]

In summary, we see that the spin correlation is reduced by a factor of about two (from its maximum value one) as soon as we inject the two electrons (entangled or not) into separate leads consisting of *interacting* Fermi liquids in their ground state. These findings are quite encouraging in view of experimental investigations, as they demonstrate that the spin correlations of a pair of electrons in a Fermi liquid will indeed be preserved in time (albeit with a reduced amplitude) as long as we can neglect spin-dependent forces such as spin-orbit interaction and spin flips induced by spin impurities or nuclear spins etc. Given the high purity of present-day GaAs 2DEG's and the possibility of suppressing the dephasing effects of nuclear spins by dynamical spin polarization [50], it looks promising to use mobile electrons in nanostructures as a means for quantum communication. Similar investigations [32] of such spin correlations are under way for non-equilibrium transport situations, as well as for leads containing impurities or consisting of superconducting or non-Fermi liquid materials, etc.

In conclusion, we believe that various aspects of quantum communication have a high chance of being realized in the not-too-distant future. As we have seen, all that is needed is one single quantum gate which is attached to leads and which can be used as a source of entanglement for mobile qubits along the lines proposed here. Although the realization of such a device is still an experimental challenge at present we are optimistic that it is within technological reach.

ACKNOWLEDGMENTS

We would like to thank G. Burkard, C. Bennett, R. Cleve, and E. Sukhorukov for useful discussions.

[1] We are indebted to an excellent monograph, G. Brassard,

Modern Cryptology, A Tutorial (Springer-Verlag, Lecture Notes in Computer Science, eds. G. Goos and J. Hartmanis, Vol. 325, New York, 1988), for knowledge about much of the subject matter of this section. Our discussion of information processing with quantum-mechanical tools is an update of Brassard's discussion in the light of the many new advances in quantum information processing in the last ten years. See also A. J. Menezes, P. C. van Oorschot, and S. A. Vanstone, *Handbook of Applied Cryptography* (CRC Press, 1996).

[2] C. H. Papadimitriou, *Computational Complexity*, (Addison-Wesley, 1994); M. Sipser, *Introduction to the Theory of Computation* (PWS Pub. Co., 1997).

[3] E. Kushelevitz and N. Nisan, *Communication Complexity* (Cambridge University Press, 1997). The concept of communication complexity was introduced in A. C.-C. Yao, "Some complexity questions related to distributive computing," *Proc. of the 11th ACM Symp. on the Theory of Computing* (ACM Press, 1979), p. 209.

[4] N. Abramson, *Information Theory and Coding* (McGraw-Hill, New York, 1963).

[5] A. Shamir, "How to share a secret," Comm. of the ACM, **22**, 612 (1979).

[6] J. von Neumann and O. Morgenstern, *Theory of Games and Economic Behavior*, 3rd ed. (Princeton University Press, Princeton, 1953).

[7] P. W. Shor, "Polynomial time algorithms for prime factorization and discrete logarithms on a quantum computer," SIAM J. Comput. **26**, 1484 (1997), and references therein.

[8] R. Beals, H. Buhrman, R. Cleve, M. Mosca, and R. de Wolf, "Quantum lower bounds by polynomials," *Proc. of the 39th Annual Symposium on the Foundations of Computer Science* (IEEE Press, Los Alamitos, 1998), p. 352; quant-ph/9802049.

[9] Y. Ozhigov, "Quantum computer cannot speed up iterated applications of a black box," quant-ph/9712051.

[10] Y. Ozhigov, "Quantum computers speed up classical with probability zero," quant-ph/9803064; E. Farhi, J. Goldstone, S. Gutmann, and M. Sipser, "A limit on the speed of quantum computation in determining parity," Phys. Rev. Lett. **81**, 5442 (1998), quant-ph/9802045; "A limit on the speed of quantum computation for insertion into an ordered list," quant-ph/9812057; "How many functions can be distinguished with k quantum queries?" quant-ph/9901012.

[11] H. Buhrman, R. Cleve, and A. Wigderson, "Quantum vs. Classical Communication and Computation," in *Proc. of the 30th Ann. ACM Symp. on the Theory of Computing* (ACM Press, 1998), p. 63; eprint quant-ph/9802040.

[12] L. K. Grover, "Quantum mechanics helps in searching for a needle in a haystack," Phys. Rev. Lett. **79**, 325 (1997).

[13] First done by R. Cleve and H. Buhrman, "Substituting quantum entanglement for communication," Phys. Rev. A **56**, 1201 (1997); quant-ph/9704026.

[14] A. Ambainis, L. Schulman, A. Ta-Shma, U. Vazirani, and A. Wigderson, "The quantum communication complexity of sampling," *Proc. of the 39th Annual Symposium on the Foundations of Computer Science* (IEEE Press, Los Alamitos, 1998); see http://www.icsi.berkeley.edu/~amnon/Papers/qcc.ps.

[15] R. Raz, "Exponential separation of classical and quantum communication complexity," *Proc. of the 31th Ann. ACM Symp. on the Theory of Computing* (ACM Press, 1999), to be published.

[16] D. Mayers, "Unconditionally secure quantum bit commitment is impossible," Phys. Rev. Lett. **78**, 3414 (1997); H.-K. Lo and H. F. Chau, "Is quantum bit commitment really possible?" *ibid.* **78**, 3410 (1997); H.-K. Lo and H. F. Chau, "Why quantum bit commitment and ideal quantum coin tossing are impossible," Physica D **120**, 177 (1998); quant-ph/9711065.

[17] C. A. Fuchs, "Nonorthogonal quantum states maximize classical information capacity," Phys. Rev. Lett. **79**, 1163 (1997) (see quant-ph/9703043); C. H. Bennett, C. A. Fuchs, and J. A. Smolin, "Entanglement-enhanced classical communication on a noisy quantum channel," in *Quantum Communication, Computing, and Measurement*, eds. O. Hirota, A. S. Holevo, and C. M. Caves (Plenum Press, NY, 1997), p. 79; quant-ph/9611006.

[18] C. H. Bennett and S. J. Wiesner, "Communication via one- and two-particle operators on Einstein-Podolsky-Rosen states," Phys. Rev. Lett. **69**, 2881 (1992).

[19] C. H. Bennett, P. Shor, and J. A. Smolin, private communication

[20] S. Wiesner, "Conjugate coding," SIGACT News **15**, (1), 78 (1983). [Manuscript prepared in 1970.]

[21] C. M. Caves, K. S. Thorne, R. W. P. Drever, V. D. Sandberg, and M. Zimmermann, "On the measurement of a weak classcial force coupled to a quantum-mechanical oscillator. I. Issues of principle," Rev. Mod. Phys. **52**, 341 (1980).

[22] C. H. Bennett and G. Brassard, "Quantum Cryptography: Public Key Distribution and Coin Tossing," in *Proceedings of the IEEE International Conference on Computers, Systems and Signal Processing, Bangalore, India* (IEEE, New York, 1984), p. 175.

[23] D. Mayers, "Unconditional security in quantum cryptography," quant-ph/9802025.

[24] H.-K. Lo and H. F. Chau, "Quantum computers render quantum key distribution unconditionally secure over arbitrarily long distance," quant-ph/9803006.

[25] M. N. Wegman and J. L. Carter, "New hash functions and their use in authentication and set equality," J. of Computer and System Sciences **22**, 265 (1981).

[26] C. H. Bennett, D. P. DiVincenzo, C. A. Fuchs, T. Mor, E. Rains, P. W. Shor, J. A. Smolin, and W. K. Wootters, "Quantum nonlocality without entanglement," Phys. Rev. A **59**, 1070 (1999); quant-ph/9804053); M. Hillery, V. Buzek, and A. Berthiaume, "Quantum secret sharing," quant-ph/9806063; A. Karlsson, M. Koashi, and N. Imoto, "Quantum entanglement for secret sharing and secret splitting," Phys. Rev. A **59**, 162 (1999).

[27] D. A. Meyer, "Quantum strategies," quant-ph/9804010; J. Eisert, M. Wilkens, and M. Lewenstein, "Quantum games and quantum strategies," quant-ph/9806088; L. Goldenberg, L. Vaidman, and S. Wiesner, "Quantum gambling," quant-ph/9808001.

[28] J. Preskill, "Reliable quantum computers," Proc. R. Soc. London **454**, 385 (1998); quant-ph/9705031.

[29] R. Cleve, D. Gottesman, and H.-K. Lo, "How to share a quantum secret," quant-ph/9901025.

[30] C. H. Bennett, G. Brassard, S. Popescu, B. Schumacher, J. A. Smolin, and W. K. Wootters, "Purification of noisy entanglement and faithful teleportation via noisy channels," Phys. Rev. Lett. **76**, 722 (1996), eprint quant-ph/9511027; C. H. Bennett, D. P. DiVincenzo, J. A. Smolin, and W. K. Wootters, "Mixed state entanglement and quantum error-correction," Phys. Rev. A **54**, 3824 (1996); e-print quant-ph/9604024.

[31] C. H. Bennett, G. Brassard, C. Crepeau, R. Jozsa, A. Peres, W. K. Wootters, "Teleporting an Unknown Quantum State via Dual Classical and Einstein-Podolsky-Rosen Channels," Phys. Rev. Lett. **70**, 1895 (1993).

[32] D. Loss, G. Burkard, and E. Sukhorukov, "Quantum communication with electrons", unpublished.

[33] A. Imamoglu *et al.*, "Quantum dot electron spin manipulation using cavity QED," unpublished.

[34] C. Monroe, D. M. Meekhof, B. E. King, W. M. Itano, and D. J. Wineland, "Demonstration of a fundamental quantum logic gate," Phys. Rev. Lett. **75**, 4714 (1995).

[35] Q. A. Turchette, C. S. Wood, B. E. King, C. J. Myatt, D. Liebfried, W. M. Itano, C. Monroe, and D. J. Wineland, "Deterministic entanglement of two trapped ions," Phys. Rev. Lett. **81**, 3631 (1998).

[36] Q. A. Turchette, C. J. Hood, W. Lange, H. Mabuchi, and H. J. Kimble, "Measurement of conditional phase shifts for quantum logic," Phys. Rev. Lett. **75**, 4710 (1995).

[37] P. D. Townsend, J. G. Rarity, and P. R. Tapster, Electronics Letters **29** (14), 1291 (1993); R. J. Hughes, D. M. Alde, P. Dyer, G. G. Luther, G. L. Morgan, and M. Schauer, "Quantum Cryptography," Contemp. Phys. **36**, 149 (1995); A. Muller, T. Herzog, B. Huttner, W. Tittel, H. Zbinden, and N. Gisin, "'Plug and Play' systems for quantum cryptography," Appl. Phys. Lett. **70**, 793 (1997).

[38] K. Mattle, H. Weinfurter, P. G. Kwiat, and A. Zeilinger, "Dense coding in experimental quantum communcation," Phys. Rev. Lett. **76**, 4656 (1996).

[39] D. Boschi, S. Branca, F. De Martini, L. Hardy, and S. Popescu, "Experimental Realization of Teleporting an Unknown Pure Quantum State via Dual Classical and Einstein-Podolsky-Rosen Channels," Phys. Rev. Lett. **80**, 1121 (1998).

[40] D. Bouwmeester, Jian Wei Pan, K. Mattle, M. Eibl, H. Weinfurter, and A. Zeilinger, "Experimental Quantum Teleportation," Nature **390**, 575 (1997).

[41] L. Vaidman, Phys. Rev. A **49** 1473 (1994); A. Furusawa, J. L. Sorensen, S. L. Braunstein, C. A. Fuchs, H. J. Kimble, and E. S. Polzik, Science **282**, 706 (1988).

[42] D. P. DiVincenzo, "Two-bit gates are universal for quantum computation," Phys. Rev. A **51**, 1015 (1995), cond-mat/9407022.

[43] N. Gershenfeld and I. Chuang, Science **275**, 350 (1997); D. Cory, A. Fahmy, and T. Havel, Proc. Nat. Acad. Sci. **94** (5), 1634 (1997).

[44] I. L. Chuang, L. Vandersypen, D. Leung, X. Zhou, and S. Lloyd, Nature **393**, 143 (1998); I. L. Chuang, N. Gershenfeld, and M. G. Kubinec, Phys. Rev. Lett. **80**, 3408 (1998).

[45] D. G. Cory, M. D. Price, W. Maas, E. Knill, R. Lafamme, W. H. Zurek, T. F. Havel, and S. S. Somaroo, "Experimental quantum error correction," Phys. Rev. Lett. **81**, 2152 (1998).

[46] M. A. Nielsen, E. Knill, and R. Laflamme, Nature **396**, 52 (1998).

[47] D. P. DiVincenzo, "Topics in Quantum Computers," in *Mesoscopic Electron Transport*, ed. L. Sohn, L. Kouwenhoven, and G. Schoen (NATO Advanced Study Institute, Series E (Applied Sciences), Vol. 345, Kluwer, 1997), p. 657, eprint (1996) cond-mat/9612126.

[48] D. Loss and D. P. DiVincenzo, "Quantum computation with quantum dots," Phys. Rev. A **57**, 120 (1998), eprint (1997) cond-mat/9701055.

[49] D. P. DiVincenzo and D. Loss, "Quantum information is physical," Superlattices and Microstructures **23**, 419 (1998), eprint (1997) cond-mat/9710259.

[50] G. Burkard, D. Loss, and D. P. DiVincenzo, "Coupled quantum dots as quantum gates," Phys. Rev. B, in press (Jan. 15, 1999); eprint cond-mat/9808026.

[51] L. Jacak, P. Hawrylak, and A. Wójs, *Quantum Dots* (Springer, Berlin, 1997).

[52] L. P. Kouwenhoven, C. M. Marcus, P. L. McEuen, S. Tarucha, R. M. Westervelt, and N. S. Wingreen, Proceedings of the Advanced Study Institute on *Mesoscopic Electron Transport*, edited by L. L. Sohn, L. P. Kouwenhoven, G. Schön (Kluwer, 1997).

[53] R. C. Ashoori, Nature **379**, 413 (1996).

[54] R. J. Luyken, A. Lorke, M. Haslinger, B. T. Miller, M. Fricke, J. P. Kotthaus, G. Medeiros-Ribiero, and P. M. Petroff, preprint.

[55] S. Tarucha, D. G. Austing, T. Honda, R. J. van der Hage, and L. P. Kouwenhoven, Phys. Rev. Lett. **77**, 3613 (1996); L. P. Kouwenhoven, T. H. Oosterkamp, M. W. S. Danoesastro, M. Eto, D. G. Austing, T. Honda, and S. Tarucha, Science **278**, 1788 (1997).

[56] F. R. Waugh, M. J. Berry, D. J. Mar, R. M. Westervelt, K. L. Chapman, and A. C. Gossard, Phys. Rev. Lett. **75**, 705 (1995); C. Livermore, C. H. Crouch, R. M. Westervelt, K. L. Chapman, and A. C. Gossard, Science **274**, 1332 (1996).

[57] T. H. Oosterkamp, S. F. Godijn, M. J. Uilenreef, Y. V. Nazarov, N. C. van der Vaart, and L. P. Kouwenhoven, Phys. Rev. Lett. **80**, 4951 (1998).

[58] J. M. Kikkawa, I. P. Smorchkova, N. Samarth, and D. D. Awschalom, "Room-temperature spin memory in two-dimensional electron gases," Science **277**, 1284 (1997).

[59] J. A. Gupta, D. D. Awschalom, X. Peng, and A. P. Alivisatos, "Spin coherence in semiconductor quantum dots," Phys. Rev. B, in press.

[60] D. P. DiVincenzo, "Two-bit gates are universal for quantum computation," Phys. Rev. A **51**, 1015 (1995), cond-mat/9407022.

[61] R. H. Blick, D. Pfannkuche, R. J. Haug, K. v. Klitzing, and K. Eberl, Phys. Rev. Lett. **80**, 4032 (1998).

[62] T. H. Oosterkamp, J. W. Janssen, L. P. Kouwenhoven, D. G. Austing, T. Honda, S. Tarucha, cond-mat/9809142.

[63] A. Barenco, C. H. Bennett, R. Cleve, D. P. DiVincenzo, N. Margolus, P. Shor, T. Sleator, J. A. Smolin, and H.

Weinfurter, "Elementary gates for quantum computation," Phys. Rev. A **52**, 3457 (1995), quant-ph/9503016.
[64] G. A. Prinz, Science **282** 1660 (1998).
[65] M. Kleiber, F. Kuemmerlen, M. Loehndorf, A. Wadas, D. Weiss, R. Wiesendanger "Magnetization switching of submicrometer Co dots induced by a magnetic force tip," Phys. Rev. B **58**, 5563 (1998).
[66] D. Gottesman, private communication.
[67] J. I. Cirac and P. Zoller, "Quantum computations with cold trapped ions," Phys. Rev. Lett. **74**, 4091 (1995).
[68] D. P. DiVincenzo, "Quantum computing and single-spin measurements using the spin-filter effect," J. Appl. Phys, in press, cond-mat/9810295.
[69] B. E. Kane, "A Si-based nuclear-spin quantum computer," Nature **393**, 133 (1998).
[70] L. Viola and S. Lloyd, "Dynamical suppression of decoherence in two-state quantum systems," Phys. Rev. A **58**, 2733 (1998), quant-ph/9803057.
[71] A. Aspect, J. Dalibard, and G. Roger, Phys. Rev. Lett. **49**, 1804 (1982).
[72] W. Tittel, J. Brendel, H. Zbinden, and N. Gisin, Phys. Rev. Lett. **81**, 3563 (1998).
[73] A. Einstein, B. Podolsky, and N. Rosen, Phys. Rev.**47**, 777 (1935).
[74] R. Hanbury Brown and R. Q. Twiss, Nature (London) **177**, 27 (1956). M. Büttiker, Phys. Rev. B**46**, 12485 (1992). R. C. Liu, B. Odom, Y. Yamamoto, and S. Tarucha, Nature **391**, 263 (1998).
[75] A.L. Fetter and J.D. Walecka, *Quantum Theory of Many-Particle Systems* (New York, McGraw-Hill, 1971).
[76] G. F. Guiliani and J. J. Quinn, Phys. Rev. B **26**, 4421 (1982).
[77] G. D. Mahan, *Many Particle Physics*, 2nd Ed. (Plenum, New York, 1993).

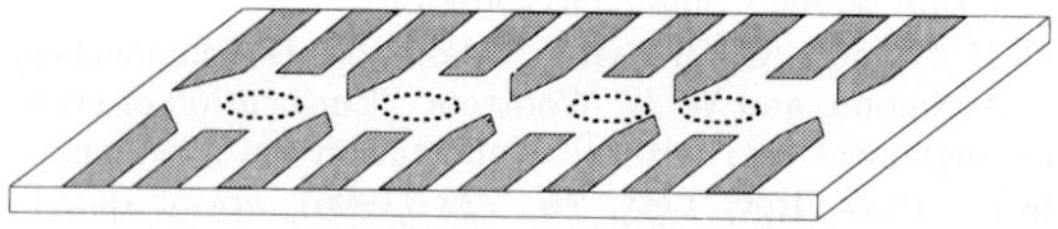

FIG. 1. A schematic of the quantum-dot array quantum computer. Single electrons are confined in a two-dimensional electron gas, and to dot regions in between the electrodes. Electrodes are shown shaded, dots are shown as dashed circles. The electrode potentials can be varied so as to push pairs of electrons into contact (see the third and fourth dots), which results in the execution of a two-bit quantum gate. One-bit gates are accomplished by the action of inhomogeneous magnetic fields (or effective fields). Readout is accomplished by tunneling the electrons through a spin-selective barrier. The magnetic elements of the device are not shown.

NMR Quantum Computing

NMR Quantum Computing

Quantum Computation with NMR

Jonathan A. Jones
Center for Quantum Computation, Oxford

A novice seeking to understand Nuclear Magnetic Resonance Quantum Computation (NMR QC) faces a formidable task. Not only is it necessary to master two new fields (NMR and QC), but many of the early papers in the field are unusually difficult to understand. Part of this opacity is, no doubt, due to the speed with which the early manuscripts were written, but an even more important cause was the lack of a common language which would allow NMR experimentalists and QC theoreticians to talk to one another. This barrier is now beginning to be crossed, and more considered pedagogical works are starting to appear.

It is, perhaps, surprising how little known NMR is within the physics community, as it provides a remarkable system for investigating a variety of topics. The name refers to spectroscopic studies of transitions between the Zeeman levels of an atomic nucleus in a magnetic field, and the subject was initially developed by physicists[1, 2] as a method for determining the size of nuclear magnetic moments, and thus testing models of nuclear structure. Within a few years, however, it became clear that measured NMR frequencies were not simply determined by nuclear properties, but also showed a subtle dependence on the properties of the surrounding electronic clouds, and thus on the chemical environment of the nucleus. The messy nature of these "chemical shift" interactions caused most physicists to lose interest, leaving the field to be developed by chemists.

NMR was of obvious interest to chemists as the chemical shift, together with the J-coupling interaction discovered a few years later, provides a powerful method of gaining insights into molecular structure. While other spectroscopic techniques can, perhaps, provide more detailed descriptions of small molecules, NMR is unique in the ease with which it can be applied to complex systems and the remarkably close relationship between the information available from NMR and the mental pictures of molecules used by most chemists. Furthermore, NMR experiments soon reached an extraordinary level of sophistication, involving the systematic generation and interconversion of single and multi spin coherent states. This sophistication was possible as a result of the long coherence times of NMR superpositions, and the exact experimental control possible with RF radiation: similar experiments involving optical transitions have only become possible in the last few years. These techniques underly modern multidimensional experiments, which have enabled NMR to become one of the most important techniques in the molecular sciences.

A vast number of introductory NMR texts are available, but most of these are aimed at chemists or biochemists; such texts typically concentrate on the applications of NMR while avoiding much of the underlying theory. Those texts which are aimed at physicists largely consider NMR studies in the solid state, which have little immediate relevance to current NMR QCs. Fortunately a small number of reasonable texts do exist: the famous text by Ernst *et al.*[3] covers most of modern NMR, and more gentle introductions[4, 5] are also available. It is particularly important to become familiar with the "product operator" description[6], which plays a central role in modern NMR theory.

While NMR has largely remained the preserve of chemists, it has occasionally been used to investigate fundamental topics in physics. One notable example is the use of NMR and its close cousin Nuclear Quadrupole Resonance (NQR) to study geometric phases[7, 8]. It has long been known that NMR is in many ways well suited to quantum computation, but early proposals foundered on the difficulty of generating an initial pure state. Most QC schemes rely on cooling to the thermodynamic ground state as an initialisation mechanism, but this approach is not practical within NMR as the energy gap between the Zeeman levels is small compared to kT at any reasonable temperature.

The great breakthrough in NMR QC was the realisation by Cory *et al.* [A] that it is not strictly necessary to form a pure state to implement an NMR QC, as NMR is an ensemble technique in which very large numbers of spins are detected simultaneously (it is not practical to detect the signal from a single spin as a result of the tiny energies involved). Instead it suffices to generate a "pseudo pure" state, that is an ensemble comprising a mixture of the desired pure state and the maximally mixed state. Assembling such a mixture is a fairly conventional problem in NMR, and Cory *et al.* [A] demonstrated all the basic elements required to build an NMR QC; subsequent papers[9, 10] have expanded and clarified many of their ideas. An alternative approach to this problem was subsequently described by Gershenfeld and Chuang[11, 12]; this approach is elegant in principle but complex in practice and has not been widely used. More recently a variety of new approaches have been suggested, among which the method of temporal averaging[13] has proved particularly popular.

Once the initialisation problem had been overcome, progress in the implementation of NMR QCs was rapid. The first algorithm was an implementation of Deutsch's algorithm [B] on a two qubit NMR QC based on the small molecule cytosine; a second implementation based on chloroform[14] was published soon afterwards. These were swiftly followed by two implementations of Grover's quantum search routine; this time the chloroform implementation [C] came first, with the cytosine implementation following behind[15].

While these two early NMR QCs share some common features, there are also some significant differences. Many of these can be traced back to the fact that the cytosine QC uses two ^{1}H nuclei (a *homonuclear* system), while the chloroform QC uses one ^{1}H nucleus and a ^{13}C nucleus (a *heteronuclear* system). In heteronuclear systems the NMR transition frequencies of the two nuclei are very different, which makes it very easy to distinguish them; in homonuclear systems the frequencies are quite similar, making discrimination between the spins a more challenging problem.

Homonuclear systems do, however, have the advantage that it is possible to address both nuclei simultaneously, and to observe both in the same spectrum, while heteronuclear systems require two completely separate RF channels. This permits a particularly simple readout scheme in homonuclear NMR QCs: it suffices to simply examine the NMR spectrum, in which state $|0\rangle$ is indicated by an absorption, while state $|1\rangle$ appears as an emissive transition. With a heteronuclear system it would be possible to perform two separate detection experiments; alternatively a more detailed analysis of the multiplet structure within the NMR spectrum of one nucleus may be used to characterise the state of the other nucleus. In fact Chuang *et al.* used a total of nine different readout experiments in order to fully characterise the final state of their NMR QC (that is, they performed full quantum state tomography). This gives an excellent idea of the experimental errors involved in their implementation, but tomography is not a practical approach for more general problems, as the number of readout experiments required rises exponentially with the number of nuclei in the system. Furthermore, their implementation uses temporal averaging to produce the initial pseudo pure state, which requires that every experiment be repeated three times. Thus their results represent the combined analysis of 27 separate experiments (although not all these experiments are strictly necessary).

Implementing these small NMR QCs is experimentally quite straightforward, and a number of new systems have been described. Most of these are broadly similar to the two systems described above; in particular multiple pulse techniques are used to modulate the nuclear Hamiltonian, creating an effective Hamiltonian[3] which implements the desired logic gate (this approach has been described in some detail in two recent papers[16, 17]). One three qubit implementation [D], however, adopts a different approach, based on the use of simultaneous line selective pulses. In this method extremely weak RF pulses are used which will excite a nucleus only when the neighbouring nuclei have specific spin states. By this means it is possible to implement several two and three qubit gates directly. This approach is experimentally challenging when applied to homonuclear systems, but may prove useful with heteronuclear implementations.

In addition to these demonstrations of quantum algorithms using NMR QCs, there has also been significant interest in using NMR to demonstrate other phenomena in quantum information processing, such as GHZ states[18, 19], state teleportation[20] and quantum error correction protocols[21, 22]. Some of these topics are related to more conventional NMR ideas, such as multiple quantum coherence and coherence transfer sequences[3], but the language of QC provides an intriguing new view of old concepts.

From the beginning there has been a strong current of concern regarding the usefulness of NMR QCs; indeed there has been some debate as to whether NMR QCs are in fact *real* QCs. Initial criticism focussed on the question of scalability[23, 24], and it is now widely accepted that current NMR implementations are probably not scalable for a variety of reasons[25, 26], including the exponential inefficiency in the preparation of pseudo pure states, the limited number of operations which can be carried out before decoherence sets in, and the experimental difficulties involved in implementing logic gates in multispin systems. Despite this somewhat depressing list it should be remembered that NMR is currently well in the lead in implementing small QCs, and that demonstrations involving five qubits[27] and hundreds of gates[28] have been performed. Furthermore, one recent proposal incorporating NMR techniques within a solid state system[29] appears to sidestep these problems.

More recently, it has been suggested[30] that NMR might not be a quantum mechanical technique at all! When assessing this comment, it should be remembered that "quantum mechanical" is used here with a technical meaning of "provably non-classical". As NMR experiments are conducted at temperatures such that kT is large compared with the splitting between the energy levels, the density matrix describing a nuclear spin system is always close to the maximally mixed state, and it can be shown that such high temperature states can always be decomposed as a mixture of product states (that is, states containing no entanglement between different nuclei). As NMR states appear to be describable without invoking entanglement, they can therefore be described using classical models (although these classical models may be somewhat contrived). However, while such classical models can be used to describe an individual NMR state, it is not clear that such models can be used to describe the evolution of the state during an NMR experiment[31]. The significance of these conclusions remains contentious and unclear.

1 Papers

A. **5 pages:** *Nuclear magnetic resonance: an experimentally accessible paradigm for quantum computing,* D. G. Cory, A. F. Fahmy, and T. F. Havel, Proceedings of PhysComp '96 (T. Toffoli, M. Biafore, and J. Leão Eds), 87–91 (1996).

B. **6 pages:** *Implementation of a quantum algorithm on a nuclear magnetic resonance quantum computer,* J. A. Jones and M. Mosca, Journal of Chemical Physics **109,** 1648–1653 (1998).

C. **4 pages:** *Experimental Implementation of Fast Quantum Searching,* I. L. Chuang, N. Gershenfeld and M. Kubinec, Physical Review Letters **80,** 3408–3411 (1998).

D. **7 pages:** *An implementation of the Deutsch–Jozsa algorithm on a three-qubit NMR quantum computer,* N. Linden, H. Barjat, and R. Freeman, Chemical Physics Letters **296,** 61–67 (1998).

References

[1] E. M. Purcell, H. C. Torrey, and R. V. Pound, *Phys. Rev.* **69,** 37 (1946).

[2] F. Bloch, W. W. Hansen, and M. E. Packard, *Phys. Rev.* **69,** 127 (1946).

[3] R. R. Ernst, G. Bodenhausen, and A. Wokaun, *Principles of Nuclear Magnetic Resonance in One and Two Dimensions* (Oxford University Press, Oxford, 1987).

[4] M. Goldman, *Quantum Description of High- Resolution NMR in Liquids* (Clarendon Press, Oxford, 1988).

[5] M. Munnowitz, *Coherence and NMR* (John Wiley, 1988).

[6] O. W. Sørensen, G. W. Eich, M. H. Levitt, G. Bodenhausen, and R. R. Ernst, *Prog. NMR Spectrosc.* **16,** 163 (1983).

[7] D. Suter, K. T. Mueller, and A. Pines, *Phys. Rev. Lett.* **60,** 1218 (1988).

[8] R. Tycko, *Phys. Rev. Lett.* **58** 2281 (1987).

[9] D. G. Cory, A. F. Fahmy and T. F. Havel, *PNAS USA* **94,** 1634 (1997).

[10] D. G. Cory, M. D. Price and T. F. Havel, *Physica D* **120,** 82 (1998).

[11] N. Gershenfeld, I. Chuang and S. Lloyd, Proceedings of PhysComp '96 (T. Toffoli, M. Biafore, and J. Leão Eds), 134 (1996).

[12] N. A. Gershenfeld and I. L. Chuang, *Science* **275,** 350 (1997).

[13] E. Knill, I. Chuang and R. Laflamme, *Phys. Rev. A* **57,** 3348 (1998).

[14] I. L. Chuang, L. M. K. Vandersypen, X. Zhou, D. W. Leung and S. Lloyd, *Nature* **393,** 143 (1998).

[15] J. A. Jones, M. Mosca, and R. H. Hansen, *Nature* **393,** 344 (1998).

[16] J. A. Jones, R. H. Hansen, and M. Mosca, *J. Magn. Reson.* **135,** 353 (1998).

[17] N. Linden, H. Barjat, R. J. Carbajo, and R. Freeman, *Chem. Phys. Lett.* **305,** 28 (1999).

[18] R. Laflamme, E. Knill, W. H. Zurek, P. Catasti, and S. V. S. Mariappan, *Phil. Trans. R. Soc. Lond. A* **356,** 1941 (1998).

[19] R. J. Nelson, D. G. Cory and S. Lloyd, `quant-ph/9905028`.

[20] M. A. Nielsen, E. Knill and R. Laflamme, *Nature* **396,** 52 (1998).

[21] D. G. Cory, M. D. Price, W. Maas, E. Knill, R. Laflamme, W. H. Zurek, T. F. Havel, and S. S. Somaroo, *Phys. Rev. Lett.* **81,** 2152 (1998).

[22] D. Leung, L. Vandersypen, X. Zhou, M. Sherwood, N. Yannoni, M. Kubinec and I. Chuang, `quant-ph/9811068`.

[23] W. S. Warren, *Science* **277,** 1688 (1997).

[24] N. Gershenfeld and I. Chuang, *Science* **277,** 1689 (1997).

[25] J. A. Jones, *Nuclear Magnetic Resonance Experiments* in *The Physics of Quantum Information* (D. Bouwmeester, A. Ekert and A. Zeilinger, Eds.), Springer-Verlag (1999, in press).

[26] J. A. Jones, *Quantum Computing and NMR* in *Quantum Computing and Communication* (M. Brooks, Ed.), Springer-Verlag (1999).

[27] R. Marx, A. F. Fahmy, J. M. Myers, W. Bermel, S. J. Glaser, Conference on Quantum Information Processing and NMR, February 22–24th, 1999, Cambridge, USA (`http://qso.lanl.gov/qc/nmrconf/abstract.html`).

[28] J. A. Jones and M. Mosca, *Phys. Rev. Lett.* (1999, in press), `quant-ph/9808056`.

[29] B. E. Kane, *Nature* **393**, 133 (1998).

[30] S. L. Braunstein, C. M. Caves, R. Jozsa, N. Linden, S. Popescu, and R. Schack, quant-ph/9811018.

[31] R. Schack and C. M. Caves, quant-ph/9903101.

Fourth Workshop on
Physics and Computation
Boston University, 22–24 Nov. 1996.

In: T. Toffoli, M. Biafore, and J. Leão (ed.), *PhysComp96*, New England Complex Systems Institute (1996), 87–91.
Also on-line in the *InterJournal*.

submitted: 12 May 1996
revised: 25 Oct., 17 Nov. 1996

Nuclear magnetic resonance spectroscopy: an experimentally accessible paradigm for quantum computing*

David G. Cory†
Nuclear Eng. Dept., M.I.T.
Amr F. Fahmy‡
Div. of Applied Sciences, Harvard University
Timothy F. Havel‡
B.C.M.P., Harvard Medical School

1 Introduction

The theory of quantum computing is advancing at a rate which vastly outstrips its experimental realization (for recent accounts, see [1, 2, 3]). Most attempts to implement a quantum computer have utilized submicroscopic assemblies of quantum spins, which are difficult to prepare, isolate, manipulate and observe. A "homologous" system which exhibits many of the same properties, but is easier to work with, would clearly be very useful both as a means of testing theoretical predictions, and for demonstration and educational purposes. Such a system is provided by weakly polarized macroscopic ensembles of spins, which are readily manipulated and observed by *nuclear magnetic resonance spectroscopy*, or NMR.

The spins of a molecule in solution are largely isolated from their surroundings by simple surface to volume considerations, and from spins in neighboring molecules by rotational averaging, which reduces dipole-dipole coupling to a second-order effect[4]. This fact enables us to work with a *reduced density matrix* $\mathbf{D}$ of size 2^n, where n is the number of spin 1/2 nuclei in the molecule, rather than 2^N where N is the total number of such spins in the sample[5]. It is also customary in NMR spectroscopy to shift the reduced density matrix by subtraction of 2^{-n} times the trace of the equilibrium reduced density matrix, since only the traceless part undergoes unitary evolution, and to scale it to have integral elements[6]. In the next paragraph, we define a manifold of statistical spin states with a reduced density matrix whose traceless part is proportional to the traceless part of the usual density matrix of a pure state.

In the following, whenever we use the term "density matrix", we mean "reduced, shifted and scaled density matrix" unless otherwise stated. When such a density matrix has rank equal to one (after adding an appropriate multiple of the unit matrix to it), it can be factored into a dyadic product of the coordinates of a "spinor" and its conjugate versus the usual $\mathbf{I}_z$ basis, and this factorization is unique up to an overall phase factor. This mapping between spinor coordinates and density matrices which can be shifted to a signature of $[1, 0, \ldots, 0]$ is *covariant*, in the sense that if we apply a unitary matrix to a spinor's coordinates, the corresponding density matrix transforms by conjugation with the *same* unitary matrix. As a result, we can regard such a density matrix as a kind of spinor, and perform essentially arbitrary unitary transformations on it via NMR spectroscopy, thereby "emulating" a quantum computer. We shall call the states described by density matrices with one positive and $2^n - 1$ equal negative eigenvalues "pseudo-pure" states, and the corresponding spinors "pseudo-spinors".

Of course, some things are lost in translation. For example, the density matrix is not changed on rotation by 2π, although spinors change sign. Since these sign changes cannot easily be observed, this seems to be of little consequence for quantum computing. More important is the fact that the coherence which is observed by NMR spectroscopy is always an ensemble average over an astronomical number of microscopic systems. As a consequence, the NMR spectrum of a pseudo-pure state yields the expectation values of certain observables relative to the corresponding pseudo-spinor, rather than a random eigenvalue of one of them. In particular, *wave function collapse does not occur*. A wide variety of more easily controlled "filtering" mechanisms are available in NMR spectroscopy, however, and we have shown that for most computational purposes the ability to measure expectation values directly is actually a great advantage[8]. NMR experiments on liquid samples possess a number of other highly desirable features as well; in particular, the decoherence times are typically on the order of seconds.

We claim that NMR spectroscopy in fact provides a means of building a nonconventional computer that can be programmed much like a quantum computer, and is capable of the same exponential speed-ups, but is much easier to implement. In some respects, this approach also resembles DNA computing, in that it can use the parallelism inherent in ensembles of molecules to efficiently count the number of solutions to combinatorial problems. A detailed account of the theory may be found in [8]; this paper will describe how basic quantum logic gates can be implemented via NMR spectroscopy, and present experimental results to validate our claims. After submitting the revised version of this abstract, we learned that an analogous approach has also been submitted to this workshop[7].

*Correspondence should be directed to havel@menelaus.med.harvard.edu.
†Thanks NSF/DMR 9357603.
‡Thanks NSF/MCB 9527181.

2 Basic results from NMR

Let us begin with the simplest nontrivial case, namely a molecule in solution containing exactly two coupled spins. The dipolar coupling averages to zero in solution, but the so-called *scalar* coupling remains, which is mediated by electron correlation in the chemical bonds linking the atoms. It will simplify our presentation if we assume *weak coupling*, i.e. that the coupling constant J is small compared to the difference $|\omega_1 - \omega_2|$ in the resonance frequencies of the two spins. With the convention that the magnetic field is along the z-axis, the Hamiltonian of this system is $\mathbf{H} = \omega_1 \mathbf{I}_z^1 + \omega_2 \mathbf{I}_z^2 + 2\pi J\, \mathbf{I}_z^1 \mathbf{I}_z^2$, where $\mathbf{I}_z^k$ $(k = 1, 2)$ are the usual matrices for the z-component of the angular momentum of each spin. Because the energy level differences are small compared to kT at room temperature, the equilibrium density matrix of this system is given to an excellent approximation by $\mathbf{Exp}(-\mathbf{H}/kT) \approx \mathbf{1} - \mathbf{H}/kT$. On taking account of the fact that $J \ll \omega_1 \approx \omega_2$, removing the trace and scaling, the equilibrium density matrix becomes

$$\mathbf{D}_{eq} = \mathbf{I}_z^1 + \mathbf{I}_z^2 = \begin{pmatrix} 1 & 0 & 0 & 0 \\ 0 & 0 & 0 & 0 \\ 0 & 0 & 0 & 0 \\ 0 & 0 & 0 & -1 \end{pmatrix} . \tag{1}$$

Rather than writing them out explicitly, NMR spectroscopists typically represent their density matrices as linear combinations of "product operators", i.e. the products of the usual matrices of one-spin operators $\mathbf{I}_x$, $\mathbf{I}_y$ and $\mathbf{I}_z$ (as above). This makes it very easy to describe the unitary transformations effected by applying RF (radio-frequency) pulses to the sample. For example, a $\pi/2$ pulse about the y-axis (in the rotating frame[6]) by definition rotates the equilibrium density matrix to

$$\mathbf{D}_{eq} \xrightarrow{[\pi/2]_y} \mathbf{I}_x^1 + \mathbf{I}_x^2 . \tag{2}$$

This rotated magnetization precesses about the applied field and generates an oscillatory magnetic moment in the xy-plane, according to the density matrix obtained from the time-dependent unitary transformation:

$$\mathbf{Exp}(-it\mathbf{H})(\mathbf{I}_x^1 + \mathbf{I}_x^2)\mathbf{Exp}(it\mathbf{H}) \tag{3}$$

The complex-valued, time-dependent signal that is detected is calculated by taking the trace of the product of this density matrix with

$$\mathbf{I}_+^1 + \mathbf{I}_+^2 = \begin{pmatrix} 0 & 1 & 1 & 0 \\ 0 & 0 & 0 & 1 \\ 0 & 0 & 0 & 1 \\ 0 & 0 & 0 & 0 \end{pmatrix} , \tag{4}$$

where $\mathbf{I}_+ = \mathbf{I}_x + i\,\mathbf{I}_y$ as usual (see e.g. [4]). Since this matrix contains only four nonzero elements, the spectrum contains direct information on only four of the ten independent elements of the density matrix, namely $D_{12} = D_{21}^*, D_{13} = D_{31}^*, D_{24} = D_{42}^*$ and $D_{34} = D_{43}^*$. These elements are called *single-quantum coherences*. After Fourier transformation,

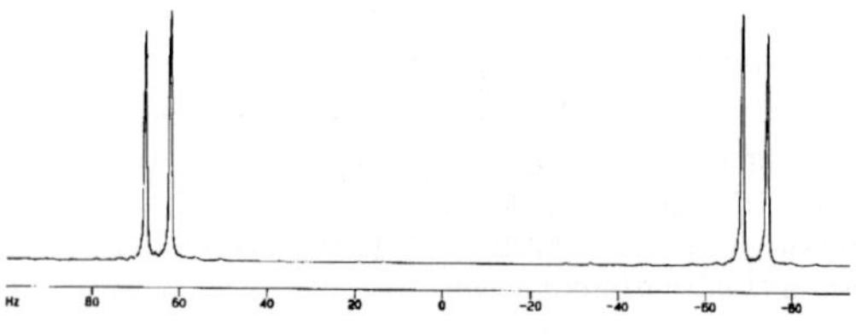

Figure 1: The experimental NMR spectrum of liquid 2,3-dibromo-thiophene (a molecule with two nonequivalent, weakly coupled spin 1/2 hydrogen atoms) after applying a nonselective $\pi/2$ y-pulse to the equilibrium state and Fourier transforming the resulting signal (see text).

we obtain a spectrum containing a pair of doublets as in Fig. 1.

If the *chemical shifts* (i.e. resonance frequencies relative to some standard) of the two spins differ by several times the sum of their line widths and coupling constant, such a pulse can be made selective for only one of the two spins, e.g. the second, which yields

$$\mathbf{D}_{eq} \xrightarrow{[\pi/2]_y^2} \mathbf{I}_z^1 + \mathbf{I}_x^2 . \tag{5}$$

A *gradient pulse*, which produces a transient field inhomogeneity along the z-axis, can then be used to selectively quench the x-coherence, giving

$$\mathbf{I}_z^1 = \begin{pmatrix} 1/2 & 0 & 0 & 0 \\ 0 & 1/2 & 0 & 0 \\ 0 & 0 & -1/2 & 0 \\ 0 & 0 & 0 & -1/2 \end{pmatrix} . \tag{6}$$

In a similar manner, we can generate the state

$$\mathbf{I}_z^2 = \begin{pmatrix} 1/2 & 0 & 0 & 0 \\ 0 & -1/2 & 0 & 0 \\ 0 & 0 & 1/2 & 0 \\ 0 & 0 & 0 & -1/2 \end{pmatrix} . \tag{7}$$

A further $\pi/2$ pulse will convert this to $\mathbf{I}_x^2$, which evolves under the influence of a coupling constant of J Hz. to

$$\mathbf{I}_x^2 \cos(\pi J t) + 2\mathbf{I}_z^1 \mathbf{I}_y^2 \sin(\pi J t) . \tag{8}$$

Thus after a period of $t = 1/(2J)$, we obtain $2\mathbf{I}_z^1 \mathbf{I}_y^2$. A final selective $\pi/2$ pulse can then be used to obtain the correlated state

$$2\mathbf{I}_z^1 \mathbf{I}_z^2 = \begin{pmatrix} 1/2 & 0 & 0 & 0 \\ 0 & -1/2 & 0 & 0 \\ 0 & 0 & -1/2 & 0 \\ 0 & 0 & 0 & 1/2 \end{pmatrix} . \tag{9}$$

3 Pseudo-spinors in NMR

It is possible to generate all three of the states (6), (7) and (9) simultaneously in the same sample, yielding:

$$\mathbf{I}_z^1 + \mathbf{I}_z^2 + 2\mathbf{I}_z^1 \mathbf{I}_z^2 = \begin{pmatrix} 3/2 & 0 & 0 & 0 \\ 0 & -1/2 & 0 & 0 \\ 0 & 0 & -1/2 & 0 \\ 0 & 0 & 0 & -1/2 \end{pmatrix} \tag{10}$$

This matrix shifts to

$$\begin{pmatrix} 2 & 0 & 0 & 0 \\ 0 & 0 & 0 & 0 \\ 0 & 0 & 0 & 0 \\ 0 & 0 & 0 & 0 \end{pmatrix}, \tag{11}$$

which can in turn be factorized into the product of a pseudo-spinor with its conjugate

$$2\,|00\rangle\langle 00| \;\equiv\; 2\begin{pmatrix} 1 \\ 0 \\ 0 \\ 0 \end{pmatrix}\begin{pmatrix} 1 & 0 & 0 & 0 \end{pmatrix}. \tag{12}$$

Therefore, the matrix $\mathbf{I}_z^1 + \mathbf{I}_z^2 + 2\mathbf{I}_z^1\mathbf{I}_z^2$ in Eq. (10) represents a pseudo-pure state in product operator notation.

The following RF and gradient pulse sequence accomplishes this task:

$$\begin{array}{l} \mathbf{I}_z^1 + \mathbf{I}_z^2 \\ \xrightarrow{[\pi/3]_x^2} \mathbf{I}_z^1 + \mathbf{I}_z^2/2 - \mathbf{I}_y^2\sqrt{3}/2 \\ \xrightarrow{[z-\text{grad}]} \mathbf{I}_z^1 + \mathbf{I}_z^2/2 \\ \xrightarrow{[\pi/4]_x^1} \mathbf{I}_z^1/\sqrt{2} + \mathbf{I}_z^2/2 - \mathbf{I}_y^1/\sqrt{2} \\ \xrightarrow{[1/(2J)]} \mathbf{I}_z^1/\sqrt{2} + \mathbf{I}_z^2/2 + \sqrt{2}\mathbf{I}_x^1\mathbf{I}_z^2 \\ \xrightarrow{[-\pi/4]_y^1} \mathbf{I}_z^1/2 + \mathbf{I}_z^2/2 - \mathbf{I}_x^1/2 - \mathbf{I}_x^1\mathbf{I}_z^2 + \mathbf{I}_z^1\mathbf{I}_z^2 \\ \xrightarrow{[z-\text{grad}]} \mathbf{I}_z^1/2 + \mathbf{I}_z^2/2 + \mathbf{I}_z^1\mathbf{I}_z^2 \end{array} \tag{13}$$

This same pseudo-pure state can also be obtained by a variety of other methods. Indeed, all four *basic* pseudo-pure states (with density matrices shifting to a single positive element on the diagonal and zeros elsewhere) can be obtained. In terms of product operators, these can be written as:

$$\begin{array}{rcl} 2\,|00\rangle\langle 00| - \frac{1}{2}\mathbf{1} &\equiv& \mathbf{I}_z^1 + \mathbf{I}_z^2 + 2\mathbf{I}_z^1\mathbf{I}_z^2 \\ 2\,|01\rangle\langle 01| - \frac{1}{2}\mathbf{1} &\equiv& \mathbf{I}_z^1 - \mathbf{I}_z^2 - 2\mathbf{I}_z^1\mathbf{I}_z^2 \\ 2\,|10\rangle\langle 10| - \frac{1}{2}\mathbf{1} &\equiv& -\mathbf{I}_z^1 + \mathbf{I}_z^2 - 2\mathbf{I}_z^1\mathbf{I}_z^2 \\ 2\,|11\rangle\langle 11| - \frac{1}{2}\mathbf{1} &\equiv& -\mathbf{I}_z^1 - \mathbf{I}_z^2 + 2\mathbf{I}_z^1\mathbf{I}_z^2 \end{array} \tag{14}$$

Given a basic pseudo-pure state, a "coherent superposition" of pseudo-spinors can be prepared by a simple nonselective $\pi/2$ pulse, for example:

$$|00\rangle\langle 00| \xrightarrow{[\pi/2]_y} \frac{1}{4}\begin{pmatrix} 1 & 1 & 1 & 1 \\ 1 & 1 & 1 & 1 \\ 1 & 1 & 1 & 1 \\ 1 & 1 & 1 & 1 \end{pmatrix} \tag{15}$$

This matrix in turn is proportional to the dyadic product of the pseudo-spinor

$$\begin{array}{l} |00\rangle + |01\rangle + |10\rangle + |11\rangle \\ = (|0\rangle + |1\rangle)(|0\rangle + |1\rangle) \end{array} \tag{16}$$

with its conjugate. Since this can be further factorized into a product of one-spin pseudo-spinors (as shown), it represents an "unentangled" state.

An entangled state can be generated with a standard pulse sequence consisting of nonselective $\pi/2$ y-pulse followed by a delay of $1/(2J)$ and a second $\pi/2$ y-pulse*, i.e.

$$|00\rangle\langle 00| \xrightarrow{[\pi/2]_y - [1/(2J)] - [\pi/2]_y} \begin{pmatrix} 1/2 & 0 & 0 & -i/2 \\ 0 & 0 & 0 & 0 \\ 0 & 0 & 0 & 0 \\ i/2 & 0 & 0 & 1/2 \end{pmatrix} \tag{17}$$

In NMR circles, the nonzero off-diagonal locations in this matrix are referred to as a *double-quantum coherence*, since they correspond to a two-spin transition. The matrix itself is proportional to the dyadic product of

$$|00\rangle + i|11\rangle \tag{18}$$

with its conjugate, which is clearly nonfactorizable. More generally, entangled pseudo-spinors appear to be associated with the multiple quantum coherences of NMR spectroscopy.

In order to simplify the experiments as well as their analysis and validation, the spectra shown below were obtained by performing the experiments on the $\mathbf{I}_z^1$, $\mathbf{I}_z^2$ and $2\mathbf{I}_z^1\mathbf{I}_z^2$ states separately, and adding the corresponding signals on the spectrometer computer. General methods of generating pseudo-pure states in arbitrary spin systems are under development.

4 An XOR gate via NMR

As is well-known, a quantum XOR gate can be implemented by selective inversion at one of the four peaks in the spectrum, just as in the original ENDOR experiment[1]; this is an example of *Pound-Overhauser double resonance*[4]. While this is often possible in NMR spectroscopy, it can also be difficult if the coupling constant is small. We have therefore developed the following pulse sequence, which constitutes an example of *spin-coherence double resonance*[4], and was easier to use for our initial experiments:

$$\begin{array}{l} [XOR]^2 \equiv [\pi/2]_y^2 - [1/(2J)] - [\pi/2]_x^2 \quad \longleftrightarrow \\ \mathbf{U}_{XOR}^2 = \frac{1}{\sqrt{2}}\begin{pmatrix} 1+i & 0 & 0 & 0 \\ 0 & 1-i & 0 & 0 \\ 0 & 0 & 0 & 1+i \\ 0 & 0 & -1+i & 0 \end{pmatrix} \end{array} \tag{19}$$

The unitary matrix $\mathbf{U}_{XOR}^2$ has the same pattern of zero and nonzero elements as the usual quantum XOR gate (with its output in the second bit), but with phases giving a determinant of 1 instead of -1 as in Pound-Overhauser implementation. We shall call this the *XOR pulse sequence.*

The effect of the XOR pulse sequence on $|10\rangle\langle 10|$, of course, is to convert it to $|11\rangle\langle 11|$. In order to demonstrate this experimentally, however, we must apply further pulses

*In the following, we do not include the effect of the Zeeman Hamiltonian on the density matrix during the delay; equivalently, we implicitly assume that we have placed a π x-pulse in the middle of the $1/(2J)$ period, which refocuses the magnetization at the end of the period.

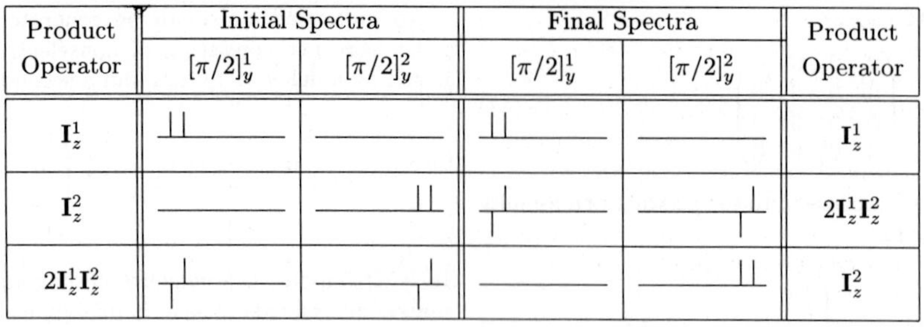

Table 1: Table of "stick" spectra that would be observed for each of the three diagonal product operators after selective $[\pi/2]_y^k$ observation pulses on the k-th spin, before (initial) and after (final) an XOR pulse sequence (see text).

to convert the unobservable diagonal elements into single quantum coherences. To this end, it is useful to evaluate the action of the XOR sequence on the diagonal product operators, as follows:

$$\begin{array}{l} \mathbf{I}_z^1 \xrightarrow{[\pi/2]_y^2} \mathbf{I}_z^1 \xrightarrow{[1/(2J)]} \mathbf{I}_z^1 \xrightarrow{[\pi/2]_x^2} \mathbf{I}_z^1 \\ \mathbf{I}_z^2 \xrightarrow{[\pi/2]_y^2} \mathbf{I}_x^2 \xrightarrow{[1/(2J)]} 2\mathbf{I}_z^1\mathbf{I}_y^2 \xrightarrow{[\pi/2]_x^2} 2\mathbf{I}_z^1\mathbf{I}_z^2 \\ 2\mathbf{I}_z^1\mathbf{I}_z^2 \xrightarrow{[\pi/2]_y^2} 2\mathbf{I}_z^1\mathbf{I}_x^2 \xrightarrow{[1/(2J)]} \mathbf{I}_y^2 \xrightarrow{[\pi/2]_x^2} \mathbf{I}_z^2 \end{array} \qquad (20)$$

The density matrices of the basic pseudo-pure states given in Eq. (14) are all sums of these three product operators, and thus the spectra that are observed after any pulse sequence is applied to a basic pseudo-pure state are sums of the spectra that would be observed if the same pulse sequence were applied to each of the states $\mathbf{I}_z^1$, $\mathbf{I}_z^2$ and $2\mathbf{I}_z^1\mathbf{I}_z^2$ separately. These spectra are depicted as "stick figures" in Table 1, in which we have separately excited spins 1 and 2 to obtain a complete "readout".

The sum of the inverted initial spectra in the first column therefore shows what we would see after applying a selective $[\pi/2]_y^1$ pulse to the state $|10\rangle\langle 10| \leftrightarrow -\mathbf{I}_z^1 + \mathbf{I}_z^2 - 2\mathbf{I}_z^1\mathbf{I}_z^2$, and consists of a single negative peak with the same total intensity as the positive peak that would be observed after applying a $[\pi/2]_y^2$ pulse. The result of the XOR sequence $|11\rangle\langle 11| \leftrightarrow -\mathbf{I}_z^1 - \mathbf{I}_z^2 + 2\mathbf{I}_z^1\mathbf{I}_z^2$ behaves similarly under the readout pulses, except that the peak observed after a $[\pi/2]_y^1$ pulse is now in phase with the peak obtained after a $[\pi/2]_y^2$ pulse, *and* shifted to the left by the coupling constant J (see Fig. 2).

As a final check, we have also applied the XOR pulse sequence to the superposition $\mathbf{I}_x^1 + \mathbf{I}_x^2 + 2\mathbf{I}_x^1\mathbf{I}_x^2$ that is obtained after applying a nonselective $\pi/2$ y-pulse to the basic pseudo-pure state $\mathbf{I}_z^1 + \mathbf{I}_z^2 + 2\mathbf{I}_z^1\mathbf{I}_z^2$. The superposition gives four lines of equal intensity and phase, just as in Fig. 1. The result of applying the XOR sequence to it is shown in Fig. 3, and consists of four lines of equal intensity as before, but shifted in phase by $\pm\pi/2$ to give four *dispersive* peaks. Thus, our XOR pulse sequence changes the spectrum only by phase factors. The effect on the actual state is to give the superposition obtained by applying a nonselective $\pi/2$ x-pulse to the state $|01\rangle\langle 01| \leftrightarrow \mathbf{I}_z^1 - \mathbf{I}_z^2 - 2\mathbf{I}_z^1\mathbf{I}_z^2$.

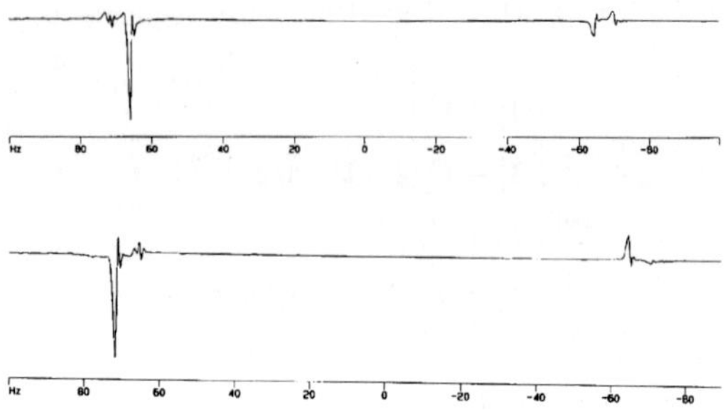

Figure 2: Two experimental NMR spectra of 2,3-dibromothiophene, which illustrate the effect of applying an XOR to the pseudo-pure state $-\mathbf{I}_z^1+\mathbf{I}_z^2-2\mathbf{I}_z^1\mathbf{I}_z^2$ (above) to get $-\mathbf{I}_z^1-\mathbf{I}_z^2+2\mathbf{I}_z^1\mathbf{I}_z^2$ (below). The change in state is indicated by the shift in the position of the peak that is observed following a selective $\pi/2$ y-pulse applied to the first spin.

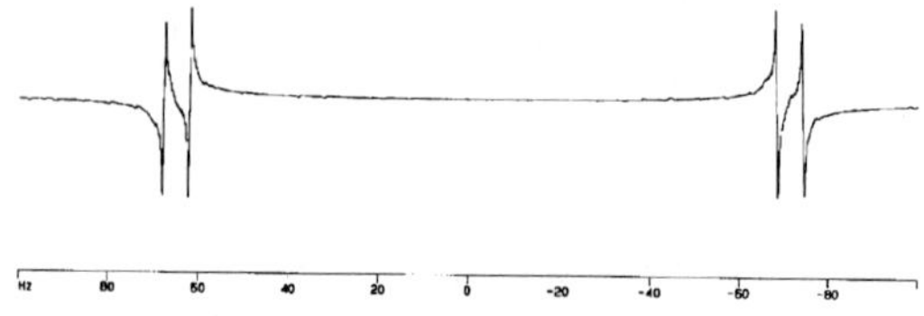

Figure 3: The result of applying the XOR pulse sequence to the superposition $\mathbf{I}_x^1+\mathbf{I}_x^2+2\mathbf{I}_x^1\mathbf{I}_x^2$ is $-\mathbf{I}_y^1+\mathbf{I}_y^2-2\mathbf{I}_y^1\mathbf{I}_y^2$, which is identical to the superposition obtained by applying a $\pi/2$ x-pulse to the basic pseudo-pure state $\mathbf{I}_z^1 - \mathbf{I}_z^2 - 2\mathbf{I}_z^1\mathbf{I}_z^2$.

5 The Toffoli gate

We have also developed a pulse sequence, analogous to the above XOR sequence, which transforms the basic pseudo-spinors of a three-spin system according to the truth table of the well-known Toffoli gate[9]. We call this the *Toffoli pulse sequence*:

$$\begin{aligned} [TOF]^3 \equiv\ & [\pi/2]_y^3 - [1/(4J)] - [\pi/2]_y^3 - [1/(4J)] \\ & - [-\pi/2]_x^3 - [1/(4J)] - [-\pi/2]_x^3 \end{aligned} \qquad (21)$$

Although this pulse sequence assumes that the coupling constants J_{13} and J_{23} have the same value J, it can modified to work with unequal coupling constants as well. Since the Toffoli gate is universal, the existence of this sequence implies that any *special* unitary transformation of pseudo-spinors can be implemented via NMR pulse sequences selective for single spins. Other unitary transformations require the use of pulses that are selective for the individual components in the multiplet of a single spin, as in Pound-Overhauser double resonance.

The matrix of the above Toffoli pulse sequence is easily shown to be $\frac{-1}{\sqrt{2}}$ times:

$$\begin{pmatrix} 1+i & 0 & 0 & 0 & 0 & 0 & 0 & 0 \\ 0 & 1-i & 0 & 0 & 0 & 0 & 0 & 0 \\ 0 & 0 & i & 0 & 0 & 0 & 0 & 0 \\ 0 & 0 & 0 & -i & 0 & 0 & 0 & 0 \\ 0 & 0 & 0 & 0 & i & 0 & 0 & 0 \\ 0 & 0 & 0 & 0 & 0 & -i & 0 & 0 \\ 0 & 0 & 0 & 0 & 0 & 0 & 0 & 1-i \\ 0 & 0 & 0 & 0 & 0 & 0 & -1-i & 0 \end{pmatrix} \tag{22}$$

Hence the output of the gate is in the third bit; the output could easily have been placed in the first or second bit by applying the $\pi/2$ pulses to the corresponding spin.

Once again, the effect of the Toffoli sequence on the actual NMR spectra (after one-spin "readout" pulses) of the eight basic pseudo-pure states is most easily established by determining how it affects the seven diagonal product operators (see Table 2). Unlike the XOR sequence, however, the Toffoli sequence does not simply permute these product operators; instead, it results in a sum of such diagonal operators, e.g.

$$\begin{aligned} \mathbf{I}_z^3 &\xrightarrow{[\pi/2]_y^3} \mathbf{I}_x^3 \\ &\xrightarrow{[1/(4J)]} \tfrac{1}{2}\mathbf{I}_x^3 + \mathbf{I}_z^1\mathbf{I}_y^3 + \mathbf{I}_z^2\mathbf{I}_y^3 - 2\mathbf{I}_z^1\mathbf{I}_z^2\mathbf{I}_x^3 \\ &\xrightarrow{[\pi/2]_y^3} \tfrac{-1}{2}\mathbf{I}_z^3 + \mathbf{I}_z^1\mathbf{I}_y^3 + \mathbf{I}_z^2\mathbf{I}_y^3 + 2\mathbf{I}_z^1\mathbf{I}_z^2\mathbf{I}_z^3 \\ &\xrightarrow{[1/(4J)]} \tfrac{-1}{2}\mathbf{I}_z^3 - \tfrac{1}{2}\mathbf{I}_x^3 + 2\mathbf{I}_z^1\mathbf{I}_z^2\mathbf{I}_x^3 + 2\mathbf{I}_z^1\mathbf{I}_z^2\mathbf{I}_z^3 \\ &\xrightarrow{[-\pi/2]_x^3} \tfrac{-1}{2}\mathbf{I}_y^3 - \tfrac{1}{2}\mathbf{I}_x^3 + 2\mathbf{I}_z^1\mathbf{I}_z^2\mathbf{I}_x^3 + 2\mathbf{I}_z^1\mathbf{I}_z^2\mathbf{I}_y^3 \\ &\xrightarrow{[1/(4J)]} \tfrac{-1}{2}\mathbf{I}_y^3 - \mathbf{I}_z^1\mathbf{I}_x^3 - \mathbf{I}_z^2\mathbf{I}_y^3 + 2\mathbf{I}_z^1\mathbf{I}_z^2\mathbf{I}_y^3 \\ &\xrightarrow{[-\pi/2]_x^3} \tfrac{1}{2}\mathbf{I}_z^3 + \mathbf{I}_z^1\mathbf{I}_z^3 + \mathbf{I}_z^2\mathbf{I}_z^3 - 2\mathbf{I}_z^1\mathbf{I}_z^2\mathbf{I}_z^3 \ . \end{aligned} \tag{23}$$

Similar calculations lead to the following complete list:

$$\begin{aligned} \mathbf{I}_z^1 &\xrightarrow{[TOF]^3} \mathbf{I}_z^1 \\ \mathbf{I}_z^2 &\xrightarrow{[TOF]^3} \mathbf{I}_z^2 \\ \mathbf{I}_z^3 &\xrightarrow{[TOF]^3} \tfrac{1}{2}\mathbf{I}_z^3 + \mathbf{I}_z^1\mathbf{I}_z^3 + \mathbf{I}_z^2\mathbf{I}_z^3 - 2\mathbf{I}_z^1\mathbf{I}_z^2\mathbf{I}_z^3 \\ 2\mathbf{I}_z^1\mathbf{I}_z^2 &\xrightarrow{[TOF]^3} 2\mathbf{I}_z^1\mathbf{I}_z^2 \\ 2\mathbf{I}_z^1\mathbf{I}_z^3 &\xrightarrow{[TOF]^3} \tfrac{1}{2}\mathbf{I}_z^3 - \mathbf{I}_z^2\mathbf{I}_z^3 + \mathbf{I}_z^1\mathbf{I}_z^3 + 2\mathbf{I}_z^1\mathbf{I}_z^2\mathbf{I}_z^3 \\ 2\mathbf{I}_z^2\mathbf{I}_z^3 &\xrightarrow{[TOF]^3} \tfrac{1}{2}\mathbf{I}_z^3 - \mathbf{I}_z^1\mathbf{I}_z^3 + \mathbf{I}_z^2\mathbf{I}_z^3 + 2\mathbf{I}_z^1\mathbf{I}_z^2\mathbf{I}_z^3 \\ 4\mathbf{I}_z^1\mathbf{I}_z^2\mathbf{I}_z^3 &\xrightarrow{[TOF]^3} \tfrac{-1}{2}\mathbf{I}_z^3 + \mathbf{I}_z^1\mathbf{I}_z^3 + \mathbf{I}_z^2\mathbf{I}_z^3 + 2\mathbf{I}_z^1\mathbf{I}_z^2\mathbf{I}_z^3 \end{aligned} \tag{24}$$

Spinor	$\mathbf{I}_z^1$	$\mathbf{I}_z^2$	$\mathbf{I}_z^3$	$2\mathbf{I}_z^1\mathbf{I}_z^2$	$2\mathbf{I}_z^1\mathbf{I}_z^3$	$2\mathbf{I}_z^2\mathbf{I}_z^3$	$4\mathbf{I}_z^1\mathbf{I}_z^2\mathbf{I}_z^3$
$\|000\rangle$	1	1	1	1	1	1	1
$\|001\rangle$	1	1	−1	1	−1	−1	−1
$\|010\rangle$	1	−1	1	−1	1	−1	−1
$\|011\rangle$	1	−1	−1	−1	−1	1	1
$\|100\rangle$	−1	1	1	−1	−1	1	−1
$\|101\rangle$	−1	1	−1	−1	1	−1	1
$\|110\rangle$	−1	−1	1	1	−1	−1	1
$\|111\rangle$	−1	−1	−1	1	1	1	−1

Table 2: Table of coefficients of the seven diagonal product operators corresponding to each basic pseudo-spinor for a three-spin system.

The spectra that are expected from each of these sums, after appropriate one-spin readout pulses, can be worked out in a straightforward fashion, and the experiments needed to validate this, as well as the usual Pound-Overhauser, implementation of the Toffoli gate are underway.

References

[1] DIVINCENZO, David P., "Quantum computation", *Science* **270** (1995), 255–261.

[2] BRASSARD, Gilles, "A quantum jump in computer science", *Computer Science Today* (J. VAN LEEUWEN ed.), Springer-Verlag (1995), 1–14.

[3] BRASSARD, Gilles, "New trends in quantum computing", *13th Symp. Theor. Aspects Comput. Sci.* (February 1996); (quant-ph/9602014).

[4] SLICHTER, Charles P., *Principles of Magnetic Resonance* (3rd. ed.), Springer-Verlag (1990).

[5] GOLDMAN, Maurice, *Quantum Description of High-Resolution NMR in Liquids*, Clarendon Press (1988).

[6] ERNST, Richard R., Geoffrey BODENHAUSEN, and Alexander WOKAUN, *Principles of Nuclear Magnetic Resonance in One and Two Dimensions*, Oxford Univ. Press (1987).

[7] GERSHENFELD, Neil, Isaac CHUANG, and Seth LLOYD, "Bulk Quantum Computation", in this volume.

[8] CORY, David G., Amr F. FAHMY, and Timothy F. HAVEL, "Ensemble quantum computing by nuclear magnetic resonance spectroscopy", submitted for publication (available as Harvard University Center for Research in Computing Technology Technical Report TR-10-96 from `ftp:deas-ftp.harvard.edu/techreports/tr.html`).

[9] TOFFOLI, Tommaso, "Reversible computing", *Automata, Languages and Programming* (J. W. DE BAKKER and J. VAN LEEUWEN ed.), Springer-Verlag (1980), 632–644.

JOURNAL OF CHEMICAL PHYSICS VOLUME 109, NUMBER 5 1 AUGUST 1998

Implementation of a quantum algorithm on a nuclear magnetic resonance quantum computer

J. A. Jones[a)]
Oxford Centre for Molecular Sciences, New Chemistry Laboratory, South Parks Road, Oxford OX1 3QT, United Kingdom and Centre for Quantum Computation, Clarendon Laboratory, Parks Road, Oxford OX1 3PU, United Kingdom

M. Mosca
Centre for Quantum Computation, Clarendon Laboratory, Parks Road, Oxford OX1 3PU, United Kingdom and Mathematical Institute, 24-29 St Giles', Oxford, OX1 3LB, United Kingdom

(Received 16 January 1998; accepted 22 April 1998)

Quantum computing shows great promise for the solution of many difficult problems, such as the simulation of quantum systems and the factorization of large numbers. While the theory of quantum computing is fairly well understood, it has proved difficult to implement quantum computers in real physical systems. It has recently been shown that nuclear magnetic resonance (NMR) can be used to implement small quantum computers using the spin states of nuclei in carefully chosen small molecules. Here we demonstrate the use of a NMR quantum computer based on the pyrimidine base cytosine, and the implementation of a quantum algorithm to solve Deutsch's problem (distinguishing between constant and balanced functions). This is the first successful implementation of a quantum algorithm on any physical system. © *1998 American Institute of Physics.* [S0021-9606(98)00729-6]

I. INTRODUCTION

In 1982 Feynman pointed out that it appears to be impossible to efficiently simulate the behavior of a quantum mechanical system with a computer.[1] This problem arises because the quantum system is not confined to its eigenstates, but can exist in any superposition of them, and so the space needed to describe the system is very large. To take a simple example, a system comprising N two-level subsystems, such as N spin-$\frac{1}{2}$ particles, inhabits a Hilbert space of dimension 2^N, and evolves under a series of transformations described by matrices containing 4^N elements. For this reason it is impractical to simulate the behavior of spin systems containing more than about a dozen spins.

The difficulty of simulating quantum systems using classical computers suggests that quantum systems have an information processing capability much greater than that of corresponding classical systems. Thus, it might be possible to build quantum mechanical computers,[1–4] which utilize this information processing capability in an effective way to achieve a computing power well beyond that of a classical computer. Such a quantum computer could be used to efficiently simulate other quantum mechanical systems,[1,3] or to solve conventional mathematical problems,[4] which suffer from a similar exponential growth in complexity, such as factoring.[5]

Considerable progress in this direction has been made in recent years. The basic logic elements necessary to carry out quantum computing are well understood, and quantum algorithms have been developed, both for simple demonstration problems[6–8] and for more substantial problems such as factoring.[4,5] Experimental implementation of a quantum computer has, however, proved difficult. Much effort has been directed toward implementing quantum computers using ions trapped by electric and magnetic fields,[9] and while this approach has shown some success,[10] it has proved difficult to progress beyond computers containing a single two-level system (corresponding to a single quantum bit, or qubit).

Recently two separate approaches have been described[11,12] for the implementation of a quantum computer using nuclear magnetic resonance[13] (NMR). These approaches show great promise, as it has proved relatively simple to investigate quantum systems containing two or three qubits.[11,12,14] Here we describe our implementation of a simple quantum algorithm for solving Deutsch's problem,[6–8] on a two qubit NMR quantum computer.

II. QUANTUM COMPUTERS

All current implementations of quantum computers are built up from a small number of basic elements. The first of these is the qubit, which plays the same role as that of the bit in a classical computer. A classical bit can be in one of two states, 0 or 1, and similarly a qubit can be represented by any two-level system with eigenstates labeled $|0\rangle$ and $|1\rangle$. One obvious implementation is to use the two Zeeman levels of a spin-$\frac{1}{2}$ particle in a magnetic field, and we shall assume this implementation throughout the rest of this paper. Unlike a bit, however, a qubit is not confined to these two eigenstates, but can, in general, exist in some superposition of the two states. It is this ability to exist in superpositions that makes quantum systems so difficult to simulate and that gives quantum computers their power.

The second requirement is a set of logic gates, corresponding to gates such as AND, OR, and NOT in conventional

[a)]Author to whom correspondence should be addressed at the New Chemistry Laboratory. Electronic mail: jones@bioch.ox.ac.uk computers.[15] Quantum gates differ from their classical counterparts in one very important way: they must be reversible.[15,16] This is because the evolution of any quantum system can be described by a series of unitary transformations, which are themselves reversible. This need for reversibility has many consequences for the design of quantum gates. Clearly, for a gate to be reversible it must be possible to reconstruct the input bits knowing only the design of the gate and the output bits, and so every input bit must be in some sense preserved in the outputs. One trivial consequence of this is that the gate must have exactly as many outputs as inputs. For this reason it is obvious that gates such as AND and OR are not reversible. It is, however, possible to construct reversible equivalents of AND and OR, in which the input bits are preserved.

Just as it can be shown that one or more gates (such as the NAND gate) are universal for classical computing[15] (that is, any classical gate can be constructed using only wires and NAND gates), it can be shown that certain gates or combinations of gates are universal for quantum computing. In particular, it can be shown[17] that the combination of a general single qubit rotation with the two bit "controlled-NOT" gate (CNOT) is universal. Furthermore, it is possible to build a reversible equivalent of the NAND gate, and thus to implement any classical logic operation using reversible logic.

Single qubit rotations are easily implemented in NMR, as they correspond to rotations within the subspace corresponding to a single spin, and such rotations can be achieved using radiofrequency (rf) fields. One particularly important single bit gate is the Hadamard gate, which performs the rotational transformation,

$$|0\rangle \xrightarrow{H} \frac{|0\rangle+|1\rangle}{\sqrt{2}}, \qquad |1\rangle \xrightarrow{H} \frac{|0\rangle-|1\rangle}{\sqrt{2}}. \tag{1}$$

The Hadamard operator can thus be used to convert eigenstates into superpositions of states. Similarly, as the Hadamard is self-inverse, it can be used to convert superpositions of states back into eigenstates for later analysis.

Two-bit gates correspond to rotations within subspaces corresponding to two spins, and thus require some kind of spin–spin interaction for their implementation. In NMR the scalar spin–spin coupling (J coupling) has the correct form, and is ideally suited for the construction of controlled gates, such as CNOT. This gate operates to invert the value of one qubit when another qubit (the control qubit) has some specified value, usually $|1\rangle$; its truth table is shown in Table I.

Finally, it is necessary to have some way of reading out information about the final quantum state of the system, and thus obtaining the result of the calculation. In most implementations of quantum computers, this process is equivalent to determining which of two eigenstates a two-level system is in, but this is not a practical approach in NMR. It is, however, possible to obtain equivalent information by exciting the spin system and observing the resulting NMR spectrum. Different qubits correspond to different spins, and thus give rise to signals at different resonance frequencies, while the eigenstate of a spin before the excitation can be determined from the relative phase (absorption or emission) of the NMR signals.

TABLE I. The truth table for the CNOT gate. The first qubit (the control qubit) is unchanged by the gate, while the second qubit is flipped if the control qubit is in state 1, effectively implementing an XOR gate.

Input		Output	
0	0	0	0
0	1	0	1
1	0	1	1
1	1	1	0

III. THE DEUTSCH ALGORITHM

Deutsch's problem in its simplest form concerns the analysis of single-bit binary functions:

$$f(x):B\mapsto B, \tag{2}$$

where $B=\{0,1\}$ is the set of possible values for a single bit. Such functions take a single bit as input, and return a single bit as their result. Clearly there are exactly four such functions, which may be described by their truth tables, as shown in Table II. These four functions can be divided into two groups: the two "constant" functions, for which $f(x)$ is independent of x (f_{00} and f_{11}), and the two "balanced" functions, for which $f(x)$ is zero for one value of x and unity for the other (f_{01} and f_{10}). Given some unknown function f (known to be one of these four functions), it is possible to determine which of the four functions it is by applying f to two known inputs: 0 and 1. This procedure also provides enough information to determine whether the function is constant or balanced. However, knowing whether the function is constant or balanced corresponds to only one bit of information, and so it might be possible to answer this question using only one evaluation of the function f. Equivalently, it might be possible to determine the value of $f(0) \oplus f(1)$ using only one evaluation of f. (The symbol $\oplus$ indicates addition modulo 2, and for two one bit numbers, a and b, $a\oplus b$ equals 0 if a and b are the same, and 1 if they are different.) In fact, this can be achieved as long as the calculation is performed using a quantum computer rather than a classical one.

Quantum computers of necessity use reversible logic, and so it is not possible to implement the binary function f directly. It is, however, possible to design a propagator, U_f, which captures f within a reversible transformation by using a system with two input qubits and two output qubits as follows:

TABLE II. The four possible binary functions mapping one bit to another.

x	$f_{00}(x)$	$f_{01}(x)$	$f_{10}(x)$	$f_{11}(x)$
0	0	0	1	1
1	0	1	0	1

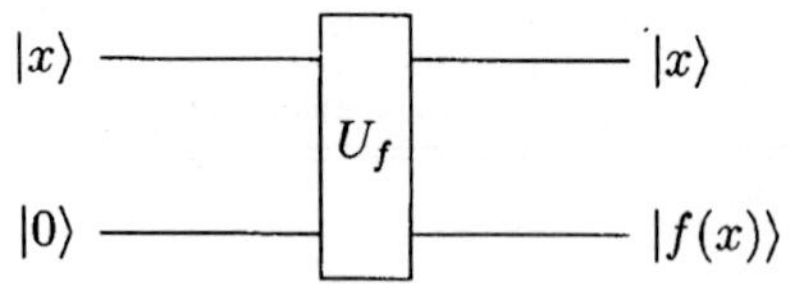

FIG. 1. Quantum circuit for the analysis of a binary function f.

$$|x\rangle|y\rangle \xrightarrow{U_f} |x\rangle|y\oplus f(x)\rangle. \tag{3}$$

The two input bits are preserved [x is preserved directly, while y is preserved by combining it with $f(x)$, the desired result], and so U_f corresponds to a reversible transformation. Note that for any one bit number a, $0\oplus a=a$, and so values of $f(x)$ can be determined by setting the second input bit to 0. Using this propagator and appropriate input states, it is possible to evaluate $f(0)$ and $f(1)$ using

$$|0\rangle|0\rangle \xrightarrow{U_f} |0\rangle|f(0)\rangle \tag{4}$$

and

$$|1\rangle|0\rangle \xrightarrow{U_f} |1\rangle|f(1)\rangle. \tag{5}$$

The approach outlined above, in which the state of a quantum computer is described explicitly, swiftly becomes unwieldy, and it is useful to use more compact notations. One particularly simple approach is to use quantum circuits,[18] which may be drawn by analogy with classical electronic circuits. In this approach lines are used to represent "wires" down which qubits "flow," while boxes represent quantum gates that perform appropriate unitary transformations. For example, the analysis of f can be summarised by the circuit shown in Fig. 1.

So far, this is simply using a quantum computer to simulate a classical computer implementing classical algorithms. With a quantum computer, however, it is not necessary to start with the system in some eigenstate; instead, it is possible to begin with a superposition of states. Suppose the calculation begins with the second qubit in the superposition $(|0\rangle-|1\rangle)/\sqrt{2}$. Then

$$|x\rangle\left(\frac{|0\rangle-|1\rangle}{\sqrt{2}}\right) \xrightarrow{U_f} |x\rangle\left(\frac{|0\oplus f(x)\rangle-|1\oplus f(x)\rangle}{\sqrt{2}}\right),$$
$$=\begin{cases} |x\rangle\dfrac{|0\rangle-|1\rangle}{\sqrt{2}}, & \text{if } f(x)=0,\\ |x\rangle\dfrac{|1\rangle-|0\rangle}{\sqrt{2}}, & \text{if } f(x)=1,\end{cases}$$
$$=(-1)^{f(x)}|x\rangle\frac{|0\rangle-|1\rangle}{\sqrt{2}}. \tag{6}$$

(We have used the fact that $0\oplus a=a$, as before, while $1\oplus a=1$ if $a=0$ and 0 if $a=1$.) The value of $f(x)$ is now encoded in the overall phase of the result, with the qubits left otherwise unchanged. While this is not particularly useful, suppose the calculation begins with the first qubit also in a superposition of states, namely $(|0\rangle+|1\rangle)/\sqrt{2}$. Then[8]

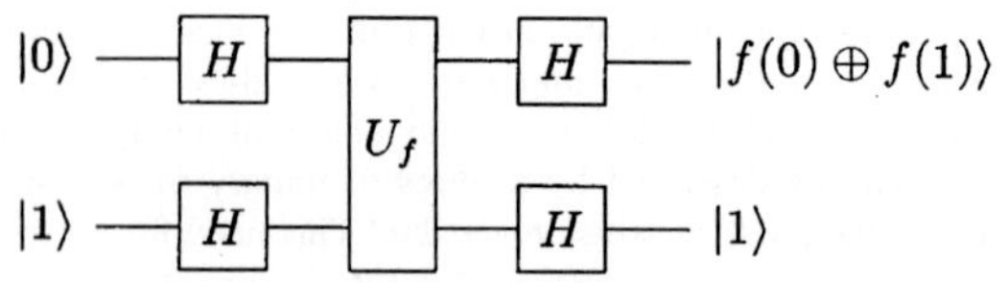

FIG. 2. A quantum circuit for solving Deutsch's problem.

$$\left(\frac{|0\rangle+|1\rangle}{\sqrt{2}}\right)\left(\frac{|0\rangle-|1\rangle}{\sqrt{2}}\right) \xrightarrow{U_f} \left(\frac{(-1)^{f(0)}|0\rangle+(-1)^{f(1)}|1\rangle}{\sqrt{2}}\right)$$
$$\times\left(\frac{|0\rangle-|1\rangle}{\sqrt{2}}\right)=(-1)^{f(0)}\left(\frac{|0\rangle+(-1)^{f(0)\oplus f(1)}|1\rangle}{\sqrt{2}}\right)$$
$$\times\left(\frac{|0\rangle-|1\rangle}{\sqrt{2}}\right), \tag{7}$$

with the first qubit ending up in the superposition $(|0\rangle \pm|1\rangle)/\sqrt{2}$, with the desired answer $[f(0)\oplus f(1)]$ encoded as the *relative* phase of the two states contributing to the superposition. This relative phase can be measured, and so the value of $f(0)\oplus f(1)$ (that is, whether f is constant or balanced) has been determined using only one application of the propagator U_f, that is, only one evaluation of the function f.

This approach can be easily implemented using a quantum circuit, as shown in Fig. 2. This circuit starts off from appropriate eigenstates, uses Hadamard transformations to convert these into superpositions, applies the propagator U_f to these superpositions, and finally uses another pair of Hadamard transforms to convert the superpositions back into eigenstates that encode the desired result.

IV. IMPLEMENTING THE DEUTSCH ALGORITHM IN NMR

The Deutsch algorithm can be implemented on a quantum computer with two qubits, such as a NMR quantum computer based on two coupled spins. First, it is necessary to show that the individual components of the quantum circuit can be built. It is convenient to begin by writing down the necessary states and operators using the product operator basis set[13,19] normally used in describing NMR experiments (this basis is formed by taking outer products between Pauli matrices describing the individual spins, together with the scaled unit matrix, 1/2 E).

The initial state, $|\psi_{01}\rangle=|0\rangle|1\rangle$, can be written as a vector in Hilbert space,

$$|\psi_{01}\rangle=\begin{pmatrix}0\\1\\0\\0\end{pmatrix}, \tag{8}$$

but this description is not really appropriate. Unlike other implementations, a NMR quantum computer comprises not just a single set of spins but rather an ensemble of spins in a statistical mixture of states. Such a system is most conveniently treated using a density matrix, which can describe either a mixture or a pure state; for example,

$$\rho_{01}=|\psi_{01}\rangle\langle\psi_{01}|=\begin{pmatrix}0&0&0&0\\0&1&0&0\\0&0&0&0\\0&0&0&0\end{pmatrix}. \tag{9}$$

This density matrix can be decomposed in the product operator basis as $\rho_{01}=(I_z-S_z-2I_zS_z+1/2\,E)/2$. Ignoring multiples of the unit matrix (which give rise to no observable effects in any NMR experiment), this can be reached from the thermal equilibrium density matrix (I_z+S_z) by a series of rf and field gradient pulses.[11]

The unitary transformation matrix corresponding to the Hadamard operator on a single spin can be written as

$$H=\frac{1}{\sqrt{2}}\begin{pmatrix}1&1\\1&-1\end{pmatrix}. \tag{10}$$

This corresponds to a 180° rotation around an axis tilted at 45° between the z and x axes. Such a rotation can be achieved directly using an off resonance pulse,[13] or using a three pulse sandwich[13] such as $45^\circ_y-180^\circ_x-45^\circ_{-y}$. Even more simply, the Hadamard can be approximated by a 90°_y pulse. While this is clearly not a true Hadamard operator (for example, it is not self-inverse), its behavior is similar, and it can be used in some cases: for example, it is possible to replace the first pair of Hadamard gates in the circuit for the Deutsch Algorithm (Fig. 2) by 90°_y pulses and the second pair of gates by 90°_{-y} pulses. Clearly, it is possible to apply the Hadamard operator either to just one of the two spins (using selective soft rf pulses[20]) or to both spins simultaneously (using nonselective hard pulses).

The unitary transformations corresponding to the four possible propagators U_f are also easily derived. Each propagator corresponds to flipping the state of the second qubit under certain conditions as follows: U_{00}, never flip the second qubit; U_{01}, flip the second qubit when the first qubit is in state one; U_{10}, flip the second qubit when the first qubit is in state zero; U_{11}, always flip the second qubit. The first and last cases are particularly simple, as U_{00} corresponds to doing nothing (the identity operation), while U_{11} corresponds to inverting the second spin (a conventional NOT gate, or, equivalently, a 180° pulse). The second and third propagators correspond to controlled-NOT gates, which can be implemented using spin–spin couplings. For example, U_{01} is described by the matrix

$$U_{01}=\begin{pmatrix}1&0&0&0\\0&1&0&0\\0&0&0&1\\0&0&1&0\end{pmatrix}, \tag{11}$$

which can be achieved using the pulse sequence

$$90S_y-\textit{couple}-90I_z-90S_{-z}-90S_{-y}, \tag{12}$$

where $90S_y$ indicates a 90° pulse on the second spin, *couple* indicates evolution under the scalar coupling Hamiltonian, $\pi J_{IS}2I_zS_z$, for a time $1/2J_{IS}$, and $90I_z$ and $90S_{-z}$ indicate

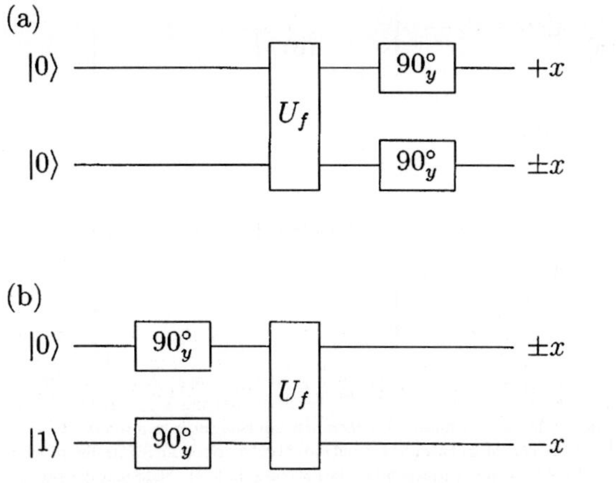

FIG. 3. Modified quantum circuits for the analysis of binary functions on a NMR quantum computer. (a) A circuit for the classical analysis of $f(0)$; the normal circuit (see Fig. 1) is followed by 90°_y pulses to excite the NMR spectrum. Clearly $f(1)$ can be obtained in a similar way. (b) A circuit for the implementation of the Deutsch algorithm, with Hadamard operators replaced by $90^\circ_{\pm y}$ pulses. The final 90°_y excitation pulses cancel out the 90°_{-y} pulses, and thus all four pulses can be omitted.

either periods of free precession under Zeeman Hamiltonians or the application of composite z pulses.[20,21] Similarly, U_{10} can be achieved using the pulse sequence

$$90S_y-\textit{couple}-90I_z-90S_z-90S_{-y}. \tag{13}$$

The pulse sequences described above can be implemented in many different ways, as different composite z pulses can be used, the order of some of the pulses can be varied, and in some cases different pulses can be combined together. We chose to use the implementation

$$90S_y-1/4J_{IS}-180_x-1/4J_{IS}-180_x-90I_y-90I_x-90_{-y}$$
$$-90S_{\pm x}, \tag{14}$$

where pulses not marked as either I or S were applied to both nuclei. The phase of the final pulse distinguishes U_{01} (for which the final pulse is S_{+x}) from U_{10} (for which it is S_{-x}).

Finally, it is necessary to consider an analysis of the final state, which could, in general, be one of the four states ρ_{00}, ρ_{01}, ρ_{10}, or ρ_{11}. In order to distinguish these states it is necessary to apply a 90°_y pulse and observe the NMR spectrum. The final NMR signal observed from spin I is I_x if the spin is in state 0, and $-I_x$ if it is in state 1. For a computer implementing the Deutsch algorithm the final detection 90°_y pulses cancel out the two final pseudo-Hadamard 90°_{-y} pulses, and thus all four pulses can be omitted (see Fig. 3). The final NMR signal observed is either $1/2\,I_x-1/2\,S_x$ (corresponding to ρ_{01}) or $-1/2\,I_x-1/2\,S_x$ (corresponding to ρ_{11}). Hence, it is simple to determine the value of $f(0)\oplus f(1)$ (that is, determine whether the function is constant or balanced) by determining the relative phase of the signals from the two spins.

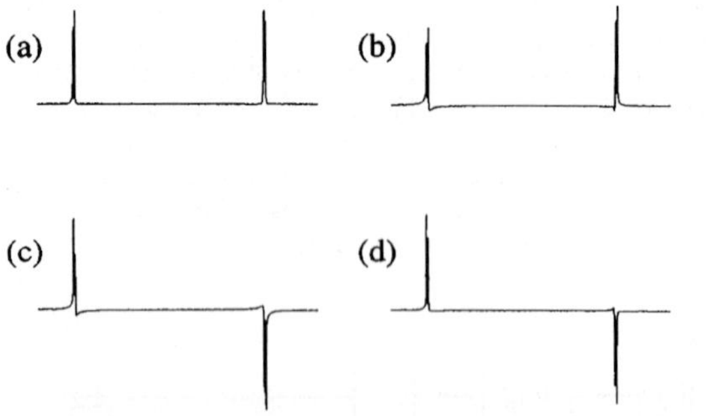

FIG. 4. Experimental implementation of an algorithm to determine $f(0)$ on a NMR quantum computer. (a) The result of applying U_{f00}; as this propagator is the identity matrix this spectrum can also serve as a reference. The left-hand pair of signals corresponds to the first spin (I), while the pair on the right-hand side correspond to the second spin (S). Note that the signals from both spins (which are in state $|0\rangle$, the ground state) are in absorption. (b) The result of applying U_{f01}; both sets of signals are still in absorption, as $f(0)=0$ for this function. (c) The result of applying U_{f10}; the signals from spin S are now in emission, since $f(0)=1$ for this function. (d) The result of applying U_{f11}; the signals from spin S are once again in emission as expected.

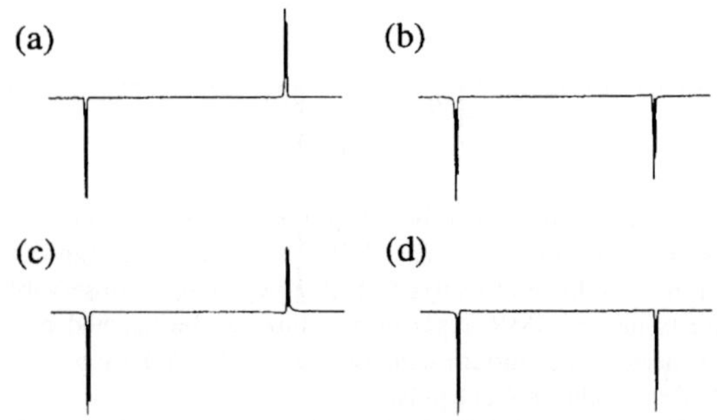

FIG. 5. Experimental implementation of an algorithm to determine $f(1)$ on a NMR quantum computer; in this case the algorithm starts with spin I in the excited state, $|1\rangle$, and so signals from spin I are in emission. For details of the labeling see Fig. 4.

V. EXPERIMENT

In order to demonstrate the results described above, we have constructed a NMR quantum computer capable of implementing the Deutsch algorithm. For our two-spin system we chose to use a 50 mM solution of the pyrimidine base cytosine in D_2O; a rapid exchange of the two amine protons and the single amide proton with the deuterated solvent leaves two remaining protons forming an isolated two-spin system. All NMR experiments were conducted at 20 °C and $p\mathrm{H}^*=7$ on a home-built NMR spectrometer at the Oxford Centre for Molecular Sciences, with a ^{1}H operating frequency of 500 MHz. The observed J coupling between the two protons was 7.2 Hz, while the difference in resonance frequencies was 763 Hz. Selective excitation was achieved using Gaussian[22] soft pulses incorporating a phase ramp[23,24] to allow excitation away from the transmitter frequency. During a selective pulse the other (unexcited) spin continues to experience the main Zeeman interaction, resulting in a rotation around the z axis, but the length of the selective pulses can be chosen such that the net rotation experienced by the other spin is zero. The residual HOD resonance was suppressed by low-power saturation during the relaxation delays.

This system can be used both for the implementation of classical algorithms to analyze $f(0)$ and $f(1)$ and for the implementation of the Deutsch algorithm; as shown in Fig. 3 the pulse sequences differ only in the placement of the 90°_y pulses. The results for the classical algorithm to determine $f(0)$ are shown in Fig. 4. The left-hand pair of signals corresponds to the first spin (I), while the pair on the right-hand side correspond to the second spin (S); the (barely visible) splitting in each pair arises from the scalar coupling J_{IS}. In this experiment the value of $f(0)$ is determined by setting both spins I and S into state $|0\rangle$, performing the calculation, and then measuring the final state of spin S; spin I should not be affected, and so should remain in state $|0\rangle$. The phase of the reference spectrum (a) was adjusted so that signals from spin I appear in absorption, and the same phase correction was applied to the other three spectra. The state of a spin after a calculation can then be determined by determining whether the corresponding signals in the spectrum are in absorption (state $|0\rangle$) or emission (state $|1\rangle$). As expected, spin I does indeed remain in state $|0\rangle$, while the value of $f(0)$ (determined from spin S) is 0 for $U_{f_{00}}$ and $U_{f_{01}}$, but 1 for $U_{f_{10}}$ and $U_{f_{11}}$.

Clearly, our NMR quantum computer is capable of implementing this classical algorithm, as it is simple to determine $f(0)$. The other value, $f(1)$, can be determined in a very similar way (see Fig. 5). In this case spin I remains in state $|1\rangle$, while $f(1)$ equals 0 for $U_{f_{00}}$ and $U_{f_{10}}$ and equals 1 for $U_{f_{01}}$ and $U_{f_{11}}$. There are, however, several imperfections visible in the results.

First, the signals are not perfectly phased: rather than exhibiting pure absorption or pure emission lineshapes, the signals have more complex shapes, including dispersive components. These arise from the difficulty of implementing perfect selective pulses, which effect the desired rotation at one spin while leaving the other spin entirely unaffected. Similarly, the selective pulses will not perfectly suppress J couplings during the excitation, leading to the appearance of antiphase contributions to the lineshape. Any practical selective pulse will be imperfect, and so will result in systematic distortions in the final result. Note that these distortions are most severe in cases (b) and (c), where the propagator is complex, containing a large number of selective pulses. Interestingly, the distortions are also more severe for the measurement of $f(0)$ (Fig. 4) than for $f(1)$ (Fig. 5); there is no simple explanation for this effect, which is due to the complex interplay of many selective pulses. We are currently seeking ways to minimize these effects.

Second, the signal intensities vary in different cases; as before, the signal loss is most severe in cases (b) and (c), corresponding to complex propagators. This is in part a consequence of imperfect selective pulses, as discussed above, but may also indicate the effects of spin relaxation, that is, decoherence of the states involved in the calculation. Decoherence is a fundamental problem, and may ultimately limit

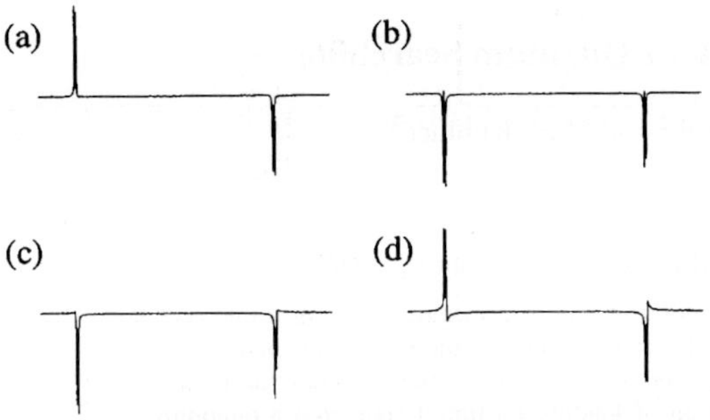

FIG. 6. Experimental implementation of a quantum algorithm to determine $f(0)\oplus f(1)$ on a NMR quantum computer. In this case the result can be read out on spin I, that is, using the signals on the left of the spectrum. For details of the labeling, see Fig. 4. As expected, the I spin is inverted when the function is balanced (f_{01} or f_{10}), but not when the function is constant (f_{00} or f_{11}).

the size of practical quantum computers,[25–27] although a variety of error correction techniques[28–30] have been devised to overcome it.

These imperfections are not a major problem in our NMR quantum computer, as it is still easy to determine the state of a spin. However, our computer is small, and the programs run on it are short (that is, they contain a small number of logic gates); if more complex programs are to be run on larger computers then these imperfections must be addressed.

The results of implementing the Deutsch quantum algorithm are shown in Fig. 6. In this case the result $[f(0)\oplus f(1)]$ can be read from the final state of spin I, while spin S remains in state $|1\rangle$. As expected, spin I is in state $|0\rangle$ for the two constant functions (f_{00} and f_{11}), but in state $|1\rangle$ for the two balanced functions (f_{01} and f_{10}). Once again a number of imperfections are visible, though in this case they appear to be most severe in the case of $U_{f_{11}}$.

VI. SUMMARY

We have demonstrated that the isolated pair of ^{1}H nuclei in partially deuterated cytosine can be used to implement a two qubit NMR quantum computer. This computer can be used to run both classical algorithms and quantum algorithms, such as that for solving Deutsch's problem (distinguishing between constant and balanced functions). This is the first successful implementation of a quantum algorithm on any physical system.[31–35]

This result confirms that NMR shows great promise as a technology for the implementation of small quantum computers. Difficulties do exist, largely as a result of the large number of selective pulses involved in the implementation of quantum gates, but we are currently seeking ways to overcome these problems. Even with the current level of errors it should be possible to build a three qubit computer capable of implementing more complex logic gates and algorithms.

ACKNOWLEDGMENTS

We are indebted to R. H. Hansen (Clarendon Laboratory) for invaluable advice and assistance. We thank N. Soffe and J. Boyd (OCMS) for assistance with implementing the NMR pulse sequences. We are grateful to A. Ekert (Clarendon Laboratory) and R. Jozsa (University of Plymouth) for helpful conversations. J.A.J. thanks C. M. Dobson (OCMS) for his encouragement and support. This is a contribution from the Oxford Centre for Molecular Sciences, which is supported by the UK EPSRC, BBSRC, and MRC. MM thanks CESG (U.K.) for their support.

[1] R. P. Feynman, Int. J. Theor. Phys. **21**, 467 (1982).
[2] D. Deutsch, Proc. R. Soc. London, Ser. A **400**, 97 (1985).
[3] S. Lloyd, Science **273**, 1073 (1996).
[4] A. Ekert and R. Jozsa, Rev. Mod. Phys. **68**, 733 (1996).
[5] P. W. Shor, in *Proceedings of the 35th Annual Symposium on the Foundations of Computer Science*, edited by S. Goldwasser (IEEE Computer Society, Los Alamitos, CA, 1994).
[6] D. Deutsch, in *Quantum Concepts in Space and Time*, edited by R. Penrose and C. J. Isham (Clarendon, Oxford, 1986).
[7] D. Deutsch and R. Jozsa, Proc. R. Soc. London, Ser. A **439**, 553 (1992).
[8] R. Cleve, A. Ekert, C. Macchiavello, and M. Mosca, Proc. R. Soc. London, Ser. A **454**, 339 (1998).
[9] J. I. Cirac and P. Zoller, Phys. Rev. Lett. **74**, 4091 (1995).
[10] C. Monroe, D. M. Meekhof, B. E. King, W. M. Itano, and D. J. Wineland, Phys. Rev. Lett. **75**, 4714 (1995).
[11] D. G. Cory, A. F. Fahmy, and T. F. Havel, Proc. Natl. Acad. Sci. USA **94**, 1634 (1997).
[12] N. A. Gershenfeld and I. L. Chuang, Science **275**, 350 (1997).
[13] R. R. Ernst, G. Bodenhausen, and A. Wokaun, *Principles of Nuclear Magnetic Resonance in One and Two Dimensions* (Oxford University Press, Oxford, 1987).
[14] See, for example, R. Laflamme, E. Knill, W. H. Zurek, P. Catasti, and S. V. S. Mariappan, *NMR GHZ*, available at the xxx.lanl.gov e-Print archive as quant-ph/9709025.
[15] R. P. Feynman, *Feynman Lectures on Computation*, edited by A. J. G. Hey and R. W. Allen (Addison-Wesley, Reading, MA, 1996).
[16] C. H. Bennett, IBM J. Res. Dev. **17**, 525 (1973).
[17] A. Barenco, C. H. Bennett, R. Cleve, D. P. DiVincenzo, N. Margolus, P. Shor, T. Sleator, J. A. Smolin, and H. Weinfurter, Phys. Rev. A **52**, 3457 (1995).
[18] D. Deutsch, Proc. R. Soc. London, Ser. A **425**, 73 (1989).
[19] O. W. Sørensen, G. W. Eich, M. H. Levitt, G. Bodenhausen, and R. R. Ernst, Prog. NMR Spectrosc. **16**, 163 (1983).
[20] R. Freeman, *Spin Choreography* (Spektrum, Oxford, 1997).
[21] R. Freeman, T. A. Frenkiel, and M. H. Levitt, J. Magn. Reson. **44**, 409 (1981).
[22] C. J. Bauer, R. Freeman, T. Frenkiel, J. Keeler, and A. J. Shaka, J. Magn. Reson. **58**, 442 (1984).
[23] H. Green, X. Wu, P. Xu, J. Friedrich, and R. Freeman, J. Magn. Reson. **81**, 646 (1989).
[24] Ē. Kupče and R. Freeman, J. Magn. Reson., Ser. A **105**, 234 (1993).
[25] I. L. Chuang, R. Laflamme, P. W. Shor, and W. H. Zurek, Science **270**, 1633 (1995).
[26] M. B. Plenio and P. L. Knight, Phys. Rev. A **53**, 2986 (1996).
[27] M. B. Plenio and P. L. Knight, Proc. R. Soc. London, Ser. A **453**, 2017 (1997).
[28] P. W. Shor, Phys. Rev. A **52**, R2493 (1995).
[29] A. Steane, Proc. R. Soc. London, Ser. A **452**, 2551 (1996).
[30] A. Steane, Phys. Rev. Lett. **78**, 2252 (1997).
[31] Since initial submission of this manuscript there has been considerable progress in this field, including another implementation of an algorithm to solve Deutsch's problem,[32] and two implementations of Grover's quantum search algorithm.[33–35]
[32] I. L. Chuang, L. M. K. Vandersypen, X. Zhou, D. W. Leung, and S. Lloyd, Nature (London) **393**, 1443 (1998).
[33] I. L. Chuang, N. Gershenfeld, and M. Kubinec, Phys. Rev. Lett. **80**, 3406 (1998).
[34] J. A. Jones, M. Mosca, and R. H. Hansen, Nature (London) **393**, 344 (1998).
[35] J. A. Jones, Science **280**, 229 (1998).

VOLUME 80, NUMBER 15 PHYSICAL REVIEW LETTERS 13 APRIL 1998

Experimental Implementation of Fast Quantum Searching

Isaac L. Chuang,[1,*] Neil Gershenfeld,[2] and Mark Kubinec[3]
[1]*IBM Almaden Research Center K10/D1, 650 Harry Road, San Jose, California 95120*
[2]*Physics and Media Group, MIT Media Lab, Cambridge, Massachusetts 02139*
[3]*College of Chemistry, D7 Latimer Hall, University of California, Berkeley, Berkeley, California 94720-1460*
(Received 21 November 1997; revised manuscript received 29 January 1998)

Using nuclear magnetic resonance techniques with a solution of chloroform molecules we implement Grover's search algorithm for a system with four states. By performing a tomographic reconstruction of the density matrix during the computation good agreement is seen between theory and experiment. This provides the first complete experimental demonstration of loading an initial state into a quantum computer, performing a computation requiring fewer steps than on a classical computer, and then reading out the final state. [S0031-9007(98)05850-5]

PACS numbers: 89.70.+c, 03.65.-w

The study of computation in quantum systems began with the recognition of the theoretical possibility [1–3]. This was followed by a series of results leading up to proofs that a quantum computer requires fewer operations than a classical computer for problems including factoring [4] and searching [5,6]. Appreciation of the power of quantum computing was quickly tempered by the realization that preserving quantum coherence made the implementation of practical quantum computers appear to be unlikely [7–9].

Two recent developments have changed that conclusion. The first is the recognition that quantum error correction can be used to compute with imperfect computers [10,11]. And the second is that it is possible to decrease the influence of decoherence by computing with mixed-state ensembles rather than isolated systems in a pure state. This can be done by introducing extra degrees of freedom [12] using quantum spins [13], space [14], or time [15] to embed within the overall system a subsystem which transforms like a pure state. We apply these ideas here in the first experimental realization of a significant quantum computing algorithm, using nuclear magnetic resonance (NMR) techniques to perform Grover's quantum search algorithm [5,6].

Classically, searching for a particular entry in an unordered list of N elements requires $\mathcal{O}(N)$ attempts. The list could be stored as a table, such as finding a name to go along with a phone number in a phone book, or computed as needed, like testing possible combinations to unlock a padlock. Grover's surprising result is that a quantum computer can obtain the result with certainty in $\mathcal{O}(\sqrt{N})$ attempts.

The simplest interesting application of Grover's algorithm is the $N = 4$ case, which can be posed as follows: on the set $x = \{0, 1, 2, 3\}$ a function $f(x) = 1$ except at some x_0, where $f(x_0) = -1$. How many evaluations of f are required to determine x_0? In the worst case, x_0 has a uniform probability of being either 0, 1, 2, or 3, and so the average number of evaluations required classically is $9/4 = 2.25$. With a quantum computer using Grover's algorithm, this is reduced to a *single* evaluation. We have experimentally implemented this case using molecules of chloroform as a quantum computer, and confirmed the periodic behavior expected of the algorithm.

The algorithm works by representing x as a pair of two-state quantum systems. We take these to be the spins of the carbon and hydrogen nuclei, writing $|\uparrow\rangle = |1\rangle$ and $|\downarrow\rangle = |0\rangle$. The function $f(x)$ is implemented as a unitary transform that flips the phase of the x_0 element. If the operator corresponding to $x_0 = 3$ is applied to the superposition $|\psi_0\rangle = (|00\rangle + |01\rangle + |10\rangle + |11\rangle)/2$ the result is $(|00\rangle + |01\rangle + |10\rangle - |11\rangle)/2$. Measurement of this state is not useful because each answer occurs with equal probability. Grover's algorithm amplifies the correct answer by following the conditional flip with a second operation that inverts each state about the mean. Applied to a superposition $\sum_k \alpha_k|k\rangle$ this step gives a new state $\sum_k \beta_k|k\rangle$ with $\beta_k = -\alpha_k + 2\langle\alpha\rangle$, where $\langle\alpha\rangle$ is the mean value of α_k. For $N = 4$ and $x_0 = 3$ the result of the conditional flip followed by the inversion about the mean is the state $|\psi_1\rangle = |11\rangle$, providing the answer immediately. For general N, about $\pi\sqrt{N}/4$ repetitions of these two steps are required to find x_0 [16].

Further iteration of the flip and inversion operations leads to a periodicity in the state. Let U be the unitary transform which does these two operations, so that $|\psi_n\rangle = U^n|\psi_0\rangle$ is the state after the nth iteration. Boyer *et al.* have shown that the amplitude $\langle x_0 | \psi_n\rangle \approx \sin[(2n + 1)\theta]$, where $\theta = \arcsin(1/\sqrt{N})$; this periodicity arises from the finite size of the system and the unitarity of U. For $N = 4$ the theoretical expectation is the sequence $|11\rangle = |\psi_1\rangle = -|\psi_4\rangle = |\psi_7\rangle = -|\psi_{10}\rangle \ldots$, a period of 6 (or 3 if the overall sign is disregarded).

Our experiments used a 0.5 milliliter, 200 millimolar sample of Carbon-13 labeled chloroform (Cambridge Isotopes) in d_6 acetone. Data were taken at room temperature with a Bruker DRX 500 MHz spectrometer. The coherence times were measured to be $T_1 = 20$ sec

 0031-9007/98/80(15)/3408(4)$15.00

and $T_2 = 0.4$ sec for the proton, and $T_1 = 21$ sec and $T_2 = 0.3$ sec for the carbon (the large ratio is due to C-Cl relaxation), and the coupling was measured to be $J = 215$ Hz. All resonance lines from other nuclei and the solvent were far from the region of interest in this experiment. In the rotating frame of the proton (at about 500 MHz) and carbon (at about 125 MHz), the Hamiltonian for this system can be approximated as [17]

$$\mathcal{H} = 2\pi\hbar J I_{zA} I_{zB} + P_{\phi A}(t) I_{\phi A} + P_{\phi B}(t) I_{\phi B} + \mathcal{H}_{\text{env}}, \tag{1}$$

where $I_{\phi A}$ and $I_{\phi B}$ are the angular momentum operators in the ϕ direction for the proton (A) and carbon (B), and $\mathcal{H}_{\text{env}}$ represents the coupling to the environment, responsible for the decoherence. $P_{\phi A}$ and $P_{\phi B}$ describe the strength of radio-frequency (rf) pulses which are applied on resonance to perform single-spin rotations to each of the two spins. These rotations will be denoted as $X \equiv \exp(i\pi I_x/2)$ for a 90° rotation about the $\hat{x}$ axis, and $\bar{Y} \equiv \exp(-i\pi I_y/2)$ for a 90° rotation about $-\hat{y}$, with a subscript specifying the affected spin.

We used temporal labeling [15] to obtain the signal from the pure initial state

$$|\psi_{\text{in}}\rangle = |00\rangle = \begin{bmatrix} 1 \\ 0 \\ 0 \\ 0 \end{bmatrix} \tag{2}$$

by repeating the experiment three times, cyclically permuting the $|01\rangle$, $|10\rangle$, and $|11\rangle$ state populations before the computation and then summing the results.

The calculation starts with a Walsh-Hadamard transform W, which rotates each quantum bit (qubit) from $|0\rangle$ to $(|0\rangle + |1\rangle)/\sqrt{2}$, to prepare the uniform superposition state

$$|\psi_0\rangle = W|\psi_{\text{in}}\rangle = \frac{1}{2}\begin{bmatrix} 1 & 1 & 1 & 1 \\ 1 & -1 & 1 & -1 \\ 1 & 1 & -1 & -1 \\ 1 & -1 & -1 & 1 \end{bmatrix}\begin{bmatrix} 1 \\ 0 \\ 0 \\ 0 \end{bmatrix} = \frac{1}{2}\begin{bmatrix} 1 \\ 1 \\ 1 \\ 1 \end{bmatrix}. \tag{3}$$

Note that $W = H_A \otimes H_B$, where $H = X^2\bar{Y}$ (pulses applied from right to left) is a single-spin Hadamard transform.

The operator corresponding to the application of $f(x)$ for $x_0 = 3$ is as

$$C = \begin{bmatrix} 1 & 0 & 0 & 0 \\ 0 & 1 & 0 & 0 \\ 0 & 0 & 1 & 0 \\ 0 & 0 & 0 & -1 \end{bmatrix}. \tag{4}$$

This conditional sign flip, testing for a Boolean string that satisfies the AND function, is implemented by using the coupled-spin evolution which occurs when no rf power is applied. During a time t the system undergoes the unitary transformation $\exp(2\pi i J I_{zA} I_{zB} t)$ in the doubly rotating frame. Denoting a $t = 1/2J$ (2.3 millisecond) period evolution as the operator τ, we find that $C = Y_A \bar{X}_A \bar{Y}_A Y_B \bar{X}_B \bar{Y}_B \tau$ (up to an irrelevant overall phase factor).

An arbitrary logical function can be tested by a network of controlled-NOT and rotation gates [13,18], leaving the result in a scratch pad qubit. This qubit can then be used as the source for a controlled phase-shift gate to implement the conditional sign flip, if necessary reversing the test procedure to erase the scratch pad. In our experiment these operations could be collapsed into a single step without requiring an extra qubit.

The operator D that inverts the states about their mean can be implemented by a Walsh-Hadamard transform W, a conditional phase shift P, and another W:

$$D = WPW = W\begin{bmatrix} 1 & 0 & 0 & 0 \\ 0 & -1 & 0 & 0 \\ 0 & 0 & -1 & 0 \\ 0 & 0 & 0 & -1 \end{bmatrix} W = \frac{1}{2}\begin{bmatrix} -1 & 1 & 1 & 1 \\ 1 & -1 & 1 & 1 \\ 1 & 1 & -1 & 1 \\ 1 & 1 & 1 & -1 \end{bmatrix} \tag{5}$$

This corresponds to the pulse sequence $P = Y_A X_A \bar{Y}_A Y_B X_B \bar{Y}_B \tau$.

Let $U = DC$ be the complete iteration. The state after one cycle is

$$|\psi_1\rangle = UW|\psi_0\rangle = |11\rangle = \begin{bmatrix} 0 \\ 0 \\ 0 \\ 1 \end{bmatrix}. \tag{6}$$

A measurements of the system's state will now give with certainty the correct answer, $|11\rangle$. For further iterations, $|\psi_n\rangle = U^n|\psi_0\rangle$,

$$|\psi_2\rangle = \frac{1}{2}\begin{bmatrix} -1 \\ -1 \\ -1 \\ 1 \end{bmatrix} \quad |\psi_3\rangle = \frac{1}{2}\begin{bmatrix} -1 \\ -1 \\ -1 \\ -1 \end{bmatrix} \quad |\psi_4\rangle = \begin{bmatrix} 0 \\ 0 \\ 0 \\ -1 \end{bmatrix}. \tag{7}$$

We see that a maximum in the amplitude of the x_0 state $|11\rangle$ recurs every third iteration.

Like any computer program that is compiled to a microcode, the rf pulse sequence for U can be optimized to eliminate unnecessary operations. In a quantum computer this is essential to make the best use of the available coherence. Ignoring irrelevant overall phase factors, and noting that $H = \bar{X}^2\bar{Y}$ also works, we can simplify U

by removing sequential rotations which cancel each other out, to get

$$U = \bar{X}_A\bar{Y}_A\bar{X}_B\bar{Y}_B\tau X_A\bar{Y}_A X_B\bar{Y}_B\tau \qquad (x_0 = 3)\,. \quad (8)$$

The other possible cases are obtained by changing the signs of the first two X rotations,

$$U = \begin{cases} \bar{X}_A\bar{Y}_A\bar{X}_B\bar{Y}_B\tau X_A\bar{Y}_A\bar{X}_B\bar{Y}_B\tau & (x_0 = 2)\,, \\ \bar{X}_A\bar{Y}_A\bar{X}_B\bar{Y}_B\tau\bar{X}_A\bar{Y}_A X_B\bar{Y}_B\tau & (x_0 = 1)\,, \\ \bar{X}_A\bar{Y}_A\bar{X}_B\bar{Y}_B\tau\bar{X}_A\bar{Y}_A\bar{X}_B\bar{Y}_B\tau & (x_0 = 0)\,. \end{cases} \quad (9)$$

Because the magnetization that is detected in an NMR experiment is the result of a weak measurement on the ensemble, the signal strength gives the fraction of the population with the measured magnetization rather than collapsing the wave function into a measurement eigenstate. The readout can be preceded by a sequence of single spin rotations to allow all terms in the deviation density matrix $\rho_\Delta = \rho - \mathrm{tr}(\rho)/N$ to be measured [19]. Nine experiments—no rotation, rotation about $\hat{x}$, and about $\hat{y}$, for each of the two spins—were performed to do this reconstruction of the density matrix to facilitate comparison between theory and experiment.

Figure 1 shows the theoretical and measured deviation density matrices $\rho_{\Delta n} = |\psi_n\rangle\langle\psi_n| - \mathrm{tr}(|\psi_n\rangle\langle\psi_n|)/4$ for the first seven iterations of U. As expected, $\rho_{\Delta 1}$ clearly reveals the $|11\rangle$ state corresponding to $x_0 = 3$. Analogous results were obtained for experiments repeated for

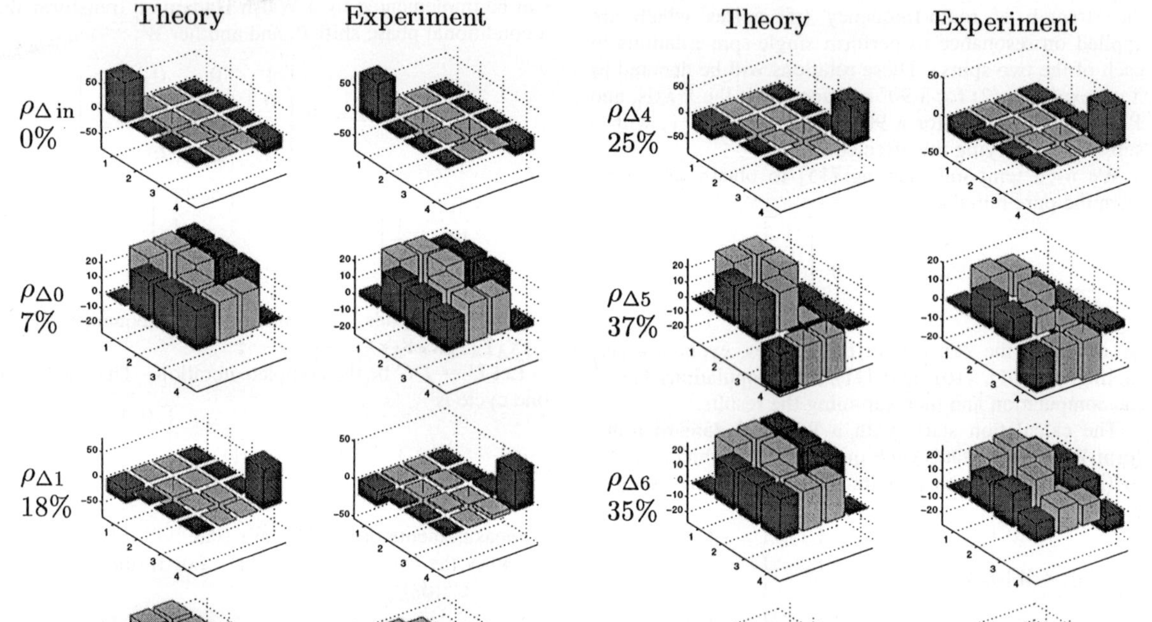

FIG. 1. Theoretical and experimental deviation density matrices (in arbitrary units) for seven steps of Grover's algorithm performed on the hydrogen and carbon spins in chloroform. Three full cycles, with a periodicity of three iterations are clearly seen. Only the real component is plotted (the imaginary portion is theoretically zero and was found to contribute less than 12% to the experimental results). Relative errors $||\rho_{\mathrm{theory}} - \rho_{\mathrm{expt}}||/||\rho_{\mathrm{theory}}||$ are shown as percentages.

the other possible values of x_0. Measuring each density matrix required $9 \times 3 = 27$ experimental repetitions, nine for the tomographic reconstruction and three for the pure state preparation. Both of these operations were performed as tests of the computation, but neither was necessary. In our experiment, starting from the thermal state the maximum population can be identified in a single iteration, with the result obtained from a single output spectrum. In the general N case, readout of $\log N$ expectation value measurements would be required, and good inputs for Grover's algorithm can be distilled in a number of steps polynomial in $\log(N)$ [15].

The longest computation, for $n = 7$, took less than 35 milliseconds, which was well within the coherence time. The periodicity of Grover's algorithm is clearly seen in Fig. 1, with good agreement between theory and experiment. The large signal-to-noise ratio (typically better than 10^4 to 1) was obtained with just single-shot measurements. Numerical simulations indicate that the 7%–44% errors are primarily due to inhomogeneity of the magnetic field, magnetization decay during the measurement, and imperfect calibration of the rotations (in order of importance).

These experimental results demonstrate the operation of a simple quantum computer that can load an initial state, perform a computation, and read out the answer. While there is a long way to go from such a demonstration to a system that can exceed the performance of the fastest classical computers, the experimental study of quantum computation has already come much farther in its short life than either early theoretical predictions or the history of mature computing technologies would have suggested. While scaling up to much larger systems poses daunting challenges, many optimizations remain to be taken advantage of, including increasing the sample size, using coherence transfer to and from electrons, and optical pumping to cool the spin system [19]. Furthermore, Grover's algorithm can be matched to convenient physical operations by performing generalized rapid search, which uses transforms other than the Walsh-Hadamard [20].

The NMR system that we have described already has all of the components of a complete computer architecture, including the rudiments of compiler optimizations. It can implement a nontrivial quantum computation; the challenge now is to accomplish a useful one.

We gratefully acknowledge the support of DARPA under the NMRQC Initiative, Contract No. DAAG55-97-1-0341, and the MIT Media Lab's Things That Think consortium. We thank Gilles Brassard, Lov Grover, and Alex Pines for helpful comments.

*Electronic address: ichuang@almaden.ibm.com

[1] P. A. Benioff, Int. J. Theor. Phys. **21**, 177 (1982).
[2] R. P. Feynman, Int. J. Theor. Phys. **21**, 467 (1982).
[3] D. Deutsch, Proc. R. Soc. London A **400**, 97 (1985).
[4] P. W. Shor, in *Proceedings of the 35th Annual Symposium on Foundations of Computer Science, Santa Fe, NM, 1994,* edited by Shafi Goldwasser (IEEE Computer Society Press, Los Alamitos, CA, 1994), pp. 124–134; SIAM J. Comput. **26**, 1484–1509 (1997).
[5] L. Grover, in *Proceedings of the 28th Annual ACM Symposium on the Theory of Computation* (ACM Press, New York, 1996), pp. 212–219.
[6] L. K. Grover, Phys. Rev. Lett. **79**, 325 (1997).
[7] W. G. Unruh, Phys. Rev. A **51**, 992 (1995).
[8] I. L. Chuang, R. Laflamme, P. Shor, and W. H. Zurek, Science **270**, 1633 (1995).
[9] G. M. Palma, K.-A. Suominen, and A. Ekert, Proc. R. Soc. London A **452**, 567 (1996).
[10] A. Steane, Proc. R. Soc. London A **452**, 2551 (1996).
[11] A. R. Calderbank and P. W. Shor, Phys. Rev. A **54**, 1098 (1996).
[12] I. Chaung and N. Gershenfeld, "State Labeling for Bulk Quantum Computation" (unpublished).
[13] N. Gershenfeld and I. L. Chuang, Science **275**, 350 (1997).
[14] D. Cory, A. Fahmy, and T. Havel, Proc. Nat. Acad. Sci. U.S.A. **94**, 1634 (1997).
[15] E. Knill, I. Chuang, and R. Laflamme, "Effective Pure States for Bulk Quantum Computation," Phys. Rev. A (to be published).
[16] M. Boyer, G. Brassard, P. Høyer, and A. Tapp, LANL e-print quant-ph/9605034; Fortschr. Phys. (to be published).
[17] R. R. Ernst, G. Bodenhausen, and A. Wokaun, *Principles of Nuclear Magnetic Resonance in One and Two Dimensions* (Oxford University Press, Oxford, 1994).
[18] A. Barenco *et al.*, Phys. Rev. A **52**, 3457 (1995).
[19] I. L. Chuang, N. Gershenfeld, M. Kubinec, and D. Leung, Proc. R. Soc. London A **454**, 447 (1998).
[20] L. Grover, LANL e-print quant-ph/9711043 (1997).

30 October 1998

ELSEVIER

CHEMICAL PHYSICS LETTERS

Chemical Physics Letters 296 (1998) 61–67

An implementation of the Deutsch–Jozsa algorithm on a three-qubit NMR quantum computer

Noah Linden [a,1], Hervé Barjat [b,2], Ray Freeman [b,*]

[a] *Isaac Newton Institute for Mathematical Sciences, 20 Clarkson Road, Cambridge, CB3 0EH, UK and Department of Applied Mathematics and Theoretical Physics, Silver Street, Cambridge CB3 9EW, UK*

[b] *Department of Chemistry, Lensfield Rd, Cambridge CB2 1EW, UK*

Received 24 August 1998

Abstract

A new approach to the implementation of a quantum computer by high-resolution nuclear magnetic resonance (NMR) is described. The key feature is that two or more line-selective radio-frequency pulses are applied simultaneously. A three-qubit quantum computer has been investigated using the 400 MHz NMR spectrum of the three coupled protons in 2,3-dibromopropanoic acid. It has been employed to implement the Deutsch–Jozsa algorithm for distinguishing between constant and balanced functions. The extension to systems containing more coupled spins is straightforward and does not require a more protracted experiment. © 1998 Elsevier Science B.V. All rights reserved.

1. Introduction

While there has long been theoretical interest in the notion of a quantum computer, it was the series of recent results leading to the remarkable algorithm of Shor [1] for finding prime factors in polynomial time which led to the recent explosion of interest in the subject. These theoretical results have led many groups to try to realise a quantum computer experimentally. Nuclear magnetic resonance offers a particularly attractive implementation of quantum computers because nuclear spins are relatively weakly coupled to the environment, and there is a long history of development of experimental techniques for manipulating the spins using radio frequency pulses.

A number of groups have already demonstrated the use of NMR computers [2–10]. One of the key challenges is to try to increase the size of the system used. Previous work on implementing quantum algorithms has focused on two algorithms in particular, the Deutsch–Jozsa [11] algorithm for distinguishing between balanced and constant functions and Grover's algorithm [12] for searching a database. Previous work on both of these algorithms has used NMR computers with two qubits. In this Letter we take the study further by implementing the Deutsch–Jozsa algorithm for a system of three qubits. A particularly notable feature of the experiments we describe is the use of *simultaneous line-selective*

* Corresponding author. E-mail: rf110@cus.cam.ac.uk
[1] E-mail: nl101@newton.cam.ac.uk
[2] E-mail: hb232@cus.cam.ac.uk

PII: S0009-2614(98)01015-X

pulses to implement the key stage of the algorithm, quantum gates which are closely related to the controlled-controlled-not gate.

The Deutsch–Jozsa algorithm which we will implement is to distinguish between two classes of two-bit binary functions:

$$f:\{0,1\} \times \{0,1\} \mapsto \{0,1\}. \qquad (1)$$

The two classes are the *constant* functions, in which all input values get mapped to the same output value, and the *balanced* functions in which exactly two of the inputs get mapped to 0. The eight balanced or constant functions are given in Table 1.

The point of the Deutsch–Jozsa algorithm is that it is possible to decide whether a function is constant or balanced with only one evaluation of the function f.

The theoretical steps of the quantum algorithm are as follows:

[1] Preparation: Prepare the system in the (pure) state $\psi_1 = |0\rangle|0\rangle|0\rangle$.

[2] Excitation: Perform rotations of the spins about the y-axis so that the state becomes $\psi_2 = (|0\rangle + |1\rangle)(|0\rangle + |1\rangle)(|0\rangle - |1\rangle)$.

[3] Evaluation: This is done by implementing the unitary transformation

$$|i\rangle|j\rangle|k\rangle \mapsto |i\rangle|j\rangle|k+f(i,j)\rangle, \qquad (2)$$

where the addition is performed modulo two. The three qubits are now in the state

$$\sum_{i,j=1}^{2} (-1)^{f(i,j)}|i\rangle|j\rangle(|0\rangle - |1\rangle). \qquad (3)$$

For example in the case of the function f_4, the state is $\psi_3 = -(|0\rangle - |1\rangle)(|0\rangle + |1\rangle)(|0\rangle - |1\rangle)$. The function f_4 is implemented by applying the unitary operator

$$U_4 = \begin{pmatrix} 0&1&0&0&0&0&0&0\\ 1&0&0&0&0&0&0&0\\ 0&0&0&1&0&0&0&0\\ 0&0&1&0&0&0&0&0\\ 0&0&0&0&1&0&0&0\\ 0&0&0&0&0&1&0&0\\ 0&0&0&0&0&0&1&0\\ 0&0&0&0&0&0&0&1 \end{pmatrix} \qquad (4)$$

to the state. We note that this may be written as

$$U_4 = \begin{pmatrix} 2\sigma_x & 0 & 0 & 0\\ 0 & 2\sigma_x & 0 & 0\\ 0 & 0 & E & 0\\ 0 & 0 & 0 & E \end{pmatrix} \qquad (5)$$

where σ_x is the Pauli matrix normalized so that $\mathrm{tr}(\sigma_x^2) = 1/2$ and E is the 2×2 identity matrix. For convenience we will denote such block diagonal matrices by the symbol Δ, so that we write

$$U_4 = \Delta(2\sigma_x, 2\sigma_x, E, E). \qquad (6)$$

The complete list of unitary operators corresponding to the eight balanced or constant functions is given in Table 2.

[4] Observation: By rotating back by the inverse of the transformation applied in the excitation stage, it may be noted that the outputs from the unitary operators corresponding to the constant functions f_1 and f_2 have a component proportional to $|0\rangle|0\rangle|0\rangle$, whereas the outputs from the balanced functions $f_3 \ldots f_8$ are orthogonal to this vector so that the constant and balanced functions may be distin-

Table 1
The eight possible balanced or constant binary functions mapping two bits to one bit

x	$f_1(x)$	$f_2(x)$	$f_3(x)$	$f_4(x)$	$f_5(x)$	$f_6(x)$	$f_7(x)$	$f_8(x)$
00	0	1	0	1	1	0	1	0
01	0	1	0	1	0	1	0	1
10	0	1	1	0	1	0	0	1
11	0	1	1	0	0	1	1	0

Table 2
The unitary operators corresponding to the eight constant or balanced binary functions mapping two bits to one bit

f	U
f_1	$\Delta(E,E,E,E)$
f_2	$\Delta(2\sigma_x,2\sigma_x,2\sigma_x,2\sigma_x)$
f_3	$\Delta(E,E,2\sigma_x,2\sigma_x)$
f_4	$\Delta(2\sigma_x,2\sigma_x,E,E)$
f_5	$\Delta(2\sigma_x,E,2\sigma_x,E)$
f_6	$\Delta(E,2\sigma_x,E,2\sigma_x)$
f_7	$\Delta(2\sigma_x,E,E,2\sigma_x)$
f_8	$\Delta(E,2\sigma_x,2\sigma_x,E)$

guished from each other with probability one by a von Neumann measurement.

The ensemble nature of an NMR quantum computer means that the implementation of the algorithm differs somewhat from the theoretical version. In particular the preparation stage differs since the states of the system are not pure and the observation stage does not use a von Neumann measurement but measures the amplitudes of spectral lines. Nonetheless the key goal of the algorithm remains to determine whether the unitary operator which acts on the system in the third step corresponds to a constant or balanced function.

One way to proceed would be to follow Ref. [3] and produce a pseudo-pure state in the preparation stage. In terms of product operators, the state corresponding to the pure state $\psi_1 = |0\rangle|0\rangle|0\rangle$ is written

$$\rho(\psi_1) = \mathbf{I}_z + \mathbf{S}_z + \mathbf{R}_z + 2\mathbf{I}_z\mathbf{S}_z + 2\mathbf{I}_z\mathbf{R}_z + 2\mathbf{S}_z\mathbf{R}_z + 4\mathbf{I}_z\mathbf{S}_z\mathbf{R}_z . \quad (7)$$

I refers to the first spin, **S**, the second and **R**, the third.

This state is excited to $\rho(\psi_2) = \mathbf{I}_x + \mathbf{S}_x - \mathbf{R}_x + 2\mathbf{I}_x\mathbf{S}_x - 2\mathbf{I}_x\mathbf{R}_x - 2\mathbf{S}_x\mathbf{R}_x - 4\mathbf{I}_x\mathbf{S}_x\mathbf{R}_x$ in stage **[2]**.

In the evaluation stage, the state to which the spins evolve depends on which function is being implemented. For example the unitary operator corresponding to f_4 produces $\rho(\psi_3) = -\mathbf{I}_x + \mathbf{S}_x - \mathbf{R}_x - 2\mathbf{I}_x\mathbf{S}_x + 2\mathbf{I}_x\mathbf{R}_x - 2\mathbf{S}_x\mathbf{R}_x + 4\mathbf{I}_x\mathbf{S}_x\mathbf{R}_x$. The full list of output states is given in Table 3.

It should be noted that, of the observable terms (i.e. those terms linear in $\mathbf{I}_x$, $\mathbf{S}_x$ and $\mathbf{R}_x$), the term in $\mathbf{R}_x$ always has the same phase, but the balanced functions have altered signs of $\mathbf{I}_x$ or $\mathbf{S}_x$, or both.

Table 3
The output states from a (pseudo) pure initial state after the evaluation stage

f	Output
f_1	$+\mathbf{I}_x+\mathbf{S}_x-\mathbf{R}_x+2\mathbf{I}_x\mathbf{S}_x-2\mathbf{I}_x\mathbf{R}_x-2\mathbf{S}_x\mathbf{R}_x-4\mathbf{I}_x\mathbf{S}_x\mathbf{R}_x$
f_2	$+\mathbf{I}_x+\mathbf{S}_x-\mathbf{R}_x+2\mathbf{I}_x\mathbf{S}_x-2\mathbf{I}_x\mathbf{R}_x-2\mathbf{S}_x\mathbf{R}_x-4\mathbf{I}_x\mathbf{S}_x\mathbf{R}_x$
f_3	$-\mathbf{I}_x+\mathbf{S}_x-\mathbf{R}_x-2\mathbf{I}_x\mathbf{S}_x+2\mathbf{I}_x\mathbf{R}_x-2\mathbf{S}_x\mathbf{R}_x+4\mathbf{I}_x\mathbf{S}_x\mathbf{R}_x$
f_4	$-\mathbf{I}_x+\mathbf{S}_x-\mathbf{R}_x-2\mathbf{I}_x\mathbf{S}_x+2\mathbf{I}_x\mathbf{R}_x-2\mathbf{S}_x\mathbf{R}_x+4\mathbf{I}_x\mathbf{S}_x\mathbf{R}_x$
f_5	$+\mathbf{I}_x-\mathbf{S}_x-\mathbf{R}_x-2\mathbf{I}_x\mathbf{S}_x-2\mathbf{I}_x\mathbf{R}_x+2\mathbf{S}_x\mathbf{R}_x+4\mathbf{I}_x\mathbf{S}_x\mathbf{R}_x$
f_6	$+\mathbf{I}_x-\mathbf{S}_x-\mathbf{R}_x-2\mathbf{I}_x\mathbf{S}_x-2\mathbf{I}_x\mathbf{R}_x+2\mathbf{S}_x\mathbf{R}_x+4\mathbf{I}_x\mathbf{S}_x\mathbf{R}_x$
f_7	$-\mathbf{I}_x-\mathbf{S}_x-\mathbf{R}_x+2\mathbf{I}_x\mathbf{S}_x+2\mathbf{I}_x\mathbf{R}_x+2\mathbf{S}_x\mathbf{R}_x-4\mathbf{I}_x\mathbf{S}_x\mathbf{R}_x$
f_8	$-\mathbf{I}_x-\mathbf{S}_x-\mathbf{R}_x+2\mathbf{I}_x\mathbf{S}_x+2\mathbf{I}_x\mathbf{R}_x+2\mathbf{S}_x\mathbf{R}_x-4\mathbf{I}_x\mathbf{S}_x\mathbf{R}_x$

Thus if one observes that the $\mathbf{I}_x$ or $\mathbf{S}_x$ (or both) multiplets are inverted one knows that the function is balanced.

We note, however, that the same goal can be achieved by starting with thermal rather than pure initial states. This is because, as we will show below, similar effects are observed from the outputs starting with thermal initial states as were visible starting from pure initial states. This is not the first time that it has been noted that in NMR quantum computers, thermal initial states are sufficient to implement the algorithms of interest [6].

Thus in the NMR implementation that we will use, the theoretical steps **[1]** to **[4]** are replaced with

[1★] Preparation: One starts with the thermal initial state

$$\mathbf{I}_z + \mathbf{S}_z + \mathbf{R}_z . \quad (8)$$

[2★] Excitation: Apply a hard $\pi/2$ pulse along the y-axis to arrive at

$$\mathbf{I}_x + \mathbf{S}_x + \mathbf{R}_x . \quad (9)$$

[3★] Evaluation: Now evolve the system with one of the unitary operators given in Table 2. This is achieved by using simultaneous line selective pulses (see below).

For example under f_4 the state evolves to

$$2\mathbf{I}_x\mathbf{R}_x + \mathbf{S}_x + \mathbf{R}_x . \quad (10)$$

The list of states to which each of $\mathbf{I}_x$, $\mathbf{S}_x$ and $\mathbf{R}_x$ evolve is given in Table 4.

However (see below) the line selective pulses produce evolution by a unitary operator which is close to that required but differs by a controlled phase shift. For example, in the case of f_4, the line selective pulse produces the unitary transformation

$$\Delta(2i\sigma_x, 2i\sigma_x, E, E) , \quad (11)$$

whereas the unitary given in Table 2 is

$$U_4 = \Delta(2\sigma_x, 2\sigma_x, E, E) . \quad (12)$$

The relation between these two matrices is

$$U_4 = \Delta(2\sigma_x, 2\sigma_x, E, E) = \Delta(2i\sigma_x, 2i\sigma_x, E, E) \times \Delta(-iE, -iE, E, E) . \quad (13)$$

Table 4
The effect on input product operators $\mathbf{I}_x$, $\mathbf{S}_x$ and $\mathbf{R}_x$ of the unitary operators in the second column

f	U	$\mathbf{I}_x$	$\mathbf{S}_x$	$\mathbf{R}_x$
f_1	$\Delta(E,E,E,E)$	$\mathbf{I}_x$	$\mathbf{S}_x$	$\mathbf{R}_x$
f_2	$\Delta(2\sigma_x,2\sigma_x,2\sigma_x,2\sigma_x)$	$\mathbf{I}_x$	$\mathbf{S}_x$	$\mathbf{R}_x$
f_3	$\Delta(E,E,2\sigma_x,2\sigma_x)$	$2\mathbf{I}_x\mathbf{R}_x$	$\mathbf{S}_x$	$\mathbf{R}_x$
f_4	$\Delta(2\sigma_x,2\sigma_x,E,E)$	$2\mathbf{I}_x\mathbf{R}_x$	$\mathbf{S}_x$	$\mathbf{R}_x$
f_5	$\Delta(2\sigma_x,E,2\sigma_x,E)$	$\mathbf{I}_x$	$2\mathbf{S}_x\mathbf{R}_x$	$\mathbf{R}_x$
f_6	$\Delta(E,2\sigma_x,E,2\sigma_x)$	$\mathbf{I}_x$	$2\mathbf{S}_x\mathbf{R}_x$	$\mathbf{R}_x$
f_7	$\Delta(2\sigma_x,E,E,2\sigma_x)$	$2\mathbf{I}_x\mathbf{R}_x$	$2\mathbf{S}_x\mathbf{R}_x$	$\mathbf{R}_x$
f_8	$\Delta(E,2\sigma_x,2\sigma_x,E)$	$2\mathbf{I}_x\mathbf{R}_x$	$2\mathbf{S}_x\mathbf{R}_x$	$\mathbf{R}_x$

For example, under $\Delta(E,2\sigma_x,E,2\sigma_x)$, the input $\mathbf{S}_x$ evolves to $2\mathbf{S}_x\mathbf{R}_x$.

The second matrix on the right hand side of this equation is a z rotation on the first spin by the angle $\pi/2$. Thus if one wants to implement U_4, it would be necessary to follow the line-selective pulse by a phase shift. One finds that similar phase shifts are required for all functions except f_1 and f_2.

[4★] Observation: Under evolution by the unitary operators corresponding to any of the balanced functions, either the I response or the S response (or both) disappears. Had we started with a pure initial state the equivalent line would have been inverted.

We note that the disappearance or otherwise of the I or S response is not affected by the final phase shift. This is because the state

$$\mathbf{I}_x+\mathbf{S}_x+\mathbf{R}_x \tag{14}$$

still evolves to states in which the same line disappears even if this last phase shift is not implemented. This may be appreciated by looking at the product operators to which the state evolves, as given in Table 5.

2. Experimental realization

One possible way to implement the evaluation stage of the algorithm would be to make use of the fact [13] that any unitary transformation can be built up from combinations of the *controlled not* operation and operations on a single qubit. The implementation of a controlled not operation by magnetic resonance involves the preparation of nuclear magnetization vectors of a given spin aligned in opposite directions in the transverse plane. This 'anti-phase' condition, which may be represented in the product operator formalism as (say) $2\mathbf{I}_y\mathbf{S}_z$, can be generated in a coupled two-spin system through the initial

Table 5
The effect on input product operators $\mathbf{I}_x$, $\mathbf{S}_x$ and $\mathbf{R}_x$ of the unitary operators in the second column

f	U	$\mathbf{I}_x$	$\mathbf{S}_x$	$\mathbf{R}_x$
f_1	$\Delta(E,E,E,E)$	$\mathbf{I}_x$	$\mathbf{S}_x$	$\mathbf{R}_x$
f_2	$\Delta(2i\sigma_x,2i\sigma_x,2i\sigma_x,2i\sigma_x)$	$\mathbf{I}_x$	$\mathbf{S}_x$	$\mathbf{R}_x$
f_3	$\Delta(E,E,2i\sigma_x,2i\sigma_x)$	$2\mathbf{I}_y\mathbf{R}_x$	$\mathbf{S}_x$	$\mathbf{R}_x$
f_4	$\Delta(2i\sigma_x,2i\sigma_x,E,E)$	$-2\mathbf{I}_y\mathbf{R}_x$	$\mathbf{S}_x$	$\mathbf{R}_x$
f_5	$\Delta(2i\sigma_x,E,2i\sigma_x,E)$	$\mathbf{I}_x$	$-2\mathbf{S}_y\mathbf{R}_x$	$\mathbf{R}_x$
f_6	$\Delta(E,2i\sigma_x,E,2i\sigma_x)$	$\mathbf{I}_x$	$2\mathbf{S}_y\mathbf{R}_x$	$\mathbf{R}_x$
f_7	$\Delta(2i\sigma_x,E,E,2i\sigma_x)$	$-4\mathbf{I}_y\mathbf{S}_z\mathbf{R}_x$	$-4\mathbf{I}_z\mathbf{S}_y\mathbf{R}_x$	$\mathbf{R}_x$
f_8	$\Delta(E,2i\sigma_x,2i\sigma_x,E)$	$4\mathbf{I}_y\mathbf{S}_z\mathbf{R}_x$	$4\mathbf{I}_z\mathbf{S}_y\mathbf{R}_x$	$\mathbf{R}_x$

For example, under $\Delta(E,2i\sigma_x,E,2i\sigma_x)$, the input $\mathbf{S}_x$ evolves to $2\mathbf{S}_y\mathbf{R}_x$.

stages of the INEPT pulse sequence [14], relying on (refocused) evolution under the $2\mathbf{I}_z\mathbf{S}_z$ operator for a fixed interval $1/(2J_{IS})$. However, the extension of this procedure to more than two coupled spins is complicated and not easy to implement. A more direct approach, and the one we have employed, is through the use of high-selectivity radio-frequency pulses designed to perturb transverse magnetization one line at a time. For example applying a π pulse with Hamiltonian of the form [15]

$$\mathbf{R}_x + 2\mathbf{I}_z\mathbf{R}_x + 2\mathbf{S}_z\mathbf{R}_x + 4\mathbf{I}_z\mathbf{S}_z\mathbf{R}_x \tag{15}$$

causes the system to evolve by the unitary operator

$$\Delta(2i\sigma_x, E, E, E). \tag{16}$$

The key observation from the point of view of our work is that more than one such line-selective perturbation may be applied simultaneously [16]. Thus any of the unitary operators in Table 5 (and indeed a very wide class of controlled rotations about more general axes) may be produced in the same time that is required to produce the perturbation given in (16). It is worth noting that this time is of the same order as that required to implement the INEPT sequence. We feel that as well as being helpful for the present work, the method of manipulating spins via simultaneous line selective pulses may well prove advantageous in NMR quantum computers with more spins.

The experimental task is to shape the radio-frequency pulse envelope so as to achieve sufficient selectivity in the frequency domain that there is negligible perturbation of the next-nearest neighbour of the spin multiplet. In this sense the technique resembles that used in pseudo-two-dimensional spectroscopy [17] where the frequency of a soft radio-frequency pulse is stepped through the spectrum of interest in very small frequency increments, exciting the transitions one by one. We investigated several possible pulse shapes for this purpose, including rectangular, Gaussian, sine-bell, and triangular, before settling on the Gaussian as the most suitable for the task.

In a weakly coupled three-spin *ISR* system the *R* spectrum is a doublet of doublets with splittings J_{IR} and J_{SR}. Application of π pulses to all four transverse *R*-spin magnetization components corresponds to a constant function in the sense of the Deutsch–Jozsa algorithm, and the 'do nothing' experiment represents the other constant function. The balanced functions may be implemented by application of soft π pulses to the individual lines two at a time, for example $[0,0,\pi,\pi]$, $[0,\pi,0,\pi]$, or $[0,\pi,\pi,0]$, where 0 denotes no soft pulse. These cases, corresponding to functions f_3, f_6 and f_8 have Hamiltonians proportional to $\mathbf{R}_x - 2\mathbf{I}_z\mathbf{R}_x$, $\mathbf{R}_x - 2\mathbf{S}_z\mathbf{R}_x$ and $\mathbf{R}_x - 4\mathbf{I}_z\mathbf{S}_z\mathbf{R}_x$, respectively. One way to calculate the effect of these Hamiltonians is to use standard product operator manipulations [15]. For example one finds that a π pulse with Hamiltonian of the form $\mathbf{R}_x - 2\mathbf{I}_z\mathbf{R}_x$ leaves $\mathbf{R}_x$ and $\mathbf{S}_x$ unchanged and changes $\mathbf{I}_x$ to $2\mathbf{I}_y\mathbf{R}_x$ as in Table 5.

The practical implementation is deceptively simple. Starting with a thermal state, a hard $\pi/2$ pulse about the *y*-axis (denoted $[\pi/2]_y$) excites transverse magnetizations $\mathbf{I}_x$, $\mathbf{S}_x$ and $\mathbf{R}_x$. The evaluation step is the application of line-selective $[\pi]_x$ pulses to the individual components of the *R* multiplet. We may choose to apply soft $[\pi]_x$ pulses to all four magnetization components, any two of the four, or none at all. In all cases the soft pulses are applied simultaneously, while the remaining transitions are simply left to evolve freely for the same period of time. However, the perturbed magnetization components lose intensity only through spin-spin relaxation during the relatively long interval of the soft pulse, *T*, because the effects of spatial inhomogeneity of the magnetic field are refocused, whereas the freely precessing components decay more rapidly, with a shorter time constant T_2^*. This difference in intensities serves to confirm which *R* transitions were perturbed.

Experiments were carried out at 400 MHz on a Varian VXR-400 spectrometer equipped with a waveform generator which controlled the shaped radiofrequency pulses. The three-spin proton system chosen for study was 2,3-dibromopropanoic acid in $CDCl_3$. The three splittings are $J_{IR} = +11.3$ Hz, $J_{IS} = -10.1$ Hz, and $J_{RS} = +4.3$ Hz. (The negative sign of the geminal coupling J_{IS} [18] has no particular significance in these experiments.) Strong coupling effects are evident between spins *I* and *S*, with $J_{IS}/\delta_{IS} = 0.12$.

Each soft $[\pi]_x$ pulse can be thought of as acting on one of the four *R*-spin magnetizations in a rotating frame at the exact resonance frequency of that particular *R* line. These four reference frames rotate at four different frequencies $(\pm J_{SR} \pm J_{IR})/2$ with respect to the transmitter frequency centred on the R chemical shift. The x-axes of all four frames must be coincident at the beginning of the soft pulse interval T. The duration of the soft pulse, T, may be chosen in such a way as to optimize the frequency selectivity.

The predicted result (Table 5), is to convert I- or S-spin magnetization into various forms of multiple-quantum coherence in the six cases where the R magnetization components are perturbed in pairs (the balanced functions) but to leave the I- and S-spin magnetizations unaffected in the remaining two cases where the four R magnetization components are all perturbed or all left alone (the constant functions). These predictions are clearly borne out by the experimental specta shown in the Fig. 1. In principle,

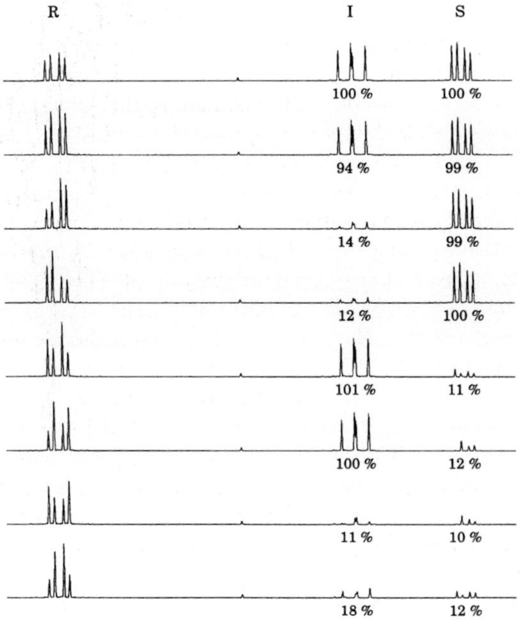

Fig. 1. Eight absolute-value 400 MHz spectra of 2,3-dibromopropanoic acid obtained with the eight different perturbations set out in Table 5. The soft pulses were applied simultaneously with a pulse duration $T = 0.65$ s. Reading from top to bottom, these spectra correspond to the functions $f_1 \dots f_8$ of Table 1. Integrals of the I- and S-spin responses are shown as percentages of those in the top trace. After the evaluation of these integrals, the line shapes were improved by pseudo-echo weighting. Note the suppression of the appropriate I- and S-spin responses by about an order of magnitude.

complete conversion into unobservable multiple-quantum coherence would be detected by the disappearance of the appropriate I- or S-spin response. In practice, owing to non-idealities of the system (for example strong coupling effects between I and S) this is observed as a roughly eightfold loss of intensity rather than complete suppression.

Eight experiments were performed to test the eight cases of Table 5. The transmitter frequency was centred on the R-spin multiplet. Note that the R spectrum remains unperturbed throughout the series, except for the intensity perturbation mentioned above, a result of the refocusing effect of the soft π pulses. The phases of the I- and S-signals will be determined by the scalar coupling and chemical shift evolution during the period T. These complex phase patterns do not interfere with the Deutsch–Jozsa test because this involves only the observation of the 'disappearance' of certain signals. These signal losses are made clearly evident by displaying absolute-value spectra, which may then be integrated. The integrated intensities are shown as percentages of the corresponding intensities in the top spectrum (no soft-pulse perturbation). Creation of multiple-quantum coherence is indicated by the roughly eightfold decrease in intensity in the appropriate places; all other I- and S-spin intensities remain essentially at 100%. This interpretation was confirmed in a second experiment with a multiplet-selective soft $\pi/2$ pulse applied to the R spins at the end of the sequence. This has the effect of restoring the 'lost' intensities by reconverting IR and SR multiple-quantum coherence into observable magnetization.

Thus, in a single measurement, a distinction can be made between constant and balanced functions simply on the grounds of the 'disappearance' of I- or S-spin lines. The fact that further details can be gleaned about the pattern of soft-pulse perturbation is irrelevant to the Deutsch–Jozsa algorithm. The extension to systems of more than three coupled spins is clear. Because the soft pulses are applied simultaneously, this involves no increase in the duration of the perturbation stage. The main limitation would be the magnitude of the smallest splitting, for this sets the frequency selectivity requirement. Extension to more qubits would most likely invoke the introduction of heteronuclear spins such as ^{13}C and ^{19}F.

Acknowledgements

The authors are indebted to Dr. Ēriks Kupče of Varian Associates for invaluable advice on the generation of shaped selective radio-frequency pulses.

References

[1] P. Shor, Proc. 35th Ann. Symp. on Found. of Computer Science, IEEE Comp. Soc. Press, Los Alamitos, CA, 1994, pp. 124–134.
[2] D.G. Cory, A.F. Fahmy, T.F. Havel, Proc. Natl. Acad. Sci. 94 (1997) 1634.
[3] D.G. Cory, M.D. Price, T.F. Havel, Physica D 120 (1998) 82.
[4] N. Gershenfeld, I.L. Chuang, Science 275 (1997) 350.
[5] I.L. Chuang, N. Gershenfeld, M.G. Kubinec, D.W. Leung, Proc. R. Soc. Lond. A 454 (1998) 447.
[6] I.L. Chuang, L.M.K. Vandersypen, X. Zhou, D.W. Leung, S. Lloyd, Nature 393 (1998) 143.
[7] I.L. Chuang, N. Gershenfeld, M. Kubinec, Phys. Rev. Lett. 80 (1998) 3408.
[8] J.A. Jones, M. Mosca, J. Chem. Phys. 109 (1998) 1648.
[9] J.A. Jones, M. Mosca, R.H. Hansen, Nature 393 (1998) 344.
[10] D.G. Cory, W. Mass, M.D. Price, E. Knill, R. Laflamme, W.H. Zurek, T.F. Havel, S.S. Somaroo, LANL E-print quant-ph/9802018 (1998).
[11] D. Deutsch, R. Jozsa, Proc. R. Soc. Lond. A 439 (1992) 553.
[12] L.K. Grover, Phys. Rev. Lett. 79 (1997) 4709.
[13] A. Barenco, C.H. Bennett, R. Cleve, D.P. DiVincenzo, N. Margolus, P. Shor, T. Sleator, J. Smolin, H. Weinfurter, Phys. Rev. A 52 (1995) 3457.
[14] G.A. Morris, R. Freeman, J. Am. Chem. Soc. 101 (1979) 760.
[15] O.W. Sorensen, G.W. Eich, M.H. Levitt, G. Bodenhausen, R.R. Ernst, Prog. NMR Spectrosc. 16 (1983) 163.
[16] Ē. Kupče, R. Freeman, J. Magn. Reson. A 105 (1993) 234.
[17] S. Davies, J. Friedrich, R. Freeman, J. Magn. Reson. 75 (1987) 540.
[18] R. Freeman, K.A. McLauchlan, J.I. Musher, K.G.R. Pachler, Mol. Phys. 5 (1962) 321.

Selected bibliography

1. L. Adleman, C. Pomerance, R. Rumely, *On distinguishing prime numbers from composite numbers*, Annals of Mathematics **117**, 173 (1983).

2. D. Aharonov and M. Ben-Or, *Fault-tolerant quantum computation with constant error*, Proc. 29th Ann. ACM Symp. on Theory of Computing (STOC '97), p. 176 (1997).

3. M.H. Anderson, J.R. Ensher, M.R. Matthews, C.E. Wieman and E.A. Cornell, *Observation of Bose-Einstein condensation in a diluite atomic vapor*, Science **269**, 198 (1995).

4. R. C. Ashoori, *Electrons in artificial atoms*, Nature **379**, 413 (1996).

5. A. Aspect, J. Dalibard, and G. Roger, *Experimental test of Bell's inequalities using time-varying analizers*, Phys. Rev. Lett. **49**, 1804 (1982).

6. D. V. Averin, *Adiabatic quantum computation with Cooper pairs*, Solid State Commun. **105**, 659 (1998).

7. A. Barenco, C.H. Bennett, R. Cleve, D.P. DiVincenzo, N. Margolus, P. Shor, T. Sleator, J. Smolin, and H. Weinfurter, *Elementary gates for quantum computation*, Phys. Rev. A **52**, 3457 (1995).

8. A. Barenco, D. Deutsch, A. Ekert, and R. Josza, *Conditional quantum dynamics and logic gates*, Phys. Rev. Lett. **74**, 4083 (1995).

9. A.Barenco, A. Ekert, C. Macchiavello, and A. Sanpera, *Un Saut Quantique Pour Les Calculateurs*, La Recherche **292**, 52 (November 1996).

10. A. Barenco, A. Berthiaume, D. Deutsch, A. Ekert, R. Jozsa and C. Macchiavello, *Stabilisation of quantum computation by symmetrisation*, SIAM J. Comput. **26**, 1541 (1997).

11. H. Barnum, C. M. Caves, C. A. Fuchs, R. Jozsa, and B. Schumacher, *Noncommuting Mixed States Cannot Be Broadcast*, Phys. Rev. Lett. **76**, 2818 (1996).

12. R. Beals, H. Buhrman, R. Cleve, M. Mosca, R. de Wolf, *Quantum lower bounds by polynomials*, Proc. 39th Ann. IEEE Symp. on Foundations of Computer Science (FOCS '98), p. 352 (1998).

13. H. Bechmann-Pasquinucci and N. Gisin *Incoherent and Coherent Eavesdropping in the 6-state Protocol of Quantum Cryptography.* quant-ph/9807041 (1998).

14. J.S. Bell, *On the Einstein-Podolsky-Rosen paradox*, Physics **1**, 195 (1964), reprinted in J. S. Bell, *Speakable and Unspeakable in Quantum Mechanics*, Cambridge U.P., Cambridge (1987).

15. C.H. Bennett, *Logical reversibility of computation*, IBM J. of Research and Development **17**, p. 525 (1973).

16. C. H. Bennett and G. Brassard, *Quantum cryptography: public key distribution and coin tossing*, Proc. Internat. Conf. Computer Systems and Signal Processing, Bangalore, 175 (1984).

17. C.H. Bennett, G. Brassard, and A.K. Ekert, *Quantum cryptography*, Scientific American, Oct. 1992, p. 50.

18. C. Bennett, *Quantum cryptography using any two nonorthogonal states*, Phys. Rev. Lett. **68**, 3121 (1992).

19. C. H. Bennett, F. Bessette, G. Brassard, L. Salvail, and J. Smolin, *Experimental quantum cryptography*, J. Cryptology **5**, 3 (1992).

20. C. H. Bennett, G. Brassard, G. Crepeau, R. Josza, A. Peres, and W. Wootters, *Teleporting an unknown quantum state via dual classical and Einstein-Podolski-Rosen channels*, Phys. Rev. Lett. **70**, 1895 (1993).

21. C. Bennett, G. Brassard, C. Crepeau, U. Maurer, *Generalized privacy amplification*, IEEE Trans. Inf. Theory **41**, 1915 (1994).

22. C.H. Bennett, *Quantum information and computation*, Physics Today **48**, 24 (1995).

23. C.H. Bennett, E. Bernstein, G. Brassard, and U. Vazirani, *Strengths and weaknesses of quantum computing*, SIAM J. Computing **26**, 1510 (1997).

24. C. H. Bennett, G. Brassard, S. Popescu, B. Schumacher, J. A. Smolin, and W. K. Wootters, *Purification of noisy entanglement and faithful teleportation via noisy channels*, Phys. Rev. Lett. **76**, 722 (1996).

25. C. H. Bennett, D. P. DiVincenzo, J. A. Smolin, and W. K. Wootters, *Mixed-state entanglement and quantum error correction*, Phys. Rev. A **54**, 3825 (1996).

26. C.H. Bennett, *Quantum information*, Physica Scripta **T76**, 210 (1998).

27. J. C. Bergquist, R. G. Hulet, W. M. Itano, and D. J. Wineland, *Observation of quantum jumps in a single atom*, Phys. Rev. Lett. **57**, 1699 (1986).

28. E. Bernstein and U.V. Vazirani, *Quantum complexity theory*, SIAM J. Comput. **26**, 1411 (1997).

29. E. Biham, M. Boyer, G. Brassard, J. van de Graaf, T. Mor, *Security of Quantum Key Distribution Against All Collective Attacks*, quant-ph/9801022 (1998).

30. G. Birkl, S. Kassner, and H. Walther, *Multiple-shell structures of laser-cooled $^{24}Mg^{+}$ ions in a quadrupole storage ring*, Nature **357**, 310 (1992).

31. R. Blatt, in *Atomic Physics*, edited by D. J. Wineland, C. E. Wieman, and S. J. Smith (AIP, New York, 1995), Vol. 14, p. 219.

32. R. H. Blick, D. Pfannkuche, R. J. Haug, K. v. Klitzing, and K. Eberl, *Formation of a coherent mode in a double quantum dot*, Phys. Rev. Lett. **80**, 4032 (1998).

33. R. H. Blick, D. W. van der Weide, R. J. Haug, K. Eberl, *Complex broad-band millimeter wave response of a double quantum dot: Rabi oscillations in an artificial molecule*, Phys. Rev. Lett. **81**, 689 (1998).

34. F. Bloch, W. W. Hansen, and M. E. Packard, *Nuclear induction*, Phys. Rev. **69,** 127 (1946).

35. J. J. Bollinger, W. M. Itano, D. J. Wineland, and D. J. Heinzen, *Optimal frequency measurements with maximally correlated states*, Phys. Rev. A **54**, R4649 (1996).

36. D. Boschi, S. Branca, F. De Martini, L. Hardy, and S. Popescu, *Experimental realization of teleporting an unknown pure quantum state via dual classical and Einstein-Podolsky-Rosen channels*, Phys. Rev. Lett. **80**, 1121 (1998).

37. S. Bose, V. Vedral, and P.L. Knight, *A multiparticle generalization of entanglement swapping*, Phys. Rev. A **57**, 822 (1998).

38. D. Bouwmeester, J-W Pan, K. Mattle, M. Eibl, H. Weinfurter, and A. Zeilinger, *Experimental quantum teleportation*, Nature **390**, 575 (1997).

39. D. Bouwmeester, J-W Pan, M. Daniell, H. Weinfurter and A.Zeilinger, *Observation of three-photon Greenberger-Horne-Zeilinger Entanglement*, Phys. Rev. Lett. **82**, 1345 (1999).

40. D. Bouwmeester, J.-W. Pan, H. Weinfurter, and A. Zeilinger, *High-fidelity teleportation of qubits*, to appear in J. Mod. Opt.

41. M. Boyer, G. Brassard, P. Høyer, and A. Tapp, *Tight bounds on quantum searching*, Fortschritte der Physik **46**, 493 (1998).

42. C.C. Bradley, C.A. Sackett, J.J. Tollett and R.G. Hulet, *Evidence of Bose-Einstein condensation in an atomic gas with attractive interactions*, Phys. Rev. Lett. **75**, 1687 (1995).

43. G. Brassard and L. Salvail *Secret-key reconciliation by public discussion.* In *Proceedings of Eurocrypt '93, held in Lofthus, Norway, 1993*, (1993).

44. G. Brassard, P. Høyer, and A. Tapp, *Quantum counting*, Proc. 25th ICALP **1443** of *Lecture Notes in Computer Science*, p. 820 (Springer), 1998.

45. G. Brassard and P. Høyer, *An exact quantum polynomial-time algorithm for Simon's problem*, Proc. 5th Israeli Symp. on Theory of Computing and Systems (ISTCS '97), p. 12 (1997).

46. S. L. Braunstein, C. M. Caves, R. Jozsa, N. Linden, S. Popescu, and R. Schack, *Separability of very noisy mixed states and implications for NMR quantum computing*, quant-ph/9811018 (1998).

47. S.L. Braunstein and H.J. Kimble, *Teleportation of Continuous Variables*, Phys. Rev. Lett. **80**, 869 (1998).

48. S.L. Braunstein and H.J. Kimble, *A posteriori teleportation*, D. Bouwmeester, J-W. Pan, M. Daniell, H. Weinfurter, M. Zukowski, and A. Zeilinger, Reply, *A posteriori teleportation - Reply*, Nature (London) **394**, 840 (1998).

49. G.K. Brennen, C.M. Caves, P.S. Jessen and I.H. Deutsch, *Quantum logic gates in optical lattices*, Phys. Rev. Lett. **82**, 1060 (1999).

50. H.-J. Briegel, W. Dür, J. I. Cirac, and P. Zoller, *Quantum repeaters: The role of imperfect local operations in quantum communication*, Phys. Rev. Lett. **81**, 5932 (1998).

51. H.-J. Briegel, T. Calarco, D. Jaksch, J.I. Cirac and P. Zoller, *Quantum computing with neutral atoms*, 1999, submitted for publication in the Special Issue on Quantum Computing of the Journal of Modern Optics (preprint quant-ph/9904010).

52. J. A. Brum and P. Hawrylak, *Coupled quantum dots as quantum exclusive-OR gate*, Superlattices and Microstructures **22**, 431 (1997).

53. M. Brune, F. Schmidt-Kaler, A. Maali, J. Dreyer, E. Hagley, J. M. Raimond, and S. Haroche, *Quantum Rabi Oscillation: a Direct Test of Field Quantization in a Cavity*, Phys. Rev. Lett. **76**, 1800 (1996).

54. M. Brune, E. Hagley, J. Dreyer, X. Matre, A. Maali, C. Wunderlich, J. M. Raimond, and S. Haroche, *Observing the Progressive Decoherence of the 'Meter' in a Quantum Measurement,* Phys. Rev. Lett. **77**, 4887 (1996).

55. D. Bruß, *Optimal eavesdropping in quantum cryptography with six states*, Phys. Rev. Lett. **81**, 3018 (1998).

56. D. Bruß, D. DiVincenzo, A. Ekert, C. Fuchs, C. Macchiavello, J. Smolin, *Optimal universal and state-dependent quantum cloning*, Phys. Rev. A **57**, 2368 (1998).

57. D.Bruß, L.Faoro C.Macchiavello and G.Massimo Palma, *Quantum entanglement and classical communication through a depolarising channel* to appean in the special issue of Journal of Modern Optics on *The Physics of Quantum Information.* also quant-ph/9903033

58. D. Brußand N. Lutkenhaus, *Quantum Key Distribution: from Principles to Practicalities*, quant-ph/9901061 (1999).

59. H. Buhrman, R. Cleve, and A. Wigderson, *Quantum vs. classical communication and computation*, Proc. 30th Ann. ACM Symp. on Theory of Computing (STOC '98), p. 63 (1998).

60. H. Buhrman, R. Cleve, and W. van Dam, *Quantum entanglement and communication complexity*, preprint quant-ph/9705033 (1997).

61. H. Buhrman, R. de Wolf, *Lower bounds for quantum search and derandomization*, preprint quant-ph/9811046 (1998).

62. G. Burkard, D. Loss, and D.P. DiVincenzo, *Coupled quantum dots as quantum gates*, Phys. Rev. B **59**, 2070 (1999).

63. W. T. Butler, R. J. Hughes, P. G. Kwiat, S. K. Lamoreaux, G. G. Luther, G. L. Morgan, J. E. Nordholt, C. G. Peterson, and C. M. Simmons, *Practical free-space quantum key distribution over 1 km*, Phys. Rev. Lett. **81**, 3283 (1998).

64. M. Büttiker, *Scattering theory of current and intensity noise correlations in conductors and wave guides*, Phys. Rev. B**46**, 12485 (1992).

65. A. R. Calderbank, E. M. Rains, P. W. Shor, and N. J. A. Sloane, *Quantum error correction and orthogonal geometry*, Phys. Rev. Lett. **78**, 405 (1997).

66. A. R. Calderbank, E. M. Rains, P. W. Shor, and N. J. A. Sloane, *Quantum error correction via codes over GF(4)*, IEEE Trans. Inf. Theory **44**, 1369 (1998).

67. M. Celio and D. Loss, *Comparison between different Markov approximations for open spin systems*, Physica **A 150**, 769 (1989).

68. M.-D. Choi, *Completely Positive Linear Maps on Complex Matrices*, Lin. Alg. Appl. **10**, 285 (1975).

69. B. Chor and O. Goldreich, *Unbiased bits from sources of weak randomness and probabilistic communication complexity*, SIAM J. Comput. **17**, 230 (1988).

70. S. Chu, *The manipulation of neutral particles*, Rev. Mod. Phys. **70**, 686 (1998).

71. I. L. Chuang and M. A. Nielsen, *Prescription for Experimental Determination of the Dynamics of a Quantum Black Box*, J. Mod. Opt. **44**, 2455 (1997).

72. I. L. Chuang, N. A. Gershenfeld, and M. Kubinec, *Experimental implementation of fast quantum searching*, Phys. Rev. Lett. **80**, 3408 (1998).

73. I. L. Chuang, L. M. K. Vandersypen, X. Zhou, D. W. Leung and S. Lloyd, *Experimental realization of a quantum algorithm*, Nature **393,** 143 (1998).

74. J.I. Cirac, P. Zoller, *Preparation of macroscopic superpositions in many-atom systems*, Phys. Rev. A **50**, R2799 (1994).

75. J. I. Cirac and P. Zoller, *Quantum computations with cold trapped ions*, Phys. Rev. Lett. **74**, 4091 (1995).

76. J. I. Cirac, A. S. Parkins, R. Blatt, and P. Zoller, *Nonclassical states of motion in ion traps*, Adv. At. Molec. Opt. Physics **37**, 237 (1996).

77. J. I. Cirac, T. Pellizzari, and P. Zoller, *Enforcing coherent evolution in dissipative quantum dynamics*, Science **273**, 1207 (1996).

78. J.I. Cirac and N. Gisin, *Coherent eavesdropping strategies for the four state quantum cryptography protocol*, Phys. Lett. A **229**, 1 (1997).

79. J. I. Cirac, P. Zoller, H. J. Kimble, and H. Mabuchi, *Quantum state transfer and entanglement distribution among distant nodes in a quantum network*, Phys. Rev. Lett. **78**, 3221 (1997).

80. J.I. Cirac, A. Ekert, S.F. Huelga and C. Macchiavello, *Distributed quantum computation over noisy channels*, Phys. Rev. A **59**, 4249 (1999).

81. J.I. Cirac, A. Ekert and C. Macchiavello, *Optimal purification of single qubits*, Phys. Rev. Lett. **82**, 4344 (1999).

82. R. Cleve and H. Buhrman, *Substituting quantum entanglement for communication*, Phys. Rev. A **56**, 1201 (1997).

83. R. Cleve, W. van Dam, M. Nielsen, and A. Tapp, *Quantum entanglement and the communication complexity of the inner product function*, to appear in Proc. 1st NASA Intl. Conf. on Quantum Computing and Quantum Communications **1509** of *Lecture Notes in Computer Science* (Springer), 1998; preprint quant-ph/9708019 (1997).

84. R. Cleve, A. Ekert, C. Macchiavello, and M. Mosca, *Quantum algorithms revisited*, Proc. R. Soc. London A **454**, 339 (1998).

85. R. Cleve, A. Ekert, L. Henderson, C. Macchiavello, M. Mosca, *On quantum algorithms*, Complexity **4**, 33 (1998).

86. C. Cohen-Tannoudji, *Manipulating atoms with photons*, Rev. Mod. Phys. **70**, 707 (1998).

87. S.A. Cook, *The complexity of theorem proving procedures*, Proc. 3rd Ann. ACM Symp. on Theory of Computing (STOC '71), p. 151 (1971).

88. D. Coppersmith, *An Approximate Fourier Transform Useful in Quantum Factoring*, IBM Research Report No. RC19642 (1994).

89. D. Cory, A. Fahmy, and T. Havel, *Ensemble quantum computing by NMR spectroscopy*, Proc. Nat. Acad. Sci. U.S.A. **94**, 1634 (1997).

90. D. G. Cory, M. D. Price, W. Maas, E. Knill, R. Laflamme, W. H. Zurek, T. F. Havel, and S. S. Somaroo, *Experimental quantum error correction*, Phys. Rev. Lett. **81**, 2152 (1998).

91. D. G. Cory, M. D. Price and T. F. Havel, *Nuclear magnetic resonance spectroscopy: An experimentally accessible paradigm for quantum computing*, Physica D **120,** 82 (1998).

92. M. Czachor and M. Kuna, *Complete Positivity of Nonlinear Evolution: A Case Study*, Phys. Rev. A **58**, 128 (1998).

93. W. van Dam, P. Høyer, A. Tapp, *Multiparty quantum communication complexity*, preprint quant-ph/9710054 (1997).

94. L. Davidovich, A. Maali, M. Brune, J. M. Raimond, and S. Haroche, *Quantum Switches and Nonlocal Microwave Fields*, Phys. Rev. Lett. **71**, 2360 (1993).

95. E. B. Davies, *Quantum Theory of Open Systems*, Academic Press, New York (1976).

96. K.B. Davies, M.-O. Mewes, M.R. Andrews, N.J. van Druten, D.S. Durfee, D.M. Kurn, and W. Ketterle, *Bose-Einstein condensation in a gas of sodium atoms*, Phys. Rev. Lett. **75**, 3969 (1995).

97. F. De Martini,*Teleportation: who was first?*, and A. Zeilinger, *Teleportation: who was first? Reply*, Physics World **11**, 23 (1998).

98. M.T. DePue, C. McCormick, S.L. Winoto, S. Oliver and D.W. Weiss, *Unity occupation of sites in a 3D optical lattice*, Phys. Rev. Lett. **82** 2262 (1999).

99. D. Deutsch, *Quantum theory, the Church-Turing principle and the universal quantum computer*, Proc. R. Soc. London A **400**, 96 (1985).

100. D. Deutsch, *Quantum computational networks*, Proc. R. Soc. London A **425**, 73 (1989).

101. D. Deutsch and R. Jozsa, *Rapid Solution of Problems by Quantum Computation*, Proc. R. Soc. London A **439**, 553 (1992).

102. D. Deutsch, A. Ekert, R. Jozsa, C. Macchiavello, S. Popescu and A. Sanpera, *Quantum privacy amplification and the security of quantum cryptography over noisy channels*, Phys. Rev. Lett. **77**, 2818 (1996).

103. D. Deutsch, *The Fabric of Reality* (Allen Lane, The Penguin Press, 1997).

104. D. Deutsch and A. Ekert, *Quantum computation*, Physics World **11**, 47 (1998).

105. I.H. Deutsch and P.S. Jessen, *Quantum-state control in optical lattices*, Phys. Rev. A **57**, 1972 (1998).

106. R. G. DeVoe, *Elliptical ion traps and trap arrays for quantum computation*, Phys. Rev. A **58**, 910 (1998).

107. C. Dhelon and G. J. Milburn, *Correcting the effects of spontaneous emission on cold-trapped ions*, Phys. Rev. A **56**, 640 (1997).

108. F. Diedrich, J. C. Bergquist, W. M. Itano, and D. J. Wineland, *Laser cooling to the zero-point energy of motion*, Phys. Rev. Lett. **62**, 403 (1989).

109. W. Diffie, and M. Hellman, *New directions in cryptography*, IEEE Trans. Inf. Theory IT-22, 644 (1976).

110. D. P. DiVincenzo, *Two-bit gates are universal for quantum computation*, Phys. Rev. A **51**, 1015 (1995).

111. D. P. DiVincenzo and P. Shor, *Fault-tolerant error correctionwith efficient quantum codes*, Phys. Rev. Lett. **77**, 3260 (1996).

112. D.P. DiVincenzo, *Quantum gates and circuits*, Proc. R. Soc. London A **454**, 261 (1998).

113. D. P. DiVincenzo and D. Loss, *Quantum information is physical*, Superlattices and Microstructures **23**, 419 (1998).

114. D. P. DiVincenzo and B. Terhal, *Decoherence: the obstacle to quantum computation*, Physics World **11**, 41 (1998).

115. D.P. DiVincenzo and D. Loss, *Quantum Computers and Quantum Coherence*, To appear in special issue of J. Mag. Magn. Matl. (vol. 200), "Magnetism beyond 2000". See cond-mat/9901137.

116. P. Domokos, J. M. Raimond, M. Brune, and S. Haroche, *Simple Cavity-QED 2-Bit Universal Quantum Logic Gate – the Principle and Expected Performances*, Phys. Rev. A **52**, 3554 (1995).

117. W. Dür, H.-J. Briegel, J. I. Cirac, and P. Zoller, *Quantum repeaters based on entanglement purification*, Phys. Rev. A **59**, 169 (1999).

118. A. Ekert, *Quantum cryptography based on Bell's theorem*, Phys. Rev. Lett. **67**, 661 (1991).

119. A. K. Ekert, B. Huttner, G. M. Palma and A. Peres *Eavesdropping on quantum-cryptographical systems.* Phys. Rev. A **50,** 1047–1056 (1994).

120. A. Ekert, J. Rarity, P. Tapster, and G.M. Palma, *Practical quantum cryptography based on two-photon interferometry*, Phys. Rev. Lett. **69**, 1293 (1992).

121. A. Ekert and R. Josza, *Quantum Computation and Shor's Factoring Algorithm*, Rev. Mod. Phys. **68**, 733 (1996).

122. A. Ekert and C. Macchiavello, *Quantum error correction for communication*, Phys. Rev. Lett. **77**, 2585 (1996).

123. A.Ekert and C. Macchiavello, *An Overview of Quantum Computing*, in Unconventional Models of Computation. Edited by C.S.Calude, J.Casti and M.J.Dinneen, p. 19, Springer Series in Discrete Mathematics and Theoretical Computer Science Springer, Singapore (1998).

124. A. Einstein, B. Podolsky, and N. Rosen, *Can quantum-mechanical description of physical reality be considered complete?*, Phys. Rev.**47**, 777 (1935).

125. S. J. van Enk, J.-I. Cirac, and P. Zoller, *Ideal Quantum Communication over Noisy Channels: a Quantum Optical Implementation*, Phys. Rev. Lett. **78**, 4293 (1997).

126. S. J. van Enk, J. I. Cirac, and P. Zoller, *Photonic channels for quantum communication*, Science **279**, 205 (1998).

127. R. R. Ernst, G. Bodenhausen, and A. Wokaun, *Principles of Nuclear Magnetic Resonance in One and Two Dimensions* (Oxford University Press, Oxford, 1987).

128. E. Farhi, J. Goldstone, S. Gutmann, and M. Sipser, *A limit on the speed of quantum computation in determining parity*, Phys. Rev. Lett. **81**, 5442 (1998).

129. R.Fazio, G.M.Palma, and J.Siewert, *Fidelity and Leakage of Josephson qubits*.Phys.Rev.Lett.(to appear) also cond-mat/9906292

130. R. Fazio and G. Schön, *Mesoscopic effects in superconductivity* , in *Mesoscopic Electron Transport* NATO ASI series E, Vol, 345, p. 407, Kluver (1997).

131. R.P. Feynman, R.B. Leighton, and M. Sands, *The Feynman Lectures on Physics*, Vol. III Addison-Wesley, Reading, MA 1966.

132. R.P. Feynman, *Simulating Physics with Computers*, Int. J. Theor. Phys. **21**, 467 (1982).

133. R.P. Feynman, *Feynman Lectures on Computation*, (Edited by A.J.G. Hey and R.W. Allen)Addison-Wesley, 1996.

134. E. Fick and G. Sauermann, *The Quantum Statistics of Dynamic Processes*, Springer Series in Solid-State Sciences 86 (Springer-Verlag, Berlin, 1990).

135. L. Fortnow and J. Rogers, *Complexity limitations on quantum computation*, Proc. 13th IEEE Conf. on Computational Complexity, p. 202 (1998).

136. J. D. Franson, and B. C. Jacobs, *Operational system for quantum cryptography*, Electron. Lett. **31**, 232 (1995).

137. S.J. Freedman, J.S. Clauser, *Experimental test of local hidden-variable theories*, Phys. Rev. Lett. **28**, 938 (1972).

138. S. Friebel, C. D'Andrea, J. Walz, M. Weitz and T.W. Hänsch, *CO2-laser optical lattice with cold rubidium atoms*, Phys. Rev. A **57**, R20 (1998).

139. C. A. Fuchs, *Nonorthogonal Quantum States Maximize Classical Information Capacity*, Phys. Rev. Lett. **79**, 1163 (1997).

140. C. A. Fuchs and A. Peres: *Quantum State Disturbance vs. Information Gain: Uncertainty Relations for Quantum Information.* Phys. Rev. A **53,** 2038–2045 (1996).

141. C. Fuchs, N. Gisin, R. Griffiths, C.-S. Niu, and A. Peres, *Optimal eavesdropping in quantum cryptography. I. Information bound and optimal strategy*, Phys. Rev. A **56**, 1163 (1997).

142. A. Furusawa, J.L. Sørenson, S.L. Braunstein, C.A. Fuchs, H.J. Kimble, and E.S. Polzik, *Unconditional quantum teleportation*, Science **282**, 706 (1998).

143. M.R. Garey and D.S. Johnson, "Computers and Intractibility: A Guide to the Theory of NP-Completeness", W.H. Freeman, 1979.

144. A. Garg, *Decoherence in ion trap quantum computers*, Phys. Rev. Lett. **77**, 964 (1996).

145. N. Gershenfeld and I. Chuang, *The usefulness of NMR quantum computing - Response*, Science **277**, 1689 (1997).

146. N. Gershenfeld and I. Chuang, *Bulk spin-resonance quantum computation*, Science **275,** 5298 (1997).

147. N. Gershenfeld, I. Chuang and S. Lloyd, Proceedings of PhysComp '96 (T. Toffoli, M. Biafore, and J. Leão Eds), 134 (1996).

148. G. Giedke, H.-J. Briegel, J. I. Cirac, and P. Zoller, *Lower bounds for attainable fidelities in entanglement purification*, Phys. Rev. A **59**, 2641 (1999).

149. N. Gisin, *Hidden quantum nonlocality revealed by local filters*, Phys. Lett. A **210**, 151 (1996).

150. N. Gisin and S. Wolf, *Quantum cryptography on noisy channels: quantum versus classical key-agreement protocols*, quant-ph/9902048 (1999).

151. L. Goldenberg and L. Vaidman: Quantum Cryptography based on Orthogonal States. Phys. Rev. Lett. **75**, 1239 (1995);

152. L. Goldenberg and L. Vaidman: Reply to Comment *Quantum Cryptography based on Orthogonal States.* Phys. Rev. Lett. **77**, 3265 (1996).

153. M. Goldman, *Quantum Description of High- Resolution NMR in Liquids* (Clarendon Press, Oxford, 1988).

154. P. K. Gosh, *Ion Traps* (Clarendon, Oxford, 1995).

155. D. Gottesman, *Class of quantum error-correcting codes saturating the quantum Hamming bounds*, Phys. Rev. A **54**, 1862 (1996).

156. D.M. Greenberger, M. Horne and A. Zeilinger in *Going beyond Bell's theorem*, Bell's theorem, Quantum Theory and the Conception of the Universe, M.Kafatos Ed. , Kluwer, Dodrecht (1989).

157. D.M. Greenberger, M.A. Horne,A. Shimony, A. Zeilinger, *Bell's theorem without inequalities*, Am. J. Phys. **58**, 1131 (1990).

158. D.M. Greenberger, M.A. Horne and A. Zeilinger, *Multiparticle Interferometry and the Superposition Principle*, Physics Today **46**, 8 (Aug. 93), p. 22.

159. R. Grimm, M. Weidenmüller and Y.B. Ovchinnikov, *Optical dipole traps for neutral atoms*, e-print quant-ph/9902072 (1999).

160. L.K. Grover, *A fast quantum mechanical algorithm for database search*, Proc. 28th Ann. ACM Symp. on Theory of Computing (STOC '96), p. 212 (1996).

161. E. Hagley, X. Maître, G. Nogues, C. Wunderlich, M. Brune, J.M. Raimond, S. Haroche, *Generation of Einstein-Podolsky-Rosen pairs of atoms*, Phys. Rev. Lett. **79**, 1 (1997).

162. D.S. Hall, M.R. Matthews, J.R. Ensher, C.E. Wieman and E.A. Cornell, *Dynamics of component separation in a binary mixture of Bose-Einstein condensates*, Phys. Rev. Lett. **81**, 1539 (1998).

163. S.E. Hamann, D.L. Haycock, G. Klose, P.H. Pax, I.H. Deutsch and P.S. Jessen, *Resolved-sideband Raman cooling to the ground state of an optical lattice*, Phys. Rev. Lett. **80**, 4149 (1998).

164. R. Hanbury Brown and R. Q. Twiss, *Correlations between photons in two coherent beams of light*, Nature (London) **177**, 27 (1956).

165. S. Haroche, *Tests of quantum mechanics with single atoms in high Q cavities*, Hyperfine Interact. **114**, 87 (1998).

166. S.Haroche *Cavity Quantum Electrodynamics*, in *Foundamental Systems in Quantum Optics*, Les Houches Summer School, J.Dalibard, J.M.Raimond and J.Zinn-Justin Eds., Elsevier 1992 .

167. S.Haroche and J.M.Raimond, *Cavity Quantum Electrodynamics*, Scientific American, (April 1993), p.26.

168. S. Haroche, Ann. N. Y. Acad. Sci. **755**, 73 (1995).

169. K. Hellwig and K. Kraus, *Pure Operations and Measurements*, Comm. Math. Phys. **11**, 214 (1969).

170. K. Hellwig and K. Kraus, *Operations and Measurements. II*, Comm. Math. Phys. **16**, 142 (1970).

171. K.-E. Hellwig, *General Scheme of Measurement Processes*, Int. J. Theor. Phys. **34**, 1467–1479 (1995).

172. C. W. Helstrom, *Quantum detection and estimation theory* (Academic Press, New York, 1976).

173. C. J. Hood, M.S.Chapman, T.W. Lynn and H.J.Kimble, *Real-Time Cavity QED with Single Atoms*, Phys. Rev. Lett. **80**, 4157 (1998).

174. A.S. Holevo, *Some estimates of the information transmitted by quantum communication channels*, Problemy Peredachi Informatsii **9**, 3 (1973). English translation in Problems of Information Transmission (USSR) **9**, 177 (1973).

175. S.F. Huelga, C. Macchiavello, T. Pellizzari, A.K. Ekert, M.B. Plenio and I.J. Cirac, *Improvement of frequency standards with quantum entanglement*, Phys. Rev. Lett. **79**, 3865 (1997).

176. R. J. Hughes, D. F. V. James, E. H. Knill, R. Laflamme, and A. G. Petschek, *Decoherence bounds on quantum computation with trapped ions*, Phys. Rev. Lett. **77**, 3240 (1996).

177. R. J. Hughes, G. G. Luther, G. L. Morgan, C. G. Peterson, and C. Simmons, *Quantum cryptography over underground optical fibers*, Lect. Notes in Comp. Sci. **1109**, 329 (1996).

178. R. J. Hughes, D. F. V. James, J. J. Gomez, M. S. Gulley, M. H. Holzscheiter, P. G. Kwiat, S. K. Lamoreaux, C. G. Peterson, V. D. Sandberg, M. M. Schauer, C. M. Simmons, C. E. Thorburn, D. Tupa, P. Z. Wang, and A. G. White, *The Los Alamos trapped ion quantum computer experiment*, Fortschritte Phys.-Prog. Phys. **46**, 329 (1998).

179. A. G. Huibers, M. Switkes, C. M. Marcus, K. Campman, and A. C. Gossard, *Dephasing in open quantum dots*, Phys. Rev. Lett. **81**, 200 (1998).

180. B. Huttner and A. K. Ekert, *Information gain in quantum eavesdropping.* J. Mod. Opt. **41,** 2455 (1994).

181. B. Huttner, N. Imoto, N. Gisin, and T. Mor, *Quantum Cryptography with Coherent States*, Phys. Rev. A **51**, 1863 (1995).

182. L. Jacak, P. Hawrylak, and A. Wójs, *Quantum Dots* (Springer, Berlin, 1997).

183. B. C. Jacobs, and J. D. Franson, *Quantum cryptography in free space*, Opt. Lett. **21**, 1854 (1996).

184. D. Jaksch, C. Bruder, J.I. Cirac, C.W. Gardiner and P. Zoller, *Cold bosonic atoms in optical lattices*, Phys. Rev. Lett. **81**, 3108 (1998).

185. D. Jaksch, H.-J. Briegel, J.I. Cirac, C.W. Gardiner and P. Zoller, *Entanglement of atoms via cold controlled collisions*, Phys. Rev. Lett. **82,** 1975 (1999).

186. D. F. V. James, *Quantum dynamics of cold trapped ions with application to quantum computation*, Appl. Phys. B-Lasers Opt. **66**, 181 (1998).

187. D. F. V. James, *Theory of heating of the quantum ground state of trapped ions*, Phys. Rev. Lett. **81**, 317 (1998).

188. S. R. Jefferts, C. Monroe, E. W. Bell, and D. J. Wineland, *Coaxial-resonator-driven rf (Paul) trap for strong confinement*, Phys. Rev. A **51**, 3112 (1995).

189. J. A. Jones and M. Mosca, *Approximate quantum counting on an NMR ensemble quantum computer*, quant-ph/9808056 (1999).

190. J. A. Jones, M. Mosca, and R. H. Hansen, *Implementation of a quantum search algorithm on a quantum computer*, Nature **393,** 344 (1998).

191. J. A. Jones, R. H. Hansen, and M. Mosca, *Quantum logic gates and nuclear magnetic resonance pulse sequences*, J. Magn. Reson. **135,** 353 (1998).

192. J. A. Jones,*Quantum Computing and NMR* in *Quantum Computing and Communication* (M. Brooks, Ed.), Springer-Verlag (1999, in press).

193. J. A. Jones, *Nuclear Magnetic Resonance Experiments* in *The Physics of Quantum Information* (D. Bouwmeester, A. Ekert and A. Zeilinger, Eds.), Springer-Verlag (1999, in press).

194. B. Kalyanasundaram and G. Schnitger, *The probabilistic communication complexity of set intersection*, Proc. 2nd Conf. on Structure in Complexity Theory, p. 41 (1987).

195. B. E. Kane, *A silicon-based nuclear spin quantum computer*, Nature **393,** 133 (1998).

196. J. M. Kikkawa and D. D. Awschalom, *Resonant spin amplification in n-type GaAs*, Phys. Rev. Lett. **80**, 4313 (1998).

197. H. J. Kimble, *Strong interactions of single atoms and photons in cavity QED*, Phys. Scripta **T76**, 127 (1998).

198. B. E. King, C. S. Wood, C. J. Myatt, Q. A. Turchette, D. Leibfried, W. M. Itano, C. Monroe, and D. J. Wineland, *Cooling the collective motion of trapped ions to initialize a quantum register*, Phys. Rev. Lett. **81**, 1525 (1998).

199. A. Yu. Kitaev, *Quantum measurements and the Abelian stabilizer problem*, quant-ph/9511026 (1995).

200. E. Knill, R. Laflamme, W. Zurek, *Threshold accuracy for quantum computation*, preprint quant-ph/9610011 (1996).

201. E. Knill and R. Laflamme, *Theory of quantum error-correcting codes*, Phys. Rev. A **55**, 900 (1997).

202. E. Knill, I. Chuang and R. Laflamme, *Effective pure states for bulk quantum computation*, Phys. Rev. A **57,** 3348 (1998).

203. M. Koashi and N. Imoto: *Quantum Cryptography Based on Split Transmission of One-Bit Information in Two Steps.* Phys. Rev. Lett. **79**, 2383 (1997).

204. K. Kraus, *General State Changes in Quantum Theory*, Ann. Phys. **64**, 311 (1971).

205. K. Kraus, *States, Effects, and Operations: Fundamental Notions of Quantum Theory*, Lecture Notes in Physics **190**, Berlin: Springer-Verlag (1983).

206. I. Kremer, *Quantum Communication*, MSc Thesis, Computer Science Department, The Hebrew University, 1995.

207. L. P. Kouwenhoven, C. M. Marcus, P. L. McEuen, S. Tarucha, R. M. Westervelt, and N. S. Wingreen, Proceedings of the Advanced Study Institute on *Mesoscopic Electron Transport*, edited by L. L. Sohn, L. P. Kouwenhoven, G. Schön (Kluwer, 1997).

208. L. P. Kouwenhoven, T. H. Oosterkamp, M. W. S. Danoesastro, M. Eto, D. G. Austing, T. Honda, and S. Tarucha, *Excitation spectra of circular, few-electron quantum dots*, Science **278**, 1788 (1997).

209. E. Kushilevitz and N. Nisan, *Communication Complexity*, (Cambridge University Press), 1998.

210. P.G. Kwiat, K. Mattle, H. Weinfurter, A. Zeilinger, A.V. Sergienko, Y.H. Shih, *New high-intensity source of polarization-entangled photon pairs*, Phys. Rev. Lett. **75**, 4337 (1995).

211. W. M. Itano, J. C. Bergquist, J. J. Bollinger, J. M. Gilligan, D. J. Heinzen, F. L. Moore, M. G. Raizen, and D. J. Wineland, *Quantum projection noise - population fluctuations in 2-level systems*, Phys. Rev. A **47**, 3554 (1993).

212. R. Laflamme, C. Miquel, J. P. Paz, and W. H. Zurek, *Perfect quantum error correcting code*, Phys. Rev. Lett. **77**, 198 (1996).

213. R. Laflamme, E. Knill, W. H. Zurek, P. Catasti, and S. V. S. Mariappan, *NMR Greenberger-Horne-Zeilinger states*, Phil. Trans. R. Soc. Lond. A **356,** 1941 (1998).

214. L. J. Landau and R. F. Streater, *On Birkhoff's Theorem for Doubly Stochastic Completely Positive Maps of Matrix Algebras*, Lin. Alg. Appl. **193**, 107 (1993).

215. R. Landauer, *Minimal energy requirements in communication*, Science **272**, 1914 (1996).

216. *Laser manipulation of Atoms and Ions,*1992, edited by E. Arimondo, W. D. Phillips, and F. Strumia,(North Holland, Amsterdam).

217. C. K. Law and H. J. Kimble, *Deterministic generation of a bit-stream of single-photon pulses*, J. Mod. Opt. **44** 2067 (1997).

218. H. Lenstra and C. Pomerance, *A rigorous time bound for factoring integers*, J. of the AMS **5**,482 (1992).

219. C.-K. Li and H. J. Woerdeman, *Special Classes of Positive and Completely Positive Maps*, Lin. Alg. Appl. **255**, 247 (1997).

220. D. A. Lidar, I. L. Chuang, and K. B. Whaley, *Decoherence free subspaces for quantum computation*, Phys. Rev. Lett. **81**, 2594 (1998).

221. E. H. Lieb, *Convex Trace Functions and the Wigner-Yanase-Dyson Conjecture*, Adv. Math. **11**, 267 (1973).

222. E. H. Lieb and M. B. Ruskai, *Proof of the Strong Subadditivity of Quantum Mechanical Entropy*, J. Math. Phys. **14**, 1938 (1973).

223. E. H. Lieb and M. B. Ruskai, *Some Operator Inequalities of the Schwarz Type*, Adv. Math. **12**, 269 (1974).

224. G. Lindblad, *Completely Positive Maps and Entropy Inequalities*, Comm. Math. Phys. **40**, 147 (1975).

225. N. Linden, H. Barjat, R. J. Carbajo, and R. Freeman, *Pulse Sequences for NMR Quantum Computers: How to Manipulate Nuclear Spins While Freezing the Motion of Coupled Neighbours*, Chem. Phys. Lett. (1999, in press), quant-ph/9811043.

226. C. Livermore, C. H. Crouch, R. M. Westervelt, K. L. Chapman, and A. C. Gossard, *The Coulomb blockade in coupled quantum dots*, Science **274**, 1332 (1996).

227. R.C. Liu, B. Odom, Y. Yamamoto, and S. Tarucha, *Quantum interference in electron collision*, Nature **391**, 263 (1998).

228. S. Lloyd, *Quantum-Mechanical Computers*, Scientific American, October 1995, p. 140.

229. S. Lloyd, *Universal Quantum Simulators*, Science **273**, 1073 (1996).

230. S. Lloyd, *Microscopic analogs of the Greenberger-Horne-Zeilinger experiment*, Phys. Rev. A **57**, R1473 (1998).

231. H.-K. Lo and H. F. Chau, *Unconditional security of quantum key distribution over arbitrarily long distances*, Science **283**, 2050 (1999).

232. D. Loss and D. P. DiVincenzo, *Quantum computation with quantum dots*, Phys. Rev. A **57**, 120 (1998).

233. N. Lütkenhaus *Security against eavesdropping in quantum cryptography*, Phys. Rev. A **54**, 97 (1996).

234. N. Lütkenhaus, *Estimates for practical quantum cryptography*, Phys. Rev. A **59**, 3301 (1999).

235. H. Mabuchi, Q. A. Turchette, M. S. Chapman, and H. J. Kimble, *Real-time detection of individual atoms falling through a high finesse optical cavity*, Opt. Lett. **21**, 1393 (1996).

236. H. Mabuchi and P. Zoller, *Inversion of Quantum Jumps in a Quantum Optical System under Continuous Observation*, Phys. Rev. Lett. **76**, 3108 (1996).

237. C. Macchiavello, *On the analytical convergence of the QPA procedure*, Phys. Lett. A **246**, 385 (1998).

238. Y. Makhlin, G. Schön, and A. Shnirman, *Josephson-Junction Qubits with Controlled Couplings*, NATURE,**398**, 305-307 (1999).

239. X. Maitre, E.Hagley, G.Nogues, C.Wunderlich, P.Goy, M.Brune, J.M.Raimond and S.Haroche *Quantum Memory with a Single Photon in a Cavity*, Phys. Rev. Lett. **79**, 769 (1997).

240. C. Marand, and P. D. Townsend, *Quantum key distribution over distances as long as 30 km*, Opt. Lett. **20**, 1695 (1995).

241. D.C. Mattis, *The Theory of Magnetism*, Vol. I, Sec. 4.5, Springer Series in Solid-State Sciences **17** (Springer, New York, 1988).

242. D. Mayers, *Unconditional security in Quantum Cryptography*, quant-ph/9802025 (1998).

243. D. M. Meekhof, C. Monroe, B. E. King, W. M. Itano, and D. J. Wineland, *Generation of nonclassical motional states of a trapped atom*, Phys. Rev. Lett. **76**, 1796 (1996).

244. C. Mennerat-Robilliard, D. Lukas, S. Guibal, J. Tabosa, C. Jurczak, J.-Y. Courtois and G. Grynberg, *Ratchet for cold rubidium atoms: The asymmetric optical lattice*, Phys. Rev. Lett. **82**, 85 (1999).

245. N.D. Mermin, *What is wrong with these elements of reality*, Physics Today, 9, June 1990.

246. N.D. Mermin, *Quantum mysteries revisited*, Am. J. Phys. **58**, 731 (1990)

247. J.-M- Mrolla, Y. Mazurenko, J.-P. Goedgebuer, H. Porte, W. T. Rhodes, *Phase-modulation transmission system for quantum cryptography*, Opt. Lett. **24** 104 (1999).

248. K. Mølmer and A. Sørensen, *Multiparticle entanglement of hot trapped ions*, Phys. Rev. Lett. **82**, 1835 (1999).

249. C. Monroe, D. M. Meekhof, B. E. King, S. R. Jefferts, W. M. Itano, D. J. Wineland, and P. Gould, *Resolved-side-band raman cooling of a bound atom to the 3D zero-point energy*, Phys. Rev. Lett. **75**, 4011 (1995).

250. C. Monroe, D. M. Meekhof, B. E. King, W. M. Itano, and D. J. Wineland, *Demonstration of a fundamental quantum logic gate*, Phys. Rev. Lett. **75**, 4714 (1995).

251. C. Monroe, D. M. Meekhof, B. E. King, and D. J. Wineland, *A "Schrödinger cat" superposition state of an atom*, Science **272**, 1131 (1996).

252. C. Monroe, D. Leibfried, B. E. King, D. M. Meekhof, W. M. Itano, and D. J. Wineland, *Simplified quantum logic with trapped ions*, Phys. Rev. A **55**, R2489 (1997).

253. G. Morigi, J. I. Cirac, M. Lewenstein, and P. Zoller, *Ground-state laser cooling beyond the Lamb-Dicke limit*, Europhys. Lett. **39**, 13 (1997).

254. M. Mosca and A. Ekert, *The hidden subgroup problem and eigenvalue estimation on a quantum computer*, to appear in Proc. 1st NASA Intl. Conf. on Quantum Computing and Quantum Communications **1509** of *Lecture Notes in Computer Science* (Springer), 1998; preprint quant-ph/9903071 (1998).

255. A. Muller, J. Breguet, and N. Gisin, *Experimental demonstration of quantum cryptography using polarized photons in optical fibre over more than 1 km*, Europhys. Lett. **23**, 383 (1993).

256. A. Muller, T. Herzog, B. Huttner, W. Tittel, H. Zbiden, and N. Gisin, *Plug and Play systems for quantum cryptography*, Appl. Phys. Lett. **70**, 793 (1997).

257. M. Munnowitz, *Coherence and NMR* (John Wiley, 1988).

258. H. C. Nägerl, W. Bechter, J. Eschner, F. Schmidt-Kaler, and R. Blatt, *Ion strings for quantum gates*, Appl. Phys. B-Lasers Opt. **66**, 603 (1998).

259. H. C. Nägerl, D. Leibfried, F. Schmidt-Kaler, J. Eschner, and R. Blatt, *Coherent excitation of normal modes in a string of Ca+ ions*, Opt. Express **3**, 89 (1998).

260. W. Nagourney, J. Sandberg, and H. Dehmelt, *Shelved optical electron amplifier - observation of quantum jumps*, Phys. Rev. Lett. **56**, 2797 (1986).

261. M. A. Nielsen, E. Knill and R. Laflamme, *Complete quantum teleportation using nuclear magnetic resonance*, Nature **396,** 52 (1998).

262. K. Nomoto, R. Ugajin, T. Suzuki, and I. Hase, *Single electron-photon logic device using coupled-quantum dots: Computation with the Fock ground state*, J. Appl. Phys. **79**, 291 (1996).

263. T. H. Oosterkamp, S. F. Godijn, M. J. Uilenreef, Y. V. Nazarov, N. C. van der Vaart, and L. P. Kouwenhoven, *Changes in the magnetization of a double quantum dot*, Phys. Rev. Lett. **80**, 4951 (1998).

264. A. O. Orlov, I. Amlani, G. H. Bernstein, C. S. Lent, and G. L. Snider, *Realization of a functional cell for quantum-dot cellular automata*, Science **277**, 928 (1997).

265. P. Owens, J. Rarity, P. Tapster, D. Knight, and P. D. Townsend, *Photon counting with passively quenched germanium avalanche photodiodes*, Appl. Opt. **33**, 6895 (1994).

266. Y. Ozhigov, *Fast quantum verification for the formulas of predicate calculus*, preprint quant-ph/9809015 (1998).

267. G.M. Palma, K.-A. Suominen and A.K. Ekert, *Quantum Computers and Dissipation*, Proc. Roy. Soc. London A **452**, 567, (1996).

268. J. W. Pan, D. Bouwmeester, H. Weinfurter, and A. Zeilinger, *Experimental entanglement swapping: Entangling photons that never interacted*, Phys. Rev. Lett. **80**, 3891 (1998).

269. C.H. Papadimitriou, *Computational Complexity*, Addison-Wesley, 1994.

270. A. S. Parkins, P. Marte, P. Zoller, and H. J. Kimble, *Synthesis of Arbitrary Quantum States via Adiabatic Transfer of Zeeman Coherence*, Phys. Rev. Lett. **71**, 3095 (1993).

271. W. Paul, *Electromagnetic traps for charged and neutral particles*, Rev. Mod. Phys. **62**, 531 (1990).

272. T. Pellizzari, S. A. Gardiner, J.-I. Cirac, and P. Zoller, *Decoherence, Continuous Observation, and Quantum Computing – a Cavity QED Model*, Phys. Rev. Lett. **75**, 3788 (1995).

273. T. Pellizzari, *Quantum networking with optical fibres*, Phys. Rev. Lett. **79**, 5242 (1997).

274. A. Peres, "Quantum theory: concepts and methods", Kluwer, Dordrecht (1993).

275. A. Peres, *Quantum entanglement: criteria and collective tests*, Physica Scripta **T76**, 52 (1998).

276. A. Peres *Quantum Cryptography with Orthogonal States?* Phys. Rev. Lett. **77**, 3264 (1996);

277. W.D. Phillips, *Laser cooling and trapping of neutral atoms*, Rev. Mod. Phys. **70**, 721 (1998).

278. M. B. Plenio, V. Vedral, and P. L. Knight, *Quantum error correction in the presence of spontaneous emission*, Phys. Rev. A **55**, 67 (1997).

279. J. A. Poluikis and R. D. Hill, *Completely Positive and Hermitian-Preserving Linear Transformations*, Lin. Alg. Appl. **35**, 1 (1981).

280. C. Pomerance, *Fast rigorous factorization and discrete logarithm algorithms*, Discrete Algorithms and Complexity (Proc. Japan-US Joint Seminar on Discrete Algorithms and Complexity theory), p. 119 (1987).

281. S. Popescu, *An optical method for teleportation*, e-print quant-ph/9501020.

282. J. F. Poyatos, J. I. Cirac, and P. Zoller, *Quantum gates with "hot" trapped ions*, Phys. Rev. Lett. **81**, 1322 (1998).

283. J. Preskill, *Reliable quantum computers*, Proc. R. Soc. Lond. A **454**, 385 (1998).

284. J. Preskill, *Fault-tolerant quantum computation*, in "Introduction to Quantum Computation and Information", edited by H.-K. Lo, S. Popescu, and T.P. Spiller (World Scientific), p. 213 (1998).

285. J. D. Prestage, G. J. Dick, and L. Maleki, *New ion trap for frequency standard applications*, J. Appl. Phys. **66**, 1013 (1989).

286. E. M. Purcell, H. C. Torrey, and R. V. Pound, *Resonance absorption by nuclear magnetic moments in a solid*, Phys. Rev. **69,** 37 (1946).

287. J. G. Rarity, P. C. M. Owens, and P. R. Tapster *Quantum random-number generation and key sharing.* J. Mod. Opt. **41,** 2435–2444 (1994).

288. R. Raz, *Exponential separation of quantum and classical communication complexity*, to appear in Proc. 31st Ann. ACM Symp. on Theory of Computing (STOC '99), 1999.

289. G. Ribordy, J.-D. Gautier, N. Gisin, O. Guinnard, and H. Zbiden, *Automated 'plug & play' quantum key distribution*, Electron. Lett. **34**, 2116 (1998).

290. G. Ribordy, J. D. Gautier, H. Zbinden, and N. Gisin, *Performance of InGaAs/InP avalanche photodiodes as gated-mode photon counters*, Appl. Opt. **37**, 2272 (1998).

291. R. Rivest, A. Shamir, and L. Adleman, *On digital signatures and public key cryptosystems*, MIT Lab. For Comp. Science, Technical Report, MIT/LCS/TR-212 (January 1979).

292. T. Sauter, W. Neuhauser, R. Blatt, and P. E. Toschek, *Observation of quantum jumps*, Phys. Rev. Lett. **57**, 1696 (1986).

293. R. Schack and C. M. Caves, *Classical model for bulk-ensemble NMR quantum computation*, quant-ph/9903101 (1999).

294. S. Schneider and G. J. Milburn, *Decoherence in ion traps due to laser intensity and phase fluctuations*, Phys. Rev. A **57**, 3748 (1998).

295. G. Schön, A. Shnirman, and Yu. Makhlin, *Josephson-Junction Qubits and the Readout Process by Single-Electron Transistors*, to be published in: *Exploring the Quantum - Classical Frontier: Recent Advances in Macroscopic and Mesoscopic Quantum Phenomena*, Eds. J.R. Friedman and S. Han, Nova Science Publishers, Commack, NY (cond-mat/9811029).

296. B. Schumacher, *Sending Entanglement Through Noisy Quantum Channels*, Phys. Rev. A **54**, 2614 (1996).

297. B. Schumacher and M. D. Westmoreland, *Sending Classical Information via Noisy Quantum Channels*, Phys. Rev. A **56**, 131 (1997).

298. A. Shnirman, G. Schön, and Z. Hermon, *Quantum manipulations of small Josephson junctions*, Phys. Rev. Lett. **79**, 2371 (1997).

299. P. Shor, *Algorithms for quantum computation: discrete logarithms and factoring*, Proc. 35th Annual Symposium on Foundations of Computer Science, Santa Fe, NM, USA, 20-22 Nov. 1994, IEEE Comput. Soc. Press (1994) p. 124.

300. P. W. Shor, *Scheme for reducing decoherence in quantum computer memory*, Phys. Rev. A **52**, 2493 (1995).

301. P.W. Shor, *Fault-tolerant quantum computation*, Proc. 37th Ann. IEEE Symp. on Foundations of Computer Science (FOCS '96), p. 56 (1996).

302. P.W. Shor, *Polynomial-time algorithms for prime factorization and discrete logarithms on a quantum computer*, SIAM J. Computing **26**, 1484 (1997).

303. D. Simon, *On the power of quantum computation*, SIAM J. Computing **26**, 1474 (1997).

304. T. Sleator and H. Weinfurter, *Realizable universal quantum logic gates*, Phys. Rev. Lett. **74**, 4087 (1995).

305. B. Slutsky, R. Rao, P.-C. Sun, and Y. Fainman, *Security of quantum cryptography against individual attacks*, Phys. Rev. A **57**, 2383 (1998).

306. R. Solovay and V. Strassen, *A fast Monte Carlo test for primality*, SIAM J. Computing **6**, 84 (1977). (Erratum in Vol. 7, p. 118, 1978.)

307. A. Sørensen and K. Mølmer, *Quantum computation with ions in thermal motion*, Phys. Rev. Lett. **82**, 1971 (1999).

308. O. W. Sørensen, G. W. Eich, M. H. Levitt, G. Bodenhausen, and R. R. Ernst, Prog. NMR Spectrosc. **16,** 163 (1983).

309. D.M. Stamper-Kurn, M.R. Andrews, A.P. Chikkatur, S. Inouye, H.-J. Miesner, J. Stenger and W. Ketterle, *Optical confinement of a Bose-Einstein condensate*, Phys. Rev. Lett. **80**, 2027 (1998).

310. A. M. Steane, *Error correcting codes in quantum theory*, Phys. Rev. Lett. **77**, 793 (1996).

311. A. M. Steane, *Multiple-particle interference and quantum error correction*, Proc. R. Soc. Lond. **A 452**, 2551 (1996).

312. A. M. Steane, *Simple quantum error correcting codes*, Phys. Rev. A **54**, 4741 (1996).

313. A. Steane, *The ion trap quantum information processor*, Appl. Phys. B-Lasers Opt. **64**, 623 (1997).

314. J. Steinbach and C. C. Gerry, *Efficient scheme for the deterministic maximal entanglement of N trapped ions*, Phys. Rev. Lett. **81**, 5528 (1998).

315. S. Stenholm, *The semiclassical theory of laser cooling*, Rev. Mod. Phys. **58**, 699 (1986).

316. D. Stevens, J. Brochard, and A. M. Steane, *Simple experimental methods for trapped-ion quantum processors*, Phys. Rev. A **58**, 2750 (1998).

317. E. Sukhorukov and D. Loss, *Universality of shot noise in multiterminal diffusive conductors*, Phys. Rev. Lett. **80**, 4959 (1998).

318. E. Sukhorukov and D. Loss, *Universality of shot noise in multiterminal diffusive conductors*, To appear in Phys. Rev. B (May 15, 1999). See cond-mat/9809239.

319. D. Suter, K. T. Mueller, and A. Pines, *Study of the Aharonov-Anandan quantum phase by NMR interferometry*, Phys. Rev. Lett. **60,** 1218 (1988).

320. S. Tarucha, D. G. Austing, T. Honda, R. J. van der Hage, and L. P. Kouwenhoven, *Shell filling and spin effects in a few electron quantum dot*, Phys. Rev. Lett. **77**, 3613 (1996).

321. B. M. Terhal, I. L. Chuang, D. P. DiVincenzo, M. Grassl, and J. A. Smolin, *Simulating Quantum Operations with Mixed Environments*, quant-ph/9806095.

322. W. Tittel, G. Ribordy, and N. Gisin, *Quantum cryptography*, Physics World **11**, 41 (1998).

323. W. Tittel, J. Brendel, H. Zbinden, and N. Gisin, *Violation of bell inequalities by photons more than 10 km apart*, Phys. Rev. Lett. **81**, 3563 (1998).

324. P. D. Townsend, *Experimental investigation of the performance limits for first telecommunication-window quantum cryptography systems*, IEEE Photon. Technol. Lett. **10**, 1048 (1998).

325. Q. A. Turchette, C. S. Wood, B. E. King, C. J. Myatt, D. Leibfried, W. M. Itano, C. Monroe, and D. J. Wineland, *Deterministic entanglement of two trapped ions*, Phys. Rev. Lett. **81**, 3631 (1998).

326. R. Tycko, *Adiabatic rotational splittings and berry's phase in nuclear quadrupole resonance*, Phys. Rev. Lett. **58** 2281 (1987).

327. W. G. Unruh, *Mantaining coherence in quantum computers*, Phys. Rev. A **51**, 992 (1995).

328. L. Vaidman, *Teleportation of quantum states*, Phys. Rev. A **49**, 1473 (1994).

329. U.V. Vazirani, *Strong communication complexity or generating quasi-random sequences from two communicating slightly-random sources*, Combinatorica **7**, 375 (1987).

330. V. Vedral and M. B. Plenio, *Basics of quantum computation*, Prog. Quantum Electron. **22**, 1 (1998).

331. L. Viola and S. Lloyd, *Dynamical suppression of decoherence in two-state quantum systems*, Phys. Rev. A **58**, 2733 (1998).

332. L. Viola, E. Knill, and S. Lloyd, *Dynamical decoupling of open quantum systems*, Phys. Rev. Lett. **82**, 2417 (1999).

333. A. C.-C. Yao, *Some complexity questions related to distributive computing*, Proc. 11th Ann. ACM Symp. on Theory of Computing (STOC '79), p. 209 (1979).

334. A. C.-C. Yao, *Quantum circuit complexity*, Proc. 34th Ann. IEEE Symp. on Foundations of Computer Science (FOCS '93), p. 352 (1993).

335. I. Waki, S. Kassner, G. Birkl, and H. Walther, *Observation of ordered structures of laser-cooled ions in a quadrupole storage ring*, Phys. Rev. Lett. **68**, 2007 (1992).

336. W. S. Warren, *The usefulness of NMR quantum computing*, Science **277**, 1688 (1997).

337. F. R. Waugh, M. J. Berry, D. J. Mar, R. M. Westervelt, K. L. Chapman, and A. C. Gossard, *Single-electron changing in double and triple quantum dots with tunable coupling*, Phys. Rev. Lett. **75**, 705 (1995).

338. G. Weish, T. Jennewein, C. Simon, H. Weinfurter, and A. Zeilinger, *Violation of Bell's inequality under strict Einstein locality conditions*, Phys. Rev. Lett. **81**, 5039 (1998).

339. D. Welsh, "Codes & Cryptography", Clarendon Press, Oxford (1988).

340. S. Wiesner, *Simulation of Many-Body Quantum Systems by a Quantum Computer*, quant-ph/9603028.

341. D. Wineland and H. Dehmelt, *Proposed 1014 delta upsilon less than upsilon laser fluorescence spectroscopy on Tl+ mono-ion oscillator III*, Bull. Amer. Phys. Soc. **20**, 637 (1975).

342. D. J. Wineland, R. E. Drullinger, and F. L. Walls, *Radiation-pressure cooling of bound resonant absorbers*, Phys. Rev. Lett. **40**, 1639 (1978).

343. D. J. Wineland and W. M. Itano, *Laser cooling of atoms*, Phys. Rev. A **20**, 1521 (1979).

344. D. J. Wineland, W. M. Itano, and R. S. Vandyck, *High-resolution spectroscopy of stored ions*, Adv. At. Mol. Phys. **19**, 135 (1983).

345. D. J. Wineland, W. M. Itano, J. C. Bergquist, and R. G. Hulet, *Laser-cooling limits and single-ion spectroscopy*, Phys. Rev. A **36**, 2220 (1987).

346. D. J. Wineland, J. J. Bollinger, W. M. Itano, and D. J. Heinzen, *Squeezed atomic states and projection noise in spectroscopy*, Phys. Rev. A **50**, 67 (1994).

347. D. J. Wineland, C. Monroe, W. M. Itano, D. Leibfried, B. E. King, and D. Meekhof, *Experimental issues in coherent quantum-state manipulation of trapped atomic ions*, J. Res. Natl. Inst. Stand. Technol. **103**, 259 (1998), quant-ph/9710025.

348. D. J. Wineland, C. Monroe, W. M. Itano, B. E. King, D. Leibfried, D. M. Meekhof, C. Myatt, and C. Wood, *Experimental primer on the trapped ion quantum computer*, Fortschritte Phys.-Prog. Phys. **46**, 363 (1998).

349. D. J. Wineland, C. Monroe, W. M. Itano, B. E. King, D. Leibfried, C. Myatt, and C. Wood, *Trapped-ion quantum simulator*, Phys. Scr. **T76**, 147 (1998).

350. W.K. Wootters and W.H. Zurek, *A single quantum cannot be cloned*, Nature **299**, 802 (1982).

351. C.S. Wu, I. Shaknov, *The angular correlation of scattered annihilation radiation*, Phys. Rev. **77**, 136 (1950).

352. C. Zalka, *Efficient Simulation of Quantum Systems by Quantum Computers*, quant-ph/9603026 (1996).

353. C. Zalka, *Grover's quantum searching algorithm is optimal*, preprint quant-ph/9711070 (1997).

354. C. Zalka, *A Grover-based quantum search of optimal order for an unknown number of marked elements*, preprint quant-ph/9902049 (1999).

355. P. Zanardi and M. Rasetti, *Noiseless quantum codes*, Phys. Rev. Lett. **79**, 3306 (1997).

356. P. Zanardi and F. Rossi, *Quantum Information in Semiconductors: Noiseless Encoding in a Quantum-Dot Array*, Phys. Rev. Lett. **81**, 4752 (1998). quant-ph/9804016.

357. H. Zbinden, J. D. Gautier, N. Gisin, B. Huttner, A. Muller, and W. Tittel, *Interferometry with Faraday mirrors fo rquantum cryptography*, Electron. Lett. **33**, 586 (1997).

358. H. Zbinden, H. Bechmann- Pasquinucci, N. Gisin, G. Ribordy, *Quantum Cryptography*, Appl. Phys. B **67**, 743 (1998).

359. H. Zbinden, N. Gisin, B. Huttner, A. Muller, and W. Tittel, *Practical aspects of quantum key distribution*, J. Cryptology **11**, 1 (1998).

360. A. Zeilinger, M.A. Horne, H. Weinfurter, M. Zukowski, *Three-particle entanglements from two entangled pairs*, Phys. Rev. Lett. **78**, 3031 (1997).

361. A. Zeilinger, *Fundamentals of quantum information*, Physics World **11**, 24 (1998).

362. A. Zeilinger, *Experiment and the Foundations of Quantum Physics*, Rev. Mod. Phys. **71**, 288 (1999).

363. M. Zukowski, A. Zeilinger, M. A. Horne, and A. Ekert, *"Event ready detectors" Bell experiment via entanglement swapping*, Phys. Rev. Lett. **71**, 4287 (1993).